QUESTION BANK
in
ENGINEERING MATHEMATICS

(For B. Tech. Students)

Dr. V. P. Mishra

QUESTION BANK
in
ENGINEERING MATHEMATICS

(For B.E., B.Tech. and A.M.I.E. students)

Dr. V.P. Mishra
Formerly, Head, Department of Applied Mathematics,
Delhi College of Engineering, Delhi.
Presently with M.A.I.T, (G.G.S. Indraprastha University, Delhi)

Mrs. Pratibha Mishra
(M.Sc.)

S.K. KATARIA & SONS®

Publisher of Engineering Books
4885/109, Prakash Mahal, Dr. Subhash Bhargav Lane,
Opposite Delhi Medical Association, Daryaganj, New Delhi–110002 (India)
Phone: +91-11-43551243; *Mobile:* +91-9871775858, 7678425209
e-mail: katariabook@gmail.com, katariabooks@yahoo.com
Website: www.skkatariaandsons.com
(e-books, subscriptions and bulk buying also available on our website)

Published by:

S.K. Kataria & Sons®
4885/109, Prakash Mahal, Dr. Subhash Bhargav Lane,
Opposite Delhi Medical Association, Daryaganj, New Delhi-110002 (India)
Phone: +91-11-43551243; *Mobile:* +91-9871775858, 7678425209
e-mail: katariabook@gmail.com, katariabooks@yahoo.com
Website: www.skkatariaandsons.com

First Edition: 2009, Second Edition: 2011, Reprint: 2026

Price : **Rs. 995/-**

Laser Type Setting by:

SARA ASSIGNMENTS, Patparganj, Delhi-110092.

Printed at:

A.S. OFFSET, Delhi.

Preface to the Second Edition

I feel immense pleasure and great satisfaction in presenting a second edition of Question Bank/Solution manual to the readers of the "Concepts of Engineering Mathematics" to the B.Tech. students in general and B.Tech. students of G.G.S.I.P. University, Delhi in particular.

Many additional problems with hints and solutions have been added in each chapter. I am really thankful to the cooperation received from all corners in bringing out this book in such a less time.

It is earnestly hoped that the valued readers would extend their suggestions for the improvement of the book and these will be gratefully acknowledged.

Date: 15 February 2011

Dr. V.P. Mishra
Mrs. Pratibha Mishra
(M.Sc.)

Preface to the First Edition

It gives me pleasure and great satisfaction in presenting a Question Bank/Solution manual to the reade
of the "Concepts of Engineering Mathematics" to the B.Tech. students in general and B.Tech. studen
of G.G.S.I.P. University, Delhi in particular.

The demand for such a book has been there and the need for it has been felt in the recent past. Man
additional problems with hints and solutions have been added. I am really thankfull to the cooperatio
received from all corners in bringing out this book in such a less time.

Any suggestions for improvement in the book will be gratefully acknowledged by the author.

Date: 1 August 2009

Dr. V.P. Mishr

Mrs. Pratibha Mishr

(M.Sc.

Contents

CHAPTER

1 Complex Numbers — de Moivre's Theorem

IMPORTANT FORMULAE AND USEFUL RESULTS

1. $z = x + iy$ then $\bar{z} = x - iy$ and $z\bar{z} = x^2 + y^2 = |z|^2$

$\overline{z_1 \pm z_2} = \bar{z}_1 \pm \bar{z}_2$, $\overline{z_1 \cdot z_2} = \bar{z}_1 \cdot \bar{z}_2$, $\overline{(z_1 / z_2)} = \bar{z}_1 / \bar{z}_2$ if $z_2 \neq 0$.

2. $|z_1 + z_2| \le |z_1| + |z_2|$ and $|z_1 - z_2| \ge |z_1| - |z_2|$.

3. $|z_1 z_2| = |z_1|\,|z_2|$ and $|z_1/z_2| = |z_1|/|z_2|$ provided $z_2 \neq 0$.

$\arg(z_1 z_2) = \arg(z_1) + \arg(z_2)$ and $\arg(z_1/z_2) = \arg(z_1) - \arg(z_2)$.

4. **De Moivre's Theorem:**

If n is an integer or a fraction then atleast one of the values of $(\cos\theta + i\sin\theta)^n$ is $\cos n\theta + i\sin n\theta$.

5. Cube roots of unity.

$$(1)^{1/3} = 1, \omega, \omega^2 \quad \text{where} \quad \omega = \frac{-1+i\sqrt{3}}{2}, \quad \omega^2 = \frac{-1-i\sqrt{3}}{2}, \quad \omega^3 = 1.$$

6. $\cos n\theta = \cos^n\theta - {}^nC_2 \cos^{n-2}\theta \sin^2\theta + {}^nC_4 \cos^{n-4}\theta \sin^4\theta - \ldots$

and $\sin n\theta = {}^nC_1 \cos^{n-1}\theta \sin\theta - {}^nC_3 \cos^{n-3}\theta \sin^3\theta + {}^nC_5 \cos^{n-5}\theta \sin^5\theta - \ldots$

7. $$\cos^n\theta = \frac{1}{2^{n-1}}[\cos n\theta - {}^nC_1 \cos(n-2)\theta + {}^nC_2 \cos(n-4)\theta + \ldots]$$

8. $$\sin^n\theta = \frac{1}{2^{n-1}(-1)^{n/2}}[\cos n\theta - {}^nC_1 \cos(n-2)\theta + {}^nC_2 (n-4)\theta - \ldots] \quad \text{when } n \text{ is an even integer}$$

$$= \frac{1}{2^{n-1} i^{n-1}}[\sin n\theta - {}^nC_1 \sin(n-2)\theta + {}^nC_2 \sin(n-4)\theta - \ldots] \text{ when } n \text{ is an odd integer.}$$

9. **Euler's Theorem:** $e^{ix} = \cos x + i\sin x$

$$\therefore \quad \cos x = \frac{1}{2}[e^{ix} + e^{-ix}] \quad \text{and} \quad \sin x = \frac{1}{2i}[e^{ix} - e^{-ix}]$$

10. Hyperbolic Functions

$$\cosh x = \frac{1}{2}(e^x + e^{-x}), \quad \sinh x = \frac{1}{2}(e^x - e^{-x})$$

$\cos(ix) = \cosh x$ and $\sin ix = i \sinh x$

$$\tanh x = \frac{e^x - e^{-x}}{e^x + e^{-x}}, \quad \coth x = \frac{1}{\tanh x}, \quad \text{sech}\, x = \frac{1}{\cosh x}, \quad \text{cosech}\, x = \frac{1}{\sinh x}.$$

$\cosh^2 x - \sinh^2 x = 1$, $\text{sech}^2 x = 1 - \tanh^2 x$, $\text{cosech}^2 x = \coth^2 x - 1$.

$\sinh(2x) = 2\sinh(x)\cosh(x)$, $\cosh(2x) = \cosh^2 x + \sinh^2 x = 2\cosh^2 x - 1 = 1 + 2\sinh^2 x$

$$\tanh(2x) = \frac{2\tanh x}{1 + \tanh^2 x}$$

$\sinh(3x) = 3\sinh x + 4\sinh^3 x$ and $\cosh(3x) = 4\cosh^3 x - 3\cosh x$

$$\tanh(3x) = \frac{3\tanh x + \tanh^3 x}{1 + 3\tanh^2 x}$$

$$\cosh x = 1 + \frac{x^2}{2!} + \frac{x^4}{4!} + \ldots$$

$$\sinh x = x + \frac{x^3}{3!} + \frac{x^5}{5!} + \ldots$$

11. Inverse Hyperbolic Functions

$$\sinh^{-1} x = \log(x + \sqrt{x^2+1}), \quad \cosh^{-1} x = \log(x + \sqrt{x^2-1})$$

$$\tanh^{-1} x = \frac{1}{2}\log\left|\frac{1+x}{1-x}\right|$$

12. Logarithm of Complex Numbers.

$$\text{Log}\, z = \log(x + iy) = \frac{1}{2}\log(x^2 + y^2) + i\tan^{-1}(y/x)$$

called the principal value of $\log z$.

$\log z = 2n\pi i + \log z$ is called the general value of $\log z$.

13. $C + iS$ – method for summation of trigonometric series

Let sine and cosine series be given by

$S = a_0 \sin\alpha + a_1 \sin(\alpha + \beta) + a_2 \sin(\alpha + 2\beta) + \ldots$

$C = a_0 \cos\alpha + a_1 \cos(\alpha + \beta) + a_2 \cos(\alpha + 2\beta) + \ldots$

Given one series we select another auxiliary series

and then $C + iS = a_0 e^{i\alpha} + a_1 e^{i(\alpha+\beta)} + a_2 e^{i(\alpha+2\beta)} + \ldots$

can be summed up and C and S can be obtained.

SOLUTIONS TO EXERCISE 1A

1. *(a) If* $z = x + iy$ *find* $|z|^4$ *and* $|z^4|$. *(b) Evaluate* $\left|\frac{(1+i)^6}{i^3(1+4i)^2}\right|$.

SOLUTION: (*a*) $|z|^4 = [|z|^2]^2 = (x^2 + y^2)^2$ and $|z^4| = |z|^4 = (x^2 + y^2)^2$ **Ans.**

(*b*) $$\left|\frac{(1+i)^6}{i^3(1+4i)^2}\right| = \frac{|(1+i)^6|}{|i^3|\,|1+4i|^2} = \frac{(\sqrt{2})^6}{1(\sqrt{17})^2} = \frac{8}{17}$$ **Ans.**

2. *Express in trigonometric form*

(*i*) $\frac{i\sqrt{2}}{4+4i}$ (*ii*) $\frac{2+3i}{5+4i}$.

SOLUTION: (*i*) $$\frac{i\sqrt{2}}{4+4i} = \frac{i}{2\sqrt{2}(1+i)} = \frac{1}{2\sqrt{2}\left(1+\frac{1}{i}\right)} = \frac{1}{2\sqrt{2}(1-i)} = \frac{1}{4\left(\frac{1}{\sqrt{2}} - \frac{i}{2}\right)}$$

$$= \frac{1}{4\left(\cos\frac{\pi}{4} - i\sin\frac{\pi}{4}\right)} = \frac{\cos\frac{\pi}{4} + i\sin\frac{\pi}{4}}{4\left(\cos^2\frac{\pi}{4} + \sin^2\frac{\pi}{4}\right)} = \frac{1}{4}\left[\cos\frac{\pi}{4} + i\sin\frac{\pi}{4}\right]$$ **Ans.**

(*ii*) $$\frac{2+3i}{5+4i} = \frac{(2+3i)(5-4i)}{25+16} = \frac{22+7i}{41} = r(\cos\theta + i\sin\theta)$$

where $r = \frac{1}{41}\sqrt{22^2 + 7^2} = 0.563$ and $\theta = \tan^{-1}\frac{7}{22} = 0.308$ rad **Ans.**

3. *Show that* $|z_1 + z_2|^2 + |z_1 - z_2|^2 = 2(|z_1|^2 + |z_2|^2)$.

SOLUTION: $$|z_1 + z_2|^2 + |z_1 - z_2|^2 = (z_1 + z_2)(\bar{z}_1 + \bar{z}_2) + (z_1 - z_2)(\bar{z}_1 - \bar{z}_2)$$

$$= z_1\bar{z}_1 + z_2\bar{z}_2 + z_1\bar{z}_2 + \bar{z}_1 z_2 + z_1\bar{z}_1 + z_2\bar{z}_2 - z_1\bar{z}_2 - \bar{z}_1 z_2$$

$$= 2(z_1\bar{z}_1 + z_2\bar{z}_2) = 2(|z_1|^2 + |z_2|^2).$$ **Hence Proved.**

4. *Express the complex number* $\frac{x-1+iy}{x+1+iy}$ *in the form* $a + ib$ *and then obtain the condition for this number to be purely imaginary.*

SOLUTION: $$\frac{x-1+iy}{x+1+iy} = \frac{x-1+iy}{x+1+iy}\cdot\frac{x+1-iy}{x+1-iy} = \frac{x^2 - 1 + y^2 + i(xy + y - xy + y)}{(x+1)^2 + y^2}$$

$$= \frac{x^2 + y^2 - 1 + 2iy}{x^2 + y^2 + 1 + 2x}$$

For this number to be purely imaginary we have $x^2 + y^2 - 1 = 0$ **Ans.**

5. *Express* $\dfrac{3}{2+\cos\theta+i\sin\theta}$ *in the form* $x+iy$ *and prove that* $x^2+y^2-4x+3=0$.

SOLUTION: $$\frac{3}{2+\cos\theta+i\sin\theta} = \frac{3\{2+\cos\theta-i\sin\theta\}}{(2+\cos\theta)^2+\sin^2\theta}$$

$$= \frac{6+3\cos\theta-3i\sin\theta}{5+4\cos\theta} = x+iy$$

$$\Rightarrow \qquad x = \frac{6+3\cos\theta}{5+4\cos\theta} \text{ and } y = \frac{-3\sin\theta}{5+4\cos\theta}.$$

Next x^2+y^2-4x+3

$$= \frac{(6+3\cos\theta)^2+9\sin^2\theta}{(5+4\cos\theta)^2} - \frac{24+12\cos\theta}{5+4\cos\theta} + 3$$

$$= \frac{1}{(5+4\cos\theta)^2}[36+9\cos^2\theta+36\cos\theta+9\sin^2\theta-12(2+\cos\theta)(5+4\cos\theta)+3(5+4\cos\theta)^2]$$

$$= \frac{1}{(5+4\cos\theta)^2}[45+36\cos\theta-120-48\cos^2\theta-156\cos\theta+75+120\cos\theta+48\cos^2\theta] = 0$$

Hence $x^2+y^2-4x+3=0$ is **Proved.**

6. *Find the locus of the point* $P(z)$ *on the Argand plane so that* $|z-z_1|+|z-z_2|=2c$ *where* 'z_1' *and* 'z_2' *are fixed points and* c *is constant.*

SOLUTION: Sum of the distances of 'z' from two fixed points 'z_1' and 'z_2' is equal to $2c$ which is the property of an ellipse where sum of the focal distances is equal to length of major axis. Hence locus of 'z' is ellipse.

Ans.

7. *Determine the region represented by the locus of* 'z' *if* $\left|\dfrac{z-i}{z+i}\right| \le 2$.

SOLUTION: $\left|\dfrac{z-i}{z+i}\right| \le 2$, to find the locus of '$z$'.

$\Rightarrow \qquad |z-i| \le 2\,|z+i|$ hence $|x+i(y-1)| \le 2\,|x+i+iy|$

or $\qquad x^2+(y-1)^2 \le 4(y+1)^2+4x^2$

or $\qquad 3x^2+3y^2+x+10y+3 \ge 0$

or $\qquad x^2+y^2+\dfrac{10}{3}y+1 \ge 0$ which is the region outside the circle with centre at $\left(0, \dfrac{-5}{3}\right)$ and radius = 4/3

Ans.

8. *If z_1, z_2, z_3 are the vertices of an isoscles triangle, right angled at z_2, show that* $z_1^2 + 2z_2^2 + z_3^2 = 2z_2(z_1 + z_3)$.

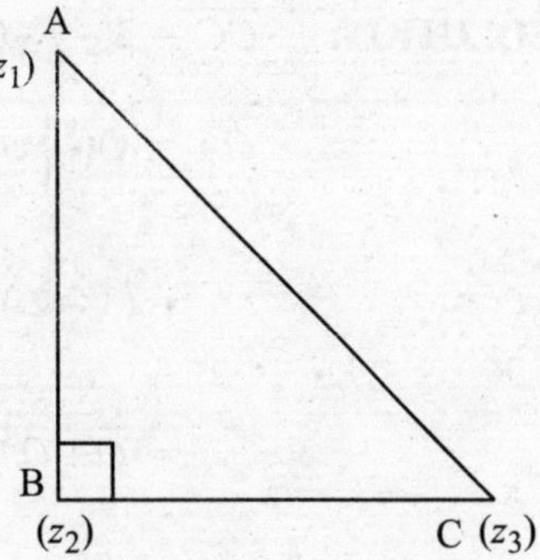

SOLUTION: $|z_1 - z_2| = |z_3 - z_2|$ given, and

$$\arg \frac{z_1 - z_2}{z_3 - z_2} = \frac{\pi}{2}$$

$$\therefore \quad \frac{z_1 - z_2}{z_3 - z_2} = 1\left(\cos\frac{\pi}{2} + i\sin\frac{\pi}{2}\right) = i$$

or $(z_1 - z_2)^2 = i^2(z_3 - z_2)^2$

or $z_1^2 + z_2^2 - 2z_1z_2 = -z_3^2 - z_2^2 + 2z_2z_3$

or $z_1^2 + 2z_2^2 + z_3^2 = 2z_2(z_1 + z_3)$ **Hence Proved.**

9. *Find the locus of the point 'z' in the complex plane such that*

(*i*) $|z - 2i| = |z + 3|$ (*ii*) $|3z - 1| = |z - 3|$.

SOLUTION: (*i*) $|z - 2i| = |z + 3|$ Hence z is equidistant from (0, 2) and (– 3, 0). Therefore the locus of 'z' is the right bisector of the line segment joining (0, 2) and (–3, 0).

(*ii*) $|3z - 1| = |z - 3|$ Here $z = x + iy$.

$\Rightarrow$ $|3x - 1 + 3iy| = |x - 3 + iy|$

or $(3x - 1)^2 + 9y^2 = (x - 3)^2 + y^2$ or $8x^2 + 8y^2 - 6x + 6x = 9 - 1$

or $x^2 + y^2 = 1$ which is circle. **Ans.**

10. *If $z = x + iy$ and x and y are real, find the locus of 'z', if*

(*i*) $\dfrac{z+i}{z+2}$ is real (*ii*) $\dfrac{z+i}{z+2}$ is imaginary.

SOLUTION: (*i*) $$\frac{z+i}{z+2} = \frac{x + (y+1)i}{x + 2 + iy} = \frac{[x + (y+1)i][(x+2) - iy]}{(x+2)^2 + y^2}$$

$$= \frac{x(x+2) + y(y+1) + i[(x+2)(y+1) - xy]}{(x+2)^2 + y^2}.$$

For this to be real $(x + 2)(y + 1) - xy = 0$ or $x + 2y + 2 = 0$

Hence locus of 'z' is the line $x + 2y + 2 = 0$ **Ans.**

(*ii*) For $\dfrac{z+i}{z+2}$ to be imaginary, $x(x + 2) + y(y + 1) = 0$

or $x^2 + y^2 + 2x + y = 0$

Hence locus of z is the circle with centre $\left(-1, \dfrac{-1}{2}\right)$ and radius $= \sqrt{5}/2$ **Ans.**

11. *The centre of a regular hexagon is at the point '2 – i' and one of the vertices is at '– 1+ i'. Find the two adjacent vertices.*

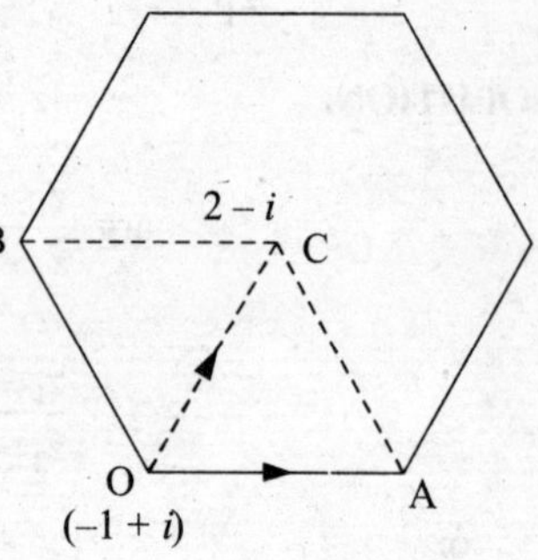

SOLUTION: $\overrightarrow{OC} = 2 - i - (-1 + i) = 3 - 2i$

$$\overrightarrow{OA} = \overrightarrow{OC}\left(\cos\frac{\pi}{3} - i\sin\left(\frac{\pi}{3}\right)\right)$$

$$= (3-2i)\left(\frac{1}{2} - \frac{i\sqrt{3}}{2}\right) = \frac{3}{2} - \sqrt{3} - i\left(1 + \frac{3\sqrt{3}}{2}\right)$$

$$\therefore \quad A = \vec{O} + \overrightarrow{OA} = \frac{3}{2} - \sqrt{3} - i\left(1 + \frac{3\sqrt{3}}{2}\right) - 1 + i$$

$$= \frac{1}{2} - \sqrt{3} - \frac{i\,3\sqrt{3}}{2}$$

$$\overrightarrow{OB} = \overrightarrow{OC}\left(\cos\frac{\pi}{3} + i\sin\frac{\pi}{3}\right) = (3-2i)\left(\frac{1}{2} + \frac{i\sqrt{3}}{2}\right) = \frac{3}{2} + \sqrt{3} + i\left(\frac{3\sqrt{3}}{2} - 1\right)$$

$$\therefore \quad \vec{B} = \vec{O} + \overrightarrow{OB} = -1 + i + \frac{3}{2} + \sqrt{3} + i\left(\frac{3\sqrt{3}}{2} - 1\right) = \frac{1}{2} + \sqrt{3} + i\frac{3\sqrt{3}}{2}$$ **Ans.**

12. *If the complex numbers $9 + 12i$ and $-5 + 10i$ reprcsent two opposite vertices of a square, find the complex numbers representing the other two vertices.*

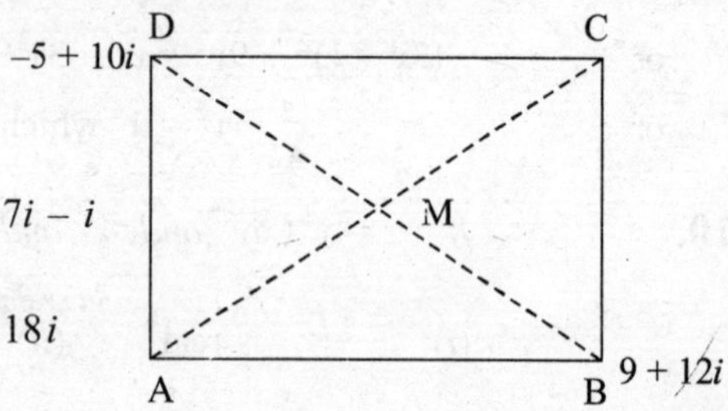

SOLUTION: Centre $M = 2 + 11i$

$\therefore \quad \overrightarrow{MB} = 9 + 12i - (2 + 11i) = 7 + i$

$\therefore \quad \overrightarrow{MC} = \overrightarrow{MB}\left(\cos\frac{\pi}{2} + i\sin\frac{\pi}{2}\right) = (7 + i)\,i = 7i - i$

$\therefore \quad \vec{C} = \overrightarrow{MC} + \overrightarrow{M} = 7i - 1 + 2 + 11i = +1 + 18i$

M is the mid point of AC $\therefore \quad A = 3 + 4i$

$\therefore$ Other vertices are $3 + 4i$ and $1 + 18i$. **Ans.**

13. *Show that, if* $\arg\left[\frac{(z_1 - z_3)(z_2 - z_4)}{(z_1 - z_4)(z_2 - z_3)}\right] = 0$, *the four points* z_1, z_2, z_3, z_4 *in the Argand diagram lie on a circle or on a straight line.*

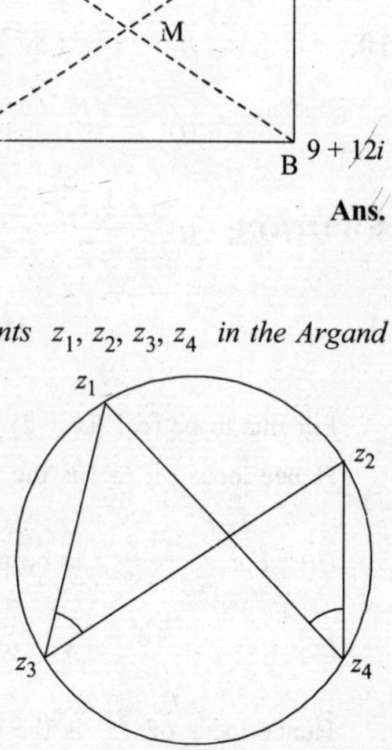

SOLUTION:
$$\arg\frac{(z_1 - z_3)(z_2 - z_4)}{(z_1 - z_4)(z_2 - z_3)} = 0$$

$$\Rightarrow \quad \arg\left(\frac{z_1 - z_3}{z_2 - z_3}\right) + \arg\left(\frac{z_2 - z_4}{z_1 - z_4}\right) = 0$$

$$\Rightarrow \quad \arg\left(\frac{z_1 - z_3}{z_2 - z_3}\right) = \arg\left(\frac{z_1 - z_4}{z_2 - z_4}\right)$$

Hence z_1, z_2, z_3, z_4 are concylic otherwise they lie on a straight line. **Hence Proved.**

14. *Find* $|z|$ *and* $\arg(z)$ *when* $z = \dfrac{(2-3i)(1+i)}{2+i}$ **(GGSIPU I Sem End Term 2008)**

SOLUTION: $|z| = \dfrac{|2-3i||1+i|}{|2+i|} = \dfrac{\sqrt{13}\sqrt{2}}{\sqrt{5}} = \sqrt{26/5}$

and $\arg(z) = \arg(2-3i) + \arg(1+i) - \arg(2+i) = \tan^{-1}\left(\dfrac{-3}{2}\right) + \tan^{-1} 1 - \tan^{-1}\dfrac{1}{2}$

$$= \tan^{-1} 1 - \tan^{-1} 8 = \tan^{-1}\left(\frac{-7}{9}\right)$$

$\therefore$ $|z| = \sqrt{\dfrac{26}{5}}$ and $\arg(z) = \tan^{-1}\left(\dfrac{-7}{9}\right)$ **Ans.**

SOLUTIONS TO EXERCISE 1B

1. *Show that* $(a+ib)^{m/n} + (a-ib)^{m/n} = 2(a^2+b^2)^{m/2n}\cos\left(\dfrac{m}{n}\tan^{-1}\dfrac{b}{a}\right)$.

SOLUTION: Let us write $a = r\cos\theta$, $b = r\sin\theta$ where $r = \sqrt{a^2+b^2}$ and $\theta = \tan^{-1} b/a$ then

$$(a+ib)^{m/n} + (a-ib)^{m/n} = r^{m/n}\left[(\cos\theta + i\sin\theta)^{m/n} + (\cos\theta - i\sin\theta)^{m/n}\right]$$

$$= r^{m/n}\left[\left(\cos\frac{m}{n}\theta + i\sin\frac{m}{n}\theta\right) + \left(\cos\frac{m}{n}\theta - i\sin\frac{m}{n}\theta\right)\right]$$

$$= 2r^{m/n}\cos\frac{m}{n}\theta = 2(a^2+b^2)^{m/2n}\cos\left[\frac{m}{n}\tan^{-1}\frac{b}{a}\right].$$ **Hence Proved.**

2. *If* $2\cos\theta = x + \dfrac{1}{x}$ *find the values of*

(i) $x^n + \dfrac{1}{x^n}$ *(ii)* $\dfrac{x^{2n}+1}{x^{2n-1}+x}$.

SOLUTION: Given that $2\cos\theta = x + \dfrac{1}{x}$ or $x^2 - 2x\cos\theta + 1 = 0$

$$\Rightarrow \quad x = \frac{2\cos\theta \pm \sqrt{4\cos^2\theta - 4}}{2} = \cos\theta \pm i\sin\theta$$

(i) $x^n + \dfrac{1}{x^n} = (\cos\theta + i\sin\theta)^n + (\cos\theta + i\sin\theta)^{-n}$

$= \cos n\theta + i\sin n\theta + \cos n\theta - i\sin n\theta = 2\cos n\theta$ **Ans.**

(ii) $\dfrac{x^{2n}+1}{x^{2n-1}+x} = \dfrac{x^n + \dfrac{1}{x^n}}{x^{n-1} + \dfrac{1}{x^{n-1}}} = \dfrac{2\cos n\theta}{2\cos(n-1)\theta} = \dfrac{\cos n\theta}{\cos(n-1)\theta}$ **Ans.**

3. *Simplify* $\dfrac{(\cos 2\theta - i\sin 2\theta)^7 (\cos 3\theta + i\sin 3\theta)^{-5}}{(\cos 4\theta - i\sin 4\theta)^{-4} (\cos 6\theta + i\sin 6\theta)^6}$.

SOLUTION: $$\frac{(\cos 2\theta - i\sin 2\theta)^7 (\cos 3\theta + i\sin 3\theta)^{-5}}{(\cos 4\theta - i\sin 4\theta)^{-4} (\cos 6\theta + i\sin 6\theta)^6}$$

$$= \frac{(\cos\theta + i\sin\theta)^{-14} (\cos\theta + i\sin\theta)^{-15}}{(\cos\theta + i\sin\theta)^{16} (\cos\theta + i\sin\theta)^{36}} = \frac{(\cos\theta + i\sin\theta)^{-29}}{(\cos\theta + i\sin\theta)^{52}}$$

$$= (\cos\theta + i\sin\theta)^{-81} = \cos 81\theta - i\sin 81\theta.$$ **Ans.**

4. If $x_r = \cos\dfrac{\pi}{2^r} + i\sin\dfrac{\pi}{2^r}$, show that $x_1 x_2 x_3 \ldots \infty = -1$.

(GGSIPU I Sem III Term Spl. 2007 – 2.5 Marks)

SOLUTION: $x_r = \cos\dfrac{\pi}{2^r} + i\sin\dfrac{\pi}{2^r} \quad r = 1, 2, 3, \ldots$

$$\therefore \quad x_1 x_2 x_3 \ldots \infty = \left(\cos\frac{\pi}{2} + i\sin\frac{\pi}{2}\right)\left(\cos\frac{\pi}{2^2} + i\sin\frac{\pi}{2^2}\right)\left(\cos\frac{\pi}{2^3} + i\sin\frac{\pi}{2^3}\right)\ldots\infty$$

$$= \cos\left(\frac{\pi}{2} + \frac{\pi}{2^2} + \frac{\pi}{2^3} + \ldots\right) + i\sin\left(\frac{\pi}{2} + \frac{\pi}{2^2} + \frac{\pi}{2^3} + \ldots\right)$$

$$= \cos\left(\frac{\pi/2}{1-1/2}\right) + i\sin\left(\frac{\pi/2}{1-1/2}\right) = \cos\pi + i\sin\pi = -1$$ **Hence Proved.**

5. *Show that the values of* $(1+i\sqrt{3})^{1/3}$ *are given by*

$$2^{1/3}\left(\cos\frac{r\pi}{9} + i\sin\frac{r\pi}{9}\right), \quad r = 1, 7, 13.$$

SOLUTION: $$(1 + i\sqrt{3})^{1/3} = 2^{1/3}\left(\frac{1}{2} + \frac{i\sqrt{3}}{2}\right)^{1/3} = 2^{1/3}\left(\cos\frac{\pi}{3} + i\sin\frac{\pi}{3}\right)^{1/3}$$

$$= 2^{1/3}\left[\cos\left(2r\pi + \frac{\pi}{3}\right) + i\sin\left(2r\pi + \frac{\pi}{3}\right)\right]^{1/3}$$

$$= 2^{1/3}\left[\cos\left(\frac{2r\pi}{3} + \frac{\pi}{9}\right) + i\sin\left(\frac{2r\pi}{3} + \frac{\pi}{9}\right)\right]$$

$$= 2^{1/3}\left[\cos(6r+1)\frac{\pi}{9} + i\sin(6r+1)\frac{\pi}{9}\right], \quad r = 0\ 1, 2, \ldots$$

(Putting $6r + 1 = n$ we get)

$$= 2^{1/3}\left[\cos\frac{n\pi}{9} + i\sin\frac{n\pi}{9}\right], \quad n = 1, 7, 13, \ldots$$ **Hence Proved.**

6. *Find the general value of* θ *satisfying the equation*

$$(\cos\theta + i\sin\theta)(\cos 2\theta + i\sin 2\theta)(\cos 3\theta + i\sin 3\theta)\ldots(\cos n\theta + i\sin n\theta) = 1.$$

SOLUTION: $(\cos\theta + i\sin\theta)(\cos 2\theta + i\sin 2\theta)(\cos 3\theta + i\sin 3\theta)\ldots(\cos n\theta + i\sin n\theta) = 1$

$\Rightarrow$ $\cos(\theta + 2\theta + 3\theta + \ldots + n\theta) + i\sin(\theta + 2\theta + 3\theta + \ldots + n\theta) = 1$

or $\cos\dfrac{n(n+1)\theta}{2} + i\sin\dfrac{n(n+1)\theta}{2} = 1$

$\Rightarrow$ $\cos\dfrac{n(n+1)\theta}{2} = 1$ and $\sin\dfrac{n(n+1)\theta}{2} = 0$

$\therefore$ $\dfrac{n(n+1)\theta}{2} = 2k\pi$, where k is an integer.

or $\theta = \dfrac{4k\pi}{n(n+1)}$ **Ans.**

7. *Solve the equation* $x^4 - x^3 + x^2 - x + 1 = 0$

SOLUTION: $x^4 - x^3 + x^2 - x + 1 = 0$

or $\dfrac{x^5+1}{x+1} = 0 \Rightarrow x^5 + 1 = 0$ where $x \neq -1$

$$x = (-1)^{1/5} = [\cos(2n+1)\pi + i\sin(2n+1)\pi)^{1/5}$$

$$= \cos\frac{(2n+1)\pi}{5} + i\sin\frac{(2n+1)\pi}{5}, \quad n = 0, 1, 2, 3, 4.$$

or $x = \cos\dfrac{\pi}{5} + i\sin\dfrac{\pi}{5}, \cos\dfrac{3\pi}{5} + i\sin\dfrac{3\pi}{5}, -1, \cos\dfrac{7\pi}{5} + i\sin\dfrac{7\pi}{9}, \cos\dfrac{9\pi}{5} + i\sin\dfrac{9\pi}{5}$

Deleting $x = -1$, we have $x = \cos\dfrac{\pi}{5} \pm i\sin\dfrac{\pi}{5}$ and $\cos\dfrac{3\pi}{5} \pm i\sin\dfrac{3\pi}{5}$. **Ans.**

8. *If* α *denotes any* nth *root of unity then show that* $1 + \alpha + \alpha^2 + \ldots + \alpha^{n-1} = 0$.

SOLUTION: $x^n = 1$ has one root $x = \alpha$, hence $\alpha^n - 1 = 0$

or $(\alpha - 1)(\alpha^{n-1} + \alpha^{n-2} + \ldots + 1) = 0, \quad \alpha \neq 1$

$\therefore$ $\alpha^{n-1} + \alpha^{n-2} + \ldots + \alpha + 1 = 0.$ **Hence Proved.**

9. *If* $x = \cos\theta + i\sin\theta$ *and* $\sqrt{1-c^2} = nc - 1$ *show that*

$$\frac{c}{2n}(1 + nx)\left(1 + \frac{n}{x}\right) = 1 + c\cos\theta.$$

SOLUTION: $x = \cos\theta + i\sin\theta$ and $\sqrt{1-c^2} = nc - 1$ or $1 - c^2 = n^2c^2 + 1 - 2nc$

or $n^2c^2 - 2nc + c^2 = 0$ or $c = \dfrac{2n}{n^2+1}$

Now $$\frac{c}{2n}(1+nx)\left(1+\frac{n}{x}\right)=\frac{c}{2n}\left[1+n\left(x+\frac{1}{x}\right)+n^2\right]$$

$$=\frac{c}{2n}[1+2n\cos\theta+n^2]=\frac{c}{2n}\left[2n\cos\theta+\frac{2n}{c}\right]$$

$$=c\left[\cos\theta+\frac{1}{c}\right]=1+c\cos\theta.$$ **Hence Proved.**

10. *If* $(1+x)^n=p_0+p_1x+p_2x^2+\ldots+p_nx^n$ *where* n *is a postive integer, find the values of*

(*i*) $p_0-p_2+p_4-p_6+\ldots$ (*ii*) $p_0+p_4+p_8+p_{12}+\ldots$

SOLUTION: $$(1+x)^n=p_0+p_1x+p_2x^2+\ldots+p_nx^n \qquad \ldots(1)$$

Putting $x=1$, we get $2^n=p_0+p_1+p_2+p_3+\ldots+p_n$

Putting $x=-1$ gives $0=p_0-p_1+p_2-p_3+p_4-\ldots+(-1)^np_n$

On adding these, we get $2^n+0=2p_0+2p_2+2p_4+\ldots$

or $$p_0+p_2+p_4+p_6+\ldots=2^{n-1} \qquad \ldots(2)$$

Putting $x=i$ in (1), gives

$$(1+i)^n=p_0+p_1i-p_2-p_3i+p_4+ip_5-p_6\ldots+i^np_n$$

$$\Rightarrow \quad p_0-p_2+p_4-p_6+\ldots=\text{Real part of }(1+i)^n=\text{Re }2^{n/2}\left(\frac{1}{\sqrt{2}}+\frac{i}{\sqrt{2}}\right)^n$$

(*i*) $$=\text{Re }2^{n/2}\left(\cos\frac{n\pi}{4}+i\sin\frac{n\pi}{4}\right)=2^{n/2}\cos\frac{n\pi}{4} \qquad \textbf{Ans.} \qquad \ldots(3)$$

Adding (2) and (3) gives

$$2p_0+2p_4+2p_8+\ldots=2^{n-1}+2^{n/2}\cos\frac{n\pi}{4}$$

(*ii*) $$p_0+p_4+p_8+\ldots=2^{n-2}+2^{\frac{n-2}{2}}\cos\frac{n\pi}{4}$$ **Ans.**

11. *Show that the sum of the infinite series*

$$1-\frac{1}{2}\cdot\frac{3}{4}+\frac{1}{2}\cdot\frac{3}{4}\cdot\frac{5}{6}\cdot\frac{7}{8}-\frac{1}{2}\cdot\frac{3}{4}\cdot\frac{5}{6}\cdot\frac{7}{8}\cdot\frac{9}{10}\cdot\frac{11}{12}+\ldots,\ \text{is}\ \frac{1}{2}\left(1+\sqrt{2}\right)^{1/2}.$$

SOLUTION: Consider the expansion of $(1+i)^{-1/2}$

$$(1+i)^{-1/2}=1+\left(\frac{-1}{2}\right)i+\frac{\left(\frac{-1}{2}\right)\left(\frac{-3}{2}\right)i^2}{2!}+\frac{\left(\frac{-1}{2}\right)\left(\frac{-3}{2}\right)\left(\frac{-5}{2}\right)i^3}{3!}+\ldots$$

$$=\left[1-\frac{1}{2}\cdot\frac{3}{4}+\frac{1}{2}\cdot\frac{3}{4}\cdot\frac{5}{6}\cdot\frac{7}{8}-\ldots\right]+i\left[-\frac{1}{2}+\frac{1}{2}\cdot\frac{3}{4}\cdot\frac{5}{6}-\ldots\right]$$

But $$(1+i)^{-1/2} = \left(\sqrt{2}\right)^{-1/2}\left(\frac{1}{\sqrt{2}} + \frac{i}{\sqrt{2}}\right)^{-1/2} = 2^{-1/4}\left(\cos\frac{\pi}{4} + i\sin\frac{\pi}{4}\right)^{-1/2}$$

$$= 2^{-1/4}\left(\cos\frac{\pi}{8} - i\sin\frac{\pi}{8}\right)$$

Equating real parts on both sides, we get

$$1 - \frac{1}{2}\cdot\frac{3}{4} + \frac{1}{2}\cdot\frac{3}{4}\cdot\frac{5}{6}\cdot\frac{7}{8} - ... = 2^{-1/4}\cos\frac{\pi}{8} = \left[2^{-1/2}\cos^2\frac{\pi}{8}\right]^{1/2}$$

$$= \left[\frac{1}{\sqrt{2}}\,\frac{1+\cos\pi/4}{2}\right]^{1/2} = \left[\frac{1}{2\sqrt{2}}\left(1+\frac{i}{\sqrt{2}}\right)\right]^{1/2}$$

$$= \frac{1}{2}\left(\sqrt{2}+1\right)^{1/2}$$

Hence Proved.

12. *Show that the roots of the equation* $(1+x)^{2n} + (1-x)^{2n} = 0$ *are given by*

$$\pm\ i\tan\frac{(2k-1)\pi}{4n}, \quad k = 1, 2, 3, ..., n.$$

SOLUTION: $(1+x)^{2n} + (1-x)^{2n} = 0$ or $\left(\dfrac{1+x}{1-x}\right)^{2n} = -1$ or $\dfrac{1+x}{1-x} = (-1)^{1/2n}$

or $$\frac{1+x}{1-x} = (\cos\pi + i\sin\pi)^{1/2n} = \cos\frac{(2m+1)\pi}{2n} + i\sin\frac{(2m+1)\pi}{2n}$$

$m = 0, 1, 2, 3, ..., 2n.$

Applying Componendo and dividendo, we get

$$\frac{2x}{2} = \frac{\cos(2m+1)\frac{\pi}{2n} - 1 + i\sin(2m+1)\frac{\pi}{2n}}{\cos(2m+1)\frac{\pi}{2n} + 1 + i\sin(2m+1)\frac{\pi}{2n}}$$

$$= \frac{-2\sin^2(2m+1)\frac{\pi}{4n} + 2i\sin(2m+1)\frac{\pi}{4n}\cos(2m+1)\frac{\pi}{4n}}{2\cos^2(2m+1)\frac{\pi}{4n} + 2i\sin(2m+1)\frac{\pi}{4n}\cos(2m+1)\frac{\pi}{4n}}$$

or $$x = \frac{i\sin(2m+1)\frac{\pi}{4n}}{\cos(2m+1)\frac{\pi}{4n}} = i\tan(2m+1)\frac{\pi}{4n}, \ m = 0, 1, 2, ... 2n-1$$

$$= i\tan(2r-1)\frac{\pi}{4n}, \ r = 1, 2, ... , n, n+1, n+2, ..., 2n.$$

When $r = 2n$, $x = i\tan(4n-1)\dfrac{\pi}{4n} = i\tan\left(\pi - \dfrac{\pi}{4n}\right) = -\,i\tan\dfrac{\pi}{4n}$

and when $r = 2n-1$, x has same value as for $r = 2$ but with – sign. and so on.

Hence $x = \pm\ i\tan(2r-1)\dfrac{\pi}{4n}$, $r = 1, 2, ..., n$ **Hence the result.**

13. *If* $\sin\alpha + 2\sin\beta + 3\sin\gamma = 0 = \cos\alpha + 2\cos\beta + 3\cos\gamma$, *show that*

$\sin 3\alpha + 8\sin 3\beta + 27\sin 3\gamma = 18\sin(\alpha+\beta+\gamma)$

and $\cos 3\alpha + 8\cos 3\beta + 27\cos 3\gamma = 18\cos(\alpha+\beta+\gamma)$.

SOLUTION: Let $a = \cos\alpha + i\sin\alpha,\ b = 2(\cos\beta + i\sin\beta),\ c = 3(\cos\gamma + i\sin\gamma)$

$\Rightarrow$ $a+b+c = 0$ hence $a^3+b^3+c^3 = 3\,abc$

Thus $(\cos\alpha + i\sin\alpha)^3 + 8(\cos\beta + i\sin\beta)^3 + 27(\cos\gamma + i\sin\gamma)^3$

$$= 3\cdot 2\cdot 3(\cos\alpha + i\sin\alpha)(\cos\beta + i\sin\beta)(\cos\gamma + i\sin\gamma)$$

or $(\cos 3\alpha + i\sin 3\alpha) + 8(\cos 3\beta + i\sin 3\beta) + 27(\cos 3\gamma + i\sin 3\gamma)$

$$= 18[\cos(\alpha+\beta+\gamma) + i\sin(\alpha+\beta+\gamma)]$$

$\Rightarrow$ $\cos 3\alpha + 8\cos 3\beta + 27\cos 3\gamma = 18\cos(\alpha+\beta+\gamma)$

and $\sin 3\alpha + 8\sin 3\beta + 27\sin 3\gamma = 18\sin(\alpha+\beta+\gamma)$ **Hence the result.**

14. *Find the equation whose roots are* $2\cos\frac{\pi}{7}, 2\cos\frac{3\pi}{7}, 2\cos\frac{5\pi}{7}$.

SOLUTION: Let $z = \cos\theta + i\sin\theta,\ \frac{1}{z} = \cos\theta - i\sin\theta \quad \therefore\ z + \frac{1}{z} = 2\cos\theta$

Consider $z^7 = -1$ or $(z+1)(z^6 - z^5 + z^4 - z^3 + z^2 - z + 1) = 0$

$$z = (-1)^{1/7} = \cos(2n+1)\frac{\pi}{7} + i\sin(2n+1)\frac{\pi}{7} \qquad n = 0, 1, 2, ..., 6$$

$$= \cos\frac{\pi}{7} \pm i\sin\frac{\pi}{7},\ \cos\frac{3\pi}{7} \pm i\sin\frac{3\pi}{7},\ \cos\frac{5\pi}{7} \pm i\sin\frac{5\pi}{7} \quad \text{and} \quad z = -1.$$

$\therefore$ $z + \frac{1}{z} = 2\cos\frac{\pi}{7},\ 2\cos\frac{3\pi}{7},\ 2\cos\frac{5\pi}{7}$

Consider $z^6 - z^5 + z^4 - z^3 + z^2 - z + 1 = 0$

or $z^3 + \frac{1}{z^3} - \left(z^2 + \frac{1}{z^2}\right) + \left(z + \frac{1}{z}\right) - 1 = 0$

or $(x^3 - 3x) - (x^2 - 2) + x - 1 = 0$ where $x = z + \frac{1}{z}$

or $x^3 - x^2 - 2x + 1 = 0$ is the required equation whose roots are $2\cos\frac{\pi}{7}, 2\cos\frac{3\pi}{7}, 2\cos\frac{5\pi}{7}$. **Ans.**

15. *Show that* z_1, z_2, z_3 *are the vertices of an equilateral triangle if and only if*

$$\frac{z_3 - z_1}{z_2 - z_1} = \frac{z_1 - z_2}{z_3 - z_2}.$$

SOLUTION: $\frac{z_3 - z_1}{z_2 - z_1} = \frac{|z_3 - z_1|}{|z_2 - z_1|} \arg\frac{z_3 - z_1}{z_2 - z_1} = 1 \arg\left(\frac{z_3 - z_1}{z_2 - z_1}\right)$

$$\frac{z_1 - z_2}{z_3 - z_2} = \frac{|z_1 - z_2|}{|z_3 - z_2|} \arg\left(\frac{z_1 - z_2}{z_3 - z_2}\right) = 1 \arg\left(\frac{z_1 - z_2}{z_3 - z_2}\right)$$

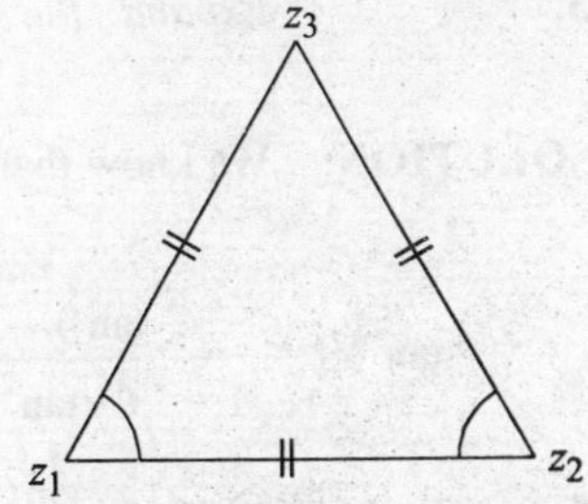

Since $\frac{z_3 - z_1}{z_2 - z_1} = \frac{z_1 - z_2}{z_3 - z_2}$ we have $\arg\frac{z_3 - z_1}{z_2 - z_1} = \arg\frac{z_1 - z_2}{z_3 - z_2}$

Hence it is an equilateral triangle. **Ans.**

16. *Prove that* $(1 + i)^n + (1 - i)^n = 2^{\frac{n}{2}+1} \cos\frac{n\pi}{4}$. **(GGSIPU I Sem End Term 2008)**

SOLUTION: We put $1 = r\cos\theta$ and $1 = r\sin\theta$ then

$$\text{L.H.S.} = r^n(\cos\theta + i\sin\theta)^n + r^n(\cos\theta - i\sin\theta)^n = 2r^n\cos n\theta$$

Here $r = \sqrt{2}$ and $\theta = \tan^{-1} 1 = \frac{\pi}{4}$ $\therefore$ L.H.S $= 2(\sqrt{2})^n \cos\frac{n\pi}{4} = 2^{\frac{n}{2}+1}\cos\frac{n\pi}{4}$

Hence Proved.

SOLUTIONS TO EXERCISE 1C

1. *Expand* $\sin 5\theta$ *in powers of* $\sin\theta$ *and* $\cos\theta$.

SOLUTION: By de Moivre's theorem.

$$\cos 5\theta + i\sin 5\theta = (\cos\theta + i\sin\theta)^5$$
$$= \cos^5\theta + {}^5C_1\cos^4\theta\,(i\sin\theta) + {}^5C_2\cos^3\theta\,(i\sin\theta)^2$$
$$+ {}^5C_3\cos^2\theta\,(i\sin\theta)^3 + {}^5C_4\cos\theta\,(i\sin\theta)^4 + (i\sin\theta)^5$$

Equating imaginary parts on both sides, we get

$$i\sin 5\theta = 5\,i\cos^4\theta\sin\theta - {}^5C_3\cos^2\theta\sin^3\theta + i\sin^5\theta$$

or $$\sin 5\theta = 5\cos^4\theta\sin\theta - 10\cos^2\theta\sin^3\theta + \sin^5\theta.$$ **Ans.**

2. *Show that* $\cos 7\theta = \cos^7\theta - 21\cos^5\theta\sin^2\theta + 35\cos^3\theta\sin^4\theta - 7\cos\theta\sin^6\theta$.

SOLUTION: $\cos 7\theta + i\sin 7\theta = (\cos\theta + i\sin\theta)^7 = \cos^7\theta + {}^7C_1\cos^6\theta\,(i\sin\theta) + {}^7C_2\cos^5\theta\,(i\sin\theta)^2$

$$+ {}^7C_3\cos^4\theta\,(i\sin\theta)^3 + {}^7C_4\cos^3\theta\,(i\sin\theta)^4 + {}^7C_5\cos^2\theta\,(i\sin\theta)^5$$
$$+ {}^7C_6\cos\theta\,(i\sin\theta)^6 + (i\sin\theta)^7$$

Equating real parts on both sides, we get

$$\cos 7\theta = \cos^7\theta - {}^7C_2\cos^5\theta\sin^2\theta + {}^7C_4\cos^3\theta\sin^4\theta + {}^7C_6\cos\theta\sin^6\theta$$
$$= \cos^7\theta - 21\cos^5\theta\sin^2\theta + 35\cos^3\theta\sin^4\theta - 7\cos\theta\sin^6\theta.$$ **Ans.**

3. *Expand $\tan 6\theta$ in powers of $\tan\theta$.*

SOLUTION: We know that $\tan n\theta = \dfrac{{}^nC_1 \tan\theta - {}^nC_3 \tan^3\theta + {}^nC_5 \tan^5\theta - ...}{1 - {}^nC_2 \tan^2\theta + {}^nC_4 \tan^4\theta - {}^nC_6 \tan^6\theta + ...}$

$$\therefore \quad \tan 6\theta = \frac{{}^6C_1 \tan\theta - {}^6C_3 \tan^3\theta + {}^6C_5 \tan^5\theta}{1 - {}^6C_2 \tan^2\theta + {}^6C_4 \tan^4\theta - \tan^6\theta} = \frac{6\tan\theta - 20\tan^3\theta + 6\tan^5\theta}{1 - 15\tan^2\theta + 15\tan^4\theta - \tan^6\theta}$$ **Ans.**

4. *Expand $\dfrac{\sin 6\theta}{\sin\theta}$ as a polynominal in $\cos\theta$.*

SOLUTION: $\cos 6\theta + i \sin 6\theta = (\cos\theta + i\sin\theta)^6$

$$= \cos^6\theta + {}^6C_1 \cos^5\theta\,(i\sin\theta) + {}^6C_2 \cos^4\theta\,(i\sin\theta)^2 + {}^6C_3 \cos^3\theta\,(i\sin\theta)^3$$
$$+ {}^6C_4 \cos^2\theta\,(i\sin\theta)^4 + {}^6C_5 \cos\theta\,(i\sin\theta)^5 + (i\sin\theta)^6$$

Equating imaginary parts on both sides, we get

$$\sin 6\theta = 6\cos^5\theta\sin\theta - 20\cos^3\theta\sin^3\theta + 6\cos\theta\sin^5\theta$$

$$\therefore \quad \frac{\sin 6\theta}{\sin\theta} = 6\cos^5\theta - 20\cos^3\theta\sin^2\theta + 6\cos\theta\sin^4\theta$$

$$= 6\cos^5\theta - 20\cos^3\theta\,(1 - \cos^2\theta) + 6\cos\theta\,(1 - \cos^2\theta)^2$$
$$= 6\cos^5\theta - 20\cos^3\theta + 20\cos^5\theta + 6\cos\theta - 12\cos^3\theta + 6\cos^5\theta$$
$$= 32\cos^5\theta - 32\cos^3\theta + 6\cos\theta.$$ **Ans.**

5. *If $\tan^{-1}a + \tan^{-1}b + \tan^{-1}c = \dfrac{\pi}{2}$, find the value of $ab + bc + ca$.*

SOLUTION: Given $\tan^{-1}a + \tan^{-1}b + \tan^{-1}c = \pi/2$

$$\Rightarrow \quad \tan^{-1}\frac{a+b}{1-ab} + \tan^{-1}c = \pi/2$$

$$\text{or} \quad \tan^{-1}\left[\frac{\left(\dfrac{a+b}{1-ab}\right) + c}{1 - \left(\dfrac{a+b}{1-ab}\right)c}\right] = \tan^{-1}\left(\frac{a + b + c(1-ab)}{1 - ab - (a+b)c}\right) = \frac{\pi}{2}$$

$$\Rightarrow \quad 1 - ab - (a+b)c = 0 \quad \therefore \quad ab + bc + ca = 1$$ **Ans.**

6. *Prove that $\dfrac{1+\cos 7\theta}{1+\cos\theta} = (x^3 - x^2 - 2x + 1)^2$ where $x = 2\cos\theta$.*

SOLUTION:
$$\frac{1+\cos 7\theta}{1+\cos\theta} = \frac{2\cos^2\dfrac{7\theta}{2}}{2\cos^2\dfrac{\theta}{2}} = \frac{2\cos^2\dfrac{7\theta}{2}}{2\cos^2\dfrac{\theta}{2}}\cdot\frac{2\sin^2\dfrac{\theta}{2}}{2\sin^2\dfrac{\theta}{2}} = \frac{\left(2\cos\dfrac{7\theta}{2}\sin\dfrac{\theta}{2}\right)^2}{\left(2\cos\dfrac{\theta}{2}\sin\dfrac{\theta}{2}\right)^2} = \left[\frac{\sin 4\theta - \sin 3\theta}{\sin\theta}\right]^2$$

Now, we know that $\sin 4\theta = {}^4C_1 \cos^3\theta\sin\theta - {}^4C_3 \cos\theta\sin^3\theta = 4\sin\theta\cos\theta\,(\cos^2\theta - \sin^2\theta)$
and $\sin 3\theta = 3\sin\theta - 4\sin^3\theta$

$$\therefore \quad \frac{\sin 4\theta - \sin 3\theta}{\sin\theta} = 4\cos\theta\,(\cos^2\theta - \sin^2\theta) - 3 + 4\sin^2\theta$$

$$= 4\cos\theta\,(2\cos^2\theta - 1) - 3 + 4 - 4\cos^2\theta$$

$$= 8\cos^3\theta - 4\cos^2\theta - 4\cos\theta + 1$$

$$\therefore \quad \left[\frac{1+\cos 7\theta}{1+\cos\theta}\right] = (8\cos^3\theta - 4\cos^2\theta - 4\cos\theta + 1)^2$$

$$= (x^3 - x^3 - 2x + 1)^2 \quad \text{where} \quad x = 2\cos\theta.$$ **Hence Proved.**

7. *Expand* $\cos^8\theta$ *in a series of cosines of multiples of* θ.

SOLUTION: We know that $\cos^n\theta = \frac{1}{2^{n-1}}\,[\cos n\theta + {}^nC_1 \cos(n-2)\theta + {}^nC_2\cos(n-4)\theta + \ldots]$

$$\therefore \quad \cos^8\theta = \frac{1}{2^7}\left[\cos 8\theta + {}^8C_1\cos 6\theta + {}^8C_2\cos 4\theta + {}^8C_3\cos 2\theta + \frac{1}{2}\,{}^8C_4\right]$$

$$= \frac{1}{128}\,[\cos 8\theta + 8\cos 6\theta + 28\cos 4\theta + 56\cos 2\theta + 35]$$ **Ans.**

8. *Expand* $\sin^7\theta$ *in a series of sines of multiples of* θ.

SOLUTION: We know that $\sin^n\theta = \frac{1}{2^{n-1}\,i^{n-1}}\,[\sin n\theta - {}^nC_1\sin(n-2)\theta + {}^nC_2\sin(n-4)\theta - \ldots]$

where n is odd integer.

$$\therefore \quad \sin^7\theta = \frac{1}{2^6 i^6}\,[\sin 7\theta - {}^7C_1\sin 5\theta + {}^7C_2\sin 3\theta - {}^7C_3\sin\theta]$$

$$= -\frac{1}{64}\,[\sin 7\theta - 7\sin 5\theta + 21\sin 3\theta - 35\sin\theta]$$

or $$\sin^7\theta = \frac{1}{64}\,[35\sin\theta - 21\sin 3\theta + 7\sin 5\theta - \sin 7\theta]$$ **Ans.**

9. *Expand* $\sin^5\theta\cos^2\theta$ *in a series of sines of multiples of* θ.

SOLUTION: $\sin^5\theta\cos^2\theta = \sin^5\theta - \sin^7\theta.$

Now $$\sin^5\theta = \frac{1}{2^4 i^4}\,[\sin 5\theta - {}^5C_1\sin 3\theta + {}^5C_2\sin\theta] = \frac{1}{16}\,[\sin 5\theta - 5\sin 3\theta + 10\sin\theta]$$

$$\therefore \quad \sin^5\theta\cos^2\theta = \frac{1}{16}\,[\sin 5\theta - 5\sin 3\theta + 10\sin\theta] - \frac{1}{64}\,[35\sin\theta - 21\sin 3\theta + 7\sin 5\theta - \sin 7\theta]$$

$$= \frac{1}{64}\,[\sin 7\theta - 3\sin 5\theta + \sin 3\theta + 5\sin\theta]$$ **Ans.**

10. *If* $\sin^4\theta \cos^3\theta = C_1\cos\theta + C_2 \cos 3\theta + C_3 \cos 5\theta + C_4 \cos 7\theta$, *find the value of* $C_1 + 9 \cdot C_2 + 25 \cdot C_3 + 49 \cdot C_4$.

SOLUTION: Given
$$\sin^4\theta \cos^3\theta = C_1 \cos\theta + C_2 \cos 3\theta + \cos 5\theta + C_4 \cos 7\theta$$
$$\sin^4\theta \cos^3\theta = (1 - \cos^2\theta)^2 \cos^3\theta = \cos^3\theta - 2\cos^5\theta + \cos^7\theta$$

We know that
$$\cos^n\theta = \frac{1}{2^{n-1}} [\cos n\theta + {}^nC_1 \cos(n-2)\theta + {}^nC_2 \cos(n-4)\theta + ...]$$

$\therefore$
$$\cos^3\theta = \frac{1}{2^2} [\cos 3\theta + {}^3C_1 \cos\theta] = \frac{1}{4} (\cos 3\theta + 3\cos\theta)$$
$$\cos^5\theta = \frac{1}{2^4} [\cos 5\theta + {}^5C_1 \cos 3\theta + {}^5C_2 \cos\theta]$$
$$= \frac{1}{16} (\cos 5\theta + 5\cos 3\theta + 10\cos\theta)$$

and
$$\cos^7\theta = \frac{1}{2^6} [\cos 7\theta + {}^7C_1 \cos 5\theta + {}^7C_2 \cos 3\theta + {}^7C_3 \cos\theta]$$
$$= \frac{1}{64} [\cos 7\theta + 7\cos 5\theta + 21\cos 3\theta + 35\cos\theta]$$

$\therefore$
$$\sin^4\theta \cos^3\theta = \frac{1}{4} (\cos 3\theta + 3\cos\theta) - \frac{2}{16} (\cos 5\theta + 5\cos 3\theta + 10\cos\theta)$$
$$+ \frac{1}{64} [\cos 7\theta + 7\cos 5\theta + 21\cos 3\theta + 35\cos\theta)$$
$$= \frac{1}{64} [\cos 7\theta - \cos 5\theta - 3\cos 3\theta + 3\cos\theta]$$

$\Rightarrow$
$$C_1 = \frac{3}{64}, C_2 = \frac{-3}{64}, C_3 = \frac{-1}{64} \text{ and } C_4 = \frac{1}{64}$$

$\therefore$
$$C_1 + 9C_2 + 25C_3 + 49C_4 = \frac{1}{64} [3 - 27 - 25 + 49] = 0$$
Ans.

11. *Using De Moivre's theorem prove that*
$$\cos^6\theta + \sin^6\theta = \frac{1}{8} (3\cos 4\theta + 5)$$

SOLUTION: To prove $\cos^6\theta + \sin^6\theta = \frac{1}{8} (3\cos 4\theta + 5)$ using De Moivre's theorem

$$\cos^6\theta = \frac{1}{2^5} (\cos 6\theta + {}^6C_1 \cos 4\theta + {}^6C_2 \cos 2\theta + \frac{1}{2}\, {}^6C_3 \cos 0)$$
$$= \frac{1}{32} (\cos 6\theta + 6\cos 4\theta + 15\cos 2\theta + 10)$$
$$\sin^6\theta = \frac{1}{2^5(-1)^{6/2}} [\cos 6\theta - {}^6C_1 \cos 4\theta + {}^6C_2 \cos 2\theta - {}^6C_3 \cos 0)$$
$$= \frac{-1}{32} [\cos 6\theta - 6\cos 4\theta + 15\cos 2\theta - 10]$$

$\therefore \quad \cos^6\theta + \sin^6\theta = \frac{1}{32}[0 + 12\cos 4\theta + 20] = \frac{1}{8}(3\cos 4\theta + 5).$

Aliter.

Let $x = \cos\theta + i\sin\theta,\ \frac{1}{x} = \cos\theta - i\sin\theta$ there $x + \frac{1}{x} = 2\cos\theta,\ x - \frac{1}{x} = 2i\sin\theta$

$$x^n + \frac{1}{x^n} = 2\cos n\theta \quad \text{and} \quad x^n - \frac{1}{x^n} = 2i\sin n\theta$$

Now $\quad 2^6\cos^6\theta = \left(x+\frac{1}{x}\right)^6 = x^6 + \frac{1}{x^6} + {}^6C_1\left(x^4+\frac{1}{x^4}\right) + {}^6C_2\left(x^2+\frac{1}{x^2}\right) + {}^6C_3$

$$= 2\cos 6\theta + 2\cdot 6\cos 4\theta + 2\cdot 15\cos 2\theta + 20$$

$\therefore \quad \cos^6\theta = \frac{1}{2^5}(\cos 6\theta + 6\cos 4\theta + 15\cos 2\theta + 10)$

Next $\quad (2i\sin\theta)^6 = \left(x-\frac{1}{x}\right)^6 = x^6 + \frac{1}{x^6} - {}^6C_1\left(x^4+\frac{1}{x^4}\right) + {}^6C_2\left(x^2+\frac{1}{x^2}\right) - {}^6C_3$

$$-2^6\sin^6\theta = 2\cos 6\theta - 2\cdot 6\cos 4\theta + 2\cdot 15\cos 2\theta - 20$$

$\therefore \quad \sin^6\theta = \frac{-1}{32}(\cos 6\theta - 6\cos 4\theta + 15\cos 2\theta - 10)$

$\therefore \quad \cos^6\theta + \sin^6\theta = \frac{1}{32}[0 + 12\cos 4\theta + 20] = \frac{1}{8}(3\cos 4\theta + 5).$ **Hence Proved.**

12. *If $\theta_1, \theta_2, \theta_3$, are theree values of θ which satisfy the relation $\tan 2\theta = \lambda\tan(\theta+\alpha)$ and are such that no two values of these differ by a multiple of π, prove that*

$$\theta_1 + \theta_2 + \theta_3 + \alpha \text{ is a multiple of } \pi.$$

SOLUTION: $\tan 2\theta = \lambda\tan(\theta+\alpha)$ is satisfied by $\theta = \theta_1, \theta_2, \theta_3$

$$\frac{2\tan\theta}{1-\tan^2\theta} = \frac{\lambda(\tan\theta + \tan\alpha)}{1-\tan\theta\tan\alpha}$$

or $\quad \lambda\tan^3\theta + (\lambda-2)\tan\alpha\tan^2\theta - (\lambda-2)\tan\theta - \lambda\tan\alpha = 0$

or $\quad \tan^3\theta + \frac{(\lambda-2)}{\lambda}\tan\alpha\tan^2\theta - \frac{\lambda-2}{\lambda}\tan\theta - \tan\alpha = 0 \quad$ ...(1)

Its roots are $\tan\theta_1$, $\tan\theta_2$ and $\tan\theta_3$.

We know that $\tan(\theta_1+\theta_2+\theta_3) = \frac{S_1 - S_3}{1-S_2}$

where $S_1 = \tan\theta_1 + \tan\theta_2 + \tan\theta_3$,

$S_2 = \tan\theta_1\tan\theta_2 + \tan\theta_2\tan\theta_3 + \tan\theta_3\tan\theta_1$ and $S_3 = \tan\theta_1\tan\theta_2\tan\theta_3$.

From equation (1) $S_1 = -\frac{(\lambda-2)}{\lambda}\tan\alpha,\ S_2 = \frac{-(\lambda-2)}{\lambda}$ and $S_3 = \tan\alpha$

$$\therefore \quad \tan(\theta_1+\theta_2+\theta_3) = -\frac{\frac{-(\lambda-2)}{\lambda}\tan\alpha - \tan\alpha}{1+\frac{\lambda-2}{\lambda}} = -\tan\alpha = \tan(-\alpha)$$

$\Rightarrow \quad \theta_1 + \theta_2 + \theta_3 + \alpha$ is a multiple of π. **Hence Proved.**

SOLUTIONS TO EXERCISE 1D

1. *If* $\tan\frac{x}{2} = \tanh\frac{u}{2}$, *show that*

(*i*) $\tan x = \sinh u$ and $\cos x \cosh u = 1$ (*ii*) $u = \log\tan\left(\frac{\pi}{4}+\frac{x}{2}\right)$.

SOLUTION: Given $\tan\frac{x}{2} = \tanh\frac{u}{2}$

(*i*) $$\tan x = \frac{2\tan\frac{x}{2}}{1-\tan^2\frac{x}{2}} = \frac{2\tanh\frac{u}{2}}{1-\tanh^2\frac{u}{2}} = \frac{2\tanh\frac{u}{2}}{\text{sech}^2\frac{u}{2}} = 2\sinh\frac{u}{2}\cosh\frac{u}{2} = \sinh u$$

and $$\cos x = \frac{\cos^2\frac{x}{2}-\sin^2\frac{x}{2}}{\cos^2\frac{x}{2}+\sin^2\frac{x}{2}} = \frac{1-\tan^2\frac{x}{2}}{1+\tan^2\frac{x}{2}} = \frac{1-\tanh^2\frac{u}{2}}{1+\tanh^2\frac{u}{2}}$$

$$= \frac{\cosh^2\frac{u}{2}-\sinh^2\frac{u}{2}}{\cosh^2\frac{u}{2}+\sinh^2\frac{u}{2}} = \frac{1}{\cosh u}$$

$\therefore$ $\cos x \cosh u = 1$.

(*ii*) $$\tanh\frac{u}{2} = \tan\frac{x}{2}$$

or $$\frac{e^{u/2}-e^{-u/2}}{e^{u/2}+e^{-u/2}} = \frac{\sin\frac{x}{2}}{\cos\frac{x}{2}}$$ Applying *c & d*, here we get

$$\frac{2e^{u/2}}{2e^{-u/2}} = \frac{\cos\frac{x}{2}+\sin\frac{x}{2}}{\cos\frac{x}{2}-\sin\frac{x}{2}} = \frac{1+\tan\frac{x}{2}}{1-\tan\frac{x}{2}} = \tan\left(\frac{\pi}{4}+\frac{x}{2}\right)$$

$\therefore$ $e^u = \tan\left(\frac{\pi}{4}+\frac{x}{2}\right)$ or $u = \log\tan\left(\frac{\pi}{4}+\frac{x}{2}\right)$. **Hence Proved.**

2. *Prove the following identities*

(*i*) $\sinh^{-1}(\tan\theta) = \log\tan\left(\frac{\pi}{4}+\frac{\theta}{2}\right)$ (*ii*) $\tanh^{-1}(\cos\theta) = \cosh^{-1}(\text{cosec}\,\theta)$.

SOLUTION: (*i*) We know that $\sinh^{-1}x = \log\left(x+\sqrt{1+x^2}\right)$

hence $\quad \sinh^{-1}(\tan\theta) = \log\left(\tan\theta + \sqrt{1+\tan^2\theta}\right) = \log(\tan\theta + \sec\theta)$

$$= \log\frac{\sin\theta+1}{\cos\theta} = \log\frac{1-\cos\left(\dfrac{\pi}{2}+\theta\right)}{\sin\left(\dfrac{\pi}{2}+\theta\right)}$$

$$= \log\frac{2\sin^2\left(\dfrac{\pi}{4}+\dfrac{\theta}{2}\right)}{2\sin\left(\dfrac{\pi}{4}+\dfrac{\theta}{2}\right)\cos\left(\dfrac{\pi}{4}+\dfrac{\theta}{2}\right)} = \log\frac{\sin\left(\dfrac{\pi}{4}+\dfrac{\theta}{2}\right)}{\cos\left(\dfrac{\pi}{4}+\dfrac{\theta}{2}\right)} = \log\tan\left(\frac{\pi}{4}+\frac{\theta}{2}\right).$$

(*ii*) We know that $\tanh^{-1}x = \dfrac{1}{2}\log\dfrac{1+x}{1-x}$

$$\therefore \quad \tanh^{-1}(\cos\theta) = \frac{1}{2}\log\left(\frac{1+\cos\theta}{1-\cos\theta}\right) = \frac{1}{2}\log\frac{2\cos^2\dfrac{\theta}{2}}{2\sin^2\dfrac{\theta}{2}} = \log\left(\cot\frac{\theta}{2}\right)$$

and since $\quad \cosh^{-1}x = \log\left(x+\sqrt{x^2-1}\right)$

we have $\cosh^{-1}(\operatorname{cosec}\theta) = \log\left(\operatorname{cosec}\theta + \sqrt{\operatorname{cosec}^2\theta - 1}\right)$

$$= \log(\operatorname{cosec}\theta + \cot\theta) = \log\frac{1+\cos\theta}{\sin\theta}$$

$$= \log\frac{2\cos^2\dfrac{\theta}{2}}{2\sin\dfrac{\theta}{2}\cos\dfrac{\theta}{2}} = \log\left(\cot\frac{\theta}{2}\right).$$

Hence Proved.

3. *Prove the following identities*

(*i*) $\sinh 3x = 3\sinh x + 4\sinh^3 x$ (*ii*) $\tanh 3x = \dfrac{3\tanh x + \tanh^3 x}{1+3\tanh^2 x}$

SOLUTION: (*i*) We know that $\sin 3x = 3\sin x - 4\sin^3 x$ Replacing x by ix we get

$$i\sinh 3x = 3i\sinh x - 4i^3\sinh^3 x \quad \text{or} \quad \sinh 3x = 3\sinh x + 4\sinh^3 x.$$

(*ii*) We know that $\tan 3x = \dfrac{3\tan x - \tan^3 x}{1-3\tan^2 x}$

Replacing x by ix throughout, we get

$$\tan 3ix = \frac{3\tan ix - \tan^3 ix}{1-3\tan^2 ix} \quad \text{or} \quad i\tanh 3x = \frac{3i\tanh x - i^3\tanh^3 x}{1-3i^2\tanh^2 x}$$

or $$\tanh 3x = \frac{3\tanh x + \tanh^3 x}{1+3\tanh^2 x}$$

Hence Proved.

4. *If* $\log \sin(\theta + i\phi) = x + iy$, *show that*

(*i*) $2\cos 2\theta = e^{2\phi} + e^{-2\phi} - 4e^{2x}$ (*ii*) $\cos(\theta - y) = e^{2\phi}\cos(\theta + y)$

SOLUTION: (*i*) $\log \sin(\theta + i\phi) = x + iy \Rightarrow \sin(\theta + i\phi) = e^{x+iy} = e^x e^{iy}$

or $\sin\theta(\cosh\phi) + \cos\theta(i\sinh\phi) = e^x(\cos y + i\sin y)$.

Equating real and imaginary parts on both sides, gives

$$\sin\theta \cosh\phi = e^x \cos y \quad ...(1)$$

and $$\cos\theta \sinh\phi = e^x \sin y \quad ...(2)$$

Eliminating y in (1) and (2), we get

$$e^{2x}(\cos^2 y + \sin^2 y) = \sin^2\theta \cosh^2\phi + \cos^2\theta \sinh^2\phi$$

or $$e^{2x} = \left(\frac{1-\cos 2\theta}{2}\right)\left(\frac{1+\cosh 2\phi}{2}\right) + \left(\frac{1+\cos 2\theta}{2}\right)\left(\frac{\cosh 2\phi - 1}{2}\right)$$

or $$4e^{2x} = 1 + \cosh 2\phi - \cos 2\theta - \cos 2\theta\cosh 2\phi + \cosh 2\phi - 1 + \cos 2\theta \cosh 2\phi - \cos 2\theta$$

$$= 2\cosh 2\phi - 2\cos 2\theta = e^{2\phi} + e^{-2\phi} - 2\cos 2\theta$$

or $$2\cos 2\theta = e^{2\phi} + e^{-2\phi} - 4e^{2x}$$ **Hence Proved.**

(*ii*) Eliminating x in (1) and (2) above, we get

$$\frac{\sin\theta\cosh\phi}{\cos\theta\sinh\phi} = \frac{\cos y}{\sin y} \quad \text{or} \quad \frac{\cosh\phi}{\sinh\phi} = \frac{\cos\theta\cos y}{\sin\theta\sin y} = \frac{e^{\phi}+e^{-\phi}}{e^{\phi}-e^{-\phi}}$$

Applying componendo and dividendo here, we get

$$\frac{2e^{\phi}}{2e^{-\phi}} = \frac{\cos\theta\cos y + \sin\theta\sin y}{\cos\theta\cos y - \sin\theta\sin y} \quad \text{or} \quad e^{2\phi} = \frac{\cos(\theta - y)}{\cos(\theta + y)}$$

or $$\cos(\theta - y) = e^{2\phi}\cos(\theta + y)$$ **Hence Proved.**

5. *If* $\sin(x + iy) = e^{i\alpha}$, *show that* $\cos^2 x = \sinh^2 y = \pm\sin\alpha$

SOLUTION: Given $\sin(x + iy) = e^{i\alpha}$

$\therefore$ $\sin x \cos iy + \cos x \sin iy = \cos\alpha + i\sin\alpha$

or $\sin x \cosh y + i\cos x \sinh y = \cos\alpha + i\sin\alpha$

$\Rightarrow$ $$\sin x \cosh y = \cos\alpha \quad ...(1)$$

and $$\cos x \sinh y = \sin\alpha \quad ...(2)$$

Eliminating y in (1) and (2), we get

$$\frac{\cos^2\alpha}{\sin^2 x} - \frac{\sin^2\alpha}{\cos^2 x} = 1 \quad \text{or} \quad \cos^2\alpha\cos^2 x - \sin^2 x\sin^2\alpha = \sin^2 x\cos^2 x$$

or $\cos^2 x(1 - \sin^2\alpha) - (1 - \cos^2 x)\sin^2\alpha = (1 - \cos^2 x)\cos^2 x$ or $\cos^4 x = \sin^2\alpha$

$\Rightarrow$ $\cos^2 x = \pm\sin\alpha$.

Equation (2) can be written as $\cos^2 x \sinh^2 y = \sin^2\alpha$,

hence $$\sinh^2 y = \frac{\sin^2\alpha}{\pm\sin\alpha} = \pm\sin\alpha$$ **Hence Proved.**

6. *Split into real and imaginary parts*

(*i*) $\cos(x+iy)$ (*ii*) $\operatorname{cosec}(x+iy)$

(*iii*) $\operatorname{sech}(\alpha+i\beta)$ (*iv*) $\tanh(\alpha+i\beta)$

SOLUTION: (*i*) $\cos(x+iy) = \cos x \cos iy - \sin x \sin iy = \cos x \cosh y - i \sin x \sinh y$

(*ii*) $$\operatorname{cosec}(x+iy) = \frac{1}{\sin(x+iy)} = \frac{1}{\sin x \cosh y + i\cos x \sinh y}$$

$$= \frac{\sin x \cosh y - i\cos x \sinh y}{\sin^2 x \cosh^2 y + \cos^2 x \sinh^2 y} = \frac{2\sin x \cosh y - 2i\cos x \sinh y}{\cosh 2y - \cos 2x}.$$

(*iii*) $$\operatorname{sech}(\alpha+i\beta) = \frac{1}{\cosh(\alpha+i\beta)} = \frac{1}{\cos i(\alpha+i\beta)} = \frac{1}{\cos(i\alpha-\beta)}$$

$$= \frac{1}{\cos i\alpha \cos\beta + \sin i\alpha \sin\beta} = \frac{1}{\cosh\alpha\cos\beta + i\sinh\alpha\sin\beta}$$

$$= \frac{\cosh\alpha\cos\beta - i\sinh\alpha\sin\beta}{\cosh^2\alpha\cos^2\beta + \sinh^2\alpha\sin^2\beta}$$

(*iv*) $$\tanh(\alpha+i\beta) = \frac{1}{i}\tan i(\alpha+i\beta) = -i\tan(i\alpha-\beta)$$

$$= \frac{-i(\tan i\alpha - \tan\beta)}{1+\tan i\alpha \tan\beta} = \frac{-i(i\tanh\alpha - \tan\beta)}{1+i\tanh\alpha\tan\beta} = \frac{\tanh\alpha + i\tan\beta}{1+i\tanh\alpha\tan\beta}$$

$$= \frac{(\tanh\alpha + i\tan\beta)(1-i\tanh\alpha\tan\beta)}{1+\tanh^2\alpha\tanh^2\beta}$$

$$= \frac{\tanh\alpha(1+\tan^2\beta) + i\tan\beta(1-\tanh^2\alpha)}{1+\tanh^2\alpha\tan^2\beta}$$

$$= \frac{\tanh\alpha\sec^2\beta + i\tan\beta\operatorname{sech}^2\alpha}{1+\tanh^2\alpha\tan^2\beta}$$ **Ans.**

7. *If* $\cos^{-1}(x+iy) = \alpha + i\beta$, *establish the following*

(*i*) $x^2\sec^2\alpha - y^2\operatorname{cosec}^2\alpha = 1$ (*ii*) $x^2\operatorname{sech}^2\beta + y^2\operatorname{cosech}^2\beta = 1.$

SOLUTION: $\cos^{-1}(x+iy) = \alpha + i\beta \Rightarrow x + iy = \cos(\alpha+i\beta) = \cos\alpha\cosh\beta - i\sin\alpha\sinh\beta.$

Equating real and imaginary parts on both sides, we get

$$\cos\alpha\cosh\beta = x \quad ...(1)$$

and $$\sin\alpha\sinh\beta = -y \quad ...(2)$$

Eliminating β in (1) and (2), gives

$$\frac{x^2}{\cos^2\alpha} - \frac{y^2}{\sin^2\alpha} = 1 \quad\text{or}\quad x^2\sec^2\alpha - y^2\operatorname{cosec}^2\alpha = 1$$

and eliminating α in (1) and (2), gives

$$\frac{x^2}{\cosh^2\beta} + \frac{y^2}{\sinh^2\beta} = 1 \quad\text{or}\quad x^2\operatorname{sech}^2\beta + y^2\operatorname{cosech}^2\beta = 1.$$ **Hence Proved.**

8. *Establish the following* $(\cosh\theta \pm \sinh\theta)^n = \cosh n\theta \pm \sinh n\theta$ *and hence, or otherwise, show that* $\left(\dfrac{1+\tanh\theta}{1-\tanh\theta}\right)^3 = \cosh 6\theta + \sinh 6\theta.$

SOLUTION: By de Moivre's theorem we can write

$$(\cos i\theta \pm i\sin i\theta)^n = \cos ni\theta \pm i\sin ni\theta$$

or $(\cosh\theta \mp \sinh\theta)^n = \cosh n\theta \mp \sinh n\theta$ **Hence the result.**

Now $$\left(\frac{1+\tanh\theta}{1-\tanh\theta}\right)^3 = \left(\frac{\cosh\theta+\sinh\theta}{\cosh\theta-\sinh\theta}\right)^3 = \left[\frac{(\cosh\theta+\sinh\theta)^2}{\cosh^2\theta-\sinh^2\theta}\right]^3$$

$$= \frac{(\cosh\theta+\sinh\theta)^6}{1} = \cosh 6\theta + \sinh 6\theta$$ **Hence Proved.**

9. *Find the principal value of* $\text{sech}^{-1}(\sin\theta)$ *and show that the general value of* $\sin^{-1}(\text{cosec}\,\theta)$ *can be written as:*

$$\{2n+(-1)^n\}\frac{\pi}{2}+(-1)^n i\log\cot\frac{\theta}{2}.$$

SOLUTION: Let $\text{sech}^{-1}(\sin\theta) = x+iy$ then $\sin\theta = \text{sech}(x+iy)$

or $\text{cosec}\,\theta = \cosh(x+iy)$

or $\text{cosec}\,\theta = \cos i(x+iy) = \cos(ix-y) = \cosh x\cos y + i\sinh x\sin y$

$\Rightarrow$ $\cosh x\cos y = \text{cosec}\,\theta$ and $\sinh x\sin y = 0$

Hence $\sin y = 0$ or $y = n\pi$ where $n\in I$ then $\cos y = 1$

Thus, $\cosh x = \text{cosec}\,\theta$ $\therefore$ $x = \cosh^{-1}(\text{cosec}\,\theta) = \log\left(\text{cosec}\,\theta+\sqrt{\text{cosec}^2\theta-1}\right)$

or $$x = \log(\text{cosec}\,\theta+\cot\theta) = \log\frac{1+\cos\theta}{\sin\theta}$$

$$= \log\frac{2\cos^2\frac{\theta}{2}}{2\sin\frac{\theta}{2}\cos\frac{\theta}{2}} = \log\cot\frac{\theta}{2}.$$

$\therefore$ $\text{sech}^{-1}(\sin\theta) = \log\cot\dfrac{\theta}{2}.$

Next, let $\sin^{-1}(\text{cosec}\,\theta) = \alpha+i\beta$ then $\sin(\alpha+i\beta) = \text{cosec}\,\theta$

or $\sin\alpha\cosh\beta + i\cos\alpha\sinh\beta = \text{cosec}\,\theta$

$\Rightarrow$ $\sin\alpha\cosh\beta = \text{cosec}\,\theta$ and $\cos\alpha\sinh\beta = 0$

$\Rightarrow$ $\cos\alpha = 0$ $\therefore$ $\alpha = (2n\pm 1)\dfrac{\pi}{2}$

Hence $\sin\alpha = \pm 1$ so $\cosh\beta = \pm\,\text{cosec}\,\theta$ or $\beta = \cosh^{-1}(\pm\,\text{cosec}\,\theta)$

or $\beta = \pm\log\cot\dfrac{\theta}{2}$ as established earlier. **Hence Proved.**

10. *Show that* $\log \sin (x + iy) = \frac{1}{2}\log\left(\frac{\cosh 2y - \cos 2x}{2}\right) + i \tan^{-1} (\cot x \tanh y)$.

SOLUTION: $\log \sin (x + iy) = \log (\sin x \cosh y + i \cos x \sinh y)$.

Since $\log (a + ib) = \frac{1}{2} \log (a^2 + b^2) + i \tan^{-1} \frac{b}{a}$, we have

$$\log \sin (x + iy) = \frac{1}{2} \log (\sin^2 x \cosh^2 y + \cos^2 x \sinh^2 y) + i \tan^{-1} \left(\frac{\cos x \sinh y}{\sin x \cosh y}\right)$$

$$= \frac{1}{2} \log \left[\left(\frac{1-\cos 2x}{2}\right)\left(\frac{1+\cosh 2y}{2}\right)+\left(\frac{1+\cos 2x}{2}\right)\left(\frac{\cosh 2y-1}{2}\right)\right]$$
$$+ i \tan^{-1} (\cot x \tanh y)$$

$$= \frac{1}{2} \log [(1 + \cosh 2y - \cos 2x - \cos 2x \cosh 2y + \cosh 2y - 1$$
$$+ \cos 2x \cosh 2y - \cos 2x)/4] + i \tan^{-1} (\cot x \tanh y)$$

$$= \frac{1}{2} \log \left(\frac{\cosh 2y - \cos 2x}{2}\right) + i \tan^{-1} (\cot x \tanh y)$$ **Hence Proved.**

11. If $i^{i^{i^{\cdots\infty}}} = a + ib$ then show that $\tan \frac{\pi a}{2} = \frac{b}{a}$ and $a^2 + b^2 = e^{-\pi b}$

SOLUTION: $i^{i^{i^{\cdots\infty}}} = a + ib \Rightarrow i^{a+ib} = a + ib$ or $(a + ib) \log i = \log (a + ib)$

or $(a + ib) \log (e^{i\pi/2}) = \log (a + ib)$

or $(a + ib)\, i\frac{\pi}{2} = \frac{1}{2} \log (a^2 + b^2) + i \tan^{-1} \frac{b}{a}$

$\Rightarrow$ $\frac{1}{2} \log (a^2 + b^2) = -\frac{b\pi}{2}$ and $\tan^{-1} \frac{b}{a} = a\frac{\pi}{2}$.

or $a^2 + b^2 = e^{-b\pi}$ and $\tan\left(\frac{a\pi}{2}\right) = \frac{b}{a}$ **Hence Proved.**

12. *Split into real and imaginary parts the principal value of* $i^{\log (1 + i)}$

SOLUTION: Let $i^{\log(1+i)} = x + iy$ then $\log (1 + i) \log i = \log (x + iy)$

Here $\log (1 + i) = \frac{1}{2} \log (1 + 1) + i \tan^{-1} 1 = \frac{1}{2} \log 2 + i\frac{\pi}{4}$

and $\log i = \log e^{i\pi/2} = i\frac{\pi}{2}$

$\therefore$ $\log (1 + i) \log i = \left[\frac{1}{2}\log 2 + \frac{i\pi}{4}\right]\frac{i\pi}{2} = \frac{-\pi^2}{8} + \frac{i\pi}{4}\log 2$

Hence $\log (x + iy) = \frac{-\pi^2}{8} + \frac{i\pi}{4} \log 2$

$\therefore$ $x + iy = e^{-\pi^2/8}\, e^{i\frac{\pi}{4}\log 2} = e^{-\pi^2/8} \left[\cos\left(\frac{\pi}{4}\log 2\right) + i\sin\left(\frac{\pi}{4}\log 2\right)\right]$ **Ans.**

13. *If* $\cos^{-1}(p + iq) = \alpha + i\beta$, find the equation whose roots are $\cos^2\alpha$ and $\cosh^2\beta$.

SOLUTION: $\cos^{-1}(p + iq) = \alpha + i\beta$ hence $\cos(\alpha + i\beta) = p + iq$

or $\cos\alpha \cosh\beta - i\sin\alpha \sinh\beta = p + iq$

$\Rightarrow$ $\cos\alpha \cosh\beta = p$ and $\sin\alpha \sinh\beta = -q.$

$\Rightarrow$ $\cos^2\alpha \cosh^2\beta = p^2$ and $(1 - \cos^2\alpha)(\cosh^2\beta - 1) = q^2$ ...(1)

or $-1 + \cos^2\alpha + \cosh^2\beta - \cos^2\alpha \cosh^2\beta = q^2$

Using (1) here we get $\cos^2\alpha + \cosh^2\beta = q^2 + 1 + p^2$

$\therefore$ Equation whose roots are $\cos^2\alpha$ and $\cosh^2\beta$, is

$$x^2 - (p^2 + q^2 + 1)x + p^2 = 0$$ **Ans.**

14. *If* $m^{x+iy} = (a + ib)^p$ *then find the principal value of* y/x.

SOLUTION: $m^{x+iy} = (a + ib)^p \Rightarrow (x + iy)\log m = p\log(a + ib)$

or $x\log m + iy\log m = \dfrac{p}{2}\log(a^2 + b^2) + pi\tan^{-1}\dfrac{b}{a}$

$\Rightarrow$ $x\log m = \dfrac{p}{2}\log(a^2 + b^2)$ and $y\log m = p\tan^{-1}\dfrac{b}{a}$

Taking their ratio, we get

$$\frac{y}{x} = \frac{p\tan^{-1}\frac{b}{2}}{\frac{p}{2}\log(a^2 + b^2)} = \frac{2\tan^{-1}\frac{b}{a}}{\log(a^2 + b^2)}.$$ **Ans.**

15. *If* $\tan(\theta + i\phi) = e^{i\alpha}$, *show that*

$$\theta = \left(n + \frac{1}{2}\right)\frac{\pi}{2} \quad \text{and} \quad \phi = -\frac{1}{2}\log\tan\left(\frac{\pi}{4} - \frac{\alpha}{2}\right).$$

SOLUTION: $\tan(\theta + i\phi) = e^{i\alpha}$ hence $\tan(\theta - i\phi) = e^{-i\alpha}$

$$\tan(2\theta) = \tan(\theta + i\phi + \theta - i\phi)$$

$$= \frac{\tan(\theta + i\phi) + \tan(\theta - i\phi)}{1 - \tan(\theta + i\phi)\tan(\theta - i\phi)} = \frac{e^{i\alpha} + e^{-i\alpha}}{1 - 1}$$

$\Rightarrow$ $2\theta = n\pi + \dfrac{\pi}{2}$ or $\theta = \left(n + \dfrac{1}{2}\right)\dfrac{\pi}{2}$, $n \in I.$

Next, $\tan(2i\phi) = \tan[(\theta + i\phi - (\theta - i\phi)] = \dfrac{\tan(\theta + i\phi) - \tan(\theta - i\phi)}{1 + \tan(\theta + i\phi)\tan(\theta - i\phi)}$

$$= \frac{e^{i\alpha} - e^{-i\alpha}}{1 + 1} = i\sin\alpha$$

But $\tan(2i\phi) = i\tanh 2\phi$, therefore

$$\tanh 2\phi = \sin\alpha \quad \text{or} \quad 2\phi = \tanh^{-1}(\sin\alpha)$$

$$= \frac{1}{2}\log\left(\frac{1+\sin\alpha}{1-\sin\alpha}\right)$$

or
$$\phi = \frac{1}{4}\log\frac{1+\cos\left(\frac{\pi}{2}-\alpha\right)}{1-\cos\left(\frac{\pi}{2}-\alpha\right)} = \frac{1}{4}\log\frac{2\cos^2\left(\frac{\pi}{4}-\frac{\alpha}{2}\right)}{2\sin^2\left(\frac{\pi}{4}-\frac{\alpha}{2}\right)}$$

$$= \frac{1}{4}\log\cot^2\left(\frac{\pi}{4}-\frac{\alpha}{2}\right) = \frac{1}{2}\log\cot\left(\frac{\pi}{4}-\frac{\alpha}{2}\right)$$

$$= -\frac{1}{2}\log\tan\left(\frac{\pi}{4}-\frac{\alpha}{2}\right)$$ **Hence Proved.**

SOLUTIONS TO EXERCISE 1E

1. *Sum the series* $x\cos\alpha - \frac{x^2}{2}\cos 2\alpha + \frac{x^3}{3}\cos 3\alpha - \ldots$

SOLUTION: Let $C = x\cos\alpha - \frac{x^2}{2}\cos 2\alpha + \frac{x^3}{3}\cos 3\alpha - \ldots$

and $S = x\sin\alpha - \frac{x^2}{2}\sin 2\alpha + \frac{x^3}{3}\sin 3\alpha - \ldots$

then $C + iS = xe^{i\alpha} - \frac{x^2}{2}e^{2i\alpha} + \frac{x^3}{3}e^{3i\alpha} - \ldots$

$$= \log(1 + xe^{i\alpha}) = \log(1 + x\cos\alpha + ix\sin\alpha)$$

or
$$C + iS = \frac{1}{2}\log[(1 + x\cos\alpha)^2 + x^2\sin^2\alpha] + i\tan^{-1}\frac{x\sin\alpha}{1+x\cos\alpha}$$

$\Rightarrow$
$$C = \frac{1}{2}\log[(1+x\cos\alpha)^2 + x^2\sin^2\alpha] = \frac{1}{2}\log[1 + x^2 + 2x\cos\alpha]$$ **Ans.**

2. *Sum the series* $\sin\theta + \frac{1}{3}\sin 3\theta + \frac{1}{3^2}\sin 5\theta + \ldots\ldots$

SOLUTION: Let $S = \sin\theta + \frac{1}{3}\sin 3\theta + \frac{1}{3^2}\sin 5\theta + \ldots$

and $C = \cos\theta + \frac{1}{3}\cos 3\theta + \frac{1}{3^2}\cos 5\theta + \ldots$

Then $C + iS = e^{i\theta} + \frac{1}{3}e^{3i\theta} + \frac{1}{3^2}e^{5i\theta} + \ldots$

$$= \frac{e^{i\theta}}{1-\frac{1}{3}e^{2i\theta}} \quad \text{as sum of an infinite geometric series.}$$

$$= \frac{\cos\theta + i\sin\theta}{1-\frac{1}{3}\cos 2\theta - \frac{i}{3}\sin 2\theta} = \frac{(\cos\theta + i\sin\theta)\left(1-\frac{1}{3}\cos 2\theta + \frac{i}{3}\sin 2\theta\right)}{\left(1-\frac{1}{3}\cos 2\theta\right)^2 + \frac{1}{9}\sin^2 2\theta}$$

$$= \frac{\cos\theta\left(1-\frac{1}{3}\cos 2\theta\right) - \frac{1}{3}\sin\theta\sin 2\theta + i\left[\sin\theta\left(1-\frac{1}{3}\cos 2\theta\right) + \frac{1}{3}\cos\theta\sin 2\theta\right]}{1-\frac{2}{3}\cos 2\theta + \frac{1}{9}(\cos^2\theta + \sin^2 2\theta)}$$

$$\therefore \quad S = \frac{\sin\theta - \frac{1}{3}\sin\theta\cos 2\theta + \frac{1}{3}\cos\theta\sin 2\theta}{\frac{10}{9} - \frac{2}{3}\cos 2\theta}$$

or
$$S = \frac{\frac{4}{3}\sin\theta}{\frac{10}{9} - \frac{2}{3}\cos 2\theta} = \frac{6\sin\theta}{5 - 3\cos 2\theta}$$ **Ans.**

3. *Sum the series* $1 + n\cos\alpha + \frac{n(n+1)}{2!}\cos 2\alpha + \frac{n(n+1)(n+2)}{3!}\cos 3\alpha + \ldots$

SOLUTION: Let $C = 1 + n\cos\alpha + \frac{n(n+1)}{2!}\cos 2\alpha + \frac{n(n+1)(n+2)}{3!}\cos 3\alpha + \ldots$

and $S = n\sin\alpha + \frac{n(n+1)}{2!}\sin 2\alpha + \frac{n(n+1)(n+2)}{3!}\sin 3\alpha + \ldots$

$$\therefore \quad C + iS = 1 + ne^{i\alpha} + \frac{n(n+1)}{2!}e^{2i\alpha} + \frac{n(n+1)(n+2)}{3!}e^{3i\alpha} + \ldots$$

$$= (1 - e^{i\alpha})^{-n} = (1 - \cos\alpha - i\sin\alpha)^{-n} = \left[2\sin^2\frac{\alpha}{2} - 2i\sin\frac{\alpha}{2}\cos\frac{\alpha}{2}\right]^{-n}$$

$$= \left(2\sin\frac{\alpha}{2}\right)^{-n}\left(\sin\frac{\alpha}{2} - i\cos\frac{\alpha}{2}\right)^{-n} = \left(2\sin\frac{\alpha}{2}\right)^{-n}\left[\cos\left(\frac{\pi}{2}-\frac{\alpha}{2}\right) - i\sin\left(\frac{\pi}{2}-\frac{\alpha}{2}\right)\right]^{-n}$$

$$= \left(2\sin\frac{\alpha}{2}\right)^{-n}\left[\cos\left(\frac{n\pi}{2}-\frac{n\alpha}{2}\right) + i\sin\left(\frac{n\pi}{2}-\frac{n\alpha}{2}\right)\right]$$

$$\therefore \quad C = \left(2\sin\frac{\alpha}{2}\right)^{-n}\cos\left(\frac{n\pi}{2}-\frac{n\alpha}{2}\right)$$ **Ans.**

4. *Sum the series* $c\sin\alpha + \frac{c^3}{3}\sin 3\alpha + \frac{c^5}{5}\sin 5\alpha + \ldots$

SOLUTION: Let $$S = c\sin\alpha + \frac{c^3}{3}\sin 3\alpha + \frac{c^5}{5}\sin 5\alpha + \ldots$$

and $$C = c\cos\alpha + \frac{c^3}{3}\cos 3\alpha + \frac{c^5}{5}\cos 5\alpha + \ldots$$

then $$C + iS = c\,e^{i\alpha} + \frac{c^3}{3}e^{3i\alpha} + \frac{c^5}{5}e^{5i\alpha} + \ldots$$

we know that $$\log(1+x) = x - \frac{x^2}{2} + \frac{x^3}{3} - \frac{x^4}{4} + \ldots$$

and $$\log(1-x) = -x - \frac{x^2}{2} - \frac{x^3}{3} - \frac{x^4}{4} - \ldots$$

or $$\frac{1}{2}\log\frac{1+x}{1-x} = x + \frac{x^3}{3} + \frac{x^5}{5} + \ldots$$

Therefore $$C + iS = \frac{1}{2}\log(1 + ce^{i\alpha}) - \frac{1}{2}\log(1 - ce^{i\alpha})$$

$$= \frac{1}{2}\log(1 + c\cos\alpha + ic\sin\alpha) - \frac{1}{2}\log(1 - c\cos\alpha - ic\sin\alpha)$$

$$= \frac{1}{4}\log[(1 + c\cos\alpha)^2 + c^2\sin^2\alpha] + \frac{i}{2}\tan^{-1}\frac{c\sin\alpha}{1+c\cos\alpha}$$

$$-\frac{1}{4}\log[(1 - c\cos\alpha)^2 + c^2\sin^2\alpha)] - \frac{i}{2}\tan^{-1}\left(\frac{-c\sin\alpha}{1-c\cos\alpha}\right)$$

$$= \frac{1}{4}\log(1 + c^2 + 2c\cos\alpha) - \frac{1}{4}\log(1 + c^2 - 2c\cos\alpha)$$

$$+ \frac{i}{2}\left[\tan^{-1}\frac{c\sin\alpha}{1+c\cos\alpha} + \tan^{-1}\frac{c\sin\alpha}{1-c\cos\alpha}\right]$$

$$= \frac{1}{4}\log\frac{1+c^2+2c\cos\alpha}{1+c^2-2c\cos\alpha} + \frac{i}{2}\tan^{-1}\left[\frac{\frac{c\sin\alpha}{1+c\cos\alpha} + \frac{c\sin\alpha}{1-c\cos\alpha}}{1 - \frac{c^2\sin^2\alpha}{(1+c\cos\alpha)(1-c\cos\alpha)}}\right]$$

$$= \frac{1}{4}\log\frac{1+c^2+2c\cos\alpha}{1+c^2-2c\cos\alpha} + \frac{i}{2}\tan^{-1}\frac{c\sin\alpha\,(1 - c\cos\alpha + 1 + c\cos\alpha)}{1 - c^2\cos^2\alpha - c^2\sin^2\alpha}$$

$$\therefore \quad = \frac{1}{4}\log\frac{1+c^2+2c\cos\alpha}{1+c^2-2c\cos\alpha} + \frac{i}{2}\tan^{-1}\left(\frac{2c\sin\alpha}{1-c^2}\right)$$

$$\therefore \quad C = \frac{1}{4}\log\frac{1+c^2+2c\cos\alpha}{1+c^2-2c\cos\alpha} \quad \text{and} \quad S = \frac{1}{2}\tan^{-1}\left(\frac{2c\sin\alpha}{1-c^2}\right)$$ **Ans.**

5. *Sum the series* $x \sin\theta - x^2 \sin 2\theta + x^3 \sin 3\theta - \ldots$ *upto n terms*

SOLUTION: Let $S = x \sin\theta - x^2 \sin 2\theta + x^3 \sin 3\theta - \ldots$ upto n terms.

and $C = x \cos\theta - x^2 \cos 2\theta + x^3 \cos 3\theta - \ldots$ upto n terms.

$\therefore$ $C + iS = x e^{i\theta} - x^2 e^{2i\theta} + x^3 e^{3i\theta} - \ldots$ upto n terms.

it is a geometric series with common ratio $-x e^{i\theta}$

$$\therefore \quad C + iS = xe^{i\theta} \frac{[(1-(-xe^{i\theta})^n]}{1-(-xe^{i\theta})} = xe^{i\theta} \frac{[1-(-1)^n x^n e^{ni\theta}}{1+x e^{i\theta}}$$

$$= \frac{x[1-(-x)^n \cos n\theta - i(-1)^n x^n \sin n\theta]}{(x+\cos\theta - i\sin\theta)}$$

$$= \frac{x[1-(-x)^n \cos n\theta - i(-x)^n \sin n\theta]\,[x+\cos\theta + i\sin\theta]}{(x+\cos\theta)^2 + \sin^2\theta}$$

$$= \frac{[x\{(x+\cos\theta)[1-(-x)^n \cos n\theta] + (-x)^n \sin n\theta \sin\theta\} + ix\{\sin\theta(1-(-x)^n \cos n\theta) - (-x)^n \sin n\theta\,(x+\cos\theta)\}]}{x^2+1+2x\cos\theta}$$

$$\therefore \quad S = \frac{x\sin\theta[1-(-x)^n \cos n\theta] - x(-x)^n \sin n\theta\,(x+\cos\theta)}{x^2+1+2x\cos\theta}$$

$$= \frac{x\sin\theta + (-1)^{n+1} x^{n+1} \{x\sin n\theta + \sin(n+1)\theta\}}{x^2+2x\cos\theta+1}$$ **Ans.**

6. *Sum the series* $n \sin\alpha + \dfrac{n(n+1)}{1.2} \sin 2\alpha + \dfrac{n(n+1)(n+2)}{3!} \sin 3\alpha + \ldots$

SOLUTION: In Question 3, we have had

$$S = n\sin\alpha + \frac{n(n+1)}{2!}\sin 2\alpha + \frac{n(n+1)(n+2)}{3!}\sin 3\alpha + \ldots$$

$$= \left(2\sin\frac{\alpha}{2}\right)^{-n} \sin\left(\frac{n\pi}{2} - \frac{n\alpha}{2}\right)$$ **Ans.**

7. *Sum the series* $\sin\alpha\cos\alpha + \sin^2\alpha\cos 2\alpha + \sin^3\alpha\cos 3\alpha + \ldots$

SOLUTION: Let $C = \sin\alpha\cos\alpha + \sin^2\alpha\cos 2\alpha + \sin^3\alpha\cos 3\alpha + \ldots$

and $S = \sin\alpha\sin\alpha + \sin^2\alpha\sin 2\alpha + \sin^3\alpha\sin 3\alpha + \ldots$

Then $C + iS = \sin\alpha\,(e^{i\alpha}) + \sin^2\alpha\, e^{2i\alpha} + \sin^3\alpha\, e^{3i\alpha} + \ldots$

$$= \frac{\sin\alpha\, e^{i\alpha}}{1-\sin\alpha\, e^{i\alpha}} = \frac{\sin\alpha}{e^{-i\alpha} - \sin\alpha} = \frac{\sin\alpha}{\cos\alpha - \sin\alpha - i\sin\alpha}$$

$$= \frac{\sin\alpha[\cos\alpha - \sin\alpha + i\sin\alpha]}{(\cos\alpha-\sin\alpha)^2 + \sin^2\alpha} = \frac{\sin\alpha(\cos\alpha-\sin\alpha) + i\sin^2\alpha}{1+\sin^2\alpha - 2\sin\alpha\cos\alpha}$$

$$\Rightarrow \quad C = \frac{\sin\alpha(\cos\alpha - \sin\alpha)}{1+\sin^2\alpha - \sin 2\alpha}$$ **Ans.**

8. Sum the series $e^{\alpha}\cos\beta - \dfrac{e^{3\alpha}}{3}\cos 3\beta + \dfrac{e^{5\alpha}}{5}\cos 5\beta - ...$

SOLUTION: Let $C = e^{\alpha}\cos\beta - \dfrac{e^{3\alpha}}{3}\cos 3\beta + \dfrac{e^{5\alpha}}{5}\cos 5\beta - ...$

and $S = e^{\alpha}\sin\beta - \dfrac{e^{3\alpha}}{3}\sin 3\beta + \dfrac{e^{5\alpha}}{5}\sin 5\beta - ...$

then $C + iS = e^{\alpha}e^{i\beta} - \dfrac{e^{3\alpha}}{3}e^{3i\beta} + \dfrac{e^{5\alpha}}{5}e^{5i\beta} - ...$

We know that $\tan^{-1}x = x - \dfrac{x^3}{3} + \dfrac{x^5}{5} - \dfrac{x^7}{7} + ...$

$\therefore$ $C + iS = \tan^{-1}(e^{\alpha}e^{i\beta}) = \tan^{-1}e^{\alpha}(\cos\beta + i\sin\beta)$

Next, consider $\tan^{-1}(x + iy) = \alpha + i\beta$ so $\tan^{-1}(x - iy) = \alpha - i\beta$

$\therefore$ $2\alpha = \tan^{-1}(x + iy) + \tan^{-1}(x - iy) = \tan^{-1}\dfrac{x + iy + x - iy}{1 - (x + iy)(x - iy)} = \tan^{-1}\dfrac{2x}{1 - x^2 - y^2}$

and $2i\beta = \tan^{-1}(x + iy) - \tan^{-1}(x - iy) = \tan^{-1}\dfrac{x + iy - (x - iy)}{1 + (x + iy)(x - iy)} = \tan^{-1}\dfrac{2iy}{1 + x^2 + y^2}$

or $\tan(2i\beta) = i\tanh 2\beta = \dfrac{2iy}{1 + x^2 + y^2}$ $\therefore$ $2\beta = \tanh^{-1}\dfrac{2y}{1 + x^2 + y^2}$

Therefore, we can wite

$$C + iS = \frac{1}{2}\tan^{-1}\frac{2e^{\alpha}\cos\beta}{1 - e^{2\alpha}\cos^2\beta - e^{2\alpha}\sin^2\beta} + i\frac{1}{2}\tanh^{-1}\frac{2e^{\alpha}\sin\beta}{1 + e^{2\alpha}\cos^2\beta + e^{2\alpha}\sin^2\beta}$$

$\therefore$ $$C = \frac{1}{2}\tan^{-1}\frac{2e^{\alpha}\cos\beta}{1 - e^{2\alpha}} = \frac{1}{2}\tan^{-1}\frac{2\cos\beta}{e^{-\alpha} - e^{\alpha}} = \frac{1}{2}\tan^{-1}\frac{\cos\beta}{-\sin\alpha}$$

$$= \frac{-1}{2}\tan^{-1}(\cos\beta\,\mathrm{cosec}\,\alpha)$$ **Ans.**

9. *Sum the series* $\sin\alpha - \dfrac{\sin(\alpha + 2\beta)}{2!} + \dfrac{\sin(\alpha + 4\beta)}{4!} - ...$

SOLUTION: Let $S = \sin\alpha - \dfrac{\sin(\alpha + 2\beta)}{2!} + \dfrac{\sin(\alpha + 4\beta)}{4!} - ...$

and $C = \cos\alpha - \dfrac{1}{2!}\cos(\alpha + 2\beta) + \dfrac{1}{4!}\cos(\alpha + 4\beta) - ...$

$\therefore$ $$C + iS = e^{i\alpha} - \frac{1}{2!}e^{i(\alpha + 2\beta)} + \frac{1}{4!}e^{i(\alpha + 4\beta)} - ... = e^{i\alpha}\left[1 - \frac{1}{2!}e^{2i\beta} + \frac{1}{4!}e^{4i\beta} - \frac{1}{6!}e^{6i\beta} + ...\right]$$

$$= e^{i\alpha}[\cos(e^{i\beta})] = (\cos\alpha + i\sin\alpha)\cos(\cos\beta + i\sin\beta)$$

$$= (\cos\alpha + i\sin\alpha)[\cos(\cos\beta)\cos(i\sin\beta) - \sin(\cos\beta)\sin(i\sin\beta)]$$

$$= (\cos\alpha + i\sin\alpha)[\cos(\cos\beta)\cosh(\sin\beta) - i\sin(\cos\beta)\sinh(\sin\beta)]$$

Equating imaginary parts on both sides, we get

$$S = \sin\alpha\cos(\cos\beta)\cosh(\sin\beta) - \cos\alpha\sin(\cos\beta)\sinh(\sin\beta)$$ **Ans.**

10. *Find the following sums*

(*i*) $\cos\dfrac{2\pi}{n} + \cos\dfrac{4\pi}{n} + \cos\dfrac{6\pi}{n} + \ldots + \cos\dfrac{2(n-1)\pi}{n}$

(*ii*) $\sin\left(\dfrac{2\pi}{n}\right) + \sin\left(\dfrac{4\pi}{n}\right) + \sin\dfrac{6\pi}{n} + \ldots + \sin\dfrac{2(n-1)\pi}{n}$.

SOLUTION: Let $C = \cos\dfrac{2\pi}{n} + \cos\dfrac{4\pi}{n} + \cos\dfrac{6\pi}{n} + \ldots + \cos\dfrac{2(n-1)\pi}{n}$

and $S = \sin\dfrac{2\pi}{n} + \sin\dfrac{4\pi}{n} + \sin\dfrac{6\pi}{n} + \ldots + \sin\dfrac{2(n-1)\pi}{n}$

$\therefore$ $C + iS = e^{2i\pi/n} + e^{4\pi i/n} + e^{6\pi i/n} + \ldots + e^{2\pi i(n-1)/n}$ which is a geometric series of $(n-1)$ terms with common ratio $e^{2\pi i/n}$ and first term $e^{2\pi i/n}$.

$$\therefore \quad C + iS = e^{2\pi i/n}\frac{[1 - e^{2\pi i(n-1)/n}]}{1 - e^{2\pi i/n}} = \frac{e^{2\pi i/n} - e^{2\pi i}}{1 - e^{2\pi i/n}} = \frac{e^{2\pi i/n} - 1}{1 - e^{2\pi i/n}} = -1$$

$\therefore$ $C = -1$ and $S = 0$

$\therefore$ (*i*) **Ans.** -1 and (*ii*) **Ans.** 0

ADDITIONAL IMPORTANT PROBLEMS

1. *Find the modulus and argument of the complex number* $A = \dfrac{(3+4i)(2-i)}{(2+3i)^2}$

SOLUTION: $$A = \frac{(3+4i)(2-i)}{(2+3i)^2} = \frac{10+5i}{-5+12i} = \frac{(10+5i)(5+12i)}{-(5^2+12^2)} = \frac{10-145i}{169}$$

Next, $$|A| = \frac{1}{169}\sqrt{10^2+145^2} = \frac{5\sqrt{845}}{169}$$

and $$\arg(A) = \tan^{-1}\frac{145}{-10} = \tan^{-1} -\frac{29}{2}$$ **Ans.**

2. *Find the modulus and argument of* $i^{\ln(1+i)}$.

SOLUTION: If r is the modulus and θ the argument of $i^{\ln(1+i)}$, then $re^{i\theta} = i^{\ln(1+i)}$

$$\Rightarrow \quad \ln(re^{i\theta}) = \ln(1+i)\ln(i) = \ln(1+i)\ln(e^{i\pi/2})$$

$$= \left(\frac{1}{2}\log 2 + \frac{i\pi}{4}\right)\left(\frac{i\pi}{2}\right)$$

$$= -\frac{\pi^2}{8} + \frac{i\pi}{4}\ln 2$$

$$\Rightarrow \quad re^{i\theta} = e^{-\pi^2/8}\cdot e^{i\pi/4}\ln 2$$

which gives $r = e^{-\pi^2/8}$ and $\theta = \dfrac{\pi}{4}\ln 2.$

3. *Discuss the Equation* $|z-a| = |z-b|$.

SOLUTION: If a and b are distinct complex numbers, the equation $|z-a| = |z-b|$ means the distance between z and a must be equal to the distance between z and b. It implies that z must lie on the perpendicular bisector of the line segment joining a and b.

For example the equation $|z+6i| = |z-1+3i|$ is satisfied by all points on the right bisector of the segment between $-6i$ and $1-3i$, that is, the line segment joining the points $(0, -6)$ and $(1, -3)$.

On taking $z = x+iy$ the equation of this line can be obtained as $y = \dfrac{-1}{3}(x+13)$. **Ans.**

4. *Find the equations of circles and disks in the complex plane.*

SOLUTION: If a is a complex number and r is a positive real number, the equation $|z-a| = r$ represents all the points on the circle whose centre is at 'a' and radius is r.

If $a = 0$, then any point on the circle $|z| = r$ has polar form $z = re^{i\theta}$ where θ is the angle from the positive real axis to the line drawn from the origin through z. As θ varies from 0 to 2π, the point $z = re^{i\theta}$ moves once counterclockwise around this circle, starting at $z = r$ on the positive real axis when $\theta = 0$, reaching ri when $\theta = \pi/2$, $-r$ when $\theta = \pi$, $-ri$ when $\theta = 3\pi/2$ and returning to r when $\theta = 2\pi$.

Any point on the circle $|z - a| = r$ has the from $z = a + re^{i\theta}$.

For example, the equation $|z - 3 + 7i| = 4$ defines a circle of radius 4 and centre at the point (3, –7). Any point on this circle has the form $z = 3 - 7i + 4e^{i\theta}$ $(0 \le \theta \le 2\pi)$.

An inequility $|z - a| < r$, specifies all points within a disk of radius r about a. Such a set is called an *open disk*, which means that points on the circumference of the circle bounding this disk are not in this set. A *closed disk* of radius r and centre 'a' consists of all points on or within the circle of radius r about a.

Ans.

5. *Find the locus of the point z which satisfies* $|z-1| = \frac{1}{2}|z-i|$.

SOLUTION: If $z = x + iy$, we have

$$|x - 1 + iy| = \frac{1}{2}|x + i(y-1)|$$

or $$(x-1)^2 + y^2 = \frac{1}{4}[x^2 + (y-1)^2] \quad \text{or} \quad 3x^2 - 8x + 3y^2 + 2y + 3 = 0$$

or $$3\left(x - \frac{4}{3}\right)^2 + 3\left(y + \frac{1}{3}\right)^2 = \frac{8}{3} \quad \text{that is,} \quad \left(x - \frac{4}{3}\right)^2 + \left(y + \frac{1}{3}\right)^2 = \frac{8}{9}$$

which represents a circle of radius $\frac{\sqrt{8}}{3}$ centred at $\left(\frac{4}{3}, \frac{-1}{3}\right)$. **Ans.**

6. *Show that the equation of the ellipse having foci at z_1 and z_2 and major axis of length 2a is $|z - z_1| + |z - z_2| = 2a$. What is its eccentricity?*

SOLUTION: We know that for any ellipse, the sum of the focal distances is equal to the length of the major axis. Let $P(z)$ be any point on the given ellipse with foci at $S_1(z_1)$ and $S_2(z_2)$, then

$$PS_1 = |z - z_1| \quad \text{and} \quad PS_2 = |z - z_2|.$$

hence $$PS_1 + PS_2 = 2a, \quad \text{gives} \quad |z - z_1| + |z - z_2| = 2a.$$

If e is the eccentricity of the ellipse then $S_1S_2 = 2ae$ or $|z_1 - z_2| = 2ae$

$$\therefore \quad e = \frac{|z_1 - z_2|}{2a}.$$

Ans.

7. *Show that the equation of the circle on the line segment joining z_1 and z_2 as diameter, is $|z - z_1|^2 + |z - z_2|^2 = |z_1 - z_2|^2$.*

SOLUTION: Let the points P, A, B represent the complex numbers z, z_1, z_2. Since $\angle APB$ is a right angle, hence $PA^2 + PB^2 = AB^2$.

or $$|z - z_1|^2 + |z - z_2|^2 = |z_1 - z_2|^2.$$

Which is the locus of P as circle with AB as diameter.

8. *If $b + ic = (1 + a)z$ and $\dfrac{a+ib}{1+c} = \dfrac{1+iz}{1-iz}$, then prove that $a^2 + b^2 + c^2 = 1$ where a, b, c are real and z complex.*

SOLUTION: Given $z = \dfrac{b+ic}{1+a}$ and $\dfrac{a+ib}{1+c} = \dfrac{1+iz}{1-iz}$

Eliminating z in the above relations, we get

$$\frac{a+ib}{1+c} = \frac{1+a+i(b+ic)}{1+a-i(b+ic)} = \frac{1+a-c+ib}{1+a+c-ib}$$

or $$(a+ib)(1+a+c-ib) = (1+c)(1+a-c+ib)$$

or $$a(1+a+c) + b^2 + ib(1+a+c-a) = (1+c)(1+a-c) + ib(1+c)$$

or $$a^2 + b^2 + a + ac + ib(1+c) = 1 + a - c + c + ac - c^2 + ib(1+c)$$

$\Rightarrow$ $$a^2 + b^2 + a + ac = 1 + a - c + c + ac - c^2$$

or $$a^2 + b^2 + c^2 = 1$$ **Hence Proved.**

9. *Prove that $\tan^{-1} i\left(\dfrac{x-a}{x+a}\right) = -\dfrac{i}{2}\log\left(\dfrac{a}{x}\right)$*

SOLUTION: Let $\tan^{-1} i\left(\dfrac{x-a}{x+a}\right) = \theta$ $\therefore$ $i\left(\dfrac{x-a}{x+a}\right) = \tan\theta$

or $$\frac{x-a}{x+a} = -i\frac{\sin\theta}{\cos\theta}$$

Applying componendo and dividendo, we get

$$\frac{x}{a} = \frac{\cos\theta - i\sin\theta}{\cos\theta + i\sin\theta} \quad \therefore \quad \frac{a}{x} = \frac{(\cos\theta + i\sin\theta)^2}{\cos^2\theta + \sin^2\theta} = \cos 2\theta + i\sin 2\theta$$

or $$\frac{a}{x} = e^{2i\theta} \quad \Rightarrow \quad \log\frac{a}{x} = 2i\theta$$

$$\therefore \quad \theta = \frac{1}{2i}\log\left(\frac{a}{x}\right) = \frac{-i}{2}\log\left(\frac{a}{x}\right).$$

10. *If $\tan(x + iy) = \sin(u + iv)$ show that $\dfrac{\sin 2x}{\sinh 2y} = \dfrac{\tan u}{\tanh v}$.*

SOLUTION: $$\tan(x+iy) = \frac{2\sin(x+iy)\cos(x-iy)}{2\cos(x+iy)\cos(x-iy)}$$

$$= \frac{\sin 2x + \sin 2iy}{\cos 2x + \cos 2iy} = \frac{\sin 2x + i\sinh 2y}{\cos 2x + \cosh 2y}$$

Since $\tan(x + iy) = \sin(u + iv) = \sin u \cos hv + i \cos u \sin hv$, we have

$$\frac{\sin 2x + i\sinh 2y}{\cos 2x + \cosh 2y} = \sin u \cosh v + i\cos u \sin hv$$

Comparing real and imaginary parts on both the sides gives

$$\frac{\sin 2x}{\cos 2x + \cosh 2y} = \sin u \cosh v \quad \text{and} \quad \frac{\sinh 2y}{\cos 2x + \cosh 2y} = \cos u \sinh v$$

$$\Rightarrow \quad \frac{\sin 2x}{\sinh 2y} = \frac{\tan u}{\tanh v}$$ **Hence Proved.**

11. *Taking only the principal value, split into real and imaginary parts* $(1+i\sqrt{3})^{1+i\sqrt{3}}$.

SOLUTION: Let $(1+i\sqrt{3})^{1+i\sqrt{3}} = a+ib$. Taking logarithm on both sides, gives

$$\log(a+ib) = (1+i\sqrt{3})\log(1+i\sqrt{3}) = (1+i\sqrt{3})\left[\frac{1}{2}\log(1+3) + i\tan^{-1}\sqrt{3}\right]$$

$$= (1+i\sqrt{3})\left(\log 2 + i\frac{\pi}{3}\right) = \log 2 - \frac{\pi\sqrt{3}}{3} + i\left(\sqrt{3}\log 2 + \frac{\pi}{3}\right)$$

$$\therefore \quad a+ib = e^{\log 2 - \frac{\pi\sqrt{3}}{3}} \cdot e^{i\left(\sqrt{3}\log 2 + \frac{\pi}{3}\right)}$$

$$= e^{\log 2 - \pi\sqrt{3}/3}\left[\cos\left(\sqrt{3}\log 2 + \frac{\pi}{3}\right) + i\sin\left(\sqrt{3}\log 2 + \frac{\pi}{3}\right)\right]$$

$$\Rightarrow \quad a = e^{\log 2 - \pi\sqrt{3}/3}\cos\left(\sqrt{3}\log 2 + \frac{\pi}{3}\right) \quad \text{and} \quad b = e^{\log 2 - \pi\sqrt{3}/3}\sin\left(\sqrt{3}\log 2 + \frac{\pi}{3}\right)$$ **Ans.**

12. *Find* z *if* $\arg(z+1) = \frac{\pi}{6}$ *and* $\arg(z-1) = \frac{2\pi}{3}$.

SOLUTION: If $z = x + iy$ then $\arg(x+1+iy) = \frac{\pi}{6}$

$$\text{or} \quad \tan^{-1}\frac{y}{x+1} = \frac{\pi}{6} \quad \text{or} \quad \frac{y}{x+1} = \tan\frac{\pi}{6} = \frac{1}{\sqrt{3}} \qquad ...(1)$$

$$\text{and} \quad \arg(x-1+iy) = \frac{2\pi}{3} \quad \text{or} \quad \tan^{-1}\frac{y}{x-1} = \frac{2\pi}{3}$$

$$\text{or} \quad \frac{y}{x-1} = \tan\frac{2\pi}{3} = -\sqrt{3} \qquad ...(2)$$

Solving (1) and (2) for x and y, gives

$$x = \frac{1}{2}, \quad y = \frac{\sqrt{3}}{2}$$ **Ans.**

13. *If* $a = \cos\alpha + i\sin\alpha$, $b = \cos\beta + i\sin\beta$, $c = \cos\gamma + i\sin\gamma$ *then prove that*

$$\frac{(b+c)(c+a)(a+b)}{abc} = 8\cos\left(\frac{\beta-\gamma}{2}\right)\cos\left(\frac{\gamma-\alpha}{2}\right)\cos\left(\frac{\alpha-\beta}{2}\right).$$

SOLUTION: Since $a = \cos\alpha + i\sin\alpha = e^{i\alpha}$, $b = \cos\beta + i\sin\beta = e^{i\beta}$, $c = \cos\gamma + i\sin\gamma = e^{i\gamma}$

we have $$\frac{(b+c)(c+a)(a+b)}{abc} = \frac{e^{i\beta}+e^{i\gamma}}{e^{i\beta/2}\cdot e^{i\gamma/2}}\cdot\frac{e^{i\gamma}+e^{i\alpha}}{e^{i\gamma/2}\cdot e^{i\alpha/2}}\cdot\frac{e^{i\alpha}+e^{i\beta}}{e^{i\alpha/2}\cdot e^{i\beta/2}}$$

$$= \left[e^{i(\beta-\gamma)/2}+e^{-i(\beta-\gamma)/2}\right]\left[e^{i(\gamma-\alpha)/2}+e^{-i(\gamma-\alpha)/2}\right]\left[e^{i(\alpha-\beta)/2}+e^{-i(\alpha-\beta)/2}\right]$$

$$= 2\cos\frac{\beta-\gamma}{2}\cdot 2\cos\frac{\gamma-\alpha}{2}\cdot 2\cos\frac{\alpha-\beta}{2}$$

$$= 8\cos\frac{\beta-\gamma}{2}\cos\frac{\gamma-\alpha}{2}\cos\frac{\alpha-\beta}{2}.$$ **Hence Proved**

14. *If* $x+iy=\sqrt[3]{a+ib}$ *prove that* $\frac{a}{x}+\frac{b}{y}=4(x^2-y^2)$.

SOLUTION: $(a+ib)^{1/3}=x+iy$ hence $a+ib=(x+iy)^3$

or $$a+ib = x^3-iy^3+3x^2\,iy-3xy^2=(x^3-3xy^2)+i(3x^2y-y^3)$$

$\Rightarrow$ $$a = x^3-3xy^2 \quad\text{and}\quad b=3x^2y-y^3$$

or $$\frac{a}{x} = x^2-3y^2 \quad\text{and}\quad \frac{b}{y}=3x^2-y^2$$

Hence $$\frac{a}{x}+\frac{b}{y} = 4(x^2-y^2).$$ **Hence Proved**

15. *Prove that* $\frac{1+\sin\theta+i\cos\theta}{1+\sin\theta-i\cos\theta}=\sin\theta+i\cos\theta$ *and hence show that*

$$\left(1+\sin\frac{\pi}{5}+i\cos\frac{\pi}{5}\right)^5+i\left(1+\sin\frac{\pi}{5}-i\cos\frac{\pi}{5}\right)^5=0.$$

SOLUTION: $$\frac{1+\sin\theta+i\cos\theta}{1+\sin\theta-i\cos\theta}=\frac{1+\cos\left(\frac{\pi}{2}-\theta\right)+i\sin\left(\frac{\pi}{2}-\theta\right)}{1+\cos\left(\frac{\pi}{2}-\theta\right)-i\sin\left(\frac{\pi}{2}-\theta\right)}$$

$$=\frac{2\cos^2\left(\frac{\pi}{4}-\frac{\theta}{2}\right)+2i\sin\left(\frac{\pi}{4}-\frac{\theta}{2}\right)\cos\left(\frac{\pi}{4}-\frac{\theta}{2}\right)}{2\cos^2\left(\frac{\pi}{4}-\frac{\theta}{2}\right)-2i\sin\left(\frac{\pi}{4}-\frac{\theta}{2}\right)\cos\left(\frac{\pi}{4}-\frac{\theta}{2}\right)}$$

$$=\frac{\cos\left(\frac{\pi}{4}-\frac{\theta}{2}\right)+i\sin\left(\frac{\pi}{4}-\frac{\theta}{2}\right)}{\cos\left(\frac{\pi}{4}-\frac{\theta}{2}\right)-i\sin\left(\frac{\pi}{4}-\frac{\theta}{2}\right)}=\frac{e^{i\left(\frac{\pi}{4}-\frac{\theta}{2}\right)}}{e^{-\left(\frac{\pi}{4}-\frac{\theta}{2}\right)}}$$

$$=e^{2i\left(\frac{\pi}{4}-\frac{\theta}{2}\right)}=e^{i\left(\frac{\pi}{2}-\theta\right)}=\cos\left(\frac{\pi}{2}-\theta\right)+i\sin\left(\frac{\pi}{2}-\theta\right)$$

$$=\sin\theta+i\cos\theta.$$

Replacing θ by $\frac{\pi}{5}$, we get $\frac{1+\sin\frac{\pi}{5}+i\cos\frac{\pi}{5}}{1+\sin\frac{\pi}{5}-i\cos\frac{\pi}{5}} = \sin\left(\frac{\pi}{5}\right)+i\cos\left(\frac{\pi}{5}\right)$.

$$\therefore \quad \left(\frac{1+\sin\frac{\pi}{5}+i\cos\frac{\pi}{5}}{1+\sin\frac{\pi}{5}-i\cos\frac{\pi}{5}}\right)^5 = \left(\sin\frac{\pi}{5}+i\cos\frac{\pi}{5}\right)^5 = \left[\cos\left(\frac{\pi}{2}-\frac{\pi}{5}\right)+i\sin\left(\frac{\pi}{2}-\frac{\pi}{5}\right)\right]^5$$

$$= \left(\cos\frac{3\pi}{10}+i\sin\frac{3\pi}{10}\right)^5 = \cos\frac{3\pi}{2}+i\sin\frac{3\pi}{2} = -i$$

or $$\left(1+\sin\frac{\pi}{5}+i\cos\frac{\pi}{5}\right)^5 + i\left(1+\sin\frac{\pi}{5}-i\cos\frac{\pi}{5}\right)^5 = 0.$$ **Hence Proved**

16. *Show that* $\left(\frac{1+\tanh x}{1-\tanh x}\right)^3 = \cosh 6x + \sinh 6x.$

SOLUTION: $$\left(\frac{1+\tanh x}{1-\tanh x}\right)^3 = \left(\frac{\cosh x+\sinh x}{\cosh x-\sinh x}\right)^3 = \left(\frac{\cos ix - i\sin ix}{\cos ix + i\sin ix}\right)^3$$

$$= \left[\frac{\cos ix - i\sin x}{\cos ix + i\sin x}\cdot\frac{\cos ix - i\sin x}{\cos ix - i\sin ix}\right]^3$$

$$= \left[\frac{(\cos ix - i\sin ix)^2}{\cos^2 ix + \sin^2 ix}\right]^3 = (\cos ix - i\sin ix)^6 = \cos 6ix - i\sin 6ix$$

$$= \cosh 6x + \sinh 6x.$$

17. *Show that the roots of the equation* $(x + 1)^7 = (x - 1)^7$ *are given by* $\pm i\cot\frac{k\pi}{7}$, *k = 1, 2, 3.*

SOLUTION: Given equation can be written as $\left(\frac{x+1}{x-1}\right)^7 = 1 = \cos 2k\pi + i\sin 2k\pi$

$$\Rightarrow \quad \frac{x+1}{x-1} = \cos\frac{2k\pi}{7}+i\sin\frac{2k\pi}{7}, \quad k = 0, 1, 2, 3, 4, 5, 6.$$

For $k = 0$ we have $\frac{x+1}{x-1} = \cos 0 + i\sin 0 = 1$

or $x + 1 = x - 1$ which is not possible.

$$\therefore \quad \frac{x+1}{x-1} = \cos\frac{2k\pi}{7}+i\sin\frac{2k\pi}{7}, \ k = 1, 2, 3, 4, 5, 6.$$

Hence using componendo and dividendo, we get

$$x = \frac{\cos\frac{2k\pi}{7} + i\sin\frac{2k\pi}{7} + 1}{\cos\frac{2k\pi}{7} + i\sin\frac{2k\pi}{7} - 1} = \frac{2\cos^2\frac{k\pi}{7} + 2i\sin\frac{k\pi}{7}\cos\frac{k\pi}{7}}{-2\sin^2\frac{k\pi}{7} + 2i\sin\frac{k\pi}{7}\cos\frac{k\pi}{7}}$$

$$= \frac{\cos\frac{k\pi}{7}}{\sin\frac{k\pi}{7}}\left[\frac{\cos\frac{k\pi}{7} + i\sin\frac{k\pi}{7}}{i^2\sin\frac{k\pi}{7} + i\cos\frac{k\pi}{7}}\right] = -i\cot\frac{k\pi}{7}, \quad k = 1, 2, 3, 4, 5, 6$$

$\Rightarrow \qquad x = -i\cot\frac{k\pi}{7}, \quad k = 1, 2, 3$

and $\qquad x = i\cot\frac{k\pi}{7} \quad$ for $\quad k = 4, 5, 6$

$\therefore \qquad x = \pm i\cot\frac{k\pi}{7}, \quad k = 1, 2, 3.$ **Hence Proved.**

CHAPTER

2 Infinite Series — Convergence and Divergence

IMPORTANT FORMULAE AND USEFUL RESULTS

1. Let $\{u_n\}$ $n = 1, 2, \ldots$ be a given sequence. Form a sequence $\{S_n\}$ where $S_1 = u_1$, $S_2 = u_1 + u_2$, $S_3 = u_1 + u_2 + u_3, \ldots$ $S_n = u_1 + u_2 + \ldots + u_n$. S_n is called nth partial sum. The series Σu_n is convergent if $\underset{n \to \infty}{\text{Lt}} S_n$ is unique and finite.

2. Necessary condition for a series Σu_n to be convergent, is $\underset{n \to \infty}{\text{Lt}} u_n = 0$.

3. **Comparision Test:**

   (*i*) If Σv_n is convergent and $u_n \le v_n$ for all $n \ge M$ then Σu_n is convergent.

   (*ii*) If Σv_n is divergent and $u_n \ge v_n$ for all $n \ge M$ then Σu_n is also divergent.

   (*iii*) If Σu_n and Σv_n are positive term series such that $\underset{n \to \infty}{\text{Lt}} \dfrac{u_n}{v_n}$ is unique and finte, then Σu_n behaves as Σv_n does.

4. **Harmonic Series (or *p*-series)**

   $\Sigma \dfrac{1}{n^p}$ is known as harmonic series or simply p-series. It converges for $p > 1$ and diverges for $p \le 1$.

5. **D'Alemberts' Ratio Test:** If Σu_n is a positve term series such that $\underset{n \to \infty}{\text{Lt}} \dfrac{u_{n+1}}{u_n} = K$ then Σu_n converges if $K < 1$ and diverges if $K > 1$.

6. **Integral Test:** If $f(x)$ is positive, continuous and monotonically decreasing and is such that $f(x) = u_n$ then Σu_n converges or diverges according as the integral $\int_1^\infty f(x)\,dx$ converges or diverges.

7. **Raabe's Test (or Higher Ratio Test):**

If Σu_n is a positive term series such that $\underset{n\to\infty}{\text{Lt}}\ n\left(\frac{u_n}{u_{n+1}} - 1\right) = \lambda$ then Σu_n converges when $\lambda > 1$ and diverges when $\lambda < 1$.

8. **Logarithmic Test:** If Σu_n is a postive term series such that $\underset{n\to\infty}{\text{Lt}}\ n \log \frac{u_n}{u_{n+1}} = \lambda$ then Σu_n converges or diverges according as $\lambda > 1$ or $\lambda < 1$.

9. **Cauchy's Root Test:** For a positive term series Σu_n, if $\underset{n\to\infty}{\text{Lt}}\ u_n^{1/n} = \lambda$ then Σu_n converges for $\lambda < 1$ and diverges for $\lambda > 1$.

10. **Lebneitz's Test:** The alternating series $u_1 - u_2 + u_3 - u_4 + \ldots$ is convergent if each term is numerically less than its preceding term and $\lim_{n\to\infty} u_n = 0$.

11. **Absolute Convergence and Conditional Convergence:** A Series Σu_n is called absolutely convergent if $\sum_{1}^{\infty} |u_n|$ is convergent.

A series which converges but not absolutely is called *conditionally convergent.*

12. **Uniform Convergence:** A series $\Sigma u_n(x)$ is said to be uniformly convergent in $[a, b]$ if the sequence of partial sums $\{S_n(x)\}$ is convergent in $[a, b]$.

13. **Weirstrass *M*-Test:** If a convergent series ΣM_n of positive constants $M_1, M_2, \ldots$ is such that $|u_n(x)| \le M_n$ for all x in the interval (a, b) then $\Sigma u_n(x)$ converges uniformaly in $[a, b]$.

SOLUTIONS TO EXERCISE 2A

1. *Test for convergence, the series* $\frac{1}{1.3}+\frac{1}{3.5}+\frac{1}{5.7}+...$

SOLUTION: $u_n = \frac{1}{(2n-1)(2n+1)} = \frac{1}{4n^2-1}$, Take $v_n = \frac{1}{n^2}$

Then $\underset{n\to\infty}{\text{Lt}} \frac{u_n}{v_n} = \underset{n\to\infty}{\text{Lt}} \frac{n^2}{4n^2-1} = \frac{1}{4}$ which is non-zero and fnite. Hence by comparison test $\Sigma\, v_n$ behaves as $\Sigma\, v_n$ does. But $\Sigma\, \frac{1}{n^2}$ is convergent as here $p = 2 > 1$. Hence $\Sigma\, u_n$ is convergent. **Ans.**

2. *Test for convergence, the series* $\sqrt{\frac{1}{2}}+\sqrt{\frac{2}{3}}+\sqrt{\frac{3}{4}}+...$

SOLUTION: $u_n = \sqrt{\frac{n}{n+1}}$, $\underset{n\to\infty}{\text{Lt}}\, u_n = \underset{n\to\infty}{\text{Lt}} \sqrt{\frac{1}{1+\frac{1}{n}}} = 1 \neq 0$. Hence $\Sigma\, u_n$ is divergent. **Ans.**

3. *Test for convergence, the series* $\frac{1.2}{3.4.5}+\frac{2.3}{4.5.6.}+\frac{3.4.}{5.6.7}+...$

SOLUTION: $u_n = \frac{n(n+1)}{(n+2)(n+3)((n+4)}$ · Take $v_n = \frac{1}{n}$

Then $\underset{n\to\infty}{\text{Lt}} \frac{u_n}{v_n} = \underset{n\to\infty}{\text{Lt}} \frac{n(n+1)\cdot n}{(n+2)(n+3)(n+4)} = 1 \neq 0$ and is finite

By comparison test $\Sigma\, u_n$ behaves as $\Sigma\, v_n$ does. But $\Sigma\, \frac{1}{n}$ is known to be divergent hence $\Sigma\, u_n$ is divergent. **Ans.**

4. *Test for convergence, the series* $\frac{1}{2}+\frac{3!}{2.4}+\frac{5!}{2.4.6}+...$

SOLUTION: Here $u_n = \frac{(2n-1)!}{2\cdot 4\cdot 6...(2n)}$ and $\underset{n\to\infty}{\text{Lt}} \frac{u_{n+1}}{u_n} = \underset{n\to\infty}{\text{Lt}} \frac{(2n+1)!\,2\cdot 4\cdot 6\cdot ...(2n)}{2\cdot 4\cdot 6\cdot ...(2n)(2n+2)(2n-1)!}$

Hence $\underset{n\to\infty}{\text{Lt}} \frac{u_{n+1}}{u_n} = \underset{n\to\infty}{\text{Lt}} \frac{(2n+1)2n}{2n+2} = \infty$

Therefore, by Ratio test $\Sigma\, u_n$ is divergent. **Ans.**

5. *Test for convergence, the series* $\frac{1}{1^2+x^2}+\frac{1}{2^2+x^2}+\frac{1}{3^2+x^2}+...$

SOLUTION: Here $u_n = \frac{1}{n^2+x^2}$. Take $v_n = \frac{1}{n^2}$ then

$$\underset{n\to\infty}{\text{Lt}} \frac{u_n}{v_n} = \underset{n\to\infty}{\text{Lt}} \frac{1\cdot n^2}{n^2+x^2} = \underset{n\to\infty}{\text{Lt}} \frac{1}{1+\frac{x^2}{n^2}} = 1 \neq 0 \text{ and finite.}$$

Hence Σu_n behaves as Σv_n does, by comparison test. But $\sum \frac{1}{n^2}$ is convergent being p-series with $p = 2 > 1$. $\therefore$ Σu_n is convergent for all x. **Ans.**

6. *Test for convergence, the series* $\frac{1}{1+x}+\frac{1}{1+2x^2}+\frac{1}{1+3x^3}+...$ $(x > 1)$

SOLUTION: $u_n = \frac{1}{1+nx^n}$. Here $\underset{n\to\infty}{\text{Lt}} u_n = 0$.

$$\therefore \quad \underset{n\to\infty}{\text{Lt}} \frac{u_{n+1}}{u_n} = \underset{n\to\infty}{\text{Lt}} \frac{1+nx^n}{1+(n+1)x^{n+1}} = \underset{n\to\infty}{\text{Lt}} \frac{1+\frac{1}{nx^n}}{\frac{1}{nx^n}+\frac{n+1}{n}\cdot x} = \underset{n\to\infty}{\text{Lt}} \frac{0+1}{0+1\cdot x}$$

$= \frac{1}{x} < 1$ as $x > 1$ Hence Σu_n is convergent by ratio test. **Ans.**

7. *Test for convergence, the series* $\frac{x}{1^2+1}+\frac{2^3x^2}{2^2+1}+\frac{3^3x^3}{3^2+1}+\frac{4^3x^4}{4^2+1}+...$ (for $x > 0$)

Solution: Here $u_n = \frac{n^3x^n}{n^2+1}$. $\quad \therefore \underset{n\to\infty}{\text{Lt}} \frac{u_{n+1}}{u_n} = \underset{n\to\infty}{\text{Lt}} \frac{(n+1)^3x^{n+1}}{(n+1)^2+1}\cdot\frac{(n^2+1)}{n^3x^n}$

or
$$\underset{n\to\infty}{\text{Lt}} \frac{u_{n+1}}{u_n} = \underset{n\to\infty}{\text{Lt}} \frac{(n+1)^3\cdot x}{n^3}\cdot\frac{n^2+1}{(n+1)^2+1} = \underset{n\to\infty}{\text{Lt}} \left[\left(1+\frac{1}{n}\right)^3\cdot x \frac{1+\frac{1}{n^2}}{\left(1+\frac{1}{n}\right)^2+\frac{1}{n^2}}\right] = x$$

By ratio test Σu_n is convergent when $x < 1$ and divergent for $x > 1$,

when $x = 1$, $\quad u_n = \frac{n^3}{n^2+1} = \frac{n}{1+\frac{1}{n^2}}$, $\underset{n\to\infty}{\text{Lt}} u_n = \infty$ hence divergent.

Thus Σu_n is convergent for $x < 1$ and divergent for $x \geq 1$. **Ans.**

8. *Test for convergence, the series* $\frac{1}{2}+\frac{\sqrt[3]{2}}{3\sqrt{2}}+\frac{\sqrt[3]{3}}{4\sqrt{3}}+...$

SOLUTION: $u_n = \frac{n^{1/3}}{(n+1)\sqrt{n}} = \frac{1}{(n+1)\,n^{1/6}}$. Here $u_n \to 0$ as $n \to \infty$.

Taking $v_n = \frac{1}{n^{7/6}}$ we have $\underset{n\to\infty}{\text{Lt}} \frac{u_n}{v_n} = \underset{n\to\infty}{\text{Lt}} \frac{n^{7/6}}{(n+1)\,n^{1/6}} = \underset{n\to\infty}{\text{Lt}} \frac{1}{1+\frac{1}{n}} = 1$, which in non-zero and finite. Hence by comparison test Σu_n behaves as Σv_n does. Σv_n is convergent being a p-series $p = \frac{7}{6} > 1$ $\therefore$ Σu_n is convergent. **Ans.**

9. *Test for convergence, the series* $1+3x+5x^2+7x^3+...$

SOLUTION: Here $u_n = (2n-1)\,x^{n-1}$ $\therefore$ $\underset{n\to\infty}{\text{Lt}} \frac{u_{n+1}}{u_n} = \underset{n\to\infty}{\text{Lt}} \frac{(2n+1)x^n}{(2n-1)x^{n-1}} = \underset{n\to\infty}{\text{Lt}} \left(\frac{2+\frac{1}{n}}{2-\frac{1}{n}}\right)x = x$

$\therefore$ By ratio test Σu_n is convergent for $x < 1$ and divergent for $x > 1$

When $x = 1$, $u_n = 2n-1$ and $\underset{n\to\infty}{\text{Lt}}\, u_n = \infty$. Hence divergent.

Thus, Σu_n is convergent for $x < 1$ and divergent $x \geq 1$. **Ans.**

10. *Test for convergence, the series* $1+\frac{2^p}{2!}+\frac{3^p}{3!}+\frac{4^p}{4!}+...$

SOLUTION: $u_n = \frac{n^p}{n!}$ $\therefore$ $\underset{n\to\infty}{\text{Lt}} \frac{u_{n+1}}{u_n} = \underset{n\to\infty}{\text{Lt}} \frac{(n+1)^p}{(n+1)!}\frac{n!}{n^p} = \underset{n\to\infty}{\text{Lt}} \left(\frac{n+1}{n}\right)^p \frac{1}{(n+1)}$

$$= \underset{n\to\infty}{\text{Lt}} \left(1+\frac{1}{n}\right)^p \cdot \frac{1}{(n+1)} = 0 < 1$$

$\therefore$ By ratio test the given series is convergent. **Ans.**

11. *Test for convergence, the series* $1+\frac{1^2\cdot 2^2}{1\cdot 3\cdot 5}+\frac{1^2\cdot 2^2\cdot 3^2}{1\cdot 3\cdot 5\cdot 7\cdot 9}+...$

SOLUTION: Here $u_n = \frac{1^2\cdot 2^2\cdot 3^2\cdot ...\cdot n^2}{1\cdot 3\cdot 5\cdot 7\cdot 9 ... (2n-1)\,(2n+1)}$

$$\underset{n\to\infty}{\text{Lt}} \frac{u_{n+1}}{u_n} = \underset{n\to\infty}{\text{Lt}} \frac{1^2\cdot 2^2\cdot 3^2 ... n^2\cdot (n+1)^2 \qquad 1\cdot 3\cdot 5\cdot 7\cdot ...(2n-1)\,(2n+1)}{[1\cdot 3\cdot 5\cdot 7 ... (2n-1)(2n+1)\,(2n+3)\,(2n+5)]\cdot 1^2\cdot 2^2\cdot 3^2 ...\cdot n^2}$$

$$= \underset{n\to\infty}{\text{Lt}} \frac{(n+1)^2}{(2n+3)\,(2n+5)} = \underset{n\to\infty}{\text{Lt}} \frac{\left(1+\frac{1}{n}\right)^2}{\left(2+\frac{3}{n}\right)\left(2+\frac{5}{n}\right)} = \frac{1}{4} < 1$$

Hence Σu_n is convergent by ratio test. **Ans.**

12. *Test for convergence, the series* $\frac{x}{x+1}+\frac{x^2}{x+2}+\frac{x^3}{x+3}+\ldots$

SOLUTION: $u_n = \frac{x^n}{x+n} \quad \therefore \quad \underset{n\to\infty}{\text{Lt}} \frac{u_{n+1}}{u_n} = \underset{n\to\infty}{\text{Lt}} \frac{x^{n+1}}{x+n+1}\cdot\frac{x+n}{x^n} = \underset{n\to\infty}{\text{Lt}} \frac{x\left(1+\frac{x}{n}\right)}{1+\frac{1}{n}+\frac{x}{n}} = x$

$\therefore$ By ratio test Σu_n is convergent for $x < 1$ and divergent for $x > 1$.

For $x = 1$ $\quad u_n = \frac{1^n}{n+1} = \frac{1}{n+1}$ which is divergent as $\Sigma \frac{1}{n}$.

Thus, the given series is convergent for $x < 1$ and divergent for $x \geq 1$. **Ans.**

13. *Test for convergence, the series* $\sum_{n=1}^{\infty} \frac{\log n}{2n^3-1}$.

SOLUTION: $\Sigma u_n = \Sigma \frac{\log n}{2n^3-1}$

$$\underset{n\to\infty}{\text{Lt}} \frac{u_{n+1}}{u_n} = \underset{n\to\infty}{\text{Lt}} \frac{\log(n+1)}{2(n+1)^3-1}\cdot\frac{2n^3-1}{\log n} = \underset{n\to\infty}{\text{Lt}} \left[\frac{\log n\left(1+\frac{1}{n}\right)}{\log n}\cdot\frac{2-\frac{1}{n^3}}{2\left(1+\frac{1}{n}\right)^3-\frac{1}{n^3}}\right]$$

$$= \underset{n\to\infty}{\text{Lt}} \frac{\log n+\log\left(1+\frac{1}{n}\right)}{\log n}\cdot 1 = \underset{n\to\infty}{\text{Lt}}\, 1+\frac{\log\left(1+\frac{1}{n}\right)}{\log n} = 1$$

Since $\log n < n \quad \therefore \quad \frac{\log n}{2n^3-1} < \frac{n}{2n^3-1} = t_n$

Σu_n is convergent if Σt_n is convergent.

For Σt_n consider $v_n = \frac{1}{n^2} \quad \therefore \quad \underset{n\to\infty}{\text{Lt}} \frac{t_n}{v_n} = \underset{n\to\infty}{\text{Lt}} \frac{n\cdot n^2}{2n^3-1} = \frac{1}{2}$

Hence Σt_n behaves as Σv_n does, but Σv_n is convergent being a p-series $(p > 2)$.

Hence Σu_n is convergent. **Ans.**

14. *Test for convergence, the series* $\sum_{n=1}^{\infty} \frac{1}{n^2\cdot 3^n}$

SOLUTION: $\sum_{n=1}^{\infty} \frac{1}{n^2\cdot 3^n}$. $\quad u_n = \frac{1}{n^2\cdot 3^n} \quad \therefore \quad \underset{n\to\infty}{\text{Lt}} \frac{u_{n+1}}{u_n} = \underset{n\to\infty}{\text{Lt}} \frac{n^2\cdot 3^n}{(n+1)^2\cdot 3^{n+1}} = \frac{1}{3}\underset{n\to\infty}{\text{Lt}} \frac{1}{\left(1+\frac{1}{n}\right)} = \frac{1}{3}$

Since $\underset{n\to\infty}{\text{Lt}} \frac{u_{n+1}}{u_n} = \frac{1}{3} < 1$ Hence Σu_n is convergent by ratio test. **Ans.**

15. *Test for convergence, the series* $\frac{1}{3}+\frac{3}{5}+\frac{7}{9}+\frac{15}{17}+...$

SOLUTION: $\Sigma u_n \frac{2-1}{2+1}+\frac{2^2-1}{2^2+1}+\frac{2^3-1}{2^3+1}+...$ $\therefore$ $u_n = \frac{2^n-1}{2^n+1} = \frac{1-\frac{1}{2^n}}{1+\frac{1}{2^n}}$ $\therefore$ $\underset{n\to\infty}{\text{Lt}}\ u_n = 1$

Σu_n is divergent since $\underset{n\to\infty}{\text{Lt}}\ u_n \neq 0$ **Ans.**

16. *Test for convergence, the series* $\sum^{\infty} \frac{n^2+2^n}{2^n \cdot n^2}$

SOLUTION: $\sum_1^{\infty} \frac{n^2+2^n}{2^n \cdot n^2} = \Sigma\left(\frac{1}{2^n}+\frac{1}{n^2}\right) = \Sigma\frac{1}{2^n}+\Sigma\frac{1}{n^2}$

$\Sigma\frac{1}{2^n}$ is convergent being a geometric series with coommon ratio = $\frac{1}{2} < 1$ and $\Sigma\frac{1}{n^2}$ is convergent being a p-series with $p = 2 > 1$

$\therefore$ Given series is convergent. **Ans.**

17. *Discuss the convergence of the series* $\Sigma\frac{x^n}{n}$, $x > 0$

SOLUTION: $\sum_1^{\infty}\frac{x^n}{n}$, $x > 0$. Here $u_n = \frac{x^n}{n}$

$$\therefore \quad \underset{n\to\infty}{\text{Lt}}\frac{u_{n+1}}{u_n} = \underset{n\to\infty}{\text{Lt}}\frac{x^{n+1}}{n+1}\cdot\frac{n}{x^n} = x\ \underset{n\to\infty}{\text{Lt}}\frac{1}{1+\frac{1}{n}} = x$$

By ratio test Σu_n is convergent for $x < 1$ and divergent for $x > 1$.

When $x = 1$ $\quad \Sigma u_n = \Sigma\frac{1}{n}$ which is known to be divergent.

Thus, Σu_n is convergent for $x < 1$ and divergent for $x \geq 1$. **Ans.**

18. *Discuss the convergence of the series* $\Sigma\frac{n^2(n+1)^2}{n!}$

SOLUTION: $\Sigma u_n\ \Sigma\frac{n^2(n+1)^2}{n!}$ hence $u_n = \frac{n^2(n+1)^2}{n!}$

$$\therefore \quad \underset{n\to\infty}{\text{Lt}}\frac{u_{n+1}}{u_n} = \underset{n\to\infty}{\text{Lt}}\frac{(n+1)^2(n+2)^2}{(n+1)!}\frac{n!}{n^2(n+1)^2} = \underset{n\to\infty}{\text{Lt}}\frac{(n+2)^2}{n^2(n+1)}$$

$$= \underset{n\to\infty}{\text{Lt}}\left[\left(1+\frac{2}{n}\right)^2\frac{1}{n+1}\right] = 0 \text{ which is less than 1}$$

Hence by ratio test Σu_n is convergent. **Ans.**

19. *Test the convergence of the series* $\frac{1}{\sqrt{1\cdot 2}}+\frac{1}{\sqrt{2\cdot 3}}+\frac{1}{\sqrt{3\cdot 4}}+\dots$

SOLUTION: $u_n=\frac{1}{\sqrt{n(n+1)}}=\frac{1}{\sqrt{n^2+n}}$. Let us take $v_n=\frac{1}{n}$.

then $\underset{n\to\infty}{\text{Lt}}\frac{u_n}{v_n}=\underset{n\to\infty}{\text{Lt}}\frac{n}{\sqrt{n^2+n}}=\underset{n\to\infty}{\text{Lt}}\frac{1}{\sqrt{1+\frac{1}{n}}}=1\neq 0$ and is finte.

$\therefore\ \Sigma u_n$ behaves as Σv_n does. But $\Sigma\frac{1}{n}$ is divergent. Hence Σu_n is divergent. **Ans.**

SOLUTIONS TO EXERCISE 2B

1. *Discuss the convergence and divergence of the harmonic series* $\Sigma\frac{1}{n^p}$ *using Integral test.*

SOLUTION: By integral test $\sum_{1}^{\infty} f(x_n)$ is convergent or divergent according as $\int_1^{\infty} f(x)\,dx$ is finite or infinite. The given series is $\sum_{1}^{\infty}\frac{1}{n^p}$. Here $f(x)=\frac{1}{x^p}$.

Now $\int_1^{\infty}\frac{1}{x^p}dx=\left[\frac{x^{-p+1}}{-p+1}\right]_1^{\infty}=\left[\frac{1}{(1-p)x^{p-1}}\right]_1^{\infty}=\frac{1}{1-p}$ if $p>1$

$=\infty$ if $p\leq 1$.

Therefore $\sum\frac{1}{n^p}$ is convergent for $p>1$ and divergent for $p\leq 1$. **Ans.**

2. *Test the convergence of the series* $\sum_{1}^{\infty} ne^{-n^2}$

SOLUTION: Let us use the integral test.

$$\int_1^{\infty} xe^{-x^2}dx=\frac{-1}{2}\left[e^{-x^2}\right]_1^{\infty}=\frac{-1}{2}\left[0-\frac{1}{e}\right]=\frac{1}{2e}$$

Hence, the given series is convergent. **Ans.**

3. *Test the convergence of the series* $\frac{2^2}{3\cdot 4}+\frac{2^2\cdot 4^2}{3\cdot 4\cdot 5\cdot 6}+\frac{2^2\cdot 4^2\cdot 6^2}{3\cdot 4\cdot 5\cdot 6\cdot 7\cdot 8}+\dots$

SOLUTION: $\frac{2^2}{3\cdot 4}+\frac{2^2\cdot 4^2}{3\cdot 4\cdot 5\cdot 6}+\frac{2^2\cdot 4^2\cdot 6^2}{3\cdot 4\cdot 5\cdot 6\cdot 7\cdot 8}+\dots=\Sigma u_n$

then $u_n=\frac{2^2\cdot 4^2\cdot 6^2\cdot\dots(2n)^2}{3\cdot 4\cdot 5\cdot 6\cdot 7\cdot 8\cdot\dots(2n+1)(2n+2)}$

$$\therefore \quad \underset{n\to\infty}{\text{Lt}} \frac{u_{n+1}}{u_n} = \underset{n\to\infty}{\text{Lt}} \frac{2^2 \cdot 4^2 \cdot 6^2 \ldots (2n)^2 \, (2n+2)^2 \cdot 3 \cdot 4 \cdot 5 \cdot 6 \cdot \ldots (2n+1)\,(2n+2)}{3 \cdot 4 \cdot 5 \cdot 6 \cdot \ldots (2n+1)\,(2n+2)\,(2n+3)\,(2n+4) \cdot 2^2 \cdot 4^2 \cdot 6^2 \ldots (2n)^2}$$

$$= \underset{n\to\infty}{\text{Lt}} \frac{(2n+2)^2}{(2n+3)\,(2n+4)} = 1, \text{ hence ratio test fails.}$$

Now we apply higher ratio test

$$\underset{n\to\infty}{\text{Lt}}\, n\left(\frac{u_n}{u_{n+1}} - 1\right) = \underset{n\to\infty}{\text{Lt}}\, n\left[\frac{(2n+3)\,(2n+4)}{(2n+2)^2} - 1\right] = \frac{3}{2} > 1$$

Hence Σu_n is convergent. **Ans.**

4. *Test the convergence of the series* $\frac{1}{3}x + \frac{1 \cdot 2}{3 \cdot 5}x^2 + \frac{1 \cdot 2 \cdot 3}{3 \cdot 5 \cdot 7}x^3 + \ldots \quad (x > 0)$

SOLUTION: Here, $$u_n = \frac{1 \cdot 2 \cdot 3 \ldots n \quad x^n}{3 \cdot 5 \cdot 7 \ldots (2n+1)}$$

$$\therefore \quad \underset{n\to\infty}{\text{Lt}} \frac{u_{n+1}}{u_n} = \underset{n\to\infty}{\text{Lt}} \frac{1 \cdot 2 \cdot 3 \ldots n(n+1)x^{n+1}}{3 \cdot 5 \cdot 7 \ldots (2n+1)\,(2n+3)} \cdot \frac{3 \cdot 5 \cdot \ldots (2n+1)}{1 \cdot 2 \cdot 3 \ldots n \cdot x^n}$$

$$= \underset{n\to\infty}{\text{Lt}} \frac{(n+1)x}{2n+3} = x \underset{n\to\infty}{\text{Lt}} \left(\frac{1+\frac{1}{n}}{2+\frac{3}{n}}\right) = \frac{x}{2}$$

$\therefore$ By ratio test Σu_n is convergent for $x < 2$ and divergent for $x > 2$.

When $x = 2$, $$u_n = \frac{1 \cdot 2 \cdot 3 \cdot \ldots \cdot n \cdot 2^n}{3 \cdot 5 \cdot 7 \cdot \ldots \cdot (2n+1)} = \frac{2 \cdot 4 \cdot 6 \cdot \ldots \cdot 2n}{3 \cdot 5 \cdot 7 \cdot \ldots \cdot (2n+1)}$$

Now let as apply Raabe's test, then

$$\underset{n\to\infty}{\text{Lt}} \left[n\left(\frac{u_n}{u_{n+1}} - 1\right)\right] = \underset{n\to\infty}{\text{Lt}} \left[\left(\frac{2n+3}{2n+2} - 1\right)n\right] = \underset{n\to\infty}{\text{Lt}} \frac{n}{2n+2} = \frac{1}{2} < 1$$

Hence Σu_n is divergent for $x = 2$.
Thus given series is convergent for $x < 2$ and divergent for $x \geq 2$. **Ans.**

5. *Test for convergence, the series* $2 + \left(\frac{3}{2}\right)^4 + \left(\frac{4}{3}\right)^9 + \left(\frac{5}{4}\right)^{16} + \ldots$

SOLUTION: $2 + \left(\frac{3}{2}\right)^4 + \left(\frac{4}{3}\right)^9 + \left(\frac{5}{4}\right)^{16} + \ldots = \Sigma u_n$

$$\therefore \quad u_n = \left(\frac{n+1}{n}\right)^{n^2}. \text{ We apply here Cauchy's root test.}$$

$$\therefore \quad \underset{n\to\infty}{\text{Lt}}\, u_n^{1/n} = \underset{n\to\infty}{\text{Lt}} \left[\left(\frac{n+1}{n}\right)^{n^2}\right]^{1/n} = \underset{n\to\infty}{\text{Lt}} \left(\frac{n+1}{n}\right)^n = \underset{n\to\infty}{\text{Lt}} \left(1+\frac{1}{n}\right)^n = e$$

which is greater than 1 hence Σu_n is divergent. **Ans.**

6. *Test the convergence of the series* $\frac{1}{4}+\frac{4}{27}+\frac{27}{256}+\ldots+\frac{(n-1)^{n-1}}{n^n}+\ldots$

SOLUTION: $\frac{1}{4}+\frac{4}{27}+\frac{27}{256}+\ldots+\frac{(n-1)^{n-1}}{n^n}+\ldots=\Sigma u_n \quad \therefore \quad u_n=\frac{n^n}{(n+1)^{n+1}}$

$\therefore \quad \underset{n\to\infty}{\text{Lt}}\ u_n=\underset{n\to\infty}{\text{Lt}}\ \frac{1}{n+1}\left(1+\frac{1}{n}\right)^{-n}=0$. Let us apply Cauchy's root test here.

$$\underset{n\to\infty}{\text{Lt}}\ u_n^{1/n}=\underset{n\to\infty}{\text{Lt}}\ \frac{1}{(n+1)^{1/n}}\left(\frac{n}{n+1}\right)=\frac{1}{e}<1$$

Hence given series is convergent by root test. **Ans.**

7. *Test the convergence of the series* $x+\frac{2^2\,x^2}{2!}+\frac{3^3\,x^3}{3!}+\frac{4^4\,x^4}{4!}+\ldots$

SOLUTION: $x+\frac{2^2\,x^2}{2!}+\frac{3^3\,x^3}{3!}+\frac{4^4\,x^4}{4!}+\ldots=\Sigma u_n$

Here $$u_n=\frac{n^n\,x^n}{n!}$$

$\therefore$ $$\frac{u_{n+1}}{u_n}=\frac{(n+1)^{n+1}\,x^{n+1}}{(n+1)!}\cdot\frac{n!}{n^n\cdot x^n}=x\left(\frac{n+1}{n}\right)^n=x\left(1+\frac{1}{n}\right)^n$$

and $$\underset{n\to\infty}{\text{Lt}}\ \frac{u_{n+1}}{u_n}=x\ \underset{n\to\infty}{\text{Lt}}\left(1+\frac{1}{n}\right)^n=xe$$

$\therefore \quad \Sigma u_n$ is convergent when $x<\frac{1}{e}$ and divergent when $x>\frac{1}{e}$. **Ans.**

8. *Test the convergence of the series* $\frac{4}{1!}x+\frac{4.7}{2!}x^2+\frac{4.7.10}{3!}x^3+\ldots$

SOLUTION: $\frac{4}{1!}x+\frac{4\cdot 7}{2!}x^2+\frac{4\cdot 7\cdot 10}{3!}x^3+\ldots=\Sigma u_n$

$\therefore$ $$u_n=\frac{4\cdot 7\cdot 10\ldots(3n+1)}{n!}x^n$$

and $$\frac{u_{n+1}}{u_n}=\frac{4\cdot 7\cdot 10\ldots(3n+1)\,(3n+4)x^{n+1}}{(n+1)!}\cdot\frac{n!}{4\cdot 7\cdot 10\ldots(3n+1)x^n}$$

$\therefore$ $$\underset{n\to\infty}{\lim}\ \frac{u_{n+1}}{u_n}=\underset{n\to\infty}{\text{Lt}}\ \frac{(3n+4)x}{(n+1)}=x\ \underset{n\to\infty}{\text{Lt}}\ \frac{3+\frac{4}{n}}{1+\frac{1}{n}}=3x$$

Hence by ratio test Σu_n is convergent for $x < \frac{1}{3}$ and divergent for $x > \frac{1}{3}$.

Next, for $x = \frac{1}{3}$, $n\left(\frac{u_n}{u_{n+1}} - 1\right) = n\left(\frac{3n+3}{3n+4} - 1\right) = \frac{-n}{3n+4}$

$$\therefore \quad \underset{n\to\infty}{\text{Lt}}\, n\left(\frac{u_n}{u_{n+1}} - 1\right) = \underset{n\to\infty}{\text{Lt}} \frac{-n}{3n+4} = -\frac{1}{3} \quad \text{which is less than 1.}$$

Hence for $x = \frac{1}{3}$ Σu_n is divergent.

Thus Σu_n is convergent for $x < \frac{1}{3}$ and divergent for $x \geq \frac{1}{3}$. **Ans.**

9. *Test the convergence of the series* $\sum\left(1+\frac{1}{\sqrt{n}}\right)^{-n^{3/2}}$

SOLUTION: $\sum\left(1+\frac{1}{\sqrt{n}}\right)^{-n^{3/2}} = \Sigma u_n \quad \therefore \quad u_n = \left(1+\frac{1}{\sqrt{n}}\right)^{-n\sqrt{n}}$

Applying root test here, we have

$$\underset{n\to\infty}{\text{Lt}}\, u_n^{1/n} = \underset{n\to\infty}{\text{Lt}} \left(1+\frac{1}{\sqrt{n}}\right)^{-\sqrt{n}} = \underset{n\to\infty}{\text{Lt}} \frac{1}{\left(1+\frac{1}{\sqrt{n}}\right)^{\sqrt{n}}} = \frac{1}{e} \quad \text{which is less than 1.}$$

Therefore, by root test Σu_n is convergent. **Ans.**

10. *Test the convergence of the series* $\frac{x}{1} + \frac{1}{2}\cdot\frac{x^3}{3} + \frac{1\cdot 3}{2\cdot 4}\cdot\frac{x^5}{5} + \frac{1\cdot 3\cdot 5}{2\cdot 4\cdot 6}\cdot\frac{x^7}{7} + \ldots\ (x > 0)$

SOLUTION: Let us quit the first term here, then

$$u_n = \frac{1\cdot 3\cdot 5\ldots(2n-1)}{2\cdot 4\cdot 6\ldots(2n)}\cdot\frac{x^{2n+1}}{2n+1}$$

$$\therefore \quad \frac{u_{n+1}}{u_n} = \frac{1\cdot 3\cdot 5\ldots(2n-1)\,(2n+1)}{2\cdot 4\cdot 6\ldots(2n)\,(2n+2)}\cdot\frac{x^{2n+3}}{2n+3}\cdot\frac{2\cdot 4\cdot 6\ldots(2n)}{1\cdot 3\cdot 5\ldots(2n-1)}\cdot\frac{(2n+1)}{x^{2n+1}}$$

$$= \frac{(2n+1)}{(2n+2)}\cdot\frac{2n+1}{2n+3}\cdot x^2$$

and $\underset{n\to\infty}{\text{Lt}} \frac{u_{n+1}}{u_n} = x^2$. Therefore, by Ratio test Σu_n is convergent for $x^2 < 1$ and divergent for $x^2 > 1$.

Next, when $x^2 = 1$ we apply the higher ratio test and get

$$\underset{n\to\infty}{\text{Lt}}\ n\left(\frac{u_n}{u_{n+1}}-1\right) = \underset{n\to\infty}{\text{Lt}}\ n\left[\frac{(2n+2)(2n+3)}{(2n+1)^2}-1\right] = \underset{n\to\infty}{\text{Lt}}\ \frac{n(6n+5)}{(2n+1)^2} = \frac{6}{4} = \frac{3}{2} > 1$$

Hence Σu_n is convergent for $x^2 = 1$

Thus the given series is convergent for $|x| \le 1$ and divergent for $|x| > 1$. **Ans.**

11. *Test the convergence of the series* $\frac{1}{2}+\left(\frac{2}{3}\right)x+\left(\frac{3}{4}\right)^2 x^2+\left(\frac{4}{5}\right)^3 x^3+...$

SOLUTION: $\frac{1}{2}+\left(\frac{2}{3}\right)x+\left(\frac{3}{4}\right)^2 x^2+\left(\frac{4}{5}\right)^3 x^3+... = \Sigma u_n$

$$u_n = \left(\frac{n+1}{n+2}\right)^n x^n \quad \left(\text{not considering the first term } \frac{1}{2}\right)$$

$$\underset{n\to\infty}{\text{Lt}}\ u_n^{1/n} = \underset{n\to\infty}{\text{Lt}}\ \frac{n+1}{n+2}x = x\ \underset{n\to\infty}{\text{Lt}}\ \frac{1+\frac{1}{n}}{1+\frac{2}{n}} = x$$

Hence by Cauchy's root test Σu_n is convergent for $x < 1$ and divergent for $x > 1$.

When $x = 1$,
$$u_n = \left(\frac{n+1}{n+2}\right)^n = \frac{\left(1+\frac{1}{n}\right)^n}{\left(1+\frac{2}{n}\right)^n} = \frac{\left(1+\frac{1}{n}\right)^n}{\left[\left(1+\frac{2}{n}\right)^{\frac{n}{2}}\right]^2}$$

and
$$\underset{n\to\infty}{\text{Lt}}\ u_n = \frac{e}{e^2} = \frac{1}{e} \ne 0 \quad \text{hence divergent.}$$

Thus Σu_n is convergent for $x < 1$ and divergent for $x \ge 1$. **Ans.**

12. *Test the convergence of the series*

$$1+\frac{a\cdot b}{1\cdot c}x+\frac{a(a+1)\,b(b+1)}{1\cdot 2\cdot c\cdot(c+1)}x^2+\frac{a(a+1)(a+2)\,b(b+1)(b+2)}{1\cdot 2\cdot 3\cdot c\cdot(c+1)(c+2)}x^3+...$$

SOLUTION: Omiting the first term,

Here,
$$u_n = \frac{a(a+1)(a+2)...(a+n-1)\,b(b+1)(b+2)...(b+n-1)x^n}{1\cdot 2\cdot 3...n\cdot c(c+1)(c+2)...(c+n-1)}$$

$$\underset{n\to\infty}{\text{Lt}}\ \frac{u_{n+1}}{u_n} = \underset{n\to\infty}{\text{Lt}}\ \frac{(a+n)(b+n)x}{(n+1)(c+n)} = x\ \underset{n\to\infty}{\text{Lt}}\ \frac{\left(1+\frac{a}{n}\right)\left(1+\frac{b}{n}\right)}{\left(1+\frac{1}{n}\right)\left(1+\frac{c}{n}\right)} = x$$

$\therefore$ By ratio test Σu_n is convergent for $x < 1$ and divergent for $x > 1$.

When $x = 1$, $$n\left(\frac{u_n}{u_{n+1}}-1\right) = n\left[\frac{(n+1)(c+n)}{(a+n)(b+n)}-1\right]=\frac{n[n(c+1)+c-ab-n(a+b)]}{(a+n)(b+n)}$$

$$\therefore \quad \text{Lt}\, n\left(\frac{u_n}{u_{n+1}}-1\right) = \underset{n\to\infty}{\text{Lt}} \frac{n^2(c+1-a-b)+n(c-ab)}{n^2+n(a+b)+ab}=c+1-a-b$$

$\Rightarrow$ by Raabe's test, for $x = 1$ Σu_n is convergent for $c > a + b$ divergent for $c \le a + b$.

Thus, Σu_n is convergent for $x < 1$, divergent for $x > 1$.

When $x = 1$, convergent for $c > a + b$ and divergent for $c \le a + b$. **Ans.**

13. *Test the convergence of the series* $\displaystyle\sum_{n=2}^{\infty}\frac{1}{n(\log n)^2}$

SOLUTION: $\displaystyle\sum_{n=2}^{\infty}\frac{1}{n(\log n)^2} = \Sigma u_n$, hence $u_n = \dfrac{1}{n(\log n)^2}$, $n \ge 2$

Consider $$\int_2^{\infty}\frac{1\,dx}{x(\log x)^2} = \left[\frac{-1}{\log x}\right]_2^{\infty} = \frac{1}{\log 2} = \text{finite}$$

therefore, by integral test the given series is convergent. **Ans.**

14. *Test the convergence of the series* $1+\frac{2}{5}x+\frac{6}{9}x^2+\frac{14}{17}x^3+\ldots \quad (x>0)$

SOLUTION: $1+\frac{2}{5}x+\frac{6}{9}x^2+\frac{14}{17}x^3+\ldots \quad (x>0) = \Sigma u_n$

$$u_n = \frac{(n^2-2)x^n}{n^2+1} \quad \text{(not considering the first term 1)} \qquad \text{where } n = 2, 3, 4,\ldots$$

$$\therefore \quad \frac{u_{n+1}}{u_n} = \frac{[(n+1)^2-2]\,x^{n+1}}{(n+1)^2+1}\cdot\frac{n^2+1}{(n^2-2)x^n} = \frac{x(n^2+1)(n^2+2n-1)}{(n^2+2n+2)(n^2-2)}$$

$$\underset{n\to\infty}{\text{Lt}}\frac{u_{n+1}}{u_n} = x\;\underset{n\to\infty}{\text{Lt}}\frac{\left(1+\frac{1}{n^2}\right)\left(1+\frac{2}{n}-\frac{1}{n^2}\right)}{\left(1+\frac{2}{n}+\frac{2}{n^2}\right)\left(1-\frac{2}{n^2}\right)} = x$$

$\therefore$ Σu_n is convergent for $x < 1$ and divergent for $x > 1$ by ratio test.

When $x = 1$ $\quad \underset{n\to\infty}{\text{Lt}}\, u_n = 1 \neq 0$ hence divergent.

Thus the given series is convergent for $x < 1$ and divergent for $x \ge 1$. **Ans.**

15. *Test the convergence of the series* $\frac{a}{x+a}+\frac{a^2}{x^2+a^2}+\frac{a^3}{x^3+a^3}+\ldots$

SOLUTION: $\frac{a}{x+a}+\frac{a^2}{x^2+a^2}+\frac{a^3}{x^3+a^3}+\ldots=\Sigma u_n$

here
$$u_n=\frac{a^n}{x^n+a^n}=\frac{1}{1+\left(\frac{x}{a}\right)^n}$$

If $x<a$ $\underset{n\to\infty}{\text{Lt}}\ u_n=1\neq 0$ and $\underset{n\to\infty}{\text{Lt}}\ u_n=\frac{1}{2}\neq 0$ when $x=a$.

Therefore Σu_n is divergent when $x\leq a$.

When $x>a$, $\underset{n\to\infty}{\text{Lt}}\ u_n=0$. Further

$$\underset{n\to\infty}{\text{Lt}}\frac{u_{n+1}}{u_n}=\underset{n\to\infty}{\text{Lt}}\frac{a^{n+1}}{x^{n+1}+a^{n+1}}\cdot\frac{x^n+a^n}{a^n}=a\ \underset{n\to\infty}{\text{Lt}}\frac{x^n+a^n}{x^{n+1}+a^{n+1}}$$

$$=a\ \underset{n\to\infty}{\text{Lt}}\frac{1+\left(\frac{a}{x}\right)^n}{x+a\left(\frac{a}{x}\right)^n}=\frac{a(1+0)}{x+0}=\frac{a}{x}$$

which is less than 1 as $x>a$, hence Σu_n is convergent. Thus Σu_n is convergent for $x>a$ and divergent for $x\leq a$.

Ans.

SOLUTIONS TO EXERCISE 2C

1. *Test the series for convergence and absolute convergence* $1-\frac{1}{5}+\frac{1}{9}-\frac{1}{13}+\ldots$

SOLUTION: $1-\frac{1}{5}+\frac{1}{9}-\frac{1}{13}+\ldots=\Sigma u_n$

Here $u_n=\frac{(-1)^{n-1}}{4n-3}$. $|u_n+1|=\frac{1}{4n+1}<|u_n|=\frac{1}{4n-3}$ for all n.

Also $\underset{n\to\infty}{\text{Lt}}\ u_n=0$, therefore, by Lebneitz test the given alternating series is convergent.

It is not absolutely convergent since $\Sigma|u_n|=\Sigma\frac{1}{4n+1}$ is divergent.

2. *Test the series for convergence and absolute convergence* $1-\frac{3}{2}+\frac{5}{4}-\frac{7}{8}+\frac{9}{16}-...$

SOLUTION: $1-\frac{3}{2}+\frac{5}{4}-\frac{7}{8}+\frac{9}{16}-... = \Sigma u_n$

Here $u_n = \frac{(2n-1)(-1)^{n-1}}{2^{n-1}}$, hence $|u_n| = \frac{2n-1}{2^{n-1}}$ and $|u_{n+1}| = \frac{n+\frac{1}{2}}{2^{n-1}}$.

Let us check if $n+\frac{1}{2} < 2n-1$ or $n>\frac{3}{2}$ which is true when $n > 2$.

$\therefore \; |u_{n+1}| < |u_n|$. Next $\underset{n\to\infty}{\text{Lt}} |u_n| = \underset{n\to\infty}{\text{Lt}} \frac{2n-1}{2^{n-1}} = \underset{n\to\infty}{\text{Lt}} \left(\frac{2}{2^{n-2}} - \frac{1}{2^{n-1}}\right) = 0.$

Therefore by Lebneitz test the given alternating series is convergent. For absolute convergence

$\underset{n\to\infty}{\text{Lt}} \left|\frac{u_{n+1}}{u_n}\right| = \frac{1}{2} < 1$ hence $\Sigma |u_n|$ is absolutely convergent. **Ans.**

3. *Test the series for convergence and absolute convergence* $1-\frac{\sqrt{2}}{2}+\frac{\sqrt{3}}{3}-\frac{\sqrt{4}}{4}+...$

SOLUTION: $1-\frac{\sqrt{2}}{2}+\frac{\sqrt{3}}{3}-\frac{\sqrt{4}}{4}+... = 1-\frac{1}{\sqrt{2}}+\frac{1}{\sqrt{3}}-\frac{1}{\sqrt{4}}+...$

It is an alternating series in which, obviously

$$\underset{n\to\infty}{\text{Lt}} \; u_n = \underset{n\to\infty}{\text{Lt}} \; \frac{(-1)^{n-1}}{\sqrt{n}} = 0 \text{ and } |u_{n+1}| < |u_n|$$

Therefore, by Lebneitz theorem Σu_n is convergent. Also, $\Sigma \frac{1}{\sqrt{n}}$ is divergent as a p-series $p = \frac{1}{2} < 1$. Hence Σu_n is not absolutely convergent. **Ans.**

4. *Test the series for convergence and absolute convergence* $1-\frac{1}{2!}+\frac{1}{3!}-\frac{1}{4!}+...$

SOLUTION: $1-\frac{1}{2!}+\frac{1}{3!}-\frac{1}{4!}+... = \Sigma u_n$

Here $u_n = \frac{(-1)^{n-1}}{n!}$. Obviously $|u_{n+1}| < |u_n|$

and $\underset{n\to\infty}{\text{Lt}} \; u_n = 0$. Hence by Lebneitz's theorem the given alternating series in convergent.

Here, $\underset{n\to\infty}{\text{Lt}} \left|\frac{u_{n+1}}{u_n}\right| = \frac{1}{n+1} = 0 < 1$ hence absolutely convergent. **Ans.**

5. *Test the series for convergence and absolute convergence* $\sum_1^{\infty}(-1)^n\left(\sqrt{n+1}-\sqrt{n}\right)$.

SOLUTION: $\sum_1^{\infty}(-1)^n\left(\sqrt{n+1}-\sqrt{n}\right) = \Sigma u_n$. It is an alternating series.

$$|u_n| = \sqrt{n+1}-\sqrt{n} = \frac{n+1-n}{\sqrt{n+1}+\sqrt{n}} = \frac{1}{\sqrt{n+1}+\sqrt{n}}$$

Obviously $|u_{n+1}| < |u_n|$ and $\underset{n\to\infty}{\text{Lt}}\ u_n = 0$. Hence by Lebneitz's test the given alternating series is convergent.

Next $\Sigma\,|u_n|$ is of the order of $\Sigma\frac{1}{\sqrt{n}}$ which is divergent hence $\Sigma\,u_n$ is not absolutely convergent. **Ans.**

6. *Test the series for convergence and absolute convergence* $\sum_1^{\infty}(-1)^n\frac{1+n^2}{1+n^3}$.

SOLUTION: $\sum_{n=1}^{\infty}(-1)^n\left(\frac{1+n^2}{1+n^3}\right) = \Sigma u_n$. It is an alternating series.

$$\therefore \qquad |u_n| = \frac{1+n^2}{1+n^3} \quad \text{and} \quad |u_{n+1}| = \frac{1+(n+1)^2}{1+(n+1)^3}$$

Here $\underset{n\to\infty}{\text{Lt}}|u_n| = \underset{n\to\infty}{\text{Lt}}\frac{\frac{1}{n^2}+1}{\frac{1}{n^2}+n} = 0$ Let us check if $|u_n+1| < |u_n|$

i.e., if $\frac{n^2+2n+2}{n^3+3n^2+3n+2} < \frac{1+n^2}{1+n^3}$

or if $x^5 + 2n^4 + 2n^3 + n^2 + 2n + 2 < n^5 + 3n^4 + 3n^3 + 2n^2 + n^3 + 3n^2 + 3n + 2$

or $n^4 + 2n^3 + 4n^2 + n > 0$ which is true, hence $|u_{n+1}| < |u_n|$

Thus by Lebneitz's test Σu_n is convergent. However $\Sigma|u_n| = \Sigma\frac{1+n^2}{1+n^3}$ is divergent hence not absolutely convergent.

7. *Test the series for convergence and absolute convergence* $\sum_1^{\infty}(-1)^{n-1}\frac{n}{n^2+1}$.

SOLUTION: $\Sigma(-1)^{n-1}\frac{n}{n^2+1} = \Sigma u_n$ where $u_n = \frac{(-1)^{n-1}n}{n^2+1}$. It is an alternating series.

Here $\underset{n\to\infty}{\text{Lt}}|u_n| = \underset{n\to\infty}{\text{Lt}}\frac{1}{n+\frac{1}{n}} = 0.$

$|u_{n+1}| < |u_n|$. Let us check, if it is true.

$$\frac{n+1}{(n+1)^2+1} < \frac{n}{n^2+1} \quad \text{or} \quad n^3 + n^2 + n + 1 < n^3 + 2n^2 + 2n$$

or $n^2 + n - 1 > 0$ or $n^2 + n > 1$ which is obviously true.

Therefore, the given series is **convergent** by Lebneitz test. $\Sigma |u_n|$ is of the order of $\Sigma \frac{1}{n}$ hence is not absolutely convergent. **Ans.**

8. *Test the series for convergence and absolute convergence*

$$\frac{1}{2} - \frac{1}{2\cdot 3} + \frac{1}{2\cdot 3\cdot 4} - \frac{1}{2\cdot 3\cdot 4\cdot 5} + \ldots$$

SOLUTION: $\frac{1}{2} - \frac{1}{2\cdot 3} + \frac{1}{2\cdot 3\cdot 4} - \frac{1}{2\cdot 3\cdot 4\cdot 5} + \ldots = \Sigma u_n$

It is alternating series where $u_n = \frac{(-1)^{n-1}}{2\cdot 3\cdot 4\cdot 5\ldots(n+1)}$

Here $\underset{n\to\infty}{\text{Lt}}\ u_n = 0$. $\left|\frac{u_{n+1}}{u_n}\right| = \frac{1}{n+2}$ which is less than 1.

$\therefore$ The given series is **convergent** by Lebneitz test. It is absolutely convergent by ratio test since

$$\underset{n\to\infty}{\text{Lt}} \left|\frac{u_{n+1}}{u_n}\right| = 0 < 1.$$ **Ans.**

9. *Test the series for convergence and absolute convergence*

$$\frac{1}{\sqrt{2}-1} - \frac{1}{\sqrt{2}+1} + \frac{1}{\sqrt{3}-1} - \frac{1}{\sqrt{3}+1} + \ldots$$

SOLUTION: $\frac{1}{\sqrt{2}-1} - \frac{1}{\sqrt{2}+1} + \frac{1}{\sqrt{3}-1} - \frac{1}{\sqrt{3}+1} + \ldots$. It is an alternating series. Here $|u_{n+1}|$ not less than $|u_n|$. Let us group the terms pairwise since it does not affect the convergence or divergence of a series. So the given series can be written as:

$$\left(\frac{1}{\sqrt{2}-1} - \frac{1}{\sqrt{2}+1}\right) + \left(\frac{1}{\sqrt{3}-1} - \frac{1}{\sqrt{3}+1}\right) + \left(\frac{1}{\sqrt{4}-1} - \frac{1}{\sqrt{4}+1}\right) + \ldots$$

or $$\frac{2}{2-1} + \frac{2}{3-1} + \frac{2}{4-1} + \ldots = 2\left(\frac{1}{1} + \frac{1}{2} + \frac{1}{3} + \ldots\right) = 2\,\Sigma\,\frac{1}{n}$$

which is **divergent.** **Ans.**

10. *State with reason, the value of x for which the series* $x-\frac{x^2}{2}+\frac{x^3}{3}-\frac{x^4}{4}+\ldots$ *converges.*

SOLUTION: $x-\frac{x^2}{2}+\frac{x^3}{3}-\frac{x^4}{4}+\ldots=\Sigma u_n$

Case I : $0<x\le 1$, Σu_n is an alternating series

$$\underset{n\to\infty}{\text{Lt}}|u_n| = \underset{n\to\infty}{\text{Lt}}\frac{x^n}{n}=0 \text{ and } \left|\frac{u_{n+1}}{u_n}\right| = \frac{nx}{n+1} = \frac{x}{1+\frac{1}{n}}<1$$

Hence Σu_n is convergent when $0\le x\le 1$.

Case II : $x>1$, Σu_n is alternating series

$$\underset{n\to\infty}{\text{Lt}}|u_n| = \underset{n\to\infty}{\text{Lt}}\frac{x^n}{n}\left(\text{it is of } \frac{\infty}{\infty} \text{ form}\right).$$

$$= \underset{n\to\infty}{\text{Lt}}\frac{x^n\log x}{1}=\infty$$

Hence Σu_n is divergent when $x>1$.

Case III : $-1<x<0$, Σu_n is a positive term series.

then $$\underset{n\to\infty}{\text{Lt}}\frac{u_{n+1}}{u_n} = \text{Lt}\frac{x^{n+1}}{n+1}\cdot\frac{n}{x^n} = x\,\text{Lt}\frac{n}{n+1}=x$$

Σu_n is convergent for $-1<x<0$ by ratio test.

Case IV : $x\le -1$, Σu_n is *a +ve* term series.

then $$\underset{n\to\infty}{\text{Lt}}\frac{u_{n+1}}{u_n}=|x|>1$$

$\therefore$ Σu_n is divergent for $x<-1$

Thus Σu_n is **convergent for $|x|\le 1$**

11. *Prove that the series* $\frac{\sin x}{1^3}-\frac{\sin 2x}{2^3}+\frac{\sin 3x}{3^3}-\ldots$ *converges absolutely.*

SOLUTION: $\frac{\sin x}{1^3}-\frac{\sin 2x}{2^3}+\frac{\sin 3x}{3^3}-\ldots=\Sigma u_n$

Σu_n is an alternating series. To test for absolute convergence

$$\underset{n\to\infty}{\text{Lt}}|u_n| = \underset{n\to\infty}{\text{Lt}}\left|\frac{\sin nx}{n^3}\right|=0$$

and $$|u_n| = \left|\frac{\sin nx}{n^3}\right|<\frac{1}{n^3}$$

and $\Sigma\frac{1}{n^3}$ is convergent hence $\Sigma\left|\frac{\sin nx}{n^3}\right|$ is also convergent by comparison test.

Thus given series is **absolutely convergent.**

12. *Test for convergence and absolute convergence of the series* $\sum_{2}^{\infty} \frac{(-1)^n}{n(\log n)^2}$

SOLUTION: Given series is convergent by Lebnitz's test and is absolutely convergent also by integral test. **Ans.**

SOLUTIONS TO EXERCISE 2D

1. *Test for uniform convergence, the series* $\frac{\cos x}{1^3}+\frac{\cos 2x}{2^3}+\frac{\cos 3x}{3^3}+\frac{\cos 4x}{4^3}+\ldots$

SOLUTION: $\frac{\cos x}{1^3}+\frac{\cos 2x}{2^3}+\frac{\cos 3x}{3^3}+\ldots=\Sigma u_n$

then $$u_n=\frac{\cos nx}{n^3}.$$

Since $\left|\frac{\cos nx}{n^3}\right|<\frac{1}{n^3}$ and $\Sigma\frac{1}{n^3}$ is convergent as a *p*-series with $p=3$ which is greater than 1.

Hence $\Sigma\frac{\cos nx}{n^3}$ is uniformly convergent for all x. **Ans.**

2. *Test for uniform convergence, the series* $\frac{\sin x}{2}+\frac{\sin 2x}{2^2}+\frac{\sin 3x}{2^3}+\frac{\sin 4x}{2^4}+\ldots$

SOLUTION: $\frac{\sin x}{2}+\frac{\sin 2x}{2^2}+\frac{\sin 3x}{2^3}+\frac{\sin 4x}{2^4}+\ldots=\Sigma u_n$

Here $u_n=\frac{\sin nx}{2^n}$. Further $\frac{\sin nx}{2^n}<\frac{1}{2^n}$

and $\Sigma\frac{1}{2^n}$ is convergent being a geometric series with common ratio less than $\frac{1}{2}$.

$\therefore$ Σu_n is uniformly convergent for all x. **Ans.**

3. *Test for uniform convergence, the series* $\frac{1}{1^3+1^4x^2}+\frac{1}{2^3+2^4x^2}+\frac{1}{3^3+3^4x^2}+\ldots$

SOLUTION: $\frac{1}{1^3+1^4x^2}+\frac{1}{2^3+2^4x^2}+\frac{1}{3^3+3^4x^2}+\ldots=\Sigma u_n$

where $$u_n=\frac{1}{n^3+n^4x^2}.$$

Since $\frac{1}{n^3+n^4+x^2}<\frac{1}{n^3}$ for all x and $\Sigma\frac{1}{n^3}$ is convergent therefore Σu_n is uniformly convergent for all x. **Ans.**

4. *Test for uniform convergence, the series* $\Sigma \dfrac{1}{n^2 + x^2}$.

SOLUTION: $\Sigma \dfrac{1}{n^2 + x^2} = \Sigma u_n$ where $u_n = \dfrac{1}{n^2 + x^2}$

Since $\dfrac{1}{n^2 + x^2} < \dfrac{1}{n^2}$ for all x and $\Sigma \dfrac{1}{n^2}$ is convergent, therefore $\Sigma\, u_n$ is uniformly convergent for all x. **Ans.**

5. *Prove that the series* $\dfrac{1}{1+x^2} - \dfrac{1}{2+x^2} + \dfrac{1}{3+x^2} - \dfrac{1}{4+x^2} + \ldots$ *is uniformly convergent for all x but not absolutely.*

SOLUTION: $\dfrac{1}{1+x^2} - \dfrac{1}{2+x^2} + \dfrac{1}{3+x^2} - \dfrac{1}{4+x^2} + \ldots = \Sigma u_n$

It is an alternating series. Hence $\underset{n\to\infty}{\text{Lt}} |u_n| = \underset{n\to\infty}{\text{Lt}} \dfrac{1}{n + x^2} = 0$ and $\left|\dfrac{u_n + 1}{u_n}\right| = \dfrac{n + x^2}{n + 1 + x^2} < 1.$

Hence $|u_n + 1| < |u_n|$ $\quad\therefore\quad$ The given series converges uniformly for all x.

However $\Sigma\, |u_n| = \Sigma \dfrac{1}{n + x^2}$. Here $\dfrac{1}{n + x^2} < \dfrac{1}{n}$

and $\Sigma \dfrac{1}{n}$ is divergent hence $\Sigma |u_n|$ is not convergent that $\Sigma\, u_n$ is **not absolutely convergent. Ans.**

6. *Show that the series* $\dfrac{x}{1\,(1+x^2)} + \dfrac{x}{2\,(1+2x^2)} + \dfrac{x}{3\,(1+3x^2)} + \ldots$ *converges uniformly for all x.*

SOLUTION. $\dfrac{x}{1\,(1+x^2)} + \dfrac{x}{2\,(1+2x^2)} + \dfrac{x}{3\,(1+3x^2)} + \ldots = \Sigma u_n$

$$u_n = \frac{x}{n(1 + nx^2)} = \frac{x}{n^2\left(x^2 + \dfrac{1}{n}\right)} = \frac{1}{xn^2\left(1 + \dfrac{1}{nx^2}\right)} < \frac{1}{n^2 x}$$

But $\Sigma \dfrac{1}{xn^2}$ is convergent for all x hence $\Sigma\, u_n$ is uniformly convergent for all x. **Ans.**

ADDITIONAL IMPORTANT PROBLEMS

1. *Test the convergence of the series*

$$\frac{1}{x}+\frac{1}{x-1}+\frac{1}{x+1}+\frac{1}{x-2}+\frac{1}{x+2}+\ldots \quad \text{where} \quad x>0.$$

SOLUTION: We know that by addition and deletion of a finite number of terms in an infinite series the nature of the series does not change. Let us delete the first term then

$$\sum u_n = \left(\frac{1}{x-1}+\frac{1}{x+1}\right)+\left(\frac{1}{x-2}+\frac{1}{x+2}\right)+\left(\frac{1}{x-3}+\frac{1}{x+3}\right)+\ldots$$

$$= \frac{2x}{x^2-1^2}+\frac{2x}{x^2-2^2}+\frac{2x}{x^2-3^2}+\ldots$$

$$= \sum_{1}^{\infty}\frac{2x}{x^2-n^2} \qquad \therefore \qquad u_n = \frac{2x}{x^2-n^2}$$

By comparison test, let $u_n = \dfrac{1}{n^2}$

$$\text{then} \quad \underset{n\to\infty}{\text{Lt}}\frac{u_n}{v_n} = \underset{n\to\infty}{\text{Lt}}\frac{2x}{x^2-n^2}\cdot\frac{n^2}{1} = \underset{n\to\infty}{\text{Lt}}\frac{2x}{\dfrac{x^2}{n^2}-1}$$

$= -2x$ which is non-zero and finite.

But $\sum u_n$ is convergent as p-series since $p = 2 > 1$.

Hence by comparison test $\sum u_n$ is convergent.

2. *Test for convergence the series* $\displaystyle\sum_{n=1}^{\infty}\frac{x^{n-1}}{n\cdot 3^n}$.

SOLUTION: Here $u_n = \dfrac{x^{n-1}}{n\cdot 3^n}$ and $u_{n+1} = \dfrac{x^n}{(n+1)\,3^{n+1}}$

$$\lim_{n\to\infty}\frac{u_{n+1}}{u_n} = \lim_{n\to\infty}\frac{x^{n+1}}{(n+1)\,3^{n+1}}\cdot\frac{n\cdot 3^n}{x^n}$$

$$= \lim_{n\to\infty}\left[\frac{x}{3}\cdot\frac{n}{n+1}\right] = \frac{x}{3}.$$

By ratio test Σu_n is convergent if $\dfrac{x}{3} < 1$ and divergent if $\dfrac{x}{3} > 1$. Thus, Σu_n is convergent for $x < 3$ and divergent for $x > 3$.

When $x = 3$ we have $u_n = \dfrac{3^{n-1}}{n\cdot 3^n} = \dfrac{1}{3n}$.

This is divergent as $\sum\dfrac{1}{n}$ is divergent.

Thus, Σu_n is convergent for $x < 3$ and divergent for $x \geq 3$. **Ans.**

3. *Test the nature of the series* $\sum \frac{\sqrt{n}}{\sqrt{n^2+1}} x^n$.

SOLUTION: $u_n = \sqrt{\frac{n}{n^2+1}}\, x^n \quad \therefore \quad u_{n+1} = \sqrt{\frac{n+1}{(n+1)^2+1}}\, x^{n+1}$

$$\therefore \quad \underset{n\to\infty}{\text{Lt}} \frac{u_{n+1}}{u_n} = \underset{n\to\infty}{\text{Lt}} \sqrt{\frac{n+1}{n} \cdot \frac{n^2+1}{(n+1)^2+1}} \frac{x^{n+1}}{x^n} = x$$

Hence by the ratio test Σu_n is convergent for $x < 1$ and divergent for $x > 1$.

When $x = 1$, $u_n = \sqrt{\frac{n}{n^2+1}} = \sqrt{\frac{1}{n+\frac{1}{n}}}$ and $\underset{n\to\infty}{\text{Lt}}\, u_n = 0$

Take $v_n = \frac{1}{\sqrt{n}}$. Hence $\underset{n\to\infty}{\text{Lt}} \frac{u_n}{v_n} = \underset{n\to\infty}{\text{Lt}} \sqrt{\frac{n \cdot n}{n^2+1}} = 1$ which is non-zero and finite. But $\sum v_n = \sum \frac{1}{n^{1/2}}$ is divergent as a *p*-series, $p = \frac{1}{2} < 1$.

Therefore, by comparison test Σu_n is divergent as Σv_n is divergent.

Thus, Σu_n is convergent for $x < 1$ and divergent for $x \geq 1$.

4. *Test the convergence the series* $\frac{1}{2\sqrt{1}} + \frac{x^2}{3\sqrt{2}} + \frac{x^4}{4\sqrt{3}} + \frac{x^6}{6\sqrt{5}} + \ldots$

SOLUTION: $u_n = \frac{x^{2n-2}}{(n+1)\sqrt{n}}$ and $u_{n+1} = \frac{x^{2n}}{(n+2)\sqrt{n+1}}$

$$\underset{n\to\infty}{\text{Lt}} \frac{u_{n+1}}{u_n} = \underset{n\to\infty}{\text{Lt}} \left[\frac{x^{2n}}{(n+2)\sqrt{n+1}} \cdot \frac{(n+1)\sqrt{n}}{x^{2n-2}}\right] = \underset{n\to\infty}{\text{Lt}} \left(\frac{n+1}{n+2}\sqrt{\frac{n}{n+1}} \cdot x^2\right) = x^2$$

Hence by ratio test, Σu_n is convergent for $x^2 < 1$ and divergent for $x^2 > 1$.

Next, for $x^2 = 1$, $u_n = \frac{1}{(n+1)\sqrt{n}}$.

We take $v_n = \frac{1}{n^{3/2}}$ then $\underset{n\to\infty}{\text{Lt}} \frac{u_n}{v_n} = \underset{n\to\infty}{\text{Lt}} \frac{n\sqrt{n}}{(n+1)\sqrt{n}} = 1$

$\therefore$ Σu_n behaves as Σv_n does.

But $\sum v_n = \sum \frac{1}{n^{3/2}}$ is convergent as a *p*-series, $p = \frac{3}{2} > 1$.

Therefore, Σu_n is convergent for $x^2 \leq 1$ and divergent for $x^2 > 1$.

5. *Test for convergence, the series* $\sum_{1}^{\infty} \frac{4 \cdot 7 \cdot 10 \ldots (3n+1)\, x^n}{n!}$.

SOLUTION: $u_n = \frac{4 \cdot 7 \cdot 10 \ldots (3n+1)\, x^n}{n!}$ and $u_{n+1} = \frac{4 \cdot 7 \cdot 10 \ldots (3n+1)\,(3n+4) x^{n+1}}{(n+1)!}$

$$\therefore \quad \underset{n\to\infty}{\text{Lt}} \frac{u_{n+1}}{u_n} = \underset{n\to\infty}{\text{Lt}} \frac{3n+4}{n+1} \cdot x = x \underset{n\to\infty}{\text{Lt}} \frac{3+4/n}{1+1/n} = 3x$$

Therefore by ratio test Σu_n is convergent for $3x < 1$, i.e., $x < \frac{1}{3}$ and divergent for $x > \frac{1}{3}$.

However, for $x = \frac{1}{3}$, let us apply the Raabe's test.

$$\underset{n\to\infty}{\text{Lt}}\, n\left(\frac{u_n}{u_{n+1}} - 1\right) = \underset{n\to\infty}{\text{Lt}}\, n\left[\frac{(n+1)\,(3)}{3n+4} - 1\right]$$

$$= \underset{n\to\infty}{\text{Lt}} \frac{-n}{3n+4} = -\frac{1}{3} \quad \text{which is} < 1$$

therefore Σu_n diverges for $x = \frac{1}{3}$.

Thus, Σu_n converges for $x < \frac{1}{3}$ and diverges for $x \geq \frac{1}{3}$.

6. *Test for convergence, the series* $\sum \frac{1 \cdot 3 \cdot 5 \ldots (2n-1)}{2 \cdot 4 \cdot 6 \ldots 2n} \frac{x^{2n+1}}{(2n+1)}$.

SOLUTION: Here $u_n = \frac{1 \cdot 3 \cdot 5 \ldots (2n-1)}{2 \cdot 4 \cdot 6 \ldots 2n} \frac{x^{2n+1}}{2n+1}$ and $u_{n+1} = \frac{1 \cdot 3 \cdot 5 \ldots (2n-1)\,(2n+1)\, x^{2n+3}}{2 \cdot 4 \cdot 6 \ldots 2n\,(2n+2)\,(2n+3)}$

$$\therefore \quad \underset{n\to\infty}{\text{Lt}} \frac{u_{n+1}}{u_n} = \underset{n\to\infty}{\text{Lt}} \frac{2n+1}{2n+2} \frac{2n+1}{2n+3} x^2 = x^2$$

By ratio test Σu_n converges for $x^2 < 1$ and diverges for $x^2 > 1$.

Next, for $x^2 = 1$, we apply Raabe's test (or higher ratio test)

$$\text{then} \quad \underset{n\to\infty}{\text{Lt}}\, n\left(\frac{u_n}{u_{n-1}} - 1\right) = \underset{n\to\infty}{\text{Lt}}\, n\left[\frac{(2n+2)\,(2n+3)}{(2n+1)\,(2n+1)} - 1\right]$$

$$= \underset{n\to\infty}{\text{Lt}}\, n \frac{(6n+5)}{(2n+1)^2} = \frac{3}{2} \quad \text{which is greater than 1}$$

$\therefore$ By Raabe's test Σu_n is convergent for $x^2 = 1$.

Thus, the given series is convergent for $x^2 \leq 1$ and divergent for $x^2 > 1$.

7. *Find the convergence and divergence of the series*

$$\frac{1^2}{4^2}+\frac{5^2}{8^2}+\frac{9^2}{12^2}+\frac{13^2}{16^2}+\ldots$$

SOLUTION: $u_n=\frac{(4n-3)^2}{(4n)^2}$. This series Σu_n is a positive term series.

The necessary condition for its convergence is $\underset{n\to\infty}{\text{Lt}}\ u_n = 0$.

Here $\underset{n\to\infty}{\text{Lt}}\ u_n = \underset{n\to\infty}{\text{Lt}}\frac{(4n-3)^2}{(4n)^2}=\underset{n\to\infty}{\text{Lt}}\left(1-\frac{3}{4n}\right)^2=1\neq 0.$

Therefore the given series is divergent.

Alternatively, we can apply the logarithmic test here.

$$\frac{u_n}{u_n+1}=\left(\frac{4n-3}{4n}\right)^2\cdot\left(\frac{4n+4}{4n+1}\right)^2=\frac{\left(1-\frac{3}{4n}\right)^2\left(1+\frac{1}{n}\right)^2}{\left(1+\frac{1}{4n}\right)^2}=\left[\frac{\left(1-\frac{3}{4n}\right)\left(1+\frac{1}{n}\right)}{\left(1+\frac{1}{4n}\right)}\right]^2$$

$$\therefore\quad n\log\frac{u_n}{u_{n+1}} = 2n\left[\log\left(1-\frac{3}{4n}\right)+\log\left(1+\frac{1}{n}\right)-\log\left(1+\frac{1}{4n}\right)\right]$$

$$= 2n\left[\left\{\frac{-3}{4n}-\frac{1}{2}\left(\frac{3}{4n}\right)^2-\ldots\right\}+\left(\frac{1}{n}-\frac{1}{2n^2}+\ldots\right)-\left\{\frac{1}{4n}-\frac{1}{2}\left(\frac{1}{4n}\right)^2+\ldots\right\}\right]$$

$$=\left(\frac{-3}{2}-\frac{9}{16n}-\ldots\right)+\left(2-\frac{1}{n}+\ldots\right)-\left(\frac{1}{2}-\frac{1}{16n^2}+\ldots\right)$$

$$\therefore\quad \text{Lt}\ n\log\frac{u_n}{u_{n+1}} = \frac{-3}{2}+2-\frac{1}{2}=0<1.$$

Hence the given series is divergent by logarithmic test.

8. *Test the convergence of the series* $1+\frac{x}{2}+\frac{2!}{3^2}x^2+\frac{3!}{4^3}x^3+\frac{4!}{5^4}x^4+\ldots$

SOLUTION: Deleting the first term, we have $u_n=\frac{n!}{(n+1)^n}x^n$

Hence $$\frac{u_n}{u_{n+1}}=\frac{n!\,x^n}{(n+1)^n}\,\frac{(n+2)^{n+1}}{(n+1)!\,x^{n+1}}=\left(\frac{n+2}{n+1}\right)^{n+1}\frac{1}{x}=\frac{1}{x}\left(1+\frac{1}{n+1}\right)^{n+1}$$

$\therefore\quad \underset{n\to\infty}{\text{Lt}}\frac{u_n}{u_{n+1}}=\frac{e}{x}$. Therefore, by ratio test

series is convergent if $\frac{e}{x}>1$ or $x<e$

and it is divergent if $\frac{e}{x}<1$ or $x>e$

Next, when $x=e$, let us apply logarithmic test.

$$\log\frac{u_n}{u_{n+1}} = (n+1)\log\left(1+\frac{1}{n+1}\right)-\log e = (n+1)\left[\frac{1}{n+1}-\frac{1}{2(n+1)^2}+\frac{1}{3(n+1)^3}-\ldots\right]-1$$

$$= \frac{-1}{2(n+1)}+\frac{1}{3(n+1)^2}-\ldots$$

$$\therefore \quad \operatorname*{Lt}_{n\to\infty} n\log\frac{u_n}{u_{n+1}} = \operatorname*{Lt}_{n\to\infty}\left[\frac{-n}{2(n+1)}+\frac{n}{3(n+1)^2}-\ldots\right] = -\frac{1}{2} < 1$$

Hence Σu_n is divergent for $x = e$.

Thus, Σu_n is convergent for $x < e$ and divergent for $x \geq e$.

9. *Test for convergence the series* $\dfrac{(a+x)}{1!}+\dfrac{(a+2x)^2}{2!}+\dfrac{(a+3x)^3}{3!}+\ldots$

SOLUTION: $u_n = \dfrac{(a+nx)^n}{n!}$ and $u_{n+1} = \dfrac{[a+(n+1)x]^{n+1}}{(n+1)!}$

$$\therefore \quad \frac{u_n}{u_{n+1}} = \frac{(n+1)!}{n!}\,\frac{1}{a+(n+1)x}\left[\frac{a+nx}{a+(n+1)x}\right]^n$$

$$= \frac{n+1}{a+(n+1)x}\left[\frac{1+\dfrac{a}{nx}}{1+\dfrac{a+x}{nx}}\right]^n = \left(\frac{1}{x+\dfrac{a}{n+1}}\right)\frac{\left(1+\dfrac{a}{nx}\right)^n}{\left(1+\dfrac{a+x}{nx}\right)^n}$$

$$\therefore \quad \operatorname*{Lt}_{n\to\infty}\frac{u_n}{u_{n+1}} = \frac{1}{x}\,\frac{e^{a/x}}{e^{a/x+1}} = \frac{1}{ex}.$$

$\therefore$ By ratio test Σu_n is convergent for $\dfrac{1}{ex} > 1$ or $x < \dfrac{1}{e}$

and is divergent for $\dfrac{1}{ex} < 1$ or $x > \dfrac{1}{e}$.

When $x = \dfrac{1}{e}$ then $\dfrac{u_n}{u_{n+1}} = \dfrac{\left(1+\dfrac{ae}{n}\right)^n\left(1+\dfrac{1}{n}\right)e}{\left(1+\dfrac{ae+1}{n}\right)^{n+1}}$

Applying logarithmic test

$$\log\frac{u_n}{u_{n+1}} = n\log\left(1+\frac{ae}{n}\right)+\log\left(1+\frac{1}{n}\right)-(n+1)\log\left(1+\frac{ae+1}{n}\right)+\log e$$

$$= n\left(\frac{ae}{n}-\frac{a^2e^2}{2n^2}+\frac{a^3e^3}{3n^3}-\ldots\right)+\left(\frac{1}{n}-\frac{1}{2n^2}+\frac{1}{3n^3}-\ldots\right)$$

$$-(n+1)\left[\frac{ae+1}{n}-\frac{(ae+1)^2}{2n^2}+\frac{(ae+1)^3}{3n^3}-\ldots\right]+1$$

$$= \left(ae - \frac{a^2e^2}{2n} + \frac{a^3e^3}{3n^2} - ...\right) + \left(\frac{1}{n} - \frac{1}{2n^2} + \frac{1}{3n^3} - ...\right)$$

$$- \left[(ae+1) - \frac{1}{2}\frac{(ae+1)^2}{n} + \frac{1}{3}\frac{(ae+1)^3}{n^2} - ... + \frac{1+ae}{n} - \frac{1}{2}\left(\frac{1+ae}{n}\right)^2 + ...\right] + 1$$

$$\therefore \quad \underset{n\to\infty}{\text{Lt}}\ n \log \frac{u_n}{u_{n+1}} = \lim_{n\to\infty}\left[\left(-\frac{a^2e^2}{2} + \frac{1}{3}\frac{a^3e^3}{n} - ...\right) + \left(1 - \frac{1}{2n} + ...\right)\right.$$

$$\left. - \left\{\frac{-1}{2}(ae+1)^2 + \frac{1}{3}\frac{(ae+1)^3}{n} + (ae+1) - \frac{1}{2}\frac{(ae+1)^2}{n} + ...\right\}\right]$$

$$= -\frac{a^2e^2}{2} + 1 + \frac{1}{2}(a^2e^2 + 1 + 2ae) - (ae+1) = \frac{1}{2} < 1$$

Hence the series is divergent for $x = \frac{1}{e}$.

Thus, Σu_n is convergent for $x < \frac{1}{e}$ and divergent for $x \geq \frac{1}{e}$.

10. *Test the convergence of the series* $\left(\frac{2^2}{1^2} - \frac{2}{1}\right)^{-1} + \left(\frac{3^3}{2^3} - \frac{3}{2}\right)^{-2} + \left(\frac{4^4}{3^4} - \frac{4}{3}\right)^{-3} + ...$

SOLUTION: $u_n = \left[\left(\frac{n+1}{n}\right)^{n+1} - \frac{n+1}{n}\right]^{-n}$. Let us apply Cauchy's root test.

$$u_n^{1/n} = \left[\left(\frac{n+1}{n}\right)^{n+1} - \frac{n+1}{n}\right]^{-1} = \left[\left(1+\frac{1}{n}\right)^{n+1} - \left(1+\frac{1}{n}\right)\right]^{-1} = \left[\left(1+\frac{1}{n}\right)^n \cdot \left(1+\frac{1}{n}\right) - \left(1+\frac{1}{n}\right)\right]^{-1}$$

$\therefore$ $\text{Lt}\ u_n^{1/n} = (e-1)^{-1} = \frac{1}{e-1}$ which is less than 1.

Therefore, by root test the given series is convergent.

11. *Test the convergence of the series* $\sum_{1}^{\infty} \frac{(n - \log n)^n}{2^n \cdot n^n}$

SOLUTION: $u_n = \frac{(n-\log n)^n}{2^n \cdot n^n}$. Let us apply the root test here.

$$u_n^{1/n} = \frac{n - \log n}{2n} = \frac{1}{2} - \frac{\log n}{2n}$$

$$\therefore \quad \underset{n\to\infty}{\text{Lt}}\ u_n^{1/n} = \underset{n\to\infty}{\text{Lt}}\left[\frac{1}{2} - \frac{1}{2}\frac{\log n}{n}\right] = \frac{1}{2} - 0 \qquad \text{(using } l\text{'Hospital's rule)}$$

$$= \frac{1}{2} < 1$$

hence the given series is convergent by root test.

12. *Test for convergence the series* $\sum_{1}^{\infty} \frac{(n+1)^n x^n}{n^{n+1}}$.

SOLUTION: $u_n = \frac{(n+1)^n x^n}{n^{n+1}}$. Applying Cauchy's root test here, we get

$$u_n^{1/n} = \frac{(n+1)x}{n^{\frac{n+1}{n}}} = \frac{(n+1)x}{n \cdot n^{1/n}} = \left(1+\frac{1}{n}\right)\frac{x}{n^{1/n}}$$

$\therefore$ $\lim_{n\to\infty} u_n^{1/n} = (1+0)\frac{x}{1} = x$ since $\lim_{n\to\infty} n^{1/n} = 1.$

$\therefore$ By root test given series is convergent for $x < 1$ and divergent for $x > 1$.

13. *Test for convergence, the series* $\frac{1}{2}+\frac{2}{3}x+\left(\frac{3}{4}\right)^2 x^2+\left(\frac{4}{5}\right)^3 x^3+\ldots$

SOLUTION: Here let us delete the first term which does not affect the nature of the series, then

$$u_n = \left(\frac{n+1}{n+2}\right)^n x^n$$

Applying Cauchy's root test, we have

$$\lim_{n\to\infty} u_n^{1/n} = \lim_{n\to\infty}\left(\frac{n+1}{n+2}\cdot x\right) = 1 \cdot x$$

Therefore, by root test the given series converges for $x < 1$ and diverges for $x > 1$.
The case of $x = 1$.

$$u_n = \left(\frac{n+1}{n+2}\right)^n$$

We first find $\lim_{n\to\infty} u_n = k$, say, then

$$\log k = \lim_{n\to\infty} n \log\left(\frac{n+1}{n+2}\right) = \lim_{n\to\infty} \frac{\log(n+1)-\log(n+2)}{\frac{1}{n}} \qquad \left(\frac{0}{0}\text{ form}\right)$$

$$= \lim_{n\to\infty} \frac{\frac{1}{n+1}-\frac{1}{n+2}}{\frac{-1}{n^2}} = \lim_{n\to\infty} \frac{-n^2}{(n+1)(n+2)} = -1 \neq 0$$

$\therefore$ $k = e^{-1} < 1$ hence Σu_n is divergent for $x = 1$.

Thus, Σu_n is convergent for $x < 1$ and divergent for $x \geq 1$.

14. *Test for convergence of the series* $\sum_{2}^{\infty} \frac{1}{n \log n}$.

SOLUTION: $u_n = \frac{1}{n\log(n)} = f(n)$

$\therefore$ $f(x) = \frac{1}{x \log x}$. Let us apply the integral test here.

$$\int_2^\infty f(x)\,dx = \int_2^\infty \frac{dx}{x\log x} = \lim_{m\to\infty}\int_2^m \frac{dx}{x\log x}$$

$$= \lim_{m\to\infty}\left[\log\log x\right]_2^m = \lim_{m\to\infty}(\log\log m - \log\log 2)$$

$$= \lim_{m\to\infty}\log\left(\log\frac{m}{2}\right) = \infty$$

Hence the series is divergent.

15. *Show that the harmonic series of order p, $\sum\frac{1}{n^p} = \frac{1}{1^p}+\frac{1}{2^p}+\frac{1}{3^p}+\ldots$ is convergent for $p > 1$ and is divergent for $p \le 1$; using integral test.*

SOLUTION: $u_n = \frac{1}{n^p} = f(n)$ then $f(x) = \frac{1}{x^p}$. We apply the integral test.

$$\therefore \quad \int_1^\infty \frac{dx}{x^p} = \lim_{m\to\infty}\left[\frac{x^{-p+1}}{-p+1}\right]_1^m = \lim_{m\to\infty}\left[\frac{m^{1-p}}{1-p}-\frac{1}{1-p}\right] \qquad (p \ne 1)$$

$$= \begin{cases} -\dfrac{1}{1-p} & \text{when } p>1 \\ \infty & \text{when } p<1 \end{cases}$$

When $p = 1$

$$\int_1^\infty f(x)dx = \int_1^\infty \frac{1}{x}dx = \lim_{m\to\infty}[\log x]_1^m = \lim_{m\to\infty}(\log m - \log 1)$$

$\therefore \quad \sum\frac{1}{n^p}$ is convergent for $p > 1$ and divergent for $p \le 1$.

16. *Test the convergence of the alternating series*

$$x - \frac{x^2}{2} + \frac{x^3}{3} - \frac{x^4}{4} + \ldots \qquad (0 < x < 1)$$

SOLUTION: $|u_n| = \frac{x^n}{n}$

$$|u_n| - |u_{n+1}| = \left|\frac{x^n}{n} - \frac{x^{n+1}}{n+1}\right| = \frac{x^n(n+1) - nx^{n+1}}{n(n+1)}$$

$$= \frac{x^n(n+1-nx)}{n(n+1)} = \frac{x^n[1+(1-x)n]}{n(n+1)} > 0 \qquad (\text{as } 0 < x < 1)$$

$\Rightarrow \quad u_n > u_{n+1}$

Also, $$\lim_{n\to\infty}|u_n| = \lim_{n\to\infty}\frac{x^n}{n} = 0 \qquad (\text{as } 0 < x < 1)$$

Hence the given series is convergent.

17. *Test the nature of the series* $\frac{1}{1^p}-\frac{1}{2^p}+\frac{1}{3^p}-\frac{1}{4^p}+\ldots$

SOLUTION: It is an alternating series. $|u_n| = \frac{1}{n^p}$

Case I: $p > 0$. $u_n - u_{n+1} = \frac{1}{n^p} - \frac{1}{(n+1)^p} = \frac{(n+1)^p - n^p}{n^p (n+1)^p} > 0$ (as $p > 0$)

$\therefore$ $|u_n| > |u_{n+1}|$. Also $\lim\limits_{n\to\infty} |u_n| = \lim\limits_{n\to\infty} \frac{1}{n^p} = 0$ (as $p > 0$)

Hence Σu_n is convergent alternating series for $p > 0$.

Case II: $p < 0$. In this case, the two conditions of Lebreitz's test are not satisfied hence oscillatory when $p < 0$.

18. *Test the series* $1-\frac{2}{3}+\frac{3}{3^2}-\frac{4}{3^3}+\ldots$ *for absolute convergence and conditional convergence.*

SOLUTION: $u_n = (-1)^{n-1}\frac{n}{3^{n-1}}$

$$\sum|u_n| = 1+\frac{2}{3}+\frac{3}{3^2}+\frac{4}{3^3}+\ldots = \sum\frac{n}{3^{n-1}}$$

$$\underset{n\to\infty}{\text{Lt}} \frac{|u_{n+1}|}{|u_n|} = \underset{n\to\infty}{\text{Lt}}\left(\frac{n+1}{3^n}\cdot\frac{3^{n-1}}{n}\right) = \frac{1}{3} < 1$$

Therefore the given series is absolutely convergent. The given series as such is convergent alternating series.

19. *Test the series* $\frac{2}{3}-\frac{3}{4}\cdot\frac{1}{2}+\frac{4}{5}\cdot\frac{1}{3}-\frac{5}{6}\cdot\frac{1}{4}+\ldots$ *for absolute or conditional convergence.*

SOLUTION: $u_n = (-1)^{n-1}\frac{n+1}{n+2}\cdot\frac{1}{n}$

$\therefore$ $|u_n| = \frac{n+1}{(n+2)\,n}$. We apply the comparison test.

Take $v_n = \frac{1}{n}$ then $\underset{n\to\infty}{\text{Lt}} \frac{|u_n|}{v_n} = \underset{n\to\infty}{\text{Lt}} \frac{n+1}{n+2} = 1$ (non-zero and finite)

Since $\sum\frac{1}{n}$ is divergent hence by comparison test $\sum|u_n|$ is also divergent. The $\sum u_n$ is not absolutely convergent.

Further $|u_{n+1}| - |u_n| = \frac{n+2}{(n+3)(n+1)} - \frac{n+1}{(n+2)\,n}$

$$= \frac{n(n+2)^2 - (n+3)(n+1)^2}{n(n+1)(n+2)(n+3)} = -\frac{(n^2+3n+3)}{n(n+1)(n+2)(n+3)} < 0$$

Hence $|u_{n+1}| < |u_n|$.

and $\underset{n\to\infty}{\text{Lt}} |u_n| = \lim\limits_{n\to\infty} \frac{n+1}{n(n+2)} = 0$

Therefore the given series is convergent but not absolutely hence is conditionally convergent.

CHAPTER

3 Successive Differentiation, Expansion of Functions and Approximate Calculations

IMPORTANT FORMULAE AND USEFUL RESULTS

1. (*i*) $\dfrac{d^n}{dx^n} e^{ax} = a^n e^{ax}$ (*ii*) $\dfrac{d^n}{dx^n} a^x = a^x (\log a)^n$.

2. (*i*) $\dfrac{d^n}{dx^n}(ax+b)^m = \dfrac{m!}{(m-n)!} a^n (ax+b)^{m-n}$

   (*ii*) $\dfrac{d^n}{dx^n}(ax+b)^n = n!\, a^n$ (*iii*) $\dfrac{d^n}{dx^n} \dfrac{1}{(ax+b)} = \dfrac{(-1)^n\, n!\, a^n}{(ax+b)^{n+1}}$

   (*iv*) $\dfrac{d^n}{dx^n} \log(ax+b) = \dfrac{(-1)^{n-1}(n-1)!\, a^n}{(ax+b)^n}$

3. $\dfrac{d^n}{dx^n} \begin{matrix}\sin\\ \cos\end{matrix}(ax+b) = a^n \begin{matrix}\sin\\ \cos\end{matrix}\left(ax+b+\dfrac{n\pi}{2}\right)$.

4. $\dfrac{d^n}{dx^n} e^{ax} \begin{matrix}\sin\\ \cos\end{matrix}(bx+c) = r^n e^{ax} \begin{matrix}\sin\\ \cos\end{matrix}(bx+c+n\alpha)$

   where $r = \sqrt{a^2+b^2}$ and $\alpha = \tan^{-1}(b/a)$.

5. **Leibnitz's theorem:** If $y = uv = u(x)\, v(x)$ then

   $y_n = (uv)_n = u_n v + {}^nC_1\, u_{n-1} v_1 + {}^nC_2\, u_{n-2} v_2 + \ldots + {}^nC_r\, u_{n-r} v_r + \ldots + uv_n$

   where $u_r = \dfrac{d^r}{dx^r} u$ and $v_r = \dfrac{d^r}{dx^r} v$

6. **Maclaurin's theorem:** If $f(x)$ is continuously differentiable any number of times and can be expanded in powers of x then $f(x) = f(0) + xf'(0) + \dfrac{x^2}{2!} f''(0) + \ldots + \dfrac{x^n}{n!} f^{(n)}(0) + \ldots$

7. **Taylors Theorem:**

Let $f(x)$ be a function of x and h be very small. If $f(x + h)$ is capable of being expanded in a convergent series of terms of positive integral powers of h, then

$$f(x + h) = f(x) + hf'(x) + \frac{h^2}{2!}f''(x) + \frac{h^3}{3!}f'''(x) + \ldots$$

or
$$f(x) = f(a) + (x - a)f'(a) + \frac{(x-a)^2}{2!}f''(a) + \ldots$$

8. **Approximate Error:** Let $y = f(x)$. If x suffers a change of δx and we intend to compute the corresponding change in y as, say δy,

then
$$\delta y = f(x + \delta x) - f(x)$$

and
$$\delta y = f(x) + f'(x)\,\delta x + \frac{f''(x)}{2!}(\delta x)^2 + \frac{f'''(x)}{3!}(\delta x)^3 + \ldots$$

SOLUTIONS TO EXERCISE 3A

1. *Find the nth derivative of the following functions*

(*i*) $\dfrac{x^2}{(x+2)(2x+3)}$ (*ii*) $\dfrac{x^4}{(x-1)(x-2)}$

SOLUTION: (*i*) $f(x) = \dfrac{x^2}{(x+2)(2x+3)} = \dfrac{1}{2} - \dfrac{(7x+6)}{2(x+2)(2x+3)} = \dfrac{1}{2} - \dfrac{4}{x+2} + \dfrac{9}{2(2x+3)}$

(On resolving into partial fractions)

$$\therefore \quad \frac{d^n}{dx^n} f(x) = 0 - 4\frac{(-1)^n n!}{(x+2)^{n+1}} + \frac{9}{2}\frac{(-1)^n 2^n n!}{(2x+3)^{n+1}}, \; n \geq 1$$

$$= \frac{4(-1)^{n+1} n!}{(x+2)^{n+1}} + \frac{9(-1)^n 2^n n!}{(2x+3)^{n+1}}.$$ **Ans.**

(*ii*) $$f(x) = \frac{x^4}{(x-1)(x-2)} = \frac{x^4}{x^2-3x+2} = x^2+3x+7+\frac{15x-14}{(x-1)(x-2)}$$

$$= x^2+3x+7-\frac{1}{x-1}+\frac{16}{x-2}$$

$$\therefore \quad f^{(n)}(x) = -\frac{(-1)^n n!}{(x-1)^{n+1}} + \frac{16(-1)^n n!}{(x-2)^{n+1}} = (-1)^n n!\left[\frac{16}{(x-2)^{n+1}} - \frac{1}{(x-1)^{n+1}}\right]$$

for $n = 3, 4, 5, \ldots$ **Ans.**

2. *Obtain the nth derivative of*

(*i*) $\sin^4 x$ (*ii*) $\dfrac{4x}{(x-1)^2(x+1)}$

SOLUTION: (*i*) $f(x) = \sin^4 x = (\sin^2 x)^2 = \left[\dfrac{1}{2}(1-\cos 2x)\right]^2 = \dfrac{1}{4}(1 - 2\cos 2x + \cos^2 2x)$

$$= \frac{1}{4} - \frac{1}{2}\cos 2x + \frac{1}{8}(1+\cos 4x) = \frac{3}{8} - \frac{1}{2}\cos 2x + \frac{1}{8}\cos 4x$$

$$\therefore \quad f^{(n)}(x) = 0 - \frac{1}{2}\cdot 2^n \cos\left(2x+\frac{n\pi}{2}\right) + \frac{1}{8}\cdot 4^n \cos\left(4x+\frac{n\pi}{2}\right)$$

$$= -2^{n-1}\cos\left(2x+\frac{n\pi}{2}\right) + 2^{2n-3}\cos\left(4x+\frac{n\pi}{2}\right)$$ **Ans.**

(*ii*) $$f(x) = \frac{4x}{(x-1)^2(x+1)} = \frac{A}{x-1} + \frac{B}{(x-1)^2} + \frac{C}{x+1}$$

$\Rightarrow \quad 4x = A(x-1)(x+1) + B(x+1) + C(x-1)^2$

or $\quad A + C = 0, \quad B - 2c = 4 \quad$ and $\quad -A + B + C = 0$

which give $\quad A = 1, \; B = 2, \; C = -1.$

$$\therefore \quad f(x) = \frac{1}{x-1} + \frac{2}{(x-1)^2} - \frac{1}{x+1}$$

$$\therefore \quad f^{(n)}(x) = \frac{(-1)^n n!}{(x-1)^{n+1}} - \frac{(-1)^n n!}{(x+1)^{n+1}} + \frac{2(-2)(-3)(-4)\ldots(-2-n+1)\,n!}{(x-1)^{n+2}}$$

$$= (-1)^n n! \left[\frac{1}{(x-1)^{n+1}} - \frac{1}{(x+1)^{n+1}} + \frac{2(n+1)}{(x-1)^{n+2}}\right] \qquad \textbf{Ans.}$$

3. *Determine* $\frac{d^2y}{dx^2}$ *if* $x = a(\theta + \sin\theta),\ y = a(1-\cos\theta)$

SOLUTION: $x = a(\theta + \sin\theta), \quad y = a(1-\cos\theta)$

$$\therefore \quad \frac{dy}{dx} = \frac{y'}{x'} = \frac{a\sin\theta}{a(1+\cos\theta)} = \frac{2\sin\frac{\theta}{2}\cos\frac{\theta}{2}}{2\cos^2\frac{\theta}{2}} = \tan\frac{\theta}{2}$$

Next
$$\frac{d^2y}{dx^2} = \frac{d}{dx}\left(\tan\frac{\theta}{2}\right) = \frac{d}{d\theta}\left(\tan\frac{\theta}{2}\right)\cdot\frac{d\theta}{dx}$$

$$= \frac{1}{2}\frac{\sec^2\frac{\theta}{2}}{x'} = \frac{\sec^2\frac{\theta}{2}}{2a(1+\cos\theta)} = \frac{\sec^2\frac{\theta}{2}}{4a\cos^2\frac{\theta}{2}} = \frac{1}{4}\sec^4\left(\frac{\theta}{2}\right) \qquad \textbf{Ans.}$$

4. *If* $y = e^{-kt}\cos(bt+c)$ *then prove that*

$$\frac{d^2y}{dt^2} + 2k\frac{dy}{dt} + n^2y = 0 \quad \text{where} \quad n^2 = k^2 + b^2.$$

SOLUTION:
$$y = e^{-kt}\cos(bt+c)$$

$$\therefore \quad \frac{dy}{dt} = -ke^{-kt}\cos(bt+c) - b\,e^{-kt}\sin(bt+c)$$

or
$$\frac{dy}{dt} = -ky - b\,e^{-kt}\sin(bt+c) \qquad \ldots(1)$$

and
$$\frac{d^2y}{dt^2} = -k\frac{dy}{dt} - b[e^{-kt}\,b\cos(bt+c) - ke^{-kt}\sin(bt+c)]$$

$$= -k\frac{dy}{dt} - b^2y + bk\,e^{-kt}\sin(bt+c)$$

$$= -k\frac{dy}{dt} - b^2y + k\left[-ky - \frac{dy}{dt}\right] = -2k\frac{dy}{dt} - (b^2+k^2)\,y$$

or
$$\frac{d^2y}{dt^2} + 2k\frac{dy}{dt} + n^2y = 0 \quad \text{where} \quad n^2 = b^2 + k^2 \qquad \textbf{Hence Proved.}$$

5. $y = \sin ax + \cos ax$, *show that*

$$y_n = a^n \sqrt{1+(-1)^n \sin 2ax}.$$

SOLUTION: $y = \sin ax + \cos ax$

$$\therefore \quad y_n = a^n \sin\left(ax + \frac{n\pi}{2}\right) + a^n \cos\left(ax + \frac{n\pi}{2}\right)$$

$$= a^n \sqrt{\left[\sin^2\left(ax+\frac{n\pi}{2}\right) + \cos^2\left(ax+\frac{n\pi}{2}\right) + 2\sin\left(ax+\frac{n\pi}{2}\right)\cos\left(ax+\frac{n\pi}{2}\right)\right]}$$

$$= a^n \sqrt{1+\sin(2ax+n\pi)} = a^n \sqrt{1+\sin 2ax \cos n\pi + \cos 2ax \sin n\pi}$$

$$= a^n \sqrt{1+(-1)^n \sin 2ax + 0} \quad \text{since} \quad \sin n\pi = 0 \quad \text{and} \quad \cos n\pi = (-1)^n$$ **Hence Proved.**

6. *If* $y = e^{-ax}\cos(bx+c)$, *show that*

$$y_n = (-1)^n (a^2+b^2)^{n/2} e^{-ax} \cos(bx + c + n\tan^{-1} b/a)$$

SOLUTION: $y = e^{-ax}\cos(bx+c)$

$$\therefore \quad y_n = r^n e^{-ax}\cos(bx+c+n\alpha) \quad \text{where} \quad r = \sqrt{a^2+b^2} \quad \text{and} \quad \alpha = \tan^{-1}\left(\frac{b}{-a}\right)$$

$$\therefore \quad y_n = (a^2+b^2)^{n/2} e^{-ax} \cos\left[bx + c + n\tan^{-1}\left(\frac{b}{-a}\right)\right]$$

We know that $\tan^{-1}\left(\frac{-b}{a}\right) = \pi - \tan^{-1}\frac{b}{a}$

$$\therefore \quad \cos\left[bx+c+n\tan^{-1}\left(\frac{b}{a}\right)\right] = \cos\left(bx+c+n\pi - n\tan^{-1}\frac{b}{a}\right) = \cos n\pi \cos\left(bx+c-n\tan^{-1}\frac{b}{a}\right)$$

Thus $$y_n = (-1)^n (a^2+b^2)^{n/2} e^{-ax} \cos\left(bx+c-n\tan^{-1}\frac{b}{a}\right) + 0$$ **Hence Proved.**

7. *Find the* nth *derivative of* $\tan^{-1}\dfrac{1-x}{1+x}$.

SOLUTION: $$y = \tan^{-1}\frac{1-x}{1+x} = \tan^{-1}\frac{1-\tan\alpha}{1+\tan\alpha} \quad \text{where} \quad x = \tan\alpha$$

$$= \tan^{-1}\tan\left(\frac{\pi}{4}-\alpha\right) = \frac{\pi}{4} - \alpha = \frac{\pi}{4} - \tan^{-1} x$$

$$\therefore \quad y_1 = 0 - \frac{1}{1+x^2} = \frac{-1}{(x-i)(x+i)} = \frac{1}{2i}\left[\frac{1}{x+i} - \frac{1}{x-i}\right]$$

and $$y_n = \frac{1}{2i}\left[\frac{(-1)^{n-1}(n-1)!}{(x+i)^n} - \frac{(-1)^{n-1}(n-1)!}{(x-i)^n}\right] = \frac{(-1)^{n-1}(n-1)!}{2i\,(x^2+1)^n}\left[(x-i)^n - (x+i)^n\right]$$

Now, putting $x = \cot\theta$, we have

$$y_n = \frac{(-1)^{n-1}(n-1)!}{2i\,\text{cosec}^{2n}\theta}\left[\left(\frac{\cos\theta - i\sin\theta}{\sin\theta}\right)^n - \left(\frac{\cos\theta + i\sin\theta}{\sin\theta}\right)^n\right]$$

$$= \frac{(-1)^{n-1}(n-1)!}{2i}\,\frac{\sin^{2n}\theta}{\sin^n\theta}\,[(\cos n\theta - i\sin n\theta) - (\cos n\theta + i\sin n\theta)]$$

$$= (-1)^n (n-1)!\,\sin^n\theta\,\sin n\theta \quad \text{where } \theta = \cot^{-1} x.$$

Ans.

8. *If* $y = \tan^{-1}x$, *prove that*

$$y_n = (n-1)!\cos\{ny + (n-1)\pi/2\}\cos^n y.$$

SOLUTION: $y = \tan^{-1} x$ hence $y_1 = \dfrac{1}{1+x^2} = \dfrac{1}{(x+i)(x-i)} = \dfrac{1}{2i}\left[\dfrac{1}{x-i} - \dfrac{1}{x+i}\right]$

$\therefore$
$$y_n = \frac{1}{2i}\frac{d^{n-1}}{dx^{n-1}}\left[\frac{1}{x-i} - \frac{1}{x+i}\right] = \frac{1}{2i}\left[\frac{(-1)^{n-1}(n-1)!}{(x-i)^n} - \frac{(-1)^{n-1}(n-1)!}{(x+i)^n}\right]$$

$$= \frac{(-1)^{n-1}(n-1)!}{2i}\left[\frac{\cos^n y}{(\sin y - i\cos y)^n} - \frac{\cos^n y}{(\sin y + i\cos y)^n}\right] \qquad \text{(Since } x = \tan y.\text{)}$$

$$= \frac{(-1)^{n-1}(n-1)!\cos^n y}{2i}\left[\frac{1}{\left\{\cos\left(\frac{\pi}{2} - y\right) - i\sin\left(\frac{\pi}{2} - y\right)\right\}^n} - \frac{1}{\left\{\cos\left(\frac{\pi}{2} - y\right) + i\sin\left(\frac{\pi}{2} - y\right)\right\}^n}\right]$$

$$= \frac{(-1)^{n-1}(n-1)!\cos^n y}{2i}\left[\frac{1}{\cos n\left(\frac{\pi}{2} - y\right) - i\sin n\left(\frac{\pi}{2} - y\right)} - \frac{1}{\cos n\left(\frac{\pi}{2} - y\right) + i\sin n\left(\frac{\pi}{2} - y\right)}\right]$$

(using de Moivre's theorem.)

$$= \frac{(-1)^{n-1}(n-1)!\cos^n y}{2i}\left[\cos\left(\frac{n\pi}{2} - ny\right) + i\sin\left(\frac{n\pi}{2} - ny\right) - \cos\left(\frac{n\pi}{2} - ny\right) + i\sin\left(\frac{n\pi}{2} - ny\right)\right]$$

$$= (-1)^{n-1}(n-1)!\cos^n y\sin\left(\frac{n\pi}{2} - ny\right)$$

Now
$$\sin\left(\frac{n\pi}{2} - ny\right) = \cos\left[\frac{\pi}{2} - \left(\frac{n\pi}{2} - ny\right)\right] = \cos\left[ny - (n-1)\frac{\pi}{2}\right] = \cos\left[ny + (n-1)\frac{\pi}{2} - (n-1)\pi\right]$$

$$= \cos\left\{ny + (n-1)\frac{\pi}{2}\right\}\cos(n+1)\pi + \sin\left\{ny + (n-1)\frac{\pi}{2}\right\}\sin(n-1)\pi$$

$$= (-1)^{n-1}\cos\left[ny + (n-1)\frac{\pi}{2}\right] + 0$$

$\therefore$
$$y_n = (-1)^{n-1}(n-1)!\cos^n y\,(-1)^{n-1}\cos\left\{ny + (n-1)\frac{\pi}{2}\right\} = (n-1)!\cos^n y\cos\left\{ny + (n-1)\frac{\pi}{2}\right\}.$$

Hence Proved.

9. *If* $(1-x^2)\tan y = 2x$, *show that*
$y_n = 2(-1)^{n-1}(n-1)!\sin^n\theta\sin n\theta$, where $\cot\theta = x$.

SOLUTION: $(1-x)^2\tan y = 2x$ or $\tan y = \dfrac{2x}{1-x^2}$ $\quad\therefore\ y = \tan^{-1}\dfrac{2x}{1-x^2} = 2\tan^{-1}x$

$$\therefore\quad y_1 = \frac{2}{1+x^2} = \frac{2}{(x+i)(x-i)} = \frac{1}{i}\left[\frac{1}{x-i}-\frac{1}{x+i}\right]$$

$$\therefore\quad y_n = \frac{d^{n-1}}{dx^{n-1}}y_1 = \frac{1}{i}\frac{d^{n-1}}{dx^{n-1}}\left[\frac{1}{x-i}-\frac{1}{x+i}\right] = \frac{1}{i}\left[\frac{(-1)^{n-1}(n-1)!}{(x-1)^n}-\frac{(-1)^{n-1}(n-1)!}{(x+i)^n}\right]$$

$$= \frac{(-1)^{n-1}}{i}(n-1)!\left[\frac{1}{(\cot\theta-i)^n}-\frac{1}{(\cot\theta+i)^n}\right] \quad (\text{taking } x = \cot\theta)$$

$$= \frac{(-1)^{n-1}}{i}(n-1)!\left[\frac{\sin^n\theta}{(\cos\theta-i\sin\theta)^n}-\frac{\sin^n\theta}{(\cos\theta+i\sin\theta)^n}\right]$$

$$= \frac{(-1)^{n-1}}{i}(n-1)!\ \sin^n\theta\,[\cos n\theta + i\sin n\theta - (\cos n\theta - i\sin n\theta)]$$

$$= \frac{(-1)^{n-1}}{i}(n-1)!\ \sin^n\theta\ 2i\sin n\theta = 2(-1)^{n-1}(n-1)!\sin^n\theta\sin n\theta \quad \text{where } \theta = \cot^{-1}x$$

Hence Proved.

10. *Find the* n^{th} *derivative of* $e^x\sin 4x\cos 6x$.

SOLUTION: Let $y = e^x\sin 4x\cos 6x$ then $y = \dfrac{e^x}{2}(\sin 10x - \sin 2x)$

$$\therefore\quad y_n = \frac{d^n y}{dx^n} = \frac{1}{2}\frac{d^n}{dx^n}(e^x\sin 10x) - \frac{1}{2}\frac{d^n}{dx^n}(e^x\sin 2x)$$

$$= \frac{1}{2}e^x\sqrt{1^2+10^2}\ \sin(10x + n\tan^{-1}10) - \frac{1}{2}e^x\sqrt{1^2+2^2}\ \sin(2x+n\tan^{-1}2)$$

$$= \frac{1}{2}e^x[(\sqrt{101})^n\sin(10x+n\tan^{-1}10)-(\sqrt{5})^n\sin(2x+n\tan^{-1}2)]$$ **Ans.**

SOLUTIONS TO EXERCISE 3B

1. *Find the* n^{th} *derivative of*

(*i*) $x^2 \sin x$ (*ii*) $e^x \log x$.

SOLUTION: (*i*) $\dfrac{d^n}{dx^n}(x^2 \sin x) = x^2 \dfrac{d^n}{dx^n} \sin x + {}^nC_1\, 2x \dfrac{d^{n-1}}{dx^{n-1}} \sin x + {}^nC_2 \cdot 2 \dfrac{d^{n-2}}{dx^{n-2}} \sin x$

(using Lebneitz's Theorem)

$$= x^2 \sin\left(x + \frac{n\pi}{2}\right) + 2nx \sin\left[x + (n-1)\frac{\pi}{2}\right] + \frac{n(n-1)}{2} \cdot 2\sin\left[x + (n-2)\frac{\pi}{2}\right]$$

$$= x^2 \sin\left(x + \frac{n\pi}{2}\right) + 2nx \sin\left(x + \frac{n\pi}{2} - \frac{\pi}{2}\right) + n(n-1) \sin\left(x + \frac{n\pi}{2} - \pi\right)$$

$$= x^2 \sin\left(x + \frac{n\pi}{2}\right) - 2nx \cos\left(x + \frac{n\pi}{2}\right) - n(n-1) \sin\left(x + \frac{n\pi}{2}\right)$$

$$= (x^2 - n^2 + n) \sin\left(x + \frac{n\pi}{2}\right) - 2nx \cos\left(x + \frac{n\pi}{2}\right) \qquad \textbf{Ans.}$$

(*ii*) $\dfrac{d^n}{dx^n}(e^x \log x)$ (Using Lebneitz's theorem)

$$= \log x \frac{d^n}{dx^n} e^x + {}^nC_1 \frac{1}{x} \frac{d^{n-1}}{dx^{n-1}} e^x + {}^nC_2 \left(\frac{-1}{x^2}\right) \frac{d^{n-2} e^x}{dx^{n-2}} + {}^nC_3 \left(\frac{2}{x^3}\right) \frac{d^{n-3}}{dx^{n-3}} e^x + \ldots + (-1)^{n-1} \frac{(n-1)!}{x^n} e^x$$

$$= e^x \left[\log x + \frac{{}^nC_1}{x} - \frac{{}^nC_2}{x^2} + {}^nC_3 \frac{2!}{x^3} - \ldots + (-1)^{n-1} \frac{(n-1)!}{x^n}\right] \qquad \textbf{Ans.}$$

2. *If* $y = x \log(x+1)$, *show that* $y_n = \dfrac{(-1)^{n-2}\,(n-2)!\,(x+n)}{(x+1)^n}$

SOLUTION: $y = x \log(x + 1)$. Using Lebneitz's theorem here, gives

$$y_n = x \frac{d^n}{dx^n} \log(x+1) + {}^nC_1 \cdot 1 \frac{d^{n-1}}{dx^{n-1}} \log(x+1) + 0$$

$$= x \frac{(-1)^{n-1}(n-1)!}{(x+1)^n} + n \frac{(-1)^{n-2}\,(n-2)!}{(x+1)^{n-1}}$$

$$= \frac{(-1)^{n-2}(n-2)!}{(x+1)^n} [-(n-1)x + n(x+1)]$$

$$= \frac{(-1)^{n-2}(n-2)!}{(x+1)^n} (x+n) \qquad \textbf{Hence Proved.}$$

3. *If* $y = a\cos(\log x) + b\sin(\log x)$, *show that*

$$x^2 y_{n+2} + (2n+1)x\, y_{n+1} + (n^2+1)\, y_n = 0$$

SOLUTION: $y = a\cos(\log x) + b\sin(\log x)$

$$\therefore \quad y_1 = -\frac{a}{x}\sin(\log x) + \frac{b}{x}\cos(\log x) \quad \text{or} \quad xy_1 = -a\sin(\log x) + b\cos(\log x)$$

and hence $xy_2 + 1.\, y_1 = -\frac{a}{x}\cos(\log x) - \frac{b}{x}\sin(\log x)$ or $x^2 y_2 + xy_1 = -y$

Differentiating both sides *w.r.t.* x, n times, and using Lebneitz's theorem, gives

$$x^2 y_{n+2} + {}^nC_1 \cdot 2x\, y_{n+1} + {}^nC_2 \cdot 2y_n + xy_{n+1} + {}^nC_1 \cdot 1 \cdot y_n = -y_n$$

or $$x^2 y_{n+2} + (2n+1)\, xy_{n+1} + \left(\frac{n(n-1)}{2}\cdot 2 + n + 1\right) y_n = 0$$

or $$x^2 y_{n+2} + (2n+1)\, xy_{n+1} + (n^2+1)\, y_n = 0$$ **Hence Proved.**

4. *If* $f(x) = x^2 e^{-x/a}$, *show that* $f_n(0) = \dfrac{n(n-1)(-1)^n}{a^{n-2}}$.

SOLUTION: $f(x) = x^2 e^{-x/a}$. Differentiating n times using Lebneitz theorem, we get

$$f^{(n)}(x) = x^2 \cdot \left(\frac{-1}{a}\right)^n e^{-x/a} + {}^nC_1 \cdot 2x \left(\frac{-1}{a}\right)^{n-1} e^{-x/a} + {}^nC_2 \cdot 2\left(\frac{-1}{a}\right)^{n-2} e^{-x/a}$$

$$\therefore \quad f^{(n)}(0) = 0 + 0 + \frac{n(n-1)}{2}\cdot 2\left(\frac{-1}{a}\right)^{n-2} = \frac{n(n-1)}{a^{n-2}}(-1)^{n-2}$$

$$= \frac{n(n-1)(-1)^n}{a^{n-2}} \quad (\text{since } (-1)^{n-2} = (-1)^n (-1)^{-2} = (-1)^n).$$ **Hence Proved.**

5. *If* $y = \tan^{-1}\left(\dfrac{a+x}{a-x}\right)$; *prove that* $(a^2+x^2)\, y_{n+2} + 2(n+1)\, x\, y_{n+1} + n(n+1)\, y_n = 0$

SOLUTION: $y = \tan^{-1}\dfrac{a+x}{a-x} = \tan^{-1}\dfrac{1+\frac{x}{a}}{1-\frac{x}{a}}$ Put $x = a\tan\theta$, then

$$y = \tan^{-1}\frac{1+\tan\theta}{1-\tan\theta} = \tan^{-1}\tan\left(\frac{\pi}{4}+\theta\right) = \frac{\pi}{4}+\theta = \frac{\pi}{4} + \tan^{-1}\left(\frac{x}{a}\right)$$

$$\therefore \quad y_1 = \frac{1}{1+\frac{x^2}{a^2}}\cdot\frac{1}{a} = \frac{a}{x^2+a^2} \quad \text{or} \quad (x^2+a^2)\, y_1 = a$$

Differentiating both sides $(n+1)$ times, using Lebneitz's theorem, we get

$$(x^2+a^2)\, y_{n+2} + {}^{n+1}C_1 \cdot 2x\, y_{n+1} + {}^{(n+1)}C_2 \cdot 2 \cdot y_n = 0$$

or $$(x^2+a^2)\, y_{n+2} + 2(n+1)x\, y_{n+1} + n(n+1)\, y_n = 0$$ **Hence Proved.**

6. *Given that* $y = \left\{\log\left(x+\sqrt{a^2+x^2}\right)\right\}^2$, *show that*

(*i*) $(a^2+x^2)y_2+xy_1=2$ (*ii*) $(a^2+x^2)\,y_{n+2}+(2n+1)\,xy_{n+1}+n^2\,y_n=0.$

SOLUTION: $y=\left[\log\left(x+\sqrt{x^2+a^2}\right)\right]^2$

$$\therefore \quad y_1 = \frac{2\log\left(x+\sqrt{x^2+a^2}\right)}{x+\sqrt{x^2+a^2}}\left[1+\frac{1\cdot 2x}{2\sqrt{x^2+a^2}}\right] = \frac{2\log\left(x+\sqrt{x^2+a^2}\right)}{\sqrt{x^2+a^2}}$$

or $$y_1^2\,(x^2+a^2) = 4\left[\log\left(x+\sqrt{x^2+a^2}\right)\right]^2 = 4y$$

Differentiating again, we get

$$(x^2+a^2)\,2y_1\,y_2 + 2xy_1^2 = 4y_1$$

or $$(x^2+a^2)\,y_2 + xy_1 = 2$$ **Hence Proved.**

(*ii*) Differentiating n times on both sides, using Lebneitz's theorem

$$(x^2+a^2)\,y_{n+2} + {}^nC_1\cdot 2xy_{n+1} + {}^nC_2\cdot 2y_n + xy_{n+1} + {}^nC_1\cdot 1\,y_n = 0$$

or $$(x^2+a^2)\,y_{n+2} + (2n+1)y_{n+1} + n\,(n-1)y_n + ny_n = 0$$

or $$(x^2+a^2)\,y_{n+2} + (2n+1)\,xy_{n+1} + n^2\,y_n = 0$$ **Hence Proved.**

7. *Given that* $y=\left(x-\sqrt{x^2-1}\right)^m$, *show that*

$$(x^2-1)\,y_{n+2}+(2n+1)\,xy_{n+1}+(n^2-m^2)\,y_n=0.$$

SOLUTION: $y=\left(x-\sqrt{x^2-1}\right)^m$ then $y_1 = m\left(x-\sqrt{x^2-1}\right)^{m-1}\left(1-\frac{x}{\sqrt{x^2-1}}\right)$

or $$y_1\sqrt{x^2-1} = m\left(x-\sqrt{x^2-1}\right)^{m-1}\left(\sqrt{x^2-1}-x\right) = -m\left(x-\sqrt{x^2-1}\right)^m = -my$$

or $y_1^2(x^2-1) = m^2\,y^2$. Again differenting both sides, gives

$$(x^2-1)\,2y_1\,y_2 + 2x\,y_1^2 = 2m^2\,y\,y_1 \quad \text{or} \quad (x^2-1)\,y_2 + xy_1 = m^2y$$

Now differentiating it n times using Lebneitz theorem, we get

$$(x^2-1)\,y_{n+2} + {}^nC_1\cdot 2x\,y_{n+1} + {}^nC_2\cdot 2\,y_n + xy_{n+1} + {}^nC_1\,1\cdot y_n = m^2\,y_n$$

or $$(x^2-1)\,y_{n+2} + (2n+1)\,xy_{n+1} + (n\,(n-1)+n-m^2)\,y_n$$

or $$(x^2-1)\,y_{n+2} + (2n+1)\,xy_{n+1} + (n^2-m^2)\,y_n = 0$$ **Hence Proved.**

8. *If* $f(x) = \cot x$, *prove that*

$$^{n}C_1\, f_{n-1}(0) - {}^{n}C_3\, f_{n-3}(0) + {}^{n}C_5\, f_{n-5}(0) - \ldots = \cos n\frac{\pi}{2}$$

SOLUTION: $f(x) = \cot x$ or $\sin x\, f(x) = \cos x$

Differentiating both sides n times, using Lebneitz's theorem, we get

$$\sin x\, f^{n}(x) + {}^{n}C_1 \sin\left(x+\frac{\pi}{2}\right) f^{n-1}(x) + {}^{n}C_2 \sin\left(x+\frac{2\pi}{2}\right) f^{n-2}(x) + {}^{n}C_3 \sin\left(x+\frac{3\pi}{2}\right) f^{n-3}(x) + \ldots\ldots$$

$$+ 1 \sin\left(x+\frac{n\pi}{2}\right) f(x) = \cos\left(x+\frac{n\pi}{2}\right)$$

At $x = 0$, we get

$$0 + {}^{n}C_1 \cdot 1\, f^{n-1}(0) + {}^{n}C_2 \times 0\, f^{n-2}(0) - {}^{n}C_3 f^{n-3}(0) + \ldots + \sin\frac{n\pi}{2} f(0) = \cos\frac{n\pi}{2}$$

Hence the result.

9. *If* $x = \sin t,\quad y = \cos at,$ *show that*

$(1 - x^2)\, y_1{}^2 = a^2\,(1 - y^2)$ *and hence deduce that*

$(1 - x^2)\, y_{n+2} - (2n+1)\, x y_{n+1} - (n^2 - a^2)\, y_n = 0$

SOLUTION: $x = \sin t,\ y = \cos at$

$$y_1 = \frac{y'}{x'} = \frac{-a\sin at}{\cos t}$$

or $$y_1^2 = a^2\,\frac{(1-y^2)}{1-x^2} \quad \text{or} \quad (1 - x^2)\, y_1^2 = (1 - y^2)\, a^2$$

Differentiating both sides again, we get

$$(1 - x^2)\, 2y_1 y_2 - 2xy_1^2 = -\, 2a^2\, yy_1$$

or $$(1 - x^2)\, y_2 - xy_1 + a^2 y = 0$$

Now differentiating n times, using Lebneitz's theorem, we get

$$(1 - x^2)\, y_{n+2} + {}^{n}C_1\, (-2x)\, y_{n+1} + {}^{n}C_2\, (-2)\, y_n - xy_{n+1} - {}^{n}C_1 \cdot 1 \cdot y_n + a^2\, y_n = 0$$

or $$(1 - x^2)\, y_{n+2} - (2n + 1)\, xy_{n+1} + y_n\left[\frac{n(n-1)}{2}(-2) - n + a^2\right] = 0$$

or $$(1 - x^2)\, y_{n+2} - (2n + 1)\, xy_{n+1} - (n^2 - a^2)\, y_n = 0$$ **Hence the result.**

10. *If* $\cos^{-1}(y/b) = \log(x/n)^n$, *show that* $x^2 y_{n+2} + (2n+1) x y_{n+1} + 2n^2 y_n = 0.$

SOLUTION: $$\cos^{-1}\left(\frac{y}{b}\right) = \log\left[\left(\frac{x}{n}\right)^n\right] = n\log\left(\frac{x}{n}\right)$$

or $$y = b\cos\left[n\log\left(\frac{x}{n}\right)\right]$$

$$\therefore \quad y_1 = -b\sin\left[n\log\left(\frac{x}{n}\right)\right]\cdot\frac{n}{x} \quad \text{or} \quad xy_1 = -nb\sin\left(n\log\frac{x}{n}\right)$$

Differentiating again w.r.t. x, we get

$$\therefore \quad xy_2 + 1\cdot y_1 = -nb\cos\left(n\log\frac{x}{n}\right)\cdot\frac{n}{x}$$

or $\quad x^2y_2 + xy_1 = -n^2y$

Differentiating both sides n times, using Lebneitz theorem, we get

$$x^2y_{n+2} + {}^nC_1\,2xy_{n+1} + {}^nC_2\cdot 2\cdot y_n + xy_{n+1} + {}^nC_1\cdot y_n\cdot 1 = -n^2y_n$$

or $\quad x^2y_{n+2} + (2n+1)xy_{n+1} + (n(n-1) + n + n^2)y_n = 0$

or $\quad x^2y_{n+2} + (2n+1)xy_{n+1} + 2n^2y_n = 0.$ **Hence the result.**

11. *If* $x = \tan(\log y)$ *show that*

$$(1+x^2)y_{n+1} + (2nx-1)y_n + n(n-1)y_{n-1} = 0.$$

SOLUTION: $x = \tan(\log y) \quad \therefore \quad \log y = \tan^{-1}x$

and $\quad \frac{1}{y}y_1 = \frac{1}{1+x^2} \quad$ or $\quad (1+x^2)y_1 = y$

Differentiating both sides n times, using Lebneitz's theorem

$$(1+x^2)y_{n+1} + {}^nC_1\cdot 2x\,y_n + {}^nC_2\cdot 2\cdot y_{n-1} = y_n$$

or $\quad (1+x^2)y_{n+1} + (2nx-1)y_n + n(n-1)y_{n-1} = 0.$ **Hence the result.**

12. *If* $x = \tan y$, *prove that* $1(1+x^2)y_{n+1} + 2nxy_n + n(n-1)y_{n-1} = 0.$

SOLUTION: $x = \tan y$ hence $y = \tan^{-1}x$ and $y_1 = \frac{1}{1+x^2}$

or $\quad (1+x^2)y_1 = 1.$ Differentiating both sides w.r.t. x, n times, we get

$$(1+x^2)y_{n+1} + {}^nC_1\cdot 2xy_n + {}^nC_2\cdot 2y_{n-1} = 0$$

or $\quad (1+x^2)y_{n+1} + 2nx\,y_n + n(n-1)y_{n-1} = 0$ **Hence the result.**

13. *If* $y = (x^2-1)^n$, *prove that* $(x^2-1)y_{n+2} + 2xy_{n+1} - n(n+1)y_n = 0.$

SOLUTION: $y = (x^2-1)^n$ hence $\quad y_1 = n(x^2-1)^{n-1}\cdot 2x$

or $\quad (x^2-1)y_1 = 2nxy$

Differentiating both sides w.r.t. x, $(n+1)$ times, we get

$$(x^2-1)y_{n+2} + {}^{n+1}C_1\cdot 2x\cdot y_{n+1} + {}^{n+1}C_2\cdot 2\cdot y_n = 2nxy_{n+1} + {}^{n+1}C_1\cdot 2ny_n$$

or $\quad (x^2-1)y_{n+2} + [2(n+1)x - 2nx]y_{n+1} + (n+1)ny_n - 2n(n+1)y_n = 0$

or $\quad (x^2-1)y_{n+2} + 2xy_{n+1} - n(n+1)y_n = 0.$ **Hence the result.**

SOLUTIONS TO EXERCISE 3C

11. *Use Maclaurin's theorem to show that* $e^x \cos x = 1 + x - \frac{2x^3}{3!} - \frac{2^2 x^4}{4!} - \frac{2^2 x^5}{5!} + \frac{2^3 x^7}{7!} + \ldots$

SOLUTION: $f(x) = e^x \cos x$, $f^{(n)}(x) = r^n e^x \cos(x + n \tan^{-1} a)$ where $r = \sqrt{1^2 + 1^2} = \sqrt{2}$

and $\tan^{-1} \alpha = \frac{1}{1}$ so $\alpha = \frac{\pi}{4}$

$$\therefore \quad f^{(n)}(x) = 2^{n/2} e^x \cos\left(x + \frac{n\pi}{4}\right) \quad \text{and} \quad f^{(n)}(0) = 2^{n/2} \cdot 1 \cdot \cos \frac{n\pi}{4}.$$

By Maclaurins theorem, we have

$$f(x) = f(0) + xf'(0) + \frac{x^2}{2!} f''(0) + \ldots + \frac{x^n}{n!} f^{(n)}(0) + \ldots\ldots, \quad f(0) = 1$$

$$f'(0) = 2^{1/2} \cos\frac{\pi}{4} = 1, \quad f''(0) = 2^{2/2} \cos\frac{\pi}{2} = 0$$

$$f'''(0) = 2^{3/2} \cos\frac{3\pi}{4} = -2, f^{iv}(0) = 2^2 \cos\frac{4\pi}{4} = 4 \cos \pi = -4, \ldots \ldots$$

$$\therefore \quad e^x \cos x = 1 + x \cdot 1 + x^2 \cdot 0 + \frac{x^3(-2)}{3!} + \frac{x^4(-4)}{4!} + \ldots \ldots$$

$$= 1 + x - \frac{x^3}{3} + \frac{x^4}{6} - \ldots\ldots$$

Hence Proved.

2. *Show that* $\sqrt{1 + x + 2x^2} = 1 + \frac{x}{2} + \frac{7}{8}x^2 - \frac{7}{16}x^3 + \ldots$

SOLUTION: $f(x) = \sqrt{1 + x + 2x^2}$, here $f(0) = 1$

$$f'(x) = \frac{1}{2}(1 + x + 2x^2)^{-1/2}(1 + 4x) \quad \therefore \quad f'(0) = \frac{1}{2}$$

$$f''(x) = \frac{1}{2}(1 + x + 2x^2)^{-1/2} \cdot 4 - \frac{1}{4}(1 + x + 2x^2)^{-3/2}(1 + 4x)^2, \quad f''(0) = 2 - \frac{1}{4} = \frac{7}{4}$$

$$f'''(x) = 2\left(\frac{-1}{2}\right)(1 + x + 2x^2)^{-3/2}(1 + 4x) - \frac{1}{4}(1 + x + 2x^2)^{-3/2} 8(1 + 4x) + \frac{3}{8}(1 + x + 2x^2)^{-5/2}(1 + 4x)^3$$

$$\therefore \quad f'''(0) = -1 - 2 + \frac{3}{8} = \frac{-21}{8}, \text{ etc.}$$

By Maclaurins theorem $\quad f(x) = f(0) + x f'(0) + \frac{x^2}{2!} f''(0) + \frac{x^3}{3!} f'''(0) + \ldots$

$$= 1 + \frac{x}{2} + \frac{7}{8}x^2 - \frac{7}{16}x^3 + \ldots$$

Hence Proved.

3. *Expand* $\sin x$ *in powers of* $(x-\pi/2)$.

SOLUTION: By Taylor's theorem $f(x)=f(a)+(x-a)f'(a)+\dfrac{(x-a)^2}{2!}f''(a)+...$

Here $f(x)=\sin x$ and $a=\dfrac{\pi}{2}$.

$$\therefore \quad \sin x=\sin\left(\frac{\pi}{2}\right)+\left(x-\frac{\pi}{2}\right)\cos\left(\frac{\pi}{2}\right)+\frac{\left(x-\frac{\pi}{2}\right)^2}{2!}\left(-\sin\frac{\pi}{2}\right)+\frac{\left(x-\frac{\pi}{2}\right)^3}{3!}\left(-\cos\frac{\pi}{2}\right)+...$$

or
$$\sin x=1-\frac{1}{2!}\left(x-\frac{\pi}{2}\right)^2+\frac{1}{4!}\left(x-\frac{\pi}{2}\right)^4+...$$
Ans.

4. *Employing Maclaurin's theorem, show that*

$$e^{a\sin^{-1}x}=1+ax+\frac{(ax)^2}{2!}+\frac{a(1^2+a^2)x^3}{3!}+\frac{a^2(2^2+a^2)x^4}{4!}+...$$

and hence deduce that

$$e^{\theta}=1+\sin\theta+\frac{\sin^2\theta}{2!}+\frac{2}{3!}\sin^3\theta+\frac{5}{4!}\sin^4\theta+...$$

SOLUTION: $y=e^{a\sin^{-1}x}$ hence $y(0)=1$. Next $y_1=\dfrac{e^{a\sin^{-1}x}\cdot a}{\sqrt{1+x^2}}$ and $y_1(0)=a$

$\therefore \quad (1-x^2)\,y_1^2=a^2y^2$. Differentiating again w.r.t., x, we get

$$(1-x^2)\,2\,y_1y_2-2xy_1^2=2a^2yy_1 \quad \text{or} \quad (1-x^2)\,y_2-xy_1-a^2y=0 \qquad ...(1)$$

At $x=0$ we get $y_2(0)-0-a^2\,y(0)=0 \quad \therefore\ y_2(0)=a^2$.

Differentiating (1) n times w.r.t. x, using Lebneitz's theorem

$$(1-x^2)\,y_{n+2}+{}^nC_1\,(-2x)\,y_{n+1}+{}^nC_2\,(-2)\,y_n-xy_{n+1}-n\cdot 1\cdot y_n-a^2\,y_n=0$$

At $x=0$ we have $y_{n+2}(0)=[n\,(n-1)+n+a^2]\,y_n(0)=(n^2+a^2)\,y_n(0)$

For $n=1$, $y_3(0)=(1^2+a^2)\,y_1(0)=a\,(1^2+a^2)$

For $n=2$, $y_4(0)=(2^2+a^2)\,y_2(0)=a^2(2^2+a^2)$, and so on.

$$\therefore \quad y(x)=y(0)+xy_1(0)+\frac{x^2}{2!}y_2(0)+\frac{x^3}{3!}y_3(0)+\frac{x^4}{4!}y_n(0)+...$$

or
$$e^{a\sin^{-1}x}=1+ax+\frac{a^2x^2}{2!}+\frac{a(1^2+a^2)}{3!}x^3+\frac{a^2(2^2+a^2)}{4!}x^4+...$$

Putting here $x=\sin\theta$ and $a=1$, we get

$$e^{\theta}=1+\sin\theta+\frac{1}{2!}\sin^2\theta+\frac{1}{3!}(2)\sin^3\theta+\frac{1\cdot 5}{4!}\sin^4\theta+...$$
Hence Proved.

5. *Show that* sin (x + h) *differs from* sin x + h cos x *by not more than* $h^2/2$.

SOLUTION: By Taylor's theorem $f(x+h) = f(x) + hf'(x) + \frac{h^2}{2!}f''(x) + ...$

$$\sin(x+h) = \sin x + h\cos x - \frac{h^2}{2!}\sin x - \frac{h^3}{3!}\cos x + \frac{h^4}{4!}\sin x + ...$$

$$\therefore \quad \sin(x+h) - (\sin x + h\cos x) = -\frac{h^2}{2!}\sin x - \frac{h^3}{3!}\cos x + ...$$

Thus, $|\sin(x+h) - (\sin x + h\cos x)| = \left|\frac{-h^2}{2}\sin x - \frac{h^3}{3!}\cos x + ...\right| \le \frac{h^2}{2}$ (as $|\sin x| < 1$).

Hence Proved.

6. *If* $|x| < a$, *then show that within* $(x/a)^2$ *we have*

$$e^{x/a} = \sqrt{\frac{a+x}{a-x}} \text{ approximately.}$$

SOLUTION: $e^{x/a} = 1 + \frac{x}{a} + \frac{1}{2!}\frac{x^2}{a^2} + \frac{1}{3!}\frac{x^3}{a^3} + \frac{1}{4!}\frac{x^4}{a^4} + ...$

$$\sqrt{\frac{a+x}{a-x}} = \sqrt{\frac{1+\frac{x}{a}}{1-\frac{x}{a}}} = \left(1+\frac{x}{a}\right)^{1/2}\left(1-\frac{x}{a}\right)^{-1/2}$$

$$= \left[1 + \frac{1}{2}\frac{x}{a} + \frac{\frac{1}{2}\left(\frac{-1}{2}\right)}{2!}\frac{x^2}{a^2} + \frac{\frac{1}{2}\left(\frac{-1}{2}\right)\left(\frac{-3}{2}\right)}{3!}\frac{x^3}{a^3} + ...\right]$$

$$\left[1 + \frac{x}{2a} + \frac{\left(\frac{-1}{2}\right)\left(\frac{-3}{2}\right)}{2!}\frac{x^2}{a} + \frac{\left(\frac{-1}{2}\right)\left(\frac{-3}{2}\right)\left(\frac{-5}{2}\right)}{3!}\frac{x^3}{a^3} + ...\right]$$

$$= 1 + \left(\frac{x}{2a} + \frac{x}{2a}\right) + \frac{x^2}{a^2}\left(\frac{1}{4} - \frac{1}{8} + \frac{3}{8}\right) + \frac{x^3}{a^3}\left(\frac{1}{16} - \frac{1}{16} + \frac{3}{16} + ...\right) + \frac{x^4}{a^4}(\ ) + ...$$

$$= 1 + \frac{x}{a} + \frac{x^2}{2a^2} + \frac{x^3}{a^3}(\ ) + ...$$

Hence $e^{x/a} = \sqrt{\frac{a+x}{a-x}}$ approximately within $\frac{x^2}{a^2}$ since two expansions are identical upto second degree terms.

Hence Proved.

7. *The pressure p and volume v of a gas are connected by the relation $pv^{1.4} = c$ where c is constant. Find the percentage increase in the pressure corresponding to a decrease of 1/2 % in the volume.*

SOLUTION: $pv^{1.4} = c$ hence $p = cv^{-1.4}$. When volume increases pressure decreases.

Now $$\delta p = \frac{dp}{dv}\delta v = -1.4\, cv^{-2.4}\,\delta v$$

$$\Rightarrow \quad \frac{\delta p}{p}\times 100 = -1.4c\frac{v^{-2.4}}{p}\cdot\delta v\times 100$$

$$= -1.4\frac{v^{-2.4}}{cv^{-1.4}}\delta v\times 100 = -1.4\frac{\delta v}{v}\times 100$$

$$= -1.4\left(\frac{1}{2}\right) = -0.7 \quad \text{since} \quad \frac{\delta v}{v}\times 100 = \frac{1}{2}$$

Hence % decrease in pressure = 0.7%. **Ans.**

8. *A soap bubble of radius 2 cm shrinks to radius 1.9 cm. Estimate the decrease in (i) Volume (ii) Surface area.*

SOLUTION: (*i*) Volume of soap bubble $V = \frac{4}{3}\pi r^3$, $r = 2$, $\delta r = -0.1$

$$\therefore \quad \delta V = \frac{dV}{dr}\delta r = 4\pi r^2\,\delta r = 4\pi\cdot 2^2(-0.1) = -1.6\pi$$

$$= -1.6\times 3.14 = -5.024 \text{ cm}^3$$ **Ans.**

(*ii*) Surface area $S = 4\pi r^2$

$$\therefore \quad \delta S = \frac{dS}{dr}\delta r = 8\pi r\,\delta r = 8\pi(2)(-0.1) = -1.6\pi = -5.024 \text{ cm}^2.$$ **Ans.**

9. *If ABCD is a rectangular protector in which AB = 6 cm, BC = 2 cm and O is the mid point of AB. An angle BOP is indicated by a mark P on the edge CD. If, in setting off an angle q degrees, the mark is made $\frac{1}{100}$ of a cm away along the edge. Show that the error in the angle is $\frac{9}{10\pi}\sin^2\theta$ degrees.*

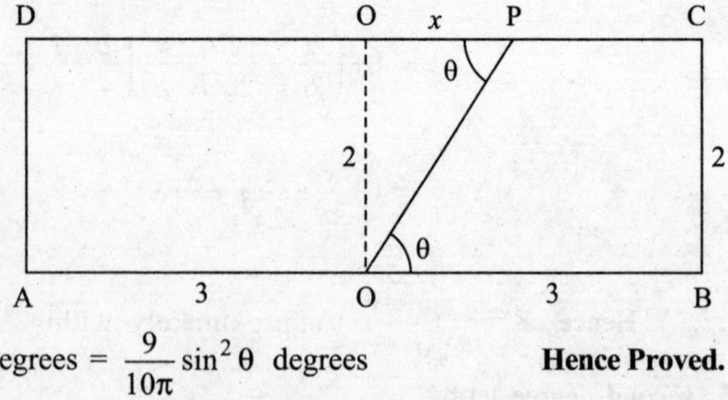

SOLUTION: $\cot\theta = \frac{x}{2}$, $\delta x = \frac{1}{100}$cm

$$-\operatorname{cosec}^2\theta\,\delta\theta = \frac{1}{2}\delta x$$

$$\delta\theta = \frac{-1}{2}\sin^2\theta\cdot\delta x \text{ in radians}$$

or $$\delta\theta = \frac{-1}{2}\sin^2\theta\cdot\frac{1}{100}\cdot\frac{180}{\pi} \text{ degrees} = \frac{9}{10\pi}\sin^2\theta \text{ degrees}$$ **Hence Proved.**

10. *Prove that* $f(mx) = f(x) + (m-1)x f'(x) + \dfrac{(m-1)^2}{2!} x^2 f''(x) + \ldots$

SOLUTION: To prove $f(mx) = f(x) + (m-1)\, xf'(x) + \dfrac{(m-1)^2}{2!} x^2 f''(x) + \ldots$ That means $f(mx)$ is to be expanded about the point x, in powers of $(mx - x)$, i.e., $(m-1)x$.

$$\therefore \quad f(mx) = f(x) + (mx - x) f'(x) + \frac{(mx-x)^2}{2!} f''(x) + \ldots$$

$$= f(x) + (m-1)\, xf'(x) + \frac{(m-1)^2}{2!} x^2 f''(x) + \ldots$$

Hence Proved.

11. *If* $x = y - \dfrac{1}{2}y^2 + \dfrac{1}{3}y^3 - \dfrac{1}{4}y^4 \ldots \infty$, *show that* $y = x + \dfrac{x^2}{2!} + \dfrac{x^3}{3!} + \dfrac{x^4}{4!} + \ldots \infty$.

SOLUTION: $x = y - \dfrac{1}{2}y^2 + \dfrac{1}{3}y^3 - \dfrac{1}{4}y^4 + \ldots \infty = \log(1+y)$. Hence $1 + y = e^x$

or $$y = e^x - 1 = \left(1 + x + \frac{x^2}{2!} + \frac{x^3}{3!} + \ldots\right) - 1 = x + \frac{x^2}{2!} + \frac{x^3}{3!} + \ldots$$

Hence Proved.

12. *Show that* $\log \sec x = \dfrac{x^2}{2} + \dfrac{1}{3}\dfrac{x^4}{4} + \dfrac{2}{15}\dfrac{x^6}{6} + \ldots$

SOLUTION: Let $f(x) = \log \sec x = -\log \cos x, \quad f(0) = 0$

$$f'(x) = -\frac{1}{\cos x}(-\sin x) = \tan x, \qquad f'(0) = 0$$

$$f''(x) = \sec^2 x \quad \therefore\ f''(0) = 1.$$

$$f'''(x) = 2 \sec x\, (\sec x \tan x) = 2 \sec^2 x \tan x, \quad \text{hence } f'''(0) = 0$$

$$f^{iv}(x) = 2 \sec^2 x \sec^2 x + 4 \sec x\, (\sec x \tan x) \tan x$$

$$= 2 \sec^4 x + 4 \sec^2 x\, (\sec^2 x - 1) = 6 \sec^4 x - 4 \sec^2 x \qquad \therefore\ f^{iv}(0) = 6 - 4 = 2$$

$$f^{(v)}(x) = 24 \sec^3 x\, (\sec x \tan x) - 8 \sec x \sec x \tan x$$

$$= 24 \sec^4 x \tan x - 8 \sec^2 x \tan x = 8 \tan x\, (3 \sec^4 x - \sec^2 x) \quad \therefore\ f^{(v)}(0) = 0.$$

$$\therefore \quad f^{(vi)}(x) = 8 \sec^2 x\, (3 \sec^4 x - \sec^2 x) + 8 \tan x\, (12 \sec^3 x \sec x \tan x - 2 \sec x \sec x \tan x)$$

$$\therefore \quad f^{vi}(0) = 8\,(3-1) + 0 = 16, \quad \text{and so on.}$$

Hence $$f(x) = f(0) + x f'(0) + \frac{x^2}{2!} f''(0) + \frac{x^3}{3!} f'''(0) + \frac{x^4}{4!} f^{iv}(0) + \frac{x^5}{5!} f^{(v)}(0) + \frac{x^6}{6!} f^{vi}(0) + \ldots$$

$$= 0 + 0 + \frac{x^2}{2!} \cdot 1 + 0 + \frac{x^4}{4!}(2) + \frac{x^5}{5!}(0) + \frac{x^6}{6!}(16) + \ldots = \frac{x^2}{2} + \frac{1}{3}\frac{x^4}{4} + \frac{2}{15}\frac{x^6}{6} + \ldots$$

Hence Proved.

13. *Compute to four decimal places the value of* $\cos 32^\circ$ *using Taylor's theorem.*

SOLUTION: To find the value of $\cos 32^\circ$. Let us take $f(x) = \cos x$

$$\cos 32^\circ = \cos (30^\circ + 2^\circ) = f(a + h) \quad \text{where } a = 30^\circ \text{ and } h = 2^\circ.$$

$$f(a + h) = f(a) + h f'(a) + \frac{h^2}{2!} f''(a) + \ldots$$

$$= \cos 30^\circ - 2^\circ \sin 30^\circ + \frac{(2^\circ)^2}{2}(-\cos 30^\circ) + \ldots$$

$$= \frac{\sqrt{3}}{2} - \frac{\pi}{180} - \frac{\sqrt{3}}{4}\left(\frac{2\pi}{180}\right)^2 - \ldots = 0.85 \text{ approx.}$$ **Ans.**

14. *Using Taylor's series compute* $\sin 31^\circ$ *to four decimal places.*

SOLUTION: To evaluate $\sin 31^\circ$, we take $f(x) = \sin x$

$$\sin 31^\circ = \sin (30^\circ + 1^\circ) = f(a + h) \quad \text{where } a = 30^\circ \text{ and } h = 1^\circ.$$

$$\therefore \quad f(a + h) = f(a) + h f'(a) + \frac{h^2}{2!} f''(a) + \frac{h^3}{3!} f'''(a) + \ldots$$

$$= \sin 30^\circ + 1^\circ \cos 30^\circ + \frac{(1^\circ)^2}{2}\sin 30^\circ - \frac{(1^\circ)^3}{6}\cos 30^\circ + \ldots$$

$$= \frac{1}{2} + \frac{\pi}{180}\frac{\sqrt{3}}{2} - \left(\frac{\pi}{180}\right)^2 \frac{1}{4} - \left(\frac{\pi}{180}\right)^3 \cdot \frac{1}{6}\frac{\sqrt{3}}{2} + \ldots$$

$$= 0.515 \text{ approx.}$$ **Ans.**

15. *Prove that* $\dfrac{e^x}{e^x + 1} = \dfrac{1}{2} + \dfrac{x}{4} - \dfrac{x^3}{48} + \ldots$

SOLUTION: $f(x) = \dfrac{e^x}{e^x + 1} = 1 - \dfrac{1}{e^x + 1}$ $\qquad f(0) = 1 - \dfrac{1}{1+1} = \dfrac{1}{2}$

$$f'(x) = \frac{e^x}{(e^x + 1)^2} = \frac{1}{e^x + 1} - \frac{1}{(e^x + 1)^2}, \quad f'(0) = \frac{1}{2} - \frac{1}{4} = \frac{1}{4}$$

$$f''(x) = \frac{-e^x}{(e^x + 1)^2} + \frac{2e^x}{(e^x + 1)^3},$$

or

$$f''(x) = -\frac{e^x + 1 - 1}{(e^x + 1)^2} + \frac{2(e^x + 1 - 1)}{(e^x + 1)^3} = \frac{-1}{e^x + 1} + \frac{1}{(e^x + 1)^2} + \frac{2}{(e^x + 1)^2} - \frac{2}{(e^x + 1)^3}$$

$$= \frac{-1}{e^x + 1} + \frac{3}{(e^x + 1)^2} - \frac{2}{(e^x + 1)^3} \qquad f''(0) = \frac{-1}{4} + \frac{2}{8} = 0$$

$$\therefore \quad f'''(x) = \frac{-e^x}{(e^x + 1)^2} - \frac{6e^x}{(e^x + 1)^3} + \frac{6e^x}{(e^x + 1)^4}, \quad f'''(0) = \frac{-1}{4} - \frac{6}{8} + \frac{6}{16} = \frac{-5}{8}$$

$$\therefore \quad f(x) = \frac{1}{2} + \frac{x}{4} + \frac{x^2}{2!}(0) + \frac{x^3}{3!}\left(\frac{-5}{8}\right) + \ldots = \frac{1}{2} + \frac{x}{4} - \frac{5x^3}{48} + \ldots$$ **Hence Proved.**

16. *Given* $\log_{10} 4 = 0.6021$ *calculate apprriximately* $\log_{10} 404$.

SOLUTION: Given $\log_{10} 4 = 0.6021$. To calculate $\log_{10}(404) = \log_{10}(400 + 4)$

$$f(x) = \log_{10} x = \log_e x \cdot \log_{10} e$$

$$f'(x) = \frac{1}{x}\log_{10} e,\ f''(x) = \frac{-1}{x^2}\log_{10} e,\ f'''(x) = \frac{2}{x^3}\log_{10} e,\ \ldots$$

$$\log(a+h) = \log a + h\cdot\frac{1}{a} + \frac{h^2}{2!}\left(\frac{-1}{a^2}\right) + \frac{h^3}{3!}\left(\frac{2}{a^3}\right) + \ldots \quad \text{Here, } a = 400,\ h = 4$$

$$\therefore \quad \log(404) = \left[\log 400 + \frac{4}{400} - \frac{4^2}{2}\frac{1}{(400)^2} + \frac{4^3}{6}\cdot\frac{2}{(400)^3} + \ldots\right]\log_{10} e$$

$$= \left[0.6021 + 0.01 - \frac{1}{20000}\right]\log_{10} e$$

$$= 2.612 \text{ approx.}$$

Ans.

ADDITIONAL IMPORTANT PROBLEMS

1. *Prove that* $\dfrac{d^{2n}}{dx^{2n}}(x^2-1)^n = (2n)!$

SOLUTION: We know that $(x^2-1)^n = (x^2)^n - {}^nc_1(x^2)^{n-1} + {}^nc_2(x^2)^{n-2} - \ldots (-1)^{n-1}$

Differentiating on both sides $2n$ times w.r.t. x, we get

$$\frac{d^{2n}}{dx^{2n}}(x^2-1)^n = \frac{d^{2n}}{dx^{2n}}[x^{2n} - {}^nc_1x^{2n-2} + {}^nc_2x^{2n-4} - \ldots + (-1)^{n-1}]$$

$$= \frac{d^{2n}}{dx^{2n}}x^{2n} + 0 + 0 + \ldots$$

$$= (2n)!$$

Hence Proved.

2. *If* $y = \dfrac{1}{x^4-a^4}$ *find* $\dfrac{d^ny}{dx^n}$.

SOLUTION:

$$y = \frac{1}{x^4-a^4} = \frac{1}{(x^2+a^2)(x^2-a^2)} = \frac{1}{2a^2}\left[\frac{1}{x^2-a^2} - \frac{1}{x^2+a^2}\right]$$

$$= \frac{1}{2a^2}\left[\frac{1}{(x+a)(x-a)} - \frac{1}{(x+ai)(x-ai)}\right]$$

$$= \frac{1}{2a^2}\left[\frac{1}{2a}\left(\frac{1}{x-a} - \frac{1}{x+a}\right) - \frac{1}{2ai}\left(\frac{1}{x-ai} - \frac{1}{x+ai}\right)\right]$$

or

$$y = \frac{1}{4a^3}\left(\frac{1}{x-a} - \frac{1}{x+a}\right) - \frac{1}{4a^3i}\left(\frac{1}{x-ai} - \frac{1}{x+ai}\right)$$

$\therefore$

$$y_n = \frac{(-1)^n n!}{4a^3}\left[\frac{1}{(x-a)^{n+1}} - \frac{1}{(x+a)^{n+1}}\right] - \frac{(-1)^n n!}{4a^3!}\left[\frac{1}{(x-ai)^{n+1}} - \frac{1}{(x+ai)^{n+1}}\right]$$

3. *Find* $\dfrac{d^ny}{dx^n}$ *if* $y = \dfrac{x^4}{(x-1)(x-2)}$

SOLUTION:

$$y = \frac{x^4}{(x-1)(x-2)} = \frac{x^4}{x^2-3x+2} = x^2 + 3x + 7 + \frac{15x-14}{x^2-3x+2}$$

$$= x^2 + 3x + 7 + \frac{15x-14}{(x-1)(x-2)} = x^2 + 3x + 7 + \left(\frac{-1}{x-1}\right) + \frac{16}{x-2}$$

Now differentiating y, w.r.t. x, n times, we get

$$y_n = -\frac{(-1)^n n!}{(x-1)^{n+1}} + \frac{16(-1)^n}{(x-2)^{n+1}} = (-1)^n n!\left[\frac{16}{(x-2)^{n+1}} - \frac{1}{(x-1)^{n+1}}\right] \quad n > 2.$$

Ans.

4. *If* $y = \tan^{-1}\left(\dfrac{x}{a}\right)$ *show that* $y_n = (-1)^{n-1}\dfrac{(n-1)!}{a^n}\sin n\theta \sin^n\theta$

where $\theta = \tan^{-1}\left(\dfrac{a}{x}\right)$.

SOLUTION: $y = \tan^{-1}\dfrac{x}{a}$. Differentiating w.r.t. x, gives

$$y_1 = \frac{a}{x^2+a^2} = \frac{a}{(x+ai)(x-ai)} = \frac{1}{2i}\left[\frac{1}{x-ai} - \frac{1}{x+ai}\right]$$

Now differentiating $(n - 1)$ times w.r.t. x, we get

$$y_n = \frac{(-1)^{n-1}(n-1)!}{2i}\left[\frac{1}{(x-ai)^n} - \frac{1}{(x+ai)^n}\right]$$

Let us write $x = r\cos\theta$, $a = r\sin\theta$ so $r = \sqrt{x^2+a^2}$, $\theta = \tan^{-1}\dfrac{a}{x}$, then

$$y_n = \frac{(-1)^{n-1}(n-1)!}{2i}\left[\frac{1}{r^n(\cos\theta - i\sin\theta)^n} - \frac{1}{r^n(\cos\theta + i\sin\theta)^n}\right]$$

$$= \frac{(-1)^{n-1}(n-1)!}{2i\cdot r^n}[(\cos\theta - i\sin\theta)^{-n} - (\cos\theta + i\sin\theta)^{-n}] \text{ (now using de Moivre's theorem)}$$

$$= \frac{(-1)^{n-1}(n-1)!}{2i.r^n}[(\cos n\theta + i\sin n\theta) - (\cos n\theta - i\sin n\theta) = \frac{(-1)^{n-1}(n-1)!}{r^n}\sin n\theta$$

$$= (-1)^{n-1}(n-1)!\frac{\sin n\theta}{(a/\sin\theta)^n} = \frac{(-1)^{n-1}(n-1)!}{a^n}\sin n\theta \sin^n\theta.$$ **Ans.**

5. *Find* n^{th} *order derivative of* $y = \cos^{-1}\left(\dfrac{x-x^{-1}}{x+x^{-1}}\right)$.

SOLUTION: We can write $y = \cos^{-1}\dfrac{x^2-1}{x^2+1}$. Putting $x = \tan\phi$,

$$y = \cos^{-1}\left(\frac{\tan^2\phi - 1}{\tan^2\phi + 1}\right) = \cos^{-1}(-\cos 2\phi) = \cos^{-1}\cos(\pi - 2\phi)$$

$$= \pi - 2\phi = \pi - 2\tan^{-1}x$$

$$\therefore \quad y_n = \pi - 2\frac{d^n}{dx^n}\tan^{-1}x. \text{ (Using the result of Problem 4 last)}$$

$$= \pi - 2\cdot(-1)^{n-1}(n-1)!\frac{\sin n\theta}{\sin^n\theta} \quad \text{where } \theta = \tan^{-1}\left(\frac{1}{x}\right)$$ **Ans.**

6. *If* $y = x\log(1+x)$ *show that* $y_n = \dfrac{(-1)^{n-1}(n-2)!(x+n)}{(x+1)^n}$.

SOLUTION: $y = x\log(1+x)$. $\therefore$ $y_1 = \dfrac{x}{x+1} + \log(1+x) = 1 - \dfrac{1}{x+1} + \log(x+1)$

Now differentiating $(n-1)$ times w.r.t. x, we get

$$y_n = 0+(-1)^{n-2}\frac{(n-2)!}{(x+1)^{n-1}} - \frac{(-1)^{n-1}(n-1)!}{(x+1)^n}$$

$$= \frac{(-1)^{n-2}(n-2)!}{(x+1)^n}\left[\frac{1}{(x+1)^{-1}} - \frac{(-1)(n-1)}{1}\right]$$

$$= \frac{(-1)^{n-2}(n-2)!}{(x+1)^n}(x+1+n-1) = \frac{(-1)^{n-2}(n-2)!(x+n)}{(x+1)^n}.$$ **Ans.**

7. *Prove that* $\frac{d^n}{dx^n}(x^{n-1}e^{1/x}) = \frac{(-1)^n e^{1/x}}{x^{n+1}}$.

SOLUTION: $$\frac{d^n}{dx^n}(x^{n-1}e^{1/x}) = \frac{d^n}{dx^n}\left[x^{n-1}\left(1+\frac{1}{x}+\frac{1}{2!x^2}+\frac{1}{3!x^3}+\ldots\right) - \frac{1}{(n-1)x^{n-1}}+\frac{1}{n!x^n}+\ldots\right]$$

$$= \frac{d^n}{dx^n}\left[x^{n-1}+x^{n-2}+\frac{x^{n-3}}{2!}+\ldots+\frac{1}{(n-1)!}+\frac{1}{n!\,x}+\frac{1}{(n+1)!\,x^2}+\ldots\right]$$

$$= 0+\frac{(-1)^n n!}{n!\,x^{n+1}}+\frac{(-1)^n(n+1)!}{(n+1)!\,x^{n+2}}+\frac{(-1)^n(n+2)}{2!(n+2)!\,x^{n+3}}+\ldots$$

$$= \frac{(-1)^n}{x^{n+1}}\left[1+\frac{1}{x}+\frac{1}{2!\,x^2}+\frac{1}{3!\,x^3}+\ldots\right] = \frac{(-1)^n}{x^{n+1}}e^{1/x}.$$ **Ans.**

8. *If* $y = (x^2-1)^n$, *prove that* $(x^2-1)y_{n+2} + 2xy_{n+1} - n(n+1)y_n = 0.$

SOLUTION: $y = (x^2-1)^n \quad \therefore \quad y_1 = n(x^2-1)^{n-1}(2x)$

or $(x^2-1)y_1 = 2n\,xy$. Differentiating w.r.t. x again we get

$(x^2-1)y_2 + 2xy_1 = 2nxy_1 + 2ny$

or $(x^2-1)y_2 + 2x(1-n)y_1 = 2ny$

Differentiating on both sides n times gives

or $(x^2-1)y_{n+2} + {}^nc_1 y_{n+1} 2x + {}^nc_2 y_n \cdot 2 + 2(1-n)[xy_{n+1} + n\cdot y_n] = 2ny_n$

or $(x^2-1)y_{n+2} + (2n+2-2n)xy_{n+1} + y_n[n(n-1)+2n(1-n)-2n] = 0$

or $(x^2-1)y_{n+2} + 2xy_{n+1} - n(n+1)y_n = 0$ **Hence Proved.**

9. *If* $y = \sec^{-1}x$ *prove that*
$x(x^2-1)y_{n+2} + [(2+3n)x^2 - (n+1)]y_{n+1} + n(3n+1)xy_n + n^2(n-1)y_{n-1} = 0$

SOLUTION: $y = \sec^{-1}x \quad \therefore \quad y_1 = \frac{+1}{x\sqrt{x^2-1}}$ or $x^2(x^2-1)y_1^2 = 1$

Differentiating again w.r.t. x, we get

$$2x(x^2-1)y_1^2 + x^2.2x\,y_1^2 + x^2(x^2-1)\,2y_1y_2 = 0$$

or $x(x^2-1)y_2 + (x^2-1)y_1 + x^2y_1 = 0$ or $(x^3-x)y_2 + (2x^2-1)y_1 = 0$

Differentiating n times w.r.t. x throughout, we get

$$(x^3-x)y_{n+2}+{}^nc_1(3x^2-1)y_{n+1}+{}^nc_2(6x)y_n+{}^nc_3\cdot 6y_{n-1}+(2x^2-1)y_{n+1}+{}^nc_1\,4xy_n+{}^nc_2\cdot 4y_{n-1}=0$$

or $\quad x(x^2-1)y_{n+2}+y_{n+1}\,\{n(3x^2-1)+2x^2-1)$

$$+\,y_n\left\{\frac{n(n-1)}{2}6x+4nx\right\}+y_{n-1}\left\{\frac{6n(n-1)(n-2)}{3!}+n(n-1)\right\}=0$$

or $\quad x(x^2-1)y_{n+2}+\{(3n+2)x^2-(n+1)\}y_{n+1}+n(3n+1)xy_n+n^2(n-1)y_{n-1}=0$

10. *If $y=\sin(m\sin^{-1}x)$, prove that*

$(y_n)_0=[(n-2)^2-m^2]\,[(n-4)^2-m^2]\ldots(1-m^2)m$, *if n is odd*

$= 0$ *if n is even.*

SOLUTION: $y=\sin(m\sin^{-1}x)$. Differentiating w.r.t. x, we get

$$y_1=\cos(m\sin^{-1}x)\cdot\frac{m}{\sqrt{1-x^2}}\quad\text{or}\quad(1-x^2){y_1}^2=m^2\cos^2(m\sin^{-1}x).$$

or $\quad(1-x^2){y_1}^2=m^2(1-y^2)$. Differentiating again w.r.t. x we get

$(1-x^2)2y_1y_2-2x{y_1}^2=-2m^2yy_1$ or $(1-x^2)y_2-xy_1=-m^2y$

Now differentiating the above equation n times using Lebrentz theorem

$$(1-x^2)y_{n+2}+n(-2x)y_{n+1}+\frac{n(n-1)}{2!}\cdot(-2)y_n-xy_{n+1}-ny_n=-m^2y_n$$

or $\quad(1-x^2)y_{n+2}-(2n+1)xy_{n+1}-(n^2-m^2)y_n=0$

Putting $x=0$ we get

$$(y_{n+2})_0=(n^2-m^2)\,(y_n)_0$$

$$=(n^2-m^2)\,[(n-2)^2-m^2]\,(y_{n-2})_0$$

$$=\begin{cases}(n^2-m^2)[(n-2)^2-m^2][(n-4)^2-m^2]\ldots(1^2-m^2)\,(y_1)_0 \text{ if } n \text{ is odd}\\(n^2-m^2)\{(n-2)^2-m^2\}\{(n-4)^2-m^2\}\ldots(0^2-m^2)\,(y)_0 \text{ if } n \text{ is even}\end{cases}$$

$$=\begin{cases}(n^2-m^2)[(n-2)^2-m^2][(n-4)^2-m^2]\ldots(1-m^2).\,m \text{ if } n \text{ is odd}\\0 \text{ if n is even.}\end{cases}$$

11. *Prove that* $f(mx)=f(x)+(m-1)x\,f'(x)+\dfrac{(m-1)^2}{2!}x^2f''(x)+\ldots$

SOLUTION: $f(mx)=f[x+(m-1)x]$. Expanding it by Taylor's series

$$g(x+h)=g(x)\;\;hg'(x)+\frac{x^2}{2!}g''(x)+\ldots\qquad\text{(take } h=(m-1)x)$$

then $\quad f(mx)=f(x)+(m-1)x\,f'(x)+\dfrac{(m-1)^2x^2}{2!}f''(x)+\ldots$ **Hence Proved.**

12. *Prove, using Taylor's theorem, that*

$$\frac{1}{1-x} = \frac{1}{3}+\frac{(x+2)}{3^2}+\frac{(x+2)^2}{3^3}+\frac{(x+2)^3}{3^4}+...$$

SOLUTION: Let $f(x) = \dfrac{1}{1-x}$. By Taylor's theorem

$$f(x) = f(a) + (x-a)f'(a) + \frac{(x-a)^2}{2!}f''(a)+\frac{(x-a)^3}{3!}f'''(a)+...$$

Here $a = -2$, therefore

$$\frac{1}{1-x} = f(-2) + (x+2)f'(-2) + \frac{(x+2)^2}{2!}f''(-2)+\frac{(x+2)^3}{3!}f'''(-2)+...$$

$$f(-2) = \frac{1}{1-(-2)} = \frac{1}{3},\ f'(x) = \frac{1}{(1-x)^2} \quad \therefore\ f'(-2) = \frac{1}{9}$$

$$f''(x) = \frac{2!}{(1-x)^3} \quad \therefore\ f''(-2) = \frac{2!}{27},\ f'''(x) = \frac{2\cdot 3}{(1-x)^4} \quad \therefore\ f'''(-2) = \frac{3!}{8!},...$$

Therefore $\dfrac{1}{1-x} = \dfrac{1}{3}+\dfrac{(x+2)}{3^2}+\dfrac{(x+2)^2}{3^3}+\dfrac{(x+2)^3}{3^4}+...$ **Hence Proved.**

13. *Find the value of tan 43°.*

SOLUTION: Let $f(x) = \tan x$. By Taylor's series **about** $x = a$

$$f(a+h) = f(a) + hf'(a) + \frac{h^2}{2!}f''(a) + ...$$

Here $a = 45°,\ h = -2° = -\dfrac{2\pi}{180} = -\dfrac{\pi}{180} = -0.0349$ radians

$\therefore$ $\tan 43° = \tan(45° - 2°) = f(45°) - 0.0349 f'(45°) + \dfrac{(-0.0359)^2}{2!}f''(45) + ...$

Here $f(45°) = \tan 45° = 1,\ f'(x) = \sec^2 x \quad \therefore\ f'(45°) = \sec^2 45 = 2$

$f''(x) = 2\sec^2 x \tan x \tan x \quad \therefore\ f''(45°) = 2\cdot 2\cdot 1 = 4$

$$\therefore\ \tan 43° = 1 - 0.0349(2) + \frac{(-0.0349)^2}{2!}\cdot(4) + ...$$

$$= 1 - 0.0698 + 0.0698\,(0.0349) + ...$$

$= 0.9326$ approximately. **Ans.**

14. *Expand* $\sqrt{1+x+2x^2}$ *in powers of* $(x-1)$.

SOLUTION: $\sqrt{1+x+2x^2} = \sqrt{4+2(x-1)^2+5(x-1)}$ [in terms of $(x-1)$]

By Taylor's theorem

$$f(a+h) = f(a) + hf'(a) + \frac{h^2}{2!}f''(a)+\frac{h^3}{3!}f'''(a)+...$$

Here $a = 4$ and $h = 2(x-1)^2 + 5(x-1)$ Let $f(x) = \sqrt{x}$

$$\therefore \quad f(4) = 2. \quad f'(x) = \frac{1}{2\sqrt{x}} \quad \therefore \quad f'(4) = \frac{1}{4}.$$

$$f''(x) = \frac{1}{2}\left(\frac{-1}{2}\right)\frac{1}{x\sqrt{x}} \quad \therefore \quad f''(4) = -\frac{1}{32}, \text{ etc.}$$

$$\therefore \quad \sqrt{4+2(x-1)^2+5(x-1)} = 2+\frac{1}{4}\left[2(x-1)^2+5(x-1)\right]+\frac{1}{2!}\left(\frac{-1}{32}\right)[2(x-1)^2+5(x-1)]^2+\ldots$$

$$= 2+\frac{5}{4}(x-1)+\frac{7}{64}(x-1)^2+\ldots$$

Ans.

15. *Prove that* $\dfrac{1}{1-x} = \dfrac{1}{3}+\dfrac{x+2}{3^2}+\dfrac{(x+2)^2}{3^2}+\dfrac{(x+2)^3}{3^4}+\ldots$

SOLUTION: Let $f(x) = \dfrac{1}{1-x}$. By Taylor's series

$$f(x) = f(a) + (x-a)f'(a) + \frac{(x-a)^2}{2!}f''(a)+\frac{(x-a)^3}{3!}f'''(a)+\ldots$$

Here $a = -2$

$$\therefore \quad f(x) = f(-2) + (x+2)f'(-2) + \frac{(x+2)^2}{2!}f''(-2)+\frac{(x+2)^3}{3!}f'''(-2)+\ldots$$

$$f(-2) = \frac{1}{3}, \quad f'(x) = \frac{1}{(1-x)^2} \quad \therefore \quad f'(-2) = \frac{1}{9},$$

$$f''(x) = \frac{2}{(1-x)^3} \quad \therefore \quad f''(-2) = \frac{2\cdot 1}{3^3} = \frac{2!}{3^3}.$$

$$f'''(x) = \frac{2\cdot 3}{(1-x)^4} \quad \therefore \quad f'''(-2) = \frac{3!}{(1-x)^4}, \text{ etc.}$$

$$\therefore \quad f(x) = \frac{1}{1-x} = \frac{1}{3}+\frac{x+2}{3^2}+\frac{(x+2)^2}{3^3}+\frac{(x+2)^3}{3^4}+\ldots$$

Hence Proved.

CHAPTER

4 Asymptotes and Curve Tracing

IMPORTANT FORMULAE AND USEFUL RESULTS

1. **Asymptotes.** The nth degree algebraic equation can be written as,

$$x^n \phi_n(y/x) + x^{n-1}\phi_{n-1}(y/x) + \dots + K_n = 0$$

Let the asymptote be $y = mx + c$ $\therefore$ $\lim_{x\to\infty} y/c = m$.

Thus m is obtained from $\phi_n(m) = 0$ and

'c' is obtained from $c\phi'_n(m) + \phi_{n-1}(m) = 0$, or $c = \dfrac{-\phi_{n-1}(m)}{\phi'_n(m)}$.

For parallel asymptotes, let $m_1 = m_2$ and 'c' for these is obtained from

$$\frac{c^2}{2!}\phi''_n(m) + \frac{c}{1}\phi'_{n-1}(m) + \phi_{n-2}(m) = 0, \text{ etc.}$$

2. **Asymptotes Parallel to coordinate Axes.**

Asymptotes parallel to y-axis are obtained by equating to zeroth coefficient of the highest degree term in y and asymptotes parallel to x-axis are obtained by equating to zero the coefficient of the highest degree term in x.

3. **Asymptotes of Polar Curves**

Let the asymptote of the polar curve be $x \cos \alpha + y \sin \alpha = p$.

Let $r \to \infty$ so as to get $f_n(\theta) = 0$, $\theta = \alpha_1, \alpha_2, \dots$.

Asymptote is $r \sin(\alpha_1 - \theta) = \dfrac{f_{n-1}(\alpha_1)}{f'_n(\alpha_1)}$, similarly, for $\alpha = \alpha_2, \alpha_3, \dots$.

4. Tracing of Curves in Cartesian form, Parametric form and Polar form.

SOLUTIONS TO EXERCISE 4A

1. *Find the asymptotes of the curve*

$$x^3+2x^2y-xy^2-2y^3+4y^2+2xy+y-1=0.$$

SOLUTION: Here $\phi_3(m) = 1 + 2m - m^2 - 2m^3$, $\phi_2(m) = 4m^2 + 2m$, $\phi_3(m) = m$.

For slope m, we have $\phi_3(m) = 0$, that is,

$$2m^3 + m^2 - 2m - 1 = 0 \quad \text{or} \quad m^2(2m + 1) - (2m + 1) = 0$$

or $(2m + 1)(m + 1)(m - 1) = 0 \quad \therefore m = 1, -1, -1/2.$

To find 'c' corresponding to 'm', we have

$$c = \frac{-\phi_2(m)}{\phi_3'(m)} = \frac{-(4m^2+2m)}{2-2m-6m^2} = \frac{-(2m^2+m)}{1-m-3m^2}$$

When $m = 1$, $\quad c = \dfrac{-(2-1)}{1-1-3} = 1$

$m = -1$, $\quad c = \dfrac{-(2-1)}{1+1-3} = 1$

$m = -1/2$, $\quad c = \dfrac{-\left(\frac{2}{4}-\frac{1}{2}\right)}{1+\frac{1}{2}-\frac{3}{4}} = 0$

Therefore the asymptotes are $\quad y = x + 1, \quad y = -x + 1, \quad y = -\dfrac{1}{2}x$ **Ans.**

2. *Find the asymptotes of the curve*

$$x^3-4y^3+3x^2y+y-x+3=0.$$

SOLUTION: $\phi_3(m) = 1 - 4m^3 + 3m$, $\phi_2(m) = 0$, $\phi_1(m) = m - 1$, $\phi'_3(m) = 3 - 12m^2$, $\phi''_3(m) = -24m$

For the slopes of the asymptotes we have $\phi_3(m) = 0$

or $\quad 4m^3 - 3m - 1 = 0$ or $(m - 1)(4m^2 + 4m + 1) = 0$ which gives $m = 1, -1/2, -1/2.$

'c' for $m = 1$ is obtained from $\quad c = \dfrac{-\phi_2(m)}{\phi_3'(m)} = 0$

'c' for the repeated root $-1/2$, are given by

$$\frac{c^2}{2}\phi_3''(m) + C\phi_2'(m) + \phi_1(m) = 0$$

or $\quad \dfrac{c^2}{2}(-24m) + 0 + m - 1 = 0$

At $m = -1/2$ we have $\quad 6c^2 - \dfrac{1}{2} - 1 = 0, \quad c^2 = \dfrac{1}{4}, \quad c = \pm 1/2$

Therefore the asymptotes are $y = x, \quad y = -\dfrac{1}{2}x \pm \dfrac{1}{2} \quad$ or $\quad 2y = -x \pm 1$ **Ans.**

3. *Find the asymptotes of the curve*

$$y^3+2xy^2+x^2y-y+1=0.$$

SOLUTION: $\phi_3(m)=m^3+2m^2+m, \quad \phi_2(m)=0, \quad \phi_1(m)=-m, \quad \phi'_3(m)=3m^2+4m+1$

$\phi_3(m)=0$ gives $m^3+2m^2+m=0$ hence $m=0,-1,-1$

'c' for $m=0$ is obtained by $c=\dfrac{-\phi_2(m)}{\phi'_3(m)}=0$ as $\phi_2(m)=0$.

'c' for $m=-1,-1$ is obtained from

$$\frac{c^2}{2}\phi''_3(m)+c\phi'_2(m)+\phi_1(m)=0 \quad \text{or} \quad \frac{c^2}{2}(6m+4)+0-m=0$$

$\therefore$ for $m=-1$ we have $-c^2+1=0$, hence $c=\pm 1$

Thus the asymptotes are $y=0$, $y=-x\pm 1$ **Ans.**

4. *Find the asymptotes of the curve*

$$\frac{x^2}{a^2}-\frac{y^2}{b^2}=1.$$

SOLUTION: Equation of the curve is $b^2x^2-a^2y^2-a^2b^2=0.$

Here $\phi_2(m)=b^2-a^2m^2$, $\phi_1(m)=0$

For slope of the asymptote, $\phi_2(m)=0$ which gives $m=\pm\dfrac{b}{a}$

For 'c' of the asymptote we have $c\,\phi'_2(m)+\phi_1(m)=0$ $\therefore c=0$, as $\phi_1(m)=0$.

Therefore, two asymptotes are $y=\dfrac{+b}{a}x$ or $ay\pm bx=0$ **Ans.**

5. *Find the asymptotes of the curve* $(x^3+a^3)\,y=bx^3$.

SOLUTION: Let us first find the asymptotes parallel to coordinate axes. For asymptotes parallel to y-axis we equate to zero the coefficient of the highest degree term in y and get $x^3+a^3=0$ or $x+a=0$. For asymptote parallel to x-axis, we equate to zero the coefficient of the highest degree term in x and get $y-b=0$.

Therefore there are only two asymptotes of the curve, namely $x+a=0$ and $y-b=0$ **Ans.**

6. *Find the asymptotes of the curve* $y\,(x-y)^3=y\,(x-y)+2$.

Solution: Here $\phi_4(m)=m(1-m)^3$, $\phi_3(m)=0$, $\phi_2(m)=-m(1-m)$, $\phi_1(m)=0$

The slope 'm' of the asymptotes are obtained from $\phi_4(m)=0$, therefore we have $m=0, 1, 1, 1$.

To obtain 'c' for $m=0$ we have $c=\dfrac{-\phi_3(m)}{\phi'_4(m)}=0$ as $\phi_3(m)=0$

Therefore one asymptotes is $y=0$ the x-axis itself.

For other three parallel asymptotes, 'c' is obtained from

$$\frac{c^3}{3!}\phi_4'''(m)+\frac{c^2}{2!}\phi_3''(m)+\frac{c}{1!}\phi_2'(m)+\phi_1(m)=0 \qquad ...(1)$$

$$\phi'_4(m) = 1(1-m)^3 + 3m(1-m)^2(-1) = (1-m)^2(1-4m)$$

$$\phi''_4(m) = -4(1-m)^2 - 2(1-m)(1-4m) = -2(1-m)[2(1-m)+1-4m]$$

$$= -6(1-m)(1-2m)$$

$$\phi_4'''(m) = 12(1-m) + 6(1-2m) = 18 - 24m$$

$\therefore$ (1) becomes $\frac{c^3}{6}(18-24m)+\frac{c^2}{2}(0)+c(-1-2m)+0 = 0$

or $c^3(3-4m) + c(-1+2m) = 0$

For $m = 1$ we have $c^3(-1) + c = 0$ which gives $c = 0, 1, -1$.

Thus there are four asymptotes $y = 0$, $y = x$, $y = x + 1$, $y = x - 1$ **Ans.**

7. *Find asymptotes of the curve* $x^3-5x^2y+8xy^2-4y^3+x^2-3xy+2y^2-1=0$.

SOLUTION: Here $\phi_3(m) = 1 - 5m + 8m^2 - 4m^3$, $\phi_2(m) = 1 - 3m + 2m^2$, $\phi_1(m) = 0$

To obtain slopes of the asymptotes, $\phi_3(m) = 0$ or $1 - 5m + 8m^2 - 4m^3 = 0$

or $(1-m)(4m^2-4m+1) = 0$ hence $m = 1, \frac{1}{2}, \frac{1}{2}$.

For $m = 1$, $c = \frac{-\phi_2(m)}{\phi_3'(m)} = \frac{-(1-3m+2m^2)}{-5+16m-12m^2} = 0$

To obtain 'c' for $m = \frac{1}{2}$, we have $\frac{c^2}{2}\phi_3''(m)+c\phi_2'(m)+\phi_1(m)=0$

or $\frac{c^2}{2}(16-24m)+\frac{c}{1}(-3+4m)+0=0$ or $c^2(8-6)-c=0$ $\therefore$ $c = 0, 1/2$

Therefore the asymptotes are $y = x$, $y = \frac{1}{2}x$, $y = \frac{1}{2}x+\frac{1}{2}$ **Ans.**

8. *Find the asymptotes of the curve* $r\theta = a$.

SOLUTION: Equation of curve in polar form is

$$r^n f_n(\theta) + r^{n-1} f_{n-1}(\theta) + ... + rf_1(\theta) + f_0(\theta) = 0$$

Writing $u = \frac{1}{r}$ then $u^n f_0(\theta) + u^{n-1} f_1(\theta) + ... + uf_{n-1}(\theta) + f_n(\theta) = 0$

If $u \to 0$ as $\phi \to \alpha$ then the equation of the asymptote is

$$r\sin(\alpha-\theta) = \frac{f_{n-1}(\alpha)}{f_n'(\alpha)}$$

In the given curve $f_1(\theta) = \theta$ and $f_0(\theta) = -a$ and $u \to 0$ as $\theta \to 0$

hence $\alpha = 0$ and asymptote is

$$r\sin(0-\theta) = \frac{f_0(\alpha)}{f_1'(\alpha)} = \frac{-a}{1} \quad \text{or} \quad r\sin\theta = a$$ **Ans.**

9. *Find the asymptote of the curve* $r = a \operatorname{cosec} \theta + b$.

SOLUTION: Curve is $r \sin \theta = a + b \sin \theta$

Here $f_1(\theta) = \sin \theta$ and $f_0(\theta) = -a - b \sin \theta$.

As $\theta \to 0$, $r \to \infty$ hence $\alpha = 0$.

Equation of asymptote is $r \sin(\alpha - \theta) = \dfrac{f_0(\alpha)}{f_1'(\alpha)}$

or $r \sin(-\theta) = \dfrac{-a - b\sin 0}{\cos 0}$ or $r \sin \theta = a$ **Ans.**

10. *Find the asymptotes of the curve* $r \sin 2\theta = a \cos 3\theta$.

SOLUTION: In the given curve $f_1(\theta) = \sin 2\theta$ and $f_0(\theta) = -a \cos 3\theta$.

Here $r \to \infty$ as $\theta \to 0$ or $\dfrac{\pi}{2}$, hence $\alpha = 0$ and $\pi/2$

Equation of asymptote is $r \sin(\alpha - \theta) = \dfrac{f_0(\alpha)}{f_1'(\alpha)}$

or $r \sin(\alpha - \theta) = \dfrac{-a \cos 3\alpha}{2 \cos(2\alpha)}$

For $\alpha = 0$, we have $r \sin(-\theta) = -\dfrac{a \cos 0}{2 \cos 0}$ or $r \sin \theta = \dfrac{a}{2}$

and for $\alpha = \dfrac{\pi}{2}$, the asymptote is

$$r \sin\left(\frac{\pi}{2} - \theta\right) = \frac{-a \cos\left(\frac{3\pi}{2}\right)}{2 \cos \pi} = 0 \quad \text{or} \quad r \cos \theta = 0$$

$\therefore$ Asymptotes are $r \sin \theta = \dfrac{a}{2}$ and $r \cos \theta = 0$ **Ans.**

11. *Find the asymptotes of the curve* $r = 4(1 + \sec 2\theta)$.

SOLUTION: Curve is $r \cos 2\theta - 4 \cos 2\theta - 4 = 0$

Here $f_1(\theta) = \cos 2\theta$ and $f_0(\theta) = -4 \cos 2\theta - 4$.

Here $f_1(\theta) = 0$ gives $\theta = \dfrac{\pi}{4}$ and $\dfrac{3\pi}{4}$ hence $\alpha = \pi/4, 3\pi/4$.

$\therefore$ Equation of asymptote is $r \sin(\alpha - \theta) = \dfrac{f_0(\alpha)}{f_1'(\alpha)}$

For $\alpha = \dfrac{\pi}{4}$, we have $r \sin(\pi/4 - \theta) = \left[\dfrac{-4\cos 2\alpha - 4}{-2 \sin 2\alpha}\right]_{\alpha = \frac{\pi}{4}}$

and for $\alpha = \dfrac{3\pi}{4}$, we have $r \sin\left(\dfrac{3\pi}{4} - \theta\right) = \left[\dfrac{-4\cos 2\alpha - 4}{-2 \sin 2\alpha}\right]_{\alpha = \frac{3\pi}{4}} = -2$ or $r \cos\left(\theta - \dfrac{\pi}{4}\right) = 2$

$\therefore$ Asymptotes are $r \sin\left(\theta - \dfrac{\pi}{4}\right) = 2$ and $r \cos\left(\theta - \dfrac{\pi}{4}\right) = -2$ **Ans.**

12. *Find all the asymptotes of the curve* $y^2(x-2a)=x^3-a^3$.

SOLUTION: Curve is $y^2x - x^3 - 2ay^2 + a^3 = 0$

Here $\theta_3(m) = m^2 - 1, \quad \phi_2(m) = -2am^2, \quad \phi_1(m) = 0$

For slope of the asymptote $\phi_3(m) = 0 \quad \therefore \quad m = \pm 1$

'c' of the asymptote is obtained from $c = \dfrac{-\phi_2(m)}{\phi_3'(m)} = \dfrac{2am^2}{2m} = am$

$\therefore c = a$ for $m = 1$ and $c = -a$ for $m = -1$.

Thus, two oblique asymptotes are $y = x + a, \quad y = -x - a$

Asymptote parallel to y-axis is obtained by equating to zero the coefficient of highest degree term in y, as $x = 2a$.

$\therefore$ Asymptotes are $x = 2a, \quad y = x + a, \quad y = -x - a$. **Ans.**

13. *Find all the asymptotes of the curve*

$$x^2y^2 - x^2y - xy^2 + x + y + 1 = 0.$$

SOLUTION: First we find asymptotes parallel to coordinate axes. Equating to zero the coefficient of the highest degree term in x we get $y^2 - y = 0$ hence $y = 0, y = 1$ are two asymptotes parallel to X-axis. Next equating to zero the coefficient of the highest degree term in y, gives $x^2 - x = 0$ hence $x = 0, \ x = 1$ are two asymptotes parallel to y-axis.

Thus, the asymptotes are $x = 0, \quad y = 0, \quad x = 1, \quad y = 1$ **Ans.**

14. *Find the asymptotes of the curve* $y^3 - xy^2 - x^2y + x^3 + x^2 - y^2 = 0$.

SOLUTION: Here $\phi_3(m) = m^3 - m^2 - m + 1, \quad \phi_2(m) = 1 - m^2, \quad \phi_1(m) = 0$

Slope 'm' of asymptotes is given by $\phi_3(m) = 0$

or $m^2(m-1) - (m-1) = 0$ hence $m = -1, 1, 1$.

'c' of the asymptote for $m = -1$, is given by $c = \dfrac{-\phi_2(m)}{\phi_3'(m)} = 0$ at $m = -1$

'c' of the two asymptotes for $m = 1$, is given by

$$\frac{c^2}{2}\theta_3''(m) + c\phi_2'(m) + \phi_1(m) = 0$$

or $\dfrac{c^2}{2}(6m-2) + (-2m)c + 0 = 0$ or $2c^2 - 2c = 0 \quad \therefore c = 0, 1$

Asymptotes are $y = -x, \quad y = x, \quad y = x + 1$ **Ans.**

SOLUTIONS TO EXERCISE 4B

Trace the following curves:

1. $y = x^3$ (*cubical parabola*)

SOLUTION:

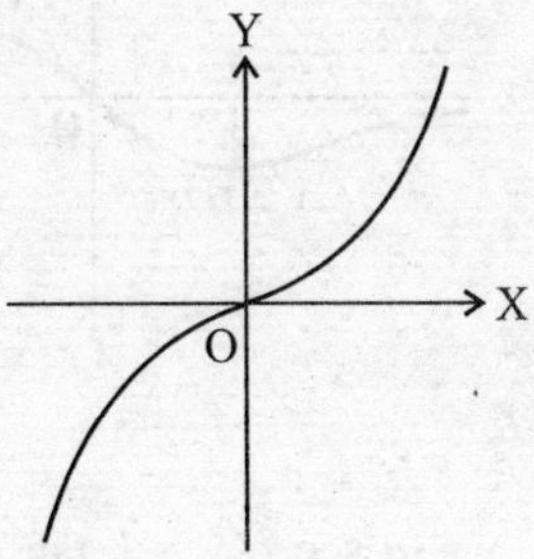

2. $y^2 = (x - a)^3$ (*Semi cubical parabola*)

SOLUTION:

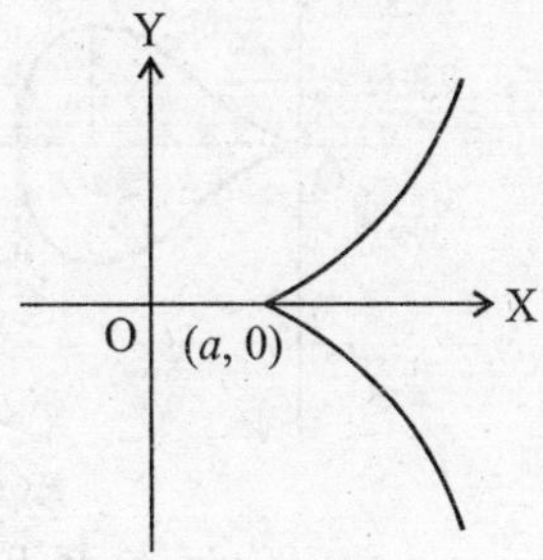

3. $3ay^2 = x^2 (a - x)$

SOLUTION:

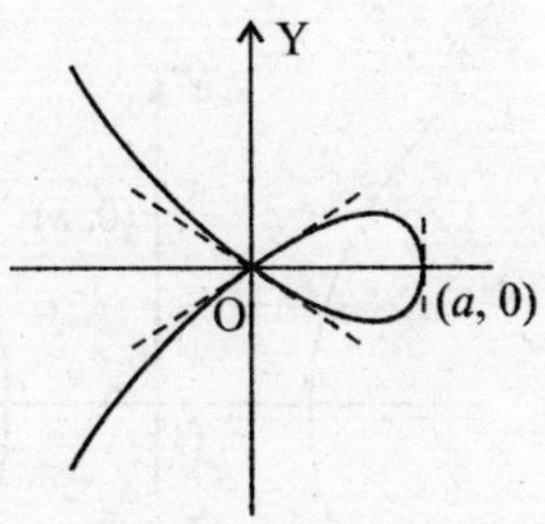

4. $ay^2 = x (a^2 + x^2)$

SOLUTION:

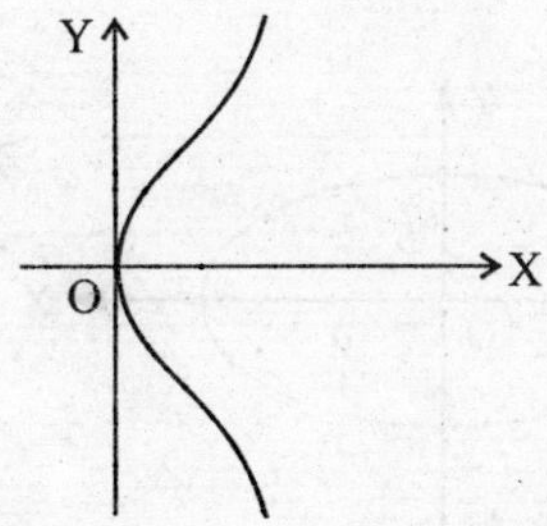

5. $a^2y^2 = x^2 (2a - x) (x - a)$

SOLUTION:

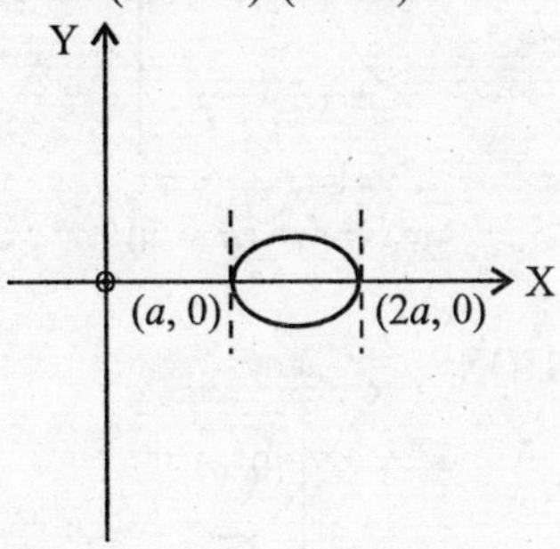

6. $y^2x = a^2 (a - x)$

SOLUTION:

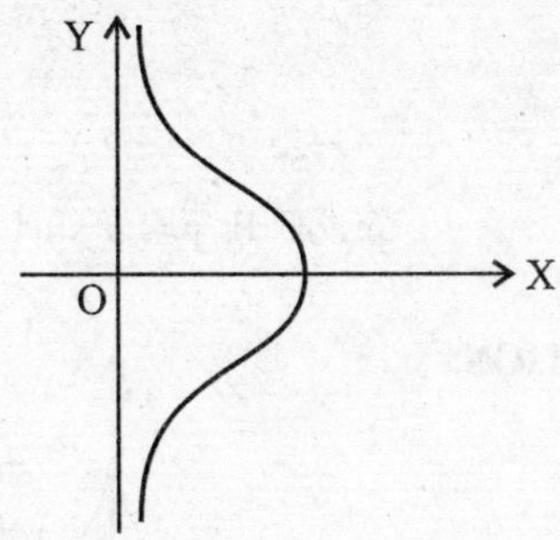

7. $y(x^2 + 4a^2) = 8a^3$

SOLUTION:

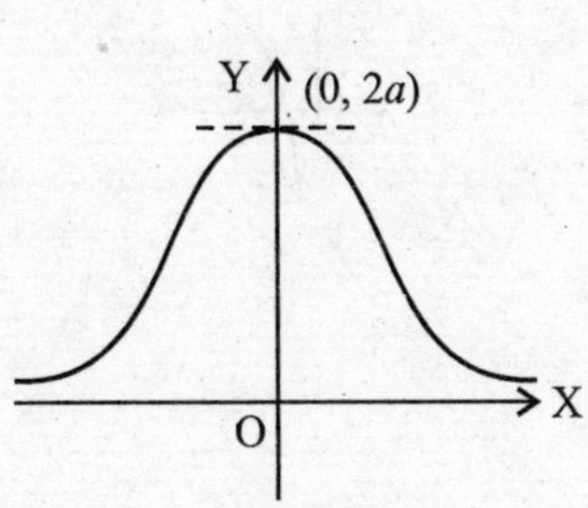

8. $x^2y^2 = a^2 (y^2 - x^2)$

SOLUTION:

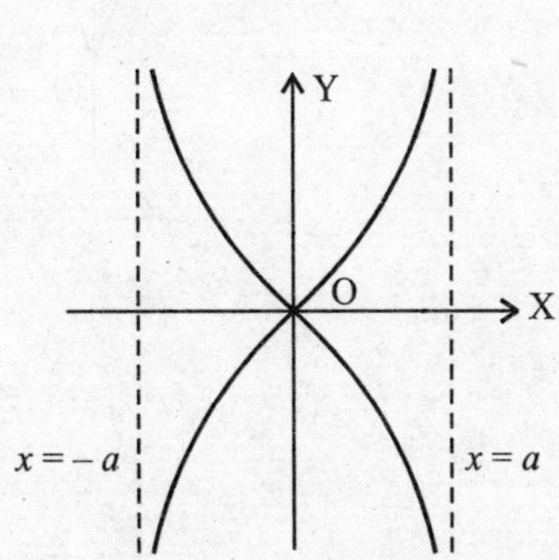

9. $a^2y^2 = x^3(2a - x)$

SOLUTION:

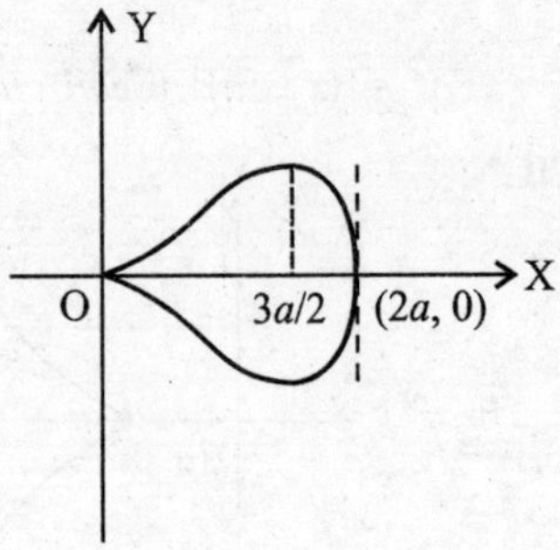

10. $y = \dfrac{x}{1+x^2}$

SOLUION:

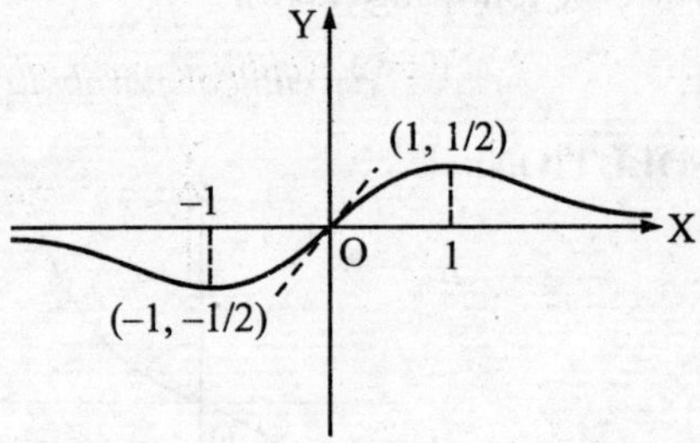

11. $x = a \cos \theta, y = b \sin \theta$

SOLUTION:

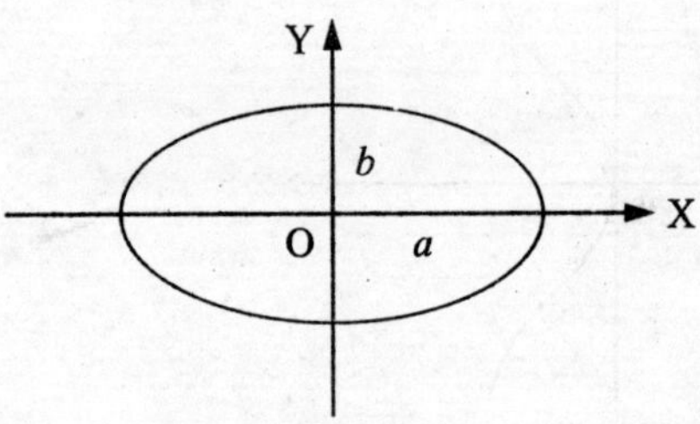

12. $x = a \sec t, y = b \tan t$

SOLUTION:

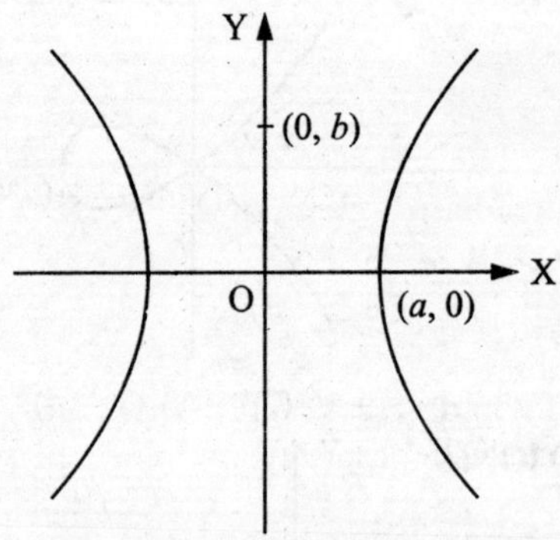

13. $x = a \cos^3 \theta, y = a \sin^3 \theta.$

SOLUTION:

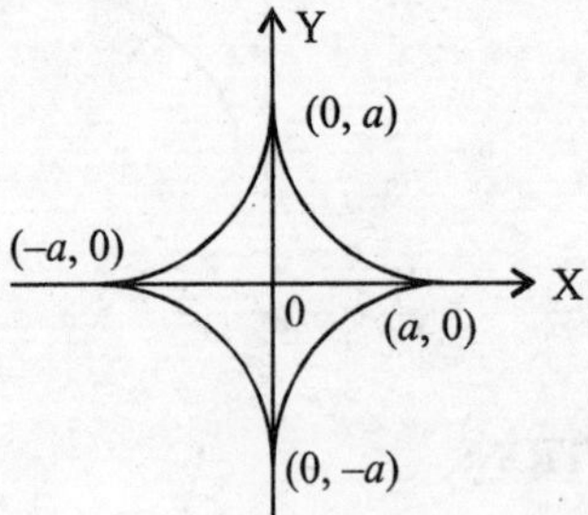

14. $y^2x = a^2(a - y)$

SOLUTION:

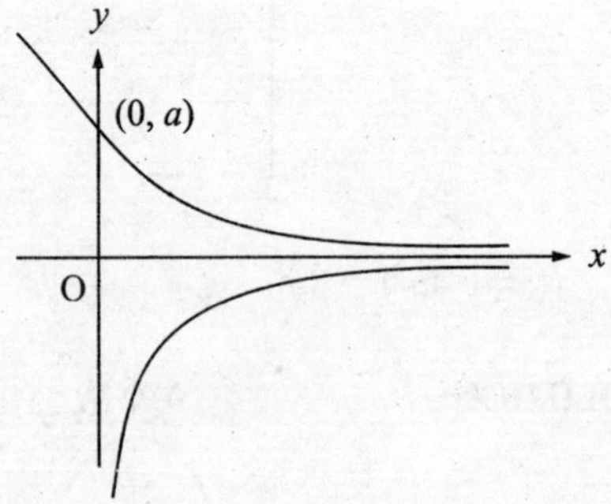

15. *Trace the curve* $y^2(2a - x) = x^3$.

SOLUTION:

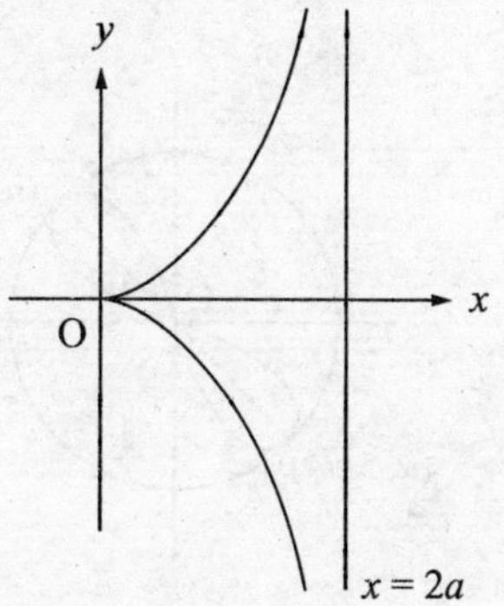

16. *Trace the curve* $x^2y^2 = x^2 + a^2$.

SOLUTION:

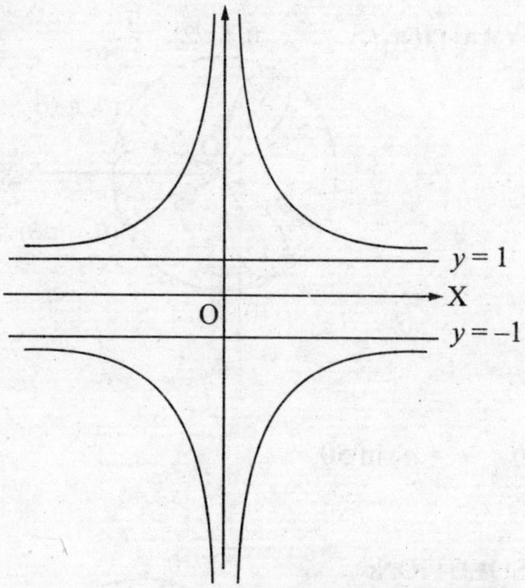

SOLUTIONS TO EXERCISE 4C

Trace the following curves:

1. (*i*) $r = 2a\cos\theta$ (*ii*) $r = 2a\cos(\theta - \alpha)$

SOLUTION: (*i*)

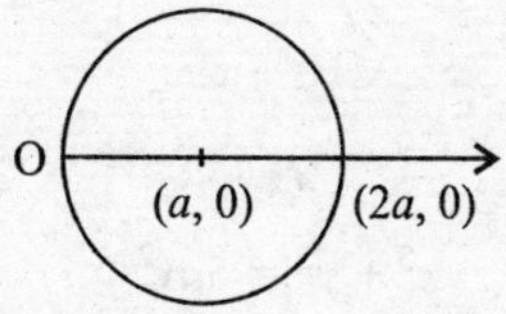

(*ii*)

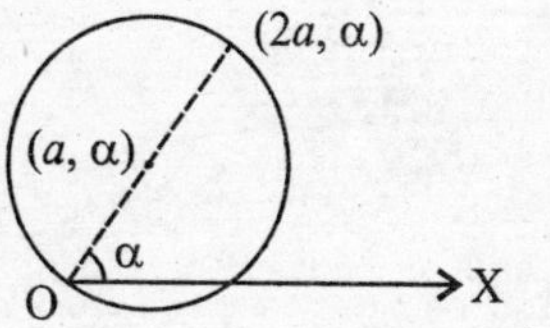

2. $r = a\theta$ (*spiral of Archamedes*).

SOLUTION:

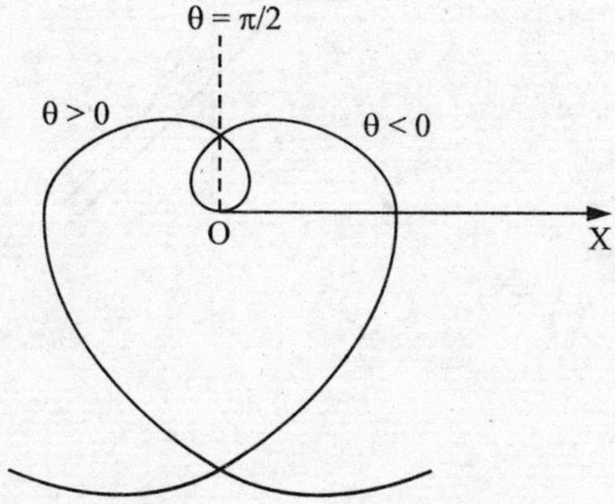

3. $r = a\cos 2\theta$.

SOLUTION:

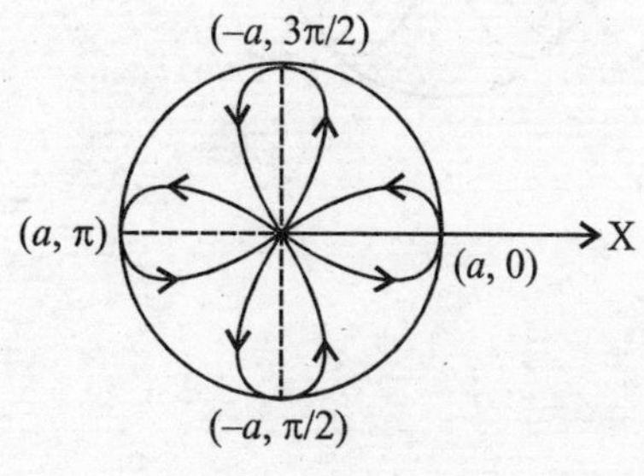

4. $r = a \cos 3\theta$.

SOLUTION:

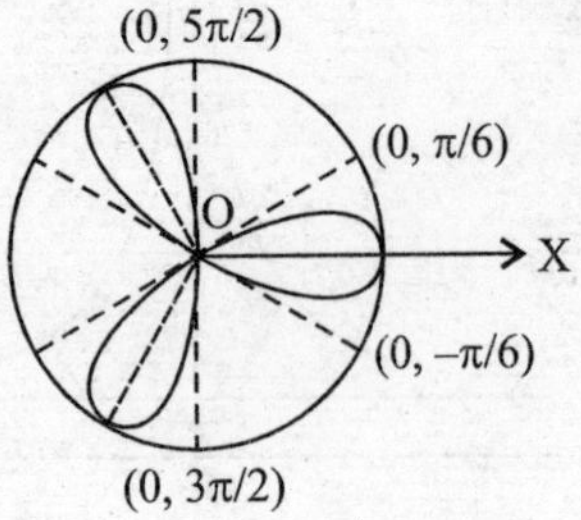

5. $r^2 = a^2 \sin 2\theta$.

SOLUTION:

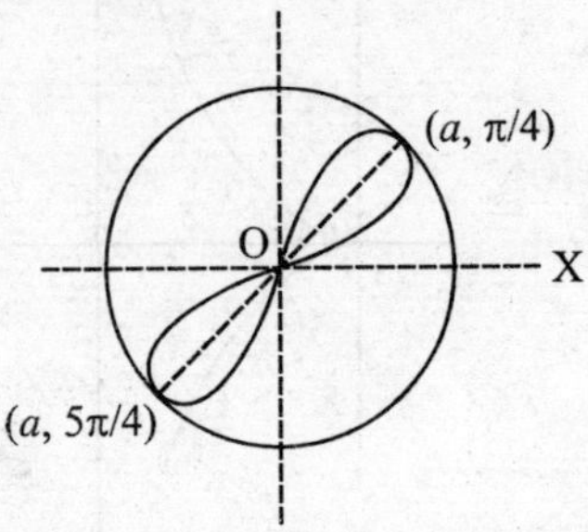

6. $r = a \sin 5\theta$.

SOLUTION:

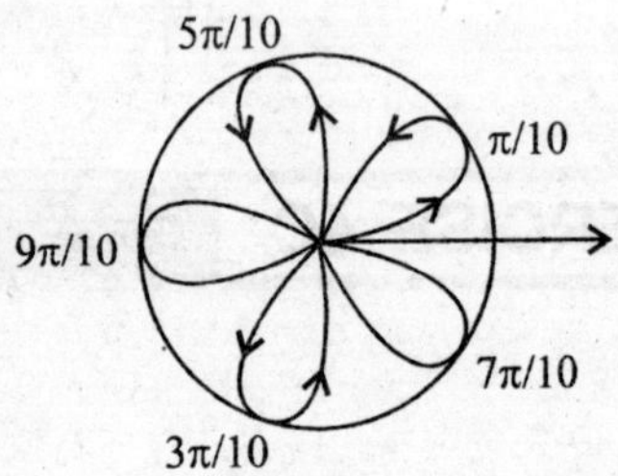

7. $r^2 \cos 2\theta = a^2$.

SOLUTION:

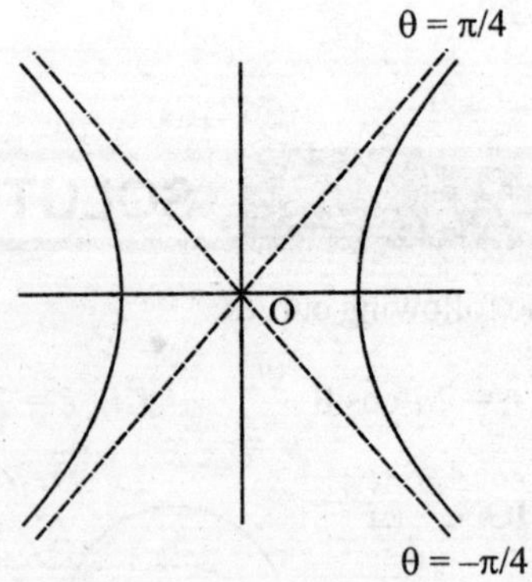

8. $x^6 + y^6 = a^2 x^2 y^2$.

SOLUTION:

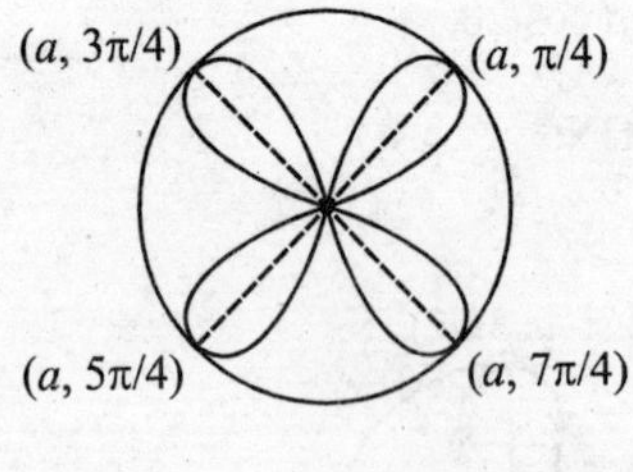

9. $x^5 + y^5 = 5ax^2 y^2$.

SOLUTION:

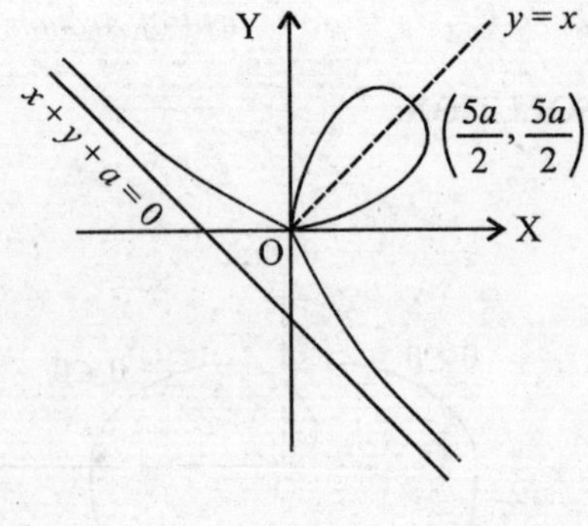

ADDITIONAL IMPORTANT PROBLEMS

1. *Find the asymptotes of the curve*

$$y^3 - 6xy^2 + 11x^2y - 6x^3 + x + y = 0$$

SOLUTION: Here $\varphi_3(m) = m^3 - 6m^2 + 11m - 6$, $\varphi_2(m) = 0$, $\varphi_1(m) = 1 + m$

For the slopes 'm' of the asymptotes $\varphi_3(m) = 0$, that is, $m^3 - 6m^2 + 11m - 6 = 0$

or $(m-1)(m-2)(m-3) = 0 \quad \therefore \quad m = 1, 2, 3.$

Next, for 'c' we have $c = -\dfrac{\varphi_2(m)}{\varphi_3'(m)} = \dfrac{0}{3m^2+12m+11} = 0$

Hence the asymptotes are $y = x$, $y = 2x$ and $y = 3x$. **Ans.**

2. *Find all the asymptotes of the curve*

$$x^3 + 4x^2y + 4xy^2 + 5x^2 + 15xy + 10y^2 - 2y + 1 = 0$$

SOLUTION: The coefficient of x^3, the highest degree term in x is constant hence no asymptote parallel to x-axis. Further equating to zero the coefficient of y^2, the highest degree term in y, we get

$4x + 10 = 0$ or $x = -\dfrac{5}{2}$ is the asymptote parallel to y-axis.

Next, $\varphi_3(m) = 1 + 4m + 4m^2$, $\varphi_2(m) = 5 + 15m + 10m^2$, $\varphi_1(m) = -2m$.

For slope 'm' of the asmptote we have $\varphi_3(m) = 0$

or $1 + 4m + 4m^2 = 0$ which gives $m = \dfrac{-1}{2}, \dfrac{-1}{2}$.

For 'C' of the parallel asymptotes, we have

$$\frac{C^2}{2!}\varphi_3''(m) + C\varphi_2'(m) + \varphi_1(m) = 0$$

or $$\frac{C^2}{2}(8) + C(15 + 20m) + 2m = 0$$

At $m = \dfrac{-1}{2}$ we have $4C^2 + 5C + 1 = 0$ or $(4C + 1)(C + 1) = 0$

hence $C = -1$ and $\dfrac{-1}{4}$.

$\therefore$ The asymptotes are $y = \dfrac{-1}{2}x - 1$ and $y = \dfrac{-1}{2}x - \dfrac{1}{4}$. **Ans.**

3. *Obtain the equation of a third degree algebraic equation of a curve whose asymptotes are the same as that of the curve $x^3 - 6x^2y + 11xy^2 - 6y^3 + 4x + 5y + 7 = 0$ and which passes through the points $(0, 0)$, $(0, 2)$ and $(2, 0)$.*

SOLUTION: The equation of the given curve is

$$x^3 - 6x^2y + 11xy^2 - 6y^3 + 4x + 5y + 7 = 0$$

or $$(x - y)(x - 2y)(x - 3y) + 4x + 5y + 7 = 0$$

which has asymptotes as $x - y = 0$, $x - 2y = 0$ and $x - 3y = 0$.

The required equation of the curve which has above cited asympotes, will be of the form

$$(x - y)(x - 2y)(x - 3y) + ax + by + c = 0$$

where a, b, c are arbitrary constants.

Since it is passes through the points $(0, 0)$, $(2, 0)$ and $(0, 2)$.

We have $c = 0$, $a = -4$ and $b = 24$. Hence the curve is

$$x^3 - 6x^2y + 11xy^2 - 6y^3 - 4x + 24y = 0.$$ **Ans.**

4. *Obtain the asymptotes of the curve*

(i) $r = \dfrac{a\theta}{\theta - 1}$. *(ii)* $r(\theta + \sin\theta) = \theta + \cos\theta$.

SOLUTION: (*i*) The given curve in polar form is $r = \dfrac{a\theta}{\theta - 1} = f(\theta)$, say.

Here $$\lim_{\theta\to\infty} f(\theta) = \lim_{\theta\to\infty} \frac{a\theta}{\theta - 1} = \lim_{\theta\to\infty} \frac{a}{1 - \frac{1}{\theta}} = a$$

Hence the required asymptote is $r = a$ in the form of a circle. **Ans.**

(*ii*) The given curve is $r = \dfrac{\theta + \cos\theta}{\theta + \sin\theta} = f(\theta)$, say.

Here $$\lim_{\theta\to\infty} \frac{\theta + \cos\theta}{\theta + \sin\theta} = \lim_{\theta\to\infty} \frac{1 + \frac{\cos\theta}{\theta}}{1 + \frac{\sin\theta}{\theta}} = \frac{1 + 0}{1 + 0} = 1$$

Hence the required asymptote is the circle $r = 1$. **Ans.**

5. *Trace the curve $ay^2 = x^3$ known as semi-cubical parabola.*

SOLUTION: $ay^2 = x^3$ is symmetrical about x-axis. It passes through the origin and tangent at the origin is x-axis. It shape is as follows. At $x = a$, we have $y = \pm a$.

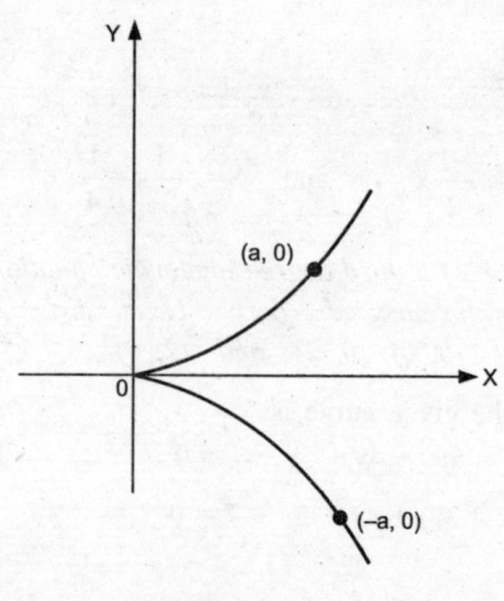

6. *Trace the curve* $y = (x^2 - x - 6)(x - 7)$

SOLUTION: $y = (x - 3)(x + 2)(x - 7)$. It meets the x-axis at the points $(3, 0), (-2, 0)$ and $(7, 0)$. Tangents to the curve at these points can be obtained by shifting the origin to these points. Shape of the curve is as per the adjoining figure.

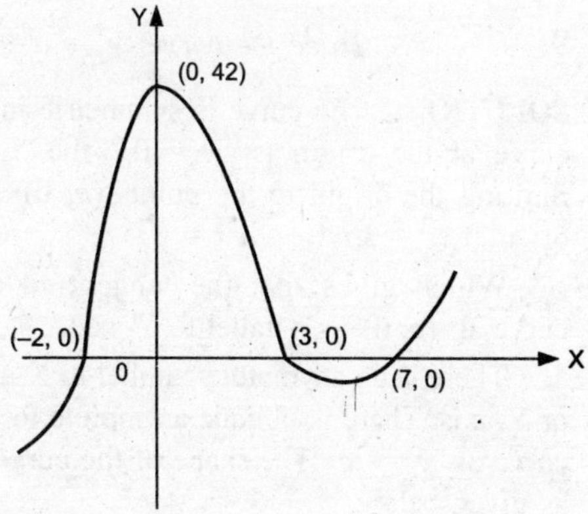

7. *Trace the curve* $y^2x^2 = x^2 - a^2$.

SOLUTION: The curve meets the x-axis at $(a, 0)$ and $(-a, 0)$. Tangent to the curve at $(a, 0)$ is parallel to y-axis. Similarly tangent at $(-a, 0)$ is parallel to y-axis.

It has two asymptotes parallel to x-axis as $y = \pm a$.

Shape of the curve is given in the adjoining figure.

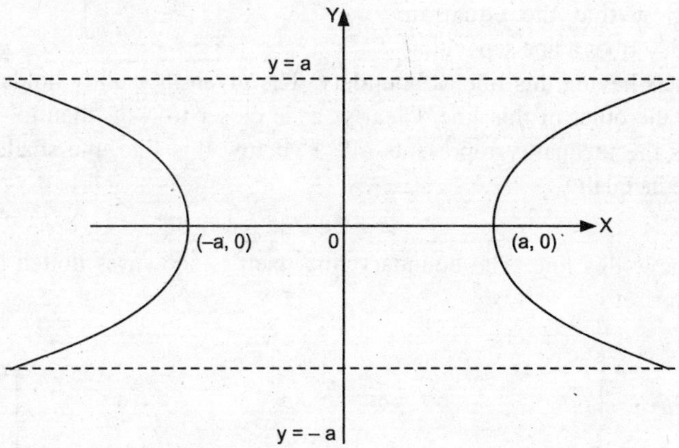

8. *Trace the curve* $y = \dfrac{x^2 + 2x}{x^2 + 4}$.

SOLUTION: The curve $y(x^2 + 4) - x^2 - 2x = 0$ is not symmetric. It passes through the origin and tangent to the curve at the origin is $4y - 2x = 0$ or $y = x/2$ by equating to zero the lowest degree term.

The curve meets the x-axis at $(-2, 0)$. Shifting the origin to $(-2, 0)$ by replacing x by $x - 2$ we get the tangent at the new origin as $8y = 6x$ or $y = \dfrac{3x}{4}$.

For asyptote parallel to x-axis we equate to zero the coefficient of the highest degree term in x and get $y = 1$. Shape of the curve is as follows.

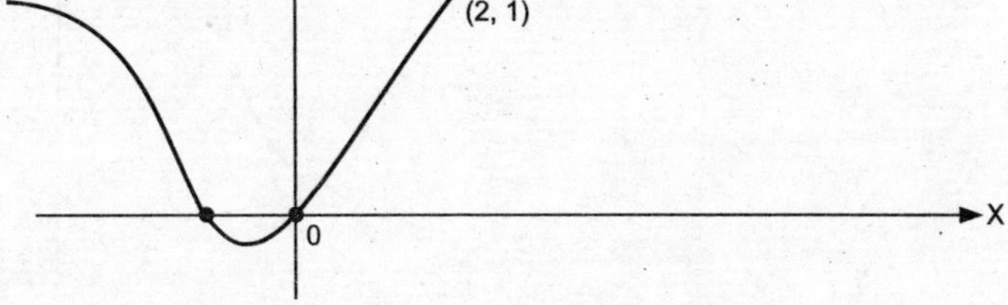

9. *Trace the curve* $y^3 = a^2x - x^3$.

SOLUTION: The curve is symmetric in opposite quadrants. It passes through the origin. Tangent to the curve at the origin is $x = 0$, the Y-axis. The curve meets the X-axis at $x = a$ and $x = -a$. Shifting the origin to the point $(a, 0)$ the equation of the curve becomes $y^3 = a^2(x + a) - (x + a)^3$ or $y^3 + x^3 - 3ax^2 + 2a^2x = 0$.

Which gives that the tangent to the curve at $(a, 0)$ is parallel to Y-axis.

There is no asymptote parallel to X-axis or Y-axis. There is oblique asymptote to the curve as $y = -x$. The shape of the curve is as given below.

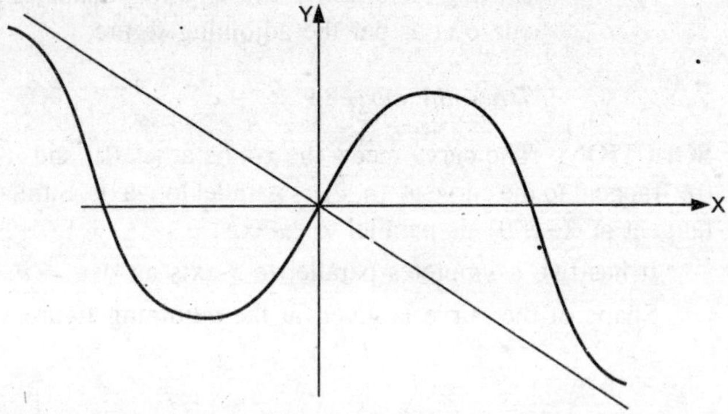

Now consider the inequality

$$|z + 6i| < |z - 1 + 3i|.$$

We already know that the equation $|z + 6i| = |z - 1 + 3i|$ describes a line separating the plane into two sets having this line as boundary. The given inequality holds for points in one of these sets, on one side or the other of this line. Clearly z is closer to $-6i$ than to $1 - 3i$ if z is below the boundary line. Thus the inequality represents all z below this line, the shaded region in the adjoining figure. Similarly, the inequality

$$|z + 6i| > |z - 1 + 3i|$$

represents all z above this line. The boundary line itself is shown as dotted because points on the line do not belong to the set.

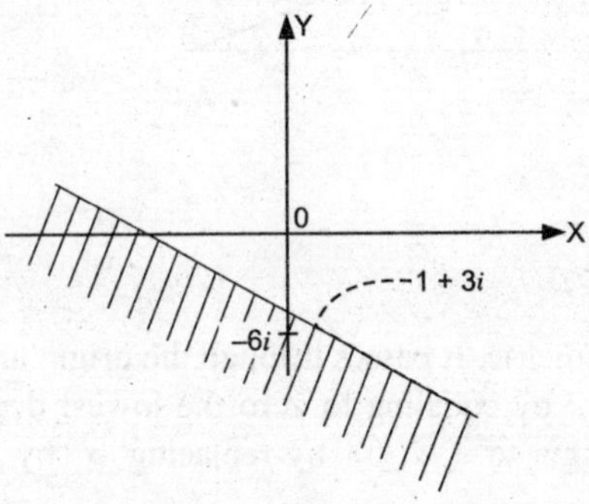

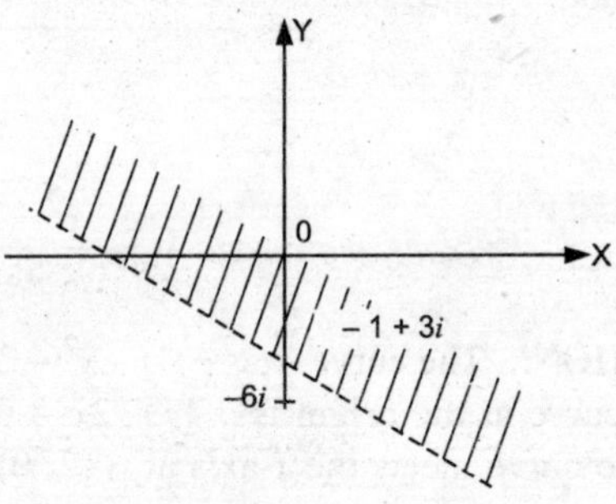

As another example, consider the equation

$$|z|^2 + 3\ \text{Re}(z^2) = 4$$

Taking $z = x + iy$ the equation becomes

$$x^2 + y^2 + 3\ (x^2 - y^2) = 4 \quad \text{or} \quad 4x^2 - 2y^2 = 4$$

which represents the hyperbola as shown below.

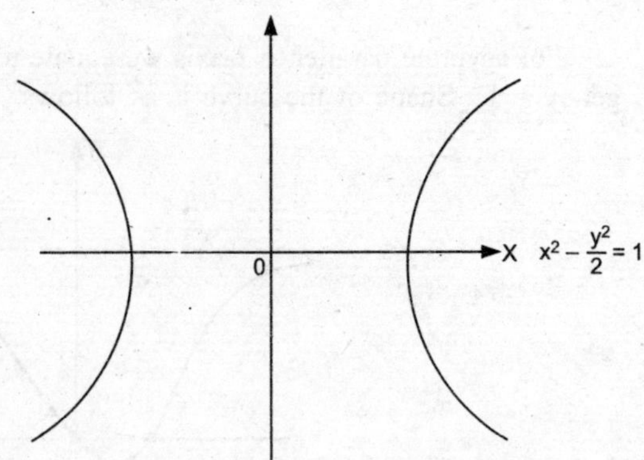

10. *Trace the luminiscate* $r^2 = a^2 \sin 2\theta$.

SOLUTION: The curve is symmetrical about the pole as the power of r is even. The equation remains unaltered on replacing θ by $\frac{\pi}{2} - \theta$ hence the curve is symmetrical about the line $\theta = \frac{\pi}{4}$.

The curve meets the line $\theta = \frac{\pi}{4}$ at $\left(a, \frac{\pi}{4}\right)$.

For tangent at this point, $r^2 = a^2 \sin 2\theta$.

$$\therefore \quad 2r\frac{dr}{d\theta} = 2a^2 \cos 2\theta \quad \text{and} \quad \tan\varphi = r\frac{d\theta}{dr} = \frac{r^2}{a^2 \cos 2\theta} = \tan 2\theta$$

$\Rightarrow \varphi = 2\theta$. Therefore, at $\theta = \frac{\pi}{4}$, $\varphi = \frac{\pi}{2}$. Therefore the tangent is perpendicular to the line $\theta = \frac{\pi}{4}$.

Following table gives values of r for various values of θ.

θ	0	$\frac{\pi}{8}$	$\frac{\pi}{4}$	$\frac{3\pi/8}{8}$	$\frac{\pi}{2}$
r	0	$\frac{a}{2^{1/4}}$	a	$\frac{a}{2^{1/4}}$	0

Shape of the curve is as follows.

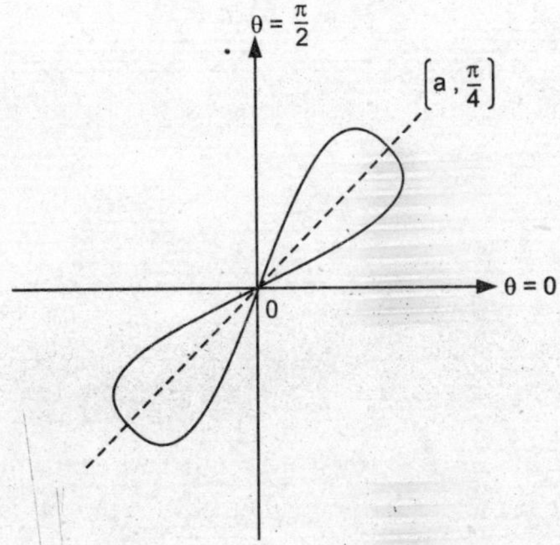

11. *Trace the four leaves rose* $r = a \cos 2\theta$.

SOLUTION: Curve is symmetric about $\theta = 0$ and about $\theta = \frac{\pi}{2}$. Since $r = 0$ at $\theta = \frac{\pi}{4}$ hence tangent to the curve at the pole is $\theta = \frac{\pi}{4}$. The curve meets the initial line at the point $(a, 0)$.

For tangent to the curve $r = a \cos 2\theta$

$$\therefore \quad \frac{dr}{d\theta} = -2a \sin 2\theta \quad \text{and} \quad \tan\varphi = r\frac{d\theta}{dr} = \frac{a \cos 2\theta}{-2a \sin^2 \theta} = \frac{-1}{2} \cot 2\theta.$$

At the point $(a, 0)$ $\tan\varphi = \infty$ hence $\varphi = \frac{\pi}{2}$.

$\therefore$ tangent at $(a, 0)$ is perpendicular to the initial line. Maximum value of r is 'a' hence it lies inside the circle with centre at the pole and radius a.

θ	0	$\frac{\pi}{8}$	$\frac{\pi}{4}$	$\frac{3\pi}{8}$	$\frac{\pi}{2}$
r	a	$\frac{a}{\sqrt{2}}$	0	$\frac{-a}{\sqrt{2}}$	$-a$

Shape of the curve is given here as four leaves.

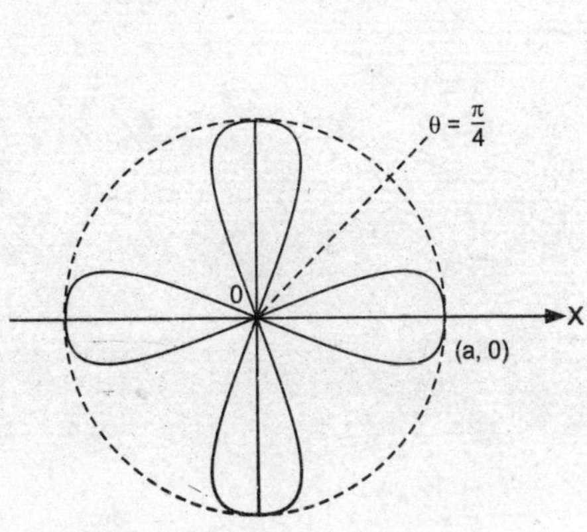

CHAPTER

5 Mean Value Theorems and Curvature

IMPORTANT FORMULAE AND USEFUL RESULTS

1. **Rolle's Theorem.** If $f(x)$ is continuous in a closed internal $[a, b]$ and is differentiable in the open interval (a, b) and if $f(a) = f(b)$ then there exists atleast one point $x = c$ in (a, b) such that $f'(c) = 0$.

2. **Lagrange's Mean Value Theorem.** If $f(x)$ is continuous in a closed interval $[a, b]$ and is differentiable in the open interval (a, b) then there exists atleast one point $x = c$ in (a, b) such that

$$f'(c) = \frac{f(b) - f(a)}{b - a}.$$

3. **Cauchy's Mean Value Theorem.** If two functions $f(x)$ and $g(x)$ are continuous in $[a, b]$ and differentiable in (a, b) and $g'(x) \neq 0$ for any x in (a, b) then there exists atleast one point $x = c$ in (a, b) such that

$$\frac{f(b) - f(a)}{g(b) - g(a)} = \frac{f'(c)}{g'(c)}.$$

4. **Curvature:** Radius of curvature $\rho = \dfrac{ds}{d\psi}$ and Curvature $= \dfrac{d\psi}{ds} = \dfrac{1}{\rho}$.

For Certesian Curves: $\rho = \dfrac{\left[1+\left(\dfrac{dy}{dx}\right)^2\right]^{3/2}}{\dfrac{d^2y}{dx^2}}$ or $\rho = \dfrac{\left[1+\left(\dfrac{dx}{dy}\right)^2\right]^{3/2}}{\dfrac{d^2x}{dy^2}}$.

For parametric curve

$$\rho = \frac{(x'^2 + y'^2)^{3/2}}{x'y'' - y'x''} \quad \text{where } x' = \frac{dx}{dt},\ y' = \frac{dy}{dt},\ x'' = \frac{d^2x}{dt^2},\ y'' = \frac{d^2y}{dt^2}$$

Newton's Method for radius of curvature at the origin

Case I: Curve passes through the origin and has x-axis as tangent there, then

$$\rho = \frac{1}{\dfrac{d^2y}{dx^2}} = \underset{x\to 0}{\text{Lt}}\left(\frac{x^2}{2y}\right)$$

Case II: Curve passes through the orign and has y-axis as tangent there, then

$$\rho = \underset{y\to 0}{\text{Lt}} \frac{y^2}{2x}.$$

Radius of curvature for polar curves

$$\rho = \frac{(r^2 + r_1^2)^{3/2}}{2r_1^2 + r^2 - rr_2}.$$

Centre of Curvature: For Cartesian curves the centre of curveture (α, β) is given by

$$\alpha = x - \frac{y_1(1+y_1^2)}{y_2}, \quad \beta = y + \frac{1+y_1^2}{y_2}$$

For parametric curves

$$\alpha = x - \frac{y'(x'^2 + y'^2)}{x'y'' - y'x''}, \quad \beta = y + \frac{x'(x'^2 + y'^2)}{x'y'' - y'x''}$$

Concavity, Convexity and Point of Inflexion

At the point of inflexion the curve changes its course either from concave downwards to concave upwards or from concave upwards to concave downwards.

In general, at the point of inflexion the even order derivative should be zero and next odd order deriative there, should not be zero.

SOLUTIONS TO EXERCISE 5A

1. *Verify Rolle's theorem for* $f(x) = \dfrac{x^2 - 4x}{x+2}$ in $[0, 4]$.

SOLUTION: $f(x)$ is continuous in $[0, 4]$ and differentiable in $(0, 4)$ since $x = -2$ lies outside $[0, 4]$. $f(0) = 0$ and $f(4) = 0$.

$$f'(x) = \frac{(x+2)(2x-4) - (x^2 - 4x)\cdot 1}{(x+2)^2} = \frac{x^2 + 4x - 8}{(x+2)^2}$$

$$f'(x) = 0 \text{ means } x^2 + 4x - 8 = 0 \text{ hence } x = -2 \pm 2\sqrt{3}$$

Out of these values of x for which $f'(x) = 0$, $x = -2 + 2\sqrt{3}$ lies in $(0, 4)$

Hence 'c' of Rolle's theorem is $2(\sqrt{3} - 1)$. **Hence Theorem is verified.**

2. *If* $f(x)$, $g(x)$ *and* $h(x)$ *are differentiable in* (a, b), *show that there exists a value* $x = c$ *in* (a, b) *satisfying*

$$\begin{vmatrix} f(a) & g(a) & h(a) \\ f(b) & g(b) & h(b) \\ f(x) & g(x) & h(x) \end{vmatrix} = 0.$$

SOLUTION: Let $F(x) = \begin{vmatrix} f(a) & g(a) & h(a) \\ f(b) & g(b) & h(b) \\ f(x) & g(x) & h(x) \end{vmatrix}$

$F(x)$ is differentiable in (a, b) as $f(x)$, $g(x)$, $h(x)$ are all differentiable in (a, b). Here $F(a) = 0$ and $F(b) = 0$ hence conditions of Rolles theorem are satisfied. Therefore there must exist $x = c$ lying in (a, b) so that $F'(c) = 0$.

i.e., $\begin{vmatrix} f(a) & g(a) & h(a) \\ f(b) & g(b) & h(b) \\ f'(x) & g'(x) & h'(x) \end{vmatrix} = 0$ for some $x = c$ in (a, b) **Hence Proved.**

3. *Prove that between any two real roots of the equation* $e^x \sin x = 1$, *there exists at least one real root of the equation* $e^x \cos x + 1 = 0$.

SOLUTION: Let $f(x) = e^x \sin x - 1$ and let its two roots be $x = a$ and $x = b$. So $e^a \sin a = 1$ and $e^b \sin b = 1$. $f(x)$ is continuous and differentiable in (a, b). Conditions of Rolle's Theorem are satisfied.

$f'(x) = e^x \cos x + e^x \sin x = e^x \cos x + 1$. Hence by Rolle's theorem, there must exist at least one $x = c$ for which $f'(x) = 0$

that is, for $x = c$, $e^x \cos x + 1 = 0$ **Hence Proved.**

4. *Verify Lagrange's Mean Value theorem for the function* $f(x) = 2x^2 - 7x - 10$ *over* (2, 5) *and find 'c' of the* L.M.V. *theorem.*

SOLUTION: $f(x)$ is continuous in [2, 5] and differentiable in (2, 5). Also $f(2) = 8 - 14 - 10 = -16$ and $f(5) = 50 - 35 - 10 = 5$. Here, $f'(x) = 4x - 7$

By L.M.V. Th. $f'(c) = \dfrac{f(5) - f(2)}{5-2}$

or $4c - 7 = \dfrac{5-(-16)}{5-2}$ hence $c = \dfrac{7}{2}$ which lies in (2, 5). **Ans.**

5. *Show that* $\log(x+1) = \dfrac{x}{1+\theta x}$ *where* $x > 0$ *and* $0 < \theta < 1$.

SOLUTION: We take $f(x) = \log(x+1)$ and interval as $(0, x)$. $\therefore f'(x) = \dfrac{1}{x+1}$.

By L.M.V. Th $f'(c) = \dfrac{f(x) - f(0)}{x-0} = \dfrac{\log(x+1) - \log 1}{x-0} = \dfrac{\log(x+1)}{x}$

or $\dfrac{\log(x+1)}{x} = \dfrac{1}{c+1}$

since $0 < c < x$ we can write $c = \theta x$ where $0 < \theta < 1$.

Hence $\dfrac{\log(x+1)}{x} = \dfrac{1}{1+\theta x}$ or $\log(x+1) = \dfrac{1}{1+\theta x}$. **Hence Verified.**

6. *Using Cauchy's Mean Value theorem, show that*

$$\frac{\sin b - \sin a}{\cos a - \cos b} = \cot c$$

where $a < c < b$, $a > 0$, $b > 0$.

SOLUTION: Let $f(x) = \sin x$ and $g(x) = \cos x$ in (a, b). Then, by Cauchy's Mean Value theorem,

$\dfrac{\sin b - \sin a}{\cos b - \cos a} = \dfrac{\cos c}{-\sin c}$ where $a < c < b$

or $\dfrac{\sin b - \sin a}{\cos a - \cos b} = \cot c$. **Hence Proved.**

7. *Verify Cauchy's Mean Value theorem for* $f(x) = 2x^3$ *and* $g(x) = x^6$ *in* $[a, b]$ *and find 'c'.*

SOLUTION: Since $f(x)$ and $g(x)$ satisfy the conditions of Cauchy's mean value theorem, in (a, b).

We have $\dfrac{f(b) - f(a)}{g(b) - g(a)} = \dfrac{f'(c)}{g'(c)}$, $a < c < b$.

$\dfrac{2b^3 - 2a^3}{b^6 - a^6} = \dfrac{6c^2}{6c^5}$ or $\dfrac{2}{b^3 + a^3} = \dfrac{1}{c^3}$

$\therefore$ $c = \left(\dfrac{a^3 + b^3}{2}\right)^{1/3}$ which lies in (a, b) **Ans.**

8. *If* $f(x) = \sqrt{x}$ *and* $g(x) = \frac{1}{\sqrt{x}}$, *prove that the 'c' of the Cauchy's Mean Value theorem is the geometric mean between* a *and* b *where* $a > 0, b > 0$.

SOLUTION: Since $f(x)$ and $g(x)$ are continuous in $[a, b]$ and differentiable in (a, b) by Cauchy's Mean Value theorem,

$$\frac{f(a)-f(b)}{g(a)-g(b)} = \frac{f'(c)}{g'(c)} \quad \text{where } a < c < b$$

or
$$\frac{\sqrt{a}-\sqrt{b}}{\frac{1}{\sqrt{a}}-\frac{1}{\sqrt{b}}} = \frac{\frac{1}{2\sqrt{c}}}{-\frac{1}{2}\frac{1}{c\sqrt{c}}} = -c \quad \text{or} \quad c = \sqrt{ab},$$

that is, c = geometric mean of a and b. **Hence Proved.**

9. *For the functions* $f(x) = x^3 + 2, g(x) = x^2 - 1$ *investigate whether Cauchy's theorem holds in the interval* $(1, 2)$ *and if so, find 'c'.*

SOLUTION: Since $f(x)$ and $g(x)$ satisfy the conditions of Cauchy's mean value theorem, we have by the theorem

$$\frac{f(2)-f(1)}{g(2)-g(1)} = \frac{f'(c)}{g'(c)} \quad \text{where } 1 < c < 2$$

$$\frac{2^3+2-(1^3+2)}{(2^2-1)-(1^2-1)} = \frac{3c^2}{2c} \quad \text{or} \quad \frac{7}{3} = \frac{3}{2}c \quad \therefore\ c = \frac{14}{9}$$ which belongs to $(1, 2)$ **Ans.**

SOLUTIONS TO EXERCISE 5B

1. *Find the radius of curvature at the point* (x, y) *of the curves*

(*i*) $y = c\cosh(x/c)$ (*ii*) $y = a\log\sec(x/a)$.

Solution: (*i*) $$y = c\cosh\left(\frac{x}{c}\right) \quad \therefore\ y_1 = \frac{dy}{dx} = \frac{c}{c}\sinh\frac{x}{c} \text{ and } y_2 = \frac{d^2y}{dx^2} = \frac{1}{c}\cosh\frac{x}{c}$$

Hence ρ at (x, y) $$= \frac{(1+y_1^2)^{3/2}}{y_2} = \frac{\left[1+\sinh^2\frac{x}{c}\right]^{3/2}}{\frac{1}{c}\cosh\frac{x}{c}} = \frac{\left[\cosh^2\left(\frac{x}{c}\right)\right]^{3/2}}{\frac{1}{c}\cosh\frac{x}{c}} = c\cosh^2\frac{x}{c}$$

or $$\rho = \frac{1}{c}y^2.$$ **Ans.**

(ii) $$y = a\log\sec\left(\frac{x}{a}\right) \quad \therefore \quad y_1 = \frac{dy}{dx} = \frac{a\sec\frac{x}{a}\tan\frac{x}{a}\cdot\frac{1}{a}}{\sec\left(\frac{x}{a}\right)} = \tan\frac{x}{a}$$

and $$y_2 = \frac{d^2y}{dx^2} = \frac{1}{a}\sec^2\frac{x}{a}.$$

Therefore, $$\rho = \frac{(1+y_1^2)^{3/2}}{y_2} = \frac{\left(1+\tan^2\frac{x}{a}\right)^{3/2}}{\frac{1}{a}\sec^2\frac{x}{a}} = a\sec\frac{x}{a} = ae^{y/a}$$ **Ans.**

2. *Obtain the radius of curvature at the point* (x, y) *of the curves*

(i) $x^{2/3} + y^{2/3} = a^{2/3}$. (ii) $\sqrt{x}+\sqrt{y} = \sqrt{a}$.

SOLUTION: $x^{2/3} + y^{2/3} = a^{2/3}$. It's parametric equation can be written as $x = a\cos^3 t, y = a\sin^3 t$.

$$\therefore \quad x' = \frac{dx}{dt} = -3a\cos^2 t\sin t, \quad y' = 3a\sin^2 t\cos t$$

and $$x'' = \frac{d^2x}{dt^2} = -3a\cos^3 t + 6a\cos t\sin^2 t = 3a\cos t\,(2\sin^2 t - \cos^2 t)$$

$$y'' = \frac{d^2y}{dt^2} = 6a\sin t\cos^2 t - 3a\sin^3 t = 3a\sin t\,(2\cos^2 t - \sin^2 t)$$

$$\therefore \quad \rho = \frac{(x'^2+y'^2)^{3/2}}{x'y''-y'x''} = \frac{(9a^2\cos^4 t\sin^2 t + 9a^2\sin^4 t\cos^2 t)^{3/2}}{-9a^2\cos^2 t\sin^2 t\,(2-3\sin^2 t) - 9a^2\sin^2 t\cos^2 t\,(2-3\cos^2 t)}$$

$$= -3a\sin t\cos t$$

$$= 3a\left(\frac{x}{a}\right)^{1/3}\left(\frac{y}{a}\right)^{1/3} = 3(axy)^{1/3}. \quad \text{Thus } \rho = 3(axy)^{1/3}$$ **Ans.**

(ii) $\sqrt{x}+\sqrt{y} = \sqrt{a}$.

$$\therefore \quad \frac{1}{2\sqrt{x}} + \frac{1}{2\sqrt{y}}\frac{dy}{dx} = 0 \quad \therefore \quad y_1 = \frac{dy}{dx} = \frac{-\sqrt{y}}{\sqrt{x}}$$

and $$y_2 = \frac{d^2y}{dx^2} = -\left[\frac{\sqrt{x}\cdot\frac{1}{2\sqrt{y}}y_1 - \sqrt{y}\cdot\frac{1}{2\sqrt{x}}}{x}\right] = \frac{-1}{2x}\left[-\frac{\sqrt{x}}{\sqrt{y}}\cdot\frac{\sqrt{y}}{\sqrt{x}} - \frac{\sqrt{y}}{\sqrt{x}}\right]$$

$$= \frac{1}{2x\sqrt{x}}\left(\sqrt{x}+\sqrt{y}\right) = \frac{\sqrt{a}}{2x\sqrt{x}}$$

Hence $$\rho = \frac{(1+y_1^2)^{3/2}}{y_2} = \frac{\left(1+\frac{y}{x}\right)^{3/2}.2x\sqrt{x}}{\sqrt{a}} = \frac{2(x+y)^{3/2}}{\sqrt{a}}$$ **Ans.**

3. *Find the radius of curvature at any point 't' on the curve*

$$x = a\left(\cos t + \log\tan\frac{t}{2}\right), \quad y = a\sin t$$

SOLUTION: $x = a\left(\cos t + \log\tan\frac{t}{2}\right), \quad y = a\sin t.$

$$\therefore \quad x' = \frac{dx}{dt} = a\left(-\sin t + \frac{1}{2}\frac{\sec^2\left(\frac{t}{2}\right)}{\tan\frac{t}{2}}\right) = a\left(-\sin t + \frac{1}{\sin t}\right) = \frac{a\cos^2 t}{\sin t}, \quad y' = a\cos t$$

and $$x'' = \frac{d^2x}{dt^2} = \frac{a(-2\sin^2 t\cos t - \cos^3 t)}{\sin^2 t} = \frac{-a\cos t}{\sin^2 t}(1+\sin^2 t), \quad y'' = -a\sin t$$

$$\therefore \quad \rho = \frac{(x'^2+y'^2)^{3/2}}{x'y''-y'x''} = \frac{\left[a^2\frac{\cos^4 t}{\sin^2 t} + a^2\cos^2 t\right]^{3/2}}{\frac{a\cos^2 t}{\sin t}(-a\sin t) + a\cos t\frac{a\cos t\,(1+\sin^2 t)}{\sin^2 t}}$$

$$= \frac{a\cos t}{\sin^3 t\,(-1+1+\operatorname{cosec}^2 t)} = a\cot t \quad \therefore \quad \rho = a\cot t$$ **Ans.**

4. *The tangents at two points P and Q on the cycloid*

$$x = a(\theta - \sin\theta), \quad y = a(1-\cos\theta)$$

are at right angles. If ρ_1 and ρ_2 are the radii of curvature at these points then show that

$$\rho_1^2 + \rho_2^2 = 16a^2.$$

SOLUTION: $x = a(\theta - \sin\theta), \quad y = a(1-\cos\theta) \quad \therefore \quad x' = a(1-\cos\theta), \quad y' = a\sin\theta$

and $x'' = a\sin\theta$, $y'' = a\cos\theta$. Hence ρ at (x, y) is given by

$$\rho = \frac{(x'^2+y'^2)}{x'y''+y'x''} = \frac{[a^2(1-\cos\theta)^2 + a^2\sin^2\theta)^{3/2}}{a(1-\cos\theta)\cdot a\cos\theta - a\sin\theta\,(a\sin\theta)} = \frac{a(2-2\cos\theta)^{3/2}}{\cos\theta - 1}$$

$$= a\cdot 2\sqrt{2}\sqrt{1-\cos\theta} = 4a\sin\frac{\theta}{2} = \rho \text{ at the point '}\theta\text{'}.$$

Let the points P and Q on the cycloide be 'θ_1' and 'θ_2',

then $\rho_1 = 4a\sin\left(\frac{\theta_1}{2}\right)$ and $\rho_2 = 4a\sin\left(\frac{\theta_2}{2}\right)$

Slopes of the tangents at any point 'θ' is $\dfrac{y'}{x'} = \dfrac{a\sin\theta}{a(1-\cos\theta)} = \cot\dfrac{\theta}{2}$

$\therefore$ Slopes of tangents at P and Q are $\cot\left(\frac{\theta_1}{2}\right)$ and $\cot\left(\frac{\theta_2}{2}\right)$. Since these are perpendicular to each other we have $\cot\frac{\theta_1}{2}\cot\frac{\theta_2}{2} = -1$

or $\quad \cos\frac{\theta_1}{2}\cos\frac{\theta_2}{2} + \sin\frac{\theta_1}{2}\sin\frac{\theta_2}{2} = 0 \quad$ or $\quad \cos\left(\frac{\theta_1}{2} - \frac{\theta_2}{2}\right) = 0$

$\Rightarrow \quad \frac{|\theta_1 - \theta_2|}{2} = \pm\frac{\pi}{2}$ or $\theta_2 = \pi - \theta_1$. Hence $\rho_2 = 4a\sin\left(\frac{\pi}{2} - \frac{\theta_1}{2}\right) = 4a\cos\frac{\theta_1}{2}$

Thus $\quad \rho_1^2 + \rho_2^2 = \left(4a\sin\frac{\theta_1}{2}\right)^2 + \left(4a\cos\frac{\theta_1}{2}\right)^2 = 16a^2.$ **Hence Proved.**

5. *Apply Newton's method to find the radius of curveture at the origin for the curve* $x^3 + y^3 - 2x^2 + 6y = 0.$

SOLUTION: $x^3 + y^3 - 2x^2 + 6y = 0$. It passes through the origin and tangent at the origin is $y = 0$, the x-axis.

$\therefore \quad \rho$ at the origin $= \underset{x\to 0}{\text{Lt}}\frac{x^2}{2y}$

Dividing the equation by $4y$ we get $\quad x\left(\frac{x^2}{4y}\right) + \frac{y^2}{4} - \frac{x^2}{2y} + \frac{3}{2} = 0$

Taking limit as $x \to 0$ we get $\quad 0 + 0 - \rho + \frac{3}{2} = 0 \quad$ or $\quad \rho = \frac{3}{2}$ **Ans.**

6. *Find the radius of curvature at the origin for the curve* $x^3 + y^3 = 3axy.$

SOLUTION: $x^3 + y^3 = 3axy$. It passes through the origin and x-axis as well as y-axis are tangents to the curve there. Dividing the equation by $(2xy)$ we get $\frac{x^2}{2y} + \frac{y^2}{2x} = \frac{3a}{2}$

Taking limit in (1) as $x \to 0$ and using $\rho = \underset{x\to 0}{\text{Lt}}\frac{x^2}{2y}$, we get

$$\rho + \underset{x\to 0}{\text{Lt}}\left(\frac{2y}{x^2}\cdot\frac{xy}{4}\right) = \frac{3a}{2}$$

or $\quad \rho + \underset{x\to 0}{\text{Lt}}\frac{2y}{x^2}\cdot\underset{x\to 0}{\text{Lt}}\frac{xy}{4} = \frac{3a}{2} \quad$ or $\quad \rho + \frac{1}{\rho}(0) = \frac{3a}{2}$

or $\quad \rho = \frac{3a}{2}$ **Ans.**

7. *Show that the radius of curvature at the end of the major axis of the ellipse*

$$\frac{x^2}{a^2}+\frac{y^2}{b^2}=1$$

is equal to the semi-latus rectum of the ellipse.

SOLUTION: End of the major axis of the ellipse is $(a, 0)$. Shift the origin to $(a, 0)$ and the equation becomes $\frac{(x+a)^2}{a^2}+\frac{y^2}{b^2}=1$ or $\frac{x^2}{a^2}+\frac{2x}{a}+\frac{y^2}{b^2}=0$. To find ρ at the origin at which y-axis is tangent, we divide throughout by $2x$ and get $\frac{x}{2a^2}+\frac{1}{a}+\frac{y^2}{2x}\cdot\frac{1}{b^2}=0$. Taking limit as $y \to 0$, we get

$$0+\frac{1}{a}+\frac{1}{b^2}\underset{x\to 0}{\text{Lt}}\frac{y^2}{2x}=0. \quad \text{or} \quad \underset{y\to 0}{\text{Lt}}\frac{y^2}{2x}=\frac{-b^2}{a}$$

$\therefore$ By Newtons method ρ at the origin $=\frac{b^2}{a}=$ semi latus rectum. **Hence Proved.**

8. *Apply Newton's method to obtain the radius of curvature of the curve* $y^2=(2a-x)/x$ *at the point* $(2a, 0)$.

SOLUTION: $y^2=\frac{(2a-x)}{x}$. It passes through $(2a, 0)$, shifting the origin to $(2a, 0)$ the equation becomes

$$y^2=\frac{2a-(x+2a)}{x+2a} \quad \text{or} \quad y^2(x+2a)=-x.$$

The tangent to the curve at the new origin is $x=0$, that is, y-axis. Therefore by Newton's method, ρ at the origin, is given by

$$\rho=\lim_{y\to 0}\frac{y^2}{2x}=\lim_{y\to 0}\frac{-x}{2x(x+2a)}=\lim_{x\to 0}\frac{-1}{2(x+2a)}=\frac{-1}{4a}=\frac{1}{4a}$$ **Ans.**

9. *Find the radius of curvature of the curve* $y^2=x^2(a+x)/(a-x)$ *at the origin.*

SOLUTION:
$$y^2=x^2\frac{(a+x)}{(a-x)} \quad \text{or} \quad (a-x)y^2=x^2(a+x)$$

Hence $\quad (a-x)\,2y\frac{dy}{dx}-y^2=2ax+3x^2$

or $\quad 2y(a-x)\frac{dy}{dx}=\frac{x^2(a+x)}{a-x}+2ax+3x^2=\frac{ax^2+x^3+2a^2x+3ax^2-2ax^2-3x^3}{(a-x)}$

or $\quad 2y\frac{dy}{dx}=\frac{2x(ax+a^2-x^2)}{(a-x)^2}$ or $\frac{dy}{dx}=\frac{x}{y}\left(\frac{a^2+ax-x^2}{(a-x)^2}\right)$

of $\quad \frac{dy}{dx}=\sqrt{\frac{a-x}{a+x}}\cdot\frac{a^2+ax-x^2}{(a-x)^2} \quad \therefore \quad \left(\frac{dy}{dx}\right)_{(0,0)}=1.$

Further $\quad \dfrac{dy}{dx} = y_1 = \dfrac{a^2 + ax - x^2}{(a-x)\sqrt{a^2 - x^2}}$

$$\therefore \quad y_2 = \frac{d^2y}{dx^2} = \frac{(a-x)\sqrt{a^2-x^2}\,(a-2x) - (a^2+ax-x^2)\left[-\sqrt{a^2-x^2} + \dfrac{(a-x)(-2x)}{2\sqrt{a^2-x^2}}\right]}{(a-x)^2(a^2-x^2)}$$

$$\therefore \quad (y_2)_{(0,\,0)} = \frac{a^3 + a^3}{a^4} = \frac{2}{a} \quad \text{at } (0, 0)$$

Hence, $\quad (\rho)_{(0,\,0)} = \dfrac{(1+1^2)^{3/2}}{\dfrac{2}{a}} = a\sqrt{2}$ **Ans.**

10. *For the curve* $s = ce^{x/c}$ *show that* $c\rho = s\,(s^2 - c^2)^{1/2}$

SOLUTION: $\quad s = c\,e^{x/c} \quad \therefore \quad \dfrac{ds}{dx} = c \cdot e^{x/c} \cdot \dfrac{1}{c} = e^{x/c} = \dfrac{s}{c}$

We know that $\quad \left(\dfrac{ds}{dx}\right)^2 = 1 + \left(\dfrac{dy}{dx}\right)^2 \quad \therefore \quad \left(\dfrac{dy}{dx}\right)^2 = \dfrac{s^2}{c^2} - 1$

or $\quad \dfrac{dy}{dx} = \dfrac{1}{c}\sqrt{s^2 - c^2} \quad \therefore \quad \dfrac{d^2y}{dx^2} = \dfrac{1}{2c}\dfrac{2s}{\sqrt{s^2-c^2}} \cdot \dfrac{ds}{dx} = \dfrac{(s/c)^2}{\sqrt{s^2-c^2}}$

$$\therefore \quad \rho = \frac{(1+y_1^2)^{3/2}}{y_2} = \frac{\left(\dfrac{s}{c}\right)^3}{\left(\dfrac{s}{c}\right)^2 \dfrac{1}{\sqrt{s^2-c^2}}} = \frac{s}{c}\sqrt{s^2-c^2}$$

or $\quad \rho c = s\sqrt{s^2 - c^2}$ **Hence Proved.**

11. *For the curve* $y = \dfrac{ax}{a+x}$, *prove that*

$$\left(\frac{2\rho}{a}\right)^{2/3} = \left(\frac{y}{x}\right)^2 + \left(\frac{x}{y}\right)^2$$

where ρ *is the radius of curvature of the curve at its point* (x, y).

SOLUTION: $\quad y = \dfrac{ax}{a+x} = \dfrac{a(a+x) - a^2}{a+x} = a - \dfrac{a^2}{a+x} \quad \therefore \quad y_1 = \dfrac{a^2}{(a+x)^2}, \quad y_2 = \dfrac{-2a^2}{(a+x)^3}$

$$\rho = \frac{(1+y_1^2)^{3/2}}{y_2} = \frac{(1+y_1^2)^{3/2}(a+x)^3}{2a^2}$$

$$\therefore \quad \rho^{2/3} = \frac{(1+y_1^2)(a+x)^2}{(2a^2)^{2/3}} = \left[1 + \frac{a^4}{(a+x)^4}\right]\frac{(a+x)^2}{2^{2/3} \cdot a^{4/3}}$$

$$\therefore \quad \left(\frac{2\rho}{a}\right)^{2/3} = \frac{1}{a^2}\left[(a+x)^2 + \frac{a^4}{(a+x)^2}\right] = \left(\frac{a+x}{a}\right)^2 + \left(\frac{a}{a+x}\right)^2$$

$$= \left(\frac{x}{y}\right)^2 + \left(\frac{y}{x}\right)^2$$

Hence Proved.

12. *Show that the ratio of radii of curvature at the points on the curves $xy = c^2$, $x^3 = 3c^2y$ which have same abscissa, varies as the square root of the ratio of the ordinates.*

SOLUTION: Curves are $y = \dfrac{c^2}{x}$ ...(1) and $y = \dfrac{x^3}{3c^2}$ ...(2)

With same abscissa points on the curves (1) and (2) are $P_{(x,\, y_1)}$ and $Q_{(x,\, y_2)}$ where $y_1 = \dfrac{c^2}{x}$ and $y_2 = \dfrac{x^3}{3c^2}$. If ρ_1 and ρ_2 are radii of curvatures at P and Q then

$$\rho_1 = \left[\frac{\left\{1+\left(\frac{dy}{dx}\right)^2\right\}^{3/2}}{\frac{d^2y}{dx^2}}\right]_{(x,\, y_1)} \quad \text{and} \quad \rho_2 = \left[\frac{\left\{1+\left(\frac{dy}{dx}\right)^2\right\}^{3/2}}{\frac{d^2y}{dx^2}}\right]_{(x,\, y_2)}$$

At P, $\quad \dfrac{dy}{dx} = \dfrac{-c^2}{x^2}$ and $\dfrac{d^2y}{dx^2} = \dfrac{2c^2}{x^3}$

and at Q, $\quad \dfrac{dy}{dx} = \dfrac{3x^2}{3c^2}$ and $\dfrac{d^2y}{dx^2} = \dfrac{2x}{c^2}$.

$$\therefore \quad \rho_1 = \frac{\left(1+\frac{c^4}{x^4}\right)^{3/2}}{\frac{2c^2}{x^3}} = \frac{(c^4+x^4)^{3/2}}{2c^2x^3} \quad \text{and} \quad \rho_2 = \frac{\left(1+\frac{x^4}{c^4}\right)^{3/2}}{\frac{2x}{c^2}} = \frac{(x^4+c^4)^{3/2}}{2xc^4}.$$

Thus $\quad \dfrac{\rho_1}{\rho_2} = \dfrac{2xc^4}{2c^2x^3} = \dfrac{c^2}{x^2}$. Also $\dfrac{y_1}{y_2} = \dfrac{c^2 \cdot 3c^2}{x \cdot x^3} = \dfrac{3c^4}{x^4}$

Therefore $\quad \dfrac{\rho_1}{\rho_2} \propto \sqrt{\dfrac{y_1}{y_2}}$ **Hence Proved.**

13. *Find the radius of curvature at the origin for the curve*

$$x^3y - xy^3 + 2x^2y + xy - y^2 + 2x = 0$$

SOLUTION: Curve $x^3y - xy^3 + 2x^2y + xy - y^2 + 2x = 0$. Passes through the origin and tangent at the origin is $x = 0$, i.e., y-axis.

By Newton's method $\rho = \underset{y \to 0}{\text{Lt}} \dfrac{y^2}{2x}$.

Dividing the given equation throughout by $2x$ and taking limit as $y \to 0$ and hence $x \to 0$, we get

$$\underset{y\to 0}{\text{Lt}}\left[\frac{x^2}{2}y - \frac{y^2}{2x}xy + xy + \frac{1}{2}y - \frac{y^2}{2x} + 1\right] = 0$$

or $\quad 0 - \rho \cdot 0 + 0 + 0 - \rho + 1 = 0 \qquad$ **Hence** $\rho = 1$ at $(0, 0)$ **Ans.**

14. *For the ellipse* $\dfrac{x^2}{a^2} + \dfrac{y^2}{b^2} = 1$, *prove that* $\rho = \dfrac{a^2b^2}{p^3}$, p *being the perpendicular distance from the centre on the tangent at* (x, y).

SOLUTION: $\dfrac{x^2}{a^2} + \dfrac{y^2}{b^2} = 1$. Any point on it is '$\theta$' so that $x = a\cos\theta$, $y = b\sin\theta$.

$\therefore \qquad x' = -a\sin\theta, \quad y' = b\cos\theta, \quad x'' = -a\cos\theta, \quad y'' = -b\sin\theta$

Hence ρ at the point 'θ' is equal to $\dfrac{(x'^2 + y'^2)^{3/2}}{x'y'' - y'x''}$

or
$$\rho = \frac{(a^2\sin^2\theta + b^2\cos^2\theta)^{3/2}}{ab\sin^2\theta + ab\cos^2\theta} = \frac{1}{ab}(a^2\sin^2\theta + b^2\cos^2\theta)^{3/2}$$

Tangent to the curve at $(a\cos\theta, b\sin\theta)$ is

$$y - b\sin\theta = \frac{b\cos\theta}{-a\sin\theta}(x - a\cos\theta) \qquad \text{or} \qquad bx\cos\theta + ay\sin\theta - ab = 0$$

$\therefore$ Perpendicular distance on this tangent from the origin, is

$$p = \frac{ab}{\sqrt{b^2\cos^2\theta + a^2\sin^2\theta}} \qquad \text{or} \qquad (b^2\cos^2\theta + a^2\sin^2\theta)^{3/2} = \frac{a^3b^3}{p^3}$$

Hence
$$\rho = \frac{1}{ab}\frac{a^3b^3}{p^3} = \frac{a^2b^2}{p^3}$$
Hence Proved.

15. *If* ρ_1 *and* ρ_2 *are the radii of curvature at the ends of a focal chord of the parabola* $y^2 = 4ax$ *then show that* $\rho_1^{-2/3} + \rho_2^{-2/3} = (2a)^{-2/3}$.

SOLUTION: $y^2 = 4ax$. Any point on it is P('t') so that $x = at^2$, $y = 2at$.

The other end of the focal chord from 't' will be $Q\left(\dfrac{-1}{t}\right)$, that is, $x = a\left(\dfrac{-1}{t}\right)^2$ and $y = \dfrac{-2a}{t}$

Thus
$$Q = \left(\frac{a}{t^2}, \frac{-2a}{t}\right)$$

Here $\qquad x' = 2at, \quad y' = 2a, \quad x'' = 2a, \quad y'' = 0$

Hence radius vector
$$= \frac{(x'^2 + y'^2)^{3/2}}{x'y'' - y'x''} = \frac{(4a^2t^2 + 4a^2)^{3/2}}{0 - 4a^2}$$

or $\qquad \rho_1 = \rho\,(\text{at } P) = 2a\,(1 + t^2)^{3/2} \quad$ and $\quad \rho_2 = \rho\,(\text{at } Q) = \dfrac{2a}{t^3}(1 + t^2)^{3/2}$

$$\therefore \qquad \rho_1^{-2/3} + \rho_2^{-2/3} = \frac{(2a)^{-2/3}}{1+t^2} + \frac{(2a)^{-2/3}t^2}{1+t^2} = (2a)^{-2/3}$$
Hence Proved.

SOLUTIONS TO EXERCISE 5C

1. *Find the radius of curvature at the point (r,θ) of the following curves*

(*i*) $r = a\theta$ (*ii*) $r = a\sin\theta$.

SOLUTION: (*i*) $r = a\,\theta$. Here $r_1 = \dfrac{dr}{d\theta} = a$ and $r_2 = \dfrac{d^2r}{d\theta^2} = 0$

Hence $$\rho = \frac{(r^2+r_1^2)^{3/2}}{2r_1^2+r^2-rr_2} = \frac{(a^2\theta^2+a^2)^{3/2}}{2a^2+a^2\theta^2-0} = \frac{a\,(1+\theta^2)^{3/2}}{2+\theta^2} = \frac{a}{2} \text{ at } (0,0).$$ **Ans.**

(*ii*) $r = a\sin\theta$. Here $r_1 = \dfrac{dr}{d\theta} = a\cos\theta$ and $r_2 = \dfrac{d^2r}{d\theta^2} = -a\sin\theta$.

$$\therefore \quad \rho = \frac{(r^2+r_1^2)^{3/2}}{2r_1^2+r^2-rr_2} = \frac{(a^2\sin^2\theta+a^2\cos^2\theta)^{3/2}}{2a^2\cos^2\theta+a^2\sin^2\theta-a\sin\theta(-a\sin\theta)}$$

$$= \frac{a}{2\,(\cos^2\theta+\sin^2\theta)} = \frac{a}{2} \text{ at } (r,\theta) \quad \text{and} = \frac{a}{2} \text{ at } (0,0)$$ **Ans.**

2. *Find the radius of curvature at the point (r,θ) of the parabola*

$\dfrac{2l}{r} = 1+\cos\theta$.

SOLUTION: $\dfrac{2l}{r} = 1+\cos\theta$ or $r = \dfrac{2l}{1+\cos\theta} = l\sec^2\dfrac{\theta}{2}$

$$r_1 = \frac{dr}{d\theta} = 2l\sec\frac{\theta}{2}\cdot\sec\frac{\theta}{2}\tan\frac{\theta}{2}\frac{1}{2} = l\sec^2\frac{\theta}{2}\tan\frac{\theta}{2} = r\tan\frac{\theta}{2}$$

$$\therefore \quad r_2 = \frac{d^2r}{d\theta^2} = r_1\tan\frac{\theta}{2} + \frac{r}{2}\sec^2\frac{\theta}{2} = r\tan^2\frac{\theta}{2}+\frac{r}{2}\sec^2\frac{\theta}{2}$$

Now $$\rho = \frac{(r^2+r_1^2)^{3/2}}{2r_1^2+r^2-rr_2} = \frac{\left(r^2+r^2\tan^2\frac{\theta}{2}\right)^{3/2}}{2r^2\tan^2\frac{\theta}{2}+r^2-r\left(r\tan^2\frac{\theta}{2}+\frac{r}{2}\sec^2\frac{\theta}{2}\right)}$$

$$= \frac{r\sec^3\frac{\theta}{2}}{2\tan^2\frac{\theta}{2}+1-\tan^2\frac{\theta}{2}-\frac{1}{2}\sec^2\frac{\theta}{2}} = \frac{r\sec^3\frac{\theta}{2}}{\frac{1}{2}\sec^2\frac{\theta}{2}} = 2\,r\sec\frac{\theta}{2}$$

or $$\rho = 2r\sqrt{\frac{r}{l}} = \frac{2}{\sqrt{l}}\,r^{3/2}.$$ **Ans.**

3. *Show that the radius of curvature at the point* (r, θ) *of the curve* $r^2 \cos 2\theta = a^2$ *is* $\dfrac{r^3}{a^2}$.

SOLUTION: $r^2 = a^2 \sec 2\theta$. $\therefore$ $2rr_1 = 2a^2 \sec 2\theta \tan 2\theta = 2r^2 \tan 2\theta$ or $r_1 = r \tan 2\theta$

and $r_2 = 2r \sec^2 2\theta + r_1 \tan 2\theta = 2r \sec^2 2\theta + r \tan^2 2\theta$

$$\therefore \quad \rho = \frac{(r^2 + r_1^2)^{3/2}}{2r_1^2 + r^2 - rr_2} = \frac{(r^2 + r^2 \tan^2 2\theta)^{3/2}}{2r^2 \tan^2 2\theta + r^2 - r^2(2\sec^2 2\theta + \tan^2 2\theta)}$$

$$= \frac{(r^2 \sec^2 2\theta)^{3/2}}{r^2(2\tan^2 2\theta + 1 - 2\sec^2 2\theta - \tan^2 2\theta)} = \frac{r^3 \sec^3 2\theta}{r^2 \sec^2 2\theta}$$

$$= r \sec 2\theta = \frac{r \cdot r^2}{a^2} = \frac{r^3}{a^2}$$ (neglecting the –ve sign). **Hence Proved.**

4. *Find the radius of curvature at the point* (r, θ) *of the curve*

$$\theta = \frac{\sqrt{r^2 - a^2}}{a} - \cos^{-1}(a/r).$$

SOLUTION:

$$\theta = \frac{\sqrt{r^2 - a^2}}{a} - \cos^{-1}\left(\frac{a}{r}\right)$$

$$\therefore \quad \frac{d\theta}{dr} = \frac{1 \cdot 2r}{2a\sqrt{r^2 - a^2}} - \frac{a/r^2}{\sqrt{1 - \dfrac{a^2}{r^2}}} = \frac{r}{a\sqrt{r^2 - a^2}} - \frac{a}{r\sqrt{r^2 - a^2}} = \frac{r^2 - a^2}{ar\sqrt{r^2 - a^2}} = \frac{\sqrt{r^2 - a^2}}{ar}$$

$$\therefore \quad \frac{dr}{d\theta} = r_1 = \frac{ar}{\sqrt{r^2 - a^2}} \text{ and } r_2 = \frac{d^2 r}{d\theta^2} = \frac{d}{dr}\left(\frac{ar}{\sqrt{r^2 - a^2}}\right) \cdot \frac{dr}{d\theta}$$

or

$$r_2 = \left(\frac{a\sqrt{r^2 - a^2} - ar \cdot \dfrac{r}{\sqrt{r^2 - a^2}}}{r^2 - a^2}\right) \frac{ar}{\sqrt{r^2 - a^2}} = \frac{ar\,[a\,(r^2 - a^2) - ar^2]}{(r^2 - a^2)^2} = \frac{-a^4 r}{(r^2 - a^2)^2}$$

Therefore

$$\rho = \frac{(r^2 + r_1^2)^{3/2}}{2r_1^2 + r^2 - rr_2} = \frac{\left[r^2 + \dfrac{a^2 r^2}{r^2 - a^2}\right]^{3/2}}{\dfrac{2a^2 r^2}{r^2 - a^2} + r^2 + \dfrac{a^4 r^2}{(r^2 - a^2)^2}}$$

or

$$\rho = \frac{r^6 \sqrt{r^2 - a^2}}{r^2(r^2 - a^2)(2a^2 + r^2 - a^2) + a^4 r^2} = \frac{r^6 (r^2 - a^2)^{1/2}}{r^2(r^4 - a^4) + a^4 r^2} = \sqrt{r^2 - a^2}.$$ **Ans.**

5. *Show that the radius of curvature at any point 't' of the curve*

$$x = a(\cos t + t\sin t),\quad y = a(\sin t - t\cos t) \quad \text{is given by} \quad \rho = at.$$

SOLUTION: $x = a(\cos t + t\sin t),\quad y = a(\sin t - t\cos t)$

$\therefore$ $x' = a(-\sin t + t\cos t + \sin t) = at\cos t,\quad y' = a(\cos t - \cos t + t\sin t) = at\sin t$

and $x'' = a(\cos t - t\sin t),\quad y'' = a(\sin t + t\cos t)$

$$\therefore \quad \rho = \frac{(x'^2 + y'^2)^{3/2}}{x'y'' - y'x''} = \frac{[a^2t^2\cos^2 t + a^2t^2\sin^2 t]^{3/2}}{a^2t\cos t(\sin t + t\cos t) - a^2t\sin t(\cos t - t\sin t)}$$

$$= \frac{at^3}{t\sin t\cos t + t^2\cos^2 t - t\sin t\cos t + t^2\sin^2 t} = at \quad \text{or} \quad p = \text{at}$$ **Hence Proved.**

6. *Find the radius of curvature at the origin for the curve* $r = a\sin n\theta$.

SOLUTION: $r = a\sin n\theta$, hence $r_1 = an\cos n\theta$ and $r_2 = -an^2\sin n\theta$.

$$\therefore \quad \rho = \frac{(r^2 + r_1^2)^{3/2}}{2r_1^2 + r^2 - rr_2} = \frac{(a^2\sin^2 n\theta + a^2n^2\cos^2 n\theta)^{3/2}}{2a^2n^2\cos^2 n\theta + a^2\sin^2 n\theta + a^2n^2\sin^2 n\theta}$$

$$= \frac{a[\sin^2 n\theta + n^2\cos^2 n\theta]^{3/2}}{2n^2\cos^2 n\theta + n^2\sin^2 n\theta + \sin^2 n\theta}$$

At $\theta = 0$, $$\rho = \frac{a(0 + n^2)^{3/2}}{2n^2} = \frac{an}{2}.$$ **Ans.**

7. *If* ρ_1 *and* ρ_2 *are the radii of curvature at the extremities of a chord of the cardioide* $r = a(1 + \cos\theta)$ *passing through the pole, prove that* $\rho_1^2 + \rho_2^2 = \frac{16}{9}a^2$.

SOLUTION: $r = a(1 + \cos\theta)$ hence $r_1 = -a\sin\theta$ and $r_2 = -a\cos\theta$

$$\rho = \frac{(r^2 + r_1^2)^{3/2}}{2r_1^2 + r^2 - rr_2} = \frac{[a^2(1+\cos\theta)^2 + a^2\sin^2\theta]^{3/2}}{2a^2\sin^2\theta + a^2(1+\cos\theta)^2 + a^2\cos\theta(1+\cos\theta)}$$

$$= \frac{a(2 + 2\cos\theta)^{3/2}}{2\sin^2\theta + 1 + 2\cos\theta + \cos^2\theta + \cos\theta + \cos^2\theta} = \frac{a\left(4\cos^2\frac{\theta}{2}\right)^{3/2}}{3(1+\cos\theta)} = \frac{4}{3}a\cos\frac{\theta}{2}$$

If one end of the chord of the cardiode through pole is θ then the other end is $\theta + \pi$.

$$\therefore \quad \rho_1 = \frac{4a}{3}\cos\frac{\theta}{2},\quad \rho_2 = \frac{4a}{3}\cos\frac{(\theta+\pi)}{2} = \frac{-4a}{3}\sin\frac{\theta}{2}$$

Therefore $$\rho_1^2 + \rho_2^2 = \left(\frac{4a}{3}\right)^2\left(\cos^2\frac{\theta}{2} + \sin^2\frac{\theta}{2}\right) = \frac{16a^2}{9}$$ **Hence Proved.**

8. *Find the centre of curvature at the point (x, y) of the hyperbola $\frac{x^2}{a^2}-\frac{y^2}{b^2}=1$ and also obtain its evolute.*

SOLUTION: $\frac{x^2}{a^2}-\frac{y^2}{b^2}=1$. It's parametric form is $x = a\sec\theta,\ y = b\tan\theta$.

$\therefore$ $x' = a\sec\theta\tan\theta,\ x'' = a(\sec^3\theta + \sec\theta\tan^2\theta),\ y' = b\sec^2\theta,\ y'' = 2b\sec^2\theta\tan\theta.$

We know that the centre of curvature (α, β) is given by

$$\alpha = x - \frac{y_1(1+y_1^2)}{y_2},\quad \beta = y + \frac{1+y_1^2}{y_2}.$$

or $$\alpha = x - \frac{y'(x'^2+y'^2)}{x'y''-y'x''},\quad \beta = y + \frac{x'(x'^2+y'^2)}{x'y''-y'x''}$$ in case of parametric curve.

$$\alpha = \frac{x'^2+y'^2}{x'y''-y'x''}\ \frac{(a^2\sec^2\theta\tan^2\theta + b^2\sec^4\theta)}{a\sec\theta\tan\theta\cdot 2b\sec^2\theta\tan\theta - b\sec^2\theta\cdot a(\sec^3\theta+\sec\theta\tan^2\theta)}$$

$$= \frac{\sec^2\theta(a^2\tan^2\theta+b^2\sec^2\theta)}{2ab\sec^3\theta\tan^2\theta - ab(\sec^3\theta+\sec\theta\tan^2\theta)\sec^2\theta}$$

$$= \frac{a^2\tan^2\theta+b^2\sec^2\theta}{ab\sec\theta(2\tan^2\theta-\sec^2\theta-\tan^2\theta)} = \frac{a^2\tan^2\theta+b^2\sec^2\theta}{-ab\sec\theta}$$

$\therefore$ $$\alpha = \alpha\sec\theta - \frac{b\sec^2\theta(a^2\tan^2\theta+b^2\sec^2\theta)}{-ab\sec\theta},$$

$$\beta = b\tan\theta + \frac{a\sec\theta\tan\theta(a^2\tan^2\theta+b^2\sec^2\theta)}{-ab\sec\theta}$$

or $$\alpha = a\sec\theta + \frac{\sec\theta}{a}(a^2\tan^2\theta+b^2\sec^2\theta),$$

and $$\beta = b\tan\theta - \frac{\tan\theta}{b}(a^2\tan^2\theta+b^2\sec^2\theta)$$

or $$\alpha = \frac{\sec\theta}{a}[a^2+a^2\tan^2\theta+b^2\sec^2\theta),\quad \beta = \frac{\tan\theta}{b}[b^2-a^2\tan^2\theta-b^2\sec^2\theta]$$

or $$\alpha = \frac{\sec^3\theta}{a}(a^2+b^2),\quad \beta = \frac{-\tan^3\theta}{b}(a^2+b^2) \qquad ...(1)$$

Thus the centre of curvature is $\left(\frac{a^2+b^2}{a}\sec^3\theta, -\frac{(a^2+b^2)}{b}\tan^3\theta\right)$. **Ans.**

Eleminating θ in (1) gives

$$\left(\frac{a\alpha}{a^2+b^2}\right)^{2/3} - \left(\frac{b\beta}{a^2+b^2}\right)^{2/3} = 1 \quad \text{or} \quad (a\alpha)^{2/3} - (b\beta)^{2/3} = (a^2+b^2)^{2/3}$$

The evolute of the curve is the locus of the centre of curvature which is

$(ax)^{2/3} - (by)^{2/3} = (a^2+b^2)^{2/3}$. **Ans.**

9. *Find the radius of curvature and the centre of curvature for the curve* $y = \tan x$ *at the point where* $x = \frac{\pi}{4}$.

SOLUTION: $y = \tan x$ hence $y_1 = \sec^2 x$ and $y_2 = 2\sec^2 x \tan x$

$$\therefore \quad \frac{1+y_1^2}{y_2} = \frac{1+\sec^4 x}{2\sec^2 x \tan x} = \frac{1+\sec^4 \frac{\pi}{4}}{2\sec^2 \frac{\pi}{4} \tan \frac{\pi}{4}} = \frac{1+4}{4} = \frac{5}{4} \left(\text{at } x = \frac{\pi}{4}\right)$$

$$\therefore \quad \rho = \frac{(1+y_1^2)^{3/2}}{y_2} = \frac{(1+\sec^4 x)^{3/2}}{2\sec^2 x \tan x} = \frac{(1+4)^{3/2}}{4} = \frac{5\sqrt{5}}{4}. \left(\text{at } x = \frac{\pi}{4}\right)$$

The centre of curvarture (α, β) is given by

$$\alpha = x - y_1 \frac{(1+y_1^2)}{y_2}, \quad \beta = y + \frac{1+y_1^2}{y_2}$$

or
$$\alpha = x - \frac{\sec^2 x\,(1+\sec^4 x)}{2\sec^2 x \tan x} = x - \frac{1+\sec^4 x}{2\tan x}$$

and
$$\beta = y + \frac{1+y_1^2}{y_2} = \tan x + \frac{1+\sec^4 x}{2\sec^2 x \tan x}$$

At $x = \frac{\pi}{4}$,
$$\alpha = \frac{\pi}{4} - \frac{1+4}{2} = \frac{\pi - 10}{4}, \quad \beta = 1 + \frac{1+4}{2\cdot 2\cdot 1} = \frac{9}{4}$$
Ans.

10. *Show that the centre of curvature at any point on the ellipse*

$$\frac{x^2}{a^2} + \frac{y^2}{b^2} = 1 \quad \text{is} \quad \left(\frac{a^2-b^2}{a^4}x^3, \frac{a^2-b^2}{-b^4}y^3\right).$$

SOLUTION: $x = a\cos\theta$, $y = b\sin\theta$ hence $x' = -a\sin\theta$, $y' = b\cos\theta$ and $x'' = -a\cos\theta$, $y'' = -b\sin\theta$.

The centre of curvature (α, β) is given by

$$\alpha = x - \frac{y'(x'^2+y'^2)}{x'y''+y'x''}, \quad \beta = y + \frac{x'(x'^2+y'^2)}{x'y''-y'x''}$$

Here
$$\frac{x'^2+y'^2}{x'y''-y'x''} = \frac{a^2\sin^2\theta + b^2\cos^2\theta}{-a\sin\theta\,(-b\sin\theta) - b\cos\theta\,(-a\cos\theta)} = \frac{a^2\sin^2\theta + b^2\cos^2\theta}{ab\,(\cos^2\theta + \sin^2\theta)}$$

$$\therefore \quad \alpha = a\cos\theta - \frac{b\cos\theta\,(a^2\sin^2\theta + b^2\cos^2\theta)}{ab},$$

and
$$\beta = b\sin\theta = \frac{a\sin\theta\,(a^2\sin^2\theta + b^2\cos^2\theta)}{ab}$$

or
$$\alpha = \frac{1}{a}\cos\theta\,(a^2 - a^2\sin^2\theta - b^2\sin^2\theta),$$

$$\beta = \frac{1}{b}\sin\theta\,(b^2 - a^2\sin^2\theta - b^2\cos^2\theta)$$

or $$\alpha = \frac{(a^2-b^2)}{a}\cos^3\theta \quad \text{and} \quad \beta = \frac{-(a^2-b^2)}{b}\sin^3\theta o,$$

Thus, the centre of curvature is $\left\{\dfrac{(a^2-b^2)x^3}{a^4}, \dfrac{(a^2-b^2)y^3}{-b^4}\right\}$, **Ans.**

11. *Find the chord of curvature through the pole of the cardioide $r = a\,(1 + \cos\theta)$.*

SOLUTION: $r = a\,(1 + \cos\theta)$, hence $r_1 = -a\sin\theta$ and $r_2 = -a\cos\theta$.

$$\therefore \quad \rho = \frac{(r^2+r_1^2)^{3/2}}{2r_1^2+r^2-rr_2} = \frac{[a^2(1+\cos\theta)^2+a^2\sin^2\theta)^{3/2}}{2a^2\sin^2\theta + a^2(1+\cos\theta)^2 + a^2(\cos\theta+\cos^2\theta)}$$

$$= \frac{a\,(2+2\cos\theta)^{3/2}}{3+3\cos\theta} = \frac{a\left(4\cos^2\dfrac{\theta}{2}\right)^{3/2}}{3\left(2\cos^2\dfrac{\theta}{2}\right)} = \frac{4a}{3}\cos\frac{\theta}{2}.$$

Next, we know that $\tan\phi = r\dfrac{d\theta}{dr} = \dfrac{a\,(1+\cos\theta)}{-a\sin\theta} = -\cot\dfrac{\theta}{2} = \tan\left(\dfrac{\pi}{2}+\dfrac{\theta}{2}\right)$

hence $$\phi = \frac{\pi}{2}+\frac{\theta}{2}$$

Also the length of the chord of curvature $= 2\rho\sin\phi = 2\cdot\dfrac{4a}{3}\cos\dfrac{\theta}{2}\sin\left(\dfrac{\pi}{2}+\dfrac{\theta}{2}\right)$

$$= \frac{4a}{3}(1+\cos\theta) = \frac{4r}{3}$$ **Ans.**

12. *For the curve $r^2 = a^2\cos 2\theta$ show that*

$$\rho = \frac{a^2}{3r}$$

SOLUTION: $r^2 = a^2\cos 2\theta$ hence $2rr_1 = -2a^2\sin 2\theta$ or $r_1 = \dfrac{-a^2\sin 2\theta}{a\sqrt{\cos 2\theta}}$

and $rr_2 + r_1^2 = -2a^2\cos 2\theta \quad \therefore rr_2 = -2a^2\cos 2\theta - \dfrac{-a^2\sin^2 2\theta}{\cos 2\theta}$

or $$rr_2 = \frac{-a^2}{\cos 2\theta}(2\cos^2 2\theta + \sin^2 2\theta) = -\frac{a^2}{\cos 2\theta}(1+\cos^2 2\theta).$$

Hence $$\rho = \frac{(r^2+r_1^2)^{3/2}}{2r_1^2+r^2-rr_2} = \frac{\left[a^2\cos 2\theta + \dfrac{a^2\sin^2 2\theta}{\cos 2\theta}\right]^{3/2}}{\dfrac{2a^2\sin^2 2\theta}{\cos 2\theta} + a^2\cos 2\theta + \dfrac{a^2}{\cos 2\theta}(1+\cos^2 2\theta)}$$

$$= \frac{a}{\sqrt{\cos 2\theta}}\cdot\left(\frac{1}{2\sin^2 2\theta + \cos^2 2\theta + 1 + \cos^2 2\theta}\right) = \frac{a}{3\sqrt{\cos 2\theta}} = \frac{a^2}{3r}$$ **Hence Proved.**

SOLUTIONS TO EXERCISE 5D

1. *Find the points of inflexion of the curve* $y = \dfrac{2x}{1+x^2}$.

SOLUTION: $y = \dfrac{2x}{1+x^2}$ hence $\dfrac{dy}{dx} = \dfrac{2(1+x^2)-4x^2}{(1+x^2)^2} = \dfrac{-2(x^2+1-2)}{(x^2+1)^2} = \dfrac{-2}{x^2+1} + \dfrac{4}{(x^2+1)^2}$

Hence $\dfrac{d^2y}{dx^2} = \dfrac{4x}{(x^2+1)^2} - \dfrac{16x}{(x^2+1)^3} = \dfrac{4x(x^2+1-4)}{(x^2+1)^3} = \dfrac{4x(x^2-3)}{(x^2+1)^3}$

For points of inflexion $\dfrac{d^2y}{dx^2} = 0$ and $\dfrac{d^3y}{dx^3} \neq 0$

which gives $x = 0, \sqrt{3}, -\sqrt{3}$. When $x = 0$ we have $y = 0$

when $x = \pm\sqrt{3}$, $y = \pm\dfrac{\sqrt{3}}{2}$. Hence the points of inflexion are (0, 0), $\left(\sqrt{3}, \dfrac{\sqrt{3}}{2}\right)$ and $\left(-\sqrt{3}, \dfrac{-\sqrt{3}}{2}\right)$.

It is easy to verify that $\dfrac{d^3y}{dx^3} \neq 0$ at these points. **Ans.**

2. *Find the points of inflexion of the curve* $y = \dfrac{x}{x^2+2x+2}$.

SOLUTION: $y = \dfrac{x}{x^2+2x+2}$ hence $\dfrac{dy}{dx} = \dfrac{(x^2+2x+2)-x(2x+2)}{(x^2+2x+2)^2} = \dfrac{2-x^2}{(x^2+2x+2)^2}$

and $\dfrac{d^2y}{dx^2} = \dfrac{-2x(x^2+2x+2)^2-(2-x^2)^2(x^2+2x+2)(2x+2)}{(x^2+2x+2)^4}$

$= \dfrac{-2x(x^2+2x+2)-4(x+1)(2-x^2)}{(x^2+2x+2)3} = \dfrac{2x^3-12x-8}{(x^2+2x+2)^2}$

Now $\dfrac{d^2y}{dx^2} = 0$ when $x^3 - 6x - 4 = 0$ or $(x + 2)(x^2 - 2x - 2) = 0$

Hence $x = -2,\ 1 \pm \sqrt{3}$. Points of inflexion are $(-2, -1)$, $\left(1+\sqrt{3}, \dfrac{-1+\sqrt{3}}{4}\right)$ and $\left[1-\sqrt{3}, \dfrac{-1-\sqrt{3}}{4}\right]$.

It can be verified that at these points $\dfrac{d^3y}{dx^3} \neq 0$. **Ans.**

3. *Obtain the angle subtended by the line joining two points of inflexion of the curve* $y^2(x-a) = x^2(x+a)$ *at the origin.*

SOLUTION: $y^2(x-a) = x^2(x+a) = x^3 + ax^2$

$\therefore$ $2(x-a)yy' + y^2 = 3x^2 + 2ax$

or $2(x-a)^2yy' + y^2(x-a) = (x-a)(3x^2+2ax)$

$\Rightarrow$ $2(x-a)^2yy' = (x-a)(3x^2+2ax) - x^3 - ax^2 = 2x^3 - 2ax^2 - 2a^2x$

or $(x-a)^4 y^2 y'^2 = x^2(x^2 - ax - a^2)^2$

or $(x-a)^3 y'^2 x^2 (x+a) = x^2(x^2 - ax - a^2)^2$

or $y'^2(x+a)(x-a)^3 = (x^2 - ax - a^2)^2$ ...(1)

Differentiating again we get

$$2y'y''(x+a)(x-a)^3 + y'^2[(x-a)^3 + 3(x+a)(x-a)^2] = 2(x^2 - ax - a^2)(2x - a)$$

At the point of inflexion $y'' = 0$ and putting for y'^2 from (1), gives

$$\frac{(x^2 - ax - a^2)}{(x+a)(x-a)}[x - a + 3(x+a)] = 2(x^2 - ax - a^2)(2x - a)$$

or $(x^2 - ax - a^2)(2x + a) = (2x - a)(x^2 - a^2) \Rightarrow x = -2a$

For $x = -2a$, $y^2 = \dfrac{-4a^2}{-3a}$ $\therefore$ $y = \pm\dfrac{2a}{\sqrt{3}}$. Hence two points of inflexion are $\left(-2a, \dfrac{2a}{\sqrt{3}}\right)$ and $\left(-2a, \dfrac{-2a}{\sqrt{3}}\right)$. The two points make an angle α at the origin

$\therefore$ $\tan\alpha = \sqrt{3}$ Hence $\alpha = \dfrac{\pi}{3}$ **Ans.**

4. *Show that the curve $y = e^{-x}$ is concave upwards at every point of it.*

SOLUTION: $y = e^{-x}$ $\therefore$ $\dfrac{dy}{dx} = -e^{-x}$ and $\dfrac{d^2y}{dx^2} = e^{-x} > 0$ for all x. Hence curve is concave upwards at every point. **Ans.**

5. *Find the points of inflexion of the curve $x = a(2\theta - \sin\theta)$, $y = a(2 - \cos\theta)$.*

SOLUTION: $x = a(2\theta - \sin\theta)$, $y = a(2 - \cos\theta)$

$$\frac{dy}{dx} = \frac{dy}{d\theta}\Big/\frac{dx}{d\theta} = \frac{a\sin\theta}{a(2-\cos\theta)} = \frac{\sin\theta}{2-\cos\theta}$$

$$\frac{d^2y}{dx^2} = \frac{d}{d\theta}\left(\frac{\sin\theta}{2-\cos\theta}\right)\frac{d\theta}{dx} = \frac{(2-\cos\theta)\cos\theta - \sin^2\theta}{(2-\cos\theta)^2 a(2-\cos\theta)} = \frac{2\cos\theta - 1}{a(2-\cos\theta)^3}$$

For points of inflexion $\dfrac{d^2y}{dx^2} = 0$ hence $2\cos\theta - 1 = 0$

or $\cos\theta = \dfrac{1}{2}$ hence $\theta = 2n\pi \pm \dfrac{\pi}{3}$. $n \in I$.

$$\frac{d^3y}{dx^3} = \frac{d}{d\theta}\frac{2\cos\theta - 1}{a(2-\cos\theta)^3}\cdot\frac{d\theta}{dx}$$

$$= \frac{1}{a^2}\left[\frac{-4\sin\theta(2-\cos\theta) - 3\sin\theta(2\cos\theta - 1)}{(2-\cos\theta)^5}\right] \neq 0 \text{ at } \theta = 2n\pi \pm \frac{\pi}{\sqrt{3}}$$

Hence the points of inflexion are at $\theta = 2n\pi \pm \dfrac{\pi}{3}$, $n \in I$. **Ans.**

6. *Obtain the points of inflexion, if any, of the curve* $x^2 y = a^2 (x - y)$.

SOLUTION: $x^2 y = a^2 (x - y)$ or $y = \dfrac{a^2 x}{x^2 + a^2}$

$$\therefore \quad \frac{dy}{dx} = \frac{(x^2+a^2)a^2 - a^2 x \cdot 2x}{(x^2+a^2)^2} = \frac{a^2(a^2-x^2)}{(x^2+a^2)^2}.$$

and hence $\dfrac{d^2y}{dx^2} = \dfrac{(x^2+a^2)^2(-2a^2x) - a^2(a^2-x^2)2(x^2+a^2)\cdot 2x}{(x^2+a^2)^4} = \dfrac{-2a^2x(x^2+a^2) - 4a^2x(a^2-x^2)}{(x^2+a^2)^3}$

$$= \frac{2ax(x^2-3a^2)}{(x^2+a^2)^3}$$

For points of inflexion, $\dfrac{d^2y}{dx^2} = 0$ hence $x = 0, \pm a\sqrt{3}$.

It can be verified that at these points $\dfrac{d^3y}{dx^3} \neq 0$. **Ans.**

7. *Show that every point in which the sine curve* $y = k \sin\left(\dfrac{x}{a}\right)$ *meets the x-axis is a point of inflexion.*

SOLUTION: $y = k \sin\left(\dfrac{x}{a}\right)$ hence $\dfrac{dy}{dx} = k \cos\left(\dfrac{x}{a}\right)$ and $\dfrac{d^2y}{dx^2} = -k \sin\left(\dfrac{x}{a}\right)$

or $\dfrac{d^2y}{dx^2} = -y.$ For point of inflexion $\dfrac{d^2y}{dx^2} = 0$ and $\dfrac{d^3y}{dx^3} \neq 0.$

$\dfrac{d^2y}{dx^2} = 0$ whenever $y = 0$, i.e., where the curve meets the x-axis. **Hence Proved.**

8. *Find the range of values of x in which the curve* $y = e^{-x^2}$ *is concave upwards or concave downwards.*

SOLUTION: $y = e^{-x^2}$ hence $\dfrac{dy}{dx} = -2xe^{-x^2}$

and $\dfrac{d^2y}{dx^2} = (-2x)^2 e^{-x^2} - 2e^{-x^2} = e^{-x^2}(4x^2 - 2) = 4e^{-x^2}\left(x^2 - \dfrac{1}{2}\right).$

when $|x| < \dfrac{1}{\sqrt{2}}$, $\dfrac{d^2y}{dx^2} < 0$ hence concave upwards.

and when $|x| < \dfrac{1}{\sqrt{2}}$, $\dfrac{d^2y}{dx^2} < 0$ hence concave downwards. **Ans.**

ADDITIONAL IMPORTANT PROBLEMS

1. *Prove that the equation $2x^3 - 3x^2 - x + 1 = 0$ has at least one real root between 1 and 2.*

SOLUTION: Let $f(x) = \frac{x^4}{2} - x^3 - \frac{x^2}{2} + x$ so that $f'(x) = 2x^3 - 3x^2 - x + 1$. Here $f(x)$ is continuous in [1, 2] and differentiable in (1, 2). Also $f(1) = 0$ and $f(2) = 0$, as such conditions of Rolle's theorem are satisfied. Therefore there must exist at least one real root c in (1, 2) such that $f'(c) = 0$, i.e., $2c^3 - 3c^2 - c + 1 = 0$

$\Rightarrow$ $x = c$ is a root of $2x^3 - 3x^2 - x + 1 = 0$ lying in (1, 2). **Hence Proved.**

2. *If a, b are real numbers, show that there exists at least one real number c such that $b^3 + ab^2 + a^2b + a^3 = 4c^3$, $a < c < b$.*

SOLUTION: Let $f(x) = x^4$ defined in (a, b). It is continuous in $[a, b]$ and $f'(x) = 4x^3$ exists for every x in (a, b), hence $f(x)$ satisfies all the conditions of L.M.V. theorem. Thus there exists at least one point c in (a, b) such that $\frac{f(b) - f(a)}{b - a} = f'(c)$.

or $\frac{b^4 - a^4}{b - a} = 4c^3$ or $\frac{(b^2 + a^2)(b - a)(b + a)}{b - a} = 4c^3$.

or $(b^2 + a^2)(b + a) = 4c^3$ or $b^3 + b^2a + a^2b + a^3 = 4c^3$, $a < c < b$. **Hence Proved.**

3. *Apply Lagrange's mean value theorem, show that*

$$\frac{\cos a\theta - \cos b\theta}{\theta} \le (b - a) \quad \text{where } \theta \ne 0.$$

SOLUTION: Let $f(x) = \cos x\theta$ defined in $[a, b]$. It is continuous in $[a, b]$ and $f'(x) = -\theta \sin x\theta$ exists for all x in (a, b).

Thus, $f(x)$ satisfies all the conditions of L.M.V. theorem.

Therefore there exists at least one point c in (a, b) such that

$$\frac{f(b) - f(a)}{b - a} = f'(c) \quad \text{or} \quad \frac{\cos b\theta - \cos a\theta}{b - a} = -\theta \sin c\,\theta$$

or $$\frac{\cos b\theta - \cos a\theta}{(b - a)\theta} = -\sin c\theta$$

or $$\frac{\cos a\theta - \cos b\theta}{(b - a)\theta} = \sin c\,\theta \le 1 \text{ if } \theta \ne 0$$

$\Rightarrow$ $$\frac{\cos a\theta - \cos b\theta}{\theta} \le (b - a), \quad \theta \ne 0.$$ **Hence Proved.**

4. *If $f(x) = \frac{1}{x^2}$ and $g(x) = \frac{1}{x}$ then show that the 'c' of the Cauchy's mean value theorem is the harmonic mean between a and b, $a > 0$, $b > 0$.*

SOLUTION: Clearly $f(x)$ and $g(x)$ are continuous in any closed interval $[a, b]$ for $a > 0, b > 0$.

$f'(x) = \frac{-2}{x^3}$ and $g'(x) = \frac{-1}{x^2}$ both exist in (a, b), $a > 0$, $b > 0$.

Hence all conditions of Cauchy's mean value theorem are satisfied.

$\therefore$ There exists at least one point c in (a, b) such that

$$\frac{f(b)-f(a)}{g(b)-g(a)} = \frac{f'(c)}{g'(c)} \quad \text{or} \quad \frac{\frac{1}{b^2}-\frac{1}{a^2}}{\frac{1}{b}-\frac{1}{a}} = \frac{\frac{-2}{c^3}}{\frac{-1}{c^2}}$$

or $$\frac{(a^2-b^2)\,ab}{(a-b)\,a^2b^2} = \frac{2}{c} \quad \text{or} \quad \frac{a+b}{ab} = \frac{2}{c}$$

or $c = \dfrac{2ab}{a+b}$ hence C is the harmonic mean between a and b. **Hence Proved.**

5. *If* $f(x) = \sqrt{x}$ *and* $g(x) = \dfrac{1}{\sqrt{x}}$ *show that the 'c' of the Cauchy's mean value theorem is the geometric mean between* a *and* b, $a > 0$, $b > 0$.

SOLUTION: $f(x) = \sqrt{x}$ and $g(x) = \dfrac{1}{\sqrt{x}}$ are continuous in any $[a, b]$, $a > 0$, $b > 0$ and $f'(x) = \dfrac{1}{2\sqrt{x}}$

and $g'(x) = \dfrac{-1}{2x\sqrt{x}}$ exist for all x in (a, b) and $g'(x) \neq 0$ for any x in (a, b).

Hence $f(x)$ and $g(x)$ satisfy all the conditions of Cauchy's mean value theorem and then there exists at least one point C (a, b) such that

$$\frac{f(b)-f(a)}{g(b)-g(a)} = \frac{f'(c)}{g'(c)}$$

or $$\frac{\sqrt{b}-\sqrt{a}}{\frac{1}{\sqrt{b}}-\frac{1}{\sqrt{a}}} = \frac{\frac{1}{2\sqrt{c}}}{\frac{-1}{2c\sqrt{c}}} \quad \text{or} \quad \frac{(\sqrt{a}-\sqrt{b})\sqrt{ab}}{\sqrt{a}-\sqrt{b}} = c$$

or $c = \sqrt{ab}$ hence c is the geometric mean between a and b. **Hence Proved.**

6. *Find the radius of curvature at any point* (x, y) *of the catenary* $y = c\cosh\left(\dfrac{x}{c}\right)$.

SOLUTION: $y = c\cosh\dfrac{x}{c}$ $\therefore$ $\dfrac{dy}{dx} = \sinh\dfrac{x}{c}$ and $\dfrac{d^2y}{dx^2} = \dfrac{1}{c}\cosh\dfrac{x}{c}$.

$$\therefore \quad \rho = \frac{\left\{1+\left(\frac{dy}{dx}\right)^2\right\}^{3/2}}{\frac{d^2y}{dx^2}} = \frac{\left(1+\sin^2 h\frac{x}{c}\right)^{3/2}}{\frac{1}{c}\cosh\frac{x}{c}} = \frac{c}{\cosh\frac{x}{c}}\cosh^3\frac{x}{c} = \frac{y^2}{c}. \qquad \textbf{Ans.}$$

7. *Find the radius of curvature of the curve* $x^2y = a(x^2 + y^2)$ *at its point* $(-2a, 2a)$.

SOLUTION: $x^2y = a(x^2 + y^2)$ ...(1)

Differentiating w.r.t. x, we get

$$2xy + x^2\frac{dy}{dx} = 2ax + 2ay\frac{dy}{dx} \quad \text{or} \quad \frac{dy}{dx} = \frac{2ax-2xy}{x^2-2ay}$$

At $(-2a, 2a)$ $\quad \dfrac{dy}{dx} = \dfrac{-4a^2+8a^2}{4a^2-4a^2} = \infty.$

Hence differentiating (1) w.r.t. y, we get

$$2xy\frac{dx}{dy}+x^2 = 2ax\frac{dx}{dy}+2ay \quad \text{or} \quad \frac{dx}{dy} = \frac{2ay-x^2}{2xy-2ax} \qquad ...(2)$$

At the point $(-2a, 2a)$, $\dfrac{dx}{dy} = 0.$

Differentiating (2) w.r.t. y, gives

$$(2xy-2ax)\frac{d^2x}{dy^2}+\frac{dx}{dy}\left[2(y-a)\frac{dx}{dy}+2x\frac{dx}{dy}\right] = 2a - 2x\frac{dx}{dy}$$

At the point $(-2a, 2a)$, we get

$$\frac{d^2x}{dy^2} = \left[\frac{a}{x(y-a)}\right]_{(-2a,\,2a)} = \frac{-1}{2a}$$

Therefore, $$\rho = \frac{\left[1+\left(\dfrac{dx}{dy}\right)^2\right]^{3/2}}{\dfrac{d^2x}{dy^2}} = -2a \quad \text{at} \quad (-2a, 2a).$$ **Ans.**

8. *Find the radius of curvature of the curve* $x = a \log(\sec\theta + \tan\theta)$, $y = a\sec\theta$.

SOLUTION: $\dfrac{dx}{d\theta} = \dfrac{a(\sec\theta\tan\theta+\sec^2\theta)}{\sec\theta+\tan\theta} = a\sec\theta$ and $\dfrac{dy}{d\theta} = a\sec\theta\tan\theta$

$$\therefore \quad \frac{d^2x}{d\theta^2} = a\sec\theta\tan\theta, \quad \frac{d^2y}{d\theta^2} = a\sec^3\theta + a\sec\theta\tan^2\theta$$

Hence $$\rho = \frac{[a^2\sec^2\theta+a^2\sec^2\theta\tan^2\theta]^{3/2}}{a\sec\theta\, a\sec\theta\,(\sec^2\theta+\tan^2\theta)-a^2\sec^2\theta\tan^2\theta}$$

$$= \frac{a\sec^6\theta}{\sec^2\theta\,(\sec^2\theta+\tan^2\theta-\tan^2\theta)} = a\sec^2\theta$$ **Ans.**

9. *Find the radius of curvature of the curve* $r = a\,e^{\theta\cot\alpha}$

SOLUTION: $r = ae^{\theta\cot\alpha}$ hence $\dfrac{dr}{d\theta} = a\cot\alpha\, e^{\theta\cot\alpha} = r\cot\alpha.$

and $\dfrac{d^2r}{d\theta^2} = \cot\alpha\dfrac{dr}{d\theta} = r\cot^2\alpha.$

$$\therefore \quad \rho = \frac{\left[r^2+\left(\frac{dr}{d\theta}\right)^2\right]^{3/2}}{r^2+2\left(\frac{dr}{d\theta}\right)^2-r\frac{d^2r}{d\theta^2}} = \frac{(r^2+r^2\cot^2\alpha)^{3/2}}{r^2+2\cdot r^2\cot^2\alpha-r\cdot r\cot^2\alpha}$$

$$= \frac{r^3\operatorname{cosec}^3\alpha}{r^2(1+\cot^2\alpha)} = r\operatorname{cosec}\alpha$$ **Ans.**

10. *Show that the radius of curvature of the hyperbolic spiral $r\theta = a$ is $\frac{r(a^2+r^2)^{3/2}}{a^3}$.*

SOLUTION: $r = \frac{a}{\theta} \quad \therefore \quad \frac{dr}{d\theta} = \frac{-a}{\theta^2}$ and $\frac{d^2r}{d\theta^2} = \frac{2a}{\theta^3}$. Therefore

$$\rho = \frac{\left[r^2+\left(\frac{dr}{d\theta}\right)^2\right]^{3/2}}{r^2+2\left(\frac{dr}{d\theta}\right)^2-r\frac{d^2r}{d\theta^2}} = \frac{\left(\frac{a^2}{\theta^2}+\frac{a^2}{\theta^4}\right)^{3/2}}{\frac{a^2}{\theta^2}+2\frac{a^2}{\theta^4}-\frac{a}{\theta}\cdot\frac{2a}{\theta^3}}$$

$$= \frac{\left(\frac{a^2}{\theta^2}+\frac{r^2}{\theta^2}\right)^{3/2}}{\frac{a^2}{\theta^2}} = \frac{(a^2+r^2)^{3/2}}{a^2\theta} = \frac{r(a^2+r^2)^{3r}}{a^3}$$ **Ans.**

11. *Find the radius of curvature at the origin for the curve*
$x^3 - 2x^2y - 4y^3 + 5x^2 - 6xy + 7y^2 - 8y = 0$

SOLUTION: Equating to zero the lowest degree term in the given equation we get $y = 0$, i.e., the X-axis as tangent to the curve at the origin.

$$\therefore \quad \rho = \lim_{x\to 0}\frac{x^2}{2y}.$$

Dividing throughout by y, the given equation, gives

$$x\cdot\frac{x^2}{y}-2x^2-4y^2+5\cdot\frac{x^2}{y}-6x+7y-8 = 0$$

when $x \to 0$, $y \to 0$ $\;0(2\rho) - 2(0) - 4(0) + 5(2\rho) - 6(0) + 7(0) - 8 = 0 \quad \Rightarrow \quad \rho = \frac{4}{5}$. **Ans.**

12. *Find the radius of curvature at the origin for the curve*

$x = a(\theta + \sin\theta),\ y = a(1 - \cos\theta)$.

SOLUTION: We know that for the cycloid $x = a(\theta + \sin\theta)$, $y = a(1 - \cos\theta)$ the tangent at the origin is X-axis, hence

$$\rho = \lim_{x\to 0}\frac{x^2}{2y} = \lim_{\theta\to 0}\frac{a^2(\theta+\sin\theta)^2}{2a(1-\cos\theta)} \qquad \left(\frac{0}{0}\text{ form}\right)$$

$$= \frac{a}{2} \lim_{\theta \to 0} \frac{2(\theta + \sin\theta)(1 + \cos\theta)}{\sin\theta}$$

$$= \frac{a}{2} \cdot 2[1 + \cos\theta]_{\theta=0} \lim_{\theta \to 0} \frac{\theta + \sin\theta}{\sin\theta} \qquad \text{again } \left(\frac{0}{0} \text{ form}\right)$$

$$= 2a \lim_{\theta \to 0} \frac{1 + \cos\theta}{\cos\theta} = 4a$$

Ans.

13. *Show that the circle of curvature at the origin of the curve* $y = mx + \frac{x^2}{a}$ *is* $x^2 + y^2 = a\ (1 + m^2)\ (y - mx)$.

SOLUTION: $y = mx + \frac{x^2}{a}$. $\therefore$ $\frac{dy}{dx} = m + \frac{2x}{a}$ and $\frac{d^2y}{dx^2} = \frac{2}{a}$.

At (0, 0), $\frac{dy}{dx} = m$ and $\frac{d^2y}{dx^2} = \frac{2}{a}$.

If (X, Y) is the centre of curvature at the origin, then

$$X = x - \frac{\frac{dy}{dx}\left[1 + \left(\frac{dy}{dx}\right)^2\right]}{\frac{d^2y}{dx^2}} = 0 - \frac{m(1+m^2)}{2/a} = -\frac{am\,(1+m^2)}{2}.$$

and
$$Y = y + \frac{\left[1 + \left(\frac{dy}{dx}\right)^2\right]}{\frac{d^2y}{dx^2}} = 0 + \frac{1+m^2}{2/a} = \frac{a(1+m^2)}{2}.$$

And ρ at the origin, is given by

$$\rho = \frac{\left[1 + \left(\frac{dy}{dx}\right)^2\right]^{3/2}}{\frac{d^2y}{dx^2}} = \frac{(1+m^2)^{3/2}}{2/a} = \frac{a\,(1+m^2)^{3/2}}{2}$$

Hence the equation of the circle of curvature at the origin, is

$$(x - X)^2 + (y - Y)^2 = \rho^2$$

or
$$\left[x + \frac{am}{2}(1+m^2)\right]^2 + \left[y - \frac{a(1+m^2)}{2}\right]^2 = \frac{a^2(1+m^2)^3}{4}$$

or
$$x^2 + y^2 = -ma\,(1 + m^2)\,x + a\,(1 + m^2)y = a\,(1 + m^2)\,(y - mx).$$

Ans.

14. *Find the evolute of the parabola* $y^2 = 4ax$.

SOLUTION: $y^2 = 4ax$ hence $\dfrac{dy}{dx} = \dfrac{2a}{y}$ and $\dfrac{d^2y}{dx^2} = \dfrac{-2a}{y^2}\dfrac{dy}{dx} = \dfrac{-4a^2}{y^3}$.

If (X, Y) are the centre of curvature, then

$$X = x - \frac{\dfrac{dy}{dx}\left[1+\left(\dfrac{dy}{dx}\right)^2\right]}{\dfrac{d^2y}{dx^2}} = x - \frac{\dfrac{2a}{y}\left(1+\dfrac{4a^2}{y^2}\right)}{\dfrac{-4a^2}{y^3}}$$

$$= x + \frac{y^2+4a^2}{2a} = \frac{2ax+4ax+4a^2}{2a} = 3x + 2a.$$

and
$$Y = y + \frac{\left[1+\left(\dfrac{dy}{dx}\right)^2\right]}{\dfrac{d^2y}{dx^2}} = y + \frac{1+\dfrac{4a^2}{y^2}}{\dfrac{-4a^2}{y^3}} = y - \frac{y(y^2+4a^2)}{4a^2} = \frac{y^3}{4a^2}$$

Therefore $x = \dfrac{X-2a}{3}$ and $y^6 = (4a^2)^2\, Y^2$. Putting these in $y^2 = 4ax$, we get

$$\therefore \quad Y^2 = \frac{4x^3}{a} = \frac{4}{a}\left(\frac{X-2a}{3}\right)^3$$

or $\quad 27aY^2 = 4(X-2a)^3$. **Ans.**

CHAPTER

6 Definite Integrals and Reduction Formulae

IMPORTANT FORMULAE AND USEFUL RESULTS

1. Property of Definite Integrals:

(i) $\int_0^a f(x)\,dx = \int_0^a f(a-x)\,dx$

(ii) $\int_0^a f(x)\,dx = \begin{cases} 2\int_0^a f(x)\,dx & \text{if } f(-x) = f(x) \\ 0 & \text{if } f(-x) = -f(x) \end{cases}$

(iii) $\int_0^{2a} f(x)\,dx = \begin{cases} 2\int_0^a f(x)\,dx & \text{if } f(2a-x) = f(x) \\ 0 & \text{if } f(2a-x) = -f(x) \end{cases}$

(iv) If m and M are the least and the greatest values of $f(x)$ in $[a, b]$

then $m \le \frac{1}{b-a}\int_a^b f(x) \le M$

(v) If $f(x)$ is continuous over $[a, b]$ and $F(x) = \int_a^x f(x)\,dx$

then $\frac{d}{dx}F(x) = f(x)$

2. Reduction Formulae:

(i) If $I_n = \int x^n e^{ax}\,dx$ then $I_n = \frac{x^n}{a}e^{ax} - \frac{n}{a}I_{n-1}$

(ii) Gamma function $\overline{|n+1} = \int_0^\infty e^{-x}x^{n-1}\,dx$ and $\overline{|n+1} = n\overline{|n}$ and $\overline{|\tfrac{1}{2}} = \sqrt{\pi}$

(iii) If $I_n = \int \sin^n x\,dx\ (n \ge 2)$ then $I_n = \frac{-1}{n}\sin^{n-1}x\cos x + \frac{n-1}{n}I_{n-2}$

$$\text{and} \int_0^{\pi/2} \sin^n x \, dx = \frac{n-1}{n} \int_0^{\pi/2} \sin^{n-2} x \, dx .$$

(*iv*) If $I_n = \int \cos^n x \, dx$ then $I_n = \frac{1}{n} \cos^{n-1} x \sin x + \frac{n-1}{n} I_n$

$$\text{and} \int_0^{\pi/2} \cos^n x \, dx = \frac{n-1}{n} \int_0^{\pi/2} \cos^{n-2} x \, dx .$$

$$\text{Thus} \int_0^{\pi/2} \begin{matrix} \sin^n \\ \cos \end{matrix} x \, dx = \begin{cases} \dfrac{(n-1)(n-3)(n-5)\ldots 3\cdot 1}{n(n-2)(n-4)\ldots 4\cdot 2} \cdot \dfrac{\pi}{2} & \text{if } n \text{ is even} \\ \dfrac{(n-1)(n-3)(n-5)\ldots 4\cdot 2}{n(n-2)(n-4)\ldots 5\cdot 3} \cdot 1 & \text{if } n \text{ is odd} \end{cases}$$

(*v*) If $I_{m,\,n} = \int \sin^m x \cos^n x \, dx = -\frac{\sin^{m-1} x \cos^{n-1} x}{m+n} + \frac{m-1}{m+n} I_{m-2,\,n}$

$$\text{and} \int_0^{\pi/2} \sin^m x \cos^n x \, dx = \frac{m-1}{m+n} \int_0^{\pi/2} \sin^{m-2} x \cos^n x \, dx .$$

(*vi*) If $I_n = \int \tan^n x \, dx$ then $I_n = \frac{\tan^{n-1} x}{n-1} - I_{n-2}$

(*vii*) If $I_n = \int \sec^n x \, dx$ then $I_n = \frac{1}{n-1} \sec^{n-2} x \tan x + \frac{n-2}{n-1} I_{n-2}$

SOLUTIONS TO EXERCISE 6A

1. *Evaluate* $\int_0^2 |1-x|\,dx$.

SOLUTION: $\int_0^2 |1-x|\,dx = \int_0^1 |1-x|\,dx + \int_1^2 |1-x|\,dx = \int_0^1 (1-x)\,dx + \int_1^2 (x-1)\,dx$

$$= \left[x-\frac{x^2}{2}\right]_0^1 + \left[\frac{x^2}{2}-x\right]_1^2 = 1-\frac{1}{2}+0-\left(\frac{1}{2}-1\right) = 1 \qquad \textbf{Ans.}$$

2. *Evaluate* $\int_0^\pi \sqrt{\frac{1+\cos 2x}{2}}\,dx.$

SOLUTION: $\int_0^\pi \sqrt{\frac{1+\cos 2x}{2}}\,dx = \int_0^\pi \sqrt{\frac{2\cos^2 x}{2}}\,dx = \int_0^\pi |\cos x|\,dx$

$$= \int_0^{\pi/2} \cos x\,dx + \int_{\pi/2}^{\pi} -\cos x\,dx = [\sin x]_0^{\pi/2} - [\sin x]_{\pi/2}^{\pi} = 1-0-(0-1) = 2. \qquad \textbf{Ans.}$$

3. *Evaluate* $\int_0^\pi \sin^5 x\,dx$

SOLUTION: $\int_0^\pi \sin^5 x\,dx = \int_0^\pi (1-\cos^2 x)^2 \sin x\,dx$ Putting $\cos x = t$

$$= -\int_1^{-1} (1-t^2)^2\,dt = 2\int_0^1 (1-t^2)^2\,dt = 2\left[t-\frac{2t^3}{3}+\frac{t^5}{5}\right]_0^1 = \frac{16}{15} \qquad \textbf{Ans.}$$

4. *Evaluate* $\int_0^\pi \frac{dx}{a^2\cos^2 x + b^2\sin^2 x}$.

SOLUTION: $\int_0^\pi \frac{dx}{a^2\cos^2 x + b^2\sin^2 x}\,dx = 2\int_0^{\pi/2} \frac{dx}{a^2\cos^2 x + b^2\sin^2 x}$ (since $a^2\cos^2(\pi-x) + b^2\sin^2(\pi-x)$

$= a^2\cos^2 x + b^2\sin^2 x$)

$$= 2\int_0^{\pi/2} \frac{\sec^2 x\,dx}{a^2+b^2\tan^2 x}$$ Putting $b\tan x = t$, so $b\sec^2 x\,dx = dt$

$$= \frac{2}{b}\int_0^\infty \frac{dt}{a^2+t^2} = \frac{2}{b}\frac{1}{a}\left[\tan^{-1}\frac{t}{a}\right]_0^\infty = \frac{2}{ab}\left(\frac{\pi}{2}-0\right) = \frac{\pi}{ab} \qquad \textbf{Ans.}$$

5. *Evaluate* $\int_0^{\pi} \frac{x\,dx}{a^2 - \cos^2 x}\ (a > 1)$.

SOLUTION: Let $I = \int_0^{\pi} \frac{x\,dx}{a^2 - \cos^2 x}$ $(a > 1)$. By property of definite integrals,

$$I = \int_0^{\pi} \frac{(\pi - x)\,dx}{a^2 - \cos^2(\pi - x)} = \int_0^{\pi} \frac{(\pi - x)\,dx}{a^2 - \cos^2 x}.$$ Adding these, gives

$$2I = \int_0^{\pi} \frac{\pi\,dx}{a^2 - \cos^2 x} = \pi \int_0^{\pi} \frac{\sec^2 x\,dx}{a^2 \sec^2 x - 1}$$

or $$I = 2\frac{\pi}{2} \int_0^{\pi/2} \frac{\sec^2 x\,dx}{a^2 \sec^2 x - 1} \qquad \left[\text{since } \frac{\sec^2(a - x)}{a^2 \sec^2(\pi - x) - 1} = \frac{\sec^2 x}{a^2 \sec^2 x - 1}\right]$$

or $$I = \pi \int_0^{\pi/2} \frac{\sec^2 x\,dx}{a^2(1 + \tan^2 x) - 1} \qquad \text{Putting } \tan x = t \ \therefore\ \sec^2 x\,dx = dt$$

$$= \pi \int_0^{\infty} \frac{dt}{a^2 - 1 + a^2 t^2} = \frac{\pi}{a^2} \int_0^{\infty} \frac{dt}{t^2 + \frac{a^2 - 1}{a^2}} = \frac{\pi}{a^2} \cdot \frac{a}{\sqrt{a^2 - 1}} \left[\tan^{-1} \frac{ta}{\sqrt{a^2 - 1}}\right]_0^{\infty}$$

$$= \frac{\pi}{a\sqrt{a^2 - 1}} \left[\frac{\pi}{2} - 0\right] = \frac{\pi^2}{2a\sqrt{a^2 - 1}}$$ **Ans.**

6. $\int_{-\pi/2}^{\pi/2} \frac{dx}{5 + 7\cos x + \sin x}$.

SOLUTION: $$\int_{-\pi/2}^{\pi/2} \frac{dx}{5 + 7\cos x + \sin x} = \int_{-\pi/2}^{\pi/2} \frac{dx}{5 + 7\left(\cos^2 \frac{x}{2} - \sin^2 \frac{x}{2}\right) + 2\sin \frac{x}{2} \cos \frac{x}{2}}$$

$$= \int_{-\pi/2}^{\pi/2} \frac{\sec^2 \frac{x}{2}\,dx}{5\sec^2 \frac{x}{2} + 7 - 7\tan^2 \frac{x}{2} + 2\tan \frac{x}{2}} \qquad \left(\text{Putting } \tan \frac{x}{2} = t\right)$$

$$= \int_{-1}^{1} \frac{2dt}{5(1 + t^2) + 7 - 7t^2 + 2t} = -\int_{-1}^{1} \frac{dt}{t^2 - t - 6}$$

$$= -\int_{-1}^{1} \frac{dt}{\left(t - \frac{1}{2}\right)^2 - \left(\frac{5}{2}\right)^2} = -\frac{1}{2 \cdot \frac{5}{2}} \left[\log \left|\frac{t - \frac{1}{2} - \frac{5}{2}}{t - \frac{1}{2} + \frac{5}{2}}\right|\right]_{-1}^{1}$$

$$= \frac{-1}{5}\left[\log \frac{2}{3} - \log \frac{4}{1}\right] = \frac{1}{5} \log 6.$$ **Ans.**

7. *Show that* $\int_0^{\pi} \frac{dx}{13+3\cos x+4\sin x} = \frac{1}{6}\left[\frac{\pi}{2} - \tan^{-1}\frac{1}{3}\right]. = \frac{1}{6}\tan^{-1} 3.$

SOLUTION: Let $I = \int_0^{\pi} \frac{dx}{13+3\cos x+4\sin x} = \int_0^{\pi} \frac{dx}{13+3\left(\cos^2\frac{x}{2} - \sin^2\frac{x}{2}\right)+8\sin\frac{x}{2}\cos\frac{x}{2}}$

$$= \int_0^{\pi} \frac{\sec^2\frac{x}{2}\,dx}{13\sec^2\frac{x}{2}+3-3\tan^2\frac{x}{2}+8\tan\frac{x}{2}},\quad \text{Putting } \tan\frac{x}{2} = t, \text{ gives}$$

or $$I = \int_0^{\infty} \frac{2dt}{10t^2+8t+16} = \int_0^{\infty} \frac{dt}{5\left(t^2+\frac{4}{5}t+\frac{8}{5}\right)} = \frac{1}{5}\int_0^{\infty} \frac{dt}{\left(t+\frac{2}{5}\right)^2+\left(\frac{6}{5}\right)^2}$$

$$= \frac{1}{5}\cdot\frac{5}{6}\left[\tan^{-1}\frac{\left(t+\frac{2}{5}\right)}{\frac{6}{5}}\right]_0^{\infty} = \frac{1}{6}\left(\frac{\pi}{2} - \tan^{-1}\frac{1}{3}\right) = \frac{1}{6}\tan^{-1} 3$$ **Hence Proved.**

8. *Evaluate* $\int_0^{\pi} \frac{2+3\cos x\ dx}{\sin x+2\cos x+3}$.

SOLUTION: $I = \int_0^{\pi} \frac{(2+3\cos x)\,dx}{\sin x+2\cos x+3}$

Let us write $2 + 3\cos x = A(\sin x + 2\cos x + 3) + B(\cos x - 2\sin x) + C$

$\Rightarrow 2A + B = 3,\quad A - 2B = 0,\quad 3A + C = 2$

which gives $A = \frac{6}{5},\quad B = \frac{3}{5},\quad C = 2 - \frac{18}{5} = \frac{-8}{5}$

$\therefore$ $$I = \int_0^{\pi} \frac{6}{5}\,dx + \frac{3}{5}\int_0^{\pi} \frac{\cos x - 2\sin x}{\sin x+2\cos x+3}\,dx - \frac{8}{5}\int_0^{\pi} \frac{dx}{\sin x+2\cos x+3}.$$

$$= \frac{6\pi}{5} + \frac{3}{5}[\log(\sin x+2\cos x+3)]_0^{\pi} - \frac{8}{5}\int_0^{\pi} \frac{dx}{2\sin\frac{x}{2}\cos\frac{x}{2}+2\left(\cos^2\frac{x}{2}-\sin^2\frac{x}{2}\right)+3}$$

$$= \frac{6\pi}{5} + \frac{3}{5}\log\frac{1}{5} - \frac{8}{5}\int_0^{\infty} \frac{\sec^2\frac{x}{2}\,dx}{2\tan\frac{x}{2}+2-2\tan^2\frac{x}{2}+3+3\tan^2\frac{x}{2}}$$

Now $$\int \frac{\sec^2\frac{x}{2}\,dx}{\tan^2\frac{x}{2}+2\tan\frac{x}{2}+5} = \int_0^{\infty} \frac{2dt}{t^2+2t+5} \quad \left(\text{On putting } t = \tan\frac{x}{2}\right)$$

$$= 2\int_0^{\infty} \frac{dt}{(t+1)^2+2^2} = \left[\tan^{-1}\frac{t+1}{2}\right]_0^{\infty}$$

$\therefore$ $$I = \frac{6\pi}{5} - \frac{3}{5}\log 5 - \frac{8}{5}\left(\frac{\pi}{2} - \tan^{-1}\frac{1}{2}\right) = \frac{2\pi}{5} - \frac{3}{5}\log 5 - \frac{8}{5}\tan^{-1} 2$$ **Ans.**

9. *Show that* $\int_0^{\pi} \frac{d\theta}{a^2 - 2a\cos\theta + 1} = \frac{\pi}{1-a^2}$ or $\frac{\pi}{a^2-1}$ according as $a < 1$ *or* $a > 1$.

SOLUTION: $\int_0^{\pi} \frac{d\theta}{a^2 - 2a\cos\theta + 1} = \int_0^{\pi} \frac{d\theta}{a^2 - 2a\left(\cos^2\frac{\theta}{2} - \sin^2\frac{\theta}{2}\right) + 1}$

$$= \int_0^{\pi} \frac{\sec^2\frac{\theta}{2}\, d\theta}{(a^2+1)\left(1+\tan^2\frac{\theta}{2}\right) - 2a\left(1-\tan^2\frac{\theta}{2}\right)} \qquad \left(\text{Put } \tan\frac{\theta}{2} = t\right)$$

$$= \int_0^{\infty} \frac{2\,dt}{(a+1)^2 t^2 + (a-1)^2} = \frac{2}{(a+1)^2}\int_0^{\infty} \frac{dt}{t^2 + \left(\frac{a-1}{a+1}\right)^2}$$

$$= \frac{2}{(a+1)^2\left(\frac{a-1}{a+1}\right)}\left[\tan^{-1}\frac{t(a+1)}{a-1}\right]_0^{\infty}$$

$$\begin{cases} = \dfrac{2}{a^2-1}\left(\dfrac{\pi}{2} - 0\right) & \text{if } a > 1 \\[2ex] = \dfrac{2}{1-a^2}\left(\dfrac{\pi}{2} - 0\right) & \text{if } a < 1 \end{cases}$$

Ans.

10. *Evaluate* $\int_a^b \frac{dx}{\sqrt{(x-a)(b-x)}}$.

SOLUTION: $\int_a^b \frac{dx}{\sqrt{(x-a)(b-x)}}$. Put $x = a\cos^2 t + b\sin^2 t$ $\therefore$ $x - a = (b-a)\sin^2 t$ and $b - x = (b-a)\cos^2 t$

$$= \int_0^{\pi/2} \frac{2(b-a)\sin t\cos t\, dt}{\sqrt{(b-a)\sin^2 t\,(b-a)\cos^2 t}} = 2\int_0^{\pi/2} \frac{(b-a)}{b-a}\, dt = \pi.$$

Ans.

11. *Evaluate* $\int_0^{\pi} \sin^6 x \cos^7 x\, dx$.

SOLUTION: $\int_0^{\pi} \sin^6 x\cos^7 x\, dx = 0$ since $\sin^6(\pi - x)\cos^7(\pi - x) = -\sin^6 x\cos^7 x$

Using the property $\int_0^{2a} f(x)\,dx = \begin{cases} 2\int_0^{\pi/2} f(x)\,dx & \text{if } f(2a-x) = f(x) \\ 0 & \text{if } f(2a-x) = -f(x) \end{cases}$

Hence value of the integral = 0. **Ans.**

12. *Evaluate* $\int_0^{\pi} \log(1+\cos x)\, dx$.

SOLUTION: Let $I = \int_0^{\pi} \log(1+\cos x)\, dx$ ...(1)

By the property $\int_0^{a} f(x)\, dx = \int_0^{a} f(a-x)\, dx$

We have $I = \int_0^{\pi} \log[1+\cos(\pi - x)]\, dx = \int_0^{\pi} \log(1-\cos x)\, dx$ ... (2)

Adding (1) and (2), we get $2I = \int_0^{\pi} [\log(1+\cos x) + \log(1-\cos x)]\, dx = \int_0^{\pi} \log(1-\cos^2 x)\, dx$

or $2I = \int_0^{\pi} \log\sin^2 x\, dx = 2\int_0^{\pi} \log\sin x\, dx$ or $I = \int_0^{\pi} \log\sin x\, dx$ ... (3)

By the property of definite integral $I = 2\int_0^{\pi/2} \log\sin x\, dx$ since $\log\sin(\pi - x) = \log\sin x$. ... (4)

Also $I = 2\int_0^{\pi/2} \log\sin\left(\frac{\pi}{2} - x\right) dx = 2\int_0^{\pi/2} \log\cos x\, dx$ ... (5)

Adding (4) and (5) we get $2I = 2\int_0^{\pi/2} (\log\sin x + \log\cos x)\, dx$

or $I = \int_0^{\pi/2} \log(\sin x\cos x)\, dx = \int_0^{\pi/2} \log\frac{\sin 2x}{2}\, dx = \int_0^{\pi/2} \log\sin 2x\, dx - \int_0^{\pi/2} \log 2\, dx$

or $I = \int_0^{\pi/2} \log\sin 2x\, dx - \frac{\pi}{2}\log 2$ ... (6)

By the property of definite integrals $\int_0^{\pi/2} \log\sin 2x\, dx = \int_0^{\pi} \log\sin t \cdot \frac{dt}{2}$

reverting back $= \frac{1}{2}\int_0^{\pi} \log\sin x\, dx = \frac{I}{2}$

$\therefore$ (6) becomes $I = \frac{1}{2} I - \frac{\pi}{2}\log 2$

$\frac{1I}{2} = -\frac{\pi}{2}\log 2$ or $I = -\pi\log 2 = \pi\log\left(\frac{1}{2}\right)$ **Ans.**

SOLUTIONS TO EXERCISE 6B

1. *Evaluate* $\displaystyle\int_0^{\pi/6} \cos^4 3x \sin^3 6x \; dx$

SOLUTION: $\displaystyle\int_0^{\pi/6} \cos^4 3x \sin^3 6x\, dx = 2^3 \int_0^{\pi/6} \cos^4 3x \sin^3 3x \cos^3 3x\, dx$

$$= 8 \int_0^{\pi/6} \cos^7 3x \sin^3 3x\, dx \qquad (\text{put } 3x = t \text{ so } 3\, dx = dt)$$

$$= \frac{8}{3}\int_0^{\pi/2} \cos^7 t \sin^3 t\, dt = \frac{8}{3}\,\frac{\Gamma\left(\frac{7+1}{2}\right)\Gamma\left(\frac{3+1}{2}\right)}{2\,\Gamma\left(\frac{12}{2}\right)} = \frac{8}{3}\,\frac{\Gamma(4)\,\Gamma(2)}{2\,\Gamma(6)} = \frac{4}{3}\,\frac{1}{5\cdot 4} = \frac{1}{15}$$ **Ans.**

2. *Show that* $\displaystyle\int_0^{\infty} \frac{x^3}{\left(1+x^2\right)^{\frac{9}{2}}}\, dx = \frac{2}{35}$

SOLUTION: Let $\displaystyle I = \int_0^{\infty} \frac{x^3\, dx}{(1+x^2)^{9/2}}$ (Put $x = \tan\theta$ so $dx = \sec^2\theta\, d\theta$)

then $$I = \int_0^{\pi/2} \frac{\tan^3\theta\cdot\sec^2\theta\, d\theta}{(1+\tan^2\theta)^{9/2}} = \int_0^{\pi/2} \sin^3\theta\cos^4\theta\, d\theta$$

$$= \frac{\Gamma\left(\frac{3+1}{2}\right)\Gamma\left(\frac{4+1}{2}\right)}{2\,\Gamma\left(\frac{4+3+2}{2}\right)} = \frac{1\cdot\Gamma\left(\frac{5}{2}\right)}{2\cdot\frac{7}{2}\cdot\frac{5}{2}\cdot\Gamma\left(\frac{5}{2}\right)} = \frac{2}{35}$$ **Hence Proved.**

3. *Show that* $\displaystyle\int_0^{a} \frac{x^n\, dx}{\sqrt{ax-x^2}} = \frac{1\cdot 3\cdot 5 \ldots (2n-1)}{2\cdot 4\cdot 6 \ldots 2n}\pi a^n$

Solution: Let $\displaystyle I = \int_0^{a} \frac{x^n\, dx}{\sqrt{ax-x^2}} = \int_0^{a} \frac{x^{n-\frac{1}{2}}\, dx}{\sqrt{a-x}}$ (Put $x = a\sin^2\theta$ so $dx = 2a\sin\theta\cos\theta\, d\theta$)

$\therefore$ $$I = \int_0^{\pi/2} \frac{(a\sin^2\theta)^{n-\frac{1}{2}}\, 2a\sin\theta\cos\theta\, d\theta}{\sqrt{a-a\sin^2\theta}} = \int_0^{\pi/2} \frac{a^{n-\frac{1}{2}}\sin^{2n-1}\theta\, 2a\sin\theta\cos\theta\, d\theta}{\sqrt{a}\cos\theta}$$

$$= 2a^n \int_0^{\pi/2} \sin^{2n}\theta \, d\theta = 2a^n \frac{\Gamma\left(\frac{2n+1}{2}\right)\Gamma\left(\frac{1}{2}\right)}{2\Gamma(n+1)} = \sqrt{\pi}\, a^n \frac{\Gamma\left(n+\frac{1}{2}\right)}{n!}$$

$$= \frac{\left(n-\frac{1}{2}\right)\left(n-\frac{3}{2}\right)\ldots\frac{3}{2}\cdot\frac{1}{2}\Gamma\left(\frac{1}{2}\right)}{n(n-1)(n-2)\ldots 3\cdot 2\cdot 1}\cdot\sqrt{\pi}\cdot a^n = \frac{\pi a^n \cdot 1\cdot 3\cdot 5\ldots(2n-1)}{2^n\cdot(1\cdot 2\cdot 3\ldots n)}$$

$$= \pi a^n \cdot \frac{1\cdot 3\cdot 5\ldots(2n-1)}{2\cdot 4\cdot 6\ldots 2n}.$$

Hence Proved.

4. *Evaluate* $\displaystyle\int_{-\pi/2}^{\pi/2} \cos^3\theta\,(1+\sin\theta)^2\, d\theta$

SOLUTION: $\displaystyle\int_{-\pi/2}^{\pi/2} \cos^3\theta(1+\sin\theta)^2\, d\theta = \int_{-\pi/2}^{\pi/2} \cos^3\theta(1+2\sin\theta+\sin^2\theta)\, d\theta$

$$= \int_{-\pi/2}^{\pi/2} \cos^3\theta\, d\theta + 2\int_{-\pi/2}^{\pi/2} \cos^3\theta\sin\theta\, d\theta + \int_{-\pi/2}^{\pi/2} \cos^3\theta\sin^2\theta\, d\theta$$

$$= 2\int_0^{\pi/2} \cos^3\theta\ d\theta + 0 + 2\int_0^{\pi/2} \cos^3\theta\sin^2\theta\, d\theta \text{ (by property of definite intagrals.)}$$

$$= \frac{2\Gamma\left(\frac{3+1}{2}\right)\Gamma\left(\frac{1}{2}\right)}{2\Gamma\left(\frac{3+2}{2}\right)} + \frac{2\Gamma\left(\frac{3+1}{2}\right)\Gamma\left(\frac{2+1}{2}\right)}{2\Gamma\left(\frac{3+2+2}{2}\right)} = \frac{1}{\frac{3}{2}\cdot\frac{1}{2}} + \frac{1}{\frac{5}{2}\cdot\frac{3}{2}} = \frac{8}{5}$$

Ans.

5. *Evaluate* $\displaystyle\int_0^{\pi/2} \frac{x\, dx}{\sin x + \cos x}$

SOLUTION: Let $\displaystyle I = \int_0^{\pi/2} \frac{x\, dx}{\sin x + \cos x}$. (by property of definite integrals.)

$$I = \int_0^{\pi/2} \frac{\left(\frac{\pi}{2}-x\right)dx}{\sin\left(\frac{\pi}{2}-x\right)+\cos\left(\frac{\pi}{2}-x\right)} = \int_0^{\pi/2} \frac{\left(\frac{\pi}{2}-\pi\right)dx}{\cos x + \sin x}.$$

Adding the above two, we get

$$2I = \int_0^{\pi/2} \frac{\frac{\pi}{2}\, dx}{\sin x + \cos x} = \frac{\pi}{2}\int_0^{\pi/2} \frac{\frac{1}{\sqrt{2}}\, dx}{\frac{1}{\sqrt{2}}\sin x + \frac{1}{\sqrt{2}}\cos x} = \frac{\pi}{2\sqrt{2}}\int_0^{\pi/2} \frac{dx}{\cos\left(x-\frac{\pi}{4}\right)}$$

or $$I=\frac{\pi}{4\sqrt{2}}\int_0^{\pi/2}\sec\left(x-\frac{\pi}{4}\right)dx=\frac{\pi}{4\sqrt{2}}\left[\log\tan\left(\frac{\pi}{4}+\frac{x}{2}-\frac{\pi}{8}\right)\right]_0^{\pi/2}$$

$$=\frac{\pi}{4\sqrt{2}}\left[\log\tan\left(\frac{\pi}{4}+\frac{\pi}{4}-\frac{\pi}{8}\right)-\log\tan\left(\frac{\pi}{4}-\frac{\pi}{8}\right)\right]=\frac{\pi}{4\sqrt{2}}\log\frac{\cot\frac{\pi}{8}}{\tan\frac{\pi}{8}}$$

$$=\frac{\pi}{4\sqrt{2}}\log\cot^2\frac{\pi}{8}=\frac{\pi}{2\sqrt{2}}\log\cot\frac{\pi}{8}.$$ **Ans.**

6. *Evaluate* $\displaystyle\int_0^{\pi}\frac{\sin^4\theta\sqrt{1-\cos\theta}}{(1+\cos\theta)^2}\,d\theta$

SOLUTION: Let $$I=\int_0^{\pi}\frac{\sin^4\theta\sqrt{1-\cos\theta}\,d\theta}{(1+\cos\theta)^2}=\int_0^{\pi}\frac{2^4\sin^4\frac{\theta}{2}\cos^4\frac{\theta}{2}\sqrt{2\sin^2\frac{\theta}{2}}\,d\theta}{\left(2\cos^2\frac{\theta}{2}\right)^2}$$

$$=4\sqrt{2}\int_0^{\pi}\sin^5\frac{\theta}{2}\,d\theta \qquad \left(\text{Putting } \frac{\theta}{2}=t\right)$$

$$=4\sqrt{2}\int_0^{\pi/2}\sin^5 t\cdot 2dt=8\sqrt{2}\,\frac{\Gamma\left(\frac{5+1}{2}\right)\Gamma\left(\frac{1}{2}\right)}{2\,\Gamma\left(\frac{5+2}{2}\right)}=\frac{8\sqrt{2}\cdot\Gamma(3)\,\Gamma\left(\frac{1}{2}\right)}{2\cdot\frac{5}{2}\cdot\frac{3}{2}\cdot\frac{1}{2}\,\Gamma\left(\frac{1}{2}\right)}=\frac{64\sqrt{2}}{15}$$ **Ans.**

7. *Show that* $\displaystyle\int_0^{\frac{\pi}{4}}(\cos 2\theta)^{\frac{3}{2}}\cos\theta\,d\theta=\frac{3\pi}{16\sqrt{2}}.$

SOLUTION: Let $$I=\int_0^{\pi/4}(\cos 2\theta)^{3/2}\cos\theta\,d\theta=\int_0^{\pi/4}(1-2\sin^2\theta)^{3/2}\cos\theta\,d\theta \qquad (\text{Put } \sqrt{2}\sin\theta=\sin\phi)$$

then $$I=\int_0^{\pi/2}(1-\sin^2\phi)^{3/2}\frac{1}{\sqrt{2}}\cos\phi\,d\phi=\frac{1}{\sqrt{2}}\int_0^{\pi/2}\cos^4\phi\,d\phi=\frac{1}{\sqrt{2}}\,\frac{\Gamma\left(\frac{5}{2}\right)\Gamma\left(\frac{1}{2}\right)}{2\,\Gamma(3)}=\frac{3\pi}{16\sqrt{2}}$$

Hence proved.

8. *Show that* $\displaystyle\int_0^\infty \frac{dx}{\left(1+x^2\right)^n} = \frac{(2n-3)(2n-5)\ldots 3\cdot 1}{(2n-2)(2n-4)\ldots 4\cdot 2}\,\frac{\pi}{2} = \frac{(2n-2)!}{2^{2n-2}\left((n-1)!\right)^2}\,\frac{\pi}{2}$

SOLUTION: Let $\displaystyle I = \int_0^\infty \frac{dx}{(1+x^2)^n}$ Putting $x = \tan\theta$, we get

$$I = \int_0^{\pi/2} \frac{\sec^2\theta\, d\theta}{\sec^{2n}\theta} = \int_0^{\pi/2} \cos^{2n-2}\theta\, d\theta = \frac{\Gamma\left(\frac{2n-2+1}{2}\right)\Gamma\left(\frac{1}{2}\right)}{2\,\Gamma\left(\frac{2n-2+2}{2}\right)} = \frac{\Gamma\left(\frac{2n-1}{2}\right)\Gamma\left(\frac{1}{2}\right)}{2\,\Gamma(n)}$$

$$= \frac{\frac{2n-3}{2}\cdot\frac{2n-5}{2}\cdot\frac{2n-7}{2}\cdots\frac{3}{2}\left(\frac{1}{2}\right)\left(\Gamma\left(\frac{1}{2}\right)\right)^2}{2\cdot(n-1)(n-2)(n-3)\ldots 3\cdot 2\cdot 1} = \frac{(2n-3)(2n-5)(2n-7)\ldots 3\cdot 1}{(2n-2)(2n-4)(2n-6)\ldots 4\cdot 2}\cdot\frac{\pi}{2}$$

$$= \frac{(2n-2)!}{2^{2n-2}[(n-1)!]^2}\cdot\frac{\pi}{2}$$

Hence Proved.

9. *Evaluate* $\displaystyle\int_0^\infty \left(1+x^2\right)^{-n-\frac{1}{2}} dx$

SOLUTION: Let $\displaystyle I = \int_0^\infty (1+x^2)^{-n-\frac{1}{2}} dx$ (Putting $x = \tan\theta$)

$$I = \int_0^{\pi/2} (\sec^2\theta)^{-n-\frac{1}{2}} \sec^2\theta\, d\theta = \int_0^{\pi/2} \sec^{-2n+1}\theta\, d\theta = \int_0^{\pi/2} \cos^{2n-1}\theta\, d\theta$$

$$= \frac{\Gamma\left(\frac{2n-1+1}{2}\right)\Gamma\left(\frac{1}{2}\right)}{2\,\Gamma\left(\frac{2n-1+2}{2}\right)} = \frac{\Gamma(n)\,\Gamma\left(\frac{1}{2}\right)}{2\,\Gamma\left(\frac{2n+1}{2}\right)} = \frac{(n-1)(n-2)(n-3)\ldots 3\cdot 2\cdot 1\,\Gamma\left(\frac{1}{2}\right)}{2\left(\frac{2n-1}{2}\right)\left(\frac{2n-3}{2}\right)\left(\frac{2n-5}{2}\right)\cdots\frac{3}{2}\cdot\frac{1}{2}\,\Gamma\left(\frac{1}{2}\right)}$$

$$= \frac{(2n-2)(2n-4)(2n-6)\ldots 6\cdot 4\cdot 2}{(2n-1)(2n-3)(2n-5)\ldots 5\cdot 3\cdot 1}$$

Hence Proved.

10. *Show that* $\displaystyle\int_0^1 x^{4n+1}\sqrt{\frac{1-x^2}{1+x^2}}\, dx = \frac{1}{2}\left[\frac{2n-1}{2n}\cdot\frac{2n-3}{2n-2}\cdots\cdots\frac{1}{2}\frac{\pi}{2} - \frac{2n}{2n+1}\,\frac{2n-2}{2n-1}\cdots\frac{2}{3}\right]$

SOLUTION: Let $\displaystyle I = \int_0^1 x^{4n+1}\sqrt{\frac{1-x^2}{1+x^2}}\, dx$ Putting $x^2 = \sin\theta$ we get $2x\, dx = \cos\theta\, d\theta$

$$I = \frac{1}{2}\int_0^{\pi/2} \sin^{2n}\theta\,\sqrt{\frac{1-\sin\theta}{1+\sin\theta}}\cdot\cos\theta\, d\theta$$

$$= \frac{1}{2}\int_0^{\pi/2} \sin^{2n}\theta \frac{1-\sin\theta}{\sqrt{1-\sin^2\theta}} \cos\theta \, d\theta = \frac{1}{2}\int_0^{\pi/2} \sin^{2n}\theta \, d\theta - \frac{1}{2}\int_0^{\pi/2} \sin^{2n+1}\theta \, d\theta$$

$$= \frac{1}{2}\frac{\left|\frac{2n+1}{2}\right|\frac{1}{2}}{2\left|\frac{2n+2}{2}\right.} - \frac{1}{2}\frac{\left|\frac{2n+2}{2}\right|\frac{1}{2}}{2\left|\frac{2n+3}{2}\right.} = \frac{\sqrt{\pi}}{4}\frac{\left|\frac{2n+1}{2}\right.}{|n+1} - \frac{\sqrt{\pi}}{4}\frac{|n+1}{\left|\frac{2n+3}{2}\right.}$$

$$= \frac{\sqrt{\pi}}{4}\frac{\frac{2n-1}{2}\cdot\frac{2n-3}{2}\cdot\frac{2n-5}{2}\cdots\frac{3}{2}\cdot\frac{1}{2}\left|\frac{1}{2}\right.}{n(n-1)(n-2)\ldots 3\cdot 2\cdot 1} - \frac{\sqrt{\pi}}{4}\frac{n(n-1)(n-2)\ldots 3\cdot 2\cdot 1}{\frac{2n+1}{2}\cdot\frac{2n-1}{2}\cdot\frac{2n-3}{2}\cdots\frac{3}{2}\cdot\frac{1}{2}\left|\frac{1}{2}\right.}$$

$$= \frac{\pi}{4}\frac{(2n-1)(2n-3)(2n-5)\ldots 3\cdot 1}{2n(2n-2)(2n-4)\ldots 4\cdot 2} - \frac{2n(2n-2)(2n-4)\ldots 4\cdot 2}{(2n+1)(2n-1)\ldots 3}$$ **Hence Proved.**

11. *Show that* $\int_0^{\pi} \theta \sin^3\theta \cos\theta \; d\theta = \frac{-3\pi}{32}$.

SOLUTION: Let $I = \int_0^{\pi} \theta \sin^3\theta \cos\theta \, d\theta$. Integrating it by parts, we get

$$I = \left[\theta \frac{\sin^4\theta}{4}\right]_0^{\pi} - \int_0^{\pi} 1\cdot\frac{\sin^4\theta}{4} d\theta = 0 - \frac{1}{4}\int_0^{\pi} \sin^4\theta \, d\theta$$

$$= \frac{-2}{4}\int_0^{\pi/2} \sin^4\theta \, d\theta. \quad \left[\text{using } \int_0^{2a} f(x)\,dx = \begin{cases} 2\int_0^a f(x)\,dx & \text{if } f(2a-x) = f(x) \\ 0 & \text{if } f(2a-x) = -f(x) \end{cases}\right]$$

$$= \frac{-1}{2}\frac{\left|\frac{5}{2}\right|\frac{1}{2}}{2\,|3} = \frac{-1}{4\cdot 2}\cdot\frac{3}{2}\cdot\frac{1}{2}\left(\left|\frac{1}{2}\right.\right)^2 = \frac{-3\pi}{32}$$ **Hence Proved.**

12. *If* $I_n = \int_0^{\pi/2} x\cos^n x \; dx$, *show that*

$$I_n = \frac{n-1}{n} I_{n-2} - \frac{1}{n^2} \text{ and hence find } I_4.$$

SOLUTION: $I_n = \int_0^{\pi/2} \cos^{n-1} x\,(x\cos x)\,dx$. We integrate it by parts taking $\cos^{n-1} x$ as first function and taking into account $\int x\cos x \, dx = x\sin x + \cos x$, we get

$$I_n = [\cos^{n-1} x\,(x\sin x + \cos x)]_0^{\pi/2} - \int_0^{\pi/2} (n-1)\cos^{n-2} x\,(-\sin x)\,(x\sin x + \cos x)\,dx$$

$$= 0 - 1 + (n-1)\int_0^{\pi/2} x\sin^2 x\cos^{n-2} x\,dx + (n-1)\int_0^{\pi/2} \cos^{n-1} x\sin x\,dx$$

$$= -1 + (n-1)\int_0^{\pi/2} x(1-\cos^2 x)\cos^{n-2} x\,dx - \frac{(n-1)}{n}[\cos^n x]_0^{\pi/2}$$

$$= -1 + (n-1)\int_0^{\pi/2} x\cos^{n-2} x\,dx - (n-1)\int_0^{\pi/2} x\cos^n x\,dx - \frac{n-1}{n}(0-1)$$

$$= -1 + (n-1)I_{n-2} - (n-1)I_n + \frac{n-1}{n}$$

or $\quad (1+n-1)\,I_n = -1 + (n-1)I_{n-2} + 1 - \frac{1}{n}$

or $\quad I_n = \frac{n-1}{n} I_{n-2} - \frac{1}{n^2}.$

Next, $I_4 = \frac{3}{4}I_2 - \frac{1}{4^2}$ and $I_2 = \frac{1}{2}I_0 - \frac{1}{2^2}$ and $I_0 = \int_0^{\pi/2} x\,dx = \frac{\pi^2}{8}$ hence $I_2 = \frac{\pi^2}{16} - \frac{1}{4}$ and in turn,

$$I_4 = \frac{3}{4}\left(\frac{\pi^2}{16} - \frac{1}{4}\right) - \frac{1}{16} = \frac{3\pi^2}{64} - \frac{1}{4}.$$

Ans.

SOLUTIONS TO EXERCISE 6C

1. *Show that* $\int_0^1 x^5 \sin^{-1} x\,dx = \frac{11\pi}{192}.$

SOLUTION: Let $I = \int_0^1 x^5 \sin^{-1} x\,dx$ Putting $x = \sin\theta$, so $dx = \cos\theta\,d\theta$, we get

$$I = \int_0^{\pi/2} \theta\sin^5\theta\cos\theta\,d\theta.$$ (Integrating by parts we get)

$$I = \left[\theta\frac{\sin^6\theta}{6}\right]_0^{\pi/2} - \int_0^{\pi/2} 1\cdot\frac{\sin^6\theta}{6}\,d\theta = \frac{\pi}{12} - 0 - \frac{1}{6}\int_0^{\pi/2}\sin^6\theta\,d\theta$$

$$= \frac{\pi}{12} - \frac{1}{6}\,\frac{\Gamma\left(\frac{7}{2}\right)\Gamma\left(\frac{1}{2}\right)}{2\,\Gamma\left(\frac{6+2}{2}\right)} = \frac{\pi}{2}\cdot\frac{\frac{5}{2}\cdot\frac{3}{2}\cdot\frac{1}{2}\left(\Gamma\left(\frac{1}{2}\right)\right)^2}{12\,\Gamma 4} = \frac{\pi}{12} - \frac{5\pi}{32\cdot 3\cdot 2} = \frac{11\pi}{192}$$

Hence Proved.

2. *Evaluate* $\int_0^{2\pi} \sin^2\theta\,(1+\cos\theta)^4\,d\theta$

SOLUTION: Let $I = \int_0^{2\pi} \sin^2\theta\,(1+\cos\theta)^4\,d\theta$. Since $\sin^2(2\pi-\theta)\,[1+\cos(2\pi-\theta)] = \sin^2\theta\,(1+\cos\theta)$ by property of definite integrals, we have

$$I = 2\int_0^{\pi} \sin^2\theta\,(1+\cos\theta)^4\,d\theta = 2\int_0^{\pi}\left(2\sin\frac{\theta}{2}\cos\frac{\theta}{2}\right)^2\left(2\cos^2\frac{\theta}{2}\right)^4 d\theta$$

$$= 32\cdot 4\int_0^{\pi/2}\sin^2\frac{\theta}{2}\cos^{10}\frac{\theta}{2} \qquad \left(\text{Putting } \frac{\theta}{2} = t \text{ so } \theta = 2t\right)$$

$$I = 128\int_0^{\pi/2}\sin^2 t\cos^{10} t\cdot 2dt = 256\,\frac{\left|\frac{3}{2}\right.\left|\frac{11}{2}\right.}{2\left|\frac{14}{2}\right.} = \frac{128\cdot\frac{1}{2}\left|\frac{1}{2}\right.\cdot\frac{9}{2}\cdot\frac{7}{2}\cdot\frac{5}{2}\cdot\frac{3}{2}\cdot\frac{1}{2}\left|\frac{1}{2}\right.}{6\cdot5\cdot4\cdot3\cdot2\cdot1}$$

$$= \frac{21\pi}{8}$$ **Ans.**

3. *Evaluate* $\int_0^{\pi/2}\frac{\cos^2\theta\;d\theta}{\cos^2\theta+4\sin^2\theta}$

SOLUTION: Let $I = \int_0^{\pi/2}\frac{\cos^2\theta\,d\theta}{\cos^2\theta+4\sin^2\theta} = \int_0^{\pi/2}\frac{d\theta}{1+4\tan^2\theta}$

Putting $\tan\theta = x$ so $\sec^2\theta\,d\theta = dx$

$$\therefore \quad I = \int_0^{\infty}\frac{dx}{(4x^2+1)(x^2+1)} = \frac{1}{3}\int_0^{\infty}\left(\frac{-1}{x^2+1}+\frac{4}{4x^2+1}\right)dx$$

$$= \frac{1}{3}[2\tan^{-1}2x - \tan^{-1}x]_0^{\infty} = \frac{\pi}{6}$$ **Ans.**

4. *If* $I_n = \int_0^{\pi/2}\cos^n x\sin nx\;dx$, *prove that*

$I_n = \frac{1}{2n}+\frac{1}{2}I_{n-1}$ *and hence, show that* $I_n = \frac{1}{2^{n+1}}\left[2+\frac{2^2}{2}+\frac{2^3}{3}+\ldots\frac{2^n}{n}\right]$

SOLUTION: $I_n = \int_0^{\pi/2}\cos^n x\sin nx\,dx$ Integrating it by parts, we get

$$I_n = \left[\cos^n x\,\frac{(-\cos nx)}{n}\right]_0^{\pi/2} - \int_0^{\pi/2} n\cos^{n-1}x\,(-\sin x)\,\frac{(-\cos nx)}{n}\,dx$$

$$= 0+\frac{1}{n}-\int_0^{\pi/2}\cos^{n-1}x\,(\cos nx\sin x)\,dx$$

$$= \frac{1}{n}-\frac{1}{2}\int_0^{\pi/2}\cos^{n-1}x\,[\sin(n+1)x-\sin(n-1)x]\,dx$$

$$= \frac{1}{n}+\frac{1}{2}\int_0^{\pi/2}\cos^{n-1}x\sin(n-1)x\,dx-\frac{1}{2}\int_0^{\pi/2}\cos^{n-1}x\sin(n+1)x\,dx$$

$$= \frac{1}{n}+\frac{1}{2}I_{n-1}-\frac{1}{2}\int_0^{\pi/2}\cos^{n-1}x\,(\sin nx\cos x+\cos nx\sin x)\,dx$$

$$= \frac{1}{n}+\frac{1}{2}I_{n-1}-\frac{1}{2}I_n-\frac{1}{2}\int_0^{\pi/2}\cos nx\cdot\cos^{n-1}x\sin x\,dx.$$

$$= \frac{1}{n}+\frac{1}{2}I_{n-1}-\frac{1}{2}I_n+\frac{1}{2}\left[\cos nx\frac{\cos^n x}{n}\right]_0^{\pi/2}+\frac{1}{2}\int_0^{\pi/2}n\sin nx\frac{\cos^n x}{n}\,dx$$

$$= \frac{1}{n}+\frac{1}{2}I_{n-1}-\frac{1}{2}I_n+0-\frac{1}{2n}+\frac{1}{2}I_n.$$ Thus, we have $I_n=\frac{1}{2n}+\frac{1}{2}I_{n-1}$.

$$\therefore\quad I_n=\frac{1}{2n}+\frac{1}{2}\left[\frac{1}{2n-2}+\frac{1}{2}I_{n-2}\right]$$

$$= \frac{1}{2n}+\frac{1}{2^2(n-1)}+\frac{1}{2^2}\left(\frac{1}{2(n-2)}+\frac{1}{2}I_{n-3}\right)$$

$$= \frac{1}{2n}+\frac{1}{2^2(n-1)}+\frac{1}{2^3(n-2)}+\frac{1}{2^n(n-3)}+\ldots+\frac{1}{2^{n-2}\cdot 3}+\frac{1}{2^{n-1}\cdot 2}+\frac{1}{2^n\cdot 2^\circ}$$

$$= \frac{1}{2^{n+1}}\left[2+\frac{2^2}{2}+\frac{2^3}{3}+\ldots+\frac{2^n}{n}\right]$$ **Hence Proved.**

5. *If* $I_n=\int_0^{\infty}e^{-x}x^n\,dx$, *show that* $I_n=n\,I_{n-1}=n!$ *where* n *is a natural number.*

SOLUTION: $I_n=\int_0^{\infty}e^{-x}\cdot x^n\,dx$ Integrating it by parts, we get

$$I_n=[-x^n e^{-x}]_0^{\infty}+\int_0^{\infty}n x^{n-1}e^{-x}\,dx=-\operatorname*{Lt}_{x\to\infty}\frac{x^n}{e^x}+0+n\,I_{n-1}$$

$$= 0+0+nI_{n-1}\quad\text{or}\quad I_n=nI_{n-1}$$

$$\Rightarrow\quad I_n=nI_{n-1}=n(n-1)\,I_{n-2}=n(n-1)\,(n-2)\,I_{n-3}=\ldots=n!\,I_0$$

Now $$I_0=\int_0^{\infty}e^{-x}x^\circ\,dx=\int_0^{\infty}e^{-x}dx=[-e^{-x}]_0^{\infty}=0-(-1)=1$$

Therefore $$I_n=nI_{n-1}=n!$$ **Hence Proved.**

6. *If* $I_n = \int \frac{x^n\, dx}{\left(x^2+a^2\right)^{3/2}}$ $(n \in \text{N})$, *show that* $(2-n)\, I_n = -\frac{x^{n-1}}{\sqrt{x^2+a^2}} = +(n-1)\, a^2\, I_{n-1}$

and hence evaluate $\int_0^1 \frac{x^5\, dx}{\left(x^2+4\right)^{3/2}}$.

SOLUTION: $$I_n = \int \frac{x^n\, dx}{(x^2+a^2)\frac{3}{2}} (n \in N) = \int x^{n-1} \cdot \frac{x}{(x^2+a^2)^{3/2}}\, dx$$

Integrating it by parts taking x^{n-1} as first function, we get

$$I_n = -x^{n-1} \frac{1}{\sqrt{(x^2+a^2)}} + \int (n-1) x^{n-2} \cdot \frac{dx}{\sqrt{x^2+a^2}}$$

$$= \frac{-x^{n-1}}{\sqrt{x^2+a^2}} + (n-1) \int \frac{x^{n-2}(x^2+a^2)}{(x^2+a^2)^{3/2}}\, dx$$

$$= \frac{-x^{n-1}}{\sqrt{x^2+a^2}} + (n-1) \int \frac{x^n\, dx}{(x^2+a^2)^{3/2}} + (n-1) a^2 \int \frac{x^{n-2} dx}{(x^2+a^2)^{3/2}}$$

$$= \frac{-x^{n-1}}{\sqrt{x^2+a^2}} + (n-1)\, I_n + (n-1) a^2\, I_{n-2}$$

or $$(2-n)\, I_n = \frac{-x^{n-1}}{\sqrt{x^2+a^2}} + (n-1) a^2\, I_{n-2}$$ **Hence Proved.**

Taking $x = 5$, we get $$-3\, I_5 = \frac{-x^4}{\sqrt{x^2+a^2}} + 4a^2\, I_3$$

and for $n = 3$, we have $$-I_3 = \frac{-x^2}{\sqrt{x^2+a^2}} + 2a^2 I_1$$

$$\therefore \quad -3\, I_5 = \frac{-x^4}{\sqrt{x^2+a^2}} + 4a^2 \left[\frac{x^2}{\sqrt{x^2+a^2}} - 2a^2 I_1\right]$$

Clearly $$I_1 = \int \frac{x\, dx}{(x^2+a^2)^{3/2}} = -\frac{1}{\sqrt{x^2+a^2}}, \text{ hence}$$

$$I_5 = \frac{1}{3} \frac{x^4}{\sqrt{x^2+a^2}} - \frac{4}{3} a^2 \frac{x^2}{\sqrt{x^2+a^2}} - \frac{8a^4}{3} \frac{1}{\sqrt{x^2+a^2}} = \frac{1}{3\sqrt{x^2+a^2}} [x^4 - 4a^2 x^2 - 8a^4]$$

$$\therefore \int_0^1 \frac{x^5\, dx}{(x^2+4)^{3/2}} = \left[\frac{1}{3\sqrt{x^2+4}} (x^4 - 16x^2 - 128)\right]_0^1 = \frac{-143}{3\sqrt{5}} + \frac{128}{6} = \frac{64\sqrt{5} - 143}{3\sqrt{5}}$$ **Ans.**

7. *If* $I_n = \int_0^{\pi/2} x^n \cos ax \; dx$, *show that* $I_n = \frac{1}{a}\left(\frac{\pi}{2}\right)^n\left[\sin\frac{a\pi}{2} + \frac{2n}{a\pi}\cos\frac{a\pi}{2}\right] - \frac{n(n-1)}{a^2} I_{n-2}$

SOLUTION: $I_n = \int_0^{\pi/2} x^n \cos ax \, dx = \int_0^{\pi/2} x^{n-1}(x\cos ax)\, dx$ Integrating by parts and taking into account

$\int x\cos ax\, dx = \frac{x}{a}\sin ax - \int \frac{1\cdot \sin ax}{a}\, dx = \frac{x}{a}\sin ax + \frac{\cos ax}{a^2}$, we get

$$I_n = \left[x^{n-1}\left(\frac{x}{a}\sin ax + \frac{\cos ax}{a^2}\right)\right]_0^{\pi/2} - \int_0^{\pi/2}(n-1)x^{n-2}\left(\frac{x}{a}\sin ax + \frac{\cos ax}{a^2}\right)dx$$

$$= \left(\frac{\pi}{2}\right)^{n-1}\left[\frac{\pi}{2a}\sin\frac{a\pi}{2} + \frac{1}{a^2}\cos\frac{a\pi}{2}\right] - \frac{n-1}{a}\int_0^{\pi/2} x^{n-1}\sin ax\, dx - \frac{n-1}{a^2}\int_0^{\pi/2} x^{n-2}\cos ax\, dx$$

$$= \frac{1}{a}\left(\frac{\pi}{2}\right)^n\left[\sin\frac{a\pi}{2} + \frac{2}{a\pi}\cos\frac{a\pi}{2}\right] - \frac{n-1}{a^2}I_{n-2} - \frac{n-1}{a}\left[\sin ax\cdot\frac{x^n}{n}\right]_0^{\pi/2} + \frac{n-1}{a}\int_0^{\pi/2} a\cos ax\cdot\frac{x^n}{n}\, dx$$

$$= \frac{1}{a}\left(\frac{\pi}{2}\right)^n\left[\sin\frac{a\pi}{2} + \frac{2}{a\pi}\cos\frac{a\pi}{2}\right] - \frac{n-1}{a^2}I_{n-2} - \frac{n-1}{an}\left(\frac{\pi}{2}\right)^n\sin\frac{a\pi}{2} + \frac{n-1}{n}I_n$$

or $$\left(1 - \frac{n-1}{n}\right)I_n = \frac{1}{a}\left(\frac{\pi}{2}\right)^n\left[\left(1-\frac{n-1}{n}\right)\sin\frac{a\pi}{2} + \frac{2}{a\pi}\cos\frac{a\pi}{2}\right] - \frac{n-1}{a^2}I_{n-2}$$

or $$I_n = \frac{1}{a}\left(\frac{\pi}{2}\right)^n\left(\sin\frac{a\pi}{2} + \frac{2n}{a\pi}\cos\frac{a\pi}{2}\right) - \frac{n(n-1)}{a^2}I_{n-2}$$ **Hence Proved.**

8. *If* $I_n = \int_0^{\pi/4} \sin^{2n} \; dx$, *show that* $I_n = \left(1 - \frac{1}{2n}\right)I_{n-1} - \frac{1}{n\cdot 2^{n+1}}$.

SOLUTION: $I_n = \int_0^{\pi/4} \sin^{2n} x\, dx = \int_0^{\pi/4}\sin^{2n-1} x\sin x\, dx$ Integrating by parts

$$I_n = \left[\sin^{2n-1}x\,(-\cos x)\right]_0^{\pi/4} - \int_0^{\pi/4}(2n-1)\sin^{2n-2}x\cos x(-\cos x)\, dx$$

$$= -\left(\frac{1}{\sqrt{2}}\right)^{2n-1}\left(\frac{1}{\sqrt{2}}\right) + (2n-1)\int_0^{\pi/4}\sin^{2n-2}x(1-\sin^2 x)\, dx = -\frac{1}{2^n} + (2n-1)I_{n-1} - (2n-1)I_n$$

or $$[1 + (2n-1)]\, I_n = \frac{-1}{2^n} + (2n-1)\, I_{n-1}$$

or $$I_n = -\frac{1}{2^{n+1}\cdot n} + \frac{2n-1}{2n}I_{n-1}$$ **Hence Proved.**

9. *If* $I_n = \int_0^{\pi/2} x^n \sin(2a+1)x \; dx$, *show that* $(2a+1)^2 I_n + n(n-1) I_{n-2} = (-1)^a n \left(\frac{\pi}{2}\right)^{n-1}$. *where a and n are positive integers.*

SOLUTION: $I_n = \int_0^{\pi/2} x^n \sin(2a+1)x\, dx = -\left[\frac{x^n \cos(2a+1)x}{2a+1}\right]_0^{\pi/2} + \int_0^{\pi/2} \frac{nx^{n-1}\cos(2a+1)x}{2a+1}\, dx$

or $$I_n = \frac{-1}{2a+1}\left(\frac{\pi}{2}\right)^n \cos(2a+1)\frac{\pi}{2} + \frac{n}{2a+1}\int_0^{\pi/2} x^{n-1}\cos(2a+1)x\, dx.$$

$$= \frac{-1}{2a+1}\left(\frac{\pi}{2}\right)^n \cos(2a+1)\frac{\pi}{2} + \frac{n}{2a+1}\left[x^{n-1}\frac{\sin(2a+1)x}{2a+1}\right]_0^{\pi/2}$$

$$-\frac{n}{2a+1}\int_0^{\pi/2} \frac{(n-1)x^{n-2}\sin(2a+1)x}{2a+1}\, dx$$

$$= -\frac{1}{2a+1}\left(\frac{\pi}{2}\right)^n \cos(2a+1)\frac{\pi}{2} + \frac{n}{(2a+1)^2}\left(\frac{\pi}{2}\right)^{n-1}\sin(2a+1)\frac{\pi}{2} - \frac{n(n-1)}{(2a+1)^2} I_{n-2}$$

or $$(2a+1)^2 I_n + n(n+1) I_{n-2} = \left(\frac{\pi}{2}\right)^{n-1}\left[-(2a+1)\frac{\pi}{2}\cos(2a+1)\frac{\pi}{2} + n\sin(2a+1)\frac{\pi}{2}\right]$$

$$= \left(\frac{\pi}{2}\right)^{n-1}\left[\left(a\pi + \frac{\pi}{2}\right)\sin a\pi + n\cos a\pi\right]$$

$$= \left(\frac{\pi}{2}\right)^{n-1} n(-1)^a \quad \text{since } a \text{ is positive integers.}$$

Hence Proved.

10. *If* $I_n = \int_{\pi/4}^{\pi/2} \cot^n x \; dx \; (n > 2)$, *show that* $I_n = \frac{1}{n-1} - I_{n-2}$ *and hence evaluate* $\int_{\pi/4}^{\pi/2} \cot^6 x \; dx$.

SOLUTION: $I_n = \int_{\pi/4}^{\pi/2} \cot^n x\; dx = \int_{\pi/4}^{\pi/2} \cot^{n-2} x\,(\text{cosec}^2 x - 1)\, dx$

$$I_n = \int_{\pi/4}^{\pi/2} \cot^{n-2} x\, \text{cosec}^2 x\, dx - I_{n-2} = -\left[\frac{\cot^{n-1} x}{n-1}\right]_{\pi/4}^{\pi/2} - I_{n-2}$$

$$= 0 + \frac{1}{n-1} - I_{n-2}$$

Hence Proved.

$$I_6 = \int_{\pi/4}^{\pi/2} \cot^6 x\, dx = \frac{1}{6-1} - I_4 = \frac{1}{5} - \left[\frac{1}{4-1} - I_2\right]$$

$$= \frac{1}{5} - \frac{1}{3} + I_2 = \frac{-2}{15} + \int_{\pi/4}^{\pi/2} \cot^2 x\, dx = \frac{-2}{15} \int_{\pi/4}^{\pi/2} (\text{cosec}^2 x - 1)\, dx$$

$$= \frac{-2}{15} + [-\cot x - x]_{\pi/4}^{\pi/2} = \frac{13}{15} - \frac{\pi}{4}$$ **Ans.**

11. *By considering the integral* $\int_0^1 \left(1-x^2\right)^n dx$, *show that*

$$1 - \frac{n}{1!\,3} + \frac{n(n-1)}{2!\,5} + \frac{n(n-1)(n-2)}{3!\,7} + \ldots = \frac{2\cdot 4\cdot 6 \ldots 2n}{3\cdot 5\cdot 7 \ldots (2n+1)}.$$

SOLUTION: Let $I = \int_0^1 (1-x^2)^n\, dx$. On expanding $(1-x^2)^n$ in powers of x, we get

$$I = \int_0^1 \left[1 - nx^2 + \frac{n(n-1)}{2!} x^4 - \frac{n(n-1)(n-2)}{3!} x^6 + \ldots\right] dx$$

$$= \left[x - \frac{nx^3}{3} + \frac{n(n-1)}{2!} \frac{x^5}{5} - \frac{n(n-1)(x-2)}{3!} \frac{x^7}{7} + \ldots\right]_0^1$$

$$= 1 - \frac{n}{1!3} + \frac{n(n-1)}{2!5} - \frac{n(n-1)(n-2)}{3!7} + \ldots$$

Next, on putting $x = \sin\theta$, we get

$$I = \int_0^{\pi/2} (1-\sin^2\theta)^n \cos\theta\, d\theta = \int_0^{\pi/2} \cos^{2n+1}\theta\, d\theta = \frac{\Gamma\left(\frac{2n+2}{2}\right)\Gamma\left(\frac{1}{2}\right)}{2\,\Gamma\left(\frac{2n+3}{2}\right)} = \frac{n!\,\Gamma\left(\frac{1}{2}\right)}{2\,\Gamma\left(\frac{2n+3}{2}\right)}$$

$$= \frac{n(n-1)(n-2)\ldots 3\cdot 2\cdot 1\,\Gamma\left(\frac{1}{2}\right)}{2\cdot \frac{2n+1}{2}\cdot\frac{2n-1}{2}\cdot\frac{2n-3}{2}\ldots\frac{3}{2}\cdot\frac{1}{2}\,\Gamma\left(\frac{1}{2}\right)} = \frac{2^n\cdot n(n-1)(n-2)\ldots 3\cdot 2\cdot 1}{(2n+1)(2n-1)(2n-3)\ldots 5\cdot 3\cdot 1}$$

$$= \frac{2\cdot 4\cdot 6\ldots 2n}{3\cdot 5\cdot 7\ldots (2n+1)}.$$ **Hence Proved.**

ADDITIONAL IMPORTANT PROBLEMS

1. *Evaluate (i)* $\int_0^{\pi/2} \sin^8 x\, dx$ *(ii)* $\int_0^{\pi/2} \cos^7 x\, dx$

SOLUTION: (*i*) Using the reduction formula

$$\int_0^{\pi/2} \sin^n x\, dx = \frac{n-1}{n}\int_0^{\pi/2} \sin^{n-2} x\, dx, \quad \text{we have}$$

$$\int_0^{\pi/2} \sin^8 x\, dx = \frac{7}{8}\int_0^{\pi/2} \sin^6 x\, dx = \frac{7}{8}\cdot\frac{5}{6}\int_0^{\pi/2} \sin^4 x\, dx$$

$$= \frac{7}{8}\cdot\frac{5}{6}\cdot\frac{3}{4}\int_0^{\pi/2} \sin^2 x\, dx = \frac{7}{8}\cdot\frac{5}{6}\cdot\frac{3}{4}\cdot\frac{1}{2}\int_0^{\pi/2} \sin^0 x\, dx.$$

$$= \frac{7}{8}\cdot\frac{5}{6}\cdot\frac{3}{4}\cdot\frac{1}{2}\cdot\frac{\pi}{2} = \frac{35\pi}{256}$$ **Ans.**

(*ii*) Using the reduction formula

$$\int_0^{\pi/2} \cos^n x\, dx = \frac{n-1}{n}\int_0^{\pi/2} \cos^{n-2} x\, dx, \quad \text{we have}$$

$$\int_0^{\pi/2} \cos^7 x\, dx = \frac{6}{7}\int_0^{\pi/2} \cos^5 x\, dx = \frac{6}{7}\cdot\frac{4}{5}\int_0^{\pi/2} \cos^3 x\, dx = \frac{6}{7}\cdot\frac{4}{5}\cdot\frac{2}{3}\int_0^{\pi/2} \cos x\, dx$$

$$= \frac{6\cdot 4\cdot 2}{7\cdot 5\cdot 3}\cdot 1 = \frac{16}{35}$$ **Ans.**

2. *Evaluate (i)* $\int_0^{\pi/2} \sin^5 x \cos^6 x\, dx$ *(ii)* $\int_0^{\pi/2} \sin^4 x \cos^8 x\, dx$

SOLUTION: (*i*) $\int_0^{\pi/2} \sin^5 x \cos^6 x\, dx = \frac{4}{11}\int_0^{\pi/2} \sin^3 x \cos^6 x\, dx = \frac{4}{11}\cdot\frac{2}{9}\int_0^{\pi/2} \sin x \cos^6 x\, dx$

$$= \frac{4}{11}\cdot\frac{2}{9}\left[\frac{-\cos^7 x}{7}\right]_0^{\pi/2} = \frac{8}{693}.$$

(*ii*) $\int_0^{\pi/2} \sin^4 x \cos^8 x\, dx = \frac{3}{12}\int_0^{\pi/2} \sin^2 x \cos^8 x\, dx = \frac{3.1}{12.10}\int_0^{\pi/2} \cos^8 x\, dx.$

$$= \frac{3\cdot 1}{12\cdot 10}\cdot\frac{7}{8}\cdot\frac{5}{6}\cdot\frac{3}{4}\cdot\frac{1}{2}\cdot\frac{\pi}{2} = \frac{7\pi}{2048}$$ **Ans.**

3. *Evaluate* $\int \sin^4 x \cos^4 x\, dx$

SOLUTION: Using the reduction formula

$$\int \sin^m x \cos^n x\, dx = -\frac{1}{m+n}\sin^{m-1} x \cos^{n+1} x + \frac{m-1}{m+n}\int \sin^{m-2} x \cos^n x\, dx, \text{ we get}$$

$$\int \sin^4 x \cos^4 x\, dx = -\frac{1}{8}\sin^3 x \cos^5 x + \frac{3}{8}\int \sin^2 x \cos^4 x\, dx$$

$$= -\frac{1}{8}\sin^3 x \cos^5 x + \frac{3}{8}\left[-\frac{1}{6}\sin x \cos^5 x + \frac{1}{6}\int \cos^4 x\, dx\right]$$

$$= -\frac{1}{8}\sin^3 x \cos^5 x - \frac{1}{16}\sin x \cos^5 x + \frac{1}{16}\left[\frac{1}{4}\cos^3 x \sin x + \frac{3}{4}\int \cos^2 x\, dx\right]$$

$$= -\frac{1}{8}\sin^3 x \cos^5 x - \frac{1}{16}\sin x \cos^5 x + \frac{1}{64}\cos^3 \sin x + \frac{3}{64}\left[\frac{1}{2}\cos x \sin x + \frac{1}{2}\int 1\, dx\right]$$

or $$\int \sin^4 x \cos^4 x\, dx = -\frac{1}{8}\sin^3 x \cos^5 x - \frac{1}{16}\sin x \cos^5 x + \frac{1}{64}\cos^3 x \sin x + \frac{3}{128}[x + \sin x \cos x]$$ **Ans.**

4. *Evaluate* $\int_{\pi/4}^{\pi/2} \cot^4 x\, dx$

SOLUTION: Applying the reduction formula $\int \cot^n dx = \frac{-\cot^{n-1} x}{n-1} - \int \cot^{n-2} x\, dx$

For $n = 4$, we get $$\int \cot^4 x\, dx = -\frac{1}{3}\cot^3 x - \int \cot^2 x\, dx$$

$$= -\frac{1}{3}\cot^3 x - \left[-\frac{1}{1}\cot x - \int \cot^0 x\, dx\right]$$

$$= -\frac{1}{3}\cot^3 x + \cot x + x$$

$$\therefore \quad \int_{\pi/4}^{\pi/2} \cot^4 x\, dx = \left[-\frac{1}{3}\cot^3 x + \cot x + x\right]_{\pi/4}^{\pi/2}$$

$$= 0 + 0 + \frac{\pi}{2} - \left[-\frac{1}{3} + 1 + \frac{\pi}{4}\right] = \frac{\pi}{4} - \frac{2}{3}$$ **Ans.**

5. *Evaluate* $\int_{\pi/6}^{\pi/2} \operatorname{cosec}^5 x\, dx$

SOLUTION: Using the reduction formula

$$\int \operatorname{cosec}^n x\, dx = -\frac{1}{n-1}\operatorname{cosec}^{n-2} x \cot x + \frac{n-2}{n-1}\int \operatorname{cosec}^{n-2} x\, dx, \quad n > 1$$

gives $\int \text{cosec}^5 x\,dx = -\frac{1}{4}\text{cosec}^3 x \cot x + \frac{3}{4}\int \text{cosec}^3 x\,dx \quad (n > 1).$

$$= -\frac{1}{4}\text{cosec}^3 x \cot x + \frac{3}{4}\left[-\frac{1}{2}\text{cosec } x \cot x + \frac{1}{2}\int \text{cosec } x\,dx\right]$$

$$= -\frac{1}{4}\text{cosec}^3 x \cot x - \frac{3}{8}\text{cosec } x \cot x + \frac{3}{8}\log(\text{cosec } x - \cot x)$$

$$\therefore \int_{\pi/6}^{\pi/2} \text{cosec}^5 x\,dx = \left[-\frac{1}{4}\cos ec^3 x \cot x - \frac{3}{8}\cos ecx \cot x + \frac{3}{8}\log(\text{cosec } x - \cot x)\right]_{\pi/6}^{\pi/2}$$

$$= 0 + 0 + \frac{3}{8}\log(1-0) + \frac{1}{4}\cdot 8 \cdot \sqrt{3} + \frac{3}{8}\cdot 2\cdot\sqrt{3} - \frac{3}{8}\log(2-\sqrt{3})$$

$$= \frac{11}{4}\sqrt{3} - \frac{3}{8}\log(2-\sqrt{3}).$$ **Ans.**

6. *Evaluate* $\int \sec^6 x\,dx$

SOLUTION: Using the reduction formula

$$\int \sec^n x\,dx = \frac{\tan x \sec^{n-2} x}{n-1} + \frac{n-2}{n-1}\int \sec^{n-2} x\,dx, \quad \text{we have}$$

$$\int \sec^6 x\,dx = \frac{1}{5}\tan x \sec^4 x + \frac{4}{5}\int \sec^4 x\,dx$$

$$= \frac{1}{5}\tan x \sec^4 x + \frac{4}{5}\left[\frac{1}{3}\tan x \sec^2 x + \frac{2}{3}\int \sec^2 x\,dx\right]$$

$$= \frac{1}{5}\tan x \sec^4 x + \frac{4}{15}\tan x \sec^2 x + \frac{8}{15}\tan x$$ **Ans.**

7. *Evaluate* $\int_0^{\pi} \frac{\sin^2\theta\sqrt{1-\cos\theta}}{1+\cos\theta}\,d\theta$

SOLUTION: Given integral $= I = \int_0^{\pi} 4\sin^2\frac{\theta}{2}\cos^2\frac{\theta}{2}\frac{\sqrt{2\sin^2\frac{\theta}{2}}}{2\cos^2\frac{\theta}{2}}\,d\theta$

$$= 2\sqrt{2}\int_0^{\pi}\sin^3\frac{\theta}{2}\,d\theta$$

Now putting $\theta = 2t$ we get

$$I = 2\sqrt{2}\int_0^{\pi/2}\sin^3 t\cdot 2dt = 4\sqrt{2}\left[\left(\frac{-1}{3}\sin^2 t\cos t\right)_0^{\pi/2} + \frac{2}{3}\int_0^{\pi/2}\sin t\,dt\right]$$

$$= 4\sqrt{2}\left[0 + \frac{2}{3}\{-\cos t\}_0^{\pi/2}\right] = \frac{8\sqrt{2}}{3}.$$ **Ans.**

8. *Evaluate* $\int_0^{\pi/4} \cos^3 2x \sin^4 4x \, dx$.

SOLUTION: Let $I = \int_0^{\pi/4} \cos^3 2x \sin^4 4x \, dx = \int_0^{\pi/4} \cos^3 2x \, (2\sin 2x \cos 2x)^4 \, dx$

$$= 16 \int_0^{\pi/4} \sin^4 2x \cos^7 2x \, dx \qquad \text{(Putting } 2x = t)$$

$$= 8 \int_0^{\pi/2} \sin^4 t \cos^7 t \cdot dt = 8 \cdot \frac{6 \cdot 4 \cdot 2 \cdot 3}{11 \cdot 9 \cdot 7 \cdot 5 \cdot 3}$$

or $$I = \frac{128}{1135}.$$ **Ans.**

9. *Evaluate* $\int_0^{\infty} \frac{dx}{(1+x^2)^5}$.

SOLUTION: Let $I = \int_0^{\infty} \frac{dx}{(1+x^2)^5}$ (Putting $x = \tan\theta$ so $dx = \sec^2\theta \, d\theta$)

$$I = \int_0^{\pi/2} \frac{\sec^2\theta \, d\theta}{(1+\tan^2\theta)^5} = \int_0^{\pi/2} \frac{d\theta}{\sec^8\theta} = \int_0^{\pi/2} \cos^8\theta \, d\theta$$

$$= \frac{7}{8} \cdot \frac{5}{6} \cdot \frac{3}{4} \cdot \frac{1}{2} \int_0^{\pi/2} \cos^0\theta \, d\theta = \frac{7}{8} \cdot \frac{5}{6} \cdot \frac{3}{4} \cdot \frac{1}{2} \cdot \frac{\pi}{2}$$

$$= \frac{35\pi}{256}$$ **Ans.**

10. *Evaluate* $\int_0^4 x^3 \sqrt{4x - x^2} \, dx$.

SOLUTION: Let $I = \int_0^4 x^3 \sqrt{4x - x^2} \, dx$ (Putting $x = 4\sin^2\theta$)

$$I = \int_0^{\pi/2} (4\sin^2\theta)^3 \cdot (16\sin^2\theta - 16\sin^4\theta)^{1/2} \cdot 8\sin\theta\cos\theta \, d\theta$$

$$= \int_0^{\pi/2} 4^4 \sin^7\theta \, (1-\sin^2\theta)^{1/2} \cdot 8\sin\theta\cos\theta \, d\theta$$

$$= 16 \cdot 16 \cdot 8 \int_0^{\pi/2} \sin^8\theta \cos^2\theta \, d\theta = 16 \cdot 128 \cdot \frac{1}{10} \int_0^{\pi/2} \sin^8\theta \, d\theta$$

$$= \frac{16 \cdot 128}{10} \cdot \frac{7}{8} \cdot \frac{5}{6} \cdot \frac{3}{4} \cdot \frac{1}{2} \int_0^{\pi/2} \sin^0\theta \, d\theta$$

$$= 28\pi$$ **Ans.**

11. *Evaluate* (*i*) $\int_0^1 x^2(1-x)^{3/2}\,dx$ (*ii*) $\int_0^2 \frac{x^4\,dx}{\sqrt{4-x^2}}$

SOLUTION: (*i*) $I = \int_0^1 x^2(1-x)^{3/2}\,dx$ Put $x = \sin^2\theta$, so $dx = 2\sin\theta\cos\theta\,d\theta$

$$I = \int_0^{\pi/2} \sin^4\theta\cos^3\theta\, 2\sin\theta\cos\theta\, d\theta = 2\int_0^{\pi/2} \sin^5\theta\cos^4\theta\, d\theta$$

$$= 2\cdot\frac{4}{9}\cdot\frac{2}{7}\int_0^{\pi/2} \sin\theta\cos^4\theta\, d\theta = \frac{16}{63}\left[-\frac{\cos^5\theta}{5}\right]_0^{\pi/2}$$

$$= \frac{16}{315}$$ **Ans.**

(*ii*) $I = \int_0^2 \frac{x^4\,dx}{\sqrt{4-x^2}}$ Putting $x = 2\sin\theta$, so $dx = 2\cos\theta\,d\theta$

$$\therefore \quad I = \int_0^{\pi/2} \frac{(2)^4\sin^4\theta\cdot 2\cos\theta\, d\theta}{2\sqrt{1-\sin^2\theta}} = \int_0^{\pi/2} 16\sin^4\theta\, d\theta$$

$$= 16\cdot\frac{3}{4}\cdot\frac{1}{2}\cdot\frac{\pi}{2} = 3\pi$$ **Ans.**

CHAPTER

7

Applications of Integration

IMPORTANT FORMULAE AND USEFUL RESULTS

1. Area under the curve $y = f(x)$ is $A = \int_a^b y\,dx = \int_a^b f(x)\,dx$ or $A = \int_c^d x\,dy = \int_c^d f(y)\,dx$.

2. Area under polar curve is $A = \int_\alpha^\beta \frac{1}{2} r^2\,d\theta$.

3. Length along a curve $= S = \int_a^b \frac{ds}{dx}\,dx = \int_a^b \sqrt{1+\left(\frac{dy}{dx}\right)^2}\,dx$ or $S = \int_c^d \frac{ds}{dy}\,dy = \int_c^d \sqrt{1+\left(\frac{dx}{dy}\right)^2}\,dy$

   For parametric curves, $S = \int_{t_1}^{t_2} \sqrt{\left(\frac{dx}{dt}\right)^2 + \left(\frac{dy}{dt}\right)^2}\,dt$ and for polar curves, $S = \int_\alpha^\beta \sqrt{r^2 + \left(\frac{dr}{d\theta}\right)^2}\,d\theta$

4. Mean value of $y = f(x)$ between $x = a$ and $x = b$, is

$$\frac{\int_a^b f(x)\,dx}{\int_a^b dx} = \frac{1}{b-a}\int_a^b f(x)\,dx$$

   and R.M.S. value of $f(x) = \sqrt{\frac{\int_a^b \{f(x)^2\}\,dx}{b-a}}$.

5. Volume of the solid of revolution of an area about x-axis $= \int_a^b \pi y^2\,dx$

   and volume of solid of revolution of area about y-axis $= \int_c^d \pi y^2\,dy$.

   and the volume of solid of revolution of a polar curve about x-axis $= \frac{2\pi}{3}\int_\alpha^\beta r^2 \sin\theta\,d\theta$.

6. The surface of solid of revolution about x-axis $= 2\pi\int_a^b y\sqrt{1+\left(\frac{dy}{dx}\right)^2}\,dx$

and that about y-axis $= 2\pi\int_c^d x\sqrt{1+\left(\frac{dy}{dt}\right)^2}\,dx.$

For parametric curves, the surface $S = \int 2\pi\, y\frac{ds}{dt}\,dt = 2\pi\int y\sqrt{\left(\frac{dx}{dt}\right)^2+\left(\frac{dy}{dt}\right)^2}\,dt$

and for polar curves, the surface $S = \int 2\pi\, y\frac{ds}{d\theta}\,d\theta = 2\pi\int r\sin\theta\sqrt{r^2+\left(\frac{dr}{d\theta}\right)^2}\,d\theta$ (about initial line)

and about $\theta = \frac{\pi}{2}$, $S = \int 2\pi x\frac{ds}{d\theta}\,d\theta = \int_\alpha^\beta 2\pi r\cos\theta\sqrt{r^2+\left(\frac{dr}{d\theta}\right)^2}\,d\theta$

7. **Centre of Gravity (C.G.).**

C.G. of an arc is given by $\bar{x} = \frac{\int x\rho\,ds}{\int \rho\,ds}$, $\bar{y} = \frac{\int y\rho\,ds}{\int \rho\,ds}$ where ρ is the density at (x, y).

C.G. for a plane lamina $\bar{x} = \frac{\int x\rho\,dA}{\int \rho\,dA}$, $\bar{y} = \frac{\int y\rho\,dA}{\int \rho\,dA}$

C.G. of the area bounded by the curve $y = f(x)$ the x-axis and ordinates at $x = a$, $x = b$, is

$$\bar{x} = \frac{\int xy\,dx}{\int y\,dx},\quad \bar{y} = \frac{\int \frac{y}{2}y\,dx}{\int y\,dx}.$$

C.G.. of the area bounded by the curve $x = f(y)$, the y-axis and the abscissa at

$y = c$ and $y = d$, is $\bar{x} = \frac{\int \frac{x}{2}x\,dy}{\int x\,dy}$, $\bar{y} = \frac{\int y\cdot x\,dy}{\int x\,dy}$.

For polar curves, C.G. is $\bar{x} = \frac{\int_\alpha^\beta \frac{2}{3}r\cos\theta\cdot\frac{1}{2}r^2\,d\theta}{\int_\alpha^\beta \frac{1}{2}r^2\,d\theta}$, $\bar{y} = \frac{\int_\alpha^\beta \frac{2}{3}r\sin\theta\cdot\frac{1}{2}r^2\,d\theta}{\int_\alpha^\beta \frac{1}{2}r^2\,d\theta}$.

C.G. of the volume of solid of revolution is given by

$$\bar{x} = \frac{\int x\cdot\pi y^2\,dx}{\int \pi y^2\,dx},\quad \bar{y} = \frac{\int y\cdot\pi y^2\,dx}{\int \pi y^2\,dx}$$ (when revolution is about x-axis)

and $$\bar{x} = \frac{\int x\cdot\pi x^2\,dy}{\int \pi x^2\,dy},\quad \bar{y} = \frac{\int y\cdot\pi y^2\,dy}{\int \pi x^2\,dy}$$ (when revolution is about y-axis)

Pappus's First Theorem: The volume of the solid of revolution of a closed curve about any axis is obtained by multiplying the area of the closed curve by the total path traversed by C.G. of the area.

Pappus's Second Theorem: The surface area of the solid of revolution of a closed curve about any axis is equal to the parameter of the curve multiplied by the path traversed by the C.G. of the parameter of the closed curve.

SOLUTIONS TO EXERCISE 7A

1. *Find the area enclosed by the ellipse* $\dfrac{x^2}{a^2}+\dfrac{y^2}{b^2}=1$.

SOLUTION: Here $y=\pm\dfrac{b}{a}\sqrt{a^2-x^2}$. The required area A is given by

$$A = 2\int_{-a}^{a}\frac{b}{a}\sqrt{a^2-x^2}\,dx = \frac{4b}{a}\int_{0}^{a}\sqrt{a^2-x^2}\,dx = \frac{4b}{a}\left[\frac{x}{a}\sqrt{a^2-x^2}+\frac{a^2}{2}\sin^{-1}\left(\frac{x}{a}\right)\right]_0^a$$

$$= \frac{4b}{a}\left[0+\frac{a^2}{2}\left(\frac{\pi}{2}-0\right)\right] = \pi ab$$ **Ans.**

2. *Show that the curve* $a^2y^2 = x^2(a^2-x^2)$ *consists of two loops and find the area of each loop.*

SOLUTION: Curve $a^2y^2=x^2(a^2-x^2)$ has two symmetrical loops in $[-a, a]$. Area A of one loop is given by

$$A = 2\int_0^a y\,dx = 2\int_0^a \frac{x}{a}\sqrt{a^2-x^2}\,dx$$

$$= \frac{2}{a}\left[\frac{(a^2-x^2)^{3/2}}{\frac{3}{2}}\left(\frac{-1}{2}\right)\right]_0^a = \frac{2}{3}a^2$$ **Ans.**

3. *Find the area bounded by the parabolas* $y^2 = 9x$ *and* $x^2 = 9y$.

SOLUTION: Required area A is given by

$$A = \int_0^9 (y_1-y_2)\,dx = \int_0^9\left(3\sqrt{x}-\frac{x^2}{9}\right)dx$$

$$= \left[3\cdot\frac{x^{3/2}}{\frac{3}{2}}-\frac{x^3}{9\cdot 3}\right]_0^9 = 27 \text{ square units}$$ **Ans.**

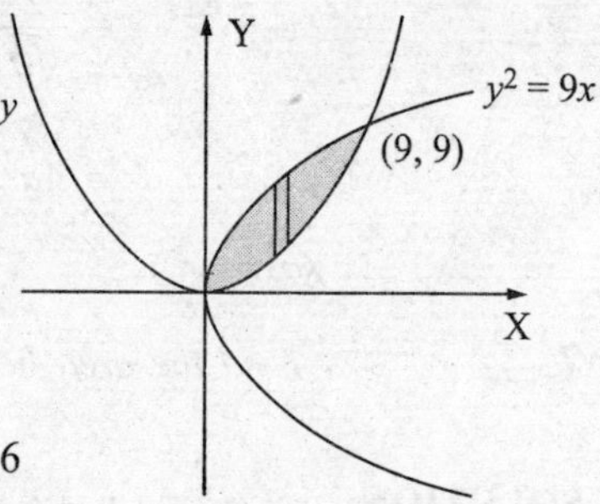

4. *Find the area bounded by the parabolas* $y^2 = 5x + 6$ *and* $x^2 = y$.

SOLUTION: $y^2 = 5x + 6$ and $x^2 = y$

For points of intersection, eliminating y, gives $x^4 - 5x - 6 = 0$ which is satisfied by $x = -1$ and $x = 2$.

$$\text{Required area} = \int_{-1}^{2}(y_1-y_2)\,dx = \int_{-1}^{2}(\sqrt{5x+6}-x^2)\,dx$$

$$= \left[\frac{(5x+6)^{3/2}}{5\cdot\frac{3}{2}}-\frac{x^3}{3}\right]_{-1}^{2} = \frac{128}{15}-\frac{8}{3}-\frac{2}{15}-\frac{1}{3} = \frac{27}{5} \text{ sq. units}$$ **Ans.**

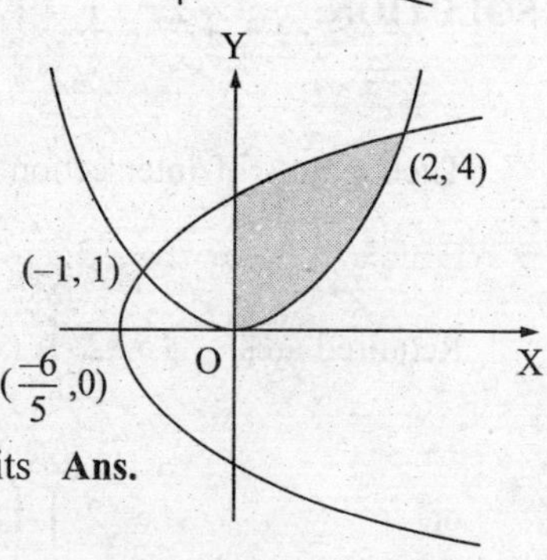

5. *Find the area between the curve* $y^2 = \dfrac{x^3}{a-x}$ *and its asymptote.*

SOLUTION: $y^2 = \dfrac{x^3}{a-x}$. Area A between the curve and the asymptote is $x = a$, given by

$$A = 2\int_0^a y\,dx = 2\int_0^a \frac{x^{3/2}dx}{\sqrt{a-x}} \quad \text{putting } x = a\sin^2\theta$$

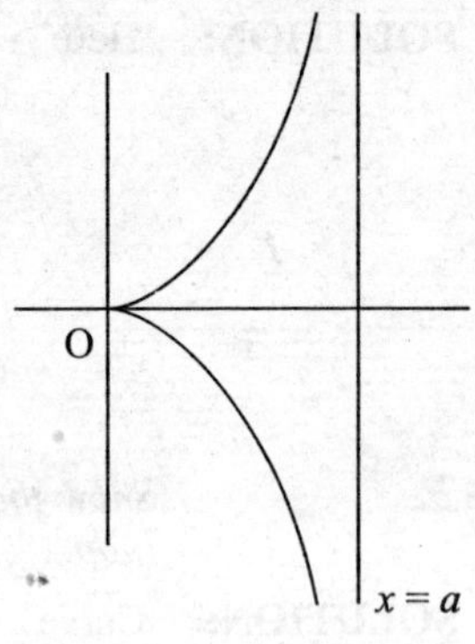

or
$$A = 2\int_0^{\pi/2} \frac{a\sqrt{a}\sin^3\theta \cdot 2a\sin\theta\cos\theta\,d\theta}{\sqrt{a}\cos\theta}$$

$$= 4a^2\int_0^{\pi/2} \sin^4\theta\,d\theta = \frac{4a^2\,\Gamma\left(\frac{5}{2}\right)\Gamma\left(\frac{1}{2}\right)}{2\,\Gamma(3)} = \frac{3\pi}{4}a^2. \quad \textbf{Ans.}$$

6. *Show that the area of the loop of the curve* $a\,y^2 = x\,(x-a)^2$ *is* $\dfrac{8a^2}{15}$.

SOLUTION: $ay^2 = x\,(x-a)^2$. Area A of the loop, is given by

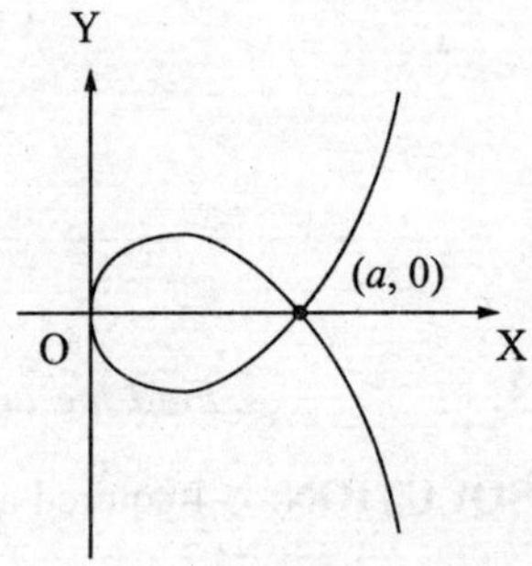

$$A = 2\int_0^a y\,dx = 2\int_0^a \frac{\sqrt{x}}{\sqrt{a}}(a-x)\,dx$$

$$= \frac{2}{\sqrt{a}}\int_0^a (x^{1/2} - ax^{3/2})\,dx$$

$$= \frac{2}{\sqrt{a}}\left[\frac{2}{3}x^{3/2} - \frac{2a}{5}x^{5/2}\right]_0^a$$

$$= 4a^2\left(\frac{1}{3} - \frac{1}{5}\right) = \frac{8a^2}{15}$$

Hence Proved.

7. *Find the area between the curves* $y = 2x^2$ *and* $y = 1 - \dfrac{x^2}{2}$.

SOLUTION: $y = 2x^2,\ y = 1 - \dfrac{x^2}{2}$ or $x^2 = \dfrac{1}{2}y,\ x^2 = -2\left(y - \dfrac{1}{2}\right)$.

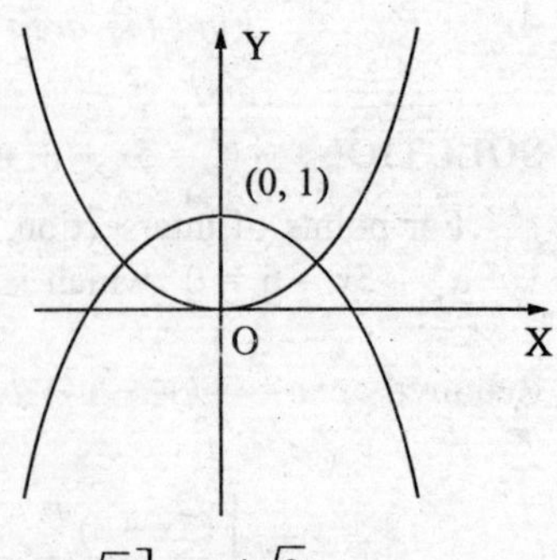

Their points of intersection are $\left(\pm\sqrt{\dfrac{2}{5}}, \dfrac{4}{5}\right)$.

Required area $A = 2\displaystyle\int_0^{\sqrt{2/5}} (y_1 - y_2)\,dx = 2\int_0^{\sqrt{2/5}} \left[1 - \frac{x^2}{2} - 2x^2\right]dx$

or
$$A = 2\int_0^{\sqrt{2/5}} \left(1 - \frac{5}{2}x^2\right)dx = 2\left[x - \frac{5x^3}{6}\right]_0^{5/2} = 2\left[\sqrt{\frac{2}{5}} - \frac{5}{6}\cdot\frac{2\sqrt{2}}{5\sqrt{5}}\right] = \frac{4\sqrt{2}}{3\sqrt{5}}$$

square units. **Ans.**

8. *Show that the area enclosed by the curve* $|x| + |y| = 2a, (a > 0)$ is $8a^2$.

SOLUTION: $|x| + |y| = 2a\ (a > 0)$ gives rectange $ABCD$

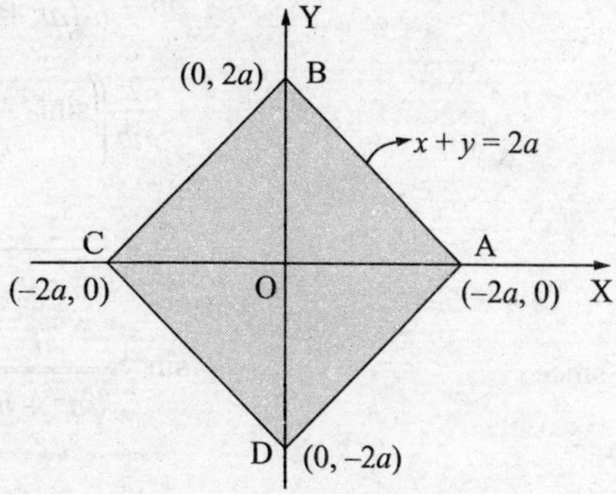

Area $A = 4\int_0^{2a} y\,dx = 4\int_0^{2a} (2a - x)\,dx$

$$= 4\left[2ax - \frac{x^2}{2}\right]_0^{2a} = 8a^2 \text{ Square units.}$$ **Ans.**

9. *Determine the area common to the two ellipses* $a^2x^2 + b^2y^2 = 1$ *and* $b^2x^2 + a^2y^2 = 1$.

SOLUTION: Curves are $\dfrac{x^2}{\left(\frac{1}{a}\right)^2} + \dfrac{y^2}{\left(\frac{1}{b}\right)^2} = 1,$ ...(1), $\dfrac{x^2}{\left(\frac{1}{b}\right)^2} + \dfrac{y^2}{\left(\frac{1}{a}\right)^2} = 1$ ...(2)

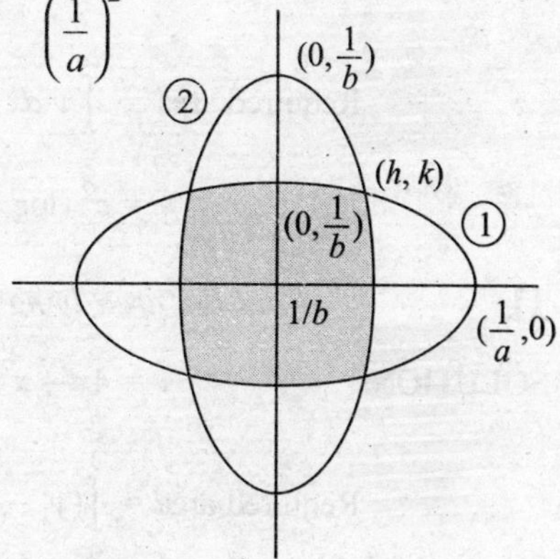

Let $a < b$ so $\dfrac{1}{a} > \dfrac{1}{b}$.

Area A common to two curves is given by

$$A = 4\left[\int_0^h y_1\,dx + \int_h^{1/b} y_2\,dx\right]$$

where $x = h$ is their point of intersection.

Eliminating y in (1) and (2), gives

$$(a^4 - b^4)\,x^2 = a^2 - b^2 \quad \text{or} \quad x^2 = \frac{1}{a^2 + b^2} \quad \text{or} \quad x = h = \frac{1}{\sqrt{a^2 + b^2}}$$

$$\therefore \quad A = 4\int_0^h \frac{1}{b}\sqrt{1 - a^2x^2}\,dx + 4\int_h^{1/b} \frac{1}{a}\sqrt{1 - b^2x^2}\,dx$$

$$= \frac{4a}{b}\int_0^h \sqrt{\left(\frac{1}{a}\right)^2 - x^2}\,dx + \frac{4b}{a}\int_h^{1/b}\sqrt{\left(\frac{1}{b}\right)^2 - x^2}\,dx \quad \text{where } h = \frac{1}{\sqrt{a^2 + b^2}}$$

$$= \frac{4a}{b}\left[\frac{x}{2}\sqrt{\left(\frac{1}{a}\right)^2 - x^2} + \frac{1}{2a^2}\sin^{-1} ax\right]_0^h + \frac{4b}{a}\left[\frac{x}{2}\sqrt{\left(\frac{1}{b}\right)^2 - x^2} + \frac{1}{2b^2}\sin^{-1} bx\right]_h^{1/b}$$

$$= \frac{2a}{b}\left[h\sqrt{\frac{1}{a^2} - h^2} + \frac{1}{a^2}\sin^{-1} ah\right] + \frac{2b}{a}\left[0 + \frac{1}{b^2}\frac{\pi}{2} - h\sqrt{\frac{1}{b^2} - h^2} - \frac{1}{b^2}\sin^{-1} bh\right]$$

$$= \frac{\pi}{ab} + \frac{2h}{b}\sqrt{1 - a^2h^2} - \frac{2h}{a}\sqrt{1 - b^2h^2} + \frac{2}{ab}\sin^{-1} ah - \frac{2}{ab}\sin^{-1} bh$$

$$= \frac{\pi}{ab} + \frac{2}{b\sqrt{a^2+b^2}}\sqrt{\left(1-\frac{a^2}{a^2+b^2}\right)} - \frac{2}{a\sqrt{a^2+b^2}}\sqrt{\left(1-\frac{b^2}{a^2+b^2}\right)}$$

$$+ \frac{2}{ab}\left(\sin^{-1}\frac{a}{\sqrt{a^2+b^2}} - \sin^{-1}\frac{b}{\sqrt{a^2+b^2}}\right)$$

$$= \frac{\pi}{ab} + \frac{2}{a^2+b^2} - \frac{2}{a^2+b^2} + \frac{2}{ab}\left[\alpha - \left(\frac{\pi}{2} - \alpha\right)\right] = \frac{4\alpha}{ab} = \frac{4}{ab}\tan^{-1}\frac{a}{b}$$

since $\alpha = \sin^{-1}\frac{a}{\sqrt{a^2+b^2}} = \tan^{-1}\frac{a}{b}$ **Ans.**

10. *Obtain the area bounded by the rectangular hyperbola $xy = c^2$, X-axis and the ordinates $x = c$, $x = 2c$.*

SOLUTION: $xy = c^2$ or $y = \frac{c^2}{x}$.

$\therefore$ Required area $= \int_c^{2c} y\,dx = \int_c^{2c} \frac{c^2}{x}\,dx = [c^2 \log x]_c^{2c}$

$= c^2 (\log 2c - \log c) = c^2 \log 2$ **Ans.**

11. *Find the area lying between the parabola $y = 4x - x^2$ and the line $y = x$.*

SOLUTION: $y = 4x - x^2$ or $(x-2)^2 = (y-4)$

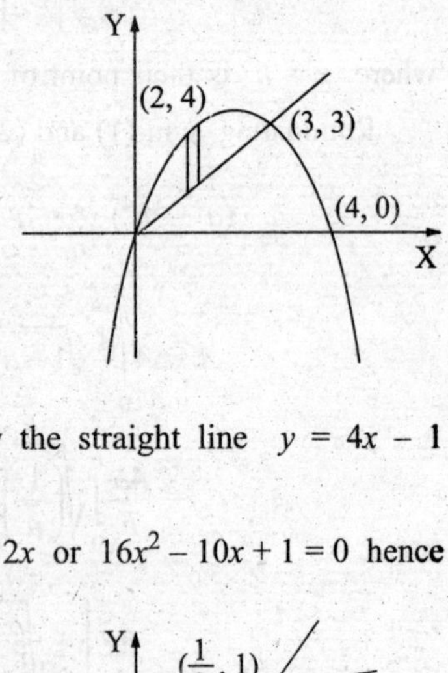

$\therefore$ Required area $= \int_0^3 (y_1 - y_2)\,dx = \int_0^3 (4x - x^2 - x)\,dx$

$= \left[\frac{3x^2}{2} - \frac{x^3}{3}\right]_0^3 = \frac{9}{2}$ sq. units. **Ans.**

12. Show that the area cut off the parabola $y^2 = 2x$ by the straight line $y = 4x - 1$ is 9/32 sq. units.

SOLUTION: $y^2 = 2x$, $y = 4x - 1$. For points of intersection $(4x-1)^2 = 2x$ or $16x^2 - 10x + 1 = 0$ hence $x = \frac{1}{8}, \frac{1}{2}$. Hence points of intersection are $\left(\frac{1}{8}, \frac{1}{2}\right)$ and $\left(\frac{1}{2}, 1\right)$.

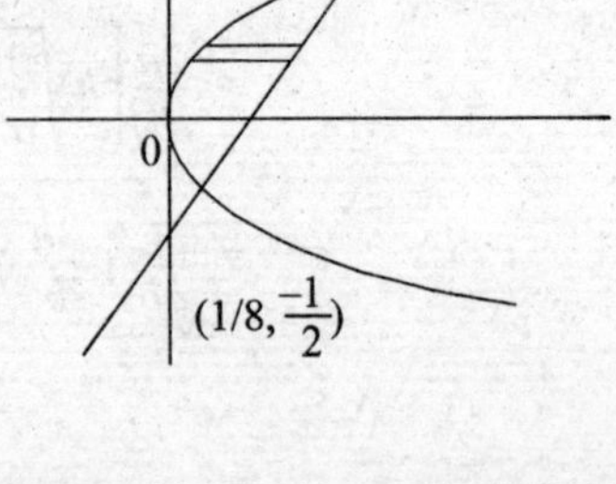

$\therefore$ Required area $= \int (x_1 - x_2)\,dy = \int_{-1/2}^{1} \left(\frac{y+1}{4} - \frac{y^2}{2}\right) dy$

$= \left[\frac{y^2}{8} + \frac{y}{4} - \frac{y^3}{6}\right]_{-1/2}^{1} = \frac{9}{32}$ **Ans.**

13. *Find the area under one arc of the curve* $y = \sin x$.

SOLUTION: One arc of $y = \sin x$ is in $(0, \pi)$

$$\text{Area} = \int_0^{\pi} y\,dx = \int_0^{\pi} \sin x\,dx = [-\cos x]_0^{\pi} \text{ sq. units}$$ **Ans.**

14. *Find the area bounded by* $y = x^3$ *and the lines* $x + 2 = 0$, $x = 1$ *and* $y = 0$.

SOLUTION: $y = x^3$, $x = -2$, $x = 1$, $y = 0$

$$\text{Required area} = \left|\int_{-2}^{0} x^3\,dx\right| + \int_0^1 x^3\,dx$$

$$= \left|\left[\frac{x^4}{4}\right]_{-2}^{0}\right| + \left[\frac{x^4}{4}\right]_0^1 = \frac{17}{4} \text{ sq. units.}$$ **Ans.**

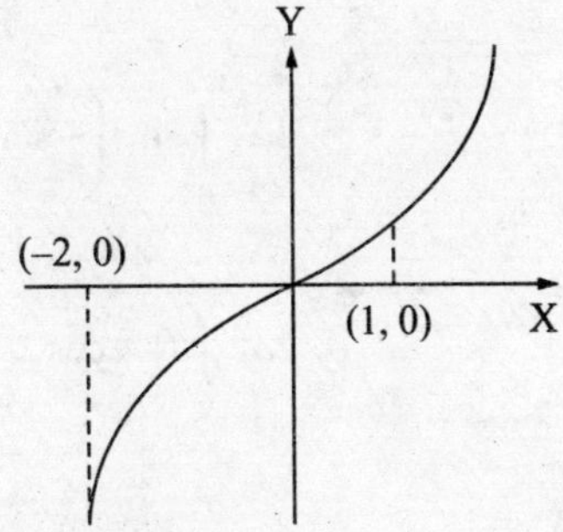

15. *Find the area of the astroid* $x^{2/3} + y^{2/3} = a^{2/3}$.

SOLUTION: $x^{2/3} + y^{2/3} = a^{2/3}$. Its parametric form is $x = a\cos^3\theta$, $y = a\sin^3\theta$

$$\text{Required area} = 4\int_0^a y\,dx = -4\int_0^{\pi/2} a\sin^3\theta \cdot 3a\cos^2\theta\sin\theta\,d\theta$$

$$= -12a^2\int_0^{\pi/2} \sin^4\theta\cos^2\theta\,d\theta = \frac{12a^2\,\Gamma\frac{5}{2}\,\Gamma\frac{3}{2}}{2\,\Gamma 4}$$

$$= \frac{3}{8}\pi a^2 \text{ sq. units.}$$ **Ans.**

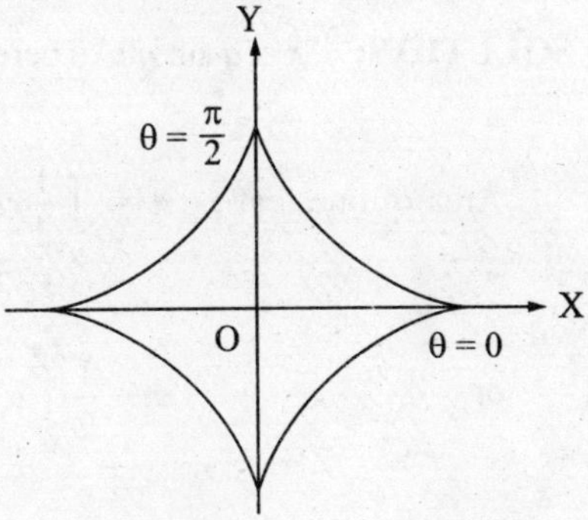

16. *Find the area included between the cycloid* $x = a(\theta - \sin\theta)$, $y = a(1 - \cos\theta)$ *and its base.*

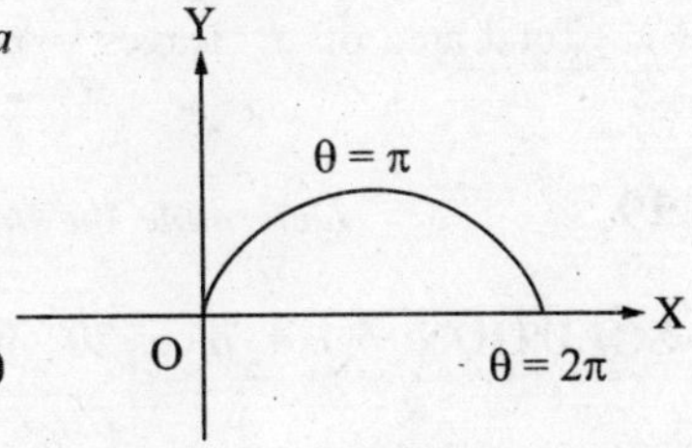

SOLUTION: $x = a(\theta - \sin\theta)$, $y = a(1 - \cos\theta)$

$$\text{Area of one cycloid is } A = \int_0^{2\pi a} y\,dx = \int_0^{2\pi} a(1-\cos\theta)\,a(1-\cos\theta)\,d\theta$$

$$= a^2\int_0^{\pi}\left(2\sin^2\frac{\theta}{2}\right)^2 d\theta = 4a^2\int_0^{\pi}\sin^4\frac{\theta}{2}\,d\theta \qquad \left(\text{Put } \frac{\theta}{2} = t\right)$$

$$= 8a^2\int_0^{\pi/2}\sin^4 t \cdot dt = 8a^2 \cdot \frac{\Gamma\frac{5}{2}\,\Gamma\frac{1}{2}}{2\,\Gamma 3} = 3/2\pi a^2$$ **Ans.**

17. *Trace the tractrix* $x = a\,(\cos t + \log|\tan t/2|)$, $y = a \sin t$ *and find the area enclosed.*

SOLUTION: $x = a\left(\cos t + \left|\log \tan\frac{t}{2}\right|\right)$, $y = a \sin t$.

It is symmetrical about both axes hence

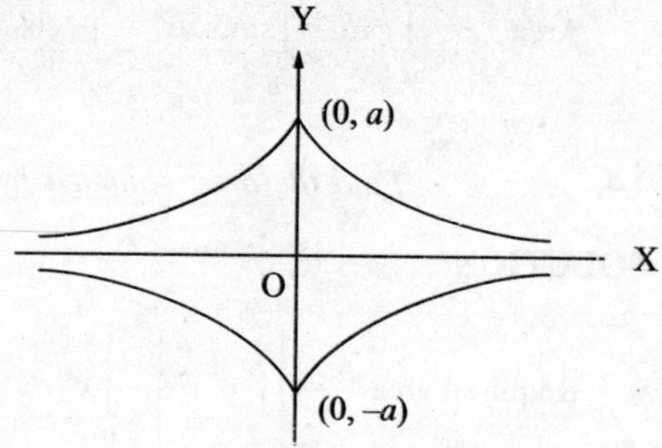

$$\text{Area} = 4\int_0^\infty y\,dx = 4\int_0^{\pi/2} a\sin t\, a\left(-\sin t + \frac{\sec^2\frac{t}{2}}{2\tan\frac{t}{2}}\right)dt$$

$$= 4a^2\int_0^{\pi/2} \sin t\left(-\sin t + \frac{1}{\sin t}\right)dt = 4a^2\int_0^{\pi/2}\cos^2 t\,dt$$

$$= 2a^2\int_0^{\pi/2}(1+\cos 2t)\,dt = 2a^2\left[t + \frac{1}{2}\sin 2t\right]_0^{\pi/2} = \pi a^2$$ **Ans.**

18. *Find the area of one loop of the curve* $r = a \sin n\theta$, $n \in N$. *What is the total area of all the loops ?*

SOLUTION: $r = a \sin n\theta$. There are n leaves from the origin, one leaf being from $\theta = 0$ to $\theta = \frac{\pi}{n}$.

Area of one leaf, $A = \int_0^{\pi/n} \frac{1}{2}r^2\,d\theta$

or $$A = \frac{1}{2}\int_0^{\pi/n} a^2\sin^2 n\theta\,d\theta = \frac{a^2}{4}\int_0^{\pi/n}(1-\cos 2n\theta)\,d\theta = \left[\frac{a^2}{4}\left[\frac{2n\theta - \sin 2n\theta}{2n}\right]\right]_0^{\pi/n} = \frac{\pi a^2}{4n}$$

Total area of n leaves $= n\frac{\pi a^2}{4n} = \frac{\pi}{4}a^2$. **Ans.**

19. *Determine the area of one loop of the curve* $r = a \cos 2\theta$.

SOLUTION: $r = a\cos 2\theta$. It has 4 leaves and on leaf is from $\theta = \frac{-\pi}{4}$ to $\theta = \frac{\pi}{4}$.

$\therefore$ $$\text{Area of one leaf} = \int_{-\pi/4}^{\pi/4}\frac{1}{2}r^2\,d\theta = \int_{-\pi/4}^{\pi/4}\frac{1}{2}a^2\cos^2 2\theta\,d\theta = a^2\int_0^{\pi/4}\cos^2 2\theta\,d\theta$$

$$= \frac{a^2}{2}\int_0^{\pi/4}(1+\cos 4\theta)\,d\theta = \frac{a^2}{2}\left[\theta + \frac{1}{4}\sin 4\theta\right]_0^{\pi/4} = \frac{\pi a^2}{8}$$ **Ans.**

20. *Show that the area bounded by the spiral $r = ae^{m\theta}$ and two radii vectors is proportional to the difference of the squares of these radii.*

SOLUTION: $r = a\,e^{m\theta}$. For $\theta = \theta_1$ and θ_2 we have $r = r_1$ and r_2 respectively.

$$\text{Area between two radii vectors} = \int_{\theta_1}^{\theta_2} \frac{1}{2} r^2\, d\theta = \frac{1}{2}\int_{\theta_1}^{\theta_2} a^2 e^{2m\theta}\, d\theta$$

$$= \frac{a^2}{2}\left[\frac{e^{2m\theta}}{2m}\right]_{\theta_1}^{\theta_2} = \frac{a^2}{4m}\left[e^{2m\theta_2} - e^{2m\theta_1}\right] = \frac{1}{4m}(r_2^2 - r_1^2)$$

Hence this area is proportional to the difference of the squares of the radii vectors. **Hence Proved.**

21. *Find the area included between two loops of the curve $r = a\left(2\cos\theta + \sqrt{3}\right)$.*

SOLUTION: $r = a\left(2\cos\theta + \sqrt{3}\right)$. Curve is symmetrical about the initial line. One loop is from $\theta = \frac{5\pi}{6}$ to $\theta = \frac{7\pi}{6}$ above the initial line the other is below the initial line.

$$\text{Area of two loops} = 2\int_{5\pi/6}^{7\pi/6} \frac{1}{2} r^2\, d\theta = \int_{5\pi/6}^{7\pi/6} a^2\left(2\cos\theta + \sqrt{3}\right)^2 d\theta = a^2 \int_{5\pi/6}^{7\pi/6} \left(4\cos^2\theta + 3 + 4\sqrt{3}\cos\theta\right) d\theta$$

$$= a^2 \int_{5\pi/6}^{7\pi/6} 2\left[(1+\cos 2\theta) + 3 + 4\sqrt{3}\cos\theta\right] d\theta = a^2\left[5\theta + \sin 2\theta + 4\sqrt{3}\sin\theta\right]_{5\pi/6}^{7\pi/6}$$

$$= a^2\left[5\left(\frac{7\pi}{6} - \frac{5\pi}{6}\right) + \sin\frac{7\pi}{3} - \sin\frac{5\pi}{3} + 4\sqrt{3}\left(\sin\frac{7\pi}{6} - \sin\frac{5\pi}{6}\right)\right]$$

$$= a^2\left[\frac{5\pi}{3} + \frac{2\sqrt{3}}{2} - 4\sqrt{3}\right] = \frac{a^2}{3}(5\pi - 9\sqrt{3})$$ **Ans.**

22. *Trace the curve $r = a\,(\sec\theta + \cos\theta)$ and find the area between the curve and its asymptote.*

SOLUTION: $r = a\,(\sec\theta + \cos\theta) = \dfrac{a(1+\cos^2\theta)}{\cos\theta}$

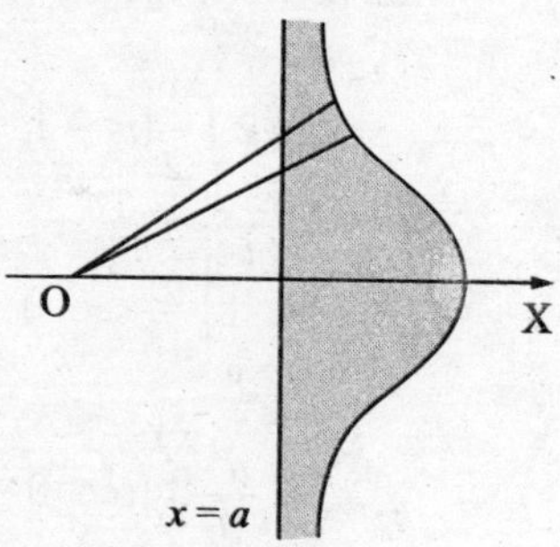

$$\therefore \quad \frac{1}{r} = \frac{\cos\theta}{a(1+\cos^2\theta)} = f(\theta)$$

and $$f'(\theta) = \frac{-\sin^3\theta}{a(1+\cos^2\theta)^2}$$

as $\theta \to \frac{\pi}{2}$, $r \to \infty$ hence $\alpha = \frac{\pi}{2}$

Asymptote is $r\sin(\theta-\alpha)=\dfrac{1}{f'(\alpha)}$ or $-r\cos\theta=\dfrac{1}{f'\left(\dfrac{\pi}{2}\right)}=-a$ or $x=a$

Required shaded area $=2\displaystyle\int_0^{\pi/2}\frac{1}{2}(r_2^2-r_1^2)\,d\theta$

$$=\int_0^{\pi/2}[a^2(\sec\theta+\cos\theta)^2-a^2\sec^2\theta]\,d\theta=a^2\int_0^{\pi/2}(\cos^2\theta+2)\,d\theta$$

$$=\frac{a^2}{2}\int_0^{\pi/2}(4+1+\cos 2\theta)=\frac{5\pi a^2}{4}$$ **Ans.**

23. *Trace the curve* $r^2=a^2\sin 2\theta$ *and find the area of one loop of the curve.*

SOLUTION: $r^2=a^2\sin 2\theta=a^2\cos 2\left(\dfrac{\pi}{4}-\theta\right)$.

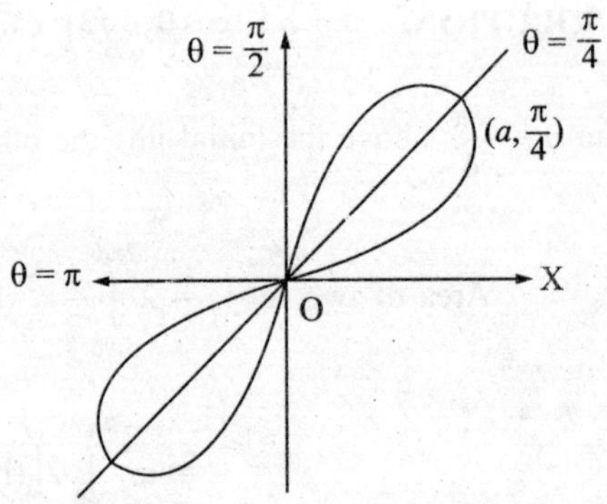

Area of one loop $=\displaystyle\int_0^{\pi/2}\frac{1}{2}r^2\,d\theta=2\int_0^{\pi/4}\frac{1}{2}r^2\,d\theta$

$$=\int_0^{\pi/4}a^2\sin 2\theta\,d\theta=-\left[\frac{a^2\cos 2\theta}{2}\right]_0^{\pi/4}$$

$$=\frac{a^2}{2}$$ **Ans.**

24. *Find the area lying between the cardioide* $r=a(1-\cos\theta)$ *and its double tangent.*

SOLUTION: $r=a(1-\cos\theta)$.

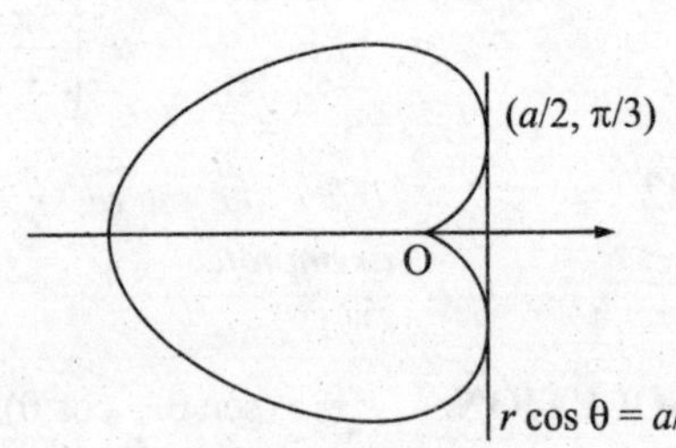

Its double tangent is $x=\dfrac{a}{4}$ or $r\cos\theta=\dfrac{a}{4}$.

Area between the curve and double tangent

$$=2\int_0^{\pi/3}\frac{1}{2}(r_2^2-r_1^2)\,d\theta$$

$$=\int_0^{\pi/3}\left[\frac{a^2}{16}\sec^2\theta-a^2(1-\cos\theta)^2\right]d\theta=\frac{a^2}{16}\int_0^{\pi/3}\left[\sec^2\theta-16(1-2\cos\theta+\cos^2\theta)\right]d\theta$$

$$=\frac{a^2}{16}\int_0^{\pi/3}[\sec^2\theta-16+32\cos\theta-8(1+\cos 2\theta)]\,d\theta=\frac{a^2}{16}[\tan\theta-24\theta+32\sin\theta-4\sin 2\theta]_0^{\pi/3}$$

$$=\frac{a^2}{16}\left[\sqrt{3}-24\frac{\pi}{3}+32\cdot\frac{\sqrt{3}}{2}-\frac{4\sqrt{3}}{2}\right]=\frac{a^2}{16}(15\sqrt{3}-8\pi)$$ **Ans.**

25. *Find the area of the loop of the curve $x^4 + y^4 = 2a^2 xy$.*

SOLUTION: $x^4 + y^4 = 2a^2 xy$. Converting into polar form $r^4 (\cos^4 \theta + \sin^4 \theta) = 2a^2 r^2 \sin\theta \cos\theta$

or $$r^2 = \frac{2a^2 \sin\theta\cos\theta}{1 - 2\sin^2\theta\cos^2\theta}$$

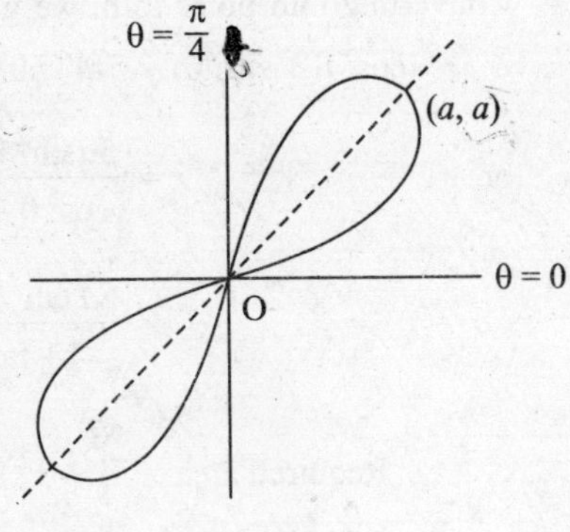

Required area of a loop $= \int_0^{\pi/2} \frac{1}{2} r^2 \, d\theta$

$$= a^2 \int_0^{\pi/2} \frac{\sin\theta\cos\theta \, d\theta}{1 - 2\sin^2\theta\cos^2\theta}$$

$$= a^2 \int_0^{\pi/2} \frac{\tan\theta\sec^2\theta \, d\theta}{\sec^4\theta - 2\tan^2\theta} = a^2 \int_0^{\pi/2} \frac{\tan\theta\sec^2\theta \, d\theta}{1 + 2\tan^2\theta + \tan^4\theta - 2\tan^2\theta}$$

$$= a^2 \int_0^{\pi/2} \frac{\tan\theta\sec^2\theta \, d\theta}{1 + \tan^4\theta}.$$ Putting $\tan^2\theta = t \;\therefore\; 2\tan\theta\sec^2\theta \, d\theta = dt$

$$= \frac{a^2}{2} \int_0^{\infty} \frac{dt}{1+t^2} = \frac{a^2}{2} [\tan^{-1} t]_0^{\infty} = \frac{\pi a^2}{4}$$ **Ans.**

26. *Find the area outside the circle $r = 2a\cos\theta$ and inside the cardioide $r = a(1 + \cos\theta)$.*

SOLUTION: $r = 2a\cos\theta$ and $r = a(1 + \cos\theta)$

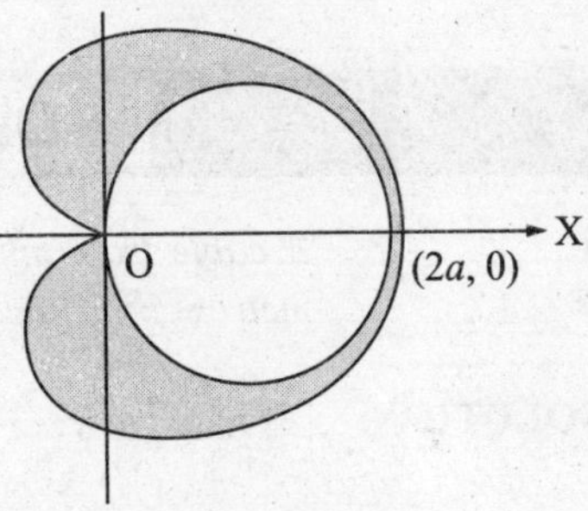

Required area $= 2\int_0^{\pi/2} \frac{1}{2} r^2 \, d\theta = 2\int_0^{\pi/2} \frac{1}{2}[a^2(1+\cos\theta)^2 - 4a^2\cos^2\theta] \, d\theta$

$$+ 2\int_{\pi/2}^{\pi} \frac{1}{2} a^2 (1+\cos\theta)^2 \, d\theta$$

$$= a^2 \int_0^{\pi/2} (1 + 2\cos\theta - 3\cos^2\theta) \, d\theta + a^2 \int_{\pi/2}^{\pi} (1 + 2\cos\theta + \cos^2\theta) \, d\theta$$

$$= \frac{a^2}{2} \int_0^{\pi/2} 2 + 4\cos\theta - 3(1 + \cos 2\theta) \, d\theta + \frac{a^2}{2} \int_{\pi/2}^{\pi} (2 + 4\cos\theta + 1 + \cos 2\theta) \, d\theta$$

$$= \frac{a^2}{2} \left[-\theta + 4\sin\theta - \frac{3}{2}\sin 2\theta \right]_0^{\pi/2} + \frac{a^2}{2} \left[3\theta + 4\sin\theta + \frac{1}{2}\sin 2\theta \right]_{\pi/2}^{\pi}$$

$$= \frac{a^2}{2} \left[-\frac{\pi}{2} + 4 - 0 \right] + \frac{a^2}{2} \left[3\pi - \frac{3\pi}{2} - 4 \right] = \frac{\pi a^2}{2}$$ **Ans.**

27. *Determine the area of the loop of the curve* $x^5 + y^5 = 5ax^2 y^2$.

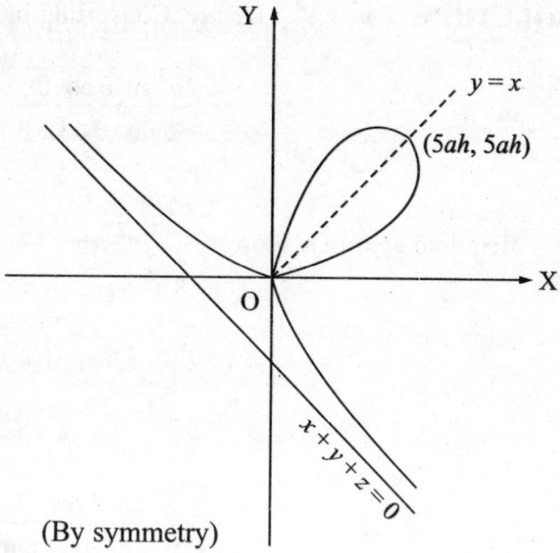

SOLUTION: $x^5 + y^5 = 5ax^2y^2$

Converting into polar form we get

$$r^5 (\cos^5\theta + \sin^5\theta) = 5ar^4 \sin^2\theta \cos^2\theta$$

or
$$r = \frac{5a\sin^2\theta\cos^2\theta}{\cos^5\theta + \sin^5\theta}$$

$$= \frac{5a\tan^2\theta\sec\theta}{1+\tan^5\theta}$$

$\therefore$ Required area $= \int_0^{\pi/2} \frac{1}{2} r^2 \, d\theta$

$$= 2\cdot\frac{1}{2}\int_0^{\pi/4} \frac{(5a)^2 \tan^4\theta\sec^2\theta\, d\theta}{(1+\tan^5\theta)^2} \quad \text{(By symmetry)}$$

$$= 25a^2 \int_1^2 \frac{dt}{5t^2} \quad (\text{On putting } 1+\tan^5\theta = t \ \therefore\ 5\tan^4\theta\sec^2\theta\, d\theta = dt)$$

$$= -5a^2\left[\frac{1}{t}\right]_1^2 = \frac{5a^2}{2}$$ **Ans.**

SOLUTIONS TO EXERCISE 7B

1. *A cable from an antenna tower has the equation* $3y = 4x^{3/2}$ *from* $x = 0$ to $x = 20$ *metres. Find the total length of the cable.*

SOLUTION: $3y = 4x^{3/2} \ \therefore\ 3\frac{dy}{dx} = 4\cdot\frac{3}{2}x^{1/2}$ or $\frac{dy}{dx} = 2x^{1/2}$

Length of the curve is $\int_0^{20} \sqrt{1+\left(\frac{dy}{dx}\right)^2}\, dx = \int_0^{20} \sqrt{1+4x}\, dx = \frac{2}{3\cdot 4}(1+4x)^{3/2}\Big|_0^{20}$

$$= \frac{1}{6}(81^{3/2}-1) = \frac{1}{6}(81\cdot 9 - 1) = 121.33 \text{ metres.}$$ **Ans.**

2. *Find the length of the arc of the parabola* $y^2 = 4ax$ *cut off by the line* $3y = 8x$.

SOLUTION: $y^2 = 4ax$, $3y = 8x$. Points of intersection are given by

$$\left(\frac{8}{3}x\right)^2 = 4ax \quad \therefore\ x = 0 \text{ and } x = \frac{9a}{16} = x_1, \text{ say.}$$

$$L = \text{Required length} = \int_0^{x_1} \sqrt{1+\left(\frac{dy}{dx}\right)^2}\, dx$$

From the equation of the curve, we have $2y\,\dfrac{dy}{dx} = 4a \;\therefore\; \left(\dfrac{dy}{dx}\right)^2 = \dfrac{4a^2}{y^2} = \dfrac{a}{x}$

$\therefore$ $$L = \int_0^{x_1} \sqrt{1+\frac{a}{x}}\, dx = \int_0^{x_1} \frac{\sqrt{x+a}}{\sqrt{x}}\, dx$$ Putting $x = t^2$ so $dx = 2t\,dt$

$$= 2\int_0^{\sqrt{x_1}} \sqrt{t^2+a}\, dt = 2\left[\frac{t}{2}\sqrt{t^2+a} + \frac{a}{2}\log(t+\sqrt{t^2+a})\right]_0^{\frac{3\sqrt{a}}{4}}$$

$$= \frac{3\sqrt{a}}{4}\sqrt{a+\frac{9a}{16}} + a\left[\log\left(\frac{3\sqrt{a}}{4}+\sqrt{a+\frac{9a}{16}}\right) - \log\sqrt{a}\right]$$

$$= \frac{15a}{16} + a\log\left(\frac{3}{4}+\frac{5}{4}\right) a\left(\log 2 + \frac{15}{16}\right)$$ **Ans.**

3. *Find the length of the arc of the curve* $4ax = y^2 - 2a^2 \log\left(\dfrac{y}{a}\right) - a^2$ *from the point* $(0, a)$ *to the point* (x, y).

SOLUTION: $4ax = y^2 - 2a^2 \log\left(\dfrac{y}{a}\right) - a^2$. Length of curve from $(0, a)$ to (x, y) is to be obtained.

$\therefore$ $$4a\frac{dx}{dy} = 2y - \frac{2a^2}{\frac{y}{a}}\cdot\frac{1}{a} \quad \text{or} \quad \frac{dx}{dy} = \frac{y^2-a^2}{2ay}$$

$\therefore$ $$\text{Required length} = \int_a^y \sqrt{1+\left(\frac{dx}{dy}\right)^2}\, dy = \int_a^y \sqrt{1+\frac{(y^2-a^2)^2}{4a^2y^2}}\, dy$$

$$= \int_a^y \frac{y^2+a^2}{2ay}\, dy = \int_a^y \left(\frac{y}{2a}+\frac{a}{2y}\right) dy = \left[\frac{y^2}{4a}+\frac{a}{2}\log y\right]_a^y = \frac{y^2-a^2}{4a} + \frac{a}{2}\log\frac{y}{a}$$

$$= \frac{y^2-a^2}{4a} + \frac{y^2-a^2}{4a} - x \quad \text{(using the equation of curve)}$$

$$= \frac{y^2-a^2}{2a} - x$$ **Ans.**

4. *If* S *is the length of the arc of the catenary* $y = c\cosh\left(\dfrac{x}{c}\right)$ *from the vertex to any point* (x, y) *on it, then prove that* $y^2 = c^2 + s^2$.

SOLUTION: Length S of catenary from vertex $(0, c)$ to its point (x, y), is given by

$$S = \int_0^x \sqrt{1+\left(\frac{dy}{dx}\right)^2}\, dx = \int_0^x \sqrt{1+\sinh^2\frac{x}{c}}\, dx = \int_0^x \cosh\frac{x}{c}\, dx = c\sinh\frac{x}{c}$$

$\therefore$ $$S^2 = c^2\sinh^2\frac{x}{c} = c^2\left(\cosh^2\frac{x}{c}-1\right) = c^2\left(\frac{y^2}{c^2}-1\right)$$

or $$y^2 = c^2 + S^2.$$ **Hence Proved.**

5. *The parabolic chain of a suspension bridge has the form* $x^2 = \dfrac{b^2 y}{h}$ *where* $2b$ *is the span and* h *is the sag. Show that the length of the chain when* h/b *is small, is* $2b + (4h^2/3b)$ *approximately.*

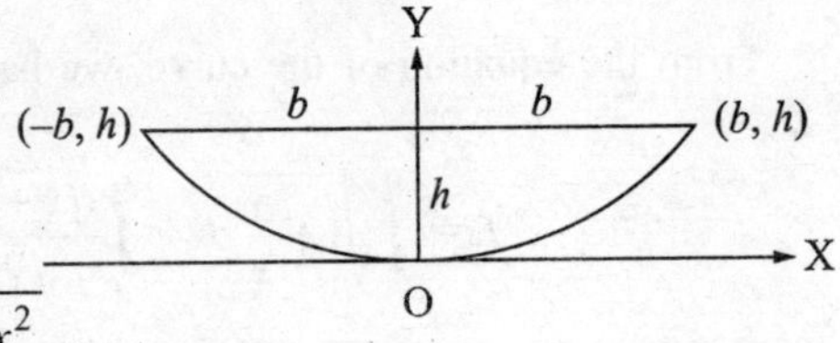

SOLUTION: $x^2 = \dfrac{b^2 y}{h} \quad \therefore \quad \dfrac{dy}{dx} = \dfrac{2hx}{b^2}$

$$\text{Length of chain} = \int_{-b}^{b} \sqrt{1+\left(\frac{dy}{dx}\right)^2}\, dx = \int_{-b}^{b} \sqrt{1+\frac{4h^2x^2}{b^4}}\, dx$$

$$= 2\int_{0}^{b} \sqrt{1+\frac{4h^2x^2}{b^4}}\, dx \qquad \left(\text{expanding the integrand in powers of } \frac{h}{b}\right)$$

$$= 2\int_{0}^{b} 1+\frac{1}{2}\cdot\frac{4h^2x^2}{b^4}+\frac{\frac{1}{2}\left(\frac{-1}{2}\right)}{2!}\left(\frac{4h^2x^2}{b^4}\right)^2+\ldots\, dx$$

$$= 2\left[x+\frac{2h^2x^3}{3b^4}-\frac{1}{8}\frac{16h^4x^5}{b^8\cdot 5}+\ldots\right]_0^b = 2b+\frac{4h^2b^3}{3b^4}+\ldots = 2b+\frac{4h^2}{3b} \quad \text{Approx.}$$

Ans.

6. *Find the perimeter of the loop of the curve* $9ay^2 = (x-2a)(x-5a)^2$.

SOLUTION: $9ay^2 = (x-2a)(x-5a)^2 \quad \therefore \quad 18\,ay\dfrac{dy}{dx} = 2(x-2a)(x-5a)+(x-5a)^2$

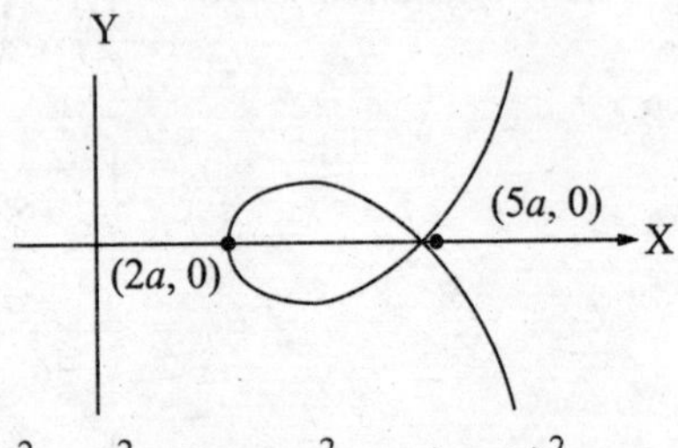

$$1+\left(\frac{dy}{dx}\right)^2 = 1+\frac{(x-3a)^2(x-5a)^2}{36a^2y^2}$$

$$= 1+\frac{(x-3a)^2(x-5a)^2}{4a(x-2a)(x-5a)^2}$$

$$= 1+\frac{1}{4a}\cdot\frac{(x-3a)^2}{(x-2a)} = \frac{4a(x-2a)+(x-3a)^2}{4a(x-2a)} = \frac{x^2-2ax+a^2}{4a(x-2a)} = \frac{(x-a)^2}{4a(x-2a)}$$

$$\therefore \quad \text{Perimeter} = \int_{2a}^{5a} \sqrt{1+\left(\frac{dy}{dx}\right)^2}\, dx = \int_{2a}^{5a} \frac{x-a}{\sqrt{4a(x-2a)}}\, dx = \frac{1}{2\sqrt{a}}\int_{2a}^{5a}\frac{x-2a+a}{\sqrt{x-2a}}\, dx$$

$$= \frac{1}{2\sqrt{a}}\int_{2a}^{5a}\left(\sqrt{x-2a}+\frac{a}{\sqrt{x-2a}}\right)dx = \frac{1}{2\sqrt{a}}\left[\frac{2}{3}(x-2a)^{3/2}+2a\sqrt{x-2a}\right]_{2a}^{5a}$$

$$= \frac{1}{2\sqrt{a}}\left[\frac{2}{3}\cdot 3a\sqrt{3a}+2a\sqrt{3a}\right] = 2a\sqrt{3}$$

By symmetry the total permeter $= 2 \times 2a\sqrt{3} = 4a\sqrt{3}$. **Ans.**

7. *Find the length of the arc of the curve* $x^2 = a^2\left(1-e^{\frac{y}{a}}\right)$ *measured from the origin to any point* (x, y) *on it.*

SOLUTION: $x^2 = a^2(1 - e^{y/a})$. Perimeter from $(0, 0)$ to (x, y) is to be obtained

or $\quad e^{y/a} = 1-\dfrac{x^2}{a^2} \quad$ or $\quad y = a\log\left(1-\dfrac{x^2}{a^2}\right) \therefore \dfrac{dy}{dx} = -\dfrac{2ax}{a^2-x^2}$

$$\therefore \text{ Perimeter} = \int_0^x \sqrt{1+\left(\frac{dy}{dx}\right)^2}\, dx = \int_0^x \sqrt{1+\frac{4a^2x^2}{(a^2-x^2)^2}}\, dx = \int_0^x \frac{a^2+x^2}{a^2-x^2}\, dx$$

$$= \int_0^x \left(-1+\frac{2a^2}{a^2-x^2}\right) dx = -x+\frac{2a^2}{2a}\log\frac{a+x}{a-x}$$

$$= a\log\frac{a+x}{a-x} - x.$$ **Ans.**

8. *Find the length of the curve* $y^2 = (2x-1)^3$ *cut off by the line* $x = 4$.

SOLUTION: $y^2 = (2x-1)^3$ and $x = 4$

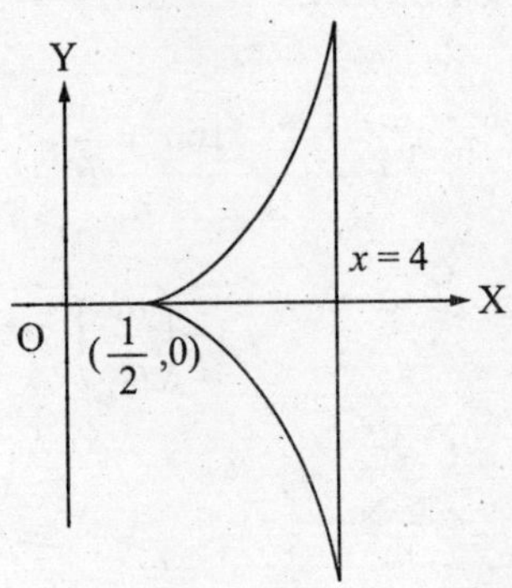

$$\therefore \quad 2y\frac{dy}{dx} = 6(2x-1)^2 \text{ or } \left(\frac{dy}{dx}\right)^2 = \frac{9(2x-1)^4}{(2x-1)^3} = 9(2x-1)$$

$$\therefore \text{ length of the curve} = \int \sqrt{1+\left(\frac{dy}{dx}\right)^2}\, dx = 2\int_{1/2}^{4} \sqrt{1+9(2x-1)}\, dx$$

$$= \frac{4}{18\cdot 3}[(18x-8)]_{1/2}^{4} = \frac{2}{27}[64^{3/2}-1]$$

$$= \frac{1022}{27}.$$ **Ans.**

9. *Obtain the length of the arc of the curve* $x = e^\theta \sin\theta$, $y = e^\theta\cos\theta$, *from* $\theta = 0$ *to* $\theta = \pi/2$.

SOLUTION: $x = e^\theta\sin\theta, \quad y = e^\theta\cos\theta, \quad 0 \le \theta \le \dfrac{\pi}{2}$.

$$\therefore \quad x' = e^\theta(\sin\theta+\cos\theta),\ y' = e^\theta(\cos\theta-\sin\theta)$$

Length of the curve

$$= \int_0^{\pi/2} \sqrt{\left(\frac{dx}{d\theta}\right)^2+\left(\frac{dy}{d\theta}\right)^2}\, d\theta = \int_0^{\pi/2} \sqrt{e^{2\theta}(\sin\theta+\cos\theta)^2+e^{2\theta}(\cos\theta-\sin\theta)^2}\, d\theta$$

$$= \int_0^{\pi/2} e^\theta\sqrt{2}\, d\theta = \sqrt{2}\,(e^{\pi/2}-1)$$ **Ans.**

10. *Find the length of one cycloid of the curve* $x = a(\theta - \sin\theta)$, $y = a(1 - \cos\theta)$.

SOLUTION: $x = a(\theta - \sin\theta)$, $y = a(1 - \cos\theta)$, $0 \le \theta \le 2\pi$.

$$\frac{dx}{d\theta} = a(1 - \cos\theta), \quad \frac{dy}{d\theta} = a\sin\theta$$

$$\text{Required length of one cycloid} = \int_0^{2\pi}\left[\left(\frac{dx}{d\theta}\right)^2 + \left(\frac{dy}{d\theta}\right)^2\right]^{1/2} d\theta$$

$$= \int_0^{2\pi}\sqrt{a^2(1-\cos\theta)^2 + a^2\sin^2\theta}\, d\theta = a\int_0^{2\pi}\sqrt{2 - 2\cos\theta}\, d\theta$$

$$= 4a\int_0^{\pi}\sin\frac{\theta}{2}\, d\theta = 4a\left[2\cos\frac{\theta}{2}\right]_0^{\pi} = 8a$$

Ans.

11. *Determine the perimeter of both the loops of the curve* $8a^2y^2 = x^2(a^2 - x^2)$.

SOLUTION: $8a^2y^2 = x^2(a^2 - x^2)$

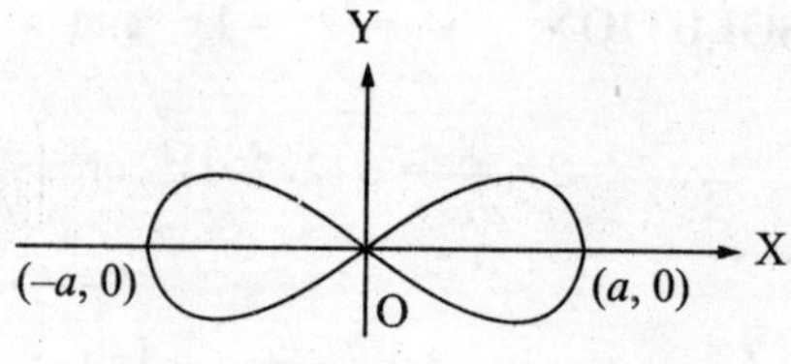

$$\therefore \quad 16a^2y\frac{dy}{dx} = 2a^2x - 4x^3 \quad \text{or} \quad \frac{dy}{dx} = \frac{a^2x - 2x^3}{8a^2y}$$

$$\therefore \quad 1 + \left(\frac{dy}{dx}\right)^2 = 1 + \frac{x^2(a^2 - 2x^2)^2}{8a^2x^2(a^2 - x^2)}$$

$$= \frac{8a^4 - 8a^2x^2 + a^4 - 4a^2x^2 + 4x^4}{8a^2(a^2 - x^2)} = \frac{(3a^2 - 2x^2)^2}{8a^2(a^2 - x^2)}$$

$$\therefore \quad \text{Required length of two loops} = 4\int_0^a\sqrt{1 + \left(\frac{dy}{dx}\right)^2}\, dx$$

$$= 4\int_0^a \frac{3a^2 - 2x^2}{2\sqrt{2}a\sqrt{a^2 - x^2}}\, dx \qquad \text{(Putting } x = a\sin\theta, \text{ we get)}$$

$$= \frac{\sqrt{2}}{a}\int_0^{\pi/2}\frac{3a^2 - 2a^2\sin^2\theta}{a\cos\theta}\cdot a\cos\theta\, d\theta = a\sqrt{2}\int_0^{\pi/2}[3 - (1 - \cos 2\theta)]\, d\theta$$

$$= a\sqrt{2}\left[2\theta + \frac{1}{2}\sin 2\theta\right]_0^{\pi/2} = \pi a\sqrt{2}$$

Ans.

12. *Show that the intrinsic equation of the curve* $x = a\,(3\cos\theta + \cos 3\theta)$, $y = a\,(3\sin\theta - \sin 3\theta)$, *length being measured from the point* $\theta = \pi/2$ *on it, is* $S = 6a\cos^2\psi$.

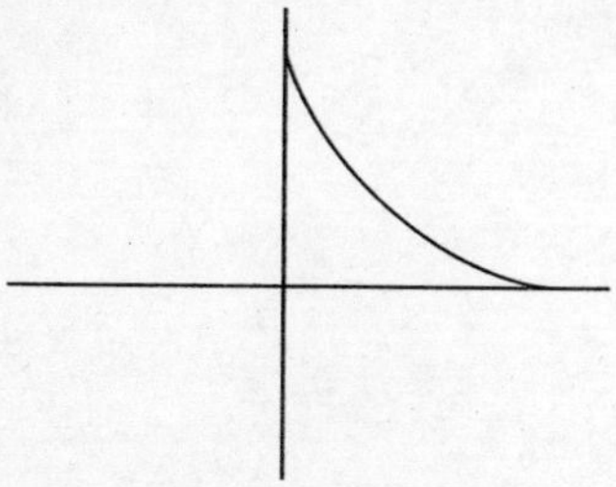

SOLUTION: $x = a\,(3\cos\theta + \cos 3\theta),\ y = a\,(3\sin\theta - \sin 3\theta)$

$$\frac{dx}{d\theta} = a\,(-3\sin\theta - 3\sin 3\theta)$$

and
$$\frac{dy}{d\theta} = a\,(3\cos\theta - 3\cos 3\theta)$$

$\therefore$
$$\tan\psi = \frac{dy}{dx} = \frac{\cos\theta - \cos 3\theta}{-\sin\theta - \sin 3\theta} = \frac{2\sin 2\theta\sin\theta}{-2\sin 2\theta\cos\theta} = \tan(\pi - \theta) \Rightarrow \psi = \pi - \theta.$$

Required length
$$S = \int_{\pi/2}^{\theta}\sqrt{\left(\frac{dx}{d\theta}\right)^2 + \left(\frac{dy}{d\theta}\right)^2}\,d\theta = a\int_{\pi/2}^{\theta}\sqrt{9(\sin\theta + \sin 3\theta)^2 + 9(\cos\theta - \cos 3\theta)^2}\,d\theta$$

$$= 3a\int_{\pi/2}^{\theta}\sqrt{1 + 1 + 2(\sin\theta\sin 3\theta - \cos\theta\cos 3\theta)}\,d\theta = 3a\sqrt{2}\int_{\pi/2}^{\theta}\sqrt{1 - \cos 4\theta}\,d\theta$$

$$= 3a\sqrt{2}\int_{\pi/2}^{\theta}\sqrt{2\sin^2 2\theta}\,d\theta = 6a\int_{\pi/2}^{\theta}\sin 2\theta\,d\theta = 6a\left[\left(\frac{-1}{2}\right)\cos 2\theta\right]_{\pi/2}^{\theta}$$

$$= -3a\,[\cos 2\theta + 1] = -6a\cos^2\theta$$

$$= 6a\cos^2(\pi - \psi) = 6a\cos^2\psi.$$ **Ans.**

13. *Show that the length of the curve* $y = \log\sec x$ *from* $x = 0$ to $x = \frac{\pi}{3}$, is $\log\left(2 + \sqrt{3}\right)$.

SOLUTION: $y = \log\sec x$ Its length from $x = 0$ to $x = \frac{\pi}{3}$ $\frac{dy}{dx} = \tan x$

$\therefore$ Length of curve
$$= \int_0^{\pi/3}\sqrt{1 + \left(\frac{dy}{dx}\right)^2}\,dx = \int_0^{\pi/3}\sqrt{1 + \tan^2 x}\,dx = \int_0^{\pi/3}\sec x\,dx$$

$$= [\log(\sec x + \tan x)]_0^{\pi/3} = \log\left(\sec\frac{\pi}{3} + \tan\frac{\pi}{3}\right) = \log(2 + \sqrt{3})$$ **Ans.**

14. *Find the length of the arc of the parabola* $y^2 = 4a\,(a - x)$ *cut off by the y-axis.*

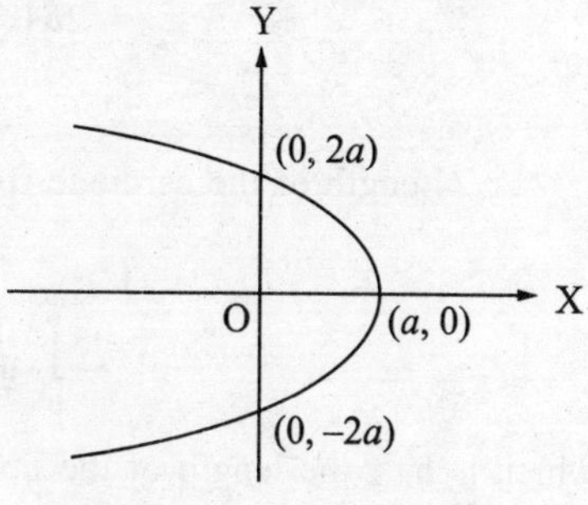

SOLUTION: $y^2 = 4a\,(a - x)\quad \therefore\quad 2y\frac{dy}{dx} = -4a$

$\therefore$
$$\left(\frac{dy}{dx}\right)^2 = \frac{4a^2}{4a(a - x)} = \frac{a}{a - x}$$

$\therefore$ Required length
$$= 2\int_0^a\sqrt{1 + \left(\frac{dy}{dx}\right)^2}\,dx = 2\int_0^a\sqrt{1 + \frac{a}{a - x}}\,dx$$

$$= 2\int_0^a \sqrt{\frac{2a-x}{a-x}}\,dx \qquad \text{Put } a - x = t^2$$

$$= 2\int_a^0 \frac{\sqrt{a+t^2}}{t}(-2t\,dt) = 4\int_0^{\sqrt{a}} \sqrt{t^2+a}\,dt$$

$$= 4\left[\frac{t}{2}\sqrt{t^2+a}+\frac{a}{2}\log(t+\sqrt{t^2+a})\right]_0^{\sqrt{a}}$$

$$= 2\left[\sqrt{a}\sqrt{2a}+\frac{a}{2}\log\frac{\sqrt{a}+\sqrt{2a}}{\sqrt{a}}\right] = 2a\left[\sqrt{2}+\log(1+\sqrt{2})\right] \qquad \textbf{Ans.}$$

15. *Find the length of the arc of the curve* $x = a(\cos\theta + \theta\sin\theta)$, $y = a(\sin\theta - \theta\cos\theta)$ *from* $\theta = 0$ *to* $\theta = 2\pi$.

SOLUTION: $x = a(\cos\theta + \theta\sin\theta)$, $y = a(\sin\theta - \theta\cos\theta)$. Its length from $\theta = 0$ to $\theta = 2\pi$

$$\frac{dx}{d\theta} = a(-\sin\theta + \sin\theta + \theta\cos\theta) = a\theta\cos\theta, \quad \frac{dy}{d\theta} = a(\cos\theta + \theta\sin\theta - \cos\theta) = a\theta\sin\theta.$$

$$\text{Length of the curve} = \int_0^{2\pi}\sqrt{\left(\frac{dx}{d\theta}\right)^2+\left(\frac{dy}{d\theta}\right)^2}\,d\theta = \int_0^{2\pi}\sqrt{a^2\theta^2\cos^2\theta + a^2\theta^2\sin^2\theta}\,d\theta$$

$$= \int_0^{2\pi} a\theta\,d\theta = 2\pi^2 a \qquad \textbf{Ans.}$$

16. *Find the full length of the cardioid* $r = a(1-\cos\theta)$ *and prove that the upper half of the curve is bisected by the line* $\theta = 2\pi/3$. *Also find the intrinsic equation of the curve.*

SOLUTION: $r = a(1-\cos\theta)$, $\dfrac{dr}{d\theta} = a\sin\theta$

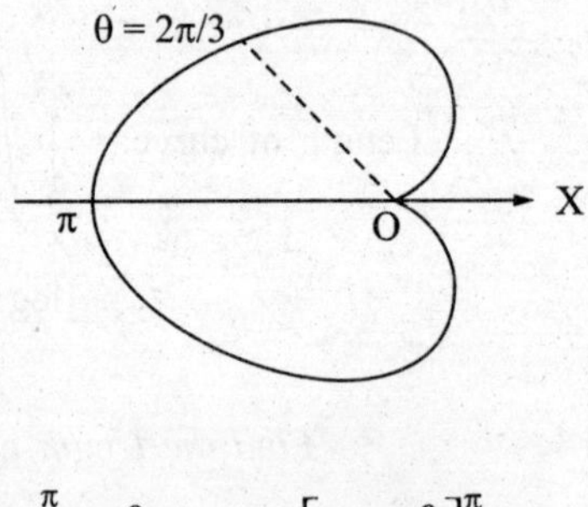

$$\text{Length of cardiode} = 2\int_0^{\pi}\sqrt{r^2+\left(\frac{dr}{d\theta}\right)^2}\,d\theta$$

$$= 2\int_0^{\pi}\sqrt{a^2(1-\cos\theta)^2+a^2\sin^2\theta}\,d\theta$$

$$= 2a\int_0^{\pi}\sqrt{2-2\cos\theta}\,d\theta = 2a\int_0^{\pi}\sqrt{4\sin^2\frac{\theta}{2}}\,d\theta = 4a\int_0^{\pi}\sin\frac{\theta}{2}\,d\theta = 8a\left[-\cos\frac{\theta}{2}\right]_0^{\pi} = 8a$$

$\therefore$ Length of the cardiode from $\theta = 0$ to $\theta = \dfrac{2\pi}{3}$

$$= \int_0^{2\pi/3}\sqrt{r^2+\left(\frac{dr}{d\theta}\right)^2}\,d\theta = 2a\int_0^{2\pi/3}\sin\frac{\theta}{2}\,d\theta = 4a\left[-\cos\frac{\theta}{2}\right]_0^{2\pi/3} = -4a\left(\frac{1}{2}-1\right) = 2a$$

which is half the length of the upperhalf of the curve. **Hence Proved.**

To get the intrinsic equation of the curve, we have

$$S = \int_0^\theta \sqrt{r^2 + \left(\frac{dr}{d\theta}\right)^2}\, d\theta = a\int_0^\theta 2\sin\frac{\theta}{2}\, d\theta = -4a\left[\cos\frac{\theta}{2}\right]_0^\theta = 4a\left(1-\cos\frac{\theta}{2}\right)$$

And $\qquad \tan\phi = r\dfrac{d\theta}{dr} = \dfrac{a(1-\cos\theta)}{a\sin\theta} = \tan\dfrac{\theta}{2} \qquad \therefore \quad \phi = \dfrac{\theta}{2}$

since $\qquad \psi = \theta + \phi = \theta + \dfrac{\theta}{2} = 3\dfrac{\theta}{2} \qquad \therefore \quad \dfrac{\theta}{2} = \dfrac{\psi}{3}$

$\therefore \qquad S = 4a\left(1-\cos\dfrac{\psi}{3}\right)$ **Ans.**

17. *Show that the length of one loop of the leminiscate* $r^2 = a^2\cos 2\theta$ *is* $\dfrac{a\sqrt{\pi}}{2}\dfrac{\Gamma(1/4)}{\Gamma(3/4)}$.

SOLUTION: $\qquad r^2 = a^2\cos 2\theta \quad \therefore \quad 2r\dfrac{dr}{d\theta} = -2a^2\sin 2\theta$

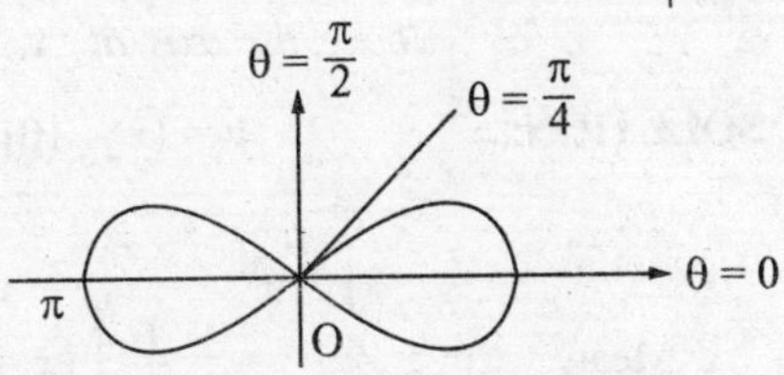

$$\therefore \quad r^2 + \left(\frac{dr}{d\theta}\right)^2 = a^2\cos 2\theta + \frac{a^4\sin^2 2\theta}{a^2\cos 2\theta} = \frac{a^2}{\cos 2\theta}$$

$$\therefore \text{Length of one loop} = 2\int_0^{\pi/4}\sqrt{r^2+\left(\frac{dr}{d\theta}\right)^2}\, d\theta = 2\int_0^{\pi/4}\frac{a}{\sqrt{\cos 2\theta}}\, d\theta$$

Put $2\theta = \phi$

$$= \frac{2a}{2}\int_0^{\pi/2}\frac{d\phi}{\sqrt{\cos\phi}} = a\frac{\Gamma\left(\frac{-\frac{1}{2}+1}{2}\right)\Gamma\left(\frac{1}{2}\right)}{2\Gamma\left(\frac{-\frac{1}{2}+2}{2}\right)} = \frac{a\sqrt{\pi}\,\Gamma\left(\frac{1}{4}\right)}{2\Gamma\left(\frac{3}{4}\right)}$$

Hence Proved.

SOLUTIONS TO EXERCISE 7C

1. *Find the RMS value of the function* $a\sin pt + b\cos qt$ *over the interval* $[0, 2\pi]$ *where* $p, q \in I$.

SOLUTION: $f(t) = a\sin pt + b\cos qt$ over $[0, 2\pi]$ where p and q are integers.

$$\text{R.M.S. Value of } f(t) = \sqrt{\frac{1}{2\pi}\int_0^{2\pi} f(t)^2\, dt} = \sqrt{\frac{1}{2\pi}\int_0^{2\pi}[a^2\sin^2 pt + b^2\cos^2 qt + 2ab\sin pt\cos qt]\, dt}$$

$$= \sqrt{\frac{1}{2\pi}\int_0^{2\pi}\frac{a^2}{2}(1-\cos 2pt)+\frac{b^2}{2}(1+\cos 2qt)+ab\sin(p+q)t+ab\sin(p-q)t\,dt}$$

$$= \sqrt{\frac{1}{4\pi}[(a^2+b^2)t-\frac{a^2}{2p}\sin 2pt+\frac{b^2}{2q}\sin 2qt-\frac{ab}{p+q}\cos(p+q)t-\frac{ab}{p-q}\cos(p+q)t]_0^{2\pi}}$$

$$= \sqrt{\frac{1}{4\pi}\left[(a^2+b^2)2\pi-\frac{a^2}{2p}\sin 4p\pi+\frac{b^2}{2q}\sin 4q\pi-\frac{ab}{(p+q)}[\cos(p+q)2\pi-1]-\frac{ab}{(p-q)}[\cos(p-q)2\pi-1]\right]}$$

$$= \sqrt{\frac{a^2+b^2}{2}}$$ **Ans.**

2. *Find the mean height of the portion of the parabola $y = (x-10)(20-x)$ which lies above the axis of x.*

SOLUTION: $y = (x-10)(20-x)$. Here height is y

$$\text{Mean} \quad y = \frac{\int_{10}^{20} y\,dx}{20-10} = \frac{1}{10}\int_{10}^{20}(x-10)(20-x)\,dx = \frac{1}{10}\int_{10}^{20}(-x^2+30x-200)\,dx$$

$$= \frac{1}{10}\left[\frac{-x^3}{3}+\frac{30}{2}x^2-200x\right]_{10}^{20} = \frac{-1}{3}(400+200+100)+15(40-10)-200 = \frac{50}{3}$$ **Ans.**

3. *If the density ρ at a distance 'r' from the centre of the earth is given by $\rho = \rho_0\left(1-\frac{kr^2}{R^2}\right)$ where k and ρ_0 are constants and R is the radius of the earth, show that the mean density is $\rho_0\left(1-\frac{k}{3}\right)$.*

SOLUTION: $\rho = \rho_0\left(1-\frac{kr^2}{R^2}\right)$. Mean $\rho = \frac{\int_0^R \rho\,dr}{R-0} = \frac{\rho_0}{R}\int_0^R\left(1-\frac{kr^2}{R^2}\right)dr$

$$= \frac{\rho_0}{R}\left[r-\frac{kr^3}{3R^2}\right]_0^R = \frac{\rho_0}{R}\left[R-\frac{k}{3}R\right] = \rho_0\left(1-\frac{k}{3}\right)$$ **Ans.**

4. *Show that the RMS value of the function $f(t) = a_1\sin p_1t + a_2\sin p_2t + ... + a_n\sin p_nt$ is equal to $\sqrt{\frac{1}{2}\left(a_1^2+a_2^2+...+a_n^2\right)}$ where $p_1, p_2, ..., p_n$ are integers.*

SOLUTION: $f(t) = a_1\sin p_1 t + a_2\sin p_2 t + ... + a_n\sin p_n t = \sum_{r=1}^{n} a_r \sin p_r t$

$$\text{R.M.S. Value of } f(t) = \left[\sum_{r=1}^{n} \frac{a_r^2}{2\pi}\int_0^{2\pi} \sin^2 p_r t\, dt + \frac{2}{2\pi}\sum_{i,\, j=1}^{n} a_i\, a_j \int_0^{2\pi} \sin p_i t \sin p_j t\, dt\right]^{1/2}$$

$$= \left[\sum_{r=1}^{n} \frac{a_r^2}{4\pi}\int_0^{2\pi} (1-\cos 2p_r t)\, dt + \frac{1}{2\pi}\sum_{v,\, j=1}^{n} a_i\, a_j \int_0^{2\pi} (\cos(p_i - p_j)t - \cos(p_i + p_j)\, t\, dt\right]^{1/2}$$

$$= \left[\sum_{r=1}^{n} \frac{a_r^2}{4\pi}\left\{t - \frac{1}{2p_r}\sin 2p_r t\right\}_0^{2\pi} + \frac{1}{2\pi}\sum_{i,\, j=1}^{n} a_i\, a_j \{\sin(p_i - p_j)t - \sin(p_i + p_j)\, t\}_0^{2\pi}\right]^{1/2}$$

$$= \left[\sum_{r=1}^{n} \frac{a_r{}^2}{2} + 0\right]^{1/2} = \sqrt{\frac{1}{2}(a_1^2 + a_2^2 + \ldots + a_n^2)}$$

Hence Proved.

5. *Find the RMS value of* $\log x$ *over the range* $x = 1$ *to* $x = e$.

SOLUTION: $f(x) = \log x$

$$\text{Required RMS value} = \left[\frac{\int_1^e (\log x)^2\, dx}{e-1}\right]^{1/2}$$

$$\int_1^e (\log x)^2\, dx = \int_1^e (\log x)^2 \cdot 1\, dx = \left[(\log x)^2 \cdot x\right]_1^e - \int_1^e 2\log x \cdot \frac{1}{x}\cdot x\, dx = \cdot 1^2 \cdot e - 0 - 2\int_1^e \log x\, dx$$

$$= e - 2\left[\{\log x \cdot x\}_1^e - \int_1^e \frac{1}{x}\cdot x\, dx\right] = e - 2[e - (e-1)] = e - 2$$

$$\therefore \quad \text{RMS value} = \sqrt{\frac{e-2}{e-1}}$$

Ans.

6. *Find the mean value of* $\cos(\omega t + \alpha)\cos(\omega t + \beta)$

SOLUTION: $f(t) = \cos(\omega t + \alpha)\cos(\omega t + \beta)$

$$\text{Mean of } f(t) \text{ over the period } 2\pi = \frac{1}{2\pi}\int_0^{2\pi} f(t)\, dt = \frac{1}{2\pi}\int_0^{2\pi} \cos(\omega t + \alpha)\cos(\omega t + \beta)\, dt$$

$$= \frac{1}{4\pi}\int_0^{2\pi} \cos(2\omega t + \alpha + \beta) + \cos(\alpha - \beta)\, dt = \frac{1}{4\pi}\left[\frac{\sin(2\omega t + \alpha + \beta)}{2\omega} + t\cos(\alpha - \beta)\right]_0^{2\pi}$$

$$= \frac{1}{4\pi}\left[\frac{\sin(\alpha + \beta)}{2\omega} + 2\pi\cos(\alpha - \beta)\right] = \frac{1}{2}\cos(\alpha - \beta) + \frac{1}{8\pi\omega}\sin(\alpha + \beta)$$

Ans.

7. *Find the mean value of* $\sin 2\omega t \cos(3\omega t + \alpha)$.

SOLUTION: $f(t) = \sin 2\omega t \cos(3\omega t + \alpha)$

Mean of $f(t)$ over the period $\dfrac{2\pi}{\omega} = \dfrac{1}{\frac{2\pi}{\omega}} \displaystyle\int_0^{2\pi/\omega} \sin 2\omega t \cos(3\omega t + \alpha)\, dt$

$$= \frac{\omega}{4\pi}\int_0^{2\pi/\omega} [\sin(5\omega t + \alpha) - \sin(\omega t + \alpha)]\, dt = \frac{\omega}{4\pi}\left[\frac{-\cos(5\omega t + \alpha)}{5\omega} + \frac{\cos(\omega t + \alpha)}{\omega}\right]_0^{2\pi/\omega}$$

$$= \frac{\omega}{4\pi}\left[\frac{-\cos\left(5\omega\cdot\frac{2\pi}{\omega} + \alpha\right)}{5\omega} + \frac{\cos\left(\omega\cdot\frac{2\pi}{\omega} + \alpha\right)}{\omega} + \frac{\cos\alpha}{5\omega} - \frac{\cos\alpha}{\omega}\right]$$

$$= \frac{1}{4\pi}\left[\frac{-1}{5}\cos(10\pi + \alpha) + \frac{\cos\alpha}{5} + \frac{\cos(2\pi + \alpha)}{\omega} - \frac{\cos\alpha}{\omega}\right]$$

$$= \frac{1}{4\pi}\left[\frac{-\cos\alpha}{5} + \frac{\cos\alpha}{5} + \frac{\cos\alpha}{\omega} - \frac{\cos\alpha}{\omega}\right] = 0$$

Ans.

8. *Find the mean ordinate of the semi circle* $x^2 + y^2 = a^2$ *above the X-axis.*

SOLUTION: $x^2 + y^2 = a^2 \quad \therefore\ y = \sqrt{a^2 - x^2}$ above the X-axis

$$\text{Mean } y = \frac{\int_{-a}^{a} \sqrt{a^2 - x^2}\, dx}{a - (-a)} = \frac{2}{2a}\int_0^a \sqrt{a^2 - x^2}\, dx = \frac{1}{a}\left[\frac{x}{2}\sqrt{a^2 - x^2} + \frac{1}{2}a^2 \sin^{-1}\frac{x}{a}\right]_0^a = \frac{a\pi}{4}$$

Ans.

9. *Calculate, by how much percentage the RMS value of the function* $a \sin bt$ *over half its period, exceeds its mean value over the same half period.*

SOLUTION: Period of $f(t) = a \sin bt$ is $\dfrac{2\pi}{b}$, hence half period $= \dfrac{\pi}{b}$.

Mean of $f(t)$ over period $\dfrac{\pi}{b} = \dfrac{1}{\frac{\pi}{b}} \displaystyle\int_0^{\pi/b} a \sin bt\, dt = \frac{ab}{\pi}\left[\frac{-1\cos bt}{b}\right]_0^{\pi/b} = \frac{-a}{\pi}(\cos\pi - 1) = \frac{2a}{\pi}$.

R.M.S. of $f(t)$ over period $\dfrac{\pi}{b} = \sqrt{\dfrac{1}{\frac{\pi}{b}} \displaystyle\int_0^{\pi/b} a^2 \sin^2 bt\, dt} = \sqrt{\dfrac{a^2 b}{2\pi}\displaystyle\int_0^{\pi/b} 2\sin^2 bt\, dt}$

$$= \sqrt{\frac{a^2 b}{2\pi}\int_0^{\pi/b} (1 - \cos 2bt)\, dt} = \sqrt{\frac{a^2 b}{2\pi}\left(t - \frac{\sin 2bt}{2b}\right)_0^{\pi/b}}$$

$$= \sqrt{\frac{a^2 b}{2\pi}\left(\frac{\pi}{b} - \frac{\sin 2b\frac{\pi}{b}}{2b}\right) - 0} = \frac{a}{\sqrt{2}}$$

Next, $\dfrac{\text{RMS} - \text{Mean}}{\text{Mean}} \times 100 = \dfrac{\frac{a}{\sqrt{2}} - \frac{2a}{\pi}}{\frac{2a}{\pi}} \times 100 = \dfrac{\pi - \sqrt{2}}{2\sqrt{2}} \times 100 = 11\%$ Approx. **Ans.**

10. *Determine the average value and the RMS value of the radius vector for the cardioide $r = a\,(1 - \cos\theta)$.*

SOLUTION: Curve is $r = a\,(1 - \cos\theta)$

$$\text{Average }(r) = \frac{1}{2\pi}\int_0^{2\pi} a(1-\cos\theta)\,d\theta = \frac{a}{2\pi}[\theta - \sin\theta]_0^{2\pi} = a$$

$$\text{RMS value of } x = \sqrt{\frac{1}{2\pi}\int_0^{2\pi} a^2(1-\cos\theta)^2\,d\theta} = \sqrt{\frac{a^2}{2\pi}\int_0^{2\pi} 1 - 2\cos\theta + \frac{1}{2}(1+\cos 2\theta)\,d\theta}$$

$$= \frac{a}{\sqrt{2\pi}}\sqrt{\left[\frac{3\theta}{2} - 2\sin\theta + \frac{1}{4}\sin 2\theta\right]_0^{2\pi}} = \frac{a}{\sqrt{2\pi}}\sqrt{3\pi} = a\sqrt{\frac{3}{2}}.$$ **Ans.**

11. *Find the mean and r.m.s. value of the ordinate y of the cycloid $x = a\,(\theta + \sin\theta)$, $y = a\,(1 - \cos\theta)$ over the range $\theta = -\pi$ to $\theta = \pi$. Hence show that the value of latter is $\sqrt{2}$ times that of the former.*

SOLUTION: $\text{Mean }(y) = \dfrac{\int y\,dx}{\int dx}$ and $\text{RMS}(y) = \sqrt{\dfrac{\int y^2\,dx}{\int dx}}$.

$$\int dx = \int \frac{dx}{d\theta}\,d\theta = \int_{-\pi}^{\pi} a(1+\cos\theta)\,d\theta = 2a\int_0^{\pi}(1+\cos\theta)\,d\theta = 2a\pi$$

$$\int y\,dx = \int \frac{y\,dx}{d\theta}\,d\theta = \int_{-\pi}^{\pi} a(1-\cos\theta)\,a(1+\cos\theta)\,d\theta = 2a^2\int_0^{\pi}\sin^2\theta\,d\theta = a^2\int_0^{\pi}(1-\cos 2\theta)\,d\theta = \pi a^2.$$

$$\therefore \quad \text{Mean }(y) = \frac{\pi a^2}{2a\pi} = \frac{a}{2}.$$

$$\int y^2\,dx = \int_{-\pi}^{\pi} a^2(1-\cos\theta)^2\,a(1+\cos\theta)\,d\theta = 2a^3\int_0^{\pi}\sin^2\theta(1-\cos\theta)\,d\theta$$

$$= a^3\int_0^{\pi}(1-\cos 2\theta)\,d\theta - 2a^3\int_0^{\pi}\sin^2\theta\cos\theta\,d\theta$$

$$= a^3\left[\left(\theta - \frac{1}{2}\sin 2\theta\right)\right]_0^{\pi} - 2a^3\,\frac{1}{3}[\sin^3\theta]_0^{\pi} = \pi a^3$$

$$\therefore \quad \text{RMS }(y) = \sqrt{\frac{\pi a^3}{2a\pi}} = \frac{a}{\sqrt{2}} = \sqrt{2}\ \text{Mean }(y)$$ **Hence Proved.**

12. *Find the mean value of r^2 over the area of the cardioide $r = a(1 + \cos\theta)$.*

SOLUTION: $r = a(1 + \cos\theta)$, If A is the area and $dA = \frac{1}{2}r^2\,d\theta$ then

$$\text{Mean of } r^2 \text{ over the area} = \frac{1}{A}\int r^2\,dA = \frac{\int_0^{2\pi} r^2 \frac{1}{2} r^2\,d\theta}{\int_0^{2\pi} \frac{1}{2} r^2\,d\theta}$$

$$= \frac{\int_0^{2\pi} a^4(1+\cos\theta)^4\,d\theta}{\int_0^{2\pi} a^2(1+\cos\theta)^2\,d\theta} = a^2\frac{\int_0^{2\pi} 16\cos^8\frac{\theta}{2}\,d\theta}{\int_0^{2\pi} 4\cos^4\frac{\theta}{2}\,d\theta} \qquad \left(\text{Putting } \frac{\theta}{2} = t\right)$$

$$= \frac{4a^2\int_0^{\pi}\cos^8 t\cdot 2dt}{\int_0^{\pi}\cos^4 t\cdot 2dt} = \frac{4a^2\int_0^{\pi/2}\cos^8 t\,dt}{\int_0^{\pi/2}\cos^4 t\,dt} = 4a^2\frac{\Gamma\left(\frac{9}{2}\right)\cdot\Gamma\left(\frac{1}{2}\right)}{2\Gamma(5)}\cdot\frac{2\Gamma(3)}{\Gamma\left(\frac{5}{2}\right)\cdot\Gamma\left(\frac{1}{2}\right)}$$

$$= \frac{4a^2}{4\cdot 3}\cdot\frac{7}{2}\cdot\frac{5}{2} = \frac{35a^2}{12}$$ **Ans.**

13. *A point moves from rest with uniform acceleration. Show that in any interval from the starting point, the space average of the velocity is 4/3 of the time average.*

SOLUTION: If a is uniform acceleration, $v = at$ and $S = \frac{1}{2}at^2 \quad \therefore \quad \frac{ds}{dt} = at$

$$\text{Mean } (V) \text{ w.r.t. } S = \frac{\int v\,ds}{\int ds} = \frac{\int at\cdot at\,dt}{\int at\,dt} = \frac{a\frac{t^3}{3}}{\frac{t^2}{2}} = \frac{2}{3}at$$

$$\text{Mean } (V) \text{ w.r.t. time } t = \frac{\int v\,dt}{\int dt} = \frac{\int at\,dt}{\int dt} = \frac{at}{2}$$

Hence Mean (V) w.r.t. space $S = \frac{4}{3}$ Mean (V) w.r.t. time t. **Hence Proved.**

SOLUTIONS TO EXERCISE 7D

1. *Find the volume of the solid generated by revolving the area included between the parabola $y^2 = 4ax$ and its latus rectum about the axis of X.*

SOLUTION: Volume of the solid generated $= \int_0^a \pi y^2\, dx$

$$= \pi\int_0^a 4ax\, dx = \pi 4a\left[\frac{x^2}{2}\right]_0^a = 2\pi a^3.$$ **Ans.**

2. *Find the volume of the solid generated by the revolution of the area of the ellipse $\frac{x^2}{16}+\frac{y^2}{9} = 1$ about the X-axis.*

SOLUTION: $\frac{x^2}{16}+\frac{y^2}{9} = 1$. Required volume of the solid generated

$$= 2\int_0^4 \pi y^2\, dx = 2\pi\int_0^4 9\left(1-\frac{x^2}{16}\right)dx = 18\pi\left[x-\frac{x^3}{48}\right]_0^4 = 18\pi\left[4-\frac{64}{48}\right] = 48\pi$$ **Ans.**

3. *Find the volume of the solid formed when the area bounded by the curve $y^2(a-x) = a^2x$ and its asymptote, is revolved about the asymptote.*

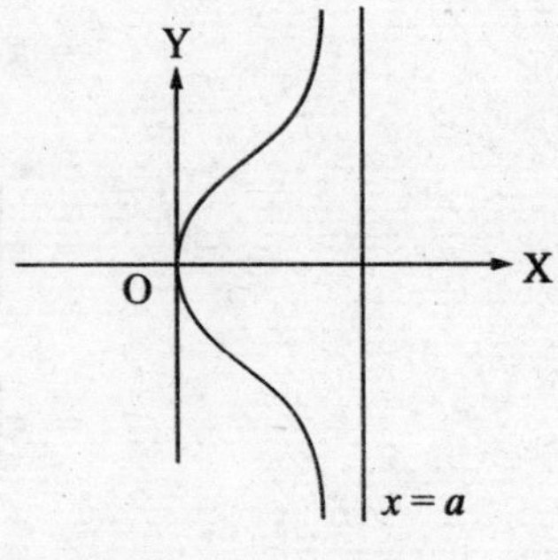

SOLUTION: $y^2(a-x) = a^2 x$. Its asymptote is $x = a$.

Required volume $= \int_{-\infty}^{\infty} \pi(a-x)^2\, dy$

$$= \pi\int_{-\infty}^{\infty}\left[a-\frac{ay^2}{a^2+y^2}\right]^2 dy \quad \left(\text{Since } x = \frac{ay^2}{a^2+y^2} \text{ from curve}\right)$$

$$= \pi\int_{-\infty}^{\infty}\left(\frac{a^3}{a^2+y^2}\right)^2 dy = 2\pi a^6\int_0^{\infty}\frac{dy}{(a^2+y^2)^2} \quad (\text{Put } y = a\tan\theta)$$

$$= 2\pi a^6\int_0^{\pi/2}\frac{a\sec^2\theta\, d\theta}{(a^2\sec^2\theta)^2} = 2\pi a^3\int_0^{\pi/2}\cos^2\theta\, d\theta = \pi a^3\int_0^{\pi/2}(1+\cos 2\theta)\, d\theta$$

$$= \pi a^3\left[\theta+\frac{1}{2}\sin 2\theta\right]_0^{\pi/2} = \frac{\pi^2 a^3}{2}$$ **Ans.**

4. *Find the volume of the solid generated by revolving the area included between the curves $ay^2 = x^3$ and $ax^2 = y^3$ about the X-axis.*

SOLUTION: Required volume $= \int_0^a \pi(y_1^2 - y_2^2)\, dx$

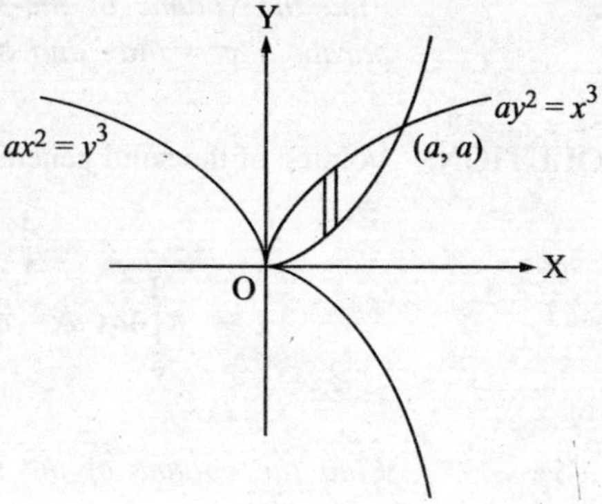

$$= \pi \int_0^a \left[-\frac{x^3}{a} + (ax^2)^{2/3} \right] dx$$

$$= \pi \left[-\frac{x^4}{4a} + a^{2/3} \frac{x^{7/3}}{\frac{7}{3}} \right]_0^a$$

$$= \pi a^3 \left(\frac{3}{7} - \frac{1}{4} \right) = \frac{5\pi a^3}{28}.$$ **Ans.**

5. *Find the volume of the spherical cap of height h cut off a sphere of radius a.*

SOLUTION: $x^2 + y^2 = a^2$

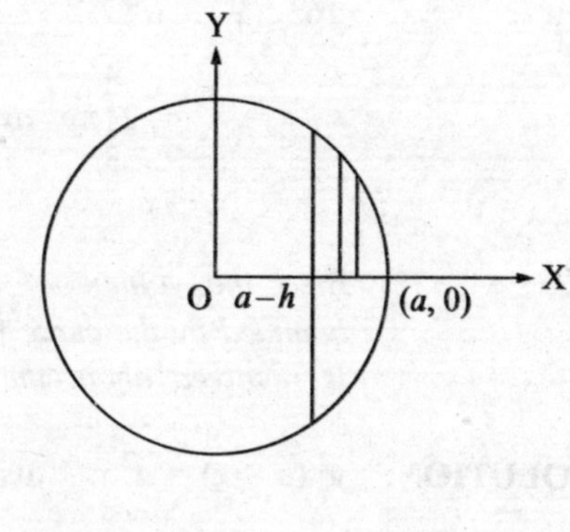

Required volume of the cap $= \int_{a-h}^{a} \pi y^2\, dx$

$$= \pi \int_{a-h}^{a} (a^2 - x^2)\, dx = \pi \left[a^2 x - \frac{x^3}{3} \right]_{a-h}^{a}$$

$$= \pi \left[a^3 - \frac{a^3}{3} - a^2(a-h) + \frac{(a-h)^3}{3} \right]$$

$$= \pi \left[\frac{2a^3}{3} - a^3 + a^2 h + \frac{1}{3}(a^3 - h^3 + 3ah^2 - 3a^2 h) \right] = \pi h^2 \left(a - \frac{h}{3} \right)$$ **Ans.**

6. *Find the volume of the solid formed by the revolution of the area enclosed between the curve $27ay^2 = 4\,(x - 2a)^3$, the axis of X and the parabola $y^2 = 4ax$ about X-axis.*

SOLUTION: $27ay^2 = 4\,(x - 2a)^3$, $y^2 = 4\,ax$, $y = 0$.

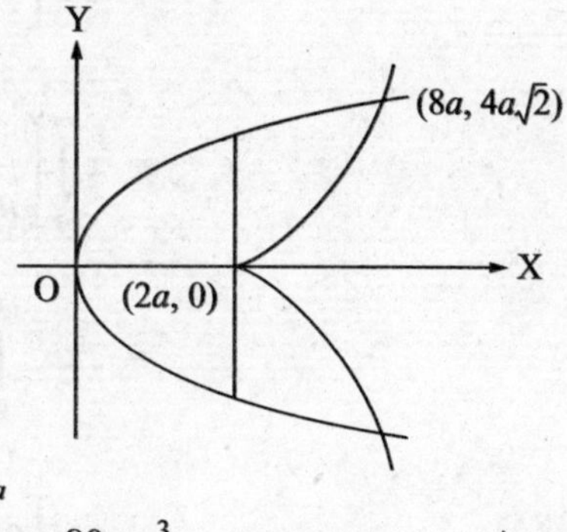

Rquired volume $= \int_0^{2a} \pi y_1^2\, dx + \int_{2a}^{8a} \pi(y_1^2 - y_2^2)\, dx$

$$= \pi \int_0^{2a} 4ax\, dx + \pi \int_{2a}^{8a} \left[4ax - \frac{4(x-2a)^3}{27a} \right] dx$$

$$= 4a\pi \left[\frac{x^2}{2} \right]_0^{2a} + \pi \left[2ax^2 - \frac{4}{27a} \frac{(x-2a)^4}{4} \right]_{2a}^{8a} = 80\, \pi a^3$$ **Ans.**

7. *Find the volume generated by revolving the area cut off the parabola* $9y = 4(9 - x^2)$ *by the line* $4x + 3y = 12$ *about X-axis*

SOLUTION: $9y = 4(9 - x^2),\ 4x + 3y = 12$

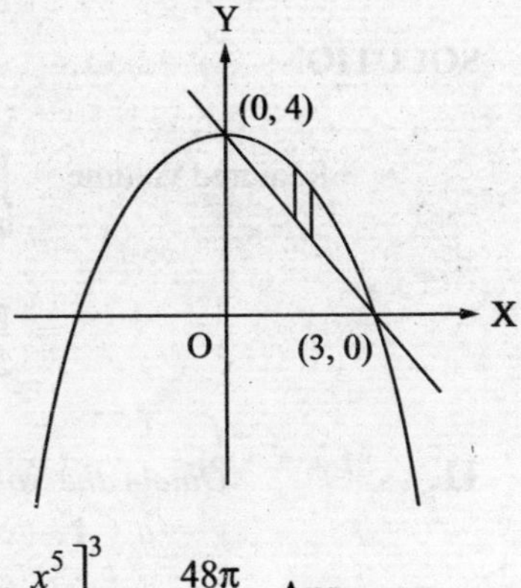

$$\text{Required volume} = \int_0^3 \pi(y_1^2 - y_2^2)\,dx$$

$$= \pi\int_0^3 \left[\frac{16}{81}(9 - x^2)^2 - \left(\frac{12 - 4x}{3}\right)^2\right]dx$$

$$= \pi\int_0^3 \left[\frac{16}{81}(81 - 18x^2 + x^4) - \frac{16}{9}(9 - 6x + x^2)\right]dx$$

$$= \frac{16\pi}{81}\int_0^3 (-27x^2 + 54x + x^4)\,dx = \frac{16\pi}{81}\left[-9x^3 + 27x^2 + \frac{x^5}{5}\right]_0^3 = \frac{48\pi}{5} \quad \textbf{Ans.}$$

8. *Find the volume generated by revolving the area bounded by the circle* $x^2 + y^2 = 25$ *cut off by the lines* $3x - 4y = 0$ *and* $y = 0$ *about the axis of* X.

SOLUTION: $x^2 + y^2 = 25, \qquad 3x = 4y,\ y = 0$

Points of intersection of line and circle are $(4, 3)$ and $(-4, -3)$.

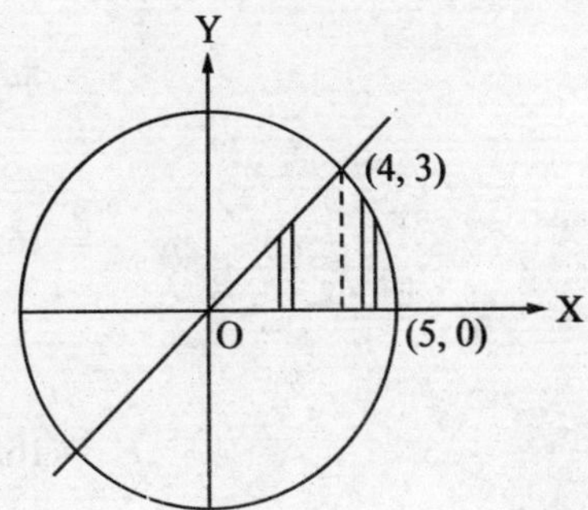

$$\text{Required volume} = \int_0^4 \pi\left(\frac{3x}{4}\right)^2 dx + \int_4^5 \pi(25 - x^2)\,dx$$

$$= \pi\left[\frac{9}{16}\frac{x^3}{3}\right]_0^4 + \pi\left[25x - \frac{x^3}{3}\right]_4^5$$

$$= 12\pi + \pi\left(25 - \frac{125 - 64}{3}\right) = \frac{50\pi}{3}$$ **Ans.**

9. *The area included between the X-axis and the upper half of the circle* $x^2 + y^2 - 4x = 0$ *is rotated about Y-axis. Find the volume thus generated.*

SOLUTION: $x^2 + y^2 - 4x = 0 \;\therefore\; x = \dfrac{4 \pm \sqrt{16 - 4y^2}}{2} = 2 \pm \sqrt{4 - y^2}$

Let $\qquad x_2 = 2 + \sqrt{4 - y^2}\,,\ x_1 = 2, -\sqrt{4 - y^2}\,.$

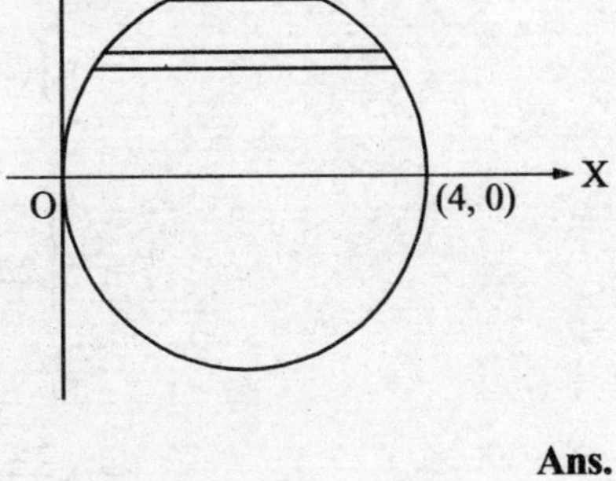

$$\therefore \quad \text{Required volume} = \int_0^2 \pi(x_2^2 - x_1^2)\,dy.$$

$$= \pi\int_0^2 (x_2 + x_1)(x_2 - x_1)\,dx$$

$$= \pi\int_0^2 4 \cdot 2\sqrt{4 - y^2}\,dy$$

$$= 8\pi\left[\frac{y}{2}\sqrt{4 - y^2} + \frac{4}{2}\sin^{-1}\frac{y}{2}\right]_0^2 = 8\pi^2$$ **Ans.**

10. *The area bounded by one arc of the sine wave $y = \sin x$ and the X-axis is revolved about the axis of X. Find the volume of the solid thus generated.*

SOLUTION: $y = \sin x, \quad 0 \le x \le \pi$

$$\text{Required volume} = \int_0^{\pi} \pi y^2\, dx = \pi \int_0^{\pi} \sin^2 x\, dx$$

$$= \frac{\pi}{2}\int_0^{\pi} (1 - \cos 2x)\, dx = \frac{\pi}{2}\left(x - \frac{1}{2}\sin 2x\right)_0^{\pi} = \frac{\pi^2}{2}$$ **Ans.**

11. *Obtain the volume generated by revolving one arch of the cycloid $x = a(\theta - \sin\theta)$, $y = a(1 - \cos\theta)$ about its base.*

SOLUTION: $x = a(\theta - \sin\theta),\ y = a(1 - \cos\theta)$

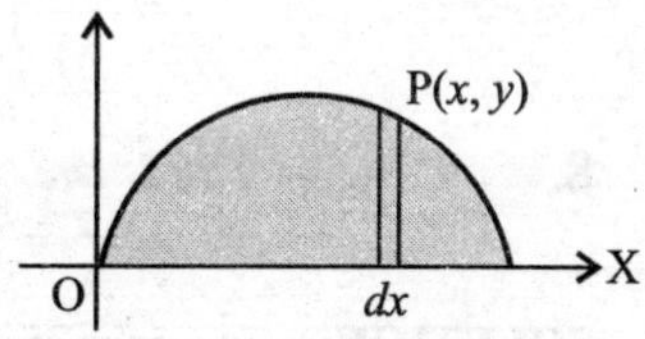

$$\text{Required volume} = \int_0^{a\pi} \pi y^2\, dx = \pi \int_0^{2\pi} a^2 (1 - \cos\theta)^2\, a(1 - \cos\theta)\, d\theta$$

$$= \pi a^3 \int_0^{2\pi} \left(2\sin^2\frac{\theta}{2}\right)^3 d\theta = 8\pi a^3 \int_0^{2\pi} \sin^6\frac{\theta}{2}\, d\theta$$

$$= 16\pi a^3 \int_0^{\pi} \sin^6\frac{\theta}{2}\, d\theta \qquad \text{Putting } \frac{\theta}{2} = t$$

$$= 16\pi a^3 \int_0^{\pi/2} \sin^6 t \cdot 2dt = 32\pi a^3 \frac{\Gamma\frac{7}{2}\,\Gamma\frac{1}{2}}{2\,\Gamma 4} = \frac{16\pi a^3}{6}\cdot\frac{5}{2}\cdot\frac{3}{2}\cdot\frac{1}{2}\left(\Gamma\frac{1}{2}\right)^2 = 5\pi^2 a^3$$ **Ans.**

12. *The area bounded by a quadrant of a circle of radius a and the tangents at the extremities revolves about one of the tangents. Find the volume of the solid generated.*

SOLUTION: $x^2 + y^2 = a^2$

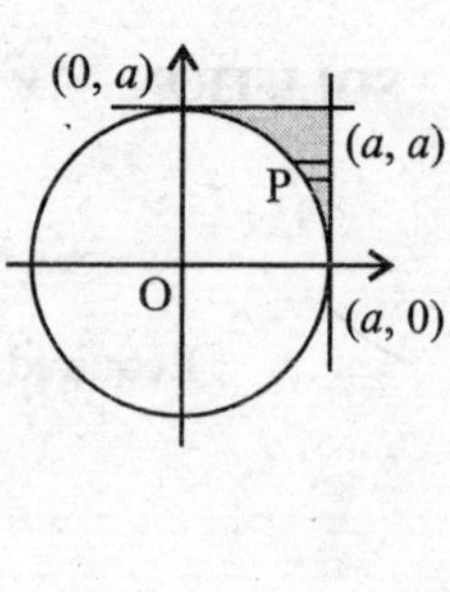

$$\text{Required volume} = \int_0^a \pi\left[a - \sqrt{a^2 - y^2}\right]^2 dy$$

$$= \pi \int_0^a (a^2 + a^2 - y^2 - 2a\sqrt{a^2 - y^2})\, dy$$

$$= \pi\left[2a^2 y - \frac{y^3}{3}\right]_0^a - 2a\pi \int_0^a \sqrt{a^2 - y^2}\, dy$$

$$= \pi\frac{5a^3}{3} - 2a\pi\left[\frac{y}{2}\sqrt{a^2 - y^2} + \frac{a^2}{2}\sin^{-1}\frac{y}{a}\right]_0^a = \pi a^3\left(\frac{5}{3} - \frac{\pi}{2}\right)$$ **Ans.**

13. *The area enclosed by the curves $y = \sin x$, $y = \cos x$ and the axis of X between $x = 0$ and $x = \frac{\pi}{2}$, is revolved about X-axis. Find the volume of the solid so formed.*

SOLUTION: $y = \sin x$, $y = \cos x$ Point of intersection $x = \frac{x}{4}$.

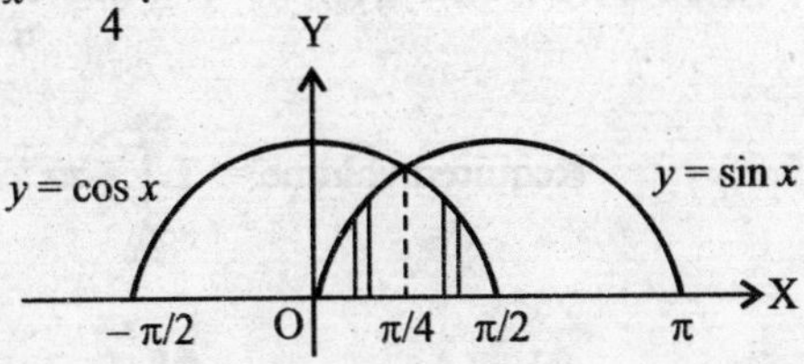

$$\text{Required volume} = \int_0^{\pi/4} \pi \sin^2 x\, dx + \int_{\pi/4}^{\pi/2} \pi \cos^2 x\, dx$$

$$= \frac{\pi}{2}\int_0^{\pi/4} (1 - \cos 2x)\, dx + \frac{\pi}{2}\int_{\pi/4}^{\pi/2} (1 + \cos 2x)\, dx$$

$$= \frac{\pi}{2}\left[x - \frac{1}{2}\sin 2x\right]_0^{\pi/4} + \frac{\pi}{2}\left[x + \frac{1}{2}\sin 2x\right]_{\pi/4}^{\pi/2}$$

$$= \frac{\pi}{2}\left[\frac{\pi}{4} - \frac{1}{2}\right] + \frac{\pi}{2}\left[\frac{\pi}{2} - \frac{\pi}{4} + \frac{1}{2}\sin \pi - \frac{1}{2}\sin\frac{\pi}{2}\right] = \frac{\pi}{4}(\pi - 2)$$ **Ans.**

14. *If the area bounded by the parabolas $y^2 = 4x$ and $y^2 = 5 - x$ is revolved about each of the co-ordinate axes, calculate the respective volumes.*

SOLUTION: $y^2 = 4x$, $y^2 = 5 - x$

Case I. Common area is revolved about X-axis, then

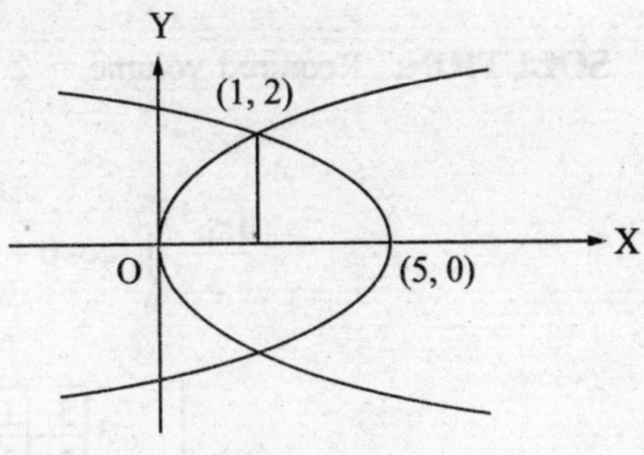

$$\text{Volume of the solid} = \pi\int_0^1 y_1^2\, dx + \pi\int_1^5 y_2^2\, dx$$

$$= \pi\int_0^1 4x\, dx + \pi\int_1^5 (5 - x)\, dx$$

$$= \pi[2x^2]_0^1 + \pi\left[5x - \frac{x^2}{2}\right]_1^5$$

$$= \left(2 + 25 - \frac{25}{2} - 5 + \frac{1}{2}\right)\pi = 10\pi$$ **Ans.**

Case II. Common area is revolved by y-axis

$$\text{Volume of the solid} = 2\pi\int_0^2 \left(x_2^2 - x_1^2\right) dy = 2\pi\int_0^2 (5 - y^2)^2 - \left(\frac{y^2}{4}\right)^2 dy$$

$$= 2\pi\int_0^2 \left(25 - 10y^2 + y^4 - \frac{y^4}{16}\right) dy = 2\pi\left[25y - \frac{10y^3}{3} + \frac{15}{16}\cdot\frac{y^5}{5}\right]_0^2$$

$$= 2\pi\left[50 - \frac{80}{3} + 6\right] = \frac{176\pi}{3}$$ **Ans.**

15. *The area of the loop of the curve* $r = a \cos \theta$ *lying between* $\theta = -\dfrac{\pi}{6}$ to $\theta = \dfrac{\pi}{6}$, *revolves about the initial line. Find the volume of the solid of revolution.*

SOLUTION: $r = a\cos\theta, \quad 0 \le \theta \le \dfrac{\pi}{6}$

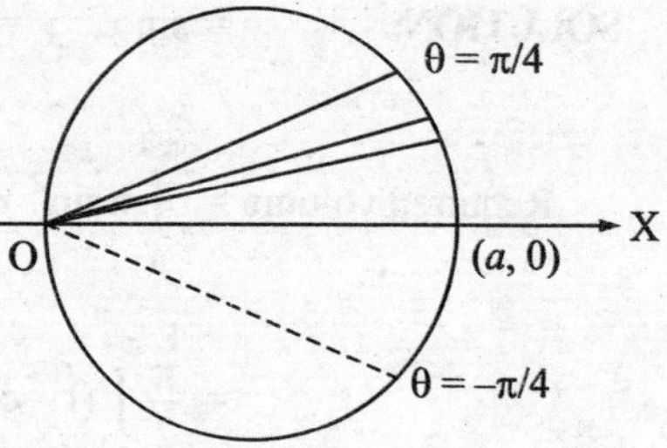

$$\text{Required volume} = 2\int_0^{\pi/6} \frac{2}{3}\pi r^2 \cdot r\sin\theta\, d\theta$$

$$= \frac{4\pi}{3}\int_0^{\pi/6} a^3\cos^3\theta\sin\theta\, d\theta$$

$$= \frac{4\pi a^3}{3}\left[\frac{-1}{4}\cos^4\theta\right]_0^{\pi/6} = \frac{\pi a^3}{3}\left[-\cos^4\frac{\pi}{6}+1\right] = \frac{\pi a^3}{3}\left[1-\left(\frac{\sqrt{3}}{2}\right)^4\right] = \frac{7\pi a^3}{48} \quad \textbf{Ans.}$$

16. *The arc of the cardioid* $r = a(1 + \cos\theta)$ *included between* $\theta = -\dfrac{\pi}{2}$ and $\theta = \dfrac{\pi}{2}$, *is rotated about the line* $\theta = \dfrac{\pi}{2}$. *Show that the volume so generated is* $2\left(2+\dfrac{5\pi}{8}\right)\pi a^3$.

SOLUTION: $\text{Required volume} = 2\cdot\dfrac{2\pi}{3}\displaystyle\int_0^{\pi/2} r^3\cos\theta\, d\theta = \dfrac{4\pi}{3}\displaystyle\int_0^{\pi/2} a^3(1+\cos\theta)^3\cos\theta\, d\theta$

$$= \frac{4\pi a^3}{3}\int_0^{\pi/2}(\cos\theta + 3\cos^2\theta + 3\cos^3\theta + \cos^4\theta)\, d\theta$$

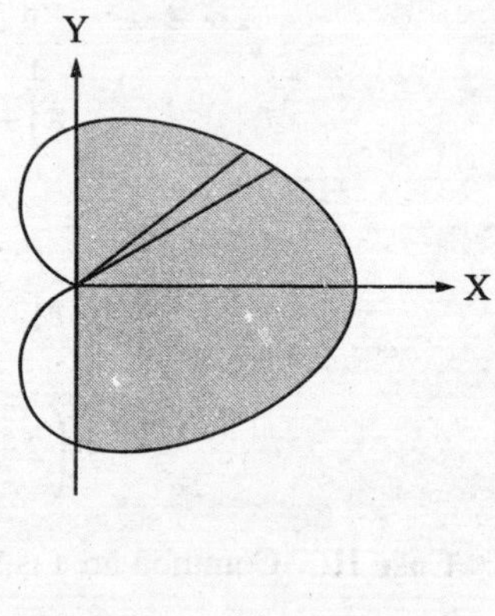

$$= \frac{4\pi a^3}{3}\left[1 + \frac{3\,\Gamma\left(\frac{3}{2}\right)\cdot\Gamma\left(\frac{1}{2}\right)}{2\,\Gamma(2)} + \frac{3\,\Gamma(2)\,\Gamma\left(\frac{1}{2}\right)}{2\,\Gamma\left(\frac{5}{2}\right)} + \frac{\Gamma\left(\frac{5}{2}\right)\cdot\Gamma\left(\frac{1}{2}\right)}{2\,\Gamma(3)}\right]$$

$$= \frac{4\pi a^3}{3}\left[1 + \frac{3}{2}\cdot\frac{1}{2}\pi + \frac{3}{2}\cdot\frac{\Gamma\left(\frac{1}{2}\right)}{\frac{3}{2}\cdot\frac{1}{2}\Gamma\left(\frac{1}{2}\right)} + \frac{\frac{3}{2}\cdot\frac{1}{2}\pi}{2\cdot 2}\right]$$

$$= \frac{4\pi a^3}{3}\left[3+\pi\left(\frac{3}{4}+\frac{3}{16}\right)\right] = 4\pi a^3\left(1+\frac{5\pi}{16}\right) \quad \textbf{Ans.}$$

17. *Find the volume of the solid generated by revolving the ellipse $\frac{x^2}{a^2}+\frac{y^2}{b^2}=1$ about the tangent at the end of (i) major axis (ii) minor axis.*

SOLUTION: $\frac{x^2}{a^2}+\frac{y^2}{b^2}=1$ when revolved about $x=a$ generates a solid whose volume is to be calculated.

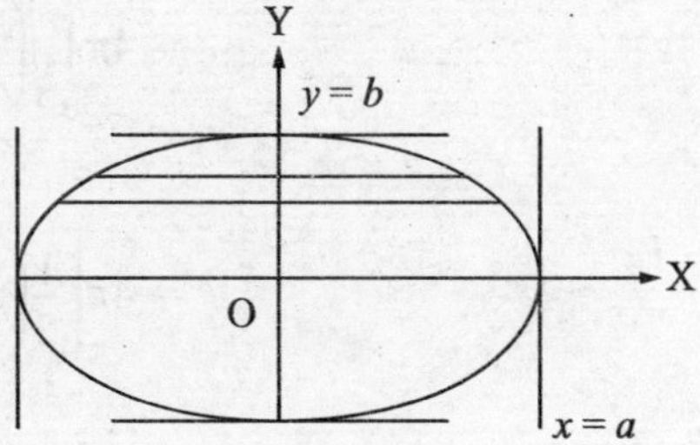

Shift the origin to $(-a, 0)$, equation of ellipise becomes $\frac{(x-a)^2}{a^2}+\frac{y^2}{b^2}=1$ which gives

$$x = a \pm \frac{a}{b}\sqrt{b^2-y^2}$$

$$\therefore \quad \text{Required volume} = \pi\int_{-b}^{b}\left[\left(a+\frac{a}{b}\sqrt{b^2-y^2}\right)^2-\left(a-\frac{a}{b}\sqrt{b^2-y^2}\right)^2\right]dy$$

$$= 4\pi\frac{a^2}{b}\int_{-b}^{b}\sqrt{b^2-y^2}\,dy = \frac{8\pi a^2}{b}\left[\frac{y}{2}\sqrt{b^2-y^2}+\frac{1}{2}b^2\sin^{-1}\frac{y}{b}\right]_0^b = 4\pi\frac{a^2}{b}\left[0+b^2\frac{\pi}{2}\right] = 2\pi^2a^2b \quad \textbf{Ans.}$$

Similarly when the curve is revolved about the line $y = b$ the volume of the solid generated $= 2\pi^2 ab^2$. **Ans.**

SOLUTIONS TO EXERCISE 7E

1. *Determine the surface area of the cone generated by revolving the line $y = 2x$ from $x = 0$ to $x = 2$ about X-axis.*

SOLUTION: Required surface area $= \int_0^2 2\pi y\,ds = 2\pi\int_0^2 y\frac{ds}{dx}\,dx$

Since $y = 2x$ $\therefore$ $\frac{dy}{dx} = 2$ and $\frac{ds}{dx} = \sqrt{1+\left(\frac{dy}{dx}\right)^2} = \sqrt{1+2^2} = \sqrt{5}$

$$\therefore \quad \text{Required surface area} = 2\pi\int_0^2 2x\sqrt{5}\,dx = 4\pi\sqrt{5}\cdot\left(\frac{x^2}{2}\right)_0^2 = 8\pi\sqrt{5} \quad \textbf{Ans.}$$

2. *The portion of the parabola $y^2 = 4ax$ cut off the latus rectum revolves about the tangent at the vertex. Find the surface area of the reel thus generated.*

SOLUTION: $y^2 = 4ax$ $\therefore$ $2y\frac{dy}{dx} = 4a$

Hence $\frac{ds}{dx} = \sqrt{1+\left(\frac{dy}{dx}\right)^2} = \sqrt{1+\frac{4a^2}{y^2}} = \sqrt{1+\frac{a}{x}}$

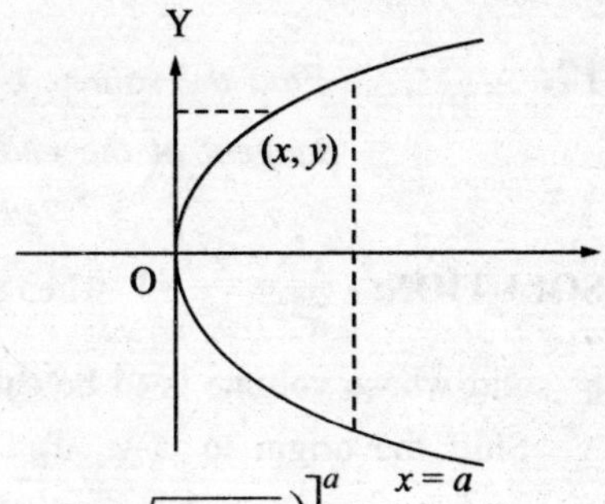

$$\therefore \quad \text{Required surface area} = 2\int_0^a 2\pi x \frac{ds}{dx}\,dx = 4\pi\int_0^a \sqrt{x^2 + ax}\,dx$$

$$= 4\pi\int_0^a \sqrt{\left(x + \frac{a}{2}\right)^2 - \left(\frac{a}{2}\right)^2}\,dx$$

$$= 4\pi\left[\frac{1}{2}\left(x + \frac{a}{2}\right)\sqrt{x^2 + ax} - \frac{1}{2}\left(\frac{a}{2}\right)^2 \log\left(x + \frac{a}{2} + \sqrt{x^2 + ax}\right)\right]_0^a$$

$$= 4\pi\left[\frac{3a}{4}\sqrt{2a^2} - \frac{a^2}{8}\log\left(\frac{\frac{3a}{2} + \sqrt{2a^2}}{\frac{a}{2}}\right)\right] = \pi\left[3a^2\sqrt{2} - \frac{a^2}{2}\log(3 + 2\sqrt{2})\right]$$

$$= \pi a^2\left[3\sqrt{2} - \frac{1}{2}\log(1 + \sqrt{2})^2\right] = \pi a^2[3\sqrt{2} - \log(1 + \sqrt{2})]$$ **Ans.**

3. *Find the curved surface area generated by revolving the cubical parabola $x = y^3$ from $y = 0$ to $y = 2$ about the axis of Y.*

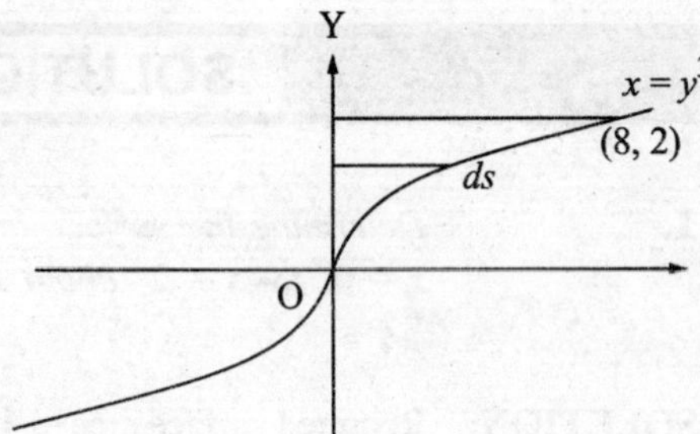

SOLUTION: $x = y^3 \therefore \dfrac{dx}{dy} = 3y^2$

$$\frac{ds}{dy} = \sqrt{1 + \left(\frac{dx}{dy}\right)^2} = \sqrt{1 + 9y^4}$$

$$\text{Required surface area} = \int_0^2 2\pi x \frac{ds}{dy}\,dy$$

$$= 2\pi\int_0^2 y^3\sqrt{1 + 9y^4}\,dy = \frac{2\pi}{36}\left[\frac{2}{3}(1 + 9y^4)^{3/2}\right]_0^2 = \frac{\pi}{27}[145^{3/2} - 1]$$ **Ans.**

4. *Find the area of the surface generated by the revolution of the arc of the catenary $y = c \cosh x/c$ from the vertex to any point (x, y) of the curve, about the X-axis.*

SOLUTION: $y = c\cosh\left(\dfrac{x}{c}\right) \quad \therefore \quad \dfrac{dy}{dx} = \sinh\left(\dfrac{x}{c}\right)$

and
$$\frac{ds}{dx} = \sqrt{1 + \left(\frac{dy}{dx}\right)^2} = \sqrt{1 + \sinh^2\left(\frac{x}{c}\right)} = \cosh\left(\frac{x}{c}\right)$$

$\therefore$ Required surface area $= \int_0^x 2\pi y \frac{ds}{dx}\, dx$

$$= 2\pi \int_0^x c \cosh^2\left(\frac{x}{c}\right) dx$$

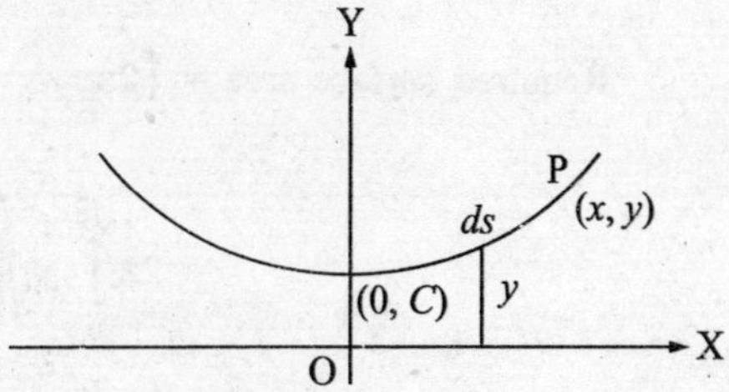

$$= \pi c \int_0^x \left[1 + \cosh\left(\frac{2x}{c}\right)\right] dx = \pi c \left[x + \frac{c}{2}\sinh\left(\frac{2x}{c}\right)\right]_0^x$$

$$= \pi c \left[x + \frac{c}{2}\sinh\left(\frac{2x}{c}\right)\right]$$ **Ans.**

5. *Show that the area of the curved surface generated when one loop of the curve $x^2(a^2 - x^2) = 8a^2y^2$ is revolved about X-axis, is $\pi a^2/4$.*

Solution: $x^2(a^2 - x^2) = 8a^2y^2$. For one loop $0 \le x \le a$

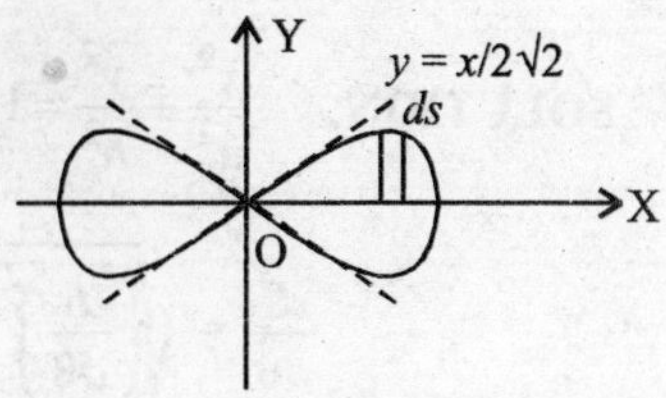

$\therefore$ $16a^2 y\frac{dy}{dx} = 2a^2x - 4x^3$ or $\frac{dy}{dx} = \frac{x(a^2 - 2x^2)}{8a^2 y}$

Required surface area $= \int_0^a 2\pi y \frac{ds}{dx} dx = 2\pi \int_0^a y\sqrt{1 + \left(\frac{dy}{dx}\right)^2}\, dx$

$$= 2\pi \int_0^a y\sqrt{1 + \frac{x^2(a^2 - 2x^2)^2}{64a^4y^2}}\, dx = 2\pi \int_0^a \sqrt{y^2 + \frac{x^2(a^2 - 2x^2)^2}{64a^4}}\, dx$$

$$= 2\pi \int_0^a \sqrt{\frac{x^2(a^2 - x^2)}{8a^2} + \frac{x^2(a^2 - 2x^2)^2}{64a^4}}\, dx = \frac{2\pi}{8a^2}\int_0^a x\sqrt{8a^2(a^2 - x^2) + (a^2 - 2x^2)^2}\, dx$$

$$= \frac{\pi}{4a^2}\int_0^a x\sqrt{9a^4 + 4x^4 - 12a^2x^2}\, dx = \frac{\pi}{4a^2}\int_0^a [x(3a^2 - 2x^2)]\, dx$$

$$= \frac{\pi}{4a^2}\left[\frac{3a^2x^2}{2} - \frac{2x^4}{4}\right]_0^a = \frac{\pi}{4}a^2$$ **Ans.**

6. *Find the surface area of the solid generated by revolving the loop of the curve $x = t^2$, $y = t - \frac{t^3}{3}$ about the axis of X.*

SOLUTION: $x = t^2$, $y = t - \frac{t^3}{3}$. It passes through origin. It is symmetrical about X-axis and $(3, 0)$ lies on the curve, loop lies in $t = 0$ and $t = \sqrt{3}$.

$$\text{Required surface area} = \int 2\pi y\, ds = 4\pi \int_0^{\sqrt{3}} y \frac{ds}{dt}\, dt$$

$$= 2\pi \int_0^{\sqrt{3}} y \sqrt{\left(\frac{dx}{dt}\right)^2 + \left(\frac{dy}{dt}\right)^2}\, dt = 2\pi \int_0^{\sqrt{3}} \left(t - \frac{t^3}{3}\right) \sqrt{(2t)^2 + (1-t^2)^2}\, dt$$

$$= 2\pi \int_0^{\sqrt{3}} t\left(1 - \frac{t^2}{3}\right)(1+t^2)\, dt = \pi \int_0^3 \left(1 - \frac{z}{3}\right)(1+z)\, dz \qquad \text{(On putting } t^2 = z)$$

$$= \pi \int_0^3 \left(1 + \frac{2}{3}z - \frac{z^2}{3}\right) dz = 3\pi.$$ **Ans.**

7. *The area under the ellipse of eccentricity e is revolved about its major axis. Find the ratio of the surface area of the spheroid generated to the area of the ellipse.*

SOLUTION: $\frac{x^2}{a^2} + \frac{y^2}{b^2} = 1$ or $x = a\cos\theta,\ y = b\sin\theta,\ e = \sqrt{1 - \frac{b^2}{a^2}}$ or $ae = \sqrt{a^2 - b^2}$

$$\frac{ds}{d\theta} = \sqrt{\left(\frac{dx}{d\theta}\right)^2 + \left(\frac{dy}{d\theta}\right)^2} = \sqrt{a^2\sin^2\theta + b^2\cos^2\theta}$$

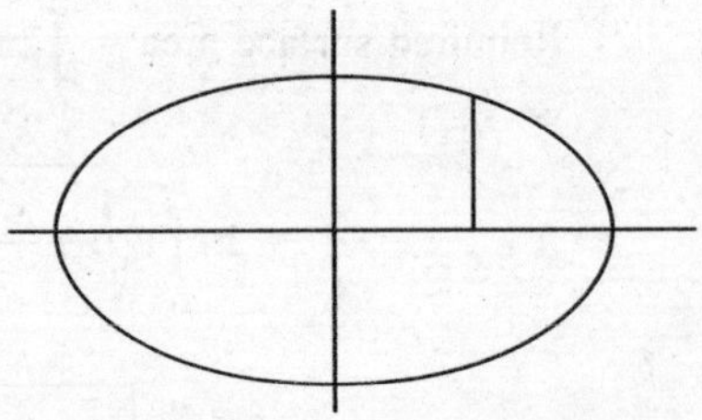

$$\therefore \quad \text{Surface area} = 2\int_0^{\pi/2} 2\pi y \frac{ds}{d\theta}\, d\theta$$

$$= 4\pi \int_0^{\pi/2} b\sin\theta \sqrt{a^2\sin^2\theta + b^2\cos^2\theta}\, d\theta$$

$$= -4\pi b \int_1^0 \sqrt{a^2 - (a^2-b^2)t^2}\, dt = 4\pi b\sqrt{a^2-b^2} \int_0^1 \sqrt{\left(\frac{a}{\sqrt{a^2-b^2}}\right)^2 - t^2}\, dt \qquad \text{(On putting } \cos\theta = t)$$

$$= 4\pi b\sqrt{a^2-b^2}\left[\frac{t}{2}\sqrt{\frac{a^2}{a^2-b^2} - t^2} + \frac{1}{2}\frac{a^2}{a^2-b^2}\sin^{-1}\left(\frac{t\sqrt{a^2-b^2}}{a}\right)\right]_0^1$$

$$= 2\pi b\sqrt{a^2-b^2}\left[1\sqrt{\frac{a^2}{a^2-b^2} - 1} + \frac{a^2}{a^2-b^2}\left(\sin^{-1}\sqrt{1 - \frac{b^2}{a^2}}\right)\right]$$

$$= 2\pi b \cdot ae\left[\sqrt{\frac{1}{e^2} - 1} + \frac{1}{e^2}\sin^{-1} e\right] = 2\pi ab\left[\sqrt{1-e^2} + \frac{1}{e}\sin^{-1} e\right]$$

Area of the ellipse $= \pi ab$

$$\therefore \quad \frac{\text{Surface area of the spheroid}}{\text{Area of the ellipse}} = 2\left[\sqrt{1-e^2} + \frac{1}{e}\sin^{-1} e\right]$$ **Ans.**

8. *A quadrant of a circle of radius a revolves about its chord so that a solid is formed. Find the surface area of the spindle generated.*

SOLUTION: $x^2 + y^2 = a^2$. AMB is the quadrant. Chord $AB = a\sqrt{2}$

Arc AMB revolves about the chord AB.

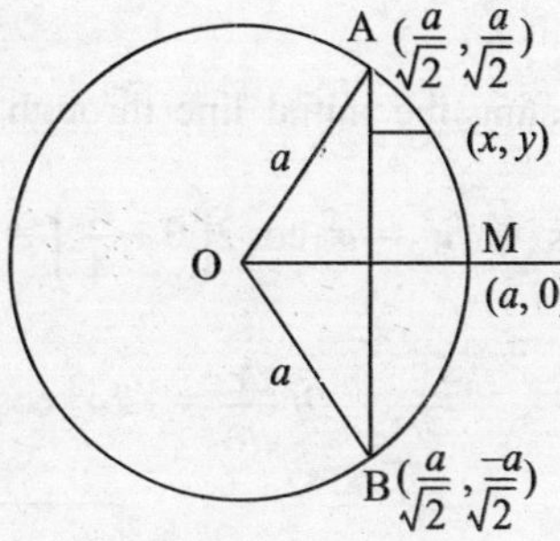

Required surface area $= 2\int 2\pi\left(x - \frac{a}{\sqrt{2}}\right)\frac{ds}{dy}\,dy$

From the curve $2y + 2x\frac{dx}{dy} = 0$ hence $\frac{dx}{dy} = \frac{-y}{x}$

or $$\frac{ds}{dy} = \sqrt{1+\left(\frac{dx}{dy}\right)^2} = \sqrt{1+\frac{y^2}{x^2}} = \sqrt{\frac{x^2+y^2}{x^2}} = \frac{a}{x}$$

$\therefore$ Surface area $$= 4\pi\int_{a/\sqrt{2}}^{a}\left(x - \frac{a}{\sqrt{2}}\right)\frac{a}{x}\,dy = 4a\pi\int_{a/\sqrt{2}}^{a}\left(1 - \frac{a}{x\sqrt{2}}\right)dy$$

$$= 4a\pi\int_{a/\sqrt{2}}^{a}\left(1 - \frac{a}{\sqrt{2}\sqrt{a^2-y^2}}\right)dy = 4a\pi\left[y - \frac{a}{\sqrt{2}}\sin^{-1}\frac{y}{a}\right]_{a/\sqrt{2}}^{a}$$

$$= 4a\pi\left[a - \frac{a}{\sqrt{2}} - \frac{a}{\sqrt{2}}\left(\frac{\pi}{2} - \frac{\pi}{4}\right)\right] = 2\sqrt{2}\pi a^2\left[\sqrt{2} - 1 - \frac{\pi}{4}\right]$$ **Ans.**

9. *Find the surface area of the solid generated by revolving the area bounded by the circle $x^2 + y^2 = a^2$ about the line $y = a$.*

SOLUTION: $x^2 + y^2 = a^2$ $\therefore 2x + 2y\frac{dy}{dx} = 0$ or $\frac{dy}{dx} = \frac{-x}{y}$.

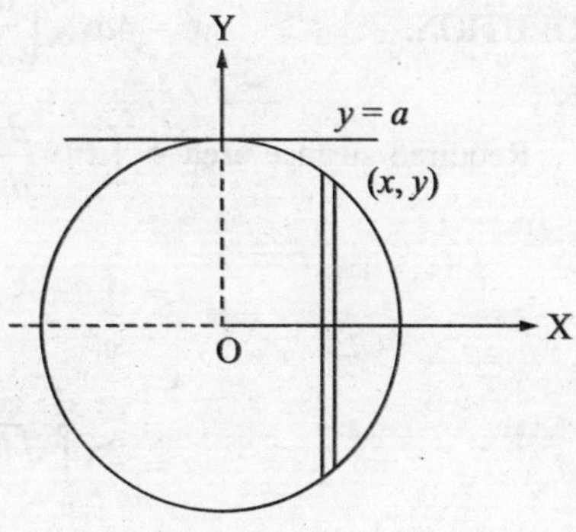

$$\therefore \quad \frac{ds}{dx} = \sqrt{1+\left(\frac{dy}{dx}\right)^2} = \sqrt{1+\frac{x^2}{y^2}} = \frac{a}{\sqrt{a^2-x^2}}$$

Required surface area of the solid generated on revolution of area

of circle about $y = a$, $= 2\int_0^a 2\pi(y_2 - y_1)\frac{ds}{dx}\,dx$

$$= 4\pi\int_0^a\left[(a+\sqrt{a^2-x^2}) - (a-\sqrt{a^2-x^2})\right]\frac{a\,dx}{\sqrt{a^2-x^2}} = 8\pi\int_0^a a\,dx$$

$= 8\pi a^2$ **Ans.**

10. *The leminiscate of Bernoulli is revolved about the tangent at the pole. Find the surface area of the solid generated.*

SOLUTION: $r^2 = a^2 \cos 2\theta$ is revolved about $\theta = \frac{\pi}{4}$

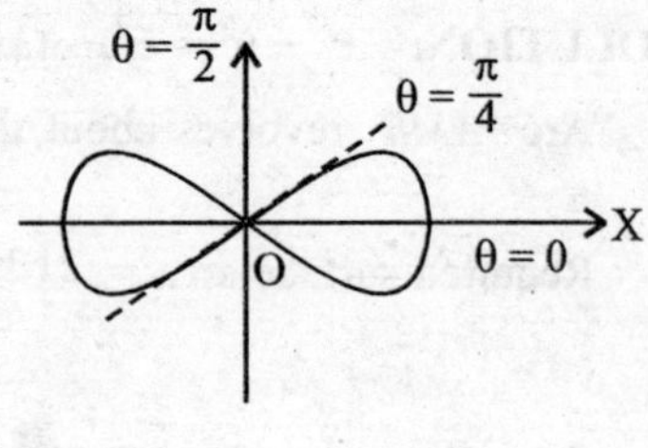

Rotating the initial line through an angle $\frac{\pi}{4}$ the above equation becomes $r^2 = a^2 \cos 2\left(\theta + \frac{\pi}{4}\right) = -a^2 \sin 2\theta.$

$$\therefore \qquad 2r\frac{dr}{d\theta} = -2a^2 \cos 2\theta \quad \text{or} \quad \frac{dr}{d\theta} = \frac{-a^2 \cos 2\theta}{r}$$

and
$$\frac{ds}{d\theta} = \sqrt{r^2 + \left(\frac{dr}{d\theta}\right)^2} = \sqrt{r^2 + \frac{a^4 \cos^2 2\theta}{r^2}} = \frac{1}{r}\sqrt{r^4 + a^4 \cos^2 2\theta}$$

$$= \frac{1}{r}\sqrt{a^4 \sin^2 2\theta + a^4 \cos^2 2\theta} = \frac{a^2}{r}$$

$$\therefore \quad \text{Required surface} = 2\int_0^{\pi/2} 2\pi\, r \sin\theta \frac{ds}{d\theta}\, d\theta$$

$$= 4\pi \int_0^{\pi/2} r\sin\theta \cdot \frac{a^2}{r}\, d\theta = 4\pi a^2 \int_0^{\pi/2} \sin\theta\, d\theta = 4\pi a^2(-\cos\theta)_0^{\pi/2} = 4\pi a^2 \qquad \textbf{Ans.}$$

11. *Find the area of the curved surface of the cup formed by the revolution about its axis of the smaller part of the parabola $y^2 = 4ax$ cut-off by the line $x = 3a$.*

SOLUTION: $y^2 = 4ax, \quad \left(\frac{dy}{dx}\right)^2 = \frac{4a^2}{y^2} = \frac{a}{x}$

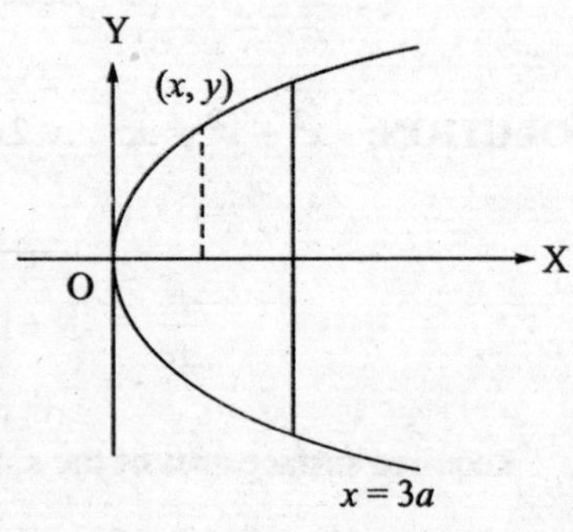

$$\text{Required surface area} = \int 2\pi y \frac{ds}{dx}\, dx$$

$$= \int_0^{3a} 2\pi y \cdot \sqrt{1 + \left(\frac{dy}{dx}\right)^2}\, dx = 2\pi \int_0^{3a} y\sqrt{1 + \frac{a}{x}}\, dx$$

$$= 2\pi \int_0^{3a} \sqrt{4ax}\sqrt{1 + \frac{a}{x}}\, dx$$

$$= 2\pi \cdot 2\sqrt{a} \int_0^{3a} \sqrt{x+a}\, dx = 4\pi\sqrt{a}\left[\frac{(x+a)^{3/2}}{\frac{3}{2}}\right]_0^{3a}$$

$$= 4\pi\sqrt{a}\,\frac{2}{3}\left[(4a)^{3/2} - a^{3/2}\right] = \frac{8\pi a^2}{3}(8-1) = \frac{56\pi a^2}{3} \qquad \textbf{Ans.}$$

12. *Find the area of the surface formed by revolving a catenary* $y = a \cosh (x/a)$ *about X-axis from* $x = 0$ to $x = a$.

SOLUTION: Refer to Question No. 4. The curved surface generated by revolving the arc of the catenary $y = c \cosh \dfrac{x}{c}$ from the vertex to any point (x, y) of the curve, about X-axis was obtained as

$$\pi c \left[x + \frac{c}{2}\sinh \frac{2x}{c}\right].$$

This, when obtained from vertex to $x = c$, is equal to

$$= \pi c^2 \left[1 + \frac{1}{2}\sinh 2\right].$$

Replacing a by c, the result is $\pi a^2 \left[1 + \dfrac{1}{2}\sinh 2\right]$. **Ans.**

13. *Show that the volume and the surface generated by the revolution of an arc of the cycloid*

$x = a\,(\theta + \sin \theta)$, $y = a\,(1 - \cos \theta)$

about the Y-axis, are respectively $\pi a^3 \left(\dfrac{3}{2}\pi^2 - \dfrac{8}{3}\right)$, $4\pi a^2 \left(2\pi - \dfrac{8}{3}\right)$.

SOLUTION: $x = a\,(\theta + \sin \theta)$, $y = a\,(1 - \cos \theta)$. Its one arc is revolved about y-axis. Volume of the solid of revolution

$$= \int \pi x^2 dy = \pi \int_0^{\pi} a^2 (\theta + \sin\theta)^2\, a \sin\theta\, d\theta$$

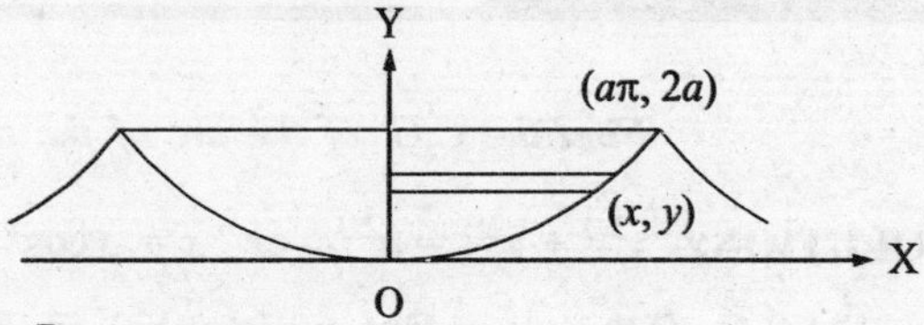

$$= \pi a^3 \int_0^{\pi} \theta^2 \sin\theta\, d\theta + 2\pi a^3 \int_0^{\pi} \theta \sin^2\theta\, d\theta + \pi a^3 \int_0^{\pi} \sin^3\theta\, d\theta$$

Here $\displaystyle\int_0^{\pi} \theta^2 \sin\theta\, d\theta = -\theta^2 \cos\theta\Big|_0^{\pi} + \int_0^{\pi} 2\theta \cos\theta\, d\theta = \pi^2 + 2\theta \sin\theta\Big|_0^{\pi} - \int_0^{\pi} 2\sin\theta\, d\theta = \pi^2 - 4$

$$\int_0^{\pi} \theta \sin^2\theta\, d\theta = \frac{1}{2}\int_0^{\pi} \theta(1 - \cos 2\theta)\, d\theta = \frac{1}{2}\left[\theta\left(\theta - \frac{1}{2}\sin 2\theta\right)\right]_0^{\pi} - \frac{1}{2}\int_0^{\pi} 1\left(\theta - \frac{1}{2}\sin 2\theta\right) d\theta$$

$$= \frac{1}{2}\pi^2 - \frac{1}{2}\left[\frac{\theta^2}{2} + \frac{1}{4}\cos 2\theta\right]_0^{\pi} = \frac{\pi^2}{4}$$

and $$\int_0^{\pi} \sin^3\theta\, d\theta = 2\int_0^{\pi/2} \sin^3\theta\, d\theta = \frac{2\Gamma 2\, \Gamma\frac{1}{2}}{2\Gamma\frac{5}{2}} = \frac{4}{3}.$$

$\therefore$ Required volume $= \pi a^3 (\pi^2 - 4) + 2\pi a^3 \cdot \dfrac{\pi^2}{4} + \pi a^3 \cdot \dfrac{4}{3} = \pi a^3 \left(\dfrac{3}{2}\pi^2 - \dfrac{8}{3}\right)$

Surface area generated on revolving one arc about y-axis $= \int 2\pi x \frac{ds}{dy}\, dy$

Here $$\frac{ds}{dy} = \sqrt{1+\left(\frac{dx}{dy}\right)^2} = \sqrt{1+\frac{\left(\frac{dx}{d\theta}\right)^2}{\left(\frac{dy}{d\theta}\right)^2}} = \sqrt{1+\frac{a^2(1+\cos\theta)^2}{a^2\sin^2\theta}} = \text{cosec}\frac{\theta}{2}$$

$$\therefore \quad \int 2\pi x \frac{ds}{dy}\, dy = 2\pi \int_0^{\pi} a(\theta+\sin\theta)\,\text{cosec}\frac{\theta}{2}\cdot a\sin\theta\, d\theta$$

$$= 2\pi a^2 \int_0^{\pi} (\theta+\sin\theta)\, 2\cos\frac{\theta}{2}\, d\theta = 4\pi a^2 \int_0^{\pi} \theta\cos\frac{\theta}{2}\, d\theta + 8\pi a^2 \int_0^{\pi} \cos^2\frac{\theta}{2}\sin\frac{\theta}{2}\, d\theta$$

$$= 4\pi a^2 \left[\left\{\theta\, 2\sin\frac{\theta}{2}\right\}_0^{\pi} - \int_0^{\pi} 2\sin\frac{\theta}{2}\, d\theta\right] - 8\pi a^2 \cdot \frac{2}{3}\left[\cos^3\frac{\theta}{2}\right]_0^{\pi}$$

$$= 4\pi a^2 \left[2\pi + \left\{4\cos\frac{\theta}{2}\right\}_0^{\pi}\right] - \frac{16\pi a^2}{3}(0-1) = 4\pi a^2\left(2\pi - \frac{8}{3}\right)$$ **Hence Proved.**

SOLUTIONS TO EXERCISE 7F

1. *Find the C.G. of the arc of the astroid $x^{2/3} + y^{2/3} = a^{2/3}$ lying in the first quadrant.*

SOLUTION: $x^{2/3} + y^{2/3} = a^{2/3}$ or $x = a\cos^3\theta,\ y = a\sin^3\theta$

Let C.G. of the arc in first quadrant be $(\bar{x}, \bar{y})$

$$\bar{x} = \frac{\int x\, ds}{\int ds} = \frac{\int x \frac{ds}{d\theta}\, d\theta}{\int \frac{ds}{d\theta}\, d\theta}, \quad \bar{y} = \frac{\int y\, ds}{\int ds} = \frac{\int y \frac{ds}{d\theta}\, d\theta}{\int \frac{ds}{d\theta}\, d\theta}$$

Here $$\frac{ds}{d\theta} = \sqrt{\left(\frac{dx}{d\theta}\right)^2 + \left(\frac{dy}{d\theta}\right)^2} = \sqrt{(-3a\cos^2\theta\sin\theta)^2 + (3a\sin^2\theta\cos\theta)^2} = 3a\sin\theta\cos\theta$$

$$\therefore \quad \int_0^{\pi/2} \frac{ds}{d\theta}\, d\theta = \int_0^{\pi/2} 3a\sin\theta\cos\theta\, d\theta = \frac{3a}{2}[\sin^2\theta]_0^{\pi/2} = \frac{3a}{2}.$$

$$\int_0^{\pi/2} x\frac{ds}{d\theta}\, d\theta = \int_0^{\pi/2} a\cos^3\theta\cdot 3a\sin\theta\cos\theta\, d\theta = -\frac{3a^2}{5}[\cos^5\theta]_0^{\pi/2} = \frac{3a^2}{5}$$

and $$\int_0^{\pi/2} y\frac{ds}{d\theta}\, d\theta = \int_0^{\pi/2} a\sin^3\theta\cdot 3a\sin\theta\cos\theta\, d\theta = 3a^2\int_0^{\pi/2} \sin^4\theta\cos\theta\, d\theta = \frac{3a^2}{5}[\sin^5\theta]_0^{\pi/2} = \frac{3a^2}{5}$$

$$\therefore \quad \bar{x} = \frac{3a^2}{5}\cdot\frac{2}{3a} = \frac{2a}{5}, \quad \bar{y} = \frac{3a^2}{5}\cdot\frac{2}{3a} = \frac{2a}{5} \quad \therefore \text{C.G.} = \left(\frac{2a}{5}, \frac{2a}{5}\right)$$ **Ans.**

2. *Find the centroid of the arc of the cardioide* $r = a(1 + \cos\theta)$ *lying above the initial line.*

SOLUTION: $r = a(1+\cos\theta)$. Let $(\bar{x}, \bar{y})$ be C.G. of the arc above the initial line.

$$\frac{ds}{d\theta} = \sqrt{r^2 + \left(\frac{dr}{d\theta}\right)^2} = \sqrt{a^2(1+\cos\theta)^2 + a^2\sin^2\theta} = a\sqrt{2+2\cos\theta} = 2a\cos\frac{\theta}{2}$$

$$\bar{x} = \frac{\int x\frac{ds}{d\theta}\,d\theta}{\int \frac{ds}{d\theta}\,d\theta}, \quad \bar{y} = \frac{\int y\frac{ds}{d\theta}\,d\theta}{\int \frac{ds}{d\theta}\,d\theta}, \quad 0 \le \theta \le \pi.$$

$$\int_0^\pi \frac{ds}{d\theta}\,d\theta = \int_0^\pi 2a\cos\frac{\theta}{2}\,d\theta = 4a\left[\sin\frac{\theta}{2}\right]_0^\pi = 4a$$

Now $$\int_0^\pi x\frac{ds}{d\theta}\,d\theta = \int_0^\pi r\cos\theta\cdot 2a\cos\frac{\theta}{2}\,d\theta = 2a\int_0^\pi a(1+\cos\theta)\cos\theta\cos\frac{\theta}{2}\,d\theta$$

$$= 2a^2\int_0^\pi 2\cos^2\frac{\theta}{2}\left(2\cos^2\frac{\theta}{2}-1\right)\cos\frac{\theta}{2}\,d\theta = 8a^2\int_0^\pi \cos^5\frac{\theta}{2}\,d\theta - 4a^2\int_0^\pi \cos^3\frac{\theta}{2}\,d\theta$$

$$= 16a^2\int_0^{\pi/2}\cos^5 t\,dt - 8a^2\int_0^{\pi/2}\cos^3 t\,dt \qquad \text{(On putting } \theta = 2t)$$

$$= 16a^2\frac{\Gamma 3\,\Gamma\frac{1}{2}}{2\Gamma\frac{7}{2}} - 8a^2\frac{\Gamma 2\,\Gamma\frac{1}{2}}{2\Gamma\frac{5}{2}} = \frac{16a^2}{\frac{5}{2}\cdot\frac{3}{2}\cdot\frac{1}{2}} - \frac{4a^2}{\frac{3}{2}\cdot\frac{1}{2}} = \frac{16a^2}{5}$$

and $$\int_0^\pi y\frac{ds}{d\theta}\,d\theta = \int_0^\pi r\sin\theta\cdot 2a\cos\frac{\theta}{2}\,d\theta = 2a\int_0^\pi a(1+\cos\theta)\sin\theta\cos\frac{\theta}{2}\,d\theta$$

$$= 2a^2\int_0^\pi 2\cos^2\frac{\theta}{2}\cdot 2\sin\frac{\theta}{2}\cos^2\frac{\theta}{2}\,d\theta = 8a^2\int_0^\pi \cos^4\frac{\theta}{2}\sin\frac{\theta}{2}\,d\theta$$

$$= -8a^2 2\left[\frac{\cos^5\frac{\theta}{2}}{5}\right]_0^\pi = \frac{16a^2}{5}$$

$\therefore$ $$\bar{x} = \frac{\frac{16a^2}{5}}{4a} = \frac{4a}{5}, \quad \bar{y} = \frac{4a}{5} \quad \therefore \text{ C.G. is } \left(\frac{4a}{5}, \frac{4a}{5}\right)$$ **Ans.**

3. *Find the C.G. of the segment of a sphere of radius a cut off by a plane at a distance c from the centre and hence, show that the C.G. of a hemisphere is at a distance of $\frac{3a}{8}$ from its centre.*

SOLUTION: C.G. of the given solid spherical segment is $(\bar{x}, \bar{y})$ where $\bar{y} = 0$

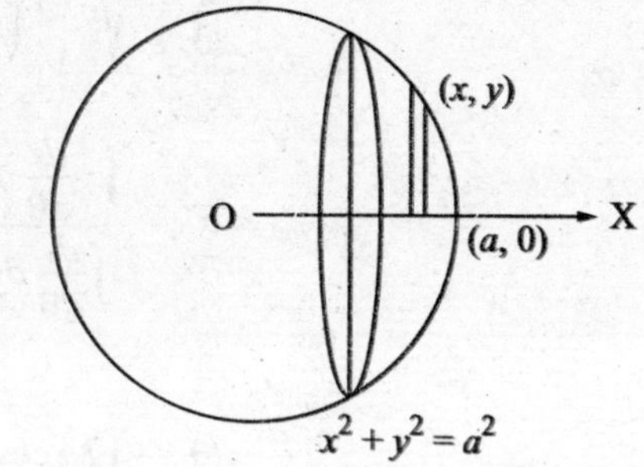

then $$\bar{x} = \frac{\int x\,\pi y^2\,dx}{\int \pi y^2\,dx} = \frac{\int_c^a x(a^2 - x^2)\,dx}{\int_c^a (a^2 - x^2)\,dx}$$

$$= \frac{\left[\frac{a^2x^2}{2} - \frac{x^4}{4}\right]_c^a}{\left[a^2x - \frac{x^3}{3}\right]_c^a} = \frac{\frac{a^4}{2} - \frac{a^4}{4} - \frac{a^2c^2}{2} + \frac{c^4}{4}}{a^3 - \frac{a^3}{3} - a^2c + \frac{c^3}{3}}$$

$$= \frac{\left(\frac{a^4}{4} - \frac{a^2c^2}{2} + \frac{c^4}{4}\right)}{\frac{2a^3}{3} - a^2c + \frac{c^3}{3}} = \frac{3}{4}\,\frac{a^4 - 2a^2c^2 + c^4}{2a^3 - 3a^2c + c^3} = \frac{3(a^2 - c^2)^2}{4(a - c)(2a^2 - ca - c^2)}$$

$$= \frac{3}{4}\,\frac{(a + c)^2(a - c)}{2a^2 - ca - c^2} = \frac{3}{4}\,\frac{(a + c)^2}{2a + c} \qquad \therefore \quad \text{C.G.} = \left(\frac{3}{4}\,\frac{(a + c)^2}{2a + c}, 0\right).$$

when $c = 0$ the C.G. of the hemisphere is $\left(\frac{3a}{8}, 0\right)$ **Ans.**

4. *Find the C.G. of the area bounded by the parabola $y^2 = 4ax$ and the double ordinate $x = h$.*

SOLUTION: By symmetry the C.G. of the area = $(\bar{x}, 0)$

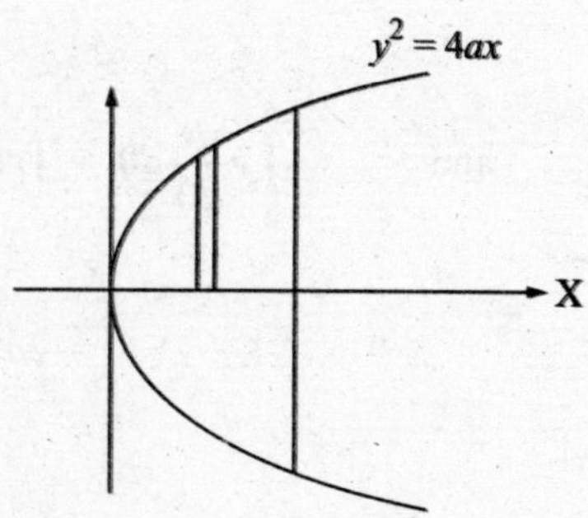

$$\therefore \quad \bar{x} = \frac{\int x \cdot y\,dx}{\int y\,dx} = \frac{\int_0^h x\sqrt{4ax}\,dx}{\int_0^h \sqrt{4ax}\,dx} = \frac{\int_0^h x^{3/2}dx}{\int_0^h x^{1/2}\,dx}$$

$$= \frac{\left[\frac{2}{5} \cdot x^{5/2}\right]_0^h}{\left[\frac{2}{3}x^{3/2}\right]_0^h} = \frac{3}{5}h$$

$\therefore$ Required C.G. $= \left(\frac{3h}{5}, 0\right)$ **Ans.**

5. *Find the centroid of the area of the loop of the curve* $r = a\sin 2\theta$ *in the first quadrant.*

SOLUTION: $r = a\sin 2\theta$, loop of the curve is from $\theta = 0$ to $\theta = \pi$.

When $\theta \to \pi + \theta$, r is unchanged so symmetry about the pole. By symmetry $\bar{x} = \bar{y}$.

$$\bar{x} = \frac{\int_0^{\pi} \frac{2}{3} r\cos\theta, \frac{1}{2} r^2\, d\theta}{\int_0^{\pi} \frac{1}{2} r^2\, d\theta} = \frac{2}{3} \frac{\int_0^{\pi} a^3 \sin^3 2\theta \cos\theta\, d\theta}{\int_0^{\pi} a^2 \sin^2 2\theta\, d\theta}$$

$$= \frac{2a}{3} \frac{\int_0^{\pi} 8\sin^3\theta\cos^3\theta\cos\theta\, d\theta}{\int_0^{\pi} \frac{1}{2}(1-\cos 4\theta)} = \frac{64a}{3} \frac{\int_0^{\pi/2} \sin^3\theta\cos^4\theta\, d\theta}{\left[\theta - \frac{1}{4}\sin 4\theta\right]_0^{\pi}} = \frac{64a}{3\pi} \frac{\Gamma 2\, \Gamma \frac{5}{2}}{2\, \Gamma \frac{9}{2}} = \frac{128a}{105\pi}$$

$\therefore$ C.G. is $\dfrac{128a}{105\pi}, \dfrac{128a}{105\pi}$ **Ans.**

6. *Locate the C.G. of an arch of the cycloid* $x = a(\theta - \sin\theta)$, $y = a(1 - \cos\theta)$.

SOLUTION: $x = a(\theta - \sin\theta)$, $y = a(1 - \cos\theta)$. Arc is symmetrical about $x = a\pi$

Hence C.G. of the arc is $(a\pi, \bar{y})$

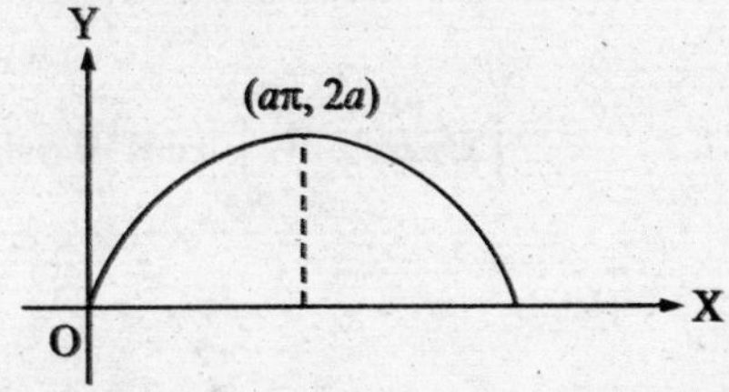

$$\bar{Y} = \frac{\int y\, ds}{\int ds} = \frac{\int_0^{2\pi} y \frac{ds}{d\theta}\, d\theta}{\int_0^{2\pi} \frac{ds}{d\theta}\, d\theta}$$

Here $\dfrac{ds}{d\theta} = \sqrt{\left(\dfrac{dx}{d\theta}\right)^2 + \left(\dfrac{dy}{d\theta}\right)^2} = \sqrt{a^2(1-\cos\theta)^2 + a^2\sin^2\theta}$

$$= a\sqrt{2 - 2\cos\theta} = 2a\sin\frac{\theta}{2}$$

Here $\bar{X} = a\pi$, $\quad \bar{Y} = \dfrac{\int_0^{2\pi} a(1-\cos\theta)\, 2a\sin\frac{\theta}{2}\, d\theta}{\int_0^{2\pi} 2a\sin\frac{\theta}{2}\, d\theta}$

$$= \frac{4a^2 \int_0^{2\pi} \sin^3\frac{\theta}{2}\, d\theta}{2a\int_0^{2\pi} \sin\frac{\theta}{2}\, d\theta} = \frac{2\cdot 4a^2 \int_0^{\pi} \sin^3 t\, dt}{-4a\left[\cos\frac{\theta}{2}\right]_0^{2\pi}} \qquad \left(\text{On putting } \frac{\theta}{2} = t\right)$$

$$= 2a\int_0^{\pi/2} \sin^3 t\, dt = 2a \frac{\Gamma\frac{3+1}{2}\, \Gamma\frac{1}{2}}{2\,\Gamma\frac{5}{2}} = \frac{4a}{3}$$

$\therefore$ C.G. of the arc is $\left(a\pi, \dfrac{4a}{3}\right)$ **Ans.**

7. *Determine the position of the centroid of the area under the curve* $x^{2/3} + y^{2/3} = a^{2/3}$ *lying in the first quadrant.*

SOLUTION: $x^{2/3} + y^{2/3} = a^{2/3}$ or $x = a\cos^3\theta,\ y = a\sin^3\theta$.

Let $(\bar{X}, \bar{Y})$ be the C.G. of the area under the curve in first quadrant.

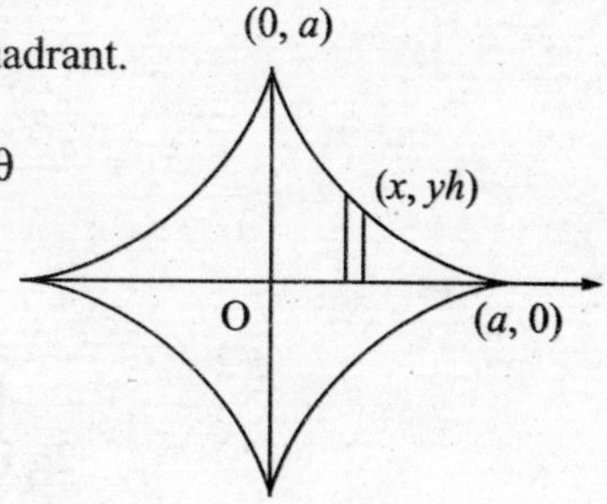

$$\frac{dx}{d\theta} = -3a\cos^2\theta\sin\theta, \qquad \frac{dy}{d\theta} = 3a\sin^2\theta\cos\theta$$

$$\bar{X} = \frac{\int_a^0 x\cdot y\,dx}{\int_a^0 y\,dx}, \quad \bar{Y} = \frac{\int_a^0 \frac{y}{2}y\,dx}{\int_a^0 y\,dx}$$

$$\int_a^0 y\,dx = \int_0^{\pi/2} a\sin^3\theta(-3a)\cos^2\theta\sin\theta\,d\theta = 3a^2\int_0^{\pi/2}\sin^4\theta\cos^2\theta\,d\theta$$

$$= 3a^2\frac{\Gamma\frac{5}{2}\Gamma\frac{3}{2}}{2\Gamma 4} = \frac{3a^2}{2\cdot 6}\cdot\frac{3}{2}\cdot\left(\frac{1}{2}\Gamma\frac{1}{2}\right)^2 = \frac{3\pi a^2}{32}$$

$$\int_a^0 xy\,dx = -\int_0^{\pi/2} a\cos^3\theta\, a\sin^3\theta(-3a\cos^2\theta\sin\theta)\,d\theta = 3a^3\int_0^{\pi/2}\cos^5\theta\sin^4\theta\,d\theta = 3a^3\frac{\Gamma 3\,\Gamma\frac{5}{2}}{2\Gamma\frac{11}{2}} = \frac{8a^3}{105}$$

$$\int_a^0 \frac{y^2}{2}\,dx = \frac{-1}{2}\int_0^{\pi/2} a^2\sin^6\theta(-3a\cos^2\theta\sin\theta)\,d\theta = \frac{3a^3}{2}\int_0^{\pi/2}\sin^7\theta\cos^2\theta\,d\theta = \frac{3a^3}{2}\frac{\Gamma 4\,\Gamma\frac{3}{2}}{2\Gamma\frac{11}{2}} = \frac{8a^3}{105}$$

$$\therefore \quad \bar{X} = \frac{8a^3}{105}\cdot\frac{32}{3\pi a^2} = \frac{256a}{315\pi}, \quad \bar{Y} = \frac{8a^3}{105}\cdot\frac{32}{3\pi a^2} = \frac{256a}{315\pi}$$

Otherwise also, since curve is symmetrical about $y = x$ hence $\bar{Y} = \bar{X}$. **Ans.**

8. *Find the C.G. of the area enclosed by the two parabolas* $y^2 = 4ax$ *and* $x^2 = 4ay$.

SOLUTION: $y^2 = 4ax, x^2 = 4ay$

Let $(\bar{X}, \bar{Y})$ be the C.G. of the common area then

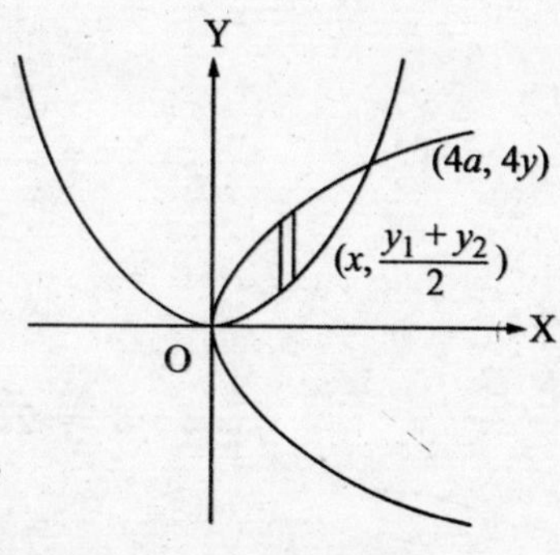

$$\bar{X} = \frac{\int_0^{4a} x(y_2 - y_1)\,dx}{\int_0^{4a}(y_2 - y_1)\,dx}, \quad \text{By symmetry } \bar{Y} = \bar{X}.$$

$$\int_0^{4a}(y_2 - y_1)\,dx = \int_0^{4a}\left(\sqrt{4ax} - \frac{x^2}{4a}\right)dx = \left[2\sqrt{a}\,x^{3/2}\cdot\frac{2}{3} - \frac{x^3}{12a}\right]_0^{4a}$$

$$= \frac{2}{3}\sqrt{a}\,4a \cdot 4\sqrt{a} - \frac{64a^3}{12a} = \frac{a^2}{3}[32-16] = \frac{16a^2}{3}$$

$$\int_0^{4a} x(y_2 - y_1)\,dx = \int_0^{4a} x\left(\sqrt{4ax} - \frac{x^2}{4a}\right)dx = \left[2\sqrt{a}\,\frac{2x^{5/2}}{5} - \frac{x^4}{4\cdot 4a}\right]_0^{4a} = \frac{48a^3}{5}$$

$\therefore$ $$\bar{X} = \frac{48a^3}{5}\cdot\frac{3}{16a^2} = \frac{9a}{5} \quad \therefore \quad \text{C.G.} = \left(\frac{9a}{5}, \frac{9a}{5}\right)$$ **Ans.**

9. *Obtain the position of the C.G. of the area bounded by the parabola $y^2 = 4x$ and the line $y = 2x - 4$.*

SOLUTION: $y^2 = 4x$ ...(1) $y = 2x - 4$ ...(2) Their points of intersection are given by

$$y^2 = 2y + 8 \quad \text{or} \quad (y-1)^2 = 3 \quad \therefore \quad y = 1 \pm \sqrt{3} = \sqrt{3} + 1, -(\sqrt{3} - 1)$$

C.G. of the elementary area strip is $\left(\frac{x_1 + x_2}{2}, y\right)$

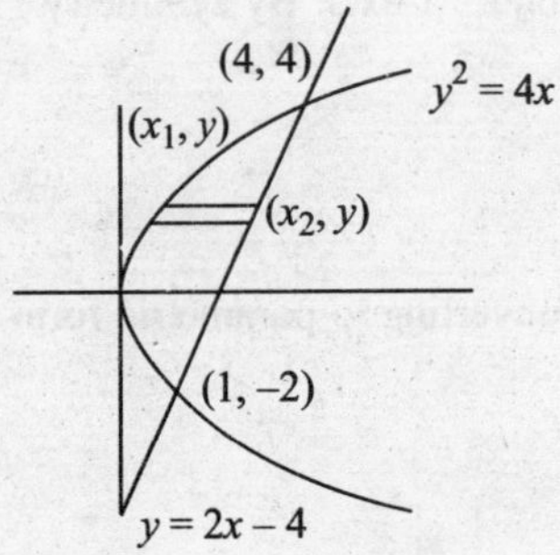

$\therefore$ $$\bar{X} = \frac{\int \left(\frac{x_1 + x_2}{2}\right)(x_2 - x_1)\,dy}{\int (x_2 - x_1)\,dy},$$

and $$\bar{Y} = \frac{\int y(x_2 - x_1)\,dy}{\int (x_2 - x_1)\,dy}$$

where $x_2 = \frac{y+4}{2}, \quad x_1 = \frac{y^2}{4}$

$$\int_{-2}^{4} (x_2 - x_1)\,dy = \int_{-2}^{4}\left(\frac{y+4}{2} - \frac{y^2}{4}\right)dy = \left[\frac{y^2}{4} + 2y - \frac{y^3}{12}\right]_{-2}^{4}$$

$$= \frac{1}{4}(16-4) + 12 - \frac{1}{12}(64+8) = 9$$

Next $$\frac{1}{2}\int_{-2}^{4}(x_2^2 - x_1^2)\,dy = \frac{1}{2}\int_{-2}^{4}\left(\left(\frac{4+y}{2}\right)^2 - \left(\frac{y^2}{4}\right)^2\right)dy = \frac{1}{2}\int_{-2}^{4}\left(4 + \frac{y^2}{4} - \frac{y^4}{16} + 2y\right)dy$$

$$= \frac{1}{2}\left[4y + \frac{y^3}{12} - \frac{y^5}{80} + y^2\right]_{-2}^{4} = \frac{1}{2}\left[16 + \frac{64}{12} - \frac{1024}{80} + 16 - \left(-8 - \frac{8}{12} + \frac{32}{80} + 4\right)\right]$$

$$= 21 - \frac{33}{5} = \frac{72}{5}$$

$\therefore$ $$\bar{X} = \frac{72}{5(9)} = \frac{8}{5}$$

and $$\int_{-2}^{4} y(x_2 - x_1)\,dy = \int_{-2}^{4} y\left(\frac{4+y}{2} - \frac{y^2}{4}\right)dy = \int_{-2}^{4}\left(2y + \frac{y^2}{2} - \frac{y^3}{4}\right)dy$$

$$= \left[y^2 + \frac{y^3}{6} - \frac{y^4}{16}\right]_{-2}^{4} = 9$$

$$\therefore \quad \bar{Y} = \frac{9}{9} = 1 \qquad \therefore \quad \text{C.G. is } \left(\frac{8}{5}, 1\right)$$ **Ans.**

10. *Find the centroid of the solid generated by revolving the part of the area of the ellipse $\frac{x^2}{a^2} + \frac{y^2}{b^2} = 1$ lying in the first quadrant, about the axis of X.*

SOLUTION: $\frac{x^2}{a^2} + \frac{y^2}{b^2} = 1$, Area in first quadrant is revolved about X-axis. By symmetry $\bar{Y} = 0$.

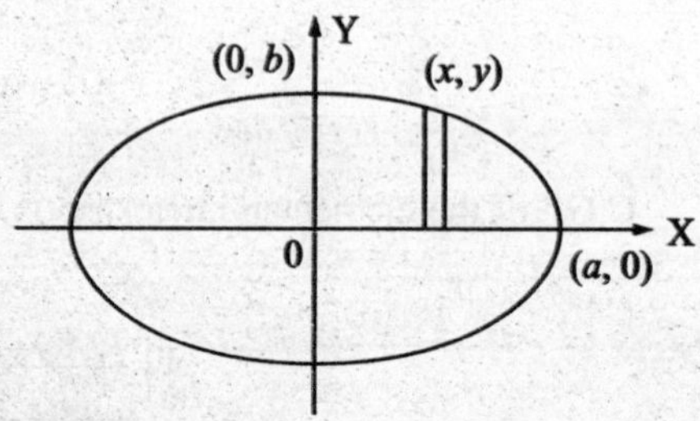

$$\bar{X} = \frac{\int x\pi y^2\,dx}{\int \pi y^2\,dx}$$

converting to parametric form $x = a\cos\theta,\ y = b\sin\theta$

$$\bar{X} = \frac{\int_0^{\pi/2} a\cos\theta\, b^2\sin^2\theta(-a\sin\theta)\,d\theta}{\int_0^{\pi/2} b^2\sin^2\theta(-a\sin\theta)\,d\theta} = \frac{a\int_0^{\pi/2}\sin^2\theta\cos\theta\,d\theta}{\int_0^{\pi/2}\sin^2\theta\,d\theta}$$

$$= a\,\frac{\dfrac{\Gamma 2\,\Gamma 1}{2\,\Gamma 3}}{\dfrac{\Gamma 2\,\Gamma\frac{1}{2}}{2\,\Gamma\frac{5}{2}}} = \frac{3a}{8} \qquad \therefore \quad \text{C.G. is } \left(\frac{3a}{8}, 0\right)$$ **Ans.**

11. *Determine the position of the C.G. of the volume of the solid generated when the area bounded by the curve $a^2 y = x^3$, the X-axis and the line $x = 2a$, is revolved about the line, $x = 2a$.*

SOLUTION: $a^2y = x^3$, $y = 0$, $x = 2a$ Area included is revolved about $x = 2a$.

The C.G. of the volume of the solid will lie on $x = 2a$, hence $\bar{X} = 2a$,

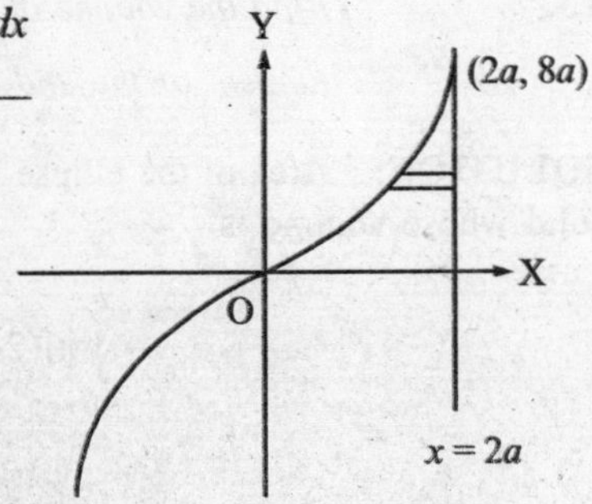

Now
$$\bar{Y} = \frac{\int_0^{8a} y(2a-x)^2\, dy}{\int_0^{8a} (2a-x)^2\, dy} = \frac{\int_0^{2a} \frac{x^3}{a^2}(2a-x)^2 \cdot \frac{3x^2}{a^2}\, dx}{\int_0^{2a} (2a-x)^2 \frac{3x^2}{a^2}\, dx}$$

$$= \frac{1}{a^2} \frac{\int_0^{2a} x^5(4a^2-4ax+x^2)\, dx}{\int_0^{2a} x^2(4a^2-4ax+x^2)\, dx}$$

$$= \frac{1}{a^2} \frac{\left[4a^2\frac{x^6}{6} - 4a\frac{x^7}{7} + \frac{x^8}{8}\right]_0^{2a}}{\left[4a^2\frac{x^3}{3} - 4a\frac{x^4}{4} + \frac{x^5}{5}\right]_0^{2a}} = \frac{1}{a^2} \frac{\left[\frac{2a^2}{3}\cdot 64a^6 - \frac{4a}{7}128a^7 + \frac{256a^8}{8}\right]}{\left[4a^2\cdot\frac{8a^3}{3} - a16a^4 + \frac{32a^5}{5}\right]}$$

$$= \frac{a\left[\frac{8}{3} - \frac{32}{7} + 2\right]}{\frac{2}{3} - 1 + \frac{2}{5}} = \frac{10a}{7} \qquad \therefore \text{ C.G. is } \left(2a, \frac{10a}{7}\right)$$ **Ans.**

12. *The loop of the curve $2ay^2 = x(a-x)^2$ revolves about the line $y = a$. Find the volume of the solid thus generated using the theorem of Pappus.*

SOLUTION: $2ay^2 = x(a-x)^2$. Its loop revolves about $y = a$

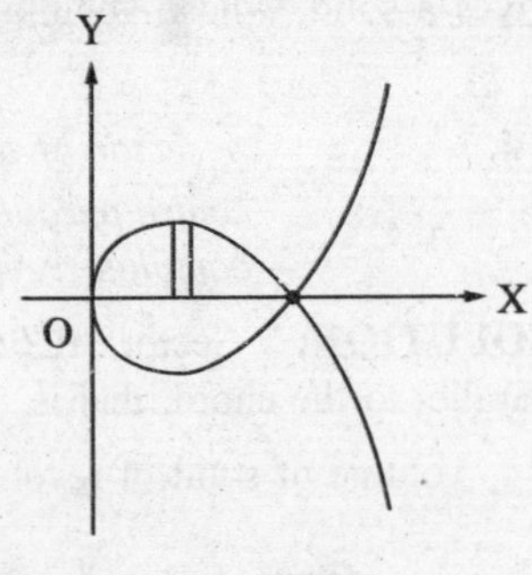

$$\text{Area of the loop} = 2\int_0^a y\, dx = 2\int_0^a \sqrt{\frac{x}{2a}}\,(a-x)\, dx$$

$$= \sqrt{\frac{2}{a}}\left[a\cdot\frac{2}{3}x\sqrt{x} - \frac{2}{5}x^2\sqrt{x}\right]_0^a$$

$$= a^2\left[\frac{2\sqrt{2}}{3} - \frac{2\sqrt{2}}{5}\right] = \frac{4\sqrt{2}}{15}a^2$$

By Pappus theorem the volume of the solid of revolution = Area of the loop × circumference of the circle of revolution by the C.G. of the area

The C.G. of the area of the loop lies on x-axis

∴ C.G. will describe a circle of circumference $2\pi a$ when revolved about $y = a$.

∴ Volume of the solid of revolution $= \frac{4\sqrt{2}}{15}a^2 \cdot 2\pi a = \frac{8\sqrt{2}}{15}\pi a^3$ **Ans.**

13. *Find the volume of the solid generated on revolving the ellipse* $\frac{x^2}{a^2}+\frac{y^2}{b^2}=1$ *about the tangent at the end of (i) the major axis, (ii) the minor axis.*

SOLUTION: Area of the ellipse when revolved about $x = a$, the tangent at one of major axis gives a solid whose volume is

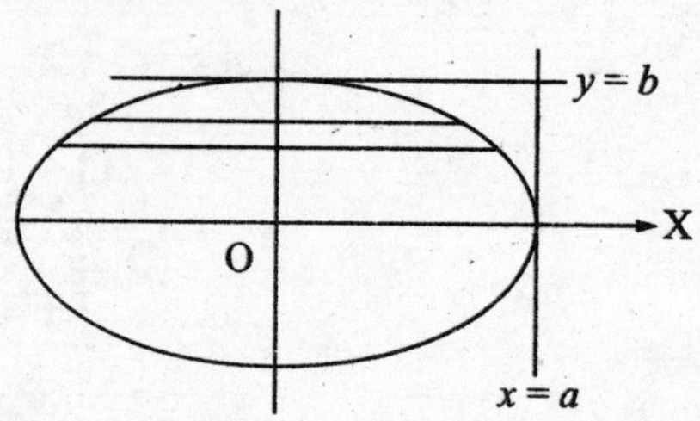

$$= \int_{-b}^{b} \pi[(2a-x)^2 - x^2]\,dy$$

$$= \pi \int_{-a}^{a} (4a^2 - 4ax)\frac{dy}{dx}\,dx$$

$$= 4a\pi \int_{-a}^{a} \frac{(a-x)(-bx)}{a\sqrt{a^2-x^2}}\,dx \qquad \left(\text{since } y = \frac{b}{a}\sqrt{a^2-x^2}\ \therefore \frac{dy}{dx} = \frac{-bx}{a\sqrt{a^2-x^2}}\right)$$

$$= 4\pi b\left[\int_{-a}^{a} \frac{-ax\,dx}{\sqrt{a^2-x^2}} + \int_{-a}^{a} \frac{bx^2\,dx}{\sqrt{a^2-x^2}}\right] = 0 + 8\pi b \int_0^a \frac{x^2\,dx}{\sqrt{a^2-x^2}}$$

(by property of definite integrals)

$$= -8\pi b \int_0^{\pi/2} \frac{a^2\sin^2\theta \cdot a\cos\theta\,d\theta}{a\cos\theta} \qquad \text{(On putting } x = a\sin\theta)$$

$$= -4\pi a^2 b \int_0^{\pi/2} (1-\cos 2\theta)\,d\theta = -4\pi a^2 b\left[\theta - \frac{1}{2}\sin 2\theta\right]_0^{\pi/2} = 2\pi^2 a^2 b$$ **Ans.**

Similarly when the area is revolved about $y = b$, the tangent at one end of the minor axis of the ellipse, gives a solid whose volume is $2\pi^2 ab^2$. **Ans.**

14. *A sector of a circle whose central angle in* 2α *is rotated about the line through the centre parallel to the chord of the arc. Find the surface and the volume generated by the complete revolution.*

SOLUTION: Sector *ACBOA* has the chord *AB* and it revolves about *a* line through its centre and parallel to the chord, that is, *X*-axis.

Volume of solid of revolution *V* is equal to

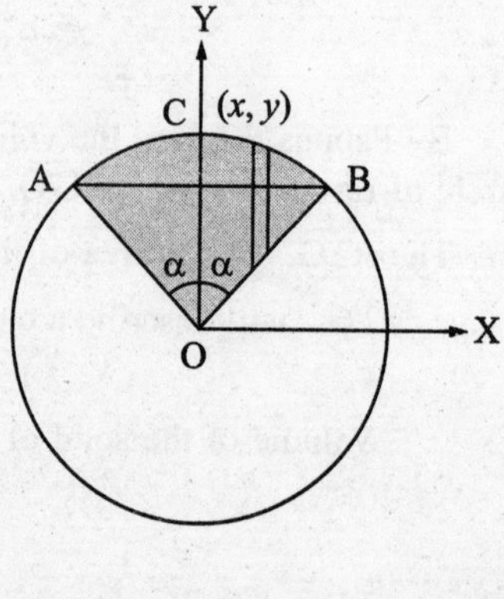

$$\int_{-a\sin\alpha}^{a\sin\alpha} \pi(y^2 - x^2\cot^2\alpha)\,dx = \pi \int_{-a\sin\alpha}^{a\sin\alpha} (a^2 - x^2 - x^2\cot^2\alpha)\,dx$$

$$= 2\pi \int_0^{a\sin\alpha} (a^2 - x^2\text{cosec}^2\alpha)\,dx$$

$$= 2\pi\left[a^2x - \frac{x^3}{3}\text{cosec}^2\alpha\right]_0^{a\sin\alpha}$$

$$= 2\pi\left[a^3 \sin\alpha - \frac{a^3 \sin^3\alpha}{3}\,\text{cosec}^2\alpha\right] = 2\pi a^3 \sin\alpha\left(1-\frac{1}{3}\right) = \frac{4\pi a^3}{3}\sin\alpha.$$ **Ans.**

Further the surface of solid of revolution is equal to

$$\int_{-a\sin\alpha}^{a\sin\alpha} 2\pi(y - x\cot\alpha)\frac{ds}{dx}\,dx$$

$$\left[x^2+y^2=a^2 \;\therefore\; \frac{dy}{dx}=\frac{-x}{y} \text{ and } \frac{ds}{dx}=\sqrt{1+\left(\frac{x}{y}\right)^2}=\frac{a}{y}=\frac{a}{\sqrt{a^2-x^2}}\right]$$

$$= 2\pi \int_{-a\sin\alpha}^{a\sin\alpha}\left(\sqrt{a^2-x^2} - x\cot\alpha\right)\frac{a}{\sqrt{a^2-x^2}}\,dx = 4\pi\int_0^{a\sin\alpha} a\,dx - 0$$

(by property of definite integrals.)

$$= 4a\pi \cdot a\sin\alpha = 4\pi a^2 \sin\alpha$$ **Ans.**

15. *The figure bounded by a quadrant of a circle of radius a and the tangents at its extremities, revolves about one of the tangents. Find the volume of solid generated.*

SOLUTION: The volume V of solid of revolution about AC, is given by

$$V = \int_0^a \pi(a-x)^2\,dy = \pi\int_0^a (a-x)^2 \frac{(-x)\,dx}{\sqrt{a^2-x^2}} \qquad \left[\text{as } y=\sqrt{a^2-x^2}\right]$$

$$= \pi\int_0^{\pi/2}(a-a\sin\theta)^2\frac{a\sin\theta}{a\cos\theta}\,a\cos\theta\,d\theta \qquad (\text{On putting } x = a\sin\theta)$$

$$= \pi a^3\int_0^{\pi/2}(\sin\theta - 2\sin^2\theta + \sin^3\theta)\,d\theta$$

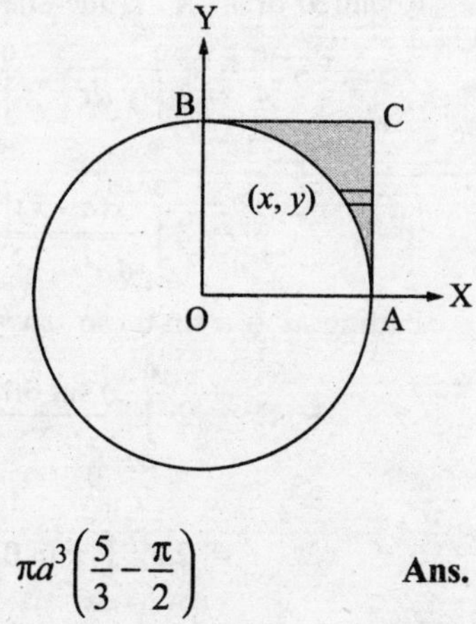

$$= \pi a^3\left[\frac{\Gamma 1\,\Gamma\frac{1}{2}}{2\,\Gamma\frac{3}{2}} - \frac{2\,\Gamma\frac{3}{2}\,\Gamma\frac{1}{2}}{2\,\Gamma 2} + \frac{\Gamma 2\,\Gamma\frac{1}{2}}{2\,\Gamma\frac{5}{2}}\right]$$

$$= \pi a^3\left[\frac{\Gamma\frac{1}{2}}{2\cdot\frac{1}{2}\Gamma\frac{1}{2}} - \frac{1}{2}\left(\Gamma\frac{1}{2}\right)^2 + \frac{1\cdot\Gamma\frac{1}{2}}{2\cdot\frac{3}{2}\cdot\frac{1}{2}\Gamma\frac{1}{2}}\right] = \pi a^3\left(1-\frac{\pi}{2}+\frac{2}{3}\right) = \pi a^3\left(\frac{5}{3}-\frac{\pi}{2}\right)$$ **Ans.**

ADDITIONAL IMPORTANT PROBLEMS

1. *Find the area included between the curve $y^2(a - x) = x^3$ and its asymptote.*

SOLUTION: Equation of the curve is $y = \dfrac{\pm x^{3/2}}{\sqrt{a-x}}$ whose asymptote is the line $x = a$. The shape of the curve is as follows.

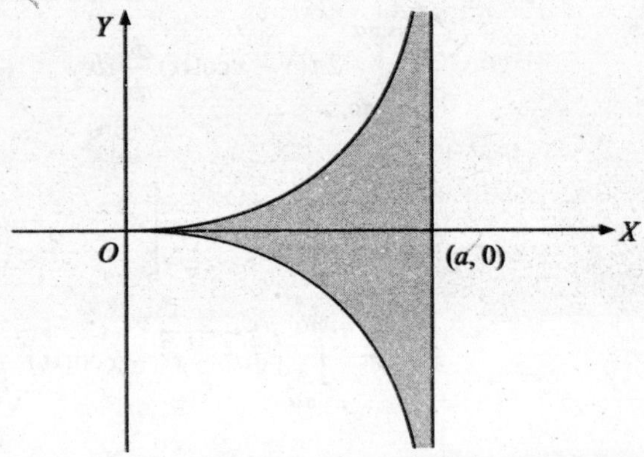

The required area A (shaded) is given by

$$A = 2\int_0^a y\,dx = 2\int_0^a \frac{x^{3/2}}{\sqrt{a-x}}dx$$

$$= 2\int_0^{\pi/2} \frac{(a\sin^2\theta)^{3/2}\cdot 2a\sin\theta\cos\theta\,d\theta}{\sqrt{a-a\sin^2\theta}}$$

(on putting $x = a\sin^2\theta$)

$$= 4a^2\int_0^{\pi/2}\sin^4\theta\,d\theta = 4a^2\cdot\frac{3}{4}\cdot\frac{1}{2}\cdot\frac{\pi}{2} = \frac{3\pi a^2}{4}.$$ **Ans.**

2. *Find the area enclosed by the curve $x(x^2 + y^2) = a(x^2 - y^2)$ and its asymptote.*

SOLUTION: Equation of the curve can be written as

$y = \pm\, x\sqrt{\dfrac{a-x}{a+x}}$ whose asymptote is the line $x = -a$.

Shape of the curve is given below.

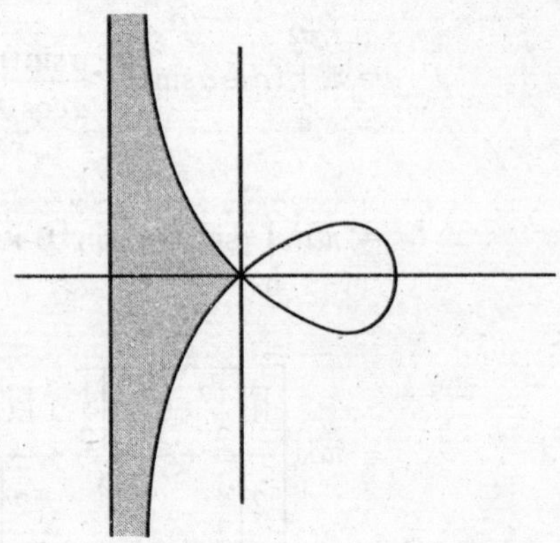

Required area A is the shaded area equal to

$$A = 2\int_{-a}^{0} y\,dx = 2\int_{-a}^{0} x\sqrt{\frac{a-x}{a+x}}\,dx$$

$$= 2\int_{-a}^{0}\frac{x(a-x)}{\sqrt{a^2-x^2}}dx$$

Putting $x = a\sin\theta$ so $dx = a\cos\theta\,d\theta$

$$\therefore \quad A = 2\int_{-\pi/2}^{0}\frac{a\sin\theta(a-a\sin\theta)\,a\cos\theta\,d\theta}{a\cos\theta}$$

$$= 2a^2\int_{-\pi/2}^{0}(\sin\theta-\sin^2\theta)\,d\theta = 2a^2\int_{-\pi/2}^{0}\left[\sin\theta-\frac{1-\cos 2\theta}{2}\right]d\theta$$

$$= 2a^2\left[-\cos\theta-\frac{\theta}{2}+\frac{1}{4}\sin 2\theta\right]_{-\pi/2}^{0} = 2a^2\left(-1-\frac{\pi}{4}\right)$$

$\therefore$ Required area $= |A| = a^2\left(2+\dfrac{\pi}{2}\right)$. **Ans.**

3. *Find the area enclosed between the curve $y(x^2 + 4a^2) = 8a^3$ and its asymptote.*

SOLUTION: Equation of the curve can be written as

$x^2 = \dfrac{4a^2(2a-y)}{y}$. Its asymptote is the line $y = 0$, i.e., X-axis shape of the curve is as follows.

Required area A is shown as shaded.

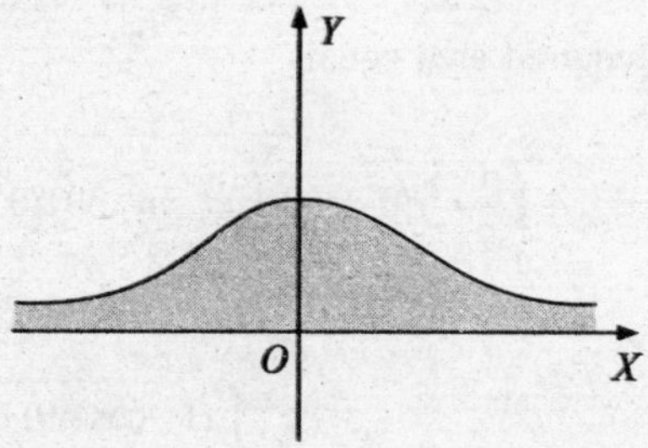

$$\therefore \quad A = 2\int_0^{\infty} y\,dx = 2\int_0^{\infty} \frac{8a^3\,dx}{x^2+4a^2}$$

$$= 16a^3 \frac{1}{2a}\left[\tan^{-1}\frac{x}{2a}\right]_0^{\infty}$$

$$= 8a^2 \cdot \frac{\pi}{2} = 4\pi a^2.$$

Ans.

4. *Find the area enclosed between the parabola $x^2 = 4y$ and the line $x = 4y - 2$.*

SOLUTION: Required area A is the shaded portion in the adjoining figure. Points of intersection of the parabola and the line are $M\left(-1, \frac{1}{4}\right)$ and $N(2, 1)$

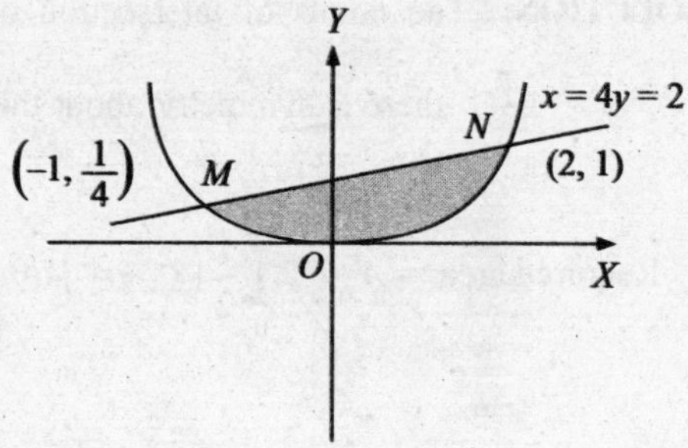

$$A = \int_{-1}^{2} (y_1 - y_2)\,dx = \int_{-1}^{2}\left(\frac{x+2}{4} - \frac{x^2}{4}\right)dx$$

$$= \left[\frac{x^2}{8} + \frac{x}{2} - \frac{x^3}{12}\right]_{-1}^{2} = \frac{4}{8} + \frac{2}{2} - \frac{8}{12} - \frac{1}{8} + \frac{1}{2} - \frac{1}{12}$$

$$= \frac{9}{8}.$$

5. *Find the area bounded by the curve $r = a \cos 3\theta$.*

SOLUTION: It is a three leaved rose as shown in the figure. In the first quadrant there is half-leaf, θ varying from $\theta = 0$ to $\theta = \dfrac{\pi}{6}$.

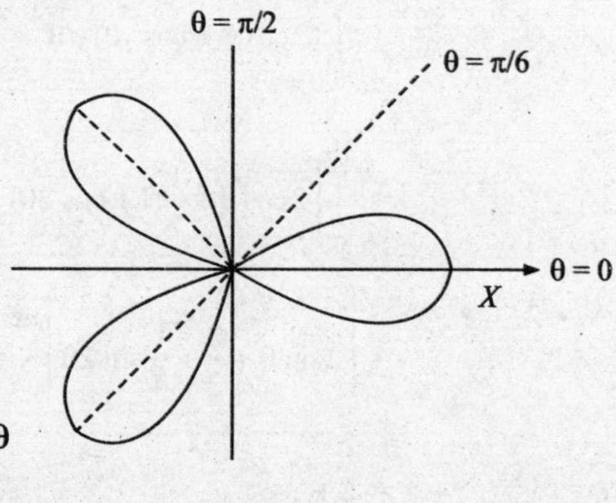

$$\text{Required area} = 6\int_0^{\pi/6} \frac{1}{2} r^2 d\theta$$

$$= 3\int_0^{\pi/6} a^2 \cos^2 3\theta\, d\theta = \frac{3a^2}{2}\int_0^{\pi/6} (1 + \cos 6\theta)\, d\theta$$

$$= \frac{3a^2}{2}\left[\theta + \frac{1}{6}\sin 6\theta\right]_0^{\pi/6} = \frac{3a^2}{2}\left[\frac{\pi}{6} + 0\right] = \frac{\pi a^2}{4}.$$

Ans.

6. *Find the area of the curve* $r = a \sin 2\theta$.

SOLUTION: It is a four leaved rose as shown in the figure.

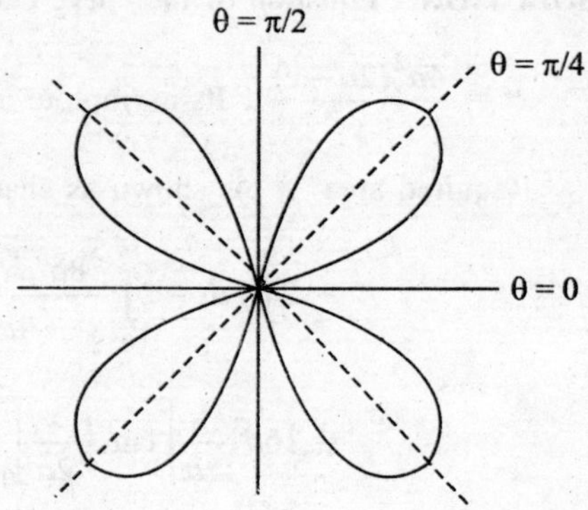

One leaf is in first quadrant, θ ranging from 0 to $\frac{\pi}{2}$.

Required area equals

$$4\int_0^{\pi/2} \frac{1}{2} r^2 d\theta = 2\int_0^{\pi/2} a^2 \sin^2 2\theta \, d\theta$$

$$= a^2 \int_0^{\pi/2} (1 - \cos 4\theta)\, d\theta = a^2 \left[\theta - \frac{1}{4}\sin 4\theta\right]_0^{\pi/2}$$

$$= a^2\left(\frac{\pi}{2} + 0\right) = \frac{\pi a^2}{2}.$$

Ans.

7. *Find the area lying inside the cardioid* $r = 1 + \cos\theta$ *and outside the circle* $r = 1$.

SOLUTION: The points of intersection of $r = 1 + \cos\theta$ and $r = 1$ are given by $\cos\theta = 0$, i.e., $\theta = \pm\frac{\pi}{2}$, there is symmetry about the initial line.

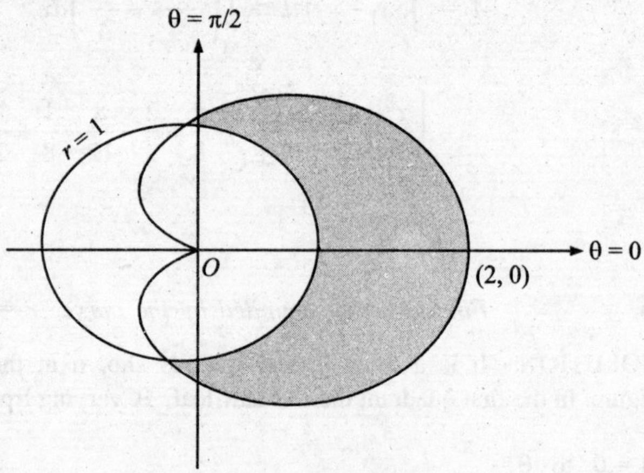

$$\text{Required area} = A = 2\int_0^{\pi/2} \frac{1}{2}\left(r_1^2 - r_2^2\right) d\theta$$

$$= \int_0^{\pi/2} [(1 + \cos\theta)^2 - 1^2]\, d\theta$$

$$= \int_0^{\pi/2} (2\cos\theta + \cos^2\theta)\, d\theta$$

$$= \int_0^{\pi/2} \left[2\cos\theta + \frac{1}{2}(1 + \cos 2\theta)\right] d\theta$$

$$= \left[2\sin\theta + \frac{\theta}{2} + \frac{1}{4}\sin 2\theta\right]_0^{\pi/2} = 2 + \frac{1}{2}\cdot\frac{\pi}{2} + 0$$

$$= 2 + \frac{\pi}{4}.$$

Ans.

8. *Prove that the length of an arc of the curve* $y^2 = x\left(1-\frac{x}{3}\right)^2$ *from the origin to the point* (x, y) *is given by* $s^2 = y^2 + \frac{4}{3}x^2$. *Hence rectify the loop.*

SOLUTION: $y = \sqrt{x}\left(1-\frac{x}{3}\right) = \sqrt{x} - \frac{1}{3}x^{3/2}$

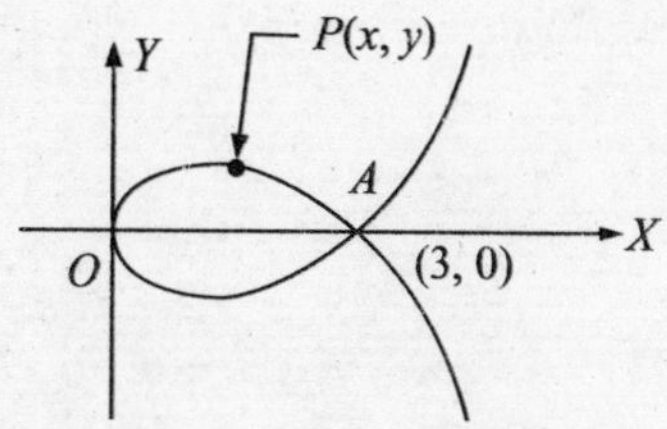

$\therefore \quad \frac{dy}{dx} = \frac{1}{2}x^{-1/2} - \frac{1}{3}\frac{3}{2}x^{1/2} = \frac{1-x}{2\sqrt{x}}$

Length S of the arc OP, is given by

$$S = \int_0^x \sqrt{1+\left(\frac{dy}{dx}\right)^2}\,dx = \int_0^x \sqrt{1+\frac{(1-x)^2}{4x}}\,dx$$

$$= \int_0^x \sqrt{\frac{(1+x)^2}{4x}}\,dx = \int_0^x \frac{1+x}{2\sqrt{x}}\,dx = \frac{1}{2}\int_0^x (x^{-1/2}+x^{1/2})\,dx = \frac{1}{2}\left[\frac{x^{1/2}}{1/2}+\frac{x^{3/2}}{3/2}\right]_0^x = \sqrt{x}\left(1+\frac{x}{3}\right)$$

or $$S^2 = x\left(1+\frac{x}{3}\right)^2 = x\left(1-\frac{x}{3}\right)^2 + \frac{4}{3}x^2 = y^2 + \frac{4}{3}x^2$$ **Hence Proved.**

Length of the upperhalf of the loop from O to $A = \left[\sqrt{y^2+\frac{4}{3}x^2}\right]_{(x=3,\, y=0)} = \sqrt{0+\frac{4}{3}\cdot 9} = 2\sqrt{3}$

$\therefore$ Length of the complete loop $= 4\sqrt{3}$ **Ans.**

9. *Obtain the length of the parabola* $y^2 = 4ax$ *from the vertex to the end of the latus return. Also obtain the length of the arc cut off by the line* $3y = 8x$.

SOLUTION: $x = \frac{y^2}{4a} \quad \therefore \quad \frac{dx}{dy} = \frac{y}{2a}$

Length S of the arc from vertex $(0, 0)$ to $(a, 2a)$, is given by

$$S = \int_0^{2a} \sqrt{1+\left(\frac{dx}{dy}\right)^2}\,dy = \int_0^{2a} \sqrt{1+\frac{y^2}{4a^2}}\,dy = \frac{1}{2a}\int_0^{2a} \sqrt{y^2+4a^2}\,dy$$

$$= \frac{1}{2a}\left[\frac{y}{2}\sqrt{y^2+4a^2} + \frac{4a^2}{2}\log\left(y+\sqrt{y^2+4a^2}\right)\right]_0^{2a}$$

$$= \frac{1}{2a}\left[a\cdot 2a\sqrt{2} + 2a^2\log(2a+2a\sqrt{2}) - 2a^2\log 2a\right]$$

$$= a[\sqrt{2}+\log(1+\sqrt{2})]$$ **Ans.**

The line $3y = 8x$ meets the parabola $y^2 = 4ax$ at O $(0, 0)$ and $A\left(\frac{9a}{16}, \frac{3a}{2}\right)$

$$\therefore \quad \text{Required length of arc} = \frac{1}{2a}\int_0^{3a/2} \sqrt{y^2+4a^2}\,dy$$

$$= \frac{1}{2a}\left[\frac{y}{2}\sqrt{y^2+4a^2} + \frac{4a^2}{2}\log\left(y+\sqrt{y^2+4a^2}\right)\right]_0^{3a/2}$$

$$= \frac{1}{2a}\left[\frac{3a}{4}\sqrt{\frac{9a^2}{4}+4a^2} + 2a^2\left\{\log\left(\frac{3a}{2}+\sqrt{\frac{9a^2}{4}+4a^2}\right) - \log 2a\right\}\right]$$

$$= \frac{1}{2a}\left[\frac{3a}{4}\cdot\frac{5a}{2} + 2a^2\log\left\{\left(\frac{3a}{2}+\frac{5}{a}\right)2a\right\}\right]$$

$$= \frac{1}{2a}\left[\frac{15a^2}{8} + 2a^2\log 2\right] = a\left[\log 2 + \frac{15}{16}\right]$$ **Ans.**

10. *Find the length of the arc of the equiangular spiral $r = ae^{\theta \cot \alpha}$ from the point corresponding to $\theta = 0$ to the point at $\theta = \tan \alpha$.*

SOLUTION: $r = ae^{\theta \cot \alpha}$ $\quad \therefore \quad \dfrac{dr}{d\theta} = a \cot \alpha\, e^{\theta \cot \alpha} = r \cot \alpha$

$$\text{Length of the arc} = \int_0^{\alpha} \sqrt{r^2 + \left(\frac{dr}{d\theta}\right)^2}\, d\theta = \int_0^{\alpha} \sqrt{r^2 + r^2 \cot^2 \alpha}\, d\theta$$

$$= \int_0^{\alpha} r \operatorname{cosec} \alpha\, d\theta = \operatorname{cosec} \alpha \int_0^{\alpha} ae^{\theta \cot \alpha}\, d\theta$$

$$= a \operatorname{cosec} \alpha \left[\frac{e^{\theta \cot \alpha}}{\cot \alpha}\right]_0^{\alpha} = \frac{a \operatorname{cosec} \alpha}{\cot \alpha}(e-1)$$

$$= a\,(e-1) \sec \alpha.$$ **Ans.**

11. *Find the volume of the solid generated by revolving the region bounded by the curve $y = \log x$ and $x = 2$ about the X-axis.*

SOLUTION: Solid is generated by revolving the region about X-axis, x varying from $x = 1$ to $x = 2$, the volume V is given by

$$V = \int_1^2 \pi y^2 dx$$

or $$V = \pi \int_1^2 (\log x)^2 dx \quad \text{(integrating by parts)}$$

$$= \pi\left[(\log x)^2 \cdot x - \int 2 \log x \cdot \frac{1}{x} \cdot x\, dx\right]_1^2$$

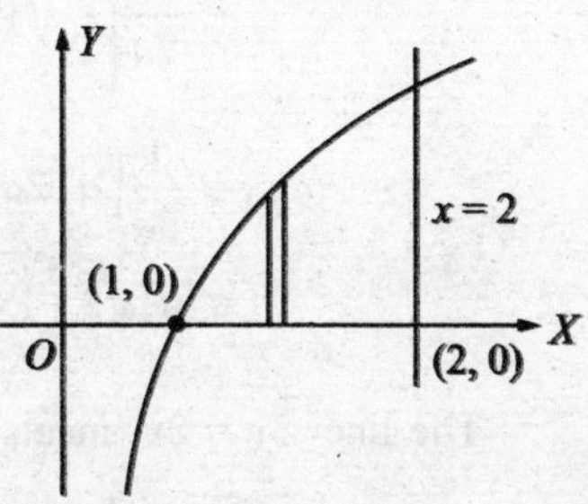

$$= \pi\left[2(\log 2)^2 - 2\int_1^2 \log x\,dx\right]$$

$$= \pi\left[2(\log 2)^2 - 2\left\{\log x\cdot x - \int \frac{1}{x}x\,dx\right\}_1^2\right]$$

$$= 2\pi\ [(\log)^2 - 2\log 2 + 1] = 2\pi(1-\log 2)^2$$ **Ans.**

12. *Find the volume of the solid generated by revolving the curve* $y^2 = \dfrac{x^3}{2a-x}$ *about its asymptote.*

SOLUTION: $y^2 = \dfrac{x^3}{2ax}$ its asymptote is $x = 2a$.

The region bounded by the curve and its asymptote is revolved about the asymptote. The volume V of the solid generated is

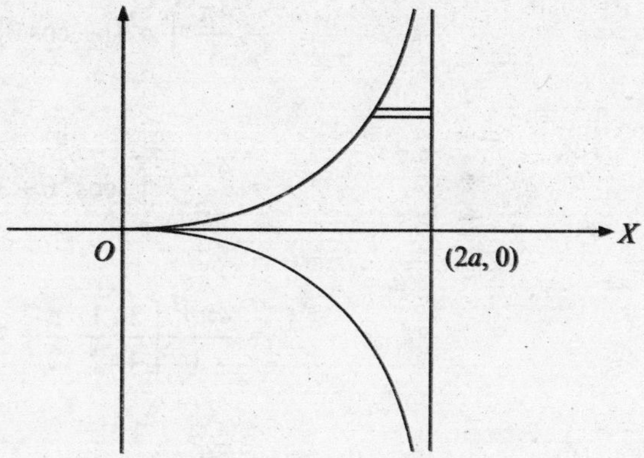

$$V = 2\int_0^\infty \pi(2a-x)^2\,dy$$

$$= 2\pi\int_0^{2a}(2a-x)^2\frac{dy}{dx}dx$$

Since $y = \dfrac{x^{3/2}}{\sqrt{2a-x}}$ $\therefore$ $\dfrac{dy}{dx} = \dfrac{\sqrt{2a-x}\cdot\frac{3}{2}x^{1/2} - x^{3/2}\dfrac{(-1)}{2\sqrt{2a-x}}}{(2a-x)}$

or $\dfrac{dy}{dx} = \dfrac{3\sqrt{x}\,(2a-x)+x\sqrt{x}}{2(2a-x)^{3/2}} = \dfrac{\sqrt{x}(3a-x)}{(2a-x)^{3/2}}$

$$\therefore\quad V = 2\pi\int_0^{2a}\frac{(2a-x)^2\sqrt{x}(3a-x)}{(2a-x)^{3/2}}dx = 2\pi\int_0^{2a}(3a-x)\sqrt{x}\,\sqrt{2a-x}\,dx$$

Putting $x = 2a\sin^2\theta$ so that $dx = 4a\sin\theta\cos\theta\,d\theta$

$$\therefore\quad V = 2\pi\int_0^{\pi/2}(3a-2a\sin^2\theta)\sqrt{2a}\sin\theta\cdot\sqrt{2a}\cos\theta\cdot 4a\sin\theta\cos\theta\,d\theta$$

$$= 16\pi a^3\int_0^{\pi/2}(3-2\sin^2\theta)\sin^2\theta\cos^2\theta\,d\theta$$

$$= 16\pi a^3\left[3\int_0^{\pi/2}\sin^2\theta\cos^2\theta\,d\theta - 2\int_0^{\pi/2}\sin^4\theta\cos^2\theta\,d\theta\right]$$

$$= 16\pi a^3\left[3\cdot\frac{1}{4}\cdot\frac{1}{2}\cdot\frac{\pi}{2} - 2\cdot\frac{3}{6}\cdot\frac{1}{4}\cdot\frac{1}{2}\frac{\pi}{2}\right] = 2\pi^2a^3$$ **Ans.**

13. *The area of the cardiode $r = a(1 + \cos\theta)$ included between $\theta = \frac{-\pi}{2}$ and $\theta = \frac{\pi}{2}$ is rotated about the line $\theta = \frac{\pi}{2}$. Find the volume of the solid of revolution.*

SOLUTION: Required volume V is given by

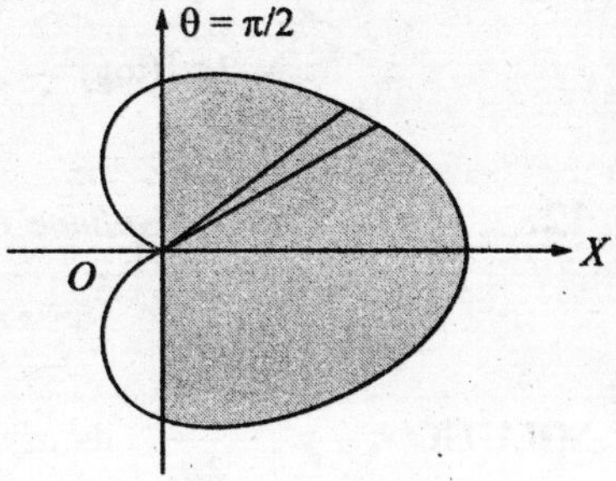

$$\int_{-\pi/2}^{\pi/2} \frac{2}{3}\pi r^2 \cdot r\cos\theta\, d\theta = 2\int_0^{\pi/2} \frac{2}{3}\pi r^3 \cos\theta\, d\theta$$

$$= \frac{4\pi}{3}\int_0^{\pi/2} a^3(1+\cos\theta)^3 \cos\theta\, d\theta$$

$$= \frac{4\pi a^3}{3}\int_0^{\pi/2} (\cos^4\theta + 3\cos^3\theta + 3\cos^2\theta + \cos\theta)\, d\theta$$

$$= \frac{4\pi a^3}{3}\left[\frac{3}{4}\cdot\frac{1}{2}\cdot\frac{\pi}{2} + 3\cdot\frac{2}{3}\cdot 1 + 3\cdot\frac{1}{2}\cdot\frac{\pi}{2} + 1\right]$$

$$= \frac{\pi a^3}{4}(5\pi + 16).$$ **Ans.**

14. *Find the area of the surface of revolution generated by revolving the cubic parabola $x = y^3$ from $y = 0$ to $y = 2$ about Y-axis.*

SOLUTION: $x = y^3 \quad \therefore \quad \frac{dx}{dy} = 3y^2$ and $\frac{ds}{dy} = \sqrt{1 + \left(\frac{dx}{dy}\right)^2} = \sqrt{1 + 9y^4}$.

The area S of the surface generated on revolving the region about Y-axis when y varies from $y = 0$ to $y = 2$, is given by

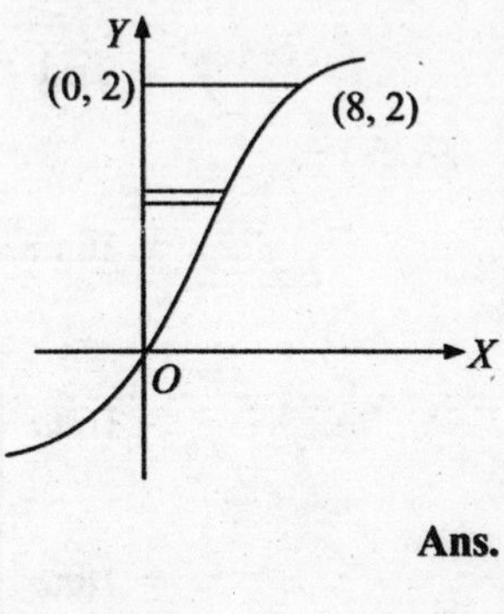

$$S = \int_0^2 2\pi x\, dy \int_0^2 2\pi\sqrt{1 + \left(\frac{dx}{dy}\right)^2}\, dy = \int_0^2 2\pi\, y^3\sqrt{1 + 9y^4}\, dy$$

$$= \frac{2\pi}{36}\int_0^2 (1 + 9y^4)^{1/2}(36y^3)\, dy = \frac{\pi}{18}\left[\frac{(1 + 9y^4)^{3/2}}{3/2}\right]_0^2$$

$$= \frac{\pi}{27}(145\sqrt{145} - 1).$$ **Ans.**

15. *Find the surface area generated by revolving the loop of the curve $9ay^2 = x(3a - x)^2$ about the axis of X.*

SOLUTION: Curve $9ay^2 = x(3a - x)^2$ is symmetrical about X-axis and meets the X-axis at (0, 0) and (3*a*, 0).

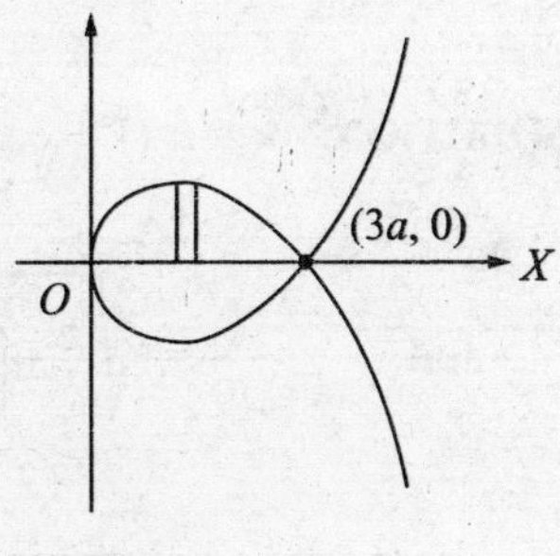

From the curve, $18ay\dfrac{dy}{dx} = (3a - x)^2 - 2x(3a - x)$

$$\therefore \quad \frac{dy}{dx} = \frac{(3a-x)(a-x)}{6ay}$$

and $$\frac{ds}{dx} = \sqrt{1+\left(\frac{dy}{dx}\right)^2} = \sqrt{1-\frac{(3a-x)^2(a-x)^2}{36a^2y^2}}$$

$$= \frac{1}{6ay}\sqrt{36a^2y^2-(3a-x)^2(a-x)^2} = \frac{1}{6ay}\sqrt{4ax(3a-x)^2+(3a-x)^2(a-x)^2}$$

$$= \frac{1}{6ay}\sqrt{(3a-x)^2(a+x)^2} = \frac{(3a-x)(a+x)}{6ay}$$

$$\therefore \quad \text{Required surface area} = \int_0^{3a} 2\pi y\frac{ds}{dy}dy = 2\pi\int_0^{3a} y\cdot\frac{(3a-x)(a+x)}{6ay}dy$$

$$= \frac{\pi}{3a}\int_0^{3a}(3a^2+2ax-x^2)\,dx = \frac{\pi}{3a}\left[3a^2x+ax^2-\frac{x^3}{3}\right]_0^{3a}$$

$$= 3\pi a^2$$

Ans.

16. *The curve $r = e^{\theta/2}$ is revolved about the initial line. Find the surface of revolution when its part between $\theta = 0$ to $\theta = \pi$ is revolved.*

SOLUTION: $r = e^{\theta/2} \quad \therefore \quad \dfrac{dr}{d\theta} = \dfrac{1}{2}e^{\theta/2}$

$$\therefore \quad \frac{ds}{d\theta} = \sqrt{r^2+\left(\frac{dr}{d\theta}\right)^2} = \sqrt{e^\theta+\frac{1}{4}e^\theta} = \frac{\sqrt{5}}{2}e^{\theta/2}$$

When the part of the curve from $\theta = 0$ to $\theta = \pi$ is revolved about the initial line the surface area S thus generated is given by

$$S = \int_0^\pi 2\pi y\frac{ds}{d\theta}d\theta = \int_0^\pi 2\pi r\sin\theta\cdot\frac{\sqrt{5}}{2}e^{\theta/2}d\theta$$

$$= \pi\sqrt{5}\int_0^\pi e^{\theta/2}\sin\theta\, e^{\theta/2}d\theta = \pi\sqrt{5}\int_0^\pi e^\theta\sin\theta\, d\theta$$

$$= \pi\sqrt{5}\left[\frac{e^\theta}{2}(1\sin\theta-\cos\theta)\right]_0^\pi = \frac{\pi}{2}\sqrt{5}(e^\pi+1)$$

17. *The arc of the cardiode $r = a(1 + \cos\theta)$ included between $\theta = \dfrac{-\pi}{2}$ and $\theta = \dfrac{\pi}{2}$ is revolved about the line $\theta = \dfrac{\pi}{2}$. Find the area of the surface generated.*

SOLUTION: $r = a(1 + \cos\theta)$ $\therefore$ $\dfrac{dr}{d\theta} = -a\sin\theta$

and $$\frac{ds}{d\theta} = \sqrt{r^2 + \left(\frac{dr}{d\theta}\right)^2} = \sqrt{a^2(1+\cos\theta)^2 + a^2\sin^2\theta} = a\sqrt{2+2\cos\theta} = 2a\cos\frac{\theta}{2}$$

When cardiode between $\theta = \dfrac{-\pi}{2}$ to $\theta = \dfrac{\pi}{2}$ is revolved about the line $\theta = \dfrac{\pi}{2}$ the surface area S thus generated, is given by

$$S = 2\int_0^{\pi/2} 2\pi y \frac{ds}{d\theta}\, d\theta = 4\pi\int_0^{\pi/2} r\cos\theta \cdot 2a\cos\frac{\theta}{2}\, d\theta$$

$$= 8\pi a^2 \int_0^{\pi/2} (1+\cos\theta)\cos\theta\cdot\cos\frac{\theta}{2} d\theta$$

$$= 8\pi a^2 \int_0^{\pi/2} 2\cos^2\frac{\theta}{2}\cdot\left(1-2\sin^2\frac{\theta}{2}\right)\cos\frac{\theta}{2} d\theta$$

$$= 16\pi a^2 \int_0^{\pi/2} \left(1-\sin^2\frac{\theta}{2}\right)\left(1-2\sin^2\frac{\theta}{2}\right)\cos\frac{\theta}{2}\, d\theta \qquad \left(\text{now put } \sin\frac{\theta}{2} = t\right)$$

$$S = 32\pi a^2 \int_0^{1/\sqrt{2}} (1-t^2)(1-2t^2)\, dt$$

$$= 32\pi a^2 \int_0^{1/\sqrt{2}} (1-3t^2+2t^4)\, dt = 32\pi a^2\left[t - t^3 + \frac{2}{5}t^5\right]_0^{1/\sqrt{2}}$$

$$= 32\pi a^2\left[\frac{1}{\sqrt{2}} - \frac{1}{2\sqrt{2}} + \frac{2}{5}\cdot\frac{1}{4\sqrt{2}}\right]$$

$$= \frac{48\sqrt{2}}{5}\pi a^2.$$ **Ans.**

18. *Find the area of the surface of revolution of a quadrant of a circular arc obtained by revolving it about a tangent at one of its ends.*

SOLUTION: Let the circle be $x^2 + y^2 = a^2$.

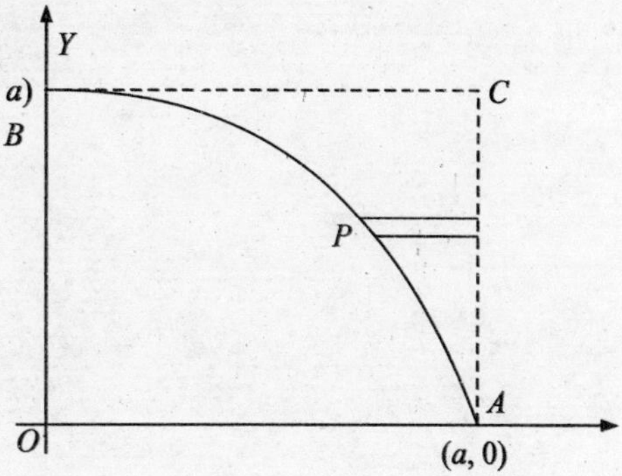

$$\therefore \quad 2x + 2y\frac{dy}{dx} = 0 \quad \text{or} \quad \frac{dy}{dx} = \frac{-x}{y}$$

and
$$\frac{ds}{dx} = \sqrt{1+\left(\frac{dy}{dx}\right)^2} = \sqrt{1+\frac{x^2}{y^2}} = \frac{a}{y}$$

If $P(x, y)$ is any point on the arc, then revolving the arc APB about the line AC, the surface S of revolution is

$$S = \int_0^a 2\pi\,(a-x)\frac{ds}{dx}dx = 2\pi\int_0^a (a-x)\frac{a}{y}dx$$

$$= 2\pi a\int_0^a \frac{(a-x)\,dx}{\sqrt{a^2-x^2}} = 2\pi a\left[\int_0^a\left(\frac{a}{\sqrt{a^2-x^2}} - \frac{x}{\sqrt{a^2-x^2}}\right)dx\right]$$

$$= 2\pi a\left[a\sin^{-1}\frac{x}{a} + \sqrt{a^2-x^2}\right]_0^a = 2\pi a\left(a\cdot\frac{\pi}{2} - a\right) = \pi a^2(\pi - 2).$$ **Ans.**

19. *Show that the surface generated by the revolution of the tractrix* $x = a\cos t + \frac{a}{2}\log\tan^2\frac{t}{2}$, $y = a\sin t$ *about its asymptote, is equal to the surface area of the sphere of radius* a.

SOLUTION: $x = a\cos t + a\log\tan\frac{t}{2}$, $y = a\sin t$

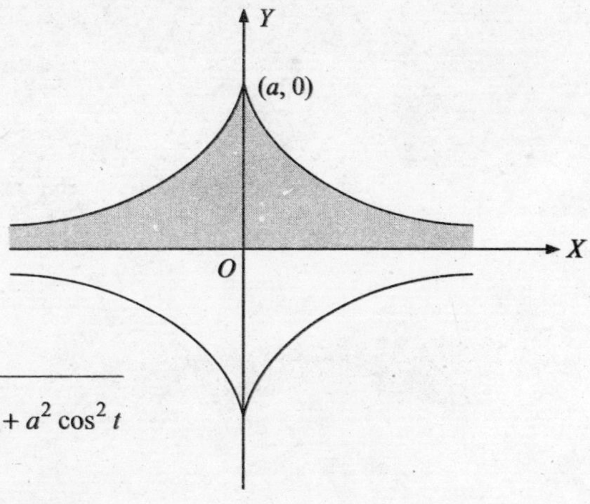

$$\therefore \quad \frac{dx}{dt} = -a\sin t + \frac{a}{2\tan\frac{t}{2}}\sec^2\frac{t}{2}$$

$$= -a\sin t + \frac{a}{\sin t} = \frac{a\cos^2 t}{\sin t}$$

and
$$\frac{dy}{dt} = a\cos t$$

$$\therefore \quad \frac{ds}{dt} = \sqrt{\left(\frac{dx}{dt}\right)^2 + \left(\frac{dy}{dt}\right)^2} = \sqrt{\frac{a^2\cos^4 t}{\sin^2 t} + a^2\cos^2 t}$$

$$= a\frac{\cos t}{\sin t}.$$

The tractrix is symmetrical about Y-axis and X-axis. When it revolves about its asymptote, the X-axis, x varies from $-\infty$ to ∞ hence t varies from 0 to π.

$\therefore$ Required surface area

$$= 2\int_0^{\pi/2} 2\pi y\frac{ds}{dt}dt = 4\pi\int_0^{\pi/2} a\sin t\cdot a\frac{\cos t}{\sin t}dt$$

$$= 4\pi a^2\int_0^{\pi/2}\cos t\,dt = 4\pi a^2.$$ **Ans.**

CHAPTER

8 Matrices and Their Applications

IMPORTANT FORMULAE AND USEFUL RESULTS

1. **Inverse of a Matrix:** If A is a non-singular matrix of order $n \times n$ and B is another matrix of same order such that $BA = I = AB$ then $B = A$.

$$A^{-1} = \frac{\text{adj}(A)}{|A|}$$

Properties of square matrices:

(*i*) $(A')^{-1} = (A^{-1})'$ (*ii*) $|AB| = |A|\,|B|$

(*iii*) $|\text{adj}\,(A)| = |A|^{n-1}$ (*iv*) $(AB)^{-1} = B^{-1}A^{-1}$

(*v*) **Orthogonal** matrix is one whose transpose is its inverse.

Thus if $AA^{-1} = A'A = I$ then A is called orthogonal.

The solution of a system of simultaneous equations $AX = B$ is $X = A^{-1}B$.

2. **Linear Transformation:** It is of the form $Y = AX$

where
$$Y = \begin{bmatrix} y_1 \\ y_2 \\ \vdots \\ y_n \end{bmatrix}, \quad A = \begin{bmatrix} a_{11} & a_{12} & \cdots & a_{1n} \\ a_{21} & a_{22} & \cdots & a_{2n} \\ \cdots & \cdots & \cdots & \cdots \\ a_{n1} & a_{n2} & \cdots & a_{nn} \end{bmatrix}, \quad X = \begin{bmatrix} x_1 \\ x_2 \\ \vdots \\ x_n \end{bmatrix}$$

If $|A| = 0$ the transformation is **singular** otherwise **non-singular.**

Elementary Matrix: When an elementary operation is applied to a unit matrix, resulting matrix is called elementary matrix.

A Deduction: Every row (column) transformation on a matrix can be effected by pre (post) multiplication by corresponding elementary matrix.

3. **Rank of a Matrix:** A matrix is said to be of rank r, if

(*i*) It has at least one non-zero minor of order r and

(*ii*) every minor of order greater than r, vanishes.

4. **Normal Form of a Matrix:** Every non-zero matrix A of rank r can be reduced by a sequence of elementary operations to the form $\begin{bmatrix} I_r & 0 \\ 0 & 0 \end{bmatrix}$ which is called the normal form of A. Corresponding to every matrix A of rank r there exist non-singular matrices P and Q such that

$$PAQ = \begin{bmatrix} I_r & 0 \\ 0 & 0 \end{bmatrix}.$$

Here if A is a matrix of order $m \times n$ then P and Q are square matrixes of order $m \times m$ and $n \times n$ respectively.

5. **Linear Depeandence and Independence of Vectors:**

The set of vectors $X_1, X_2, ..., X_n$ is said to be **linearly dependent** if there exist scalars $c_1, c_2, ..., c_n,$ not all zero, such that $c_1 X_1, + c_2 X_2 + ... + c_n X_n = 0$ and, if no such scalars other than all zero exist, the vectors $X_1, X_2, ..., X_n$ are called **linearly independent.**

6. **Consisttency of Simultaneous Equations.**

For the matrix equation system $AX = B$ we form the augmented matrix $[A : B]$. The rule for its consistency is

(*i*) If rank $(A) \neq$ rank $[A : B]$ the system is **inconsistent** having no solution

(*ii*) If rank $(A) =$ rank $[A : B]$ the system is **consistent,** and two cases arise

(*a*) If rank $(A) =$ rank $[A : B] = n$ the number of unknows then we have **unique** solution.

(*b*) If rank $(A) =$ rank $[A : B] < n$ then infinitely **many** solutions.

7. **Hermitian and Skew-Hermitian Matrix:**

If $Z = (z_{mn})$ is a matrix of complex numbers and $\overline{Z} = (\overline{z}_{mn})$ and $Z^* = (\overline{Z})'$ then Z is Hermitian if $Z^* = Z$ and Z is **Skew Hermitian** if $Z^* = -Z$.

In a Hermitian matrix the leading diagonal elements are all real and in a Skew-Hermitian matrix the leading diagonal elements are all parely imaginary.

8. **Eigen Values and Eigen Vectors:**

Characterstic equation of a square matrix A is $|A - \lambda I| = 0$. Roots of the characterstic equation are **eigen values.** or characterstic roots.

The vectors X corresponding to $(A - \lambda I)X = 0$ are called **eigen vectors** or characterstic vectors.

Caley-Hamilton Theorem: Every square matrix satisfies its own characterstic equation.

9. **Diagonaliting a Square Matrix:** If $\lambda_1, \lambda_2, \lambda_3$ are eigen values of a matrix A and $X_1 = \begin{bmatrix} x_1 \\ y_1 \\ z_1 \end{bmatrix}$ $X_2 = \begin{bmatrix} x_2 \\ y_2 \\ z_2 \end{bmatrix}$, $X_3 = \begin{bmatrix} x_3 \\ y_3 \\ z_3 \end{bmatrix}$ are the corresponding eigen vectors then $P = [X_1\ X_2\ X_3] = \begin{bmatrix} x_1 & x_2 & x_3 \\ y_1 & y_2 & y_3 \\ z_1 & z_2 & z_3 \end{bmatrix}$ is the **modal** matrix.

then $P^{-1}AP = D$ is the **diagonal** matrix called **Spectral** matrix.

SOLUTIONS TO EXERCISE 8A

1. *If* $A=\begin{bmatrix}1 & 2\\ 3 & 4\end{bmatrix}$ *show that* $A^2-5A=2I$ *where* I *is the unit matrix of order* 2, *and hence find* A^4.

SOLUTION: $A=\begin{bmatrix}1 & 2\\ 3 & 4\end{bmatrix}$ hence $A^2=\begin{bmatrix}1 & 2\\ 3 & 4\end{bmatrix}\begin{bmatrix}1 & 2\\ 3 & 4\end{bmatrix}=\begin{bmatrix}7 & 10\\ 15 & 22\end{bmatrix}$

$\therefore$ $A^2-5A=\begin{bmatrix}7 & 10\\ 15 & 22\end{bmatrix}-\begin{bmatrix}5 & 10\\ 15 & 20\end{bmatrix}=\begin{bmatrix}2 & 0\\ 0 & 2\end{bmatrix}=2I$

Also $A^4=(5A+2I)^2=25A^2+20A+4I=25\begin{bmatrix}7 & 10\\ 15 & 22\end{bmatrix}+20\begin{bmatrix}1 & 2\\ 3 & 4\end{bmatrix}+\begin{bmatrix}4 & 0\\ 0 & 4\end{bmatrix}$

$=\begin{bmatrix}175+20+4 & 250+40+0\\ 375+60+0 & 550+80+4\end{bmatrix}=\begin{bmatrix}199 & 290\\ 435 & 634\end{bmatrix}$ **Ans.**

2. *If* $A=\begin{bmatrix}1 & 1 & 1\\ 1 & 1 & 1\\ 1 & 1 & 1\end{bmatrix}$ *and* $B=\begin{bmatrix}1 & 2 & 3\\ 4 & 5 & 6\\ 7 & 8 & 9\end{bmatrix}$ *verify that* $(AB)'=B'A'$.

SOLUTION: $AB=\begin{bmatrix}1 & 1 & 1\\ 1 & 1 & 1\\ 1 & 1 & 1\end{bmatrix}\begin{bmatrix}1 & 2 & 3\\ 4 & 5 & 6\\ 7 & 8 & 9\end{bmatrix}=\begin{bmatrix}12 & 15 & 18\\ 12 & 15 & 18\\ 12 & 15 & 18\end{bmatrix}$

$B'A'=\begin{bmatrix}1 & 4 & 7\\ 2 & 5 & 8\\ 3 & 6 & 9\end{bmatrix}\begin{bmatrix}1 & 1 & 1\\ 1 & 1 & 1\\ 1 & 1 & 1\end{bmatrix}=\begin{bmatrix}12 & 12 & 12\\ 15 & 15 & 15\\ 18 & 18 & 18\end{bmatrix}=(AB)'$ **Hence Proved.**

3. *Let* $\mathrm{A}=\begin{bmatrix}-1 & 2 & 3\\ 1 & 3 & -1\\ 4 & 5 & 1\end{bmatrix}$ *show that* $(A^2)'=(A')^2$.

SOLUTION: $A^2=\begin{bmatrix}-1 & 2 & 3\\ 1 & 3 & -1\\ 4 & 5 & 1\end{bmatrix}\begin{bmatrix}-1 & 2 & 3\\ 1 & 3 & -1\\ 4 & 5 & 1\end{bmatrix}=\begin{bmatrix}15 & 19 & -2\\ -2 & 6 & -1\\ 5 & 28 & 8\end{bmatrix}$

and $(A')^2=\begin{bmatrix}-1 & 1 & 4\\ 2 & 3 & 5\\ 3 & -1 & 1\end{bmatrix}\begin{bmatrix}-1 & 1 & 4\\ 2 & 3 & 5\\ 3 & -1 & 1\end{bmatrix}=\begin{bmatrix}15 & -2 & 5\\ 19 & 6 & 28\\ -2 & -1 & 8\end{bmatrix}=(A^2)'$ **Hence Proved.**

4. *Express the matrix* $\begin{bmatrix} 2 & 1 & 3 \\ -1 & 4 & 1 \\ 0 & 2 & -2 \end{bmatrix}$ *as the sum of a symmetric and a skew-symmetric matrix.*

SOLUTION: We know that $A = \frac{1}{2}(A + A') + \frac{1}{2}(A - A')$

Here $A = \begin{bmatrix} 2 & 1 & 3 \\ -1 & 4 & 1 \\ 0 & 2 & -2 \end{bmatrix}$ and $A' = \begin{bmatrix} 2 & -1 & 0 \\ 1 & 4 & 2 \\ 3 & 1 & -2 \end{bmatrix}$

$$\frac{1}{2}(A + A') = \begin{bmatrix} 2 & 0 & 3/2 \\ 0 & 4 & 3/2 \\ 3/2 & 3/2 & -2/1 \end{bmatrix} \text{ which is symmetric}$$

$$\frac{1}{2}(A - A') = \begin{bmatrix} 0 & 1 & 3/2 \\ -1 & 0 & -1/2 \\ -3/2 & 1/2 & 0 \end{bmatrix} \text{ which is skew symmetric}$$

Hence Proved.

5. *Let* $A = \begin{bmatrix} a_1 & b_1 \\ a_2 & b_2 \end{bmatrix}$ *show that* $A^{-1} = \frac{1}{a_1b_2 - a_2b_1}\begin{bmatrix} b_2 & -b_1 \\ -a_2 & a_1 \end{bmatrix}$ *and hence find the inverse of the matrix*

$$B = \begin{bmatrix} \cos\theta & -\sin\theta \\ \sin\theta & \cos\theta \end{bmatrix}. \quad \textit{Is } B \textit{ orthogonal?}$$

SOLUTION: $A = \begin{bmatrix} a_1 & b_1 \\ a_2 & b_2 \end{bmatrix}$ then $A^{-1} = \frac{1}{|A|}\begin{bmatrix} b_2 & -a_2 \\ -b_1 & a_1 \end{bmatrix} = \frac{1}{a_1 b_2 - a_2 b_1}\begin{bmatrix} b_2 & -b_1 \\ -a_2 & a_1 \end{bmatrix}$

$$B = \begin{bmatrix} \cos\theta & -\sin\theta \\ \sin\theta & \cos\theta \end{bmatrix} \text{ hence } B^{-1} = \frac{1}{\cos^2\theta + \sin^2\theta}\begin{bmatrix} \cos\theta & \sin\theta \\ -\sin\theta & \cos\theta \end{bmatrix} = \begin{bmatrix} \cos\theta & \sin\theta \\ -\sin\theta & \cos\theta \end{bmatrix} = B'$$

Since $B^{-1} = B'$, B is orthogonal matrix **Ans.**

6. *Determine the inverse of the matrix* $\begin{bmatrix} 5 & -2 & 4 \\ -2 & 1 & 1 \\ 4 & 1 & 0 \end{bmatrix}$.

SOLUTION: $A = \begin{bmatrix} 5 & -2 & 4 \\ -2 & 1 & 1 \\ 4 & 1 & 0 \end{bmatrix}$

$$\therefore \quad \text{adj}(A) = \begin{bmatrix} -1 & 4 & -6 \\ 4 & -16 & -13 \\ -6 & -13 & 1 \end{bmatrix} = \begin{bmatrix} -1 & 4 & -6 \\ 4 & -16 & -13 \\ -6 & -13 & 1 \end{bmatrix}$$

and $$|A| = 5(-1) + (-2)\,4 + 4(-6) = -5 - 8 - 24 = -37$$

$$\therefore \quad A^{-1} = \frac{\text{adj}(A)}{|A|} = \frac{-1}{37}\begin{bmatrix} -1 & 4 & -6 \\ 4 & -16 & -13 \\ -6 & -13 & 1 \end{bmatrix}$$ **Ans.**

7. *Given that* $A = \begin{bmatrix} 0 & 1 & -1 \\ 2 & -3 & 4 \\ 3 & -2 & 3 \end{bmatrix}$ *and* $B = \begin{bmatrix} 1 & -1 & 1 \\ -3 & 2 & -1 \\ -2 & 1 & -1 \end{bmatrix}$,

verify (*i*) $(A^2)' = (A')^2$ (*ii*) $(AB)^{-1} = B^{-1}A^{-1}$.

SOLUTION: (*i*)
$$A^2 = \begin{bmatrix} 0 & 1 & -1 \\ 2 & -3 & 4 \\ 3 & -2 & 3 \end{bmatrix}\begin{bmatrix} 0 & 1 & -1 \\ 2 & -3 & 4 \\ 3 & -2 & 3 \end{bmatrix} = \begin{bmatrix} -1 & -1 & 1 \\ 6 & 3 & -2 \\ 5 & 3 & -2 \end{bmatrix}$$

$$(A')^2 = \begin{bmatrix} 0 & 2 & 3 \\ 1 & -3 & -2 \\ -1 & 4 & 3 \end{bmatrix}\begin{bmatrix} 0 & 2 & 3 \\ 1 & -3 & -2 \\ -1 & 4 & 3 \end{bmatrix} = \begin{bmatrix} -1 & 6 & 5 \\ -1 & 3 & 3 \\ 1 & -2 & -2 \end{bmatrix} = (A^2)'$$

(*ii*)
$$AB = \begin{bmatrix} 0 & 1 & -1 \\ 2 & -3 & 4 \\ 3 & -2 & 3 \end{bmatrix}\begin{bmatrix} 1 & -1 & 1 \\ -3 & 2 & -1 \\ -2 & 1 & -1 \end{bmatrix} = \begin{bmatrix} -1 & 1 & -2 \\ 3 & -4 & 1 \\ 3 & -4 & 2 \end{bmatrix}$$

$$(AB)^{-1} = \frac{1}{|AB|}\text{adj}(AB) = \frac{1}{(4-3-0)}\begin{bmatrix} -4 & -3 & 0 \\ 6 & 4 & -1 \\ -7 & -5 & 1 \end{bmatrix} = \frac{1}{1}\begin{bmatrix} -4 & 6 & -7 \\ -3 & 4 & -5 \\ 0 & -1 & 1 \end{bmatrix}$$

$$A^{-1} = \frac{1}{|A|}\text{adj}(A) = \frac{1}{0+6-5}\begin{bmatrix} -1 & 6 & 5 \\ -1 & 3 & 3 \\ 1 & -2 & -2 \end{bmatrix}' = \begin{bmatrix} -1 & -1 & 1 \\ 6 & 3 & -2 \\ 5 & 3 & -2 \end{bmatrix}$$

$$B^{-1} = \frac{1}{|B|}\text{adj}(B) = \frac{1}{-1-1+1}\begin{bmatrix} -1 & -1 & 1 \\ 0 & 1 & 1 \\ -1 & -2 & -1 \end{bmatrix}' = -\begin{bmatrix} -1 & 0 & -1 \\ -1 & 1 & -2 \\ 1 & 1 & -1 \end{bmatrix} = \begin{bmatrix} 1 & 0 & 1 \\ 1 & -1 & 2 \\ -1 & -1 & 1 \end{bmatrix}$$

Now
$$B^{-1}A^{-1} = \begin{bmatrix} 1 & 0 & 1 \\ 1 & -1 & 2 \\ -1 & -1 & 1 \end{bmatrix}\begin{bmatrix} -1 & -1 & 1 \\ 6 & 3 & -2 \\ 5 & 3 & -2 \end{bmatrix} = \begin{bmatrix} -4 & 6 & -7 \\ -3 & 4 & -5 \\ 0 & -1 & 1 \end{bmatrix} (AB)^{-1}$$ **Hence Proved.**

8. *If* $A = \begin{bmatrix} -1 & -2 & -2 \\ 2 & 1 & -2 \\ 2 & -2 & 1 \end{bmatrix}$ *show that* adj $A = 3A'$ *and find* adj (adj A).

SOLUTION:
$$\text{adj}(A) = \begin{bmatrix} -3 & -6 & -6 \\ 6 & 3 & -6 \\ 6 & -6 & 3 \end{bmatrix}' = \begin{bmatrix} -3 & 6 & 6 \\ -6 & 3 & -6 \\ -6 & -6 & 3 \end{bmatrix} = 3\begin{bmatrix} -1 & 2 & 2 \\ -2 & 1 & -2 \\ -2 & -2 & 1 \end{bmatrix} = 3A'$$

Also $|A| = (-1)(-5) + (-2)(-6) + (-2)(-6) = 3 + 12 + 12 = 27$

Next
$$\text{adj}(\text{adj } A) = \text{adj}\begin{bmatrix} -3 & 6 & 6 \\ -6 & 3 & -6 \\ -6 & -6 & 3 \end{bmatrix} = \begin{bmatrix} -27 & 54 & 54 \\ -54 & 27 & -54 \\ -54 & -54 & 27 \end{bmatrix}'$$

$$= \begin{bmatrix} -27 & -54 & -54 \\ 54 & 27 & -54 \\ 54 & -54 & 27 \end{bmatrix} = 27\begin{bmatrix} -1 & -2 & -2 \\ 2 & 1 & -2 \\ 2 & -2 & 1 \end{bmatrix} = |A|\,A$$ **Ans.**

9. *Find* $\begin{bmatrix} 1 & -\tan\theta \\ \tan\theta & 1 \end{bmatrix}\begin{bmatrix} 1 & \tan\theta \\ -\tan\theta & 1 \end{bmatrix}^{-1}$.

SOLUTION:
$$\begin{bmatrix} 1 & \tan\theta \\ -\tan\theta & 1 \end{bmatrix}^{-1} = \frac{1}{1+\tan^2\theta}\begin{bmatrix} 1 & -\tan\theta \\ \tan\theta & 1 \end{bmatrix}$$

$\therefore$
$$\begin{bmatrix} 1 & -\tan\theta \\ \tan\theta & 1 \end{bmatrix}\begin{bmatrix} 1 & \tan\theta \\ -\tan\theta & 1 \end{bmatrix}^{-1} = \frac{1}{\sec^2\theta}\begin{bmatrix} 1 & -\tan\theta \\ \tan\theta & 1 \end{bmatrix}\begin{bmatrix} 1 & -\tan\theta \\ \tan\theta & 1 \end{bmatrix}$$

$$= \cos^2\theta\begin{bmatrix} 1-\tan^2\theta & -2\tan\theta \\ 2\tan\theta & -\tan^2\theta+1 \end{bmatrix} = \begin{bmatrix} \cos 2\theta & -\sin 2\theta \\ \sin 2\theta & \cos 2\theta \end{bmatrix}$$ **Ans.**

10. *Solve by matrix inversion method, the following set of equations*

$x + 3y + 3z = 1$

$x + 4y + 3z = 0$

$x + 3y + 4z = 2.$

SOLUTION: Given set of equation can be written as $AX = B \Rightarrow X = A^{-1}B$

$$A^{-1} = \frac{1}{|A|}\text{adj}(A) = \frac{1}{7-3-3}\begin{bmatrix} 7 & -1 & -1 \\ -3 & 1 & 0 \\ -3 & 0 & 1 \end{bmatrix}' = \frac{1}{1}\begin{bmatrix} 7 & -3 & -3 \\ -1 & 1 & 0 \\ -1 & 0 & 1 \end{bmatrix}$$

$\Rightarrow$
$$X = \begin{bmatrix} 7 & -3 & -3 \\ -1 & 1 & 0 \\ -1 & 0 & 1 \end{bmatrix}\begin{bmatrix} 1 \\ 0 \\ 2 \end{bmatrix} = \begin{bmatrix} 1 \\ -1 \\ 1 \end{bmatrix} \quad \therefore \quad x = 1,\ y = -1,\ z = 1$$ **Ans.**

SOLUTIONS TO EXERCISE 8B

1. *Using the rank concept investigate the consistency of the system of equations*

$$4x-2y+6z=8$$
$$x+y-3z=-1$$
$$15x-3y+9z=21.$$

and, if consistent, solve it.

SOLUTION: The system of equations in $AX = B$

The augmental metrix $D=[A:B]=\begin{bmatrix}4 & -2 & 6 & \vdots & 8\\ 1 & 1 & -3 & \vdots & -1\\ 15 & -3 & 9 & \vdots & 21\end{bmatrix}$

Applying $R_1 \leftrightarrow R_2$ and $R_3 \rightarrow \frac{1}{3}R_3$, we get

$D \approx \begin{bmatrix}1 & 1 & -3 & \vdots & -1\\ 4 & -2 & 6 & \vdots & 8\\ 5 & -1 & 3 & \vdots & 7\end{bmatrix}$ Applying now $R_2 - 4R_1$ and $R_3 - 5R_1$, we get

$D \approx \begin{bmatrix}1 & 1 & -3 & \vdots & -1\\ 0 & -6 & 18 & \vdots & 12\\ 0 & -6 & 18 & \vdots & 12\end{bmatrix}$ Applying $R_3 - R_2$

$\approx \begin{bmatrix}1 & 1 & -3 & \vdots & -1\\ 0 & -6 & 18 & \vdots & 12\\ 0 & 0 & 0 & \vdots & 0\end{bmatrix}$ now applying $-\frac{1}{6}R_2$

$\sim \begin{bmatrix}1 & 1 & -3 & \vdots & -1\\ 0 & 1 & -3 & \vdots & -2\\ 0 & 0 & 0 & \vdots & 0\end{bmatrix}$ Hence rank (A) = 2 and rank (D) = 2

Hence system is consistent with many solutions.

$\therefore$ Equations can be written as $x + y - 3z = -1$ and $y - 3z = -2$

$\Rightarrow$ $x = 1$, $z = k$, $y = -2 + 3k$ where k is arbitrary **Ans.**

2. *Show that the system of equations*

$$x+y+3z=6$$
$$2x+3y+z=8$$
$$x+5y+7z=20$$
$$x+z=10$$

has no solution.

SOLUTION: Here we have four equations and three unknows. $AX = B$ where

$$A = \begin{bmatrix} 1 & 1 & 3 \\ 2 & 3 & 1 \\ 1 & 5 & 7 \\ 1 & 0 & 1 \end{bmatrix} \quad \text{and} \quad B = \begin{bmatrix} 6 \\ 8 \\ 20 \\ 10 \end{bmatrix} \quad \text{and} \quad X = \begin{bmatrix} x \\ y \\ z \end{bmatrix}.$$

$\therefore$ Augmented matrix $D = \begin{bmatrix} 1 & 1 & 3 & : & 6 \\ 2 & 3 & 1 & : & 8 \\ 1 & 5 & 7 & : & 20 \\ 1 & 0 & 1 & : & 10 \end{bmatrix}$ applying $R_2 - 2R_1$, $R_3 - R_1$ and $R_4 - R_1$, we get

$$D = \begin{bmatrix} 1 & 1 & 3 & : & 6 \\ 0 & 1 & -5 & : & -4 \\ 0 & 4 & 4 & : & 12 \\ 0 & -1 & -2 & : & 4 \end{bmatrix}$$ Now applying $R_3 - 4R_2$ and $R_3 + R_2$, gives

$$D = \begin{bmatrix} 1 & 1 & 3 & : & 6 \\ 0 & 1 & -5 & : & -4 \\ 0 & 0 & 24 & : & 28 \\ 0 & 0 & -7 & : & 0 \end{bmatrix} \quad \Rightarrow \quad 24z = 28 \text{ and } -7z = 0$$

Hence system is inconsistent.

3. *Investigate whether the set of equations*

$$2x - y - z = 2$$
$$x + 2y + z = 2$$
$$4x - 7y - 5z = 2.$$

is consistent or not and, if consistent, solve it.

SOLUTION: System is $AX = B$

$$D = \begin{bmatrix} 2 & -1 & -1 & : & 2 \\ 1 & 2 & 1 & : & 2 \\ 4 & -7 & -5 & : & 2 \end{bmatrix}$$ Applying $R_1 \leftrightarrow R_2$ then $R_3 - 2R_2$ and $R_2 - 2R_1$, we get

$$D \approx \begin{bmatrix} 1 & 2 & 1 & : & 2 \\ 0 & -5 & -3 & : & -2 \\ 0 & -5 & -3 & : & -2 \end{bmatrix}$$ then applying $R_3 - R_2$, gives

$$\approx \begin{bmatrix} 1 & 2 & 1 & : & 2 \\ 0 & -5 & -3 & : & -2 \\ 0 & 0 & 0 & : & 0 \end{bmatrix} \quad \therefore \quad \text{Rank (D)} = 2 \text{ and Rank (A)} = 2$$

hence consistent and have many solutions.

Equations are $x + 2y + z = 2$

and $5y + 3z = 2$

If $z = k$ then $y = \dfrac{2-3k}{5}$ and hence $x = 2 - k - \dfrac{4-6k}{5} = \dfrac{6+k}{5}$ k being arbitrary. **Ans.**

4. *Investigate if the following system of equations is consistent or not.*

$$\begin{aligned} x_1 + x_2 + x_3 &= 3 \\ 2x_1 - x_2 + 3x_3 &= 1 \\ 4x_1 + x_2 + 5x_3 &= 3 \\ 3x_1 - 2x_2 + x_3 &= 4. \end{aligned}$$

SOLUTION: Augmented Matrix $D = \begin{bmatrix} 1 & 1 & 1 & \vdots & 3 \\ 2 & -1 & 3 & \vdots & 1 \\ 4 & 1 & 5 & \vdots & 3 \\ 3 & -2 & 1 & \vdots & 4 \end{bmatrix}$. Applying $R_3 - 2R_2$ and $R_2 - R_1$ and $R_4 - 3R_1$

$$\sim \begin{bmatrix} 1 & 1 & 1 & \vdots & 3 \\ 0 & -3 & 1 & \vdots & -5 \\ 0 & -2 & -1 & \vdots & 1 \\ 0 & -5 & -2 & \vdots & -5 \end{bmatrix} \text{ Now apply } R_2 - R_3 \text{ we get}$$

$$\sim \begin{bmatrix} 1 & 1 & 1 & \vdots & 3 \\ 0 & -1 & 2 & \vdots & -6 \\ 0 & -2 & -1 & \vdots & 1 \\ 0 & -5 & -2 & \vdots & -5 \end{bmatrix} \text{ Now applying } R_3 - 2R_2 \text{ and } R_4 - 5R_2$$

$$\sim \begin{bmatrix} 1 & 1 & 1 & \vdots & 3 \\ 0 & 1 & 3 & \vdots & -7 \\ 0 & -2 & -1 & \vdots & 1 \\ 0 & -5 & -2 & \vdots & -5 \end{bmatrix} \text{ Now applying } R_3 + 2R_2 \text{ and } R_4 - 5R_2$$

$$\sim \begin{bmatrix} 1 & 1 & 1 & \vdots & 3 \\ 0 & 1 & 3 & \vdots & -7 \\ 0 & 0 & 5 & \vdots & -13 \\ 0 & 0 & 13 & \vdots & -40 \end{bmatrix} \text{ Applying } R_4 - \frac{13}{5} R_3$$

$$= \begin{bmatrix} 1 & 1 & 1 & 3 \\ 0 & 1 & 3 & -7 \\ 0 & 0 & 5 & -13 \\ 0 & 0 & 0 & -40 + \dfrac{169}{5} \end{bmatrix}$$

$\therefore$ $\rho(D) = 4$ and $\rho(A) = 3$ **Hence Inconsistent.**

5. *Determine the value(s) of λ for which the following equations*

$$2x-3y+z = \lambda x$$
$$2x-3y+2z = \lambda y$$
$$-x+2y+0z = \lambda z$$

can possess a non-trivial solution.

SOLUTION: This is a homogeneous system of simultaneous equations $AX = 0$

where $A = \begin{bmatrix} 2-\lambda & -3 & 1 \\ 2 & -3-\lambda & 2 \\ -1 & 2 & -\lambda \end{bmatrix}$ and $X = \begin{bmatrix} x \\ y \\ z \end{bmatrix}$

Here $|A| = \begin{vmatrix} 2-\lambda & -3 & 1 \\ 2 & -3-\lambda & 2 \\ -1 & 2 & -\lambda \end{vmatrix} = (2-\lambda)(3\lambda+\lambda^2-4)+3(-2\lambda+2)+1(4-3-\lambda)$

$$= (2-\lambda)(\lambda^2+3\lambda-4)+3(-2\lambda+2)-\lambda+1$$
$$= -\lambda^3-\lambda^2+3\lambda-1 = (\lambda-1)(-\lambda^2-2\lambda+1)$$

For given equations to have non-trivial solutions $|A| = 0$

which gives $\lambda = 1$ and $\lambda^2+2\lambda-1 = 0$ or $\lambda = -1, \pm\sqrt{2}$

Thus $\lambda = 1, \ -1 \pm \sqrt{2}$ **Ans.**

6. *Determine the values of λ and μ so that the equations*

$$2x+3y+5z = 9$$
$$7x+3y-2z = 8$$
$$2x+3y+\lambda z = \mu$$

have (i) no solution, (ii) a unique solution and (iii) infinitely many solutions.

SOLUTION: $AX = B$

Augmented matrix $D\,[A:B] = \begin{bmatrix} 2 & 3 & 5 & \vdots & 9 \\ 7 & 3 & -2 & \vdots & 8 \\ 2 & -3 & \lambda & \vdots & \mu \end{bmatrix}$ Applying $R_3 - R_1$, we get

$$D \sim \begin{bmatrix} 2 & 3 & 5 & \vdots & 9 \\ 7 & 3 & -2 & \vdots & 8 \\ 0 & 0 & \lambda-5 & \vdots & \mu-9 \end{bmatrix}$$

Case I : $\lambda \neq 5 \ \therefore \ \rho(A) = \rho(D) = 3$ hence consistent and unique solution.

Case II : $\lambda = 5, \ \mu \neq 9$ then $\rho(A) = 2, \ \rho(D) = 3$ hence no solution or inconsistent.

Case III : $\lambda = 5, \ \mu = 9$ then $\rho(A) = \rho(D) = 2 <$ number of unknows, hence many solutions. **Ans.**

7. *Solve by matrix method, the system of equations*

$$x+3y-z-\omega=0$$
$$3x+y+2z-2\omega=4$$
$$4x-3y-4z-3\omega=-13.$$

SOLUTION: $AX=B$ where $A=\begin{bmatrix}1 & 3 & -1 & -1\\ 3 & 1 & 2 & -2\\ 4 & -3 & -4 & -3\end{bmatrix}$, $X=\begin{bmatrix}x\\ y\\ z\\ w\end{bmatrix}$, $B=\begin{bmatrix}0\\ 4\\ -13\end{bmatrix}$

Augmented matrix $D=\left[\begin{array}{cccc:c}1 & 3 & -1 & -1 & 0\\ 3 & 1 & 2 & -2 & 4\\ 4 & -3 & -4 & -3 & -13\end{array}\right]$ Applying R_2-3R_1 and R_4-4R_1

$D\sim\left[\begin{array}{cccc:c}1 & 3 & -1 & -1 & 0\\ 0 & -8 & 5 & 1 & 4\\ 0 & -15 & 0 & 1 & -13\end{array}\right]$ Applying $R_3-\frac{15}{8}R_2$, gives

$$\sim\left[\begin{array}{cccc:c}1 & 3 & -1 & -1 & 0\\ 0 & -8 & 5 & 1 & 4\\ 0 & 0 & \frac{-75}{8} & \frac{-7}{8} & \frac{-41}{2}\end{array}\right]$$

Here Rank $(A)=3$, Rank (D) = 3 which is less than the number of unknowns.
Hence system is consistent having many solutions.

Applying $R_3\to\frac{-8}{75}R_3$, $R_2\to\frac{-1}{8}R_2$, we get

$$D\sim\left[\begin{array}{cccc:c}1 & 3 & -1 & -1 & 0\\ 0 & 1 & -5/8 & -1/8 & -1/2\\ 0 & 0 & 1 & \frac{7}{75} & \frac{164}{75}\end{array}\right]$$

Thus, the given equations become (taking $w=k$) as

$$z+\frac{7k}{75}=\frac{164}{75} \qquad \therefore\ z=\frac{-7k+164}{75}$$

$$y-\frac{5z}{8}-\frac{w}{8}=\frac{-1}{2} \qquad \therefore\ y=\frac{5}{8}\left(\frac{-7k+164}{75}\right)+\frac{k}{8}-\frac{1}{2}=\frac{k+13}{15}$$

and $\quad x\neq 3y-z-w=0 \qquad \therefore\ x=\frac{-3}{15}(k+13)+\frac{-7k+164}{75}+k=\frac{53k}{75}-\frac{31}{75}$

Thus the solution is $x=\frac{53k-31}{75}$, $y=\frac{k+13}{15}$, $z=\frac{164-7k}{75}$, $w=k$. **Ans.**

8. *Test for consistency the following system of equations*

$$\begin{aligned} x+y+z &= 3 \\ x+2y+3z &= 4 \\ x+4y+9z &= 4 \\ 2x+3y+9z &= 6. \end{aligned}$$

SOLUTION: The augmented matrix D ($\Leftarrow$ $A : B$) is given by

$$D = \begin{bmatrix} 1 & 1 & 1 & \vdots & 3 \\ 1 & 2 & 3 & \vdots & 4 \\ 1 & 4 & 9 & \vdots & 4 \\ 2 & 3 & 9 & \vdots & 6 \end{bmatrix}$$ Applying $R_4 - 2R_1,\ R_3 - R_2,\ R_2 - R_1,$ we get

$$\sim \begin{bmatrix} 1 & 1 & 1 & \vdots & 3 \\ 0 & 1 & 2 & \vdots & 1 \\ 0 & 2 & 6 & \vdots & 0 \\ 0 & 1 & 7 & \vdots & 0 \end{bmatrix}$$ Then applying $R_4 - R_2,\ R_3 - 2R_2$

$$\sim \begin{bmatrix} 1 & 1 & 1 & \vdots & 3 \\ 0 & 1 & 2 & \vdots & 1 \\ 0 & 0 & 2 & \vdots & -2 \\ 0 & 0 & 5 & \vdots & -1 \end{bmatrix}$$ Now applying $R_4 - \dfrac{5}{2} R_3$

$$\sim \begin{bmatrix} 1 & 1 & 1 & \vdots & 3 \\ 0 & 1 & 2 & \vdots & 1 \\ 0 & 0 & 2 & \vdots & -2 \\ 0 & 0 & 0 & \vdots & 4 \end{bmatrix}$$

Hence rank of $D = 4$ and Rank of $A = 3$

Since $\rho(D) \neq \rho(A)$ **system is Inconsistent.** **Ans.**

9. *Solve the following system of equations using elementary transformations*

$$\begin{aligned} 2x+y-z+3\omega &= 11 \\ x-2y+z+\omega &= 8 \\ 4x+7y+2z+\omega &= 0 \\ 3x+5y+4z+4\omega &= 17. \end{aligned}$$

SOLUTION: $AX = B$. The augmented matrix $D \Leftarrow [A : B]$ is given by

$$D = \begin{bmatrix} 2 & 1 & -1 & 3 & \vdots & 11 \\ 1 & -2 & 1 & 1 & \vdots & 8 \\ 4 & 7 & 2 & 1 & \vdots & 6 \\ 3 & 5 & 4 & 4 & \vdots & 17 \end{bmatrix}$$ Applying $R_1 \leftrightarrow R_2$ then $R_2 - 2R_1, R_3 - 2R_2, R_4 - 3R_1$

$$\sim \begin{bmatrix} 1 & -2 & 1 & 1 & \vdots & 8 \\ 0 & 5 & -3 & 1 & \vdots & -5 \\ 0 & 5 & 4 & -5 & \vdots & -16 \\ 0 & 11 & 1 & 1 & \vdots & -7 \end{bmatrix}$$ Now applying $R_3 - R_2$

$$\sim \begin{bmatrix} 1 & -2 & 1 & 1 & \vdots & 8 \\ 0 & 5 & -3 & 1 & \vdots & -5 \\ 0 & 0 & 7 & -6 & \vdots & -11 \\ 0 & 11 & 1 & 1 & \vdots & -7 \end{bmatrix} \quad \text{Next, apply } R_4 - \frac{11}{5} R_2$$

$$\sim \begin{bmatrix} 1 & -2 & 1 & 1 & \vdots & 8 \\ 0 & 5 & -3 & 1 & \vdots & -5 \\ 0 & 0 & 7 & -6 & \vdots & -11 \\ 0 & 0 & \frac{38}{5} & \frac{-6}{5} & \vdots & 4 \end{bmatrix} \quad \text{Next applying } R_4 \to 5R_4, \text{ gives}$$

$$\sim \begin{bmatrix} 1 & -2 & 1 & 1 & \vdots & 8 \\ 0 & 5 & -3 & 1 & \vdots & -5 \\ 0 & 0 & 7 & -6 & \vdots & -11 \\ 0 & 0 & 38 & -6 & \vdots & 20 \end{bmatrix} \quad \text{Apply } R_4 - \frac{38}{7} R_3 \text{ then } R_4 \to 7R_4$$

$$\sim \begin{bmatrix} 1 & -2 & 1 & 1 & \vdots & 8 \\ 0 & 5 & -3 & 1 & \vdots & -5 \\ 0 & 0 & 7 & -6 & \vdots & -11 \\ 0 & 0 & 0 & 186 & \vdots & 558 \end{bmatrix} \quad \text{Next } R_4 \to \frac{1}{186} R_4$$

$$\sim \begin{bmatrix} 1 & -2 & 1 & 1 & \vdots & 8 \\ 0 & 5 & -3 & 1 & \vdots & -5 \\ 0 & 0 & 7 & -6 & \vdots & -11 \\ 0 & 0 & 0 & 1 & \vdots & 3 \end{bmatrix} \quad \text{Hence } w = 3, 7z - 6w = -11 \therefore z = 1 \text{ and } y = -1.$$

Thus $x = 2, \quad y = -1, \quad z = 1, \quad w = 3.$ **Ans.**

10. *Find the values of a and b such that the system of equations*

$$3x - 2y + z = 6$$
$$5x - 8y + 9z = 3$$
$$2x + y + az = -b$$

may have unique solution.

SOLUTION: $AX = B$. Augmented matrix D ($\Leftarrow$ $[A : B]$), is given by

$$D = \begin{bmatrix} 3 & -2 & 1 & \vdots & 6 \\ 5 & -8 & 9 & \vdots & 3 \\ 2 & 1 & a & \vdots & -b \end{bmatrix} \quad \text{Applying } R_2 - R_1 \text{ and then } R_3 - R_2, \text{ gives}$$

$$\sim \begin{bmatrix} 3 & -2 & 1 & \vdots & 6 \\ 2 & -6 & 8 & \vdots & -3 \\ 0 & 7 & a-8 & \vdots & -b+3 \end{bmatrix} \quad \text{Now applying } R_1 - R_2 \text{ and then } R_2 - 2R_1, \text{and } R_3 \to R_3 + \frac{1}{2} R_2, \text{ gives}$$

$$\sim \begin{bmatrix} 1 & 4 & -7 & \vdots & 9 \\ 0 & -14 & 22 & \vdots & -21 \\ 0 & 0 & a+3 & \vdots & -b-\frac{15}{2} \end{bmatrix}$$

If $a \neq -3$ $\rho(A) = 3$ and $\rho(D) = 3$ = number of unknowns, hence unique solution.

Thus $a \neq -3$ and b may have any value. **Ans.**

11. *Test whether the vectors (1, 1, 1, 3), (1, 2, 3, 4) and (2, 3, 4, 7) are linearly dependent or not. If dependent, find the relation between them.*

SOLUTION: Vectors are $X = (1, 1, 1, 3)$, $Y = (1, 2, 3, 4)$, $Z = (2, 3, 4, 7)$ which are linearly dependent if $Z = a\ X + bY$ where a and b are not both zero.

$\Rightarrow$ $2 = a + b$ ...(1), $3 = a + 2b,$ ...(2), $4 = a + 3b,$ ...(3), $7 = 3a + 4b$ ...(4)

Solving (1) and (2) gives $a = 1$, $b = 1$ and this satisfies (3) and (4) hence X, Y, Z are linearly dependent.

And $Z = aX + bY$ becomes $Z = X + Y$ **Ans.**

12. *Show that the square matrix*

$$\begin{bmatrix} \cos\phi\cos\theta & \sin\phi & \cos\phi\sin\theta \\ -\sin\phi\cos\theta & \cos\phi & -\sin\phi\sin\theta \\ -\sin\theta & 0 & \cos\theta \end{bmatrix} \text{ is orthogonal.}$$

SOLUTION:

$$A = \begin{bmatrix} \cos\phi\cos\theta & \sin\phi & \cos\phi\sin\theta \\ -\sin\phi\cos\theta & \cos\phi & -\sin\phi\sin\theta \\ -\sin\theta & 0 & \cos\theta \end{bmatrix}, \ A' \begin{bmatrix} \cos\phi\cos\theta & -\sin\phi\cos\theta & -\sin\theta \\ \sin\phi & \cos\phi & 0 \\ \cos\phi\sin\theta & -\sin\phi\sin\theta & \cos\theta \end{bmatrix}$$

Now $$AA' = \begin{bmatrix} 1 & 0 & 0 \\ 0 & 1 & 0 \\ 0 & 0 & 1 \end{bmatrix} = I, \text{ hence } A \text{ is orthogonal}$$ **Hence Proved.**

13. *Consider the linear transformations*

$$x_1 = y_1 - 2y_2 + y_3$$

$$x_2 = 2y_1 + y_2 - 3y_3,$$

and

$$y_1 = z_1 + 2z_2$$

$$y_2 = 2z_1 - z_2$$

$$y_3 = 2z_1 + 3z_2.$$

Using matrix method find the linear transformation from (z_1, z_2) to (x_1, x_2).

SOLUTION: Let $X = \begin{bmatrix} x_2 \\ x_2 \end{bmatrix}, \quad Y = \begin{bmatrix} y_1 \\ y_2 \\ y_3 \end{bmatrix}, \quad Z = \begin{bmatrix} z_1 \\ z_2 \end{bmatrix}$.

Given equations are $X = AY$ where $A = \begin{bmatrix} 1 & -2 & 1 \\ 2 & 1 & -3 \end{bmatrix}$

and $Y = BZ$ where $B = \begin{bmatrix} 1 & 2 \\ 2 & -1 \\ 2 & 3 \end{bmatrix} \quad \Rightarrow \quad X = A\,(BZ) = (AB)\,Z$

Now
$$AB = \begin{bmatrix} 1 & -2 & 1 \\ 2 & 1 & -3 \end{bmatrix} \begin{bmatrix} 1 & 2 \\ 2 & -1 \\ 2 & 3 \end{bmatrix} = \begin{bmatrix} -1 & 7 \\ -2 & -6 \end{bmatrix}$$

or
$$\begin{bmatrix} x_1 \\ x_2 \end{bmatrix} = \begin{bmatrix} -1 & 7 \\ -2 & -6 \end{bmatrix} \begin{bmatrix} z_1 \\ z_2 \end{bmatrix}$$

$\Rightarrow \quad x_1 = -z_1 + 7z_2 \quad$ and $\quad x_2 = -2z_1 - 6z_2$ **Ans.**

14. *Given the linear transformation* $Y = AX$ *where*

$$Y = [y_1, y_2, y_3]', \; X = [x_1, x_2, x_3]' \; \textit{and } A = \begin{bmatrix} 1 & 1 & 2 \\ 1 & 2 & 5 \\ 1 & 3 & 3 \end{bmatrix},$$

find the coordinates (x_1, x_2, x_3) *corresponding to* $(2, 0, 5)$ *in* Y.

SOLUTION:
$$Y = AX, \quad \begin{bmatrix} y_1 \\ y_2 \\ y_3 \end{bmatrix} = \begin{bmatrix} 1 & 1 & 2 \\ 1 & 2 & 5 \\ 1 & 3 & 3 \end{bmatrix} \begin{bmatrix} x_1 \\ x_2 \\ x_3 \end{bmatrix} \quad \therefore \; X = A^{-1}Y$$

To find A^{-1} we write $A = A\;I = \begin{bmatrix} 1 & 1 & 2 \\ 1 & 2 & 5 \\ 1 & 3 & 3 \end{bmatrix} \begin{bmatrix} 1 & 0 & 0 \\ 0 & 1 & 0 \\ 0 & 0 & 1 \end{bmatrix}$

We apply elementary row operations on A and I simultaneously.

Applying $R_3 - R_2$ and $R_2 - R_1$ gives

$$\sim \begin{bmatrix} 1 & 1 & 2 \\ 0 & 1 & 3 \\ 0 & 1 & -2 \end{bmatrix} \begin{bmatrix} 1 & 0 & 0 \\ -1 & 1 & 0 \\ 0 & -1 & 1 \end{bmatrix}$$

Next applying $R_3 - R_2$ and $R_1 - R_2$

$$\sim \begin{bmatrix} 1 & 0 & -1 \\ 0 & 1 & 3 \\ 0 & 0 & -5 \end{bmatrix} \begin{bmatrix} 2 & -1 & 0 \\ -1 & 1 & 0 \\ 1 & -2 & 1 \end{bmatrix}$$

Now applying $R_2 + \frac{3}{5}R_3$ and $R_1 - \frac{1}{5}R_3 - \frac{1}{5}R_3$, gives

$$\sim \begin{bmatrix} 1 & 0 & 0 \\ 0 & 1 & 0 \\ 0 & 0 & 1 \end{bmatrix} \begin{bmatrix} \frac{9}{5} & \frac{-3}{5} & \frac{-1}{5} \\ \frac{-2}{5} & \frac{-1}{5} & \frac{3}{5} \\ \frac{-1}{5} & \frac{+2}{5} & \frac{-1}{5} \end{bmatrix} = I A^{-1}$$

$$\therefore \quad X = A^{-1}Y = \begin{bmatrix} \frac{9}{5} & \frac{-3}{5} & \frac{-1}{5} \\ \frac{-2}{5} & \frac{-1}{5} & \frac{3}{5} \\ \frac{-1}{5} & \frac{2}{5} & \frac{-1}{5} \end{bmatrix} \begin{bmatrix} 2 \\ 0 \\ 5 \end{bmatrix} = \begin{bmatrix} 13/5 \\ 11/5 \\ -7/5 \end{bmatrix}$$

$\therefore \; x_1 = \frac{13}{5}, \quad x_2 = \frac{11}{5}, \quad x_3 = \frac{-7}{5}$ **Ans.**

15. *Show that the transformation*

$$y_1 = x_1 - x_2 + x_3$$
$$y_2 = 3x_1 - x_2 + 2x_3$$
$$y_3 = 2x_1 - 2x_2 + 3x_3$$

is non-singular. Find the inverse transformation.

SOLUTION: $Y = AX, \; Y' = (y_1, y_2, y_3), \; X' = (x_1, x_2, x_3), \; A = \begin{bmatrix} 1 & -1 & 1 \\ 3 & -1 & 2 \\ 2 & -2 & 3 \end{bmatrix}$

$|A| = -3 + 4 + 9 - 4 - 6 + 2 = 2$ hence A is non-singular

To find A^{-1}, we write $A = \begin{bmatrix} 1 & -1 & 1 \\ 3 & -1 & 2 \\ 2 & -2 & 3 \end{bmatrix} \begin{bmatrix} 1 & 0 & 0 \\ 0 & 1 & 0 \\ 0 & 0 & 1 \end{bmatrix} = AI$

Applying $R_3 - 3R_1$ and $R_3 - 2R_1$ $\sim \begin{bmatrix} 1 & -1 & 1 \\ 0 & 2 & -1 \\ 0 & 0 & 1 \end{bmatrix} \begin{bmatrix} 1 & 0 & 0 \\ -3 & 1 & 0 \\ -2 & 0 & 1 \end{bmatrix}$

Next, applying $R_1 - R_3$ and $R_2 + R_3$ $\sim \begin{bmatrix} 1 & -1 & 0 \\ 0 & 2 & 0 \\ 0 & 0 & 1 \end{bmatrix} \begin{bmatrix} 3 & 0 & -1 \\ -5 & 1 & 1 \\ -2 & 0 & 1 \end{bmatrix}$

Next $R_2 \rightarrow \frac{1}{2}R_2$ and $R_1 + R_3$ gives $\sim \begin{bmatrix} 1 & 0 & 0 \\ 0 & 1 & 0 \\ 0 & 0 & 1 \end{bmatrix} \begin{bmatrix} 1/2 & 1/2 & -1/2 \\ -5/2 & 1/2 & 1/2 \\ -2 & 0 & 1 \end{bmatrix} = A^{-1}$ **Ans.**

16. *Find the rank of the matrix* $\begin{bmatrix} 2 & 3 & 4 & -1 \\ 5 & 2 & 0 & -1 \\ -4 & 5 & 12 & -1 \end{bmatrix}$.

SOLUTION: Applying $R_2 - R_1$ and $R_3 - R_1$, we get

$$A \sim \begin{bmatrix} 2 & 3 & 4 & -1 \\ 3 & -1 & -4 & 0 \\ -6 & 2 & 8 & 0 \end{bmatrix}.$$ Now applying $R_3 + 2R_2$ gives

$$\sim \begin{bmatrix} 2 & 3 & 4 & -1 \\ 3 & -1 & -4 & 0 \\ 0 & 0 & 0 & 0 \end{bmatrix}$$ Hence its rank is 2. **Ans.**

17. *Find the rank of the matrix* $\begin{bmatrix} 5 & 6 & 7 & 8 \\ 6 & 7 & 8 & 9 \\ 11 & 12 & 13 & 14 \\ 16 & 17 & 18 & 19 \end{bmatrix}$.

SOLUTION: Applying $C_4 - C_3$, $C_3 - C_2$ and $C_2 - C_1$, we get

$$A \sim \begin{bmatrix} 5 & 1 & 1 & 1 \\ 6 & 1 & 1 & 1 \\ 11 & 1 & 1 & 1 \\ 16 & 1 & 1 & 1 \end{bmatrix}$$ Now applying $R_4 - R_3$, $R_3 - R_2$ and $R_2 - R_1$

$$A \sim \begin{bmatrix} 5 & 1 & 1 & 1 \\ 1 & 0 & 0 & 0 \\ 5 & 0 & 0 & 0 \\ 5 & 0 & 0 & 0 \end{bmatrix}$$ Hence its rank is two. **Ans.**

18. *Compute the inverse of the following matrix by using elementary transformations*

$$\begin{bmatrix} 2 & -6 & -2 & -3 \\ 5 & -13 & -4 & -7 \\ -1 & 4 & 1 & 2 \\ 0 & 1 & 0 & 1 \end{bmatrix}$$

SOLUTION: $$A = AI = \begin{bmatrix} 2 & -6 & -2 & -3 \\ 5 & -13 & -4 & -7 \\ -1 & 4 & 1 & 2 \\ 0 & 1 & 0 & 1 \end{bmatrix} \begin{bmatrix} 1 & 0 & 0 & 0 \\ 0 & 1 & 0 & 0 \\ 0 & 0 & 1 & 0 \\ 0 & 0 & 0 & 1 \end{bmatrix}$$

Applying $R_2 - 2R_1$ and then $R_1 + R_3$, we get

$$\sim \begin{bmatrix} 1 & -2 & -1 & -1 \\ 1 & -1 & 0 & -1 \\ -1 & 4 & 1 & 2 \\ 0 & 1 & 0 & 1 \end{bmatrix} \begin{bmatrix} 1 & 0 & 1 & 0 \\ -2 & 1 & 0 & 0 \\ 0 & 0 & 1 & 0 \\ 0 & 0 & 0 & 1 \end{bmatrix}$$

Now applying $R_3 + R_2$ and $R_2 - R_1$

$$\sim \begin{bmatrix} 1 & -2 & -1 & -1 \\ 0 & 1 & 1 & 0 \\ 0 & 3 & 1 & 1 \\ 0 & 1 & 0 & 1 \end{bmatrix} \begin{bmatrix} 1 & 0 & 1 & 0 \\ -3 & 1 & -1 & 0 \\ -2 & 1 & 1 & 0 \\ 0 & 0 & 0 & 1 \end{bmatrix}$$

Next applying $R_4 - R_2$ and $R_3 - 3R_2$

$$\sim \begin{bmatrix} 1 & -2 & -1 & -1 \\ 0 & 1 & 1 & 0 \\ 0 & 0 & -2 & 1 \\ 0 & 0 & -1 & 1 \end{bmatrix} \begin{bmatrix} 1 & 0 & 1 & 0 \\ -3 & 1 & -1 & 0 \\ 7 & -2 & 4 & 0 \\ 3 & -1 & 1 & 1 \end{bmatrix}$$

Applying $R_3 - R_4$

$$\sim \begin{bmatrix} 1 & -2 & -1 & -1 \\ 0 & 1 & 1 & 0 \\ 0 & 0 & -1 & 0 \\ 0 & 0 & -1 & 1 \end{bmatrix} \begin{bmatrix} 1 & 0 & 1 & 0 \\ -3 & 1 & -1 & 0 \\ 4 & -1 & 3 & -1 \\ 3 & -1 & 1 & 1 \end{bmatrix}$$

Applying $R_4 - R_3$ and $R_2 + R_3$

$$\sim \begin{bmatrix} 1 & -2 & -1 & -1 \\ 0 & 1 & 0 & 0 \\ 0 & 0 & -1 & 0 \\ 0 & 0 & 0 & 1 \end{bmatrix} \begin{bmatrix} 1 & 0 & 1 & 0 \\ 1 & 0 & 2 & -1 \\ 4 & -1 & 3 & -1 \\ -1 & 0 & -2 & 2 \end{bmatrix}$$

Applying $R_1 - R_3$

$$\sim \begin{bmatrix} 1 & -2 & 0 & -1 \\ 0 & 1 & 0 & 0 \\ 0 & 0 & -1 & 0 \\ 0 & 0 & 0 & 1 \end{bmatrix} \begin{bmatrix} -3 & 1 & -2 & 1 \\ 1 & 0 & 2 & -1 \\ 4 & -1 & 3 & -1 \\ -1 & 0 & -2 & 2 \end{bmatrix}$$

Applying $R_1 + R_4$

$$\sim \begin{bmatrix} 1 & -2 & 0 & 0 \\ 0 & 1 & 0 & 0 \\ 0 & 0 & -1 & 0 \\ 0 & 0 & 0 & 1 \end{bmatrix} \begin{bmatrix} -4 & 1 & -4 & 3 \\ 1 & 0 & 2 & -1 \\ 4 & -1 & 3 & -1 \\ -1 & 0 & -2 & 2 \end{bmatrix}$$

Applying $R_1 + 2R_2 - R_3$

$$\sim \begin{bmatrix} 1 & 0 & 0 & 0 \\ 0 & 1 & 0 & 0 \\ 0 & 0 & 1 & 0 \\ 0 & 0 & 0 & 1 \end{bmatrix} \begin{bmatrix} -2 & 1 & 0 & 1 \\ 1 & 0 & 2 & -1 \\ -4 & 1 & -3 & 1 \\ -1 & 0 & -2 & 2 \end{bmatrix} = A^{-1}$$

Ans.

19. *Find the value of θ for which the system of equations*

$$(2 \sin \theta)\, x + y - 2z = 0$$
$$3x + (2 \cos 2\theta)\, y = 0$$
$$5x + 3y - z = 0$$

has a non-trivial solution.

SOLUTION: The system of homogeneous equations will be consistent if $\begin{vmatrix} 2\sin\theta & 1 & -2 \\ 3 & 2\cos 2\theta & 3 \\ 5 & 3 & -1 \end{vmatrix} = 0$

or $2 \sin \theta\, (-2 \cos 2\theta - 9) + (15 + 3) - 2(9 - 10 \cos 2\theta) = 0$

or $20 \cos 2\theta - 4 \sin \theta \cos 2\theta - 18 \sin \theta = 0$

or $10\, (1 - 2 \sin^2 \theta) - 2\sin \theta\, (1 - 2 \sin^2 \theta) - 9 \sin \theta = 0$

or $4 \sin^3 \theta - 20 \sin^2 \theta - 11 \sin \theta + 10 = 0$

$\sin \theta = \dfrac{1}{2}$ satisfies this relation hence $\theta = \dfrac{\pi}{6}$ **Ans.**

20. *For the matrix* $A = \begin{bmatrix} 1 & 2 & 3 & -2 \\ 2 & -2 & 1 & 3 \\ 3 & 0 & 4 & 1 \end{bmatrix}$, *find non- singular matrices P and Q such that PAQ is in normal form.*

SOLUTION: We have $A = \begin{bmatrix} 1 & 2 & 3 & -2 \\ 2 & -2 & 1 & 3 \\ 3 & 0 & 4 & 1 \end{bmatrix} = \begin{bmatrix} 1 & 0 & 0 \\ 0 & 1 & 0 \\ 0 & 0 & 1 \end{bmatrix} A \begin{bmatrix} 1 & 0 & 0 & 0 \\ 0 & 1 & 0 & 0 \\ 0 & 0 & 1 & 0 \\ 0 & 0 & 0 & 1 \end{bmatrix}$

When we apply row operation on A, it is to be applied on I_3 also and when column operation is applied on A it is to be applied on I_4 also. Thus $R_3 \to R_3 - 3R_1$ and $R_2 \to R_2 - 2R_1$, gives

$$\begin{bmatrix} 1 & 2 & 3 & -2 \\ 0 & -6 & -5 & 7 \\ 0 & -6 & -5 & 7 \end{bmatrix} = \begin{bmatrix} 1 & 0 & 0 \\ -2 & 1 & 0 \\ -3 & 0 & 1 \end{bmatrix} A \begin{bmatrix} 1 & 0 & 0 & 0 \\ 0 & 1 & 0 & 0 \\ 0 & 0 & 1 & 0 \\ 0 & 0 & 0 & 1 \end{bmatrix}$$

Now $R_3 \to R_3 - R_2$ and $R_2 \to -R_2$, we get

$$\begin{bmatrix} 1 & 2 & 3 & -2 \\ 0 & 6 & 5 & -7 \\ 0 & 0 & 0 & 0 \end{bmatrix} = \begin{bmatrix} 1 & 0 & 0 \\ 2 & -1 & 0 \\ -1 & -1 & 1 \end{bmatrix} A \begin{bmatrix} 1 & 0 & 0 & 0 \\ 0 & 1 & 0 & 0 \\ 0 & 0 & 1 & 0 \\ 0 & 0 & 0 & 1 \end{bmatrix}$$

Next, $C_3 \to C_3 - 3C_1$ and $C_4 \to C_4 + C_2$ we get

$$\begin{bmatrix} 1 & 2 & 0 & 0 \\ 0 & 6 & 5 & -1 \\ 0 & 0 & 0 & 0 \end{bmatrix} = \begin{bmatrix} 1 & 0 & 0 \\ 2 & -1 & 0 \\ -1 & -1 & 1 \end{bmatrix} A \begin{bmatrix} 1 & 0 & -3 & 0 \\ 0 & 1 & 0 & 1 \\ 0 & 0 & 1 & 0 \\ 0 & 0 & 0 & 1 \end{bmatrix}$$

Now apply $C_2 \to \frac{1}{2}C_2$ and $C_4 \to 5C_4 + C_3$ we get

$$\begin{bmatrix} 1 & 1 & 0 & 0 \\ 0 & 3 & 5 & 0 \\ 0 & 0 & 0 & 0 \end{bmatrix} = \begin{bmatrix} 1 & 0 & 0 \\ 2 & -1 & 0 \\ -1 & -1 & 1 \end{bmatrix} A \begin{bmatrix} 1 & 0 & -3 & -3 \\ 0 & \frac{1}{2} & 0 & 5 \\ 0 & 0 & 1 & 0 \\ 0 & 0 & 0 & 5 \end{bmatrix}$$

Then apply $C_2 \to C_2 - C_1$ and get

$$\begin{bmatrix} 1 & 0 & 0 & 0 \\ 0 & 3 & 5 & 0 \\ 0 & 0 & 0 & 0 \end{bmatrix} = \begin{bmatrix} 1 & 0 & 0 \\ 2 & -1 & 0 \\ -1 & -1 & 1 \end{bmatrix} A \begin{bmatrix} 1 & -1 & -3 & -3 \\ 0 & \frac{1}{2} & 0 & 5 \\ 0 & 0 & 1 & 1 \\ 0 & 0 & 0 & 5 \end{bmatrix}$$

Next, operate $C_3 \to 5C_2 - 3C_3$ to get

$$\begin{bmatrix} 1 & 0 & 0 & 0 \\ 0 & 3 & 0 & 0 \\ 0 & 0 & 0 & 0 \end{bmatrix} = \begin{bmatrix} 1 & 0 & 0 \\ 2 & -1 & 0 \\ -1 & -1 & 1 \end{bmatrix} A \begin{bmatrix} 1 & -1 & 4 & -3 \\ 0 & \frac{1}{2} & \frac{5}{2} & 5 \\ 0 & 0 & -3 & 1 \\ 0 & 0 & 0 & 5 \end{bmatrix}$$

Finally, operating $R_2 \to \frac{1}{3}R_2$, gives

$$\begin{bmatrix} 1 & 0 & 0 & 0 \\ 0 & 1 & 0 & 0 \\ 0 & 0 & 0 & 0 \end{bmatrix} = \begin{bmatrix} 1 & 0 & 0 \\ \frac{2}{3} & \frac{-1}{3} & 0 \\ -1 & -1 & 1 \end{bmatrix} A \begin{bmatrix} 1 & -1 & 4 & -3 \\ 0 & \frac{1}{2} & \frac{5}{2} & 5 \\ 0 & 0 & -3 & 1 \\ 0 & 0 & 0 & 5 \end{bmatrix} = PAQ.$$

Hence $$P = \begin{bmatrix} 1 & 0 & 0 \\ \frac{2}{3} & \frac{-1}{3} & 0 \\ -1 & -1 & 1 \end{bmatrix} \text{ and } Q = \begin{bmatrix} 1 & -1 & 4 & -3 \\ 0 & \frac{1}{2} & \frac{5}{2} & 5 \\ 0 & 0 & -3 & 1 \\ 0 & 0 & 0 & 5 \end{bmatrix}$$

21. *Reduce the following matrix into its normal form and hence find its rank:*

$$A = \begin{bmatrix} 2 & 3 & -1 & -1 \\ 1 & -1 & -2 & -4 \\ 3 & 1 & 3 & -2 \\ 6 & 3 & 0 & -7 \end{bmatrix}$$

SOLUTION: Interchanging R_1 and R_2 we get

$$A \sim \begin{bmatrix} 1 & -1 & -2 & -4 \\ 2 & 3 & -1 & -1 \\ 3 & 1 & 3 & -2 \\ 6 & 3 & 0 & -7 \end{bmatrix}$$

$$\sim \begin{bmatrix} 1 & -1 & -2 & -4 \\ 0 & 5 & 3 & 7 \\ 0 & 4 & 9 & 10 \\ 0 & 9 & 12 & 17 \end{bmatrix} \quad (\text{by } R_2 \to R_2 - 2R_1,\ R_3 \to R_3 - 3R_1,\ R_4 \to R_4 - 6R_1)$$

$$\sim \begin{bmatrix} 1 & 0 & 0 & 0 \\ 0 & 5 & 3 & 7 \\ 0 & 4 & 9 & 10 \\ 0 & 9 & 12 & 17 \end{bmatrix} \quad (\text{by } C_2 \to C_2 + C_1,\ C_3 \to C_3 + 2C_1,\ C_4 \to C_4 + 4C_1)$$

$$\sim \begin{bmatrix} 1 & 0 & 0 & 0 \\ 0 & 5 & 3 & 7 \\ 0 & 4 & 9 & 10 \\ 0 & 0 & 0 & 0 \end{bmatrix} \quad (\text{by } R_4 \to R_4 - R_2 - R_3)$$

$$\sim \begin{bmatrix} 1 & 0 & 0 & 0 \\ 0 & 1 & -6 & -3 \\ 0 & 4 & 9 & 10 \\ 0 & 0 & 0 & 0 \end{bmatrix} \quad (\text{by } R_2 \to R_2 - R_3)$$

$$\sim \begin{bmatrix} 1 & 0 & 0 & 0 \\ 0 & 1 & -6 & -3 \\ 0 & 0 & 33 & 22 \\ 0 & 0 & 0 & 0 \end{bmatrix} \quad (\text{by } R_3 \to R_3 - 4R_2)$$

$$\sim \begin{bmatrix} 1 & 0 & 0 & 0 \\ 0 & 1 & 0 & 0 \\ 0 & 0 & 33 & 22 \\ 0 & 0 & 0 & 0 \end{bmatrix} \quad (\text{by } C_3 \to C_3 + 6C_2,\ C_4 \to C_4 + 3C_2)$$

$$\sim \begin{bmatrix} 1 & 0 & 0 & 0 \\ 0 & 1 & 0 & 0 \\ 0 & 0 & 1 & 22 \\ 0 & 0 & 0 & 0 \end{bmatrix} \quad \left(\text{by } C_3 \to \frac{1}{33} C_1\right)$$

$$\sim \begin{bmatrix} 1 & 0 & 0 & 0 \\ 0 & 1 & 0 & 0 \\ 0 & 0 & 1 & 0 \\ 0 & 0 & 0 & 0 \end{bmatrix} \quad (\text{by } C_4 \to 2C_3)$$

$$\sim \begin{bmatrix} I_3 & 0 \\ 0 & 0 \end{bmatrix}$$

Hence the rank $(A) = 3$. **Ans.**

22. *Find two non-singular matrix P and Q such the PAQ is in the normal form, where*

$$A = \begin{bmatrix} 1 & -1 & -1 \\ 1 & 1 & 1 \\ 3 & 1 & 1 \end{bmatrix}$$

SOLUTION: We write $A = IAI$, that is

$$\begin{bmatrix} 1 & -1 & -1 \\ 1 & 1 & 1 \\ 3 & 1 & 1 \end{bmatrix} = \begin{bmatrix} 1 & 0 & 0 \\ 0 & 1 & 0 \\ 0 & 0 & 1 \end{bmatrix} A \begin{bmatrix} 1 & 0 & 0 \\ 0 & 1 & 0 \\ 0 & 0 & 1 \end{bmatrix}$$

Applying $C_2 \to C_2 + C_1$ and $C_3 \to C_3 + C_1$, we get

$$\begin{bmatrix} 1 & 0 & 0 \\ 1 & 2 & 2 \\ 3 & 4 & 4 \end{bmatrix} = \begin{bmatrix} 1 & 0 & 0 \\ 0 & 1 & 0 \\ 0 & 0 & 1 \end{bmatrix} A \begin{bmatrix} 1 & 0 & 0 \\ 0 & 1 & 0 \\ 0 & 0 & 1 \end{bmatrix}$$

Now, appling $R_2 \to R_2 - R_1$ and $R_3 \to R_3 - 3R_1$, we get

$$\begin{bmatrix} 1 & 0 & 0 \\ 0 & 2 & 2 \\ 3 & 4 & 4 \end{bmatrix} = \begin{bmatrix} 1 & 0 & 0 \\ -1 & 1 & 0 \\ -3 & 0 & 1 \end{bmatrix} A \begin{bmatrix} 1 & 1 & 1 \\ 0 & 1 & 0 \\ 0 & 0 & 1 \end{bmatrix}$$

Next, applying $C_3 \to C_3 - C_2$ and $C_2 \to \frac{1}{2} C_2$, we get

$$\begin{bmatrix} 1 & 0 & 0 \\ 0 & 1 & 0 \\ 0 & 2 & 0 \end{bmatrix} = \begin{bmatrix} 1 & 0 & 0 \\ -1 & 1 & 0 \\ -3 & 0 & 1 \end{bmatrix} A \begin{bmatrix} 1 & 1/2 & 0 \\ 0 & 1/2 & -1 \\ 0 & 0 & 1 \end{bmatrix}$$

Finally applying $R_3 \to R_3 - 2R_2$ we get

$$\begin{bmatrix} 1 & 0 & 0 \\ 0 & 1 & 0 \\ 0 & 0 & 0 \end{bmatrix} = \begin{bmatrix} 1 & 0 & 0 \\ -1 & 1 & 0 \\ -1 & -2 & 1 \end{bmatrix} A \begin{bmatrix} 1 & 1/2 & 0 \\ 0 & 1/2 & -1 \\ 0 & 0 & 1 \end{bmatrix} = PAQ$$

$$\begin{bmatrix} I_2 & 0 \\ 0 & 0 \end{bmatrix} = PAQ, \text{ where } P = \begin{bmatrix} 1 & 0 & 0 \\ -1 & 1 & 0 \\ -1 & -2 & 0 \end{bmatrix} \text{ and } Q = \begin{bmatrix} 1 & 1/2 & 0 \\ 0 & 1/2 & -1 \\ 0 & 0 & 1 \end{bmatrix}$$ **Ans.**

23. *Find the rank of the matrix* $\begin{bmatrix} 3 & -1 & 2 \\ -6 & 2 & 4 \\ -3 & 1 & 2 \end{bmatrix}$ *by reducing it in its normal form.*

SOLUTION: Let $A = \begin{bmatrix} 3 & -1 & 2 \\ -6 & 2 & 4 \\ -3 & 1 & 2 \end{bmatrix}$. Applying $C_1 \leftrightarrow C_2$, $C_2 + 3C_1$ and $-C_1$, we get

$$A \sim \begin{bmatrix} 1 & 0 & 2 \\ -2 & 0 & 4 \\ -1 & 0 & 2 \end{bmatrix}$$

Next apply $R_2 \to R_2 + 2R_1$ and $R_3 \to R_3 + R_1$, and get

$$A \sim \begin{bmatrix} 1 & 0 & 2 \\ 0 & 0 & 8 \\ 0 & 0 & 4 \end{bmatrix}$$

Then apply $C_2 \leftrightarrow C_3$, $R_1 \to R_1 - \frac{1}{4}R_2$, $R_3 \to R_3 - \frac{1}{2}R_2$ and $R_2 \to \frac{1}{8}R_2$ and get

$$A \sim \begin{bmatrix} 1 & 0 & 0 \\ 0 & 1 & 0 \\ 0 & 0 & 0 \end{bmatrix} = \begin{bmatrix} I_2 & 0 \\ 0 & 0 \end{bmatrix}.$$ Hence rank of A is 2. **Ans.**

24. *Are the vectors* $X_1 = (1, 3, 4, 2)$, $X_2 = (3, -5, 2, 2)$, $X_3 = (2, -1, 3, 2)$ *linearly dependent? If so, express one of these as a linearl combination of others.*

SOLUTION: If X_1, X_2, X_3 are linearly dependent then there exist scalars $\lambda_1, \lambda_2, \lambda_3$ such that $\lambda_1 X_1 + \lambda_2 X_2 + \lambda_3 X_3 = 0$ where not all $\lambda_1, \lambda_2, \lambda_3$ are zero.

Hence we have $\lambda_1 + 3\lambda_2 + 2\lambda_3 = 0$, $3\lambda_1 - 5\lambda_2 - \lambda_3 = 0$,

$4\lambda_1 + 2\lambda_2 + 3\lambda_3 = 0$, and $2\lambda_1 + 2\lambda_2 + 2\lambda_3 = 0$ or $\lambda_1 + \lambda_2 + \lambda_3 = 0$

Solving any two equations, say first and third of the above four, we get

$\lambda_1 = 1$, $\lambda_2 = 1$, $\lambda_3 = -2$ which do satisfy the other two equations.

Hence X_1, X_2, X_3 are linearly dependent and the relation between them is $X_1 + X_2 - 2X_3 = 0$. **Ans.**

25. *Examine the following system of vectors for linear dependence. If dependent, find the relation between them* $X_1 = (1, 1, -1, 1)$, $X_2 = (1, -1, 2, -1)$, $X_3 = (3, 1, 0, 1)$.

SOLUTION: X_1, X_2, X_3 will be dependent if there exist $\lambda_1, \lambda_2, \lambda_3$ not all zero such that

$$\lambda_1 X_1 + \lambda_2 X_2 + \lambda_3 X_3 = 0$$

$$\Rightarrow \quad \lambda_1 + \lambda_2 + 3\lambda_3 = 0, \qquad \lambda_1 - \lambda_2 + \lambda_3 = 0$$

and

$$-\lambda_1 + 2\lambda_2 + 0\lambda_3 = 0, \qquad \lambda_1 - \lambda_2 + \lambda_3 = 0.$$

Thus $\lambda_1 = 2\lambda_2$, $\lambda_2 = -\lambda_3$ or $\dfrac{\lambda_1}{2} = \dfrac{\lambda_2}{1} = \dfrac{\lambda_3}{-1}$ which are not all zero.

Hence X_1, X_2, X_3 are dependent and the relation between them is $2X_1 + X_2 - X_3 = 0$. **Ans.**

26. *Find the value of* λ *for which the equations*

$$(\lambda - 1)\,x + (3\lambda + 1)\,y + 2\lambda z = 0$$

$$(\lambda - 1)\,x + (4\lambda - 2)\,y + (\lambda + 3)\,z = 0$$

$$2x + (3\lambda + 1)\,y + 3\,(\lambda - 1)\,z = 0$$

are consistent and find the ratios of $x : y : z$ *when* λ *has the smallest of these values. What happens when* λ *has the greatest of these values.*

SOLUTION: The given equations are homogeneous equations in x, y, z and will be consistent if

$$\begin{vmatrix} \lambda-1 & 3\lambda+1 & 2\lambda \\ \lambda-1 & 4\lambda-2 & \lambda+3 \\ 2 & 3\lambda+1 & 3(\lambda-1) \end{vmatrix} = 0. \qquad \text{Applying } R_2 - R_1, \text{ gives}$$

or

$$\begin{vmatrix} \lambda-1 & 3\lambda+1 & 2\lambda \\ 0 & \lambda-3 & 3-\lambda \\ 2 & 3\lambda+1 & 3(\lambda-1) \end{vmatrix} = 0 \qquad \text{Now applying } C_3 + C_2, \text{ gives}$$

or

$$\begin{vmatrix} \lambda-1 & 3\lambda+1 & 5\lambda+1 \\ 0 & \lambda-3 & 0 \\ 2 & 3\lambda+1 & 6\lambda-2 \end{vmatrix} = 0 \qquad \text{or} \qquad (\lambda-3)\begin{vmatrix} \lambda-1 & 5\lambda+1 \\ 2 & 6\lambda-2 \end{vmatrix} = 0$$

or

$$(\lambda - 3)^2\,[2\,(3\lambda - 1) + 2] = 0 \qquad \text{or} \qquad 6\lambda\,(\lambda - 3)^2 = 0$$

Thus, the given equations are consistent when $\lambda = 0, 3, 3$. When λ is least equal to 0 the equations become $-x + y = 0$, $-x - 2y + 3z = 0$ and $2x + y - 3z = 0$ which gives $x = y = z$.

Further, when λ is greatest equal to 3, the equations are all identical as

$$2x + 10y + 6z = 0 \qquad \text{or} \qquad x + 5y + 3z = 0.$$

27. *For what values of the parameters λ and μ do the system of equations $x+y+z=6,\ x+2y+3z=10,\ x+2y+\lambda z=\mu$, have (i) no solution (ii) unique solutions (iii) more than one solution?*

SOLUTION: Given equations can be written as $AX = B$

$$\text{where } A = \begin{bmatrix} 1 & 1 & 1 \\ 1 & 2 & 3 \\ 1 & 2 & \lambda \end{bmatrix}, \quad X = \begin{bmatrix} x \\ y \\ z \end{bmatrix}, \quad B = \begin{bmatrix} 6 \\ 10 \\ \mu \end{bmatrix}$$

$$\text{Augmented matrix} = [A \mid B] = \left[\begin{array}{ccc|c} 1 & 1 & 1 & 6 \\ 1 & 2 & 3 & 10 \\ 1 & 2 & \lambda & \mu \end{array}\right]$$

Appling $R_3 \to R_3 - R_2$ and $R_2 \to R_2 - R_1$ we get

$$D \approx [A \mid B] \sim \left[\begin{array}{ccc|c} 1 & 1 & 1 & 6 \\ 0 & 1 & 2 & 4 \\ 0 & 0 & \lambda-3 & \mu-10 \end{array}\right]$$

Case I: $\lambda = 3,\ \mu \neq 10$, we have $\rho(A) = 2,\ \rho(D) = 3$. Since $\rho(A) \neq \rho(D)$ no solutions.

Case II: $\lambda \neq 3,\ \mu$ has any value then $\rho(A) = 3,\ \rho(D) = 3$ since $\rho(A) = \rho(D) = 3$ = number of unknown hence unique solution.

Case III: $\lambda = 3$ and $\mu = 10$ then $\rho(A) = 2,\ \rho(D) = 2$ since $\rho(A) = \rho(D) <$ number of unknowns hence many solutions.

SOLUTIONS TO EXERCISE 8C

1. *Verify Caley-Hamilton theorem for the matrix*

$$A = \begin{bmatrix} 8 & -8 & -2 \\ 4 & -3 & -2 \\ 3 & -4 & 1 \end{bmatrix}$$

and hence find A^{-1}.

SOLUTION: Characterstic equation of A is $|A - \lambda I| = 0$

$$\begin{vmatrix} 8-\lambda & -8 & -2 \\ 4 & -3-\lambda & -2 \\ 3 & -4 & 1-\lambda \end{vmatrix} = 0 \quad \text{or} \quad \lambda^3 - 6\lambda^2 + 11\lambda - 6 = 0.$$

By Caley Hamiltom theorem matrix A satisfies this equation hence $A^3 - 6A^2 + 11A - 6I = 0$.
Multiplying throughout by A^{-1} we get $6A^{-1} = A^2 - 6A + 11I$.

Here $$A^2 = \begin{bmatrix} 8 & -8 & -2 \\ 4 & -3 & -2 \\ 3 & -4 & 1 \end{bmatrix}\begin{bmatrix} 8 & -8 & -2 \\ 4 & -3 & -2 \\ 3 & -4 & 1 \end{bmatrix} = \begin{bmatrix} 26 & -32 & -2 \\ 14 & -15 & -4 \\ 11 & -16 & 3 \end{bmatrix}$$

$\therefore$ $$6A^{-1} = \begin{bmatrix} 26 & -32 & -2 \\ 14 & -15 & -4 \\ 11 & -16 & 3 \end{bmatrix} + \begin{bmatrix} -48 & 48 & 12 \\ -24 & 18 & 12 \\ -18 & 24 & -6 \end{bmatrix} + \begin{bmatrix} 11 & 0 & 0 \\ 0 & 11 & 0 \\ 0 & 0 & 11 \end{bmatrix}$$

or $$A^{-1} = \frac{1}{6}\begin{bmatrix} -11 & 16 & 10 \\ -10 & 14 & 8 \\ -7 & 8 & 8 \end{bmatrix}$$ **Ans.**

2. *Verify Caley-Hamilton theorem for the matrix* $\begin{bmatrix} 1 & 4 \\ 2 & 3 \end{bmatrix}$ *and use it to simplify the expression*

$$A^5 + 5A^4 - 6A^3 + 2A^2 - 4A + 7I.$$

SOLUTION: $A = \begin{bmatrix} 1 & 4 \\ 2 & 3 \end{bmatrix}$ $\therefore$ Characterstic equation is $|A - \lambda I| = 0$

or $\begin{vmatrix} 1-\lambda & 4 \\ 2 & 3-\lambda \end{vmatrix} = 0$ or $\lambda^2 - 4\lambda - 5 = 0$. By Caley Hamilton theorem, it is satisfied by A itself, hence $A^2 - 4A - 5I = 0$

Given expression is $A^5 + 5A^4 - 6A^3 + 2A^2 - 4A + 7I$

$$= A^3(A^2 - 4A - 5I) + 9A^4 - A^3 + 2A^2 - 4A + 7I$$
$$= 0 + 9A^2(A^2 - 4A - 5I) + 35A^3 + 47A^2 - 4A + 7I$$
$$= 0 + 0 + 35A(A^2 - 4A - 5I) + 187A^2 + 171A + 7I$$
$$= 0 + 0 + 0 + 187(A^2 - 4A - 5I) + 919A + 942I$$
$$= 919A + 942I$$
$$= \begin{bmatrix} 919 & 3676 \\ 1838 & 2757 \end{bmatrix} + \begin{bmatrix} 942 & 0 \\ 0 & 942 \end{bmatrix} = \begin{bmatrix} 1861 & 3676 \\ 1838 & 3699 \end{bmatrix}$$ **Ans.**

3. *Determine the eigen values and eigen vectors of the matrix*

$$\begin{bmatrix} 4 & 2 & -2 \\ -5 & 3 & 2 \\ -2 & 4 & 1 \end{bmatrix}.$$

SOLUTION: Characterstic Equation of A is $|A - \lambda I| = 0$

or $$\begin{vmatrix} 4-\lambda & 2 & -2 \\ -5 & 3-\lambda & 2 \\ -2 & 4 & 1-\lambda \end{vmatrix} = 0$$

or $(4-\lambda)[(3-\lambda)(1-\lambda)-8]-2(-1+5\lambda)-2(-14-2\lambda)=0$

or $(4-\lambda)(\lambda^2-4\lambda-5)+2-10\lambda+28+4\lambda=0$

or $-\lambda^3+8\lambda^2-17\lambda+10=0$ or $(\lambda-1)(\lambda^2-7\lambda+10)=0$ $\therefore$ $\lambda=1, 2, 5$

Thus, eigen values of A are 1, 2 and 5. Next, for eigen vectors

$$(A-\lambda I)X=0 \quad \text{or} \quad \begin{bmatrix} 4-\lambda & 2 & -2 \\ -5 & 3-\lambda & 2 \\ -2 & 4 & 1-\lambda \end{bmatrix}\begin{bmatrix} x \\ y \\ z \end{bmatrix}=0$$

For $\lambda=1$ we have $3x+2y-2z=0$, $-5x+2y+2z=0$, $-2x+4y+0z=0$

$$\Rightarrow \quad \frac{x}{8}=\frac{y}{4}=\frac{z}{16} \quad \therefore \text{ Eigen vector is } (2, 1, 4)'.$$

For $\lambda=2$ we have $2x+2y-2z=0$, $-5x+2y+2z=0$, $-2x+4y-z=0$.

$$\Rightarrow \quad \frac{x}{6}=\frac{y}{6}=\frac{z}{12} \quad \therefore \text{ Eigen Vector is } (1, 1, 2)'$$

For $\lambda=5$ we have $-x+2y-2z=0$, $-5x-2y+2z=0$, $-2x+4y-4z=0$

$$\Rightarrow \quad \frac{x}{4-4}=\frac{y}{10+2}=\frac{z}{2+10} \quad \therefore \text{ Eigen Vector is } (0, 1, 1)'.$$

Thus eigen values are 1, 2, 5 and corresponding vectors are

$$\begin{bmatrix} 2 \\ 1 \\ 4 \end{bmatrix}, \begin{bmatrix} 1 \\ 1 \\ 2 \end{bmatrix} \text{ and } \begin{bmatrix} 0 \\ 1 \\ 1 \end{bmatrix}$$

Ans.

4. *Show that any square matrix A and its transpose A' have same characteristic equation.*

SOLUTION: Characterstic equation for A is $|A-\lambda I|=0$

$\therefore$ $(A-\lambda I)X=0 \Rightarrow X'(A-\lambda I)'=0$ or $X'(A'-\lambda I)=0$

or $X'A'-\lambda X'I=0$ or $X'(A'-\lambda I)=0 \Rightarrow |A'-\lambda I|=0$.

Hence A' has same characterstic equation as of $A=0$ **Hence Proved.**

5. *If λ is an eigen value of an orthogonal matrix then show that $1/\lambda$ is also its eigen value.*

SOLUTION: If A is an orthogonal matrix then $AA'=I$ or $A'=A^{-1}$.

We know that if λ is the eigen value of A then $\frac{1}{\lambda}$ will be the eigen value of A^{-1} and

$A^{-1}=A'$ hence $\frac{1}{\lambda}$ is also the eigen value. **Hence Proved.**

6. *Show that the matrix* $A = \begin{bmatrix} -9 & 4 & 4 \\ -8 & 3 & 4 \\ -16 & 8 & 7 \end{bmatrix}$ *can be reduced to the diagonal form with diagonal elements as eigen values of A.*

SOLUTION: $A = \begin{bmatrix} -9 & 4 & 4 \\ -8 & 3 & 4 \\ -16 & 8 & 7 \end{bmatrix}$ The characterstic equation is $|A - \lambda I| = 0$

or $\begin{vmatrix} -9-\lambda & 4 & 4 \\ -8 & 3-\lambda & 4 \\ -16 & 8 & 7-\lambda \end{vmatrix} = 0$ or $\lambda^3 - \lambda^2 - 5\lambda - 3 = 0$

or $(\lambda - 3)(\lambda^2 + 2\lambda + 1) = 0$ $\therefore$ $\lambda = -1, -1, 3$ are the eigen values.

Now, for eigen vectors, $(A - \lambda I)X = 0$.

For $\lambda = 3$ we have

$$-12x + 4y + 4z = 0$$
$$-8x + 0y + 4z = 0$$
$$-16x + 8y + 4z = 0$$

$\Rightarrow$ $z = 2x$, $y = x$ $\therefore$ Eigen vector is $[1, 1, 2]'$.

For $\lambda = -1$ we have $2x = y + z$.

One eigen vector is clearly $(1, 1, 1)'$ and the other is $(0, 1, -1)'$.

$\therefore$ $P = \begin{bmatrix} 1 & 1 & 0 \\ 1 & 1 & 1 \\ 2 & 1 & -1 \end{bmatrix}$ Hence $P^{-1} = \begin{bmatrix} -2 & 1 & 1 \\ 3 & -1 & -1 \\ -1 & 1 & 0 \end{bmatrix}$

$\therefore$ The spectratum $= D = P^{-1}AP = \begin{bmatrix} -2 & 1 & 1 \\ 3 & -1 & -1 \\ -1 & 1 & 0 \end{bmatrix} \begin{bmatrix} -9 & 4 & 4 \\ -8 & 3 & 4 \\ -16 & 8 & 7 \end{bmatrix} \begin{bmatrix} 1 & 1 & 0 \\ 1 & 1 & 1 \\ 2 & 1 & -1 \end{bmatrix}$

$$= \begin{bmatrix} -6 & 3 & 3 \\ -3 & 1 & 1 \\ 1 & -1 & 0 \end{bmatrix} \begin{bmatrix} 1 & 1 & 0 \\ 1 & 1 & 1 \\ 2 & 1 & -1 \end{bmatrix} = \begin{bmatrix} 3 & 0 & 0 \\ 0 & -1 & 0 \\ 0 & 0 & -1 \end{bmatrix}$$ **Ans.**

7. *Find the eigen values and eigen vectors of the matrix A given by*

$$A = \begin{bmatrix} 1 & 1 & 3 \\ 1 & 5 & 1 \\ 3 & 1 & 1 \end{bmatrix}$$

Also, diagonalise A.

SOLUTION: $A = \begin{bmatrix} 1 & 1 & 3 \\ 1 & 5 & 1 \\ 3 & 1 & 1 \end{bmatrix}$. Characterstic equation is $|A - \lambda I| = 0$

or $\begin{vmatrix} 1-\lambda & 1 & 3 \\ 1 & 5-\lambda & 1 \\ 3 & 1 & 1-\lambda \end{vmatrix} = 0$ or $(1 - \lambda)(\lambda^2 - 6\lambda + 4) + \lambda + 2 + 3(3\lambda - 14) = 0$

or $-\lambda^3 + 7\lambda^2 + \lambda(0) - 36 = 0$ $\therefore$ $\lambda = 3,\ 6, -2.$

Hence eigen values of A are $-2, 3, 6.$

For eigen vectors, $(A - \lambda I)X = 0$ or $\begin{bmatrix} 1-\lambda & 1 & 3 \\ 1 & 5-\lambda & 1 \\ 3 & 1 & 1-\lambda \end{bmatrix}\begin{bmatrix} x \\ y \\ z \end{bmatrix} = 0$

For $\lambda = -2$ we have

$$3x + y + 3z = 0$$
$$x + 7y + z = 0$$
$$3x + y + 4z = 0$$

these give $\dfrac{x}{-20} = \dfrac{y}{0} = \dfrac{z}{20}$ $\therefore$ Eigen Vector = (1, 0, –1)

For $\lambda = 3$ we have

$$-2x + y + 3z = 0$$
$$x + 2y + z = 0$$
$$3x + y - 2z = 0$$

which give $\dfrac{x}{1-6} = \dfrac{y}{3+2} = \dfrac{z}{-4-1}$ $\therefore$ Eigen Vector = (1, –1, 1)

For $\lambda = 6$ we have

$$-5x + y + 3z = 0$$
$$x - y + z = 0$$
$$3x + y - 5z = 0$$

which give $\dfrac{x}{1+3} = \dfrac{y}{3+5} = \dfrac{z}{5-1}$ $\therefore$ Eigen Vector = (1, 2, 1)

$\therefore$ P = Modal matrix = $\begin{bmatrix} 1 & 1 & 1 \\ 0 & -1 & 2 \\ -1 & 1 & 1 \end{bmatrix}$ $\therefore$ $P^{-1} = \dfrac{1}{6}\begin{bmatrix} 3 & 0 & -3 \\ 2 & -2 & 2 \\ 1 & 2 & 1 \end{bmatrix}$

$\therefore$ Diagonal Matrix $D = P^{-1}AP$

$$= \frac{1}{6}\begin{bmatrix} 3 & 0 & -3 \\ 2 & -2 & 2 \\ 1 & 2 & 1 \end{bmatrix}\begin{bmatrix} 1 & 1 & 3 \\ 1 & 5 & 1 \\ 3 & 1 & 1 \end{bmatrix}\begin{bmatrix} 1 & 1 & 1 \\ 0 & -1 & 2 \\ -1 & 1 & 1 \end{bmatrix} = \frac{1}{6}\begin{bmatrix} 3 & 0 & -3 \\ 2 & -2 & 2 \\ 1 & 2 & 1 \end{bmatrix}\begin{bmatrix} -2 & 3 & 6 \\ 0 & -3 & 12 \\ 2 & 3 & 6 \end{bmatrix}$$

$$= \frac{1}{6}\begin{bmatrix} -12 & 0 & 0 \\ 0 & 18 & 0 \\ 0 & 0 & 36 \end{bmatrix} = \begin{bmatrix} -2 & 0 & 0 \\ 0 & 3 & 0 \\ 0 & 0 & 6 \end{bmatrix}$$ **Ans.**

8. *If* $A = \begin{bmatrix} 3 & -3 & 4 \\ 2 & -3 & 4 \\ 0 & -1 & 1 \end{bmatrix}$, *find two non-singular matrices*

P and Q such that PAQ = I and hence find A^{-1}.

SOLUTION:

$$A = \begin{bmatrix} 3 & -3 & 4 \\ 2 & -3 & 4 \\ 0 & -1 & 1 \end{bmatrix} = \begin{bmatrix} 1 & 0 & 0 \\ 0 & 1 & 0 \\ 0 & 0 & 1 \end{bmatrix} A \begin{bmatrix} 1 & 0 & 0 \\ 0 & 1 & 0 \\ 0 & 0 & 1 \end{bmatrix} \quad \text{Operating } R_1 - R_2$$

$$\begin{bmatrix} 1 & 0 & 0 \\ 2 & -3 & 4 \\ 0 & -1 & 1 \end{bmatrix} = \begin{bmatrix} 1 & -1 & 0 \\ 0 & 1 & 0 \\ 0 & 0 & 1 \end{bmatrix} A \begin{bmatrix} 1 & 0 & 0 \\ 0 & 1 & 0 \\ 0 & 0 & 1 \end{bmatrix}$$

Now operating $R_2 \to R_2 - 2R_1$

$$\begin{bmatrix} 1 & 0 & 0 \\ 0 & -3 & 4 \\ 0 & -1 & 1 \end{bmatrix} = \begin{bmatrix} 1 & -1 & 0 \\ -2 & 3 & 0 \\ 0 & 0 & 1 \end{bmatrix} A \begin{bmatrix} 1 & 0 & 0 \\ 0 & 1 & 0 \\ 0 & 0 & 1 \end{bmatrix}$$

Next applying $C_2 \to -C_2$

$$\begin{bmatrix} 1 & 0 & 0 \\ 0 & 3 & 4 \\ 0 & 1 & 1 \end{bmatrix} = \begin{bmatrix} 1 & -1 & 0 \\ -2 & 3 & 0 \\ 0 & 0 & 1 \end{bmatrix} A \begin{bmatrix} 1 & 0 & 0 \\ 0 & -1 & 0 \\ 0 & 0 & 1 \end{bmatrix}$$

Next apply $R_2 \leftrightarrow R_3$ and $R_3 - 3R_2$

$$\begin{bmatrix} 1 & 0 & 0 \\ 0 & 1 & 1 \\ 0 & 0 & 1 \end{bmatrix} = \begin{bmatrix} 1 & -1 & 0 \\ 0 & 0 & 1 \\ -2 & 3 & -3 \end{bmatrix} A \begin{bmatrix} 1 & 0 & 0 \\ 0 & -1 & 0 \\ 0 & 0 & 1 \end{bmatrix}$$

Then apply $C_3 \to C_3 - C_2$

$$\begin{bmatrix} 1 & 0 & 0 \\ 0 & 1 & 0 \\ 0 & 0 & 1 \end{bmatrix} = \begin{bmatrix} 1 & -1 & 0 \\ 0 & 0 & 1 \\ -2 & 3 & -3 \end{bmatrix} A \begin{bmatrix} 1 & 0 & 0 \\ 0 & -1 & 1 \\ 0 & 0 & 1 \end{bmatrix}$$

$\therefore \; I = PAQ$ where $P = \begin{bmatrix} 1 & -1 & 0 \\ 0 & 0 & 1 \\ -2 & 3 & -3 \end{bmatrix}$, $Q = \begin{bmatrix} 1 & 0 & 0 \\ 0 & -1 & 1 \\ 0 & 0 & 1 \end{bmatrix}$ **Ans.**

Next $A^{-1} = QP = \begin{bmatrix} 1 & 0 & 0 \\ 0 & -1 & 1 \\ 0 & 0 & 1 \end{bmatrix} \begin{bmatrix} 1 & -1 & 0 \\ 0 & 0 & 1 \\ -2 & 3 & -3 \end{bmatrix} = \begin{bmatrix} 1 & -1 & 0 \\ -2 & 3 & -4 \\ -2 & 3 & -3 \end{bmatrix}$ **Ans.**

9. *Prove that every Hermitian matrix can be written as $A + iB$ where A is real and symmetric and B is real and skew symmetric.* ***(GGSIPU Ist Sem. IInd Term 2004)***

SOLUTION: Let Z be the Hermitian matrix $= \begin{bmatrix} a & p_1 + iq_1 & p_2 + iq_2 \\ p_1 - iq_1 & b & p_3 + iq_3 \\ p_2 - iq_2 & p_3 - iq_3 & c \end{bmatrix}$

Then $$Z = \begin{bmatrix} a & p_1 & p_2 \\ p_1 & b & p_3 \\ p_2 & p_3 & c \end{bmatrix} + i\begin{bmatrix} 0 & q_1 & q_2 \\ -q_1 & 0 & q_3 \\ -q_2 & -q_3 & 0 \end{bmatrix} = A + iB$$

where $A = \begin{bmatrix} a & p_1 & p_2 \\ p_1 & b & p_3 \\ p_2 & p_3 & c \end{bmatrix}$ is real and symmetric

and $B = \begin{bmatrix} 0 & q_1 & q_2 \\ -q_1 & 0 & q_3 \\ -q_2 & -q_3 & 0 \end{bmatrix}$ is real and skew-symmetric. **Hence Proved.**

10. *Find the eigen values and eigen vectors of the following matrix*

$$A = \begin{bmatrix} -9 & 2 & 6 \\ 5 & 0 & -3 \\ -16 & 4 & 11 \end{bmatrix}$$

SOLUTION: Characterstic equation is $|A - \lambda I| = 0$

or $$\begin{vmatrix} -9-\lambda & 2 & 6 \\ 5 & -\lambda & -3 \\ -16 & 4 & 11-\lambda \end{vmatrix} = 0 \quad \text{or} \quad \lambda^3 - 2\lambda^2 - \lambda + 2 = 0$$

or $(\lambda^2 - 1)(\lambda - 2) = 0$. Thus the eigen values are 1, –1, 2.

For eigen vectors we have $(A - \lambda I)X = 0$.

or $$\begin{bmatrix} -9-\lambda & 2 & 6 \\ 5 & -\lambda & -3 \\ -16 & 4 & 11-\lambda \end{bmatrix}\begin{bmatrix} x \\ y \\ z \end{bmatrix} = 0$$

For $\lambda = 1$, we have $-10x + 2y + 6z = 0$, $5x - y - 3z = 0$, $-16x + 4y + 10z = 0$ which give the eigen vector as $[1, -1, 2]'$.

For $\lambda = 2$, we have $-11x + 2y + 6z = 0$, $5x - 2y - 3z = 0$, $-16x + 4y + 9z = 0$. which give eigen vector as $[2, -1, 4]'$.

For $\lambda = -1$ we have $-8x + 2y + 6z = 0$, $5x + y - 3z = 0$, $-16x + 4y + 12z = 0$ which gives the eigen vector as $[2, -1, 3]'$.

Thus, eigen values are 1, –1, 2 and eigen vectors are

$(1, -1, 2)'$, $(2, -1, 3)'$, $(2, -1, 4)'$ **Ans.**

11. *Find the eigen values and eigen vectors of the matrix*

$$\begin{bmatrix} -2 & 2 & -3 \\ 2 & 1 & -6 \\ -1 & -2 & 0 \end{bmatrix}$$

SOLUTION: The characterstic equation is $\begin{vmatrix} -2-\lambda & 2 & -3 \\ 2 & 1-\lambda & -6 \\ -1 & -2 & -\lambda \end{vmatrix} = 0$ or $\lambda^3 + \lambda^2 - 21\lambda - 45 = 0$

or $(\lambda + 3)^2 (\lambda - 5) = 0$ $\therefore$ Eigen values are $-3, -3, 5$.

For eigen vectors we have $(A - \lambda I) X = 0$

For $\lambda = 5$, we have $-7x + 2y - 3z = 0$, $2x - 4y - 6z = 0$, $-x - 2y - 5z = 0$ which give eigen vector as $(1, 2, -1)'$

For $\lambda = -3$, we have $x + 2y - 3z = 0$, $2x + 4y - 6z = 0$, $-x - 2y + 3z = 0$, that is, there is one equation only as $x + 2y - 3z = 0$.

We have here two eigen vectors $(1, 1, 1)$ and $(2, -1, 0)$.

Thus we have eigen values as $-3, -3, 5$

and eigen vectors as $(1, 2, -1)', (1, 1, 1)'$ and $(2, -1, 0)'$ **Ans.**

12. *Find the inverse of the matrix* $A = \begin{bmatrix} 7 & -1 & 3 \\ 6 & 1 & 4 \\ 2 & 4 & 8 \end{bmatrix}$ *using Caley Hamilton theorem.*

SOLUTION: The characterstic equation of A is $\begin{vmatrix} 7-\lambda & -1 & 3 \\ 6 & 1-\lambda & 4 \\ 2 & 4 & 8-\lambda \end{vmatrix} = 0$

or $\lambda^3 - 16\lambda^2 + 55\lambda - 50 = 0$. Hence by Caley-Hamilton theorem.

$A^3 - 16A^2 + 55A - 50I = 0$. Multiplying throughout by A^{-1}, we get

$A^2 - 16A + 55I = 50A^{-1}$ or $A^{-1} = \dfrac{1}{50}[A^2 - 16A + 55I]$

Next, $$A^2 = \begin{bmatrix} 7 & -1 & 3 \\ 6 & 1 & 4 \\ 2 & 4 & 8 \end{bmatrix}\begin{bmatrix} 7 & -1 & 3 \\ 6 & 1 & 4 \\ 2 & 4 & 8 \end{bmatrix} = \begin{bmatrix} 49 & 4 & 41 \\ 56 & 11 & 54 \\ 54 & 34 & 86 \end{bmatrix}$$

$$\therefore \quad A^{-1} = \frac{1}{50}\begin{bmatrix} 49 & 4 & 41 \\ 56 & 11 & 54 \\ 54 & 34 & 86 \end{bmatrix} - \frac{16}{50}\begin{bmatrix} 7 & -1 & 3 \\ 6 & 1 & 4 \\ 2 & 4 & 8 \end{bmatrix} + \frac{1}{50}\begin{bmatrix} 55 & 0 & 0 \\ 0 & 55 & 0 \\ 0 & 0 & 55 \end{bmatrix}$$

$$= \frac{1}{50}\begin{bmatrix} -8 & 20 & -7 \\ -40 & 50 & -10 \\ 22 & -30 & 13 \end{bmatrix} = \begin{bmatrix} -\frac{4}{25} & \frac{2}{5} & -\frac{7}{50} \\ -\frac{4}{5} & 1 & -\frac{1}{5} \\ \frac{11}{25} & -\frac{3}{5} & \frac{13}{50} \end{bmatrix}$$ **Ans.**

13. *Verify Caley-Hamilton theorem and find the inverse of the matrix*

$$\begin{bmatrix} 7 & 2 & -2 \\ -6 & -1 & 2 \\ 6 & 2 & -1 \end{bmatrix}$$

SOLUTION: Characterstic equation of the given matrix is

$$\begin{vmatrix} 7-\lambda & 2 & -2 \\ -6 & -1-\lambda & 2 \\ 6 & 2 & -1-\lambda \end{vmatrix} = 0$$

which gives $\lambda^3 - 5\lambda^2 + 7\lambda - 3 = 0$. By Caley Hamilton theorem. $A^3 - 5A^2 + 7A - 3I = 0$ ...(1)

Now $$A^2 = \begin{bmatrix} 7 & 2 & -2 \\ -6 & -1 & 2 \\ 6 & 2 & -1 \end{bmatrix}\begin{bmatrix} 7 & 2 & -2 \\ -6 & -1 & 2 \\ 6 & 2 & -1 \end{bmatrix} = \begin{bmatrix} 25 & 8 & -8 \\ -24 & -7 & 8 \\ 24 & 8 & -7 \end{bmatrix}$$

and $$A^3 = \begin{bmatrix} 25 & 8 & -8 \\ -24 & -7 & 8 \\ 24 & 2 & -7 \end{bmatrix}\begin{bmatrix} 7 & 2 & -2 \\ -6 & -1 & 2 \\ 6 & 2 & -1 \end{bmatrix} = \begin{bmatrix} 79 & 26 & -26 \\ -78 & -25 & 26 \\ 78 & 26 & -25 \end{bmatrix}$$

Here $A^3 - 5A^2 + 7A - 3I = 0$ can be easily verified.

To find A^{-1} we multiply (1) throughout by A^{-1} and get $A^2 - 5A + 7I = 3A^{-1}$

$$\therefore \quad A^{-1} = \frac{1}{3}\left[A^2 - 5A + 7I\right] = \frac{1}{3}\begin{bmatrix} 25 & 8 & -8 \\ -24 & -7 & 8 \\ 24 & 8 & -7 \end{bmatrix} - \frac{5}{3}\begin{bmatrix} 7 & 2 & -2 \\ -6 & -1 & 2 \\ 6 & 2 & -1 \end{bmatrix} + \frac{1}{3}\begin{bmatrix} 7 & 0 & 0 \\ 0 & 7 & 0 \\ 0 & 0 & 7 \end{bmatrix}$$

$$= \frac{1}{3}\begin{bmatrix} -3 & -2 & 2 \\ 6 & 5 & -2 \\ -6 & -2 & 5 \end{bmatrix}$$ **Ans.**

14. *Diagonalise the matrix* $A = \begin{bmatrix} -1 & 1 & 2 \\ 0 & -2 & 1 \\ 0 & 0 & -3 \end{bmatrix}$.

SOLUTION: The characterstic equation of A is $|A - \lambda I| = 0$

or $$\begin{vmatrix} -1-\lambda & 1 & 2 \\ 0 & -2-\lambda & 1 \\ 0 & 0 & -3-\lambda \end{vmatrix} = 0 \quad \text{or} \quad (-1-\lambda)(-2-\lambda)(-3-\lambda) = 0.$$

Hence the eigen values of A are $\lambda = -1, -2, -3$.

For $\lambda = -1$, the eigen vector X is given by $(A - \lambda I)X = 0$

$$0x + y + 2z = 0,\quad -y + z = 0,\quad -2z = 0$$

Hence the eigen vector is (1, 0, 0)

For $\lambda = -2$, we have equations $x + y + 2z = 0,\ 0y + z = 0,\ -z = 0$

Hence eigen vector is (1, –1, 0).

For $\lambda = -3$ we have equation as

$$2x + y + 2z = 0,\quad y + z = 0,\quad 0z = 0 \quad \therefore \quad \text{eigen vector is } (1, 2, -2).$$

$$\therefore \text{ Modal Matrix} = P = \begin{bmatrix} 1 & 1 & 1 \\ 0 & -1 & 2 \\ 0 & 0 & -2 \end{bmatrix}, \qquad |P| = 2$$

$$\therefore \qquad P^{-1} = \frac{1}{2}\begin{bmatrix} 2 & 2 & 3 \\ 0 & -2 & -2 \\ 0 & 0 & -1 \end{bmatrix}$$

$$\therefore \text{ Diagonal matrix} = D = P^{-1}AP = \frac{1}{2}\begin{bmatrix} 2 & 2 & 3 \\ 0 & -2 & -2 \\ 0 & 0 & -1 \end{bmatrix}\begin{bmatrix} -1 & 1 & 2 \\ 0 & -2 & 1 \\ 0 & 0 & -3 \end{bmatrix}\begin{bmatrix} 1 & 1 & 1 \\ 0 & -1 & 2 \\ 0 & 0 & -2 \end{bmatrix}$$

$$= \frac{1}{2}\begin{bmatrix} 2 & -2 & -3 \\ 0 & 4 & 4 \\ 0 & 0 & 3 \end{bmatrix}\begin{bmatrix} 1 & 1 & 1 \\ 0 & -1 & 2 \\ 0 & 0 & -2 \end{bmatrix} = \begin{bmatrix} -1 & 0 & 0 \\ 0 & -2 & 0 \\ 0 & 0 & -3 \end{bmatrix}$$ **Ans.**

15. *Verify Caley Hamilton theorem for the matrix*

$$A = \begin{bmatrix} 1 & 2 & 0 \\ -1 & 1 & 2 \\ 1 & 2 & 1 \end{bmatrix}. \text{ Also } A^{-1} \text{ and } A^3.$$

SOLUTION: Characterstic equation of the matrix A is $|A - \lambda I| = 0$

$$\begin{vmatrix} 1-\lambda & 2 & 0 \\ -1 & 1-\lambda & 2 \\ 1 & 2 & 1-\lambda \end{vmatrix} = 0 \quad \text{or} \quad (1 - 1)\,[(1 - 1)^2 - 4] - 2\,[-1 + \lambda - 2] = 0$$

or $\lambda^3 - 3\lambda^2 + \lambda - 3 = 0$

By Caley-Hamilton therome $A^3 - 3A^2 + A - 3I = 0$ ...(1)

Here
$$A^2 = \begin{bmatrix} 1 & 2 & 0 \\ -1 & 1 & 2 \\ 1 & 2 & 1 \end{bmatrix}\begin{bmatrix} 1 & 2 & 0 \\ -1 & 1 & 2 \\ 1 & 2 & 1 \end{bmatrix} = \begin{bmatrix} -1 & 4 & 4 \\ 0 & 3 & 4 \\ 0 & 6 & 5 \end{bmatrix}$$

$$\therefore \quad A^3 = \begin{bmatrix} -3 & 12 & 12 \\ 0 & 9 & 12 \\ 0 & 18 & 15 \end{bmatrix} + \begin{bmatrix} -1 & -2 & 0 \\ +1 & 1 & 2 \\ 1 & 2 & 1 \end{bmatrix} + \begin{bmatrix} 3 & 0 & 0 \\ 0 & 3 & 0 \\ 0 & 0 & 3 \end{bmatrix} = \begin{bmatrix} -1 & 10 & 12 \\ 1 & 11 & 10 \\ -1 & 16 & 17 \end{bmatrix}$$

Multiplying (1) by A^{-1} we get

$$3A^{-1} = I - 3A = \begin{bmatrix} 1 & 0 & 0 \\ 0 & 1 & 0 \\ 0 & 0 & 1 \end{bmatrix} - \begin{bmatrix} 3 & 6 & 0 \\ -3 & 3 & 6 \\ 3 & 6 & 3 \end{bmatrix}$$

$$\therefore \quad A^{-1} = \frac{1}{3}\begin{bmatrix} -2 & -6 & 0 \\ 3 & -2 & -6 \\ -3 & -6 & -2 \end{bmatrix} = \begin{bmatrix} -2/3 & -2 & 0 \\ 1 & -2/3 & -2 \\ -1 & -2 & -2/3 \end{bmatrix}$$ **Ans.**

ADDITIONAL IMPORTANT PROBLEMS

1. *Show that every square matrix can be uniquely expressed as the sum of a Hermitian matrix and a skew-Hermitian matrix, and express the matrix as the sum of a Hermitian and a skew-Hermitian matrix.*

$$\begin{bmatrix} 2+3i & 0 & 4i \\ 5 & i & 8 \\ 1-i & -3+i & 6 \end{bmatrix}$$

SOLUTION: Let A be any square matrix, then

$$A = \frac{1}{2}(A + A^*) + \frac{1}{2}(A - A^*) = P + Q$$

where $P = \frac{1}{2}(A + A^*)$ and $Q = \frac{1}{2}(A - A^*)$.

Now $P^* = \frac{1}{2}(A + A^*)^* = \frac{1}{2}(A^* + A) = P$

and $Q^* = \frac{1}{2}(A - A^*)^* = \frac{1}{2}(A^* - A) = -Q$.

Hence P is a Hermitian matrix and Q is a skew Hermitian matrix.

$$A = \begin{bmatrix} 2+3i & 0 & 4i \\ 5 & i & 8 \\ 1-i & -3+i & 6 \end{bmatrix} \quad \therefore \quad A^* = (\bar{A})^1 = \begin{bmatrix} 2-3i & 5 & 1+i \\ 0 & -i & -3-i \\ -4i & 8 & 6 \end{bmatrix}$$

$$\therefore \quad P = \frac{1}{2}(A + A^*) = \frac{1}{2}\cdot\left\{\begin{bmatrix} 2+3i & 0 & 4i \\ 5 & i & 8 \\ 1-i & -3+i & 6 \end{bmatrix} + \begin{bmatrix} 2-3i & 5 & 1+i \\ 0 & -i & -3-i \\ -4i & 8 & 6 \end{bmatrix}\right\}$$

$$= \frac{1}{2}\begin{bmatrix} 4 & 5 & 1+5i \\ 5 & 0 & 5-i \\ 1-5i & 5+i & 12 \end{bmatrix}$$

and $$Q = \frac{1}{2}(A - A^*) = \frac{1}{2}\left\{\begin{bmatrix} 2+3i & 0 & 4i \\ 5 & i & 8 \\ 1-i & -3+i & 6 \end{bmatrix} - \begin{bmatrix} 2-3i & 5 & 1+i \\ 0 & -i & -3-i \\ -4i & 8 & 6 \end{bmatrix}\right\}$$

$$= \frac{1}{2}\begin{bmatrix} 6i & -5 & -1+3i \\ 5 & 2i & 11+i \\ 1+3i & -11+i & 0 \end{bmatrix}$$

We see here that P is Hermitian and Q is skew-Hermitian matrix. **Ans.**

2. *Show that every square matrix can be uniquely expressed as* $P + iQ$ *where* P *and* Q *are Hermitian matrixes and hence express the matrix* $A = \begin{bmatrix} 2i & -3 & 1-i \\ 0 & 2+3i & 1+i \\ -3i & 3+2i & 2-5i \end{bmatrix} = P + iQ.$

SOLUTION: $A = \frac{1}{2}(A + A^*) + \frac{i}{2i}(A - A^*) = P + iQ$

where $P = \frac{1}{2}(A + A^*)$ and $Q = \frac{1}{2i}(A - A^*)$.

Here $$P^* = \frac{1}{2}(A + A^*)^* = \frac{1}{2}(A^* + A) = P \quad \text{as } (A^*)^* = A$$

and $$Q^* = \left[\frac{1}{2i}(A - A^*)\right]^* = \frac{1}{2(-i)}(A^* - A) = \frac{1}{2i}(A - A^*) = Q$$

Hence P and Q are both Hermitian.

Further $$A = \begin{bmatrix} 2i & -3 & 1-i \\ 0 & 2+3i & 1+i \\ -3i & 3+2i & 2-5i \end{bmatrix} \quad \therefore \quad A^* \begin{bmatrix} -2i & 0 & 3i \\ -3 & 2-3i & 3-2i \\ 1+i & 1-i & 2+5i \end{bmatrix}$$

$\therefore$ $$P = \frac{1}{2}(A + A^*) = \frac{1}{2}\left\{\begin{bmatrix} 2i & -3 & 1-i \\ 0 & 2+3i & 1+i \\ -3i & 3+2i & 2-5i \end{bmatrix} + \begin{bmatrix} -2i & 0 & 3i \\ -3 & 2-3i & 3-2i \\ 1+i & 1-i & 2+5i \end{bmatrix}\right\}$$

$$= \frac{1}{2}\begin{bmatrix} 0 & -3 & 1+2i \\ -3 & 4 & 4-i \\ 1-2i & 4+i & 4 \end{bmatrix}$$

and $$Q = \frac{1}{2i}(A - A^*) = \frac{1}{2i}\left\{\begin{bmatrix} 2i & -3 & 1-i \\ 0 & 2+3i & 1+i \\ -3i & 3+2i & 2-5i \end{bmatrix} - \begin{bmatrix} -2i & 0 & 3i \\ -3 & 2-3i & 3-2i \\ 1+i & 1-i & 2+5i \end{bmatrix}\right\}$$

$$= \frac{1}{2i}\begin{bmatrix} 4i & -3 & 1-4i \\ 3 & 6i & -2+3i \\ -1-4i & 2+3i & -10i \end{bmatrix}$$

Clearly P and Q are Hermitian matrices.

$$\therefore \qquad P + iQ = \frac{1}{2}\begin{bmatrix} 0 & -3 & 1+2i \\ -3 & 4 & 4-i \\ 1-2i & 4+i & 4 \end{bmatrix} + \frac{1}{2}\begin{bmatrix} 4i & -3 & 1-4i \\ 3 & 6i & -2+3i \\ -1-4i & 2+3i & -10i \end{bmatrix}$$

$$= \begin{bmatrix} 2i & -3 & 1+i \\ 0 & 2+3i & 1+i \\ -3i & 3+2i & 2-5i \end{bmatrix} = A.$$ **Ans.**

3. *Show that every Hermitian matrix can be expressed as $P + iQ$ where P is real symmetric and Q is real skew symmetric matrix. Hence express the Hermitian matrix*

$$A = \begin{bmatrix} 1 & -i & 1+i \\ i & 0 & 2-3i \\ 1-i & 2+3i & 2 \end{bmatrix} \quad \textit{as} \quad P + iQ.$$

SOLUTION: Let A be a Hermitian matrix so $A^* = (\bar{A})' = A$.

$$A = \frac{1}{2}(A+\bar{A}) + i\frac{1}{2i}(A-\bar{A}) = P + iQ$$

where $P = \frac{1}{2}(A+\bar{A})$ and $Q = \frac{1}{2i}(A-\bar{A})$ are real matrixes.

Now $$P' = \left[\frac{1}{2}(A+\bar{A})\right]' = \frac{1}{2}(A'+\bar{A}') = \frac{1}{2}(A'+A^*) = \frac{1}{2}(A'+A) = P$$

Hence P is real symmetric matrix,

and $$Q' = \left[\frac{1}{2i}(A-\bar{A})\right]' = \frac{1}{2i}(A'-\bar{A}') = \frac{1}{2i}(A'-A^*) \quad \text{as} \quad A' = \bar{A} \text{ and } (A^* = A)$$

$$= \frac{1}{2i}(\bar{A}-A) = -\frac{1}{2i}(A-\bar{A}) = -Q.$$

Hence Q is real skew symmetric matrix

$$A = \begin{bmatrix} 1 & -i & 1+i \\ i & 0 & 2-3i \\ 1-i & 2+3i & 2 \end{bmatrix} \quad \therefore \quad \bar{A} = \begin{bmatrix} 1 & i & 1-i \\ -i & 0 & 2+3i \\ 1+i & 2-3i & 2 \end{bmatrix}$$

Let $$P = \frac{1}{2}(A+\bar{A}) = \frac{1}{2}\left\{\begin{bmatrix} 1 & -i & 1+i \\ i & 0 & 2-3i \\ 1-i & 2+3i & 2 \end{bmatrix} + \begin{bmatrix} 1 & i & 1-i \\ -i & 0 & 2+3i \\ 1+i & 2-3i & 2 \end{bmatrix}\right\}$$

$$= \begin{bmatrix} 1 & 0 & 1 \\ 0 & 0 & 2 \\ 1 & 2 & 2 \end{bmatrix}$$

and $$Q=\frac{1}{2i}(A-\bar{A})=\frac{1}{2i}\left\{\begin{bmatrix}1 & -i & 1+i\\ i & 0 & 2-3i\\ 1-i & 2+3i & 2\end{bmatrix}-\begin{bmatrix}1 & i & 1-i\\ -i & 0 & 2+3i\\ 1+i & 2-3i & 2\end{bmatrix}\right\}$$

$$=\frac{1}{2i}\begin{bmatrix}0 & -2i & 2i\\ 2i & 0 & -6i\\ -2i & 6i & 0\end{bmatrix}=\begin{bmatrix}0 & -1 & 1\\ 1 & 0 & -3\\ -1 & 3 & 0\end{bmatrix}$$

Here P is symmetric and Q is skew symmetric. **Ans.**

4. *Show that the matrix* $A=\frac{1}{2}\begin{bmatrix}1+i & -1+i\\ 1+i & 1-i\end{bmatrix}$ *is a unitary matrix and hence find* A^{-1}.

SOLUTION: $A=\frac{1}{2}\begin{bmatrix}1+i & -1+i\\ 1+i & 1-i\end{bmatrix}$ $\therefore$ $A^{-1}=\frac{1}{2}\begin{bmatrix}1+i & 1+i\\ -1+i & 1-i\end{bmatrix}$ and $A^*=\frac{1}{2}\begin{bmatrix}1-i & 1-i\\ -1-i & 1+i\end{bmatrix}$

We know that A is unitary if $AA^*=I$.

$$AA^*=\frac{1}{4}\begin{bmatrix}1+i & -1+i\\ 1+i & 1-i\end{bmatrix}\begin{bmatrix}1-i & 1-i\\ -1-i & 1+i\end{bmatrix}=\frac{1}{4}\begin{bmatrix}2+2 & 2-2\\ 2-2 & 2+2\end{bmatrix}=\begin{bmatrix}1 & 0\\ 0 & 1\end{bmatrix}=I$$

Hence A is a unitary matrix.

Since $AA^*=I$ hence $A^{-1}=A^*=\frac{1}{2}\begin{bmatrix}1-i & 1-i\\ -1-i & 1+i\end{bmatrix}$.

5. *Find a, b and c and* A^{-1} *if* $A=\begin{bmatrix}0 & 2b & c\\ a & b & -c\\ a & -b & c\end{bmatrix}$ *is orthogonal.*

SOLUTION: Since A is orthogonal we have $AA'=I$

or $$\begin{bmatrix}0 & 2b & c\\ a & b & -c\\ a & -b & c\end{bmatrix}\begin{bmatrix}0 & a & a\\ 2b & b & -b\\ c & -c & c\end{bmatrix}=\begin{bmatrix}1 & 0 & 0\\ 0 & 1 & 0\\ 0 & 0 & 1\end{bmatrix}$$

or $$\begin{bmatrix}4b^2+c^2 & 2b^2-c^2 & -2b^2+c^2\\ 2b^2-c^2 & a^2+b^2+c^2 & a^2-b^2-c^2\\ -2b^2+c^2 & a^2-b^2-c^2 & a^2+b^2+c^2\end{bmatrix}=\begin{bmatrix}1 & 0 & 0\\ 0 & 1 & 0\\ 0 & 0 & 1\end{bmatrix}$$

$\Rightarrow$ $4b^2+c^2=1,\ 2b^2-c^2=0,\ a^2+b^2+c^2=1$

Hence $a^2=\frac{1}{2},\quad b^2=\frac{1}{6},\quad C^2=\pm\frac{1}{\sqrt{3}}.$

or $a=\frac{\pm 1}{\sqrt{2}},\ b=\frac{\pm 1}{\sqrt{6}},\ c=\pm\frac{1}{\sqrt{3}}$

$$\therefore\quad A^{-1}=A'=\begin{bmatrix}0 & \pm\frac{1}{\sqrt{2}} & \pm\frac{1}{\sqrt{2}}\\ \pm\frac{2}{\sqrt{6}} & \pm\frac{1}{\sqrt{6}} & \mp\frac{1}{\sqrt{6}}\\ \pm\frac{1}{\sqrt{3}} & \mp\frac{1}{\sqrt{3}} & \pm\frac{1}{\sqrt{3}}\end{bmatrix}$$ **Ans.**

6. *Using elementary row operations find the inverse of the matrix* $A = \begin{bmatrix} 1 & -1 & 0 & 2 \\ 0 & 1 & 1 & -1 \\ 2 & 1 & 2 & 1 \\ 3 & -2 & 1 & 6 \end{bmatrix}$.

SOLUTION: $A = \begin{bmatrix} 1 & -1 & 0 & 2 \\ 0 & 1 & 1 & -1 \\ 2 & 1 & 2 & 1 \\ 3 & -2 & 1 & 6 \end{bmatrix} = \begin{bmatrix} 1 & 0 & 0 & 0 \\ 0 & 1 & 0 & 0 \\ 0 & 0 & 1 & 0 \\ 0 & 0 & 0 & 1 \end{bmatrix} A$

Applying $R_3 \to R_3 - 2R_1$ and $R_4 \to R_4 - 3R_1$

$$\begin{bmatrix} 1 & -1 & 0 & 2 \\ 0 & 1 & 1 & -1 \\ 0 & 3 & 2 & -3 \\ 0 & 1 & 1 & 0 \end{bmatrix} = \begin{bmatrix} 1 & 0 & 0 & 0 \\ 0 & 1 & 0 & 0 \\ -2 & 0 & 1 & 0 \\ -3 & 0 & 0 & 1 \end{bmatrix} A$$

Now applying $R_3 \to R_3 - 3R_2$ and $R_4 \to R_4 - R_2$

$$\begin{bmatrix} 1 & -1 & 0 & 2 \\ 0 & 1 & 1 & -1 \\ 0 & 0 & -1 & 0 \\ 0 & 0 & 0 & 1 \end{bmatrix} = \begin{bmatrix} 1 & 0 & 0 & 0 \\ 0 & 1 & 0 & 0 \\ -2 & -3 & 1 & 0 \\ -3 & -1 & 0 & 1 \end{bmatrix} A$$

Next applying $R_1 \to R_1 - 2R_4$ and $R_2 \to R_2 + R_3$

$$\begin{bmatrix} 1 & -1 & 0 & 0 \\ 0 & 1 & 0 & -1 \\ 0 & 0 & -1 & 0 \\ 0 & 0 & 0 & 1 \end{bmatrix} = \begin{bmatrix} 7 & 2 & 0 & -2 \\ -2 & -2 & 1 & 0 \\ -2 & -3 & 1 & 0 \\ -3 & -1 & 0 & 1 \end{bmatrix} A$$

Now applying $R_2 \to R_2 + R_4$

$$\begin{bmatrix} 1 & -1 & 0 & 0 \\ 0 & 1 & 0 & 0 \\ 0 & 0 & -1 & 0 \\ 0 & 0 & 0 & 1 \end{bmatrix} = \begin{bmatrix} 7 & 2 & 0 & -2 \\ -5 & -3 & 1 & 1 \\ -2 & -3 & 1 & 0 \\ -3 & -1 & 0 & 1 \end{bmatrix} A$$

Finally applying $R_1 \to R_1 + R_2$ and $R_3 \to -R_3$

$$\begin{bmatrix} 1 & 0 & 0 & 0 \\ 0 & 1 & 0 & 0 \\ 0 & 0 & 1 & 0 \\ 0 & 0 & 0 & 1 \end{bmatrix} = \begin{bmatrix} 2 & -1 & 1 & -1 \\ -5 & -3 & 1 & 1 \\ 2 & 3 & -1 & 0 \\ -3 & -1 & 0 & 1 \end{bmatrix} A$$

$\therefore$ $$A^{-1} = \begin{bmatrix} 2 & -1 & 1 & -1 \\ -5 & -3 & 1 & 1 \\ 2 & 3 & -1 & 0 \\ -3 & -1 & 0 & 1 \end{bmatrix}$$ **Ans.**

7. *Determine the value of* a *so that the rank of* A *is 3 where* $A = \begin{bmatrix} 1 & 1 & -1 & 0 \\ 4 & 4 & -3 & 1 \\ a & 2 & 2 & 2 \\ 9 & 9 & a & 3 \end{bmatrix}$.

SOLUTION: Rank of A will be 3 if $|A| = 0$ and at least one minor of order 3 is non-zero. Applying the operations $C_2 \to C_2 - C_1$ and $C_3 \to C_3 + C_1$ we get

$$A \sim \begin{bmatrix} 1 & 0 & 0 & 0 \\ 4 & 0 & 1 & 1 \\ a & 2-a & a+2 & 2 \\ 9 & 0 & a+9 & 3 \end{bmatrix}$$

$$\therefore \quad |A| = \begin{vmatrix} 1 & 0 & 0 & 0 \\ 4 & 0 & 1 & 1 \\ a & 2-a & a+2 & 2 \\ 9 & 0 & a+9 & 3 \end{vmatrix} = \begin{vmatrix} 0 & 1 & 1 \\ 2-a & 2+a & 2 \\ a & a+9 & 3 \end{vmatrix} = (2-a)(a+6)$$

$|A| = 0$, for $a = 2, -6$

when $a = 2$ one minor of A of order 3 is $\begin{vmatrix} 4 & -3 & 1 \\ 2 & 2 & 2 \\ 9 & 2 & 3 \end{vmatrix} \neq 0$.

When $a = -6$, one minor of A of order 3 is $\begin{vmatrix} 4 & -3 & 1 \\ 2 & 2 & 2 \\ 9 & -6 & 3 \end{vmatrix} \neq 0$. Hence required values of a are 2, –6. **Ans.**

8. *Find the rank of the matrix* $\begin{bmatrix} 1 & 2 & 3 & -1 \\ -2 & -1 & -3 & -1 \\ 1 & 0 & 1 & 1 \\ 0 & 1 & 1 & -1 \end{bmatrix}$ *by reducing it to echelon form.*

SOLUTION: Let $A = \begin{bmatrix} 1 & 2 & 3 & -1 \\ -2 & -1 & -3 & -1 \\ 1 & 0 & 1 & 1 \\ 0 & 1 & 1 & -1 \end{bmatrix}$. Applying $R_2 + 2R_1$ and $R_3 - R_1$

$\sim \begin{bmatrix} 1 & 2 & 3 & -1 \\ 0 & 3 & 3 & -3 \\ 0 & -2 & -2 & 2 \\ 0 & 1 & 1 & -1 \end{bmatrix}$. Now applying $R_3 \to R_3 + \dfrac{2}{3} R_2$

$$\sim \begin{bmatrix} 1 & 2 & 3 & -1 \\ 0 & 3 & 3 & -3 \\ 0 & 0 & 0 & 0 \\ 0 & 1 & 1 & -1 \end{bmatrix} \text{ Now applying } R_4 \to R_4 - \frac{1}{3} R_2$$

$$\sim \begin{bmatrix} 1 & 2 & 3 & -1 \\ 0 & 3 & 3 & -3 \\ 0 & 0 & 0 & 0 \\ 0 & 0 & 0 & 0 \end{bmatrix}$$ which is in the echelon form. Clearly rank of $A = 2$. **Ans.**

9. *Find the rank of the matrix* $\begin{bmatrix} 1 & 2 & -1 & 3 \\ 3 & 4 & 0 & -1 \\ -1 & 0 & -2 & 7 \end{bmatrix}$ *by reducing to normal form.*

SOLUTION: Let $A = \begin{bmatrix} 1 & 2 & -1 & 3 \\ 3 & 4 & 0 & -1 \\ -1 & 0 & -2 & 7 \end{bmatrix}$ Applying $R_2 \to R_2 - 3R_1$ and $R_3 \to R_3 + R_1$

$$\sim \begin{bmatrix} 1 & 2 & -1 & 3 \\ 0 & -2 & 3 & -10 \\ 0 & 2 & -3 & 10 \end{bmatrix} \text{ Applying } R_3 \to R_3 + R_2$$

$$\sim \begin{bmatrix} 1 & 2 & -1 & 3 \\ 0 & -2 & 3 & -10 \\ 0 & 0 & 0 & 0 \end{bmatrix} \text{ Applying } C_2 \to C_2 - 2C_1, C_3 \to C_3 + C_1, C_4 \to C_4 - 3C_1$$

$$\sim \begin{bmatrix} 1 & 0 & 0 & 0 \\ 0 & -2 & 3 & -10 \\ 0 & 0 & 0 & 0 \end{bmatrix} \text{ Applying } C_2 \to \frac{-1}{2} C_2, C_3 \to C_3 - 3C_2, C_4 \to C_4 - 4C_2$$

$$\sim \begin{bmatrix} 1 & 0 & 0 & 0 \\ 0 & 1 & 0 & 0 \\ 0 & 0 & 0 & 0 \end{bmatrix}.$$ Therefore, the rank of $A = 2$.

10. *Find non-singular matrices P and Q such that PAQ is in normal form and hence find ρ(A) for* $A = \begin{bmatrix} 2 & 1 & 1 & 3 \\ 1 & 0 & 1 & 2 \\ 3 & 1 & 2 & 5 \end{bmatrix}$.

SOLUTION: Write $A = I_3 A I_4$, that is,

$$\begin{bmatrix} 2 & 1 & 1 & 3 \\ 1 & 0 & 1 & 2 \\ 3 & 1 & 2 & 5 \end{bmatrix} = \begin{bmatrix} 1 & 0 & 0 \\ 0 & 1 & 0 \\ 0 & 0 & 1 \end{bmatrix} A \begin{bmatrix} 1 & 0 & 0 & 0 \\ 0 & 1 & 0 & 0 \\ 0 & 0 & 1 & 0 \\ 0 & 0 & 0 & 1 \end{bmatrix} \text{ Applying } R_1 \leftrightarrow R_2$$

$$\begin{bmatrix} 1 & 0 & 1 & 2 \\ 2 & 1 & 1 & 3 \\ 3 & 1 & 2 & 5 \end{bmatrix} = \begin{bmatrix} 0 & 1 & 0 \\ 1 & 0 & 0 \\ 0 & 0 & 1 \end{bmatrix} A \begin{bmatrix} 1 & 0 & 0 & 0 \\ 0 & 1 & 0 & 0 \\ 0 & 0 & 1 & 0 \\ 0 & 0 & 0 & 1 \end{bmatrix}$$ Applying $R_2 \to R_2 - 2R_1$ and $R_3 \to R_3 - 3R_1$

$$\begin{bmatrix} 1 & 0 & 1 & 2 \\ 0 & 1 & -1 & -1 \\ 0 & 1 & -1 & -1 \end{bmatrix} = \begin{bmatrix} 0 & 1 & 0 \\ 1 & -2 & 0 \\ 0 & -3 & 1 \end{bmatrix} A \begin{bmatrix} 1 & 0 & 0 & 0 \\ 0 & 1 & 0 & 0 \\ 0 & 0 & 1 & 0 \\ 0 & 0 & 0 & 1 \end{bmatrix}$$ Applying $C_3 \to C_3 + C_2$ and $C_4 \to C_4 + C_2$

$$\begin{bmatrix} 1 & 0 & 1 & 2 \\ 0 & 1 & 0 & 0 \\ 0 & 1 & 0 & 0 \end{bmatrix} = \begin{bmatrix} 0 & 1 & 0 \\ 1 & -2 & 0 \\ 0 & -3 & 1 \end{bmatrix} A \begin{bmatrix} 1 & 0 & 0 & 0 \\ 0 & 1 & 1 & 1 \\ 0 & 0 & 1 & 0 \\ 0 & 0 & 0 & 1 \end{bmatrix}$$ Applying $C_3 \to C_3 - C_1$ and $C_4 \to C_4 - 2C_1$

$$\begin{bmatrix} 1 & 0 & 0 & 0 \\ 0 & 1 & 0 & 0 \\ 0 & 1 & 0 & 0 \end{bmatrix} = \begin{bmatrix} 0 & 1 & 0 \\ 1 & -2 & 0 \\ 0 & -3 & 1 \end{bmatrix} A \begin{bmatrix} 1 & 0 & -1 & -2 \\ 0 & 1 & 1 & 1 \\ 0 & 0 & 1 & 0 \\ 0 & 0 & 0 & 1 \end{bmatrix}$$ Finally applying $R_3 \to R_3 - R_2$

$$\begin{bmatrix} 1 & 0 & 0 & 0 \\ 0 & 1 & 0 & 0 \\ 0 & 0 & 0 & 0 \end{bmatrix} = \begin{bmatrix} 0 & 1 & 0 \\ 1 & -2 & 0 \\ -1 & -1 & 1 \end{bmatrix} A \begin{bmatrix} 1 & 0 & -1 & -2 \\ 0 & 1 & 1 & 1 \\ 0 & 0 & 1 & 0 \\ 0 & 0 & 0 & 1 \end{bmatrix}$$

or $$\begin{bmatrix} I_2 & 0 \\ 0 & 0 \end{bmatrix} = PAQ$$

where $P = \begin{bmatrix} 0 & 1 & 0 \\ 1 & -2 & 0 \\ -1 & -1 & -1 \end{bmatrix}$ and $Q = \begin{bmatrix} 1 & 0 & -1 & -2 \\ 0 & 1 & 1 & 1 \\ 0 & 0 & 1 & 0 \\ 0 & 0 & 0 & 1 \end{bmatrix}$ Hence rank of $A = 2$. **Ans.**

11. *Solve the following system of equations*

$$3x - y - z = 0$$
$$x + y + 2z = 0$$
$$5x + y + 3z = 0$$

SOLUTION: $AX = 0$

or $$\begin{bmatrix} 3 & -1 & -1 \\ 1 & 1 & 2 \\ 5 & 1 & 3 \end{bmatrix} \begin{bmatrix} x \\ y \\ z \end{bmatrix} = \begin{bmatrix} 0 \\ 0 \\ 0 \end{bmatrix}$$ Applying $R_1 \leftrightarrow R_2$, gives

$$\begin{bmatrix} 1 & 1 & 2 \\ 3 & -1 & -1 \\ 5 & 1 & 3 \end{bmatrix} \begin{bmatrix} x \\ y \\ z \end{bmatrix} = \begin{bmatrix} 0 \\ 0 \\ 0 \end{bmatrix}$$

Applying now $R_2 \to R_2 - 3R_1$ and $R_3 \to R_3 - 5R_1$

$$\begin{bmatrix} 1 & 1 & 2 \\ 0 & -4 & -7 \\ 0 & -4 & -7 \end{bmatrix} \begin{bmatrix} x \\ y \\ z \end{bmatrix} = \begin{bmatrix} 0 \\ 0 \\ 0 \end{bmatrix}$$

Now applying $R_3 \to R_3 - R_2$, gives

$$\begin{bmatrix} 1 & 1 & 2 \\ 0 & -4 & -7 \\ 0 & 0 & 0 \end{bmatrix} \begin{bmatrix} x \\ y \\ z \end{bmatrix} = \begin{bmatrix} 0 \\ 0 \\ 0 \end{bmatrix}$$

Here $\rho(A) = 2 < 3$ (the number of unknowns)

Hence the system has non-trivial solutions

Equations are $x + y + 2z = 0$

$$-4y - 7z = 0$$

Taking $z = t$ we get $y = -\frac{7}{4}t$ hence $x = \frac{7}{4}t - 2t = -\frac{t}{4}$

Thus, solutions are $x = -\frac{t}{4}$, $y = -\frac{7t}{4}$, $z = t$ where t is a parameter. **Ans.**

12. *Examine whether the following vectors are linearly independent or dependent.*
[1, 1, –1], [2, 3, –5], [2, –1, 4]

SOLUTION: Let $X_1 = [1, 1, -1]$, $X_2 = [2, 3, -5]$, $X_3 = [2, -1, 4]$.

Let $k_1X_1 + k_2X_2 + k_3X_3 = 0$ where not all k_1, k_2, k_3 are zero.

$$k_1[1, 1, -1] + k_2\,[2, 3, -5] + k_3\,[2, -1, 4] = (0, 0, 0)$$

$\Rightarrow$ $k_1 + 2k_2 + 2\,k_3 = 0,\quad k_1 + 3k_2 - k_3 = 0,\quad -k_1 - 5k_2 + 4k_3 = 0$

or
$$\begin{bmatrix} 1 & 2 & 2 \\ 1 & 3 & -1 \\ -1 & -5 & 4 \end{bmatrix} \begin{bmatrix} k_1 \\ k_2 \\ k_3 \end{bmatrix} = \begin{bmatrix} 0 \\ 0 \\ 0 \end{bmatrix}$$

Applying $R_2 \to R_2 - R_1$, $R_3 \to R_3 + R_1$ gives

$$\begin{bmatrix} 1 & 2 & 2 \\ 0 & 1 & -3 \\ 0 & -3 & 6 \end{bmatrix} \begin{bmatrix} k_1 \\ k_2 \\ k_3 \end{bmatrix} = \begin{bmatrix} 0 \\ 0 \\ 0 \end{bmatrix}.$$ Now applying $R_3 \to R_3 + 3R_2$

$$\begin{bmatrix} 1 & 2 & 2 \\ 0 & 1 & -3 \\ 0 & 0 & -3 \end{bmatrix} \begin{bmatrix} k_1 \\ k_2 \\ k_3 \end{bmatrix} = \begin{bmatrix} 0 \\ 0 \\ 0 \end{bmatrix}$$

or $\quad k_1 + 2k_2 + 2k_3 = 0, \quad k_2 - 3k_3 = 0, \quad -3k_1 = 0.$

$\Rightarrow \quad k_1 = 0, \quad k_2 = 0$ and $k_3 = 0$. Hence, vectors are independent. **Ans.**

13. *Investigate whether the following vectors are independent on dependent.*

$$X_1 = [1, 2, 4]', \ X_2 = [3, 7, 10]'$$

SOLUTION: $X_1 = [1, 2, 4]' = \begin{bmatrix} 1 \\ 2 \\ 4 \end{bmatrix}, \ X_2 = \begin{bmatrix} 3 \\ 7 \\ 10 \end{bmatrix},$

Let $\quad k_1 X_1 + k_2 X_2 = 0$

$$k_1 \begin{bmatrix} 1 \\ 2 \\ 4 \end{bmatrix} + k_2 \begin{bmatrix} 3 \\ 7 \\ 10 \end{bmatrix} = \begin{bmatrix} 0 \\ 0 \\ 0 \end{bmatrix}$$

or
$$k_1 + 3k_2 = 0$$
$$2k_1 + 7k_2 = 0$$
$$4k_1 + 10k_2 = 0.$$

In matrix form

$\Rightarrow \quad \begin{bmatrix} 1 & 3 \\ 2 & 7 \\ 4 & 10 \end{bmatrix} \begin{bmatrix} k_1 \\ k_2 \end{bmatrix} = \begin{bmatrix} 0 \\ 0 \\ 0 \end{bmatrix}$ Applying $R_2 \to R_2 - 2R_1, R_3 \to R_3 - 4R_1$

$$\begin{bmatrix} 1 & 3 \\ 0 & 1 \\ 0 & -2 \end{bmatrix} \begin{bmatrix} k_1 \\ k_2 \end{bmatrix} = \begin{bmatrix} 0 \\ 0 \\ 0 \end{bmatrix}$$

$\Rightarrow \quad k_1 + 3k_2 = 0,$ and $k_2 = 0$, hence $k_1 = 0$. Hence independent. **Ans.**

14. *Find the eigen values and eigen vectors for the matrix*

$$A = \begin{bmatrix} 1 & 2 & 2 \\ 0 & 2 & 1 \\ -1 & 2 & 2 \end{bmatrix}$$

SOLUTION: Characterstic equation is $|A - \lambda I| = 0$

or $\quad \begin{vmatrix} 1-\lambda & 2 & 2 \\ 0 & 2-\lambda & 1 \\ -1 & 2 & 2-\lambda \end{vmatrix} = 0 \quad$ or $\quad \lambda^3 - 5\lambda^2 + 8\lambda - 4 = 0$

Its roots are $\lambda = 1, 2, 2$. which are eigen values of A. Now we find corresponding eigenvectors.

For $\lambda = \lambda_1 = 1,\ (A - \lambda_1 I)X = 0$

or $$\begin{bmatrix} 0 & 2 & 2 \\ 0 & 1 & 1 \\ -1 & 2 & 1 \end{bmatrix}\begin{bmatrix} x \\ y \\ z \end{bmatrix} = \begin{bmatrix} 0 \\ 0 \\ 0 \end{bmatrix}$$

$$\Rightarrow \quad y + z = 0,\ -x + 2y + z = 0 \quad \Rightarrow \quad \frac{x}{-1} = \frac{y}{-1} = \frac{z}{1} \quad \text{or} \quad \frac{x}{1} = \frac{y}{1} = \frac{z}{-1}.$$

Hence eigen vector for $\lambda = 1$ is $X_1 = \begin{bmatrix} 1 \\ 1 \\ -1 \end{bmatrix}$

For $\lambda = \lambda_2 = 2,\ (A - \lambda_2 I)X = 0$

or $$\begin{bmatrix} -1 & 2 & 2 \\ 0 & 0 & 1 \\ -1 & 2 & 0 \end{bmatrix}\begin{bmatrix} x \\ y \\ z \end{bmatrix} = \begin{bmatrix} 0 \\ 0 \\ 0 \end{bmatrix}$$

$$\Rightarrow \quad z = 0,\ -x + 2y + 2z = 0,\ -x + 2y = 0 \Rightarrow \frac{x}{2} = \frac{y}{1} = \frac{z}{0}$$

Hence eigen vector for $\lambda = 2$ is $\begin{bmatrix} 2 \\ 1 \\ 0 \end{bmatrix}$. **Ans.**

15. *If* $A = \begin{bmatrix} 3 & -1 & 1 \\ -1 & 5 & -1 \\ 1 & -1 & 3 \end{bmatrix}$ *find eigen values and eigen vectors for* A *and also for*

(i) A', *(ii)* A^{-1}, *(iii)* A^*, *(iv)* A^2 *(v)* $A^2 - 2A + I$, *(vi)* $A^3 + 2I$, *(vii) adj* (A).

SOLUTION: $A = \begin{bmatrix} 3 & -1 & 1 \\ -1 & 5 & -1 \\ 1 & -1 & 3 \end{bmatrix}$. Characterstic equation is $|A - \lambda I| = 0$.

or $$\begin{vmatrix} 3-\lambda & -1 & 1 \\ -1 & 5-\lambda & -1 \\ 1 & -1 & 3-\lambda \end{vmatrix} = 0 \quad \text{or} \quad \lambda^3 - 11\lambda^2 + 36\lambda - 36 = 0$$

whose roots are $\lambda = 2, 3, 6$. Hence eigen values of A are $2, 3, 6$ as those of A.

and the eigen values of A' are same as $2, 3, 6$.

and the eigen values of A^{-1} are $\frac{1}{2}, \frac{1}{3}, \frac{1}{6}$

and the eigen values of A^* are $2, 3, 6$

and the eigen values of A^2 are $4, 9, 36$.

Obviously the eigen values of $A^2 - 2A + I$ are $2^2 - 2(2) + 1,\quad 3^2 - 6 + 1$ and $6^2 - 2(6) + 1$.

Obviously the eigen values of $A^3 + 2I$ or $1, 4, 25$

Obviously the eigen values of adj(A) are $\frac{|A|}{\lambda}$ or $18, 12, 6$ (since $|A| = 36$).

As done in previous problems, the eigen vectors of A are

$$X_1 = \begin{bmatrix} 1 \\ 0 \\ -1 \end{bmatrix}, \ X_2 = \begin{bmatrix} 1 \\ 1 \\ 1 \end{bmatrix}, \ X_3 = \begin{bmatrix} 1 \\ -2 \\ 1 \end{bmatrix}.$$

16. *If* $A = \begin{bmatrix} 2 & 1-2i \\ 1+2i & -2 \end{bmatrix}$ *verify whether its eigen vectors are mutually orthogonal.*

SOLUTION: Characterstic equation is $|A - \lambda I| = 0$ or $\begin{vmatrix} 2-\lambda & 1-2i \\ 1+2i & -2-\lambda \end{vmatrix} = 0$

or $\lambda^2 - 9 = 0$ hence the eigen values are $\lambda = -3, 3$.

For eigen vectors, when $\lambda = -3$, $[A - \lambda I]X = 0$ or $\begin{bmatrix} 5 & 1-2i \\ 1+2i & 1 \end{bmatrix}\begin{bmatrix} x \\ y \end{bmatrix} = \begin{bmatrix} 0 \\ 0 \end{bmatrix}$

$\Rightarrow$ $5x + (1 - 2i)\, y = 0$ and $(1 + 2i)\, x + y = 0$

$\therefore$ Eigen vector $X_1(\text{for } \lambda = -3) = \begin{bmatrix} 1-2i \\ -5 \end{bmatrix}$

For eigen vector when $\lambda = 3$, $[A - 1I]X = 0$ or $\begin{bmatrix} -1 & 1-2i \\ 1+2i & -5 \end{bmatrix}\begin{bmatrix} x \\ y \end{bmatrix} = \begin{bmatrix} 0 \\ 0 \end{bmatrix}$

$\Rightarrow$ $-x + (1 - 2i)y = 0$ and $(1 + 2i)x - 5y = 0$

$\therefore$ Eigen vector X_2 (for $\lambda = 3$) $= \begin{bmatrix} 1-2i \\ 1 \end{bmatrix}$.

Two eigen vectors are mutually orthogonal if $X_1 \cdot X'_2 = 0$

or $\begin{bmatrix} 1-2i \\ -5 \end{bmatrix}\begin{bmatrix} 1-2i \\ 1 \end{bmatrix} = -5 + 5 = 0.$ **Hence proved.**

17. *Find the characteristic roots of the matrix* $A = \begin{bmatrix} 1 & 4 \\ 2 & 3 \end{bmatrix}$ *and verify Caley-Hamilton theorem for this matrix. Find* A^{-1} *and also simplify* $A^5 - 4A^4 - 7A^3 + 11A^2 - A - 10I$.

SOLUTION: $A = \begin{bmatrix} 1 & 4 \\ 2 & 3 \end{bmatrix}$. Characteristic equation is $|A - \lambda I| = 0$

or $\begin{vmatrix} 1-\lambda & 4 \\ 2 & 3-\lambda \end{vmatrix} = 0$ or $\lambda^2 - 4\lambda - 5 = 0$ $\therefore$ $\lambda = -1, 5$ are eigen values of A.

By Caley-Hamilton theorem A must satisfy the characteristic equation, that is, $A^2 - 4A - 5I = 0$.

Here $$A^2 = \begin{bmatrix} 1 & 4 \\ 2 & 3 \end{bmatrix}\begin{bmatrix} 1 & 4 \\ 2 & 3 \end{bmatrix} = \begin{bmatrix} 9 & 16 \\ 8 & 17 \end{bmatrix}$$

$\therefore$ $$A^2 - 4A - 5I = \begin{bmatrix} 9 & 16 \\ 8 & 17 \end{bmatrix} - \begin{bmatrix} 4 & 16 \\ 8 & 12 \end{bmatrix} - \begin{bmatrix} 5 & 0 \\ 0 & 5 \end{bmatrix} = \begin{bmatrix} 0 & 0 \\ 0 & 0 \end{bmatrix}$$

Hence Caley-Hamilton theorem is verified for A.

Pre multiplying $A^2 - 4A - 5I = 0$ by A^{-1} we get $A - 4I = 5A^{-1}$

$$\therefore \quad A^{-1} = \frac{1}{5}\begin{bmatrix} -3 & 4 \\ 2 & -1 \end{bmatrix}$$

Next,

$$A^5 - 4A^4 - 7A^3 + 11A^2 - A - 10I = A^3(A^2 - 4A - 5I) - 2A^3 + 11A^2 - 10I$$
$$= 0 - 2A(A^2 - 4A - 5I) + 3A^2 - 11A - 10I = 0 + 0 + 3(A^2 - 4A - 5I) + A + 5I.$$
$$= 0 + 0 + 0 + A + 5I = A + 5I.$$

Ans.

18. *Find the modal matrix of* $A = \begin{bmatrix} 4 & 2 & -2 \\ -5 & 3 & 2 \\ -2 & 4 & 1 \end{bmatrix}$ *and diagonalize it.*

SOLUTION: Characterstic equation of A is $|A - \lambda I| = 0$

$$\text{or} \quad \begin{vmatrix} 4-\lambda & 2 & -2 \\ -5 & 3-\lambda & 2 \\ -2 & 4 & 1-\lambda \end{vmatrix} = 0 \quad \text{or} \quad \lambda^3 - 8\lambda^2 + 17\lambda - 10 = 0$$

whose roots are $\lambda = 1, 2, 5$ which are the eigen values of A.

For eigen vector for $\lambda = \lambda_1 = 1$, $[A - \lambda_1 I]X = 0$ or $\begin{bmatrix} 3 & 2 & -2 \\ -5 & 2 & 2 \\ -2 & 4 & 0 \end{bmatrix}\begin{bmatrix} x \\ y \\ z \end{bmatrix} = \begin{bmatrix} 0 \\ 0 \\ 0 \end{bmatrix}$

$\Rightarrow \quad 3x + 2y - 2z = 0, \ -5x + 2y + 2z = 0$ and $-2x + 4y = 0$

$$\therefore \quad \frac{x}{8} = \frac{y}{4} = \frac{z}{16} \quad \text{or} \quad \frac{x}{2} = \frac{y}{1} = \frac{z}{4} \qquad \therefore \ X_1 = \begin{bmatrix} 2 \\ 1 \\ 4 \end{bmatrix}.$$

For eigen vector for $\lambda = \lambda_2 = 2$, $[A - \lambda_2 I]X = 0$ or $\begin{bmatrix} 2 & 2 & -2 \\ -5 & 1 & 2 \\ -2 & 4 & -1 \end{bmatrix}\begin{bmatrix} x \\ y \\ z \end{bmatrix} = \begin{bmatrix} 0 \\ 0 \\ 0 \end{bmatrix}$

$\Rightarrow \quad 2x + 2y - 2z = 0, \ -5x + y + 2z = 0, \ -2x + 4y - z = 0$

$$\therefore \quad \frac{x}{6} = \frac{y}{6} = \frac{z}{12} \quad \text{or} \quad \frac{x}{1} = \frac{y}{1} = \frac{z}{2} \quad \therefore \ X_2 = \begin{bmatrix} 1 \\ 1 \\ 2 \end{bmatrix}$$

For eigen vector for $\lambda = \lambda_3 = 5$, $[A - \lambda_3 I]X = 0$ or $\begin{bmatrix} -1 & 2 & -2 \\ -5 & -2 & 2 \\ -2 & 4 & -4 \end{bmatrix}\begin{bmatrix} x \\ y \\ z \end{bmatrix} = \begin{bmatrix} 0 \\ 0 \\ 0 \end{bmatrix}$

$\Rightarrow \quad -x + 2y - 2z = 0, \ -5x - 2y + 2z = 0$

$$\therefore \quad \frac{x}{0} = \frac{y}{12} = \frac{7}{12} \quad \text{or} \quad \frac{x}{0} = \frac{y}{1} = \frac{z}{1} \qquad \therefore \quad X_3 = \begin{bmatrix} 0 \\ 1 \\ 1 \end{bmatrix}$$

$\therefore$ Diagonal matrix $D = \begin{bmatrix} 1 & 0 & 0 \\ 0 & 2 & 0 \\ 0 & 0 & 5 \end{bmatrix}$ and modal matrix $P = \begin{bmatrix} 2 & 1 & 0 \\ 1 & 1 & 1 \\ 4 & 2 & 1 \end{bmatrix}$. **Ans.**

CHAPTER

9 First Order Differential Equations and their Applications

IMPORTANT FORMULAE AND USEFUL RESULTS

1. **Various Forms:**

(*a*) Variable separable form

(*b*) Homogenous form

(*c*) Reducible to homogenous form

(*d*) Linear equation form

$\dfrac{dy}{dx} + P(x)y = Q(x)$, I.F. $= e^{\int p dx}$ solution is $y.I.F. = \int Q.I.F.dx + \text{constant}$.

(*e*) Reducible to linear form $\dfrac{dy}{dx} + P(x)y = Q(x)y^n$

[Divide by y^n and put $y^{1-n} = y$, etc]

2. **Exact Differential Equations**

Equations is $Mdx + Ndy = 0$. It is exact if $\dfrac{\partial M}{\partial y} = \dfrac{\partial N}{\partial x}$, then it is directly integrable.

Reducible to exact form:

Rule 1. If $Mdx + Ndy = 0$ is homogenous equation in x and y then $\dfrac{1}{Mx+Ny}$ is the Integrating factor.

Rule 2. If $Mdx + Ndy = 0$ is of the form $f_1(xy)ydx + f_2(xy)xdy = 0$ then $\dfrac{1}{Mx - Ny}$ is the Integrating factor provided $Mx \neq Ny$.

Rule 3. In the equation $Mdx + Ndy = 0$ if $\dfrac{\dfrac{\partial M}{\partial y} - \dfrac{\partial N}{\partial x}}{N}$ is a function of x only $= f(x)$ say, then I.F. $= e^{\int f(x)dx}$.

Rule 4. In the equation $Mdx + Ndy = 0$ if $\dfrac{\dfrac{\partial M}{\partial y} - \dfrac{\partial N}{\partial x}}{M}$ is a function of y alone $= f(y)$ say,

then I.F. $= e^{\int f(y)dy}$.

Rule 5. If the equation is of the form $x^a y^b (pydx + qxdy) + x^c y^d (rydx + sxdy) = 0$ then I.F. $= x^h y^k$.

Rule 6. Sometimes conversion to polar coordinates helps solve the equation.

3. Differential Equations of first order and higher degree.

(*i*) Solvable for p where $p = \dfrac{dy}{dx}$

(*ii*) Solvable for y

(*iii*) Solvable for x

(*iv*) Clairaut's equation $y = px + f(p)$ its solution is $y = cx + f(c)$.

SOLUTIONS TO EXERCISE 9A

1. *Form the differential equation whose general solution is* $ax^2 + by^2 = 1$.

SOLUTION: $ax^2 + by^2 = 1 \quad \therefore \quad 2ax + 2by\dfrac{dy}{dx} = 0 \quad$ or $\quad \dfrac{dy}{dx} = -\dfrac{ax}{by}$ ...(1)

and $\quad a + b\dfrac{d^2y}{dx^2} + b\left(\dfrac{dy}{dx}\right)^2 = 0 \quad$ or $\quad \dfrac{d^2y}{dx^2} + \left(\dfrac{dy}{dx}\right)^2 = -\dfrac{a}{b}$ ...(2)

$$\frac{d^2y}{dx^2} + \left(\frac{dy}{dx}\right)^2 = \frac{y}{x}\frac{dy}{dx} \quad \text{or} \quad x\frac{d^2y}{dx^2} + x\left(\frac{dy}{dx}\right)^2 - y\frac{dy}{dx} = 0$$ **Ans.**

2. *Form the differential equation for the relation* $y = ax + \dfrac{b}{x}$ *where* a *and* b *are arbitrary constants.*

SOLUTION: $y = ax + \dfrac{b}{x}$ ...(1) $\quad \therefore \quad \dfrac{dy}{dx} = a - \dfrac{b}{x^2}$ ...(2) $\quad$ and $\quad \dfrac{d^2y}{dx^2} = \dfrac{2b}{x^3}$ ...(3)

Eliminating a between (1) and (2), we get

$$y = x\left(\frac{dy}{dx} + \frac{b}{x^2}\right) + \frac{b}{x} = x\frac{dy}{dx} + \frac{2b}{x} \qquad \text{...(4)}$$

Eliminating b in (3) and (4), gives

$$y = x\frac{dy}{dx} + x^2\frac{d^2y}{dx^2} \quad \text{or} \quad x^2\frac{d^2y}{dx^2} + x\frac{dy}{dx} - y = 0.$$ **Ans.**

3. *Eliminate the arbitrary constants* a *and* α *by way of obtaining the differential equation for* $y = a\cos(nx + \alpha)$.

SOLUTION: $y = a\cos(nx + \alpha) \quad \therefore \quad \dfrac{dy}{dx} = -an\sin(nx + \alpha)$

and $\quad \dfrac{d^2y}{dx^2} = -an^2\cos(nx + \alpha) = -n^2y \quad$ or $\quad \dfrac{d^2y}{dx^2} + n^2y = 0.$ **Ans.**

4. *Obtain the differential equation for the general solution* $(y - k)^2 = 4a(x - h)$ *where* h *and* k *are arbitrary constants.*

SOLUTION: $(y - k)^2 = 4a(x - h) \quad \therefore \quad 2(y - k)\dfrac{dy}{dx} = 4a$

and $\quad (y - k)\dfrac{d^2y}{dx^2} + \left(\dfrac{dy}{dx}\right)^2 = 0$

Eliminating $(y-k)$ in the above two relations, we get

$$\frac{2a}{\frac{dy}{dx}}\frac{d^2y}{dx^2}+\left(\frac{dy}{dx}\right)^2 = 0 \quad \text{or} \quad 2a\frac{d^2y}{dx^2}+\left(\frac{dy}{dx}\right)^3 = 0.$$ **Ans.**

5. *Solve* $\frac{dy}{dx}+\frac{y^2+y+1}{x^2+x+1} = 0.$

SOLUTION: Given differential equation, can be written as

$$\frac{dy}{y^2+y+1}+\frac{dx}{x^2+x+1} = 0 \quad \text{or} \quad \frac{dy}{\left(y+\frac{1}{2}\right)^2+\left(\frac{\sqrt{3}}{2}\right)^2}+\frac{dx}{\left(x+\frac{1}{2}\right)^2+\left(\frac{\sqrt{3}}{2}\right)^2} = 0$$

Integrating the above relation, gives

$$\frac{2}{\sqrt{3}}\tan^{-1}\left(\frac{y+1/2}{\sqrt{3}/2}\right)+\frac{2}{\sqrt{3}}\tan^{-1}\left(\frac{x+1/2}{\sqrt{3}/2}\right) = \frac{2}{\sqrt{3}}\tan^{-1}C \quad \text{where } C \text{ is constant}$$

or $$\tan^{-1}\left[\frac{\left(\frac{2y+1}{\sqrt{3}}+\frac{2x+1}{\sqrt{3}}\right)}{1-\frac{(2y+1)(2x+1)}{3}}\right] = \tan^{-1}C \quad \text{or} \quad \frac{(2x+2y+2)\sqrt{3}}{2-4xy-2x-2y} = C$$

or $$x+y+1+C\,(2xy+x+y-1) = 0$$ **Ans.**

6. *Solve* $\frac{y}{x}\frac{dy}{dx} = \sqrt{1+x^2+y^2+x^2y^2}.$

SOLUTION: Given differential equation is $\frac{y}{x}\frac{dy}{dx} = \sqrt{(1+x^2)(1+y^2)}$

or $\frac{y\,dy}{\sqrt{1+y^2}} = x\sqrt{1+x^2}\,dx$. On integrating both sides, we get

$$\sqrt{1+y^2} = \frac{1}{3}(1+x^2)^{3/2}+C \quad \text{where } C \text{ is constant.}$$ **Ans.**

7. *Solve* $\cos x\,dy-(\sin x+x\sec x)\cot y\,dx = 0.$

SOLUTION: Equation can be written as $\tan y\,dy = (\tan x+x\sec^2 x)\,dx$

On integrating the above, we get

$$-\log\cos y = \int\tan x\,dx+x\tan x-\int 1.\tan x\,dx$$

or $$\log\sec y = x\tan x+C \quad \text{where } C \text{ is constant of integration}$$ **Ans.**

8. *Solve* $\dfrac{dy}{dx} = \dfrac{x \cos x}{2e^y \sinh y}$.

SOLUTION: The given equation can be written as

$$2\, e^y \sinh y\, dy = x \cos x\, dx \quad \text{or} \quad e^y(e^y - e^{-y})\, dy = x \cos x\, dx$$

or $(e^{2y} - 1)\, dy = x \cos x\, dx$ which on integration, gives

$$\frac{1}{2}e^{2y} - y = x \sin x - \int 1 \sin x\, dx$$

or $\dfrac{1}{2}e^{2y} - y = x \sin x + \cos x + C$ where C is constant of integration. **Ans.**

9. *Solve* $\left(\dfrac{x+y-a}{x+y+b}\right)\dfrac{dy}{dx} = \left(\dfrac{x+y+a}{x+y-b}\right)$.

SOLUTION: Putting $x + y = t$ so $1 + \dfrac{dy}{dx} = \dfrac{dt}{dx}$ the given equation becomes

$$\left(\frac{t-a}{t+b}\right)\left(\frac{dt}{dx} - 1\right) = \frac{t+a}{t-b} \quad \text{or} \quad \frac{t-a}{t+b}\frac{dt}{dx} = \frac{t+a}{t-b} + \frac{t-a}{t+b} = \frac{2(t^2+ab)}{t^2-b^2}$$

or $2\, dx = \dfrac{(t-a)(t-b)dt}{t^2+ab}$ Integrating on both insides, we get

$$2x = \int \frac{t^2-(a+b)t+ab}{t^2+ab}\, dt = \int\left[1 - \frac{(a+b)t}{t^2+ab}\right]dt$$

or $2x = t - \left(\dfrac{a+b}{2}\right)\log(t^2+ab) + C$ or $4x = 2(x + y) - (a + b) \log [(x + y)^2 + ab) + 2C$

or $2(y - x) - (a + b) \log [(x + y)^2 + ab)] = -2C = C'$ where C' is constant of integration. **Ans.**

10. *Solve* $\dfrac{dy}{dx} = (4x + y + 1)^2$

SOLUTION: Putting $4x + y + 1 = t$ the given equation becomes

$\dfrac{dt}{dx} - 4 = t^2$ or $\dfrac{dt}{t^2+4} = dx$. This, on integration gives

$$\frac{1}{2}\tan^{-1}\frac{t}{2} = x \quad \text{or} \quad \tan^{-1}\frac{4x+y+1}{2} = 2x + C$$ **Ans.**

11. *Solve* $x^3\dfrac{dy}{dx} = y^3 + y^2\sqrt{y^2 - x^2}$.

SOLUTION: The equation can be written as $\dfrac{dy}{dx} = \dfrac{y^3 + y^2\sqrt{y^2-x^2}}{x^3}$ which is of homogenous form.

Putting $y = vx$, gives $v + x\dfrac{dv}{dx} = v^3 + v^2\sqrt{v^2-1}$ or $\dfrac{dv}{v^3 - v + v^2\sqrt{v^2-1}} = \dfrac{dx}{x}$

$\therefore \quad \int \frac{dx}{x} = \int \frac{dv}{v(v^2-1)+v^2\sqrt{v^2-1}}$ Putting $v = \sec\theta$, we get

$$\therefore \quad \log x = \int \frac{\sec\theta \tan\theta \, d\theta}{\sec\theta \tan^2\theta + \sec^2\theta \tan\theta} = \int \frac{d\theta}{\sec\theta + \tan\theta} = \int \frac{\cos\theta \, d\theta}{1+\sin\theta}$$

$\therefore \quad \log x = \log(1 + \sin\theta) - \log c$ or $Cx = (1 + \sin\theta)$

or $\quad Cx = 1+\sqrt{1-\cos^2\theta} = 1+\sqrt{1-\frac{1}{v^2}} = 1 + \sqrt{1-\frac{x^2}{y^2}}$

or $\quad C\,xy = y + \sqrt{y^2-x^2}$ where C is constant of integration. **Ans.**

12. *Solve* $xdy - ydx = \sqrt{x^2+y^2}\,dx$.

SOLUTION: Equation is $x\frac{dy}{dx} - y = \sqrt{x^2+y^2}$

This is of homogeneous form. Putting $y = vx$, we get

$\therefore \quad x\left(v + x\frac{dv}{dx}\right) - vx = \sqrt{x^2+v^2x^2}$ or $x\frac{dv}{dx} = \sqrt{1+v^2}$

or $\quad \frac{dx}{x} = \frac{dv}{\sqrt{1+v^2}}$ which on integration, gives

$$\log x = \log(v+\sqrt{1+v^2}) - \log C \quad \text{where } C \text{ is arbitrary constant.}$$

$$Cx = v+\sqrt{1+v^2} = \frac{y}{x}+\sqrt{1+\frac{y^2}{x^2}}$$

or $\quad y+\sqrt{x^2+y^2} = Cx^2$ **Ans.**

13. *Solve* $x \sin\left(\frac{y}{x}\right)\frac{dy}{dx} = y \sin\left(\frac{y}{x}\right) + x$.

SOLUTION: Putting $y = vx$, the equation becomes $x \sin v\left(v + x\frac{dv}{dx}\right) = vx \sin v + x$

or $\quad x \sin v\frac{dv}{dx} = 1$ or $\sin v \, dv = \frac{dx}{x}$

On integration, we get $-\cos v = \log x$ or $\cos\left(\frac{y}{x}\right) = -\log x + \log C$

or $\quad \cos(y/x) = \log(C/x)$ where C is constant. **Ans.**

14. *Find the equation to the curve for which the tangent at any point on it cuts the x-axis at a point which is equidistant from the origin and the point on the curve.*

SOLUTION: Let P be any point (x, y) on the carve and tangent at P is

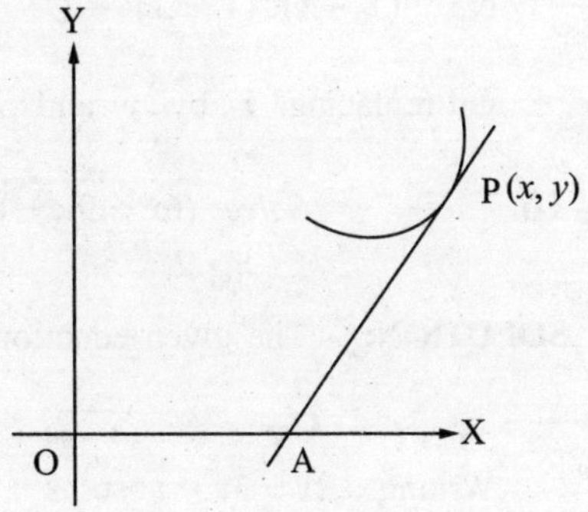

$$Y - y = \frac{dy}{dx}(X - x).$$ It meets X-axis at a point A.

Here $\quad Y = 0,$ gives $X = x - y\dfrac{dx}{dy} = OA$

$$\therefore \quad AP^2 = \left[\left(x - \frac{y}{y'}\right) - x\right]^2 + y^2 = \frac{y^2}{y'^2} + y^2$$

Since $\quad AP^2 = OA^2$ we have $\dfrac{y^2}{y'^2} + y^2 = \left(x - \dfrac{y}{y'}\right)^2$

or $\quad y^2 = x^2 - \dfrac{2xy}{y'}$ or $y' = \dfrac{2xy}{x^2 - y^2}$.

Putting $\quad y = vx$ we get $v + x\dfrac{dv}{dx} = \dfrac{2x^2 v}{x^2 - v^2 x^2}$

or $\quad x\dfrac{dv}{dx} = \dfrac{2v}{1 - v^2} - v = \dfrac{v + v^3}{1 - v^2}$

or $\quad \dfrac{dx}{x} = \dfrac{(1 - v^2)dv}{v + v^3} = \left(\dfrac{1}{v} - \dfrac{2v}{v^2 + 1}\right)dv$

$\therefore \quad \log x = \log v - \log (v^2 + 1) + \log c$

or $\quad x = \dfrac{cv}{v^2 + 1}$ or $x^2 + y^2 = cy$ **Ans.**

15. *Solve* $(3y - 7x + 7)\, dx + (7y - 3x + 3)dy = 0.$

SOLUTION: Given equation can be written as

$\dfrac{dy}{dx} = \dfrac{7x - 3y - 7}{7y - 3x + 3}$. Putting $x = X + h,\ y = Y + k,$ we get

$\dfrac{dY}{dX} = \dfrac{7X - 3Y + 7h - 3k - 7}{7Y - 3X + 7k - 3h + 3}$. Choose h and k so that $7h - 3k - 7 = 0$ and $7k - 3h + 3 = 0$

On solving for h and k, we get $h = 1,\ k = 0$

$\therefore \quad \dfrac{dY}{dX} = \dfrac{7X - 3Y}{7Y - 3X}$. Putting now $Y = v\,X$, gives

or $\quad v + X\dfrac{dv}{dX} = \dfrac{7 - 3v}{7v - 3}$ or $X\dfrac{dv}{dX} = \dfrac{7 - 7v^2}{7v - 3}$ or $\dfrac{dX}{X} = \dfrac{7v - 3}{7 - 7v^2}$

$$\therefore \quad 7 \log X = \int \frac{-7v + 3}{v^2 - 1} dv = -\frac{7}{2}\log(v^2 - 1) + \frac{3}{2}\log\frac{v - 1}{v + 1} + \log C$$

or $$X^7 = C(v^2-1)^{-7/2} \times \left(\frac{v-1}{v+1}\right)^{3/2} = \frac{C}{(v-1)^2}\frac{1}{(v+1)^5} = \frac{CX^7}{(Y-X)^2(Y+X)^5}$$

or $(Y-X)^2 (Y+X)^5 = C.$

On replacing Y by y and X by $x-1$, gives $(y-x+1)^2 (y+x-1)^5 = C.$ **Ans.**

16. *Solve* $(6x + 9y + 6)\, dy - (2x + 3y - 1)dx = 0.$

SOLUTION: The given equation is $\dfrac{dy}{dx} = \dfrac{2x+3y-1}{6x+9y+6}$

Writing $2x + 3y = t$ so $2 + 3\dfrac{dy}{dx} = \dfrac{dt}{dx}$, we get

$$\frac{dt}{dx} = 2 + 3\frac{(t-1)}{3(t+2)} = \frac{3t+3}{t+2} \text{ or } 3dx = \frac{t+2}{t++1}dt$$

This on integration, gives $3x = t + \log(t+1) - \log C$

or $3x = 2x + 3y + \log(2x + 3y + 1) - \log e$

$\therefore$ The solution is $2x + 3y + 1 = C\, e^{x-3y}$ where C is constant. **Ans.**

17. *Solve* $(1 + x^2)\dfrac{dy}{dx} + 2xy = \cos x.$

SOLUTION: The equation is $\dfrac{dy}{dx} + \dfrac{2x}{1+x^2}y = \dfrac{\cos x}{1+x^2}$ which is linear in y.

$\therefore$ I.F. $= e^{\int \frac{2x}{1+x^2}dx} = e^{\log(1+x^2)} = 1 + x^2$

$\therefore$ Solution is $y(1+x^2) = \int \cos x\, dx + C$ or $y(1+x^2) = \sin x + C.$ **Ans.**

18. *Solve* $\tan y \dfrac{dy}{dx} + \tan x = \cos y \cos^2 x.$

SOLUTION: The equation can be written as $\sec y \tan y \dfrac{dy}{dx} + \tan x \sec y = \cos^2 x.$

Putting $\sec y = t$ we get $\dfrac{dt}{dx} + t\tan x = \cos^2 x$ which is linear in t.

Hence Integrating factor $= e^{\int \tan x\, dx} = e^{\log \sec x} = \sec x$

$\therefore$ Solution is $t \sec x = \int \cos^2 x \sec x\, dx = \int \cos x\, dx$.

or $\sec x \sec y = \sin x + C$ or $\sec y = (\sin x + C)\cos x$ where C is constant **Ans.**

19. *Solve* $\left(\dfrac{e^{-2\sqrt{x}}}{\sqrt{x}}-\dfrac{y}{\sqrt{x}}\right)\dfrac{dx}{dy}=1.$

SOLUTION: The given equation can be written as

$$\frac{dy}{dx}+\frac{y}{\sqrt{x}}=\frac{e^{-2\sqrt{x}}}{\sqrt{x}} \quad \text{which is linear in } y. \quad \text{I.F.} = e^{\int\frac{1}{\sqrt{x}}dx}=e^{2\sqrt{x}}$$

$\therefore$ The solution is $y\,e^{2\sqrt{x}}=\int e^{2\sqrt{x}}\cdot\dfrac{e^{-2\sqrt{x}}}{\sqrt{x}}dx=\int\dfrac{1}{\sqrt{x}}dx=2\sqrt{x}+C$

or $y\,e^{2\sqrt{x}}=2\sqrt{x}+C$ where C is constant of integration. **Ans.**

20. *Solve* $\dfrac{dx}{dy}+\dfrac{x}{1+y^2}=\dfrac{\tan^{-1}y}{1+y^2}$

SOLUTION: $\dfrac{dx}{dy}+\dfrac{x}{1+y^2}=\dfrac{\tan^{-1}y}{1+y^2}$ is linear in x.

$\therefore$ I.F. $=e^{\int\frac{dy}{1+y^2}}=e^{\tan^{-1}y}$. Thus, the solution is

$$xe^{\tan^{-1}y}=\int\frac{\tan^{-1}y}{1+y^2}\cdot e^{\tan^{-1}y}dy=\int te^t dt \quad (\text{On putting } \tan^{-1}y=t)$$

$$=(t-1)e^t=(\tan^{-1}y-1)\,e^{\tan^{-1}y}+C$$

or $x=(\tan^{-1}y-1)+C\,e^{-\tan^{-1}y}$ where C is constant. **Ans.**

21. *Solve* $ye^y=(y^3+2x\,e^y)\dfrac{dy}{dx}$.

SOLUTION: Equation can be written as $\dfrac{dx}{dy}-\dfrac{2x}{y}=y^2\,e^{-y}$ which is linear is x

Hence I.F. $=e^{\int-\frac{2}{y}dy}=e^{-2\log y}=\dfrac{1}{y^2}$ $\therefore$ Solution is

$$x\cdot\frac{1}{y^2}=\int y^2e^{-y}\cdot\frac{1}{y^2}dy=\int e^{-y}dy=-e^{-y}+C$$

or $x+y^2\,e^{-y}=C\,y^2$ where C is constant of integration. **Ans.**

22. *Solve* $\dfrac{dy}{dx}-y\tan x=y^4\sec x.$

SOLUTION: Dividing the given equation by y^4, we get

$\dfrac{1}{y^4}\dfrac{dy}{dx}-\dfrac{1}{y^3}\tan x=\sec x$ Putting $\dfrac{-1}{y^3}=t$, gives

$$\frac{dt}{dx} + 3\tan x.t = 3 \sec x.$$ This is linear in t.

$\therefore$ I.F. $= e^{\int 3\tan x\, dx} = e^{3\log \sec x} = \sec^3 x$ $\therefore$ Solution is

$$t \sec^3 x = \int 3\sec x.\sec^3 x\, dx = 3\int (1+\tan^2 x)\sec^2 x\, dx = 3 \tan x + \tan^3 x + C$$

or $$\frac{-1}{y^3}\sec^3 x = 3 \tan x + \tan^3 x + C \quad \text{or} \quad \frac{1}{y^3} = -\sin^3 x - 3 \tan x \cos^3 x - \cos^3 x$$

or $$\frac{1}{y^3} + 3 \sin x - 2 \sin^3 x + C \cos^3 x = 0$$ where C is constant **Ans.**

23. *Solve* $\frac{dy}{dx} = e^{x-y}(e^x - e^y)$.

SOLUTION: Equation can be written as $e^y \frac{dy}{dx} + e^x e^y = e^{2x}$

Putting $e^y = t$, we get $\frac{dt}{dx} + e^x.t = e^{2x}$ which is linear in t.

$\therefore$ I.F. $= e^{\int e^x dx} = e^{e^x}$ $\therefore$ The solution is

$$te^{e^x} = \int e^{2x}\left(e^{e^x}\right)dx + e \quad \text{or} \quad t = e^x - 1 + C e^{-e^x} \quad \text{or} \quad e^y = e^x - 1 - C e^{-e^x}$$ **Ans.**

24. *Solve* $xy\frac{dy}{dx} + y^2 = 4x$ if $y = 0$ at $x = 0$.

SOLUTION: The given equation can be written as $2y\frac{dy}{dx} + 2\frac{y^2}{x} = 8.$

Putting $y^2 = t$, gives $\frac{dt}{dx} + \frac{2}{x}t = 8$ which is linear in t.

$\therefore$ I.F. $= e^{\int 2/x\, dx} = e^{2\log x} = x^2$, hence the solution is

$$x^2 t = \int 8.x^2 dx = \frac{8}{3}x^3 + C \quad \text{or} \quad x^2 y2 = \frac{8}{3}x^3 + C.$$

Given that $y = 0$ at $x = 0$, we have $C = 0$ hence $3y^2 = 8x$ is the solution. **Ans.**

25. *Solve* $\frac{dy}{dx} = \frac{x^2 + y^2 + 1}{2xy}$.

SOLUTION: Equation can be written as $2y\frac{dy}{dx} - \frac{1}{x}y^2 = x + \frac{1}{x}$

Putting here $y^2 = t$, gives $\frac{dt}{dx} - \frac{t}{x} = x + \frac{1}{x}$, linear in t.

$\therefore$ I.F. $= e^{\int -\frac{1}{x}dx} = e^{-\log x} = \frac{1}{x}$. Hence the solution is

$$t \cdot \frac{1}{x} = \int\left(x+\frac{1}{x}\right)\frac{1}{x}dx = \int\left(1+\frac{1}{x^2}\right)dx \quad \text{or} \quad y^2 = x^2 - 1 + Cx$$ **Ans.**

26. *Solve* $\frac{dy}{dx} - x^2 y = y^2 e^{-x^3/3}$

SOLUTION: Equation can be written as $\frac{1}{y^2}\frac{dy}{dx} - \frac{x^2}{y} = e^{-x^3/3}$.

Putting $\frac{-1}{y} = t$ we get $\frac{dt}{dx} + x^2 t = e^{-x^3/3}$. It is linear in t. IF. $= e^{\int x^2 dx = e^{x^3/3}}$

$\therefore$ Solution is $t e^{x^3/3} = \int e^{x^3/3} \cdot e^{-x^3/3} = x + C$ or $-\frac{1}{y}e^{x^3/3} = x + C$

or $y(x + C) + e^{x^3/3} = 0$ is the required solution where C is constant of integration. **Ans.**

27. *Solve* $y(2xy + e^x)\,dx - e^x\,dy = 0$

SOLUTION: Equation is $e^x\frac{dy}{dx} - ye^x = 2xy^2$ Dividing throughout by $y^2 \cdot e^x$, gives

$\frac{1}{y^2}\frac{dy}{dx} - \frac{1}{y} = 2x\,e^{-x}$. Putting here $\frac{-1}{y} = t$, we get

$\frac{dt}{dx} + t = 2x\,e^{-x}$. It is linear in t with I.F. $= e^x$

$\therefore$ Solution is $te^x = \int 2xe^{-x}\,e^x dx = x^2 + C$ or $-\frac{1}{y}e^x = x^2 + C$

which is the required solution. Here C is constant of integration. **Ans.**

28. *Solve* $(x^2y - 2xy^2)\,dx - (x^3 - 3x^2y)\,dy = 0$

SOLUTION: It is homogeneous hence putting $y = vx$, we get

$$v + x\frac{dv}{dx} = \frac{v - 2v^2}{1 - 3v} \quad \text{or} \quad x\frac{dv}{dx} = \frac{-2v^2 + 3v^2}{1 - 3v}$$

or $$\frac{dx}{x} = \frac{1 - 3v}{v^2}dv \Rightarrow \log x = \frac{-1}{v} - 3\log v + \log k$$

or $$xv^3 = K\,e^{-1/v} \quad \text{or} \quad \frac{xy^3}{x^3} = K\,e^{-x/y}$$

$\therefore$ Solution is $y^3 = kx^2\,e^{-x/y}$ where K is constant. **Ans.**

SOLUTIONS TO EXERCISE 9B

1. *Solve the differential equations* $(2x + y - 1)dy - (x - 2y + 5)dx = 0$

SOLUTION: $(x - 2y + 5)\, dx - (2x + y - 1)\, dy = 0$ which is of form $Mdx + Ndy = 0$

$\dfrac{\partial M}{\partial y} = -2$ and $\dfrac{\partial N}{\partial x} = -2$. The equation is exact as $\dfrac{\partial M}{\partial y} = \dfrac{\partial N}{\partial x}$. Therefore the solution is

$$\int (x - 2y + 5)\, dx - \int (y - 1)\, dy = C$$

or $$\frac{x^2}{2} - 2xy + 5x - \frac{y^2}{2} + y = C \quad \text{or} \quad x^2 - y^2 - 4xy + 10x + 2y = C$$

where C is constant of integration. **Ans.**

2. *Solve the differential equations* $(2xy + 2y^2 e^{2x})dx + (x^2 + 2y\, e^{2x})\, dy = 0$

SOLUTION: Equation is of form $Mdx + Ndy = 0$

Here $\dfrac{\partial M}{\partial y} = 2x + 4y\, e^{2x}$, $\dfrac{\partial N}{\partial x} = 2x + 4y\, e^{2x}$. It is exact as $\dfrac{\partial M}{\partial y} = \dfrac{\partial N}{\partial x}$

$\therefore$ The solution is $\int [2xy + 2y^2 e^{2x}]\, dx + \int o\, dy = C$ or $x^2 y + y^2 e^{2x} = C$

where C is arbitrary constant. **Ans.**

3. *Solve the differential equations* $(y\, e^{xy} - \tan x)\, dx + (x\, e^{xy} - \sec y)\, dy = 0.$

SOLUTION: Equation is of the form $Mdx + Ndy = 0$

Here $\dfrac{\partial M}{\partial y} = ye^{xy} \cdot x + 1 \cdot e^{xy}$, $\dfrac{\partial N}{\partial x} = xe^{xy} \cdot y + 1\, e^{xy}$ It is exact since $\dfrac{\partial M}{\partial y} = \dfrac{\partial N}{\partial x}$.

The solution is $\int [ye^{xy} - \tan x]\, dx + \int -\sec y dy = C$

or $$e^{xy} + \log \cos x - \log (\sec y + \tan y) = C$$

where C is arbitrary constant of integration. **Ans.**

4. *Solve the differential equation*

$$[y \sin (xy) + xy^2 \cos (xy)]\, dx + [x \sin (xy) + x^2 y \cos (xy)]dy = 0.$$

SOLUTION: Equation is of form $Mdx + Ndy = 0.$

Here $\dfrac{\partial M}{\partial y} = y \cos (xy) \cdot x + 1.\ \sin (xy) + 2xy \cos (xy) - xy^2 \sin (xy)$

and $\dfrac{\partial N}{\partial x} = x \cos (xy) \cdot y + 1.\ \sin (xy) + 2xy \cos (xy) - x^2 y \sin (xy)$

Since $\dfrac{\partial M}{\partial y} = \dfrac{\partial N}{\partial x}$ hence equation is exact.

$\therefore$ The solution is $\int [y\sin(xy) + xy^2\cos(xy)]\,dx + \int o\,dy = C$

or $$-y\cos(xy)\cdot\frac{1}{y} + xy^2\sin(xy)\cdot\frac{1}{y} - \int y^2\sin(xy)\cdot\frac{1}{y}\,dx = C$$

or $$-\cos(xy) + xy\sin(xy) + y\cos(xy)\cdot\frac{1}{y} = C$$

or $xy \sin xy = C$ where C is constant. **Ans.**

5. *Solve the differential equations* $(x^4 e^x - 2a\, xy^2)dx + 2a\, x^2y\, dy = 0.$

SOLUTION: Equation is of the form $Mdx + Ndy = 0$. Here

$$\frac{\partial M}{\partial y} = -4axy, \quad \frac{\partial N}{\partial x} = 4\,axy \quad \text{and} \quad \frac{\dfrac{\partial M}{\partial y} - \dfrac{\partial N}{\partial x}}{N} = \frac{-8axy}{2ax^2y} = \frac{-4}{x}$$

$\therefore$ Integrating factor $= e^{\int -4/x\,dy} = e^{-4\log x} = \dfrac{1}{x^4}$

Multiplying the given equation by $\dfrac{1}{x^4}$, we get

$$\left(e^x - \frac{2ay^2}{x^3}\right)dx + \frac{2ay}{x^2}dy = 0 \quad \text{which is exact.}$$

It solution is $\int \left[e^x - \dfrac{2ay^2}{x^3}\right]dx + \int 0\,dy = C$

or $e^x + \dfrac{ay^2}{x^2} = C$ or $x^2 e^x + ay^2 = Cx^2$ **Ans.**

6. *Solve the differential equation* $(y^4 + 2y)dx + (xy^3 + 2y^4 - 4x)dy = 0.$

SOLUTION: Equation is of the form $Mdx + Ndy = 0$

Here $$\frac{\partial M}{\partial y} = 4y^3 + 2, \quad \frac{\partial N}{\partial x} = y^3 - 4 \quad \text{and} \quad \frac{\dfrac{\partial N}{\partial x} - \dfrac{\partial M}{\partial y}}{M} = \frac{-3y^3 - 6}{y^4 + 2y} = \frac{-3}{y}$$

$\therefore$ I.F. $= e^{\int \frac{-3}{y}dy} = e^{-3\log y} = \dfrac{1}{y^3}.$

Multiplying throughout by $\dfrac{1}{y^3}$, we get

$$\left(y + \frac{2}{y^2}\right)dx + \left(x + 2y - \frac{4x}{y^3}\right)dy = 0 \quad \text{which is exact.}$$

Its solution is $\int\left[y+\frac{2}{y^2}\right]dx + \int 2y\,dy = C$ y as constant

or $xy+\frac{2x}{y^2}+y^2 = C$ or $xy^3 + 2x + y^4 = cy^2$ **Ans.**

is the solution where C is constant. **Ans.**

7. *Solve the differential equations* $y(x^2y + e^x)dx - e^x\,dy = 0.$

SOLUTION: Equation is $Mdx + Ndy = 0.$

Here $\frac{\partial M}{\partial y}=2x^2y+e^x$, $\frac{\partial N}{\partial x} = -e^x$, $\frac{\frac{\partial N}{\partial x}-\frac{\partial M}{\partial y}}{M}=-\frac{2}{y}$

$\therefore$ I.F. $= e^{\int\frac{-2}{y}dy} = e^{-2\log y} = \frac{1}{y^2}$

Multiplying the given equation throughout by $\frac{1}{y^2}$, we get

$\left(x^2+\frac{1}{y}e^x\right)dx-\frac{e^x}{y^2}dy=0$ which is exact.

It solution is

$$\int\left[x^2+\frac{1}{y}e^x\right]dx - \int 0\,dy=\frac{C}{3} \quad y \text{ as constant.}$$

or $$\frac{x^3}{3}+\frac{1}{y}e^x = \frac{C}{3} \quad \text{or} \quad x^3y + 3e^x = Cy$$

is the solution where C is constant. **Ans.**

8. *Solve the differential equations* $x^2y\,dx - (x^3 + y^3)\,dy = 0.$

SOLUTION: Equation is of the form $Mdx + Ndy = 0$

Here $\frac{\partial M}{\partial y}=x^2$ and $\frac{\partial N}{\partial x} = -3x^2$ $\therefore$ $\frac{\frac{\partial N}{\partial x}-\frac{\partial M}{\partial y}}{M}=\frac{-4x^2}{x^2y}=\frac{-4}{y}$

$\therefore$ I.F. $= e^{\int\frac{-4}{y}dy} = e^{-4\log y} = \frac{1}{y^4}$.

Then equation becomes $\frac{x^2}{y^3}dx-\frac{x^3+y^3}{y^4}dy = 0$

Its solution in $\int\frac{x^2}{y^3}dx-\int\frac{y^3}{y^4}dy = C$

or $\frac{x^3}{3y^3}-\log y = C$ **Ans.**

9. *Solve the differential equations* $(x^2y^2 + 5xy + 2)\, y\, dx + (x^2y^2 + 4xy + 2)\, x\, dy = 0$.

SOLUTION: Equation is of the form $f_1(xy)\, ydx + f_2(xy)\, xdy = 0$

$$\therefore \quad \text{I.F.} = \frac{1}{Mx - Ny} = \frac{1}{xy(x^2y^2 + 5xy + 2) - xy(x^2y^2 + 4xy + 2)} = \frac{1}{x^2y^2}$$

Multiplying throughout by $\frac{1}{x^2y^2}$, gives

$$\left(y + \frac{5}{x} + \frac{2}{x^2y}\right)dx + \left(x + \frac{4}{y} + \frac{2}{xy^2}\right)dy = 0. \text{ It is exact.}$$

Its solution is $\int\left[y + \frac{5}{x} + \frac{2}{x^2y}\right]dx + \int \frac{4}{y}\, dy = C$ where C is any constant.

or $\quad xy + 5\log x - \frac{2}{xy} + 4\log y = \log C$ or $\log\left(\frac{x^5y^4}{C}\right) = \frac{2}{xy} - xy$

or $\quad x^5y^4 = C\, e^{2/xy - xy}$ where C is constant. **Ans.**

10. *Solve the differential equations* $(3xy^2 - y^3)dx + (xy^2 - 2x^2y)dy = 0$

SOLUTION: Equation can be written as $\frac{dy}{dx} = \frac{y^3 - 3xy^2}{xy^2 - 2x^2y}$. It is homogeneous.

Putting $y = vx$ we get $v + x\frac{dv}{dx} = \frac{v^3 - 3v^2}{v^2 - 2v}$ or $x\frac{dv}{dx} = \frac{-v^2}{v^2 - 2v}$.

$$\text{or} \quad \frac{dx}{x} = \frac{v^2 - 2v}{-v^2}\, dv = \left(-1 + \frac{2}{v}\right)dv$$

On integration, it gives $\quad \log x = -v + 2\log v + \log C$

$$\log\frac{x}{cv^2} = -v \quad \text{or} \quad \frac{x^3}{cy^2} = e^{-y/x}$$

or $\quad x^3 = Cy^2 e^{-y/x}$ where C is constant. **Ans.**

11. *Solve the differential equations* $(x + 2y^3)\frac{dy}{dx} = y + 2x^3y^2$.

SOLUTION: The equation is $(y + 2x^3y^2)\, dx + (-x - 2y^3)\, dy = 0$ or $Mdx + Ndy = 0$. ...(1)

$$\therefore \quad \frac{\partial M}{\partial y} = 1 + 4x^3y, \quad \frac{\partial N}{\partial x} = -1 \quad \therefore \quad \frac{\frac{\partial N}{\partial x} - \frac{\partial M}{\partial y}}{M} = \frac{-2 - 4x^3y}{y(1 + 2x^3y)} = \frac{-2}{y}$$

$$\therefore \quad \text{I.F.} = e^{\int -\frac{2}{y}dy} = e^{-2\log y} = \frac{1}{y^2}.$$

Multiplying (1) throughout by $\frac{1}{y^2}$, gives

$$\left(\frac{1}{y}+2x^3\right)dx+\left(\frac{-x}{y^2}-2y\right)dy=0 \quad \text{which is exact.}$$

Its solution is $\int\left[\frac{1}{y}+2x^3\right]dx+\int -2y\,dy=C$ or $\frac{x}{y}+\frac{x^4}{2}-y^2=C$ C being constant. **Ans.**

12. *Solve the differential equations* $\left(xy\log\frac{x}{y}\right)dx+\left(y^2-x^2\log\frac{x}{y}\right)dy=0.$

SOLUTION: Equation can be taken as homogeneous since $\left(\frac{x}{y}\right)$ is of zero degree, however,

It is of the form $Mdx + Ndy = 0$, hence

$$\frac{\partial M}{\partial y}=x\log\left(\frac{x}{y}\right)+xy\left(\frac{y}{x}\right)\left(\frac{-x}{y^2}\right)=x\log\left(\frac{x}{y}\right)-x.$$

and

$$\frac{\partial N}{\partial x}=0-2x\log\frac{x}{y}-x^2\cdot\frac{y}{x}\cdot\frac{1}{y}=-2x\log\frac{x}{y}-x$$

Here

$$\frac{\frac{\partial N}{\partial x}-\frac{\partial M}{\partial y}}{M}=\frac{2x\log\frac{x}{y}-x\log\frac{x}{y}}{xy\log\left(\frac{x}{y}\right)}=\frac{-3}{y}.$$

$\therefore$

$$\text{I.F.}=e^{\int -\frac{3}{y}dy}=e^{-3\log y}=\frac{1}{y^3}.$$

$\therefore$ Multiplying throughout by $\frac{1}{y^3}$, equation becomes

$$\frac{x}{y^2}\log\frac{x}{y}\,dx+\left(\frac{1}{y}-\frac{x^2}{y^3}\log\frac{x}{y}\right)dy=0 \quad \text{which is exact.}$$

$\therefore$ Solution is $\int\left[\frac{x}{y^2}\log\frac{x}{y}\right]dx+\int\frac{1}{y}dy=C$ where C is arbitrary constant.

or

$$\log\frac{x}{y}\cdot\frac{x^2}{2y^2}-\int\frac{y}{x}\cdot\frac{1}{y}\cdot\frac{x^2}{2y^2}dx+\log y=C$$

or

$$\frac{x^2}{2y^2}\left[\log\left(\frac{x}{y}\right)-\frac{1}{2}\right]+\log y=C$$ **Ans.**

13. *Solve the differential equations* $(2x^2y^2 + y)\,dx - (x^3 y - 3x)\,dy = 0.$

SOLUTION: Equation is $Mdx + Ndy = 0$ and can be written as $x^2y\,(2y\,dx - xdy) + (ydx + 3x\,dy) = 0$

Here I.F. $= x^h\, y^k$ then multiplying throughout by it, we get

$(2x^{h+2}\, y^{k+2} + x^h\, y^{k+1})\, dx + (3x^{h+1}\, y^k - x^{h+3}\, y^{k+1})\, dy = 0.$ This should be exact.

Here $\dfrac{\partial M}{\partial y} = 2(k+2)\, x^{h+2} \cdot y^{k+1} + (k+1)x^h y^k,\quad \dfrac{\partial N}{\partial x} = 3(h+1)x^h y^k - (h+3)x^{h+2} y^{k+1}$

For $\dfrac{\partial M}{\partial y} = \dfrac{\partial N}{\partial x}$ we should have $2\,(k+2) = -(h+3)$ and $(k+1) = 3\,(h+1)$

which on solving for h and k, gives $h = \dfrac{-11}{7}, k = \dfrac{-19}{7}.$

$\therefore$ Equation now becomes

$$\left[2 \cdot x^{3/7} y^{-5/7} + x^{-11/7} y^{-12/7}\right] dx + \left[3x^{-4/7} y^{-19/7} - x^{-10/7}\, y^{-5/7}\right] dy = 0 \text{ which is exact.}$$

Its solution is $-2 \cdot \dfrac{7}{10} x^{10/7} y^{-5/7} - \dfrac{7}{4} x^{-4/7} y^{-12/7} + \int 0\, dy = C$

or $\dfrac{7}{5} x^{10/7} y^{-5/7} - \dfrac{7}{4} x^{-4/7} y^{-12/7} = C$ where C is arbitrary constant. **Ans.**

14. *Solve the differential equations* $\left(2x \sinh \dfrac{y}{x} + 3y \cosh \dfrac{y}{x}\right) dx - 3x \cosh \dfrac{y}{x}\, dy = 0.$

SOLUTION: Equation is of the form $Mdx + Ndy = 0.$

Here $\dfrac{\partial M}{\partial y} = \dfrac{2x}{x} \cosh \dfrac{y}{x} + \dfrac{3y}{x} \sinh \dfrac{y}{x} + 3 \cosh \dfrac{y}{x} = 5 \cosh \dfrac{y}{x} + \dfrac{3y}{x} \sinh \dfrac{y}{x}$

and $\dfrac{\partial N}{\partial x} = -3x\left(\sinh \dfrac{y}{x}\right)\left(\dfrac{-y}{x^2}\right) - 3\cosh \dfrac{y}{x} = 3\dfrac{y}{x} \sinh \dfrac{y}{x} - 3\cosh \dfrac{y}{x}$

$\therefore \quad \dfrac{\dfrac{\partial M}{\partial y} - \dfrac{\partial N}{\partial x}}{N} = \dfrac{8 \cosh \dfrac{y}{x}}{-3x \cosh \dfrac{y}{x}} = \dfrac{-8}{3x} \quad \therefore$ I.F. $= e^{\int \frac{-8}{3x} dx} = e^{-8/3 \log x} = x^{-8/3}$

Multiplying the given equation by $x^{-8/3}$, we get

$$\left(2x^{-5/3} \sinh \frac{y}{x} + 3x^{-8/3} y \cosh \frac{y}{x}\right) dx - \left(3x^{-5/3} \cosh \frac{y}{x}\right) dy = 0. \text{ It is exact.}$$

It's solution is $\displaystyle\int \left[3x^{-5/3} \cosh \frac{y}{x}\right] dy + \int 0\, dx = C$

x as constant

or $3x^{-5/3}\left(\sin \dfrac{y}{x}\right) \cdot x = C$ or $\sinh \dfrac{y}{x} = C\, x^{2/3}$

where C is arbitrary constant of integration. **Ans.**

15. Solve the differential equations $(y^2 + 2yx^2)dx + (2x^3 - xy)dy = 0$.

SOLUTION: Given equation can be written as $y(ydx - xdy) + 2x^2(ydx + xdy) = 0$, hence I.F. $= x^h y^k$ then multiplying throught by $x^h y^k$ given equation becomes

$$(x^h y^{k+2} + 2y^{k+1} x^{h+2})\,dx + (2x^{h+3} y^k - x^{h+1} y^{k+1})\,dy = 0 \qquad ...(1)$$

Now $$\frac{\partial M}{\partial y} = (k+2)x^h y^{k+1} + 2(k+1)x^{h+2} y^k,$$

and $$\frac{\partial N}{\partial x} = 2(h+3)x^{h+2} y^k - (h+1)x^h y^{k+1}$$

For $\dfrac{\partial M}{\partial y} = \dfrac{\partial N}{\partial x}$ we must have $k + 2 = -(h + 1)$ and $2(k + 1) = 2(h + 3)$

Solving h and k, we get $h = \dfrac{-5}{2}$ and $k = \dfrac{-1}{2}$

$\therefore$ (1) becomes $(x^{-5/2} y^{3/2} + 2x^{-1/2} y^{1/2})\,dx + (2x^{1/2} y^{-1/2} - x^{-3/2} y^{1/2})\,dy = 0$

This is exact equation, hence its solution is

$$\int (x^{-5/2} y^{3/2} + 2x^{-1/2} y^{1/2})\,dx + \int 0\,dy = 0$$

or $$\frac{-2}{3}x^{-3/2} y^{3/2} + 2(2)x^{1/2} y^{1/2} = C \quad \text{where } C \text{ is arbitrary constant.}$$

or $$6x^{1/2} y^{1/2} - x^{-3/2} y^{3/2} = C$$ **Ans.**

16. *Solve the differential equations* $x^2(x\,dx + y\,dy) + y(x\,dy - y\,dx) = 0$.

SOLUTION: Equation can be written as

$$xdx + ydy + y\left(\frac{xdy - ydx}{x^2}\right) = 0 \quad \text{or} \quad d\left(\frac{x^2 + y^2}{2}\right) + y\,d\left(\frac{y}{x}\right) = 0$$

Converting into polar coordinates, we get

$$d\left(\frac{r^2}{2}\right) + r\sin\theta\,d(\tan\theta) = 0 \quad \text{or} \quad rdr + r\sin\theta\sec^2\theta\,d\theta = 0$$

or $$dr + \tan\theta\sec\theta\,d\theta = 0, \quad \text{which on integration, gives}$$

$$r + \sec\theta = C \quad \text{or} \quad r\cos\theta + 1 = C\cos\theta$$

or $$x + 1 = C\frac{x}{r} \quad \text{or} \quad r(x + 1) = Cx$$

or $$(x+1)\sqrt{x^2 + y^2} = Cx \quad \text{where } C \text{ is an arbitrary constant.}$$ **Ans.**

17. *Solve the differential equations* $x\,dy - y\,dx = (x^2 + y^2)(x\,dx + y\,dy)$.

SOLUTION: Equation can be written as

$$\frac{xdy - ydx}{x^2} = \frac{(x^2 + y^2)(xdx + ydy)}{x^2}$$

or
$$d\left(\frac{y}{x}\right) = \left[1 + \left(\frac{y}{x}\right)^2\right]\frac{1}{2}\,d(x^2 + y^2)$$

Converting into polar coordinates, we get

$$d(\tan\theta) = (1 + \tan^2\theta)\frac{1}{2}d(r^2)$$

or $\sec^2\theta\, d\theta = \sec^2\theta \cdot \frac{1}{2}d(r^2)$ or $2d\theta = d(r^2)$

which on integration, gives $2\theta = r^2 + C$ where C is an arbitrary constant.

or $2\tan^{-1}\frac{y}{x} = x^2 + y^2 + C$. **Ans.**

18. *Solve the differential equations* $(2x^2 + 3y^2 - 7)x\,dx - (3x^2 + 2y^2 - 8)y\,dy = 0$.

SOLUTION: Putting $x^2 = X,\ y^2 = Y,$ the equation becomes

$$(2X + 3Y - 7)\,dX - (3X + 2Y - 8)\,dY = 0$$

or $\dfrac{dY}{dX} = \dfrac{2X + 3Y - 7}{3X + 2Y - 8}$. Writing $X = \alpha + h,\ Y = \beta + k,$ we get

$\dfrac{d\beta}{d\alpha} = \dfrac{2\alpha + 3\beta + 2h + 3k - 7}{3\alpha + 2\beta + 3h + 2k - 8}$. Choose h and k such that

$2h + 3k - 7 = 0$ and $3h + 2k - 8$ which give $h = 2,\ k = 1$

$\therefore$ $\dfrac{d\beta}{d\alpha} = \dfrac{2\alpha + 3\beta}{3\alpha + 2\beta}$. Now putting $\beta = v\alpha$ gives

$$v + \alpha\frac{dv}{d\alpha} = \frac{2 + 3v}{3 + 2v} \quad \text{or} \quad \alpha\frac{dv}{d\alpha} = \frac{2 + 3v}{3 + 2v} - v = \frac{-2v^2 + 2}{2v + 3}$$

or $\dfrac{d\alpha}{\alpha} = \dfrac{2v + 3}{-2v^2 + 2}dv$ or $-2\dfrac{d\alpha}{\alpha} = \dfrac{2v + 3}{v^2 - 1}dv$. On integrating both sides, we get

$$-2\log\alpha = \log(v^2 - 1) + \frac{3}{2}\log\frac{v - 1}{v + 1} - \log C \quad \text{or} \quad (v^2 - 1)\left(\frac{v - 1}{v + 1}\right)^{3/2}\cdot\alpha^2 = C$$

$$\alpha^2\frac{(v - 1)^{5/2}}{(v + 1)^{1/2}} = C \quad \text{or} \quad \frac{(\beta - \alpha)^{5/2}}{(\beta + \alpha)^{1/2}} = C$$ But $\alpha = x^2 - 2,\ \beta = y^2 - 1$

$$\frac{(y^2 - x^2 + 1)^{5/2}}{(y^2 + x^2 - 3)^{1/2}} = C \quad \text{or} \quad (x^2 - y^2 - 1)^5 = C\,(x^2 + y^2 - 3),$$ C being constant. **Ans.**

19. *Solve the differential equations* $(x+y)^2\left(x\frac{dy}{dx}+y\right) = xy\left(1+\frac{dy}{dx}\right)$.

SOLUTION: Equation can be written as $(x + y)^2\ (xdy + ydx) = xy\ (dx + dy)$

or $\frac{xdy + ydx}{xy} = \frac{dx + dy}{(x+y)^2}$ or $\frac{d(xy)}{xy} = \frac{d(x+y)}{(x+y)^2}$

which on integration, gives $\log(xy) + \frac{1}{x+y} = C$. **Ans.**

20. ***Solve the differential equation*** $\left(x^4\frac{dy}{dx}+x^3y\right) = -\sec(xy)$.

SOLUTION: Equation is $x^3\ (xdy + ydx) = -\sec(xy)\ dx$ or $\cos(xy)\ d(xy) = \frac{-dx}{x^3}$

On integrating both sides, we get $\sin(xy) = \frac{1}{2x^2} + C$ **Ans.**

21. *Solve the differential equations* $\left[\cos x \log(2y-8)+\frac{1}{x}\right]dx+\frac{\sin x}{y-4}dy = 0,\ \ y(1) = \frac{9}{2}$

SOLUTION: Equation is of the form $Mdx + Ndy = 0$

Here $\frac{\partial M}{\partial y} = \cos x\cdot\frac{2}{2y-8}$ and $\frac{\partial N}{\partial x} = \frac{\cos x}{y-4}$, **Hence exact.**

The solution is

$$\int\left[\cos x\log(2y-8)+\frac{1}{x}\right]dx + \int 0\,dy = C$$

y as constant.

or $\log(2y - 8)\sin x + \log x = C$. Since $y = \frac{9}{2}$ at $x = 1$ we have $C = 0$

$\therefore$ Solution is $\log(2y - 8)\sin x + \log x = 0$. **Ans.**

22. *Solve the differential equation* $(2x\log x - xy)\ dy + 2y\ dx = 0$

SOLUTION: Equation is $Mdx + Ndy = 0$ where $M = 2y,\ \ N = 2x\log x - xy$

Here $\frac{\partial M}{\partial y} = 2$, $\frac{\partial N}{\partial x} = 2x\cdot\frac{1}{x}+2\log x - y$ $\therefore$ $\frac{\frac{\partial M}{\partial y}-\frac{\partial N}{\partial x}}{N} = \frac{2-2-2\log x+y}{2x\log x - xy} = \frac{-1}{x}$

Hence I.F. $= e^{\int\frac{-1}{x}dx} = e^{-\log x} = \frac{1}{x}$.

Multiplying throughout by $\frac{1}{x}$ the equation becomes $\frac{2y}{x}dx + (2\log x - y)\,dy = 0$ which is exact.

Its solution is $\int\left[\frac{2y}{x}\right]_{y=\text{constant}} dx + \int -y\,dy = C$ or $2y\log x - \frac{y^2}{2} = C$ **Ans.**

23. *Solve the differential equation* $y(xy + 2x^2y^2)\,dx + x(xy - x^2y^2)\,dy = 0$

SOLUTION: The equation of the form $f_1(xy)\,ydx + f_2(xy)\,xdy = 0$

The I.F. $= \frac{1}{Mx - Ny} = \frac{1}{xy(xy+2x^2y^2) - xy(xy - x^2y^2)} = \frac{1}{3x^3y^3}$

Multiplying the equation throughout by I.F., we get

$$y\left(\frac{1}{x^2y^2} + \frac{2}{xy}\right)dx + x\left(\frac{1}{x^2y^2} - \frac{1}{xy}\right)dy = 0$$

or $\left(\frac{1}{x^2y} + \frac{2}{x}\right)dx + \left(\frac{1}{xy^2} - \frac{1}{y}\right)dy = 0$ which is exact.

Its solution is $\int\left[\frac{1}{x^2y} + \frac{2}{x}\right]dx + \int\frac{-1}{y}\,dy = C$

or $-\frac{1}{xy} + 2\log x - \log y = C$ is the solution. **Ans.**

24. *Solve the differential equation* $(y^3 + x^3)\,dx - xy^2\,dy = 0$

SOLUTION: Equation is of the form $Mdx + Ndy = 0$

Here $\frac{\partial M}{\partial y} = 3y^2,\ \frac{\partial N}{\partial x} = -y^2 \quad \therefore \quad \frac{\frac{\partial M}{\partial y} - \frac{\partial N}{\partial x}}{N} = \frac{4y^2}{-xy^2} = -\frac{4}{x}$

$\therefore$ Integrating Factor $= e^{\int -4/x\,dx} = e^{-4\log x} = \frac{1}{x^4}$

Multiplying the given equation by the I.F., we get

$\left(\frac{y^3}{x^4} + \frac{1}{x}\right)dx - \frac{y^2}{x^3}\,dy = 0$ which is exact.

The solution is $\int\left[\frac{y^3}{x^4} + \frac{1}{x}\right]_{y=\text{constant}} dx + 0 = C$

or $\frac{-y^3}{3x^3} + \log x = C$ where C is an arbitrary constant. **Ans.**

25. *Solve the differential equation* $(x^2 + y^2 - a^2)\,x\,dx + (x^2 - y^2 - b^2)\,y\,dy = 0.$

SOLUTION: The equation is $x(x^2 - a^2)\,dx - y(y^2 + b^2)\,dy + xy\,(ydx + xdy) = 0$

or $(x^3 - a^2x)\,dx - (y^3 + b^2y) + xy\,d(xy) = 0$

which when integrated, gives

$$\frac{x^4}{4}-\frac{a^2x^2}{2}-\frac{y^4}{4}-\frac{b^2y^2}{2}+\frac{1}{2}(xy)^2=C$$

or $x^4 - y^4 + 2(x^2y^2 - a^2x^2 - b^2y^2) = C$ where C is an arbitrary constant. **Ans.**

26. *Solve the equation* $\left[y\left(1+\frac{1}{x}\right)+\cos y\right]dx+(x+\log x - x\sin y)\,dy=0.$

SOLUTION: Given equation can be written as $y\left(1+\frac{1}{x}\right)dx+(x+\log x)\,dy+\cos y\,dx - x\sin y\,dy=0$

or $d[y(x+\log x)]+d[x\cos y]=0$ which on integration, gives

$y(x + \log x) + x\cos y = C$ where C is an arbitrary constant.

which is the requried solution. **Ans.**

27. *Solve* $(xy^3 + y)\,dx + 2(x^2y^2 + x + y^4)\,dy = 0.$

SOLUTION: It is of the form $Mdx + Ndy = 0$

Therefore, $\frac{\partial M}{\partial y}=3xy^2+1,\quad \frac{\partial N}{\partial x}=4xy^2+2.$

Here $$\frac{\frac{\partial N}{\partial x}-\frac{\partial M}{\partial y}}{M}=\frac{xy^2+1}{xy^3+y}=\frac{1}{y}$$

$\therefore$ Integrating factor $= e^{\int 1/y\,dy}=e^{\log y}=y.$

Multiplying the given equation by y, we get

$(xy^4 + y^2)\,dx + 2(x^2y^3 + xy + y^5)\,dy = 0$ which is exact.

$\therefore$ The solution is $\int[xy^4+y^2]\,dx+\int 2y^5dy=C$

or $$\frac{x^2y^4}{2}+xy^2+\frac{2y^6}{6}=C$$ where C is an arbitrary constant.

which is the required solution. **Ans.**

28. *Solve* $(x^2y^2 + xy + 1)\,y\,dx + (x^2y^2 - xy + 1)\,x\,dy = 0.$

SOLUTION: Equation is of the form $f_1(xy)\,ydx + f_2(xy)\,xdy = 0$

$$\text{I.F.}=\frac{1}{Mx-Ny}=\frac{1}{xy(x^2y^2+xy+1-x^2y^2+xy-1)}=\frac{1}{2x^2y^2}$$

Multiplying the equation throughout by $\frac{1}{x^2y^2}$, we get

$$\left(1+\frac{1}{xy}+\frac{1}{x^2y^2}\right)ydx+\left(1-\frac{1}{xy}+\frac{1}{x^2y^2}\right)xdy=0$$

or $$\left(y+\frac{1}{x}+\frac{1}{x^2y}\right)dx+\left(x-\frac{1}{y}+\frac{1}{xy^2}\right)dy=0$$ which is exact.

Its solution is

$$\int\left[y+\frac{1}{x}+\frac{1}{x^2y}\right]_{y=\text{constant}}dx+\int-\frac{1}{y}dy=0$$

or $$xy+\log x-\frac{1}{xy}-\log y=C \quad \text{or} \quad xy-\frac{1}{xy}+\log\left(\frac{x}{y}\right)=C$$

which is the required solution where C is an arbitrary constant. **Ans.**

29. *Solve* $y(2xy+1)\,dx+x(1+2xy-x^3y^3)\,dy=0$.

SOLUTION: Equation is of the form $f_1(xy)\,ydx+f_2(xy)\,xdy=0$

$\therefore$ $$\text{I.F.}=\frac{1}{Mx-Ny}=\frac{1}{xy(2xy+1-1-2xy+x^3y^3)}=\frac{1}{x^4y^4}$$

Multiplying the equation throughout by $\frac{1}{x^4y^4}$, we get

$$y\left(\frac{2}{x^3y^3}+\frac{1}{x^4y^4}\right)dx+x\left(\frac{1}{x^4y^4}+\frac{2}{x^3y^3}-\frac{1}{xy}\right)dy=0$$

or $$\left(\frac{2}{x^3y^2}+\frac{1}{x^4y^3}\right)dx+\left(\frac{1}{x^3y^4}+\frac{2}{x^2y^3}-\frac{1}{y}\right)dy=0$$ which is exact.

Therefore the solution is

$$\int\left[\frac{2}{x^3y^2}+\frac{1}{x^4y^3}\right]_{y=\text{constant}}dx+\int-\frac{1}{y}\,dy=C$$

or $$\frac{-1}{x^2y^2}-\frac{1}{3x^3y^3}-\log y-C=0$$

or $$\frac{1}{x^2y^2}+\frac{1}{3x^3y^3}+\log y+C=0$$

is the requried solution where C is constant of integration. **Ans.**

SOLUTIONS TO EXERCISE 9C

1. *Solve the differential equation* $xp - y + e^p = 0$, where $p = \frac{dy}{dx}$.

SOLTUTION: $xp - y + e^p = 0,\ p = \frac{dy}{dx}$ or $y = px + e^p$. It is of Claraut's form $y = px + f(p)$.

Hence its solution is $y = cx + e^c$ where C is constant of integration. **Ans.**

2. *Solve the differential equation* $xyp^2 + p(3x^2 - 2y^2) - 6xy = 0$, *where* $p = \frac{dy}{dx}$.

SOLUTION: $xy\,p^2 + p(3x^2 - 2y^2) - 6xy = 0$. It is solvable for p.

$$p = \frac{2y^2 - 3x^2 \pm \sqrt{(3x^2 - 2y^2)^2 + 24x^2y^2}}{2xy} = \frac{2y^2 - 3x^2 \pm (3x^2 + 2y^2)}{2xy} = \frac{2y}{x}, \frac{-3x}{y}$$

Case I. $\frac{dy}{dx} = \frac{2y}{x}$ or $\frac{dy}{y} = \frac{2dx}{x} \Rightarrow \log y = 2\log x + \log c$ or $y = Cx^2$.

Case II. $\frac{dy}{dx} = \frac{-3x}{y}$ or $ydy + 3x\,dx = 0 \Rightarrow y^2 + 3x^2 = c$

$\therefore$ Solution is $(y - Cx^2)(y^2 + 3x^2 - c) = 0.$ **Ans.**

3. *Solve the differential equation* $x + \frac{p}{\sqrt{1+p^2}} = a$ *where* $p = \frac{dy}{dx}$.

SOLUTION: It is solvable for x as $x = a - \frac{p}{\sqrt{1+p^2}}$

$$\therefore \quad \frac{dx}{dy} = 0 - \frac{\sqrt{1+p^2}\cdot 1 - p\cdot\frac{p}{\sqrt{1+p^2}}}{(1+p^2)}\frac{dp}{dy} \quad \text{or} \quad -\frac{1}{p} = \frac{\frac{dp}{dy}}{(1+p^2)^{3/2}}$$

or $dy = \frac{-pdp}{(1+p^2)^{3/2}}$ which on integration, gives $y = \frac{-1}{\sqrt{1+p^2}} + b.$

Thus $x - a = \frac{-p}{\sqrt{1+p^2}}$ and $y - b = \frac{-1}{\sqrt{1+p^2}}$

$\Rightarrow \quad (x-a)^2 + (y-b)^2 = 1$ **Ans.**

4. *Solve the differential equation* $2p^2y - p^3x + 16x^2 = 0$ *where* $p = \frac{dy}{dx}$.

SOLUTION: The equation can be written as $y = \frac{px}{2} - \frac{8x^2}{p^2}$. Differentiating w.r.t. x, we get

$$\therefore \quad p = \frac{dy}{dx} = \frac{p}{2} + \frac{x}{2}\frac{dp}{dx} - \frac{16x}{p^2} + \frac{16x^2}{p^3}\frac{dp}{dx}$$

or $$\frac{p}{2}+\frac{16x}{p^2} = \frac{dp}{dx}\left(\frac{x}{2}+\frac{16x^2}{p^3}\right)=\frac{x}{p}\left(\frac{p}{2}+\frac{16x}{p^2}\right)\frac{dp}{dx}$$

or $$\left[\frac{p}{2}+\frac{16x}{p^2}\right]\left(1-\frac{x}{p}\frac{dp}{dx}\right)=0 \quad \Rightarrow \quad \frac{xdp}{pdx}=1.$$

or $\dfrac{dp}{p} = \dfrac{dx}{x}$ which on integration, gives

$\log p = \log x + \log c$ or $p = cx$ where c is constant of integration.

Now substituting $p = cx$ in the given equation, we get

$$y=\frac{cx^2}{2}-\frac{8x^2}{c^2x^2} \quad \text{or} \quad y=\frac{cx^2}{2}-\frac{8}{c^2} \quad \text{or} \quad c^3x^2-2c^2y=16$$ **Ans.**

5. *Solve the differential equations* $p^3+mp^2=y+mx$ *where* $p=\dfrac{dy}{dx}$.

SOLUTION: Differentiating w.r.t. x, $y + mx = p^3 + mp^2$ ...(1)

we get $$p+m=(3p^2+2mp)\frac{dp}{dx} \quad \text{or} \quad dx=\frac{3p^2+2mp}{p+m}dp$$

or $$x=\int\frac{3p^2+2mp}{p+m}dp=\int\left(3p-m+\frac{m^2}{p+m}\right)dp$$

or $$x=\frac{3}{2}p^2-mp+m^2\log(p+m)+C \qquad ...(2)$$

The relations (1) and (2) together represent the solution. **Ans.**

6. *Solve the differential equation* $p=\tan\left(x-\dfrac{p}{1+p^2}\right)$ *where* $p=\dfrac{dy}{dx}$.

SOLUTION: Equation can be written as $x-\dfrac{p}{1+p^2}=\tan^{-1}p$ ...(1)

Differentiating both sides of (1) w.r.t. y, we get

$$\frac{dx}{dy}-\frac{1+p^2-2p^2}{(1+p^2)^2}\frac{dp}{dy}=\frac{1}{1+p^2}\frac{dp}{dy} \quad \text{or} \quad \frac{1}{p}=\left[\frac{1-p^2}{(1+p^2)^2}+\frac{1}{1+p^2}\right]\frac{dp}{dy}$$

or $$\frac{1}{p}=\frac{2}{(1+p^2)^2}\frac{dp}{dy} \quad \text{or} \quad dy=\frac{2p\,dp}{(1+p^2)^2}$$

$\therefore$ $$y=\int\frac{2pdp}{(p^2+1)^2}=\frac{-1}{p^2+1}+c \quad \text{or} \quad p^2+1=\frac{1}{c-y}$$

$\therefore$ $$p=\sqrt{\frac{1}{c-y}-1}=\sqrt{\frac{1-c+y}{c-y}} \qquad ...(2)$$

Using (2) in (1), gives

$$\tan\left[x-\sqrt{\frac{1-c+y}{c-y}}\,(c-y)\right]=\sqrt{\frac{1-c-y}{c-y}}$$ where C is an arbitrary constant.

or $$\tan\left[x-\sqrt{c-y-(c-y)^2}\right]=\sqrt{\frac{1-c-y}{c-y}}$$ is the requried solution. **Ans.**

7. *Solve the differential equation* $p^2(x^2-1)-2p\,xy+y^2-1=0$ *where* $p=\dfrac{dy}{dx}$.

SOLUTION: The given equation is solvable for y.

$$y^2-2pxy+p^2(x^2-1)-1=0$$

$$\therefore\quad y=\frac{2px\pm\sqrt{4p^2x^2+4-4p^2(x^2-1)}}{2}$$

or $$y=px\pm\sqrt{1+p^2}$$ which is of Claraut's form.

It's solution is $y=cx\pm\sqrt{c^2+1}$ or $(y-cx)^2=c^2+1$

which is the requried solution. **Ans.**

8. *Solve the differential equation* $y=p\tan p+\log\cos p$ *where* $p=\dfrac{dy}{dx}$.

SOLUTION: $y=p\tan p+\log\cos p$ ...(1) Differentiating it w.r.t. x, we get

$$p=p\sec^2 p\frac{dp}{dx}+\tan p\frac{dp}{dx}-\tan p\frac{dp}{dx}.\qquad\text{or}\qquad \frac{dp}{dx}=\cos^2 p$$

or $dx=\sec^2 p\,dp$ $\therefore$ $x=\tan p+c,$ ...(2)

(1) and (2) together give the solution of the given equation. **Ans.**

9. *Solve the differential equation* $ayp^2+(2x-b)p-y=0$ *where* $p=\dfrac{dy}{dx}$.

SOLUTION: $ayp^2+(2x-b)p-y=0$

or $2x-b=\dfrac{y}{p}-ayp$ or $x=\dfrac{b}{2}+\dfrac{y}{2}\left(\dfrac{1}{p}-ap\right)$

Differentiating both sides w.r.t. y, gives

$$\frac{1}{p}=\frac{1}{2}\left(\frac{1}{p}-ap\right)+\frac{y}{2}\left(\frac{-1}{p^2}-a\right)\frac{dp}{dy}\quad\text{or}\quad y\left(\frac{1}{p^2}+a\right)\frac{dp}{dy}=-ap-\frac{1}{p}$$

or $$\frac{dy}{y}=\frac{-dp}{p}\Rightarrow \log y=-\log p+\log c\quad\text{or}\quad p=\frac{c}{y}$$

Putting $p=\dfrac{c}{y}$ in (1), gives $ay\dfrac{c^2}{y^2}+(2x-b)\dfrac{c}{y}-y=0$

or $ac^2+c(2x-b)=y^2$ or $y^2=c(2x-b)+ac^2$ is the solution. **Ans.**

10. *Solve the differential equation* $3p^2y^2 - 2xy\,p + 4y^2 - x^2 = 0$ *where* $p = \frac{dy}{dx}$.

SOLUTION: Given equation is solvable for p

$$\therefore \quad p = \frac{2xy \pm \sqrt{4x^2y^2 - 12y^2(4y^2 - x^2)}}{6y^2} = \frac{2x \pm 4\sqrt{x^2 - 3y^2}}{6y} = \frac{dy}{dx}$$

Putting $y = vx$ we get $v + x\frac{dv}{dx} = \frac{2 \pm 4\sqrt{1 - 3v^2}}{6v}$

or $x\frac{dv}{dx} = \frac{2 - 6v^2 \pm 4\sqrt{1 - 3v^2}}{6y}$ or $\frac{dx}{x} = \frac{6v\,dv}{2 - 6v^2 \pm 4\sqrt{1 - 3v^2}}$

Now put $1 - 3v^2 = t^2$ to get

$$\frac{dx}{x} = \frac{-2t\,dt}{2t^2 \pm 4t} = \frac{-dt}{t \pm 2}$$

On integrating both sides, gives $\log x = -\log(t \pm 2) + \log c$ or $x(t \pm 2) = c$

or $x\left[\sqrt{1 - 3v^2} \pm 2\right] = c$ or $\sqrt{x^2 - 3y^2} = c \pm 2x$

or $x^2 - 3y^2 = c^2 + 4x^2 \pm 4Cx$ or $3(x^2 + y^2) + c^2 \pm 4Cx = 0$

which is the required solution. **Ans.**

11. *Solve the differential equation* $x^2 = 1 + p^2$ *where* $p = \frac{dy}{dx}$.

SOLUTION: $p^2 = x^2 - 1$ or $p = \pm\sqrt{x^2 - 1} = \frac{dy}{dx}$. or $dy = \pm\sqrt{x^2 - 1}\,dx$

$\therefore \quad y = \pm\frac{x}{2}\sqrt{x^2 = 1} \pm \frac{1}{2}\log\left(x + \sqrt{x^2 - 1}\right) + C$ is the required solution. **Ans.**

12. *Solve the differential equation* $y = 2px + yp^2$ *where* $p = \frac{dy}{dx}$.

SOLUTION: It is solvable for p hence we have

$$p = \frac{-2x \pm \sqrt{4x^2 + 4y^2}}{2y} = \frac{-x \pm \sqrt{x^2 + y^2}}{y} = \frac{dy}{dx}$$

Putting $y = vx$ here, gives $v + x\frac{dv}{dx} = \frac{-1 \pm \sqrt{1 + v^2}}{v}$

or $x\frac{dv}{du} = \frac{-1 \pm \sqrt{1 + v^2}}{v} - v = \frac{-(1 + v^2) \pm \sqrt{1 + v^2}}{v}$

or $\frac{dx}{x} = \frac{vdv}{-(1 + v^2) \pm \sqrt{1 + v^2}}$ Now putting $1 + v^2 = t^2$, we get

$$\frac{dx}{x} = \frac{dt}{-t \pm 1} \Rightarrow \log x = -\log\left(\sqrt{1 + v^2} \pm 1\right) + \log c$$

or $\sqrt{1 + v^2} \pm 1 = \frac{c}{x}$ or $y^2 = c^2 \pm 2cx$ which is required solution. **Ans.**

SOLUTIONS TO EXERCISE 9D

1. *A particle is projected vertically upwards in the gravitational field under a resistance equal to k times the square of its velocity per unit mass. If V_0 is the initial velocity find the maximum height attained.*

SOLUTION: By Newton's second law of motion $mv\frac{dv}{dx} = -mg - Kv^2m$ or $\frac{vdv}{-g-Kv^2} = dx$

$$\Rightarrow \quad -\int_{V_0}^{0}\frac{vdv}{g+Kv^2} = \int_0^h dx \text{ or } h = \int_0^{V_0}\frac{vdv}{g+kv^2} = \frac{1}{k}\int_0^{V_0}\frac{vdv}{v^2+\frac{g}{k}}$$

or
$$h = \frac{1}{2k}\log\left(v^2+\frac{g}{k}\right)\Bigg|_0^{V_0} = \frac{1}{2k}\log\frac{V_0^2+\frac{g}{k}}{\frac{g}{k}} = \frac{1}{2k}\log\left(1+\frac{kV_0^2}{g}\right) \quad \textbf{Ans.}$$

2. *The differential equation for the motion of a particle moving in a central orbit is*

$$\frac{d^2x}{dt^2} = -\mu\left(x+\frac{a^4}{x^3}\right).$$

If it starts from test at a distance 'a', find the time it will take to arrive at the origin.

SOLUTION: Let us write $v = \frac{dx}{dt}$ then $\frac{d}{dt}\left(\frac{dx}{dt}\right) = \frac{dv}{dx}\cdot v$

$\therefore$ Equation is $v\frac{dv}{dx} = -\mu\left(x+\frac{a^4}{x^3}\right)$. On integration, we get

$$\frac{v^2}{2} = -\mu\left(\frac{x^2}{2}-\frac{a^4}{2x^2}\right)+C. \text{ Here } C = 0 \text{ as } v = 0 \text{ at } x = a$$

$$\therefore \quad v = \frac{dx}{dt} = \sqrt{\frac{a^4}{x^2}-x^2}\,\sqrt{\mu} \text{ or } \sqrt{\mu}\,dt = \frac{-xdx}{\sqrt{a^4-x^4}}$$

or $\sqrt{\mu}\,t = \int\frac{xdx}{\sqrt{a^4-x^4}}$ Putting $x^2 = a^2 z$

$$= -\int\frac{a^2dz}{2a^2\sqrt{1-z^2}} = -\frac{1}{2}\sin^{-1}z = \frac{-1}{2}\sin^{-1}\left(\frac{x^2}{a^2}\right)$$

or $\sin(2\sqrt{\mu}\,t) = \frac{x^2}{a^2}$ $\therefore$ $x = 0$ when $t = \frac{\pi}{2\sqrt{\mu}}$. **Ans.**

3. *A car starts from rest and its acceleration at a time t is equal to* $k\left(1-\frac{t}{T}\right)$ *where k is a constant and T is the time taken to attain the highest speed which is maintained constant after that time. Find the highest speed and the distance travelled till this speed is attained.*

SOLUTION: Acceleration $= \frac{dv}{dt} = k\left(1-\frac{t}{T}\right)$, v is maximum when $\frac{dv}{dt} = 0$ so $t = T$.

On integration we get $v = k\left(t-\frac{t^2}{2T}\right)$, hence $V_{max} = k\left(1-\frac{t}{T}\right) = \frac{kT}{2}$.

As $v = \frac{ds}{dt}$ $\therefore$ Distance travelled in time t is given by

$$s = \int k\left(t-\frac{t^2}{2T}\right)dt = k\left(\frac{t^2}{2}-\frac{T^3}{6T}\right) = \frac{kT^2}{3}.$$

Thus max. Velocity $= \frac{kT}{2}$ and distance travelled to attain it $= \frac{kT^2}{3}$ **Ans.**

4. *Show that the curve for which the portion of the tangent intercepted between the co-ordinate axes is bisected at the point of contact, is a rectangular hyperbola.*

SOLUTION: Equation of tangent at (x, y) is $Y - y = \frac{dy}{dx}(X - x)$

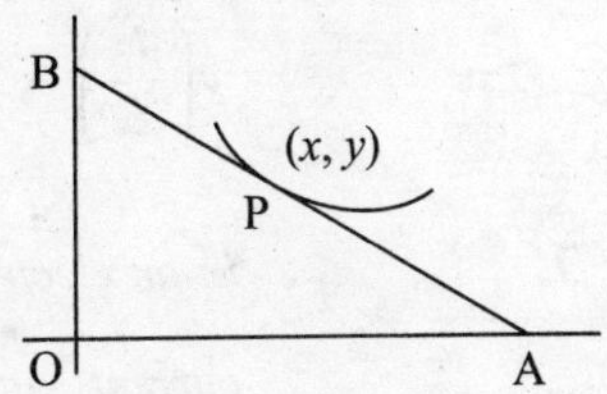

$A = \left(x - \frac{y}{\frac{dy}{dx}}, 0\right)$ and $B = \left(0, y - x\frac{dy}{dx}\right)$.

$P(x, y)$ is mid point of AB if $x = \frac{1}{2}\left(x - \frac{y}{\frac{dy}{dx}}\right)$, $y = \frac{1}{2}\left(y - x\frac{dy}{dx}\right)$

Therefore $x = \frac{-y}{\frac{dy}{dx}}$ and $y = -x\frac{dy}{dx}$ $\Rightarrow$ $x\,dy + y\,dx = 0$

or $d(xy) = 0$ $\therefore$ xy = constant, which is a rectangular hyperbola. **Hence the result.**

5. *Find the equation of the curve whose subtenant is twice the abscissa of the point of contact and passes through the point* $(1, 2)$.

SOLUTION: Subtangent $= TM = y\cot\psi = \frac{y}{\frac{dy}{dx}}$

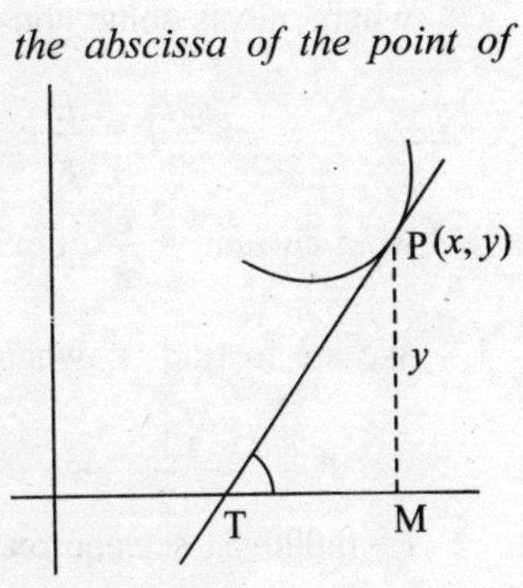

Given that $\frac{y}{\frac{dy}{dx}} = 2x$ or $\frac{2dy}{y} = \frac{dx}{x}$

$$\Rightarrow \qquad 2\log y = \log x + C.$$

It passes through (1, 2), hence $2\log 2 = 0 + C$.

Thus $\qquad 2\log y = \log x + 2\log 2 \quad \Rightarrow y^2 = 4x$ **Ans.**

6. *Find the orthogonal trajectories of a system of confocal and co-axial parabolas.*

SOLUTION: System of parabolas having same focus and same axis, say x-axis, is $y^2 = 4a(x + a)$ ...(1)

Here $2y\dfrac{dy}{dx} = 4a.$ ...(2) Eliminating 'a' in (1) and (2), we get

$$y^2 = 2xy\frac{dy}{dx} + \left(y\frac{dy}{dx}\right)^2 \qquad \text{or} \qquad y\left(\frac{dy}{dx}\right)^2 + 2x\frac{dy}{dx} - y = 0$$

To get orthogonal trajectory, we replace $\dfrac{dy}{dx}$ by $-\dfrac{dx}{dy}$ and get

$$y\left(\frac{-dx}{dy}\right)^2 - 2x\frac{dx}{dy} - y = 0$$

or
$$-y\left(\frac{dy}{dx}\right)^2 - 2x\frac{dy}{dx} + y = 0$$

or
$$y\left(\frac{dy}{dx}\right)^2 + 2x\frac{dy}{dx} - y = 0$$ which is same as above hence self orthogonal. **Hence Proved.**

7. *In an electric circuit containing an inductance L, a resistance R and a battery E, the current i builds up at a rate given by* $L\dfrac{di}{dt} + Ri = E$. *Find the current i in terms of time t. Also, calculate the time taken before the current takes half its peak value.*

SOLUTION: $L\dfrac{di}{dt} + Ri = E$ or $\dfrac{di}{dt} + \dfrac{R}{L}i = \dfrac{E}{L}$. It is linear in i.

I.F. $= e^{\int \frac{R}{L}dt} = e^{Rt/L}$ $\quad\therefore$ Solution is

$$e^{Rt/L}\, i = \int \frac{E}{L} e^{Rt/L}\, dt = \frac{E}{L}\cdot\frac{L}{R} e^{Rt/L} + C$$

where C is some constant. Since $i = 0$ at $t = 0$ $\therefore$ $C = -\dfrac{E}{R}$

$$\therefore \qquad e^{Rt/L} i = \frac{E}{R} e^{+Rt/L} - \frac{E}{R} \quad \text{or} \quad i = \frac{E}{R}\left[1 - e^{-Rt/L}\right].$$

Max current $= \dfrac{E}{R}$ obviously = peak value of current.

We are to find t when $i = \dfrac{1}{2}\dfrac{E}{R}$, i.e. $\dfrac{1}{2} = 1 - e^{-Rt/L}$

or $e^{-Rt/L} = \dfrac{1}{2}$ or $e^{Rt/L} = 2 \Rightarrow \dfrac{Rt}{L} = \log 2$ or $t = \dfrac{L}{R}\log 2$. **Ans.**

$t = 0.000931$ sec approx. **Ans.**

8. *A voltage Ee^{-at} is applied initially to an electric circuit containing inductance L and resistance R. Obtain the current developed at any time t.*

SOLUTION: $L\dfrac{di}{dt} + Ri = E\,e^{-at}$ or $\dfrac{di}{dt} + \dfrac{R}{L}i = \dfrac{E}{L}e^{-at}$. It is linear in t.

$$\text{I.F.} = e^{\int \frac{R}{L}dt} = e^{Rt/L}$$

$\therefore$ Solution is
$$i\,e^{Rt/L} = \int \frac{E}{L} e^{-at} e^{Rt/L}\,dt = \frac{E}{L}\int e^{-t(a-R/L)}\,dt$$

$$= \frac{E}{L}\frac{e^{-t(a-R/L)}}{-\left(a-\dfrac{R}{L}\right)} + C \quad \text{where } C \text{ is arbitrary constant}$$

or
$$i = \frac{E}{L}\frac{e^{-at}}{\left(\dfrac{R}{L}-a\right)} + C e^{-Rt/L}.$$

Since $i = 0$ at $t = 0$, we have $C = \dfrac{-E}{R-aL}$

$\therefore$
$$i = \frac{E}{R-aL}\left[e^{-at} - e^{-Rt/L}\right]$$ **Ans.**

9. *The temperature θ of a body satisfies the equation $\dfrac{d\theta}{dt} = -k(\theta - \theta_0)$ where θ_0 is the temperature of the surrounding medium. A thermometer which reads 75° F indoors is taken outdoors. After 5 minutes it reads 65°F and after another 5 minutes, it reads 60°F. What is the outside temperature ?*

SOLUTION: Given equation is $\dfrac{d\theta}{\theta-\theta_0} = -k\,dt \Rightarrow \log(\theta - \theta_0) = -kt + \log C$

or $\theta - \theta_0 = C e^{-kt}$. At $t = 0$, $\theta = 75°$ $\therefore$ $C = 75 - \theta_0$.

Hence $\theta - \theta_0 = (75 - \theta_0)\,e^{-kt}$. At $t = 5$, $\theta = 65°$ $\therefore$ $65 - \theta_0 = (75 - \theta_0)\,e^{-5k}$

Further at $t = 0$, $\theta = 60°$ $\therefore$ $60 - \theta_0 = (75 - \theta_0)\,e^{-10t}$

Eliminating k in the last two relations, we get

$$\frac{(65-\theta_0)^2}{(75-\theta_0)^2} = \frac{60-\theta_0}{75-\theta_0} \quad \text{or} \quad (65-\theta_0)^2 = (60-\theta_0)(75-\theta_0)$$

or $65^2 - 130\,\theta_0 = 60 \times 75 - 135\,\theta_0$ $\therefore$ $5\theta_0 = 60 \times 75 - 65^2$

Hence $\theta_0 = 55°$. **Ans.**

10. *The rate of which certain mass of radium decomposes is at any instant proportional to the mass then present and the mass decreases to half its original values in* 1500 *years. If the initial mass be* 100 mg, *in how many years will it be* 80 mg ?

SOLUTION: $\frac{dm}{dt} \propto m$ hence $\frac{dm}{dt} = km$ where k is constant of proportionality.

or $\frac{dm}{m} = k\,dt$ hence $\log m = kt + C.$ Next, at $t = 0$, $m = 100$ mg

$\therefore$ $C = \log 100.$ Thus, $\log m - \log 100 = kt$ or $m = 100\, e^{kt}$

Also, since $m = 50$ mg at $t = 1500$ years. $\therefore$ $50 = 100\, e^{1500k}$

or $1500\,k = \log\frac{1}{2}.$ We are find t so that $m = 80$ mg.

$$\therefore \quad kt = \log\left(\frac{80}{100}\right) \qquad \therefore \quad t = \frac{\log\left(\frac{8}{10}\right)}{\log\frac{1}{2}}\,1500$$ **Ans.**

11. *The equation of e.m.f. in terms of current for an electric circuit having resistance* R *and a condenser of capacity* C *in series, is* $E = Ri + \int \frac{i\,dt}{e}$. *Find the current* i *at any time* t *when* $E = E_0 \sin \omega t$.

SOLUTION: $E = Ri + \int \frac{i\,dt}{C} = E_0 \sin wt.$

$$\Rightarrow \quad R\frac{di}{dt} + \frac{i}{C} = E_0\, w \cos wt \qquad \text{or} \qquad \frac{di}{dt} + \frac{1}{RC}i = \frac{E_0 w}{R}\cos wt\,.$$

It is linear in i. Therefore, I.F. $= e^{\int \frac{1}{RC}dt} = e^{t/RC}$

$$\therefore \quad \text{Solution is } i\,e^{t/RC} = \int \frac{E_0 w}{R} e^{t/RC} \cos wt\, dt + K$$

$$\text{or} \quad i\,e^{t/RC} = \frac{E_0 w}{R}\frac{e^{t/RC}}{\sqrt{\frac{1}{R^2C^2} + w^2}} \cos\left[wt - \tan^{-1}(wRC)\right] + K$$

$$\text{or} \quad i = \frac{E_0\, wC}{\sqrt{1 + R^2C^2w^2}} \cos\left[wt - \tan^{-1} wRC\right] + Ke^{-t/RC}$$

where K is arbitrary constant. **Ans.**

ADDITIONAL IMPORTANT PROBLEMS

1. *Solve* $2x^2 y \frac{dy}{dx} = \tan(x^2 y^2) - 2xy^2$.

SOLUTION: Let us put $x^2 y^2 = t$ then $2x^2 y \frac{dy}{dx} + 2xy^2 = \frac{dt}{dx}$.

Then the given equation becomes, $\frac{dt}{dx} = \tan t$

or $\cot t \, dt = dx$. On integration we get

$\log \sin t = x + c$ or $\log \sin(x^2 y^2) = x + c$

or $\sin(x^2 y^2) = e^{x+c} = ke^x$ where k is arbitrary constant. **Ans.**

2. *Solve* $(x + y)^2 \left(x\frac{dy}{dx} + y \right) = xy\left(1 + \frac{dy}{dx}\right)$

SOLUTION: The given equation can be written as

$$\frac{x\frac{dy}{dx} + y}{xy} = \frac{1 + \frac{dy}{dx}}{(x+y)^2} \quad \text{or} \quad \frac{d(xy)}{xy} = d\left(\frac{-1}{x+y}\right)$$

which on integration gives $\log(xy) = -\frac{1}{x+y} + C$ where C is an arbitrary constant. **Ans.**

3. *Solve* $2ye^{x/y}\, dx + (y - 2xe^{x/y})\, dy = 0$.

SOLUTION: The given equation can be written as

$$\frac{dy}{dx} = \frac{-2ye^{x/y}}{y - 2xe^{x/y}}$$

we put here $y/x = v$, that is, $y = vx$

hence $\frac{dy}{dx} = v + x\frac{dv}{dx}$. The given equation becomes

$$v + x\frac{dv}{dx} = \frac{-2vx\, e^{1/v}}{vx - 2x\, e^{1/v}} = \frac{-2v\, e^{1/v}}{v - xe^{1/v}}.$$

or $$x\frac{dv}{dx} = \frac{-2ve^{1/v}}{v - 2e^{1/v}} - v = \frac{-v^2}{v - 2e^{1/v}}$$

or $\frac{v - 2e^{1/v}}{-v^2}\, dv = \frac{dx}{x}$. Integrating both sides, we get

$$\int \frac{dx}{x} = \int -\frac{1}{v}\, dv - 2\int e^{\frac{1}{v}}\left(-\frac{1}{v^2}\right) dv$$

or $\log x + c = -\log v - 2e^{1/v}$ or $-\log(y/x) - 2e^{x/y} = \log x + c$

or $\log y - 2e^{x/y} + c = 0$. **Ans.**

4. *Solve* $\frac{dy}{dx} - \frac{y}{x} + \text{cosec}\left(\frac{y}{x}\right) = 0,\ y(1) = 0.$

SOLUTION: $\frac{dy}{dx} - \frac{y}{x} = -\text{cosec}\left(\frac{y}{x}\right)$

It is of homogeneous form. Taking $y = vx$, we get $v + x\frac{dv}{dx} = v - \text{cosec } v$

or $x\frac{dv}{dx} = -\text{cosec } v$ or $\sin v\, dv = -\frac{dx}{x}$

Integrating both sides, we get

$$-\cos v = -\log x + C \quad \text{or} \quad \log x - \cos\left(\frac{y}{x}\right) = C$$ **Ans.**

5. *Solve* $\frac{dy}{dx} = \frac{2x - 5y + 3}{2x + 4y - 6}.$

SOLUTION: The equation is of reducible to homogeneous form we put $x = X + h,\ y = Y + k$

then equation becomes $\frac{dY}{dX} = \frac{2X - 5Y + 2h - 5k + 3}{2X + 4Y + 2h + 4k - 6}$

We choose h and k such that $2h - 5k + 3 = 0$ and $2h + 4k - 6 = 0$

On solving these for h and k, we get $h = 1, k = 1$.

then $\frac{dY}{dX} = \frac{2X - 5Y}{2X + 4Y}$. Now writing $Y = vX$, gives

$$v + X\frac{dv}{dX} = \frac{2 - 5v}{2 + 4v} \quad \text{or} \quad X\frac{dv}{dX} = \frac{2 - 5v}{2 + 4v} - v = \frac{-4v^2 - 7v + 2}{2 + 4v}$$

$$\therefore \quad -\int \frac{dX}{X} = \int \frac{2 + 4v}{(4v - 1)(v + 2)}\, dv$$

or $$-\int \frac{dX}{X} = \int \left[\frac{4}{3(4v - 1)} + \frac{2}{3(v + 2)}\right] dv$$

or $$-\log X + \log C = \frac{4}{3}\cdot\frac{1}{4}\log(4v - 1) + \frac{2}{3}\log(v + 2)$$

or $$\log\frac{C}{X} = \frac{1}{3}\log[(4v - 1)(v + 2)^2]$$

or $$\frac{C}{X} = (4v - 1)^{1/3}(v + 2)^{2/3} \quad \text{or} \quad \frac{C}{X} = \left(\frac{4Y - X}{X}\right)^{1/3}\left(\frac{Y + 2X}{X}\right)^{2/3}$$

or $$(4Y - X)^{1/3}(Y + 2X)^{2/3} = C \quad \text{or} \quad (4Y - X)(Y + 2X)^2 = C^3 = k, \text{ say.}$$

Replacing X by $x - 1$ and Y by $y - 1$, we get

$(4y - x - 3)(y + 2x - 3)^2 = k$ where k is arbitrary constant. **Ans.**

6. *Solve* $(x\sqrt{1-x^2y^2} - y)\,dy + (x + y\sqrt{1-x^2y^2})\,dx = 0.$

SOLUTION: Equation is $Mdx + Ndy = 0$ where

$$M = x + y\sqrt{1-x^2y^2} \quad \text{and} \quad N = x\sqrt{1-x^2y^2} - y$$

$$\therefore \quad \frac{\partial M}{\partial y} = \sqrt{1-x^2y^2} + \frac{y}{2}\frac{(-2x^2y)}{\sqrt{1-x^2y^2}} = \sqrt{1-x^2y^2} - \frac{x^2y^2}{\sqrt{1-x^2y^2}}$$

$$\text{and} \quad \frac{\partial N}{\partial x} = \sqrt{1-x^2y^2} + \frac{x(-2xy^2)}{2\sqrt{1-x^2y^2}} = \sqrt{1-x^2y^2} - \frac{x^2y^2}{\sqrt{1-x^2y^2}}$$

Since $\frac{\partial M}{\partial y} = \frac{\partial N}{\partial x}$ hence the equation is exact. It's solution is

$$\int Mdx + \int_{\text{trms not constaing } x} Ndy = C \quad \text{where } C \text{ is constant of integration.}$$

or $$\int\left(x + y\sqrt{1-x^2y^2}\right)dx + \int(0 - y)\,dy = C$$

or $$\frac{x^2}{2} + y^2\int\sqrt{\frac{1}{y^2} - x^2}\,dx - \frac{y^2}{2} = C$$

or $$\frac{x^2}{2} + y^2\left[\frac{x}{2}\sqrt{\frac{1}{y^2} - x^2} + \frac{1}{2y^2}\sin^{-1}\left(\frac{x}{1/y}\right)\right] - \frac{y^2}{2} = C$$

or $$\frac{x^2 - y^2}{2} + \frac{xy}{2}\sqrt{1-x^2y^2} + \frac{1}{2}\sin^{-1}(xy) = C$$

or $$x^2 - y^2 + xy\sqrt{1-x^2y^2} + \sin(xy) = 2C$$ **Ans.**

7. *Solve* $\frac{dy}{dx} + \frac{y\cos x + \sin y + y}{\sin x + x\cos y + x} = 0.$

SOLUTION: Equation is of the form $Mdx + Ndy = 0$

where $M = y\cos x + \sin y + y, \; N = \sin x + x\cos y + x.$

$$\therefore \quad \frac{\partial M}{\partial y} = \cos x + \cos y + 1 \quad \text{and} \quad \frac{\partial N}{\partial x} = \cos x + \cos y + 1$$

Since $\frac{\partial M}{\partial y} = \frac{\partial N}{\partial x}$ the equation is exact. The solution is

$$\int_{y-\text{constt.}} Mdx + \int_{\substack{\text{terms not}\\ \text{containg } x}} Ndy = C$$

or $$\int(y\cos x + \sin y + y)\,dx + \int 0\,dy = C$$

$$y\sin x + x\sin y + xy = C \quad \text{where } C \text{ is arbitrary constant.}$$ **Ans.**

8. *Solve $(x^2 + y^2 + 1)\,dx - 2xy\,dy = 0$.*

SOLUTION: Equation is $Mdx + Ndy = 0$ where $M = x^2 + y^2 + 1$ and $N = -2xy$.

Hence $\dfrac{\partial M}{\partial y} = 2y,\ \dfrac{\partial N}{\partial x} = -2y$. Since $\dfrac{\partial M}{\partial y} \neq \dfrac{\partial N}{\partial x}$ the equation is not exact.

Now
$$\frac{\dfrac{\partial M}{\partial y} - \dfrac{\partial N}{\partial x}}{N} = \frac{2y-(-2y)}{-2xy} = -\frac{2}{x}$$

$$\therefore \qquad \text{I.F.} = e^{\int -\frac{2}{x}dx} = e^{-2\log x} = \frac{1}{x^2}$$

Multiplying throughout by I.F. we get

$$\left(1 + \frac{y^2}{x^2} + \frac{1}{x^2}\right)dx - \frac{2y}{x}\,dy = 0 \text{ which is exact.}$$

$\therefore$ The solution is $\underset{y-\text{constant}}{\int Mdx} + \underset{\substack{\text{terms not}\\ \text{containing } x}}{\int Ndy} = C$, a constant.

$$x - \left(\frac{y^2+1}{x}\right) = C \quad \text{or} \quad x^2 - y^2 - 1 = Cx \text{ where } C \text{ is arbitrary constant.}$$ **Ans.**

9. *Solve $(2x \log x - xy)\,dy + 2y\,dx = 0$.*

SOLUTION: Here $M = 2y$ and $N = 2x \log x - xy$.

Therefore $\dfrac{\partial M}{\partial y} = 2$ and $\dfrac{\partial N}{\partial x} = 2\log x + 2 - y$

Since $\dfrac{\partial M}{\partial y} \neq \dfrac{\partial N}{\partial x}$ equation is not exact.

Next,
$$\frac{\dfrac{\partial M}{\partial y} - \dfrac{\partial N}{\partial x}}{N} = \frac{2 - 2\log x - 2 + y}{2x\log x - xy} = -\frac{1}{x}$$

$$\therefore \qquad \text{I.F.} = e^{\int -\frac{1}{x}dx} + e^{-\log x} = \frac{1}{x}.$$

Multiplying throughout by the I.F., equation becomes $\dfrac{1}{x}(2y)\,dx + (2\log x - y)\,dy = 0$ which is exact

Hence the solution is

$$\underset{y \text{ cosstant}}{\int Mdx} + \underset{\text{terms not involving } x}{\int Ndy} = C$$

or $\displaystyle\int \frac{2y}{x}dx + \int -y\,dy = C$ or $2y\log x - \dfrac{y^2}{2} = C$ **Ans.**

10. *Solve* $xe^x (dx - dy) + e^x dx + ye^y dy = 0.$

SOLUTION: Equation is $(xe^x + e^x)\, dx + (ye^y - xe^x)\, dy = 0$

Here $M = (x + 1)\, e^x$ and $N = ye^y - xe^x$

$$\therefore \quad \frac{\partial M}{\partial y} = 0 \quad \text{and} \quad \frac{\partial N}{\partial x} = -xe^x - e^x.$$

Since $\dfrac{\partial M}{\partial y} \neq \dfrac{\partial N}{\partial x}$ the equation is not exact.

$$\frac{\dfrac{\partial N}{\partial x} - \dfrac{\partial M}{\partial y}}{M} = \frac{-xe^x - e^x - 0}{xe^x + e^x} = -1 \quad \therefore \quad \text{I.F.} = e^{\int -1\, dy} = e^{-y}.$$

Multiplying throughout by I.F. we get $e^{-y}(x+1)e^x\, dx + e^{-y}(ye^y - xe^x)\, dy = 0$

or $$e^{x-y}(x+1)\, dx + (y - x^{\,x-y})\, dy = 0 \quad \text{which is exact.}$$

Solution is $$\underset{y\,\text{constt.}}{\int} Mdx + \underset{\substack{\text{terms not}\\ \text{involving } x}}{\int} Ndy = C$$

or $$\int e^{-y}(xe^x + e^x)\, dx + \int y dy = C$$

or $$e^{-y}(xe^x - e^x + e^x) + \frac{y^2}{2} = C \quad \text{or} \quad xe^{x-y} + \frac{y^2}{2} = C.$$ **Ans.**

11. *Solve* $(xy \sin xy + \cos xy)\, ydx + (xy \sin xy - \cos xy)\, xdy = 0.$

SOLUTION: Equation is of the form $f_1(xy)\, ydx + f_2(xy)\, xdy = 0$

$$\text{I.F.} = \frac{1}{Mx - Ny} = \frac{1}{xy(xy \sin xy + \cos xy) - xy(xy \sin xy - \cos xy)} = \frac{1}{2xy \cos xy}$$

Multiplying throughout by I.F., we get

$$\frac{1}{2xy \cos xy}(xy \sin xy + \cos xy)\, ydx + \frac{1}{2xy \cos xy}(xy \sin xy - \cos xy)\, x\, dy = 0$$

or $$\left(\frac{y}{2} \tan xy + \frac{1}{2x}\right) dx + \left(\frac{x \tan y}{2} - \frac{1}{2y}\right) dy = 0 \quad \text{which is exact.}$$

Solution is $$\underset{y\text{constt.}}{\int} Mdx + \underset{\text{terms not involving } x}{\int} Ndy = C$$

or $$\frac{1}{2}\int \left(y \tan xy + \frac{1}{x}\right) dx - \int \frac{1}{2y}\, dy = 0$$

or $$\frac{1}{2}\left(\frac{y}{y} \log \sec xy + \log x\right) - \frac{1}{2} \log y = C$$

or $$\log (x \sec xy) - \log y = 2C \quad \text{or} \quad \log\left(\frac{x}{y} \sec xy\right) = 2C$$

or $$\frac{x}{y} \sec xy = e^{2c} = k \quad \text{or} \quad \frac{x}{y} \sec xy = k$$

where k is constant of integration.

12. *Solve $(x^7y^2 + 3y)\,dx + (3x^8\,y - x)\,dy = 0$.*

SOLUTION: The equation is $Mdx + Ndy = 0$ where $M = x^7\,y^2 + 3y,\ N = 3x^8\,y - x$.

$$\frac{\partial M}{\partial y} = 2x^7\,y + 3 \quad \text{and} \quad \frac{\partial N}{\partial x} = 24x^7\,y - 1. \text{ Since } \frac{\partial M}{\partial y} \neq \frac{\partial N}{\partial x}, \text{ the equation not exact.}$$

Equation can be written as $x^7y\,(ydx + 3xdy) + (3ydx - xdy) = 0$

Multiplying throughout by $x^h\,y^k$ and choosing h and k so that equation becomes exact.

Then $(x^{7+h}\,y^{2+k} + 3x^hy^{k+1})dx + (3x^{8+h}\,y^{k+1} - x^{h+1}\,y^k)dy = 0$

hence $\dfrac{\partial M}{\partial y} = (2 + k)\,x^{7+h}\,y^{k+1} + 3(k + 1)x^hy^k$ and $\dfrac{\partial N}{\partial x} = 3(8 + h)x^{7+h}y^{k+1} - (h + 1)x^hy^k$

Since $\dfrac{\partial M}{\partial y} = \dfrac{\partial N}{\partial x}$ we have $2 + k = 3(8 + h)$ and $3(k + 1) = -(h + 1)$

$\Rightarrow \qquad h = -7,\ k = 1.$

and hence I.F. $= x^{-7}\,y^1$. Multiplying by $x^{-7}\,y^1$ throughout, we get

$$(y^3 + 3x^{-7}\,y^2)\,dx + (3xy^2 - x^{-6}y)\,dy = 0$$

or $\qquad M_1\,dx + N_1\,dy = 0$ where $M_1 = y^3 + 3x^{-7}\,y^2,\ N_1 = 3xy^2 - x^{-6}\,y.$

$$\frac{\partial M_1}{\partial y} = 3y^2 + 6x^{-7}y \quad \text{and} \quad \frac{\partial N_1}{\partial x} = 3y^2 + 6x^{-7}y \text{ hence exact.}$$

$\therefore$ Solution is $\displaystyle\int_{y\text{ constt.}} M_1dx + \int_{\substack{\text{terms not}\\ \text{involving } x}} N_1dy$

or $\qquad \displaystyle\int (y^3 + 3x^{-7}\,y^2)\,dx + \int 0\,dy = C$

or $\qquad xy^3 - \dfrac{1}{2}x^{-6}y^2 = C$ **Ans.**

13. *Solve $(1 + xy)\,ydx + (1 - xy)\,x\,dy = 0$.*

SOLUTION: Equation can be written as $ydx + xdy + xy^2\,dx - x^2\,y\,dy = 0$.

Dividing throughout by $x^2\,y^2$, we get

$$\frac{ydx + xdy}{x^2y^2} + \frac{dx}{x} - \frac{dy}{y} = 0 \quad \text{or} \quad d\left(\frac{-1}{xy}\right) + \frac{dx}{x} - \frac{dy}{y} = 0$$

Integrating both sides, we get

$$-\frac{1}{xy} + \log x - \log y = C$$

or $\qquad -\dfrac{1}{xy} + \log\dfrac{x}{y} = C$ **Ans.**

14. *Solve* $y(x^3 e^{xy} - y)\, dx + x(y + x^3 e^{xy})\, dy = 0.$

SOLUTION: Equation can be written as

$$x^3 y e^{xy}\, dx + x^4 e^{xy} dy - y^2 dx + xy\, dy = 0$$

Dividing throughout by x^3, we get

$$y e^{xy}\, dx + x e^{xy} dy - \frac{1}{2}\left(\frac{y^2 \cdot 2x\, dx - x^2 \cdot 2y\, dy}{x^4}\right) = 0$$

or $$d(e^{xy}) + \frac{1}{2} d\left(\frac{y^2}{x^2}\right) = 0$$

On integration, it gives $e^{xy} + \frac{1}{2}\frac{y^2}{x^2} = C.$ **Ans.**

15. *Solve* $(1 + x + xy^2)\, dy + (y + y^3)\, dx = 0.$

SOLUTION: Given equation can be written as

$$\frac{dx}{dy}(y + y^3) + x(1 + y^2) + 1 = 0$$

or $\frac{dx}{dy} + \frac{x}{y} = -\frac{1}{y(y+1)}$ which is linear in x. I.F. $= e^{\int \frac{1}{y} dy} = e^{\log y} = y$

$\therefore$ The solution is $x(\text{I.F.}) = \int -\frac{(\text{I.F.})\, dy}{y(y^2+1)}$

or $xy = \int \frac{-y\, dy}{y(y^2+1)} + C$ or $xy = -\tan^{-1} y + C$ **Ans.**

16. *Solve* $(1 + y^2)\, dx = (\tan^{-1} y - x)\, dy.$

SOLUTION: The given equation can be written as

$$\frac{dx}{dy} + \frac{x}{1+y^2} = \frac{\tan^{-1} y}{1+y^2}$$ which is linear in x.

$$\text{I.F.} = e^{\int \frac{dy}{1+y^2}} = e^{\tan^{-1} y}$$

Hence the solution is $x \cdot e^{\tan^{-1} y} = \int e^{\tan^{-1} y} \frac{\tan^{-1} y}{1+y^2} dy + C$ Putting $\tan^{-1} y = t$ we have

$$x e^{\tan^{-1} y} = \int t e^t + C = t e^t - e^t + C = e^{\tan^{-1} y}(\tan^{-1} y - 1) + C$$

or $$x = \tan^{-1} y - 1 + C e^{-\tan^{-1} y}$$ is the solution. **Ans.**

17. *Solve* $L\frac{di}{dt} + Ri = \sin \omega t, \quad t \geq 0, \quad i(0) = 0$ *where* R, ω *and* L *are constants and* i *is the current at anytime* t.

SOLUTION: Equation can be written as $\frac{di}{dt} + \frac{R}{L} i = \frac{1}{L} \sin \omega t$ which is linear in i.

I.F. $= e^{\int \frac{R}{L} dt} = e^{Rt/L}$. $\therefore$ Solution is

$$ie^{Rt/L} = \int e^{Rt/L} \cdot \frac{1}{L} \sin \omega t \, dt = \frac{1}{L} \frac{e^{Rt/L}}{\frac{R^2}{L^2} + \omega^2} \left(\frac{R}{L} \sin \omega t - \omega \cos \omega t \right)$$

$$= \frac{e^{Rt/L}}{R^2 + L^2\omega^2} \left(\frac{R}{L} \sin \omega t - \omega \cos \omega t \right) + C$$

or $\quad i = \frac{1}{R^2 + \omega^2 L^2} (R \sin \omega t - \omega L \cos \omega t) + Ce^{-Rt/L}$

Since $\quad i = 0$ at $t = 0$ we get $C = \frac{\omega L}{R^2 + \omega^2 L^2}$

$\therefore$ The solution is $\quad i = \frac{1}{R^2 + \omega^2 t^2} [R \sin \omega t - \omega L \cos \omega t] + \frac{\omega L}{R^2 + \omega^2 L^2} e^{-Rt/2}$. **Ans.**

18. *If* $\frac{dy}{dx} = y \tan x = \sin 2x, \quad y(0) = 0,$ *show that maximum value of* y *is 1/2.*

SOLUTION: Given equation is $\frac{dy}{dx} + y \tan x = \sin 2x$, which is linear in y.

I.F. $= e^{\int \tan x \, dx} = e^{\log \sec x} = \sec x$.

$\therefore$ The solution is $\quad y \sec x = \int \sec x \cdot \sin 2x \, dx = 2 \int \sin x \, dx + C = -2 \cos x + C$

or $\quad y = -2 \cos^2 x + C \cos x$.

Since $\quad y(0) = 0, \quad \therefore \quad C = 2$

$\therefore$ Solution is $\quad y = -2 \cos^2 x + 2 \cos x$

For y to be maximum or minimum $\frac{dy}{dx} = 0$.

$\therefore \quad -4 \cos x (-\sin x) - 2 \sin x = 0$ or $2 \sin x (2 \cos x - 1) = 0$

$\Rightarrow \quad x = 0$ and $x = \cos^{-1}\left(\frac{1}{2}\right) = \frac{\pi}{3}$.

Further $\qquad \dfrac{d^2y}{dx^2} = 4 \cos 2x - 2 \cos x$

At $x = 0$, $\dfrac{d^2y}{dx^2} = 4 - 2 > 0$ hence y is minimum at $x = 0$.

At $x = \pi/3$, $\dfrac{d^2y}{dx^2} = 4\cos\dfrac{2\pi}{3} - 2\cos\dfrac{\pi}{3} = -3 < 0$ hence maximum at $x = \dfrac{\pi}{3}$. **Ans.**

19. *Solve* $xy\,(1 + xy^2)\dfrac{dy}{dx} = 1.$

SOLUTION: Equation is $\dfrac{dx}{dy} - xy = x^2y^3$. It is of Bernoulli's form. Dividing throughout by x^2, we get

$$\frac{1}{x^2}\frac{dx}{dy} - \frac{y}{x} = y^3. \text{ Putting } -\frac{1}{x} = t, \text{ gives}$$

$$\frac{dt}{dy} + ty = y^3 \quad \text{which is linear in } t. \quad \therefore \quad \text{I.F.} = e^{\int y\,dy} = e^{y^2/2}$$

and solution is $\qquad e^{y^2/2} \cdot t = \int e^{y^2/2} \cdot y^3\, dy + C$ Putting $\dfrac{y^2}{2} = v$, so $ydy = dv$

We get $\qquad e^{y^2/2} . t = \int e^v \cdot 2v\, dv + C$

or $\qquad e^v \cdot t = 2ve^v - \int 2e^v\, dv = 2e^v(v - 1) + C \qquad$ or $\qquad t = 2(v - 1) + Ce^{-v}$

or $\qquad -\dfrac{1}{x} = 2\left(\dfrac{y^2}{2} - 1\right) + C\,e^{-y^2/2}.$

or $\qquad \dfrac{1}{x} = 2 - y^2 - Ce^{-y^2/2}$ is the solution. **Ans.**

20. *Solve the equation* $\dfrac{dr}{d\theta} = r \tan\theta - \dfrac{r^2}{\cos\theta}.$

SOLUTION: The equation is of Bermoulli's form. Dividing by r^2, we get

$$\frac{1}{r^2}\frac{dr}{d\theta} - \frac{1}{r}\tan\theta = -\frac{1}{\cos\theta}. \text{ Putting } -\frac{1}{r} = t, \text{ we get}$$

$$\frac{dt}{d\theta} + t\tan\theta = -\frac{1}{\cos\theta}. \text{ which is linear in } t \quad \therefore \quad \text{I.F.} = e^{\int \tan\theta d\theta} = e^{\log(\sec\theta)} = \sec\theta$$

and the solution is $\quad t\sec\theta = \int \sec\theta\left(-\dfrac{1}{\cos\theta}\right)d\theta = -\tan\theta + C$

or $\quad \dfrac{\sec\theta}{r} = \tan\theta - C$ where C is an arbitrary constant. **Ans.**

21. *Solve* $\dfrac{dy}{dx} = \dfrac{y^3}{e^{2x} + y^2}$.

SOLUTION: Equation can be written as $\dfrac{dx}{dy} = \dfrac{e^{2x}}{y^3} + \dfrac{1}{y}$ or $e^{-2x}\dfrac{dx}{dy} - \dfrac{e^{-2x}}{y} = \dfrac{1}{y^3}$

Putting $e^{-2x} = t$ so $-2e^{-2x}\dfrac{dx}{dy} = \dfrac{dt}{dy}$, equation becomes

$$\left(-\frac{1}{2}\right)\frac{dt}{dy} - \frac{t}{y} = \frac{1}{y^3} \quad \text{or} \quad \frac{dt}{dy} + \frac{2t}{y} = -\frac{2}{y^3} \quad \text{which is linear in } t.$$

$$\therefore \quad \text{I.F.} = e^{\int \frac{2}{y} dy} = e^{2\log y} = y^2$$

and the solution is $ty^2 = \displaystyle\int \frac{-2}{y^3}\cdot y^2\, dy + C = -2\log y + C$

Hence the solution is $e^{-2x}\, y^2 = -2\log y + C$ where C is an arbitrary constant. **Ans.**

22. *Solve* $\dfrac{dz}{dx} + \dfrac{z}{x}\log z = \dfrac{z}{x^2}(\log z)^2$.

SOLUTION: Equation can be rewritten as

$$\frac{1}{z(\log z)^2}\frac{dz}{dx} + \frac{1}{x\log z} = \frac{1}{x^2}$$

On putting $-\dfrac{1}{\log z} = t$ so $\dfrac{1}{z(\log z)^2}\dfrac{dz}{dx} = \dfrac{dt}{dx}$, we get

$$\frac{dt}{dx} - \frac{t}{x} = \frac{1}{x^2} \quad \text{which is linear in } t.$$

$$\text{I.F.} = e^{\int -\frac{1}{x} dx} = e^{-\log x} = 1/x, \text{ hence the solution is}$$

$$t\left(\frac{1}{x}\right) = \int \frac{1}{x}\cdot\frac{1}{x^2}\, dx = \frac{1}{-2x^2} + C$$

or $$\frac{1}{x}\left[-\frac{1}{\log z}\right] = -\frac{1}{2x^2} + C$$

or $$\frac{1}{\log z} = \frac{1}{2x} + Cx \quad \text{where } C \text{ is arbitrary constant.}$$ **Ans.**

23. *Solve* $\dfrac{dy}{dx} + x(x+y) = x^3(x+y)^3 - 1.$

SOLUTION: The equation is $\dfrac{dy}{dx} + x(x+y) = x^3(x+y)^3 - 1$

Putting $x + y = z$ $\quad \therefore \quad 1 + \dfrac{dy}{dx} = \dfrac{dz}{dx}$, the equation becomes

$$\frac{dz}{dx} - 1 + xz = x^3z^3 - 1 \quad \text{or} \quad \frac{dz}{dx} + xz = x^3z^3$$

or $\dfrac{1}{z^3}\dfrac{dz}{dx} + \dfrac{x}{z^2} = x^3$. Putting here $\dfrac{1}{z^2} = t$, we get

$$-\frac{1}{2}\frac{dt}{dx} + xt = x^3 \quad \text{or} \quad \frac{dt}{dx} - 2xt = -2x^3 \text{ which is linear in } t.$$

$$\text{I.F.} = e^{\int -2x\,dx} = e^{-x^2}$$

$\therefore$ Solution is $te^{-x^2} = \int e^{-x^2}(-2x^3)\,dx$ putting here $x^2 = v$, so $2xdx = dv$.

or $$e^{-x^2}t = -\int e^{-v} v\,dv = ve^{-v} + e^{-v} + C$$

$$= (x^2 + 1)e^{-x^2} + C$$

or $$t = x^2 + 1 + Ce^{x^2} \quad \text{but } t = \frac{1}{z^2} = \frac{1}{(x+y)^2}$$

$\therefore$ Solution is $\dfrac{1}{(x+y)^2} = x^2 + 1 + Ce^{-x^2}$ where C is constant of integration. **Ans.**

CHAPTER

10 Linear Differential Equations of Higher Order and Applications

IMPORTANT FORMULAE AND USEFUL RESULTS

1. The equation to be solved is $\dfrac{d^n y}{dx^n}+a_1\dfrac{d^{n-1}y}{dx^{n-1}}+a_2\dfrac{d^{n-2}y}{dx^{n-2}}+\ldots+a_n y = \phi(x)$

or, $(D^n + D^{n-1} + D^{n-2} + \ldots + a_n)\, y = \phi(x)$ or, $f(D)\, y = \phi(x)$ where $D \equiv \dfrac{d}{dx}$

The solution is y = complementary function + particular integral = C.F + P.I.

For C.F we write auxiliary equation $(A.E.)$ as $f(m) = 0$ whose roots are $m = m_1, m_2, \ldots, m_n$.

C.F. $= c_1\, e^{m_1 x} + c_2\, e^{m_2 x} + \ldots c_n\, e^{m_n x}$ if roots are all real and distinct

If $m_1 = m_2$ then C.F $= (c_1 x + c_2)\, e^{m_1 x}$, etc.

Next, P.I. $= \dfrac{1}{f(D)}\phi(x)$. Here $\dfrac{1}{D-a}X = e^{ax}\int X e^{-ax} dx$.

Further, $\dfrac{1}{f(D)}e^{ax} = \dfrac{1}{f(a)}e^{ax}$, where $f(a) \neq 0$

$$\frac{1}{f(D^2)}\begin{matrix}\sin\\ \cos\end{matrix}\, ax = \frac{1}{f(-a^2)}\begin{matrix}\sin\\ \cos\end{matrix}\, ax \quad \text{if } f(-a^2) \neq 0.$$

$$\frac{1}{f(D)}x^m = [f(D)]^{-1}x^m \quad \text{and} \quad \frac{1}{f(D)}e^{ax}V = e^{ax}\frac{1}{f(D+a)}V.$$

$$\frac{1}{f(D)}(xV) = x\frac{1}{f(D)}V + \left(\frac{d}{dD}\frac{1}{f(D)}\right)V.$$

2. **Method of Variation of Parameters–Explained in the Text.**

3. **Cauchys Homogeneous Linear Equation.**

Equation is of the form $x^n \frac{d^n y}{dx^n} + a_1 x^{n-1} \frac{d^{n-1} y}{dx^{n-1}} + ... + a_n y = X(x)$.

Putting $x = e^t$, we get $x \frac{dy}{dx} = Dy$, $x^2 \frac{d^2 y}{dx^2} = D(D-1)y$,

$x^3 \frac{d^3 y}{dx^3} = D(D-1)(D-2)y$ where $D \equiv \frac{d}{dt}$. Thus the given homogenous equation becomes linear differential equation with constant coefficients.

4. **Simulataneous Linear Equations.** The form is

$$f_1(D)x + f_2(D)y = \varphi_1(t)$$

$$g_1(D)x + g_2(D)y = \varphi_2(t) \text{ where } D \equiv \frac{d}{dt}.$$

By eliminating one dependent variable either x or y we get differential equation with one dependent variable and one independent variable t. Then it is solved by usual method.

SOLUTIONS TO EXERCISE 10A

1. *Solve* $(D^3 + 7D^2 + 16D + 10)\,y = 0$ *where* $D \equiv \dfrac{d}{dx}$.

SOLUTION: A.E. is $m^3 + 7m^2 + 16\,m + 10 = 0$

or $(m + 1)\,(m^2 + 6\,m + 10) = 0$. Its roots are $m = -1, -3 \pm i$

$\therefore$ solution is $y = C_1e^{-x} + e^{-3x}\,(C_2 \cos x + C_3 \sin x)$ **Ans.**

2. *Solve* $\dfrac{d^4y}{dx^4} - \dfrac{d^3y}{dx^3} - 9\dfrac{d^2y}{dx^2} - 11\dfrac{dy}{dx} - 4y = 0.$

SOLUTION: A.E. is $m^4 - m^3 - 9m^2 - 11m - 4 = 0$

or $(m + 1)\,(m^3 - 2\,m^2 - 7m - 4) = 0$ or $(m + 1)\,(m + 1)\,(m^2 - 3m - 4) = 0$

or $(m + 1)^3\,(m - 4) = 0$. Hence roots are $m = -1, -1, -1, 4$

$\therefore$ Solution is $y = (C_1x^2 + C_2x + C_3)e^{-x} + C_4e^{4x}$ **Ans.**

3. *Solve* $(D^4 + D^2 + 1)\,y = 0$ *where* $D \equiv \dfrac{d}{dx}$.

SOLUTION: A.E. is $m^4 + m^2 + 1 = 0$ or $(m^2 + 1^2) - m^2 = 0$

or $m^2 + m + 1= 0$ and $m^2 - m + 1 = 0$ which give $m = \dfrac{\pm 1 \pm i\sqrt{3}}{2}$.

$\therefore$ Solution is $y = e^{x/2}\left(C_1 \cos\dfrac{\sqrt{3}}{2}x + C_2 \sin\dfrac{\sqrt{3}}{2}x\right) + e^{-x/2}\left(C_3 \cos\dfrac{\sqrt{3}}{2}x + C_4 \sin\dfrac{\sqrt{3}}{2}x\right)$

4. *Solve* $\dfrac{d^3y}{dx^3} - \dfrac{d^2y}{dx^2} - 18y = 2^x.$

SOLUTION: Equation is $(D^3 - D^2 - 18)y = 2^x$.

The A.E. is $m^3 - m^2 - 18 = 0$ or $(m - 3)\,(m^2 + 2m + 6) = 0$ $\therefore$ $m = 3,\ -1 \pm i\sqrt{5}$

$\therefore$ $\text{C.F.} = C_1e^{3x} + e^{-x}\left(C_2 \cos\sqrt{5}\,x + C_3 \sin\sqrt{5}x\right)$

and $\text{P.I.} = \dfrac{1}{D^3 - D^2 - 18}2^x = \dfrac{1}{D^3 - D^2 - 18}e^{\log(2^x)} = \dfrac{1}{D^3 - D^2 - 18}e^{x\log 2}$

$$= \frac{1}{(\log 2)^3 - (\log 2)^2 - 18}e^{x\log 2} = \frac{1}{(\log 2)^3 - (\log 2)^2 - 18}2^x$$

$\therefore$ Complete solution is

$$y = C_1 e^{3x} + e^{-x}\left(C_2 \cos\sqrt{5}\,x + C_3 \sin\sqrt{5}x\right) + \frac{1}{(\log 2)^3 - (\log 2)^2 - 18}(2^x).$$ **Ans.**

5. *Solve* $\dfrac{d^3y}{dx^3} - y = (1 + e^x)^2.$

SOLUTION: Equation is $(D^3 - 1)\,y = (1 + e^x)^2$ where $D \equiv \dfrac{d}{dx}$.

The A.E. is $m^3 - 1 = 0$ $\therefore$ $m = 1, \dfrac{-1 \pm i\sqrt{3}}{2}$

$$\therefore \quad \text{C.F.} = C_1 e^{-x} + e^{-x/2}\left(C_2 \cos\frac{\sqrt{3}}{2}x + C_3 \sin\frac{\sqrt{3}}{2}x\right)$$

$$\text{P.I.} = \frac{1}{D^3 - 1}(1 + 2e^x + e^{2x}) = \frac{1}{D^3 - 1}e^{ox} + \frac{1}{D^3 - 1}e^{2x} + \frac{2}{D^3 - 1}e^x$$

$$= \frac{1}{0-1}e^{ox} + \frac{1}{8-1}e^{2x} + 2\frac{x}{3D^2}e^x = -1 + \frac{1}{7}e^{2x} + \frac{2x}{3}e^x$$

$\therefore$ Complete solution is

$$y = C_1 e^{-x} + e^{-x/2}\left(C_2 \cos\frac{\sqrt{3}}{2}x + C_3 \sin\frac{\sqrt{3}}{2}x\right) - 1 + \frac{2x}{3}e^x + \frac{1}{7}e^{2x}$$ **Ans.**

6. *Solve* $\dfrac{d^4y}{dx^4} - y = \cosh x \sinh x.$

SOLUTION: Equation is $(D^4 - 1)y = (1 + e^x)^2$ where $D \equiv \dfrac{d}{dx}$.

The A.E. is $m^4 - 1 = 0$. Its roots are $m = \pm 1, \pm i$

$$\therefore \quad \text{C.F.} = c_1 e^x + c_2 e^{-x} + c_3 \sin x + c_4 \cos x$$

$$\text{P.I.} = \frac{1}{D^4 - 1}\cosh x \sinh x = \frac{1}{D^4 - 1}\frac{1}{2}\sinh 2x = \frac{1}{4(D^4 - 1)}(e^{2x} - e^{-2x})$$

$$= \frac{1}{4(D^4 - 1)}e^{2x} - \frac{1}{4(D^4 - 1)}e^{-2x}$$

$$= \frac{1}{4(2^4 - 1)}e^{2x} - \frac{1}{4[(-2)^4 - 1]}e^{-2x} = \frac{1}{60}(e^{2x} - e^{-2x}) = \frac{1}{30}\sinh 2x$$

$\therefore$ Complete solution is $y = c_1 e^x + c_2{}^{-x} + c_3 \sin x + c_4 \cos x + \dfrac{1}{30}\sinh 2x$ **Ans.**

7. *Solve* $(D-1)^2 (D^2+1)^2 y = \sin^2 \frac{x}{2} + e^x$ *where* $D \equiv \frac{d}{dx}$.

SOLUTION: The A.E. is $(m-1)^2 (m^2+1)^2 = 0 \quad \therefore \quad m = 1, 1, \pm i, \pm i$

$\therefore$ C.F. $= (C_1x + C_2)e^x + (C_3x\ C_4) \cos x + (C_5x + C_6) \sin x$

and $$\text{P.I.} = \frac{1}{(D-1)^2 (D^2+1)^2}\left[\sin^2 \frac{x}{2} + e^x\right].$$

$$= \frac{1}{2(D-1)^2 (D^2+1)^2}(1-\cos x) + \frac{1}{(D-1)^2 (D^2+1)} e^x$$

$$= \frac{1}{2(D-1)^2 (D^2+1)^2} e^{ox} - \frac{1}{2(D-1)^2 (D^2+1)^2} \cos x + \frac{1}{(D-1)^2 (D^2+1)^2} e^x$$

$$= \frac{1}{2} - \frac{1}{2(D^2+1)^2}\left[\frac{1}{D^2-2D+1}\cos x\right] + \frac{1}{(D-1)^2} \cdot \frac{1}{(1^2+1)^2} e^x$$

$$= \frac{1}{2} - \frac{1}{4(D^2+1)^2} \sin x + \frac{1}{4} \frac{1}{(D-1)^2} e^x$$

$$= \frac{1}{2} - \frac{x^2}{4} \frac{1}{\frac{d^2}{dD^2}(D^2+1)^2} \sin x + \frac{1}{4} \frac{x^2}{\frac{d^2}{dD^2}(D-1)^2} e^x$$

$$= \frac{1}{2} - \frac{x^2}{\frac{d}{dD}4(D^3+D)} \sin x \frac{x^2}{4\frac{d}{dD}2(D-1)} e^x$$

$$= \frac{1}{2} - \frac{x^2}{4.4(3D^2+1)} \sin x + \frac{x^2}{8} e^x = \frac{1}{2} - \frac{x^2}{32} \sin x + \frac{x^2}{8} e^x$$

$\therefore$ Solution is $$Y = (C_1x + C_2)e^x + (C_3x + C_4) \cos x + (C_5x + C_6) \sin x + \frac{1}{2} - \frac{x^2}{32} \sin x + \frac{x^2}{8} e^x.$$

8. *Solve* $\operatorname{cosec} x \frac{d^4 y}{dx^4} + y \operatorname{cosec} x = \sin 2x$.

SOLUTION: Equation is $(D^4 + 1)y = \sin 2x \sin x$

The A.E. is $m^4 + 1 = 0$ or $(m^2+1)^2 - 2m^2 = 0$ or $(m^2 - \sqrt{2} + 1)(m^2 + \sqrt{2}m + 1) = 0$

Its roots are $$m = \frac{\sqrt{2} \pm i\sqrt{2}}{2}, \frac{-\sqrt{2} \pm i\sqrt{2}}{2} = \frac{1}{\sqrt{2}}[\pm 1 \pm i]$$

$\therefore$ $$\text{C.F.} = e^{x/\sqrt{2}}(C_1 \cos x/\sqrt{2} + C_2 \sin x/\sqrt{2}) + e^{-x/\sqrt{2}}(C_3 \cos x/\sqrt{2} + C_4 \sin x/\sqrt{2})$$

Next, $$\text{P.I.} = \frac{1}{D^4+1} \sin 2x \sin x = \frac{1}{2(D^4+1)}(\cos 3x - \cos x)$$

$$= \frac{1}{2\{(D^2)^2+1\}}\cos 3x - \frac{1}{2\{(D^2)^2+1\}}\cos x$$

$$= \frac{1}{2\{(-9^2)^2+1\}}\cos 3x - \frac{1}{2\{(-1)^2+1\}}\cos x = \frac{1}{164}\cos 3x - \frac{1}{4}\cos x$$

$\therefore$ Complete solution is

$$Y = e^{x/\sqrt{2}}(C_1 \cos x/\sqrt{2} + C_2 \sin x/\sqrt{2}) + e^{-x/\sqrt{2}}(C_3 \cos x/\sqrt{2} + C_4 \sin x/\sqrt{2}) + \frac{1}{164}\cos 3x - \frac{1}{4}\cos x. \quad \textbf{Ans.}$$

9. *Solve* $\frac{d^3y}{dx^3} - 2\frac{dy}{dx} + 4y = 3x^2 - 5x + 2.$

SOLUTION: Equation is $(D^3 - 2D + 4)y = 3x^2 - 5x + 2$

The A.E. is $m^3 - 2m + 4 = 0$ or $(m + 2)(m^2 - 2m + 2) = 0.$ Its roots are $m = -2, 1 \pm i$

$\therefore$ C.F. $= C_1e^{-2x} + e^x (C_3 \cos x + C_4 \sin x).$

$$\text{P.I.} = \frac{1}{D^3 - 2D + 4}(3x^2 - 5x + 2) = \frac{1}{4}\left[1 - \left(\frac{D}{2} - \frac{D^3}{4}\right)\right]^{-1}(3x^2 - 5x + 2)$$

$$= \frac{1}{4}\left[1 + \left(\frac{D}{2} - \frac{D^3}{4}\right) + \left(\frac{D}{2} - \frac{D^3}{4}\right)^2 + \ldots\right](3x^2 - 5x + 2)$$

$$= \frac{1}{4}\left(1 + \frac{D}{2} + \frac{D^2}{1}\right)(3x^2 - 5x + 2) = \frac{1}{4}\left[3x^2 - 5x + 2 + \frac{1}{2}(6x - 5) + \frac{6}{4}\right] = \frac{3}{4}x^2 - \frac{1}{2}x + \frac{1}{4}$$

$\therefore$ Complete solution is $\quad y = C_1e^{-2x} + e^x(C_3 \cos x + C_4 \sin x) + \frac{1}{4}(3x^2 - 2x + 1)$ **Ans.**

10. *Solve* $\frac{d^3y}{dx^3} - \frac{7dy}{dx} - 6y = e^{2x}(1 + x) + \cosh x \cos x.$

SOLUTION: Equation is $(D^3 - 7D - 6)y = e^{2x}(1 + x) + \cosh x \cos x$

For C.F. the A.E. is $m^3 - 7m - 6 = 0$ or $(m + 1)(m^2 - m - 6) = 0$

Its roots are $m = -1, -2, 3 \quad \therefore \quad$ C.F. $= C_1e^{-x} + C_2e^{-2x} + C_3e^{3x}.$

and $\quad \text{P.I.} = \frac{1}{D^3 - 7D - 6}e^{2x}(1 + x) + \frac{1}{2(D^3 - 7D - 6)}(e^x \cos x + e^{-x}\cos x)$

$$= e^{2x}\frac{1}{(D+2)^3 - 7(D+2) - 6}(1 + x) + \frac{e^x}{2}\left[\frac{1}{(D+1)^3 - 7(D+1) - 6}\right]\cos x$$

$$+ \frac{e^{-x}}{2}\left[\frac{1}{(D-1)^3 - 7(D-1) - 6}\right]\cos x$$

$$= e^{2x}\frac{1}{D^3+6D^2+12D+8-7D-20}(1+x)+\frac{e^x}{2}\frac{1}{D^3+3D^2+3D1-7D-13}\cos x$$

$$+\frac{e^{-x}}{2}\frac{1}{D^3-3D^2-4D}\cos x$$

$$=\frac{e^{2x}}{-12}\left[1-\left(\frac{5D}{12}+\frac{1}{2}D^2+\frac{D^3}{12}\right)\right]^{-1}(x+1)+\frac{e^x}{2}\frac{1}{-D-3-4D-12}\cos x+\frac{e^{-x}}{2}\frac{1}{-D+3-4D}\cos x$$

$$=\frac{e^{2x}}{-12}\left[1+\frac{5}{12}D\right](x+1)-\frac{e^x}{10}\frac{D-3}{D^2-9}\cos x-\frac{e^{-x}}{2}\frac{5D+3}{25D^2-9}\cos x$$

$$=\frac{e^{2x}}{-12}\left[x+1+\frac{5}{12}\right]-\frac{e^x}{10}\frac{(D-3)}{-1-9}\cos x-\frac{e^{-x}}{2}\frac{(5D+3)}{-25-9}\cos x$$

$$=\frac{-1}{12}e^{2x}\left(x+\frac{17}{12}\right)+\frac{e^x}{100}(-\sin x-3\cos x)+\frac{e^{-x}}{2(34)}(-5\sin x+3\cos x)$$

$\therefore$ Complete solution is

$$y=e^{-x}C_1+C_2e^{-2x}+C_3e^{3x}-\frac{e^{2x}}{12}\left(x+\frac{17}{12}\right)-\frac{e^x}{100}(\sin x+3\cos x)+\frac{e^{-x}}{68}(-5\sin x+3\cos x)$$ **Ans.**

11. *Solve* $\dfrac{d^2x}{dt^2}+2\dfrac{dx}{dt}+x=e^{-t}(\log t)$

SOLUTION: Equation is $(D^2+2D+1)x=e^{-t}\log t$ where $D\equiv\dfrac{d}{dt}$.

For C.F. the A.E. is $m^2+2m+1=0$ whose roots are $m=-1,-1$. $\therefore$ C.F. $=(C_1t+C_2)e^{-t}$.

$$\text{P.I.}=\frac{1}{D^2+2D+1}e^{-t}\log t=e^{-t}\frac{1}{(D-1)^2+2(D-1)+1}\log t$$

$$=e^{-t}\frac{1}{D^2}\log t=e^{-t}\cdot\frac{1}{D}\left[\int\log t\,dt\right]=e^{-t}\frac{1}{D}[t\log t-t]=e^{-t}\int(t\log t-t)\,dt$$

$$=e^{-t}\left[\frac{t^2}{2}\log t-\frac{1}{4}t^2\right]$$

$\therefore$ Complete solution is $x=(C_1t+C_2)e^{-t}+e^{-t}\left[\dfrac{t^2}{2}\log t-\dfrac{3}{4}t^2\right]$ **Ans.**

12. *Find the particular integral of the equation* $\dfrac{d^4y}{dx^4} - y = \cos x \cosh x$.

SOLUTION: Equation is $(D^4 - 1)y = \cos x \cosh x$

For C.F. the A.E. is $m^4 - 1 = 0$ hence $m = \pm 1, \pm i$

$\therefore$ C.F. $= C_1e^x + C_2e^{-x} + C_3 \cos x + C_4 \sin x$

$$\text{P.I.} = \frac{1}{D^4-1}\cos x\cosh x = \frac{1}{2(D^4-1)}(e^x\cos x) + \frac{1}{2(D^4-1)}(e^{-x}\cos x)$$

$$= \frac{e^x}{2}\left[\frac{1}{(D+1)^4-1}\right]\cos x + \frac{e^{-x}}{2}\left[\frac{1}{(D-1)^4-1}\right]\cos x$$

$$= \frac{e^x}{2}\frac{1}{(-1)^2-4D-6+4D}\cos x + \frac{e^{-x}}{2}\frac{1}{(-1)^2+4D-6-4D}\cos x$$

$$= \frac{1}{2}e^x\left[\frac{1}{D^4+4D^3+6D^2+4D+1-1}\right]\cos x + \frac{1}{2}e^{-x}\left[\frac{1}{D^4-4D^3+6D^2-4D+1-1}\right]\cos x$$

$$= \frac{e^x}{-10}\cos x + \frac{e^{-x}}{-10}\cos x = -\frac{1}{10}\cos x\,(e^x + e^{-x}) = -\frac{1}{5}\cos x\cosh x$$

$\therefore$ Complete solution is $\quad y = C_1e^x + C_2e^{-x} + C_3\cos x + C_4\sin x - \dfrac{1}{5}\cos x\cosh x$ **Ans.**

13. *Solve* $\dfrac{d^2y}{dx^2} + \dfrac{dy}{dx} + y = x \sin x$.

SOLUTION: Equation is $(D^2 + D + 1)y = x \sin x$ where $D \equiv \dfrac{d}{dx}$.

For C.F., the A.E. is $m^2 + m + 1 = 0$ whose roots are $m = \dfrac{-1}{2} \pm \dfrac{i\sqrt{3}}{2}$

$$\therefore \quad \text{C.F.} = e^{-x/2}\left[C_1\cos\frac{\sqrt{3}}{2}x + C_2\sin\frac{\sqrt{3}}{2}x\right]$$

and $\quad$ P.I. $= \dfrac{1}{D^2+D+1}(x\sin x)$.

We know that $\dfrac{1}{f(D)}(xV) = x\dfrac{1}{f(D)}V - \dfrac{f'(D)}{(f(D))^2}V$

$$\therefore \quad \text{P.I.} = x\cdot\frac{1}{D^2+D+1}\sin x - \frac{2D+1}{(D^2+D+1)^2}\sin x$$

$$= x\frac{1}{-1+D+1}\sin x - \frac{2D+1}{(-1+D+1)^2}\sin x$$

$$= x\int\sin x\,dx - \frac{2D+1}{D^2}\sin x = -x\cos x - (2D+1)(-\sin x) = -x\cos x + (2\cos x + \sin x)$$

Thus the complete solution is

$$y = e^{-x/2}\left(C_1 \cos\frac{\sqrt{3}\,x}{2} + C_2 \sin\frac{\sqrt{3}\,x}{2}\right) - x\cos x + 2\cos x + \sin x \qquad \textbf{Ans.}$$

14. *Find the particular Integral of the equation* $\dfrac{d^2y}{dx^2} + 4y = x\ \sin^2 x.$

SOLUTION: Equation is $(D^2 + 4)y = x\ \sin^2 x$

$$\text{P.I.} = \frac{1}{D^2+4}(x\sin^2 x) = x\frac{1}{D^2+4}\sin^2 x - \frac{2D}{(D^2+4)^2}\sin^2 x$$

$$= \frac{x}{2}\frac{1}{D^2+4}(1-\cos 2x) - \frac{D}{(D^2+4)^2}(1-\cos 2x)$$

$$= \frac{x}{2}\left[\frac{1}{D^2+4}\cdot 1 - \frac{1}{D^2+4}\cos 2x\right] - \frac{1}{(D^2+4)^2}2\sin 2x$$

$$= \frac{x}{2}\frac{1}{D^2+4}e^{0x} - \frac{x}{2}\cdot\frac{x}{2D}\cos 2x - 2x^2\frac{1}{\dfrac{d^2}{dD^2}(D^2+4)^2}\sin 2x$$

$$= \frac{x}{8} - \frac{x^2}{4}\frac{\sin 2x}{2} - 2x^2\frac{1}{12\,D^2+16}\sin 2x$$

$$= \frac{x}{8} - \frac{x^2}{8}\sin 2x - x^2\frac{1}{6\,(-4)+8}\sin 2x$$

$$= \frac{x}{8} - \frac{x^2}{8}\sin 2x + \frac{x^2}{16}\sin 2x = \frac{x}{8} - \frac{x^2}{16}\sin 2x \qquad \textbf{Ans.}$$

15. *Solve* $\dfrac{d^2y}{dx^2} + 3\dfrac{dy}{dx} + 2y = \sin(e^x).$

SOLUTION: Equation is $(D^2 + 3D + 2)y = \sin(e^x)$

For C.F. the A.E. is $m^2 + 3m + 2 = 0$ whose roots are $m = -1, -2$.

$\therefore$ C.F. $= c_1e^{-x} + c_2e^{-2x}$

and

$$\text{P.I.} = \frac{1}{D^2+3D+2}\sin(e^x) = \frac{1}{(D+2)(D+1)}\sin(e^x)$$

$$= \left(\frac{1}{D+1} - \frac{1}{D+2}\right)\sin(e^x) = \frac{1}{D+1}\sin(e^x) - \frac{1}{D+2}\sin(e^x)$$

$$= e^{-x}\int e^x \sin(e^x)\,dx - e^{-2x}\int e^{2x}\sin(e^x)\,dx$$

$$= e^{-x}\int \sin t\, dt - e^{-2x}\int t \sin t\, dt \qquad \text{(on putting } e^x = t)$$

$$= -e^{-x}\cos(e^x) - e^{-2x}\left[-t\cos t - \int 1\cdot(-\cos t)\, dt\right]$$

$$= -e^{-x}\cos(e^x) - e^{-2x}\left[-e^x \cos(e^x) + \sin(e^x)\right] = -e^{-2x}\sin(e^x)$$

$\therefore$ Complete solution is $\quad y = c_1e^{-x} + c_2e^{-2x} - e^{-2x}\sin(e^x)$ **Ans.**

16. *Find the P.I. of the equation* $\dfrac{d^2y}{dx^2} - y = (1 - e^{-x})^{-2}$.

SOLUTION: Equation is $(D^2 - 1)y = (1 - e^{-x})^{-2} = \dfrac{e^{2x}}{(e^x - 1)^2}$

$$\text{P.I.} = \frac{1}{D^2-1}(1-e^{-x})^{-2} = \frac{1}{2}\left[\frac{1}{D-1} - \frac{1}{D+1}\right]\frac{e^{2x}}{(e^x-1)^2}$$

$$= \frac{1}{2(D-1)}\frac{e^{2x}}{(e^x-1)^2} - \frac{1}{2(D+1)}\frac{e^{2x}}{(e^x-1)}$$

$$= \frac{1}{2}e^x\int\frac{e^{-x}e^{2x}}{(e^x-1)^2}dx - \frac{1}{2}e^{-x}\int\frac{e^x e^{2x}\,dx}{(e^x-1)^2} = \frac{1}{2}e^x\int\frac{dt}{t^2} - \frac{1}{2}e^{-x}\int\frac{(t+1)^2\,dt}{t^2}$$

(on putting $e^x - 1 = t$)

$$= \frac{1}{2}e^x\left(-\frac{1}{t}\right) - \frac{1}{2}e^{-x}\int\left(1+\frac{2}{t}+\frac{1}{t^2}\right)dt = \frac{-e^x}{2(e^x-1)} - \frac{e^{-x}}{2}\left[t + 2\log t - \frac{1}{t}\right]$$

$$= \frac{-e^x}{2(e^x-1)} - \frac{1}{2e^x}\left[e^x - 1 + 2\log(e^x-1) - \frac{1}{e^x-1}\right] = -1 - e^{-x}\log(e^x - 1)$$ **Ans.**

17. *Solve* $(D^2 + 3D + 2)\,y = e^{e^x}$ where $D \equiv \dfrac{d}{dx}$

SOLUTION: A.E. is $m^2 + 3m + 2 = 0$ where roots are $m = -1, -2$. $\quad\therefore\quad$ C.F. $= c_1e^{-x} + c_2e^{-2x}$

and $\quad \text{P.I.} = \dfrac{1}{D^2+3D+2}e^{e^x} = \left(\dfrac{1}{D+1} - \dfrac{1}{D+2}\right)e^{e^x} = \dfrac{1}{D+1}e^{e^x} - \dfrac{1}{D+2}e^{e^x}$

$$= e^{-x}\int e^x e^{e^x}\,dx - e^{-2x}\int e^{2x}e^{e^x}\,dx \qquad \text{(Put } e^x = t)$$

$$= e^{-x}\int e^t\,dt - e^{-2x}\int e^t\cdot t\,dt = e^{-x}e^t - e^{-2x}\cdot e^t(t-1)$$

$$= e^{-x}e^{e^x} - e^{-2x}\cdot e^{e^x}(e^x - 1) = e^{-2x}e^{e^x}$$

$\therefore$ Solution is $y = c_1e^{-x} + c_2e^{-2x} + e^{-2x}e^{e^x}$

18. *Apply the method of variation of parameters, to solve*

$$\frac{d^2y}{dx^2} - y = e^{-x}\sin(e^{-x}) + \cos(e^{-x}).$$

SOLUTION: Equation is $(D^2 - 1)y = e^{-x}\sin(e^{-x}) + \cos(e^{-x})$

For C.F., the A.E. is $m^2 = 1$ hence $m = \pm 1$.

$\therefore$ C.F. $= c_1e^x + c_2e^{-x}$. We take $y = u\,e^x + v\,e^{-x} =$ P.I. where u and v are functions of x.

$$\therefore\ Dy = u\,e^x + u'\,e^x - ve^{-x} + v'\,e^{-x}$$

Let us choose u and v such that $u'\,e^x + v'\,e^{-x} = 0$ ... (1)

Then $Dy = u\,e^x - ve^{-x}$ and $D^2y = u'\,e^x + ue^x + ve^{-x} - v'\,e^{-x}$ ... (2)

Using (2) in the given equation, we get

$$u'\,e^x + u\,e^x + v\,e^{-x} - v'\,e^{-x} - u\,e^x - v\,e^{-x} = e^{-x}\sin(e^{-x}) + \cos(e^{-x})$$

or $u'\,e^x - v'\,e^{-x} = e^{-x}\sin(e^{-x}) + \cos(e^{-x})$ ... (3)

Solving (1) and (3) for u' and v', we get

$$u' = \frac{1}{2}e^{-2x}\sin(e^{-x}) + \frac{1}{2}e^{-x}\cos(e^{-x}) \quad \text{and} \quad v' = \frac{-1}{2}\sin(e^{-x}) - \frac{1}{2}e^x\cos(e^{-x})$$

$$\therefore\quad 2u = \int[(e^{-2x}\sin(e^{-x}) + e^{-x}\cos(e^{-x})]\,dx \quad \text{and} \quad 2v = \int[-\sin(e^{-x}) - e^{+x}\cos(e^{-x})]\,dx$$

Putting $e^{-x} = t$ so that $-e^{-x}\,dx = dt$ we have

$$2u = \int(-t\sin t - \cos t)\,dt = -\sin t + t\cos t + \int -\cos t\,dt = t\cos t - 2\sin t$$

$$= e^{-x}\cos(e^{-x}) - 2\sin(e^{-x})$$

$$2v = \int\left(-\sin t - \frac{1}{t}\cos t\right)\frac{dt}{-t} = \int\left(\frac{1}{t}\sin t + \frac{1}{t^2}\cos t\right)dt$$

$$= \frac{-1}{t}\cos t - \int\frac{1}{t^2}\cos t\,dt + \int\frac{1}{t^2}\cos t\,dt$$

or $2v = -e^x\cos(e^{-x})$

$$\therefore\quad \text{P.I.} = e^x\left[\frac{1}{2}e^{-x}\cos(e^{-x}) - \sin(e^{-x})\right] + e^{-x}\left(\frac{-1}{2}\right)e^x\cos(e^{-x}) = -e^x\sin(e^{-x})$$

$\therefore$ Complete solution is $y = c_1e^x + c_2e^{-x} - e^x\sin(e^{-x})$ **Ans.**

19. *By the method of variation of parameters, solve* $\dfrac{d^2y}{dx^2} - y = \dfrac{2}{1+e^x}$.

SOLUTION: Equation is $(D^2 - 1)y = \dfrac{2}{1+e^x}$

Here C.F. $= c_1e^x + c_2e^{-x}$. We take $y =$ P.I. $= ue^x + ve^{-x}$ where u and v are functions of x.

$$\therefore\quad y' = u'e^x + ue^x + v'e^{-x} - ve^{-x}.$$

We choose u and v such that $u'e^x + v'e^{-x} = 0$ ...(1)

$\therefore \quad y' = ue^x - ve^{-x}$ and $y'' = u'e^x + ue^x - v'e^{-x} + ve^{-x}$ ...(2)

Using (2) in the given equation, gives

$$u'e^x + ue^x - v'e^{-x} + ve^{-x} - ue^{-x} - ve^{-x} = \frac{2}{1+e^x}$$

or $$u'e^x - v'e^{-x} = \frac{2}{1+e^x} \quad ...(3)$$

Solving (1) and (3) for u' and v', we get $u' = \dfrac{1}{e^x(1+e^x)}$ and $v' = \dfrac{e^x}{1+e^x}$

$$\therefore \quad u = \int \frac{dx}{e^x(1+e^x)} \quad \text{and} \quad v = \int \frac{e^x\,dx}{1+e^x} = \log(1+e^x)$$

on putting $$e^x = t, \quad u = \int \frac{dt}{t^2(1+t)} = \int\left(\frac{1}{t^2} - \frac{1}{t} + \frac{1}{1+t}\right)dt = \frac{-1}{t} - \log t + \log(1+t)$$

or $$u = \frac{-1}{t} + \log\frac{1+t}{t} = -e^{-x} + \log(1+e^{-x})$$

$$\therefore \quad \text{P.I.} = ue^x + v\,e^{-x} = -1 + e^x\log(1+e^{-x}) + e^{-x}\log(1+e^x)$$

$$= -1 - xe^x + (e^x + e^{-x})\log(1+e^x)$$

$\therefore$ Complete solution is $y = c_1e^x + c_2e^{-x} - 1 - xe^x + (e^x - e^{-x})\log(1+e^x)$ **Ans.**

20. *Solve* $(D^2 - 1)\,y = \text{sech}\,x$ *by the method of variation of parameters.*

SOLUTION: The A.E. is $m^2 - 1 = 0$ so $m = \pm 1$ $\therefore$ C.F. $= c_1e^x + c_2e^{-x}$

We take P.I. $= y = u\,e^x + ve^{-x}$ $\therefore$ $y' = ue^x - ve^{-x} + u'e^x + v'e^{-x}$

Choose u and v such that $u'e^x + v^1e^{-x} = 0$ ...(1)

then $y' = ue^x - ve^{-x}$

Hence $y'' = u\,e^x + u'e^x - v'e^{-x} + ve^{-x}$.

Putting these values of y and y'' in the given equation, we get

$$ue^x + u'e^x - v'e^{-x} + v\,e^{-x} - ue^x - ve^{-x} = \text{sech}\,x$$

or $u'e^x - v'e^{-x} = \text{sech}\,x$ ...(2)

Solving (1) and (2) for u' and v', we get

$$u' = \frac{1}{2}e^{-x}\,\text{sech}\,x = \frac{1}{e^x(e^x+e^{-x})} = \frac{1}{1+e^{2x}}$$

and $$v' = \frac{-e^x}{2}\text{sech}\,x = \frac{-e^x}{e^x+e^{-x}} = \frac{-e^{2x}}{1+e^{2x}}$$

$$\therefore \quad u = \int\frac{dx}{1+e^{2x}} = \frac{1}{2}\int\frac{dt}{t(1+t)} = \frac{1}{2}\log\frac{t}{1+t} = \frac{1}{2}\log\frac{e^{2x}}{1+e^2x} \quad \text{(on putting } t = e^{2x})$$

and $\quad v = -\frac{1}{2}\log(1+e^{2x})$

$$\therefore \quad \text{P.I.} = u\,e^x + ve^{-x} = \frac{1}{2}e^x \log\frac{e^{2x}}{1+e^{2x}} - \frac{1}{2}e^{-x}\log(1+e^{2x})$$

$$= \frac{1}{2}e^x \cdot 2x - \frac{1}{2}e^x \log(1+e^{2x}) - \frac{1}{2}e^{-x}\log(1+e^{2x})$$

$$= xe^x - \cosh x \log(1+e^{2x})$$

$\therefore$ Complete solution is $y = c_1e^x + c_2e^{-x} + x\,e^x - \cosh x \log(1+e^{2x})$. **Ans.**

21. *Solve* $y'' - 2y' + 2y = e^x \tan x$ *by the method of variation of parameters.*

SOLUTION: Equation is $(D^2 - 2D + 2)\,y = e^x \tan x$ where $D \equiv \frac{d}{dx}$

A.E. is $m^2 - 2m + 2 = 0$ whose roots are $m = 1 \pm i$

$\therefore$ C.F. $= e^x(c_1 \cos x + c_2 \sin x)$. Let us take P.I. $= u\,e^x \cos x + v\,e^x \sin x = y$

$\therefore\ y' = u'\,e^x \cos x + u\,(e^x \cos x - e^x \sin x) + v'\,e^x \sin x + v\,(e^x \sin x + e^x \cos x)$.

Let us choose u and v such that $\quad u'\,e^x \cos x + v'e^x \sin x = 0$

or $\quad u' \cos x + v' \sin x = 0$ ... (1)

then $\quad y' = u\,(e^x \cos x - e^x \sin x) + v\,(e^x \sin x + e^x \cos x)$

$\quad = e^x[u\,(\cos x - \sin x) + v\,(\cos x + \sin x)]$ ...(2)

$\therefore \quad y'' = e^x\,[u\,(\cos x - \sin x) + v'(\cos x + \sin x)]$

$\quad + e^x\,[u'(\cos x - \sin x) + v'\,(\cos x + \sin x) + u\,(-\sin x - \cos x) + v\,(-\sin x + \cos x)]$

or $\quad y'' = e^x\,[u'(\cos x - \sin x) + v'(\cos x + \sin x) - 2u \sin x + 2v \cos x]$

Substituting for y, y' and y'' in the given equation, we get

$e^x[u'(\cos x - \sin x) + v'(\cos x + \sin x) - 2u \sin x + 2v \cos x]$

$- 2e^x\,[u(\cos x - \sin x) + v(\cos x + \sin x)] + 2e^x(u \cos x + v \sin x) = e^x \tan x$

or $\quad u'(\cos x - \sin x) + v'(\cos x + \sin x) = \tan x$ ... (2)

Solving (1) and (2) for u' and v', we get

$$u' = -\frac{\sin^2 x}{\cos x} = \cos x - \sec x. \quad v' = \sin x$$

Hence $\quad v = -\cos x, \quad \therefore \quad u = \sin x - \log(\sec x + \tan x)$

$\therefore \quad$ P.I. $= e^x[\cos x\,[\sin x - \log(\sec x + \tan x)] + \sin x\,(-\cos x)]$

$\quad = -e^x \cos x \log(\sec x + \tan x)$

$\therefore$ Complete solution is

$$y = e^x(c_1 \cos x + c_2 \sin x) - e^x \cos x \log(\sec x + \tan x)$$ **Ans.**

22. *Solve* $\frac{d^2y}{dx^2}+4y=e^x+\sin 2x$.

SOLUTION: The A.E. is $m^2+4=0$ $\therefore$ $m=\pm 2i$

$\therefore$ C.F. $= c_1\cos 2x + c_2 \sin 2x$

and P.I. $=\frac{1}{D^2+4}(e^x+\sin 2x) = \frac{1}{D^2+4}e^x+\frac{1}{D^2+4}\sin 2x$

$$=\frac{1}{1+4}e^x+\frac{x}{2D}\sin 2x = \frac{1}{5}e^x-\frac{x}{4}\cos 2x$$

$\therefore$ Complete solution is $y = c_1\cos 2x + c_2\sin 2x + \frac{1}{5}e^x-\frac{x}{4}\cos 2x$. **Ans.**

23. *Solve* $(D-2)^2 y = 8(e^{2x}+\sin 2x + x^2)$ *where* $D\equiv\frac{d}{dx}$.

SOLUTION: A.E. is $(m-2)^2=0$ $\therefore$ $m=2, 2.$ $\therefore$ C.F. $=(c_1x+c_2)e^{2x}$

and P.I. $=\frac{8}{(D-2)^2}(e^{2x}+\sin 2x+x^2)$

$$=\frac{8}{(D-2)^2}e^{2x}+\frac{8}{D^2-4D+4}\sin 2x+8\frac{1}{(D-2)^2}x^2$$

$$=8x\frac{1}{2(D-2)}e^{2x}+\frac{8}{-4-4D+4}\sin 2x+\frac{8}{4}\left(1-\frac{D}{2}\right)^{-2}x^2$$

$$=8\cdot\frac{x^2}{2}e^{2x}-2\int\sin 2x\,dx+2\left(1+D+\frac{2.3}{2.4}D^2\right)x^2$$

$$=4x^2e^{2x}+\cos 2x+2\left(x^2+2x+\frac{3}{2}\right)$$

$\therefore$ Solution is $y=(c_1x+c_2)e^{2x}+4x^2e^{2x}+\cos 2x+2x^2+4x+3.$ **Ans.**

24. *Solve* $\frac{d^3y}{dx^3}+2\frac{d^2y}{dx^2}+\frac{dy}{dx}=e^{2x}+\sin 2x$.

SOLUTION: Equation is $(D^3+2D^2+D)y=e^{2x}+\sin 2x$ where $D\equiv\frac{d}{dx}$

A.E. is $m^3+2m^2+m=0$ whose roots are $m=0, -1, -1$

$\therefore$ C.F. $=c_1e^{ox}+(c_2x+c_3)e^{-x}$

and P.I. $=\frac{1}{D^3+2D^2+D}e^{2x}+\frac{1}{D^3+2D^2+D}\sin 2x$

$$= \frac{1}{8+8+2}e^{2x} + \frac{1}{-4D+2(-4)+D}\sin 2x = \frac{1}{18}e^{2x} - \frac{3D-8}{9D^2-64}\sin 2x$$

$$= \frac{1}{18}e^{2x} - \frac{3D-8}{9(-4)-64}\sin 2x = \frac{1}{18}e^{2x} + \frac{1}{100}(6\cos 2x - 8\sin 2x)$$

$\therefore$ Solution is $y = c_1 + (c_2x + c_3)e^{-x} + \frac{1}{18}e^{2x} + \frac{3}{50}\cos 2x - \frac{2}{25}\sin 2x$. **Ans.**

25. *Solve* $(D-1)^2(D+1)^2 y = \sin^2\left(\frac{x}{2}\right) + e^x + x$ *where* $D \equiv \frac{d}{dx}$

SOLUTION: A.E. is $(m^2-1)^2 = 0$ $\therefore$ $m = 1, 1, -1, -1$ and C.F. $= (c_1x + c_2)e^x + (c_3x + c_4)e^{-x}$

$$\text{P.I.} = \frac{1}{(D^2-1)^2}\left[\frac{1}{2}(1-\cos x) + e^x + x\right]$$

$$= \frac{1}{(D-1)^2(D+1)^2}e^x - \frac{1}{(D^2-1)}\cos x + \frac{1}{(1-D^2)^2}\left(x+\frac{1}{2}\right)$$

$$= \frac{1}{(D-1)^2}\frac{1}{(1+1)^2}e^x - \frac{1}{2(-1-1)^2}\cos x + (1-D^2)^{-2}\left(x+\frac{1}{2}\right)$$

$$= \frac{1}{4(D-1)^2}e^x - \frac{1}{8}\cos x + (1+2D^2)\left(x+\frac{1}{2}\right)$$

$$= \frac{1}{4}\cdot\frac{x^2}{2}e^x - \frac{1}{8}\cos x + \left(x+\frac{1}{2}+0\right)$$

$\therefore$ Complete solution is $y = (c_1x+c_2)e^x + (c_3x+c_4)e^{-x} + \frac{1}{2} + x - \frac{x^2}{8}e^x - \frac{1}{8}\cos x$. **Ans.**

26. *Solve* $(D^3 + 2D^2 + D)y = x^2e^{2x} + \sin^2 x$ *where* $D \equiv \frac{d}{dx}$.

SOLUTION: A.E. is $m^3 + 2m^2 + m = 0$ $\therefore$ $m = 0, -1, -1$ $\therefore$ C.F. $= c_1e^{ox} + (c_2x + c_3)e^{-x}$.

$$\text{P.I.} = \frac{1}{D(D+1)^2}(x^2e^{2x} + \sin^2 x)$$

$$= \frac{1}{D^3+2D^2+D}(x^2e^{2x}) + \frac{1}{D^3+2D^2+D}\frac{1-\cos 2x}{2}$$

$$= e^{2x}\frac{1}{(D+2)^3+2(D+2)^2+(D+2)}x^2 + \frac{1}{2D(D^2+2D+1)}e^{ox} - \frac{1}{2(D^3+2D^2+D)}\cos 2x$$

$$= e^{2x}\frac{1}{D^3+8D^2+21D+18}x^2 + \frac{x}{2} - \frac{1}{2(-4D-8+D)}\cos 2x$$

$$= \frac{e^{2x}}{18}\left[1+\frac{21D}{18}+\frac{8D^2}{18}+\frac{D^3}{18}\right]^{-1} x^2+\frac{x}{2}+\frac{(3D-8)}{2(-36-64)}\cos 2x$$

$$= \frac{e^{2x}}{18}\left[x^2-\frac{21}{9}x-\frac{8}{9}+\frac{21^2}{9.18}\right]+\frac{x}{2}+\frac{1}{200}(-6\sin 2x-8\cos 2x)$$

$\therefore$ Solution is $y = c_1+(c_2x+c_3)e^{-x}+\frac{x}{2}-\frac{1}{100}(3\sin 2x+4\cos 2x)+\frac{e^{2x}}{18}\left(x^2-\frac{7x}{3}+\frac{33}{18}\right)$. **Ans.**

27. *Solve* $\frac{d^2y}{dx^2}-4y = x\sinh x$.

SOLUTION: Equation is $(D^2-4)\,y = x\sinh x = \frac{x}{2}(e^x-e^{-x})$

$\therefore$ C.F. $= c_1e^{2x}+c_2\,e^{-2x}$.

$$\text{P.I.} = \frac{1}{D^2-4}(x\sinh x) = \frac{1}{D^2-4}\frac{xe^x}{2}-\frac{1}{D^2-4}\frac{xe^{-x}}{2}$$

$$= \frac{1}{2}e^x\frac{1}{(D+1)^2-4}x-\frac{e^{-x}}{2}\frac{1}{(D-1)^2-4}x = \frac{e^x}{2}\frac{1}{D^2+2D-3}x-\frac{e^{-x}}{2}\frac{1}{D^2-2D-3}x$$

$$= \frac{-e^x}{6}\left[1-\frac{2}{3}D-\frac{D^2}{3}\right]^{-1}x+\frac{e^{-x}}{6}\left[1+\frac{2}{3}D-\frac{D^2}{3}\right]^{-1}x$$

$$= \frac{-e^x}{6}\left[1+\frac{2}{3}D+\frac{D^2}{3}\right]x+\frac{e^{-x}}{6}\left[1-\frac{2}{3}D+\frac{D^2}{3}\right]x$$

$$= \frac{-e^x}{6}\left(x+\frac{2}{3}\right)+\frac{e^{-x}}{6}\left(x-\frac{2}{3}\right) = \frac{-x}{3}\sinh x-\frac{2}{9}\cosh x$$

$\therefore$ Solution is $y = c_1e^{2x}+c_2e^{-2x}-\frac{x}{3}\sinh x-\frac{2}{9}\cosh x$. **Ans.**

28. *Solve by the method of variation of parameters* $\frac{d^2y}{dx^2}-6\frac{dy}{dx}+9y = e^{3x}/x^2$.

SOLUTION: A.E. is $m^2-6m+9=0$ whose roots are $m = 3, 3$, hence C.F. $= (c_1x+c_2)e^{3x}$

For P.I. we take $y = (xu+v)e^{3x}$ where u, v are functions of x.

$\therefore$ $y' = (xu'+u+v')\,e^{3x}+3(xu+v)e^{3x}$.

Let us chose u and v such that $xu'+v' = 0$ ... (1)

Then $y' = [(1+3x)\,u+3v]\,e^{3x}$

and $y'' = [(1+3x)\,u'+3u+3v']\,e^{3x}+3\,[(1+3x)\,u+3v]\,e^{3x}$

$= e^{3x}\,[(1+3x)\,u'+3v'+3(2+3x)\,u+9v]$

Substituting for y, y' and y'' is the given equation, we get

$$e^{3x}[(1+3x)u' + 3v' + 3(2+3x)u + 9v] - 6[(1+3x)u + 3v]e^{3x} + 9e^{3x}(xu+v) = \frac{e^{3x}}{x^2}$$

or $$(1+3x)u' + 3v' = \frac{1}{x^2} \qquad ...(2)$$

Solving for u' and and v' in (1) and (2), we get

$$u' = \frac{1}{x^2} \quad \text{and} \quad v' = -\frac{1}{x} \quad \Rightarrow \quad u = -\frac{1}{x} \quad \text{and} \quad v = -\log x$$

$\therefore$ $\quad$ P.I. $= (ux + v)\,e^{3x} = (-1 - \log x)\,e^{3x}$.

Complete solution is $y = (c_1 x + c_2)e^{3x} - (1 + \log x)\,e^{3x}$. **Ans.**

29. *Solve* $\dfrac{d^2y}{dx^2} + y = x\sin x$ *by the method of variation of parameters.*

SOLUTION: A.E. is $m^2 + 1 = 0$ $\therefore m = \pm i.$ hence C.F. $= A\cos x + B\sin x$

Let us choose, for finding P.I., $y = u\cos x + v\sin x$ where u and v are unknown function of x.

$$\therefore \quad y' = u'\cos x + v'\sin x + v\cos x - u\sin x$$

Let us choose u and v such that $u'\cos x + v'\sin x = 0$ ...(1)

then, $$y' = -u\sin x + v\cos x$$

hence $$y'' = -u'\sin x + v'\cos x - u\cos x - v\sin x$$

putting these values of y, y', y'' is the given equation, we get

$$-u'\sin x + v'\cos x - u\cos x - v\sin x + u\cos x + v\sin x = x\sin x$$

or $\quad u'\sin x + v'\cos x + x\sin x = 0.$ ...(2)

Solving (1) and (2) for u' and v', we get

$$u' = -x\sin^2 x, \quad v' = x\sin x\cos x$$

$$\therefore \quad u = \int \frac{x}{2}(\cos 2x - 1)\,dx = \frac{x}{2}\left(\frac{1}{2}\sin 2x - x\right) - \int \frac{1}{2}\left(\frac{1}{2}\sin 2x - x\right)dx$$

$$= \frac{x}{4}\sin 2x - \frac{x^2}{2} + \frac{1}{8}\cos 2x + \frac{x^2}{4}$$

and $$v = \int \frac{x}{2}\sin 2x\,dx = -\frac{x}{4}\cos 2x + \frac{1}{8}\sin 2x$$

$\therefore$ $\quad$ P.I. $= u\cos x + v\sin x$

$$= \frac{x}{4}\sin 2x\cos x - \frac{x^2}{4}\cos x + \frac{1}{8}\cos x\cos 2x + \frac{x}{4}\cos 2x\sin x + \frac{1}{8}\sin 2x\sin x$$

$$\therefore \quad y = c_1\cos x + c_2\sin x + \frac{x}{4}\sin 3x - \frac{x^2}{4}\cos x + \frac{1}{8}\cos x$$ **Ans.**

30. *Find the particular integral of* $\frac{d^2y}{dx^2} - 2\frac{dy}{dx} + y = xe^x \sin x.$

SOLUTION: $\text{P.I.} = \frac{1}{D^2 - 2D + 1} xe^x \sin x = \frac{1}{(D-1)^2} xe^x \sin x$ where $D = \frac{d}{dx}$

$\therefore \quad \text{P.I.} = e^x \frac{1}{(D+1-1)^2} x \sin x = e^x \frac{1}{D}\left(\frac{1}{D} x \sin x\right)$

$= e^x \frac{1}{D} \int x \sin x \, dx = e^x \int (-x \cos x + \sin x) \, dx$

$= e^x \left[-\cos x - \int x \cos x dx\right] = e^x [-\cos x - (x \sin x + \cos x)]$

or $\quad \text{P.I.} = -e^x [2 \cos x + x \sin x]$ **Ans.**

SOLUTIONS TO EXERCISE 10B

1. *Solve* $x^2\dfrac{d^2y}{dx^2} - x\dfrac{dy}{dx} + y = \log x.$

SOLUTION: Equation is of Cauchy's homogeneous form.

On putting $x = e^t$ the given equation becomes

$$[D(D-1) - D + 1]\, y = \log e^t = t \quad \text{where } D = \frac{d}{dt}.$$

or $(D-1)^2 y = t$ Here A.E. is $(m-1)^2 = 0$ hence $m = 1, 1$

$\therefore$ C.F. $= (c_1 t + c_2)\, e^t.$

and P.I. $= \dfrac{1}{(D-1)^2} t = (1-D)^{-2} t = \left[1 + 2D + \dfrac{(-2)(-3)}{2!} D^2 + \ldots\right] t$

$\therefore$ Complete solution is $y = (c_1 t + c_2)\, e^t + t + 2$

or $y = (c_1 \log x + c_2)\, x + \log x + 2$ **Ans.**

2. *Solve* $x^2\dfrac{d^2y}{dx^2} - 4x\dfrac{dy}{dx} + 6y = -x^4 \sin x.$

SOLUTION: Equation is of Cauchy's homogeneous form.

Putting $x = e^t$, the given equation becomes

$$[D(D-1) - 4D + 6]\, y = -e^{4t} \sin(e^t) \quad \text{where } D \equiv \frac{d}{dt}.$$

The *A.E.*, is $m^2 - 5m + 6 = 0$ $\therefore$ $m = 2, 3$

$\therefore$ C.F. $= C_1 e^{2t} + C_2 e^{3t} = C_1 x^2 + C_2 x^3$

$$\text{P.I.} = \frac{1}{D^2 - 5D + 6}\{-e^{4t} \sin(e^t)\} = \left(\frac{1}{D-2} - \frac{1}{D-3}\right) e^{4t} \sin(e^t)$$

$$= e^{2t}\int e^{-2t} e^{4t} \sin(e^t)\, dt - e^{3t}\int e^{-3t} e^{4t} \sin(e^t)\, dt$$

$$= x^2 \int x \sin x\, dx - x^3 \int \sin x\, dx$$

$$= x^2\left[x(-\cos x) + \int 1 \cos x\, dx\right] + x^3 \cos x = x^2 \sin x$$

$\therefore$ Complete solution is $y = C_1 x^2 + C_2 x^3 + x^2 \sin x$ **Ans.**

3. *Solve* $x^2\dfrac{d^2y}{dx^2} + 3x\dfrac{dy}{dx} + y = \dfrac{1}{(1-x)^2}.$

SOLUTION: It is homogeneous type. Putting $x = e^t$ the given equation becomes

$$[D(D-1) + 3D + 1]\, y = \frac{1}{(1-e^t)^2} \quad \text{where } D \equiv \frac{d}{dt}$$

$\therefore$ $\quad$ $\text{C.F.} = (C_1 + C_2 t)e^{-t} = \frac{1}{x}(C_1 + C_2 \log x)$

and $\quad$ $\text{P.I.} = \frac{1}{(D+1)^2}\frac{1}{(1-e^t)^2} = \frac{1}{D+1}\left[\frac{1}{D+1}\frac{1}{(1-e^t)^2}\right] = \frac{1}{D+1}\left[e^{-t}\int\frac{e^t\,dt}{(1-e^t)^2}\right]$

$$= \frac{1}{D+1}\left[e^{-t}\left(\frac{1}{1-e^t}\right)\right] = e^{-t}\int\frac{e^t\cdot e^{-t}\,dt}{1-e^t} = e^{-t}\int\frac{dt}{1-e^t} \quad \text{But } e^t = x$$

$$= \frac{1}{x}\int\frac{dx}{x(1-x)} = \frac{1}{x}\int\left(\frac{1}{1-x}+\frac{1}{x}\right)dx = \frac{1}{x}[\log x - \log(1-x)]$$

$\therefore$ Complete solution is $y = \frac{1}{x}(C_1 + C_2 \log x) + \frac{1}{x}[\log x - \log(1-x)]$

$$= \frac{1}{x}[C_1 + C_2 \log x - \log(1-x)]$$ **Ans.**

4. *Solve* $\frac{d^2u}{dr^2} + \frac{1}{r}\frac{du}{dr} - \frac{u}{r^2} + kr = 0$ *given that* $u = 0$ *for* $r = 0,\ r = a.$

SOLUTION: Equation is of homogeneous form.

Equation can be written as $r^2\frac{d^2u}{dr^2} + r\frac{du}{dr} - u = -kr^3$

Putting $r = e^t$ the equation becomes

$$[D(D-1) + D - 1]\,u = -ke^{3t} \quad \text{where} \quad D \equiv \frac{d}{dt}$$

or $\quad$ $(D^2 - 1)\,u = -ke^{3t}.$ $\quad$ Here $\text{C.F.} = C_1 e^t + C_2 e^{-t} = C_1 r + \frac{C_2}{r}.$

$$\text{P.I.} = \frac{1}{D^2-1}(-ke^{3t}) = \frac{-k}{2}\left(\frac{1}{D-1} - \frac{1}{D+1}\right)e^{3t}$$

$$= \frac{k}{2}\left[\frac{1}{D+1}e^{3t} - \frac{1}{D-1}e^{3t}\right] = \frac{k}{2}\left[e^{-t}\int e^t e^{3t}dt - e^t\int e^{-t}e^{3t}dt\right] = \frac{k}{2}\left[\frac{1}{4}e^{-t}e^{4t} - \frac{1}{2}e^t.e^{2t}\right]$$

$$= \frac{k}{8}[e^{3t} - 2e^{3t}] = \frac{-k}{8}r^3$$

$\therefore$ Solution is $u = C_1 r + \frac{C_2}{r} - \frac{k}{8}r^3.$ Since $u = 0$ at $r = 0$ we have $C_2 = 0$

Further, $u = 0$ at $r = a$ we have $0 = C_1 a - \frac{k}{8}a^3$ $\therefore$ $C_1 = \frac{k}{8}a^2.$

Hence $u = \frac{k}{8}a^2 r - \frac{k}{8}r^3 = \frac{k}{8}(a^2 r - r^3).$ **Ans.**

5. *Solve* $x^3\dfrac{d^3y}{dx^3}+3x^2\dfrac{d^2y}{dx^2}+x\dfrac{dy}{dx}+y=x+\log x.$

SOLUTION: Equation is of homegeneous form.

Given equation can be written on putting $x=e^t$, as

$$[D(D-1)(D-2)+3D(D-1)+D+1]\,y=e^t+t \text{ where } D\equiv\frac{d}{dt}$$

or $(D^3+1)\,y=e^t+t.$ The Auxiliary equation is $m^3+1=0$

whose roots are $m=-1,\ \dfrac{1}{2}\pm i\dfrac{\sqrt{3}}{2}$

$$\therefore\quad \text{C.F.}=C_1e^{-t}+e^{t/2}\left(C_2\cos\sqrt{3}\frac{t}{2}+C_3\sin\sqrt{3}\frac{t}{2}\right)$$

$$=\frac{C_1}{x}+\sqrt{x}\left[C_2\cos\left(\frac{\sqrt{3}}{2}\log x\right)+C_3\sin\left(\frac{\sqrt{3}}{2}\log x\right)\right]$$

and $$\text{P.I.}=\frac{1}{D^3+1}e^t+\frac{1}{D^3+1}t=\frac{1}{1^3+1}e^t+(1+D^3)^{-1}t=\frac{1}{2}e^t+(1-D^3)t=\frac{1}{2}e^t+t=\frac{1}{2}x+\log x$$

$$\therefore\ \text{Complete solution is}\quad y=\frac{C_1}{x}+\sqrt{x}\left[C_2\cos\left(\frac{\sqrt{3}}{2}\log x\right)+C_3\sin\left(\frac{\sqrt{3}}{2}\log x\right)\right]+\frac{x}{2}+\log x.$$ **Ans.**

6. *Solve* $(x+a)^2\dfrac{d^2y}{dx^2}-4(x+a)\dfrac{dy}{dx}+6y=x.$

SOLUTION: Equation is of Legendre's form.

Putting $x+a=e^t$ in the given equation, it becomes

$$[D(D-1)-4D+6]\,y=e^t-a \quad\text{where}\quad D\equiv\frac{d}{dt}$$

or $(D^2-5D+6)\,y=e^t-a.$ The auxiliary equation is $m^2-5m+6=0$

whose roots are $m=2,3,$ hence $\text{C.F.}=C_1e^{2t}+C_2e^{3t}=C_1(x+a)^2+C_2(x+a)^3$

$$\text{P.I.}=\frac{1}{D^2-5D+6}(e^t-a)=\frac{1}{D^2-5D+6}e^t-\frac{1}{D^2-5D+6}ae^{0t}$$

$$=\frac{1}{1-5+6}e^t-\frac{1}{0-0+6}a=\frac{1}{2}e^t-\frac{a}{6}=\frac{1}{2}(x+a)-\frac{a}{6}=\frac{1}{6}(3x+2a)$$

$\therefore$ Complete solution is $y=C_1(x+a)^2+C_2(x+a)^3+\dfrac{1}{6}(3x+2a).$ **Ans.**

7. *Solve* $(x+1)^2 \dfrac{d^2y}{dx^2} + (x+1)\dfrac{dy}{dx} + y = 2\sin[\log(x+1)]$.

SOLUTION: The equation is of Legendre's homogeneous form. Putting $x + 1 = e^t$, it becomes

$$[D(D-1) + D + 1]\,y = 2\sin t \quad \text{where} \quad D \equiv \frac{d}{dt}$$

The A.E. is $m^2 + 1 = 0$ hence $m = \pm i$

$\therefore$ C.F. $= C_1 \cos t + C_2 \sin t = C_1 \cos(\log(x+1)) + C_2 \sin(\log(x+1))$.

$$\text{P.I.} = \frac{1}{D^2+1}\,2\sin t = \frac{2\cdot t}{2D}\sin t = -t\cos t = -\log(x+1)\cos(\log(x+1))$$

$\therefore$ Complete solution is

$$y = C_1\cos(\log(x+1)) + C_2\sin(\log(x+1)) - \log(x+1)\cos(\log(x+1)).$$ **Ans.**

8. *Solve* $(3x+2)^2 \dfrac{d^2y}{dx^2} + 3(3x+2)\dfrac{dy}{dx} - 36y = 3x^2 + 4x + 1$.

SOLUTION: Equation is of Legendre's homogeneous form.

Putting $3x + 2 = e^t$, equation becomes

$$[3^2 D(D-1) + 3.3D - 36]\,y = 3\left(\frac{e^t-2}{3}\right)^2 + 4\left(\frac{e^t-2}{3}\right) + 1$$

or $$(D^2-4)\,y = \frac{1}{27}(e^{2t} - 4e^t + 4) + \frac{4}{27}(e^t - 2) + \frac{1}{9} = \frac{1}{27}(e^{2t} - 1)$$

Here C.F. $= C_1 e^{2t} + C_2 e^{-2t} = C_1(3x+2)^2 + C_2(3x+2)^{-2}$.

and $$\text{P.I.} = \frac{1}{D^2-4}\frac{1}{27}(e^{2t}-1) = \frac{1}{27}\left[\frac{1}{D^2-4}e^{2t} - \frac{e^{ot}}{D^2-4}\right] = \frac{1}{27}\left[\frac{1}{D^2-4}e^{2t} + \frac{1}{4}\right]$$

$$\frac{1}{D^2-4}e^{2t} = \frac{1}{(D-2)}\left[\frac{1}{D+2}e^{2t}\right] = \frac{1}{D-2}\left[\frac{1}{2+2}e^{2t}\right] = \frac{1}{4}\frac{1}{D-2}e^{2t}$$

$$= \frac{1}{4}e^{2t}\int e^{-2t}\cdot e^{2t}\,dt = \frac{1}{4}e^{2t}\cdot t$$

$\therefore$ $$\text{P.I.} = \frac{1}{27}\left[\frac{t}{4}e^{2t} + \frac{1}{4}\right] = \frac{1}{108}[(3x+2)^2\log(3x+2) + 1]$$

Hence complete solution is $y = C_1(3x+2)^2 + \dfrac{C_2}{(3x+2)^2} + \dfrac{1}{108}[1 + (3x+2)^2\log(3x+2)]$ **Ans.**

9. *Solve the equation*

$$x^2\frac{d^2y}{dx^2}+x\frac{dy}{dx}+y=\sin(\log x^2)$$

SOLUTION: Putting $x = e^t$ the given equation becomes

$$[D(D-1)+D+1]\,y=\sin(\log e^{2t}) \quad \text{or} \quad (D^2+1)\,y=\sin 2t$$

$$\text{C.F.} = C_1\cos t + C_2\sin t = C_1\cos(\log x) + C_2\sin(\log x).$$

$$\text{P.I.} = \frac{1}{D^2+1}\sin 2t=\frac{1}{-4+1}\sin 2t=\frac{-1}{3}\sin(2\log x)$$

$\therefore$ Complete solution is $\quad y = C_1\cos(\log x) + C_2\sin(\log x) - \frac{1}{3}\sin[\log(x^2)]$. **Ans.**

10. *Solve the equation*

$$x^2\frac{d^3y}{dx^3}+3x\frac{d^2y}{dx^2}+\frac{dy}{dx}=x^2\log x$$

SOLUTION: Given equation can be written as $x^3\dfrac{d^3y}{dx^3}+3x^2\dfrac{d^2y}{dx^2}+x\dfrac{dy}{dx}=x^3\log x.$

Putting $x = e^t$, the equation, which is of homogeneous form, becomes

$$[D(D-1)(D-2)+3D(D-1)+D]\,y = e^{3t}\cdot t \quad \text{where } D\equiv\frac{d}{dt}.$$

or $D^3y = te^{3t}$. The C.F. $= C_1 + C_2 t + C_3 t^2 = C_1 + C_2\log x + C_3(\log x)^2$

$$\text{P.I.} = \frac{1}{D^3}e^{3t}\cdot t = e^{3t}\cdot\frac{1}{(D+3)^3}t = e^{3t}\frac{1}{27}\left(1+\frac{D}{3}\right)^{-3}t$$

$$=\frac{e^{3t}}{27}\left[1-\frac{3D}{3}+\ldots\right]t=\frac{e^{3t}}{27}[t-1]=\frac{x^3}{27}(\log x-1)$$

$\therefore$ Complete solution is $\quad y = C_1 + C_2\log x + C_3(\log x)^2 + \dfrac{x^3}{27}(\log x - 1)$ **Ans.**

11. *Show that the solution of the equation*

$$(1+x)^2\frac{d^2y}{dx^2}+(1+x)\frac{dy}{dx}+y=\sin[2\log(1+x)]$$

is $y = c_1\cos\log(1+x) + c_2\sin\log(1+x) - \dfrac{1}{3}\sin[2\log(1+x)]$

where c_1 and c_2 are arbitrary constants.

SOLUTION: Equation is of Legendre's form. Putting $x + 1 = e^t$, it becomes

$$[D(D-1)+D+1]\,y = \sin[2\log(x+1)] = \sin(2t)$$

or $(D^2 + 1)\, y = \sin (2t)$

$$\text{C.F.} = C_1 \cos t + C_2 \sin t = C_1 \cos \log (x + 1) + C_2 \sin \log (x + 1)$$

and $$\text{P.I.} = \frac{1}{D^2 + 1} \sin 2t = \frac{1}{-4 + 1} \sin 2t = \frac{-1}{3} \sin[2 \log (x + 1)]$$

$\therefore$ Complete solution is

$$y = C_1 \cos \log (x + 1) + C_2 \sin \log (x + 1) - \frac{1}{3} \sin[2 \log (x + 1)]$$ **Ans.**

12. *Solve the differential equation* $(x^2D^2 - xD + 2)\, y = x \log x$ where $D \equiv \dfrac{d}{dx}$.

SOLUTION: Putting $x = e^t$ the given equation becomes

$$[D'(D' - 1) - D' + 2]\, y = t\, e^t \text{ when } D' = \frac{d}{dt}.$$

or $(D'^2 - 2D' + 2)y = t\, e^t$. Here A.E. is $m^2 - 2m + 2 = 0$

whose roots are $m = 1 \pm 1$.

$\therefore$ $$\text{C.F.} = e^t\, (C_1 \cos t + C_2 \sin t) = x[C_1 \cos \log x + C_2 \sin \log x]$$

and $$\text{P.I.} = \frac{1}{D'^2 - 2D' + 2} t e^t = e^t \frac{1}{(D' + 1)^2 - 2(D' + 1) + 2} t$$

$$= e^t \frac{1}{D'^2 + 1} t = e^t(1 + D'^2)^{-1}\, t = e^t\, (1 - D'^2 - ...)\, t = e^t \cdot t = x \log x.$$

$\therefore$ Complete solution is $y = x[C_1 \cos \log x + C_2 \sin \log x] + x \log x$. **Ans.**

SOLUTIONS TO EXERCISE 10C

1. *Solve* $\dfrac{dx}{dt} + 5x + y = e^t,\ \dfrac{dy}{dt} + 3y - x = e^{2t}.$

SOLUTION: Equation can be written as

$$(D + 5)\,x + y = e^t \quad \text{and} \quad -x + (D + 3)\,y = e^{2t} \quad \text{where } D \equiv \frac{d}{dt}.$$

Eliminating x in the above equations, we get

$$[(D + 3)\,(D + 5) + 1]\,y = (D + 5)\,e^{2t} + e^t$$

or $\quad (D^2 + 8D + 16)\,y = 2e^{2t} + 5e^{2t} + e^t \quad$ or $\quad (D + 4)^2\,y = 7e^{2t} + e^t$

The complementary function for $y = (C_1 + C_2\,t)\,e^{-4t}$

and P.I. for $\quad y = \dfrac{1}{(D+4)^2}(7e^{2t} + e^t) = \dfrac{7}{(2+4)^2}e^{2t} + \dfrac{1}{(1+4)^2}e^t$

$\therefore \quad y = \text{C.F.} + \text{P.I.} = (C_1 + C_2\,t)\,e^{-4t} + \dfrac{7}{36}e^{2t} + \dfrac{1}{25}e^t.$

and hence $\quad x = (D + 3)\,y - e^{2t} = Dy + 3y - e^{2t}$

$$= C_2\,e^{-4t} - 4\,(C_1 + C_2\,t)\,e^{-4t} + \frac{7}{18}e^{2t} + \frac{1}{25}e^t + 3(C_1 + C_2 t)\,e^{-4t} + \frac{7}{12}e^{2t} + \frac{3}{25}e^t - e^{2t}$$

$\therefore \quad x = (C_2 - C_1 - C_2 t)e^{-4t} + \dfrac{4}{25}e^t - \dfrac{1}{36}e^{2t}$

Hence $\quad x = (C_2 - C_1 - C_2 t)e^{-4t} + \dfrac{4}{25}e^t - \dfrac{1}{36}e^{2t},$

and $\quad y = (C_1 + C_2 t)e^{-4t} + \dfrac{7}{36}e^{2t} + \dfrac{1}{25}e^t.$ **Ans.**

2. *Two coupled electrical circuits have currents* i_1 *and* i_2 *given by*

$$L\frac{di_1}{dt} + Ri_1 + R\,(i_1 - i_2) = E$$

$$L\frac{di_2}{dt} + Ri_2 - R(i_1 - i_2) = 0$$

where the inductance L, resistance R and E.M.F. E are constants. Find i_1, i_2 as functions of time t, if, initially $i_1 = i_2 = 0$ at $t = 0$.

SOLUTION: Equation can be written as

$$(LD + 2R)\,i_1 - Ri_2 = E \quad \text{and} \quad -Ri_1 + (LD + 2R)\,i_2 = 0$$

Eliminating i_1, in the above two equations, we get

$$[-R^2 + (LD + 2R)^2]\,i_2 = RE \quad \text{or} \quad (L^2 D^2 + 4LRD + 3R^2)\,i_2 = RE$$

A.E. is $L^2 m^2 + 4LRm + 3R^2 = 0 \quad \therefore \quad m = \dfrac{-4LR \pm \sqrt{16L^2R^2 - 12L^2R^2}}{2L^2} = \dfrac{-R}{L}, \dfrac{-3R}{L}$

$\therefore$ C.F. (for i_2) $= C_1 e^{-Rt/L} + C_2 e^{-3Rt/L}$

and P.I. (for i_2) $= \dfrac{1}{L^2D^2 + 4LRD + 3R^2}(ER\, e^{ot}) = \dfrac{1}{3R^2}(ER) = \dfrac{E}{3R}$

$$\therefore \quad i_2 = C_1 e^{-Rt/L} + C_2 e^{-3Rt/L} + \frac{E}{3R}$$

Given $\quad i_2 = 0$ at $t = 0$ hence $C_1 + C_2 = -\dfrac{E}{3R}$ ...(1)

Next,
$$i_1 = \frac{1}{R}\left[L\frac{di_2}{dt} + 2Ri_2\right] = \frac{L}{R}\frac{di_2}{dt} + 2i_2$$
$$= \frac{L}{R}\left[-C_1\frac{R}{L}e^{-Rt/L} - 3C_2\frac{R}{L}e^{-3Rt/L}\right] + 2C_1 e^{-Rt/L} + 2C_2 e^{-3Rt/L} + \frac{2E}{3R}$$
$$= C_1 e^{-Rt/L} - C_2 e^{-3Rt/L} + \frac{2}{3}\frac{E}{R}$$

Since $\quad i_1 = 0$ at $t = 0$ we have $C_1 - C_2 = -\dfrac{2}{3}\dfrac{E}{R}$ ...(2)

Solving (1) and (2), we get $C_1 = -\dfrac{E}{2R}$ and $C_2 = \dfrac{1}{6}\dfrac{E}{R}$

Thus,
$$i_1 = \frac{-E}{2R}e^{-Rt/L} - \frac{E}{6R}e^{-3Rt/L} + \frac{2}{3}\frac{E}{R}$$

and $\quad i_2 = \dfrac{-E}{2R}e^{-Rt/L} + \dfrac{E}{6R}e^{-3Rt/L} + \dfrac{E}{3R}$ is the solution. **Ans.**

3. *Solve* $\dfrac{dx}{dt} = x^2 + xy, \dfrac{dy}{dt} = y^2 + xy$ *given that* $x = 1, y = 2$ at $t = 0$.

SOLUTION: Equation can be written as $\dfrac{dx}{dt} = x(x+y), \quad \dfrac{dy}{dt} = y(x+y)$

Adding the above, we get $\dfrac{d}{dt}(x+y) = (x+y)^2$

or
$$\frac{d(x+y)}{(x+y)^2} = dt \quad \therefore \quad \frac{-1}{x+y} = t + C_1$$

Given that $x = 1, y = 2$ at $t = 0$, we get $-\dfrac{1}{3} = C_1 \quad \therefore \quad \dfrac{-1}{x+y} = t - \dfrac{1}{3}$ ...(1)

Using (1) in the first given equation, we get

$$\frac{dx}{dt} = \frac{-x}{t-\frac{1}{3}} \quad \text{or} \quad \frac{dx}{x} = \frac{-dt}{t-\frac{1}{3}}.$$

$$\Rightarrow \qquad \log x = -\log\left(t-\frac{1}{3}\right) + \log C_2 \quad \text{or} \quad x = \frac{C_2}{t-\frac{1}{3}}$$

Since $x = 1$ at $t = 0$ we get $1 = -3C_2$ or $C_2 = -\frac{1}{3}$

$$\therefore \qquad x = \frac{-\frac{1}{3}}{t-\frac{1}{3}} = \frac{1}{1-3t}$$

Next $$y = -x + \frac{3}{1-3t} = \frac{-1}{1-3t} + \frac{3}{1-3t} = \frac{2}{1-3t}$$

$\therefore$ Solution is $$x = \frac{1}{1-3t}, \quad y = \frac{2}{1-3t}.$$ **Ans.**

4. *Velocity components of a particle are given by*

$$\frac{dx}{dt} + 2x - y = 1 \quad \text{and} \quad \frac{dy}{dt} + 2y - x = 0.$$

If the particle was at the origin initially, find its path.

SOLUTION: The given simultaneous equations can be written as

$$(D+2)\,x - y = 1 \quad \text{and} \quad -x + (D+2)\,y = 0 \quad \text{where } D \equiv \frac{d}{dt}.$$

Eliminating x in the above equation, we get

$$[(D+2)^2 - 1]\,y = 1 \quad \text{or} \quad (D+3)\,(D+1)\,y = 1$$

The C.F. for $y = C_1\, e^{-t} + C_2\, e^{-3t}$

and $$\text{P.I.} = \frac{1}{D^2+4D+3}1 = \frac{1}{D^2+4D+3}e^{0t} = \frac{1}{0+0+3}e^{0t} = \frac{1}{3}.$$

$\therefore\; y = C_1\, e^{-t} + C_2\, e^{-3t} + \frac{1}{3}$. Since $y = 0$ at $t = 0$ we have $0 = C_1 + C_2 + \frac{1}{3}$ ...(2)

Substituting this value of y in the second given equation, we get

$$x = Dy + 2y = -C_1\, e^{-t} - 3C_2\, e^{-3t} + 2C_1 e^{-t} + 2C_2\, e^{-3t} + \frac{2}{3}$$

$$= C_1\, e^{-t} - C_2\, e^{-3t} + \frac{2}{3} \qquad \text{...(3)}$$

Since $x = 0$ at $t = 0$ hence $0 = C_1 - C_2 + \frac{2}{3}$ ...(4)

Solving (2) and (4) for C_1 and C_2, gives $C_1 = \frac{-1}{2}$ and $C_2 = \frac{1}{6}$

Hence the solution of the given equations is

$$x = \frac{-1}{2}e^{-t} - \frac{1}{6}e^{-3t} + \frac{2}{3} \quad \text{and} \quad y = \frac{-1}{2}e^{-t} + \frac{1}{6}e^{-3t} + \frac{1}{3}$$ **Ans.**

5. *The equations of motion of a particle, are given by*

$$\frac{dx}{dt} + \omega y = 0, \quad \frac{dy}{dt} - \omega x = 0$$

Find the path of the particle and show that it is a circle.

SOLUTION: The given equations can be written as

$$Dx + wy = 0 \quad ...(1), \quad -wx + Dy = 0 \quad ...(2) \qquad \text{where } D \equiv \frac{d}{dt}.$$

Eliminating x in between (1) and (2), we get $(D^2 + w^2)\, y = 0$

whose solution is $y = C_1 \cos wt + C_2 \sin wt$ ...(3)

Substituting (3) in (2), we get

$$x = \frac{1}{w}Dy = \frac{1}{w}[-C_1\, w \sin wt + C_2\, w \cos wt] = -C_1 \sin wt + C_2 \cos wt$$

Thus, the path of the particle is $x = -C_1 \sin wt + C_2 \cos wt$, $y = C_1 \cos wt + C_2 \sin wt$

or $x^2 + y^2 = C_1^2 + C_2^2$ which is a circle. **Ans.**

6. *A mechanical system satisfies the simultaneous equations*

$$2\frac{d^2x}{dt^2} + 3\frac{dy}{dt} = 4, \quad 2\frac{d^2y}{dt^2} - 3\frac{dx}{dt} = 0.$$

Obtain x and y in terms of t, given that x, y, $\frac{dx}{dt}, \frac{dy}{dt}$ all vanish at $t = 0$.

SOLUTION: The given simultaneous equations can be written as

$$2D^2x + 3Dy = 4, \quad 3Dx - 2D^2y = 0 \quad \text{where } D \equiv \frac{d}{dt}.$$

Eliminating x in the above equations we get $(9D + 4D^3)\, y = 12$

Its A.E. is $4m^3 + 9m = 0$ whose roots are $m = 0$, $m = \pm\frac{3}{2}i$

$\therefore$ C.F. (for y) $= C_1 + C_2 \cos\frac{3t}{2} + C_3 \sin\frac{3t}{2}$.

and P.I. (for y) $= \frac{1}{4D^3 + 9D}12 = \frac{12}{D}\frac{1}{4D^2 + 9}e^{0t} = \frac{12}{D}\frac{1}{0+9}e^{0t} = \frac{4}{3}\int 1\, dt = \frac{4t}{3}$

$$\therefore \qquad y = C_1 + C_2 \cos\frac{3t}{2} + C_3 \sin\frac{3t}{2} + \frac{4t}{3}$$

From $3Dx - 2D^2 y = 0$ we have $x = \frac{2}{3} Dy$

$$\therefore \qquad x = \frac{2}{3}\left[0 - \frac{3}{2} C_2 \sin\frac{3t}{2} + \frac{3}{2} C_3 \cos\frac{3t}{2} + \frac{4}{3}\right] = \frac{8}{9} - C_2 \sin\frac{3t}{2} + C_3 \cos\frac{3t}{2}$$

Next, since $x = y = 0$ at $t = 0$, we have $C_1 + C_2 = 0$ and $\frac{8}{9} + C_3 = 0$

Also since $\frac{dx}{dt} = \frac{dy}{dt} = 0$ at $t = 0$, we get

$$0 - \frac{3}{2} C_2 = 0 \text{ and } 0 = \frac{+3}{2} C_3 + \frac{4}{3}$$

$$\therefore \qquad C_1 = C_2 = 0 \text{ and } C_3 = -\frac{-8}{9}.$$

$$\therefore \qquad x = \frac{8}{9} - \frac{8}{9} \cos\frac{3t}{2}, \quad y = \frac{-8}{9} \sin\frac{3t}{2} + \frac{4t}{3}$$

$\therefore$ Solution is $x = \frac{8}{9}\left(1 - \cos\frac{3t}{2}\right)$ and $y = \frac{4}{3}\left(t - \frac{2}{3}\sin\frac{3t}{2}\right)$ **Ans.**

7. *Solve* $\frac{l\,dx}{(m-n)\,yz} = \frac{m\,dy}{(n-l)\,zx} = \frac{n\,dz}{(l-m)\,xy}$.

SOLUTION: $\frac{ldx}{(m-n)yz} = \frac{m\,dy}{(n-l)\,zx} = \frac{n\,dz}{(l-m)\,xy} = \frac{lx \cdot dx + my\,dy + nz\,dz}{xyz\,(m-n+n-l+l-m)}$

$\Rightarrow$ $l\,xdx + m\,ydy + nz\,dz = 0$ which gives $l\,x^2 + my^2 + nz^2 = C_1$

Also $\frac{ldx}{(m-n)yz} = \frac{m\,dy}{(n-l)\,zx} = \frac{n\,dz}{(l-m)\,xy} = \frac{l^2 xdx + m^2 y\,dy + n^2\,zdz}{xyz\,(lm - nl + nm - lm + \ln - mn)}$.

$\Rightarrow$ $l^2\,xdx + m^2\,y\,dy + n^2\,zdz = 0$ which gives $l^2 x^2 + m^2\,y^2 + n^2 z^2 = C_2$

$\therefore$ Solution is $lx^2 + my^2 + nz^2 = C_1$, $l^2\,x^2 + m^2\,y^2 + n^2\,z^2 = C_2$. **Ans.**

8. *Solve* $\frac{dx}{y+z} = \frac{dy}{z+x} = \frac{dz}{x+y}$.

SOLUTION: $\frac{dx}{y+z} = \frac{dy}{z+x} = \frac{dz}{x+y} = \frac{dx+dy+dz}{2(x+y+z)} = \frac{dx-dy}{y-x} = \frac{dy-dz}{z-y}$

Here $\frac{d\,(x-y)}{x-y} = \frac{d\,(y-z)}{y-z} \Rightarrow \log\,(x-y) = \log\,(y-z) + \log C_1$

or $\frac{x-y}{y-z} = C_1$.

Also $\dfrac{d(x+y+z)}{2(x+y+z)} = \dfrac{d(x-y)}{-(x-y)}$

$\Rightarrow$ $\log(x+y+z) = -2\log(x-y) + \log C_2$ or $(x+y+z)(x-y)^2 = C_2$

$\therefore$ Solution is $\dfrac{x-y}{y-z} = C_1$ and $(x-y)^2(x+y+z) = C_2$ **Ans.**

9. *Solve the simultaneous linear equations* $\dfrac{dx}{y^3x-2x^4} = \dfrac{dy}{2y^4-x^3y} = \dfrac{dz}{9z(x^3-y^3)}$.

SOLUTION: $\dfrac{dx}{y^3x-2x^4} = \dfrac{dy}{2y^4-x^3y} = \dfrac{dz}{9z(x^3-y^3)} = \dfrac{\dfrac{dx}{x}+\dfrac{dy}{y}+\dfrac{dz}{3z}}{0}$

$\Rightarrow$ $\dfrac{dx}{x}+\dfrac{dy}{y}+\dfrac{dz}{3z} = 0$ or $x^3y^3z = C_1$

Next, $\dfrac{dx}{y^3x-2x^4} = \dfrac{dy}{2y^4-x^3y} = \dfrac{x^2dx+y^2dy}{2(y^6-x^6)} = \dfrac{dz}{9z(x^3-y^3)}$

or $\dfrac{x^2dx+y^2\cdot dy}{2(x^3+y^3)} = \dfrac{-dz}{9z}$ or $\dfrac{3x^2dx+3y^2\,dy}{x^3+y^3}+\dfrac{2dz}{3z}=0$

$\Rightarrow$ $\log(x^3+y^3)+\dfrac{2}{3}\log z = \log C_2$ or $(x^3+y^3)^3z^2 = C_2$

$\therefore$ The solution are

$x^3y^3z = C_1$ and $(x^3+y^3)^3z^2 = C_2$ **Ans.**

10. *Solve* $\dfrac{dx}{2x} = \dfrac{dy}{-y} = \dfrac{dz}{4xy^2-2z}$.

SOLUTION: $\dfrac{dx}{2x} = \dfrac{dy}{-y}$ or $\dfrac{dx}{x}+\dfrac{2dy}{y}=0$ $\Rightarrow$ $\log x + 2\log y = \log C_1$

or $xy^2 = C_1$. ...(1)

The given equations can be written as

$$\frac{dx}{2x} = \frac{dy}{-y} = \frac{dz}{4C_1-2z}.$$

For second solution, consider $\dfrac{dx}{2x} = \dfrac{dz}{4C_1-2z}$ $\Rightarrow$ $\log x = -\log(2C_1-z)+\log C_2$

or $x(2C_1-z) = C_2$ or $x(2xy^2-z) = C_2$

Hence solutions are $xy^2 = C_1$ and $x(2xy^2-z) = C_2$ **Ans.**

11. *Solve* $\dfrac{dx}{x(z^2-y^2)} = \dfrac{dy}{y(x^2-z^2)} = \dfrac{dz}{z(y^2-x^2)}$.

SOLUTION: $\dfrac{dx}{x(z^2-y^2)} = \dfrac{dy}{y(x^2-z^2)} = \dfrac{dz}{z(y^2-x^2)} = \dfrac{xdx+ydy+zdz}{0} = \dfrac{\dfrac{dx}{x}+\dfrac{dy}{y}+\dfrac{dz}{z}}{0}$

Hence, $xdx + ydy + zdz = 0$ and $\dfrac{dx}{x}+\dfrac{dy}{y}+\dfrac{dz}{z}=0$

$\Rightarrow$ $x^2+y^2+z^2 = C_1$ and $\log x + \log y + \log z = \log C_2$ or $xyz = C_2$

Hence, the solution is $x^2+y^2+z^2=C_1$ and $xyz = C_2$ **Ans.**

12. *Solve* $\dfrac{dx}{x(y-z)} = \dfrac{dy}{y(z-x)} = \dfrac{dz}{z(x-y)}$.

SOLUTION: $\dfrac{dx}{x(y-z)} = \dfrac{dy}{y(z-x)} = \dfrac{dz}{z(x-y)} = \dfrac{dx+dy+dy}{0} = \dfrac{\dfrac{dx}{x}+\dfrac{dy}{y}+\dfrac{dz}{z}}{0}$

Hence, $dx + dy + dz = 0$ and $\dfrac{dx}{x}+\dfrac{dy}{y}+\dfrac{dz}{z}=0$

$\Rightarrow$ $x+y+z=C_1$ and $\log x + \log y + \log z = \log C_2$

Hence the solution is $x+y+z=C_1$ and $xyz = C_2$ **Ans.**

13. *Solve* $\dfrac{dx}{1} = \dfrac{dy}{1} = \dfrac{dz}{(x+y)(1+2xy+3x^2y^2)}$

SOLUTION: Equation is $\dfrac{dx}{1} = \dfrac{dy}{1} = \dfrac{dz}{(x+y)(1+2xy+3x^2y^2)}$

Consider $\dfrac{dx}{1} = \dfrac{dy}{1}$ which gives $x - y = C_1$

Also, we have $\dfrac{dx}{1} = \dfrac{dy}{1} = \dfrac{ydx+xdy}{y+x} = \dfrac{d(xy)}{x+y} = \dfrac{dz}{(x+y)(1+2xy+3x^2y^2)}$

$\Rightarrow$ $(1+2xy+3x^2y^2)\,d(xy) = dz$

Its solution is $xy + (xy)^2 + (xy)^3 = z - C_2$

Thus the solutions are $x - y = C_1$ and $z = xy + (xy)^2 + (xy)^3 + C_2$ **Ans.**

14. *Solve the simultaneous equations*

$$\frac{dx}{dt}+4x+3y=t, \quad \frac{dy}{dt}+2x+5y=e^t$$

SOLUTION: The equations can be written as $(D+4)x+3y=t$, $2x+(D+5)y=e^t$

Eliminating y between the above equations, we get

$$[(D+4)(D+5)-6]\,x = (D+5)\,t - 3e^t \quad \text{or} \quad (D^2+9D+14)\,x = 1+5t-3e^t$$

A.E. is $m^2 + 9m + 14 = 0 \therefore m = -2, -7 \quad \therefore$ C.F. (for x) $= C_1 e^{-2t} + C_2 e^{-7t}$

and for x, P.I. $= \dfrac{1}{D^2+9D+14}(1+5t-3e^t)$

$$= \frac{1}{D^2+9D+14}e^{0t} - \frac{3}{D^2+9D+14}e^t + \frac{5}{14+9D+D^2}t$$

$$= \frac{1}{14} - \frac{1}{8}e^t + \frac{5}{14}\left(1+\frac{9}{14}D+\frac{D^2}{14}\right)^{-1} t = \frac{1}{14} - \frac{1}{8}e^t + \frac{5}{14}\left(t - \frac{9}{14}\right)$$

$$= \frac{5}{14}t - \frac{1}{8}e^t - \frac{31}{196}$$

Therefore $\quad x = C_1 e^{-2t} + C_2 e^{-7t} + \dfrac{5}{14}t - \dfrac{1}{8}e^t - \dfrac{31}{196}$

and $\quad Dx = -2C_1 e^{-2t} - 7C_2 e^{-7t} + \dfrac{5}{14} - \dfrac{1}{8}e^t$

Substituting these values of x and Dx in the first given equation, gives

$$y = \frac{1}{3}[t - Dx - 4x] = -\frac{1}{3}\left[2C_1 e^{-2t} - 3C_2 e^{-7t} - \frac{5}{8}e^t + \frac{5}{14} - \frac{31}{49} + \frac{3}{7}t\right]$$

$\therefore$ Solution is $x = C_1 e^{-2t} + C_2 e^{-7t} + \dfrac{5t}{14} - \dfrac{1}{8}e^t - \dfrac{31}{196}$,

$$y = \frac{-2}{3}C_1 e^{-2t} + C_2 e^{-7t} + \frac{5}{24}e^t - \frac{1}{7}t + \frac{9}{98}$$ **Ans.**

15. *Solve the system of simultaneous equations*

$$\frac{d^2y}{dt^2} + \frac{dy}{dt} - 2y = \sin t, \quad \frac{dx}{dt} + x - 3y = 0.$$

SOLUTION: $(D^2 + D - 2)\,y = \sin t$ and $(D + 1)\,x - 3y = 0$

For y, A.E. is $m^2 + m - 2 = 0 \quad \therefore \quad m = 1, -2$, so C.F. for y is $= C_1 e^t + C_2 e^{-2t}$.

For y, P.I. $= \dfrac{1}{D^2+D-2}\sin t = \dfrac{1}{-1+D-2}\sin t = \dfrac{D+3}{D^2-9}\sin t = \dfrac{-1}{10}(\cos t + 3\sin t)$.

$\therefore \quad y = C_1 e^t + C_2 e^{-2t} - \dfrac{1}{10}\cos t - \dfrac{3}{10}\sin t$.

Further, $\dfrac{dx}{dt} + x = 3y = 3C_1 e^t + 3C_2 e^{-2t} - \dfrac{3}{10}\cos t - \dfrac{9}{10}\sin t$ which is linear in x.

I.F. $= e^{\int dt} = e^t$, hence the solution is

$$x\,e^t = \int e^t\left(3C_1 e^t + 3C_2 e^{-2t} - \frac{3}{10}\cos t - \frac{9}{10}\sin t\right)dt$$

$$= 3C_1\int e^{2t}dt + 3C_2\int e^{-t}dt - \frac{3}{10}\int e^t\cos t\,dt - \frac{9}{10}\int e^t\sin t\,dt$$

$$= \frac{3C_1}{2}e^{2t} - 3C_2\,e^{-t} - \frac{3}{10}\frac{e^t}{2}(\cos t + \sin t) - \frac{9}{10}\frac{e^t}{2}(\sin t - \cos t)$$

$$\therefore \quad x = \frac{3C_1}{2}e^t - \frac{3C_2}{2}e^{-2t} - \frac{3}{20}(\cos t + \sin t) - \frac{9}{20}(\sin t - \cos t)$$

Thus, $x = \frac{3C_1}{2}e^t - \frac{3C_2}{2}e^{-2t} + \frac{3}{10}(\cos t - 2\sin t)$

and $y = C_1 e^t + C_2 e^{-2t} - \frac{1}{10}(\cos t + 3\sin t)$ is the solution. **Ans.**

16. *Solve the simultaneous equations* $\frac{dx}{dt} + 5x - 2y = t$, $\frac{dy}{dt} + 2x + y = 0$ *given that* $x = y = 0$ *at* $t = 0$.

SOLUTION: Equations are $(D + 5)x - 2y = t$, $2x + (D + 1)y = 0$ where $D \equiv \frac{d}{dt}$

Eliminating x between these, we get $(D^2 + 6 + 9)y = -2t$

For C.F. of y, A.E. is $m^2 + 6m + 9 = 0$ $\therefore$ $m = -3, -3$ and C.F. $= (C_1 t + C_2)e^{-3t}$

For y, P.I. $= \frac{1}{(D+3)^2}(-2t) = \frac{-2}{9}\left(1 + \frac{D}{3}\right)^{-2} t = \frac{-2}{9}\left(1 - \frac{2D}{3}\right)t = \frac{-2t}{9} + \frac{4}{27}$

$$\therefore \quad y = (C_1 t + C_2)e^{-3t} - \frac{2}{9}t + \frac{4}{27}$$

Substituting for y and Dy in the equation $(D + 1)y + 2x = 0$, we get

$$x = \frac{1}{2}\left[(C_1 t + C_2)e^{-3t} - \frac{2}{9}t + \frac{4}{27} + C_1 e^{-3t} - 3(C_1 t + C_2)e^{-3t} - \frac{2}{9}\right]$$

or $$x = e^{-3t}\left[\left(t - \frac{1}{2}\right)C_1 + C_2\right] + \frac{t}{9} + \frac{1}{27}$$

Since $x = y = 0$ at $t = 0$, we get $C_1 = -\frac{2}{9}$ and $C_2 = -\frac{4}{27}$.

$$\therefore \quad x = \frac{-1}{27}e^{-3t}(6t + 1) + \frac{1}{27}(1 + 3t) \quad \text{and} \quad y = \frac{-2}{27}e^{-3t}(3t + 2) + \frac{2}{27}(2 - 3t)$$ **Ans.**

17. *Solve the following simultaneous equations*

$$\frac{dx}{dt} + 2x + 3y = 0, \quad 3x + \frac{dy}{dt} + 2y = 2e^{2t}.$$

SOLUTION: Equations are $(D + 2)\,x + 3y = 0, \quad 3x + (D + 2)\,y = 2e^{2t}$ where $D \equiv \dfrac{d}{dt}$.

Eliminating y in these, we get

$$[(D + 2)^2 - 9]\,x = -6\,e^{2t} \qquad \text{or} \qquad (D^2 + 4D - 5)\,x = -6\,e^{2t}$$

For x, A.E. is $m^2 + 4m - 5 = 0 \quad \therefore \; m = -5, 1 \quad \therefore$ C.F. $= C_1\,e^t + C_2\,e^{-5t}$.

For x, P.I. $= \dfrac{1}{D^2 + 4D - 5}(-6e^{2t}) = \dfrac{-6}{2^2 + 4 \cdot 2 - 5}e^{2t} = \dfrac{-6}{7}e^{2t}$

Thus, $x = C_1\,e^t + C_2\,e^{-5t} - \dfrac{6}{7}e^{2t}$. $\therefore \quad \dfrac{dx}{dt} = C_1\,e^t - 5\,C_2\,e^{-5t} - \dfrac{12}{7}e^{2t}$

Putting these in the first of the given equations, we get

$$y = \frac{-1}{3}\left[2x + \frac{dx}{dt}\right] = \frac{-1}{3}\left[2C_1\,e^t + 2C_2\,e^{-5t} - \frac{12}{7}e^{2t} + C_1\,e^t - 5C_2\,e^{-5t} - \frac{12}{7}e^{2t}\right]$$

$$= -C_1\,e^t + C_2\,e^{-5t} + \frac{8}{7}e^{2t}$$

Therefore, the solution is

$$x = C_1\,e^t + C_2\,e^{-5t} - \frac{6}{7}e^{2t}, \quad y = -C_1\,e^t + C_2\,e^{-5t} + \frac{8}{7}e^{2t}$$ **Ans.**

18. *Solve* $\dfrac{dx}{dt} = 7x - y$ and $\dfrac{dy}{dt} = 2x + 5y$

SOLUTION: Equations are $(D - 7)\,x + y = 0, \quad -2x + (D - 5)\,y = 0$ where $D \equiv \dfrac{d}{dt}$.

Eliminating x in the above equations, we get

$$[(D - 5)(D - 7) - 2]\,y = 0 \qquad \text{or} \qquad (D^2 - 12D + 33)\,y = 0$$

A.E. for y is $m^2 - 12m + 33 = 0 \qquad \therefore \qquad m = 6 \pm \sqrt{3}$

$\therefore$ Solution is $y = C_1\,e^{(6+\sqrt{3})t} + C_2\,e^{(6-\sqrt{3})t}$ ...(1)

Putting this value of y in the second given equation, we get

$$x = \frac{-5}{2}y + \frac{1}{2}\frac{dy}{dt} = \frac{-5}{2}C_1 e^{(6+\sqrt{3})t} - \frac{5}{2}C_2 e^{(6-\sqrt{3})t} + \frac{C_1}{2}(6+\sqrt{3})\,e^{(6+\sqrt{3})\,t} + \frac{C_2}{2}(6-\sqrt{3})\,e^{(6-\sqrt{3})t}$$

or $x = \dfrac{1}{2}(1+\sqrt{3})\,C_1\,e^{(6+\sqrt{3})t} + \dfrac{1}{2}(1-\sqrt{3})\,C_2\,e^{(6-\sqrt{3})t}$ ...(2)

(1) and (2) together represent the solutuion. **Ans.**

SOLUTIONS TO EXERCISE 10D

1. *Solve the equation of damped harmonic oscillations* $\dfrac{d^2x}{dt^2}+2nh\dfrac{dx}{dt}+n^2x=0$ *where* $0<h<1$ *and* $x=0, \dfrac{dx}{dt}=v_0$ *when* $t=0$. *To measure h the ratio E of amplitudes of successive swings (i.e. on opposite sides of the equilibrium position) is found. Show that*

$$h^2=\frac{(\log E)^2}{\pi^2+(\log E)^2}.$$

SOLUTION: The equation is $(D^2 + 2nh\,D + n^2)\,x = 0$ where $D \equiv \dfrac{d}{dt}$, $0 < h < 1$.

The A.E. is $m^2 + 2nh\,m + n^2 = 0$ $\quad\therefore\quad m = \dfrac{-2nh \pm \sqrt{4n^2h^2-4n^2}}{2} = nh \pm in\sqrt{1-h^2}$

$$\therefore \qquad x = e^{-nht}\left[C_1 \cos nt\sqrt{1-h^2} + C_2 \sin nt\sqrt{1-h^2}\right]$$

Now, since $x = 0$ at $t = 0$ hence $C_1 = 0$

Thus $\qquad x = C_2\, e^{-nht} \sin\left(nt\sqrt{1-h^2}\right)$

and $\qquad \dfrac{dx}{dt} = C_2\, e^{-nht}\, n\sqrt{1-h^2}\, \cos\left(nt\sqrt{1-h^2}\right) - C_2\, nhe^{-nht} \sin\left(nt\sqrt{1-h^2}\right)$

Since $\dfrac{dx}{dt} = v_0$ at $t = 0$, we have $nC_2\sqrt{1-h^2} = v_0$ $\quad\therefore\quad C_2 = \dfrac{V_0}{n\sqrt{1-h^2}}$

$\therefore$ Solution is $\qquad x = \dfrac{V_0}{n\sqrt{1-h^2}} \sin\left(nt\sqrt{1-h^2}\right) e^{-nht}$

Here the amplitude $\qquad A = \dfrac{V_0}{n\sqrt{1-h^2}}\, e^{-nht}$ and Period $= \dfrac{2\pi}{n\sqrt{1-h^2}} = T$.

If A_1 and A_2 are amplitudes of successive swings then

$$A_1 = \frac{V_0}{n\sqrt{1-h^2}}\, e^{-nht} \quad\text{and}\quad A_2 = \frac{x_0}{n\sqrt{1-h^2}}\, e^{-nh(t+T/2)}$$

$$\therefore \qquad E = \frac{A_1}{A_2} = e^{nhT/2} \quad\text{or}\quad \log E = -nh\frac{T}{2} = \frac{-nh\pi}{n\sqrt{1-h^2}} = \frac{-\pi h}{\sqrt{1-h^2}}$$

or $\qquad (1 - h^2)(\log E)^2 = \pi^2 h^2 \quad\therefore\quad h^2 = \dfrac{(\log E)^2}{\pi^2 + (\log E)^2}$ **Hence Proved.**

2. *If* $a\frac{d\theta}{dt}+2b\theta+u=0$ *and* $\theta=\frac{dx}{dt}$, $x=\text{K}u$, *find the differential equation connecting x and t and the constants a, b and K. If x denotes the displacement and t the time show that the motion will be just non-oscillatory when* $a=Kb^2$ *and express x in terms of t if* $x=x_0$ *and* $\theta=\theta_0$ *when* $t=0$.

SOLUTION: $a\frac{d\theta}{dt}+2b\theta+u=0$. Since $\theta=\frac{dx}{dt}$ and $x=ku$, we have

$$a\frac{d^2x}{dt^2}+2b\frac{dx}{dt}+\frac{x}{k}=0 \quad \text{or} \quad \frac{d^2x}{dt^2}+\frac{2b}{a}\frac{dx}{dt}+\frac{1}{ak}x=0$$

The auxiliary equation is $m^2+\frac{2b}{a}m+\frac{1}{ak}=0$

$$\therefore \qquad m=\frac{\frac{-2b}{a}\pm\sqrt{\frac{4b^2}{a^2}-\frac{4}{ak}}}{2}=\frac{b}{a}\pm\sqrt{\frac{b^2}{a^2}-\frac{1}{ak}}=\frac{-b}{a}\pm\frac{1}{a}\sqrt{b^2-\frac{a}{k}}$$

Motion will be non-oscillatory if $a=kb^2$.

Then $m=\frac{-b}{a},\frac{-b}{a}$ $\quad\therefore\quad x=(C_1t+C_2)\,e^{-b\pm/a}$

Given that $x=x_0$ and $\theta=\theta_\theta$ when $t=0$ $\therefore$ $x_0=C_2$

$\therefore$ $\qquad x=(C_1t+x_0)\,e^{-bt/a}$

and $\qquad \frac{dx}{dt}=C_1e^{-bt/a}-(C_1t+x_0)\,\frac{b}{a}e^{-bt/a}$

$\therefore$ $\qquad \theta_0=C_1-x_0\frac{b}{a}$ $\quad$ or $\quad C_1=\theta_0+\frac{bx_0}{a}$

Hence $\qquad x=\left[x_0+\left(\theta_0+\frac{bx_0}{a}\right)t\right]e^{-bt/a}$ **Ans.**

3. *For an electrical circuit with circuit constants L, R, C, the charge q on a plate of the condenser is given by* $\text{L}\ddot{q}+\text{R}\dot{q}+\frac{q}{c}=E\sin\omega t$ *and the current* $i=\dot{q}$. *The circuit is tuned to resonance so that* $\omega^2=\frac{1}{\text{LC}}$. *If* $\text{R}^2<\frac{4\text{L}}{\text{C}}$ *and* $q=i=0$ *at* $t=0$, *show that*

$$q=\frac{\text{E}}{\text{R}\omega}\left[-\cos\omega t+e^{\frac{-\text{R}t}{2\text{L}}}\left(\cos pt+\frac{\text{R}}{2\text{L}p}\sin pt\right)\right]$$

and $i=\frac{\text{E}}{\text{R}}\left[\sin\omega t-\frac{1}{p\sqrt{\text{LC}}}e^{-\frac{\text{R}t}{2\text{L}}}\sin pt\right]$ *where* $p^2=\frac{1}{\text{LC}}-\frac{\text{R}^2}{4\text{L}^2}$.

SOLUTION: The differential equation can be written as

$$\left(LD^2+RD+\frac{1}{C}\right)q=E\sin wt \quad \text{or} \quad \left(D^2+\frac{R}{L}D+\frac{1}{LC}\right)q=\frac{E}{L}\sin wt$$

Auxiliary Equation is $m^2 + \frac{R}{L}m + \frac{1}{LC} = 0$

It's roots are $m = \frac{-R}{2L} \pm \frac{1}{2}\sqrt{\frac{R^2}{L^2} - \frac{4}{LC}} = \frac{-R}{2L} \pm \frac{1}{2L}\sqrt{R^2 - \frac{4L}{C}}$

$$= \frac{-R}{2L} \pm \frac{i}{2L}\sqrt{\frac{4L}{C} - R^2} \quad \text{since} \quad R^2 < \frac{4L}{C}$$

$$= \frac{-R}{2L} \pm ip \quad \text{where} \quad p = \left(\frac{1}{LC} - \frac{R^2}{4L^2}\right)$$

$\therefore$ C.F. $= e^{\frac{-R}{2L}}(C_1 \cos pt + C_2 \sin pt)$

and P.I. $= \frac{1}{D^2 + \frac{R}{L}D + \frac{1}{LC}} \frac{E}{L} \sin wt = \frac{E}{L} \frac{1}{-w^2 + \frac{R}{L}D + \frac{1}{LC}} \sin wt$

$$= \frac{E}{L \cdot \frac{R}{L} D} \sin wt \quad \left(\text{since circuit is in resonance, } w^2 = \frac{1}{LC}\right)$$

$$= \frac{E}{R}\,\frac{1}{D} \sin wt = \frac{-E}{Rw} \cos wt.$$

$\therefore$ $$q = e^{\frac{-Rt}{2L}}(C_1 \cos pt + C_2 \sin pt) - \frac{E}{Rw} \cos wt$$

since $q = 0$ at $t = 0$ we have $C_1 = \frac{E}{Rw}$

$\therefore$ $$q = e^{\frac{-Rt}{2L}}\left[\frac{E}{Rw} \cos pt + C_2 \sin pt\right] - \frac{E}{Rw} \cos wt.$$

Next $\frac{dq}{dt} = i = e^{\frac{-Rt}{2L}}\left[\frac{-Ep}{Rw} \sin pt + C_2\, p \cos pt\right] - \frac{R}{2L} e^{\frac{-Rt}{2L}}\left(\frac{E}{Rw} \cos pt + C_2 \sin pt\right) + \frac{E}{R} \sin wt$

Since $i = 0$ at $t = 0$ we have $C_2 p - \frac{R}{2L}\,\frac{E}{Rw} = 0 \quad \therefore \quad C_2 = \frac{E}{2p\,Lw}$

Thus, $i = e^{\frac{-Rt}{2L}}\left[-\frac{Ep}{Rw} \sin pt + \frac{E}{2Lw} \cos pt\right] - \frac{R}{2L} e^{\frac{-Rt}{2L}}\left[\frac{E}{Rw} \cos pt + \frac{E}{2pLw} \sin pt\right] + \frac{E}{R} \sin wt$

$$= e^{\frac{-Rt}{2L}}\left[-\frac{Ep}{Rw} - \frac{ER}{4L^2 pw}\right] \sin pt + \frac{E}{R} \sin wt = \frac{E}{R}\left[\sin wt - \left(\frac{p}{w} + \frac{R^2}{4L^2 pw}\right) \sin pt\right]$$

$$= \frac{E}{R}\left[\sin wt - \left\{\frac{p}{w} + \left(LC - p^2\right)\frac{1}{pw}\right\} \sin pt\right] = \frac{E}{R}\left[\sin wt - \frac{1}{LC\,pw} \sin pt\right]$$

$$= \frac{E}{R}\left[\sin wt - \frac{\sqrt{LC}}{p\,LC} \sin pt\right] = \frac{E}{R}\left[\sin wt - \frac{1}{p\sqrt{LC}} \sin pt\right]$$

Hence Proved.

4. *A body executes damped forced oscillations given by* $\dfrac{d^2x}{dt^2}+2K\dfrac{dx}{dt}+b^2x=e^{-Kt}\sin\omega t.$

Obtain the solution expressing the displacement x *in terms of the time* t *for the cases when* $w^2 \neq b^2-K^2$ *and when* $w^2 = b^2-K^2$.

SOLUTION: Equation is $[D^2+2K\,D+6^2]\,x = e^{-Kt}\sin wt$, A.E. is $m^2+2km+b^2=0$

$$\therefore\ m = -K \pm \sqrt{K^2-b^2} = -K \pm i\sqrt{b^2-K^2}$$

$$\therefore\quad \text{C.F.} = e^{-Kt}\left[C_1\cos\sqrt{b^2-K^2}\,t + C_2\sin\sqrt{b^2-K^2}\,t\right] = Ae^{-Kt}\cos\left(\sqrt{b^2-K^2}\,t+\beta\right)$$

$$\text{P.I.} = \frac{1}{D^2+2KD+b^2}e^{-Kt}\sin wt = e^{-Kt}\frac{1}{(D-K)^2+2K(D-K)+b^2}\sin wt$$

$$= e^{-Kt}\frac{1}{D^2+b^2-K^2}\sin wt = \frac{e^{-Kt}\sin wt}{-w^2+b^2-K^2}$$

Case I: When $w^2 \neq b^2-K^2$ then

$$x = Ae^{-Kt}\cos\left(\sqrt{b^2-K^2}\,t+\beta\right)+\frac{e^{-Kt}\sin wt}{-w^2+b^2-K^2}$$

Case II: When $w^2 = b^2-K^2$ then $\text{P.I.} = e^{-Kt}\dfrac{t}{2D}\sin wt = \dfrac{-e^{-Kt}t\cos wt}{2w}$

$$\therefore\quad x = A\,e^{-Kt}\cos(wt+\beta) - \frac{e^{-Kt}t}{2w}\cos wt.$$ **Ans.**

5. *The deflection* y *of a strut with one end built-in and the other supported and subjected to end thrust* P, *satisfies the equation* $\dfrac{d^2y}{dx^2}+a^2y=\dfrac{a^2R}{P}(l-x)$.

Given that $\dfrac{dy}{dx}=y=0$ *at* $x=0$ *and* $y=0$ *at* $x=l$ *show that*

$$y=\frac{R}{P}\left[\frac{\sin ax}{a}-l\cos ax+l-x\right]\quad \textit{where}\quad al=\tan al.$$

SOLUTION: Equation is $(D^2+a^2)\,y = \dfrac{a^2R}{P}(l-x)$. C.F. $= C_1\cos ax + C_2\sin ax$

$$\text{P.I.} = \frac{a^2R}{P}\frac{1}{D^2+a^2}(l-x) = \frac{R}{P}\left(1+\frac{D^2}{a^2}\right)^{-1}(l-x) = \frac{R}{P}[l-x]$$

$$\therefore\quad y = C_1\cos ax + C_2\sin ax + \frac{R}{P}(l-x)$$

Since $y=0$ at $x=0$ we have $0 = C_1 + 0 + \dfrac{Rl}{P}$ $\quad\therefore\ C_1 = \dfrac{-Rl}{P}$

$\therefore \qquad y = \dfrac{-Rl}{P}\cos ax + C_2 \sin ax \dfrac{R}{P}(l - x)$

Also, $y = 0$ at $x = l$ hence $0 = \dfrac{-Rl}{p}\cos al + C_2 \sin al + 0 \quad \therefore \quad C_2 = \dfrac{Rl}{p}\cot al$

$\therefore \qquad y = \dfrac{-Rl}{p}\cos ax + \dfrac{Rl}{p}\cot al \sin x + \dfrac{R}{p}(l - x)$

Now $\qquad \dfrac{dy}{dx} = \dfrac{Rl}{P} a \sin ax + \dfrac{aRl}{P}\cot al \cos ax - \dfrac{R}{P}$

Since $\qquad \dfrac{dy}{dx} = 0$ at $x = 0$ we have $0 = 0 + \dfrac{aRl}{P}\cot al - \dfrac{R}{P} \qquad (\tan al = al)$

$\therefore \qquad y = \dfrac{-Rl}{P}\cos ax + \dfrac{Rl}{P}\dfrac{1}{al}\sin ax + \dfrac{R}{P}(l - x) = \dfrac{R}{P}\left[\dfrac{1}{a}\sin ax - l\cos ax + (l - x)\right]$ **Hence Proved.**

6. *The differential equation of a shaft which is whirling with the line bearings horizontal, is*

$$\text{EI}\,\frac{d^4 y}{dx^4} - \frac{\text{W}\omega^2}{g}\cdot y = W$$

where W is the weight of the shaft and ω is the whirling speed. Taking the shaft of length $2l$ with origin as its centre and short bearing at both ends, show that

$$y = \frac{g}{2\omega^2}\left[\frac{\cos mx}{\cos ml} + \frac{\cosh mx}{\cosh ml} - 2\right] \quad \textit{where } m^4 = \frac{W\omega^2}{EIg}.$$

SOLUTION: Equation is $\left(El\,D^4 - \dfrac{W\,w^2}{g}\right)y = W$ or $\left(D^4 - \dfrac{W\,w^2}{Elg}\right)y = \dfrac{W}{El}$

or $\qquad (D^4 - m^4)\,y = \dfrac{W}{El}$ where $m^4 = \dfrac{W\,w^2}{Elg}$

$\therefore \qquad \text{C.F} = c_1 e^{mx} + c_2 e^{-mx} + c_3 \cos mx + c_4 \sin mx$

and $\qquad \text{P.I.} = \dfrac{W}{El}\left(\dfrac{1}{D^4 - m^4}\right)\cdot e^{ox} = \dfrac{-W}{El\,m^4} = \dfrac{-g}{w^2}$

$\therefore \qquad y = c_1 e^{mx} + c_2 e^{-mx} + c_3 \cos mx + c_4 \sin x - \dfrac{g}{w^2}$

or $\qquad y = e_1 \cosh mx + c_2 \sinh mx + c_3 \cos mx + c_4 \sin mx - \dfrac{g}{w^2}$

By boundary conditions $y = 0$ at $x = \pm\, l,$ we have

$$0 = c_1 \cosh ml + c_2 \sinh ml + c_3 \cos ml + c_4 \sin ml - \frac{g}{w^2}$$

and $\quad 0 = c_1 \cosh ml - c_2 \sin h\, ml + c_3 \cos ml - c_4 \sin ml - \dfrac{g}{w^2}$

$\Rightarrow \quad 2\, c_1 \cosh ml + 2\, c_4 \sin ml = \dfrac{2g}{w^2}$ and $2\, c_2 \sinh ml + 2\, c_4 \sin \mathrm{ml} = 0$

Also $\quad \dfrac{d^2 y}{dx^2} = c_1\, m^2 \cosh mx + c_2\, m^2 \sinh mx - c_3 m^2 \cos mx - c_4 m^2 \sin mx$

Since $\dfrac{d^2 y}{dx^2} = 0$ at $x = \pm\, l$, hence

$c_1 \cosh ml + c_2 \sinh ml - c_3 \cos ml - c_4 \sin ml = 0$

and $\quad c_1 \cosh ml - c_2 \sinh ml - c_3 \cos ml + c_4 \sin ml = 0$

$\Rightarrow \quad c_1 \cosh ml - c_3 \cos ml = 0$

and $\quad c_2 \sin ml - c_4 \sin ml = 0$ Therefore $c_2 = c_4 = 0$

Further, $\quad 2c_1 \cosh ml = \dfrac{g}{w^2}$ and $2c_3 \cos ml = \dfrac{g}{w^2}$

or $\quad c_1 = \dfrac{g}{2w^2 \cosh ml}$ and $c_3 = \dfrac{g}{2w^2 \cos ml}$

$$\therefore \quad y = \frac{g \cosh mx}{2w^2 \cosh ml} + \frac{g \sin ml}{2w^2 \sin ml} - \frac{g}{w^2} = \frac{g}{2w^2}\left[\frac{\cosh mx}{\cosh ml} + \frac{\sin ml}{\sin ml} - 2\right]$$

7. *A simple pendulum of length l is oscillating through a small angle θ in a medium in which the resistance is proportional to the velocity. Show that the differential equation of its motion is* $\dfrac{d^2\theta}{dt^2} + 2\mathrm{K}\dfrac{d\theta}{dt} + \dfrac{g}{l}\theta = 0$. *Also find the period of oscillation.*

SOLUTION: Equation of motion along the tangent PT is $m\dfrac{d^2 s}{dt^2} = -mg \sin\theta - \lambda\dfrac{ds}{dt}$

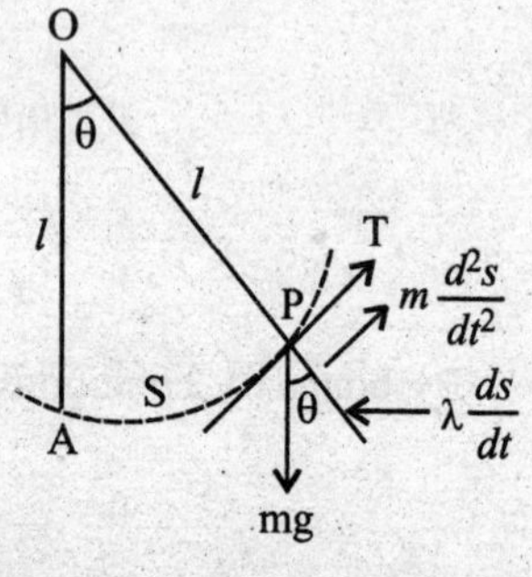

and since $s = l\theta$. $\quad \dfrac{d^2 s}{dt^2} = l\dfrac{d^2\theta}{dt^2}$.

We write $\quad \dfrac{\lambda}{m} = 2K$ and $\dfrac{g}{l} = w^2$

Hence the above equation becomes

$$ml\frac{d^2\theta}{dt^2} = -mg \sin\theta - 2mkl\frac{d\theta}{dt}$$

or $\quad \dfrac{d^2\theta}{dt^2} + 2k\dfrac{d\theta}{dt} + \dfrac{g}{l}\sin\theta = 0$ and take $\sin\theta = \theta$ as θ is small

$\therefore \quad \dfrac{d^2\theta}{dt^2} + 2k\dfrac{d\theta}{dt} + w^2\theta = 0.$ **Hence Proved.**

Next for the period of oscillation.

Equation can be written as $(D^2 + 2k\,D + w^2)\,\theta = 0$ A.E. is $p^2 + 2k\,p + w = 0$

$$\therefore \quad p = \frac{-2k \pm \sqrt{4k^2 - 4w^2}}{2} = -k \pm i\sqrt{w^2 - k^2}$$

$\therefore$ Solution is $\quad \theta = (e^{-kt}\cos(\sqrt{w^2 - k^2}\,t + c_2))$

Period of oscillation $= \dfrac{2\pi}{\sqrt{w^2 - k^2}}$ **Ans.**

8. *A strut of length l freely hinged at each end, follows the equation*

$$\text{EI}\,y'' + \text{P}y = -\frac{\text{W}l^2}{g}\sin\frac{\pi x}{l}.$$

Find the deflection of the centre of the beam and the bending moment there.

SOLUTION: Equation can be written as $(EI\,D^2 + P)\,y = -W\dfrac{l^2}{g}\sin\dfrac{\pi x}{l}$.

or $\quad \left(D^2 + \dfrac{P}{EI}\right)y = -\dfrac{W\,l^2}{g\,E\,I}\sin\dfrac{\pi x}{l}$ or $(D^2 + n^2)y = -\dfrac{W\,l^2\,n^2}{g\,P}\sin\dfrac{\pi x}{l}$ where $\dfrac{P}{EI} = n^2$.

$\therefore \quad$ C.F. $= c_1 \cos nx + c_2 \sin nx$

and $\quad$ P.I. $= \dfrac{1}{D^2 + n^2}\left(\dfrac{-Wl^2\,n^2}{g\,p}\sin\dfrac{\pi x}{l}\right) = -\dfrac{We^2n^2}{gp}\dfrac{1}{\left(-\dfrac{\pi^2}{l^2} + n^2\right)}\sin\pi\dfrac{x}{l}$.

$\therefore$ Deflection of the centre of the beam $\; y = c_1\cos nx + c_z\sin nx + \dfrac{Wl^4n^2}{gp(\pi^2 - n^2l^2)}\sin\dfrac{\pi x}{l}$

Boundary condition are $\quad y = 0$ at $x = 0$ and $x = l$.

$\therefore \quad 0 = c_1$ and $0 = 0 + c_2 \sin nl + 0 \Rightarrow \sin nl = 0$ as $c_2 \neq 0$ $\quad \therefore\; n = \dfrac{m\pi}{l}$

$$\therefore \quad y = \sin\frac{m\pi}{l}x + \frac{W\,l^4m^2\pi^2\sin(\pi x/l)}{l^2gp(\pi^2 - m^2\pi^2)} = \sin\frac{m\pi x}{l} + \frac{W\,l^2m^2}{g\,p(1-m^2)}\sin\left(\pi\frac{x}{l}\right)$$

Hence, bending moment $= EI\dfrac{d^2y}{dx^2} = EI\left[\dfrac{-m^2\pi^2}{l^2}\sin\dfrac{m\pi x}{l} + \dfrac{Wl^2w^2}{gp(1-m^2)}\left(\dfrac{-\pi^2}{l^2}\right)\sin\pi\dfrac{x}{l}\right]$ **Ans.**

9. *A resistance R in series with inductance L is shunted by an equal resistance R in series with capacity C. An alternating e.m.f. E sin pt produces currents i_1 and i_2 in the two branches. If i_1 and i_2 are zero when $t = 0$, determine i_1 and i_2 from the equations*

$$L\frac{di_1}{dt} + Ri_1 = E\sin pt$$

$$R\frac{di_2}{dt} + \frac{i_2}{C} = pE\cos pt$$

Verify that if $R^2C = L$ *the total current* $i_1 + i_2$ *will be* $\frac{E\sin pt}{R}$.

SOLUTION: Equations for i_1 and i_2 are

$$\frac{di_1}{dt} + \frac{Ri_1}{L} = \frac{E}{L}\sin pt\,,\quad \frac{di_2}{dt} + \frac{1}{RC}i_2 = \frac{pE}{R}\cos pt \text{ which are linear}$$

$$\text{I.F. for } i_1 = e^{\int\frac{Rdt}{L}} = e^{Rt/L} \text{ and I.F. for } i_2 = e^{\int\frac{dt}{RC}} = e^{t/RC}$$

$$\therefore \quad i_1e^{Rt/L} = \int e^{Rt/L}\frac{E}{L}\sin pt\,dt + C_1 = \frac{E}{L}\frac{e^{Rt/L}}{\frac{R^2}{L^2}+p^2}\left[\frac{R}{L}\sin pt - p\cos pt\right] + c_1$$

or
$$i_1 = \frac{E}{R^2+p^2L^2}[R\sin pt - pL\cos pt] + c_1\,e^{-Rt/L}$$

since $i_1 = 0$ at $t = 0$ we have $c_1 = \frac{EpL}{R^2+p^2L^2}$

$$\therefore \quad i_1 = \frac{E}{R^2+p^2L^2}[R\sin pt - pL\cos pt] + \frac{EpL}{R^2+p^2L} = e^{-Rt/L}$$

Similarly, $\text{(I.F. for } i_2) = e^{\int\frac{1}{RC}dt} = e^{t/RC}$

$$\therefore \quad i_2\,e^{t/RC} = \int e^{t/RC}\frac{pE}{R}\cos pt + c_2 \quad \text{or} \quad i_2e^{t/RC} = \frac{pE}{R}e^{t/RC}\left[\frac{1}{RC}\cos pt + p\sin pt\right] + c_2$$

or
$$i_2 = \frac{pE/R}{p^2+\frac{1}{R^2C^2}}\left[\frac{1}{RC}\cos pt + p\sin pt\right] + c_2e^{-t/RC}$$

Since $i_z = 0$ at $t = 0$ we have $c_2 = -\frac{pER^2C^2}{R^2C(1+p^2R^2C^2)} = \frac{-pEC}{1+R^2C^2}$

$$\therefore \quad i_2 = \frac{pEc}{1+p^2R^2-c^2} - \left[\frac{1}{RC}\cos pt + p\sin pt\right] - \frac{pEC}{1+R^2p^2c^2}e^{-t/RC}$$

Next, if $R^2C = L$ $\quad i_1 + i_2 = \frac{E\sin pt}{R}$ on simplification. **Ans.**

10. *A particle of mass m is projected vertically upwards with an initial velocity V_0. The resisting force per unit mass at any time is K times the velocity. Find the distance S covered by the particle at any time t.*

SOLUTION: Equation of motion is $m\dfrac{d^2s}{dt^2} = -mg - km\dfrac{ds}{dt}$

or $\dfrac{d^2s}{dt^2} + km\dfrac{ds}{dt} + g = 0$ or $(D^2 + kD)s = -g$ where $D \equiv \dfrac{d}{dt}$

A.E. is $p^2 + kp = 0 \quad \therefore \quad p = 0, -k \quad \therefore$ C.F. $= c_1 + c_2 e^{-kt}$

$$\text{P.I.} = \frac{1}{D(D+K)}(-g) = \frac{1}{K}\left(\frac{1}{D+k} - \frac{1}{D}\right)g = \frac{1}{K}\left[\frac{1}{D+K} g e^{ot} - \int g\,dt\right]$$

$$= \frac{1}{K}\left[\frac{g}{k} - gt\right] \quad \therefore \quad S = c_1 + c_2 e^{-kt} + \frac{g}{k^2} - \frac{g}{k}t.$$

Further, since $S = 0$ at $t = 0$ we have $c_1 + c_2 = \dfrac{-g}{k^2}$.

Next, $\dfrac{ds}{dt} = -kc_2 e^{-kt} - \dfrac{g}{k}$. Since $\dfrac{ds}{dt} = V_0$ at $t = 0$, we have

$$V_0 = -kc_2 - \frac{g}{K} \quad \therefore c_2 = \frac{-g}{k^2} - \frac{V_0}{k} \quad \therefore c_1 = \frac{V_0}{k}$$

$$\therefore \quad S = \frac{V_0}{k} + \left(\frac{-g}{k^2} - \frac{V_0}{k}\right)e^{-kt} + \frac{g}{k^2} - \frac{gt}{k}$$

or $$S = \left(\frac{-g}{k^2} - \frac{V_0}{k}\right)(1 - e^{-kt}) - \frac{gt}{k}$$ **Ans.**

11. *In a chemical transformation of certain substance the following equations occur*

$$\frac{dx}{dt} + ax = 0, \quad \frac{dz}{dt} = by$$

and $x + y + z = c'$ where a, b, c are constants. Obtain a differential equation for z and solve it for z, given that $z = \dfrac{dz}{dt} = 0$ when $t = 0$.

SOLUTION: Equation are $\dfrac{dx}{dt} + ax = 0$, $\dfrac{dz}{dt} = by$ and $x + y + z = c$

or $$\frac{dx}{dt} + \frac{dy}{dt} + \frac{dz}{dt} = 0 \qquad \text{... (1)}$$

From $\dfrac{dz}{dt}$ = by we get $\dfrac{d^2z}{dt^2} = b\dfrac{dy}{dt} \quad \therefore$ (1) becomes

$$-ax + \frac{1}{b}\frac{d^2z}{dt^2} + \frac{dz}{dt} = 0 \quad \text{or} \quad \frac{d^2z}{dt^2} + \frac{b\,dz}{dt} - abz = 0. \qquad \text{... (2)}$$

From $x + y + z = c$ we have $x = c - y - z$ $\therefore$ (2) becomes

$$\therefore \quad \frac{d^2 z}{dt^2} + b\frac{dz}{dt} - ab(c - y - z) = 0$$

or $$\frac{d^2 z}{dt^2} + b\frac{dz}{dt} = abc + a\,(by) + abz = 0 \quad \text{but } by = \frac{dz}{dt}$$

$$\therefore \quad \frac{d^2 z}{dt^2} + (a+b)\frac{dz}{dt} + abz = abc$$

or $(D^2 + (a + b)\,D + ab)z = abc$ or $(D + a)\,(D + b)\,z = abc$

Here C.F. $= c_1 e^{-at} + e_2\, e^{-bt}$ and P.I. $= \dfrac{1}{D^2 + (a+b)z + ab} abc\, e^{0t} = \dfrac{abc}{ab} = C$

$\therefore$ $z = c_1\, e^{-at} + c_2\, e^{-bt} + c$

Since $z = 0$ at $t = 0$ we have $c_1 + c_2 = -c$.

$\dfrac{dz}{dt} = -a\, c_1\, e^{-at} - bc_2 e^{-bt}$. Since $\dfrac{dz}{dt} = 0$ at $t = 0$ we have $ac_1 + bc_2 = 0$

$\Rightarrow$ $c_1 = \dfrac{bc}{a-b}$ and $c_2 = \dfrac{-ac}{a-b}$

$\therefore$ $z = \dfrac{bc}{a-b} e^{-at} - \dfrac{ac}{a+b} e^{-bt} + c$ or $z = c + \dfrac{c}{a-b}(be^{-at} - ae^{-pt})$. **Ans.**

12. *A resistance R of 5 Ω and an inductance of 0.1 H are connected with a battery of 10 volts. Find the current i as a function of time t using the differential equation* $Ri + L\dfrac{di}{dt} = E$.

SOLUTION: Equation can be written as $\dfrac{di}{dt} + \dfrac{R}{L}i = \dfrac{E}{L}$.

Which is linear in i. The I.F. $= e^{\int \frac{R}{L} dt} = e^{Rt/L}$.

$\therefore$ The solution is $i\, e^{Rt/L} = \displaystyle\int \frac{E}{L} e^{Rt/L} dt + K$ where K is constant.

or $i\, e^{Rt/L} = \dfrac{E}{L} \cdot \dfrac{L}{R} e^{Rt/L} + K$ or $i = \dfrac{E}{R} + Ke^{-Rt/L}$

Since $i = 0$ at $t = 0$ we have $K = -\dfrac{E}{R}$.

$\therefore$ $i = \dfrac{E}{R}\left(1 - e^{-Rt/L}\right)$. We are given that $R = 5\ \Omega$.

$L = 0.1$ H and $E = 10$ volts, therefore

$$i = \frac{10}{5}\left(1 - e^{\frac{-5t}{0.1}}\right) = 2\left(1 - e^{-50t}\right)$$ **Ans.**

ADDITIONAL IMPORTANT PROBLEMS

1. Solve the equation $\frac{d^2y}{dx^2}+6\frac{dy}{dx}+9y=5^x-\log 2.$

SOLUTION: Equation can be written as $(D^2+6D+9)\,y=5^x-\log 2$ where $D\equiv\frac{d}{dx}$.

A.E. is $m^2+6m+9=0$ or $(m+3)^2=0$ $\therefore$ $m=-3,-3.$

$\therefore$ $\quad$ C.F. $=(c_1+c_2x)\,e^{-3x}.$

and

$$\text{P.I.}=\frac{1}{D^2+6D+9}(5^x-\log 2)=\frac{1}{(D+3)^2}e^{x\log 5}-\log 2\frac{1}{(D+3)^2}e^{\circ x}$$

$$=\frac{1}{(3+\log 5)^2}e^{x\log 5}-\log 2\frac{1}{(0+3)^2}e^{\circ x}$$

$$=\frac{1}{(3+\log 5)^2}\cdot(5^x)-\frac{\log 2}{9}$$

$\therefore$ Solution is $\quad y=(c_1+c_2x)\,e^{-3x}+\frac{5^x}{(3+\log 5)^2}-\frac{1}{9}\log 2.$ **Ans.**

2. Solve $(D^3-4D)y=2\cosh^2(2x).$

SOLUTION: Auxiliary equation is $m^3-4m=0$ $\therefore$ $m=0,2,-2.$

$\therefore$ $\quad$ C.F. $=c_1e^{\circ x}+c_2\,e^{2x}+c_3\,e^{-2x}=c_1+c_2\,e^{2x}+c_3\,e^{-2x}.$

and

$$\text{P.I.}=\frac{1}{D^3-4D}2\cosh^2(2x)=\frac{2}{D^3-4D}\left(\frac{e^{2x}+e^{-2x}}{2}\right)^2$$

$$=\frac{1}{2}\frac{1}{D^3-4D}[e^{4x}+e^{-4x}+2e^{\circ x}]$$

$$=\frac{1}{2}\left[\frac{1}{4^3-16}e^{4x}+\frac{1}{(-4)^3+16}e^{-4x}+x\frac{1}{(3D^2-4)}(2e^{\circ x})\right]$$

$$=\frac{1}{2}\left[\frac{e^{4x}}{48}+\frac{1}{-48}e^{-4x}+x\cdot\frac{1}{(0-4)}2e^{\circ x}\right]$$

$$=\frac{1}{2}\left[\frac{e^{4x}-e^{-4x}}{48}-\frac{x}{2}\right]=\frac{\sinh 4x}{48}-\frac{x}{4}$$

$\therefore$ Solution is $\quad y=c_1+c_2\,e^{2x}+c_3\,e^{-2x}+\frac{1}{48}\sinh 4x-\frac{x}{4}.$ **Ans.**

3. Solve the equation $\frac{d^4y}{dx^4}-2\frac{d^3y}{dx^3}+\frac{2d^2y}{dx^2}-2\frac{dy}{dx}+y=e^x+\sin^2\left(\frac{x}{2}\right)$.

SOLUTION: The equation is $(D^4-2D^3+2D^2-2D+1)\,y=e^x+\sin^2\left(\frac{x}{2}\right)$.

A.E. is $m^4-2m^3+2m^2-2m+1=0$ or $(m^2-2m+1)(m^2+1)=0$

whose roots are $m=1, 1, i, -i$.

$\therefore$ C.F. $=(c_1+xc_2)\,e^x+c_3\cos x+c_4\sin x$

and P.I. $=\frac{1}{(D-1)^2(D^2+1)}\left\{e^x+\sin^2\left(\frac{x}{2}\right)\right\}=\frac{1}{(D-1)^2(D^2+1)}\left\{e^x+\frac{1-\cos x}{2}\right\}$

Here $\frac{1}{(D-1)^2(D^2+1)}e^x=\frac{1}{(D-1)^2}\frac{1}{1^2+1}e^x=\frac{1}{2}\frac{1}{(D-1)^2}e^x=\frac{1}{2}\left(\frac{x^2}{2}e^x\right)=\frac{x^2}{4}e^x,$

$$\frac{1}{(D-1)^2(D^2+1)}\frac{1-\cos x}{2}=\frac{1}{2(D-1)^2(D^2+1)}e^{(ox)}-\frac{1}{2(D^2-1)(D^2-2D+1)}(\cos x)$$

$$=\frac{1}{2(0-1)^2}\left(\frac{1}{0+1}e^{ox}\right)-\frac{1}{2(D^2+1)(-1-2D+1)}\cos x$$

$$=\frac{1}{2}e^{ox}+\frac{1}{4(D^2+1)}\frac{1}{D}\cos x=\frac{1}{2}+\frac{1}{4(D^2+1)}\int\cos x\,dx$$

$$=\frac{1}{2}+\frac{1}{4(D^2+1)}\sin x=\frac{1}{2}+\frac{x}{4}\frac{1}{2D}\sin x=\frac{1}{2}+\frac{x}{8}\int\sin x\,dx$$

$$=\frac{1}{2}-\frac{x}{8}\cos x$$

$\therefore$ Solution is $y=\text{C.F.}+\text{P.I.}=(c_1+c_2x)e^x+c_3\cos x+c_4\sin x+\frac{1}{2}-\frac{x}{8}\cos x.$ **Ans.**

4. Solve $\frac{d^3y}{dx^3}+3\frac{d^2y}{dx^2}-4\frac{dy}{dx}-12y=12x\,e^{-2x}$.

SOLUTION: The equation is $(D^3-3D^2-4D-12)y=12x\,e^{-2x}$ where $D\equiv\frac{d}{dx}$.

A.E. is $m^3-3m^2-4m-12=0$ or $(m+3)(m^2-4)=0$ $\therefore$ $m=-3, 2, -2$.

$\therefore$ C.F. $=c_1\,e^{-3x}+c_2\,e^{-2x}+c_3\,e^{2x}$.

and $$\text{P.I.} = \frac{1}{(D+3)(D+2)(D-2)} 12x\, e^{-2x} = 12e^{-2x} \frac{1}{(D-2+3)(D-2+2)(D-2-2)} x$$

$$= 12e^{-2x} \frac{1}{D(D+1)(D-4)} x = 12e^{-2x} \frac{1}{D(D^2-3D-4)} x$$

$$= 12e^{-2x} \frac{1}{-4D\left(1+\dfrac{3D-D^2}{4}\right)} x = -3e^{-2x} \frac{1}{D}\left[1+\frac{3D-D^2}{4}\right]^{-1} x$$

$$= -3e^{-2x} \frac{1}{D}\left[1-\frac{3D-D^2}{4}+\ldots\right] x$$

$$= -3e^{-2x} \frac{1}{D}\left[x-\frac{3}{4}\right] = -3e^{-2x} \int\left(x-\frac{3}{4}\right) dx$$

$$= -3e^{-2x}\left[\frac{x^2}{2}-\frac{3}{4}x\right]$$

$\therefore$ Solution is $y = c_1 e^{-3x} + c_2 e^{-2x} + c_3 e^{2x} - 3e^{-2x}\left(\dfrac{x^2}{2}-\dfrac{3}{4}x\right)$. **Ans.**

5. Solve $(4D^2 + 8D + 3)y = xe^{-x/2} \cos x$ where $D \equiv \dfrac{d}{dx}$.

SOLUTION: A.E. is $4m^2 + 8m + 3 = 0 \quad \therefore \quad m = -\dfrac{1}{2}, -\dfrac{3}{2}$.

$\therefore$ C.F. $= c_1 e^{-x/2} + c_2 e^{-3x/2}$.

$$\text{P.I.} = \frac{1}{4D^2+8D+3}(e^{-x/2} x \cos x) = e^{-x/2} \frac{1}{4\left(D-\dfrac{1}{2}\right)^2+8\left(D-\dfrac{1}{2}\right)+3} (x \cos x).$$

$$= e^{-x/2} \frac{1}{4D^2+4D} x \cos x = \frac{e^{-x/2}}{4} \frac{1}{D}\left[\frac{1}{D+1} x \cos x\right]$$

$$= \frac{e^{-x/2}}{4} \cdot \frac{1}{D}\left[x \frac{1}{D+1} \cos x - \frac{1}{(D+1)^2} \cos x\right]$$

$$= \frac{e^{-x/2}}{4} \frac{1}{D}\left[x \frac{D-1}{D^2-1} \cos x - \frac{1}{D^2+2D+1} \cos x\right]$$

$$= \frac{e^{-x/2}}{4D}\left[x\left(\frac{D-1}{-1-1}\right) \cos x - \frac{1}{-1+2D+1} \cos x\right]$$

$$= \frac{e^{-x/2}}{4}\frac{1}{D}\left[-\frac{x}{2}(-\sin x - \cos x) - \frac{1}{2}\int \cos x\, dx\right] = \frac{e^{-x/2}}{8}\cdot\frac{1}{D}[x(\sin x + \cos x) - \sin x]$$

$$= \frac{e^{-x/2}}{8}\int [x(\sin x + \cos x) - \sin x]\, dx$$

$$= \frac{e^{-x/2}}{8}\left[x(-\cos x + \sin x) - \int 1(-\cos x + \sin x)\, dx + \cos x\right]$$

$$= \frac{e^{-x/2}}{8}[x(-\cos x + \sin x) + \sin x + \cos x + \cos x]$$

$\therefore$ The solution is $y = c_1 e^{-x/2} + c_2 e^{-3x/2} + \frac{e^{-x/2}}{8}[x(\sin x - \cos x) + \sin x + 2\cos x]$ **Ans.**

6. Solve the equation $(D^2 + 3D + 2)y = e^{e^x}$.

SOLUTION: A.E. is $m^2 + 3m + 2 = 0$ whose roots are $-1, -2$

$\therefore$ $\text{C.F.} = c_1 e^{-x} + c_2 e^{-2x}$

and $\text{P.I.} = \frac{1}{D^2 + 3D + 2} e^{e^x} = \frac{1}{(D+1)(D+2)} e^{e^x} = \frac{1}{D+2}\left(\frac{1}{D+1} e^{e^x}\right)$

$$= \frac{1}{D+2}\left[e^{-x}\int e^x e^{e^x}\, dx\right] = \frac{1}{D+2}[e^{-x}\cdot e^{e^x}]$$

$$= e^{-2x}\int e^{2x}\cdot e^{-x} e^{e^x}\, dx = e^{-2x}\int e^x \cdot e^{e^x}\, dx.$$

$$= e^{-2x} e^{e^x}$$

$\therefore$ General solution $y = c_1 e^{-x} + c_2 e^x + e^{-2x} e^{e^x}$. **Ans.**

7. Solve the equation $4x^2\frac{d^2y}{dx^2} + y = 19\cos(\log x) + 22\sin(\log x)$.

SOLUTION: Putting $x = e^t$ in the given equation, we get

$$[4D(D-1) + 1]\, y = 19\cos t + 22\sin t \quad \text{where } D \equiv \frac{d}{dt}.$$

or $(4D^2 - 4D + 1)y = 19\cos t + 22\sin t$

A.E. is $4m^2 - 4m + 1 = 0$ or $(2m-1)^2 = 0$ $\therefore$ $m = 1/2, 1/2.$

$\therefore$ $\text{C.F.} = (C_1 + tC_2)e^{t/2} = (C_1 + C_2\log x)x^{1/2}$

and $\text{P.I.} = \frac{1}{4D^2 - 4D + 1}(19\cos t + 22\sin t)$

$$= \frac{1}{4(-1)-4D+1}(19\cos t + 22\sin t) = \frac{-1}{4D+3}\frac{(4D-3)}{(4D-3)}(19\cos t + 22\sin t)$$

$$= \frac{-(4D-3)}{16(-1)-9}(19\cos t + 22\sin t) = \frac{1}{25}[4(-19\sin t + 22\cos t) - 3(19\cos t + 22\sin t)]$$

$$= \frac{1}{25}(31\cos t - 142\sin t) = \frac{1}{25}[31\cos\log x - 142\sin\log x]$$

$\therefore$ The solution is $y = (c_1 + c_2\log x)\sqrt{x} + \frac{1}{25}[31\cos\log x - 142\sin\log x]$.

8. Solve $x\frac{d^2y}{dx^2} + \frac{dy}{dx} - \frac{y}{x} = -ax^2$

SOLUTION: Multiplying throughout by x we get

$$x^2\frac{d^2y}{dx^2} + x\frac{dy}{dx} - y = -ax^3.$$

Putting $x = e^t$ so $t = \log x$, we get

$$[D(D-1) + D - 1]y = -ae^{3t} \quad \text{or} \quad (D^2 - 1)y = -ae^{3t}$$

A.E. is $m^2 - 1 = 0 \quad \therefore \quad m = 1, -1$

$\therefore$ $$\text{C.F.} = c_1e^t + c_2e^{-t} = c_1x + \frac{c_2}{x}.$$

and $$\text{P.I.} = \frac{(-ae^{3t})}{D^2-1} = \frac{-a\,e^{3t}}{9-1} = \frac{-a}{8}x^3.$$

$\therefore$ Solution is $$y = c_1x + \frac{c_2}{x} - \frac{a}{8}x^3.$$

9. Solve $\left(\frac{d}{dx} + \frac{1}{x}\right)^2 y = \frac{1}{x^4}$

SOLUTION: $$\left(\frac{d}{dx} + \frac{1}{x}\right)\left(\frac{dy}{dx} + \frac{y}{x}\right) = \frac{1}{x^4}$$

or $$\frac{d^2y}{dx^2} + \frac{1}{x}\frac{dy}{dx} - \frac{1}{x^2}y + \frac{1}{x}\frac{dy}{dx} + \frac{y}{x^2} = \frac{1}{x^4}.$$

or $$x^2\frac{d^2y}{dx^2} + 2x\frac{dy}{dx} = \frac{1}{x^2}$$ which is Cauchy's equation form.

Putting $x = e^t$ we get

$$[D(D-1) + 2D]\, y = e^{-3t} \text{ or } (D^2 + D)\, y = e^{-2t}$$

A.E. is $m^2 + m = 0 \quad \therefore \quad m = 0, -1$

$\therefore$ $$\text{C.F.} = c_1 + c_2 e^{-t} = c_1 + \frac{c_2}{x}.$$

and $$\text{P.I.} = \frac{1}{D^2 + D} e^{-2t} = \frac{1}{(-2)^2 + (-2)} e^{-2t} = \frac{1}{2} e^{-2t} = \frac{1}{2x^2}$$

$\therefore$ Solution is $y = c_1 + \dfrac{c_2}{x} + \dfrac{1}{2x^2}.$ **Ans.**

10. Solve $x^2 \dfrac{d^2y}{dx^2} + 5x \dfrac{dy}{dx} + 3y = \dfrac{\log x}{x^2}$

SOLUTION: Putting $x = e^t$ or $t = \log x$ we get the equation as

$$[D(D-1) + 5D + 3]\, y = te^{-2t} \quad \text{where} \quad D \equiv \frac{d}{dt}.$$

or $$(D^2 + 4D + 3)\, y = te^{-2t}$$

A.E. is $m^2 + 4m + 3 = 0 \quad \therefore \quad m = -1, -3$

$\therefore$ $$\text{C.F.} = c_1 e^{-t} + c_2 e^{-3t} = \frac{c_1}{x} + \frac{c_2}{x^3}.$$

and $$\text{P.I.} = \frac{1}{D^2 + 4D + 3} te^{-2t} = e^{-2t} \frac{1}{(D-2)^2 + 4(D-2) + 3} t$$

$$= \frac{e^{-2t} 1}{D^2 - 1} t = -e^{-2t} (1 - D^2)^{-1} t = -e^{-2t} (1 + D^2 + \ldots)\, t$$

$$= -e^{-2t}\, t = -\frac{1}{x^2} \log x$$

$\therefore$ The solution is $y = \dfrac{c_1}{x} + \dfrac{c_2}{x^3} - \dfrac{1}{x^2} \log x.$ **Ans.**

11. Solve $(x-1)^3 \dfrac{d^3y}{dx^3} + 2(x-1)^2 \dfrac{d^2y}{dx^2} - 4(x-1) \dfrac{dy}{dx} + 4y = 4\log(x-1).$

SOLUTION: Putting $x - 1 = e^t$ the given equation becomes

$$[D(D-1)(D-2) + 2D(D-1) - 4D + 4]\, y = 4t$$

or $$(D^3 - D^2 - 4D + 4)\, y = 4t$$

A.E. is $m^3 - m^2 - 4m + 4 = 0$ or $(m-1)(m^2-4) = 0$ $\therefore$ $m = 1, 2, -2$.

$$\therefore \quad \text{C.F.} = c_1 e^t + c_2 e^{2t} + c_3 e^{-2t} = c_1 (x-1) + c_2 (x-1)^2 + c_3 (x-1)^{-2}$$

and
$$\text{P.I.} = \frac{1}{D^3 - D^2 - 4D + 4} 4t = \frac{1}{4}\left[1 - \frac{(4D + D^2 - D^3)}{4}\right]^{-1} 4t$$

$$= \left[1 - \frac{4D + D^2 - D^3}{4}\right]^{-1} t$$

$$= \left[1 + \frac{4D + D^2 - D^3}{4} + \frac{(4D + D^2 - D^3)^2}{16} + \ldots\right] t$$

$$= (1 + D)t = t + 1 = \log(x-1) + 1$$

$\therefore$ Solution is $\quad y = (x-1)\, c_1 + (x-1)^2\, c_2 + (x-1)^{-2}\, c_3 + \log(x-1) + 1.$ **Ans.**

12. Solve $(D^2 + 1)\, y = \operatorname{cosec} x$ where $D \equiv \dfrac{d}{dx}$, by the method of variation of parameters.

SOLUTION: A.E. is $m^2 + 1 = 0$ $\therefore$ $m = \pm i$

$\therefore$ C.F. $= c_1 \cos x + c_2 \sin x$. Let $y_1 = \cos x$, $y_2 = \sin x$

and P.I. $= u(x) \cos x + v(x) \sin x$

$$\text{The wronskian} = W = \begin{vmatrix} y_1 & y_2 \\ y_1' & y_2' \end{vmatrix} = \begin{vmatrix} \cos x & \sin x \\ -\sin x & \cos x \end{vmatrix} = 1$$

$$\therefore \quad u(x) = \int \frac{-y_2 Q}{W} dx = \int \frac{\sin x \operatorname{cosec} x}{1} dx = x$$

$$\text{and} \quad v(x) = \int \frac{y_1 Q}{W} dx = \int \frac{\cos x \cdot \operatorname{cosec} x}{1} dx = \int \cot x\, dx = \log \sin x$$

$\therefore$ P.I. $= x \cos x + (\log \sin x) \cdot \sin x$

$\therefore$ Solution is $\quad y = c_1 \cos x + c_2 \sin x + x \cos x + \sin x (\log \sin x)$. **Ans.**

13. Solve $(D^2 - 1)\, y = e^x \sin(e^{-x}) + \cos(e^{-x})$.

SOLUTION: A.E. is $m^2 - 1 = 0$ $\therefore$ $m = \pm 1$

$\therefore$ C.F. $= c_1 e^x + c_2 e^{-x}$ so $y_1 = e^x$, $y_2 = e^{-x}$

$$\text{and} \quad \text{Wronskion} = W = \begin{vmatrix} y_1 & y_2 \\ y_1' & y_2' \end{vmatrix} = \begin{vmatrix} e^x & e^{-x} \\ e^x & -e^{-x} \end{vmatrix} = -2$$

Let the P.I. $= v_1(x)\,e^{x} + v_2(x)\,e^{-x}$

where $v_1 = \int -\frac{y_2 Q}{W}\,dx = \int \frac{e^{-x}(e^{-x}\sin(e^{-x}) + \cos(e^{-x}))}{-2}\,dx$

putting here $e^{-x} = t \quad \therefore \quad dt = -e^{-x}$

$$\therefore \qquad v_1 = \frac{1}{2}\int -(t\sin t + \cos t)\,dt = -\frac{1}{2}[t(-\cos t) - (-\sin t) + \sin t]$$

$$\therefore \qquad = \frac{1}{2}\,t\cos t\sin t = \frac{1}{2}\,e^{-x}\cos(e^{-x})\sin(e^{-x})$$ **Ans.**

14. Solve $\frac{dx}{dt} - 3x - 6y = t^2,\ \frac{dy}{dt} + \frac{dx}{dt} - 3y = e^t.$

SOLUTION: Equations are $(D-3)\,x - 6y = t^2$ ...(1)

and $Dx + (D-3)\,y = e^t$ where $D \equiv \frac{d}{dt}.$ ...(2)

Eliminating y in the equations (1) and (2), we get

$$[(D-3)^2 + 6D]\,x = (D-3)\,t^2 + 6e^t$$

or $$(D^2+9)\,x = 2t - 3t^2 + 6e^t$$

Its auxiliary equation is $m^2 + 9 = 0 \quad \therefore \quad m = \pm 3i$

$\therefore$ For x C.F. $= c_1\cos 3t + c_2\sin 3t$

and $$\text{P.I.} = \frac{1}{D^2+9}(2t - 3t^2) + \frac{1}{D^2+9}6e^t$$

$$= \frac{1}{9}\left(1+\frac{D^2}{9}\right)^{-1}(2t-3t^2) + \frac{6}{1+9}e^t = \frac{1}{9}\left(1-\frac{D^2}{9}\right)(2t-3t^2) + \frac{6}{10}e^t$$

$$= \frac{1}{9}\left(2t - 3t^2 + \frac{6}{9}\right) + \frac{6}{10}e^t = \frac{-t^2}{3} + \frac{2t}{9} + \frac{2}{27} + \frac{3}{5}e^t$$

$$\therefore \qquad x = c_1\cos 3t + c_2\sin 3t - \frac{t^2}{3} + \frac{2}{9}t + \frac{2}{27} + \frac{3}{5}e^t \qquad ...(3)$$

hence $$Dx = -3c_1\sin 3t + 3c_2\cos 3t - \frac{2t}{3} + \frac{2}{9} + \frac{3}{5}e^t$$

$\therefore$ From (1) we get $6y = Dx - 3x - t^2$

or $$6y = -3c_1\sin 3t + 3c_2\cos 3t - \frac{2t}{3} + \frac{2}{9} + \frac{3}{5}e^t - 3\left[c_1\cos 3t + c_2\sin 3t - \frac{t^2}{3} + \frac{2}{9}t + \frac{2}{27} + \frac{3}{5}e^t\right] - t^2$$

$$= (-3c_1 - 3c_2)\sin 3t + (-3c_1 + 3c_2)\cos 3t + t\left(-\frac{2}{3} - \frac{2}{3}\right) + t^2 - t^2 + e^t\left(\frac{3}{5} - \frac{9}{5}\right)$$

or $$y = -\frac{1}{2}(c_1+c_2)\sin 3t - \frac{1}{2}(c_1 - c_2)\cos 3t - \frac{2t}{9} - \frac{1}{5}e^t \qquad ...(4)$$

$\therefore$ Equations (3) and (4) constitute the solution. **Ans.**

15. Solve $\dfrac{d^2 y}{dx^2} = x - 2, \quad \dfrac{d^2 x}{dt^2} = y + 2.$

SOLUTION: Equations are $D^2 y - x = -2$ and $D^2 x - y = 2$ where $D \equiv \dfrac{d}{dt}$.

Eliminating y from the above equations, we get $(D^4 - 1)\, x = -2.$

Its A.E. is $m^4 - 1 = 0 \quad \therefore \quad m = \pm 1, \pm i$

$\therefore$ For x, C.F. $= c_1 e^t + c_2 e^{-t} + c_3 \cos t + c_4 \sin t.$

and P.I. (for x) $= \dfrac{1}{D^4 - 1}(-2) = \dfrac{1}{D^4 - 1}(-2e^{ot}) = \left(\dfrac{-2}{0-1}\right) e^{ot} = 2.$

$$\therefore \quad x = c_1 e^t + c_2 e^{-t} + c_3 \cos t + c_4 \sin t + 2 \qquad \text{...(1)}$$

Substituting for x in the second given equation, we get

$$y = D^2 x - 2 = c_1 e^t + c_2 e^{-t} - c_3 \cos t - c_4 \sin t - 2 \qquad \text{...(2)}$$

(1) and (2) together represent the solution.

16. Obtain the orthogonal trajectories of the family of curves $\dfrac{x^2}{a^2} + \dfrac{y^2}{b^2 + \lambda} = 1$ where λ is a parameter.

SOLUTION: Family of curves is $\dfrac{x^2}{a^2} + \dfrac{y^2}{b^2 + \lambda} = 1.$ On differentiation, we get

$$\frac{2x}{a^2} + \frac{2y}{b^2 + \lambda}\frac{dy}{dx} = 0 \quad \text{or} \quad \frac{y^2}{b^2 + \lambda} = \frac{-xy}{a^2\left(\dfrac{dy}{dx}\right)}$$

$\therefore$ From the equation of the curve, we have

$$\frac{x^2}{a^2} - \frac{xy}{a^2\left(\dfrac{dy}{dx}\right)} = 1 \quad \text{or} \quad (x^2 - a^2)\frac{dy}{dx} = xy.$$

This is the differential equation of the given family of curves. Now replacing $\dfrac{dy}{dx}$ by $-\dfrac{dx}{dy}$ in the above equation, we get

$$(a^2 - x^2)\frac{dx}{dy} = xy \quad \text{or} \quad \frac{a^2 - x^2}{x}\,dx = y\,dy$$

which on integration gives $a^2 \log x - \dfrac{x^2}{2} = \dfrac{y^2}{2} + c'$ or $x^2 + y^2 = 2a^2 \log x + 2c$

which represents orthogonal trajectories of the given curve. **Ans.**

17. A circuit consisting of resistance R and a condenser of capacity C, is connected in series with a voltage E. Assuming that there is no charge on condenser at $t = 0$, find the current i and charge q at any time t.

SOLUTION: As per Kirchoff's law to the R-C circuit, we have

$$Ri + \frac{1}{C}\int i\,dt = E \quad \text{or} \quad R\frac{dq}{dt} + \frac{q}{C} = E \text{ since } i = \frac{dq}{dt}.$$

or $$\frac{dq}{dt} + \frac{1}{RC}q = \frac{E}{R} \text{ which is linear in } q.$$

$\therefore$ $$\text{I.F.} = e^{\int \frac{1}{RC}dt} = e^{\frac{t}{RC}}$$

and the solution is $$qe^{t/RC} = \int e^{t/RC}\cdot\frac{E}{R}\,dt = \frac{E}{R}\cdot RCe^{t/RC} + K$$

or $q = EC + Ke^{-t/RC}$

Since $q = 0$ at $t = 0$ hence $K = -EC$ $\therefore$ $q = EC\,(1 - e^{-t/RC})$

and $$i = \frac{dq}{dt} = \frac{EC}{RC}e^{-t/RC} = \frac{E}{R}e^{-t/RC}.$$

Also, the e.m.f. e is given by

$$e = \frac{1}{C}\int i dt = \frac{1}{C}\int \frac{E}{R}e^{-t/RC}\,dt = \frac{-1}{C}\frac{E}{R}\cdot RC\,e^{-t/RC} + K$$

Since $e = 0$ at $t = 0$ $\therefore$ $0 = -E + K$ or $K = E$

$\therefore$ $$e = -Ee^{-t/RC} + E = E\,(1 - e^{-t/RC}).$$ **Ans.**

18. The motion of a simple pendulum is given by $\frac{d^2x}{dt^2} + \omega^2 x = F \sin nt$ where w and n constants. If $x = 0$, $\frac{dx}{dt} = 0$ at $t = 0$, determine the motion when $n = w$.

SOLUTION: Equation is $(D^2 + w^2)\,x = F \sin nt$ where $D \equiv \frac{d}{dx}$.

A.E. is $m^2 + w^2 = 0$ $\therefore$ $m = \pm\, iw$.

hence C.F. $= c_1 \cos \omega t + c_2 \sin \omega t$

and $$\text{P.I.} = \frac{1}{D^2 + \omega^2}F \sin nt = \frac{F}{\omega^2 - n^2}\sin nt$$

Hence $$x = c_1 \cos \omega t + c_2 \sin \omega t + \frac{F}{\omega^2 - n^2}\sin nt.$$

Since $x = 0$ at $t = 0$ we have $c_1 = 0$ hence $x = c_2 \sin \omega t + \frac{F}{\omega^2 - n^2}\sin nt$

Next, $\frac{dx}{dt} = c_2\omega \cos \omega t + \frac{Fn}{\omega^2 - n^2} \cos nt$.

Since $\frac{dx}{dt} = 0$ at $t = 0$, we have $c_2 = \frac{Fn}{\omega(n^2 - \omega^2)}$

Thus, $x = \frac{Fn}{\omega(n^2 - \omega^2)} \sin \omega t - \frac{F}{n^2 - \omega^2} \sin nt = \frac{F(n \sin \omega t - \omega \sin nt)}{\omega(n^2 - \omega^2)}$ is the solution.

However, when $n = \omega$ we get $x = \underset{n \to \omega}{\text{Lt}} \frac{F(\sin \omega t - \omega t \cos nt)}{(2n\omega)} = \frac{-Ft}{2\omega} \cos \omega t + \frac{F}{2\omega^2} \sin \omega t.$ **Ans.**

CHAPTER

11 Partial Differentiation

IMPORTANT FORMULAE AND USEFUL RESULTS

1. If $u = f(x, y)$ then $\dfrac{\partial f}{\partial x} = f_x = p,\quad \dfrac{\partial f}{\partial y} = f_y = q$

$$\frac{\partial^2 f}{\partial x^2} = f_{xx} = \frac{\partial p}{\partial x} = r, \quad \frac{\partial^2 f}{\partial x \partial y} = f_{xy} = \frac{\partial q}{\partial x} = \frac{\partial r}{\partial y} = s, \quad \frac{\partial^3 f}{\partial y^2} = f_{yy} = \frac{\partial q}{\partial y} = t.$$

2. **Euler's Theorem on Homogeneous Functions.**

If u is homogeneous function in x and y of degree n, then

I $\quad x\dfrac{\partial u}{\partial y} + y\dfrac{\partial u}{\partial y} = nu$

II $\quad x^2 \dfrac{\partial^2 u}{\partial x^2} + 2xy \dfrac{\partial^2 u}{\partial x \partial y} + y^2 \dfrac{\partial^2 u}{\partial y^2} = n(n-1)\, u.$

III If z is homogeneous is x and y of degree n and z is a function of u as $z = f(u)$

then $\quad x\dfrac{\partial u}{\partial x} + y\dfrac{\partial u}{\partial y} = n\dfrac{f(u)}{f'(u)}$

and $\quad x^2 \dfrac{\partial^2 u}{\partial x^2} + 2xy \dfrac{\partial^2 u}{\partial x \partial y} + y^2 \dfrac{\partial^2 u}{\partial y^2} = g(u)\,[g'(u) - 1]$ where $g(u) = \dfrac{nf(u)}{f'(u)}$

3. **Total Derivative of $u\,(x, y)$ is**

$$du = \frac{\partial u}{\partial x} dx + \frac{\partial u}{\partial y} dy$$

If $f(x, y) = 0$ then $\dfrac{dy}{dx} = \dfrac{-\partial f}{\partial x} \Big/ \dfrac{\partial f}{\partial y} = \dfrac{-p}{q}$. and $\dfrac{d^2 y}{dx^2} = -\dfrac{1}{q^3}[q^2 r - 2pqs + p^2 t]$

4. **Change of Independent Variables**

If $u = f(x, y)$ and $x = x(t_1, t_2)$ and $y = y(t_1, t_2)$, then

$$\frac{\partial u}{\partial t_1} = \frac{\partial u}{\partial x}\frac{\partial x}{\partial t_1} + \frac{\partial u}{\partial y}\frac{\partial y}{\partial t_1} \quad \text{and} \quad \frac{\partial u}{\partial t_2} = \frac{\partial u}{\partial x}\frac{\partial x}{\partial t_2} + \frac{\partial u}{\partial y}\frac{\partial y}{\partial t_2}.$$

SOLUTIONS TO EXERCISE 11A

1. *If* $u = \dfrac{e^{x+y+z}}{e^x + e^y + e^z}$, *show that* $\dfrac{\partial u}{\partial x} + \dfrac{\partial u}{\partial y} + \dfrac{\partial u}{\partial z} = 2u.$

SOLUTION: $u\,(e^x + e^y + e^z) = e^{x+y+z}$. Differentiating w.r.t. x, gives $(e^x + e^y + e^z)\,\dfrac{\partial u}{\partial x} + 4e^x = e^{x+y+z}.$

or
$$(e^x + e^y + e^z)\,\frac{\partial u}{\partial x} = e^{x+y+z} - \frac{e^x \cdot e^{x+y+z}}{e^x + e^y + e^z} \qquad \therefore \quad \frac{\partial u}{\partial x} = \frac{e^{x+y+z}(e^y + e^z)}{(e^x + e^y + e^z)^2}.$$

Similarly
$$\frac{\partial u}{\partial y} = \frac{e^{x+y+z}\,(e^z + e^x)}{(e^x + e^y + e^z)^2} \quad \text{and} \quad \frac{\partial u}{\partial z} = \frac{e^{x+y+z}(e^x + e^y)}{(e^x + e^y + e^z)^2}$$

$$\therefore \quad \frac{\partial u}{\partial x} + \frac{\partial u}{\partial y} + \frac{\partial u}{\partial z} = \frac{e^{x+y+z}\,(2e^x + 2e^y + 2e^z)}{(e^x + e^y + e^z)^2} = \frac{2e^{x+y+z}}{e^x + e^y + e^z} = 2u$$ **Hence Proved.**

2. *If* $V = \log\,(\tan x + \tan y + \tan z)$ *find the value of*
$$\sin 2x\frac{\partial V}{\partial x} + \sin 2y\frac{\partial V}{\partial y} + \sin 2z\,\frac{\partial V}{\partial z}.$$

SOLUTION: Given equation can be written as
$e^V = \tan x + \tan y + \tan z.$ Differentiating w.r.t. x, gives
$$e^V\,\frac{\partial V}{\partial x} = \sec^2 x \quad \therefore \quad \sin 2x\,\frac{\partial V}{\partial x} = \sin 2x\,\sec^2 x\; e^{-V} = 2\tan x\; e^{-V}$$

Similarly, $\sin 2y\dfrac{\partial V}{\partial y} = 2\tan y\; e^{-V}$ and $\sin 2z\,\dfrac{\partial V}{\partial z} = 2\tan z\; e^{-V}$

$$\therefore \quad \sin 2x\,\frac{\partial V}{\partial x} + \sin 2y\,\frac{\partial V}{\partial y} + \sin 2z\,\frac{\partial V}{\partial z} = 2e^{-V}\,(\tan x + \tan y + \tan z) = 2e^{-V}\,e^V = 2.$$ **Ans.**

3. *Given that* $z = x^2 \tan^{-1}(y/x) - y^2 \tan^{-1}(x/y)$, *find the value of* $\dfrac{\partial^2 z}{\partial y\,\partial x}$

and verify that $\dfrac{\partial^2 z}{\partial x\,\partial y} = \dfrac{\partial^2 z}{\partial y\,\partial x}.$

SOLUTION:
$$z = x^2 \tan^{-1}\left(\frac{y}{x}\right) - y^2 \tan^{-1}\left(\frac{x}{y}\right)$$

$$\therefore \quad \frac{\partial z}{\partial x} = x^2\,\frac{1}{1+\dfrac{y^2}{x^2}}\left(\frac{-y}{x^2}\right) + 2x\tan^{-1}\left(\frac{y}{x}\right) - y^2 \cdot \frac{1}{1+\dfrac{x^2}{y^2}}\left(\frac{1}{y}\right)$$

$$= \frac{-x^2 y}{x^2 + \frac{y^2}{x^2}} + 2x \tan^{-1}\left(\frac{y}{x}\right) - \frac{y^3}{x^2 + y^2} = 2x \tan^{-1}\left(\frac{y}{x}\right) - y.$$

$$\therefore \quad \frac{\partial^2 z}{\partial y\, \partial x} = \frac{2x}{1 + \frac{y^2}{x^2}}\left(\frac{1}{x}\right) - 1 = \frac{y^3}{x^2 + y^2} - 1 = \frac{x^2 - y^2}{x^2 + y^2}$$

Similarly, we can obtain $\dfrac{\partial^2 z}{\partial x\, \partial y} = \dfrac{x^2 - y^2}{x^2 + y^2}$ **Hence Proved.**

4. *If* $u^2 = x^2 + y^2 + z^2$ *find the value of* $\dfrac{\partial^2 u}{\partial x^2} + \dfrac{\partial^2 u}{\partial y^2} + \dfrac{\partial^2 u}{\partial z^2}$.

SOLUTION: Differentiating $u^2 = x^2 + y^2 + z^2$ on both sides w.r.t. x, we get

$2u\dfrac{\partial u}{\partial x} = 2x$. On again differentiating w.r.t. x, gives

$$u\frac{\partial^2 u}{\partial x^2} + \left(\frac{\partial u}{\partial x}\right)^2 = 1 \quad \text{or} \quad u^3\frac{\partial^2 u}{\partial x^2} + \left(u\frac{\partial u}{\partial x}\right)^2 = u^2$$

or
$$u^3\frac{\partial^2 u}{\partial x^2} + x^2 = x^2 + y^2 + z^2. \quad \text{or} \quad \frac{\partial^2 u}{\partial x^2} = \frac{y^2 + z^2}{u^3}$$

Similarly,
$$\frac{\partial^2 u}{\partial y^2} = \frac{z^2 + x^2}{u^3} \quad \text{and} \quad \frac{\partial^2 u}{\partial z^2} = \frac{x^2 + y^2}{u^3}$$

Hence
$$\frac{\partial^2 u}{\partial x^2} + \frac{\partial^2 u}{\partial y^2} + \frac{\partial^2 u}{\partial z^2} = \frac{2(x^2 + y^2 + z^2)}{u^3} = \frac{2u^2}{u^3} = \frac{2}{u}$$ **Ans.**

5. *If* $u(x, t) = ae^{-gx}\sin(nt - gx)$ *where* a, g *and* n *are constants, and* $\dfrac{\partial u}{\partial t} = a^2\dfrac{\partial^2 u}{\partial x^2}$, *show that* $g = \dfrac{1}{a}\sqrt{\dfrac{n}{2}}$.

SOLUTION: $u = ae^{-gx}\sin(nt - gx) \quad \therefore \quad \dfrac{\partial u}{\partial t} = ae^{-gx}\cos(nt - gx)\cdot n$

and
$$\frac{\partial u}{\partial x} = -ag\, e^{-gx}[\sin(nt - gx) + \cos(nt - gx)]$$

and then
$$\frac{\partial^2 u}{\partial x^2} = ag^2 e^{-gx}\{\sin(nt - gx) + \cos(nt - gx)\} - ag\, e^{-gx}[-g\cos(nt - gx) + g\sin(nt - gx)]$$

$= a\, g^2 e^{-gx}\{2\cos(nt - gx)\}$ Given that $\dfrac{\partial u}{\partial t} = a^2\dfrac{\partial^2 u}{\partial x^2}$, we get $g^2 = \dfrac{n}{2a^2}$.

Hence Proved.

6. *If* $u = (1 - 2xy + y^2)^{-1/2}$ *prove that* $x\dfrac{\partial u}{\partial x} - y\dfrac{\partial u}{\partial y} = y^2 u^3.$

And also evaluate $\dfrac{\partial}{\partial x}\left\{(1-x^2)\dfrac{\partial u}{\partial x}\right\} + \dfrac{\partial}{\partial y}\left\{y^2\dfrac{\partial u}{\partial y}\right\}.$

SOLUTION: Given that $u^{-2} = 1 - 2xy + y^2$ ...(1). Differentiating w.r.t. x, we get

$$-2u^{-3}\frac{\partial u}{\partial x} = -2y \quad \text{or} \quad \frac{\partial u}{\partial x} = yu^3$$

Differentiating (1) w.r.t. y, we get

$$-2u^{-3}\frac{\partial u}{\partial y} = -2x + 2y \quad \text{or} \quad \frac{\partial u}{\partial y} = u^3(x - y)$$

$$\therefore \quad x\frac{\partial u}{\partial x} - y\frac{\partial u}{\partial y} = xy\,u^3 - u^3(x-y)\,y = y^2 u^3$$ **Hence Proved.**

Next
$$\frac{\partial}{\partial x}\left[(1-x^2)\frac{\partial u}{\partial x}\right] = \frac{\partial}{\partial x}[(1-x^2)\,yu^3] = y(1-x^2)\cdot 3u^2\frac{\partial u}{\partial x} - 2xy\,u^3$$
$$= 3y^3(1-x^2)\,u^5 - 2xy\,u^3$$

and
$$\frac{\partial}{\partial y}\left[y^2\frac{\partial u}{\partial y}\right] = \frac{\partial}{\partial y}\left[u^3(xy^2 - y^3)\right] = u^3(2xy - 3y^2) + 3(xy^2 - y^3)\,u^2\frac{\partial u}{\partial y}$$
$$= u^3(2xy - 3y^2) + 3u^5 y^2 (x-y)^2$$

$$\therefore \quad \frac{\partial}{\partial x}\left[(1-x^2)\frac{\partial u}{\partial y}\right] + \frac{\partial}{\partial y}\left[y^2\frac{\partial u}{\partial y}\right] = u^3(2xy - 3y^2 - 2xy) + 3y^2u^5(1 - x^2 + x^2 - 2xy + y^2)$$
$$= -3y^2u^3 + 3y^2u^5(1 - 2xy + y^2)$$
$$= -3y^2u^3 + 3y^2u^3 = 0$$ **Ans.**

7. *If* $u = e^{r\cos\theta}\cos(r\sin\theta)$ *and* $v = e^{r\cos\theta}\sin(r\sin\theta)$, *show that* $\dfrac{\partial u}{\partial r} = \dfrac{1}{r}\dfrac{\partial v}{\partial \theta}$ *and* $\dfrac{\partial v}{\partial r} = -\dfrac{1}{r}\dfrac{\partial u}{\partial \theta}.$

SOLUTION: $u = e^{r\cos\theta}\cos(r\sin\theta),\ v = e^{r\cos\theta}\sin(r\sin\theta)$

$$\therefore \quad \frac{\partial u}{\partial r} = e^{r\cos\theta}\cos(r\sin\theta)\cos\theta - e^{r\cos\theta}\sin(r\sin\theta)\sin\theta = e^{r\cos\theta}\cos(\theta + r\sin\theta)$$
$$\frac{\partial v}{\partial r} = e^{r\cos\theta}\cos\theta\sin(r\sin\theta) + e^{r\cos\theta}\cos(r\sin\theta)\sin\theta = e^{r\cos\theta}\sin(\theta + r\sin\theta)$$
$$\frac{\partial u}{\partial \theta} = -e^{r\cos\theta}\,r\sin\theta\cos(r\sin\theta) - e^{r\cos\theta}\sin(r\sin\theta)\,r\cos\theta = -re^{r\cos\theta}\sin(\theta + r\sin\theta)$$
$$\frac{\partial v}{\partial \theta} = -e^{r\cos\theta}\,r\sin\theta\sin(r\sin\theta) + e^{r\cos\theta}\cos(r\sin\theta)\,r\cos\theta = re^{r\cos\theta}\cos(\theta + r\sin\theta)$$

$$\therefore \quad \frac{\partial u}{\partial r} = \frac{1}{r}\frac{\partial v}{\partial \theta} \quad \text{and} \quad \frac{\partial v}{\partial r} = \frac{-1}{r}\frac{\partial u}{\partial \theta}.$$ **Hence Proved.**

8. *If* $u = r^n$ *and* $r^2 = x^2 + y^2 + z^2$, *prove that*

$$u_{xx} + u_{yy} + u_{zz} = n(n+1)\, r^{n-2}.$$

SOLUTION: $u = r^n \quad \therefore \dfrac{\partial u}{\partial x} = nr^{n-1}\dfrac{\partial r}{\partial x}$. Since $r^2 = x^2 + y^2 + z^2$ we have $\dfrac{\partial r}{\partial x} = \dfrac{x}{r}$.

$$\therefore \quad \frac{\partial u}{\partial x} = nr^{n-1}\cdot\frac{x}{r} = nx\cdot r^{n-2}. \text{ Further } \frac{\partial^2 u}{\partial x^2} = nx(n-2)\, r^{n-3}\frac{\partial r}{\partial x} + n\cdot r^{n-2}$$

$$\text{or} \quad u_{xx} = \frac{\partial^2 u}{\partial x^2} = n(n-2)\, x r^{n-3}\cdot\frac{x}{r} + nr^{n-2} = n(n-2)\, x^2 r^{n-4} + nr^{n-2}$$

Similarly, $u_{yy} = n(n-2)\, y^2 r^{n-4} + nr^{n-2}$ and $u_{zz} = n(n-2)\, z^2 r^{n-4} + nr^{n-2}$

$$\therefore \quad u_{xx} + u_{yy} + u_{zz} = n(n-2)\, r^{n-4}(x^2 + y^2 + z^2) + 3nr^{n-2} = n(n-2)\, r^{n-2} + 3nr^{n-2}$$

$$= n\, r^{n-2}(n - 2 + 3) = n(n+1)\, r^{n-2}$$

Hence Proved.

9. *Given that* $u = x \log(x + r) - r$ *where* $r^2 = x^2 + y^2$, *evaluate* $u_{xx} + u_{yy}$ *in the simplified form.*

SOLUTION: $u = x\log(x+r) - r$, hence $u_x = \dfrac{\partial u}{\partial x} = \dfrac{x}{x+r}\left(1 + \dfrac{\partial r}{\partial x}\right) + \log(x+r) - \dfrac{\partial r}{\partial x}$.

Since $r^2 = x^2 + y^2$ hence $2r\dfrac{\partial r}{\partial x} = 2x$ or $\dfrac{\partial r}{\partial x} = \dfrac{x}{r}$ and $\dfrac{\partial r}{\partial y} = \dfrac{y}{x}$.

$$\therefore \quad u_x = \frac{x}{x+r}\left(1 + \frac{x}{r}\right) + \log(x+r) - \frac{x}{r} = \log(x+r)$$

$$\text{and} \quad u_{xx} = \frac{1}{x+r}\left(1 + \frac{\partial r}{\partial x}\right) = \frac{1}{x+r}\left(1 + \frac{x}{r}\right) = \frac{1}{r}$$

$$\text{Next,} \quad u_y = \frac{-y}{x+r} \quad \text{and} \quad u_{yy} = \frac{-r(x+r) + y^2}{r(x+r)^2} = \frac{-x}{r(x+r)}.$$

$$\text{hence} \quad u_{xx} + u_{yy} = \frac{1}{r} - \frac{x}{r(x+r)} = \frac{1}{x+r}.$$

Ans.

10. *Let* $u = lx + my$ *and* $v = mx - ly$ *where* l *and* m *are constants, show that*

$$\left(\frac{\partial u}{\partial x}\right)_y \left(\frac{\partial x}{\partial u}\right)_v = \frac{l^2}{l^2 + m^2} = \left(\frac{\partial y}{\partial v}\right)_u \left(\frac{\partial v}{\partial y}\right)_x$$

SOLUTION: $u = lx + my$, $v = mx - ly$ hence $\left(\dfrac{\partial u}{\partial x}\right)_y = l$ and $\left(\dfrac{\partial v}{\partial y}\right)_x = -l$.

Solving the given equations for x and y, we get

$$x = \frac{lu + mv}{l^2 + m^2} \quad \text{and} \quad y = \frac{mu - lv}{l^2 + m^2}.$$

$$\therefore \quad \left(\frac{\partial x}{\partial u}\right)_v = \frac{l}{l^2+m^2} \quad \text{and} \quad \left(\frac{\partial y}{\partial v}\right)_u = \frac{-l}{l^2+m^2}$$

$$\therefore \quad \left(\frac{\partial u}{\partial x}\right)_y \left(\frac{\partial x}{\partial u}\right)_v = \frac{l \cdot l}{l^2+m^2} \quad \text{and} \quad \left(\frac{\partial v}{\partial y}\right)_x \left(\frac{\partial y}{\partial v}\right)_u = \frac{l \cdot l}{l^2+m^2}$$

Thus $$\left(\frac{\partial u}{\partial x}\right)_y \cdot \left(\frac{\partial x}{\partial u}\right)_v = \frac{l^2}{l^2+m^2} = \left(\frac{\partial v}{\partial y}\right)_x \left(\frac{\partial y}{\partial v}\right)_u$$ **Hence Proved.**

11. *If* $u = \log(\tan x + \tan y)$, *evaluate* $\sin 2x \dfrac{\partial u}{\partial x} + \sin 2y \dfrac{\partial u}{\partial y}$.

SOLUTION: $u = \log(\tan x + \tan y)$ $\quad \therefore \quad \dfrac{\partial u}{\partial x} = \dfrac{\sec^2 x + 0}{\tan x + \tan y}$ and $\dfrac{\partial u}{\partial y} = \dfrac{0 + \sec^2 y}{\tan x + \tan y}$

Hence $$\sin 2x \frac{\partial u}{\partial x} + \sin 2y \frac{\partial u}{\partial y} = \frac{2 \sin x \cos x \sec^2 x}{\tan x + \tan y} + \frac{2 \sin y \cos y \sec^2 y}{\tan x + \tan y}$$

$$= \frac{2 \tan x + 2 \tan y}{\tan x + \tan y} = 2$$ **Ans.**

12. *Let* $xu + yv = 0$ *and* $\dfrac{u}{x} + \dfrac{v}{y} = 1$, *then show that*

$$\left(\frac{\partial u}{\partial x}\right)_y - \left(\frac{\partial v}{\partial y}\right)_x = \frac{y^2+x^2}{y^2-x^2}.$$

SOLUTION: Solving $xu + yv = 0$ and $\dfrac{u}{x} + \dfrac{v}{y} = 1$ for u and v, we get

$$v = \frac{x^2 y}{x^2 - y^2} \quad \text{and} \quad u = \frac{-xy^2}{x^2 - y^2}$$

$$\therefore \quad \frac{\partial u}{\partial x} = \frac{(x^2-y^2)(-y^2) - (-xy^2)2x}{(x^2-y^2)^2} = \frac{-y^2(x^2-y^2-2x^2)}{(x^2-y^2)^2} = \frac{y^2(x^2+y^2)}{(x^2-y^2)^2}$$

and $$\frac{\partial v}{\partial y} = \frac{(x^2-y^2)x^2 - x^2 y(-2y)}{(x^2-y^2)^2} = \frac{x^2(x^2-y^2+2y^2)}{(x^2-y^2)^2} = \frac{x^2(x^2+y^2)}{(x^2-y^2)^2}$$

$$\frac{\partial u}{\partial x} - \frac{\partial y}{\partial y} = \frac{(x^2+y^2)(y^2-x^2)}{(y^2-x^2)^2} = \frac{x^2+y^2}{y^2-x^2}$$ **Ans.**

13. *If* $z = (x+y)\left[1+\phi\left(\frac{y}{x}\right)\right]$ *prove that*

$$x\left(\frac{\partial^2 z}{\partial x^2}-\frac{\partial^2 z}{\partial x\partial y}\right) = y\left(\frac{\partial^2 z}{\partial y^2}-\frac{\partial^2 z}{\partial y\partial x}\right).$$

SOLUTION: $z = (x+y)\left[1+\phi\left(\frac{y}{x}\right)\right]$

$$\therefore \quad \frac{\partial z}{\partial x} = (x+y)\left[0+\varphi'\left(\frac{y}{x}\right)\cdot\frac{-y}{x^2}\right]+1\left[1+\varphi\left(\frac{y}{x}\right)\right] = 1+\varphi\left(\frac{y}{x}\right)-\frac{y(x+y)}{x^2}\varphi'\left(\frac{y}{x}\right)$$

and
$$\frac{\partial z}{\partial y} = (x+y)\left[0+\varphi'\left(\frac{y}{x}\right)\cdot\frac{1}{x}\right]+1\left[1+\varphi\left(\frac{y}{x}\right)\right] = 1+\varphi\left(\frac{y}{x}\right)+\frac{x+y}{x}\varphi'\left(\frac{y}{x}\right)$$

and
$$\frac{\partial^2 z}{\partial x^2} = 0+\varphi'\left(\frac{y}{x}\right)\left(\frac{-y}{x^2}\right)-\frac{y(x+y)}{x^2}\varphi''\left(\frac{y}{x}\right)\left(\frac{-y}{x^2}\right)-\left(\frac{-y}{x^2}-\frac{2y^2}{x^3}\right)\varphi'\left(\frac{y}{x}\right)$$

$$= \frac{2y^2}{x^3}\varphi'\left(\frac{y}{x}\right)+\frac{y^2}{x^4}(x+y)\,\varphi''\left(\frac{y}{x}\right)$$

$$\frac{\partial^2 z}{\partial y^2} = 0+\varphi'\left(\frac{y}{x}\right)\frac{1}{x}+\frac{x+y}{x}\varphi''\left(\frac{y}{x}\right)\cdot\frac{1}{x}+\frac{1}{x}\varphi'\left(\frac{y}{x}\right) = \frac{2}{x}\varphi'\left(\frac{y}{x}\right)+\frac{x+y}{x^2}\varphi''\left(\frac{y}{x}\right)$$

$$\frac{\partial^2 z}{\partial x\partial y} = \varphi'\left(\frac{y}{x}\right)\cdot\frac{1}{x}-\frac{y(x+y)}{x^2}\varphi''\left(\frac{y}{x}\right)\cdot\frac{1}{x}-\left(\frac{1}{x}+\frac{2y}{x^2}\right)\varphi'\left(\frac{y}{x}\right)$$

$$= \frac{-y(x+y)}{x^3}\varphi''\left(\frac{y}{x}\right)-\frac{2y}{x^2}\varphi'\left(\frac{y}{x}\right)$$

$$\therefore \quad x\left(\frac{\partial^2 z}{\partial x^2}-\frac{\partial^2 z}{\partial x\partial y}\right) = x\left[\left(\frac{2y^2}{x^3}+\frac{2y}{x^2}\right)\varphi'\left(\frac{y}{x}\right)+\left\{\frac{y^2(x+y)}{x^4}+\frac{y(x+y)}{x^3}\right\}\varphi''\left(\frac{y}{x}\right)\right]$$

$$= \left(\frac{2y}{x}+\frac{2y^2}{x^2}\right)\varphi'\left(\frac{y}{x}\right)+(x+y)\left(\frac{y}{x^2}+\frac{y^2}{x^3}\right)\varphi''\left(\frac{y}{x}\right)$$

$$= 2\left(\frac{y}{x}+\frac{y^2}{x^2}\right)\varphi'\left(\frac{y}{x}\right)+(x+y)^2\,\frac{y}{x^3}\varphi''\left(\frac{y}{x}\right)$$

and
$$y\left(\frac{\partial^2 z}{\partial y^2}-\frac{\partial^2 z}{\partial y\partial x}\right) = y\left[\frac{z}{x}\varphi'\left(\frac{y}{x}\right)+\frac{x+y}{x^2}\varphi''\left(\frac{y}{x}\right)+\frac{2y}{x^2}\varphi'\left(\frac{y}{x}\right)+\frac{y(x+y)}{x^3}\varphi''\left(\frac{y}{x}\right)\right]$$

$$= 2\left(\frac{y}{x}+\frac{y^2}{x^2}\right)\varphi'\left(\frac{y}{x}\right)+\frac{(x+y)^2}{x}\left(\frac{y}{x^2}\right)\cdot\varphi''\left(\frac{y}{x}\right)$$

Hence
$$x\left(\frac{\partial^2 z}{\partial x^2}-\frac{\partial^2 z}{\partial x\partial y}\right) = y\left(\frac{\partial^2 z}{\partial y^2}-\frac{\partial^2 z}{\partial x\partial y}\right)$$ **Ans.**

14. *If* $u = e^{xyz} f\left(\frac{xz}{y}\right)$ *show that* $x\frac{\partial u}{\partial x} + y\frac{\partial u}{\partial y} = 2\,xyzu = y\frac{\partial u}{\partial y} + z\frac{\partial u}{\partial z}$

and deduce that $x\frac{\partial^2 u}{\partial x \partial z} = y\frac{\partial^2 u}{\partial y \partial z}$

SOLUTION: Since $u = e^{xyz} + f\left(\frac{xz}{y}\right)$, we have

$$\frac{\partial u}{\partial x} = e^{xyz} \cdot yz\, f\left(\frac{xz}{y}\right) + e^{xyz} f'\left(\frac{xz}{y}\right) \cdot \frac{z}{y} = z\, e^{xyz}\left[y\, f\left(\frac{xz}{y}\right) + \frac{1}{y} f'\left(\frac{xz}{y}\right)\right]$$

$$\frac{\partial u}{\partial y} = e^{xyz} x z\, f\left(\frac{xz}{y}\right) + e^{xyz} f'\left(\frac{xz}{y}\right)\left(\frac{-xz}{y^2}\right) = x z\, e^{xyz}\left[f\left(\frac{xz}{y}\right) - \frac{1}{y^2} f'\left(\frac{xz}{y}\right)\right]$$

$$\therefore \quad x\frac{\partial u}{\partial x} + y\frac{\partial u}{\partial y} = z\, e^{xyz}\left[xyf\left(\frac{xz}{y}\right) + xy\, f\left(\frac{xz}{y}\right) + \frac{x}{y} f'\left(\frac{xz}{y}\right) - \frac{x}{y} f'\left(\frac{xz}{y}\right)\right] = 2\, xyz\, e^{xyz} f\left(\frac{xz}{y}\right)$$

and
$$\frac{\partial u}{\partial z} = e^{xyz} xy\, f\left(\frac{xz}{y}\right) + e^{xyz} f'\left(\frac{xz}{y}\right) \cdot \frac{x}{y} = e^{xyz} x\left[y\, f\left(\frac{xz}{y}\right) + \frac{1}{y} f'\left(\frac{xz}{y}\right)\right]$$

$$\therefore \quad y\frac{\partial u}{\partial y} + z\frac{\partial u}{\partial z} = xyz\, e^{xyz}\left[f\left(\frac{xz}{y}\right) - \frac{1}{y^2} f'\left(\frac{xz}{y}\right)\right] + xz\, e^{xyz}\left[y\, f\left(\frac{xz}{y}\right) + \frac{1}{y} f'\left(\frac{xz}{y}\right)\right]$$

$$= 2xyz\, e^{xyz} f\left(\frac{xz}{y}\right)$$

Hence $$x\frac{\partial u}{\partial x} + y\frac{\partial u}{\partial y} = y\frac{\partial u}{\partial y} + z\frac{\partial u}{\partial z}$$ **Hence Proved**

15. *Given* $x = \cos\theta - r\sin\theta$, $y = \sin\theta + r\cos\theta$ *show that*

(*i*) $\frac{\partial \theta}{\partial x} = \frac{-\cos\theta}{r}$, $\frac{\partial r}{\partial x} = \frac{x}{r}$

(*ii*) $\frac{\partial^2 \theta}{\partial x^2} = \frac{\cos\theta}{r^3}(\cos\theta - 2r\sin\theta)$

SOLUTION: $x = \cos\theta - r\sin\theta$ ...(1), $y = \sin\theta + r\cos\theta$ ...(2) hence $x^2 + y^2 = r^2 + 1$ ...(3)

(*i*) Differentiating (3) partially w.r.t. x, we get $2x = 2r\frac{\partial r}{\partial x}$ or $\frac{\partial r}{\partial x} = \frac{x}{r}$.

Next, differentiating (1) partially w.r.t. x, gives $1 = -\sin\theta\frac{\partial \theta}{\partial x} - r\cos\theta\frac{\partial \theta}{\partial x} - \sin\theta\frac{\partial r}{\partial x}$

or
$$(\sin\theta + r\cos\theta)\frac{\partial \theta}{\partial x} = -1 - \sin\theta \cdot \frac{x}{r} = -1 - \frac{\sin\theta}{r}(\cos\theta - r\sin\theta)$$

or $$\frac{\partial \theta}{\partial x} = \frac{-r - \sin\theta\cos\theta + r\sin^2\theta}{r(\sin\theta + r\cos\theta)} = \frac{-r\cos^2\theta - \sin\theta\cos\theta}{r(\sin\theta + r\cos\theta)} = \frac{-\cos\theta}{r}$$

(*ii*) Now $$\frac{\partial^2\theta}{\partial x^2} = \frac{\partial}{\partial x}\left(\frac{-\cos\theta}{r}\right) = \frac{r\sin\theta\dfrac{\partial\theta}{\partial x} + \cos\theta\dfrac{\partial r}{\partial x}}{r^2} = \frac{1}{r^2}\left[r\sin\theta\frac{(-\cos\theta)}{r} + \cos\theta\cdot\frac{x}{r}\right]$$

$$= \frac{1}{r^2}\left[-\sin\theta\cos\theta + \frac{\cos^2\theta}{r} - \cos\theta\sin\theta\right] = \frac{\cos\theta}{r^3}(-2r\sin\theta + \cos\theta)$$ **Hence Proved.**

16. *If* $u = f(r)$ *and* $x = r\cos\theta,\ \ y = r\sin\theta,$ *prove that*

$$\frac{\partial^2 u}{\partial x^2} + \frac{\partial^2 u}{\partial y^2} = f''(r) + \frac{1}{r}f'(r).$$

SOLUTION: Since $u = f(r)$ we have $\dfrac{\partial u}{\partial x} = f'(r)\dfrac{\partial r}{\partial x}$

Also since $r^2 = x^2 + y^2$ we have $2r\dfrac{\partial r}{\partial x} = 2x$ or $\dfrac{\partial r}{\partial x} = \dfrac{x}{r}$ $\therefore$ $\dfrac{\partial u}{\partial x} = f'(r)\dfrac{x}{r}$.

Further $$\frac{\partial^2 u}{\partial x^2} = x\cdot\frac{d}{dr}\left(\frac{f'(r)}{r}\right)\cdot\frac{\partial r}{\partial x} + 1\cdot\frac{f'(r)}{r} = \frac{x^2}{r}\frac{d}{dr}\left(\frac{f'(r)}{r}\right) + \frac{f'(r)}{r}$$

$$= \frac{x^2}{r^3}[r\,f''(r) - f'(r)] + \frac{f'(r)}{r}$$

Similarly $$\frac{\partial^2 u}{\partial y^2} = \frac{y^2}{r^3}[r\,f''(r) - f'(r)] + \frac{f'(r)}{r}$$

$$\therefore \quad \frac{\partial^2 u}{\partial x^2} + \frac{\partial^2 u}{\partial y^2} = \frac{x^2 + y^2}{r^3}[r\,f''(r) - f'(r)] + \frac{2f'(r)}{r}$$

$$= \frac{r^2}{r^3}[r\,f''(r) - f'(r)] + \frac{2f'(r)}{r} = f''(r) + \frac{f'(r)}{r}.$$ **Hence Proved.**

SOLUTIONS TO EXERCISE 11B

1. *If* $u = \dfrac{x^3+y^3}{y\sqrt{x}} + \dfrac{1}{x^7}\sin^{-1}\left(\dfrac{x^2+y^2}{x^2+2xy}\right)$ *find the value of*

$$x^2 u_{xx} + 2xy\, u_{xy} + y^2 u_{yy} + x\, u_x + y\, u_y \quad \textit{at the point } (1, 2).$$

SOLUTION: $u = \dfrac{x^3+y^3}{y\sqrt{x}} + \dfrac{1}{x^7}\sin^{-1}\left(\dfrac{x^2+y^2}{x^2+2xy}\right) = f_1(x, y) + f_2(x, y)$. Here $f_1(x, y) = \dfrac{x^3+y^3}{y+\sqrt{x}}$ is hemogenous in x and y of degree $\dfrac{3}{2}$. Also, $\dfrac{x^2+y^2}{x^2+2xy}$ is homogenous in x and y of degree 0, hence can be treated as cosntant. As such $f_2(x, y) = \dfrac{1}{x^7}\sin^{-1}\left(\dfrac{x^2+y^2}{x^2+2xy}\right)$ is homogeneous of degree −7. Therefore, by Euler's theorem

$$x\frac{\partial f_1}{\partial x} + y\frac{\partial f_2}{\partial y} = \frac{3}{2}f_1 \quad \text{and} \quad x\frac{\partial f_2}{\partial x} + y\frac{\partial f_2}{\partial y} = -7f_2$$

Also $$x^2\frac{\partial^2 f_1}{\partial x^2} + 2xy\frac{\partial^2 f_1}{\partial x\partial y} + y^2\frac{\partial^2 f_1}{\partial y^2} = \frac{3}{2}\left(\frac{3}{2}-1\right)f_1 = \frac{3}{4}f_1$$

and $$x^2\frac{\partial^2 f_2}{\partial x^2} + 2xy\frac{\partial^2 f_2}{\partial x\partial y} + y^2\frac{\partial^2 f_2}{\partial y^2} = -7(-7-1)f_2 = 56f_2$$

$$\therefore \quad x^2 u_{xx} + 2xy\, u_{xy} + y^2 u_{yy} + xu_x + yu_y$$

$$= x^2 f_{1xx} + 2xy\, f_{1xy} + y^2 f_{1yy} + x f_{1x} + y f_{1y} + x^2 f_{2xx} + 2xy f_{2xy} + y^2 f_{2yy} + x f_{2x} + y f_{2y}$$

$$= \frac{3}{4}f_1 + \frac{3}{2}f_1 + 56f_2 - 7f_2 = \frac{9}{4}f_1(x, y) + 49f_2(x, y) = \frac{9}{4}f_1(1, 2) + 49f_2(1, 2)$$

$$= \frac{9}{4}\cdot\frac{1^3+2^3}{2\sqrt{1}} + 49\,\frac{1}{1}\sin^{-1}\left(\frac{1^2+2^2}{1^2+2\cdot1\cdot2}\right) \quad \text{at } (1, 2)$$

$$= \frac{9}{4}\cdot\frac{9}{2} + 49\sin^{-1}(1) = \frac{81}{8} + 49\,\frac{\pi}{2}$$ **Ans.**

2. *If* $u = \sin^{-1}\left(\dfrac{\sqrt{x}-\sqrt{y}}{\sqrt{x}+\sqrt{y}}\right)$ *show that* $\dfrac{\partial u}{\partial x} = \dfrac{-y}{x}\dfrac{\partial u}{\partial y}$.

SOLUTION: $u = \sin^{-1}\left(\dfrac{\sqrt{x}-\sqrt{y}}{\sqrt{x}+\sqrt{y}}\right)$ $\quad\therefore\quad \sin u = \dfrac{\sqrt{x}-\sqrt{y}}{\sqrt{x}+\sqrt{y}}$

Here sin u is homogenous in x and y of degree 0. Hence, by Euler's Theorem.

$$x\frac{\partial \sin u}{\partial x} + y\frac{\partial \sin u}{\partial y} = 0.\ \sin u \quad \text{or} \quad x\cos u\frac{\partial u}{\partial x} + y\cos u\frac{\partial u}{\partial y} = 0$$

or $$x\frac{\partial u}{\partial x} + y\frac{\partial u}{\partial y} = 0 \quad \text{Hence} \quad \frac{\partial u}{\partial x} = -\frac{y}{x}\frac{\partial u}{\partial y}$$ **Hence Proved.**

3. *If* $z = x^3 e^{-x/y}$ *prove that* $x^2 z_{xx} + 2xy\, z_{xy} + y^2 z_{yy} = 6z$.

SOLUTION: $z = x^3 e^{-x/y}$. Here $\dfrac{x}{y}$ is hemogeneous in x and y of degree 0. Hence z is hemogeneous in x and y of degree 3. By Euler's theorem.

$$x^2 z_{xx} + 2xy\, z_{xy} + y^2 z_{yy} = 3(3-1)z = 6z$$ **Hence Proved.**

4. *If* $u = \sin^{-1}\left(\dfrac{x^{1/4}+y^{1/4}}{x^{1/6}+y^{1/6}}\right)$, *evaluate* $x^2\dfrac{\partial^2 u}{\partial x^2} + 2xy\dfrac{\partial^2 u}{\partial x\,\partial y} + y^2\dfrac{\partial^2 u}{\partial y^2}$.

SOLUTION: $\sin u = \dfrac{x^{1/4}+y^{1/4}}{x^{1/6}+y^{1/6}} = f(u)$.

It is hemogeneous is x and y of degree $\left(\dfrac{1}{4}-\dfrac{1}{6}\right)\left(=\dfrac{1}{12}\right)$. By Euler's theorem we have

$$xu_x + yu_y = \frac{n\,f(u)}{f'(u)} = \frac{1}{12}\frac{\sin u}{\cos u} = \frac{1}{12}\tan u = g(u)$$

$\therefore$ By Euler's theorem $x^2 u_{xx} + 2xy\, u_{xy} + y^2 u_{yy} = g(u)\{g'(u) - 1\}$

$$= \frac{1}{12}\tan u\left\{\frac{1}{12}\sec^2 u - 1\right\} = \frac{1}{144}\tan u(\tan^2 u - 11)$$ **Hence Proved.**

5. *If* $y = x\sin u$, *find the value of* $x^2 u_{xx} + 2xy\, u_{xy} + y^2 u_{yy}$.

SOLUTION: $y = x\sin u$. Let $f(u) = \sin u = \dfrac{y}{x}$. Hence $\sin u$ is homogeneous in x and y of degree 0. Hence by Euler's theorm $x\dfrac{\partial \sin u}{\partial x} + y\dfrac{\partial \sin u}{\partial y} = 0$

or $\quad x \cos u \dfrac{\partial u}{\partial x} + y \cos u \dfrac{\partial u}{\partial y} = 0$ or $x \dfrac{\partial u}{\partial x} + y \times \dfrac{\partial u}{\partial x} = 0.$

Differetating it partially w.r.t. x and y, gives

$$x \frac{\partial^2 u}{\partial x^2} + 1 \cdot \frac{\partial u}{\partial x} + y \frac{\partial^2 u}{\partial x \partial y} = 0 \qquad ...(1)$$

and $\quad x \dfrac{\partial^2 u}{\partial y \partial x} + y \dfrac{\partial^2 u}{\partial y^2} + 1 \cdot \dfrac{\partial u}{\partial y} = 0 \qquad ...(2)$

Multiplying (1) by x and (2) by y and adding, we get

$$x^2 \frac{\partial^2 u}{\partial x \partial y} + 2xy \frac{\partial^2 u}{\partial x \partial y} + y^2 \frac{\partial^2 u}{\partial y^2} + x \frac{\partial u}{\partial x} + y \frac{\partial u}{\partial y} = 0$$

But $\qquad x \dfrac{\partial u}{\partial x} + y \dfrac{\partial u}{\partial y} = 0,$

hence $\qquad x^2 \dfrac{\partial^2 u}{\partial x^2} + 2xy \dfrac{\partial^2 u}{\partial x \partial y} + y^2 \dfrac{\partial^2 u}{\partial y^2} = 0 \qquad$ **Ans.**

6. *If* $f(x, y) = \dfrac{1}{x^2} + \dfrac{1}{xy} + \dfrac{\log x - \log y}{x^2 + y^2}$, *show that* $x \dfrac{\partial f}{\partial x} + y \dfrac{\partial f}{\partial y} + 2f = 0.$

SOLUTION: $\qquad f(x, y) = \dfrac{1}{x^2} + \dfrac{1}{xy} + \dfrac{\log x - \log y}{x^2 + y^2} = u + v$

where $\qquad u = \dfrac{1}{x^2} + \dfrac{1}{xy}, \quad v = \dfrac{\log (x/y)}{x^2 + y^2}.$

Hence u is homogenous in x and y of degree – 2 and v is homogenous in x and y of degree – 2, since $\dfrac{x}{y}$ is of degree 0. $\quad\therefore$ By Euler's theorm, we have

$$x \frac{\partial u}{\partial x} + y \frac{\partial u}{\partial y} = -2u \quad \text{and} \quad x \frac{\partial v}{\partial x} + y \frac{\partial v}{\partial y} = -2v$$

Adding these, we get $\qquad x \dfrac{\partial}{\partial x} (u + v) + y \dfrac{\partial}{\partial y} (u + v) = -2 (u + v)$

or $\qquad x \dfrac{\partial f}{\partial x} + y \dfrac{\partial f}{\partial y} + 2f = 0 \qquad$ **Hence Proved.**

7. *Determine the value of* $x^2 u_{xx} + 2xy\, u_{xy} + y^2 u_{yy}$ *if* $u = \sin^{-1} (x^3 + y^3)^{2/5}$.

SOLUTION: Here $f(u) = \sin u = (x^3 + y^3)^{\frac{2}{5}}$

which is a homogeneous function in x and y degree $\frac{6}{5}$. Applying Euler's theorem, we get

$$x \frac{\partial u}{\partial x} + y \frac{\partial u}{\partial y} = \frac{nf(u)}{f'(u)} = \frac{6}{5} \frac{\sin u}{\cos u} = \frac{6}{5} \tan u = g(u)$$

$$\therefore \quad x^2 \frac{\partial^2 u}{\partial x^2} + 2xy \frac{\partial^2 u}{\partial x \partial y} + y^2 \frac{\partial^2 u}{\partial y^2} = g(u)\,[g'(u) - 1] = \frac{6}{5} \tan u \left[\frac{6}{5} \sec^2 u - 1\right]$$ **Ans.**

8. *If* $u = \sec^{-1}\left(\frac{x^3 - y^3}{x - y}\right)$, *then evaluate* $x^2 u_{xx} + 2xy\, u_{xy} + y^2 u_{yy}$.

SOLUTION: Here $\sec u = f(u) = \frac{x^3 - y^3}{x - y}$. This is homogenous functions in x and y of degree 2.

Using Euler's theorem, we get

$$x \frac{\partial u}{\partial x} + y \frac{\partial u}{\partial y} = \frac{n\, f(u)}{f'(u)} = \frac{2 \sec u}{\sec u \tan u} = 2 \cot u = g(u)$$

then $x^2 u_{xx} + 2xy\, u_{xy} + y^2 u_{yy} = g(u)\,[g'(u) - 1] = 2 \cot u\,[-2 \operatorname{cosec}^2 u - 1]$

$$= -\frac{2 \cos u}{\sin^3 u} (2 + \sin^2 u).$$ **Ans.**

9. *If* $u = \frac{x^2 y^2 z^2}{x^2 + y^2 + z^2} + \cos \frac{2xy + yz + zx}{x^2 + y^2 + z^2}$ *evaluate* $x\, u_x + y\, u_y + z\, u_z$.

SOLUTION: Let us write $u = V + \cos W$.

where $V = \frac{x^2 y^2 z^2}{x^2 + y^2 + z^2}$ and $W = \frac{2xy + yz + zx}{x^2 + y^2 + z^2}$.

Here V and W are homogeous functions in x and y of degree 4 and 0 respectively.
Applying Euler's theorem to V and W, we get

$$x \frac{\partial V}{\partial x} + y \frac{\partial V}{\partial y} = 4V \quad \text{and} \quad x \frac{\partial W}{\partial x} + y \frac{\partial W}{\partial y} = 0$$

$$\therefore \quad x \frac{\partial u}{\partial x} + y \frac{\partial u}{\partial y} = x \frac{\partial}{\partial x} (V + \cos W) + y \frac{\partial}{\partial y} (V + \cos W)$$

$$= x \frac{\partial V}{\partial x} + y \frac{\partial V}{\partial y} - x \sin W \frac{\partial W}{\partial x} - y \sin W \frac{\partial W}{\partial y}$$

$$= 4V - \sin W\,(0) = 4V = \frac{4x^2 y^2 z^2}{x^2 + y^2 + z^2}$$ **Ans.**

10. *If* $u = \log \dfrac{x+y}{\sqrt{x^2+y^2}} \sin^{-1}\left(\dfrac{x+y}{\sqrt{x}+\sqrt{y}}\right)$, *show that*

$$x^2 \frac{\partial^2 u}{\partial x^2} + 2xy \frac{\partial^2 u}{\partial x \partial y} + y^2 \frac{\partial^2 u}{\partial y^2} = \frac{1}{2} \tan\theta \left(\frac{1}{2}\sec^2\theta - 1\right) \text{ where } \theta = \sin^{-1} \frac{x+y}{\sqrt{x}+\sqrt{y}}.$$

SOLUTION: $\log \dfrac{x+y}{\sqrt{x^2+y^2}}$ behaves like a costant and the degree of $\dfrac{x+y}{\sqrt{x^2+y^2}}$ is zero.

$\therefore$ we can write $u = k \sin^{-1} \dfrac{x+y}{\sqrt{x}+\sqrt{y}} = k\theta$ where $k = \log \dfrac{x+y}{\sqrt{x^2+y^2}}$ and $\sin\theta = \dfrac{x+y}{\sqrt{x}+\sqrt{y}} = f(\theta)$

Applying Euler's theorem, we have $x \dfrac{\partial\theta}{\partial x} + y \dfrac{\partial\theta}{\partial y} = \dfrac{1}{2} \tan\theta = g(\theta)$

$$\therefore \quad x^2 \frac{\partial^2\theta}{\partial x^2} + 2xy \frac{\partial^2\theta}{\partial x \partial y} + y^2 \frac{\partial^2\theta}{\partial y^2} = g(\theta)[g'(\theta) - 1] = \frac{1}{2}\tan\theta\left[\frac{1}{2}\sec^2\theta - 1\right]$$

But $\theta = \dfrac{u}{k}$, hence

$$\frac{1}{k} x^2 \frac{\partial^2 u}{\partial x^2} + \frac{1}{k} \cdot 2xy \frac{\partial^2 u}{\partial x \partial y} + \frac{1}{k} y^2 \frac{\partial^2 u}{\partial y^2} = \frac{1}{2}\tan\theta\left[\frac{1}{2}\sec^2\theta - 1\right]$$

$$\therefore \quad x^2 u_{xx} + 2xy\, u_{xy} + y^2 u_{yy} = \frac{1}{2}\tan\theta\left(\frac{1}{2}\sec^2\theta - 1\right)\log\left(\frac{x+y}{\sqrt{x^2+y^2}}\right)$$

Hence Proved.

11. *If* $u = \dfrac{1}{r} f(\theta)$ *where* $x = r\cos\theta$, $y = r\sin\theta$, *then evaluate* $x \dfrac{\partial u}{\partial x} + y \dfrac{\partial u}{\partial y}$.

SOLUTION: Since $x = r\cos\theta$, $y = \sin\theta$ we have $r = \sqrt{x^2+y^2}$ and $\theta = \tan^{-1} \dfrac{y}{x}$

Therefore
$$u = \frac{1}{r} f(\theta) = \frac{1}{\sqrt{x^2+y^2}} f\left(\tan^{-1}\frac{y}{x}\right).$$

Here $\dfrac{1}{\sqrt{x^2+y^2}}$ is a homogaeous function in x and y of degree -1 and $\dfrac{y}{x}$ of degree 0.

Hence $f\left(\tan^{-1}\dfrac{y}{x}\right)$ can be treated as costant. Therefore, we can take u as homeganeous in x and y of degree -1. Applying Eluer's theorem, we get

$$x\frac{\partial u}{\partial x} + y\frac{\partial u}{\partial y} = (-1)\, u = -u$$

Ans.

12. *If* $u = \log\sqrt{x^2+y^2}$, *evaluate* $x^2\dfrac{\partial^2 u}{\partial x^2}+2xy\dfrac{\partial^2 u}{\partial x\,\partial y}+y^2\dfrac{\partial^2 u}{\partial y^2}$.

SOLUTION: $e^u = \sqrt{x^2+y^2} = f(u)$ which is homogeneous in x and y of degree 1, hence by Euler's theorem we have

$$x\frac{\partial u}{\partial x}+y\frac{\partial u}{\partial y} = \frac{1\,f(u)}{f'(u)} = \frac{e^u}{e^u} = 1 = g(u)$$

and $$x^2\frac{\partial^2 u}{\partial x^2}+2xy\frac{\partial^2 u}{\partial x\partial y}+y^2\frac{\partial^2 u}{\partial y^2} = g(u)[g'(u)-1]=1(0-1) = -1$$ **Ans.**

13. *If* $x = e^u \tan v$, $y = e^u \sec v$, *find the value of*

$$\left(x\frac{\partial u}{\partial x}+y\frac{\partial u}{\partial y}\right)\left(x\frac{\partial v}{\partial x}+y\frac{\partial v}{\partial y}\right)$$

SOLUTION: Since $x = e^u \tan v$, $y = e^u \sec v$, we have $y^2 - x^2 = e^{2u}$ and $\dfrac{x}{y} = \sin v$.

Since e^{2u} is homogeneous function is x and y of degree 2 and $\sin v$ is homogeneous function of degree 0, applying Euler's theorem, we get

$$x\frac{\partial}{\partial x}e^{2u}+y\frac{\partial}{\partial y}e^{2u} = 2\,e^{2u} \quad \text{and} \quad x\frac{\partial}{\partial x}\sin v + y\frac{\partial}{\partial y}\sin v = 0 \sin v = 0$$

or $$2x\,e^{2u}\frac{\partial u}{\partial x}+2ye^{2u}\frac{\partial u}{\partial y} = 2\,e^{2u} \quad \text{and} \quad x\cos v\frac{\partial v}{\partial x}+y\cos v\frac{\partial v}{\partial y} = 0$$

$\Rightarrow$ $$x\frac{\partial u}{\partial x}+y\frac{\partial u}{\partial y} = 1 \quad \text{and} \quad x\frac{\partial v}{\partial u}+y\frac{\partial v}{\partial y}=0$$

$\therefore$ $$\left(x\frac{\partial u}{\partial x}+y\frac{\partial u}{\partial y}\right)\left(x\frac{\partial v}{\partial x}+y\frac{\partial v}{\partial y}\right)=0$$ **Ans.**

14. *If* $u = \log\dfrac{x^4+y^4}{x+y}$, *show that* $x\dfrac{\partial u}{\partial x}+y\dfrac{\partial u}{\partial y}=3.$

SOLUTION: $e^u = \dfrac{x^4+y^4}{x+y}$. Here e^u is homogeneous function in x and y of degree 3 hence

applying Euler's theorem on e^u we get $x\dfrac{\partial}{\partial x}e^u + y\dfrac{\partial}{\partial y}e^u = 3e^u$

or $$x\,e^u\frac{\partial u}{\partial x}+ye^u\frac{\partial u}{\partial y} = 3\,e^x \quad \text{or} \quad x\frac{\partial u}{\partial x}+y\frac{\partial u}{\partial y} = 3$$ **Hence Proved.**

15. *If* $u(x, y) = \sin^{-1}\left(\frac{x}{y}\right) + \tan^{-1}\left(\frac{y}{x}\right)$, *prove that* $x\frac{\partial u}{\partial x} + y\frac{\partial u}{\partial y} = 0.$

SOLUTION: Let $u = V + W$ where $V = \sin^{-1}\left(\frac{x}{y}\right)$ and $W = \tan^{-1}\left(\frac{y}{x}\right)$

Here $\sin V = \frac{x}{y}$ and $\tan W = \frac{y}{x}$. Here $\frac{x}{y}$ and $\frac{y}{x}$ are both homogeneous function in x and y of degree. 0, hence applying Euler's theorem on $\sin V$ and $\tan W$, we get

$$x\frac{\partial}{\partial x}\sin V + y\frac{\partial}{\partial y}\sin V = 0 \sin V \quad \text{and} \quad x\frac{\partial}{\partial x}\tan W + y\frac{\partial}{\partial y}\tan W = 0 \tan W$$

$$\Rightarrow \quad x\cos V\frac{\partial V}{\partial x} + y\cos V\frac{\partial V}{\partial y} = 0 \quad \text{and} \quad x\sec^2 W\frac{\partial W}{\partial x} + y\sec^2 W\frac{\partial W}{\partial y} = 0$$

or $$x\frac{\partial V}{\partial x} + y\frac{\partial V}{\partial y} = 0 \text{ and } x\frac{\partial W}{\partial x} + y\frac{\partial W}{\partial y} = 0$$

$$\Rightarrow \quad x\frac{\partial}{\partial x}(V+W) + y\frac{\partial}{\partial y}(V+W) = 0 \text{ or } x\frac{\partial u}{\partial x} + y\frac{\partial u}{\partial y} = 0$$

Hence Proved.

16. *If* $u = \sin^{-1}\frac{x+2y+3z}{\sqrt{x^8+y^8+z^8}}$, *show that* $x\frac{\partial u}{\partial x} + y\frac{\partial u}{\partial y} + z\frac{\partial u}{\partial z} = -3\tan u.$

SOLUTION: $\sin u = \frac{x+2y+3z}{\sqrt{x^8+y^8+z^8}}$, It is a homogeneous function in x, y, z of degree -3, hence applying Euler's theorem on $\sin u$, we get

$$x\frac{\partial u}{\partial x}\sin u + y\frac{\partial}{\partial y}\sin u = -3 \sin u$$

or $$x\cos u\frac{\partial u}{\partial x} + y\cos u\frac{\partial u}{\partial y} = -3 \sin u$$

or $$x\frac{\partial u}{\partial x} + y\frac{\partial u}{\partial y} = -3 \tan u$$

Hence Proved.

SOLUTIONS TO EXERCISE 11C

1. *If* $x^3 + 3x^2y + 6xy^2 + y^3 = 1$, *find* $\dfrac{dy}{dx}$.

SOLUTION: $f(x, y) = x^3 + 3x^2y + 6xy^2 + y^3 - 1 = 0$

Hence $$p = \frac{\partial f}{\partial x} = 3x^2 + 6xy + 6y^2, \quad q = \frac{\partial f}{\partial y} = 3x^2 + 12xy + 3y^2$$

$$\therefore \quad \frac{dy}{dx} = \frac{-p}{q} = \frac{-(x^2 + 2xy + 2y^2)}{(x^2 + 4xy + y^2)} \quad \textbf{Ans.}$$

2. *If* $ax^2 + 2hxy + by^2 + 2gx + 2fv + c = 0$ *prove that* $\dfrac{d^2y}{dx^2} = \dfrac{\Delta}{(hx + by + f)^3}$

where $\Delta = \begin{vmatrix} a & h & g \\ h & b & f \\ g & h & c \end{vmatrix}$.

SOLUTION: $f(x, y) = ax^2 + 2hxy + by^2 + 2gx + 2fy + C = 0$

$$\therefore \quad p = \frac{\partial f}{\partial x} = 2(ax + hy + g), \quad \text{and} \quad q = \frac{\partial f}{\partial y} = 2(hx + by + f),$$

and $$r = \frac{\partial^2 f}{\partial x^2} = 2a, \quad s = \frac{\partial^2 f}{\partial x\,\partial y} = 2h, \quad t = \frac{\partial^2 f}{\partial y^2} = 2b,$$

$$\therefore \quad \frac{d^2y}{dx^2} = \frac{-1}{q^3}[q^2r - 2pqs + p^2t]$$

$$= \frac{-1}{(hx + by + f)^3}\left[a(hx + by + f)^2 - 2h(ax + hy + g)(hx + by + f) + b(ax + hy + g)^2\right]$$

$$= \frac{-1}{(hx + by + f)^3}\Big[x^2(ah^2 - 2ah^2 + a^2h) + y^2(ab^2 - 2h^2b + bh^2) + 2xy(abh - abh - h^3)$$
$$+ 2x(ahf - ahf - gh^2 + abg) + 2y(abf - h^2f - bgh + bgh) + af^2 - 2fgh + bg^2\Big]$$

$$= \frac{-1}{(hx + by + f)^3}[x^2a(ab - h^2) + by^2(ab - h^2) - 2h^3xy + 2gx(ab - h^2)$$
$$+ 2fy(ab - h^2) + af^2 + bg^2 - 2fgh]$$

$$= \frac{-1}{(hx + by + f)^3}[(ab - h^2)(ax^2 + by^2 + agx + 2fy) - 2h^3xy + af^2 + bg^2 - 2fgh]$$

$$= \frac{-1}{(hx + by + f)^3}[(ab - h^2)(-2hxy - c) - 2h^3xy + af^2 + bg^2 - 2fgh)]$$

(On using $ax^2 + 2hxy + by^2 + 2gx + 2fy + c = 0$)

$$= \frac{1}{(hx+by+f)^3}[abc+2fgh-af^2-bg^2-ch^2] = \frac{\Delta}{(hx+by+f)^3}$$

where $\Delta = \begin{vmatrix} a & h & g \\ h & b & f \\ g & f & c \end{vmatrix}$

3. *If* $x^4 + y^4 + 4a^2xy = 0$, *prove that* $(y^3 + a^2x)^3 \dfrac{d^2y}{dx^2} = 2a^2xy\,(x^2y^2 + 3a^4)$.

SOLUTION: $f(x, y) = x^4 + y^4 + 4a^2\,xy = 0$

hence $p = \dfrac{\partial f}{\partial x} = 4x^3 + 4a^2y,\quad q = \dfrac{\partial f}{\partial y} = 4y^3 + 4a^2x,$

and $r = \dfrac{\partial^2 f}{\partial x^2} = 12x^2,\quad s = \dfrac{\partial^2 f}{\partial x\,\partial y} = 4a^2,\quad t = \dfrac{\partial^2 f}{\partial y^2} = 12y^2$

$$\therefore \quad \frac{d^2y}{dx^2} = \frac{-1}{q^3}[q^2r - 2pqs + p^2t] = \frac{-1}{64(y^3+a^2+x)^3}\cdot$$

$$[16(y^3+a^2x)^2\cdot 12x^2 - 2.4.4.4(y^3+a^2x)(x^3+a^2y)a^2 + 16(x^3+a^2y)^2.12y^2]$$

$$= \frac{-1}{(y^3+a^2x)^3}[3x^2(y^3+a^2x)^2 - 2a^2(y^3+a^2x)(x^3+a^2y) + 3y^2(x^3+a^2y)^2]$$

or $$(y^3+a^2x)^3\frac{d^2y}{dx^2} = -3\,(xy^3 + a^2x^2)^2 + 2a^2(y^3 + a^2x)\,(x^3 + a^2y) - 3\,(x^3y + a^2y^2)^2$$

$$= -3(x^2y^6 + a^4x^4 + 2a^2x^3y^3) + 2a^2\,(x^3y^3 + a^4xy + a^2y^4 + a^2x^4) - 3(x^6y^2 + a^4y^4 + 2a^2x^3y^3)$$

$$= -3x^2y^2\,(x^4 + y^4) - 3a^4(x^4 + y^4) - 10\,a^2x^3y^3 + 2a^6xy + 2a^4\,(x^4 + y^4)$$

$$= -4a^2\,xy\,(-3x^2y^2 - a^4) - 10a^2x^3y^3 + 2a^6xy = 2a^2xy\,(x^2y^2 + 3a^4)$$ **Hence Proved.**

4. *If* $u = x\log(xy)$ *where* $x^3 + y^3 + 3xy = 1$, *find* $\dfrac{du}{dx}$ *at* $(1, 1)$.

SOLUTION: $du = \dfrac{\partial u}{\partial x}dx + \dfrac{\partial u}{\partial y}dy$ or $\dfrac{du}{dx} = \dfrac{\partial u}{\partial x} + \dfrac{\partial u}{\partial x}\dfrac{dy}{dx}$

From the relation $f(x, y) = x^3 + 3xy + y^3 - 1 = 0$, we have

$$\frac{dy}{dx} = \frac{\dfrac{-\partial f}{\partial x}}{\dfrac{\partial f}{\partial y}} = \frac{-(3x^2+3y)}{3y^2+3x} = \frac{-(x^2+y)}{y^2+x} = -1 \quad \text{at} \quad (1, 1)$$

Also we have $\dfrac{\partial u}{\partial x} = x\dfrac{1}{xy}\cdot y + 1\cdot\log(xy) = 1 + \log(xy) = 1$ at $(1, 1)$

and $\dfrac{\partial u}{\partial y} = x\dfrac{1}{xy}x = \dfrac{x}{y} = 1$ at $(1, 1)$

Hence $\dfrac{du}{dx} = 1 + 1\cdot(-1) = 0$ at $(1, 1)$ **Ans.**

5. *The relations $f(x, y, z) = constant$ and $xyz = constant$ together define y as a function of x, then show that*

$$\frac{dy}{dx} = \frac{-y\left(x\frac{\partial f}{\partial x} - z\frac{\partial f}{\partial z}\right)}{x\left(y\frac{\partial f}{\partial y} - z\frac{\partial f}{\partial z}\right)}.$$

SOLUTION: $f(x, y, z)$ = constant, hence $df = \frac{\partial f}{\partial x}dx + \frac{\partial f}{\partial y}dy + \frac{\partial f}{\partial z}dz = 0$

and xyz = constant gives $yzdx + xzdy + xydz = 0$

or $\frac{\partial f}{\partial y}\frac{dy}{dx} + \frac{\partial f}{\partial z}\frac{dz}{dx} + \frac{\partial f}{\partial x} = 0$ or $xz\frac{dy}{dx} + xy\frac{dz}{dx} + yz = 0$

Eliminating $\frac{dz}{dx}$ in the above equations, we get

$$\frac{dy}{dx}\left[xy\frac{\partial f}{\partial y} - xz\frac{\partial f}{\partial z}\right] = -xy\frac{\partial f}{\partial x} + yz\frac{\partial f}{\partial z}$$

$$\therefore \quad \frac{dy}{dx} = \frac{-y\left(x\frac{\partial f}{\partial x} - z\frac{\partial f}{\partial z}\right)}{x\left(y\frac{\partial f}{\partial y} - z\frac{\partial f}{\partial z}\right)}$$ **Hence Proved.**

6. *If* $u = \log(\tan x + \tan y + \tan z)$, find the value of $\sin 2x\frac{\partial u}{\partial x} + \sin 2y\frac{\partial u}{\partial y} + \sin 2z\frac{\partial u}{\partial z}$.

SOLUTION: $e^u = \tan x + \tan y + \tan z \quad \therefore \quad e^u\frac{\partial u}{\partial x} = \sec^2 x$ or $\frac{\partial u}{\partial x} = \frac{\sec^2 x}{e^u}$

$$\therefore \quad \sin 2x\frac{\partial u}{\partial x} = \frac{2\sin x\cos x}{\cos^2 x\, e^u} = \frac{2\tan x}{e^u}$$

Hence $\sin 2x\frac{\partial u}{\partial x} + \sin 2y\frac{\partial u}{\partial y} + \sin 2z\frac{\partial u}{\partial z} = \frac{2(\tan x + \tan y + \tan z)}{e^u} = 2$ **Ans.**

7. *If* $u = f\left(\frac{x}{y}, \frac{y}{z}, \frac{z}{x}\right)$ *prove that* $x\frac{\partial u}{\partial x} + y\frac{\partial u}{\partial y} + z\frac{\partial u}{\partial z} = 0.$

SOLUTION: Let $u = f\left(\frac{x}{y}, \frac{y}{z}, \frac{z}{x}\right) = f(p, q, r)$, where $p = \frac{x}{y},\ q = \frac{y}{z},\ r = \frac{z}{x}$.

$$\therefore \quad \frac{\partial u}{\partial x} = \frac{\partial f}{\partial p}\cdot\frac{\partial p}{\partial x} + \frac{\partial f}{\partial q}\frac{\partial q}{\partial x} + \frac{\partial f}{\partial r}\frac{\partial r}{\partial x} = \frac{1}{y}\frac{\partial f}{\partial p} + 0\cdot\frac{\partial f}{\partial q} + \frac{\partial f}{\partial r}\left(\frac{-z}{x^2}\right)$$

or $$x\frac{\partial u}{\partial x} = \frac{x}{y}\frac{\partial f}{\partial p} - \frac{z}{x}\frac{\partial f}{\partial r} \quad ...(1)$$

Similarly $$\frac{\partial u}{\partial y} = \frac{\partial f}{\partial p}\frac{\partial p}{\partial y}+\frac{\partial f}{\partial q}\cdot\frac{\partial q}{\partial y}+\frac{\partial f}{\partial r}\frac{\partial r}{\partial y} = \frac{-x}{y^2}\frac{\partial f}{\partial p}+\frac{1}{z}\frac{\partial f}{\partial q}+0\cdot\frac{\partial f}{\partial r}$$

or $$y\frac{\partial u}{\partial y} = \frac{-x}{y}\frac{\partial f}{\partial p}+\frac{y}{z}\frac{\partial f}{\partial q}+0 \qquad \text{...(2)}$$

And $$\frac{\partial u}{\partial z} = \frac{\partial f}{\partial p}\frac{\partial p}{\partial z}+\frac{\partial f}{\partial q}\cdot\frac{\partial q}{\partial z}+\frac{\partial f}{\partial r}\cdot\frac{\partial r}{\partial z} = 0\cdot\frac{\partial f}{\partial p}-\frac{y}{z^2}\frac{\partial f}{\partial q}+\frac{1}{x}\frac{\partial f}{\partial r}$$

or $$z\frac{\partial u}{\partial z} = 0-\frac{y}{z}\frac{\partial f}{\partial p}+\frac{z}{x}\frac{\partial f}{\partial r} \qquad \text{...(3)}$$

$\therefore$ $$x\frac{\partial u}{\partial x}+y\frac{\partial u}{\partial y}+z\frac{\partial u}{\partial z} = \frac{\partial f}{\partial p}\left(\frac{x}{y}-\frac{x}{y}\right)+\frac{\partial f}{\partial q}\left(\frac{y}{z}-\frac{y}{z}\right)+\frac{\partial f}{\partial r}\left(\frac{z}{x}-\frac{z}{x}\right)=0$$ **Ans.**

8. *If* $u = x + ay$ *and* $v = x + by$ *transforms the equation* $2\dfrac{\partial^2 z}{\partial x^2}-5\dfrac{\partial^2 z}{\partial x\,\partial y}+3\dfrac{\partial^2 z}{\partial y^2}=0$ *into* $\dfrac{\partial^2 z}{\partial u\,\partial v}=0$, *find the values of* a *and* b.

SOLUTION: $u = x + ay,\ v = x + by$ hence $\dfrac{\partial u}{\partial x}=1,\ \dfrac{\partial u}{\partial y}=a,\ \dfrac{\partial v}{\partial x}=1,\ \dfrac{\partial v}{\partial y}=b$

$$\frac{\partial z}{\partial x} = \frac{\partial z}{\partial u}\frac{\partial u}{\partial x}+\frac{\partial z}{\partial v}\frac{\partial v}{\partial x} = \frac{\partial z}{\partial u}+\frac{\partial z}{\partial v} \Rightarrow \frac{\partial}{\partial x}\equiv\frac{\partial}{\partial u}+\frac{\partial}{\partial v}.$$

$$\frac{\partial z}{\partial y} = \frac{\partial z}{\partial u}\frac{\partial u}{\partial y}+\frac{dz}{\partial v}\frac{\partial v}{\partial y} = a\frac{\partial z}{\partial u}+b\frac{\partial z}{\partial v} \Rightarrow \frac{\partial}{\partial y}\equiv a\frac{\partial}{\partial u}+b\frac{\partial}{\partial v}.$$

then $$\frac{\partial^2 z}{\partial x^2} = \frac{\partial}{\partial x}\left(\frac{\partial z}{\partial x}\right) = \left(\frac{\partial}{\partial u}+\frac{\partial}{\partial v}\right)\left(\frac{\partial z}{\partial u}+\frac{\partial z}{\partial v}\right) = \frac{\partial^2 z}{\partial u^2}+2\frac{\partial^2 z}{\partial u\,\partial v}+\frac{\partial^2 z}{\partial v^2},$$

$$\frac{\partial^2 z}{\partial y^2} = \frac{\partial}{\partial y}\left(\frac{\partial z}{\partial y}\right)=\left(a\frac{\partial}{\partial u}+b\frac{\partial}{\partial v}\right)\left(a\frac{\partial z}{\partial u}+b\frac{\partial z}{\partial v}\right) = a^2\frac{\partial^2 z}{\partial u^2}+2ab\frac{\partial^2 z}{\partial u\,\partial v}+b^2\frac{\partial^2 z}{\partial v^2},$$

and $$\frac{\partial^2 z}{\partial x\,\partial y} = \frac{\partial}{\partial x}\left(\frac{\partial z}{\partial y}\right)=\left(\frac{\partial}{\partial u}+\frac{\partial}{\partial v}\right)\left(a\frac{\partial z}{\partial u}+b\frac{\partial z}{\partial v}\right) = a\frac{\partial^2 z}{\partial x\,\partial y}+(a+b)\frac{\partial^2 z}{\partial u\,\partial v}+b\frac{\partial^2 z}{\partial v^2}$$

Putting for $\dfrac{\partial^2 z}{\partial x^2}, \dfrac{\partial^2 z}{\partial x\,\partial y}$ and $\dfrac{\partial^2 z}{\partial y^2}$ in $2\dfrac{\partial^2 z}{\partial x^2}-5\dfrac{\partial^2 z}{\partial x\,\partial y}+3\dfrac{\partial^2 z}{\partial y^2}=0$, we get

$$(2-5a+3a^2)\frac{\partial^2 z}{\partial u^2}+(4-5a+6ab)\frac{\partial^2 z}{\partial u\,\partial v}+(2-5b+3b^2)\frac{\partial^2 z}{\partial v^2}=0$$

But it should be $\dfrac{\partial^2 z}{\partial u\,\partial v}=0$, hence

$2-5a+3a^2=0,\ 2-5b+3b^2=0 \Rightarrow a = 1, \dfrac{2}{3}$ and $b = 1, \dfrac{2}{3}$

Thus $a = 1,\ b = \dfrac{2}{3}$ or $a = \dfrac{2}{3},\ b = 1$ **Ans.**

9. *If* $z = f(x, y)$ *where* $x = \log u,\ y = \log v$ *then show that*

(i) $$\frac{\partial^2 z}{\partial x^2} + \frac{\partial^2 z}{\partial y^2} = u^2 \frac{\partial^2 z}{\partial u^2} + v^2 \frac{\partial^2 z}{\partial v^2} + u\frac{\partial z}{\partial u} + v\frac{\partial z}{\partial v}.$$

(ii) $$\frac{\partial^2 z}{\partial x \partial y} = uv \frac{\partial^2 z}{\partial u \partial v}.$$

SOLUTION: z is a function of x and y and $u = e^x,\ v = e^y$

(i) $$\frac{\partial z}{\partial x} = \frac{\partial z}{\partial u}\cdot\frac{\partial u}{\partial x} + \frac{\partial z}{\partial v}\frac{\partial v}{\partial x} = u\frac{\partial z}{\partial u} + 0, \quad \frac{\partial z}{\partial y} = \frac{\partial z}{\partial u}\frac{\partial u}{\partial y} + \frac{\partial z}{\partial v}\frac{\partial v}{\partial y} = 0 + v\frac{\partial z}{\partial v}$$

Next, $$\frac{\partial^2 z}{\partial x^2} = \frac{\partial}{\partial x}\left(\frac{\partial z}{\partial x}\right) = \left(u\frac{\partial}{\partial u}\right)\left(u\frac{\partial z}{\partial u}\right) = u\left[u\frac{\partial^2 z}{\partial u^2} + 1\frac{\partial z}{\partial u}\right] = u^2\frac{\partial^2 z}{\partial u^2} + u\frac{\partial z}{\partial u}$$

and $$\frac{\partial^2 z}{\partial y^2} = \frac{\partial}{\partial y}\left(\frac{\partial z}{\partial y}\right) = \left(v\frac{\partial}{\partial v}\right)\left(v\frac{\partial z}{\partial v}\right) = v^2\frac{\partial^2 z}{\partial v^2} + v\frac{\partial z}{\partial v}$$

$\therefore$ $$\frac{\partial^2 z}{\partial x^2} + \frac{\partial^2 z}{\partial y^2} = u^2\frac{\partial^2 z}{\partial u^2} + v^2\frac{\partial^2 z}{\partial v^2} + u\frac{\partial z}{\partial u} + v\frac{\partial z}{\partial v}$$

(ii) $$\frac{\partial^2 z}{\partial x\,\partial y} = \frac{\partial}{\partial x}\left(\frac{\partial z}{\partial y}\right) = \left(u\frac{\partial}{\partial u}\right)\left(v\frac{\partial z}{\partial v}\right) = uv\frac{\partial^2 z}{\partial u\,\partial v}$$

Hence Proved.

10. *If* $z = f(u, v)$ *where* $u = x + y,\ v = x - y$ *show that* $\dfrac{\partial z}{\partial x} + \dfrac{\partial z}{\partial y} = 2\dfrac{\partial z}{\partial u}$.

SOLUTION: $z = f(u, v),\ u = x + y,\ v = x - y$

Therefore $$\frac{\partial z}{\partial x} = \frac{\partial z}{\partial u}\frac{\partial u}{\partial x} + \frac{\partial z}{\partial v}\frac{\partial v}{\partial x} = \frac{\partial z}{\partial u} + \frac{\partial z}{\partial v}$$

and $$\frac{\partial z}{\partial y} = \frac{\partial z}{\partial u}\cdot\frac{\partial u}{\partial y} + \frac{\partial z}{\partial v}\cdot\frac{\partial v}{\partial y} = 1\cdot\frac{\partial z}{\partial u} - \frac{\partial z}{\partial v}$$

$\therefore$ $$\frac{\partial z}{\partial x} + \frac{\partial z}{\partial y} = 2\frac{\partial z}{\partial u}$$

Hence Proved.

11. *If* $z = f(x, y)$ *where* $x = u\cosh v$ *and* $y = u\sinh v$, *then show that*

$$\left(\frac{\partial z}{\partial x}\right)^2 - \left(\frac{\partial z}{\partial y}\right)^2 = \left(\frac{\partial z}{\partial u}\right)^2 - \frac{1}{u^2}\left(\frac{\partial z}{\partial v}\right)^2.$$

SOLUTION: $x = u\cosh v,\ y = u\sinh v$

$$\frac{\partial z}{\partial u} = \frac{\partial z}{\partial x}\frac{\partial x}{\partial u} + \frac{\partial z}{\partial y}\frac{\partial y}{\partial u} = \cosh v\frac{\partial z}{\partial x} + \sinh v\frac{\partial z}{\partial y}$$

$$\frac{\partial z}{\partial v} = \frac{\partial z}{\partial x}\frac{\partial x}{\partial v} + \frac{\partial z}{\partial y}\frac{\partial y}{\partial v} = u\sinh v\frac{\partial z}{\partial x} + u\cosh v\frac{\partial z}{\partial y}$$

$$\therefore \quad \left(\frac{\partial z}{\partial u}\right) - \frac{1}{u^2}\left(\frac{\partial z}{\partial v}\right)^2 = \left(\frac{\partial z}{\partial x}\right)^2 (\cosh^2 v - \sinh^2 v) + \left(\frac{\partial z}{\partial y}\right)^2 (\sinh^2 v - \cosh^2 v)$$

$$= \left(\frac{\partial z}{\partial x}\right)^2 - \left(\frac{\partial z}{\partial y}\right)^2$$

Hence Proved.

12. *If* $z = f(x, y)$ *where* $x = uv$ *and* $y = \frac{u}{v}$ *then show that*

(i) $\frac{\partial z}{\partial x} = \frac{1}{2v}\frac{\partial z}{\partial u} + \frac{1}{2u}\frac{\partial z}{\partial v}$ (ii) $\frac{\partial z}{\partial y} = \frac{v}{2}\frac{\partial z}{\partial u} - \frac{v^2}{2u}\frac{\partial z}{\partial v}$.

SOLUTION: $x = uv,\ y = \frac{u}{v} \quad \therefore\ u = \sqrt{xy},\ v = \sqrt{\left(\frac{x}{y}\right)}$

(i) $$\frac{\partial z}{\partial x} = \frac{\partial z}{\partial u}\frac{\partial u}{\partial x} + \frac{\partial z}{\partial v}\frac{\partial v}{\partial x} = \frac{1}{2}\sqrt{\frac{y}{x}}\frac{\partial z}{\partial u} + \frac{1}{2\sqrt{xy}}\frac{\partial z}{\partial v} = \frac{1}{2v}\frac{\partial z}{\partial u} + \frac{1}{2u}\frac{\partial z}{\partial v}.$$

(ii) $$\frac{\partial z}{\partial y} = \frac{\partial z}{\partial u}\frac{\partial u}{\partial y} + \frac{\partial z}{\partial v}\frac{\partial v}{\partial y} = \frac{1}{2}\sqrt{\frac{x}{y}}\frac{\partial z}{\partial u} - \frac{1}{2}\frac{\sqrt{x}}{y\sqrt{y}}\frac{\partial z}{\partial v} = \frac{v}{2}\frac{\partial z}{\partial u} - \frac{1}{2}\frac{v}{y}\frac{\partial z}{\partial v} = \frac{v}{2}\frac{\partial z}{\partial u} - \frac{v^2}{2u}\frac{\partial z}{\partial v}$$ **Hence Proved.**

13. *A function* $f(x, y)$ *is expressed in terms of new variables* u *and* v *given by* $u = e^x \cos y$ *and* $v = e^x \sin y$ *show that*

(i) $\frac{\partial f}{\partial x} = u\frac{\partial f}{\partial u} + v\frac{\partial f}{\partial v}$

(ii) $\frac{\partial f}{\partial y} = -v\frac{\partial f}{\partial u} + u\frac{\partial f}{\partial v}$

(iii) $\frac{\partial^2 f}{\partial x^2} + \frac{\partial^2 f}{\partial y^2} = (u^2 + v^2)\left(\frac{\partial^2 f}{\partial u^2} + \frac{\partial^2 f}{\partial v^2}\right)$

SOLUTION: $u = e^x \cos y,\ v = e^x \sin y$

(i) $$\frac{\partial f}{\partial x} = \frac{\partial f}{\partial u}\frac{\partial u}{\partial x} + \frac{\partial f}{\partial v}\frac{\partial v}{\partial x} = e^x \cos y\frac{\partial f}{\partial u} + e^x \sin y\frac{\partial f}{\partial v} = u\frac{\partial f}{\partial u} + v\frac{\partial f}{\partial v}$$

(ii) $$\frac{\partial f}{\partial y} = \frac{\partial f}{\partial u}\frac{\partial u}{\partial y} + \frac{\partial f}{\partial v}\frac{\partial v}{\partial y} = -e^x \sin y\frac{\partial f}{\partial u} + e^x \cos y\frac{\partial f}{\partial v} = -v\frac{\partial f}{\partial u} + u\frac{\partial f}{\partial v}.$$

(iii) $$\frac{\partial^2 f}{\partial x^2} = \left(u\frac{\partial}{\partial u} + v\frac{\partial}{\partial v}\right)\left(u\frac{\partial f}{\partial u} + v\frac{\partial f}{\partial v}\right)$$

$$= u\left[u\frac{\partial^2 f}{\partial u^2} + \frac{\partial f}{\partial u} + v\frac{\partial^2 f}{\partial u\,\partial v}\right] + v\left[u\frac{\partial^2 f}{\partial u\,\partial v} + v\frac{\partial^2 f}{\partial v^2} + \frac{\partial f}{\partial v}\right]$$

$$= u^2\frac{\partial^2 f}{\partial u^2} + 2uv\frac{\partial^2 f}{\partial u\,\partial v} + v^2\frac{\partial^2 f}{\partial y^2} + u\frac{\partial f}{\partial u} + v\frac{\partial f}{\partial v}$$

and
$$\frac{\partial^2 f}{\partial y^2} = \left(-v\frac{\partial}{\partial u}+u\frac{\partial}{\partial v}\right)\left(-v\frac{\partial f}{\partial u}+u\frac{\partial f}{\partial v}\right)$$

$$= -v\left(-v\frac{\partial^2 f}{\partial u^2}+u\frac{\partial^2 f}{\partial u\,\partial v}+\frac{\partial f}{\partial v}\right)+u\left(-v\frac{\partial^2 f}{\partial u\,\partial v}-\frac{\partial f}{\partial u}+u\frac{\partial^2 f}{\partial v^2}\right)$$

$$= v^2\frac{\partial^2 f}{\partial u^2}-2uv\frac{\partial^2 f}{\partial u\,\partial v}+u^2\frac{\partial^2 f}{\partial v^2}-v\frac{\partial f}{\partial v}-u\frac{\partial f}{\partial u}$$

$\therefore$
$$\frac{\partial^2 f}{\partial x^2}+\frac{\partial^2 f}{\partial y^2} = (u^2+v^2)\left(\frac{\partial^2 f}{\partial u^2}+\frac{\partial^2 f}{\partial v^2}\right)$$
Hence Proved.

14. *If* $u = f(ax^2 + 2hxy + by^2)$ *and* $v = \phi(ax^2 + 2hxy + by^2)$ *prove that*

$$\frac{\partial}{\partial y}\left(u\frac{\partial v}{\partial x}\right) = \frac{\partial}{\partial x}\left(u\frac{\partial v}{\partial y}\right).$$

SOLUTION: $u = f(ax^2 + 2hxy + by^2)$, $v = \phi(ax^2 + 2hxy + by^2)$

Hence
$$\frac{\partial v}{\partial x} = \phi'\cdot(2ax+2hy) \quad\text{and}\quad u\frac{\partial v}{\partial x} = (2ax+2hy)\,\phi'\cdot f$$

$\therefore$
$$\frac{\partial}{\partial y}\left(u\frac{\partial v}{\partial x}\right) = \frac{\partial}{\partial y}[(2ax+2hy)\phi'\cdot f]$$

$$= (2ax+2hy)\,[\phi''\cdot(2hx+2by)\,f+\phi'\;f'\cdot(2hx+2by)]+2h\,\phi'\cdot f$$

Also
$$\frac{\partial v}{\partial y} = \phi'\cdot(2hx+2by) \quad\text{and}\quad u\frac{\partial v}{\partial y} = (2hx+2by)\,\phi'\cdot f$$

$\therefore$
$$\frac{\partial}{\partial x}\left(u\frac{\partial v}{\partial y}\right) = \frac{\partial}{\partial x}[(2hx+2by)\,\phi'\cdot f]$$

$$= (2hx+2by)\,[\phi''\cdot(2ax+2hy)\,f+\phi'\cdot(2ax+2hy)]+2h\,\phi'\cdot f$$

Hence
$$\frac{\partial}{\partial y}\left(u\frac{\partial v}{\partial x}\right) = \frac{\partial}{\partial x}\left(u\frac{\partial v}{\partial y}\right)$$
Hence Proved.

15. *If* $u = u\left(\dfrac{y-x}{xy},\dfrac{z-x}{xz}\right)$ *prove that* $x^2\dfrac{\partial u}{\partial x}+y^2\dfrac{\partial u}{\partial y}+z^2\dfrac{\partial u}{\partial z} = 0.$

SOLUTION: $u = u\left(\dfrac{y-x}{xy},\dfrac{z-x}{xz}\right) = u(t_1,t_2)$ where $t_1 = \dfrac{y-x}{xy} = \dfrac{1}{x}-\dfrac{1}{y}$, and $t_2 = \dfrac{z-x}{xz} = \dfrac{1}{x}-\dfrac{1}{z}$.

Now
$$\frac{\partial u}{\partial x} = \frac{\partial u}{\partial t_1}\frac{\partial t_1}{\partial x}+\frac{\partial u}{\partial t_2}\frac{\partial t_2}{\partial x} = \frac{-1}{x^2}\frac{\partial u}{\partial t_1}-\frac{1}{x^2}\frac{\partial u}{\partial t_2}$$

or $$x^2\frac{\partial u}{\partial x} = -\frac{\partial u}{\partial t_1} - \frac{\partial u}{\partial t_2}$$

Next, $$\frac{\partial u}{\partial y} = \frac{\partial u}{\partial t_1}\cdot\frac{\partial t_1}{\partial y} + \frac{\partial u}{\partial t_2}\cdot\frac{\partial t_2}{\partial y} = \frac{1}{y^2}\frac{\partial u}{\partial t_1} + 0 \quad \therefore \quad y^2\frac{\partial u}{\partial y} + \frac{\partial u}{\partial t_1}$$

and $$\frac{\partial u}{\partial z} = \frac{\partial u}{\partial t_1}\frac{\partial t_1}{\partial z} + \frac{\partial u}{\partial t_2}\frac{\partial t_2}{\partial z} = 0 + \frac{\partial u}{\partial t_2}\cdot\frac{1}{z^2} \quad \therefore \quad z^2\frac{\partial u}{\partial z} = \frac{\partial u}{\partial t_2}$$

$\Rightarrow$ $$x^2\frac{\partial u}{\partial x} + y^2\frac{\partial u}{\partial y} + z^2\frac{\partial u}{\partial z} = 0$$ **Hence Proved.**

16. *If* $x + y = 2\,e^{\theta}\cos\phi$ and $x - y = 2ie^{\theta}\sin\phi$ *then show that* $\dfrac{\partial^2 f}{\partial\theta^2} + \dfrac{\partial^2 f}{\partial\phi^2} = 4xy\dfrac{\partial^2 f}{\partial x\,\partial y}$.

SOLUTION: $x + y = 2\,e^{\theta}\cos\phi$ and $x - y = 2ie^{\theta}\sin\phi$ $\therefore$ $x = e^{\theta}(\cos\phi + i\sin\theta) = e^{\theta+i\phi}$

and $y = e^{\theta}(\cos\phi - i\sin\phi) = e^{\theta-i\phi}$

Next $$\frac{\partial f}{\partial\theta} = \frac{\partial f}{\partial x}\frac{\partial x}{\partial\theta} + \frac{\partial f}{\partial y}\frac{\partial y}{\partial\theta} = e^{\theta+i\phi}\cdot\frac{\partial f}{\partial x} + e^{\theta-i\phi}\cdot\frac{\partial f}{\partial y} = x\frac{\partial f}{\partial x} + y\frac{\partial f}{\partial y}$$

and $$\frac{\partial f}{\partial\phi} = \frac{\partial f}{\partial x}\frac{\partial x}{\partial\phi} + \frac{\partial f}{\partial y}\frac{\partial y}{\partial\theta} = i\,e^{\theta+i\phi}\frac{\partial f}{\partial x} - i\,e^{\theta-i\phi}\frac{\partial f}{\partial y} = i\left(x\frac{\partial f}{\partial x} - y\frac{\partial f}{\partial y}\right)$$

Furhter $$\frac{\partial^2 f}{\partial\theta^2} = \frac{\partial}{\partial x}\left(x\frac{\partial f}{\partial x} + y\frac{\partial f}{\partial y}\right)\cdot\frac{\partial x}{\partial\theta} + \frac{\partial}{\partial y}\left(x\frac{\partial f}{\partial x} + y\frac{\partial f}{\partial y}\right)\cdot\frac{\partial y}{\partial\theta}$$

$$= \left(x\frac{\partial^2 f}{\partial x^2} + \frac{\partial f}{\partial x} + y\frac{\partial^2 f}{\partial x\,\partial y}\right)x + \left(x\frac{\partial^2 f}{\partial x\,\partial y} + y\frac{\partial^2 f}{\partial y^2} + \frac{\partial f}{\partial y}\right)y$$

$$= x^2\frac{\partial^2 f}{\partial x^2} + 2xy\frac{\partial^2 f}{\partial x\,\partial y} + y^2\frac{\partial^2 f}{\partial y^2} + x\frac{\partial f}{\partial x} + y\frac{\partial f}{\partial y}$$

and $$\frac{\partial^2 f}{\partial\phi^2} = \frac{i\partial}{\partial x}\left(x\frac{\partial f}{\partial x} - y\frac{\partial f}{\partial y}\right)\frac{\partial x}{\partial\phi} + i\frac{\partial}{\partial y}\left(x\frac{\partial f}{\partial x} - y\frac{\partial f}{\partial y}\right)\frac{\partial y}{\partial\phi}$$

$$= i^2x\left(x\frac{\partial^2 f}{\partial x^2} + \frac{\partial f}{\partial x} - y\frac{\partial^2 f}{\partial x\,\partial y}\right) - i_y^2\left(x\frac{\partial^2 f}{\partial x\,\partial y} - y\frac{\partial^2 f}{\partial y^2} - \frac{\partial f}{\partial y}\right)$$

$$= -x^2\frac{\partial^2 f}{\partial x^2} - x\frac{\partial f}{\partial x} + xy\frac{\partial^2 f}{\partial x\,\partial y} + xy\frac{\partial f}{\partial x\,\partial y} - y^2\frac{\partial^2 f}{\partial y^2} - y\frac{\partial f}{\partial y}$$

$\therefore$ $$\frac{\partial^2 f}{\partial\theta^2} + \frac{\partial^2 f}{\partial\phi^2} = 4xy\frac{\partial^2 f}{\partial x\,\partial y}$$ **Hence Proved.**

17. *If* $u = f(r, s, t)$ *where* $r = \frac{x}{y},\ s = \frac{y}{z},\ t = \frac{z}{x}$ *prove that* $x\frac{\partial u}{\partial x} + y\frac{\partial u}{\partial y} + z\frac{\partial u}{\partial z} = 0.$

SOLUTION: $\frac{\partial u}{\partial x} = \frac{\partial u}{\partial r}\cdot\frac{\partial r}{\partial x} + \frac{\partial u}{\partial s}\cdot\frac{\partial s}{\partial x} + \frac{\partial u}{\partial t}\cdot\frac{\partial t}{\partial x} = \frac{1}{y}\frac{\partial u}{\partial r} + 0\frac{\partial u}{\partial s} - \frac{z}{x^2}\frac{\partial u}{\partial t} \quad \therefore \quad x\frac{\partial u}{\partial x} = \frac{x}{y}\frac{\partial u}{\partial r} - \frac{z}{x}\frac{\partial u}{\partial t}$

Next, $\frac{\partial u}{\partial y} = \frac{\partial u}{\partial r}\frac{\partial r}{\partial y} + \frac{\partial u}{\partial s}\frac{\partial s}{\partial y} + \frac{\partial u}{\partial t}\cdot\frac{\partial t}{\partial y} = \frac{-x}{y^2}\frac{\partial u}{\partial r} + \frac{1}{z}\frac{\partial u}{\partial s} + 0 \quad \therefore \quad y\frac{\partial u}{\partial y} = \frac{-x}{y}\frac{\partial u}{\partial r} + \frac{y}{z}\frac{\partial u}{\partial s}$

Further $\frac{\partial u}{\partial z} = \frac{\partial u}{\partial r}\frac{\partial r}{\partial z} + \frac{\partial u}{\partial s}\frac{\partial s}{\partial z} + \frac{\partial u}{\partial t}\cdot\frac{\partial t}{\partial z} = 0\ 1 - \frac{y}{z^2}\frac{\partial u}{\partial s} + \frac{1}{x}\frac{\partial u}{\partial t} \quad \therefore \quad z\frac{\partial u}{\partial z} = \frac{-y}{z}\frac{\partial u}{\partial s} + \frac{z}{x}\frac{\partial u}{\partial t}$

Hence $x\frac{\partial u}{\partial x} + y\frac{\partial u}{\partial y} + z\frac{\partial u}{\partial z} = 0$ **Hence Proved.**

18. *If* $f(x, y, z) = 0$ *show that* $\left(\frac{\partial y}{\partial z}\right)_x \left(\frac{\partial z}{\partial x}\right)_y \left(\frac{\partial x}{\partial y}\right)_z = -1.$

SOLUTION: In $f(x, y, z) = 0$, treating z as constant we have

$$\left(\frac{\partial x}{\partial y}\right)_z = \frac{-\frac{\partial f}{\partial y}}{\frac{\partial f}{\partial x}}$$

Similarly, in $f(x, y, z) = 0$, treating y as constant we have

$$\left(\frac{\partial z}{\partial x}\right)_y = \frac{-\frac{\partial f}{\partial x}}{\frac{\partial f}{\partial z}}$$

Similarly, in $f(x, y, z) = 0$, treating x as constant we have

$$\left(\frac{\partial y}{\partial z}\right)_x = \frac{-\frac{\partial f}{\partial z}}{\frac{\partial f}{\partial y}}$$

Multiplying the above three relations, we get

$$\left(\frac{\partial x}{\partial y}\right)_z \cdot \left(\frac{\partial z}{\partial x}\right)_y \left(\frac{\partial y}{\partial z}\right)_x = -1.$$

Hence Proved.

ADDITIONAL IMPORTANT PROBLEMS

1. If $u = \dfrac{e^{x+y+z}}{e^x + e^y + e^z}$ prove that $\dfrac{\partial u}{\partial x} + \dfrac{\partial u}{\partial y} + \dfrac{\partial u}{\partial z} = 2u.$

SOLUTION: Differentiating partially w.r.t. x, we get

$$\frac{\partial u}{\partial x} = \frac{e^{x+y+z}(e^x + e^y + e^z) - e^{x+y+z} \cdot e^x}{(e^x + e^y + e^z)^2} = \frac{e^{x+y+z}(e^y + e^z)}{(e^x + e^y + e^z)^2}$$

Similarly, $$\frac{\partial u}{\partial y} = \frac{e^{x+y+z}(e^z + e^x)}{(e^x + e^y + e^z)^2} \quad \text{and} \quad \frac{\partial u}{\partial z} = \frac{e^{x+y+z}(e^x + e^y)}{(e^x + e^y + e^z)^2}$$

$\therefore$ $$\frac{\partial u}{\partial x} + \frac{\partial u}{\partial y} + \frac{\partial u}{\partial z} = \frac{e^{x+y+z}\, 2(e^x + e^y + e^z)}{(e^x + e^y + e^z)^2} = 2u.$$ **Hence Proved.**

2. If $u = (3xy - y^3) - (y^2 - 2x)^{3/2}$ show that $\dfrac{\partial^2 u}{\partial x \partial y} = \dfrac{\partial^2 u}{\partial y \partial x} = 3 + \dfrac{3y}{\sqrt{y^2 - 2x}}.$

SOLUTION: Since $u = (3xy - y^3) - (y^2 - 2x)^{3/2}$ we have

$$\frac{\partial u}{\partial x} = 3y - \frac{3}{2}(y^2 - 2x)^{1/2}(-2) = 3y + 3(y^2 - 2x)^{1/2}$$

$\therefore$ $$\frac{\partial^2 u}{\partial y \partial x} = 3 + \frac{3}{2}(y^2 - 2x)^{-1/2} \cdot 2y = 3 + \frac{3y}{\sqrt{y^2 - 2x}}.$$

and $$\frac{\partial u}{\partial y} = 3x - 3y^2 - \frac{3}{2}(y^2 - 2x)^{1/2}(2y) = 3x - 3y^2 + 3y\sqrt{y^2 - 2x}$$

$\therefore$ $$\frac{\partial^2 u}{\partial x \partial y} = 3 + \frac{3y}{2}(y^2 - 2x)(-2) = 3 + \frac{3y}{\sqrt{y^2 - 2x}}$$ **Hence Proved.**

3. If $u = 3(ax + by + cz)^2 - (x^2 + y^2 + z^2)$ and $a^2 + b^2 + c^2 = 1$, then show that

$$\frac{\partial^2 u}{\partial x^2} + \frac{\partial^2 u}{\partial y^2} + \frac{\partial^2 u}{\partial z^2} = 0.$$

SOLUTION: $u = 3(ax + by + cz)^2 - (x^2 + y^2 + z^2)$ hence $\dfrac{\partial u}{\partial x} = 3.2\,(ax + by + cz)\,a - 2x$

and $$\frac{\partial^2 u}{\partial x^2} = 6a^2 - 2$$

Similarly, $$\frac{\partial^2 u}{\partial y^2} = 6b^2 - 2 \quad \text{and} \quad \frac{\partial^2 u}{\partial z^2} = 6c^2 - 2.$$

$$\therefore \quad \frac{\partial^2 u}{\partial x^2} + \frac{\partial^2 u}{\partial y^2} + \frac{\partial^2 u}{\partial z^2} = 6a^2 - 2 + 6b^2 - 2 + 6c^2 - 2$$

$$= 6 - 6 = 0, \text{ since } a^2 + b^2 + c^2 = 1.$$ **Hence Proved.**

4. If $u = (ar^n + br^{-n})(\cos n\theta + \sin n\theta)$ prove that

$$\frac{\partial^2 u}{\partial r^2} + \frac{1}{r}\frac{\partial u}{\partial r} + \frac{1}{r^2}\frac{\partial^2 u}{\partial \theta^2} = 0$$

SOLUTION: $u = (ar^n + br^{-n})(\cos n\theta + \sin n\theta)$. Differentiating partially w.r.t. r, we get

$$\frac{\partial u}{\partial r} = (nar^{n-1} - bnr^{-n-1})(\cos n\theta + \sin n\theta)$$

and $$\frac{\partial^2 u}{\partial r^2} = [n(n-1)\, ar^{n-2} + bn(n+1) r^{-n-2}](\cos n\theta + \sin n\theta)$$

Differentiating u w.r.t. θ partially, we get

$$\frac{\partial u}{\partial \theta} = (ar^n + br^{-n})(-n \sin n\theta + n \cos n\theta)$$

and $$\frac{\partial^2 u}{\partial \theta^2} = (ar^n + br^{-n})(-n^2 \cos n\theta - n^2 \sin n\theta) = -n^2 u$$

$$\therefore \quad r^2 \frac{\partial^2 u}{\partial r^2} = [(n^2 - n)\, ar^n + (n^2 + n)\, br^{-n}](\cos n\theta + \sin n\theta)$$

$$= [n^2 u - n(ar^n - br^{-n})](\cos n\theta + \sin n\theta)$$

and $$r \frac{\partial u}{\partial r} = (nar^n - bnr^{-n})(\cos n\theta + \sin n\theta)$$

$$\therefore \quad r^2 \frac{\partial^2 u}{\partial r^2} + r\frac{\partial u}{\partial r} + \frac{\partial^2 u}{\partial \theta^2} = n^2 u + [-n(ar^n - br^{-n}) + n(ar^n - br^{-n})](\cos n\theta + \sin n\theta) - n^2 u = 0$$

Ans.

5. If $u = f\left(\frac{x^2}{y}\right)$ prove that $x^2 \frac{\partial^2 u}{\partial x^2} + 3xy \frac{\partial^2 u}{\partial x \partial y} + 2y^2 \frac{\partial^2 u}{\partial y^2} = 0$

SOLUTION: Since $u = f\left(\frac{x^2}{y}\right)$ we have

$$\frac{\partial u}{\partial x} = f'\left(\frac{x^2}{y}\right) \cdot \frac{2x}{y} \quad \text{and} \quad \frac{\partial u}{\partial y} = f'\left(\frac{x^2}{y}\right)\left(\frac{-x^2}{y^2}\right)$$

and $$\frac{\partial^2 u}{\partial x^2} = \frac{2}{y} f'\left(\frac{x^2}{y}\right) + \frac{2x}{y} f''\left(\frac{x^2}{y}\right) \cdot \frac{2x}{y} = \frac{2}{y} f'\left(\frac{x^2}{y}\right) + \frac{4x^2}{y^2} f''\left(\frac{x^2}{y}\right).$$

$$\frac{\partial^2 u}{\partial x \partial y} = -\frac{2x}{y^2} f'\left(\frac{x^2}{y}\right) + \frac{2x}{y} f''\left(\frac{x^2}{y}\right)\left(\frac{-x^2}{y^2}\right)$$

$$= \frac{-2x}{y^2} f'\left(\frac{x^2}{y}\right) - \frac{2x^3}{y^3} f''\left(\frac{x^2}{y}\right).$$

$$\frac{\partial^2 u}{\partial y^2} = f'\left(\frac{x^2}{y}\right)\left(\frac{2x^2}{y^3}\right) - \frac{x^2}{y^2} f''\left(\frac{x^2}{y}\right)\left(\frac{-x^2}{y^2}\right)$$

$$= \frac{2x^2}{y^3} f'\left(\frac{x^2}{y}\right) + \frac{x^4}{y^4} f''\left(\frac{x^2}{y}\right).$$

$$x^2 \cdot \frac{\partial^2 u}{\partial x^2} + 3xy \frac{\partial^2 u}{\partial x \, \partial y} + 2y^2 \frac{\partial^2 u}{\partial y^2} = 0$$ (On simplification) **Hence Proved.**

6. If $u = f(r^2)$ where $r^2 = x^2 + y^2 + z^2$ show that

$$\frac{\partial^2 u}{\partial x^2} + \frac{\partial^2 u}{\partial y^2} + \frac{\partial^2 u}{\partial z^2} = 4r^2 f''(r^2) + 6f'(r^2).$$

SOLUTION: $u = f(r^2)$ $\therefore$ $$\frac{\partial u}{\partial x} = f'(r^2)\, 2r \frac{\partial r}{\partial x} = f'(r^2)\, 2r \cdot \frac{x}{r} = 2xf'(r^2)$$

and $$\frac{\partial^2 u}{\partial x^2} = 2x\, f''(r^2)\, 2r \frac{\partial r}{\partial x} + 2f'(r^2) = 2x \cdot 2r \cdot \frac{x}{r} f''(r^2) + 2f'(r^2)$$

$$= 4x^2 f''(r^2) + 2f'(r^2)$$

Similarly $$\frac{\partial^2 u}{\partial y^2} = 4y^2 f''(r^2) + 2f'(r^2) \quad \text{and} \quad \frac{\partial^2 u}{\partial z^2} = 4z^2 f''(r^2) + 2f'(r^2)$$

$\therefore$ $$\frac{\partial^2 u}{\partial x^2} + \frac{\partial^2 u}{\partial y^2} + \frac{\partial^2 u}{\partial z^2} = 4(x^2 + y^2 + z^2) f''(r^2) + 3(2f'(r^2))$$

$$= 4r^2 f''(r^2) + 6f'(r^2)$$ **Hence Proved.**

7. If $u = x^2 + y^2 + z^2 - 2xyz = 1$ establish the relation

$$\frac{dx}{\sqrt{1-x^2}} + \frac{dy}{\sqrt{1-y^2}} + \frac{dz}{\sqrt{1-z^2}} = 0$$

SOLUTION: We know that $du = \frac{\partial u}{\partial x} dx + \frac{\partial u}{\partial y} dy + \frac{\partial u}{\partial z} dz = 0$ (as $n = 1$).

Here $\frac{\partial u}{\partial x} = 2x - 2yz, \quad \frac{\partial u}{\partial y} = 2y - 2xz$ and $\frac{\partial u}{\partial z} = 2z - 2xy$

$\therefore \quad (x - yz)\, dx + (y - zx)\, dy + (z - xy)\, dz = 0$ ...(1)

Next, $x^2 + y^2 + z^2 - 2xyz = 1$ or $x^2 - 2xyz = 1 - y^2 - z^2$

or $x^2 - 2xyz + y^2 z^2 = 1 - y^2 - z^2 - y^2 - z^2$

or $(x - yz)^2 = (1 - y^2)(1 - z^2)$ or $x - yz = \sqrt{1-y^2}\sqrt{1-z^2}$

Similarly we can have $y - zx = \sqrt{1-z^2}\sqrt{1-x^2}$ and $z - xy = \sqrt{1-x^2}\sqrt{1-y^2}$.

Substituting these in (1) we get

$$\sqrt{1-y^2}\sqrt{1-z^2}\, dx + \sqrt{1-x^2}\sqrt{1-z^2}\, dy + \sqrt{1-x^2}\sqrt{1-y^2}\, dz = 0$$

or $$\frac{dx}{\sqrt{1-x^2}} + \frac{dy}{\sqrt{1-y^2}} + \frac{dz}{\sqrt{1-z^2}} = 0.$$ **Hence Proved.**

8. If $z = f(x, y)$ and $x = e^u + e^{-v}$ and $y = e^{-u} - e^v$

show that $\frac{\partial z}{\partial u} - \frac{\partial z}{\partial v} = x\frac{\partial z}{\partial x} - y\frac{\partial z}{\partial y}$.

SOLUTION: $z = f(x, y)$ where $x = e^u + e^{-v}, y = e^{-u} - e^v$, hence

$$\frac{\partial z}{\partial u} = \frac{\partial z}{\partial x}\cdot\frac{\partial x}{\partial u} + \frac{\partial z}{\partial y}\frac{\partial y}{\partial u} = e^u \frac{\partial z}{\partial x} - e^{-u}\frac{\partial z}{\partial y}$$

$$\frac{\partial z}{\partial v} = \frac{\partial z}{\partial x}\frac{\partial x}{\partial v} + \frac{\partial z}{\partial y}\cdot\frac{\partial y}{\partial v} = -e^{-v}\frac{\partial z}{\partial x} - e^v\frac{\partial z}{\partial y}$$

Hence $$\frac{\partial z}{\partial u} - \frac{\partial z}{\partial v} = (e^u + e^{-v})\frac{\partial z}{\partial x} - (e^{-u} - e^v)\frac{\partial z}{\partial y}$$

$$= x\frac{\partial z}{\partial x} - y\frac{\partial z}{\partial y}$$ **Hence Proved.**

9. If $z = f(u, v)$ and $u = x^2 - y^2$ and $v = 2xy$ prove that

$$\left(\frac{\partial z}{\partial x}\right)^2 + \left(\frac{\partial z}{\partial y}\right)^2 = 4(u^2 + v^2)^{1/2}\left\{\left(\frac{\partial z}{\partial u}\right)^2 + \left(\frac{\partial z}{\partial v}\right)^2\right\}$$

SOLUTION: $z = f(u, v)$ where $u = x^2 - y^2,\ v = 2xy$ we have

$$\frac{\partial z}{\partial x} = \frac{\partial z}{\partial u}\cdot\frac{\partial u}{\partial x} + \frac{\partial z}{\partial v}\frac{\partial v}{\partial x} = 2x\frac{\partial z}{\partial u} + 2y\frac{\partial z}{\partial v}$$

and $$\frac{\partial z}{\partial y} = \frac{\partial z}{\partial u}\frac{\partial u}{\partial y} + \frac{\partial z}{\partial v}\frac{\partial v}{\partial y} = -2y\frac{\partial z}{\partial u} + 2x\frac{\partial z}{\partial v}$$

Hence $$\left(\frac{\partial z}{\partial x}\right)^2 + \left(\frac{\partial z}{\partial y}\right)^2 = \left(2x\frac{\partial z}{\partial u} + 2y\frac{\partial z}{\partial v}\right)^2 + \left(-2y\frac{\partial z}{\partial u} + 2x\frac{\partial z}{\partial v}\right)^2$$

$$= 4x^2\left(\frac{\partial z}{\partial u}\right)^2 + 4y^2\left(\frac{\partial z}{\partial v}\right)^2 + 4xy\frac{\partial z}{\partial u}\frac{\partial z}{\partial v} + 4y^2\left(\frac{\partial z}{\partial u}\right)^2 + 4x^2\left(\frac{\partial z}{\partial v}\right)^2 - 4xy\frac{\partial z}{\partial u}\frac{\partial z}{\partial v}$$

$$= 4(x^2 + y^2)\left(\frac{\partial z}{\partial u}\right)^2 + 4(x^2 + y^2)\left(\frac{\partial z}{\partial v}\right)^2$$

$$= 4(x^2 + y^2)\left[\left(\frac{\partial z}{\partial u}\right)^2 + \left(\frac{\partial z}{\partial v}\right)^2\right]$$

Since $(x^2 + y^2)^2 = (x^2 - y^2)^2 + 4x^2 y^2 = u^2 + v^2$

$\therefore$ $$\left(\frac{\partial z}{\partial x}\right)^2 + \left(\frac{\partial z}{\partial y}\right)^2 = 4\sqrt{u^2 + v^2}\left[\left(\frac{\partial z}{\partial u}\right)^2 + \left(\frac{\partial z}{\partial v}\right)^2\right]$$ **Hence Proved.**

10. If $u = f(x^2 - y^2, y^2 - z^2, z^2 - x^2)$ show that $\frac{1}{x}\frac{\partial u}{\partial x} + \frac{1}{y}\frac{\partial u}{\partial y} + \frac{1}{z}\frac{\partial u}{\partial z} = 0.$

SOLUTION: Let $l = x^2 - y^2,\ m = y^2 - z^2,\ n = z^2 - x^2$ then we have

$$\frac{\partial l}{\partial x} = 2x,\ \frac{\partial l}{\partial y} = -2y,\ \frac{\partial l}{\partial z} = 0,\quad \frac{\partial m}{\partial x} = 0,\ \frac{\partial m}{\partial y} = 2y,\ \frac{\partial m}{\partial z} = -2z$$

and $$\frac{\partial n}{\partial x} = -2x,\ \frac{\partial n}{\partial y} = 0,\ \frac{\partial n}{\partial z} = 2z.$$

Now $$\frac{\partial u}{\partial x} = \frac{\partial u}{\partial l}\cdot\frac{\partial l}{\partial x} + \frac{\partial u}{\partial m}\frac{\partial m}{\partial x} + \frac{\partial u}{\partial n}\cdot\frac{\partial n}{\partial x} = 2x\frac{\partial u}{\partial l} + 0 - 2x\frac{\partial u}{\partial n}$$

or $$\frac{1}{x}\frac{\partial u}{\partial x} = 2\left(\frac{\partial u}{\partial l} - \frac{\partial u}{\partial n}\right).$$

Similarly we can arrive at

$$\frac{1}{y}\frac{\partial u}{\partial y} = 2\left(\frac{\partial u}{\partial m} - \frac{\partial u}{\partial l}\right) \quad \text{and} \quad \frac{1}{z}\frac{\partial u}{\partial z} = 2\left(\frac{\partial u}{\partial n} - \frac{\partial u}{\partial m}\right)$$

Adding these we get

$$\frac{1}{x}\frac{\partial u}{\partial x} + \frac{1}{y}\frac{\partial u}{\partial y} + \frac{1}{z}\frac{\partial u}{\partial z} = 0.$$

Hence Proved.

11. If $u = \dfrac{x^3 y^3 z^3}{x^2 + y^2 + z^2} + \cos\left(\dfrac{xy + yz + zx}{x^2 + y^2 + z^2}\right)$ then find the value of $x\dfrac{\partial u}{\partial x} + y\dfrac{\partial u}{\partial y} + z\dfrac{\partial u}{\partial z}$.

SOLUTION: Let $u = P + Q$ where $P = \dfrac{x^3 y^3 z^3}{x^2 + y^2 + z^2}$, $Q = \left(\cos\dfrac{xy + yz + zx}{x^2 + y^2 + z^2}\right)$.

Now P is a homogeneous function in x, y, z of degree 7 hence by Euler's theorem $x\dfrac{\partial P}{\partial x} + y\dfrac{\partial P}{\partial y} + z\dfrac{\partial P}{\partial z} = 7P$.

Also, Q is a homogeneous function in x, y, z of degree 0

hence $$x\frac{\partial Q}{\partial x} + y\frac{\partial Q}{\partial y} + z\frac{\partial Q}{\partial z} = 0. \quad Q = 0.$$

$\Rightarrow$ $$x\frac{\partial}{\partial x}(P+Q) + y\frac{\partial}{\partial y}(P+Q) + z\frac{\partial}{\partial z}(P+Q) = 0$$

or $$x\frac{\partial u}{\partial x} + y\frac{\partial u}{\partial y} + z\frac{\partial u}{\partial z} = 0.$$

Hence Proved.

12. If $u = x^3 \sin^{-1}(y/x) + x^4 \tan^{-1}(y/x)$, find the value of

$$x^2\frac{\partial^2 u}{\partial x^2} + 2xy\frac{\partial^2 u}{\partial x\,\partial y} + y^2\frac{\partial^2 u}{\partial y^2} + x\frac{\partial u}{\partial x} + y\frac{\partial u}{\partial y} \quad \text{at} \quad x = 1, y = 1.$$

SOLUTION: Let $u = P + Q$ where $P = x^3 \sin^{-1}(y/x)$ and $Q = x^4 \tan^{-1}(y/x)$.

Replacing x by xt and y by yt we get

$$P = t^3\{x^3 \sin^{-1}(y/x)\} \quad \text{and} \quad Q = t^4\{x^4 \tan^{-1} y/x\}$$

Hence P and Q are homogeneous functions in x and y of degreee 3 and 4 respctively. Therefore, by Euler's theorem

$$x^2\frac{\partial^2 P}{\partial x^2} + 2xy\frac{\partial^2 P}{\partial x\,\partial y} + y^2\frac{\partial^2 R}{\partial y^2} + x\frac{\partial P}{\partial x} + y\frac{\partial P}{\partial y} = 3(3-1)P + 3P$$

and $$x^2\frac{\partial^2 Q}{\partial x^2} + 2xy\frac{\partial^2 Q}{\partial x\,\partial y} + y^2\frac{\partial Q}{\partial y^2} + x\frac{\partial Q}{\partial x} + y\frac{\partial Q}{\partial y} = 4(4-1)Q + 4Q.$$

Adding the above two relations, gives

$$x^2\frac{\partial^2(P+Q)}{\partial x^2}+2xy\frac{\partial^2(P+Q)}{\partial x\,\partial y}+y^2\frac{\partial^2(P+Q)}{\partial y^2}+x\frac{\partial}{\partial x}(P+Q)+y\frac{\partial}{\partial y}(P+Q)=9P+16Q$$

or $$x^2\frac{\partial^2 u}{\partial x^2}+2xy\frac{\partial^2 u}{\partial x\,\partial y}+y^2\frac{\partial^2 u}{\partial y^2}+x\frac{\partial u}{\partial x}+y\frac{\partial u}{\partial y}=9P+16Q.$$

At $x = 1,\ y = 1$ we have $P = 1\sin^{-1}(1) = \pi/2$ and $Q = 1\tan^{-1}(1) = \pi/4$

$$\therefore\quad x^2\frac{\partial^2 u}{\partial x^2}+2xy\frac{\partial^2 u}{\partial x\,\partial y}+y^2\frac{\partial^2 u}{\partial y^2}+x\frac{\partial u}{\partial x}+y\frac{\partial u}{\partial y}=9\frac{\pi}{2}+16\frac{\pi}{4}=\frac{17\pi}{2}.$$ **Ans.**

CHAPTER

12 Applications of Partial Differentiation

IMPORTANT FORMULE AND USEFUL RESULTS

1. **Errors And Approximation:** If u is a function of two variables x and y and let δx and δy be errors in x and y then the error in u is equal to $\delta u = \frac{\partial u}{\partial y}\delta x + \frac{\partial u}{\partial y}\delta y$ approximately.

   It can be extended to $\delta u = \frac{\partial u}{\partial x}\delta x + \frac{\partial u}{\partial y}\delta y + \frac{\partial u}{\partial z}\delta z$ approx.

2. **Jacobians.**

$$J\left(\frac{u, v}{x, y}\right) = \frac{\partial(u, v)}{\partial(x, y)} = \begin{vmatrix} \frac{\partial u}{\partial x} & \frac{\partial u}{\partial y} \\ \frac{\partial v}{\partial x} & \frac{\partial v}{\partial y} \end{vmatrix}$$

and

$$J\left(\frac{u, v, w}{x, y, z}\right) = \frac{\partial(u, v, w)}{\partial(x, y, z)} = \begin{vmatrix} \frac{\partial u}{\partial x} & \frac{\partial u}{\partial y} & \frac{\partial u}{\partial z} \\ \frac{\partial v}{\partial x} & \frac{\partial v}{\partial y} & \frac{\partial v}{\partial z} \\ \frac{\partial w}{\partial x} & \frac{\partial w}{\partial y} & \frac{\partial w}{\partial z} \end{vmatrix}.$$

Chain Rule of Jacobians $\frac{\partial(u, v)}{\partial(x, y)} \cdot \frac{\partial(x, y)}{\partial(r, s)} = \frac{\partial(u, v)}{\partial(r, s)} \Rightarrow \frac{\partial(u, v)}{\partial(x, y)} \cdot \frac{\partial(x, y)}{\partial(u, v)} = 1.$

Jacobians of Implicit Functions

If $f_1(u, v, x, y) = 0$ and $f_2(u, v, x, y) = 0$

then

$$\frac{\partial(u, v)}{\partial(x, y)} = (-1)^2 \frac{\frac{\partial(f_1, f_2)}{\partial(x, y)}}{\frac{\partial(f_1, f_2)}{\partial(u, v)}}.$$

Its extention is

$$\frac{\partial(u, v, w)}{\partial(x, y, z)} = (-1)^3 \frac{\frac{\partial(f_1, f_2, f_3)}{\partial(x, y, z)}}{\frac{\partial(f_1, f_2, f_3)}{\partial(u, v, w)}}.$$

3. Functional Dependance:

Function $f(x, y)$ and $\phi(x, y)$ are functionally dependent if $\dfrac{\partial(f, \phi)}{\partial(x, y)} = 0$

4. Taylor's Theorem on Functions of Two Variables.

$$f(x+h, y+k) = f(x, y) + \left(h\frac{\partial f}{\partial x} + k\frac{\partial f}{\partial y}\right) + \frac{1}{2!}\left(h^2\frac{\partial^2 f}{\partial x^2} + 2hk\frac{\partial^2 f}{\partial x\,\partial y} + k^2\frac{\partial^2 f}{\partial y^2}\right) + \dots .$$

5. Maxima and Minima of Function of Two Variables.

For the function $f(x, y)$, the the relations $\dfrac{\partial f}{\partial x} = 0$, $\dfrac{\partial f}{\partial y} = 0$, or $p = q = 0$ give pairs of values of x and y called critical points or stationary points. Then calculate r, s and t at the ciritical points. Now if $rt - s^2 > 0$ then Maxima if r (or t) < 0 and minima if r (or t) > 0. Also if $rt - s^2 < 0$ then neither maxima nor minima and points are called saddle points.

SOLUTIONS TO EXERCISE 12A

1. *Find the percentage error in the calculated area of rectangle when an error of + 1 % is made in measuring its length and breadth.*

SOLUTION: Area of rectangle $= A = lb$. $\delta A = \dfrac{\partial A}{\partial l}\,\delta l + \dfrac{\partial A}{\partial b}\,\delta b = b.\delta P + l\delta b$

$$\therefore \quad \frac{\delta A}{A}\times 100 = \frac{b}{lb}\,\delta l \times 100 + \frac{l}{lb}\,\delta b \times 100 = \frac{\delta l}{l}\times 100 + \frac{\delta b}{b}\times 100 = 1 + 1 = 2\%$$ **Ans.**

2. *The period of a simple pendulum is given by* $T = 2\pi\sqrt{\dfrac{l}{g}}$. *Find the maximum percentage error in* T *due to possible errors up to* 1 % *in* l *and* 2.5 % *in* g.

SOLUTION: $T = 2\pi\sqrt{(l/g)}$. Given $\dfrac{\delta l}{l}\times 100 = \pm 1$ and $\dfrac{\delta g}{g}\times 100 = \pm 2.5$.

Hence $$\delta T = \frac{\partial T}{\partial l}\,\delta l + \frac{\partial T}{\partial g}\,\delta g = \frac{2\pi}{2\sqrt{(l\,g)}}\,\delta l + \frac{2\pi\sqrt{l}}{g\sqrt{g}}\left(-\frac{1}{2}\right)\delta g$$

$$\therefore \quad \frac{\delta T}{T}\times 100 = \frac{2\pi}{2\sqrt{(lg)}}\,\frac{\sqrt{g}}{2\pi\sqrt{l}}\,\delta l \times 100 - \frac{2\pi\sqrt{l}}{2g\sqrt{g}}.\frac{\sqrt{g}}{2\pi\sqrt{l}}\,\delta g \times 100$$

$$= \frac{1}{2}\,\frac{\partial l}{l}\times 100 - \frac{1}{2}\,\frac{\delta g}{g}\times 100$$

Maximum $$\frac{\delta T}{T}\times 100 = \frac{1}{2}(1) - \frac{1}{2}(-2.5) = 1.75$$

$\therefore$ Maximum possible error in $T = 1.75\%$ **Ans.**

3. *In computing the volume of a right circular cone, errors of 2 % and 1 % are made in the height and the radius of the base respectively. Calculate the resulting percentage error in the volume.*

SOLUTION: $V = \dfrac{1}{3}\pi r^2 h$. Here $\dfrac{\partial V}{\partial r} = \dfrac{2\pi}{3}\,rh$ and $\dfrac{\partial V}{\partial h} = \dfrac{1}{3}\pi r^2$

$$\therefore \quad \delta T = \frac{\partial V}{\partial r}\,\delta r + \frac{\partial V}{\partial h}\,\delta h = \frac{2\pi}{3}\,rh\,\delta r + \frac{1}{3}\pi r^2\,\delta h$$

and $$\frac{\partial V}{V}\times 100 = \left(\frac{2\pi}{3}\,\frac{3rh}{\pi r^2}\,\delta r + \frac{1}{3}\,\frac{\pi r^2.3}{\pi r^2 h}\,\delta h\right)\times 100 = \frac{2}{r}\,\delta r \times 100 + \frac{\delta h}{h}\times 100$$

$\therefore$ Percentage error in volume $= 4\%$ **Ans.**

4. *A pile of bricks measures 2 m × 20 m × 6 m and the count is 430 bricks to 1 m³. Find the error in the cost when the tape is stretched by 1% beyond its standard length, the cost of bricks being Rs. 400 per thousand.*

SOLUTION: Number of brievs $= 2 \times 20 \times 6 \times 430 = 240 \times 430$. Volume of the brick $= lbh = V$.

$$\delta V = \frac{\partial V}{\partial l}\,\delta l + \frac{\partial V}{\partial b}\,\delta b + \frac{\partial V}{\partial h}\,\delta h = bh\delta l + hl\delta b + lb\delta h$$

$$\therefore \quad \frac{\partial V}{V} \times 100 = \left(\frac{\delta l}{l} + \frac{\delta b}{b} + \frac{\delta h}{h}\right) \times 100 = 1 + 1 + 1$$

$\therefore$ Error in cost $= 240 \times 430 \times 0.40 \times 3 = 1238.4$ Rs. **Ans.**

5. *Find the approximate value of* $[0.98^2 + 2.01^2 + 1.94^2]^{\frac{1}{2}}$.

SOLUTION: Let $f(x, y, z) = \sqrt{x^2 + y^2 + z^2}$ Here $x = 1,\ y = 2,\ z = 2 \ \therefore\ f = 3$.

and $\delta x = -0.02, \quad \delta y = +0.01, \quad \delta z = -0.06$

$$\therefore \quad \delta f = \frac{\partial f}{\partial x}\delta x + \frac{\partial f}{\partial y}\delta y + \frac{\partial f}{\partial z}\delta z = \frac{2x\,\delta x}{2\sqrt{x^2+y^2+z^2}} + \frac{2y\,\delta y}{2\sqrt{x^2+y^2+z^2}} + \frac{2z\,\delta z}{2\sqrt{x^2+y^2+z^2}}$$

$$= \frac{1(-0.02) + 2(0.01) + 2(-0.06)}{\sqrt{1^2 + 2^2 + 2^2}}$$

$$= \frac{1}{3}(-0.02 + 0.02 - 0.12) = \frac{0.12}{3} = -0.04$$

Hence $\sqrt{0.98^2 + 2.01^2 + 1.94} = f + \delta f = 3 - 0.04 = 2.96$ **Ans.**

6. *The area of a triangle ABC is computed from the formula* $\Delta = \frac{1}{2}\, bc \sin A$. *Errors of* 1%, 2% *and* 3% *are made in measuring* b, c *and* A. *If* $A = 45°$, *find the percentage error in the calculated value of* Δ.

SOLUTION: $\Delta = \frac{1}{2} bc \sin A$. Hence $\delta\Delta = \frac{\partial \Delta}{\partial b}\delta b + \frac{\partial \Delta}{\partial c}\delta c + \frac{\partial \Delta}{\partial A}\delta A$

$$= \frac{c}{2}\sin A\,\delta b + \frac{b}{2}\sin A\,\delta c + \frac{1}{2} bc\cos A\,\delta A$$

$$\therefore \quad \frac{\delta\Delta}{\Delta} \times 100 = \left(\frac{\delta b}{b} + \frac{\delta c}{c} + \frac{\cos A}{\sin A}\,\delta A\right) \times 100 = 1 + 2 + 3A\cot\frac{\pi}{4} = 3\left(1 + \frac{\pi}{4}\right)$$

$\left(\text{since } \frac{\delta A}{A} \times 100 = 3 \quad \text{and} \quad \delta A = \frac{3A}{100} = \frac{3\pi}{400}.\right)$ **Ans.** $3\left(1 + \frac{\pi}{4}\right)$.

7. *The focal length of a mirror is found from the formula* $\frac{2}{f} = \frac{1}{v} - \frac{1}{u}$. *Find the percentage error in* f *if* u *and* v *are both in error by 2%.*

SOLUTION: $\frac{2}{f} = \frac{1}{v} - \frac{1}{u}$ Hence $\frac{-2}{f^2}\frac{\partial f}{\partial u} = \frac{+1}{u^2}$, $\frac{-2}{f^2}\frac{\partial f}{\partial v} = \frac{-1}{v^2}$

$$\therefore \quad \delta f = \frac{\partial f}{\partial f}\delta u + \frac{\partial f}{\partial v}\delta v = \frac{-f^2}{2u^2}\delta u + \frac{f^2}{2v^2}\delta v$$

$$\text{or} \quad \frac{\delta f}{f}\times 100 = \left(\frac{-f}{2u}\frac{\delta u}{u} + \frac{f}{2v}\frac{\delta v}{v}\right)\times 100 = \frac{-f}{2u}\cdot 2 + \frac{f}{2v}\cdot 2$$

$$= \frac{f}{2}\left(\frac{1}{v} - \frac{1}{u}\right)\cdot = \frac{f}{2}\cdot\frac{2}{f}\cdot 2 = 2$$

Ans. **2%.**

8. *The resistance* R *of a circuit is found by the formula* $I = \frac{\mathrm{E}}{\mathrm{R}}$. *If there is an error of 0.1 ampear in reading* I *and* 0.5 *volts in* E, *find the corresponding possible percentage error in* R *when* $I = 15$ amp *and* $E = 100$ *volts.*

SOLUTION: $R = \frac{E}{I}$ $\therefore$ $\delta R = \frac{\partial R}{\partial E}\delta E + \frac{\partial R}{\partial I}\delta I = \frac{1}{I}\delta E - \frac{E}{I^2}\delta I$

$$\therefore \quad \frac{\partial R}{R}\times 100 = \left(\frac{1}{IR}\delta E - \frac{E}{I^2 R}\delta I\right)100 = \left(\frac{1}{E}\delta E - \frac{\partial I}{I}\right)100$$

But $\delta I = 0.1$ and $\delta E = 0.5$ and $I = 15$, $E = 100$

$$\therefore \quad \frac{\partial R}{R}\times 100 = \left(\frac{0.5}{100} - \frac{0.1}{15}\right)\times 100 = \frac{1}{2} - \frac{2}{3} = \frac{-1}{6}$$

Ans. $= \frac{-1}{6}\%$

9. *The resonant frequency* f *in a circuit is given by* $f = \frac{1}{2\pi\sqrt{\mathrm{LC}}}$. *If the measurements of inductance* L *and capacitance* C *are in error by* + 2% and – 1% *respectively, find the percentage error in the calculated value of* f.

SOLUTION: $f = \frac{L^{-1/2}C^{-1/2}}{2\pi}$. Given that $\frac{\delta L}{L}\times 100 = 2$ and $\frac{\delta C}{C}\times 100 = -1$

$$\therefore \quad \delta f = \frac{\partial f}{\partial L}\delta L + \frac{\partial f}{\partial C}\delta C = \frac{-1}{2\cdot 2\pi}L^{-3/2}C^{-1/2}\delta L - \frac{1}{2\cdot 2\pi}C^{-3/2}L^{-1/2}\delta C$$

$$\text{and} \quad \frac{\delta f}{f}\times 100 = \frac{-1}{4\pi}\cdot\frac{2\pi}{L^{-1/2}C^{-1/2}}\left[L^{-3/2}C^{-1/2}\delta L + L^{-1/2}C^{-3/2}\delta C\right]\times 100$$

$$= -\frac{1}{2}\left[\frac{\delta L}{L} + \frac{\delta C}{C}\right]\times 100 = \frac{-1}{2}[2-1] = \frac{-1}{2} = -0.5$$

Ans.

10. *The inner dimensions of a rectangular box are measured to be* 1, 1.5 *and* 3 *metres subject to an error of* ± 0.2 *metres in each. Within what limits must the diagonal of the box lie?*

SOLUTION: Dimension of the box are $x, y, z.$ and $x = 1,\ y = 1.5,\ y = 3.$

Diagonal $D = \sqrt{x^2 + y^2 + z^2}$. Also, $\delta x = \pm 0.2,\ \delta y = \pm 0.2,\ \delta z = \pm 0.2\text{m}$

$$D = \sqrt{1^2 + \left(\frac{3}{2}\right)^2 + 3^2} = \sqrt{12.25} = 3.5$$

$$\therefore \qquad \delta D = \frac{\partial D}{\partial x}\delta x + \frac{\partial D}{\partial y}\delta y + \frac{\partial D}{\partial z}\delta z = \frac{x\,\delta x + y\delta y + z\delta z}{\sqrt{x^2 + y^2 + z^2}}$$

$$= \frac{\pm 0.2(1) \pm 0.2(1.5) \pm 0.2(3)}{\sqrt{12.25}} = \frac{1}{3.5}(\pm 0.2 \pm 0.3 \pm 0.6) = \pm\frac{1.1}{3.5} = \pm 0.314$$

Hence D would lie between $3.5 \pm 0.314 = (3.186, 3.814)$ **Ans.**

11. *At a distance of* 120 ft. *from the foot of a tower, the elevation of its top is* 60°, *if the possible error in measuring the distance and the elevation are* 1 *inch and* 1 *minute, find the approximate error in the calculated height.*

SOLUTION: $x = 120,\ \theta = 60°,\ h = x\tan\theta$

Given $\delta x = \frac{1}{12}$ ft. and $\delta\theta = \frac{1}{60}$ degrees $= \frac{\pi}{60} \times 180$

$$\therefore \qquad \delta h = \frac{\partial h}{\partial x}\delta x + \frac{\partial h}{\partial \theta}\delta\theta = \tan\theta\ \delta x + x\sec^2\theta\ \delta\theta$$

$$= \tan 60° \cdot \frac{1}{12} + 120.\sec^2 60° \cdot \frac{\pi}{60(180)}$$

$$= \frac{\sqrt{3}}{12} + \frac{2(1+3)\pi}{180} = \frac{1.732}{12} + \frac{8 \times 3.14}{180} = 0.144 + 0.14 = 0.284$$

$\therefore$ Error in the height = 0.284 ft. **Ans.**

12. *The area of a triangle is calculated from the lengths a, b, c of its sides by using the formula* $\Delta = \sqrt{s(s-a)(s-b)(s-c)}$. *If a is diminished and b is increased by the same small amount k, show that the resulting change in the area is given by*

$$\frac{\delta\Delta}{\Delta} = 2(a-b)\,k[c^2 - (a-b)^2]^{-1}$$

SOLUTION: $\Delta^2 = s(s-a)(s-b)(s-c)$ or $2\log\Delta = \log s + \log(s-a) + \log(s-b) + \log(s-c)$

where $s = (a + b + c)/2$. Differentiating partially w.r.t. a and b we get

$$\frac{2}{\Delta}\frac{\partial\Delta}{\partial a} = \frac{1}{2s} - \frac{1}{2(s-a)} + \frac{1}{2(s-b)} + \frac{1}{2(s-c)}$$

and
$$\frac{2}{\Delta}\frac{\partial\Delta}{\partial b} = \frac{1}{2s} + \frac{1}{2(s-a)} - \frac{1}{2(s-b)} + \frac{1}{2(s-c)}$$

Error in $\Delta = \delta\Delta = \frac{\partial\Delta}{\partial a}\delta a + \frac{\partial\Delta}{\partial b}\delta b$ Here $\delta a = -k$ and $\delta b = k$

$$\therefore \quad \frac{\delta\Delta}{\Delta} = \frac{-k}{2}\left[\frac{1}{2s} - \frac{1}{2(s-a)} + \frac{1}{2(s-b)} + \frac{1}{2(s-c)}\right] + \frac{k}{2}\left[\frac{1}{2s} + \frac{1}{2(s-a)} - \frac{1}{2(s-b)} + \frac{1}{2(s-c)}\right]$$

$$= \frac{k}{2}\left[\frac{1}{s-a} - \frac{1}{s-b}\right] = \frac{2k(2-b)}{(2s-2a)(2s-2b)}$$

$$= \frac{2k(a-b)}{(-a+b+c)(-b+a+c)} = \frac{2k(a-b)}{[c^2-(a-b)^2]}$$

Hence Proved.

13. *Find the approximate value of* $\sqrt{24}\ (18)^{1/4}\ (30)^{V1/3}$ *using the concept of total derivative.*

SOLUTION: Consider $f(x, y, z) = x^{1/2}\, y^{1/4}\, z^{1/3}$. Let $x = 25,\ y = 16,\ z = 27$

$$\frac{\partial f}{\partial x} = \frac{f}{2x}, \quad \frac{\partial f}{\partial y} = \frac{f}{4y}, \quad \frac{\partial f}{\partial z} = \frac{f}{3z}$$

$$\therefore \quad \delta f = \frac{\partial f}{\partial x}\delta x + \frac{\partial f}{\partial y}\delta y + \frac{\partial f}{\partial z}\delta z = f\left[\frac{\delta x}{2x} + \frac{\delta y}{4y} + \frac{\delta z}{3z}\right]$$

$$= 25^{1/2}\cdot 16^{1/4}\cdot 27^{1/3}\left[\frac{-1}{2.25} + \frac{2}{4.16} + \frac{3}{3.27}\right]$$

$$= 5\cdot 2\cdot 3\left[\frac{-1}{50} + \frac{1}{32} + \frac{1}{27}\right] = \left[\frac{2086}{288}\right] = 1.45$$

$\therefore$ Approximate Value of $(24^{1/2})(18^{1/4})(30^{1/3}) = (5)\,(2)\,(3) + 1.45 = 31.45$ **Ans.**

SOLUTIONS TO EXERCISE 12B

1. *Find the Jacobian of the following transformations*
$u = \cos x,\ v = \sin x \cos y,\ w = \sin x \sin y \cos z.$

SOLUTION: $u = \cos x,\ v = \sin x \cos y,\ w = \sin x \sin y \cos z$

$$\frac{\partial(u,v,w)}{\partial(x,y,z)} = \begin{vmatrix} \frac{\partial u}{\partial x} & \frac{\partial u}{\partial y} & \frac{\partial u}{\partial z} \\ \frac{\partial v}{\partial x} & \frac{\partial v}{\partial y} & \frac{\partial v}{\partial z} \\ \frac{\partial w}{\partial x} & \frac{\partial w}{\partial y} & \frac{\partial w}{\partial z} \end{vmatrix} = \begin{vmatrix} -\sin x & 0 & 0 \\ \cos x \cos y & -\sin x \sin y & 0 \\ \cos x \sin y \cos z & \sin x \cos y \cos z & -\sin x \sin y \sin z \end{vmatrix}$$

$$= -\sin^3 x \sin^2 y \sin z.$$

2. *If (x, y, z) and (r, θ, ϕ) are respectively the cartesian and spherical polar coordinates of a point then show that* $\dfrac{\partial (x, y, z)}{\partial (r, \theta, \phi)} = r^2 \sin^2 \theta$.

SOLUTION: Spherical polar coordinatar are

$$x = r\sin\theta\cos\phi,\ \ y = r\sin\theta\sin\phi,\ \ z = r\cos\theta$$

$$\frac{\partial(x,y,z)}{\partial(r,\theta,\phi)} = \begin{vmatrix} \frac{\partial x}{\partial r} & \frac{\partial x}{\partial \theta} & \frac{\partial x}{\partial \phi} \\ \frac{\partial y}{\partial r} & \frac{\partial y}{\partial \theta} & \frac{\partial y}{\partial \phi} \\ \frac{\partial z}{\partial r} & \frac{\partial z}{\partial \theta} & \frac{\partial z}{\partial \phi} \end{vmatrix} = \begin{vmatrix} \sin\theta\cos\phi & r\cos\theta\cos\phi & -r\sin\theta\sin\phi \\ \sin\theta\sin\phi & r\cos\theta\sin\phi & r\sin\theta\cos\phi \\ \cos\theta & -r\sin\theta & 0 \end{vmatrix}$$

$$= r^2\sin\theta \begin{vmatrix} \sin\theta\cos\phi & \cos\theta\cos\phi & -\sin\phi \\ \sin\theta\sin\phi & \cos\theta\sin\phi & \cos\phi \\ \cos\theta & -\sin\theta & 0 \end{vmatrix}$$

Expanding it w.r.t. C_3 we get

$$= -r^2\sin\theta\sin\phi(-\sin^2\theta\sin\phi - \cos^2\theta\sin\phi) - r^2\sin\theta\cos\phi\,(-\sin^2\theta\cos\phi - \cos^2\theta\cos\phi)$$

$$= r^2\sin\theta\ \sin^2\phi + r^2\sin\theta\cos^2\phi = r^2\sin\theta$$ **Ans.**

3. *For the transformation* $x = e^u \cos v$, $y = e^u \sin v$ *show that* $\dfrac{\partial (u, v)}{\partial (x, y)} = e^{-2u}$.

SOLUTION: $x = e^u\cos v,\ \ y = e^u\sin v$

$$\therefore\quad \frac{\partial(x,y)}{\partial(u,v)} = \begin{vmatrix} \frac{\partial x}{\partial u} & \frac{\partial x}{\partial v} \\ \frac{\partial y}{\partial u} & \frac{\partial y}{\partial v} \end{vmatrix} = \begin{vmatrix} e^u\cos v & -e^u\sin v \\ e^u\sin v & e^u\cos v \end{vmatrix} = e^{2u}.$$ **Ans.**

4. *Verify* $JJ' = 1$ *when* $x = v^2 + w^2$, $y = w^2 + u^2$, $z = u^2 + v^2$. *where* $J = \dfrac{\partial (x, y, z)}{\partial (u, v, w)}$.

SOLUTION:

$$J = \frac{\partial(x,y,z)}{\partial(u,v,w)} = \begin{vmatrix} 0 & 2v & 2w \\ 2u & 0 & 2w \\ 2u & 2v & 0 \end{vmatrix} = 16uvw$$

Now $x+y+z = 2\,(u^2+v^2+w^2) = 2(u^2 + x)$ $\therefore$ $u^2 = \frac{1}{2}(-x+y+z)$, $v^2 = \frac{1}{2}(x-y+z)$, $w^2 = \frac{1}{2}(x+y-z)$.

$$J' = \frac{\partial(u,v,w)}{\partial(x,y,z)} = \begin{vmatrix} \frac{\partial u}{\partial x} & \frac{\partial u}{\partial y} & \frac{\partial u}{\partial z} \\ \frac{\partial v}{\partial x} & \frac{\partial v}{\partial y} & \frac{\partial v}{\partial z} \\ \frac{\partial w}{\partial x} & \frac{\partial w}{\partial y} & \frac{\partial w}{\partial z} \end{vmatrix}$$

$$= \begin{vmatrix} \frac{-1}{4u} & \frac{1}{4u} & \frac{1}{4u} \\ \frac{1}{4v} & \frac{-1}{4v} & \frac{1}{4v} \\ \frac{1}{4w} & \frac{1}{4w} & \frac{-1}{4w} \end{vmatrix} = \frac{1}{4^3 uvw} \begin{vmatrix} -1 & 1 & 1 \\ 1 & -1 & 1 \\ 1 & 1 & -1 \end{vmatrix} = \frac{1}{64\, uvw} \begin{vmatrix} -1 & 1 & 1 \\ 0 & 0 & 2 \\ 0 & 2 & 1 \end{vmatrix} = \frac{1}{16\, uvw}$$

$\therefore \quad JJ' = 1$ **Hence Proved.**

5. *If* $x^2 + y^2 + u^2 - v^2 = 0$ and $uv + xy = 0$ *show that* $\dfrac{\partial(u, v)}{\partial(x, y)} = \dfrac{x^2 - y^2}{u^2 + v^2}$.

SOLUTION : $f_1(x, y, u, v) = x^2 + y^2 + u^2 - v^2 = 0$ and $f_2(x, y, u, v) = uv + xy = 0$

$$\therefore \quad \frac{\partial(f_1, f_2)}{\partial(x, y)} = \begin{vmatrix} 2x & 2y \\ y & x \end{vmatrix} = 2(x^2 - y^2) \text{ and } \frac{\partial(f_1, f_2)}{\partial(u, v)} = \begin{vmatrix} 2u & -2v \\ v & u \end{vmatrix} = 2(u^2 + v^2)$$

$$\therefore \quad \frac{\partial(u, v)}{\partial(x, y)} = (-1)^2 \frac{\dfrac{\partial(f_1, f_2)}{\partial(x, y)}}{\dfrac{\partial(f_1, f_2)}{\partial(u, v)}} = \frac{2(x^2 - y^2)}{2(u^2 + v^2)} = \frac{x^2 - y^2}{u^2 + v^2}$$

Hence Proved.

6. *If* $u^3 + v^3 + w^3 = x + y + z$, $u^2 + v^2 + w^2 = x^3 + y^3 + z^3$ *and* $u + v + w = x^2 + y^2 + z^2$, *find* $\dfrac{\partial(u, v, w)}{\partial(x, y, z)}$.

SOLUTION: $f_1(x, y, z, u, v, w) = u^3 + v^3 + w^3 - x - y - z$,

$f_2(x, y, z, u, v, w) = u^2 + v^2 + w^2 - x^3 - y^3 - z^3$

and $f_3(x, y, z, u, v, w) = u + v + w - x^2 - y^2 - z^2$

$$\text{Now} \frac{\partial(f_1, f_2, f_3)}{\partial(x, y, z)} = \begin{vmatrix} -1 & -1 & -1 \\ -3x^2 & -3y^2 & -3z^2 \\ -2x & -2y & -2z \end{vmatrix} = 6 \begin{vmatrix} -1 & -1 & -1 \\ x^2 & y^2 & z^2 \\ x & y & z \end{vmatrix} = 6 \begin{vmatrix} -1 & 0 & 0 \\ x^2 & y^2 - x^2 & z^2 - x^2 \\ x & y - x & z - x \end{vmatrix}$$

$$= 6(y - x)(z - x) \begin{vmatrix} -1 & 0 & 0 \\ x^2 & y + x & z + x \\ x & 1 & 1 \end{vmatrix} = 6\,(x - y)\,(y - z)\,(z - x)$$

$$\text{and} \quad \frac{\partial(f_1, f_2, f_3)}{\partial(u, v, w)} = \begin{vmatrix} 3u^2 & 3v^2 & 3w^2 \\ 2u & 2v & 2w \\ 1 & 1 & 1 \end{vmatrix} = 6 \begin{vmatrix} u^2 & v^2 - u^2 & w^2 - u^2 \\ u & v - u & w - u \\ 1 & 0 & 0 \end{vmatrix}$$

$$= 6(v - u)\,(w - u) \begin{vmatrix} u^2 & v + u & w + u \\ u & 1 & 1 \\ 1 & 0 & 0 \end{vmatrix} = -\,6(u - v)\,(w - u)\,(v - w).$$

$$\therefore \quad \frac{\partial(u,v,w)}{\partial(x,y,z)} = \frac{(-1)^3 \dfrac{\partial(f_1,f_2,f_3)}{\partial(x,y,z)}}{\dfrac{\partial(f_1,f_2,f_3)}{\partial(u,v,w)}} = \frac{(-1)^3(x-y)(y-z)(z-x)}{-(u-v)(u-w)(w-u)} = \frac{(x-y)(y-z)(z-x)}{(u-v)(u-w)(w-u)} \quad \textbf{Ans.}$$

7. *If* $u = xyz, \; v = x^2 + y^2 + z^2, \; w = x + y + z,$ *find* $\dfrac{\partial x}{\partial u}$.

SOLUTION: $f_1 \equiv u - xyz = 0, \quad f_2 \equiv v - u^2 - y^2 - z^2, \quad f_3 \equiv w - x - y - z.$

$$\therefore \quad \frac{\partial x}{\partial u} = -\frac{\dfrac{\partial(f_1,f_2,f_3)}{\partial(u,y,z)}}{\dfrac{\partial(f_1,f_2,f_3)}{\partial(x,y,z)}} = -\frac{\begin{vmatrix} \dfrac{\partial f_1}{\partial u} & \dfrac{\partial f_1}{\partial y} & \dfrac{\partial f_1}{\partial z} \\ \dfrac{\partial f_2}{\partial u} & \dfrac{\partial f_2}{\partial y} & \dfrac{\partial f_2}{\partial z} \\ \dfrac{\partial f_3}{\partial u} & \dfrac{\partial f_3}{\partial y} & \dfrac{\partial f_3}{\partial z} \end{vmatrix}}{\begin{vmatrix} \dfrac{\partial f_1}{\partial x} & \dfrac{\partial f_1}{\partial y} & \dfrac{\partial f_1}{\partial z} \\ \dfrac{\partial f_2}{\partial x} & \dfrac{\partial f_2}{\partial y} & \dfrac{\partial f_2}{\partial z} \\ \dfrac{\partial f_3}{\partial x} & \dfrac{\partial f_3}{\partial y} & \dfrac{\partial f_3}{\partial z} \end{vmatrix}} = \frac{-\begin{vmatrix} 1 & -xz & -xy \\ 0 & -2y & -2z \\ 0 & -1 & -1 \end{vmatrix}}{\begin{vmatrix} -yz & -xz & -xy \\ -2x & -2y & -2z \\ -1 & -1 & -1 \end{vmatrix}}$$

$$= \frac{-(2y-2z)}{\begin{vmatrix} -yz & z(y-x) & y(z-x) \\ -2x & 2x-2y & 2x-2z \\ -1 & 0 & 0 \end{vmatrix}} = \frac{-2(y-z)}{-[z(y-x)2(x-z) - y(z-x)2(x-y)]}$$

$$= \frac{y-z}{z(x-y)(y-x) - y(z-x)(x-y)} = \frac{-1}{(x-y)(z-x)} = \frac{1}{(x-y)(z-x)} \quad \textbf{Ans.}$$

8. *If* $x + y + z + u + v = a$ *and* $x^2 + y^2 + z^2 + u^2 + v^2 = b$ *where* a *and* b *are constants, find* $\left(\dfrac{\partial v}{\partial y}\right)_{(x,u)}$ and $\left(\dfrac{\partial y}{\partial v}\right)_{(x,z)}$.

SOLUTION: $f_1 \equiv x + y + z + u + v = a, \quad f_2 \equiv x^2 + y^2 + u^2 + v^2 = b.$

Hence $$\left(\frac{\partial v}{\partial y}\right)_{x,z} = \frac{\dfrac{\partial(f_1,f_2)}{\partial(z,y)}}{\dfrac{\partial(f_1,f_2)}{\partial(z,u)}} = -\frac{\begin{vmatrix} 1 & 1 \\ 2z & 2y \end{vmatrix}}{\begin{vmatrix} 1 & 1 \\ 2z & 2v \end{vmatrix}} = \frac{-(2y-2z)}{(2v-2z)} = \frac{z-y}{v-z}$$

and $$\frac{\partial y}{\partial v} = -\frac{\dfrac{\partial(f_1, f_2)}{\partial(v,u)}}{\dfrac{\partial(f_1, f_2)}{\partial(y,u)}} = \frac{\begin{vmatrix} 1 & 1 \\ 2v & 2u \end{vmatrix}}{\begin{vmatrix} 1 & 1 \\ 2y & 2u \end{vmatrix}} = \frac{-2(u-v)}{2(u-y)} = \frac{v-u}{u-y}$$

$$\therefore \quad \frac{\partial v}{\partial y} = \frac{z-y}{v-z}, \quad \frac{\partial y}{\partial v} = \frac{v-u}{u-y} \qquad \text{Similarly } \left(\frac{\partial v}{\partial y}\right)_{x,u} = \frac{z-y}{v-z}$$ **Ans.**

9. *If $u = x + y + z$, $v = x^2 + y^2 + z^2$, $w = xy + yz + zx$ investigate whether u, v, w are functionally dependent on x, y, z, or not. If yes, find the relation between them.*

SOLUTION: $$\frac{\partial(u,v,w)}{\partial(x,y,z)} = \begin{vmatrix} \frac{\partial u}{\partial x} & \frac{\partial u}{\partial y} & \frac{\partial u}{\partial z} \\ \frac{\partial v}{\partial x} & \frac{\partial v}{\partial y} & \frac{\partial v}{\partial z} \\ \frac{\partial w}{\partial x} & \frac{\partial w}{\partial y} & \frac{\partial w}{\partial z} \end{vmatrix} = \begin{vmatrix} 1 & 1 & 1 \\ 2x & 2y & 2z \\ y+z & z+x & x+y \end{vmatrix} = \begin{vmatrix} 1 & 0 & 0 \\ 2x & 2(y-x) & 2(z-x) \\ y+z & x-y & x-z \end{vmatrix} = 0.$$

$= 0$ Hence v, u, w are functionally dependent.

Relation between u, v, w is $u^2 = v + 2w$ **Ans.**

10. *If $u = \sin^{-1} x + \sin^{-1} y$, $v = x\sqrt{1-y^2} + y\sqrt{1-x^2}$, find out if they are functionally dependent and if so, find the relation between them.*

SOLUTION: $u = \sin^{-1} x + \sin^{-1} y$, $v = x\sqrt{1-y^2} + y\sqrt{1-x^2}$

then the Jaeobian $$\frac{\partial(u,v)}{\partial(x,y)} = \begin{vmatrix} \dfrac{1}{\sqrt{1-x^2}} & \dfrac{1}{\sqrt{1-y^2}} \\ \sqrt{1-y^2} - \dfrac{xy}{\sqrt{1-x^2}} & \dfrac{-xy}{\sqrt{1-y^2}} + \sqrt{1-x^2} \end{vmatrix}$$

$= 0$ Hence functionally dependent.

Next, let $\sin^{-1} x = \theta$ and $\sin^{-1} y = \phi$:, $u = \theta + \phi$

and $\sin u = \sin\theta\cos\phi + \cos\theta\sin\phi = x\sqrt{1-y^2} + y\sqrt{1-x^2} = v$

$\therefore v = \sin u$ which is the required relation **Ans.**

11. *Investigate if $u = \dfrac{x-y}{x+z}$, $v = \dfrac{x+z}{y+z}$ are functionally dependent and if so, find the functional relation between them.*

SOLUTION: $u = \dfrac{x-y}{x+z}$, $v = \dfrac{x+z}{y+z}$

$$\therefore \quad \frac{\partial u}{\partial x} = \frac{x+z-(x-y)}{(x+z)^2} = \frac{y+z}{(x+z)^2}, \quad \frac{\partial u}{\partial y} = \frac{-1}{x+z} \text{ and } \frac{\partial u}{\partial z} = \frac{y-x}{(x+z)^2}.$$

and $\frac{\partial v}{\partial x} = \frac{1}{y+z}, \quad \frac{\partial v}{\partial y} = \frac{-(x+z)}{(y+z)^2}, \quad \frac{\partial v}{\partial z} = \frac{(y+z)-(x+z)}{(y+z)^2} = \frac{y-x}{(y+z)^2}.$

$$\therefore \quad \frac{\partial(u,v)}{\partial(x,y)} = \begin{vmatrix} \frac{y+z}{(x+z)^2} & \frac{-1}{x+z} \\ \frac{1}{y+z} & \frac{-(x+z)}{(y+z)^2} \end{vmatrix} = \frac{1}{(x+z)^2 (y+z)^2} \begin{vmatrix} y+z & -(x+z) \\ y+z & -(x+z) \end{vmatrix} = 0$$

$$\text{and} \quad \frac{\partial(u,v)}{\partial(y,z)} = \begin{vmatrix} \frac{-1}{x+z} & \frac{y-x}{(x+z)^2} \\ \frac{-(x+z)}{(y+z)^2} & \frac{y-x}{(y+z)^2} \end{vmatrix} = \frac{1}{(x+z)^2 (y+z)^2} \begin{vmatrix} -(x+z) & (y-x) \\ -(x+z) & (y-x) \end{vmatrix} = 0$$

$$\text{and} \quad \frac{\partial(u,v)}{\partial(x,z)} = \begin{vmatrix} \frac{y+z}{(x+z)^2} & \frac{y-x}{(x+z)^2} \\ \frac{1}{y+z} & \frac{y-x}{(y+z)^2} \end{vmatrix} = \frac{1}{(x+z)^2 (y+z)^2} \begin{vmatrix} y+z & y-x \\ y+z & y-x \end{vmatrix} = 0$$

Hence u and v are functionally dependent. To get the relation between them

$$1-u = 1 - \frac{x-y}{x+z} = \frac{y+z}{x+z} = \frac{1}{v} \qquad \text{Hence} \quad v(1-u) = 1$$ **Ans.**

12. *If u and v are functions of x and y then show that* $\frac{\partial(u,v)}{\partial(x,y)} \cdot \frac{\partial(x,y)}{\partial(u,v)} = 1$

SOLUTION: We know that by chain rule, if u, v are functions of x, y and x, y are functions of r, s then

$$\frac{\partial(u,v)}{\partial(x,y)} \cdot \frac{\partial(x,y)}{\partial(r,s)} = \frac{\partial(u,v)}{\partial(r,s)}$$

Replacing r and s by u and v respectively in the above relation, gives

$$\frac{\partial(u,v)}{\partial(x,y)} \cdot \frac{\partial(x,y)}{\partial(u,v)} = \frac{\partial(u,v)}{\partial(u,v)} = 1.$$ **Hence Proved.**

13. *Given that* $x = u + v + w$, $y = u^2 + v^2 + w^2$, $z = u^3 + v^3 + w^3$ *find* $\frac{\partial u}{\partial x}$.

SOLUTION: $f_1 \equiv u+v+w-x=0, \quad f_2 \equiv u^2+v^2+w^2-y=0, \quad f_3 \equiv u^3+v^3+w^3-z=0.$

$$\frac{\partial u}{\partial x} = -\frac{\frac{\partial(f_1, f_2, f_3)}{\partial(x, v, w)}}{\frac{\partial(f_1, f_2, f_3)}{\partial(u, v, w)}}$$

$$\frac{\partial(f_1, f_2, f_3)}{\partial(x, v, w)} = \begin{vmatrix} \frac{\partial f_1}{\partial x} & \frac{\partial f_1}{\partial v} & \frac{\partial f_1}{\partial w} \\ \frac{\partial f_2}{\partial x} & \frac{\partial f_2}{\partial v} & \frac{\partial f_2}{\partial w} \\ \frac{\partial f_3}{\partial x} & \frac{\partial f_3}{\partial v} & \frac{\partial f_3}{\partial w} \end{vmatrix} = \begin{vmatrix} -1 & 1 & 1 \\ 0 & 2v & 2w \\ 0 & 3v^2 & 3w^2 \end{vmatrix} = 6\,(v^2 w - v\,w^2)$$

$$\frac{\partial(f_1, f_2, f_3)}{\partial(u, v, w)} = \begin{vmatrix} \frac{\partial f_1}{\partial u} & \frac{\partial f_1}{\partial v} & \frac{\partial f_1}{\partial w} \\ \frac{\partial f_2}{\partial u} & \frac{\partial f_2}{\partial v} & \frac{\partial f_2}{\partial w} \\ \frac{\partial f_3}{\partial u} & \frac{\partial f_3}{\partial v} & \frac{\partial f_3}{\partial w} \end{vmatrix} = \begin{vmatrix} 1 & 1 & 1 \\ 2u & 2v & 2w \\ 3u^2 & 3v^2 & 3w^2 \end{vmatrix} = 6\begin{vmatrix} 1 & 0 & 0 \\ u & v-u & w-u \\ u^2 & v^2-a^2 & w^2-u^2 \end{vmatrix}$$

$$= 6\,(v-u)\,(w-u)\begin{vmatrix} 1 & 0 & 0 \\ u & 1 & 1 \\ 3u^2 & 3v^2 & 3w^2 \end{vmatrix} = 6\begin{vmatrix} 1 & 0 & 0 \\ u & v-u & w-u \\ u^2 & v^2-u^2 & w^2-u^2 \end{vmatrix} = 6(v-u)\,(w-u)\,(w-v).$$

$$\therefore \quad \frac{\partial u}{\partial x} = \frac{-6vw\,(v-w)}{6\,(v-u)\,(w-u)\,(w-v)} = \frac{vw}{(u-v)\,(u-w)}$$ **Ans.**

14. *Show that the functions* $u = x + y + z$, $v = x^3 + y^3 + z^3 - 3xyz$ and $w = x^2 + y^2 + z^2 - xy - yz - zx$ *are functionally dependent. Also find the relation between them.* **(GGSIPU IInd Sem. End Term 2005)**

SOLUTION:

$$\frac{\partial\,(u, v, w)}{\partial\,(x, y, z)} = \begin{vmatrix} 1 & 1 & 1 \\ 3(x^2 - yz) & 3(y^2 - zx) & 3(z^2 - xy) \\ 2x - y - z & 2y - x - z & 2z - x - y \end{vmatrix}$$

$$= 3\begin{vmatrix} 1 & 0 & 0 \\ x^2 - yz & y^2 - x^2 + z(y-x) & z^2 - x^2 + y(z-x) \\ 2x - y - z & 3(y-x) & 3(z-x) \end{vmatrix}$$

$$= 3\,(y-x)\,(z-x)\begin{vmatrix} y + x + z & z + x + y \\ 3 & 3 \end{vmatrix} = 0.$$

Hence u, v, w are functionally dependent on x, y, z.

Since $x^3 + y^3 + z^3 - 3xyz = (x + y + z)\,(x^2 + y^2 + z^3 - xy - yz - zx)$

or $$v = uw$$ **Ans.**

15. *If* $u = \dfrac{y^2}{2x}$, $v = \dfrac{x^2 + y^2}{2x}$, *find* $\dfrac{\partial\,(u, v)}{\partial\,(x, y)}$.

SOLUTION: If $$u = \frac{y^2}{2x}, \quad v = \frac{x^2 + y^2}{2x}$$

$$\therefore \quad \frac{\partial\,(u, v)}{\partial\,(x, y)} = \begin{vmatrix} \frac{\partial u}{\partial x} & \frac{\partial v}{\partial x} \\ \frac{\partial u}{\partial y} & \frac{\partial v}{\partial y} \end{vmatrix} = \begin{vmatrix} \frac{-y^2}{2x^2} & \frac{1}{2} - \frac{y^2}{2x} \\ \frac{2y}{2x} & \frac{2y}{2x} \end{vmatrix} = \frac{-y}{2x}$$ **Ans.**

16. *If* $u^3 + v^3 = x + y,\ u^2 + v^2 = x^3 + y^3,$ *show that* $\dfrac{\partial(u, v)}{\partial(x, y)} = \dfrac{y^2 - x^2}{2uv(u-v)}.$

SOLUTION: Let $f_1 = u^3 + v^3 - x - y = 0,\quad f_2 = u^2 + v^2 - x^3 - y^3.$

$$\frac{\partial(u, v)}{\partial(x, y)} = \frac{\dfrac{\partial(f_1, f_2)}{\partial(x, y)}}{\dfrac{\partial(f_1, f_2)}{\partial(u, v)}} = \frac{\begin{vmatrix} -1 & -1 \\ -3x^2 & -3y^2 \end{vmatrix}}{\begin{vmatrix} 3u^2 & 3v^2 \\ 2u & 2v \end{vmatrix}} = \frac{3y^2 - 3x^2}{6u^2v - 6uv^2} = \frac{y^2 - x^2}{2uv(u-v)} \qquad \textbf{Ans.}$$

17. *If* u, v, w *are the roots of the equation* $(x - a)^3 + (x - b)^3 + (x - c)^3 = 0$ *find* $\dfrac{\partial(u, v, w)}{\partial(a, b, c)}.$

SOLUTION: Given equation can be written as

$$3x^3 - 3(a + b + c)x^2 + 3(a^2 + b^2 + c^2)x - (a^3 + b^3 + c^3) = 0$$

Since its foots are $u, v, w,$ we have

$$u + v + w = a + b + c,\quad uv + uw + wu = a^2 + b^2 + c^2 \quad \text{and} \quad uvw = \frac{a^3 + b^3 + c^3}{3}$$

Thus $\quad f_1 \equiv u + v + w - a - b - c, = 0,\quad f_2 \equiv (uv + vw + wu - a^2 - b^2 - c^2) = 0$

and $\quad f_3 \equiv uvw - \dfrac{1}{3}(a^3 + b^3 + c^3) = 0.$

$$\therefore \quad \frac{\partial(f_1, f_2, f_3)}{\partial(a, b, c)} = \begin{vmatrix} -1 & -1 & -1 \\ -2a & -2b & -2c \\ -a^2 & -b^2 & -c^2 \end{vmatrix} \begin{vmatrix} -1 & 0 & 0 \\ -2a & 2(a-b) & 2(a-c) \\ -a^2 & a^2 - b^2 & a^2 - c^2 \end{vmatrix}$$

$$= 2(a-b)(a-c)\begin{vmatrix} 1 & 1 \\ a+b & a+c \end{vmatrix} = -2(a-b)(b-c)(c-a)$$

$$\text{and} \quad \frac{\partial(f_1, f_2, f_3)}{\partial(u, v, w)} = \begin{vmatrix} 1 & 1 & 1 \\ v+w & w+u & u+v \\ vw & wu & uv \end{vmatrix} = \begin{vmatrix} 1 & 0 & 0 \\ v+w & u-v & u-w \\ vw & w(u-v) & v(u-w) \end{vmatrix}$$

$$\therefore \quad \frac{\partial(u, v, w)}{\partial(a, b, c)} = (-1)^3 \frac{\partial(f_1, f_2, f_3)}{\partial(a, b, c)} \Big/ \frac{\partial(f_1, f_2, f_3)}{\partial(u, v, w)} = \frac{2(a-b)(b-c)(c-a)}{-(u-v)(v=w)(w-u)} \qquad \textbf{Ans.}$$

SOLUTIONS TO EXERCISE 12C

1. *Find the maximum and minimum value of the function* $f(x, y) = x^4 + y^4 - 2x^2 + 4xy - 2y^2$.

SOLUTION: $p = \dfrac{\partial f}{\partial x} = 4x^3 - 4x + 4y, \quad q = \dfrac{\partial f}{\partial y} = 4y^3 + 4x - 4y$

For f to be maximum or minimum $p = q = 0$

$\therefore \quad x^3 - x + y = 0$ and $y^3 - y + x = 0$

Adding these we get $x^3 + y^3 = 0. \quad \Rightarrow \quad y = -x$

Subtracting the above equations gives

$$x^3 - y^3 - 2(x - y) = 0 \quad \text{or} \quad (x - y)[x^2 + xy + y^2 - 2] = 0$$

$\Rightarrow \quad y = x$ or $x^2 + xy + y^2 = 2.$

taking $y = -x$ gives $x^2 + x^2 - x^2 - 2 = 0 \quad x = \pm\sqrt{2}$

$\therefore$ Critical points are $(0, 0), (\sqrt{2}, \sqrt{2}), (\sqrt{2}, -\sqrt{2}), (-\sqrt{2}, \sqrt{2}), (-\sqrt{2}, -\sqrt{2})$

But $y \neq x \quad \therefore$ Points are $(0, 0), (\sqrt{2}, -\sqrt{2}), (-\sqrt{2}, \sqrt{2})$.

Further $\quad r = \dfrac{\partial^2 f}{\partial x^2} = 12x^2 - 1, \quad s = \dfrac{\partial^2 f}{\partial x\,\partial y} = 4, \quad t = \dfrac{\partial^2 f}{\partial y^2} = 12y^2 - 4$

Hence $\quad rt - s^2 = 16(3x^2 - 1)(3y^2 - 1) - 16 = 16[9x^2y^2 - 3x^2 + 1 - 1]$

At the points $(\sqrt{2}, -\sqrt{2})$ and $(-\sqrt{2}, \sqrt{2})$, $rt - s^2 = 36 - 12 - 12 > 0$

and r and t are positive hence minimum.

At (0, 0) $rt = s^2 = 0$ hence neither maximum nor minimum. **Ans.**

2. *Show that* $u = x^2y^2(x^2 + y^2 - 1)$ *has a maximum value of* $-\dfrac{1}{27}$ *for four sets of values of* x *and* y *and a minimum value at* $x = 0, \ y = 0$.

SOLUTION: $u = x^4y^2 + x^2y^4 - x^2y^2$ hence $\dfrac{\partial u}{\partial x} = 4x^3y^2 + 2xy^4 - 2xy^2 = 2xy^2(2x^2 + y^2 - 1)$

and $\quad \dfrac{\partial u}{\partial y} = 2x^4y + 4x^2y^3 - 2x^2y = 2x^2y(x^2 + 2y^2 - 1).$

For u to be maximum or minimum $\dfrac{\partial u}{\partial x} = 0$ and $\dfrac{\partial u}{\partial y} = 0.$

which gives $x = 0, \ y = 0$ and $2x^2 + y^2 = 1, \ x^2 + 2y^2 = 1.$

Thus, $\left(\dfrac{1}{\sqrt{3}}, \dfrac{1}{\sqrt{3}}\right), \left(\dfrac{-1}{\sqrt{3}}, \dfrac{-1}{\sqrt{3}}\right), \left(\dfrac{1}{\sqrt{3}}, \dfrac{-1}{\sqrt{3}}\right)$ and $\left(\dfrac{-1}{\sqrt{3}}, \dfrac{1}{\sqrt{3}}\right)$ are critical points.

$$r = \frac{\partial^2 u}{\partial x^2} = 2xy^2 \cdot 4x + 2y^2 (2x^2 + y^2 - 1) = 2y^2 (4x^2 + y^2 - 1),$$

$$t = \frac{\partial^2 u}{\partial y^2} = 2x^2 (4y^2 + x^2 - 1) \quad \text{and} \quad s = \frac{\partial^2 u}{\partial x \partial y} = 8x^3 y + 8xy^3 - 4xy$$

At (0, 0), r, s and t all are 0.

$$rt - s^2 = 4x^2 y^2 (4x^2 + y^2 - 1)(4y^2 + x^2 - 1) - 16x^2y^2 (2x^2 + 2y^2 - 1)$$
$$= 4x^2 y^2 [(4x^2 + y^2 - 1)(4y^2 + x^2 - 1) - 8x^2 - 8x^2 + 4] = 0 \quad \text{at} \quad (0, 0).$$

At the points $\left(\pm\frac{1}{\sqrt{3}}, \pm\frac{1}{\sqrt{3}}\right)$ we have

$$rt - s^2 = 4 \cdot \frac{1}{3} \cdot \frac{1}{3}\left[\left(\frac{4}{3} + \frac{1}{3} - 1\right)\left(\frac{4}{3} + \frac{1}{3} - 1\right) - \frac{8}{3} - \frac{8}{3} + 4\right] = \frac{4}{81}(4 - 48 + 36) < 0$$

Hence maximum at $\left(\frac{1}{\sqrt{3}}, \frac{1}{\sqrt{3}}\right), \left(\frac{1}{\sqrt{3}}, \frac{-1}{\sqrt{3}}\right), \left(\frac{-1}{\sqrt{3}}, \frac{1}{\sqrt{3}}\right), \left(\frac{-1}{\sqrt{3}}, \frac{-1}{\sqrt{3}}\right)$. **Ans.**

3. *Find the maximum and minimum values of* $\sin x \sin y \sin (x + y)$.

SOLUTION: $f(x, y) = \sin x \sin y \sin (x + y) = \frac{1}{2} \sin(x + y)[\cos(x - y) - \cos(x + y)]$

$$= \frac{1}{4}[\sin 2x + \sin 2y - \sin(2x + 2y)]$$

Hence $\frac{\partial f}{\partial x} = \frac{1}{4}[2\cos 2x - 2\cos(2x + 2y)] = \frac{1}{2}[\cos 2x - \cos(2x + 2y)]$

Similarly, $\frac{\partial f}{\partial y} = \frac{1}{2}[\cos 2y - \cos(2x + 2y)]$

For max. and min. $\frac{\partial f}{\partial x} = 0, \ \frac{\partial f}{\partial y} = 0.$

hence $\cos(2x + 2y) = \cos 2x$ and $\cos(2x + 2y) = \cos 2y$

$\Rightarrow$ $\cos 2x = \cos 2y$ so $y = \pm x$

$2x + 2y = \pm 2\pi \pm 2x$ or $2x + 2y = \pm 2\pi - 2x$

$\Rightarrow$ $2x + y = \pm \pi$ and $y = x$ gives $x = \pm \frac{\pi}{3}$ and $y = x$ hence $y = \pm\frac{\pi}{3}$.

$\therefore$ General values are $x = m\pi + \frac{\pi}{3}, \quad y = n\pi + \frac{\pi}{3}$

when $m = 1$, $x = \pi - \frac{\pi}{3} = \frac{2\pi}{3}$ and $n = 1, \ y = \pi - \frac{\pi}{3} = \frac{2\pi}{3}$

Thus $x = m\pi + \frac{\pi}{3}, \ y = n\pi + \frac{\pi}{3}$

and $x = m\pi + \frac{2\pi}{3}, \ y = n\pi \ \frac{2\pi}{3}.$

$\therefore$ Critical points are $\left(m\pi+\frac{\pi}{3}, n\pi+\frac{\pi}{3}\right)$ and $\left(m\pi+\frac{2\pi}{3}, n\pi+\frac{2\pi}{3}\right)$.

Next, $\frac{\partial^2 f}{\partial x^2} = -\sin 2x + \sin 2(x+y)$, $\frac{\partial^2 f}{\partial y^2} = -\sin 2y + \sin 2(x+y)$ and $\frac{\partial^2 f}{\partial x \partial y} = \sin(2x+2y)$

$$\therefore \quad \frac{\partial^2 f}{\partial x^2}\cdot\frac{\partial^2 f}{\partial y^2} - \left(\frac{\partial^2 f}{\partial x\,\partial y}\right)^2 = [\sin 2(x+y) - \sin 2x]\,[\sin 2(x+y) - \sin 2y] - \sin^2(2x+2y)$$

$$= \sin 2x \sin 2y - \sin 2(x+y)(\sin 2x + \sin 2y) > 0$$

At $\left(m\pi+\frac{\pi}{3}, n\pi+\frac{\pi}{3}\right)$, $\frac{\partial^2 f}{\partial x^2} = \sin\left(2m\pi+2n\pi+\frac{4\pi}{3}\right) - \sin\left(2m\pi+\frac{2\pi}{3}\right)$

$$= \sin\left(\pi+\frac{\pi}{3}\right) - \sin\frac{2\pi}{3} = -\sin\frac{\pi}{3} - \sin\left(\pi-\frac{\pi}{3}\right) = -\sqrt{3} < 0 \text{ hence Max.}$$

At $\left(m\pi+\frac{2\pi}{3}, n\pi+\frac{2\pi}{3}\right)$, $\frac{\partial^2 f}{\partial x^2} = \sin\left(2m\pi+2n\pi+\frac{8\pi}{3}\right) - \sin\left(2m\pi+\frac{4\pi}{3}\right)$

$$= \sin\frac{8\pi}{3} - \sin\frac{4\pi}{3} = \sin\left(3\pi-\frac{\pi}{3}\right) - \sin\left(\pi+\frac{\pi}{3}\right)$$

$$= \sin\frac{\pi}{3} + \sin\frac{\pi}{3} = \sqrt{3} > 0.$$ **hence Min.**

$\therefore$ At the points $\left(m\pi+\frac{\pi}{3}\ n\pi+\frac{\pi}{3}\right)$, $f(x, y)$ is maximum and at the points $\left(m\pi+\frac{2\pi}{3}\ n\pi+\frac{2\pi}{3}\right)$,

$f(x, y)$ is minimum at the points $\left(m\pi+\frac{\pi}{3}\ n\pi+\frac{\pi}{3}\right)$.

Max. $f(x, y) = \frac{3\sqrt{3}}{8}$ and Min $f(x, y) = \frac{-3\sqrt{3}}{8}$ **Ans.**

4. *A tent on a square base of side 2a consists of four vertical sides of height b surmounted by a regular pyramid of height h. If the volume enclosed is V show that the area of the canvas in the tent is*

$\frac{2V}{a} - \frac{8}{3}ah + 4a\sqrt{h^2+a^2}$. *Also show that if* a *and* h *can both vary, the least area of the canvas corresponding to a given volume V is given by*

$$a = \frac{\sqrt{5}}{2}h, \quad b = \frac{h}{4}.$$

SOLUTION: $V = 4a^2 b + \frac{1}{3}\cdot 3a^2 h$ ...(1)

Surface area $S = 4(2ab) + 4a\sqrt{a^2+h^2} = \frac{8a}{4a^2}\left(V - \frac{4a^2 h}{3}\right) + 4a\sqrt{a^2+h^2}$

$$= \frac{2V}{a} - \frac{8}{3}ah + 4a\sqrt{a^2+h^2}$$ **Hence Proved.**

Next, if a and h are variables and V is given, S should be least, for which $\frac{\partial S}{\partial a} = \frac{\partial S}{\partial h} = 0$

$$\frac{\partial S}{\partial a} = \frac{-2V}{a^2} - \frac{8h}{3} + \frac{4a}{2}\frac{2a}{\sqrt{a^2+h^2}} + 4\sqrt{a^2+h^2} = 0 \quad ...(2)$$

$$\frac{\partial S}{\partial h} = \frac{-8a}{3} + \frac{4a}{2}\frac{2h}{\sqrt{a^2+h^2}} = 0 \quad \text{or} \quad \frac{h}{\sqrt{a^2+h^2}} = \frac{2}{3}$$

or $\quad 9h^2 = 4a^2 + 4h^2 \quad \therefore \quad a = \frac{\sqrt{5}h}{2}$ **Hence Proved.**

From (1), $\frac{V}{a^2} = 4b + \frac{4h}{3}$ and taking $a = \sqrt{5}\frac{h}{2}$, (2) becomes

$$-2\left(4b + \frac{4h}{3}\right) - \frac{8h}{3} + \frac{4a^2 + 4a^2 + h^2}{\sqrt{a^2+h^2}} = 0$$

or $\quad 8b = \frac{22h}{3} - \frac{16h}{3} = 2h \quad \therefore \quad b = \frac{h}{4}.$ **Hence Proved.**

5. *Find the extreme values of* $x^2 + y^2 + z^2$ *subject to the constraints*

$$x^2 + \frac{y^2}{2} + \frac{z^2}{3} = 2 \quad \text{and} \quad 3x + 2y + z = 0.$$

SOLUTION: $u = x^2 + y^2 + z^2, \quad f_1 = x^2 + \frac{y^2}{2} + \frac{z^2}{3} = 2, \quad f_2 = 3x + 2y + z = 0$

Let $\quad F = u + \lambda_1 f_1 + \lambda_2 f_2$

$$\frac{\partial F}{\partial x} = \frac{\partial u}{\partial x} + \lambda_1\frac{\partial f_1}{\partial x} + \lambda_2\frac{\partial f_2}{\partial x} = 2x + \lambda_1\, 2x + \lambda_2 \cdot 3 = 0 \quad ...(1)$$

$$\frac{\partial F}{\partial y} = \frac{\partial u}{\partial y} + \lambda_1\frac{\partial f_1}{\partial y} + \lambda_2\frac{\partial f_2}{\partial y} = 2y + \lambda_1 \cdot y + 2\lambda_2 = 0 \quad ...(2)$$

$$\frac{\partial F}{\partial z} = \frac{\partial u}{\partial z} + \lambda_1\frac{\partial f_1}{\partial z} + \lambda_2\frac{\partial f_2}{\partial z} = 2z + \lambda_1 \cdot \frac{2}{3}z + 1\lambda_2 = 0 \quad ...(3)$$

Multiplying (1) by x (2) by y and (3) by z and adding, we get

$$2(x^2+y^2+z^2) + \lambda_1\left(2x^2 + y^2 + \frac{2}{3}z^2\right) + \lambda_2(3x+2y+z) = 0$$

or $\quad 2u + \lambda_1 \cdot 4 + 0\,\lambda_2 = 0 \quad \therefore \quad \lambda_1 = \frac{-u}{2}.$

Substituting for λ_1 in (1), (2), (3), we get

$$x = \frac{-3\lambda_2}{2-u} = \frac{3\lambda_2}{u-2}, \quad y = \frac{-2\lambda_2}{2-\frac{u}{2}} = +\frac{4\lambda_2}{u-4}, \quad z = \frac{-\lambda_2}{2-\frac{u}{3}} = \frac{3\lambda_2}{u-6}$$

Putting these in $3x + 2y + z = 0$, we get

$$\frac{9\lambda_2}{u-2} + \frac{8\lambda_z}{u-4} + \frac{3\lambda_z}{u-6} = 0 \quad \text{or} \quad \frac{9}{u-2} + \frac{8}{u-4} + \frac{3}{u-6} = 0$$

or $\quad 9(u-4)(u-6) + 8(u-2)(u-6) + 3(u-2)(u-4) = 0 \quad$ or $\quad 5u^2 - 43u + 84 = 0$

$\therefore \quad u = \frac{43+13}{10} = 3, \ \frac{28}{5}.$ Hence $u_{max} = \frac{28}{5}$ and $u_{min} = 3.$ **Ans.**

6. *Find the minimum value of* $x^2 + y^2$ *subject to the condition* $ax + by = c.$

SOLUTION: $u = x^2 + y^2$, condition is $\phi = ax + by = c$

Let $F = u + \lambda\phi$ Then for max. & min. $\dfrac{\partial F}{\partial x} = \dfrac{\partial u}{\partial x} + \lambda\dfrac{\partial \phi}{\partial x} = 0,\quad \dfrac{\partial F}{\partial y} = \dfrac{\partial u}{\partial y} + \dfrac{\partial \phi}{\partial y} = 0$

or $2x + \lambda a = 0$ ...(1) $2y + \lambda b = 0$ ...(2)

Multiplying (1) by x and (2) by y and adding, we get

$$2x^2 + 2y^2 + \lambda(ax + by) = 0 \quad \text{or} \quad 2u + c\lambda = 0 \quad \text{so} \quad \lambda = \frac{-2u}{c}.$$

$\therefore$ from (1) and (2), we get $x = \dfrac{au}{c}$ and $y = \dfrac{bu}{c}$.

Putting for x and y in $ax + by = c$, we get

$$\frac{a^2u}{c} + \frac{b^2u}{c} = c \quad \therefore \quad u = \frac{c^2}{a^2 + b^2}.$$ **Ans.**

7. *Find the points on the ellipse* $\dfrac{x^2}{4} + \dfrac{y^2}{1} = 1$ *whose distances from the straight line* $2x + 3y = 6$ *are the greatest and the least.*

SOLUTION: Let u = Distance of the point (x, y) from the line $2x + 3y = 6$

$\therefore \quad u = \dfrac{2x + 3y - 6}{\sqrt{2^2 + 3^2}}$ where the point (x, y) lies on $x^2 + 4y^2 = 4$ ...(1)

Consider $F = u + \lambda\phi$ so for critical points $\dfrac{\partial F}{\partial x} = \dfrac{\partial u}{\partial x} + \lambda\dfrac{\partial \phi}{\partial x} = 0$ and $\dfrac{\partial F}{\partial y} = \dfrac{\partial u}{\partial y} + \lambda\dfrac{\partial \phi}{\partial y} = 0$

or $\dfrac{2}{\sqrt{13}} + \lambda 2x = 0$ ....(2) and $\dfrac{3}{\sqrt{13}} + \lambda 8y = 0$ ...(3)

$$\therefore \quad x = \frac{-1}{\sqrt{13}\,\lambda}, \quad y = \frac{-3}{8\sqrt{13}\,\lambda}.$$

Multiplying (2) by x and (3) by y and adding, gives

$$\frac{2x + 3y}{\sqrt{13}} + \lambda(2x^2 + 8y^2) = 0$$

or $$u + \frac{6}{\sqrt{13}} + 8\lambda = 0 \quad \therefore \quad \lambda = \frac{-u}{8} - \frac{6}{8\sqrt{13}} = \frac{-(u\sqrt{13} + 6)}{8\sqrt{13}}$$

$$\therefore \quad x = -\frac{1}{\lambda\sqrt{13}} = \frac{8\sqrt{13}}{\sqrt{13}\,(u\sqrt{13} + 6)} = \frac{8}{(u\sqrt{13} + 6)},$$

and $$y = \frac{3}{8\sqrt{13}} \cdot \frac{8\sqrt{13}}{u\sqrt{13} + 6} = \frac{3}{u\sqrt{13} + 6}$$

Substituting for x and y in (1), gives

$$\frac{8^2}{(u\sqrt{13}+6)^2}+\frac{4\cdot 3^2}{(u\sqrt{13}+6)^2} = 4 \quad \text{or} \quad (u\sqrt{13}+6)^2 = 16 + 9 = 5^2$$

$$\therefore \quad u\sqrt{13}+6 = \pm 5 \quad \therefore \; u = \frac{-11}{\sqrt{13}} \text{ and } \frac{-1}{\sqrt{13}}$$

$$\therefore \quad u_{\max} = \frac{-1}{\sqrt{13}} \quad \text{and} \quad u_{\min} = \frac{-11}{\sqrt{13}}.$$

Thus $\quad$ Point of maxima $= \left(\dfrac{8}{-1+6}, \dfrac{3}{-1+6}\right) = \left(\dfrac{8}{5}, \dfrac{3}{5}\right)$

and $\quad$ Point of minima $= \left(\dfrac{8}{-11+6}, \dfrac{3}{-11+6}\right) = \left(\dfrac{-8}{5}, \dfrac{-3}{5}\right)$ **Ans.**

8. *Find the minimum value of* $f(x, y, z) = x^2 + y^2 + z^2$ *subject to the condition* $x + y + z = 1$.

SOLUTION: $f = x^2 + y^2 + z^2$ and the condition is $\phi \equiv x + y + z - 1 = 0$ ...(1)

Consider $F = f + \lambda\phi$. For maxima and minima $\dfrac{\partial F}{\partial x} = 0, \; \dfrac{\partial F}{\partial y} = 0, \; \dfrac{\partial F}{\partial z} = 0$

or $$\frac{\partial f}{\partial x}+\lambda\frac{\partial \phi}{\partial x} = 0 \text{ and } \frac{\partial f}{\partial y}+\lambda\frac{\partial \phi}{\partial y} = 0 \text{ and } \frac{\partial f}{\partial z}+\lambda\frac{\partial \phi}{\partial z} = 0 \text{ that is,}$$

$2x + \lambda \cdot 1 = 0, \;\; 2y + \lambda \cdot 1 = 0, \;\; 2z + \lambda \cdot 1 = 0 \;\;$ or $\;\; x = \dfrac{-\lambda}{2}, \; y = \dfrac{-\lambda}{2}, \; z = \dfrac{-\lambda}{2}.$

Putting these values of x, y, z in (1), we get $-\dfrac{3\lambda}{2} = 1$ or $\lambda = -\dfrac{2}{3}.$

$\therefore \quad f = 3\left(-\dfrac{\lambda}{2}\right)^2 = \dfrac{3\lambda^2}{4} = \dfrac{1}{3} \qquad$ Hence F has minimum value $= \dfrac{1}{3}.$ **Ans.**

9. *Determine the values of* x, y, z *that maximise the function* $3x + 5y + z - x^2 - y^2 - z^2$ *subject to the constraint* $x + y + z = 6$.

SOLUTION: Let $f = 3x + 5y + z - x^2 - y^2 - z^2$, constraint is $\phi \equiv x + y + z = 6$ ...(1)

Consider $F = f + \lambda f$. For maximum and minimum value of f, we must have $\dfrac{\partial F}{\partial x} = \dfrac{\partial F}{\partial y} = \dfrac{\partial F}{\partial z} = 0,$

that is, $$\frac{\partial f}{\partial x}+\lambda\frac{\partial \phi}{\partial x} = 0, \quad \frac{\partial f}{\partial y}+\lambda\frac{\partial \phi}{\partial y} = 0, \quad \frac{\partial f}{\partial z}+\lambda\frac{\partial \phi}{\partial z} = 0.$$

or $$3 - 2x + \lambda \cdot 1 = 0, \;\; 5 - 2y + \lambda \cdot 1 = 0, \;\; 1 - 2z + \lambda \cdot 1 = 0$$

or $\quad x = \dfrac{\lambda+3}{2}, \; y = \dfrac{\lambda+5}{2}, \; z = \dfrac{\lambda+1}{2}.$ These must satisfy (1).

hence $\quad \dfrac{\lambda+3}{2}+\dfrac{\lambda+5}{2}+\dfrac{\lambda+1}{2} = 6 \quad \therefore \quad \lambda = 1$

$\therefore$ Point of maxima is $x = 2, \; y = 3, \; z = 1$ or $(2, 3, 1)$. **Ans.**

10. *Divide* 120 *into three parts in such a way that the sum of their products taken two at a time, is maximum.*

SOLUTION: Let the three parts of 120 be x, y, z hence $\phi \equiv x + y + z = 120$

Here $f = xy + yz + zx$ is to be maximised.

Consider $F = f + \lambda\phi = xy + yz + zx + \lambda(x + y + z)$

For f to be maximum $\frac{\partial F}{\partial x} = \frac{\partial F}{\partial y} = \frac{\partial F}{\partial z} = 0$ is $y + z + \lambda = 0,\ z + x + \lambda = 0,\ x + y + \lambda = 0.$

Adding these, gives $2(x + y + z) + 3\lambda = 0 \quad \therefore \quad \lambda = \frac{2(120)}{3} - 80$

Above equations can be written as $120 - x + \lambda = 0,\ 120 - y + \lambda = 0,\ 120 - z + \lambda = 0$

hence $x = 40,\ y = 40,\ z = 40.$ **Ans.**

11. *Find a point within a given triangle so that the sum of the squares of its distances from the sides of the triangle, is minimum.*

SOLUTION: Let ABC be the given triangle with sides a, b, c and area Δ. Let P be any point inside the triangle whose distances from the sides BC, CA and AB are denoted by x, y, z respectively. Let $u = x^2 + y^2 + z^2$ then the relation between $x, y, z,$ is

$$\frac{a}{2}x + \frac{b}{2}y + \frac{c}{2}z = ar(\Delta PBC) + ar(\Delta PCA) + ar(\Delta PAB) = \Delta.$$

or $\phi \equiv ax + by + cz = 2\Delta.$

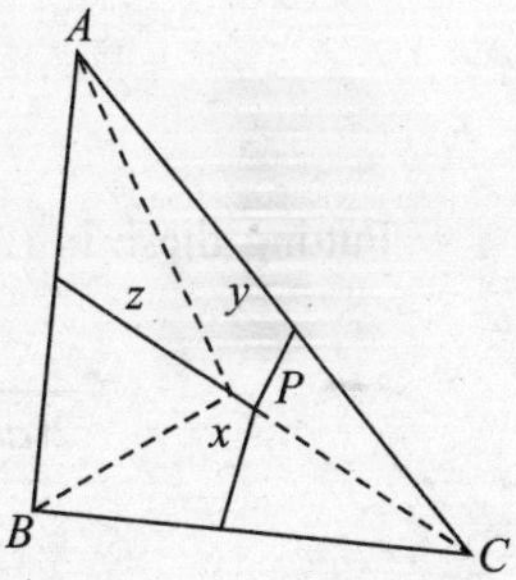

Consider $F = u + \lambda f$. For u to be max. or min. $\frac{\partial F}{\partial x} = \frac{\partial F}{\partial y} = \frac{\partial F}{\partial z} = 0$

or $\frac{\partial u}{\partial x} + \lambda\frac{\partial \phi}{\partial x} = 0, \quad \frac{\partial u}{\partial y} + \lambda\frac{\partial \phi}{\partial y} = 0, \quad \frac{\partial u}{\partial z} + \lambda\frac{\partial \phi}{\partial z} = 0$

or $2x + a\lambda = 0,\ 2y + b\lambda = 0,\ 2z + c\lambda = 0.$

Multiplying these relations by x, y, z respectively and adding, we get

$$2u + 2\Delta\lambda = 0 \quad \therefore \quad \lambda = \frac{-u}{\Delta} \quad \therefore \quad x = \frac{au}{2\Delta}, \quad y = \frac{bu}{2\Delta}, \quad z = \frac{cw}{2\Delta}$$

These satisfy $ax + by + cz = 2\Delta$ hence

$$\frac{a^2u}{2\Delta} + \frac{b^2u}{2\Delta} + \frac{c^2u}{2\Delta} = 2\Delta \quad \therefore \quad u = \frac{4\Delta^2}{a^2 + b^2 + c^2}$$

$$\therefore \quad x = \frac{2a\Delta}{a^2 + b^2 + c^2}, \quad y = \frac{2b\Delta}{a^2 + b^2 + c^2}, \quad z = \frac{2c\Delta}{a^2 + b^2 + c^2}$$ **Ans.**

12. *If r is the distance of a point on the conic $ax^2 + by^2 + cz^2 = 1$, $lx + my + nz = 0$ from the origin, then show that the stationary values of r are given by*

$$\frac{l^2}{1 - ar^2} + \frac{m^2}{1 - br^2} + \frac{n^2}{1 - cr^2} = 0$$

SOLUTION: Let any point on the conic $ax^2 + by^2 + cz^2 = 1$ ...(1) be (x, y, z)

$lx + my + nz = 0$ ...(2) then $r^2 = x^2 + y^2 + z^2$.

Let $f = x^2 + y^2 + z^2$, $\phi_1 = ax^2 + by^2 + cz^2 = 1$, $\phi_2 = lx + my + nz = 0$

Consider $F = f + \lambda_1\phi_1 + \lambda_2\phi_2$.

For f to have stationary values, $\frac{\partial F}{\partial x} = \frac{\partial F}{\partial y} = \frac{\partial F}{\partial z} = 0$

$$2x + 2ax\lambda_1 + l\lambda_2 = 0 \qquad ...(3)$$

$$2y + 2by\lambda_1 + m\lambda_2 = 0 \qquad ...(4)$$

$$2z + 2cz\lambda_1 + n\lambda_2 = 0 \qquad ...(5)$$

Multiplying (3) by x, (4) by y and (5) by z and adding, we get

$$2f + 2\lambda_1(ax^2 + by^2 + cz^2) + \lambda_2(lx + my + nz) = 0$$

or $$2f + 2\lambda_1 + 0\lambda_2 = 0 \quad \therefore \quad \lambda_1 = -f \text{ or } \lambda_1 = -r^2.$$

$\therefore$ (3), (4) and (5), give

$$x = \frac{l\lambda_2}{2(af - 1)}, \quad y = \frac{m\lambda_2}{2(bf - 1)}, \quad z = \frac{2n\lambda_2}{2(cf - 1)}.$$

Putting these in (2), we get

$$\frac{l^2\lambda_2}{2(ar^2 - 1)} + \frac{m^2\lambda_2}{2(br^2 - 1)} + \frac{n^2\lambda_2}{2(cr^2 - 1)} = 0$$

or $$\frac{l^2}{1 - ar^2} + \frac{m^2}{1 - br^2} + \frac{n^2}{1 - cr^2} = 0$$ **Hence proved.**

13. *Find the extreme value of $x^l \cdot y^m \cdot z^n$ subject to the conditions*

$$ax + by + cz = l + m + n \text{ and } \frac{a}{x} + \frac{b}{y} + \frac{c}{z} = 1.$$

SOLUTION: Let $f = \log(x^l y^m z^n) = l \log x + m \log y + n \log z$

and let $\phi = ax + by + cz - (l + m + n)$.

We construct a function $F = f + \lambda\phi$ where λ is arbitrary

Next, for f to be maximum or minimum $\frac{\partial F}{\partial x} = \frac{\partial F}{\partial y} = \frac{\partial F}{\partial z} = 0$

or $\frac{l}{x} + \lambda a = 0$ ...(1) $\frac{m}{y} + \lambda b = 0$ ...(2) $\frac{n}{z} + \lambda c = 0$ ...(3)

Multiplying (1) by x, (2) by y and (3) by z and adding, we get

$$l + m + n + \lambda\,(ax + by = cz) = 0$$

or $$l + m + n + \lambda(l + m + n) = 0 \quad \text{hence} \quad \lambda = -1$$

$\Rightarrow \quad x = l/a, \quad y = m/b \quad$ and $\quad z = n/c \qquad \therefore \qquad x^l\, y^m\, z^n = \left(\frac{l}{a}\right)^l \left(\frac{m}{b}\right)^m \left(\frac{n}{c}\right)^n.$ **Ans.**

14. *Expand* $x^2y + 3y - 2$ *in powers of* $x - 1$ *and* $y + 2$ *using Taylor's theorem.*

SOLUTION: $f(x, y) = x^2y + 3y - 2 \quad \therefore \quad \frac{\partial f}{\partial x} = 2xy, \; \frac{\partial f}{\partial y} = x^2 + 3, \; \frac{\partial^2 f}{\partial x^2} = 2y, \; \frac{\partial^2 f}{\partial x\,\partial y} = 2x, \; \frac{\partial^2 f}{\partial y^2} = 0$

Expansion of $f(x, y)$ about the point $(1, -2)$ to be obtained by Taylor's theorem, is

$$f(x,y) = f(1,-2), + (x-1)\left(\frac{\partial f}{\partial x}\right)_{(1,-2)} + (y+2)\left(\frac{\partial f}{\partial y}\right)_{(1,-2)}$$

$$+ \frac{1}{2!}\left[(x-1)^2\left(\frac{\partial^2 f}{\partial x^2}\right)_{(1,-2)} + 2(x-1)(y+2)\left(\frac{\partial^2 f}{\partial x\,\partial y}\right)_{(1,-2)} + (y+2)^2\left(\frac{\partial^2 f}{\partial y^2}\right)_{(1,-2)}\right] + \ldots$$

$$= -10 - 4(x-1) + 4(y+2) + \frac{1}{2!}[-4(x-1)^2 + 4(x-1)(y+2) + 0] + \ldots$$

$$= -10 - 4(x-1) + 4(y+2) - 2(x-1)^2 + 2(x-1)(y+2) + 0(y+2)^2 + \ldots.$$ **Ans.**

15. *Find Taylor's expansion of the function* $f(x, y) = e^{-x^2-y^2}\cos(xy)$ *about the point* $x_0 = 0,\; y_0 = 0$ *upto three terms.*

SOLUTION: By Taylor's theorem, expansion of $f(x, y)$ about (x_0, y_0), is

$$f(x,y) = f(x_0,y_0) + (x-x_0)\left(\frac{\partial f}{\partial x}\right)_{x_0/y_0} + (y-y_0)\left(\frac{\partial f}{\partial y}\right)_{(0,0)}$$

$$+ \frac{1}{2!}\left[(x_0-x_0)^2\left(\frac{\partial^2 f}{\partial x^2}\right)_{(x_0,y_0)} + 2(x-x_0)(y-y_0)\left(\frac{\partial^2 f}{\partial x\,\partial y}\right)_{(x_0,y_0)} + (y-y_0)^2\left(\frac{\partial^2 f}{\partial y^2}\right)_{(x_0,y_0)}\right] + \ldots.$$

Here $\quad x_0 = y_0 = 0 \quad$ hence $\quad f(x_0, y_0) = 1.$

$$\frac{\partial f}{\partial x} = (-2x)e^{-x^2-y^2}\cos(xy) - ye^{-x^2-y^2}\sin(xy) \quad \therefore \quad \left(\frac{\partial f}{\partial x}\right)_{(0,0)} = 0$$

$$\frac{\partial f}{\partial y} = -2ye^{-x^2-y^2}\cos(xy) - xe^{-x^2+y^2}\sin(xy) \quad \therefore \quad \left(\frac{\partial f}{\partial y}\right)_{(0,0)} = 0$$

$$\frac{\partial^2 f}{\partial x^2} = -2xe^{-x^2-y^2}[-2x\cos(xy) - y\sin(xy)] + e^{-x^2-y^2}[2xy\sin(xy) - 2\cos(xy) - y^2\sin(xy)]$$

$$\therefore \quad \left(\frac{\partial^2 f}{\partial x^2}\right)_{(0,0)} = -2$$

$$\frac{\partial^2 f}{\partial x\,\partial y} = -2ye^{-x^2-y^2}[-2x\cos xy - y\sin xy] + e^{-x^2-y^2}[2x^2\sin(xy) - xy\cos xy - \sin xy]$$

$$= 0 + 0 = 0 \quad \text{at} \quad (0, 0)$$

$$\frac{\partial^2 y}{\partial y^2} = -2ye^{-x^2-y^2}[-2y\cos(xy) - x\sin(xy)] + e^{-x^2-y^2}[-2\cos(xy) + 2xy\sin(xy) - x^2\cos(xy)]$$

$$\therefore \quad \left(\frac{\partial^2 f}{\partial y^2}\right)_{(0,0)} = -2$$

$$\therefore \quad f(x, y) = 1 + (x-0)\cdot 0 + (y-0)\cdot 0 + \frac{1}{2!}[(x-0)^2(-2) + 2(x-0)(y-0)\,0 + (y-0)^2(-2)] + \ldots$$

$$= 1 - x^2 - y^2 \quad \text{up to three terms only}$$ **Ans.**

16. *Expand $e^{a\sin^{-1}x}$ in ascending powers of x as far upto x^4 by Maclaurins series.*

SOLUTION: $y = e^{a\sin^{-1}x}$, $\quad \dfrac{dy}{dx} = y_1 = \dfrac{e^{a\sin^{-1}x}\,a}{\sqrt{1-x^2}}$. Here $y(0) = 1$ and $y_1(0) = a$

or $\quad (1-x^2)\,y_1^2 = a^2y^2 \quad$ hence $\quad (1-x^2)\,2y_1y_2 - 2x\,y_1^2 = 2a^2\,yy_1$.

or $\quad (1-x^2)\,y_2 - xy_1 = a^2\,y \quad \ldots(1) \qquad \therefore\ y_2(0) = a^2$.

Differentiating (1) n times using Lebneitz's theorem, we get

$$(1-x^2)\,y_{n+2} - n\,(2x)\,y_{n+1} + \frac{n(n-1)}{2!}(-2)\,y_n - xy_{n+1} - n\cdot y_n = a^2y_n$$

At $x = 0$ we have $\quad (y_{n+2}) = (n^2 + a^2)\,(y_n)_0$

For $n = 1 \quad (y_3)_0 = (1^2 + a^2)\,(y_1)_0 = a\,(1 + a^2)$

For $n = 2, \quad (y_4)_0 = (2^2 + a^2)\,(y_2)_0 = a^2\,(4 + a^2)$, etc.

$\therefore$ By Maclaurins theorem

$$y = e^{a\sin^{-1}x} = 1 + xa + \frac{x^2}{2!}a^2 + \frac{x^3}{3!}a(1+a^2) + \frac{x^4}{4!}a^2(4+a^2) + \ldots$$

$$= 1 + ax + \frac{a^2}{2}x^2 + \frac{a}{6}(a^2+1)\,x^3 + \frac{a^2(4+a^2)}{24}x^4 \quad \text{upto} \quad x^4$$ **Ans.**

17. *Expand x^y in powers of $(x-1)$ and $(y-1)$ upto third degree terms.*

SOLUTION: $f(x, y) = x^y$, $\quad f(1, 1) = 1$.

Hence $\quad \dfrac{\partial f}{\partial x} = yx^{y-1}$ and $\dfrac{\partial f}{\partial y} = x^y\log x \quad \therefore \left(\dfrac{\partial f}{\partial x}\right)_{(1,1)} = 1, \quad \left(\dfrac{\partial f}{\partial y}\right)_{(1,1)} = 0$

and $\quad \dfrac{\partial^2 f}{\partial x^2} = y\,(y-1)\,x^{y-2}, \quad \dfrac{\partial^2 f}{\partial x\,\partial y} = y\,x^{(y-1)}\log x + x^{y-1}, \quad \dfrac{\partial^2 f}{\partial y^2} = x^y\,(\log x)^2$.

$$\therefore \quad \left(\frac{\partial^2 f}{\partial x^2}\right)_{(1,1)} = 0, \quad \left(\frac{\partial^2 f}{\partial x\,\partial y}\right)_{(1,1)} = 1 \quad \text{and} \quad \left(\frac{\partial^2 f}{\partial y^2}\right)_{(1,1)} = 0$$

$$\frac{\partial^3 f}{\partial x^3} = y(y-1)(y-2)\,x^{y-3}, \qquad \frac{\partial^3 f}{\partial x^2\,\partial y} = y(y-1)\,x^{y-2}\log x + (2y-1)\,x^{y-2}$$

$$\frac{\partial^3 f}{\partial x\,\partial y^2} = y\,x^{y-1}(\log x)^2 + x^{y-1}\,2\log x, \qquad \frac{\partial^3 f}{\partial y^3} = x^y(\log x)^3.$$

$$\therefore \quad \left(\frac{\partial^3 f}{\partial x^3}\right)_{(1,1)} = 0, \quad \left(\frac{\partial^3 f}{\partial x^2\,\partial y}\right) = 1, \quad \left(\frac{\partial^3 f}{\partial x\,\partial y^2}\right)_{(1,1)} = 0, \quad \left(\frac{\partial^3 f}{\partial y^3}\right)_{(1,1)} = 0.$$

$$\therefore \quad f(x,y) = 1 + (x-1)(1) + (y-1)(0) + \frac{1}{2!}[(x-1)^2\cdot 0 + 2(x-1)(y-1)\cdot 1 + (y-1)^2\cdot 0]$$

$$+ \frac{1}{3!}[(x-1)^3\cdot 0 + 3(x-1)^2(y-1)\cdot 1 + 3(x-1)(y-1)^2\cdot 0 + (y-1)^3\cdot 0]$$

$$= 1 + (x-1) + (x-1)(y-1) + \frac{1}{2}(x-1)^2(y-1)$$

upto third degree terms **Ans.**

18. *Divide $2u$ into three parts such that the continued product of the first, square of the second and cube of the third may be maximum.*

SOLUTION: Let the first part be x, second be y then third is $2u - (x + y)$.

We are to maximise $f(x,y) = xy^2(2u - x - y)^3$.

$$\frac{\partial f}{\partial x} = y^2 \times 1\,(2u - x - y)^3 + 3xy^2(2u - x - y)^2(-1) = y^2(2u - x - y)^2(2u - x - y - 3x)$$

and
$$\frac{\partial f}{\partial y} = 2xy(2u - x - y)^3 + 3xy^2(2u - x - y)^2(-1) = xy(2u - x - y)^2(4u - 2x - 2y - 3y)$$

For f to be maximum $\frac{\partial f}{\partial x} = 0, \; \frac{\partial f}{\partial y} = 0$

$\Rightarrow \; 2u - 4x - y = 0$ and $4u - 2x - 5y = 0$ or $4x + y = 2u$ and $2x + 5y = 4u.$

Which on solving for x and y, give $x = \frac{u}{3}, \; y = \frac{2u}{3}$ hence the third part $= 2u - \left(\frac{u}{3} + \frac{2u}{3}\right) = u.$

It can be easily seen that at this poinht $rt > s^2$ and $r < 0$ **Ans.**

19. *Find the points on the surface $z^2 = xy + 1$ nearest to origin.*

SOLUTION: Let $f = x^2 + y^2 + z^2 = x^2 + y^2 + xy + 1$

$$\therefore \quad \frac{\partial f}{\partial x} = 2x + y, \quad \frac{\partial f}{\partial y} = 2y + x.$$

For f to be minimum $\frac{\partial f}{\partial x} = 0, \; \frac{\partial f}{\partial y} = 0$ which give $x = 0, \; y = 0$

For these $x = 0, \; y = 0, \; z^2 = xy + 1$ hence $z^2 = 1$ hence $z = \pm 1$

Thus the points on the given surface nearest to the origin are $(0, 0, 1)$ and $(0, 0, -1)$ **Ans.**

20. *Find the maximum and minimum distances of the point* (3, 4, 12) *from the sphere* $x^2 + y^2 + z^2 = 1$.

SOLUTION: Let $P(x, y, z)$ be any point on the sphere $\phi \equiv x^2 + y^2 + z^2 = 1$ ...(1)

Point A is (3, 4, 12). Distance AP is to be maximised and minimised.

$$f(x, y, z) = (x-3)^2 + (y-4)^2 + (z-12)^2 \quad \text{subject to} \quad \phi \equiv x^2 + y^2 + z^2 = 1$$

Let $F = f + \lambda f$. $\therefore \quad \frac{\partial f}{\partial x} + \lambda \frac{\partial \phi}{\partial x} = 0, \quad \frac{\partial f}{\partial y} + \lambda \frac{\partial \phi}{\partial y} = 0, \quad \frac{\partial f}{\partial z} + \lambda \frac{\partial \phi}{\partial z} = 0.$

or $2(x-3) + 2\lambda x = 0, \quad 2(y-4) + 2\lambda y = 0, \quad 2(z-12) + 2\lambda z = 0$

$\therefore \quad x = \frac{3}{1+\lambda}, \quad y = \frac{4}{1+\lambda}, \quad z = \frac{12}{1+\lambda}$ These values satisfy (1), hence

$$\left(\frac{3}{1+\lambda}\right)^2 + \left(\frac{4}{1+\lambda}\right)^2 + \left(\frac{12}{1+\lambda}\right)^2 = 1 \quad \text{or} \quad (1+\lambda)^2 = 9 + 16 + 144 = 169 = 13^2$$

$\therefore \quad \lambda + 1 = \pm 13 \quad \text{or} \quad \lambda = 12, -14.$

For $\lambda = 12, \quad x = \frac{3}{13}, \; y = \frac{4}{13}, \; z = \frac{12}{13}$

For $\lambda = -14, \quad x = \frac{3}{-13}, \; y = \frac{4}{-13}, \; z = \frac{12}{-13}$

hence at $\left(-\frac{3}{13}, -\frac{4}{13}, -\frac{12}{13}\right)$, $f = \left(\frac{3}{-13} - 3\right)^2 + \left(\frac{4}{-13} - 4\right)^2 + \left(\frac{12}{-13} - 12\right)^2 = 14^2$

and at $\left(\frac{3}{13}, \frac{4}{13}, \frac{12}{13}\right) = \frac{144}{169}[9 + 16 + 144] = 144 = 12^2$

Thus, maximum and minimum distances are 14 and 12. **Ans.**

21. *Find the maximum distance from the point* (1, 2, 0) *to the cone* $z^2 = x^2 + y^2$.

SOLUTION: $(\text{Distance})^2 = (x-1)^2 + (y-2)^2 + (z-0)^2 = (x-1)^2 + (y-2)^2 + z^2$ but $z^2 = x^2 + y^2$

$\therefore \quad f(x, y) = (x-1)^2 + (y-2)^2 + x^2 + y^2 = 2x^2 + 2y^2 - 2x - 4y + 1 + 4$

$$\frac{\partial f}{\partial x} = 4x - 2, \quad \frac{\partial f}{\partial y} = 4y - 4.$$

For f to be max. $\frac{\partial f}{\partial x} = 0, \; \frac{\partial f}{\partial y} = 0$ or $x = \frac{1}{2}, \; y = 1$

$\therefore \quad f(x, y) = \left(\frac{1}{2} - 1\right)^2 + (1-2)^2 = \frac{1}{4} + 1 = \frac{5}{4}$

$\therefore$ Min. Distance $= \sqrt{\frac{5}{4}} = \frac{\sqrt{5}}{2}$ **Ans.**

ADDITIONAL IMPORTANT PROBLEMS

1. A balloon is in the form of a right circular cylinder of radius 1.5 m and height 4 m and is surmounted by hemispherical ends. If the radius is increased by 0.01m and the height by 0.05m, find the percentage change in the volume of the balloon.

SOLUTION: Volume of the cylinder = $\pi r^2 h$. Volume of the hemisphers = $2\cdot\left(\frac{2}{3}\pi r^3\right)$

$\therefore$ Volume V of the balloon = $\pi r^2 h + \frac{4}{3}\pi r^3$

$$\delta V = \pi\,(2rh\,\delta r + r^2\delta h) + \frac{4}{3}\pi\cdot 3r^2\,\delta r$$

Given that $r = 1.5$ m, $h = 4$ m, $\delta r = 0.01$ m, $\delta h = 0.05$ m.

$$\therefore \quad \delta V = \pi(2\times 1.5\times 4\times 0.01 + (1.5)^2\times 0.05) + \frac{4}{3}\pi\cdot 3(1.5)^2\times 0.01$$

$$= 0.3225\pi$$

$$V = \pi\left[(1.5)^2\times 4 + \frac{4}{3}(1.5)^3\right] = 13.5\pi$$

$\therefore$ Percentage change in volume = $\frac{\delta V}{V}\times 100 = \frac{0.3225\pi}{13.5\pi}\times 100 = 2.389\%$.

2. Find the approximate value of $[(0.98)^2 + (2.01)^2 + (1.94)^2]^{1/2}$.

SOLUTION: Let $U = (x^2 + y^2 + z^2)^{1/2}$ or $U^2 = x^2 + y^2 + z^2$

$$\therefore \quad 2U\,\delta U = 2x\,\delta x + 2y\,\delta y + 2z\,\delta z \quad \text{or} \quad U\delta U = x\delta x + y\delta y + z\delta z.$$

Now let $x = 1,\ y = 2,\ z = 2$ then $\delta x = -0.02,\ \delta y = 0.01,\ \delta z = -0.06$

Also, $$U = \sqrt{1^2 + 2^2 + 2^2} = 3.$$

$$\therefore \quad U\,\delta U = 1(-0.02) + 2\,(0.01) + 2(-0.06) = -0.12$$

Hence $$\delta U = -\frac{0.12}{3} = -0.04.$$

Therefore, the approximate value of U is equal to $3 - 0.04 = 2.96$ **Ans.**

3. Find the percentage error in the area of an ellipse when an error of 1.5% is made in measuring its major and minor axes.

SOLUTION: Let a and b be the lengths of the semi-major axis and semi-minor axis of the ellipse respctively and A be its area, then

$A = \pi ab \quad \therefore \quad \log A = \log \pi + \log a + \log b$

$$\therefore \quad \frac{1}{A}\,\delta A = 0 + \frac{1}{a}\,\delta a + \frac{1}{b}\,\delta b$$

or $$\frac{1}{A}\,\delta A\times 100 = \frac{1}{a}\,\delta a\times 100 + \frac{1}{b}\,\delta b\times 100 = 1.5 + 1.5$$

$\therefore$ Percentage error in the area of ellipse = 3%. **Ans.**

4. If $x = uv$ and $y = \dfrac{u+v}{u-v}$ find $\dfrac{\partial(u, v)}{\partial(x, y)}$

SOLUTION: $x = uv$ and $y = \dfrac{u+v}{u-v}$ then

$$\frac{\partial(x, y)}{\partial(u, v)} = \begin{vmatrix} \dfrac{\partial x}{\partial u} & \dfrac{\partial x}{\partial v} \\ \dfrac{\partial y}{\partial u} & \dfrac{\partial y}{\partial v} \end{vmatrix} = \begin{vmatrix} v & u \\ \dfrac{-2v}{(u-v)^2} & \dfrac{2u}{(u-v)^2} \end{vmatrix}$$

$$= \frac{2}{(u-v)^2}\begin{vmatrix} v & u \\ -v & u \end{vmatrix} = \frac{4uv}{(u-v)^2}$$

Since $\dfrac{\partial(x, y)}{\partial(u, v)} \cdot \dfrac{\partial(u, v)}{\partial(x, y)} = 1 \quad \therefore \quad \dfrac{\partial(u, v)}{\partial(x, y)} = \dfrac{(u-v)^2}{4uv}$ **Ans.**

5. If $u = x^2 - y^2$, $v = 2xy$ where $x = r \cos \theta$, $y = r \sin \theta$, find $\dfrac{\partial(u, v)}{\partial(r, \theta)}$.

SOLUTION:
$$\frac{\partial(u, v)}{\partial(x, y)} = \begin{vmatrix} \dfrac{\partial u}{\partial x} & \dfrac{\partial u}{\partial y} \\ \dfrac{\partial v}{\partial x} & \dfrac{\partial v}{\partial y} \end{vmatrix} = \begin{vmatrix} 2x & -2y \\ 2y & 2x \end{vmatrix} = 4(x^2 + y^2) = 4r^2.$$

and
$$\frac{\partial(x, y)}{\partial(r, \theta)} = \begin{vmatrix} \dfrac{\partial x}{\partial r} & \dfrac{\partial x}{\partial \theta} \\ \dfrac{\partial y}{\partial r} & \dfrac{\partial y}{\partial \theta} \end{vmatrix} = \begin{vmatrix} \cos\theta & -r\sin\theta \\ \sin\theta & r\cos\theta \end{vmatrix} = r(\cos^2\theta + \sin^2\theta) = r$$

Therefore
$$\frac{\partial(u, v)}{\partial(r, \theta)} = \frac{\partial(u, v)}{\partial(x, y)} \cdot \frac{\partial(x, y)}{\partial(r, \theta)} = 4r^2 \cdot r = 4r^3.$$ **Ans.**

6. If $x = \sqrt{vw}$, $y = \sqrt{uw}$, $z = \sqrt{u\,v}$ and $u = r \sin \theta \cos \theta$, $v = r \sin \theta \sin \theta$, $w = r \cos \theta$ find $\dfrac{\partial(x, y, z)}{\partial(r, \theta, \phi)}$.

SOLUTION:
$$\frac{\partial(x, y, z)}{\partial(u, v, w)} = \begin{vmatrix} \dfrac{\partial x}{\partial u} & \dfrac{\partial x}{\partial v} & \dfrac{\partial x}{\partial w} \\ \dfrac{\partial y}{\partial u} & \dfrac{\partial y}{\partial v} & \dfrac{\partial y}{\partial w} \\ \dfrac{\partial z}{\partial u} & \dfrac{\partial z}{\partial v} & \dfrac{\partial z}{\partial w} \end{vmatrix} = \begin{vmatrix} 0 & \dfrac{1}{2}\sqrt{\dfrac{w}{v}} & \dfrac{1}{2}\sqrt{\dfrac{v}{w}} \\ \dfrac{1}{2}\sqrt{\dfrac{w}{u}} & 0 & \dfrac{1}{2}\sqrt{\dfrac{u}{w}} \\ \dfrac{1}{2}\sqrt{\dfrac{v}{u}} & \dfrac{1}{2}\sqrt{\dfrac{u}{v}} & 0 \end{vmatrix} = \frac{1}{4}$$

and $$\frac{\partial(u,v,w)}{\partial(r,\theta,\phi)}=\begin{vmatrix}\frac{\partial u}{\partial r} & \frac{\partial u}{\partial \theta} & \frac{\partial u}{\partial \phi}\\ \frac{\partial v}{\partial r} & \frac{\partial v}{\partial \theta} & \frac{\partial v}{\partial \phi}\\ \frac{\partial w}{\partial r} & \frac{\partial w}{\partial \theta} & \frac{\partial w}{\partial \phi}\end{vmatrix}=\begin{vmatrix}\sin\theta\cos\theta & r\cos\theta\cos\phi & -r\sin\theta\sin\phi\\ \sin\theta\sin\phi & r\cos\theta\sin\phi & r\sin\theta\cos\phi\\ \cos\theta & -r\sin\theta & 0\end{vmatrix}=r^2\sin\theta$$

$\therefore$ By chain rule $$\frac{\partial(x,y,z)}{\partial(r,\theta,\phi)}=\frac{\partial(x,y,z)}{\partial(u,v,w)}\cdot\frac{\partial(u,v,w)}{\partial(r,\theta,\phi)}$$

$$=\frac{1}{4}(r^2\sin\theta)=\frac{r^2}{4}\sin\theta.$$ **Ans.**

7. Determine if the following functions are functionally dependent. If dependent, find the relation between them:

(*i*) $u = e^x \sin y, \quad v = e^x \cos y$

(*ii*) $u = xy + yz + zx, \quad v = x^2 + y^2 + z^2, \quad w = x + y + z.$

SOLUTION: (*i*) $u = e^x \sin y, \; v = e^x \cos y$

$$\therefore \quad \frac{\partial(u,v)}{\partial(x,y)}=\begin{vmatrix}\frac{\partial u}{\partial x} & \frac{\partial u}{\partial y}\\ \frac{\partial v}{\partial x} & \frac{\partial v}{\partial y}\end{vmatrix}=\begin{vmatrix}e^x\sin y & e^x\cos y\\ e^x\cos y & -e^x\sin y\end{vmatrix}$$

$$= e^{2x}(-\sin^2 y - \cos^2 y) = -e^{2x} \neq 0$$

since $\frac{\partial(u,v)}{\partial(x,y)}\neq 0$, u and v are functionally independent.

(*ii*) $u = xy + yz + zx, \; v = x^2 + y^2 + z^2, \; w = x + y + z.$

$$\frac{\partial(u,v,w)}{\partial(x,y,z)}=\begin{vmatrix}\frac{\partial u}{\partial x} & \frac{\partial u}{\partial y} & \frac{\partial u}{\partial z}\\ \frac{\partial v}{\partial x} & \frac{\partial v}{\partial y} & \frac{\partial v}{\partial z}\\ \frac{\partial w}{\partial x} & \frac{\partial w}{\partial y} & \frac{\partial w}{\partial z}\end{vmatrix}=\begin{vmatrix}y+z & z+x & x+y\\ 2x & 2y & 2z\\ 1 & 1 & 1\end{vmatrix}$$

$$=2\begin{vmatrix}y+z & z+x & x+y\\ x & y & z\\ 1 & 1 & 1\end{vmatrix}=2\begin{vmatrix}x+y+z & x+y+z & x+y+z\\ x & y & z\\ 1 & 1 & 1\end{vmatrix}=0$$

Hence u, v, w are functionally dependent.

For the relation between them

$$w^2 = x^2 + y^2 + z^2 + 2(xy + yz + zx) = v + 2u.$$ **Ans.**

8. Find the extreme values of $u = x^3 + 3xy^2 - 3x^2 - 3y^2 + 7$, if any.

SOLUTION: $\dfrac{\partial u}{\partial x} = 3x^2 + 3y^2 - 6x$ and $\dfrac{\partial u}{\partial y} = 6xy - 6y$

For extreme values $\dfrac{\partial u}{\partial x} = 0, \ \dfrac{\partial u}{\partial y} = 0$

or $y(x-1) = 0$ and $x^2 + y^2 = 2x.$

when $y = 0,\ x = 0, 2\ \therefore$ stationary points are $(0, 0)$ and $(2, 0)$

when $x = 1,\ y^2 = 1\ \therefore\ y = \pm 1$ and stationary points are $(1, 1)$ and $(1, -1)$.

Next $r = \dfrac{\partial^2 u}{\partial x^2} = 6x - 6 = 6(x-1),\ \ s = \dfrac{\partial^2 u}{\partial x \partial y} = 6y,\ t = \dfrac{\partial^2 u}{\partial y^2} = 6(x-1)$

At $(0, 0)$, $r = -6,\ s = 0,\ t = -6$ and $rt - s^2 = 36 > 0$ hence maximum.

At $(2, 0)$, $r = 6,\ t = 0,\ s = 6$ and $rt - s^2 = 36 > 0$ hence minimum.

At $(1, 1)$, $r = 0,\ t = 0,\ s = 6$ and $rt - s^2 = -36 < 0$ hence neither maximum nor minimum.

At $(1, -1)$, $r = 0,\ t = 0,\ s = -6$ and $st - s^2 = -36 < 0$ hence neither maximum nor minimum.

9. Show that the rectangular solid of maximum volume that can be inscribed in a given sphere is a cube.

SOLUTION: If x, y, z are the length, breadth and height of the rectangular solid and V is its volume them $V = xyz$.

Let the given sphere be $x^2 + y^2 + z^2 = a^2$ so $z^2 = a^2 - x^2 - y^2$.

$$\therefore \quad V = xy\sqrt{a^2 - x^2 - y^2} \quad \text{or} \quad V^2 = x^2 y^2 (a^2 - x^2 - y^2) = f(x, y), \text{ say.}$$

$$\frac{\partial f}{\partial x} = y^2 (2a^2 x - 4x^3 - 2xy^2) = 2xy^2 (a^2 - 2x^2 - y^2)$$

and $$\frac{\partial f}{\partial y} = x^2 (2a^2 y - 4y^3 - 2x^2 y) = 2x^2 y (a^2 - x^2 - 2y^2)$$

For extreme values $\dfrac{\partial f}{\partial x} = 0 \Rightarrow x = 0,\ y = 0,\ 2x^2 + y^2 = a^2$

and $\dfrac{\partial f}{\partial y} = 0 \Rightarrow x = 0, y = 0,\ x^2 + 2y^2 = a^2$

But x and y being sides of the rectangular solid, cannot be zero.

Thus we have $2x^2 + y^2 = a^2$ and $x^2 + 2y^2 = a^2$

Solving these we get $x = a/\sqrt{3},\ y = a/\sqrt{3}$

then $$z = \sqrt{a^2 - x^2 - y^2} = \sqrt{a^2 - \frac{a^2}{3} - \frac{a^2}{3}} = \frac{a}{\sqrt{3}}$$

$\therefore$ Stationary point is $\left(a/\sqrt{3}, a/\sqrt{3}, a/\sqrt{3}\right)$.

Next,
$$r = \frac{\partial^2 f}{\partial x^2} = 2a^2y^2 - 12x^2y^2 - 2y^4 = -8a^4/9 \text{ at } \left(\frac{a}{\sqrt{3}}, \frac{a}{\sqrt{3}}\right)$$

$$s = \frac{\partial^2 f}{\partial x \partial y} = 4a^2xy - 8x^3y - 8xy^3 = -4a^4/9 \text{ at } \left(\frac{a}{\sqrt{3}}, \frac{a}{\sqrt{3}}\right)$$

$$t = \frac{\partial^2 f}{\partial y^2} = 2a^2x^2 - 2x^4 - 12x^2y^2 = -8a^4/9 \text{ at } \left(\frac{a}{\sqrt{3}}, \frac{a}{\sqrt{3}}\right)$$

Hence
$$rt - s^2 = \frac{64a^8}{81} - \frac{16a^8}{81} - \frac{16a^8}{81} = \frac{48}{81}a^8 > 0$$

$\therefore$ $f(x, y)$, that is, V^2 is maximum when $x = y = z$ hence V is maximum. **Ans.**

10. Find the maximum value of $f = x^2 y^3 z^4$ subject to the constraint $x + y + z = 5$.

SOLUTION: $f = x^2 y^3 z^4$, $\phi(x, y, z) = x + y + z - 5 = 0$

By Lagrang's method of undetrmined multipliers, we have

$$\frac{\partial f}{\partial x} + \lambda \frac{\partial \phi}{\partial x} = 0, \quad \frac{\partial f}{\partial y} + \lambda \frac{\partial \phi}{\partial y} = 0, \quad \frac{\partial f}{\partial z} + \lambda \frac{\partial \phi}{\partial z} = 0$$

or
$$2xy^3 z^4 + \lambda \cdot 1 = 0, \quad 3x^2 y^2 z^4 + \lambda \cdot 1 = 0, \quad 4x^2 y^3 z^3 + \lambda \cdot 1 = 0$$

$\Rightarrow$
$$2xy^3 z^4 = 3x^2 y^2 z^4 = 4x^2 y^3 z^3$$

dividing throughout by $x^2 y^3 z^4$, we get

$$\frac{2}{x} = \frac{3}{y} = \frac{4}{z} = t, \text{ say}$$

$\Rightarrow$
$$x = \frac{2}{t}, \quad y = \frac{3}{t}, \quad z = \frac{4}{t}$$

substituting $x + y + z = 5$ gives $\frac{2}{t} + \frac{3}{t} + \frac{4}{t} = 5 \quad \therefore \quad t = \frac{9}{5}$.

Hence
$$x = \frac{10}{9}, \quad y = \frac{15}{9} = \frac{5}{3}, \quad z = \frac{20}{9}.$$

then maximum
$$f = \left(\frac{10}{9}\right)^2 \left(\frac{5}{3}\right)^3 \left(\frac{20}{9}\right)^4 = \frac{2^{10} \cdot 5^3}{3^{15}}.$$ **Ans.**

11. A closed rectangular box has length twice its breadth and has constant volume V. Determine the dimensions of the box requiring least surface area.

SOLUTION: Let x be the breadth, y be the height of the box so that length be $2x$. Volume $V = x \cdot 2x \cdot y =$ constant. ...(1)

Surface area S of the box, is given by

$$S = 2(2x \cdot x + x \cdot y + y \cdot 2x) = 4x^2 + 6xy. \qquad \text{...(2)}$$

Let $\phi = 2x^2y - V = 0.$

For extreme value of S, by Lagrange's method, we have

$$\frac{\partial S}{\partial x} + \lambda\frac{\partial \phi}{\partial x} = 0 \quad \text{and} \quad \frac{\partial S}{\partial y} + \lambda\frac{\partial \phi}{\partial y} = 0$$

or $\quad 8x + 6y + \lambda \cdot 4xy = 0 \quad$ and $\quad 6x + \lambda \cdot 2x^2 = 0$

which give $x = \dfrac{-3}{\lambda}$ and $y = \dfrac{-4}{\lambda}$.

Putting these in $2x^2 y = V$, we get

$$2\left(\frac{-3}{\lambda}\right)^2\left(\frac{-4}{\lambda}\right) = V \Rightarrow \lambda = -\left(\frac{72}{V}\right)^{1/3}.$$

$$\therefore \qquad x = -\frac{3}{\lambda} = 3\left(\frac{V}{72}\right)^{1/3} \quad \text{and} \quad y = -\frac{4}{\lambda} = 4\left(\frac{V}{72}\right)^{1/3}.$$

Therefore, the dimension of the box with least surface area are $3\left(\dfrac{V}{72}\right)^{1/3}, 4\left(\dfrac{V}{72}\right)^{1/3}$. **Ans.**

12. Find the maximum and minimum distance from the point (1, 2, 2) to the sphere $x^2 + y^2 + z^2 = 36$.

SOLUTION: Let (x, y, z) is any point on the sphere $x^2 + y^2 + z^2 = 36$. It D is distance of this point from (1, 2, 2), then D^2 is given by

$$D^2 = (x-1)^2 + (y-2)^2 + (z-2)^2 = f(x, y, z). \qquad \text{...(1)}$$

where $\phi(x, y, z) = x^2 + y^2 + z^2 - 36$ ...(2)

Now the Lagrangian equations are

$$\frac{\partial f}{\partial x} + \lambda\frac{\partial \phi}{\partial x} = 0 \quad \text{or} \quad 2(x-1) + \lambda(2x) = 0 \quad \text{or} \quad x - 1 + \lambda x = 0 \qquad \text{...(3)}$$

$$\frac{\partial f}{\partial y} + \lambda\frac{\partial \phi}{\partial y} = 0 \quad \text{or} \quad 2(y-2) + \lambda(2y) = 0 \quad \text{or} \quad y - 2 + \lambda y = 0 \qquad \text{...(4)}$$

and $\frac{\partial f}{\partial z} + \lambda \frac{\partial \phi}{\partial z} = 0$ or $2(z - 2) + \lambda(2z) = 0$ or $z - 2 + \lambda z = 0$ ...(5)

Multiplying (3), (4) and (5) by x, y and z respectively and adding gives

$$x(x - 1) + y(y - 2) + z(z - 2) + \lambda(x^2 + y^2 + z^2) = 0$$

or $$x^2 + y^2 + z^2 - (x + 2y + 2z) + \lambda(x^2 + y^2 + z^2) = 0$$

or $$36(1 + \lambda) = x + 2y + 2z. \quad ...(6)$$

From (3), (4) and (5) we have $x = \frac{1}{1+\lambda}, y = \frac{2}{1+\lambda}, z = \frac{2}{1+\lambda}$ ...(7)

Substituting for x, y, z in (6) we get

$$36(1 + \lambda) = \frac{1+4+4}{1+\lambda} \quad \text{or} \quad (1 + \lambda)^2 = \frac{9}{36} = \frac{1}{4}$$

$\therefore$ $1 + \lambda = \pm 1/2$, hence from (7) we get

$$x = \pm 2, \quad y = \pm 4, \quad z = \pm 4.$$

Therefore the maximum distance $= \sqrt{(-2-1)^2 + (-4-2)^2 + (-4-2)^2} = \sqrt{9+36+36} = 9.$

and minimum distance $= \sqrt{(2-1)^2 + (4-2)^2 + (4-2)^2} = 3.$ **Ans.**

CHAPTER

13 Multiple Integrals

IMPORTANT FORMULAE AND USEFUL RESULTS

1. *Double Integral in Cartesian Coordinates* $\iint\limits_R f(x, y)\, dx\, dy = \lim\limits_{x \to \infty} \sum\limits_{i, j=1}^{n} f(x_i, y_j)\, \delta x_i\, \delta y_j$

2. *Double Integral in Polar Coordinates* $\iint\limits_R f(r\cos\theta, r\sin\theta)\, rd\, rd\theta$.

3. *Triple Integral* $\iiint f(x, y, z)\, dx\, dy\, dz = \lim\limits_{n \to \infty} \sum\limits_{i, j, k}^{n} f(x_i, y_j, z_k)\, \delta x_i\, \delta y_j\, \delta y_k$

4. *Cylindrical Polar Coordinates* are r, θ, and z. Then elementary volume $dx\, dy\, dz = rd\, rd\theta\, dz$.

5. *Spherical Polar Coordinates* are r, θ, and ϕ. Here $x = r \sin\theta \cos\phi$, $y = r \sin\theta \sin\phi$, $z = r\cos\theta$ and elementary volume $= dx\, dy\, dz = r^2 \sin\theta\, dr\, d\theta\, d\phi$.

6. *In Double Integral* Area $= \iint dx\, dy$ in Cartesian coordinates

   $= \iint r d\theta\, dr$ in polar coordinates

7. *In Triple Integrals* Volume $= \iiint dx\, dy\, dz$ in Cartesian coordinates

   $= \iiint r^2 \sin\theta\, dr\, d\theta\, d\phi$ in spherical polar coordinates.

8. *Determining Mass of a Body*

   In case of are of a curve, Mass $= \int \rho\, ds$ where ρ is density at (x, y)

   In case of plane lamina, Mass $= \iint \rho\, dx\, dy$

   and in case of solid, the Mass $= \iiint \rho\, dx\, dy\, dz$

9. *Finding Centroid or Centre of Gravity (C.G.)*

C.G. of the mass of an arc, $\bar{x} = \dfrac{\int x\rho\, ds}{\int \rho\, ds}, \quad \bar{y} = \dfrac{\int y\rho\, ds}{\int \rho\, ds}$

C.G. of the mass of a plane area, $\bar{x} = \dfrac{\iint \rho \cdot x\, dx\, dy}{\iint \rho\, dx\, dy}, \quad \bar{y} = \dfrac{\iiint \rho\, y\, dx\, dy}{\iiint \rho\, dx\, dy}$

C.G. of the mass of a solid, $\bar{x} = \dfrac{\iiint \rho\, x\, dx\, dy\, dz}{\iiint \rho\, dx\, dy\, dz}, \quad \bar{y} = \dfrac{\iiint \rho\, y\, dx\, dy\, dz}{\iiint \rho\, dx\, dy\, dz}, \quad \bar{z} = \dfrac{\iiint \rho\, z\, dx\, dy\, dz}{\iiint \rho\, dx\, dy\, dz}$

SOLUTIONS TO EXERCISE 13A

1. *Evaluate* $\displaystyle\int_0^1\int_0^{x^2} e^{\frac{y}{x}}\,dy\,dx.$

SOLUTION: $\displaystyle\int_0^1\int_0^{x^2} e^{y/x}\,dy\,dx = \int_0^1\left[x e^{y/x}\right]_{y=0}^{x^2} dx = \int_0^1 x(e^x-1)\,dx = \left[x(e^x-x)\right]_0^1 - \int_0^1 (e^x - x)\,dx$

$$= 1(e-1) - \left(e^x - \frac{x^2}{2}\right)_0^1 = e-1-e+\frac{1}{2}+1 = \frac{1}{2} \qquad \textbf{Ans.}$$

2. *Evaluate* $\displaystyle\int_0^{a\sqrt{3}}\int_0^{\sqrt{x^2+a^2}} \frac{x\,dy\,dx}{y^2+x^2+a^2}.$

SOLUTION: $\displaystyle\int_0^{a\sqrt{3}}\int_0^{\sqrt{x^2+a^2}} \frac{x\,dy\,dx}{y^2+x^2+a^2} = \int_0^{a\sqrt{3}}\left[\frac{x}{\sqrt{x^2+a^2}}\tan^{-1}\frac{y}{\sqrt{x^2+a^2}}\right]_0^{\sqrt{a^2+x^2}}$

$$= \int_0^{a\sqrt{3}} \frac{x}{\sqrt{x^2+a^2}}(\tan^{-1}1 - 0)dx = \frac{\pi}{4}\int_0^{a\sqrt{3}}\frac{x\,dx}{\sqrt{x^2+a^2}}$$

$$= \frac{\pi}{4}\left[\sqrt{x^2+a^2}\right]_0^{a\sqrt{3}} = \frac{\pi}{4}[2a-a) = \frac{a\pi}{4} \qquad \textbf{Ans.}$$

3. *Evaluate the double integral* $\displaystyle\iint\frac{dx\,dy}{x^4+y^2}$ *over the region bounded by* $x = 1$ *and* $y = x^2$.

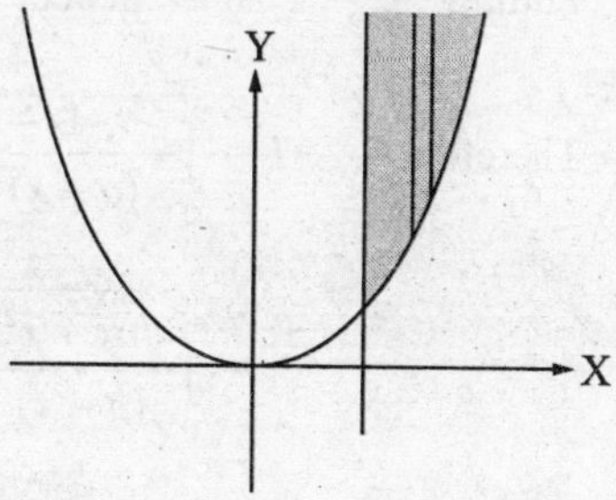

SOLUTION: The given integral over the shaded region bounded by $x = 1$ and $y = x^2$ is equal to

$$\int_{x=1}^{\infty}\int_{y=x^2}^{\infty}\frac{dy\,dx}{x^4+y^2} = \int_1^{\infty}\left[\frac{1}{x^2}\tan^{-1}\frac{y}{x^2}\right]_{x^2}^{\infty} dx$$

$$= \int_1^{\infty}\frac{1}{x^2}\left(\tan^{-1}\infty - \tan^{-1}1\right)dx$$

$$= \int_1^{\infty}\left(\frac{\pi}{2}-\frac{\pi}{4}\right)\frac{1}{x^2}dx = \frac{\pi}{4}\left[-\frac{1}{x}\right]_1^{\infty} = \frac{\pi}{4} \qquad \textbf{Ans.}$$

4. *Evaluate* $\iint \sin(ax+by)\,dx\,dy$ *over the triangular area bounded by*

$$x = 0, \quad y = 0, \quad ax + by = 1.$$

SOLUTION: Given integral is equal to

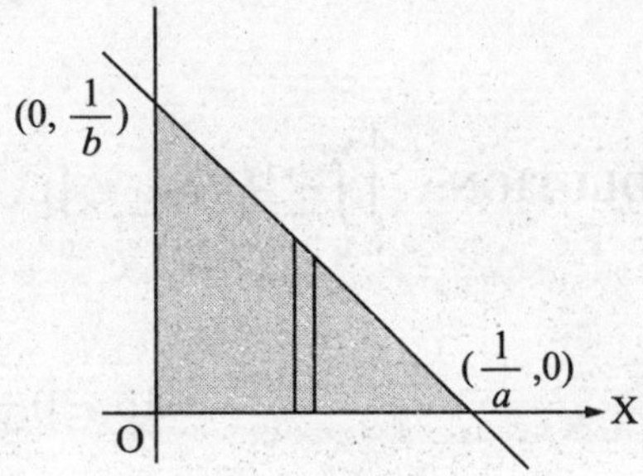

$$\int_0^{1/a}\int_0^{\frac{1-ax}{b}} \sin(ax+by)\,dy\,dx = \int_0^{1/a}\left[-\frac{1}{b}\cos(ax+by)\right]_0^{\frac{1-ax}{b}} dx$$

$$= -\frac{1}{b}\int_0^{1/a}[\cos 1 - \cos ax]\,dx$$

$$= -\frac{1}{b}\left[x\cos 1 - \frac{1}{a}\sin ax\right]_0^{1/a}$$

$$= -\frac{1}{b}\left[\frac{1}{a}\cos 1 - \frac{1}{a}\sin 1\right] = \frac{1}{ab}(\sin 1 - \cos 1) \qquad \textbf{Ans.}$$

5. *Evaluate* $\displaystyle\int_0^a\int_{\frac{y^2}{a}}^{y} \frac{y\,dx\,dy}{(a-x)\sqrt{ax-y^2}}$ *by changing the order of integration.*

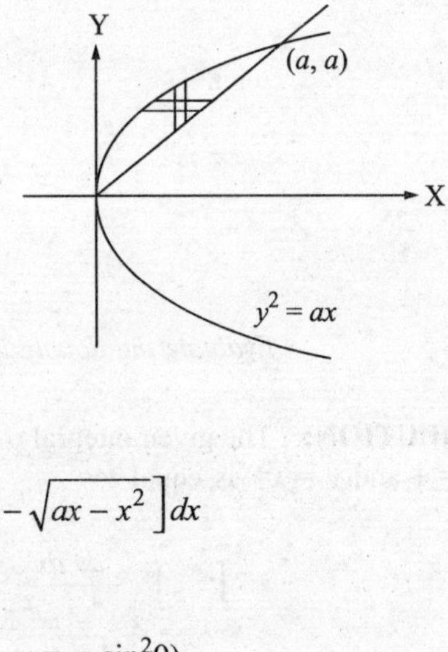

SOLUTION: $I = \displaystyle\int_0^a\int_{y^2/a}^{y} \frac{y\,dx\,dy}{(a-x)\sqrt{ax-y^2}}$

Charging the order of integration

$$I = \int_0^a\int_x^{\sqrt{ax}} \frac{y\,dy\,dx}{(a-x)\sqrt{ax-y^2}}$$

Putting $ax - y^2 = t^2$ hence $-2y\,dy = 2t\,dt$

Therefore, $I = \displaystyle\int_0^a -\frac{1}{(a-x)}\left[\sqrt{ax-y^2}\right]_{y=x}^{\sqrt{ax}} dx = -\int_0^a \frac{1}{(a-x)}\left[0-\sqrt{ax-x^2}\right]dx$

$$= \int_0^a \frac{\sqrt{ax-x^2}}{(a-x)}\,dx = \int_0^a \frac{\sqrt{x}}{\sqrt{a-x}}\,dx \qquad (\text{Putting } x = a\sin^2\theta)$$

$$= \int_0^{\pi/2} \frac{\sqrt{a}\sin\theta}{\sqrt{a}\cos\theta}\cdot 2a\sin\theta\cos\theta\,d\theta = 2a\int_0^{\pi/2}\sin^2\theta\,d\theta$$

$$= a\int_0^{\pi/2}(1-\cos 2\theta)\,d\theta = a\left(\theta - \frac{1}{2}\sin 2\theta\right)_0^{\pi/2} = \frac{a\pi}{2} \qquad \textbf{Ans.}$$

6. *Change the order of integration in the double integral* $\int_0^a \int_{\frac{x^2}{a}}^{2a-x} xy\, dy\, dx$ *and the evaluate it.*

SOLUTION: The region of integration is divided into two parts I and II as shown in figure. Changing the order of integration,

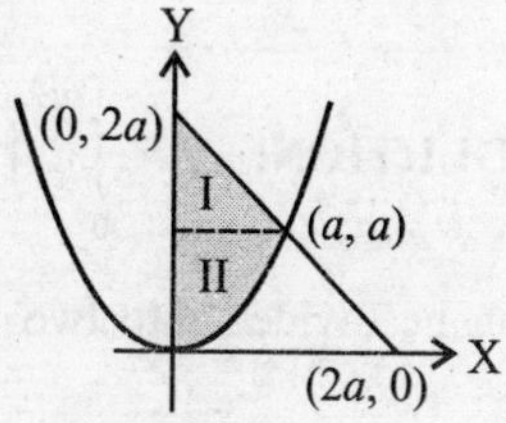

$$I = \int_0^a \int_{x^2/a}^{2a-x} xy\, dy\, dx = \int_a^{2a} \int_0^{2a-y} xy\, dx\, dy + \int_0^a \int_0^{\sqrt{ay}} xy\, dx\, dy$$

$$= \int_a^{2a} y\left[\frac{x^2}{2}\right]_{x=0}^{2a-y} dy + \int_0^a y\left[\frac{x^2}{2}\right]_0^{\sqrt{ay}} dy = \int_a^{2a} \frac{y}{2}(2a-y)^2\, dy + \int_0^a \frac{y}{2}\cdot ay\, dy$$

$$= \frac{1}{2}\int_a^{2a} (4a^2 y - 4ay^2 + y^3)\, dy + \frac{1}{2}\int_0^a ay^2 dy$$

$$= \frac{1}{2}\left[2a^2y^2 - \frac{2ay^3}{3} + \frac{y^4}{4}\right]_a^{2a} + \frac{a}{2}\left[\frac{y^3}{3}\right]_0^a = \frac{a^4}{2}\left(8 - \frac{32}{3} + \frac{16}{4} - 2 + \frac{4}{3} - \frac{1}{4}\right) + \frac{a^4}{6}$$

$$= \frac{3}{8}a^4.$$

Ans.

7. *Change the order of integration in the integral*

$$\int_{-a}^{a} \int_0^{\sqrt{a^2-y^2}} f(x,y)\, dx\, dy.$$

SOLUTION: Area of integrations is shown in figure. On changing the order of integration, we have

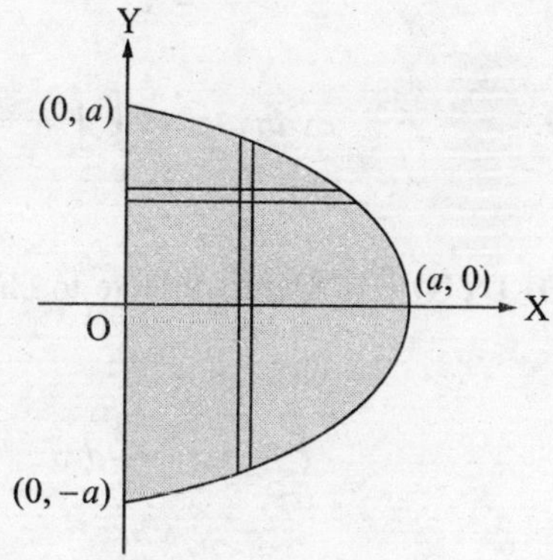

$$\int_0^a \int_{-\sqrt{a^2-x^2}}^{\sqrt{a^2-x^2}} f(x, y)\, dy\, dx$$

Ans.

8. *Evaluate* $\int_0^4 \int_0^{x^2} e^{\frac{y}{x}}\, dy\, dx.$

SOLUTION: $\int_0^4 \int_0^{x^2} e^{y/x} dy\, dx = \int_0^4 \left[xe^{y/x}\right]_{y=0}^{x^2} dx$

$$= \int_0^4 (xe^{x^2/x} - xe^0)dx = \int_0^4 xe^x\, dx - \left[\frac{x^2}{2}\right]_0^4$$

$$= \left[xe^x\right]_0^4 - \int_0^4 1\, e^x dx - 8 = 4e^4 - (e^4 - 1) - 8 = 3e^4 - 7$$

Ans.

9. *Change the order of integration in the integral* $\int_0^1 \int_x^{\sqrt{2-x^2}} \frac{x\,dy\,dx}{\sqrt{x^2+y^2}}$.

SOLUTION: $I = \int_0^1 \int_x^{\sqrt{2-x^2}} \frac{x\,dy\,dx}{\sqrt{x^2+y^2}}$. The region of integration is the shaded area in the figure. Area can be divided into two parts by the line $y = 1$. On changing the order of integration.

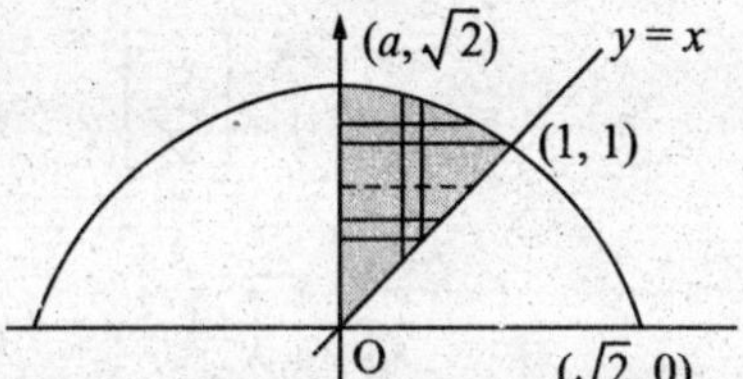

$$I = \int_0^1 \int_0^y \frac{x\,dx\,dy}{\sqrt{x^2+y^2}} + \int_1^{\sqrt{2}} \int_0^{\sqrt{2-y^2}} \frac{x\,dx\,dy}{\sqrt{x^2+y^2}}$$

$$= \int_0^1 \left[\sqrt{x^2+y^2}\right]_0^y dy + \int_1^{\sqrt{2}} \left[\sqrt{x^2+y^2}\right]_0^{\sqrt{2-y^2}} dy$$

$$= \int_0^1 (y\sqrt{2} - y)\,dy + \int_1^{\sqrt{2}} (\sqrt{2} - y)\,dy = \left[\frac{y^2}{2}(\sqrt{2}-1)\right]_0^1 + \left[\sqrt{2}\,y - \frac{y^2}{2}\right]_1^{\sqrt{2}}$$

$$= \frac{\sqrt{2}-1}{2} + 2 - 1 - \sqrt{2} + \frac{1}{2} = 1 - \frac{1}{\sqrt{2}} \quad \therefore \quad I = 1 - \frac{1}{\sqrt{2}}$$ **Ans.**

10. *Evaluate* $\int_0^\infty \int_x^\infty \frac{e^{-y}}{y}\,dy\,dx$.

SOLUTION: Here we have to change the order of integration $I = \int_0^\infty \int_x^\infty \frac{e^{-y}}{y}\,dy\,dx$.

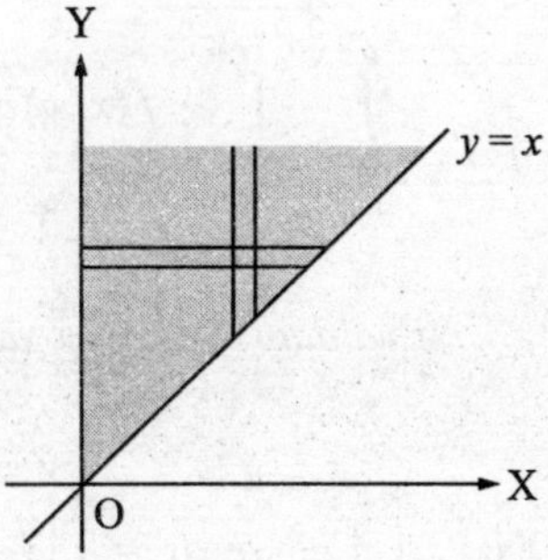

$$I = \int_0^\infty \int_0^y \frac{e^{-y}}{y}\,dy dx = \int_0^\infty \frac{e^{-y}}{y}[x]_0^y\,dx$$

$$= \int_0^\infty \frac{e^{-y}}{y} \cdot y\,dy = -\left[e^{-y}\right]_0^\infty = 1$$

Thus $I = 1$. **Ans.**

11. *Evaluate* $\int_0^1 dx \int_1^\infty e^{-y}\, y^x \log y\,dy$.

SOLUTION: $I = \int_1^\infty dx \int_1^\infty e^{-y} y^x \log y\,dy$. The area of integration is as shown in the figure. To evaluate I, we change the order of integration.

$$I = \int_1^\infty \int_{x=0}^1 e^{-y}\, y^x \log y\, dx\, dy = \int_1^\infty e^{-y} \log y \cdot \left[\frac{y^x}{\log y}\right]_{x=0}^1 dy$$

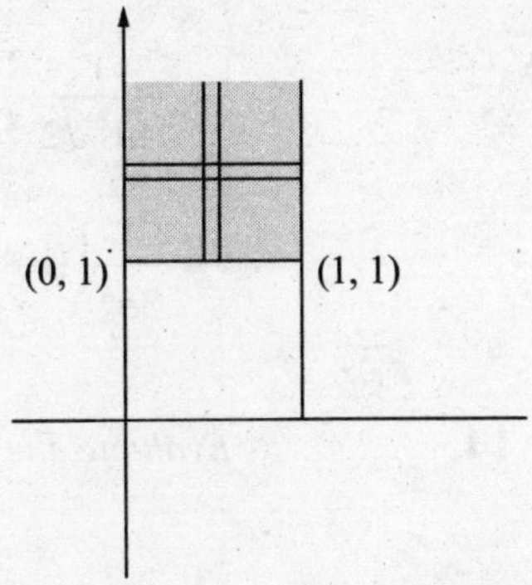

$$= \int_1^\infty e^{-y} \log y \cdot \frac{y-1}{\log y}\, dy = \int_1^\infty e^{-y}(y-1)\, dy$$

$$= -\left[(y-1)e^{-y}\right]_1^\infty + \int_1^\infty e^{-y}\, dy = 0 - \left[e^{-y}\right]_1^\infty$$

$$= \frac{1}{e} \qquad \therefore \qquad I = \frac{1}{e}$$ **Ans.**

12. *Evaluate, by changing the order of integration, the double integral* $\displaystyle\int_0^1 \int_{e^x}^e \frac{dx\, dy}{y^2 \log y}$.

SOLUTION: $\displaystyle I = \int_0^1 \int_{e^x}^e \frac{dx\, dy}{y^2 \log y}$

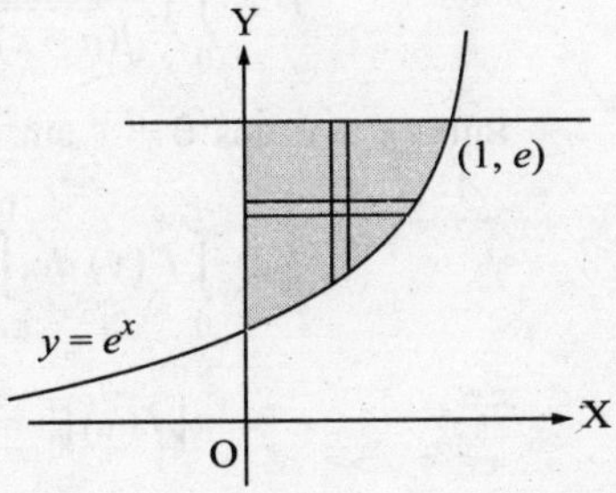

The area of integration is shown in the figure as shaded area. On changing the order of integration.

$$I = \int_1^e \int_0^{\log y} \frac{dx\, dy}{y^2 \log y} = \int_1^e \frac{1}{y^2 \log y} [x]_0^{\log y}\, dy$$

$$= \int_1^e \frac{1}{y^2}\, dy = -\left[\frac{1}{y}\right]_1^e = 1 - \frac{1}{e} \qquad \therefore \quad I = \frac{e-1}{e}.$$ **Ans.**

13. *Sketch the region of integration and evaluate* $\displaystyle\int_0^{\frac{\pi}{2}} \int_0^y \cos 2y \sqrt{1 - a^2 \sin^2 x}\; dx\, dy$.

SOLUTION: $\displaystyle I = \int_0^{\pi/2} \int_0^y \cos 2y \sqrt{1 - a^2 \sin^2 x}\, dx\, dy$

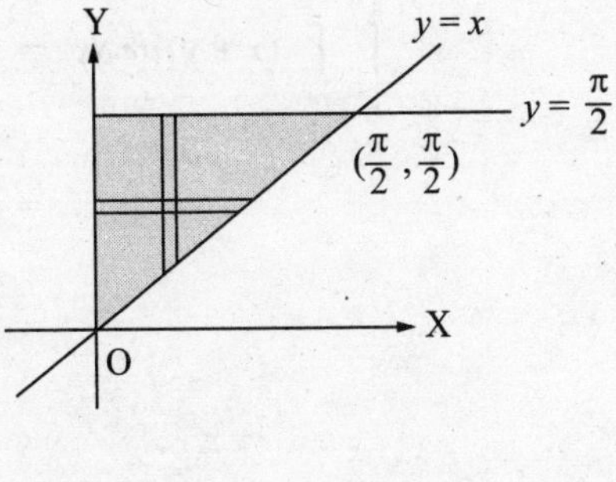

Are of integration is sketched here. Changing the order of integration

$$I = \int_0^{\pi/2} \int_x^{\pi/2} \cos 2y \sqrt{1 - a^2 \sin^2 x}\, dy\, dx$$

$$= \int_0^{\pi/2} \frac{1}{2} \sqrt{1 - a^2 \sin^2 x}\, [\sin 2y]_{y=x}^{\pi/2}\, dx$$

$$= \frac{1}{2\sqrt{2}} \int_0^{\pi/2} \sqrt{2 - a^2(1 - \cos 2x)}\, (0 - \sin 2x)\, dx = \frac{1}{2\sqrt{2}} \int_0^{\pi/2} \sqrt{2 - a^2 + a^2 \cos 2x}\, \sin 2x\, dx$$

$$= \frac{1}{4a^2\sqrt{2}}\frac{2}{3}\left[(2-a^2+a^2\cos 2x)^{3/2}\right]_0^{\pi/2} = \frac{1}{6a^2\sqrt{2}}\left[(2-2a^2)^{3/2}-2\sqrt{2}\right]$$

$$= \frac{1}{3a^2}\left[(1-a^2)^{3/2}-1\right]$$ **Ans.**

14. *Evaluate the double integral* $\int_0^a \int_0^x \frac{f'(y)\,dy\,dx}{\sqrt{(a-x)(x-y)}}$.

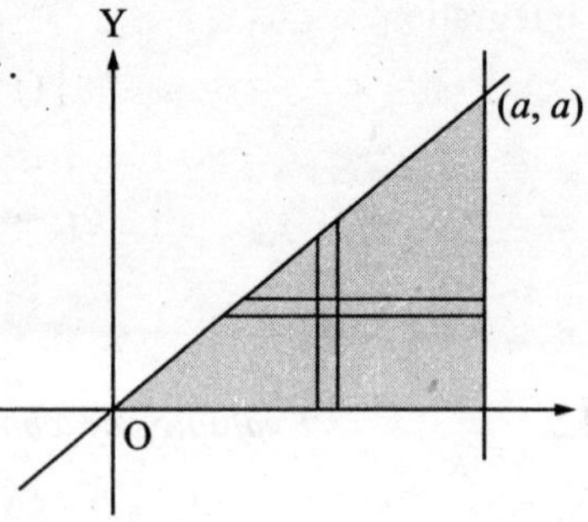

SOLUTION: $I = \int_0^a \int_0^x \frac{f'(y)\,dy\,dx}{\sqrt{(a-x)(x-y)}}$

Let us change the order of integration, then

$$I = \int_0^a \int_y^a \frac{f'(y)\,dx\,dy}{\sqrt{(a-x)(x-y)}}$$

Put $x = a\cos^2\theta + y\sin^2\theta$ so $dx = (y-a)^2 \sin\theta\cos\theta\, d\theta$

$$\therefore \quad I = \int_0^a f'(y)\,dy \int_{\pi/2}^0 \frac{-(a-y)^2\sin\theta\cos\theta\,d\theta}{(a-y)\sin\theta\cos\theta} = \frac{2\pi}{2}\int_0^a f'(y)\,dy$$

$$= \pi[f(y)]_0^a = \pi[f(a)-f(0)], \quad \therefore \quad I = \pi[f(a)-f(0)]$$ **Ans.**

15. *Evaluate the integral* $\int_0^3 \int_1^{\sqrt{4-y}} (x+y)\,dxdy$

by changing the order of integration.

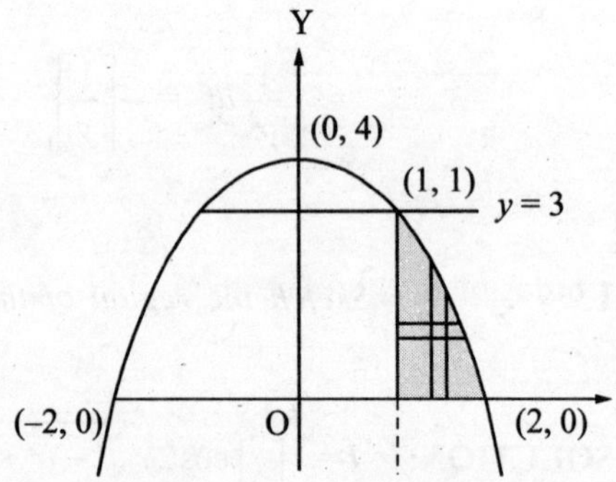

SOLUTION: Area of integration is shown as shaded into figure.

Changing the order of integration the given integral becomes

$$\int_1^2 \int_0^{4-x^2} (x+y)\,dy\,dx = \int_1^2 \left[xy+\frac{y^2}{2}\right]_0^{4-x^2} dx$$

$$= \int_1^2 \left[x(4-x^2)+\frac{1}{2}(4-x^2)^2\right]dx = \int_1^2 \left(4x-x^3+8-4x^2+\frac{1}{2}x^4\right)dx$$

$$= \left[2x^2-\frac{x^4}{4}+8x-\frac{4x^3}{3}+\frac{x^5}{10}\right]_1^2$$

$$= 2(4-1)-\frac{1}{4}(16-1)+8(2-1)-\frac{4}{3}(8-1)+\frac{1}{10}(32-1)$$

$$= \frac{241}{60} \quad \therefore \quad I = \frac{241}{60}$$ **Ans.**

16. *Evaluate the integral* $\int_0^\infty \int_0^x x e^{-x^2/y}\, dy dx$ *by changing the order of integration.*

SOLUTION: The area of integration is the shaded region in the figure. $I = \int_0^\infty \int_0^x x\, e^{-x^2/y}\, dy\, dx$. Changing the order of integration, we get

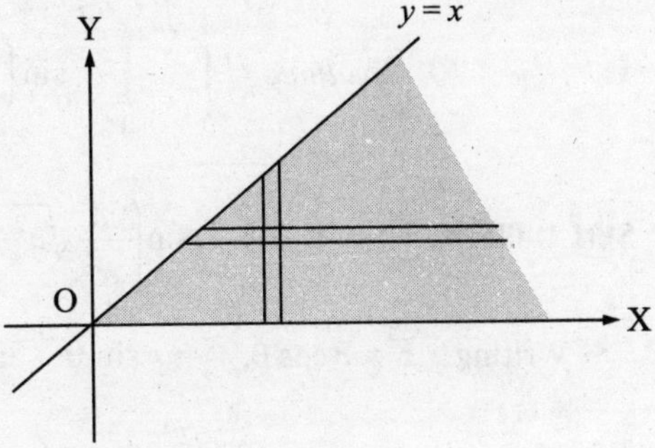

$$I = \int_0^\infty \int_y^\infty x\, e^{-x^2/y}\, dx\, dy$$

$$= \int_0^\infty -\left[\frac{y}{2} e^{-x^2/y}\right]_{x=y}^\infty dy$$

$$= -\frac{1}{2}\int_0^\infty y(0 - e^{-y})\, dy = \frac{1}{2}\int_0^\infty y e^{-y}\, dy$$

$$= \frac{1}{2}\left[-y\, e^{-y}\right]_0^\infty - \frac{1}{2}\int_0^\infty -e^{-y}\, dy = 0 - \frac{1}{2}\left[e^{-y}\right]_0^\infty = \frac{1}{2} \quad \therefore \quad I = \frac{1}{2}$$ **Ans.**

17. *By changing the order of integration in the integral* $\int_0^\infty \int_0^\infty e^{-xy} \sin px\, dx dy$ *show that* $\int_0^\infty \frac{\sin px}{x} dx = \frac{\pi}{2}$.

SOLUTION: $I = \int_0^\infty \int_0^\infty e^{-xy} \sin px\, dx\, dy$. The area of integration is the positive quadrant. Changing the order of integration, gives

$$I = \int_0^\infty \int_0^\infty e^{-xy} \sin px\, dy\, dx = \int_0^\infty \left[\frac{e^{-xy}}{-x}\right]_{y=0}^\infty \sin px\, dx = \int_0^\infty \frac{\sin px}{x} dx$$

$$\text{Originally } I = \int_0^\infty \left[\frac{e^{-xy}}{y^2 + p^2}(-y \sin px - p \cos px)\right]_{x=0}^\infty dy$$

$$= \int_0^\infty \left[0 - \frac{1}{y^2 + p^2}(0 - p)\right] dy = \int_0^\infty \frac{p}{y^2 + p^2} dy = \left[\tan^{-1}\frac{y}{p}\right]_0^\infty = \frac{\pi}{2}$$

Thus, $\int_0^\infty \frac{\sin px}{x} dx = \frac{\pi}{2}$ **Hence proved.**

SOLUTIONS TO EXERCISE 13B

1. *Evaluate* $\int_0^a \int_0^{\sqrt{a^2-x^2}} \sin\left(\frac{\pi}{a^2}\sqrt{a^2-x^2-y^2}\right) dx\, dy$.

SOLUTION: $I = \int_0^a \int_0^{\sqrt{a^2-x^2}} \sin\left(\frac{\pi}{a^2}\sqrt{a^2-x^2-y^2}\right) dx\, dy$. Area of integration is shaded in the figure.

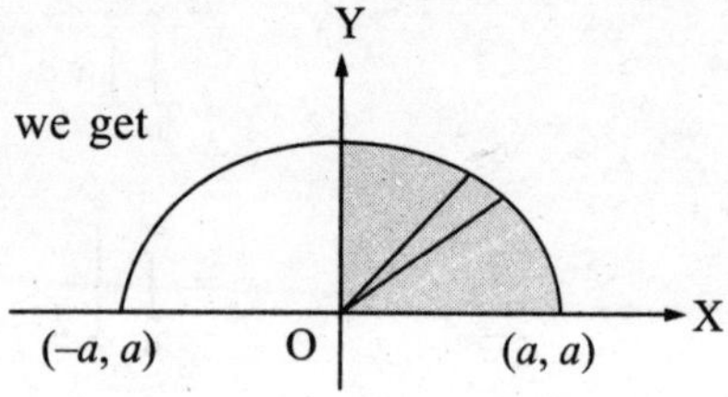

Writing $x = r\cos\theta,\ y = r\sin\theta$ and so $dx\,dy = r\,dr\,d\theta$, we get

$$I = \int_{\theta=0}^{\pi/2} \int_{r=0}^{a} \sin\left(\frac{\pi}{a^2}\sqrt{a^2-r^2}\right) r\,dr\,d\theta$$

Substituting $a^2 - r^2 = t^2$ so $-r\,dr = t\,dt$

we have $$I = \int_{\theta=0}^{\pi/2} \int_{a}^{0} \sin\left(\frac{\pi t}{a^2}\right)\cdot(-t\,dt)\,d\theta = \int_0^{\pi/2}\int_0^a t\sin\left(\frac{\pi t}{a^2}\right) dt\cdot d\theta$$

$$= \int_0^{\pi/2}\left[-\left\{t\cos\frac{\pi t}{a^2}\cdot\frac{a^2}{\pi}\right\}_0^a + \int_0^{\pi/2}\cos\left(\frac{\pi t}{a^2}\right)\cdot\frac{a^2}{\pi}dt\right]d\theta$$

$$= \int_0^{\pi/2}\left[-\frac{a^3}{\pi}\cos\left(\frac{\pi}{a}\right) + \frac{a^4}{\pi^2}\left\{\sin\frac{\pi t}{a^2}\right\}_0^a\right]d\theta$$

$$= \int_0^{\pi/2}\left[-\frac{a^3}{\pi}\cos\left(\frac{\pi}{a}\right) + \frac{a^4}{\pi^2}\sin\left(\frac{\pi}{a}\right)\right]d\theta = \frac{\pi}{2}\left[\frac{a^4}{\pi^2}\sin\frac{\pi}{a} - \frac{a^3}{\pi}\cos\frac{\pi}{a}\right]$$

$$= \frac{a^3}{2}\left(\frac{a}{\pi}\sin\frac{\pi}{a} - \cos\frac{\pi}{a}\right)$$ **Ans.**

2. *Evaluate* $\int_{-\infty}^{\infty}\int_{-\infty}^{\infty} \frac{dx\,dy}{(1+x^2+y^2)^{\frac{3}{2}}}$.

SOLUTION: $I = \int_{-\infty}^{\infty}\int_{-\infty}^{\infty} \frac{dx\,dy}{(1+x^2+y^2)^{3/2}}$. Area of integration is whole of *XOY*-plane. Inducting polar coordinates $x = r\cos\theta,\ y = r\sin\theta$ and $dx\,dy = r\,dr\,d\theta$, we get

$$I = \int_0^{2\pi}\int_0^{\infty} \frac{r\,dr\,d\theta}{(1+r^2)^{3/2}} = \int_0^{2\pi} -\left[\frac{1}{\sqrt{1+r^2}}\right]_0^{\infty} d\theta = 2\pi$$

$I = 2\pi$. **Ans.**

3. *Find the value of* $\iint \frac{x^2\, y^2\, dx\, dy}{x^2 + y^2}$ *over the annular region between*
$x^2 + y^2 = a^2, \quad x^2 + y^2 = b^2\ (b > a > 0).$

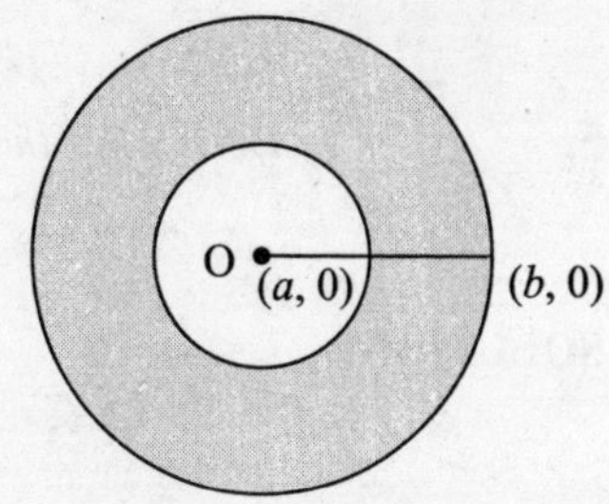

SOLUTION: $I = \iint \frac{x^2 y^2\, dx\, dy}{x^2 + y^2}$ over annular region between $x^2 + y^2 = a^2,\ x^2 + y^2 = b^2$. Converting to polar coordinates, we get

$$I = \int_0^{2\pi}\int_a^b \frac{r^4.\cos^2\theta,\ \sin^2\theta\, .\, r\, dr\, d\theta}{r^2} = \int_0^{2\pi}\left[\frac{r^4}{4}\right]_a^b \cos^2\theta \sin^2\theta\, d\theta$$

$$= \frac{1}{4}\left[b^4 - a^4\right]\int_0^{2\pi} \cos^2\theta \sin^2\theta\, d\theta = \frac{(b^4 - a^4)}{4} 4\int_0^{\pi/2} \cos^2\theta \sin^2\theta\, d\theta$$

$$= \left(b^4 - a^4\right)\frac{\overline{|3h}\,\overline{|3h}}{2\overline{|3}} = \frac{\pi(b^4 - a^4)}{16} \qquad \therefore \quad I = \frac{\pi(b^4 - a^4)}{16}$$ **Ans.**

4. *Evaluate* $\int_0^{4a}\int_{\frac{y^2}{4a}}^{y} \frac{x^2 - y^2}{x^2 + y^2}\, dy\, dx.$

SOLUTION: $I = \int_0^{4a}\int_{y^2/4a}^{y} \frac{x^2 - y^2}{x^2 + y^2} dx\, dy = \int_0^{4a}\int_{y^2/4a}^{y} \left(1 - \frac{2y^2}{x^2 + y^2}\right) dx\, dy$

$$= \int_0^{4a}\left[x - \frac{2y^2}{y}\tan^{-1}\frac{x}{y}\right]_{x=\frac{y^2}{4a}}^{y} dy = \int_0^{4a}\left[y - \frac{y^2}{4a} - 2y\tan^{-1}(1) + 2y\tan^{-1}\frac{y}{4a}\right] dy$$

$$= \left[\frac{y^2}{2} - \frac{y^3}{12a} - y^2\cdot\frac{\pi}{4}\right]_0^{4a} + 2\int_0^{4a} y\tan^{-1}\left(\frac{y}{4a}\right) dy$$

$$= 8a^2 - \frac{16a^2}{3} - 4\pi a^2 + \left[2\tan^{-1}\left(\frac{y}{4a}\right)\cdot\frac{y^2}{2}\right]_0^{4a} - \int 2\cdot\frac{y^2}{2}\left[\frac{1}{1 + \frac{y^2}{16a^2}}\cdot\frac{1}{4a}\right] dy$$

$$= 4a^2\left(2 - \frac{4}{3} - \pi\right) + 16a^2\cdot\frac{\pi}{4} - 4a\int_0^{4a}\frac{y^2}{y^2 + 16a^2} dy$$

$$= \frac{8a^2}{3} - 4a\int_0^{4a}\left(1 - \frac{16a^2}{y^2 + 16a^2}\right) dy = \frac{8a^2}{3} - 4a\left[y - \frac{16a^2}{4a}\tan^{-1}\frac{y}{4a}\right]_0^{4a}$$

$$= \frac{8a^2}{3} - 4a\left[4a - 4a\cdot\frac{\pi}{4}\right] = 8a^2\left(\frac{\pi}{2} - \frac{5}{3}\right).$$ **Ans.**

5. *Determine the value of* $\int_0^a dx \int_{\sqrt{a^2-x^2}}^{\infty} \frac{x^{\frac{1}{3}}\,dy}{(x^2+y^2)^{\frac{3}{2}}}$.

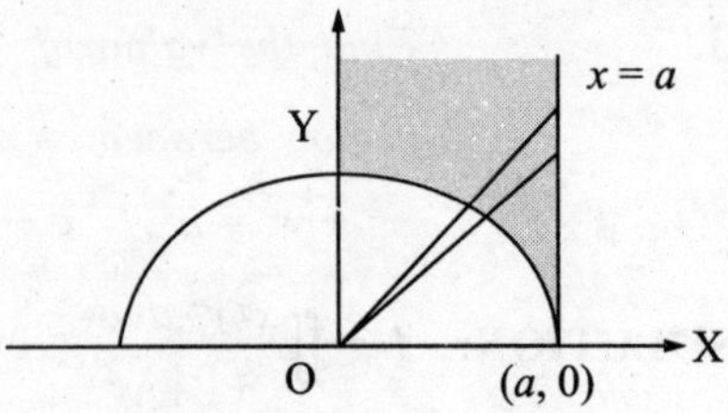

SOLUTION: $I = \int_0^a \int_{\sqrt{a^2-x^2}}^{\infty} \frac{x^{1/3}\,dx\,dy}{(x^2+y^2)^{3/2}}$

Converting to polar coordinates $x = r\cos\theta,\ y = r\sin\theta$ and $dx\,dy = r\,dr\,d\theta$, we get

$$I = \int_0^{\pi/2}\int_a^{a\sec\theta} \frac{(r\cos\theta)^{1/3}\,r\,dr\,d\theta}{r^3} = \int_0^{\pi/2}(\cos\theta)^{1/3}\int_a^{a\sec\theta} r^{-5/3}\,dr\,d\theta$$

$$= \int_0^{\pi/2}(\cos\theta)^{1/3}\left[r^{-2/3}\left(-\frac{3}{2}\right)\right]_a^{a\sec\theta} d\theta = -\frac{3}{2}\int_0^{\pi}(\cos\theta)^{1/3}a^{-2/3}(\sec^{-2/3}\theta - 1)\,d\theta$$

$$= -\frac{3}{2}a^{-2/3}\int_0^{\pi/2}(1-\cos^{1/3}\theta)\,d\theta = \frac{3}{2}a^{-2/3}\left[\frac{\overline{|2/3}\,\overline{|1/2}}{2\overline{|7/6}} - 1\right]$$

$$= \frac{3}{2}a^{-2/3}\left[\frac{\overline{|2/3}\cdot\sqrt{\pi}}{2\cdot\frac{1}{6}\overline{|\frac{1}{6}}} - 1\right] = \frac{3}{2}a^{-2/3}\left[\frac{3\overline{|2/3}\sqrt{\pi}}{\overline{|\frac{1}{6}}} - 1\right]$$ **Ans.**

6. *Evaluate* $\int_0^2 \int_{1-\sqrt{2x-x^2}}^{1+\sqrt{2x-x^2}} \frac{dy\,dx}{(x^2+y^2)^2}$.

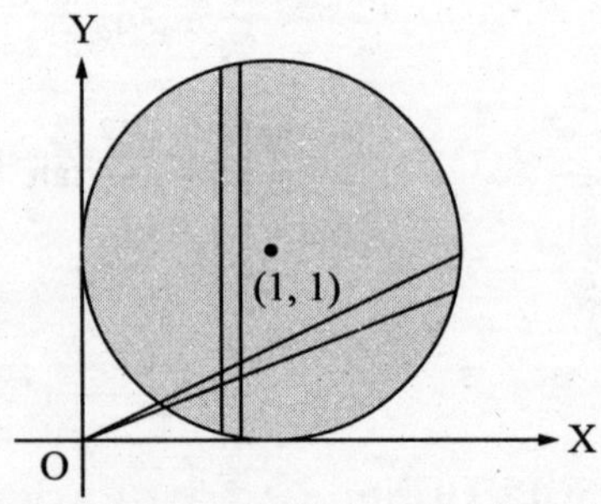

SOLUTION: $I = \int_0^2 \int_{1-\sqrt{2x-x^2}}^{1+\sqrt{2x-x^2}} \frac{dy\,dx}{(x^2+y^2)^2}$

Curve is $y = 1 \pm \sqrt{2x-x^2}$ or $(y-1)^2 = 2x - x^2$ or $(x-1)^2 + (y-1)^2 = 1$

Area of integration is the shaded area in the figure. Circle in polar coordinates is

$$(r\cos\theta - 1)^2 + (r\sin\theta - 1)^2 = 1 \quad \text{or} \quad r^2 - 2r(\cos\theta + \sin\theta) + 1 = 0$$

$$r = \frac{2(\cos\theta+\sin\theta) + \sqrt{4(\cos\theta+\sin\theta)^2 - 4}}{2} = (\cos\theta+\sin\theta) \pm \sqrt{2\sin\theta\cos\theta}$$

$$\therefore \quad I = \int_0^{\pi/2}\int_{\cos\theta+\sin\theta-\sqrt{2\sin\theta\cos\theta}}^{\cos\theta+\sin\theta+\sqrt{2\sin\theta\cos\theta}} \frac{r\,dr\,d\theta}{r^4} = \int_0^{\pi/2}\left[-\frac{1}{2r^2}\right]_{\cos\theta+\sin\theta-\sqrt{2\sin\theta\cos\theta}}^{\cos\theta+\sin\theta+\sqrt{2\sin\theta\cos\theta}} d\theta$$

$$= \frac{1}{2}\int_0^{\pi/2}\left[\frac{1}{\left(\cos\theta+\sin\theta-\sqrt{2\sin\theta\cos\theta}\right)^2} - \frac{1}{\left(\cos\theta+\sin\theta+\sqrt{2\sin\theta\cos\theta}\right)^2}\right] d\theta$$

$$= \frac{1}{2}\int_0^{\pi/2} \frac{4(\sin\theta + \cos\theta)\sqrt{2\sin\theta\cos\theta}}{1}d\theta = 2\sqrt{2}\int_0^{\pi/2}\left(\sin^{3/2}\theta\cos^{1/2}\theta + \cos^{3/2}\theta\sin^{1/2}\theta\right)d\theta$$

$$= 2\sqrt{2}\left(\frac{\overline{|5/4}\,\overline{|3/4}}{2\overline{|2}} + \frac{\overline{|5/4}\,\overline{|3/4}}{2\overline{|2}}\right) = \frac{1}{\sqrt{2}}\overline{\left|\frac{1}{4}\right.}\,\overline{\left|1-\frac{1}{4}\right.} = \frac{1}{\sqrt{2}}\cdot\pi\sqrt{2} = \pi$$

(since $\overline{|n}\,\overline{|1-n} = \dfrac{\pi}{\sin\cdot n\pi}$, taken $n = \dfrac{1}{4}$ $= \pi$ **Ans.**)

7. *Evaluate* $\iint_R xy\,dx\,dy$ *where* R *is the positive quadrant of the circle* $x^2 + y^2 = a^2$.

SOLUTION: $I = \iint_R xy\,dx\,dy$. Converting to polar coordinates

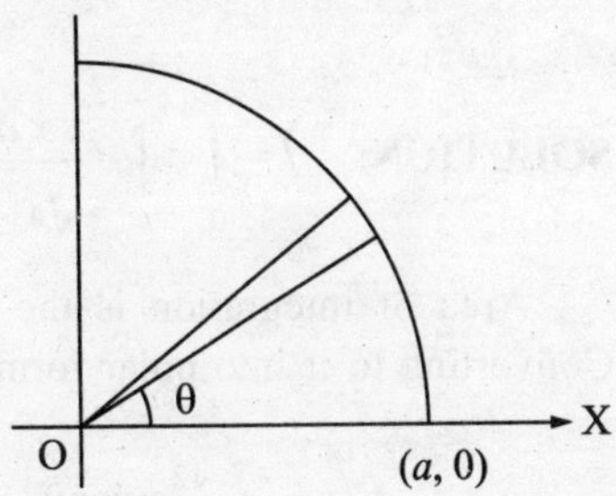

$$I = \int_0^{\pi/2}\int_0^a r\cos\theta.r\sin\theta.r\,dr\,d\theta$$

$$= \int_0^{\pi/2}\left[\frac{r^4}{4}\right]_0^a \cos\theta\sin\theta\,d\theta$$

$$= \frac{a^4}{8}\int_0^{\pi/2}\sin 2\theta\,d\theta = -\frac{a^4}{8}\left[\frac{1}{2}\cos 2\theta\right]_0^{\pi/2} = \frac{a^4}{8}$$ **Ans.**

8. *Determine the value of* $\int_0^{2a}\int_0^{\sqrt{2ax-x^2}}(x^2+y^2)\,dx\,dy$.

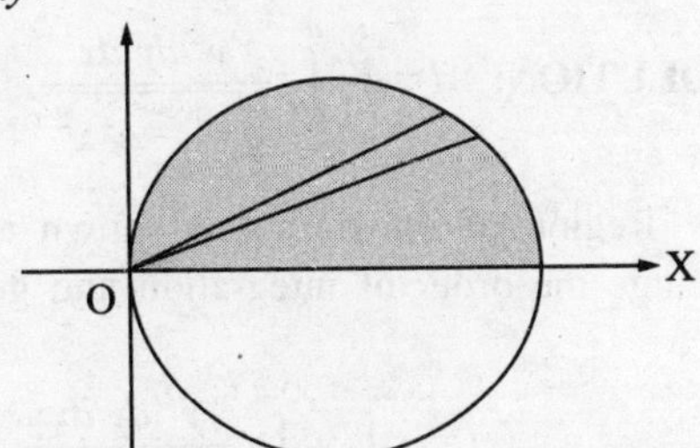

SOLUTION: $I = \int_0^{2a}\int_0^{\sqrt{2ax-x^2}}(x^2+y^2)\,dx\,dy$

$y = \sqrt{2ax - x^2}$ or $x^2 + y^2 - 2ax = 0$. or $r = 2a\cos\theta$

Area of integration is the shaded portion in the figure. Converting to polar from

$$I = \int_0^{\pi/2}\int_0^{2a\cos\theta} r^2.r\,dr\,d\theta = \int_0^{\pi/2}\left[\frac{r^4}{4}\right]_0^{2a\cos\theta} d\theta = 4a^4\int_0^{\pi/2}\cos^4\theta\,d\theta$$

$$= 4a^4\frac{\overline{|5/2}\,\overline{|1/2}}{2\overline{|3}} = \frac{4a^4\cdot\frac{3}{2}\cdot\frac{1}{2}\left(\overline{\left|\frac{1}{2}\right.}\right)^2}{2\cdot 2} = \frac{3\pi a^4}{4}$$ **Ans.**

9. *Evaluate* $\int_0^\infty \int_0^\infty e^{-(x^2+y^2)}\,dx\,dy$ *by transforming it into polar co-ordinates.*

SOLUTION: The area of integration is the first quadrant. Converting to polar form

$$I = \int_0^\infty \int_0^\infty e^{-(x^2+y^2)}\,dx\,dy = \int_0^{\pi/2} \int_0^\infty e^{-r^2}\,r\,dr\,d\theta = -\frac{1}{2}\int_0^{\pi/2}\left[e^{-r^2}\right]_0^\infty d\theta$$

$$= -\frac{\pi}{4}\left[e^{-r^2}\right]_0^\infty = \frac{\pi}{4}$$ **Ans.**

10. *Evaluate* $\int_0^1 \int_x^{\sqrt{2-x^2}} \frac{x\,dy\,dx}{\sqrt{x^2+y^2}}$.

SOLUTION: $I = \int_0^1 \int_x^{\sqrt{2-x^2}} \frac{x\,dy\,dx}{\sqrt{x^2+y^2}}$

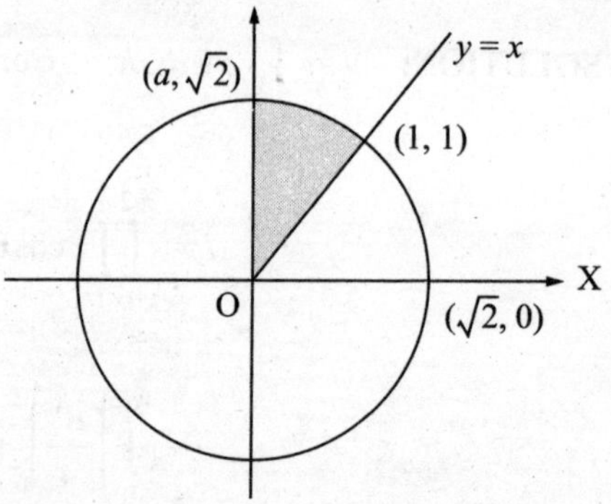

Area of integration is the shaded portion in the figure. Converting to it into polar form

$$I = \int_{\pi/4}^{\pi/2} \int_0^{\sqrt{2}} \frac{r\cos\theta \cdot r\,dr\,d\theta}{r} = \int_{\pi/4}^{\pi/2} \cos\theta\left[\frac{r^2}{2}\right]_0^{\sqrt{2}} d\theta$$

$$= \int_{\pi/4}^{\pi/2} \cos\theta\,d\theta = [\sin\theta]_{\pi/4}^{\pi/2} = 1 - \frac{1}{\sqrt{2}} \quad \therefore \quad I = 1 - \frac{1}{\sqrt{2}}$$ **Ans.**

11. *Change the order of integration and evaluate* $\int_0^a \int_{\sqrt{ax}}^a \frac{y^2\,dx\,dy}{\sqrt{y^4 - a^2x^2}}$

SOLUTION: $I = \int_0^a \int_{\sqrt{ax}}^a \frac{y^2\,dy\,dx}{\sqrt{y^4 - a^2x^2}}$.

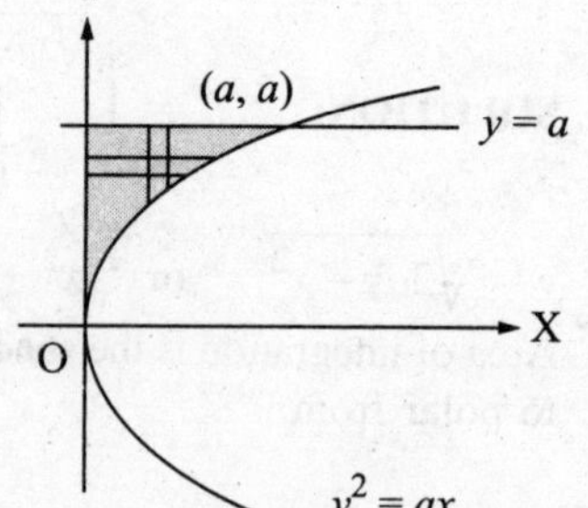

Region of integration is shown as shaded area in the figure. We change the order of integration and get

$$I = \int_{y=0}^{a} \int_{x=0}^{y^2/a} \frac{y^2\,dy\,dx}{\sqrt{y^4 - a^2x^2}} = \int_0^a \frac{y^2}{a} \int_0^{y^2/a} \frac{dx\,dy}{\sqrt{\left(\frac{y^2}{a}\right)^2 - x^2}}$$

$$= \int_0^a \frac{y^2}{a}\left[\sin^{-1}\frac{xa}{y^2}\right]_0^{y^2/a} dy = \int_0^a \frac{y^2}{a}(\sin^{-1}1 - 0)\,dy = \frac{\pi}{2a}\left[\frac{y^3}{3}\right]_0^a$$

$$= \frac{\pi a^2}{6} \quad \therefore \quad I = \frac{\pi a^2}{6}$$ **Ans.**

12. *Evaluate the integral* $\int_0^{\infty}\int_0^{x} x \exp(-x^2/y)dy\, dx$

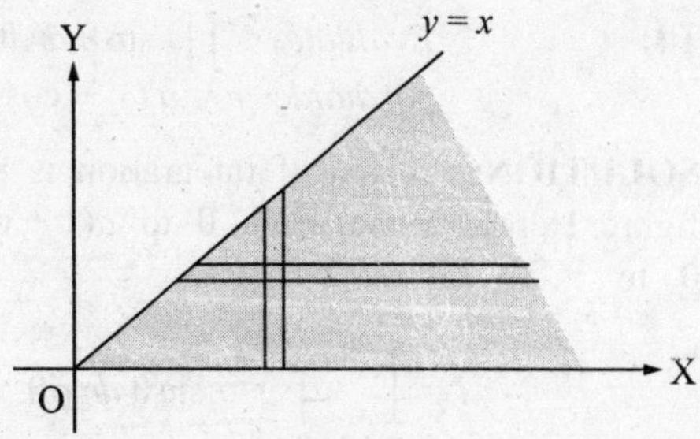

SOLUTION: $I = \int_0^{\infty}\int_0^{x} x\, e^{-x^2/y}\, dy\, dx$

Area of integration is the shaded position in the figure. We change the order of integration to get

$$I = \int_0^{\infty}\int_y^{\infty} x\, e^{-x^2/y}\, dx\, dy = \int_0^{\infty}\left[-\frac{y}{2}e^{-x^2/y}\right]_{x=y}^{\infty} dy$$

$$= \int_0^{\infty}\left(0 - \frac{y}{2}e^{-y}\right)dy = \left[\frac{y}{2}e^{-y}\right]_0^{\infty} + \int_0^{\infty}\frac{1}{2}e^{-y}\, dy = 0 - \frac{1}{2}\left[e^{-y}\right]_0^{\infty} = \frac{1}{2}$$

$\therefore \quad I = \frac{1}{2}$ **Ans.**

13. *Evaluate* $\int_0^{2}\int_0^{\sqrt{2x-x^2}} \frac{x\, dydx}{\sqrt{x^2+y^2}}$.

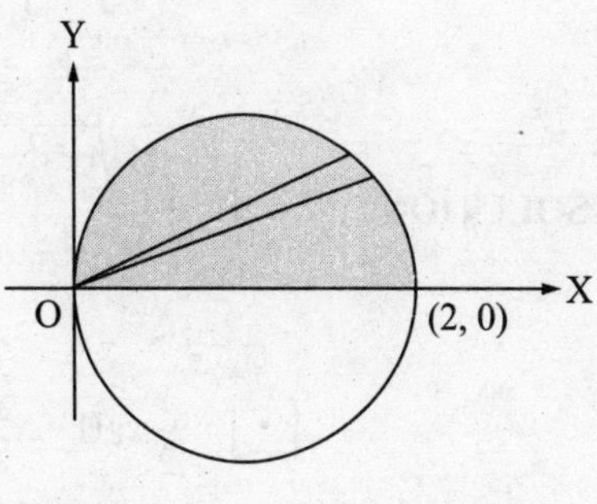

SOLUTION: $I = \int_0^{2}\int_0^{\sqrt{2x-x^2}} \frac{x\, dy\, dx}{\sqrt{x^2+y^2}}$

Area of integration is the shaded position in the figure. Circle is $y = \sqrt{2x - x^2}$ or $r = 2\cos\theta$. Converting to polar coordinates

$$I = \int_0^{\pi/2}\int_0^{2\cos\theta} \frac{r\cos\theta \cdot r\, dr\, d\theta}{r} = \int_0^{\pi/2} \cos\theta\left[\frac{r^2}{2}\right]_0^{2\cos\theta} d\theta$$

$$= \int_0^{\pi/2} \frac{1}{2}\cos\theta\cdot(2\cos\theta)^2\, d\theta = 2\int_0^{\pi/2}\cos^3\theta\, d\theta = 2\int_0^{\pi/2}(1-\sin^2\theta)\cos\theta\, d\theta$$

$$= 2\left[t - \frac{t^3}{3}\right]_0^1 \quad \text{(on putting } \sin\theta = t\text{)} = \frac{4}{3}$$

$\therefore \quad I = \frac{4}{3}$ **Ans.**

14. *Evaluate* $\iint r \sin\theta \, dr d\theta$ *over the area of the cardioide* $r = a(1 + \cos\theta)$ *above the initial line.*

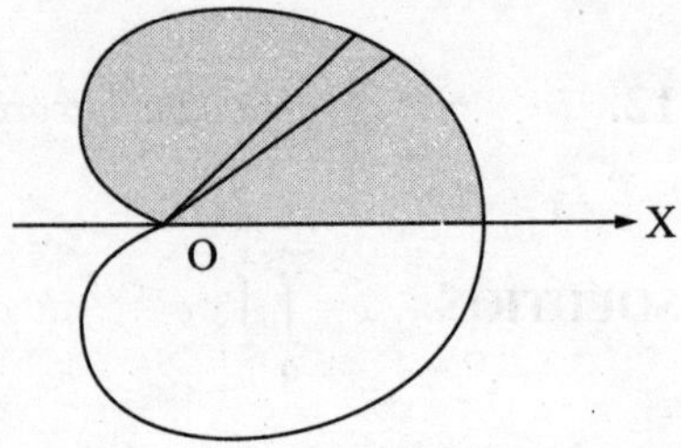

SOLUTION: Area of integration is the shaded portion in the figure. Here r varies from 0 to $a(1 + \cos\theta)$ and θ varies from 0 to π.

$$I = \int_0^{\pi} \int_0^{a(1+\cos\theta)} r \sin\theta \, dr \, d\theta = \int_0^{\pi} \left[\frac{r^2}{2}\right]_0^{a(1+\cos\theta)} \sin\theta \, d\theta$$

$$= \frac{a^2}{2}\int_0^{\pi} (1+\cos\theta)^2 \sin\theta \, d\theta$$ Put $1 + \cos\theta = t$ so $-\sin\theta \, d\theta = dt$

$$= \frac{a^2}{-2}\int_2^0 t^2 \, dt = \frac{a^2}{2}\cdot\left[\frac{t^3}{3}\right]_0^2 = \frac{4}{3}a^2$$ **Ans.**

SOLUTIONS TO EXERCISE 13C

1. *Evaluate* $\int_0^1 \int_0^{\sqrt{1-x^2}} \int_0^{\sqrt{1-x^2-y^2}} xyz \, dz \, dy \, dx$.

SOLUTION: $$I = \int_0^1 \int_0^{\sqrt{1-x^2}} \int_0^{\sqrt{1-x^2-y^2}} xyz \, dz \, dy \, dx = \int_0^1 \int_0^{\sqrt{1-x^2}} xy\left[\frac{z^2}{2}\right]_0^{\sqrt{1-x^2-y^2}} dy \, dx$$

$$= \int_0^1 \int_0^{\sqrt{1-x^2}} \frac{1}{2} xy(1-x^2-y^2) \, dy \, dx = \frac{1}{2}\int_0^1 \int_0^{\sqrt{1-x^2}} (xy - x^3y - xy^3) \, dy \, dx$$

$$= \frac{1}{2}\int_0^1 \left[(x-x^3)\frac{y^2}{2} - \frac{xy^4}{4}\right]_0^{\sqrt{1-x^2}} dx = \frac{1}{2}\int_0^1 \left[\frac{x}{2}(1-x^2)^2 - \frac{x}{4}(1-x^2)^2\right] dx = \int_0^1 \frac{x}{8}(1-x^2)^2 \, dx$$

$$= \frac{1}{8}\int_0^1 (x - 2x^3 + x^5) \, dx = \frac{1}{8}\left[\frac{x^2}{2} - \frac{x^4}{2} + \frac{x^6}{6}\right]_0^1 = \frac{1}{48} \quad \therefore \quad I = \frac{1}{48}$$ **Ans.**

2. *Evaluate* $\int_{-1}^{1} \int_0^{z} \int_{x-z}^{x+z} (x+y+z) \, dy \, dx \, dz$.

SOLUTION: $$I = \int_{-1}^{1} \int_0^{z} \int_{x-z}^{x+z} (x+y+z) \, dy \, dx \, dz = \int_{-1}^{1} \int_0^{z} \left[(x+z)y + \frac{y^2}{2}\right]_{x-z}^{x+z} dx \, dz$$

$$= \int_{-1}^{1}\int_{0}^{z}\left[(x+z)^2-(x^2-z^2)+\frac{1}{2}(x+z)^2-\frac{1}{2}(x-z)^2\right]dx\,dz = \int_{-1}^{1}\int_{0}^{z}(2z^2+4xz)\,dx\,dz$$

$$= \int_{-1}^{1}\left[2z^2x+2x^2z\right]_0^z dz = \int_{-1}^{1}(2z^3+2z^3)dz = 0 \quad \therefore \quad I = 0 \qquad \textbf{Ans.}$$

3. *Evaluate* $\displaystyle\int_1^e\int_1^{\log y}\int_1^{e^x}\log z\,dz\,dx\,dy$.

SOLUTION: $\displaystyle I = \int_1^e\int_1^{\log y}\int_1^{e^x}\log z\,dz\,dx\,dy = \int_1^e\int_1^{\log y}\left[z(\log z-1)\right]_1^{e^x}dx\,dy = \int_1^e\int_1^{\log y}\left[e^x(x-1)+1\right]dx\,dy$

$$\left[\text{since } \int \log z\cdot dz = \log z\cdot z-\int z\cdot\frac{1}{z}dz = z(\log z-1)\right]$$

$$= \int_1^e\left[(x-1)e^x-e^x+x\right]_1^{\log y}dy = \int_1^e\left[(\log y-1)y-y+\log y+e-1\right]dy$$

$$= \int_1^e\left[\log y(y+1)-2y+(e-1)\right]dy = \left[\log y\left(\frac{y^2}{2}+y\right)-\int\frac{1}{y}\left(\frac{y^2}{2}+y\right)dy-y^2+(e-1)y\right]_1^e$$

$$= 1\left(\frac{e^2}{2}+e\right)-\left[\frac{y^2}{4}+y\right]_1^e-e^2+(e-1)e+1-(e-1) = \frac{1}{4}(e^2-8e+13) \qquad \textbf{Ans.}$$

4. *Find volume of the ellipsoid* $\dfrac{x^2}{a^2}+\dfrac{y^2}{b^2}+\dfrac{z^2}{c^2}=1$.

SOLUTION: Let us take $X = x/a$, $Y = y/b$ and $Z = z/c$ then the given ellipsolid becomes $X^2+Y^2+Z^2=1$

$$\therefore \quad \text{Volume } V = \iiint dx\,dy\,dz = \iiint\frac{\partial(x,y,z)}{\partial(X,Y,Z)}dX\,dY\,dZ = abc\iiint dX\,dY\,dZ$$

since
$$\frac{\partial(x,y,z)}{\partial(X,Y,Z)} = \begin{vmatrix}\frac{\partial x}{\partial X} & \frac{\partial x}{\partial Y} & \frac{\partial x}{\partial Z}\\ \frac{\partial y}{\partial X} & \frac{\partial y}{\partial Y} & \frac{\partial y}{\partial Z}\\ \frac{\partial z}{\partial X} & \frac{\partial z}{\partial Y} & \frac{\partial z}{\partial Z}\end{vmatrix} = \begin{vmatrix}a & 0 & 0\\ 0 & b & 0\\ 0 & 0 & c\end{vmatrix} = abc$$

Next converting to spherical polar coordinates as $X = r\sin\theta\cos\phi$, $Y = r\sin\theta\sin\phi$, $Z = r\cos\theta$.

$$\frac{\partial(X,Y,Z)}{\partial(r,\theta,\phi)} = r^2\sin\theta, \quad \text{hence} \quad V = abc\iiint\frac{\partial(X,Y,Z)}{\partial(r,\theta,\phi)}dr\,d\theta\,d\phi$$

or $$V = abc \int_{\phi=0}^{2\pi} \int_{\theta=0}^{\pi} \int_{r=0}^{1} r^2 \sin\theta \, dr \, d\theta \, d\phi = abc \int_{\theta=0}^{2\pi} \int_{0}^{\pi} \left[\frac{r^3}{3}\right]_0^1 \sin\theta \, d\theta \, d\phi$$

$$= \frac{abc}{3} \int_0^{2\pi} [-\cos\theta]_0^{\pi} \, d\phi = \frac{4\pi}{3} abc$$ **Ans.**

5. *A circular hole of radius b is made centrally through a sphere of radius a. Find the volume of the remaining portion of the sphere.*

SOLUTION: Volume of the upper half of the central hole $= \iint_R z \, dx \, dy$ where Z is obtained from $x^2 + y^2 + z^2 = a^2$ and b is the radius of the circle in the XOY-plower, that is, $x^2 + y^2 = b^2$.

Volume of the sphere $= \frac{4}{3}\pi a^3$ and volume V_1 of the hole, is given by

$$V_1 = 2\iint_R \sqrt{a^2 - x^2 - y^2} \, dx \, dy = 2\int_0^{2\pi} \int_0^{b} \sqrt{a^2 - r^2} \cdot r \, d\theta \, dr$$

$$= 2 \cdot 2\pi \int_0^b \sqrt{a^2 - r^2} \, r \, dr$$ [Putting $a^2 - r^2 = t^2$ so $-r\,dr = t\,dt$]

$$= \frac{4\pi}{3}\left[a^3 - (a^2 - b^2)^{3h}\right]$$

$\therefore$ Required volume $= \frac{4}{3}\pi a^3 - \frac{4\pi}{3}\left[a^3 - (a^2 - b^2)^{3/2}\right] = \frac{4\pi}{3}(a^2 - b^2)^{3/2}$ **Ans.**

6. *Evaluate* $\displaystyle\int_0^a \int_y^{a+\sqrt{a^2-y^2}} \frac{dx\,dy}{(4a^2 + x^2 + y^2)^2}$.

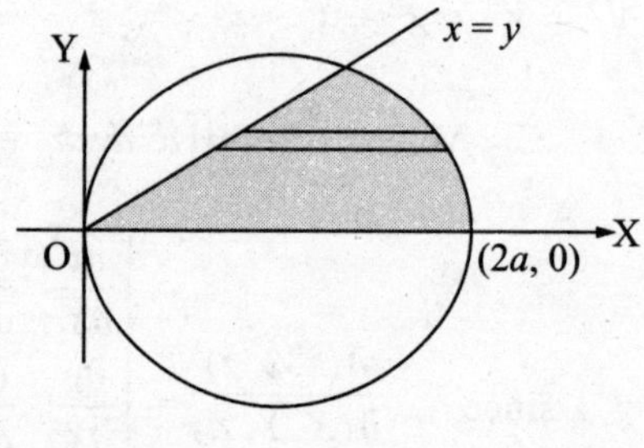

SOLUTION: $I = \displaystyle\int_0^a \int_y^{a+\sqrt{a^2-y^2}} \frac{dx\,dy}{(4a^2 + x^2 + y^2)^2}$

Shaded area is the region of integration. Converting to polar coordinates, we get

$$I = \int_{\theta=0}^{\pi/4} \int_{r=0}^{2a\cos\theta} \frac{r\,dr\,d\theta}{(4a^2 + r^2)^2} = \int_0^{\pi/4} \left[\frac{-1}{2(r^2 + 4a^2)}\right]_0^{2a\cos\theta} d\theta = \int_0^{\pi/4} \left[\frac{-1}{8a^2(1+\cos^2\theta)} + \frac{1}{8a^2}\right] d\theta$$

$$= \frac{1}{8a^2}\left[\frac{\pi}{4} - \int_0^{\pi/4} \frac{\sec^2\theta\,d\theta}{1+\sec^2\theta}\right] = \frac{1}{8a^2}\left[\frac{\pi}{4} - \frac{1}{\sqrt{2}}\tan^{-1}\left(\frac{\tan\theta}{\sqrt{2}}\right)_0^{\pi/4}\right] = \frac{1}{8a^2}\left(\frac{\pi}{4} - \tan^1\frac{1}{\sqrt{2}}\right)$$ **Ans.**

7. *Evaluate* $\int_0^{\pi}\int_0^{a(1+\cos\theta)}\int_0^{h} 2r\left\{1-\frac{r}{a(1+\cos\theta)}\right\} dz\, dr\, d\theta$.

SOLUTION: $I = \int_0^{\pi}\int_0^{a(1+\cos\theta)}\int_0^{h} 2r\left\{1-\frac{r}{a(1+\cos\theta)}\right\} dz\, dr\, d\theta$

$$= h\int_0^{\pi}\int_0^{a(1+\cos\theta)}\left[2r-\frac{2r^2}{a(1+\cos\theta)}\right] dr\, d\theta = h\int_0^{\pi}\left[r^2-\frac{2r^3}{3a(1+\cos\theta)}\right]_0^{a(1+\cos\theta)} d\theta$$

$$= h\int_0^{\pi}\left[a^2(1+\cos\theta)^2-\frac{2a^3(1+\cos\theta)^3}{3a(1+\cos\theta)}\right] d\theta = \frac{ha^2}{3}\int_0^{\pi}(1+\cos\theta)^2\, d\theta$$

$$= \frac{ha^2}{3}\int_0^{\pi}\left(1+2\cos\theta+\frac{1+\cos 2\theta}{2}\right) d\theta = \frac{ha^2}{3}\left[3\frac{\theta}{2}+2\sin\theta+\frac{1}{4}\sin 2\theta\right]_0^{\pi}$$

$$= \frac{\pi h a^2}{2}$$ **Ans.**

8. *Evaluate* $\iiint_R x^{l-1} y^{m-1} z^{n-1}\, dx\, dy\, dz$ *over the region* R *bounded by* $x > 0, \quad y > 0, \quad x + y + z < 1.$

SOLUTION: $I = \int_0^1\int_0^{1-x}\int_0^{1-x-y} x^{l-1}y^{m-1}z^{n-1}\, dx\, dy\, dz = \int_0^1\int_0^{1-x} x^{l-1}y^{m-1}\left[\frac{z^n}{n}\right]_0^{1-x-y} dy\, dx$

$$= \frac{1}{n}\int_0^1\int_0^{1-x} x^{l-1}y^{m-1}(1-x-y)^n\, dy\, dx \qquad [\text{put } y = (1-x)t]$$

$$= \frac{1}{n}\int_0^1 x^{l-1} dx\int_0^1 (1-x)^{m-1}t^{m-1}(1-x)^n(1-t)^n(1-x)\, dt$$

$$= \frac{1}{n}\int_0^1 x^{l-1}(1-x)^{m+n} dx\int_0^1 t^{m-1}(1-t)^n\, dt$$

$$= \frac{1}{n}\beta(l,\ m+n+1)\,\beta(m,\ n+1) = \frac{1}{n}\frac{\Gamma l\ \Gamma(m+n+1)}{\Gamma(l+m+n+1)}\cdot\frac{\Gamma m\ \Gamma(n+1)}{\Gamma(m+n+1)}$$

$$= \frac{\Gamma l\ \Gamma m\ n\Gamma n}{n(l+m+n)\Gamma(l+m+n)} = \frac{\Gamma l\ \Gamma m\ \Gamma n}{(l+m+n)\Gamma(l+m+n)}$$ **Ans.**

9. *Evaluate* $\iiint z^2\,dx\,dy\,dz$ *over the region common to the sphere* $x^2 + y^2 + z^2 = a^2$ *and the cylinder* $x^2 + y^2 = ax$.

SOLUTION: $I = \iiint z^2 dx\,dy\,dz$

Limits for z come from the equation of the sphere $x^2 + y^2 + z^2 = a^2$ and those for x and y from $x^2 + y^2 = ax$.

$$\therefore \quad I = \int_0^a \int_{-\sqrt{ax-x^2}}^{\sqrt{ax-x^2}} \int_{-\sqrt{a^2-x^2-y^2}}^{\sqrt{a^2-x^2-y^2}} z^2\,dz\,dy\,dz = 2\int_0^a \int_{-\sqrt{ax-x^2}}^{\sqrt{ax-x^2}} \int_0^{\sqrt{a^2-x^2-y^2}} z^2\,dz\,dy\,dz$$

$$= 2\int_0^a \int_{-\sqrt{ax-x^2}}^{\sqrt{ax-x^2}} \left[\frac{z^3}{3}\right]_0^{\sqrt{a^2-x^2-y^2}} dy\,dx = \frac{2}{3}\int_0^a \int_{-\sqrt{ax-x^2}}^{\sqrt{ax-x^2}} (a^2 - x^2 - y^2)^{3/2}\,dy\,dx$$

Since $x^2 + y^2 = ax$ is the circle $r = a\cos\theta$. Hence converting to polar coordinates, we get

$$I = \frac{2}{3}\int_{-\pi/2}^{\pi/2} \int_0^{a\cos\theta} (a^2 - r^2)^{3/2} \cdot r\,dr\,d\theta = \frac{2}{3}\int_{-\pi/2}^{\pi/2} -\frac{1}{5}\left[(a^2 - r^2)^{5/2}\right]_0^{a\cos\theta} d\theta$$

$$= -\frac{2}{15}\int_{-\pi/2}^{\pi/2} (a^5 \sin^5\theta - a^5)\,d\theta = \frac{2a^5}{15}\int_{-\pi/2}^{\pi/2} (1 - \sin^5\theta)\,d\theta$$

$$= \frac{2a^5}{15}\left[\{\theta\}_{-\pi/2}^{\pi/2} - 0\right] = \frac{2\pi a^5}{15}$$ **Ans.**

10. *Determine the value of* $\int_0^a \int_0^\alpha \int_0^\beta (x^2 + y^2 + z^2)\,dx\,dy\,dz$ *where a is constant and*

$$\alpha = \sqrt{a^2 - z^2}, \quad \beta = \sqrt{a^2 - y^2 - z^2}.$$

SOLUTION: $I = \iiint_R (x^2 + y^2 + z^2)\,dx\,dy\,dz$ *where* R *is the surface of the sphere.* $x^2 + y^2 + z^2 = a^2$.

Converting to spherical polar coordinates $x = r\sin\theta\cos\phi$, $y = r\sin\theta\sin\phi$, $z = r\cos\theta$ *and*

$$dx\,dy\,dz = \frac{\partial(x, y, z)}{\partial(r, \theta, \phi)}\,dr\,d\theta\,d\phi$$

$$\therefore \quad I = \iiint r^2 \frac{\partial(x, y, z)}{\partial(r, \theta, \phi)}\,dr\,d\theta\,d\phi = \int_0^a \int_{\phi=0}^{\pi/2} \int_{\theta=0}^{\pi/2} r^2 . r^2 \sin\theta\,dr\,d\theta\,d\phi$$

$$= \left[\frac{r^5}{5}\right]_0^a [-\cos\theta]_0^{\pi/2}\,[\phi]_0^{\pi/2} = \frac{\pi a^5}{10}$$ **Ans.**

11. *Evaluate* $\iiint_R x^2\,dx\,dy\,dz$ *where* R *encloses the volume of the ellipsoid*

$$\frac{x^2}{a^2}+\frac{y^2}{b^2}+\frac{z^2}{c^2}=1.$$

SOLUTION: $I=\iiint_R x^2\,dx\,dy\,dz$. Let us take $\frac{x}{a}=X,\ \frac{y}{b}=Y,\ \frac{z}{c}=Z.$

$$=\iiint_R a^2x^2\frac{\partial(x,y,z)}{\partial(X,Y,Z)}dX\,dY\,dZ=\iiint_{R'} a^2X^2.abc\,dX\,dY\,dZ \quad \text{where } R': X^2+Y^2+Z^2=1$$

Converting to spherical polar coordinates $X=r\sin\theta\cos\phi,\ Y=r\sin\theta\sin\phi,\ Z=r\cos\theta$, then

$$I=a^3bc\iiint X^2\frac{\partial(X,Y,Z)}{\partial(r,\theta,\phi)}dr\,d\theta\,d\phi=a^3bc\int_0^1\int_0^{2\pi}\int_0^{\pi/2} r^2\sin^2\theta\cos^2\phi\cdot(r^2\sin\theta)\,dr\,d\theta\,d\phi$$

$$=a^3bc\left[\frac{r^5}{5}\right]_0^1\int_0^{2\pi}\cos^2\phi\,d\phi\int_0^{\pi}\sin^3\theta\,d\theta=\frac{a^3bc}{5}\int_0^{2\pi}\frac{1+\cos 2\phi}{2}d\phi\int_0^{\pi}\frac{3\sin\theta-\sin 3\theta}{4}d\theta$$

$$=\frac{a^3bc}{5}\cdot\left[\frac{\theta}{2}+\frac{1}{4}\sin 2\phi\right]_0^{2\pi}\left[-\frac{3}{4}\cos\theta+\frac{1}{12}\cos 3\theta\right]_0^{\pi}=\frac{a^3bc}{5}\pi\left(\frac{3}{4}-\frac{1}{12}+\frac{3}{4}-\frac{1}{12}\right)=\frac{4}{15}\pi a^3bc. \quad \textbf{Ans.}$$

12. *Evaluate* $\iint_R (x^2+y^2)\,dx\,dy$ *where* R *is bounded by the lines*

$$x+y=0,\quad x+y=2,\quad x-y=2,\quad x-y=0.$$

SOLUTION: Let $I=\iint_R (x^2+y^2)\,dx\,dy$

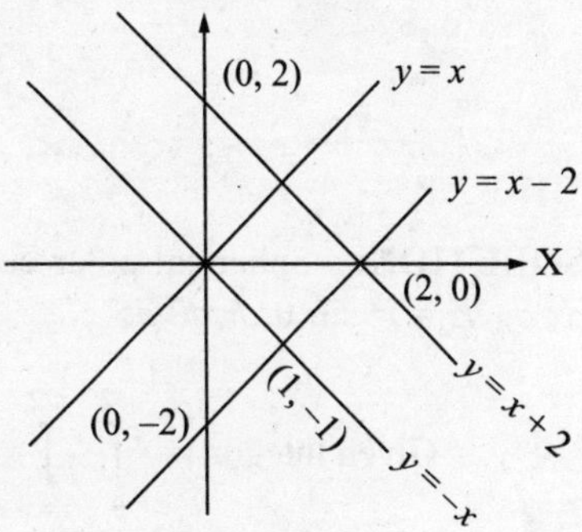

Let us write $u=x+y,\ v=x-y$, hence u varies from 0 to 2 and v varies from 0 to 2.

Here $x=\frac{u+v}{2},\ y=\frac{u-v}{2}$

$$\therefore \quad I=\iint\left[\left(\frac{u+v}{2}\right)^2+\left(\frac{u-v}{2}\right)^2\right]dx\,dy$$

and since $dx\,dy=\frac{\partial(x,y)}{\partial(u,v)}du\,dv$ and

$$\frac{\partial(x,y)}{\partial(u,v)}=\begin{vmatrix}\frac{1}{2} & \frac{1}{2}\\ \frac{1}{2} & -\frac{1}{2}\end{vmatrix}=-\frac{1}{2}$$

Hence $$I=\int_0^2\int_0^2\frac{1}{2}(u^2+v^2)\left(-\frac{1}{2}\right)du\,dv=\frac{1}{4}\int_0^2\left[u^2v+\frac{v^3}{3}\right]_{v=0}^2 du$$

$$=\frac{1}{4}\int_0^2\left(2u^2+\frac{8}{3}\right)du=\frac{1}{4}\left(2\frac{u^3}{3}+\frac{8}{3}u\right)_0^2=\frac{8}{3} \quad \textbf{Ans.}$$

13. *Evaluate* $\iiint x^2y^2z^2\,dx\,dy\,dz$ *over the volume bounded by*

$xy = 4, \quad xy = 9, \quad yz = 1, \quad yz = 4, \quad zx = 25$ *and* $zx = 49$.

SOLUTION: Let $I = \iiint x^2y^2z^2\,dx\,dy\,dz$

Let us write $xy = w, yz = u$ and $zx = v$ so $uvw = x^2y^2z^2$

$$\therefore \quad I = \iiint u\,v\,w\,dx\,dy\,dz = \iiint u\,v\,w\frac{\partial(x, y, z)}{\partial(u, v, w)}du\,dv\,dw$$

Next,
$$\frac{\partial(u, v, w)}{\partial(x, y, z)} = \begin{vmatrix} \frac{\partial u}{\partial x} & \frac{\partial u}{\partial y} & \frac{\partial u}{\partial z} \\ \frac{\partial v}{\partial x} & \frac{\partial v}{\partial y} & \frac{\partial v}{\partial z} \\ \frac{\partial w}{\partial x} & \frac{\partial w}{\partial y} & \frac{\partial w}{\partial z} \end{vmatrix} = \begin{vmatrix} 0 & z & y \\ z & 0 & x \\ y & x & 0 \end{vmatrix} = 2xyz$$

$$\therefore \quad I = \iiint \frac{u\,v\,w}{2\,xyz}du\,dv\,dw = \frac{1}{2}\iiint \frac{uvw}{\sqrt{uvw}}du\,dv\,dw = \frac{1}{2}\iiint \sqrt{u\,v\,w}\,du\,dv\,dw$$

$$= \frac{1}{2}\int_1^4 u^{1/2}du\int_{25}^{49} v^{1/2}dv\int_4^9 w^{1/2}dw = \frac{1}{2}\cdot\frac{2}{3}\cdot\frac{2}{3}\cdot\frac{2}{3}[u^{3/2}]_1^4\,[v^{3/2}]_{25}^{49}\,[w^{3/2}]_4^9$$

$$= \frac{4}{27}(8-1)(7^3-5^3)(27-8) = 4295.4$$ **Ans.**

14. *Evaluate by changing to spherical polar coordinates, the integral*

$$\int_0^1\int_0^{\sqrt{1-x^2}}\int_0^{\sqrt{1-x^2-y^2}} \frac{dx\,dy\,dz}{\sqrt{1-x^2-y^2-z^2}}$$

SOLUTION: Spherical polar coordinates are $x = r\sin\theta\cos\phi$, $y = r\sin\theta\sin\phi$, $z = r\cos\theta$ and $dx\,dy\,dz = r^2\sin\theta\,dr\,d\theta\,d\phi$

$$\therefore \quad \text{Given integral} = \int_{\theta}^{\pi/2}\int_{\phi=0}^{\pi/2}\int_{r=0}^{1} \frac{r^2\sin\theta\,dr\,d\theta\,d\phi}{\sqrt{1-r^2}}$$

$$= \int_0^{\pi/2} d\phi\int_0^{\pi/2}\sin\theta\,d\theta\int_0^1\frac{r^2\,dr}{\sqrt{1-r^2}} = \frac{\pi}{2}\cdot 1\cdot\int_0^1\frac{1-(1-r^2)}{\sqrt{1-r^2}}dr$$

$$= \frac{\pi}{2}\int_0^1\frac{dr}{\sqrt{1-r^2}} - \frac{\pi}{2}\int_0^1\sqrt{1-r^2}\,dr = \frac{\pi}{2}[\sin^{-1}r]_0^1 - \frac{\pi}{2}\left[\frac{r}{2}\sqrt{1-r^2}+\frac{1}{2}\sin^{-1}r\right]_0^1$$

$$= \frac{\pi^2}{4} - \frac{\pi^2}{8} = \frac{\pi^2}{8}$$ **Ans.**

15. *Find the area bounded by the parabolas* $y^2 = a - x$ *and* $y^2 = a - 4x$.

SOLUTION: The required area is

$$\int_{-\sqrt{a}}^{\sqrt{a}} \int_{(a-y^2)/4}^{a-y^2} dx\,dy = \int_{-\sqrt{a}}^{\sqrt{a}} \left[(a-y^2) - \frac{1}{4}(a-y^2)\right] dy$$

$$= 2\cdot\frac{3}{4}\int_0^{\sqrt{a}} (a-y^2)dy = \frac{3}{2}\left[ay - \frac{y^3}{3}\right]_0^{\sqrt{a}} = a\sqrt{a} \qquad \textbf{Ans.}$$

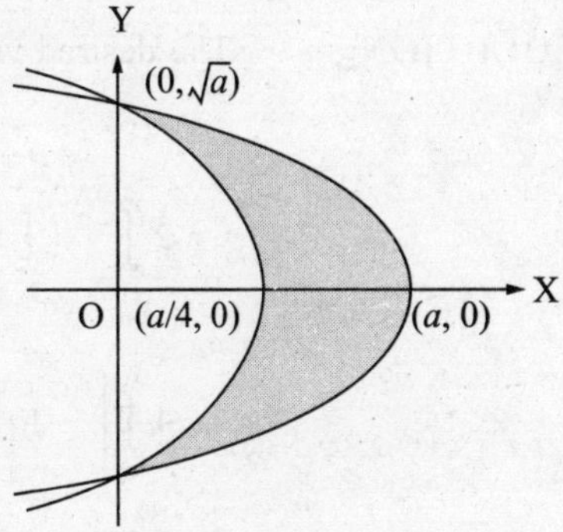

16. *Find the volume bounded by the elliptic parabolas* $z = x^2 + 9y^2$ *and* $z = 18 - x^2 - 9y^2$.

SOLUTION: Required volume = $\iiint dx\,dy\,dz$ where V is the volume bounded by the paraboloids $z = x^2 + 9y^2$ and $z = 18 - x^2 - 9y^2$.

Intersection of these paraboloids is $x^2 + 9y^2 = 18 - x^2 - 9y^2$ or $x^2 + 9y^2 = 9$ or $\frac{x^2}{9} + y^2 = 1$.

$$\therefore \quad \text{Volume} = \int_{-3}^{3} \int_{-\frac{1}{3}\sqrt{9-x^2}}^{\frac{1}{3}\sqrt{9-x^2}} \int_{x^2+9y^2}^{18-x^2-9y^2} dz\,dy\,dx = \int_{-3}^{3} \int_{-\frac{1}{3}\sqrt{9-x^2}}^{\frac{1}{3}\sqrt{9-x^2}} (18 - x^2 - 9y^2 - x^2 - 9y^2)\,dy\,dx$$

$$= 2\int_{-3}^{3}\left[18y - 2x^2y - 18\cdot\frac{y^3}{3}\right]_0^{\frac{1}{3}\sqrt{9-x^2}} dx = 2\int_{-3}^{3} \frac{1}{3}(18 - 2x^2)\sqrt{9-x^2} - \frac{18}{81}(9-x^2)^{3/2}\,dx$$

$$= 4\left(\frac{2}{3} - \frac{18}{81}\right)\int_0^3 (9-x^2)^{3/2}\,dx = \frac{16}{9}\int_0^3 (9-x^2)^{3/2}\,dx \qquad [\text{Putting } x = 3\sin\theta]$$

$$= \frac{16}{9}\cdot 27.3\int_0^{\pi/4} \cos^4\theta\theta = 27\pi \qquad \textbf{Ans.}$$

17. *Find the volume bounded by the cylinder* $x^2 + y^2 = 4$ *and the plane* $y + z = 4$. and $z = 0$.

SOLUTION: $\therefore$ Volume as triple integral = $\int_{-2}^{2} \int_{-\sqrt{4-x^2}}^{\sqrt{4-x^2}} \int_0^{4-y} dz\,dy\,dx$

$$= \int_{-2}^{2} \int_{-\sqrt{4-x^2}}^{\sqrt{4-x^2}} [z]_0^{4-y}\,dy\,dx = \int_{-2}^{2} \int_{-\sqrt{4-x^2}}^{\sqrt{4-x^2}} (4-y)\,dy\,dx = 16\int_0^2 \int_0^{\sqrt{4-x^2}} dy\,dx - 0$$

(by property of definite integrals)

$$= 16\int_0^2 \sqrt{4-x^2}\,dx = 16\left[\frac{x}{2}\sqrt{4-x^2} + \frac{1}{2}\cdot 4\sin^{-1}\frac{x}{2}\right]_0^2 = 16\cdot 2\frac{\pi}{2} = 16\pi. \qquad \textbf{Ans.}$$

18. *Find by the double integration, the volume of the sphere* $x^2 + y^2 + z^2 = 9$

SOLUTION: $\therefore$ The desired volume $= 2\int_{-a}^{a}\int_{-\sqrt{a^2-x^2}}^{\sqrt{a^2-x^2}} z\,dy\,dx$ when sphere is $x^2 + y^2 + z^2 = a^2$.

$$= 2\int_{-a}^{a}\int_{-\sqrt{a^2-x^2}}^{\sqrt{a^2-x^2}} \sqrt{a^2-x^2-y^2}\,dy\,dx = 4\int_{-a}^{a}\int_{0}^{\sqrt{a^2-x^2}} \sqrt{a^2-x^2-y^2}\,dy\,dx$$

$$= 4\int_{-a}^{a}\left[\frac{y}{2}\sqrt{a^2-x^2-y^2} + \frac{1}{2}(a^2-x^2)\sin^{-1}\frac{y}{\sqrt{a^2-x^2}}\right]_0^{\sqrt{a^2-x^2}} dx$$

$$= 4\int_{-a}^{a}\left[0 + \frac{1}{2}(a^2-x^2)(\sin^{-1}1 - 0)\right]dx = 4\int_0^a (a^2-x^2)\frac{\pi}{2}dx$$

$$= 2\pi\left[a^2x - \frac{x^3}{3}\right]_0^2 = \frac{4\pi a^3}{3}\ \frac{4\pi}{3}\cdot 3^3 = 36\pi \text{ when } a = 3. \qquad \textbf{Ans.}$$

19. *Find the volume of the portion of the sphere* $x^2 + y^2 + z^2 = a^2$ *lying inside the cylinder* $x^2 + y^2 = ax$.

SOLUTION: From the equation of the sphere we have $Z = \pm\sqrt{a^2-x^2-y^2}$

$$\therefore \text{ Required volume} = I = \iiint dx\,dy\,dz = \iint\int_{-\sqrt{a^2-x^2-y^2}}^{\sqrt{a^2-x^2-y^2}} dz\,dy\,dx = 2\iint\sqrt{a^2-x^2-y^2}\,dy\,dx$$

Now, x and y vary as per the cylinder $x^2 + y^2 = ax$ or $r = a\cos\theta$ converting into polar form.

$$I = 2\int_{-\pi/2}^{\pi/2}\int_0^{a\cos\theta}\sqrt{a^2-r^2}\,.r\,dr\,d\theta = \int_{-\pi/2}^{\pi/2}\left[-\frac{2}{3}(a^2-r^2)^{3/2}\right]_0^{a\cos\theta} d\theta$$

$$= -\frac{2}{3}\int\left[(a^2-a^2\cos^2\theta)^{3/2} - a^3\right]d\theta = -\frac{4a^3}{3}\int_0^{\pi/2}(\sin^3\theta - 1)\,d\theta$$

$$= \frac{4a^3}{3}\left[\frac{\pi}{2} - \frac{\lceil 2\,\lceil 1/2}{2\lceil 5/2}\right] = \frac{2a^3}{3}\left[\pi - \frac{4}{3}\right] = \frac{2a^3}{9}(3\pi - 4) \qquad \textbf{Ans.}$$

SOLUTIONS TO EXERCISE 13D

1. *The density at any point of a non-uniform circular lamina of radius a varies as its distance from a fixed point on the circumference of the circle. Find its mass.*

SOLUTION: Mass $= \int_{-\pi/2}^{\pi/2}\int_{0}^{2a\cos\theta} kr\cdot r\,dr\,d\theta \quad (\rho = kr).$

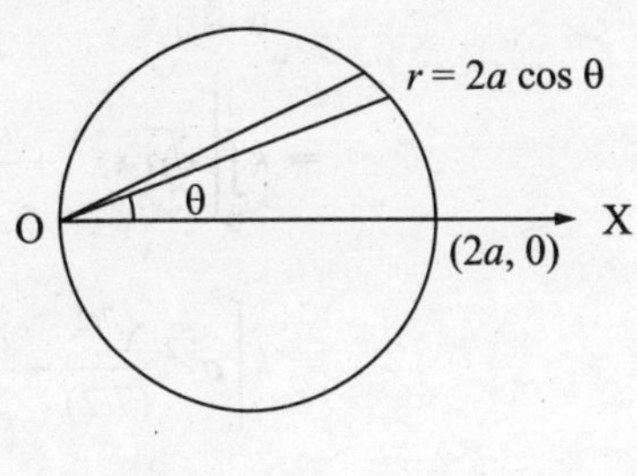

$$= k\int_{-\pi/2}^{\pi/2}\left[\frac{r^3}{3}\right]_0^{2a\cos\theta} d\theta = \frac{k}{3}\int_{-\pi/3}^{\pi/2} 8a^3\cos^3\theta\,d\theta$$

$$= \frac{16}{3}k\,a^3\int_0^{\pi/2}\cos^3\theta\,d\theta = \frac{16}{3}ka^3\frac{\lceil 2\,\lceil 1/2}{2\lceil 5/2}$$

$$= \frac{16ka^3\cdot 1}{3\cdot 2\cdot\frac{3}{2}\cdot\frac{1}{2}} = \frac{32}{9}ka^3 \quad \text{where } k \text{ is constant of proportionality.}$$ **Ans.**

2. *If the mass per unit area (i.e. density) varies as the square of the ordinate of the point, find the mass of the lamina bounded by one cycloid x = a (t + sin t), y = a (1 – cos t), the ordinates from the two cusps and the tangent at the vertex.*

SOLUTION: Mass $= \iint \rho\,dy\,dx \quad (\rho = ky^2)$

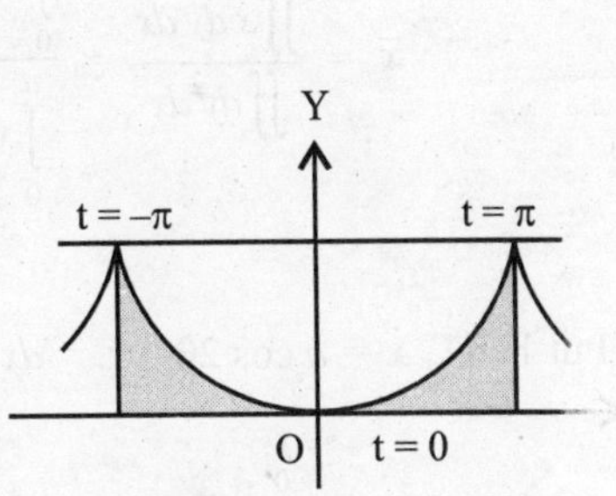

$$= \int_{-\pi}^{\pi}\int_0^{a(1-\cos t)} ky^2\,dy\,dx = k\int_{-\pi}^{\pi}\left[\frac{y^3}{3}\right]_0^{a(1-\cos t)}\frac{dx}{dt}dt$$

$$= \frac{k}{3}\int_{-\pi}^{\pi} a^3(1-\cos t)^3\,a(1+\cos t)\,dt$$

$$= 2\frac{ka^4}{3}\int_0^{\pi}(1-\cos t)^2\sin^2 t\,dt = \frac{2ka^4}{3}\int_0^{\pi}\left(2\sin^2\frac{t}{2}\right)^2\cdot 4\sin^2\frac{t}{2}\cos^2\frac{t}{2}dt \quad \left(\text{put } \frac{t}{2} = \theta\right)$$

$$= \frac{32ka^4}{3}\int_0^{\pi/2}\sin^6\theta\cos^2\theta\cdot 2d\theta = \frac{64ka^4}{3}\frac{\lceil 7/2\,\lceil 3/2}{2\lceil 5}$$

$$= \frac{32ka^4\,\frac{5}{2}\cdot\frac{3}{2}}{3\cdot 4\cdot 3\cdot 2}\cdot\left(\frac{1}{2}\lceil\frac{1}{2}\right)^2 = \frac{5\pi ka^4}{12} \quad \text{where } k \text{ is constant of proportionality.}$$ **Ans.**

3. *Find the mass of a lamina enclosed between the parabolas $y^2 = ax$ and $x^2 = ay$ when density at any point varies as the square of its distance from the origin.*

SOLUTION: Mass $= \iint \rho \, dy \, dx \qquad \rho = k(x^2 + y^2)$

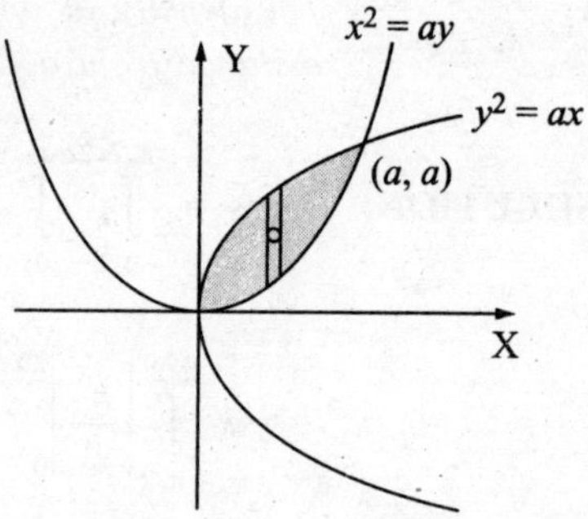

$$= \int_0^a \int_{x^2/a}^{\sqrt{ax}} k(x^2 + y^2)\, dy\, dx = k \int_0^a \left[x^2 y + \frac{y^3}{3} \right]_{y = x^2/a}^{\sqrt{ax}} dx$$

$$= k \int_0^a \left[\sqrt{a}\, x^{5/2} + \frac{a^{3/2} x^{3/2}}{3} - \frac{x^4}{a} - \frac{x^6}{3a^3} \right] dx$$

$$= k \left[a^{1/2} \frac{x^{7/2}}{(7/2)} + \frac{a^{3/2} x^{5/2}}{3 \cdot 5/2} - \frac{x^5}{5a} - \frac{x^7}{21a^3} \right]_0^a$$

$$= ka^4 \left(\frac{2}{7} + \frac{2}{15} - \frac{1}{5} - \frac{1}{21} \right) = \frac{6\,ka^4}{35}$$ where k is constant of proportionality. **Ans.**

4. *Find the centroid of the loop of the curve $y^2 (a + x) = x^2 (a - x)$*

SOLUTION: Because of symmetry, the C.G. of the loop lies on X-axis hence $\bar{y} = 0$ and

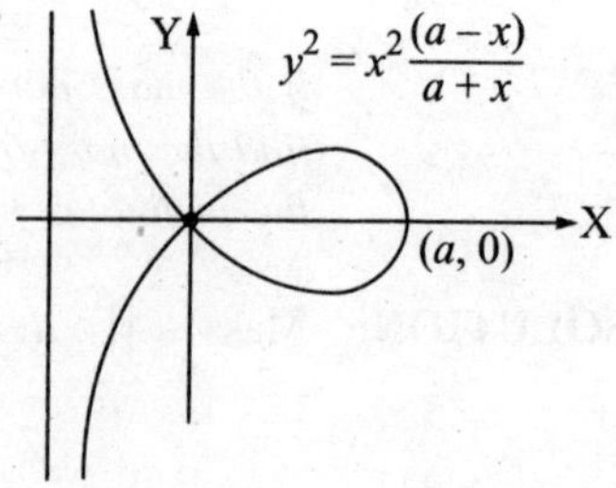

$$\bar{x} = \frac{\iint x\, dy\, dx}{\iint dy\, dx} = \frac{\int_0^a x^2 \sqrt{\frac{a-x}{a+x}}\, dx}{\int_0^a x \sqrt{\frac{a-x}{a+x}}\, dx}$$

Put here $x = a \cos 2\theta \quad \therefore \quad dx = -2a \sin 2\theta \, d\theta$ and $\sqrt{\frac{a-x}{a+x}} = \sqrt{\frac{a(1 - \cos 2\theta)}{a(1 + \cos 2\theta)}} = \tan\theta$

$$\therefore \quad \bar{x} = \frac{\int_{\pi/4}^{0} a^2 \cos^2 2\theta \cdot \frac{\sin\theta}{\cos\theta} \cdot (-4a \sin\theta \cos\theta)\, d\theta}{\int_{\pi/4}^{0} a \cos 2\theta \frac{\sin\theta}{\cos\theta} (-4a \sin\theta \cos 0)\, d\theta} = \frac{a \int_0^{\pi/4} \cos^2 2\theta\, (1 - \cos 2\theta)\, d\theta}{\int_0^{\pi/4} \cos 2\theta\, (1 - \cos 2\theta)\, d\theta} \quad (\text{put } 2\theta = t)$$

$$= \frac{a \int_0^{\pi/2} (\cos^2 t - \cos^3 t)\, dt}{\int_0^{\pi/2} (\cos t - \cos^2 t)\, dt} = a \left[\frac{\frac{\Gamma(3/2)\Gamma(1/2)}{2\Gamma(2)} - \frac{\Gamma(2)\Gamma(1/2)}{2\Gamma(5/2)}}{1 - \frac{\Gamma(3/2)\Gamma(1/2)}{2\Gamma(2)}} \right] = a \left[\frac{\frac{\pi}{2} - \frac{2}{3}}{1 - \frac{\pi}{4}} \right] = \frac{a(3\pi - 8)}{3(4 - \pi)}$$

$\therefore$ C.G. is $\left(\frac{a(3\pi - 8)}{3(4 - \pi)}, 0 \right)$ **Ans.**

5. *A plate in the form of a quadrant of the ellipse $\frac{x^2}{a^2}+\frac{y^2}{b^2}=1$ whose thickness varies from point to point and is at any point proportional to the product of the distances of the point from the axes. Find its centre of gravity.*

SOLUTION: $\frac{x^2}{a^2}+\frac{y^2}{b^2}=1$, density at $P(x, y) = \rho \propto xy$

$\therefore \quad \rho = k\,xy$ where k is constant of proportionality

$$x = \frac{\iint x\rho \cdot dy\,dx}{\iint \rho\,dy\,dx} = \frac{\iint x^2 y\,dy\,dx}{\iint xy\,dy\,dx} = \frac{\int_0^a \int_0^{b/a\sqrt{a^2-x^2}} x^2 y\,dy\,dx}{\int_0^a \int_0^{b/a\sqrt{a^2-x^2}} xy\,dy\,dx} = \frac{\int_0^a x^2\left[\frac{y^2}{2}\right]_0^{b/a\sqrt{a^2-x^2}} dx}{\int_0^a x\left[\frac{y^2}{2}\right]_0^{b/a\sqrt{a^2-x^2}} dx}$$

$$= \frac{\int_0^a x^2 \frac{b^2}{a^2}(a^2-x^2)\,dx}{\int_0^a x\frac{b^2}{x^2}(a^2-x^2)\,dx} = \frac{\int_0^a (a^2x^2-x^4)\,dx}{\int_0^a (a^2x-x^3)dx} = \frac{\left[a^2\frac{x^3}{3}-\frac{x^5}{5}\right]_0^a}{\left[a^2\frac{x^2}{2}-\frac{x^4}{4}\right]_0^a} = \frac{a\left(\frac{1}{3}-\frac{1}{5}\right)}{\left(\frac{1}{2}-\frac{1}{4}\right)} = \frac{8a}{15}$$

Similarly, we can get $\overline{Y} = \frac{8b}{15}$

$\therefore$ Required C.G. is $\left(\frac{8a}{15}, \frac{8b}{15}\right)$ **Ans.**

6. *Find the centroid of a homogeneous lamina bounded by the parabola $y = 2 - 3x^2$ and the line $3x + 2y = 1$, density being uniform throughout.*

SOLUTION: $y = 2 - 3x^2$ and $3x + 2y = 1$. These intersect at the point $\left(-\frac{1}{2}, \frac{5}{4}\right)$ and $(1, -1)$. Here ρ is constant.

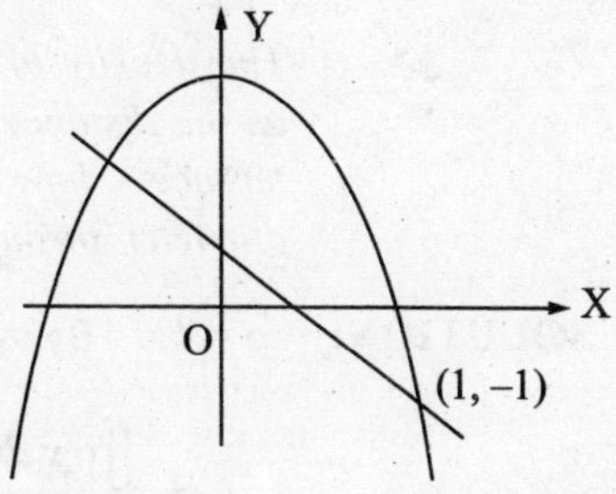

$$\overline{x} = \frac{\iint \rho\,x\,dx\,dy}{\iint \rho\,dx\,dy} = \frac{\iint x\,dx\,dy}{\iint dx\,dy},$$

$$\overline{y} = \frac{\iint \rho\,y\,dx\,dy}{\iint \rho\,dx\,dy} = \frac{\iint y\,dx\,dy}{\iint dx\,dy}$$

$$\iint dy\,dx = \int_{-1/2}^{1}\int_{\frac{1-3x}{2}}^{2-3x^2} dy\,dx = \int_{-1/2}^{1} [y]_{\frac{1-3x}{2}}^{2-3x^2} dx = \int_{-1/2}^{1}\left(2-3x^2-\frac{1}{2}+\frac{3x}{2}\right)dx$$

$$= \left[\frac{3}{2}x+\frac{3}{4}x^2-x^3\right]_{-1/2}^{1} = \frac{3}{2}+\frac{3}{4}-1+\frac{3}{4}-\frac{3}{16}-\frac{1}{8} = \frac{27}{16}$$

$$\iint x\,dy\,dx = \int_{-1/2}^{1}\int_{\frac{1-3x}{2}}^{2-3x^2} x\,dy\,dx = \int_{-1/2}^{1} x\left[y\right]_{\frac{1-3x}{2}}^{2-3x^2} ex = \int_{-1/2}^{1} x\left[2-3x^2-\frac{1}{2}+\frac{3}{2}x\right]dx$$

$$= \int_{-1/2}^{1}\left(\frac{3}{2}x+\frac{3}{2}x^2-3x^3\right)dx = \left[\frac{3}{4}x^2+\frac{x^3}{2}-\frac{3x^4}{4}\right]_{-1/2}^{1}$$

$$= \left[\frac{3}{4}\left(1-\frac{1}{4}\right)+\frac{1}{2}\left(1+\frac{1}{8}\right)-\frac{3}{4}\left(1-\frac{1}{16}\right)\right] = \frac{27}{64}$$

$$\iint y\,dy\,dx = \int_{-1/2}^{1}\int_{\frac{1-3x}{4}}^{2-3x^2} y\,dy\,dx = \int_{-1/2}^{1}\left[\frac{y^2}{2}\right]_{\frac{1-3x}{2}}^{2-3x^2} dx = \int_{-1/2}^{1}\frac{1}{2}(2-3x^2)^2-\frac{1}{2}\left(\frac{1-3x}{2}\right)^2 dx$$

$$= \int_{-1/2}^{1}\left[\frac{1}{2}(4-12x^2+9x^4)-\frac{1}{8}(1-6x+9x^2)\right]dx = \int_{-1/2}^{1}\left(\frac{15}{8}+\frac{3}{4}x-\frac{57}{8}x^2+\frac{9}{2}x^4\right)dx$$

$$= \left[\frac{15}{8}x+\frac{3}{8}x^2-\frac{19}{8}x^3+\frac{9x^5}{10}\right]_{-1/2}^{1}$$

$$= \frac{15}{8}\left(1+\frac{1}{2}\right)+\frac{3}{8}\left(1-\frac{1}{4}\right)-\frac{19}{8}\left(1-\frac{1}{8}\right)+\frac{9}{10}\left(1+\frac{1}{32}\right)$$

$$= \frac{45}{16}+\frac{9}{32}-\frac{171}{64}+\frac{297}{320} = \frac{432}{320} = \frac{27}{20}$$

$$\therefore \quad \bar{x} = \frac{27}{64}\cdot\frac{16}{27} = \frac{1}{4}, \quad \bar{y} = \frac{27}{80}\cdot\frac{16}{27} = \frac{1}{5} \quad \therefore \text{ C.G.} = \left(\frac{1}{4}, \frac{1}{5}\right)$$ **Ans.**

7. *The density of a point on a circular lamina varies as the distance from a fixed point O on its circumference. Show that the C.G. of lamina divides the diameter through O in the ratio 3:2.*

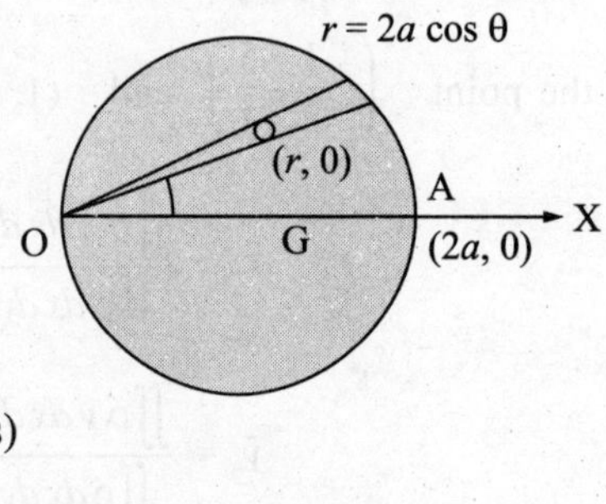

SOLUTION: $\rho = kr$. By symmetry $\bar{y} = 0$.

$$\bar{x} = \frac{\iint \rho\, x\, dx\, dy}{\iint \rho\, dx\, dy} \quad \text{(converting to polar coordinates)}$$

$$= \frac{\iint r\cdot r\cos\theta\cdot r\,dr\,d\theta}{\iint r\cdot r\,dr\,d\theta} = \frac{\int_{-\pi/2}^{\pi/2}\left[\frac{r^4}{4}\right]_0^{2a\cos\theta}\cos\theta\,d\theta}{\int_{-\pi/2}^{\pi/2}\left[\frac{r^3}{3}\right]_0^{2a\cos\theta} d\theta} = OG$$

or $$OG = \frac{3}{4}\cdot 2a\frac{\int_{-\pi/2}^{\pi/2}\cos^5\theta\,d\theta}{\int_{-\pi/2}^{\pi/2}\cos^3\theta\,d\theta} = \frac{3a}{2}\frac{\int_0^{\pi/2}\cos^5\theta\,d\theta}{\int_0^{\pi/2}\cos^3\theta\,d\theta} = \frac{3a}{2}\frac{\dfrac{\Gamma 3\,\Gamma 1/2}{2\Gamma 7/2}}{\dfrac{\Gamma 2\,\Gamma 1/2}{2\Gamma 5/2}} = \frac{6a}{5}.$$

$\therefore \quad AG = 2a - \overline{x} = \dfrac{4a}{5}.$ Thus, C.G. divides OA in the ratio 3 : 2. **Hence proved.**

8. *Find the C.G. of the area between the curve $y^2(2a - x) = x^3$ and its asymptote.*

SOLUTION: By symmetry $\overline{y} = 0$, Here $y^2 = \dfrac{x^3}{2a-x}$ $\quad\therefore\quad y = \pm\sqrt{\dfrac{x^3}{2a-x}} = \pm f(x)$

$$\overline{x} = \frac{\iint x\,dx\,dy}{\iint dx\,dy} = \frac{\int_0^{2a} x\,dx\int_{-f(x)}^{f(x)} dy}{\int_0^{2a} dx\int_{-f(x)}^{f(x)} dy} = \frac{2\int_0^{2a} x\sqrt{\dfrac{x^3}{2a-x}}\,dx}{2\int_0^{2a}\sqrt{\dfrac{x^3}{2a-x}}\,dx} \qquad (\text{putting } x = 2a\sin^2\theta)$$

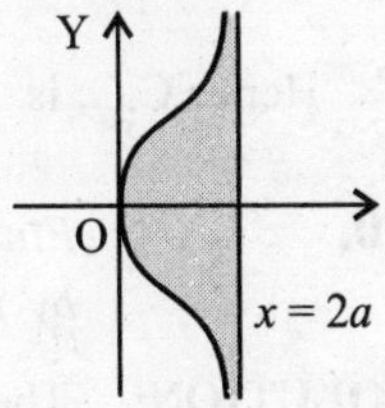

$$= \frac{\int_0^{\pi/2}(2a)^{5/2}\dfrac{\sin^5\theta}{\cos\theta}\cdot 4a\sin\theta\cos\theta\,d\theta}{\int_0^{\pi/2}(2a)^{3/2}\dfrac{\sin^3\theta}{\cos\theta}\cdot 4a\sin\theta\cos\theta\,d\theta} = \frac{2a\int_0^{\pi/2}\sin^6\theta\,d\theta}{\int_0^{\pi/2}\sin^4\theta\,d\theta}$$

$$= 2a\frac{\dfrac{\Gamma 7/2\,\Gamma 1/2}{2\Gamma 4}}{\dfrac{\Gamma 5/2\,\Gamma 1/2}{2\Gamma 3}} = 2a\cdot\frac{(5/2)}{3} = \frac{5a}{3}.\quad \text{Hence C.G.} = \left(\frac{5a}{3}, 0\right)$$ **Ans.**

9. *Find the distance from the cusp of the centre of gravity of the cardioide $r = a(1 + \cos\theta)$ if the density at a point varies as the square of its distance from the cusp.*

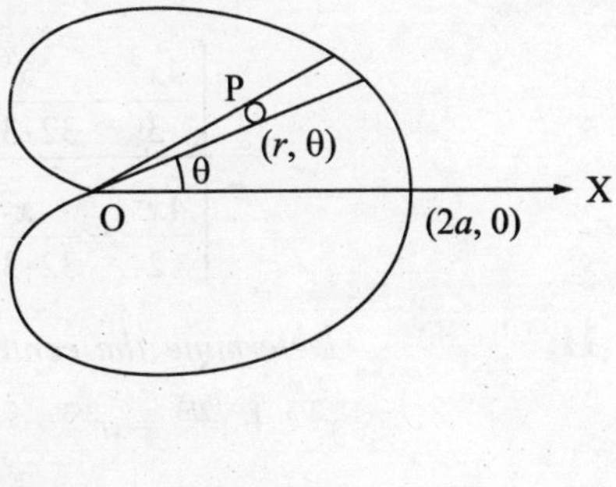

SOLUTION: $\rho = kr^2$. By symmetry $\overline{y} = 0$.

$$\overline{x} = \frac{\iint \rho\cdot x\,dx\,dy}{\iint \rho\,dx\,dy} = \frac{\iint kr^2\cdot x\,dx\,dy}{\iint kr^2\cdot dx\,dy}$$

Converting to polar form $dx\,dy = r\,dr\,d\theta$

$$\overline{x} = \frac{\iint r^2\cdot r\cos\theta\cdot r\,dr\,d\theta}{\iint r^2\cdot r\,dr\,d\theta} = \frac{\int_{-\pi}^{\pi}\int_0^{a(1+\cos\theta)} r^4\,dr\cos\theta\,d\theta}{\int_{-\pi}^{\pi}\int_0^{a(1+\cos\theta)} r^3\,dr\,d\theta}$$

$$= \frac{\int_{-\pi}^{\pi} \cos\theta \left[\frac{r^5}{5}\right]_0^{a(1+\cos\theta)} d\theta}{\int_{-\pi}^{\pi} \left[\frac{r^4}{4}\right]_0^{a(1+\cos\theta)} d\theta} = \frac{4}{5}\frac{\int_{-\pi}^{\pi} a^5 (1+\cos\theta)^5 \cos\theta \, d\theta}{\int_{-\pi}^{\pi} a^4 (1+\cos\theta)^4 \, d\theta}$$

$$= \frac{4a}{5}\frac{\int_0^{\pi}\left(2\cos^2\frac{\theta}{2}\right)^5\left(2\cos^2\frac{\theta}{2}-1\right)d\theta}{\int_0^{\pi}\left(2\cos^2\frac{\theta}{2}\right)^4 d\theta} \quad \text{Put here } \theta = 2t$$

$$= \frac{4a}{5}\cdot\frac{\left[\int_0^{\pi/2} 2^7\cos^{12} t\, dt - \int_0^{\pi/2} 2^6 \cos^{10} t\, dt\right]}{\int_0^{\pi/2} 2^5 \cos^8 t\, dt} = \frac{8a}{5}\left[\frac{2\frac{\overline{|13/2}\,\overline{|1/2}}{2\overline{|7}} - \frac{\overline{|11/2}\,\overline{|1/2}}{2\overline{|6}}}{\frac{\overline{|9/2}\,\overline{|1/2}}{2\overline{|5}}}\right] = \frac{6a}{5}.$$

Hence C.G. is $\left(\frac{6a}{5}, 0\right)$. **Ans.**

10. *Find the centroid of the volume of the solid generated by revolving the area bounded by the parabolas $y^2 = 4x$ and $x^2 = 32y$ about the axis of x.*

SOLUTION: The shaded area is revolved about the axis of X. By symmetry, the C.G. of the solid of revolution will be on X-axis, hence $\overline{Y} = 0$

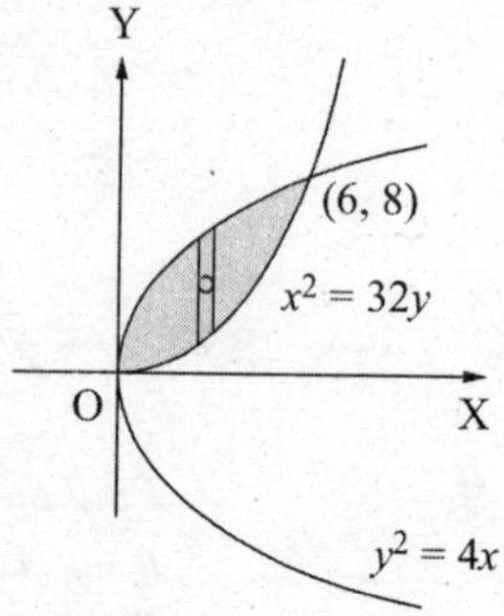

$$\overline{x} = \frac{\int_0^{16} x\,\pi(y_1^2 - y_2^2)\,dx}{\int_0^{16} \pi(y_1^2 - y_2^2)\,dx} = \frac{\int_0^{16} x\left(4x - \frac{x^4}{32\cdot 32}\right)dx}{\int_0^{16}\left(4x - \frac{x^4}{32\cdot 32}\right)dx}$$

$$= \frac{\left[\frac{4x^3}{3} - \frac{x^6}{32\cdot 32.6}\right]_0^{16}}{\left[\frac{4x^2}{2} - \frac{x^5}{32\cdot 32.5}\right]_0^{16}} = \frac{16^3\left(\frac{4}{3} - \frac{16^3}{32\cdot 32.6}\right)}{16^2\left(2 - \frac{4}{5}\right)} = \frac{80}{9} \quad \therefore \quad \text{C.G.} = \left(\frac{80}{9}, 0\right)$$ **Ans.**

11. *Determine the centre of gravity of the arc in the positive quadrant of the astroid $x^{2/3} + y^{2/3} = a^{2/3}$.*

SOLUTION: $\overline{x} = \frac{\int x\,ds}{\int ds}$ and $\overline{y} = \frac{\int y\,ds}{\int ds}$

Curve is $x^{2/3} + y^{2/3} = a^{2/3}$ or $x = a\cos^3\theta$, $y = a\sin^3\theta$ θ where varies from 0 to $\pi/2$ in the positive quadrant. By symmetry $\overline{x} = \overline{y}$.

$$\frac{ds}{d\theta} = \sqrt{\left(\frac{dx}{d\theta}\right)^2 + \left(\frac{dy}{d\theta}\right)^2} = \sqrt{(-3a\cos^2\theta\sin\theta)^2 + (3a\sin^2\theta\cos\theta)^2} = 3a(\sin\theta\cos\theta)$$

$$\int ds = \int_0^{\pi/2} 3a\sin\theta\cos\theta\, d\theta = \frac{3a}{2}\left[\sin^2\theta\right]_0^{\pi/2} = \frac{3a}{2}$$

$$\int x\, ds = \int x\frac{ds}{d\theta}d\theta = \int_0^{\pi/2} a\cos^3\theta\cdot 3a\sin\theta\cos\theta\, d\theta = 3a^2\int_0^{\pi/2}\cos^4\theta\sin\theta\, d\theta$$

$$= -3a^2\left[\frac{\cos^5\theta}{5}\right]_0^{\pi/2} = \frac{3a^2}{5}$$

Hence $\bar{x} = \frac{3a^2}{5}\cdot\frac{2}{3a} = \frac{2a}{5}$. By symmetry $\bar{x} = \bar{y} = \frac{2a}{5}$. **Ans.**

12. *Find the centroid of the arc of the cycloid $x = a(\theta - \sin\theta)$, $y = a(1 - \cos\theta)$ between its two conscutive cusps.*

SOLUTION: $\bar{x} = \frac{\int x\, ds}{\int ds}$ and $\bar{y} = \frac{\int y\, ds}{\int ds}$, $x = a(\theta - \sin\theta)$, $y = a(1 - \cos\theta)$, $0 < \theta < 2\pi$

$$\frac{ds}{d\theta} = \sqrt{\left(\frac{dx}{d\theta}\right)^2 + \left(\frac{dy}{d\theta}\right)^2} = \sqrt{a^2(1-\cos\theta)^2 + a^2\sin^2\theta} = a\sqrt{2 - 2\cos\theta} = 2a\sin\frac{\theta}{2}$$

$$\int ds = \int_0^{2\pi}\frac{ds}{d\theta}d\theta = \int_0^{2\pi} 2a\sin\frac{\theta}{2}d\theta = 8a$$

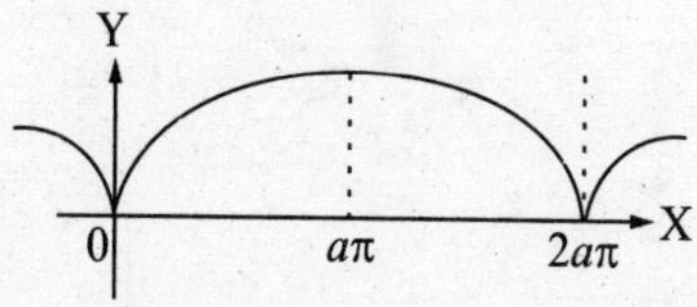

By symmetry $\bar{x} = a\pi$.

$$\int y\, ds = \int_0^{2\pi} a(1-\cos\theta)\frac{ds}{d\theta}d\theta = \int_0^{2\pi} a\, 2\sin^2\frac{\theta}{2}\cdot 2a\sin\frac{\theta}{2}d\theta = 4a^2\int_0^{2\pi}\sin^3\frac{\theta}{2}d\theta$$

$$= 8a^2\int_0^{\pi}\sin^3\frac{\theta}{2}d\theta = 8a^2\int_0^{\pi/2}\sin^3 t\cdot 2dt \qquad \text{(on putting } \theta/2 = t\text{)}$$

$$= 16a^2\frac{\Gamma(2)\,\Gamma(1/2)}{2\,\Gamma(5/2)} = \frac{16a^2}{2\cdot\frac{3}{2}\cdot\frac{1}{2}} = \frac{32a^2}{3}$$

$\therefore \quad \bar{y} = \frac{32}{3}\cdot\frac{a^2}{8a} = \frac{4a}{3}$ Hence C.G. is $\left(a\pi, \frac{4a}{3}\right)$ **Ans.**

13. *Find the C.G. of a thin hemispherical shell of radius a.*

SOLUTION: Arc AB of the circle $x^2 + y^2 = a^2$ is revolved about Y-axis to get hemispherical shell. By symmetry $\bar{x} = 0$. Circle is $x = a\cos\theta,\ y = a\sin\theta$. Hence $ds = a\,d\theta$.

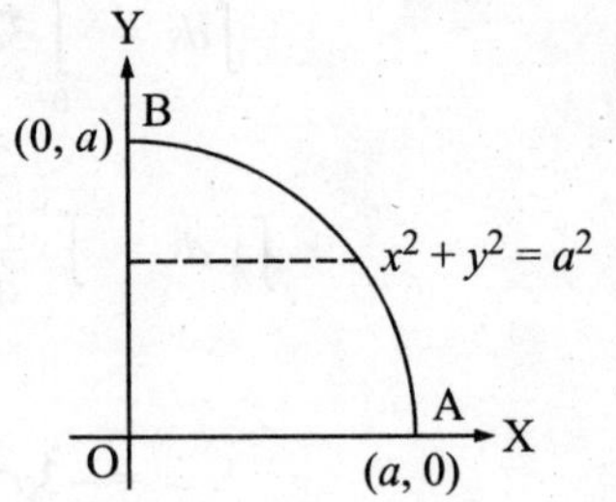

$$\bar{y} = \frac{\int y \cdot 2\pi x\, ds}{\int 2\pi x\, ds} = \frac{\int_0^{\pi/2} a\cos\theta \cdot a\sin\theta \cdot a\,d\theta}{\int_0^{\pi/2} a\cos\theta \cdot a\,d\theta}$$

$$= a\frac{\left[\frac{1}{2}\sin^2\theta\right]_0^{\pi/2}}{[\sin\theta]_0^{\pi/2}} = \frac{a}{2}$$

Hence $\quad \text{C.G.} = \left(0, \frac{a}{2}\right).$ **Ans.**

14. *Obtain the centroid of the volume generated by revolving a sector of a circle about either of its bounding radii and hence deduce the C.G. of the solid hemisphere.*

SOLUTION: The sector AOB of the circle $x^2 + y^2 = a^2$ with angle $\angle AOB = \alpha$ is revolved about its bounding radius say OA. By symmetry the C.G. of the volume of solid of revolution lies on OX hence $\bar{y} = 0$.

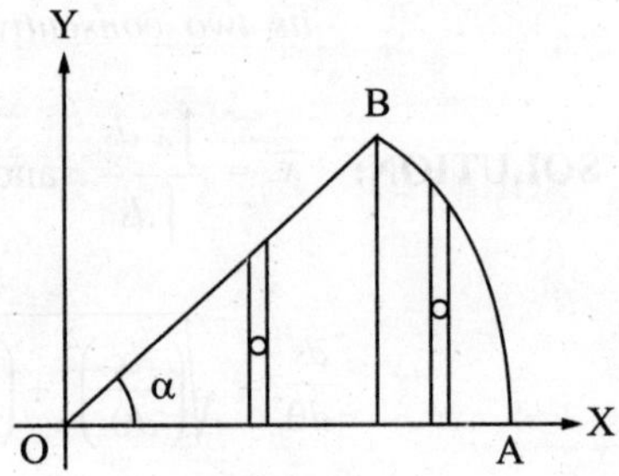

$$\bar{x} = \frac{\iint x \cdot 2\pi y\, dx\, dy}{\iint 2\pi y\, dx\, dy}$$

$$= \frac{\int_{a\cos\alpha}^{a} x \int_0^{\sqrt{a^2-x^2}} y\, dy\, dx + \int_0^{a\cos\alpha} x \int_x^{x\sec\alpha} y\, dy\, dx}{\int_{a\cos\alpha}^{a} \int_0^{\sqrt{a^2-x^2}} y\, dy\, dx + \int_0^{a\cos\alpha} \int_x^{x\sec\alpha} y\, dy\, dx}$$

$$= \frac{\int_{a\cos\alpha}^{a} x\left[\frac{y^2}{2}\right]_0^{\sqrt{a^2-x^2}} dx + \int_0^{a\cos\alpha} x\left[\frac{y^2}{2}\right]_x^{x\sec\alpha} dx}{\int_{a\cos\alpha}^{a} \left[\frac{y^2}{2}\right]_0^{\sqrt{a^2-x^2}} dx + \int_0^{a\cos\alpha} \left[\frac{y^2}{2}\right]_x^{x\sec\alpha} dx}$$

$$= \frac{\int_{a\cos\alpha}^{a} x(a^2 - x^2)dx + \int_0^{a\cos\alpha} x(x^2\sec^2\alpha - x^2)dx}{\int_{a\cos\alpha}^{a} (a^2 - x^2)dx + \int_0^{a\cos\alpha} (x^2\sec^2\alpha - x^2)dx}$$

$$= \frac{\left[a^2\frac{x^2}{2} - \frac{x^4}{4}\right]_{a\cos\alpha}^{a} + \left[\frac{x^4}{4}\tan^2\alpha\right]_0^{a\cos\alpha}}{\left[a^2x - \frac{x^3}{3}\right]_{a\cos\alpha}^{a} + \left[\frac{x^3}{3}\tan^2\alpha\right]_0^{a\cos\alpha}}$$

$$= \frac{\frac{a^4}{2} - \frac{a^4}{4} - \frac{a^4\cos^2\alpha}{2} + \frac{a^4\cos^4\alpha}{4} + \frac{a^4}{4}\cos^2\alpha\tan^2\alpha}{a^3 - \frac{a^3}{3} - a^3\cos\alpha + \frac{a^3}{3}\cos^3\alpha + \frac{a^3}{3}\cos^3\alpha\tan^2\alpha}$$

$$= \frac{a\left[\frac{1}{4} - \frac{1}{2}\cos^2\alpha + \frac{1}{4}\cos^2\alpha\right]}{\left[\frac{2}{3} - \cos\alpha + \frac{1}{3}\cos\alpha\right]} = \frac{a\left(\frac{1}{4} - \frac{1}{4}\cos^2\alpha\right)}{\frac{2}{3} - \frac{2}{3}\cos\alpha} = \frac{3a}{8}(1+\cos\alpha)$$

Hence C.G. is $\left\{\frac{3a}{8}(1+\cos\alpha),\ 0\right\}$. When $\alpha = \frac{\pi}{2}$, we get the hemisphere whose C.G. is at $\left(\frac{3a}{8},\ 0\right)$. **Ans.**

15. *Employ Pappus theorem to find the volume of the solid generated by revolving the loop of the curve $2ay^2 = x(a-x)^2$ about the line $y = a$.*

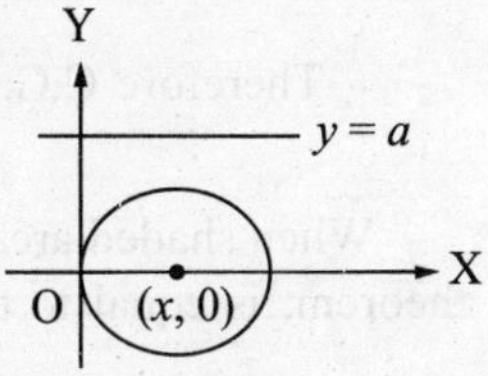

SOLUTION: By pappus theorem $V = 2\pi a\,(A)$. curve is $2ay^2 = x(a-x)^2$.
Area of the loop, A is given by

$$A = 2\int_0^a y\,dx = 2\int_0^a \frac{(a-x)\sqrt{x}}{\sqrt{2a}}\,dx = \frac{2}{\sqrt{2a}}\int_0^a \left(2\sqrt{x} - x^{3/2}\right)dx$$

$$= \frac{2}{\sqrt{2a}}\left[\frac{2a}{3}x^{3/2} - \frac{2}{5}x^{5/2}\right]_0^a = \frac{2}{\sqrt{2a}}\left[\frac{2}{3}a^{5/2} - \frac{2}{5}a^{5/2}\right] = \frac{4\sqrt{2}}{15}a^2.$$

C.G. of the loop lies on the X-axis and its distance from the line $y = a$ is 'a'

hence $\quad V = 2\pi a\,A = 2\pi a\cdot\frac{4\sqrt{2}}{15}a^2 = \frac{8\sqrt{2}}{15}\pi a^3.$ **Ans.**

16. *Find the centre of gravity of the smaller segment of the ellipse $\frac{x^2}{a^2} + \frac{y^2}{b^2} = 1$ cut-off by the line $\frac{x}{a} + \frac{y}{b} = 1$. Next, this segment is revolved about this chord, find the volume of the solid generated using Pappus theorem.*

SOLUTION: C.G. of the shaded area is $(\bar{x}, \bar{y})$

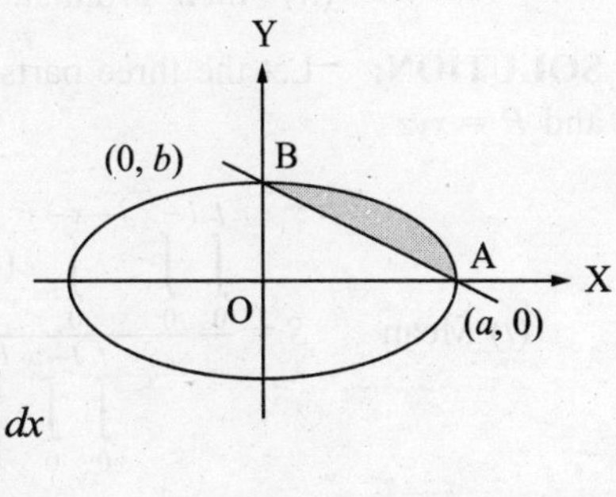

$$\bar{x} = \frac{\iint x\,dx\,dy}{\iint dx\,dy}, \quad \bar{y} = \frac{\iint y\,dx\,dy}{\iint dx\,dy}$$

$$\iint dx\,dy = \int_0^a [y]_{\frac{b}{a}(a-x)}^{\frac{b}{a}\sqrt{a^2-x^2}}\,dx = \int_0^a \left[\frac{b}{a}\sqrt{a^2-x^2} - \frac{b}{a}(a-x)\right]dx$$

$$= \frac{b}{a}\cdot\left[\frac{x}{2}\sqrt{a^2-x^2}+\frac{1}{2}a^2\sin^{-1}\frac{x}{a}-ax+\frac{x^2}{2}\right]_0^a = \frac{b}{a}\left[\frac{a^2}{2}\cdot\frac{\pi}{2}-a^2+\frac{a^2}{2}\right] = \frac{ab}{4}(\pi-2)$$

$$\iint x\,dx\,dy = \int_0^a x[y]_{\frac{b}{a}(a-x)}^{\frac{b}{a}\sqrt{a^2-x^2}}\,dx = \frac{b}{a}\int_0^a\left[x\sqrt{a^2-x^2}-x(a-x)\right]dx$$

$$= \frac{b}{a}\left[-\frac{1}{3}(a^2-x^2)^{3/2}-\frac{ax^2}{2}+\frac{x^3}{3}\right]_0^a = \frac{b}{a}\left[\frac{a^3}{3}-\frac{a^3}{2}+\frac{a^3}{3}\right] = \frac{1}{6}a^2b$$

and $$\iint y\,dx\,dy = \int_0^a\left[\left[\frac{y^2}{2}\right]_{\frac{b}{a}(a-x)}^{\frac{b}{a}\sqrt{a^2-x^2}}\right]dx = \frac{1}{2}\int_0^a\frac{b^2}{a^2}\left[(a^2-x^2)^{3/2}-(a-x)^2\right]dx$$

$$= \frac{b^2}{2a^2}\left[a^2x-\frac{x^3}{3}-a^2x+ax^2-\frac{x^3}{3}\right]_0^a = \frac{ab^2}{2}\left[1-\frac{1}{3}-1+1-\frac{1}{3}\right] = \frac{ab^2}{6}$$

Therefore C.G. of the shaded area $= (\bar{x}, \bar{y}) = \left(\frac{2a}{3(\pi-2)}, \frac{2b}{3(\pi-2)}\right)$. **Ans.**

When shaded area is revolved about the chord AB, the volume of the solid generated, by pappus theorem, is equal to the product of the area and the path traversed by the C.G.

distance of the C.G. from the chord $= \dfrac{\dfrac{2a}{3a(\pi-2)}+\dfrac{2b}{3b(\pi-2)}-1}{\sqrt{\dfrac{1}{a^2}+\dfrac{1}{b^2}}} = \dfrac{(10-3\pi)ab}{3(\pi-2)\sqrt{a^2+b^2}}$

$\therefore$ Required volume $\doteq \dfrac{ab}{4}\,\dfrac{(\pi-2)\cdot 2\pi(10-3\pi)ab}{3(\pi-2)\sqrt{a^2+b^2}} = \dfrac{\pi}{6}\,\dfrac{(10-3\pi)a^2b^2}{\sqrt{a^2+b^2}}$ **Ans.**

17. *A rod of length l is divided into three parts at random. Find the mean value of*

(i) the sum of the squares of three parts

(ii) their product.

SOLUTION: Let the three parts of the rod be x, y and z that $x+y+z=l$. $S = x^2+y^2+z^2$ and $P = xyz$

(i) Mean $$S = \frac{\int_0^l\int_0^{l-x}\int_0^{l-x-y}(x^2+y^2+z^2)\,dz\,dy\,dz}{\int_0^l\int_0^{l-x}\int_0^{l-x-y}dz\,dy\,dx} = \frac{\int_0^l\int_0^{l-x}\left[(x^2+y^2)z+\frac{z^3}{3}\right]_0^{l-x-y}dy\,dx}{\int_0^l\int_0^{l-x}[z]_0^{l-x-y}\,dy\,dx}$$

$$= \frac{\int_0^l \int_0^{l-x} \left[(x^2+y^2)(l-x-y) + \frac{1}{3}(l-x-y)^3\right] dy\, dx}{\int_0^l \int_0^{l-x} (l-x-y)\, dy\, dx}$$

$$= \frac{\int_0^l \int_0^{l-x} \left[x^2(l-x) - yx^2 + (l-x)y^2 - y^3 + \frac{1}{3}(l-x-y)^3\right] dx}{\int_0^l \left[(l-x)y - \frac{y^2}{2}\right]_0^{l-x} dx}$$

$$= \frac{\int_0^l \left[x^2(l-x)y - \frac{x^2y^2}{2} + (l-x)\frac{y^3}{3} - \frac{y^4}{4} - \frac{1}{12}(l-x-y)^4\right]_0^{l-x} dx}{\int_0^l \left[(l-x)^2 - \frac{1}{2}(l-x)^2\right] dx}$$

$$= \frac{\int_0^l \left[x^2(l-x)^2 - \frac{x^2}{2}(l-x)^2 + \frac{(l-x)^4}{3} - \frac{(l-x)^4}{4} + \frac{(l-x)^4}{12}\right] dx}{\left[-\frac{1}{6}(l-x)^3\right]_0^l}$$

$$= \frac{6}{l^3}\int_0^l \left[\frac{x^2}{2}(l-x)^2 + \frac{(l-x)^4}{6}\right] dx = \frac{1}{l^3}\int_0^l \left[3x^2(l-x)^2 + (l-x)^4\right] dx$$

$$= \frac{1}{l^3}\int_0^l \left[3x^2l^2 - 6lx^3 + 3x^4 + (l-x)^4\right] dx = \frac{1}{l^3}\left[l^2x^3 - \frac{3}{2}lx^4 + 3\frac{x^5}{5} - \frac{(l-x)^5}{5}\right]_0^l$$

$$= \frac{1}{l^3}\left[l^5 - \frac{3}{2}l^5 + \frac{3}{5}l^5 + \frac{l^5}{5}\right] = \frac{3l^2}{10}$$ **Ans.**

(*ii*) Similarly, mean of the product can be calculated as $P = \frac{6}{l^3}\int_0^l \int_0^{l-x} \int_0^{l-x-y} xyz\, dz\, dy\, dx = \frac{l^3}{120}$

18. *A quantity of a gas in a cylindrical tube under a piston follows the adiabatic law $p\, v^n = c$. Find the average value of the pressure as volume changes from $v = v_1$ to $v = v_2$.*

SOLUTION: $pv^n = c$ or $p = cv^{-n}$, $p_1 = cv_1^{-n}$, $p_2 = cv_2^{-n}$

$$\text{Average pressure} = \frac{\int p\, dv}{\int dv} = \frac{\int_{v_1}^{v_2} cv^{-n} dv}{\int_{v_1}^{v_2} dv} = \frac{c\left[\frac{v^{-n+1}}{-n+1}\right]_{v_1}^{v_2}}{v_2 - v_1} = \frac{c}{(1-n)(v_2-v_1)}\left(v_2^{-n+1} - v_1^{-n+1}\right)$$

$$= \frac{v_2p_2 - v_1p_1}{(1-n)(v_2-v_1)} = \frac{p_1v_1 - p_2v_2}{(n-1)(v_2-v_1)}$$ **Ans.**

ADDITIONAL IMPORTANT PROBLEMS

1. Evaluate $\iint xy\left(\frac{x^2}{a^2}+\frac{y^2}{b^2}\right)^{n/2} dx\, dy$ over the first quadrant of the ellipse $\frac{x^2}{a^2}+\frac{y^2}{b^2}=1$.

SOLUTION: $\frac{x^2}{a^2}+\frac{y^2}{b^2}=1 \;\therefore\; y=\pm b\sqrt{1-\frac{x^2}{a^2}}$. Hence, in the first quadrant y varies from 0 to $b\sqrt{1-\frac{x^2}{a^2}}$ and x varies from 0 to a.

$$\therefore \quad I=\iint xy\left(\frac{x^2}{a^2}+\frac{y^2}{b^2}\right)^{n/2} dx\, dy=\int_0^a x\int_0^{b\sqrt{1-x^2/a^2}} \frac{b^2}{2}\left(\frac{x^2}{a^2}+\frac{y^2}{b^2}\right)^{n/2}\frac{2y}{b^2}\, dy\, dx$$

$$=\frac{b^2}{2}\int_0^a x\left[\left(\frac{x^2}{a^2}+\frac{y^2}{b^2}\right)^{n/2+1}\frac{1}{\frac{n}{2}+1}\right]_0^{b\sqrt{1-x^2/a^2}} dx$$

$$=\frac{b^2}{n+2}\int_0^a x\left[1-\left(\frac{x}{a}\right)^{n+2}\right]dx=\frac{b^2}{n+2}\left[\frac{x^2}{2}-\frac{x^{n+4}}{(n+4)a^{n+2}}\right]_0^a$$

$$=\frac{b^2}{n+2}\left[\frac{a^2}{2}-\frac{a^2}{n+4}\right]=\frac{a^2b^2}{(n+2)}\cdot\frac{n+2}{2(n+4)}=\frac{a^2b^2}{2(n+4)}$$ **Ans.**

2. Evaluate $\iint (x^2-y^2)\, dx\, dy$ over the triangle with vertices (0, 1), (1, 1) and (1, 2).

SOLUTION: Integral is to be carried out over the area of the shaded triangle.

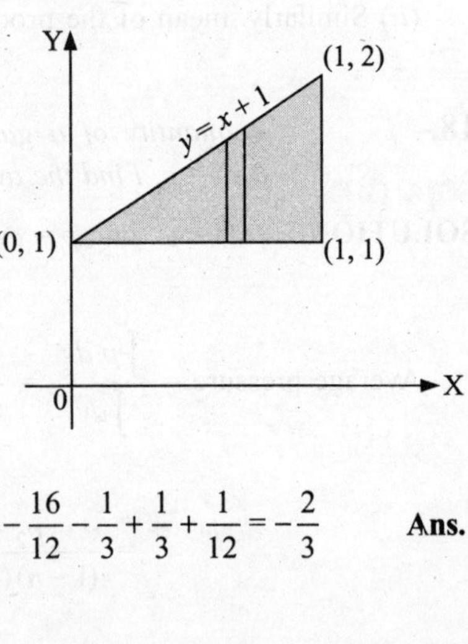

$$\text{Given integral}=\int_0^1\int_1^{x+1}(x^2-y^2)\, dy\, dx$$

$$=\int_0^1\left[x^2y-\frac{y^3}{3}\right]_1^{x+1} dx$$

$$=\int_0^1\left[x^2(x+1)-\frac{(x+1)^3}{3}-x^2+\frac{1}{3}\right]dx$$

$$=\left[\frac{x^4}{4}+\frac{x^3}{3}-\frac{(x+1)^4}{12}-\frac{x^3}{3}+\frac{x}{3}\right]_0^1=\frac{1}{4}+\frac{1}{3}-\frac{16}{12}-\frac{1}{3}+\frac{1}{3}+\frac{1}{12}=-\frac{2}{3}$$ **Ans.**

3. Evaluate $\iint xy(x+y)\,dx\,dy$ over the region bounded by the parabolas $y^2 = -x$ and $x^2 = y$.

SOLUTION:
$$I = \iint xy(x+y)\,dx\,dy = \int_0^1 y \int_{-\sqrt{y}}^{-y^2} (x^2 + xy)\,dx\,dy$$

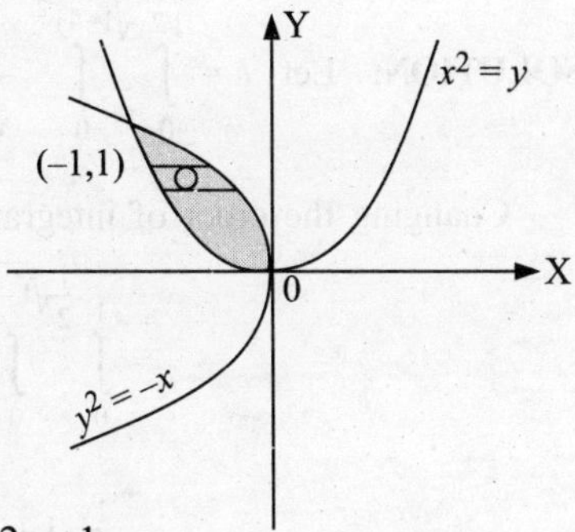

$$= \int_0^1 y\left[\frac{x^3}{3} + \frac{x^2 y}{2}\right]_{-\sqrt{y}}^{-y^2} dy$$

$$= \int_0^1 y\left[\frac{-y^6}{3} + \frac{y^5}{2} + \frac{y\sqrt{y}}{3} - \frac{y^2}{2}\right] dy$$

$$= \left[\frac{-y^8}{24} + \frac{y^7}{14} + \frac{2}{21}y^{7/2} - \frac{y^4}{8}\right]_0^1 = \frac{-1}{24} + \frac{1}{14} + \frac{2}{21} - \frac{1}{8} = 0.$$ **Ans.**

4. Evaluate $\iint x^2\,dx\,dy$ over the region in the first quadrant enclosed by the hyperbola $xy = 16$ and the lines $y = x$, $y = 0$, $x = 8$.

SOLUTION: The area of integration is the shaded region.

$$I = \iint x^2\,dx\,dy = \int_0^4\int_0^x x^2\,dy\,dx + \int_4^8\int_0^{16/x} x^2\,dy\,dx$$

$$= \int_0^4 x^2[y]_0^x\,dx + \int_4^8 x^2[y]_0^{16/x}\,dx$$

$$= \int_0^4 x^3\,dx + \int_4^8 x^2\cdot\frac{16}{x}\,dx = \left[\frac{x^4}{4}\right]_0^4 + 16\left[\frac{x^2}{2}\right]_4^8$$

$$= 64 + 8(64 - 16) = 448.$$ **Ans.**

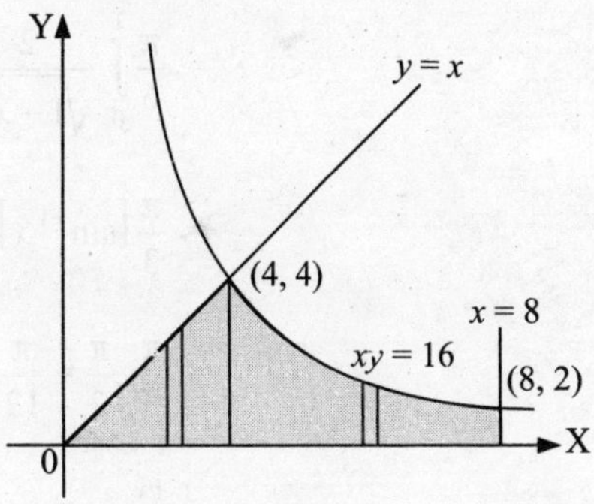

5. Evaluate $\int_0^\pi\int_x^\pi \frac{\sin y}{y}\,dy\,dx.$

SOLUTION: Area of integration is the shaded region in the figure. Integration w.r.t. y first is not possible hence we change the order of integration.

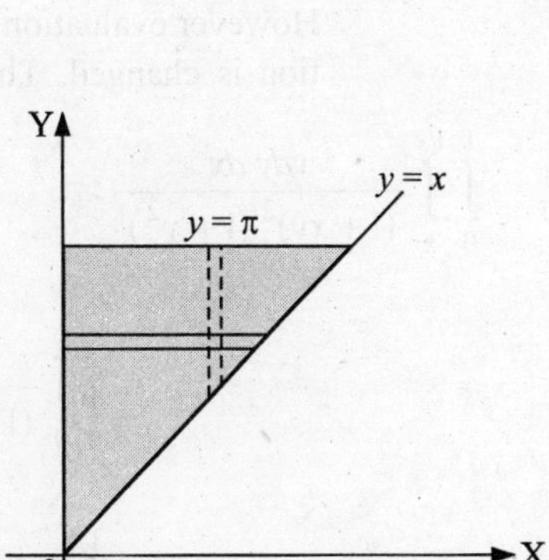

Then $$\int_0^\pi\int_x^\pi \frac{\sin y}{y}\,dy\,dx = \int_0^\pi\int_0^y \frac{\sin y}{y}\,dx\,dy$$

$$= \int_0^\pi \frac{\sin y}{y}\cdot[x]_0^y\,dy$$

$$= \int_0^\pi \sin y\,dy = [-\cos y]_0^\pi = 2.$$ **Ans.**

6. *Evaluate* $I = \int_0^{1/2} \int_0^{\sqrt{1-4y^2}} \frac{1+x^2}{\sqrt{1-x^2}\,\sqrt{1-x^2-y^2}}\,dx\,dy.$

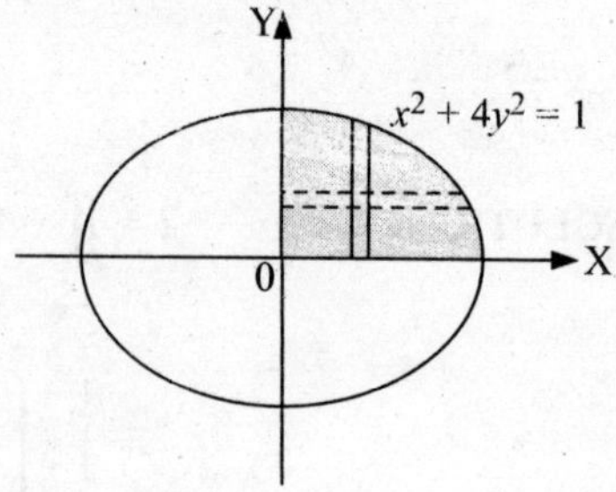

SOLUTION: Let $I = \int_0^{1/2} \int_0^{\sqrt{1-4y^2}} \frac{1+x^2}{\sqrt{1-x^2}\,\sqrt{1-x^2-y^2}}\,dx\,dy$

Changing the order of integration, we get

$$= \int_0^1 \int_0^{\frac{1}{2}\sqrt{1-x^2}} \frac{(1+x^2)}{\sqrt{1-x^2}} \frac{1}{\sqrt{1-x^2-y^2}}\,dy\,dx$$

$$= \int_0^1 \frac{1+x^2}{\sqrt{1-x^2}} \cdot \left[\sin^{-1}\frac{y}{\sqrt{1-x^2}}\right]_0^{\frac{1}{2}\sqrt{1-x^2}} dx$$

$$= \int_0^1 \frac{1+x^2}{\sqrt{1-x^2}}\left(\sin^{-1}\frac{1}{2} - 0\right)dx = \frac{\pi}{6}\int_0^1 \frac{2-(1-x^2)}{\sqrt{1-x^2}}\,dx$$

$$= \frac{\pi}{6}\int_0^1 \frac{2}{\sqrt{1-x^2}}\,dx - \frac{\pi}{6}\int_0^1 \sqrt{1-x^2}\,dx$$

$$= \frac{\pi}{3}\left[\sin^{-1}x\right]_0^1 - \frac{\pi}{6}\left[\frac{x}{2}\sqrt{1-x^2} + \frac{1}{2}\sin^{-1}x\right]_0^1$$

$$= \frac{\pi}{3}\cdot\frac{\pi}{2} - \frac{\pi}{12}\cdot\frac{\pi}{2} = \frac{\pi^2}{8}$$ **Ans.**

7. *Evaluate* $\int_0^1 \int_x^{1/x} \frac{y\,dx\,dy}{(1+xy)^2\,(1+y^2)}$

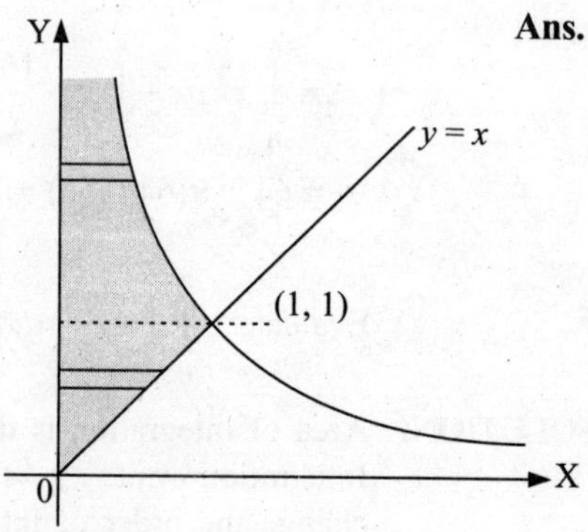

SOLUTION: Here integration is first w.r.t. y and then w.r.t. x. However evaluation is easier when order of integration is changed. Then

$$\int_0^1 \int_x^{1/x} \frac{y dy\,dx}{(1+xy)^2\,(1+y^2)}$$

$$= \int_0^1 \int_0^y \frac{y\,dx\,dy}{(1+xy)^2\,(1+y^2)} + \int_1^\infty \int_0^{1/y} \frac{y\,dx\,dy}{(1+xy)^2\,(1+y^2)}$$

$$= \int_0^1 \frac{y}{1+y^2}\left[\int_0^y \frac{dx}{(1+xy)^2}\right]dy + \int_1^\infty \frac{y}{1+y^2}\left[\int_0^{1/y} \frac{dx}{(1+xy)^2}\right]dy$$

$$= \int_0^1 \frac{y}{1+y^2}\left[\frac{-1}{y(1+xy)}\right]_0^y dy + \int_1^\infty \frac{y}{1+y^2}\left[\frac{-1}{y(1+xy)}\right]_0^{1/y} dy$$

$$= \int_0^1 \frac{y}{1+y^2}\left[\frac{-1}{y(1+y^2)}+\frac{1}{y}\right] dy + \int_1^\infty \frac{y}{1+y^2}\left[\frac{-1}{2y}+\frac{1}{y}\right] dy$$

$$= \int_0^1 \frac{1}{1+y^2}\left(1-\frac{1}{1+y^2}\right) dy + \frac{1}{2}\int_1^\infty \frac{dy}{1+y^2} = \int_0^1 \frac{dy}{1+y^2} + \frac{1}{2}\int_1^\infty \frac{dy}{1+y^2} - \int_0^1 \frac{dy}{(1+y^2)^2}$$

$$= \left[\tan^{-1} y\right]_0^1 + \frac{1}{2}\left[\tan^{-1} y\right]_1^\infty - \int_0^1 \frac{dy}{(1+y^2)^2} \quad \text{Putting } \quad y = \tan\theta$$

$$= \tan^{-1} 1 + \frac{1}{2}\left(\frac{\pi}{2}-\frac{\pi}{4}\right) - \int_0^{\pi/4} \frac{\sec^2\theta\, d\theta}{\sec^4\theta} = \frac{\pi}{4}+\frac{\pi}{8}-\frac{1}{2}\int_0^{\pi/4}(1+\cos 2\theta)\, d\theta$$

$$= \frac{3\pi}{8} - \frac{1}{2}\left[\theta + \frac{1}{2}\sin 2\theta\right]_0^{\pi/4} = \frac{3\pi}{8} - \frac{1}{2}\left[\frac{\pi}{4}+\frac{1}{2}\sin\frac{\pi}{2}\right] = \frac{-1}{4}+\frac{\pi}{4}$$

$$= \frac{\pi - 1}{4}.$$ **Ans.**

8. Evaluate $\iint r^4 \cos^3\theta\, dr\, d\theta$ over the interior of the circle $r = 2a\cos\theta$.

SOLUTION: Let $I = \iint r^4 \cos^3\theta\, dr\, d\theta$

or $$I = \int_{-\pi/2}^{\pi/2} \int_0^{2a\cos\theta} r^4 \cos^3\theta\, dr\, d\theta = \int_{-\pi/2}^{\pi/2} \cos^3\theta \left(\frac{r^5}{5}\right)_0^{2a\cos\theta} d\theta$$

$$= \int_{-\pi/2}^{\pi/2} \frac{(2a)^5}{5}\cos^3\theta\cos^5\theta\, d\theta = \frac{(2a)^5}{5}\int_{-\pi/2}^{\pi/2}\cos^8\theta\, d\theta$$

$$= 2\cdot\frac{32}{5}a^5\int_0^{\pi/2}\cos^8\theta\, d\theta$$

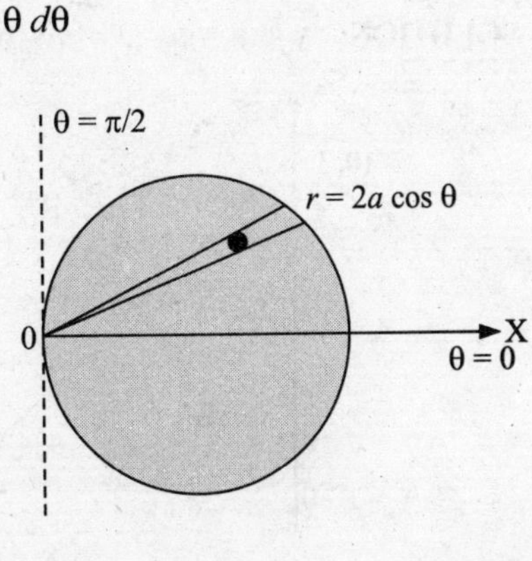

$$= a^5\,\frac{64}{5}\cdot\frac{7}{8}\cdot\frac{5}{6}\cdot\frac{3}{4}\cdot\frac{1}{2}\int_0^{\pi/2}\cos^2\theta\, d\theta$$

$$= \frac{7}{8}\pi\cdot a^5.$$ **Ans.**

9. Evaluate $\iint \frac{x^2 y^2}{(x^2+y^2)} dx\, dy$ over the region bounded by the circles $x^2 + y^2 = a^2$, $x^2 + y^2 = b^2$, $a > b$.

SOLUTION: Converting to polar coordinates

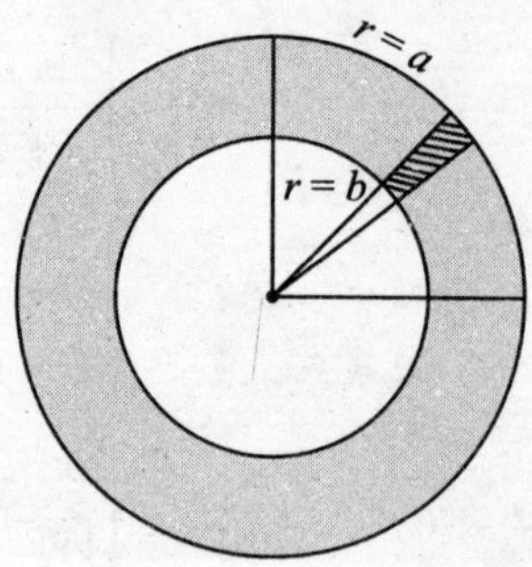

or
$$I = \iint \frac{r^2 \cos^2\theta\, r^2 \sin^2\theta}{r^2} \cdot r d\theta\, dr$$

$$= \int_0^{2\pi}\int_b^a r^3 \cos^2\theta \sin^2\theta\, dr\, d\theta$$

$$= \left[\frac{r^4}{4}\right]_b^a \int_0^{2\pi} \cos^2\theta \sin^2\theta\, d\theta$$

$$= \frac{a^4 - b^4}{16} \cdot \int_0^{2\pi} \sin^2 2\theta\, d\theta = \frac{a^4 - b^4}{32} \int_0^{2\pi} (1 + \cos 4\theta)\, d\theta$$

$$= \frac{a^4 - b^4}{32}\left[\theta + \frac{1}{4}\sin 4\theta\right]_0^{2\pi} = \frac{a^4 - b^4}{32} \cdot (2\pi) = \frac{\pi}{16}(a^4 - b^4)$$ **Ans.**

10. Evaluate $\int_0^\infty \int_0^\infty e^{-(x^2+y^2)} dx\, dy$

SOLUTION: Converting to polar coordinates

or
$$I = \int_0^{\pi/2}\int_0^\infty e^{-r^2} \cdot r\, dr\, d\theta = \int_0^{\pi/2} \left[\frac{e^{-r^2}}{-2}\right]_0^\infty d\theta$$

$$= \int_0^{\pi/2} -\frac{1}{2}(0-1)\, d\theta = \frac{\pi}{4}$$ **Ans.**

11. Using the transformation $x + y = u$, $y = uv$, prove that $\int_0^1 \int_0^{1-x} e^{\frac{y}{x+y}} dy\, dx = \frac{1}{2}(e-1)$.

SOLUTION: $x + y = u$, $y = uv$ or $x = u(1 - v)$, $y = uv$.

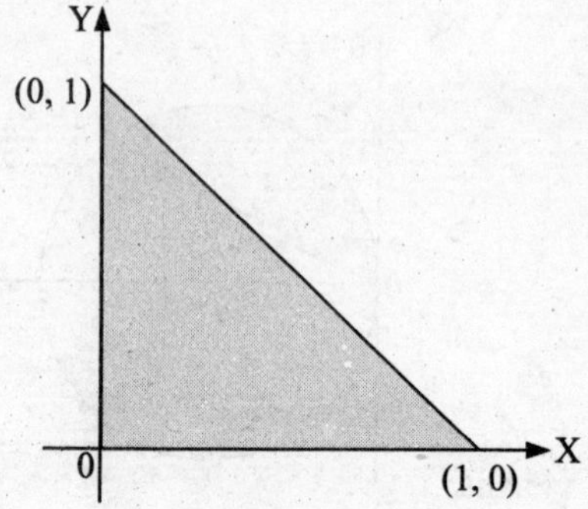

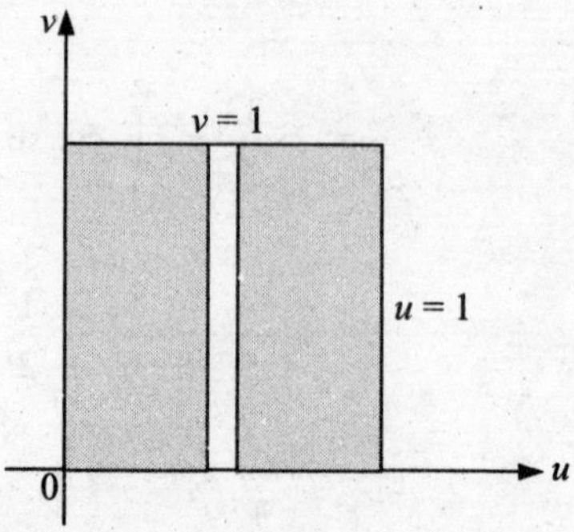

$$\therefore \qquad \frac{\partial(x, y)}{\partial(u, v)} = \begin{vmatrix} \dfrac{\partial x}{\partial u} & \dfrac{\partial x}{\partial v} \\[2mm] \dfrac{\partial y}{\partial u} & \dfrac{\partial y}{\partial v} \end{vmatrix} = \begin{vmatrix} 1-v & -u \\ v & u \end{vmatrix} = (1-v)\,u + uv = u \quad \Rightarrow \quad dxdy = u\ dudv$$

Hence $$\int_0^1 \int_0^{1-x} e^{y/(x+y)}\, dx\, dy = \int_0^1 \int_0^1 e^v\, u\, du\, dv$$

$$= \int_0^1 e^v\, dv \int_0^1 u\, du = \left[e^v\right]_0^1 \left[\frac{u^2}{2}\right]_0^1 = (e-1)\left(\frac{1}{2} - 0\right) = \frac{e-1}{2} \qquad \textbf{Ans.}$$

12. Evaluate $\displaystyle\int_0^\infty \int_0^\infty \int_0^\infty \frac{dx\, dy\, dz}{(1+x^2+y^2+z^2)^2}$

SOLUTION: The region of integration is the first octant. It is difficult to integrate in Cartesian form, so we convert to spherical polar coordinates. $x = r \sin \theta \cos \phi,\ \ y = r \sin \theta \sin \phi,\ \ z = r \cos \theta.$ $x^2 + y^2 + z^2 = r^2$

since $$dxdydz = \frac{\partial(x, y, z)}{\partial(r, \theta, \phi)}\, dr\, d\theta\, d\phi$$

and $$\frac{\partial(x, y, z)}{\partial(r, \theta, \phi)} = r^2 \sin \theta$$

$$\therefore \qquad I = \int_{r=0}^{\infty} \int_{\theta=0}^{\pi/2} \int_{\phi=0}^{\pi/2} \frac{r^2 \sin\theta\, dr\, d\theta\, d\phi}{(1+r^2)^2}$$

Putting $r = \tan t$ then t varies from 0 to $\pi/2$.

$$\therefore \qquad I = \int_{t=0}^{\pi/2} \frac{\tan^2 t\, dt}{\sec^4 t} \int_{\theta=0}^{\pi/2} \sin\theta\, d\theta \int_{\phi=0}^{\pi/2} d\phi$$

$$= \int_{t=0}^{\pi/2} \sin^2 t \cos^2 t\, dt\, [-\cos\theta]_0^{\pi/2} \cdot [\phi]_0^{\pi/2} = \frac{\pi}{2} \int_0^{\pi/2} \sin^2 t \cos^2 t\, dt = \frac{\pi^2}{32} \qquad \textbf{Ans.}$$

13. Find the area common to the circles $x^2 + y^2 - 4y = 0$ and $x^2 + y^2 - 4x - 4y + 4 = 0.$

SOLUTION: Required area is shown as shaded in the figure. Points of intersection of two circles are: $A(1, 2-\sqrt{3})$ and $B(1, 2+\sqrt{3})$.

Area is symmetrical about line AB.

$$\therefore \qquad \text{Area} = 2\int_1^2 \int_{2-\sqrt{4-x^2}}^{2+\sqrt{4-x^2}} dy\, dx$$

$$= 2\int^2 (2+\sqrt{4-x^2}) - (2-\sqrt{4-x^2})\, dx$$

$$= 4\int_1^2 \sqrt{4-x^2}\, dx = 4\left[\frac{x}{2}\sqrt{4-x^2} + \frac{4}{2}\sin^{-1}\frac{x}{2}\right]_1^2$$

$$= 4\left[0 + 2\sin^{-1}1 - \frac{\sqrt{3}}{2} - 2\sin^{-1}\left(\frac{1}{2}\right)\right]$$

$$= 4\left[2\cdot\frac{\pi}{2} - \frac{\sqrt{3}}{2} - 2\cdot\frac{\pi}{6}\right] = \frac{8\pi}{3} - 2\sqrt{3} \quad \textbf{Ans.}$$

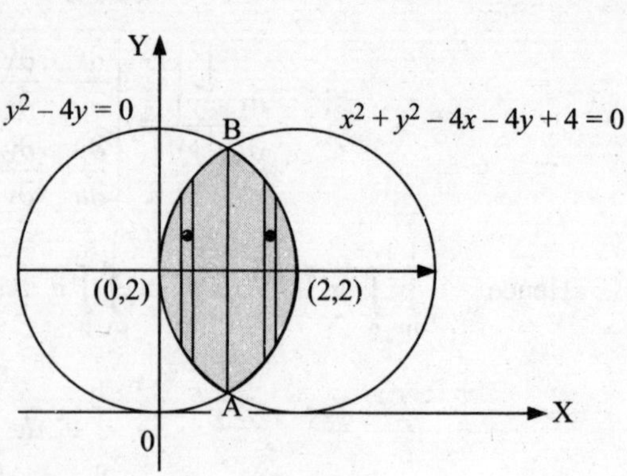

14. Find the area between the rectangular hyperbola $3xy = 2$ and the line $12x + y = 6$.

SOLUTION: The line $12x + y = 6$ meets the hyperbola $3xy = 2$. Required area is shown as shaded portion in the figure.

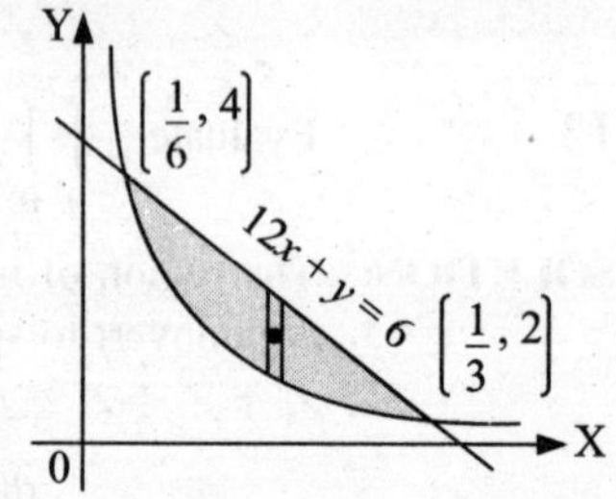

$$\text{Area} = \int_{1/6}^{1/3}\int_{2/3x}^{6-12x} dy\, dx = \int_{1/6}^{1/3} [y]_{2/3x}^{6+12x}\, dx$$

$$= \int_{1/6}^{1/3}\left(6 - 12x - \frac{2}{3x}\right)dx = \left[6x - 6x^2 - \frac{2}{3}\log x\right]_{1/6}^{1/3}$$

$$= 2 - \frac{6}{9} - \frac{2}{3}\log\frac{1}{3} - 1 + \frac{1}{6} + \frac{2}{3}\log\frac{1}{6} = \frac{1}{3} - \frac{2}{3}\log 2. \quad \textbf{Ans.}$$

15. Find the area inside the cardiode $r = 3(1 + \cos\theta)$ and outside the parabola $r = \dfrac{3}{1+\cos\theta}$.

SOLUTION: Points of intersection of the two curve are $\left(3, \frac{\pi}{2}\right)$ and $\left(3, -\frac{\pi}{2}\right)$. Area is symmetrical about $\theta = 0$.

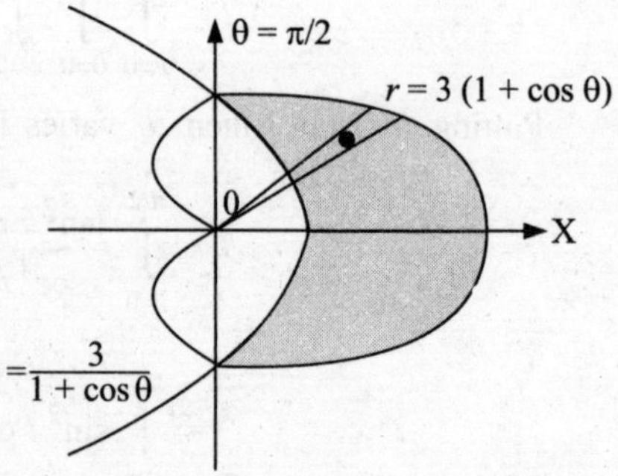

$$\text{Area} = 2\int_0^{\pi/2}\int_{\frac{3}{1+\cos\theta}}^{3(1+\cos\theta)} r\, dr\, d\theta$$

$$= 2\int_0^{\pi/2}\left[\frac{1}{2}r^2\right]_{3/(1+\cos\theta)}^{3(1+\cos\theta)} d\theta = \int_0^{\pi/2}\left[9(1+\cos\theta)^2 - \frac{9}{(1+\cos\theta)^2}\right]d\theta$$

$$= 9\int_0^{\pi/2}\left[1 + 2\cos\theta + \cos^2\theta - \frac{1}{\left(2\cos^2\frac{\theta}{2}\right)^2}\right]d\theta$$

$$= 9\int_0^{\pi/2}\left[1+2\cos\theta+\frac{1+\cos 2\theta}{2}-\frac{1}{4}\left(1+\tan^2\frac{\theta}{2}\right)\sec^2\frac{\theta}{2}\right]d\theta$$

$$= 9\left[\frac{3\theta}{2}+2\sin\theta+\frac{1}{4}\sin 2\theta-\frac{1}{4}2\tan\frac{\theta}{2}-\frac{1}{2.3},\tan^3\frac{\theta}{2}\right]_0^{\pi/2}$$

$$= 9\left[\frac{3\pi}{4}+2\sin\frac{\pi}{2}+\frac{1}{4}\sin\pi-\frac{1}{2}\tan\frac{\pi}{4}-\frac{1}{6}\tan^3\frac{\pi}{4}\right]=9\left(\frac{3\pi}{4}+\frac{4}{3}\right)$$ **Ans.**

16. Find the area common to the circle $r = a$ and the cardiode $r = a(1 + \cos\theta)$.

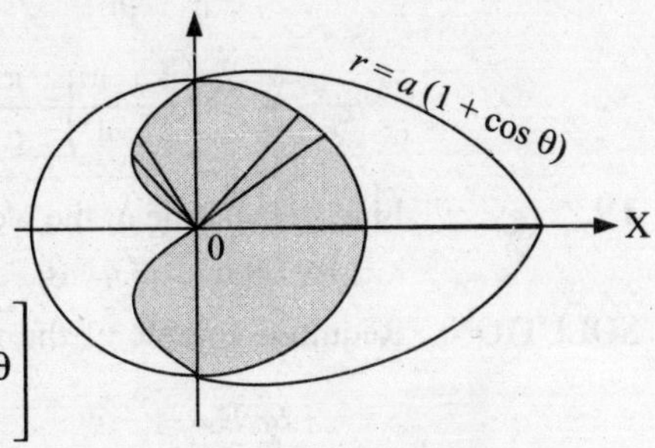

SOLUTION: For points of intersection $a = a(1 + \cos\theta)$

$\therefore\quad \theta = \dfrac{\pi}{2}.\ r = a$ consider shaded portion.

$$\text{Required area} = 2\left[\int_0^{\pi/2}\int_0^a r\,dr\,d\theta+\int_{\pi/2}^{\pi}\int_0^{a(1+\cos\theta)} r\,dr\,d\theta\right]$$

$$= 2\int_0^{\pi/2}\left[\frac{r^2}{2}\right]_0^a d\theta+2\int_{\pi/2}^{\pi}\left[\frac{r^2}{2}\right]_0^{a(1+\cos\theta)} d\theta$$

$$= a^2\cdot\frac{\pi}{2}+\int_{\pi/2}^{\pi} a^2(1+\cos\theta)^2\,d\theta=\frac{\pi a^2}{2}+a^2\int_{\pi/2}^{\pi}\left[1+2\cos\theta+\frac{1+\cos 2\theta}{2}\right]d\theta$$

$$= \frac{\pi a^2}{2}+a^2\left[\frac{3\theta}{2}+2\sin\theta+\frac{1}{4}\sin 2\theta\right]_{\pi/2}^{\pi}$$

$$= \frac{\pi a^2}{2}+a^2\left[\frac{3}{2}\left(\pi-\frac{\pi}{2}\right)+2\sin\pi-2\sin\frac{\pi}{2}+\frac{1}{4}\sin 2\pi-\frac{1}{4}\sin\pi\right]$$

$$= a^2\left[\frac{\pi}{2}+\frac{3\pi}{4}-2\right]=a^2\left(\frac{5\pi}{4}-2\right).$$ **Ans.**

17. Find the area of the loop of the curve $x^4 + y^4 = 8xy$.

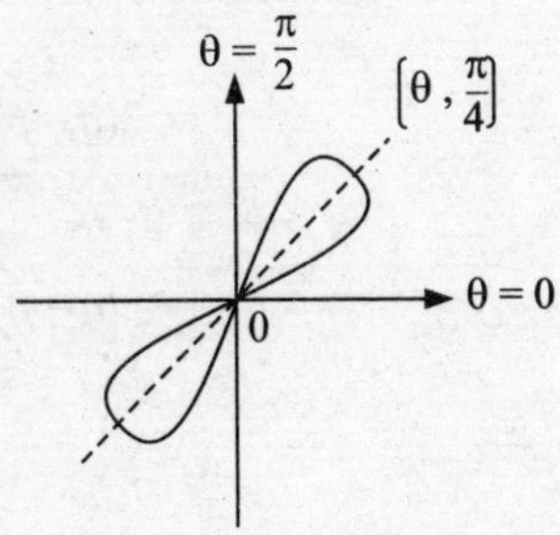

SOLUTION: Converting to polar form $r^4(\sin^4\theta+\cos^4\theta) = 8r^2\sin\theta\cos\theta$

or $$r^2 = \frac{8\sin\theta\cos\theta}{\cos^4\theta+\sin^4\theta}$$

$$\text{Area} = \int_0^{\pi/2}\int_0^{\sqrt{\frac{8\sin\theta\cos\theta}{\sin^4\theta\cos^4\theta}}} r\,dr\,d\theta$$

$$= \int_0^{\pi/2}\frac{1}{2}\cdot\frac{8\sin\theta\cos\theta}{\sin^4\theta+\cos^4\theta}d\theta=4\int_0^{\pi/2}\frac{\tan\theta\sec^2\theta}{1+\tan^4\theta}d\theta$$

$= \pi$ (on putting $\tan^2\theta = t$) **Ans.**

18. Find the volume bounded by the core $x^2 + y^2 = z^2$ and the paraboloid $x^2 + y^2 = z$.

SOLUTION: We take the cylindrical polar coordinates. Here z varies from $z = r^2$ to $z = r$, r varies from 0 to 1 and θ varies from $\theta = 0$ to $\theta = \pi/2$.

$$\text{Volume} = 4\int_0^{\pi/2}\int_0^1\int_{r^2}^{r} r\, dz.dr\, d\theta = 4\int_0^{\pi/2}\int_0^1 [z]_{r^2}^{r}\, dr\, d\theta$$

$$= 4\int_0^{\pi/2} d\theta \int_0^1 r(r-r^2)dr = \frac{4\pi}{2}\int_0^1 (r^2 - r^3)dr$$

$$= 2\pi\left(\frac{1}{3}-\frac{1}{4}\right) = \frac{\pi}{6}.$$ **Ans.**

19. Find the volume of the wedge intercepted between the cylinder $x^2 + y^2 = 2ax$ and the planes $z = px$ and $z = qx$ $(q > p)$.

SOLUTION: Required volume of the wedge

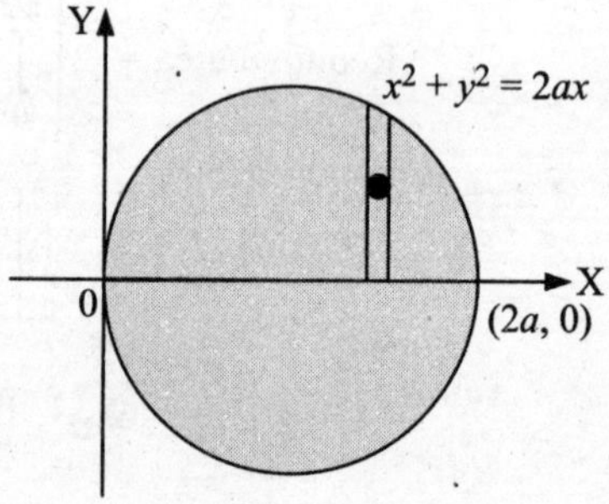

$$= 2\int_0^{2a}\int_0^{\sqrt{2ax-x^2}}\int_{z=px}^{qx} dz\, dy\, dx = 2\int_0^{2a}\int_0^{\sqrt{2ax-x^2}} [z]_{px}^{qx}\, dy\, dx$$

$$= 2\int_0^{2a}\int_0^{\sqrt{2ax-x^2}} (q-p)x\, dy\, dx$$

$$= 2(q-p)\int_0^{2a} [y]_0^{\sqrt{2ax-x^2}}\, dx = 2(q-p)\int_0^{2a} x\sqrt{2ax-x^2}\, dx$$

$$= 2(q-p)\int_0^{2a} x^{3/2}\sqrt{2a-x}\, dx \quad (\text{Putting } x = 2a\sin^2\theta \quad \therefore \quad dx = 4a\sin\theta\cos\theta\, d\theta)$$

$$= 2(q-p)\int_0^{\pi/2} (2a)^{3/2}\sin^3\theta \cdot (2a)^{1/2}\cos\theta \cdot 4a\sin\theta\cos\theta\, d\theta$$

$$= 2(q-p)\,16a^3\int_a^{\pi/2} \sin^4\theta\cos^2\theta\, d\theta$$

$$= 2(q-p)\,16a^3\,\frac{\Gamma(5/2)\,\Gamma(3/2)}{2\,\Gamma(4)} = \pi(q-p)a^3$$ **Ans.**

CHAPTER

14 Vector Differentiation

IMPORTANT FORMULAE AND USEFUL RESULTS

1. **Gradient** of $\phi(x, y, z) = \text{grad } \phi = \nabla\phi = i\frac{\partial\phi}{\partial x} + j\frac{\partial\phi}{\partial y} + k\frac{\partial\phi}{\partial x}$. It is along the normal to the level surface at P and its magnitude is the rate of change of ϕ along the normal at P. Directional Derivative of $\phi(x, y, z)$ along of vector $\vec{a} = \nabla\phi \cdot \hat{a}$.

2. **Divergence of a vector $\vec{F}$**

$= \nabla \cdot \vec{F} = \frac{\partial F_1}{\partial x} + \frac{\partial F_2}{\partial y} + \frac{\partial F_3}{\partial z}$ where $\vec{F} = F_1 i + F_2 j + F_3 k$. If $\nabla \cdot \vec{F} = 0$, $\vec{F}$ is called **solenoidal** vector. **Fluid flux** is a physical application of divergence.

3. **Curl of a vector field $\vec{F}$**

$$= \nabla \times \vec{F} = \begin{vmatrix} i & j & k \\ \frac{\partial}{\partial x} & \frac{\partial}{\partial y} & \frac{\partial}{\partial x} \\ F_1 & F_2 & F_3 \end{vmatrix} \quad \text{where } \vec{F} = F_1 i + F_2 j + F_3 k.$$

If $\nabla \times \vec{F} = 0$, $\vec{F}$ is called **irrotational** (or **conservative**) filed. As a physical application of Curl, consider a body rotating about an axis, with constant angular velocity ω then Curl of V, the liner velocity, is twice the angular velocity. $\nabla \times \vec{V} = 2\vec{\omega}$

4. **Vector Identities**

(1) $\nabla \cdot (\phi F) = \nabla\phi \cdot F + \phi\nabla \cdot F$

(2) $\nabla \times (\phi F) = \nabla\phi \times F + \phi\nabla \times F$

(3) $\nabla \cdot (F \times G) = G \cdot (\nabla \times F) - F \cdot (\nabla \times G)$.

(4) $\nabla \times (F \times G) = (\nabla \cdot G)F - (\nabla \cdot F)G + (G \cdot \nabla)F - (F \cdot \nabla)G$.

(5) $\nabla (F \cdot G) = \vec{F} \times (\nabla \times G) + \vec{G} \times (\nabla \times F) + (F \cdot \nabla)G + (G \cdot \nabla)F$.

(6) $\nabla \cdot \nabla\phi = \nabla^2\phi = \frac{\partial^2\phi}{\partial x^2} + \frac{\partial^2\phi}{\partial y^2} + \frac{\partial^2\phi}{\partial z^2}$

(7) $\nabla \times (\nabla\phi) = 0$

(8) $\nabla \cdot (\nabla \times F) = 0$

(9) $\nabla \times (\nabla \times F) = \nabla(\nabla \cdot F) - (\nabla \cdot \nabla)F$

SOLUTIONS TO EXERCISE 14A

1. *Evaluate* $\nabla\phi$ *where* $\phi = x^3 + y^3 + z^3 - 3xyz$ *at the point* $(1, -1, 2)$

SOLUTION: $\nabla\phi = i\dfrac{\partial\phi}{\partial x} + j\dfrac{\partial\phi}{\partial y} + k\dfrac{\partial\phi}{\partial z} = i(3x^2 - 3yz) + j(3y^2 - 3zx) + k(3z^2 - 3xy)$

$$[\nabla\phi]_{(1,-1,2)} = (3+6)i + (3-6)j + (12+3)k = 9i - 3j + 15k$$

$$\nabla\phi = 9i - 3j + 15k$$ **Ans.**

2. *Obtain the directional derivative of the scalar field* $\phi = xy + yz + zx$ *at the point* $(1, 2, 3)$ *in the direction of the vector* $3i + 4j + 5k$.

SOLUTION: $\nabla\phi = i\dfrac{\partial\phi}{\partial x} + j\dfrac{\partial\phi}{\partial y} + k\dfrac{\partial\phi}{\partial z} = i(y+z) + j(z+x) + k(x+y)$

$\nabla\phi$ at $(1, 2, 3) = 5i + 4j + 3k$

Directional derivative along $3i + 4j + 5k$

$$= (5i + 4j + 3k)\cdot\frac{(3i+4j+5k)}{\sqrt{3^2+4^2+5^2}} = \frac{15+16+15}{\sqrt{50}} = \frac{46}{5\sqrt{2}}$$ **Ans.**

3. *Find div* $\vec{V}$ *and Curl* $\vec{V}$ *if* $\vec{V} = \nabla(x^3 + y^3 + z^3 - 3xyz)$

SOLUTION: $\vec{V} = \nabla(x^3 + y^3 + z^3 - 3xyz) = (3x^2 - 3yz)i + (3y^2 - 3zx)j + k(3z^2 - 3xy)$

Divergence $\vec{V} = \nabla\cdot\vec{V} = \dfrac{\partial}{\partial x}(3x^2 - 3yz) + \dfrac{\partial}{\partial y}(3y^2 - 3zx) + \dfrac{\partial}{\partial z}(3z^2 - 3xy)$

$$= 6x + 6y + 6z = 6(x + y + z)$$

$$\text{Curl } \vec{V} = \nabla\times\vec{V} = \begin{vmatrix} i & j & k \\ \dfrac{\partial}{\partial x} & \dfrac{\partial}{\partial y} & \dfrac{\partial}{\partial z} \\ 3(x^2 - yz) & 3(y^2 - zx) & 3(z^2 - xy) \end{vmatrix}$$

$$= 3i(-x + x) + 3j(-y + y) + 3k(-z + z) = 0$$

Hence $\nabla\times\vec{V} = 0$ and $\nabla\cdot\vec{V} = 6(x + y + z)$ **Ans.**

4. *If* $\vec{r}$ *is the position vector of the* $P(x, y, z)$ *and* $\vec{a}$ *and* $\vec{b}$ *are constant vectors, prove that*

$$\vec{a}\cdot\nabla\left\{\vec{b}\cdot\nabla\left(\frac{1}{r}\right)\right\} = \frac{3(\vec{a}\cdot\vec{r})(\vec{b}\cdot\vec{r})}{r^5} - \frac{\vec{a}\cdot\vec{b}}{r^3}$$

SOLUTION: $$\nabla\left(\frac{1}{r}\right) = \sum i\frac{\partial}{\partial x}\left(\frac{1}{r}\right) = \sum -\frac{i}{r^2}\cdot\frac{\partial r}{\partial x} = \sum -i\frac{x}{r^3} = -\frac{\vec{r}}{r^3}$$

then $$\vec{b}\cdot\nabla\left(\frac{1}{r}\right) = -\frac{\vec{b}\cdot\vec{r}}{r^3}$$

Next $$\nabla\left[\vec{b}\cdot\nabla\left(\frac{1}{r}\right)\right] = -\nabla\left(\frac{\vec{b}\cdot\vec{r}}{r^3}\right) = -\vec{b}\cdot\vec{r}\,\nabla\left(\frac{1}{r^3}\right) - \frac{1}{r^3}\nabla(\vec{b}\cdot\vec{r})$$

$$= -\vec{b}\cdot\vec{r}\sum i\frac{\partial}{\partial x}\left(\frac{1}{r^3}\right) - \frac{1}{r^3}\nabla(\vec{b}\cdot\vec{r}) \quad \text{but} \quad \nabla(\vec{b}\cdot\vec{r}) = \vec{b}$$

$$= (\vec{b}\cdot\vec{r})\sum i\frac{3i}{r^4}\frac{\partial r}{\partial x} - \frac{\vec{b}}{r^3} = 3(b\cdot r)\sum\frac{xi}{r^5} - \frac{\vec{b}}{r^3}$$

$$= 3(\vec{b}\cdot\vec{r})\frac{\vec{r}}{r^5} - \frac{\vec{b}}{r^3}$$

$$\therefore \quad a\cdot\nabla\left[\vec{b}\cdot\nabla\left(\frac{1}{r}\right)\right] = \frac{3(\vec{b}\cdot\vec{r})}{r^5}(\vec{a}\cdot\vec{r}) - \frac{\vec{a}\cdot\vec{b}}{r^3}$$

Hence proved.

5. *If* $\phi_1 = x + y + z$, $\phi_2 = x + y$ *and* $\phi_3 = -(2xz + 2yz + z^2)$ *show that* $\nabla\phi_1\cdot[\nabla\phi_2\times\nabla\phi_3] = 0.$

SOLUTION: $$\nabla\phi_1 = i\frac{\partial\phi_1}{\partial x} + j\frac{\partial\phi_1}{\partial y} + k\frac{\partial\phi_1}{\partial z} = i + j + k, \quad \nabla\phi_2 = i\frac{\partial\phi_2}{\partial x} + \frac{\partial\phi_2}{\partial y} + \frac{\partial\phi_2}{\partial z} = i + j$$

and $$\nabla\phi_3 = i\frac{\partial\phi_3}{\partial x} + j\frac{\partial\phi_3}{\partial y} + k\frac{\partial\phi_3}{\partial z} = -(2z)i - (2z)j - (2x + 2y + 2z)k$$

$$\nabla\phi_1\cdot[\nabla\phi_2\times\nabla\phi_3] = \begin{vmatrix} 1 & 1 & 1 \\ 1 & 1 & 0 \\ -2z & -2z & -2(x+y+z) \end{vmatrix} = 0 \text{ as } C_1 \text{ and } C_2 \text{ are identical.}$$

Hence proved.

6. *Given that* $\rho\vec{F} = \nabla\phi$ *where* ρ *and* ϕ *are scalar point functions and* $\vec{F}$ *is vector point function, find* $F\cdot Curl\ F$.

SOLUTION: $\rho\vec{F} = \nabla\phi$ or $\vec{F} = \frac{1}{\rho}\nabla\phi$ where ρ and ϕ and scalar functions.

$$\text{Curl } \vec{F} = \nabla\times\vec{F} = \nabla\times\left(\frac{1}{\rho}\nabla\phi\right) = \nabla\left(\frac{1}{\rho}\right)\times\nabla\phi + \frac{1}{\rho}\nabla\times\nabla\phi$$

$$= \nabla\left(\frac{1}{\rho}\right)\times\nabla\phi \quad \text{since } \nabla\times\nabla\phi = 0$$

$$\therefore \quad \vec{F}\cdot\text{Curl }\vec{F} = \frac{1}{\rho}\nabla\phi\cdot\left(\nabla\frac{1}{\rho}\times\nabla\phi\right)$$

= 0 since scalar triple product is zero if two vectors are parallel

$$\therefore \quad \vec{F}\cdot\text{Curl }\vec{F} = 0$$

Ans.

7. *Find the angle between the two surfaces* $x^2y + z = 3$ *and* $x \log z - y^2 = -4$ *at their point of intersection* $(-1, 2, 1)$.

SOLUTION: $\phi \equiv x^2y + z = 3, \quad \psi \equiv x \log z - y^2 = 4$

$$\nabla\phi = 2xy\, i + x^2 j + 1k \quad \text{and} \quad \nabla\psi = (\log z)\, i - 2yj + \frac{x}{z}k$$

$$\therefore \quad \nabla\phi \cdot \nabla\psi = 2xy \log z - x^2 \cdot 2y + \frac{x}{z}$$

The point of intersection to the two surfaces is $(-1, 2, 1)$.

$$\therefore \quad [\nabla\phi \cdot \nabla\psi]_{(-1,2,1)} = 0 - 4 - 1 = -5.$$

Also $[\nabla\phi]_{(-1,2,1)} = -4i + 4j + k$ and $[\nabla\psi]_{(-1,2,1)} = 0i - 4j - k$

If θ is the angle between the normals to the surfaces

$$\cos\theta = \frac{-5}{\sqrt{[4^2+1^2+1^2]\,[0^2+4^2+1^2]}} = \frac{5}{\sqrt{17\times 18}} \quad \therefore \quad \theta = \cos^{-1}\cdot\frac{5}{\sqrt{17\times 18}}$$ **Ans.**

8. *Determine the values of m and n so that the vector point* function $\vec{F} = (xyz)^m (x^n i + y^n j + z^n k)$ *is irrotational.*

SOLUTION: $\vec{F}$ is irrotational if $\nabla\times\vec{F} = 0$. Here $\vec{F} = (xyz)^m (x^n i + y^n j + z^n k)$

$$\nabla\times\vec{F} = \nabla\times\{x^{m+n}y^m z^m i + x^m y^{m+n} z^m j + x^m y^m z^{m+n} k\}$$

$$= \begin{vmatrix} i & j & k \\ \dfrac{\partial}{\partial x} & \dfrac{\partial}{\partial y} & \dfrac{\partial}{\partial z} \\ x^{m+n}y^m z^m & x^m y^{m+n} z^m & x^m y^m z^{m+n} \end{vmatrix}$$

$$= i\{mx^m y^{m-1} z^{m+n} - m x^m y^{m+n} z^{m-1}\} + j\{mx^{m+n} y^m z^{m-1} - mx^{m-1} y^m z^{m+n}\}$$

$$+ k\{mx^{m-1} y^{m+n} z^m - mx^{m+n} y^{m-1} z^m\} = 0$$

or $m[ix^m y^{m-1} z^{m-1}(z^{n+1} - y^{n+1}) + jx^{m-1} y^m z^{m-1}(x^{n+1} - z^{n+1}) + kx^{m-1} y^{m-1} z^m (y^{n+1} - x^{n+1})] = 0$

$$\Rightarrow \quad m = 0, \quad n = -1$$ **Ans.**

9. *Show that* $\nabla\times\{\vec{a}\times\nabla(1/r)\} + \nabla\{\vec{a}\cdot\nabla(1/r)\} = 0$ *where* $\vec{a}$ *is constant.*

SOLUTION:

$$\nabla\left(\frac{1}{r}\right) = \Sigma i\frac{\partial}{\partial x}\left(\frac{1}{r}\right) = -\Sigma\frac{i}{r^2}\frac{\partial r}{\partial x} = -\Sigma\frac{i}{r^2}\frac{x}{r} = \frac{-\vec{r}}{r^3}$$

$$\therefore \quad a\cdot\nabla\left(\frac{1}{r}\right) = \frac{-\vec{a}\cdot\vec{r}}{r^3}$$

$$\Rightarrow \quad \nabla\cdot\left\{a\cdot\nabla\left(\frac{1}{r}\right)\right\} = -\nabla\left(\frac{1}{r^3}\vec{a}\cdot\vec{r}\right) = -\frac{1}{r^3}\nabla(\vec{a}\cdot\vec{r}) - (\vec{a}\cdot\vec{r})\nabla\left(\frac{1}{r^3}\right) = -\frac{\vec{a}}{r^3}(\vec{a}\cdot\vec{r})\left(-3\frac{\vec{r}}{r^5}\right)$$

Next $\quad a\times\nabla\left(\frac{1}{r}\right) = -\frac{(\vec{a}\times\vec{r})}{r^3}$

$$\therefore \quad \nabla\times\left\{\vec{a}\times\nabla\left(\frac{1}{r}\right)\right\} = \nabla\times\left(\frac{\vec{a}\times\vec{r}}{r^3}\right) = +\nabla\times\left(\vec{r}\times\frac{\vec{a}}{r^3}\right)$$

We know that $\nabla\times(F\times G) = (\nabla\cdot G)F - (\nabla\cdot F)G + (G\cdot\nabla)F - (F\cdot\nabla)G$

$$\therefore \quad \nabla\times\left(\vec{r}\times\frac{\vec{r}}{r^3}\right) = \left(\nabla\cdot\frac{\vec{a}}{r^3}\right)\vec{r} - (\nabla\cdot r)\frac{\vec{a}}{r^3} + \left(\frac{\vec{a}}{r^3}\cdot\nabla\right)\vec{r} - (\vec{r}\cdot\nabla)\frac{\vec{a}}{r^3}$$

$$= \left[\nabla\frac{1}{r^3}\cdot\vec{a} + \frac{1}{r^3}\nabla\cdot a\right]\vec{r} - \frac{3\vec{a}}{r^3} + \frac{1}{r^3}\left(a_1\frac{\partial}{\partial x} + a_2\frac{\partial}{\partial y} + a_3\frac{\partial}{\partial z}\right)\vec{r} - \left(x\frac{\partial}{\partial x} + y\frac{\partial}{\partial y} + z\frac{\partial}{\partial z}\right)\frac{\vec{a}}{r^3}$$

$$= \left[-\frac{3(\vec{r}\times\vec{a})}{r^5} + 0\right]\vec{r} - \frac{3\vec{a}}{r^3} + \frac{1}{r^3}\vec{a} - \left(\Sigma\frac{x(-3)}{r^4}\frac{x}{r}\right)\vec{a}$$

$$= -\frac{3}{r^5}(\vec{r}.\vec{a})\vec{r} - \frac{2\vec{a}}{r^3} + 3\frac{\vec{a}}{r^3} = \frac{-3}{r^5}(\vec{r}\cdot\vec{a})\vec{r} + \frac{\vec{a}}{r^3}$$

Therefore $\quad \nabla\left[a\cdot\nabla\left(\frac{1}{r}\right)\right] + \nabla\times\vec{a}\times\nabla\left(\frac{1}{r}\right) = 0$ **Hence proved.**

10. *If $\vec{F}$ is a solenoidal vector field, show that*

$$\text{Curl Curl Curl Curl } \vec{F} = \nabla^4\,\vec{F}.$$

SOLUTION: We know that

$$\text{Curl Curl } \vec{F} = \nabla\times(\nabla\times\vec{F}) = \nabla(\nabla\cdot\vec{F}) - \nabla^2\times\vec{F}$$

Since $\vec{F}$ is solenoidal $\nabla\cdot\vec{F} = 0$, therefore $\nabla\times(\nabla\times\vec{F}) = 0 - \nabla^2\vec{F}$

Repeating the operation, we get

Curl Curl (Curl Curl $\vec{F}$) = Curl Curl $(-\nabla^2 F) = -\nabla^2(-\nabla^2\vec{F}) = \nabla^4 F.$ **Hence proved.**

11. *If $\vec{a}$ and $\vec{b}$ are constant vectors show that*

$$\nabla\{(\vec{r}\times\vec{a})\cdot(\vec{r}\times\vec{b})\} = \vec{b}\times(\vec{r}\times\vec{a}) + \vec{a}\times(\vec{r}\times\vec{b})$$

SOLUTION: $\quad \nabla\{(\vec{r}\times\vec{a})\times(\vec{r}\times\vec{b})\} = \Sigma i\frac{\partial}{\partial x}\{(\vec{r}\times\vec{a})\cdot(\vec{r}\times\vec{b})\}$

$$= \Sigma i\left[(\vec{r}\times\vec{a})\cdot\frac{\partial}{\partial x}(\vec{r}\times\vec{b}) + (\vec{r}\times\vec{b})\cdot\frac{\partial}{\partial x}(\vec{r}\times\vec{a})\right]$$

Consider $\quad \Sigma i\left[(r\times a)\cdot\frac{\partial}{\partial x}(r\times b)\right] = \Sigma i\left[(\vec{r}\times\vec{a})\cdot\left(\frac{\partial\vec{r}}{\partial x}\times\vec{b} + \vec{r}\times\frac{\partial\vec{b}}{\partial x}\right)\right]$

$$= \Sigma i\left[(\vec{r}\times\vec{a})\cdot(i\times\vec{b}) + 0\right] \qquad \left(\text{since as } \vec{b} \text{ is constant}\right)$$

$$= \Sigma i\left[i\cdot\left\{\vec{b}+(\vec{r}\times\vec{a})\right\}\right] \quad (\text{as } a\cdot(b\times c) = b\cdot(c\times a) = c\cdot(a\times b)]$$

$$= \vec{b}\times(\vec{r}\times\vec{a}) \qquad (\text{as } \Sigma i(i\cdot\vec{V} = \Sigma iV_1 = \vec{V})$$

Similarly, $\Sigma i\left[(\vec{r}\times\vec{b})\cdot\frac{\partial}{\partial x}(\vec{r}\times\vec{a})\right] = \vec{a}\times(\vec{r}\times\vec{b})$ **Hence the result.**

12. *If the field* $\vec{F} = (x - 4y + az)\, i + (bx + 2y + z)\, j + (2x - cy + 3z)\, k$ *is conservative, find a, b, c. Also find the work done in moving an object in this field from* (1, 1, 1) *to the point* (2, 1, 3).

SOLUTION: If $\vec{F}$ is conservative filed then $\nabla\times\vec{F} = 0$

or $\begin{vmatrix} i & j & k \\ \frac{\partial}{\partial x} & \frac{\partial}{\partial y} & \frac{\partial}{\partial z} \\ x-4y+az & bx+2y+z & 2x-cy+3z \end{vmatrix} = 0$ or $i(-c-1) + j(a-2) + k(b+4) = 0$

$\Rightarrow$ $a = 2,\quad b = -4,\quad c = -1$ $\quad\therefore F = (x + 4y + 2z)i + (-4x + 2y + z)\,j + (2x + y + 3z)\,k$

For work done, we find the scalar potential ϕ so that $\nabla\phi = \vec{F}$

Thus, $\frac{\partial\phi}{\partial x} = x - 4y + 2z, \quad \frac{\partial\phi}{\partial y} = -4x + 2y + z, \quad \frac{\partial\phi}{\partial z} = 2x + y + 3z.$

$\therefore \quad \phi(x, y, z) = \frac{x^2}{2} - 4xy + 2xz + k(y, z)$

Now $\frac{\partial\phi}{\partial y} = -4x + \frac{\partial k}{\partial y} \Rightarrow \frac{\partial k}{\partial y} = 2y + z$ hence $k = y^2 + yz + A(z)$

$\therefore \quad \phi(x, y, z) = \frac{x^2}{2} - 4xy + 2xz + y^2 + yz + A(z)$

Now $\frac{\partial\phi}{\partial z} = 2x + y + \frac{dA}{dz} \Rightarrow \frac{dA}{dz} = 3z$ hence $A = \frac{3z^2}{2}.$

$\therefore \quad \phi(x, y, z) = \frac{x^2}{2} - 4xy + 2xz + y^2 + yz + \frac{3}{2}z^2$

The work done by the force in moving the object from (1, 1, 1) to (2, 1, 3)

$= \phi(2, 1, 3) - \phi(1, 1, 1) = (2 - 8 + 12 + 1 + 3 + \frac{27}{2}) - \left(\frac{1}{2} - 4 + 2 + 1 + 1 + \frac{3}{2}\right) = \frac{43}{2}$ units **Ans.**

13. *Show that the vector field* $\vec{F} = (x^2 - yz)\, i + (y^2 - zx)\, j + (z^2 - xy)\, k$ *is irrotational and find its corresponding scalar field* ϕ *so that* $\vec{F} = \nabla\phi$. *Also find the work done by the field if an object gets displaced from the origin to the point* (1, 2, 3).

SOLUTION: $$\nabla \times F = \begin{vmatrix} i & j & k \\ \dfrac{\partial}{\partial x} & \dfrac{\partial}{\partial y} & \dfrac{\partial}{\partial z} \\ x^2 - yz & y^2 - zx & z^2 - xy \end{vmatrix} = i(-x + x) + j(-y + y) + k(-z + z) = 0$$

Now, let $\vec{F} = \nabla\phi$ then $\dfrac{\partial\phi}{\partial x} = x^2 - yz,\quad \dfrac{\partial\phi}{\partial y} = y^2 - zx,\quad \dfrac{\partial\phi}{\partial z} = z^2 - xy$

$\therefore \quad \phi = \dfrac{x^3}{3} - xyz + A(y,z)$ hence $\dfrac{\partial\phi}{\partial y} = -xz + \dfrac{\partial A}{\partial y} \Rightarrow \dfrac{\partial A}{\partial y} = y^2 \quad \therefore \quad A = \dfrac{y^3}{3} + B(z)$

$\therefore \quad \phi = \dfrac{x^3}{3} - xyz + \dfrac{y^3}{3} + B(z) \quad \therefore \quad \dfrac{\partial\phi}{\partial z} = -xy + \dfrac{dB}{dz} \Rightarrow \dfrac{\partial B}{\partial z} = z^2 \quad \therefore \quad B = \dfrac{z^3}{3}$

Hence $\phi = \dfrac{x^3}{3} + \dfrac{y^3}{3} + \dfrac{z^3}{3} - xyz.$

Thus the work done by the field if an object gets displaced from the origin to the point (1, 2, 3)

$= \phi(1, 2, 3) - \phi(0, 0, 0)$

$= \dfrac{1}{3} + \dfrac{8}{3} + \dfrac{27}{3} - 6 - 0 = 6$ units **Ans.**

14. *Find the values of* λ *and* μ *so that the surfaces* $\lambda x^2 - \mu yz = (\lambda + z)x$ *and* $4x^2y + z^3 = 4$ *intersect orthogonally at the point* $(1, -1, 2)$.

(GGSIPU IInd Sem. IInd Term 2005)

SOLUTION: $\phi \equiv \lambda x^2 - \mu yz - (\lambda + z)\, x = 0$ and $\Psi \equiv 4x^2y + z^3 - 4 = 0.$

$\therefore \quad \nabla\phi = \{2\lambda x - (\lambda + z)\}\, i + (-\mu z) j + (-\mu y - x) k$ and $\nabla\psi = 8xyi + 4x^2 j + 3z^2 k$

Hence $\nabla\phi \cdot \nabla\psi = 8xy(2\lambda x - \lambda - z) - 4x^2\mu z - 3z^2(\mu y + x)$

At $(1, -1, 2)$ $\nabla\phi \cdot \nabla\psi = -8(2\lambda - \lambda - 2) - 8\mu - 12(-\mu + 1) = -8\lambda + 4\mu + 4$

Since two surfaces intersect outhogernally $-8\lambda + 4\mu + 4 = 0$

The point $(1, -1, 2)$ lies on the surface $\phi = 0$ hence $\lambda + 2\mu - (\lambda + 2) = 0$

$\therefore \quad \mu = 1, \quad \lambda = 1$ **Ans.**

15. *Find the directional derivative of* $\phi = x^2yz + 4xz^2$ at the point $(1, -2, -1)$ in the direction of the vector $2i - j - 2k$.

SOLUTION: $\phi = x^2yz + 4xz^2 \quad \therefore \quad \nabla\phi = (2xyz + 4z^2)\, i + x^2 zj + (x^2y + 8xz)\, k$

$[\nabla\phi]_{(1,-2,-1)} = 8i - j - 10k.$ Unit vector $\hat{n}$ along $2i - j - 2k$ is given by

$$\hat{n} = \frac{2i - j - 2k}{\sqrt{4+1+4}} = \frac{1}{3}(2i - j - 2k).$$ Hence the required directional derivative

$$= \nabla \phi \cdot \hat{n} = (8i - j - 10k) \cdot \frac{1}{3}(2i - j - 2k) = \frac{1}{3}(16 + 1 + 20) = \frac{37}{3}.$$ **Ans.**

16. *Find the directional derivative of* $\phi = 5x^2y - 5y^2z + \frac{5}{2}z^2x$ *at the point* (1, 1, 1) *in the direction of the line* $\frac{x-1}{2} = \frac{y-3}{-2} = \frac{z}{1}$.

SOLUTION: $\phi = 5x^2y - 5y^2z + \frac{5}{2}z^2x$

$$\therefore \quad \nabla \phi = \left(10xy + \frac{5}{2}z^2\right)i + (5x^2 - 10yz)\,j + (-5y^2 + 5zx)\,k.$$

$$[\nabla\phi]_{(1,\,1,\,1)} = \frac{25}{2}i - 5j + 0k$$

Unit vector along the line $\frac{x-1}{2} = \frac{y-3}{-2} = \frac{z}{1}$ is $\hat{n} = \frac{2i - 2j + k}{3}$

Hence the directional derivative as required

$$= \nabla\phi \cdot \hat{n} = \frac{1}{3}\left(\frac{25}{2}i - 5j + 0k\right) \cdot (2i - 2j + k)$$

$$= \frac{1}{3}(25 + 10 + 0) = \frac{35}{3} \text{ units.}$$ **Ans.**

17. *Find the directional derivative of* $\frac{1}{r^2}$ *in the direction of* $\vec{r} = xi + yj + zk$.

SOLUTION: $\phi = \frac{1}{r^2} \quad \therefore \quad \nabla\phi = \sum \frac{\partial \phi}{\partial x} i = \sum \frac{-2}{r^3}\frac{\partial r}{\partial x} i$

or $$\nabla\phi = \sum \frac{-2}{r^3}\frac{x}{r} i = \frac{-2}{r^4}\vec{r}$$

$\therefore$ Directional derivative of ϕ along $\vec{r} = \nabla\phi \cdot \vec{r} = \nabla\phi \cdot \frac{\vec{r}}{r}$

$$= -\frac{2}{r^4}\vec{r} \cdot \frac{\vec{r}}{r} = \frac{-2}{r^5}r^2 = \frac{-2}{r^3}.$$ **Ans.**

18. (*i*) If $f(x, y, z) = 3x^2y - y^2z^2$ find gradient of f at (1, –2, –1).

(*ii*) If $\vec{F} = xz^3\, i - 2x^2yz\, j + 2yz^4\, k$ find divergence and curl of $\vec{F}$.

SOLUTION: (*i*) grad $f = \nabla f = i\frac{\partial f}{\partial x} + j\frac{\partial f}{\partial y} + k\frac{\partial f}{\partial z} = 6xy\, i + (3x^2 - 2yz^2)\,j - 2y^2 zk$

$\therefore \quad \nabla f$ at $(1, -2, -1) = -12i + 7j + 8k$ **Ans.**

(*ii*) $\vec{F} = x z^3 i - 2x^2 yz j + 2yz^4 k$

$$\text{div } \vec{F} = \nabla \times \vec{F} = \frac{\partial}{\partial x}(xz^3) + \frac{\partial}{\partial y}(-2x^2yz) + \frac{\partial}{\partial z}(2yz^4) = z^3 - 2x^2z + 8yz^3$$

and
$$\text{curl } \vec{F} = \begin{vmatrix} i & j & k \\ \frac{\partial}{\partial x} & \frac{\partial}{\partial y} & \frac{\partial}{\partial z} \\ xz^3 & -2x^2yz & 2yz^4 \end{vmatrix} = i(2z^4 + 2x^2y) + j(3xz^2 - 0) + k(-4xyz - 0).$$

$$= i(2z^4 + 2x^2y) + 3xz^2 j - 4xyz\, k.$$ **Ans.**

19. *Find the directional derivative of* $\frac{1}{r}$ *in the direction of* $\vec{r} = xi + yj + zk.$

SOLUTION: $\phi = \frac{1}{r} \quad \therefore \quad \nabla\phi = \sum \frac{-i}{r^2}\frac{\partial r}{\partial x} = \sum \frac{-xi}{r^3} = \frac{-\vec{r}}{r^3}.$

Directional derivative of ϕ along $\vec{r} = \nabla\phi \cdot \hat{r}$

$$= -\frac{\vec{r}}{r^3} \cdot \frac{\vec{r}}{r} = \frac{-r^2}{r^4} = \frac{-1}{r^2}.$$ **Ans.**

20. *Find the angle between the surfaces* $x^2 + y^2 + z^2 = 9$ *and* $z = x^2 + y^2 - 3$ *at the point* (2, –1, 2).

SOLUTION: Let $\phi_1 \equiv x^2 + y^2 + z^2 - 9 = 0$, $\phi_2 = x^2 + y^2 - 3 - z = 0$ be two surfaces

$$\nabla\phi_1 = i\frac{\partial\phi_1}{\partial x} + j\frac{\partial\phi_1}{\partial y} + k\frac{\partial\phi_1}{\partial z} = 2xi + 2yj + 2zk = 4i - 2j + 4k \text{ at } (2, -1, 2).$$

$$\nabla\phi_2 = i\frac{\partial\phi_2}{\partial x} + j\frac{\partial\phi_2}{\partial y} + k\frac{\partial\phi_2}{\partial z} = 2xi + 2yj - k = 4i - 2j - k \text{ at } (2, -1, 2).$$

If ϕ is the angle between the surfaces so is between their normals, hence

$$\cos\phi = \frac{\nabla\phi_1 \cdot \nabla\phi_2}{|\nabla\phi_1||\nabla\phi_2|} = \frac{16+4-4}{\sqrt{16+4+16}\,\sqrt{16+4+1}} = \frac{16}{6\sqrt{21}} \quad \therefore \ \theta = \cos^{-1}\frac{8}{3\sqrt{21}}$$ **Ans.**

21. *For the function* $\phi(x, y) = \frac{x}{x^2 + y^2}$ find the magnitude of the directional derivative along a line making an angle 30° with the positive x-axis at (0, 2).

SOLUTION: $\nabla\phi = i\frac{\partial\phi}{\partial x} + j\frac{\partial\phi}{\partial y} = \frac{i(y^2 - x^2)}{(y^2 + x^2)^2} - \frac{2xyj}{(x^2 + y^2)^2}$

At (0, 2), $\nabla\phi = \frac{4i + 0j}{16} = \frac{i}{4}.$

Unit vector along a line making an angle 30° with X-axis

$$= i\cos\frac{\pi}{6} + j\sin\frac{\pi}{6} = \frac{i\sqrt{3}}{8} + \frac{j}{2}.$$

$\therefore$ Required directional derivative $= \frac{i}{4}\cdot\left(\frac{i\sqrt{3}}{2} + \frac{j}{2}\right) = \frac{\sqrt{3}}{8}$ **Ans.**

22. *Find the values of a, b, c for which the vector V, given by*

$\vec{V} = (x + y + az)\, i + (bx + 3y - z) j + (3x + cy + z)\, k$ is irrotational.

SOLUTION: Vector V is irrotational when $\nabla \times \vec{V} = 0$.

$$\nabla \times V = \begin{vmatrix} i & j & k \\ \dfrac{\partial}{\partial x} & \dfrac{\partial}{\partial y} & \dfrac{\partial}{\partial z} \\ x+y+az & bx+3y-z & 3x+cy+z \end{vmatrix} = i(c+1) + j(a-3) + k(b-1)$$

$\nabla \times V = 0$ when $a = 3,\ b = 1,\ c = -1.$ **Ans.**

ADDITIONAL IMPORTANT PROBLEMS

1. If $\vec{r} = (a\cos t)i + (a\sin t)j + (at\tan\alpha)k$, show that

(i) $\left|\dfrac{d\vec{r}}{dt} \times \dfrac{d^2\vec{r}}{dt^2}\right| = a^2 \sec\alpha$ (ii) $\left[\dfrac{d\vec{r}}{dt}, \dfrac{d^2\vec{r}}{dt^2}, \dfrac{d^3\vec{r}}{dt^3}\right] = a^3 \tan\alpha.$

SOLUTION: $\vec{r} = (a\cos t)i + (a\sin t)j + (at\tan\alpha)k,$

$$\therefore \quad \frac{d\vec{r}}{dt} = (-a\sin t)i + (a\cos t)j + (a\tan\alpha)k \quad \text{and} \quad \frac{d^2\vec{r}}{dt^2} = (-a\cos t)i - (a\sin t)j + ok$$

and $$\frac{d^3\vec{r}}{dt^3} = (a\sin t)i - (a\cos t)j.$$

Now (i) $$\left|\frac{d\vec{r}}{dt} \times \frac{d^2\vec{r}}{dt^2}\right| = \begin{vmatrix} i & j & k \\ -a\sin t & a\cos t & a\tan\alpha \\ -a\cos t & -a\sin t & 0 \end{vmatrix}$$

$$= i(a^2\tan\alpha\sin t) + j(-a^2\tan\alpha\cos t) + k(a^2\sin^2 t + a^2\cos^2 t)$$

$$= (a^2\tan\alpha\sin t)i + (-a^2\tan\alpha\cos t)j + a^2k$$

$$\therefore \quad \left|\frac{d\vec{r}}{dt} \times \frac{d^2\vec{r}}{dt^2}\right| = \sqrt{a^4\tan^2\alpha\sin^2 t + a^4\tan^2\alpha\cos^2 t + a^4} = a^2\sqrt{\tan^2\alpha + 1} = a^2\sec\alpha.$$

(ii) $$\left[\frac{d\vec{r}}{dt}, \frac{d^2\vec{r}}{dt^2}, \frac{d^3\vec{r}}{dt^3}\right] = \frac{d^3\vec{r}}{dt^3} \cdot \left(\frac{d\vec{r}}{dt} \times \frac{d^2\vec{r}}{dt^2}\right)$$

$$= \{(a\sin t)i - (a\cos t)j\}.\{(a^2\tan\alpha\sin t)i - (a^2\tan\alpha\cos t)j + a^2k\}$$

$$= a\sin t\,(a^2\tan\alpha\sin t) + a\cos t(a^2\tan\alpha\cos t) = a^3\tan\alpha.$$ **Hence Proved**

2. Show that a particle whose position vector $\vec{r}$ at any time t, is given by $\vec{r} = (a \cos nt)i + (b \sin nt)j$ moves in an ellipse whose centre is at the origin and whose acceleration varies directly as its distance from the centre and is directed towards it.

SOLUTION: $\vec{r} = (a \cos nt)i + (b \sin nt)j = xi + yi$

hence $x = a \cos nt, \quad y = b \sin nt \Rightarrow \dfrac{x^2}{a^2} + \dfrac{y^2}{b^2} = 1$, ellipse with centre at (0, 0).

Now Velocity $= \dfrac{d\vec{r}}{dt} = (-an \sin nt)i + (bn \cos nt)j$ is the velocity at any time t.

and

$$\text{acceleration} = \frac{d^2\vec{r}}{dt^2} = (-an^2 \cos nt)i + (-bn^2 \sin nt)j$$

$$= -n^2\{(a \cos nt)i + (b \sin nt)j\} = -n^2\vec{r}.$$

This shows that acceleration of the particle varies as its distance $|\vec{r}|$ from the origin (centre of the ellipse) and the minus sign shows that it is directed towards the centre.

3. *Show that* $\nabla\left(\dfrac{\vec{a}\cdot\vec{r}}{r^n}\right) = \dfrac{\vec{a}}{r^n} - \dfrac{n(\vec{a}.\vec{r})}{r^{n+2}}(\vec{r})$ *where* $\vec{r} = xi + yj + zk$ *and* $\vec{a}$ *is a constant vector.*

SOLUTION: Let $\vec{a} = a_1 i + a_2 j + a_3 k$, and $\phi = \dfrac{\vec{a}.\vec{r}}{r^n}$ where $\vec{r} = xi + yj + zk$

$$\therefore \quad \phi = \frac{\vec{a}.\vec{r}}{r^n} = \frac{a_1x + a_2y + a_3z}{r^n}$$

$$\therefore \quad \frac{\partial\phi}{\partial x} = \frac{1}{r^n}\frac{\partial}{\partial x}(a_1x + a_2y + a_3z) + (a_1x + a_2y + a_3z)\left(\frac{-n}{r^{n+1}}\right)\frac{\partial r}{\partial x}$$

$$= \frac{a_1}{r^n} - n(a_1x + a_2y + a_3z)\frac{1}{r^{n+1}}\frac{x}{r} = \frac{a_1}{r^n} -- n(a_1x + a_2y + a_3z)\frac{x}{r^{n+2}}$$

Similarly,

$$\frac{\partial\phi}{\partial y} = \frac{a_2}{r^n} - n(a_1x + a_2y + a_3z)\frac{y}{r^{n+2}} \text{ and } \frac{\partial\phi}{\partial z} = \frac{a_3}{r^n} - n(a_1x + a_2y + a_3z)\frac{z}{r^{n+2}}$$

$$\therefore \quad \nabla\phi = i\frac{\partial\phi}{\partial x} + j\frac{\partial\phi}{\partial x} + k\frac{\partial\phi}{\partial z}$$

$$= \frac{a_1i + a_2j + a_3k}{r^n} - \frac{n}{r^{x+2}}(a_1x + a_2y + a_3z)(xi + yj + zk)$$

$$= \frac{\vec{a}}{r^n} - n\frac{(\vec{a}.\vec{r})}{r^{n+2}}(\vec{r}) \text{ where } \vec{a} \text{ is a constant vector.}$$

Hence proved.

4. Find the directional derivative of $\phi = e^{2x}\cos yz$ at the origin in the direction of the tangent to the curve $x = a \sin t$, $y = a \cos t$, $z = at$ at $t = \pi/4$.

SOLUTION:

$$\nabla\phi = i\frac{\partial}{\partial x}\phi + j\frac{\partial\phi}{\partial y} + k\frac{\partial\phi}{\partial z}$$

$$= i(2e^x \cos yz) + j(-ze^{2x}\sin yz) + k(-ye^{2x}\sin yz)$$

At the origin $\nabla\phi = 2i$

Tangent T to the given curve $x = a \sin t$, $y = a \cos t$, $z = at$ is given by

$$T = \frac{d\vec{r}}{dt} = \frac{d}{dt}[a \sin t i + a \cos t j + at\, k]$$

$$= (a \cos t)i - a \sin t j + ak$$

At $t = \pi/4$, $\frac{d\vec{r}}{dt} = \frac{a}{\sqrt{2}} i - \frac{a}{\sqrt{2}} j + ak$ and $\left|\frac{d\vec{r}}{dt}\right| = |T| = \sqrt{\frac{a^2}{2} + \frac{a^2}{2} + a^2} = a\sqrt{2}$

$\therefore$ Required Directional derivative $= \nabla\phi \cdot \hat{T}$

$$= 2i \cdot \frac{\left(\frac{a}{\sqrt{2}} i - \frac{a}{\sqrt{2}} j + ak\right)}{a\sqrt{2}} = \frac{2\frac{a}{\sqrt{2}}}{a\sqrt{2}} = 1.$$ **Ans.**

5. Find the directional derivative of v^2 where $\vec{v} = xy^2 i + zy^2 j + xz^2 k$ at the point (2, 0, 3) in the direction of the outward normal to the sphere $x^2 + y^2 + z^2 = 14$ at the point (3, 2, 1).

SOLUTION: $v^2 = \vec{v}.\vec{v} = x^2y^4 + z^2y^4 + x^2z^4 = \phi$, say.

$$\therefore \quad \nabla\phi = i(2xy^4 + 2xz^4) + j(4x^2y^3 + 4x^2y^3) + k(2zy^4 + 4x^2z^3)$$

At the point (2, 0, 3) $\nabla\phi = (0 + 324)i + (0 + 0)j + k(0 + 432) = 324i + 432k.$

Next, let $\psi = x^2 + y^2 + z^2 - 14$ then $\nabla\psi = 2xi + 2yj + 2zk$ is along the outward drawn normal to the surface and $\nabla\psi$ at the point $(3, 2, 1) = 6i + 4j + 2k$.

$\therefore$ Directional derivative along the normal to the sphere

$$= \nabla\phi . \frac{\nabla\psi}{|\nabla\psi|} = (324i + 432k) . \frac{(6i + 4j + 2k)}{\sqrt{6^2 + 4^2 + 2^2}} = \frac{1404}{\sqrt{14}}$$ **Ans.**

6. Find the directional derivative of $\phi = x^2 - y^2 + 2z^2$ at the point $A(1, 2, 3)$ in the direction of the line AB where $B = (5, 0, 4)$. In what direction it will be maximum? Find the maximum value of it.

SOLUTION: $\phi = x^2 - y^2 + 2z^2 \quad \therefore \quad \nabla\phi = i\frac{\partial\phi}{\partial x} + j\frac{\partial\phi}{\partial y} + k\frac{\partial\phi}{\partial z} = 2xi - 2yj + 4zk$

At the point $A(1, 2, 3)$, $\nabla\phi = 2i - 4j + 12k$. Now $B = (5, 0, 4)$, therefore

$$\overrightarrow{AB} = (5 - 1)i + (0 - 2)j + (4 - 3)k = 4i - 2j + k.$$

$\therefore$ Directional derivative of ϕ along $\overrightarrow{AB}$

$$= (2i - 4j + 12k) . \frac{(4i - 2j + k)}{\sqrt{16 + 4 + 1}} = \frac{8 + 8 + 12}{\sqrt{21}} = \frac{28}{\sqrt{21}}.$$

Directional derivative at A is maximum along $\nabla\phi$.

$\therefore$ Its Maximum value $= \sqrt{2^2 + (-4)^2 + (12)^2} = \sqrt{4 + 16 + 144} = \sqrt{164} = 2\sqrt{41}.$ **Ans.**

7. Find the values of a, b, c if the directional derivative of $\phi = axy^2 + byz + cz^2x^3$ at $(1, 2, -1)$ has maximum magnitude 64 in the direction parallel to z-axis.

SOLUTION: Since $\phi = axy^2 + byz + cz^2x^3$ we have $\nabla\phi = i(ay^2 + 3cx^2z^2) + j(2axy + bz) + k(by + 2czx^3)$.

At the point $(1, 2, -1)$, $\nabla\phi = (4a + 3c)i + (4a - b)j + (2b - 2c)k$

The directional derivative is maximum along z-axis

Hence $\nabla\phi = (2b - 2c)k \Rightarrow 4a + 3c = 0$ and $4a - b = 0$.

Thus $b = 4a$ and $c = -4a/3$

$$\therefore \quad \nabla\phi = \left(8a + \frac{8a}{3}\right)k = \frac{32}{3}ak.$$

Given $|\nabla\phi| = 64 \Rightarrow \dfrac{32a}{3} = 64 \quad \therefore a = 6.$

Thus $a = 6$, $b = 24$ and $c = -8$. **Ans.**

8. Find the direction in which temperature changes most rapidly with distance from the point $(1, 1, 1)$ and determine the maximum rate of change if the temperature at any point is given by $\phi(x, y, z) = xy + yz + zx$.

SOLUTION: Rate of change of temperature with be maximum a'ong $\nabla\phi$.

$$\nabla\phi = i(y + z) + j(x + z) + k(x + y)$$

At the point $(1, 1, 1)$, $\nabla\phi = 2i + 2j + 2k$.

Maximum rate of change = maximum directional derivative

$$= |\nabla\phi| = \sqrt{4+4+4} = 2\sqrt{3}.$$ **Ans.**

9. The temperature of the points in space is given by $\phi = x^2 + y^2 - z$. A mosquito located at the point $(1, 1, 2)$ desires to fly in such a direction that it will get warm as soon as possible. In what direction should it move?

SOLUTION: $\phi = x^2 + y^2 - z$. Rate of change in temperature $= \nabla\phi$

$\nabla\phi = 2xi + 2yj - k$. At the point $(1, 1, 2)$, $\nabla\phi = 2i + 2j - k$.

Mosquito will get warm at the earliest if it moves in the direction in which rate of change in temperature is maximum, that is, along $\nabla\phi$ itself.

$$\text{Unit vector along } \nabla\phi = \frac{2i + 2j - k}{\sqrt{4+4+1}} = \frac{2i + 2j - k}{3}$$

Thus, mosquito should move in the direction of $(2i + 2j - k)/3$. **Ans.**

10. Show that the vector field $\vec{V} = \dfrac{a(xi + yj)}{\sqrt{x^2 + y^2}}$ is a source field or sink field according as $a > 0$, or $a < 0$.

SOLUTION: We know that a vector field $\vec{F}$ is a source field or sink field according as $\nabla.\vec{F} > 0$ or < 0.

Here
$$\nabla.\vec{V} = \nabla.\left(\frac{ax}{\sqrt{x^2+y^2}}i + \frac{ay}{\sqrt{x^2+y^2}}j\right)$$
$$= \frac{\partial}{\partial x}\left(\frac{ax}{\sqrt{x^2+y^2}}\right) + \frac{\partial}{\partial y}\left(\frac{ay}{\sqrt{x^2+y^2}}\right)$$
$$= \frac{a}{\sqrt{x^2+y^2}} - \frac{2ax^2}{2(x^2+y^2)^{3/2}} + \frac{a}{\sqrt{x^2+y^2}} - \frac{2ay^2}{2(x^2+y^2)^{3/2}}$$
$$= \frac{2a}{\sqrt{x^2+y^2}} - \frac{a(x^2+y^2)}{\sqrt{(x^2+y^2)^{3/2}}} = \frac{a}{\sqrt{x^2+y^2}}.$$

Therefore vector field $\vec{V}$ is a source field or sink field according as $a > 0$ or $a < 0$. **Ans.**

11. Find curl curl $\vec{F}$ if $\vec{F} = x^2yi - 2xzj + 2yzk$ at the point (1, 0, 2).

SOLUTION: Curl $\vec{F} = \nabla \times \vec{F} = \begin{vmatrix} i & j & k \\ \dfrac{\partial}{\partial x} & \dfrac{\partial}{\partial y} & \dfrac{\partial}{\partial z} \\ x^2y & -2xz & 2yz \end{vmatrix} = i(2z+2x) + j(0-0) + k(-2z-x^2)$

$\therefore$ Curl curl $\vec{F} = \nabla \times \vec{F} = \begin{vmatrix} i & j & k \\ \dfrac{\partial}{\partial x} & \dfrac{\partial}{\partial y} & \dfrac{\partial}{\partial z} \\ 2(z+x) & 0 & -(2z+x^2) \end{vmatrix}$

$= i(0-0) + j(2+2x) + k(0-0) = 2(x+1)j.$

At the point (1, 0, 2), $\nabla \times (\nabla \times \vec{F}) = 4j.$ **Ans.**

12. Show that the vector field $\vec{F} = [y^2 - z^2 + 3yz - 2x)i + (3xz + 2xy)j + (3xy - 2xz + 2z)k$ is solenoidal as well as irrotational.

SOLUTION: $\nabla.\vec{F} = \dfrac{\partial}{\partial x}(y^2 - z^2 + 3yz - 2x) + \dfrac{\partial}{\partial y}(3xz + 2xy) + \dfrac{\partial}{\partial z}(3xy - 2xz + 2z)$

$= -2 + 2x - 2x + 2 = 0$

Hence $\vec{F}$ is solenoidal.

$$\nabla \times \vec{F} = \begin{vmatrix} i & j & k \\ \dfrac{\partial}{\partial x} & \dfrac{\partial}{\partial y} & \dfrac{\partial}{\partial z} \\ y^2 - z^2 + 3yz - 2x & 3xz + 2xy & 3xy - 2xz + 2z \end{vmatrix}$$
$$= i(3x - 3x) + j(-2z + 3y - 3y + 2z) + k(3z + 2y - 2y - 3z) = 0$$

Hence $\vec{F}$ is irrotational. **Hence proved.**

13. Find the directional derivative of the divergence of the vector field $\vec{F} = xyi + xy^2 j + z^2 k$ at the point (2, 1, 2) in the direction of the outer normal to the sphere $x^2 + y^2 + z^2 = 9$.

SOLUTION:

$$\nabla.\vec{F} = \frac{\partial}{\partial x}(xy) + \frac{\partial}{\partial y}(xy^2) + \frac{\partial}{\partial z}(z^2)$$

$$= y + 2xy + 2z = \phi(x, y, z), \text{ say.}$$

$$\therefore \quad \nabla\phi = \nabla(\nabla.\vec{F}) = \left(i\frac{\partial}{\partial x} + j\frac{\partial}{\partial y} + k\frac{\partial}{\partial z}\right)(y + 2xy + 2z)$$

$$= i(2y) + j(1 + 2x) + k(2)$$

At the point (2, 1, 2), $\nabla\phi = 2i + 5j + 2k$

Normal to the given sphere $= \left(i\frac{\partial}{\partial x} + j\frac{\partial}{\partial y} + k\frac{\partial}{\partial z}\right)(x^2 + y^2 + z^2 - 9)$

$$= 2xi + 2yj + 2zk = 4i + 2j + 4k \text{ at the point } (2, 1, 2).$$

$\therefore$ Directional derivative in the direction of the normal to the sphere

$$= (2i + 5j + 2k).\frac{4i + 2j + 4k}{\sqrt{16 + 4 + 16}} = \frac{1}{6}(8 + 10 + 8) = \frac{13}{3}.$$ **Ans.**

14. Show that Div (grad r^n) = $n(n + 1)r^{n-2}$ where $\vec{r} = xi + yj + zk$.

SOLUTION: $\text{grad}(r^n) = \nabla(r^n) = i\frac{\partial}{\partial x}r^n + j\frac{\partial}{\partial y}r^n + k\frac{\partial}{\partial z}r^n$

$$= i\,n\,r^{n-1}\frac{\partial r}{\partial x} + j\,n r^{n-1}\frac{\partial r}{\partial y} + k\,n\,r^{n-1}\frac{\partial r}{\partial z}$$

$$= n\left[i r^{n-1}\frac{x}{r} + j r^{n-1}\frac{y}{r} + k r^{n-1}\frac{z}{r}\right] = n r^{n-2}(xi + yj + zk)$$

$$\therefore \text{ Div (grad } r^n) = \nabla.(\nabla r^n) = \frac{\partial}{\partial x}(nx\,r^{n-2}) + \frac{\partial}{\partial y}(ny\,r^{n-2}) + \frac{\partial}{\partial z}(nz\,r^{n-2}) + nr^{n-2} + nz\,(n-z)\,r^{n-3}\frac{\partial r}{\partial z}$$

$$= nr^{n-2} + nx(n-2)r^{n-3}\frac{\partial r}{\partial x} + nr^{n-2} + ny(n-2)r^{n-3}\frac{\partial r}{\partial y} + nr^{n-2} + nz(n-2)r^{n-3}\frac{\partial r}{\partial z}$$

$$= 3nr^{n-2} + n(n-2)\left[\frac{x^2}{r}r^{n-3} + \frac{y^2}{r}r^{n-3} + \frac{z^2}{r}r^{n-3}\right]$$

$$= 3nr^{n-2} + n(n-2)r^{n-4}(x^2 + y^2 + z^2)$$

$$= 3nr^{n-2} + n(n-2)r^{n-2} \qquad (\text{as } r^2 = x^2 + y^2 + z^2).$$

$$= (n^2 + 3n - 2n)\,r^{n-2} = n(n+1)r^{n-2}.$$ **Hence proved.**

15. Show that the vector field $\vec{F} = \frac{\vec{r}}{r^2}$ is irrotational.

SOLUTION: $\nabla \times \vec{F} = \nabla \times \left(\frac{xi}{r^2} + \frac{yj}{r^2} + \frac{zk}{r^2} \right) = \begin{vmatrix} i & j & k \\ \frac{\partial}{\partial x} & \frac{\partial}{\partial y} & \frac{\partial}{\partial z} \\ \frac{x}{r^2} & \frac{y}{r^2} & \frac{z}{r^2} \end{vmatrix}$

$$= i\left[\frac{\partial}{\partial y}\left(\frac{z}{r^2}\right) - \frac{\partial}{\partial z}\left(\frac{y}{r^2}\right)\right] + j\left[\frac{\partial}{\partial z}\left(\frac{x}{r^2}\right) - \frac{\partial}{\partial x}\left(\frac{z}{r^2}\right)\right] + k\left[\frac{\partial}{\partial x}\left(\frac{y}{r^2}\right) - \frac{\partial}{\partial y}\left(\frac{x}{r^2}\right)\right]$$

$$= i\left[z\frac{(-2)}{r^3}\frac{\partial r}{\partial y} - \frac{y(-2)}{r^3}\frac{\partial r}{\partial z}\right] + j\left[x\frac{(-2)}{r^3}\frac{\partial r}{\partial z} - \frac{z(-2)}{r^3}\frac{\partial r}{\partial x}\right] + k\left[y\frac{(-2)}{r^3}\frac{\partial r}{\partial x} - \frac{x(-2)}{r^3}\frac{\partial r}{\partial y}\right]$$

$$= \frac{-2i}{r^3}\left(z.\frac{y}{r} - y\frac{z}{r}\right) - \frac{2j}{r^3}\left(x.\frac{z}{r} - z\frac{x}{r}\right) - \frac{2k}{r^3}\left(y\frac{x}{r} - \frac{xy}{r}\right) = 0$$

Thus, $\frac{\vec{r}}{r^2}$ is irrotational. **Hence Proved.**

16. Show that $\nabla \times \left(\frac{\vec{a} \times \vec{r}}{r^n}\right) = \frac{(2-n)\,\vec{a}}{r^n} + \frac{n(\vec{a}.\vec{r})\vec{r}}{r^{n+2}}$ where $\vec{a}$ is a constant vector and $\vec{r} + xi + yj + zk$.

SOLUTION: $\nabla \times \left(\frac{\vec{a} \times \vec{r}}{r^n}\right) = \nabla \times (r^{-n}\,\vec{F})$ where $\vec{a} \times \vec{r} = \vec{F}$, say.

We know that $\nabla \times (\phi\vec{F}) = \phi(\nabla \times \vec{F}) + (\nabla\phi) \times \vec{F}$

$\therefore$
$$\nabla \times (r^{-n}\vec{F}) = r^{-n}(\nabla \times \vec{F}) + (\nabla r^{-n}) \times \vec{F}$$
$$= r^{-n}[\nabla \times (a \times \vec{r})] + \left(i\frac{\partial}{\partial x}r^{-n} + j\frac{\partial}{\partial y}r^{-n} + k\frac{\partial}{\partial z}r^{-n}\right) \times \vec{F}$$

We know that $\nabla \times (\vec{a} \times \vec{r}) = 2\vec{a}$

$\therefore$
$$\nabla \times (r^{-n} \times \vec{F}) = r^{-n}(2\vec{a}) - nr^{-n-1}\left(i\frac{\partial r}{\partial x} + j\frac{\partial r}{\partial y} + k\frac{\partial r}{\partial z}\right) \times \vec{F}$$
$$= \frac{2\vec{a}}{r^n} - nr^{-n-1}\left(i\frac{x}{r} + j\frac{y}{r} + k\frac{z}{r}\right) \times (\vec{a} \times \vec{r})$$
$$= \frac{2\vec{a}}{r^n} - \frac{n}{r^{n+2}}[\vec{r} \times (\vec{a} \times \vec{r})] = \frac{2\vec{a}}{r^n} - \frac{n}{r^{n+2}}[(r.r)\vec{a} - (a.r)\vec{r}]$$
$$= \frac{2\vec{a}}{r^n} - \frac{n}{r^{n+2}}[r^2\vec{a} - (\vec{a} \cdot \vec{r})\vec{r}] = \frac{(2-n)\vec{a}}{r^n} + \frac{n(\vec{a} \cdot \vec{r})\vec{r}}{r^{n+2}}.$$

Hence proved.

17. Find $\nabla^2\phi$ where $\phi = 3x^2z - y^2z^3 + 4x^3y + 2x - 3y$ at the point (1, 1, 0).

SOLUTION: $\nabla^2\phi = \dfrac{\partial^2\phi}{\partial x^2} + \dfrac{\partial^2\phi}{\partial y^2} + \dfrac{\partial^2\phi}{\partial z^2}$. Here $\phi = 3x^2z - y^2z^3 + 4x^3y + 2x - 3y$.

$$\therefore \quad \frac{\partial\phi}{\partial x} = 6xz + 12x^2y + 2 \quad \text{and} \quad \frac{\partial^2\phi}{\partial x^2} = 6z + 24xy$$

$$\frac{\partial\phi}{\partial y} = -2yz^3 + 4x^3 - 3 \quad \text{and} \quad \frac{\partial^2\phi}{\partial y^2} = -2z^3$$

$$\frac{\partial\phi}{\partial z} = 3x^2 - 3y^2z^2 \quad \text{and} \quad \frac{\partial^2\phi}{\partial z^2} = -6y^2z$$

$$\therefore \quad \nabla^2\phi = 6z + 24xy - 2z^3 - 6y^2z$$

$$= 0 + 24 - 0 - 0 = 24 \quad \text{at the point } (1, 1, 0)$$ **Ans.**

CHAPTER 15

Vector Integration

IMPORTANT FORMULAE AND USEFUL RESULTS

Line Integral: $\int_C \vec{F} \cdot \overline{dr} = \int_C F_1 dx + F_2 dy + F_3 dz$ where $\vec{F} = F_1 i + F_2 j + F_3 k.$

Green's Lemma: If two functions $u(x, y)$ and $v(x, y)$ and their partial derivatives are single valued and continuous over a plane region R bounded by a closed curve C, then

$$\int_C u\, dx + v\, dy = \iint_R \left(\frac{\partial v}{\partial x} - \frac{\partial u}{\partial y} \right) dx\, dy.$$

Surface Integral: If $\vec{F} = F_1 i + F_2 j = F_3 k$ then surface integral of $\vec{F}$ is given by

$$\iint_S \vec{F} \cdot \overline{ds} = \iint_S (F_1 dydz + F_2 dzdx + F_3 dxdy)$$

Stoke's Theorem: If S is an open surface bounded by a closed curve C and if a vector field $\vec{F}$ has continuous partial derivatives, then

$$\oint_C \vec{F} \cdot \overline{dr} = \iint_S (\nabla \times F) \cdot \overline{dS} = \iint_S \nabla \times F \cdot \hat{n}\, ds$$

Gauss Divergence Theorem: The surface integral of a vector field $\vec{F}$ taken over a closed surface S containing volume V, is equal to the volume integral of the divergence of $\vec{F}$ taken over the volume V bounded by the surface S.

$$\iint_S F \cdot ds = \iiint_V \nabla \cdot F\, dv$$

SOLUTIONS TO EXERCISE 15A

1. *Evaluate* $\oint_C \vec{F}\cdot d\vec{r}$ *where* $\vec{F} = (\sin z)\, i + (\cos x)\, j + (\sin y)\, k$ *and* C *is the boundary of the rectangle determined by the planes* $x = 0, x = \pi, y = 0, y = 1$ and $z = 3$.

SOLUTION: The four vertices of the rectangle *ABCD* are *A* (0, 0, 3), *B* (0, 1, 3), *C*(π, 1, 3) and *D* (π, 0, 3). On the side *AB*, $x = 0, 0 \le y \le 1, z = 3$, on side *BC*, $0 \le x \le \pi, y = 1, z = 3$; On the side *CD* we have $x = \pi, 0 \le y \le 1, z = 3$ and on side *DA*, $0 \le x \le \pi, y = 0, z = 3$.

$$\therefore \quad \int_C \vec{F}.d\vec{r} = \int_C (\sin z dx + \cos x dy + \sin y dz)$$

Where *C* is the closed curve from *A* to *B* then from *B* to *C* then from *C* to *D* and finally from *D* to *A*.

$$\therefore \quad \int_C \vec{F}.d\vec{r} = \int_{AB} (0+\cos 0\, dy+0) + \int_{BC} (\sin 3dx+0+0) + \int_{CD} (0+\cos\pi\, dy+0) + \int_{DA} (\sin 3dx+0+0)$$

$$= \int_0^1 dy + \int_0^\pi \sin 3dx + \int_1^0 -1dy + \int_\pi^0 \sin 3dx$$

$$= 1 + \pi \sin 3 + 1 - \pi \sin 3 = 2 \text{ (taken clockwise)}$$

$$= -2 \text{ if taken counter clockwise.}$$

Ans.

2. *Determine the value of the line integral* $\int_C \vec{F}\cdot dr$ *where* C *is the curve along the path* $x^2 = 4y,\ \ 3x^3 = 8z$ *from* $x = 0$ to $x = 2$ *and* $\vec{F} = 3x^2 i + (2xz - y)\, j + zk$.

SOLUTION:

$$\vec{F} = 3x^2 i + (2xz - y)\, j + zk$$

$$\therefore \quad \int F.dr = \int 3x^2 dx + (2xz - y)dy + zdz$$

Since $y = \dfrac{x^2}{4}$ and $z = \dfrac{3x^3}{8}$ we have $dy = \dfrac{xdx}{2}$, $dz = \dfrac{9x^2}{8}dx$

Therefore

$$\int F.dr = \int_0^2 \left[3x^2 dx + \left(\frac{6x^4}{8} - \frac{x^2}{4}\right)\frac{xdx}{2} + \frac{3x^3}{8}.\frac{9x^2}{8}dx\right]$$

$$= \int_0^2 \left(3x^2 + \frac{3x^5}{8} - \frac{x^3}{8} + \frac{27x^5}{64}\right)dx = \left[x^3 - \frac{x^4}{32} + \frac{(24+27)}{64}.\frac{x^6}{6}\right]_0^2$$

$$= 8 - \frac{1}{2} + \frac{17}{2} = 16$$

Ans.

3. *Find the work done in moving a particle free round the ellipse* $\dfrac{x^2}{16} + \dfrac{y^2}{9} = 1$ *in the plane* $z = 0$, *if the force field is given by* $\vec{F} = (3x - 4y + 2z)\, i + (4x + 2y - 3z^2)\, j + (2xz - 4y^2 + z^3)\, k$.

SOLUTION: Curve is $\dfrac{x^2}{16} + \dfrac{y^2}{9} = 1, z = 0$ or $x = 4\cos\theta, y = 3\sin\theta, z = 0$

$$\int \vec{F}.d\vec{r} = \int (3x - 4y + 2z)dx + (4x + 2y - 3z^2)dy + (2xz - 4y^2 + z^3)dz$$

$$= \int_0^{2\pi} (12\cos\theta - 12\sin\theta + 0)(-4\sin\theta)d\theta + (16\cos\theta + 6\sin\theta - 0)(3\cos\theta)d\theta + 0$$

$$= \int_0^{2\pi} \left[48\sin^2\theta + 48\cos^2\theta - 48\sin\theta\cos\theta + 18\sin\theta\cos\theta\right]d\theta$$

$$= \int_0^{2\pi} (48 - 30\sin\theta\cos\theta)d\theta = \left[48\theta - 15\sin^2\theta\right]_0^{2\pi} = 96\pi$$ **Ans.**

4. *Find the work done when a particle moves along the curve* $x = a\cos\theta$, $y = a\sin\theta$, $z = b\theta$ *from* $\theta = \pi/4$ *to* $\theta = \pi/2$ *under the action of the force field*

$$\vec{F} = -3a\sin^2\theta\cos\theta\, i + a(2\sin\theta - 3\sin^3\theta)\, j + (b\sin 2\theta)\, k.$$

SOLUTION: Curve is $x = a\cos\theta$, $y = a\sin\theta$, $z = b\theta$, $\frac{\pi}{4} \le \theta \le \frac{\pi}{2}$.

$$\int F.dr = \int -3a\sin^2\theta\cos\theta dx + a(2\sin\theta - 3\sin^3\theta)dy + b\sin 2\theta\, dz$$

$$= \int_{\pi/4}^{\pi/2} \left[-3a\sin^2\theta\cos\theta(-a\sin\theta)d\theta + a(2\sin\theta - 3\sin^3.\theta)(a\cos\theta)d\theta + b\sin 2\theta.bd\theta\right]$$

$$= \int_{\pi/4}^{\pi/2} (3a^2\sin^3\theta\cos\theta) + a^2(2\sin\theta - 3\sin^3\theta)\cos\theta + 2b^2\sin.\theta\cos\theta)d\theta$$

$$= \int_{1/\sqrt{2}}^{1} \left[3a^2t^3 + a^2(2t - 3t^3) + 2b^2t\right]dt \text{ (on putting } \sin\theta = t)$$

$$= \int_{1/\sqrt{2}}^{1} (2a^2t + 2b^2t)dt = (a^2 + b^2)\left[t^2\right]_{1/\sqrt{2}}^{1} = \frac{(a^2 + b^2)}{2}$$ **Ans.**

5. *Find* $\int_C \vec{F}\cdot dr$ *where* $\vec{F} = (x^2 + y^2)\, i + (x^2 - y^2)\, j + z^2k$ *along the boundary of the triangle with vertices* (1, 1, 0), (0, 1, 0), (0, 0, 0).

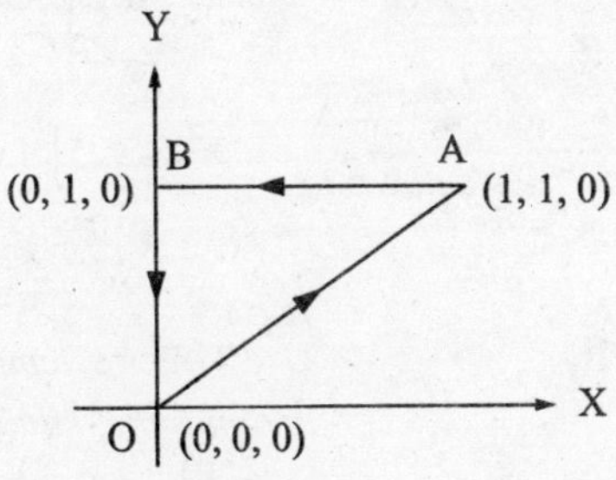

SOLUTION: $\int \vec{F}.\vec{dr} = \int (x^2 + y^2)dx + (x^2 - y^2)dy + z^2dz$

$$= \int_{AB} + \int_{BO} + \int_{OA}$$

$$= \int_1^0 (x^2 + 1)dx + \int_1^0 -y^2dy + \int_0^1 (x^2 + x^2)\, dx$$

$$= -\left[\frac{x^3}{3} + x\right]_0^1 + \left[\frac{y^3}{3}\right]_0^1 + \left[\frac{2x^3}{3}\right]_0^1 = -\frac{4}{3} + \frac{1}{3} + \frac{2}{3} = \frac{-1}{3}$$ **Ans.**

6. *Evaluate* $\int_C \vec{F}\cdot dr$ *where* C *is the rectilinear path from the origin* (0, 0, 0) *to* B (1, 0, 0) *to* C (1, 1, 1) *and* $\vec{F} = 2yi + xj + z^2k$.

SOLUTION: $\int \vec{F}.\vec{dr} = \int 2ydx + xdy + z^2dz$. The path is from O to B and then from B to C.

Along OB, $y = 0$, $z = 0$ and x varies from 0 to 1, and then along BC, $x = 1$, $y = z$ and y varies from 0 to 1.

$$\therefore \quad \int F.dr = \int_0^1 2(0)dx + 0 + 0 + \int_0^1 1dy + y^2 dy$$

$$= 0 + \left[y + \frac{y^3}{3}\right]_0^1 = 1 + \frac{1}{3} = \frac{4}{3} \text{ units.}$$

Ans.

7. *Verify Green's lemma for* $\oint_C (3x^2 - 8y^2)\, dx + (4y - 6xy)\, dy$ *where* C *is boundary of the area enclosed by the curves* $y = \sqrt{x}$ *and* $y = x^2$.

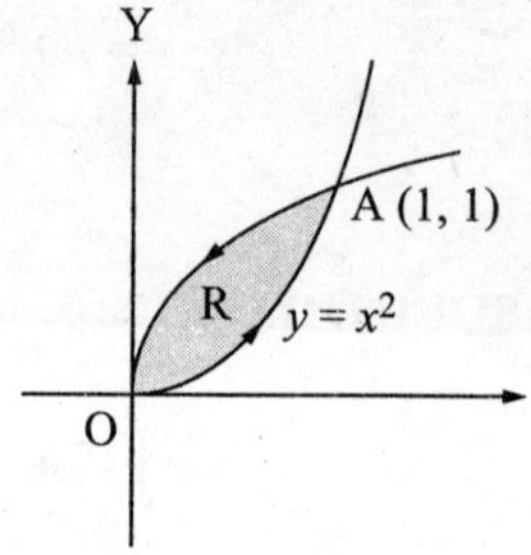

SOLUTION: By Green's theorem $\int\int_C u dx + v dy = \int_R \int \left(\frac{\partial v}{\partial x} - \frac{\partial u}{\partial y}\right) dx\, dy$

Here $u = 3x^2 - 8y^2$ and $v = 4y - 6xy$. And C consists of the curve C_1 from O to A along $y = x^2$ and the curve C_2 from A to O along $y = \sqrt{x}$.

$$\therefore \quad \int_C u\, dx + v dy = \int_{C_1} \left[3x^2 - 8y^2)dx + (4y - 6xy)\right]dy + \int_{C_2} (3x^2 - 8y^2)dx + (4y - 6xy)dy$$

$$= \int_0^1 (3x^2 - 8x^4)dx + (4x^2 - 6x^3)2xdx + \int_1^0 (3x^2 - 8x)dx + (4\sqrt{x} - 6x\sqrt{x})\frac{1}{2\sqrt{x}} dx$$

$$= \left[x^3 + 2x^4 - 4x^5\right]_0^1 - \left[x^3 - 4x^2 + 2x - \frac{3x^2}{2}\right]_0^1 = 1 + 2 - 4 - \left(1 - 4 + 2 - \frac{3}{2}\right) = \frac{3}{2}$$

$$\int_R \int \left(\frac{\partial v}{\partial x} - \frac{\partial u}{\partial y}\right) dydx = \int_0^1 \int_{x^2}^{\sqrt{x}} [-6y + 16y]dydx = \int_0^1 \left[5y^2\right]_{x^2}^{\sqrt{x}} dx$$

$$= 5\int_0^1 (x - x^4)dx = 5\left(\frac{x^2}{2} - \frac{x^5}{5}\right)_0^1 = \frac{3}{2}$$

Hence verified.

8. *Use Green's lemma to evaluate* $\int (2x\, dy - 3y\, dx)$ *around the square with vertices* (0, 2), (2, 0), (– 2, 0), (0, – 2).

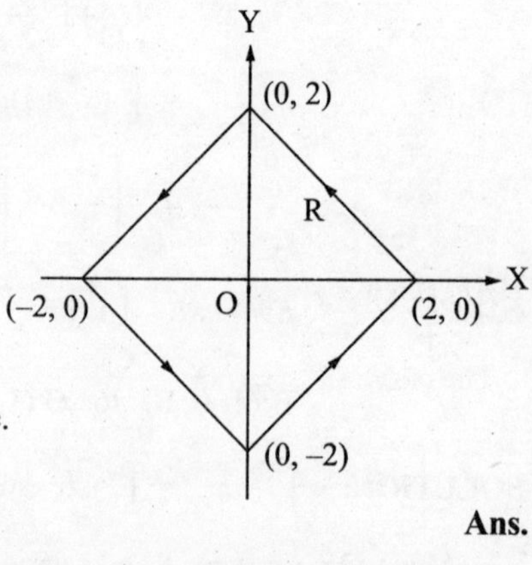

SOLUTION: By Green's theorem

$$\int_C u dx + v dy = \int\int_R \left(\frac{\partial v}{\partial x} - \frac{\partial u}{\partial y}\right) dxdy$$

$$\int_C 2xdy - 3ydx = \int\int_R (2 + 3)dxdy = 5 \times \text{Area of the square.}$$

$$= 5 \times \left(2\sqrt{2}\right)^2 = 40 \text{ units}$$

Ans.

9. *Verify Green's lemma for the line integral* $\int (x^2+y^2)\,dx + xy\,dy$ taken over $x^2+y^2=4,\ x\ge 0,\ y\ge 0$.

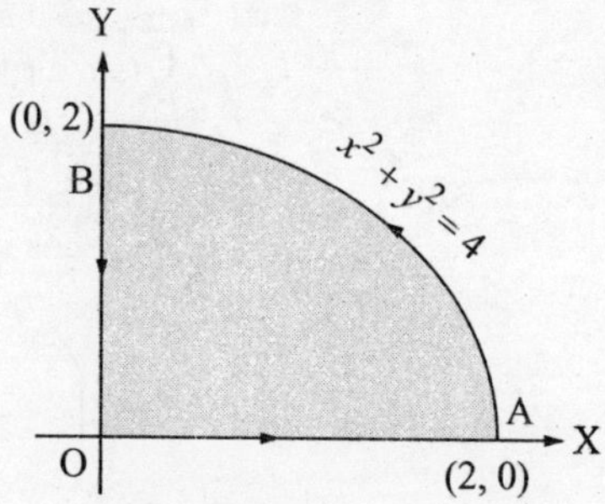

SOLUTION: By Green's lemma $\int_C u\,dx + v\,dy = \iint_R \left(\frac{\partial v}{\partial x} - \frac{\partial u}{\partial y}\right) dx\,dy$

$$\therefore \quad \text{L.H.S.} = \int_C (x^2+y^2)\,dx + xy\,dy = \int_{O \text{ to } A} + \int_{AB\text{ Curve}} + \int_{B \text{ to } A}$$

$$= \int_0^2 (x^2+0)dx + \int_2^0 4dx + \int_0^2 x(-xdx) + \int 0 = \int_0^2 (x^2-4+x^2)dx = \left[\frac{2x^3}{3} - 4x\right]_0^2 = \frac{-8}{3}$$

$$\text{and} \quad \text{R.H.S.} = \int_0^2\int_0^{\sqrt{4-x^2}} (y-2y)dydx = \int_0^2 \left[-\frac{y^2}{2}\right]_0^{\sqrt{4-x^2}} dx = \int_0^2 -\frac{1}{2}(4-x^2)dx = \left[-2x + \frac{x^3}{6}\right]_0^2$$

$$= -4 + \frac{8}{6} = \frac{-16}{6} = \frac{-8}{3} = \text{L.H.S.}$$

Hence verified.

10. *Use Green's lemma to evaluate the line integral* $\int (x^2-y^2)\,dx - 4xy\,dy$ *over the region bounded by* $y^2=4x,\ y=0,\ x=1$ *in the first quadrant.*

SOLUTION: By Green's lemma $\int_C u\,du + v\,dy = \iint_R \left(\frac{\partial v}{\partial x} - \frac{\partial u}{dy}\right) dx\,dy$

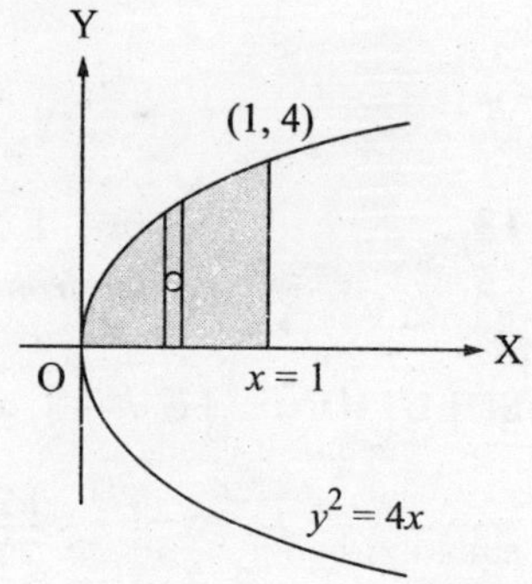

$$\therefore \quad \int_C (x^2-y^2)\,dx - 4xydy = \int_R\int (-4y+2y)\,dxdy$$

where R is the shaded region.

$$\text{RHS} = \int_0^1\int_0^{2\sqrt{x}} -2ydydx = \int_0^1 \left[-y^2\right]_0^{2\sqrt{x}} dx$$

$$= \int_0^1 -4xdx = \left[-2x^2\right]_0^1 = -2 = -2 \text{ units.}$$

Ans.

11. *Apply Green's theorem to evaluate* $\int_C (y-\sin x)\,dx + \cos x\,dy$ *where* C *is the plane triangle enclosed by the lines* $y=0,\ x=\pi/2,\ y=2x/\pi$.

SOLUTION: By Green's lemma $\int_C u\,dx + v\,dy = \iint_R \left(\frac{\partial v}{\partial x} - \frac{\partial u}{\partial y}\right) dx\,dy$

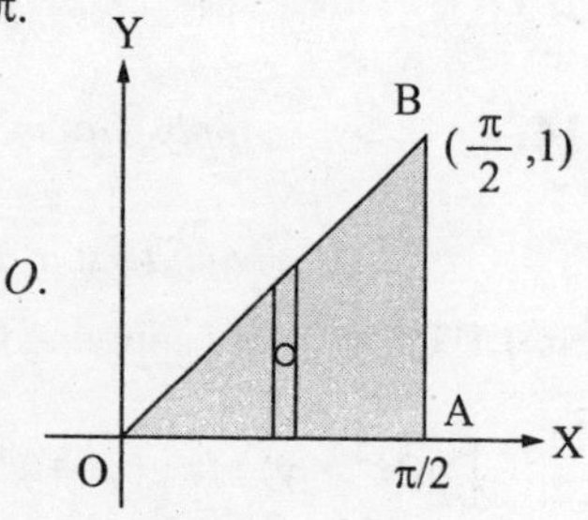

The curve C is from O to A, from A to B and finally from B to O.

$$\therefore \int_C (y-\sin x)dx + \cos xdy = \int_0^{\pi/2}\int_0^{2x/\pi} (-\sin x - 1)dydx$$

$$= -\int_0^{\pi/2} (\sin x+1)[y]_0^{2x/\pi}\,dx = -\int_0^{\pi/2}(\sin x+1)\frac{2x}{\pi}\,dx = \frac{-2}{\pi}\int_0^{\pi/2} x(\sin x+1)dx$$

$$= \frac{-2}{\pi}[x(-\cos x+x)]_0^{\pi/2} + \frac{2}{\pi}\int_0^{\pi/2}(-\cos x+x)dx = \frac{-2}{\pi}.\frac{\pi}{2}\left(0+\frac{\pi}{2}\right)+\frac{2}{\pi}\left(-\sin x+\frac{x^2}{2}\right)_0^{x/2}$$

$$= -\frac{\pi}{2}+\frac{2}{\pi}\left(-1+\frac{\pi^2}{8}\right) = -\left(\frac{2}{\pi}+\frac{\pi}{4}\right)$$ **Ans.**

12. *Apply Green's theorem to evaluate* $\int_C (2y^2dx + 3x\,dy)$ *where* C *is the boundary of the closed region bounded by* $y = x$ *and* $y = x^2$.

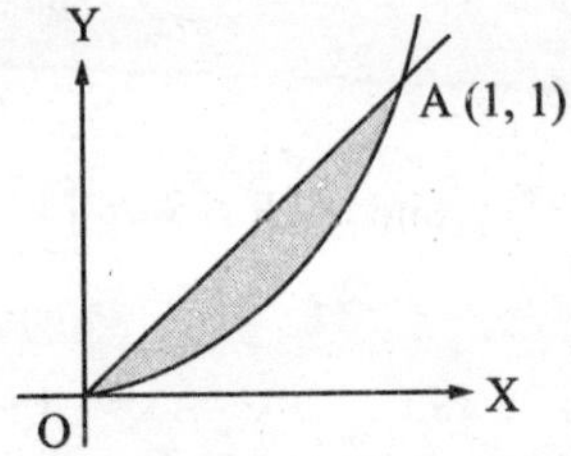

SOLUTION: By Green's lemma

$$\int_C (2y^2dx+3xdy) = \iint_R\left[\frac{\partial}{\partial x}(3x)-\frac{\partial}{\partial y}(2y^2)\right]dxdy$$

where R is the shaded region.

$$= \int_0^1\int_{x^2}^{x}(3-4y)dydx$$

$$= \int_0^1\left[3y-2y^2\right]_{x^2}^{x}dx = \int_0^1(3x-2x^2-3x^2+2x^4)dx$$

$$= \left[\frac{3x^2}{2}-\frac{2x^3}{3}-x^3+\frac{2x5}{5}\right]_0^1 = \frac{3}{2}-\frac{2}{3}-1+\frac{2}{5} = \frac{7}{30} = \frac{7}{30}$$ **Ans.**

13. *Suppose* $\vec{F}$ *is the force field as* $\vec{F} = x^3i + yj + zk$. *Find the work done by* F *along the line from* (1, 2, 3) *to* (3, 5, 7).

SOLUTION: $\int \vec{F}.\vec{dr} = \int_C x^3dx + ydy + zdz$ where C is the straight line from (1, 2, 3) to (3, 5, 7)

equation being $\frac{x-1}{3-1} = \frac{y-2}{5-2} = \frac{z-3}{7-3}$ or $y = \frac{3x}{2}+\frac{1}{2}$ and $z = 2x + 1$

$\therefore$ Work done $= \int_1^3 x^3dx+\left(\frac{3}{2}x+\frac{1}{2}\right)\frac{3}{2}dx+(2x+1)2dx = \int_1^3\left(x^3+\frac{25}{4}x+\frac{11}{4}\right)dx$

$$= \left[\frac{x^4}{4}+\frac{25}{8}x^2+\frac{11x}{4}\right]_1^3 = 50.5 \text{ units.}$$ **Ans.**

14. *Apply Green's theorem to evaluate* $\int_C (2x^2 - y^2)\,dx + (x^2 + y^2)\,dy$ *where* C *is the boundary of the area enclosed by the* X-axis *and the upper half of* $x^2 + y^2 = a^2$.

SOLUTION: By applying Green's theorem, we have

$$\int_C(2x^2-y^2)dx+(x^2+y^2)dy = \int_R\int\left[\frac{\partial}{\partial x}(x^2+y^2)-\frac{\partial}{\partial y}(2x^2-y^2)\right]dxdy$$

$$= \int_R \int (2x+2y)dydx = 2\int_{-a}^{a}\int_0^{\sqrt{a^2-x^2}} (x+y)dy\,dx$$

$$= 2\int_{-a}^{a}\left[xy+\frac{y^2}{2}\right]_0^{\sqrt{a^2-x^2}} dx = 2\int_{-a}^{a}\left[x\sqrt{a^2-x^2}+\frac{1}{2}(a^2-x^2)\right]dx$$

$$= 0 + 4\int_a^a (a^2-x^2)dx = 4\left[a^2x-\frac{x^3}{3}\right]_0^a = 2\left(a^3-\frac{a^3}{3}\right) = \frac{4a^3}{3}$$ **Ans.**

15. *Using Green's theorem evaluate* $\int_C (x^2y\,dx + x^2dy)$ *where C is the boundary described counter clockwise of the triangle with vertices* $(0, 0), (1, 0), (1, 1)$.

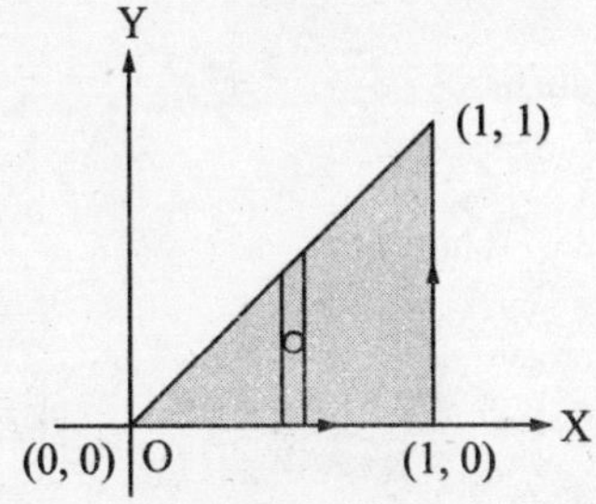

SOLUTION: Using Green's theorem

$$\int_C (x^2ydx + x^2dy) = \int_R\int\left[\frac{\partial}{\partial x}(x^2) - \frac{\partial}{\partial y}(x^2y)\right]dxdy$$

Where R is the shaded area:

$$= \int_0^1\int_{y=0}^{x}(2x-x^2)dydx = \int_0^1\left[2xy-x^2y\right]_0^x dx$$

$$= \int_0^1 (2x^2-x^3)dx = \left[\frac{2x^3}{3}-\frac{x^4}{4}\right]_0^1 = \frac{2}{3}-\frac{1}{4} = \frac{5}{12}$$ **Ans.**

16. *Find the work done in moving a particle once round the circle* $^2 + y^2 = 9,\ z = 0$ *in the XY-plane if the field of force is* $\overline{F} = (2x-y-z)\,i + (x+y-z^2)\,j + (3x-2y+4z)\,k$. *If possible, find its scalar potential.* **(GGSIPUII Sem End Term 2009)**

SOLUTION: Work done = $\int F\cdot dr = \int (2x-y-z)\,dx + (x+y-z^2)\,dy + (3x-2y+4z)\,dz.$

Particle moves round the circle $x^2 + y^2 = 9,\ z = 0,$ hence

Work done = $\int (2x-y)\,dx + (x+y)\,dy.$ (Put $x = 3\cos\theta,\ y = 3\sin\theta$)

$$= \int_0^{2\pi} [(6\cos\theta - 3\sin\theta)(-3\sin\theta) + 3(\cos\theta+\sin\theta)3\cos\theta]\,d\theta$$

$$= \int_0^{2\pi} (9\cos^2\theta + 9\sin^2\theta - 9\sin\theta\cos\theta)\,d\theta = \int_0^{2\pi}\left(9-\frac{9}{2}\sin 2\theta\right)d\theta$$

$= 18\pi.$

The scalar potential corresponding to the field $\overline{F}$ is not possible as it is not a conservative field.

Ans.

SOLUTIONS TO EXERCISE 15B

1. *Evaluate the surface integral* $\iint_S \vec{F}\cdot\vec{ds}$ *where* $\vec{F} = xi + yj + zk$ *and S is the surface of the sphere* $x^2 + y^2 + z^2 = 1$.

SOLUTION: $\iint_S \vec{F}.\vec{ds} = \iint (xi + yj + zk)\,\hat{n}\, ds$ where $\hat{n}$ in the unit vector along the normal to the surface $\phi = x^2 + y^2 = z^2 = 1$. $\nabla\phi = 2xi + 2yj + 2zk$

Therefore $$\hat{n} = \frac{\nabla\phi}{|\nabla\phi|} = \frac{2xi + 2yj + 2zk}{\sqrt{4x^2 + 4y^2 + 4z^2}} = xi + yj + zk$$

$$\therefore \quad \iint \vec{F}.\vec{ds} = \iint_S (x^2 + y^2 + z^2)\, ds$$

$$= \iint_S ds = 4\pi \cdot 1^2 = 4\pi = \text{surface of the unit sphere.}$$ **Ans.**

2. *Evaluate the surface integral* $\iint_S (yzi + zxj + xyk)\cdot\vec{ds}$ *where S is the surface of the sphere* $x^2 + y^2 + z^2 = 1$ *in the first octant.*

SOLUTION: $\iint_S \vec{F}.\vec{ds} = \iint (yzi + zxj + xyk).\hat{n}\, ds$ where $\hat{n}$ is unit vector along the normal to the surface $\phi = x^2 + y^2 + z^2 = 1$ in the positive octant $\hat{n} = \dfrac{\nabla\phi}{|\nabla\phi|} = \dfrac{2xi + 2yj + 2zk}{2\sqrt{x^2 + y^2 + z^2}} = xi + yj + zk$

Therefore $$\iint \vec{F}.\vec{ds} = \iint (yzi + zxj + xyk).(xi + yj + zk)ds = \iint 3xyz\, ds$$

We know that $dx\, dy = ds\ (\hat{n}\cdot k) = z\, ds \quad \therefore \quad ds = \dfrac{dx\, dy}{z}$

$$\therefore \quad \iiint 3xy\, zds = \int_0^1 \int_0^{\sqrt{1-x^2}} 3xy\, dx\, dy = 3\int_0^1 x\left[\frac{y^2}{2}\right]_0^{\sqrt{1-x^2}} dx$$

$$= \frac{3}{2}\int_0^1 x(1 - x^2)dx = \frac{3}{2}\left[\frac{x^2}{2} - \frac{x^4}{4}\right]_0^1 = \frac{3}{8} \text{ units}$$ **Ans.**

3. *Verify Stoke's theorem for the vector field* $\vec{F} = (x^2 - y^2)\, i + 2xyj$ *in the rectangular region in the XY-plane bounded by the lines* $x = 0,\ x = a,\ y = 0,\ y = b.$

SOLUTION: By stokes theorem $\int_C \vec{F}\cdot d\vec{r} = \iint_S curl\ F.d\vec{s}$ In the *XY*-plane Stoke's theorem is same as Green's lemma.

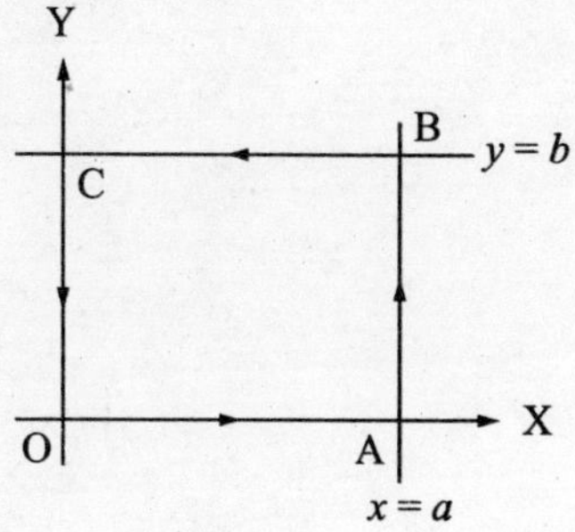

$$\text{curl } \vec{F} = \nabla \times [(x^2 - y^2)i + 2xyj] = \begin{vmatrix} i & j & k \\ \dfrac{\partial}{\partial x} & \dfrac{\partial}{\partial y} & \dfrac{\partial}{\partial z} \\ x^2 - y^2 & 2xy & 0 \end{vmatrix}$$

$$= i(0) + j(0) + k(2y + 2y)$$

$$\text{R.H.S.} = \iint 4yk.kdx\ dy = 4\iint y\ dy\ dx = 4\int_0^a \left[\frac{y^2}{2}\right]_0^b dx = 2ab^2$$

$$\text{L.H.S} \int_C \vec{F}.d\vec{r} = \int_{OA} (x^2 - y^2)dx + 2xy\ dy + \int_{AB} (x^2 - y^2)dx + 2xy\ dy + \int_{BC} (x^2 - y^2)dx + 2xy\ dy$$

$$+ \int_{CO} (x^2 - y^2)\ dx + 2xy\ dy$$

$$= \int_0^a x^2 dx + \int_0^b + 2ay\ dy \int_a^0 (x^2 - b^2)\ dx = \left[\frac{x^3}{3}\right]_0^a + [ay^2]_0^b + \left[\frac{x^2}{3} - b^2 x\right]_a^0 = 2ab^2 = \text{R.H.S.}$$

Hence Stokes theorem is verified.

4. *Evaluate the line integral* $\oint_C \text{F} \cdot dr$ *using Stoke's theorem if* C *is the boundary of the triangle* ABC *cut off the coordinate planes by the plane* $2x + 2y + z = 4$ *and the vector field*

$$\vec{F} = 2y\,(1 - x)\ i + (x - x^2 - y^2)\ j + (x^2 + y^2 + z^2)\ k.$$

SOLUTION: By Stokes theorem $\int_C F.dr = \iint_S (\nabla \times F).\hat{n}\, ds$

Here $\hat{n}$ is the unit vector along the normal to the plane $\phi = 2x + 2y + z = 4$

Therefore $$\hat{n} = \frac{\nabla\phi}{|\nabla\phi|} = \frac{2i + 2j + k}{\sqrt{2^2 + 2^2 + 1^2}} = \frac{1}{3}(2i + 2j + k).$$

and $$\text{curl } F = \nabla \times F = \begin{vmatrix} i & j & k \\ \dfrac{\partial}{\partial x} & \dfrac{\partial}{\partial y} & \dfrac{\partial}{\partial z} \\ 2y(1-x) & x - x^2 - y^2 & x^2 + y^2 + z^2 \end{vmatrix}$$

$$= i(2y - 0) + j(0 - 2x) + k(1 - 2x - 2 + 2x) = 2yi - 2xj - k$$

$$\therefore \quad \iint (\nabla \times F).\hat{n}\ ds = \iint (2yi - 2xj - k)\cdot\frac{1}{3}(2i + 2j + k)\ ds \approx \frac{1}{3}\iint (4y - 4x - 1)\ ds$$

Now $dx\,dy = ds(\hat{n}.k)$ $\therefore$ $ds = \dfrac{dx\,dy}{1/3} = 3\,dx\,dy$

$$\therefore \int_C \vec{F}.\vec{dr} = \frac{1}{3}\iint (4y-4x-1)\,3\,dx\,dy = \int_0^2\int_0^{2-x} (4y-4x-1)\,dx\,dy$$

$$= \int_0^2 \left[2y^2 - 4xy - y\right]_0^{2-x} dx = \int_0^2 [2(2-x)^2 - (4x+1)(2-x)]dx$$

$$= \int_0^2 (6x^2 - 15x + 6)dx = \left[2x^3 - 15\frac{x^2}{2} + 6x\right]_0^2 = -2$$

$$\therefore \int F.dr = -2.$$ **Ans.**

5. *Use Stoke's theorem to calculate the line integral $\int_C ydx + zdy + xdz$ where C is the intersection of the sphere $x^2 + y^2 + z^2 = a^2$ and the plane $x + z = a$.*

SOLUTION: By Stokes theorem $\int_C \vec{F}.\vec{dr} = \iint_S (\nabla \times \vec{F}).\vec{ds}$

Here we have $\vec{F} = yi + zj + xk$ Hence $\nabla \times \vec{F} = \begin{vmatrix} i & j & k \\ \dfrac{\partial}{\partial x} & \dfrac{\partial}{\partial y} & \dfrac{\partial}{\partial z} \\ y & z & x \end{vmatrix} = -i - j - k$

$$\therefore \int_C \vec{F}.\vec{dr} = \int_C ydx + zdy + xdz = -\iint_S (i + j + k).\hat{n}\,ds$$

Here S is the area of cross-section of the sphere $x^2 + y^2 + z^2 = a^2$ and plane $x + z = a$.

Also, $\hat{n} = \dfrac{i+k}{\sqrt{1+1}}$, therefore $\int_C \vec{F}.\vec{dr} = -\dfrac{1}{\sqrt{2}}\iint (i+j+k).(i+j)ds$

$$= -\frac{2}{\sqrt{2}}\iint ds$$

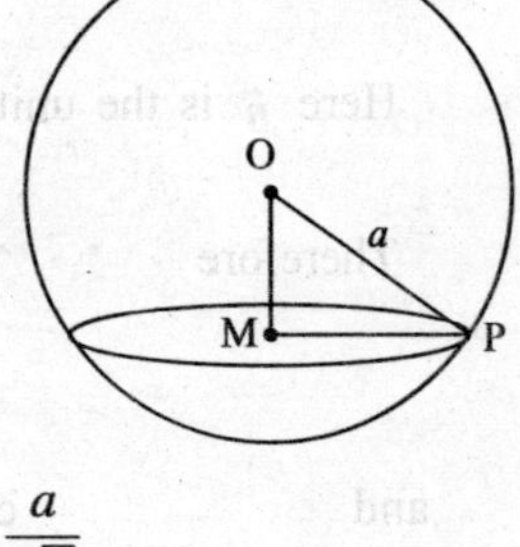

OM = Perpendicular distance of the plane $x + z = a$ from the centre of the sphere, which is origin.

$$= \left|\frac{0+0-a}{\sqrt{1^2+1^2}}\right| = \frac{a}{\sqrt{2}}$$

$\therefore$ radius of the circle of intersection $= MP = \sqrt{OP^2 - OM^2} = \sqrt{a^2 - \dfrac{a^2}{2}} = \dfrac{a}{\sqrt{2}}$

$\therefore$ Area of this circle $= \pi\left(\dfrac{a}{\sqrt{2}}\right)^2 = \pi a^2/2$

$$\therefore \int_C F.dr = -\sqrt{2}.\frac{\pi a^2}{2} = -\pi a^2/\sqrt{2}.$$ **Ans.**

6. *Evaluate* $\int_C (xy\,dx + xy^2 dy)$ *by using Stoke's theorem taking C to be a square in the xy-plane with vertices* (1, 0), (– 1, 0), (0, 1) *and* (0, – 1).

SOLUTION: Stokes theorem in the XY-plane is nothing but Green's lemma.

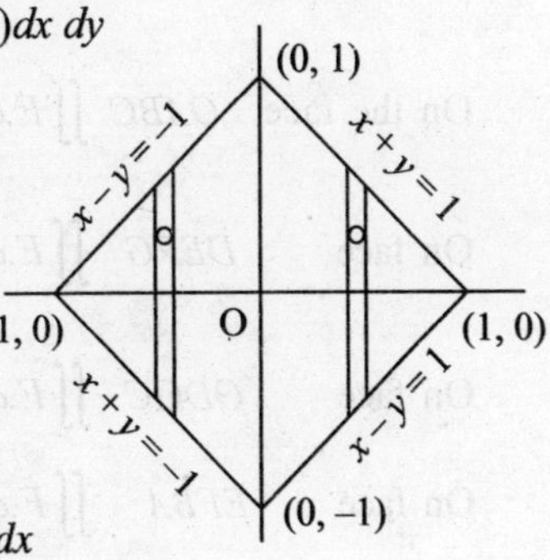

$$\therefore \quad \int_C (xy\,dx + xy^2 dy) = \iint_S \left[\frac{\partial}{\partial x}(xy^2) - \frac{\partial}{\partial y}(xy)\right] dx\,dy = \iint_S (y^2 - x)dx\,dy$$

$$= \int_{-1}^{0}\int_{-1-x}^{x+1} (y^2 - x)dy\,dx + \int_{0}^{1}\int_{x-1}^{-x+1} (y^2 - x)dy\,dx$$

$$= \int_{-1}^{0}\left[\frac{y^3}{3} - xy\right]_{-(x+1)}^{x+1} dx + \int_{0}^{1}\left[\frac{y^3}{3} - xy\right]_{x-1}^{-(x-1)} dx$$

$$= 2\int_{-1}^{0}\left[\frac{(x+1)^3}{3} - x(x+1)\right]dx + 2\int_{0}^{1}\left[-\frac{1}{3}(x-1)^3 + x(x-1)\right]dx$$

$$= \left[\frac{1}{6}(x+1)^4 - \frac{2x^3}{3} - x^2\right]_{-1}^{0} + \left[-\frac{(x-1)^4}{6} + \frac{2x^3}{3} - x^2\right]_{0}^{1} = \frac{1}{3}$$ **Ans.**

7. *Use Stoke's theorem to find* $\int_C \vec{F}\cdot\vec{dr}$ *if* $\vec{F} = -2yi + xj + zk$ *and C is the loop of the curve* $x^3 + y^3 = 3axy$.

SOLUTION: By Stoke's theorem $\int_C \vec{F}.\vec{dr} = \iint_S (\nabla \times \vec{F}).\vec{ds}$

$$\nabla \times \vec{F} = \begin{vmatrix} i & j & k \\ \frac{\partial}{\partial x} & \frac{\partial}{\partial y} & \frac{\partial}{\partial z} \\ -2y & x & z \end{vmatrix} = i(0-0) + j(0-0) + k(1+2) = 3k$$

$$\therefore \quad \iint_S \nabla \times F.\vec{ds} = \iint_S (3k)\cdot \hat{n}\,ds$$

Here $\hat{n} = k \Rightarrow 3\iint ds = 3 \times$ Area under the loop of $x^3 + y^3 = 3axy$

Polar Equation of the curve is $r = \dfrac{3a\sin\theta\cos\theta}{\sin^3\theta + \cos^3\theta}$. The loop lies in $0 \le \theta \le \dfrac{\pi}{2}$

$$\therefore \quad \text{Area of loop} = \int_0^{\pi/2} \frac{1}{2}r^2 d\theta = \frac{1}{2}\int_0^{\pi/2} \frac{9a^2 \sin^2\theta\cos^2\theta\,d\theta}{(\sin^3\theta + \cos^3\theta)^2}$$

$$= \frac{9a^2}{2}\int_0^{\pi/2} \frac{\tan^2\theta\sec^2\theta d\theta}{(1+\tan^3\theta)^2} = \frac{9a^2}{2}\int_1^{\infty} \frac{\frac{1}{3}dt}{t^2} \quad \text{In putting } 1 + \tan^3\theta = t$$

$$= \frac{3a^2}{2}\left[-\frac{1}{t}\right]_1^{\infty} = \frac{3a^2}{2} \quad \text{Hence } \int F.dr = 3 \times \frac{3a^2}{2} = \frac{9a^2}{2}$$ **Ans.**

8. *Verify Gauss divergence theorem for the function* $\vec{F} = x^2 i + zj + yzk$ *over the unit cube* $x=0,\ x=1;\ y=0,\ y=1;\ z=0,\ z=1$.

SOLUTION: $\iint_S F.ds = \iiint \nabla.F dv$ **by Gauss divergence theorem.**

$$F = x^2 i + zj + yzk$$

On the face *OABC* $\iint F.ds = \iint (x^2 i + zj + yzk)\cdot(-k)\,dx\,dy = -\iint yz\,dx\,dy = 0$ *as* $z = 0$

On face *DEFG* $\iint F.ds = \iint (x^2 i + zj + yzk).k\ dx\,dy = \iint yz\,dx\,dy = \int_0^1\int_0^1 y\,dy\,dx = \frac{1}{2}$

On face *ODGC* $\iint F.ds = \iint (x^2 i + zy + yzk).(-i) dy\,dz = -\iint x^2\,dy\,dz = 0$ as $x = 0$ there.

On face *EFBA* $\iint F.ds = \iint (x^2 i + zj + yzk).(i) dy\,dz = \iint x^2 dydz = \iint 1 dy\,dz = 1$ as $x = 1$ there

On face *DEAO* $\iint F.ds = \iint (x^2 i + zj + yzk)(-j) dx\,dz = -\iint zdx\,dz = -\int_0^1 dx\left(\frac{z^2}{2}\right)_0^1 = -\frac{1}{2}$

On face *FBCG* $\iint F.ds = \iint (x^2 i + zj + yzk).jdx\,dz = \iint z\,dx\,dz = \frac{1}{2}$

$\therefore$ L.H.S. $= \iint F\cdot ds = \iint F.ds = 0 + \frac{1}{2} + 1 - \frac{1}{2} + \frac{1}{2} = \frac{3}{2}$

Next R.H.S. $= \iiint \nabla.F dv = \iiint \nabla.(x^2 i + zj + yzk) dx\,dy\,dz$

$$= \int_0^1\int_0^1\int_0^1 (2x + 0 + y) dz\,dy\,dx = \int_0^1\int_0^1 (2x + y)[z]_0^1 dy\,dx$$

$$= \int_0^1\int_0^1 (2x + y) dy\,dx = \int_0^1\left[2xy + \frac{y^2}{2}\right]_0^1 dx = \int_0^1\left(2x + \frac{1}{2}\right)dx = \frac{3}{2}$$

Hence Gauss theorem is verified.

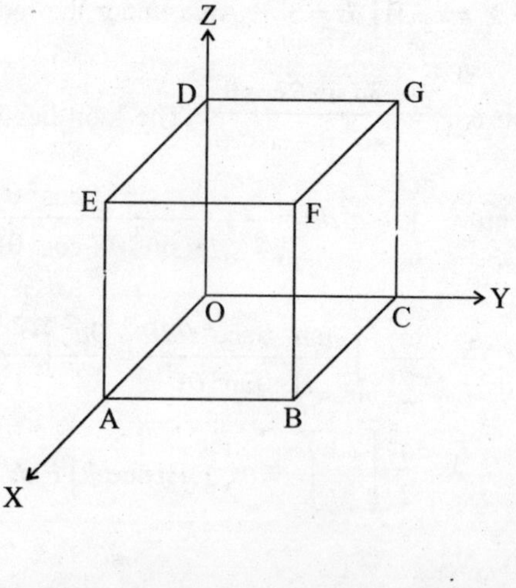

9. *Evaluate* $\iint_S (y^2z^3 i + z^2x^3 j + x^2y^3 k)\cdot \vec{ds}$ *where* S *is the curved surface of the sphere* $x^2+y^2+z^2=a^2$ *above the plane* $z=0$.

SOLUTION: By Gauss's theorem $\iint \vec{F}.\hat{N}ds = \iiint_v \nabla.\vec{F}dv$

Let S_1 be the curved surface and S_2 be in $z = 0$.

$$= \iint_{S_1} \vec{F}.\hat{n}ds_1 + \iint_{S_2} \vec{F}.\hat{n}\, ds_2 = \iiint_v \nabla.\vec{F}dv$$

But $\nabla \cdot \vec{F} = \frac{\partial}{\partial x}(y^2z^3) + \frac{\partial}{\partial y}(z^2x^3) + \frac{\partial}{\partial z}(x^2y^3) = 0$

For S_2, $\hat{n} = -k \therefore \vec{F}.\hat{n} = -x^2y^3$.

$\therefore$ $\iint_{S_2} \vec{F}.\hat{n}ds_2 = \iint_{S_2} -x^2y^4 dx\, dy$ where S_2 is $x^2+y^2=a^2, z=0$.

$$= -\int_0^{2\pi}\int_0^a r^2\cos^2\theta r^4 \sin^4\theta . r\, d\theta\, dr$$

$$= -\frac{a^8}{8}\int_0^{2\pi} \sin^4\theta\cos^2\theta d\theta = \frac{2a^8}{8}\int_0^{\pi} \sin^4\theta\cos^2\theta\, d\theta$$

$$= -\frac{a^8}{8}.4\int_0^{\pi/2} \sin^4\theta\cos^2\theta d\theta = -\frac{a^8}{2}.\frac{3}{6}.\frac{1}{4}.\frac{\pi}{2.2} = \frac{\pi a^8}{64}$$

Since $\iiint_v \nabla.\vec{F}dv = 0$, $\iint_{S_1} F.\hat{n}.ds_1 - \frac{\pi a^8}{64} = 0$

Hence $\iint_S F.\hat{n}ds = \frac{\pi a^8}{64}$. **Ans.**

10. *Evaluate by divergence theorem* $\iint_S (lx^2+my^2+nz^2)\, ds$ *over the sphere* $(x-a)^2+(y-b)^2+(z-c)^2=\rho^2$, *l, m, n being the direction cosines of the drawn normal to the sphere.*

SOLUTION: Given surface S is a closed surface hence divergence theorem is applicable. We have

$$\iint_S (lx^2+my^2+nz^2)ds = \iint_S \vec{F}.\hat{n}ds$$

If $\phi \equiv (x-a)^2+(y-b)^2+(z-c)^2$ then $\nabla\phi = 2(x-a)i + 2(y-b)j + 2(z-c)$

$$\hat{n} = \frac{(x-a)i}{\rho} + \frac{(y-b)j}{\rho} + \frac{(z-c)k}{\rho} = li + mj + nk$$

If we write $\vec{F} = F_1 i + F_2 j + F_3 k$ then $F_1 l = lx^2$, $F_2 m = my^2$, $F_3 n = nz^2$

$\therefore$ $\vec{F} = x^2 i + y^2 j + z^2 k$ and $\nabla . \vec{F} = 2(x + y + z)$

By Gauss Theorem $\iint_S F\, ds = \iiint_v 2(x+y+z)dx\, dy\, dz$

If we substitute $x - a = X,\ y - b = Y,\ z - c = Z$

then
$$\iint_S F.ds = 2\iiint (a+b+c+X+Y+Z)dXdYdZ$$
$$= 2(a+b+c)\iiint dX\, dY\, dZ + 2\iiint (X+Y+Z)\, dX\, dYdZ$$
$$= 2(a+b+c)\frac{4}{3}\pi\rho^3 + 0 \quad \text{(by property of definite integrals.)}$$
$$= \frac{8}{3}(a+b+c)\pi\rho^3$$

11. *Use Gauss divergence theorem for evaluating the surface integral* $\iint_S \mathrm{F}\cdot\vec{ds}$ *where* S *is the closed surface bounded by the cylinder* $x^2 + y^2 = a^2$ *and the planes* $z = 0, z = h$ and $\vec{F} = i\, yz + j\, zx + kz^2$.

SOLUTION: As S is a closed surface by Gauss divergence theorem $\iint_S \vec{F}.\vec{ds} = \iiint_R divF\, dv$ where R is the region enclosed by the surface S bounded by the cylinder $x^2 + y^2 = a^2$ and planes $z = 0$ and $z = h$.

$$F = iyz + jzx + kz^2 \quad \therefore\ \nabla.F = 0 + 0 + 2z$$

$$\therefore \quad \iiint_v \nabla.Fdv = \int_{-a}^{a}\int_{-\sqrt{a^2-x^2}}^{\sqrt{a^2-x^2}}\int_{z=0}^{h} 2z\, dx\, dy\, dz = \int_{-a}^{a}\int_{-\sqrt{a^2-x^2}}^{\sqrt{a^2-x^2}} [z^2]_0^h\, dy\, dz$$
$$= h^2\int_{-a}^{a}\int_{-\sqrt{a^2-x^2}}^{\sqrt{a^2-x^2}} dy\, dx = \pi a^2 h^2$$ **Ans.**

12. *Evaluate* $\iint_S xz^2 dy\, dz + (x^2y - z^3)\, dz\, dx + (2xy + y^2z)\, dx\, dy$ *where* S *is the surface enclosing the region bounded by the hemisphere* $x^2 + y^2 + z^2 = 4$ *above the XY-plane.*

SOLUTION: Given surface integral $= \iint \vec{F}.\vec{ds} = \iint [xz^2 i + (x^2y - z^3)j + (2xy + y^2z)k].\vec{ds}$

where S is the surface enclosing the region bounded by the hemisphere $x^2 + y^2 + z^2 = 4$ above the XY-plane. Since S is the closed surface, we can use Gauss theorem

$$\therefore \quad \iint \vec{F}.ds = \iiint div\, F\, dv = \iiint (z^2 + x^2 + y^2)dx\, dy\, dz$$

Converting to the spherical polar coordinates

$x = r \sin\theta \cos\phi,\ y = r \sin\theta \sin\phi,\ z = r \cos\theta,$ we get

$$\int_{\phi=0}^{2\pi}\int_{\theta=0}^{\pi/2}\int_{0}^{2} r^2 . r^2 \sin\theta \, dr\, d\theta\, d\phi = \int_{\phi=0}^{2\pi}\int_{\theta=0}^{\pi/2} \sin\theta \left[\frac{r^5}{5}\right]_0^2 d\theta\, d\phi$$

$$= \frac{32}{5}\int_{\theta=0}^{2\pi} [-\cos\theta]_{\theta=0}^{\pi/2} d\phi = \frac{32}{5}\int_0^{2\pi} d\phi = \frac{64\pi}{5}$$ **Ans.**

13. *It S is any closed surface prove that* $\iint_S \text{curl}\, \vec{F} \cdot \overrightarrow{ds} = 0.$

SOLUTION: Since S is defined over a closed surface S, Gauss theroem is applicable.

$$\therefore \quad \iint_S \text{curl}\, \vec{F} \cdot \overrightarrow{ds} = \iiint_V \nabla \cdot \text{curl}\, F\, dv = \iiint \nabla \cdot (\nabla \times F)\, dv$$

$$= 0 \text{ since } \nabla \cdot (\nabla \times F) = 0.$$ **Hence proved.**

14. *Evaluate* $\iint_S \vec{r} \cdot \hat{n}\, ds$ where S is a closed surface.

SOLUTION: Since S is a closed surface Gauss theorem is applicable here, hence $\iint_S \vec{r} \cdot \overrightarrow{ds} = \iiint_V \nabla \cdot \vec{r}\, dv$

$$= \iiint_V \left(i\frac{\partial}{\partial x} + j\frac{\partial}{\partial y} + k\frac{\partial}{\partial z}\right) \cdot (xi + yj + zk)\, dv = \iiint_V 3dv = 3V$$

where V is the volume enclosed by the surface S. **Ans.**

15. *Evaluate the surface integral* $\iint_S (xdydz + ydzdx + zdxdy)$ *where S is the portion of the plane* $x + 2y + 3z = 6$ *lying in the first octanct.*

SOLUTION: The given integral $= I_1 + I_2 + I_3$

where $I_1 = \iint_S x\, dy\, dz, \quad I_2 = \iint_S y\, dz\, dx$ and $I_3 = \iint_S z\, dx\, dy$.

For I_1, $x = 6 - 2y - 3z$, for I_2, $y = 3 - \frac{x}{2} - \frac{3z}{2}$ and for I_3, $z = 2 - \frac{x}{3} - \frac{2y}{3}$.

$$\therefore \quad I_1 = \int_0^3 \int_0^{2-\frac{2}{3}y} (6 - 2y - 3z)\, dy\, dz = \int_0^3 \left[(6-2y)z - \frac{3z^2}{2}\right]_0^{(6-2y)/3} dy$$

$$= \int_0^3 \left[\frac{1}{3}(6-2y)^2 - \frac{3}{2}\left(\frac{6-2y}{3}\right)^2\right] dy = \frac{1}{6}\int_0^3 (6-2y)^2\, dy = 6$$

$$I_2 = \int_0^6 \int_0^{(6-x)/3} \left(3 - \frac{x}{2} - \frac{3z}{2}\right) dz\, dx = \int_0^6 \left[\left(\frac{6-x}{2}\right)z - \frac{3z^2}{4}\right]_0^{\frac{6-x}{3}} dx$$

$$= \int_0^6 \left[\frac{1}{6}(6-x^2) - \frac{3}{4}\left(\frac{6-x}{3}\right)^2\right] dx = \int_0^6 \frac{1}{12}(6-x)^2\, dx = 6$$

$$I_3 = \int_0^6 \int_0^{(6-x)/2} \left(2 - \frac{x}{3} - \frac{2y}{3}\right) dy\, dx = \int_0^6 \left[\left(\frac{6-x}{3}\right)y - \frac{y^2}{3}\right]_0^{(6-x)/2} dx$$

$$= \int_0^6 \left[\frac{1(6-x)^2}{6} - \frac{1(6-x)^2}{12}\right] dx = \frac{1}{12}\int_0^6 (6-x)^2\, dx = 6 \ = I + I_1 + I_2 + I_3 = 18 \qquad \textbf{Ans.}$$

16. *Verify Stoke's theorem for $F = xzi - yj + x^2yk$ defined over the surface of the region S bounded* by $x = 0, y = 0, z = 0$ and $2x + y + 2z = 8$ *which does not include the zx plane.*

SOLUTION: $\oint_C F \cdot dr = \int_C (xzi - yj + x^2yk) \cdot (idx + jdy + kdz)$

$\int_C xz\, dx - y\, dy + x^2 y\, dz = 0$ On the open face $y = 0$ $2x + 2z = 8$

$\therefore \qquad \int_C \vec{F}.\vec{dr} = 0 \int_0^4 x(4-x)dx = \left[2x^2 - \frac{x^3}{3}\right]_0^4 = \frac{32}{3}$

Next, $\iint \text{Curl}\, \vec{F}.\vec{ds} = \int (\nabla \times F).\hat{n}\, ds$

$$\nabla \times F = \begin{vmatrix} i & j & k \\ \frac{\partial}{\partial x} & \frac{\partial}{\partial y} & \frac{\partial}{\partial z} \\ xz & -y & x^2y \end{vmatrix} = ix^2 + j(x - 2xy) + 0\ \hat{k}$$

and $\hat{n} = \frac{2i + j + 2k}{\sqrt{2^2 + 1^2 + 2^2}} = \pm\frac{1}{3}(2i + j + 2k)$. Here $ds = \frac{dx\, dy}{\hat{n}.k} = \frac{dx\, dy}{2/3}$

$\therefore \qquad \iint_S (\nabla \times F).\hat{n}\, ds = \iint [x^2 i + x(1-2y)j]\left(-\frac{1}{3}\right)(2i + j + 2k)\frac{3}{2}dx\, dy$

$$= -\int_0^4 \int_0^{8-2x} \left[x^2 + \frac{x}{2}(1-2y)\right] dx\, dy \qquad S = 2x + y + 2z = 8, z = 0$$

$$= -\int_0^4 \left[x^2 y + \frac{x}{2}y - \frac{xy^2}{2}\right]_0^{8-2x} dx = \int_0^4 \left[x^2(8-2x) + \frac{x}{2}(8-2x) - \frac{x}{2}(8-2x)^2\right] dx$$

$$= -\int_0^4 (-2x^3 + 8x^2 + 4x - x^2 - 32x + 16x^2 - 2x^3)dx = \left[x^4 - 23\frac{x^3}{3} + 14x^2\right]_0^4 = \frac{32}{3}$$

Hence verified.

17. *Verify Gauss divergence theorem evaluating the surface integral*

$$\iint (x^3 dydz + x^2 y\, dzdx + x^2 zdxdy)$$

as triple integral, where S *is the closed surface consisting of the cylinder* $x^2 + y^2 = a^2$ $(a \le z \le b)$ *and the circular discs at* $z = 0$ *and* $z = b$, $(x^2 + y^2 \le a^2)$.

SOLUTION: By Gauss divergence theorem $\iint_S F.\hat{n}ds = \iiint_V \nabla.F\, dv$. Here $\vec{F} = x^3 i + x^2 yj + x^2 zk$

$\therefore$ div $F = \nabla.F = 3x^2 + x^2 + x^2 = 5x^2$.

$\therefore$ $\iiint \nabla.F\, dv = \iint \int_0^b 5x^2\, dx\, dy\, dz$ where z varies from 0 to b, y varies from $-\sqrt{a^2 - x^2}$ to $\sqrt{a^2 - x^2}$ and x varies from $-a$ to a.

$$\therefore \quad \iiint \nabla.Fdv = 5\int_{-a}^{a}\int_{-\sqrt{a^2-x^2}}^{\sqrt{a^2-x^2}}\int_0^b x^2\, dz\, dy\, dx = 20b\int_0^a \int_0^{\sqrt{a^2=x^2}} x^2 dy\, dx.$$

$$= 20\, b\int_0^a x^2 \sqrt{a^2 - x^2}\, dx = \frac{5\pi a^4 b}{4}.$$

To calculate the surface integral, S consists of three parts: S_1, S_2, S_3 where S_1 : the circular disc in the plane $z = 0$, S_2 : The circular disc in the plane $z = b$ and S_3 : the curved surface of the cylinder $x^2 + j^2 = a^2$.

Therefore $\iint_{S_1} F.\hat{n}\, ds = 0$ (since $z = 0$ and $dz = 0$)

and $$\iint_{S_2} F.\hat{n}\, ds = \int_{-a}^{a}\int_{a-\sqrt{a^2-x^2}}^{\sqrt{a^2-x^2}} x^2 b\, dy\, dx = 4b\int_0^a \int_0^{\sqrt{a^2-x^2}} x^2\, dy\, dx = \frac{\pi a^4 b}{4}$$

For S_3, $\nabla (x^2 + y^2) = 2xi + 2yj$ $\therefore$ $\hat{n} = \dfrac{2xi + 2yj}{\sqrt{4x^2 + 4y^2}} = \dfrac{1}{a}(xi + yj)$

Hence $$F.\, \hat{n} = \frac{1}{a}(x^3 i + x^2 yj + x^2 zk).(xi + yj)$$

$$= \frac{1}{a}(x^4 + x^2 y^2) = \frac{x^2}{a}(x^2 + y^2) = ax^2 \quad (\text{since } x^2 + y^2 = a^2)$$

$\therefore$ $\iint_{S_3} \vec{F}.\hat{n}\, ds = \iint ax^2 ds$ we put $x = a\cos\theta$, $y = a\sin\theta$ hence $ds = a\, d\theta\, dz$)

$$= a\int_0^{2\pi}\int_0^b a^2 \cos^2\theta . ad\theta dz = a^4 b\int_0^{2\pi} \cos^2\theta\, d\theta = \pi a^4 b$$

Finally, $$\iint_S F.\hat{n}ds = 0 + \frac{\pi a^4 b}{4} + \pi a^4 b = \frac{5}{4}\pi a^4 b = \iiint \nabla.F\, dv$$ **Hence verified.**

ADDITIONAL IMPORTANT PROBLEMS

1. *Show that* $\int_C F.dr = 3\pi$ *if* $\vec{F} = zi + xj + yk$ *and C is the portion of the curve* $\vec{r} = (\cos t)i + (\sin t)j + tk$ *from* $t = 0$ to $t = 2\pi$.

SOLUTION: $\vec{r} = xi + yj + zk$ hence $x = \cos t,\ y = \sin t,\ z = t$

Here $\vec{F} = zi + xj + yk$ hence $\vec{F}.\vec{dr} = zdx + xdy + ydz$

or $$\vec{F}.\vec{dr} = -t \sin t\ dt + \cos t\ .\ \cos t\ dt + \sin t\ dt$$

$$\therefore \quad \int \vec{F}.\overrightarrow{dr} = \int_0^{2\pi} (-t \sin t + \cos^2 t + \sin t)\, dt$$

$$= \left[t \cos t - \int \cos t\, dt\right]_0^{2\pi} + \int_0^{2\pi} \left(\frac{1+\cos 2t}{2} - \sin t\right) dt$$

$$= 2\pi + 0 + \left[\frac{t}{2} + \frac{\sin 2t}{4} + \cos t\right]_0^{2\pi} = 3\pi.$$ **Hence proved.**

2. *If* $\vec{F} = (2x - y + 2z)i + (x + y - z)j + (3x - 2y - 5z)k$. *Calculate the circulation of F along the circle in the* xy *plane of radius 2 and centre at the origin.*

SOLUTION: Circulation of $\vec{F} = \oint_C \vec{F}.\overrightarrow{dr}$.

Here $$\vec{F} \cdot \overrightarrow{dr} = (2x - y + 2z)dx + (x + y - z)\, dy + (3x - 2y - 5z)dz$$

The path of integration is the circle $x^2 + y^2 = 4,\ z = 0 \Rightarrow x = 2 \cos \theta,\ y = 2 \sin \theta,\ z = 0.$

$$\therefore \oint_C \vec{F}.\overrightarrow{dr} = \int_0^{2\pi} (4 \cos\theta - 2\sin\theta + 0)\,(-2\sin\theta)d\theta + \int_0^{2\pi} (2 \cos\theta + 2\sin\theta + 0)\, 2\cos\theta\, d\theta + \int_0^{2\pi} (6 \cos\theta - 4\sin\theta - 0)\,(0)$$

$$= \int_0^{2\pi} (-8 \sin\theta \cos\theta + 4\sin^2\theta + 4\cos^2\theta + 4\sin\theta \cos\theta)\, d\theta = \int_0^{2\pi} (-4\sin\theta \cos\theta + 4)\, d\theta$$

$$= \left[-2\sin^2\theta + 4\theta\right]_0^{2\pi} = 0 + 8\pi = 8\pi$$ **Ans.**

3. *Evaluate* $\int_C \vec{F}.\overrightarrow{dr}$ *over the circular path* $x^2 + y^2 = a^2$ *where* $\vec{F} = \sin y i + x(1 + \cos y) j$.

SOLUTION: Here $\vec{F} = (\sin y)i + x(1 + \cos y)j$, and $\vec{r} = xi + yj$ so $\vec{dr} = idx + jdy$.

$$\therefore \quad \oint_C \vec{F}.\overrightarrow{dr} = \int \sin y\, dx + x(1 + \cos y)dy = \int \sin y\, dx + x \cos y dy + xdy$$

$$= \int d\,(x \sin y) + xdy$$

Path is $C : x^2 + y^2 = a^2$ hence $x = a \cos\theta,\ y = a \sin\theta,\ dx = -a \sin d\theta,\ dy = a \cos\theta\, d\theta$

$$\therefore \quad \oint_C \vec{F}.\overrightarrow{dr} = \int_0^{2\pi} d[a\cos\theta.\sin(a\sin\theta)] + a\cos\theta.a\cos\theta\, d\theta$$

$$= [a\cos\theta.\sin(a\sin\theta)]_0^{2\pi} + a^2\int_0^{2\pi}\frac{1+\cos 2\theta}{2}d\theta$$

$$= a\cos 2\pi\sin(a\sin 2\pi) - 0 + \frac{a^2}{2}\left[\theta + \frac{1}{2}\sin 2\theta\right]_0^{2\pi}$$

$$= 0 + \frac{a^2}{2}.2\pi = \pi a^2.$$ **Ans.**

4. *Find the work done in moving a particle along the line segments joining the points (0, 0, 0) to (1, 0, 0) then to (1, 1, 0) and then to (1, 1, 1) under the force field* $\vec{F} = (3x^2+6y)i - 14yz\, j + 20xz^2k.$

SOLUTION: Work done $\int \vec{F}.\overrightarrow{dr} = \int (3x^2+6y)dx - 14yzdy + 20xz^2dz$

Path is from 0(0, 0, 0) to $A(1, 0, 0)$ then from $A(1, 0, 0)$ to $B(1, 1, 0)$ and then from $B(1, 1, 0)$ to $C(1, 1, 1)$. Thus, $\int_C F.dr = \int_{OA} F.dr + \int_{AB} F.dr + \int_{BC} F.dr$

Along OA we have $y = 0$, $z = 0$ and x varies from 0 to 1, $dy = 0$ hence $dz = 0$.

Along AB, $x = 1$, $z = 0$ hence y varies from 0 to 1, $dx = 0$, $dz = 0$

Along BC we have $x = 1$, $y = 1$ and z varies from 0 to 1, $dx = 0$, $dy = 0$

$$\therefore \quad \int_C \vec{F}.\overrightarrow{dr} = \int_0^1 (3x^2+0)\,dx + \int_0^1 (0-0+0)dy + \int_0^1 0+0+20(1)z^2dz$$

$$= \left[x^3\right]_0^1 + 20\left[\frac{z^3}{3}\right]_0^1 = 1 + \frac{20}{3} = \frac{23}{3}.$$ **Ans.**

5. *Show that the line integral of* $\vec{F} = (y\sin z - \sin x)i + (x\sin z + 2yz)j + (xy\cos z + y^2)k$ *is independent of the path of integration and hence find its scalar potential.*

SOLUTION: $$\nabla\times F = \begin{vmatrix} i & j & k \\ \frac{\partial}{\partial x} & \frac{\partial}{\partial y} & \frac{\partial}{\partial z} \\ y\sin z - \sin x & x\sin z + 2yz & xy\cos z + y^2 \end{vmatrix}$$

$$= i(x\cos z + 2y - x\cos z - 2y) + j(y\cos z - y\cos z) + k(\sin z - \sin z) = 0$$

Thus, vector field $\vec{F}$ is conservative hence line integral is independent of the path of integration.

$$\vec{F} = \nabla\phi = i\frac{\partial\phi}{\partial x} + j\frac{\partial\phi}{\partial y} + k\frac{\partial\phi}{\partial z}$$

$$\therefore \quad \frac{\partial\phi}{\partial x} = y\sin z - \sin x, \quad \frac{\partial\phi}{\partial y} = x\sin z + 2yz, \quad \frac{\partial\phi}{\partial z} = xy\cos z + y^2$$

But $$d\phi = \frac{\partial \phi}{\partial x}dx + \frac{\partial \phi}{\partial y}dy + \frac{\partial \phi}{\partial z}dz$$

$$= (y \sin z - \sin x)dx + (x \sin z + 2yz)dy + (xy \cos z + y^2)dz$$

$$\int d\phi = \int_{y,\,z \text{ constt}} (y \sin z - \sin x) + \int_{z,\,x \text{ constt}} (x \sin z + 2yz)\,dy + \int_{x,\,y, \text{ constt.}} (xy \cos z + y^2)dz$$

$\therefore$ $\phi = xy \sin z + \cos x + y^2 z + c$ is the required scalar potential. **Ans.**

6. *Determine the constants a, b, c such that the vector field* $\vec{F} = (x+2y+az)i + (bx-3y-z)j + (4x+cy+2z)k$ *is conservative.*

Also find its scalar potential ϕ *and find the work done in moving a particle under this force field from (1, 2, –4) to (3, 3, 2).*

SOLUTION:
$$\nabla \times \vec{F} = \begin{vmatrix} i & j & k \\ \frac{\partial}{\partial x} & \frac{\partial}{\partial y} & \frac{\partial}{\partial z} \\ x+2y+az & bx-3y-z & 4x+cy+2z \end{vmatrix}$$

$= i(c + 1) + j(a - 4) + k(b - 2) = 0$ as $\vec{F}$ is a conservative field.

$\Rightarrow c = -1, \quad a = 4, \quad b = 2.$

Since $\vec{F}$ is conservative, $\vec{F} = \nabla\phi$

$$\therefore \quad \frac{\partial \phi}{\partial x} = x+2y+4z, \quad \frac{\partial \phi}{\partial y} = 2x-3y-z \quad \text{and} \quad \frac{\partial \phi}{\partial z} = 4x-y+2z.$$

But $$d\phi = \frac{\partial \phi}{\partial x}dx + \frac{\partial \phi}{\partial y}dy + \frac{\partial \phi}{\partial z}dz$$

$$\therefore \quad \phi = \int_{y,z \text{ constt}} (x+2y+4z)dx + \int_{x,z \text{ constt}} (2x-3y-z)dy + \int_{x,y \text{ constt}} (4x-y+2z)dz$$

$$= \frac{x^2}{2} + 2xy + 4xz + 2xy - \frac{3y^2}{2} - yz + 4xz - yz + z^2$$

$$= \frac{x^2}{2} - \frac{3y^2}{2} + z^2 + 2xy - yz + 4xz + k \quad \text{where } k \text{ is constant.}$$

$$\text{Work done} = \int_C \vec{F}.\vec{dr} = \int_C d\phi = \phi\Big|_{(1,\,2,\,-4)}^{(3,\,3,\,2)}$$

$$= \left[\frac{x^2}{2} - \frac{3y^2}{2} + z^2 + 2xy - yz + 4xz\right]_{(1,\,2,\,-4)}^{(3,\,3,\,2)}$$

$$= \frac{49}{2} \quad \text{(On simplification).}$$ **Ans.**

7. *Verify Green's theorem for* $\oint_C (2xy\,dx - y^2 dy)$ *where* C *is the boundary of the region bounded by the ellipse* $3x^2 + 4y^2 = 12$.

SOLUTION: $\oint_C M\,dx + N\,dy = \oint_C (2xy\,dx - y^2 dy) \quad \therefore \quad M = 2xy,\ N = -y^2$

Equation of ellipse is $3x^2 + 4y^2 = 12$ or $\frac{x^2}{4} + \frac{y^2}{3} = 1 \quad \therefore \quad x = 2\cos\theta,\ y = \sqrt{3}\sin\theta$.

$$\int_C Mdx + Ndy = \int_0^{2\pi} 2.(2\cos\theta)(\sqrt{3}\sin\theta)(-2\sin\theta)\,d\theta - 3\sin^2\theta\,(\sqrt{3}\cos\theta)\,d\theta$$

$$= \int_0^{2\pi} -11\sqrt{3}\sin^2\theta\cos\theta\,d\theta = -22\sqrt{3}\int_0^{\pi}\sin^2\theta\cos\theta\,d\theta = 0 \text{ by property of definite integrals.}$$

By Green's theorem
$$\oint_C Mdx + Ndy = \iint_R \left(\frac{\partial N}{\partial x} - \frac{\partial M}{\partial y}\right)dxdy = \int_{-2}^{2}\int_{-\frac{1}{2}\sqrt{12-3x^2}}^{\frac{1}{2}\sqrt{12-3x^2}} (0 - 2x)\,dydx$$

$$= \int_{-2}^{2} -4x\sqrt{3}\sqrt{1 - \frac{x^2}{4}}\,dx$$

$$= 0 \text{ by property of definite integrals.}$$

Hence Green's theorem is **verified.**

8. *Evaluate* $\oint_C \left[(x^2 + 2y)dx + (4x + y^2)\,dy\right]$ *by Green's theorem where* C *is the boundary of the region bounded by* $y = 0$, $y = 2x$ *and* $x + y = 3$.

SOLUTION: By Green's theorem

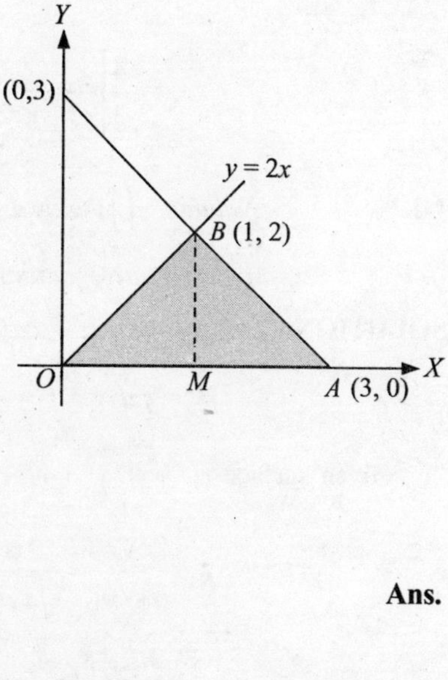

$$\oint_C Mdx + Ndy = \iint_R \left(\frac{\partial N}{\partial x} - \frac{\partial M}{\partial y}\right)dxdy$$

or $\oint_C (x^2 + 2y)dx + (4x + y^2)dy$

$$= \iint_R \left[\frac{\partial}{\partial x}(4x + y^2) - \frac{\partial}{\partial y}(x^2 + 2y)\right]dxdy$$

$$= \iint_R (4 - 2)\,dxdy = 2\,(\text{Area of } \Delta AOB)$$

$$= 2\left[\frac{1}{2}(3).(2)\right] = 6.$$ **Ans.**

9. *Evaluate* $\iint_S \vec{F}.\hat{n}\,ds$ *where* $\vec{F} = 18zi - 12j + 3yk$ *and* S *is the part of the plane* $2x + 3y + 6z = 12$ *in the first octant.*

SOLUTION: The surface S is the plane $2x + 3y + 6x = 12$ in the first octant.

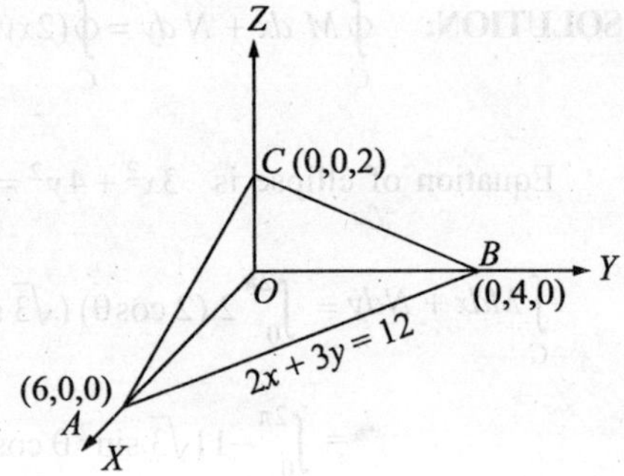

$$\phi = 2x + 3y + 6z.$$

$$\therefore \quad \hat{n} = \frac{\nabla\phi}{|\nabla\phi|} = \frac{2i+3j+6k}{\sqrt{4+9+36}} = \frac{1}{7}(2i+3j+6k)$$

$$dS = \frac{dx\,dy}{|\hat{n}.k|} = \frac{7}{6}\,dxdy.$$

$$F = 18\,zi - 12j + 3yk$$

$$\iint_S \vec{F}.\hat{n}dS = \iint_R (18zi - 12j + 3yk)\cdot\left(\frac{2i+3j+6k}{7}\right)\frac{7}{6}\,dxdy$$

$$= \frac{1}{6}\iint_R (36z - 36 + 18y)\,dxdy$$

$$= \frac{1}{6}\iint_R \left[36\left(2 - \frac{2x+3y}{6}\right) - 36 + 18y\right]dxdy$$

$$= \iint_R [12 - (2x+3y) - 6 + 3y]\,dxdy = \int_0^6\int_0^{\frac{12-2x}{3}} (6-2x)\,dydx$$

$$= 2\int_0^6\int_0^{(12-2x)} (6-2x)\,dydx = 2\int_0^6 (3-x)\left(\frac{12-2x}{3}\right)dx$$

$$= \frac{4}{3}\int_0^6 (x^2 - 9x + 18)dx = \frac{4}{3}\left[\frac{x^3}{3} - \frac{9x^2}{2} + 18x\right]_0^6$$

$$= \frac{4}{3}\left[\frac{x^3}{3} - \frac{9x^2}{2} + 18x\right]_0^6 = \frac{4}{3}\left[\frac{36\times 6}{3} - \frac{9}{2}\times 36 + 18(6)\right] = 24.$$ **Ans.**

10. *Evaluate* $\iint_S (yz\,dy\,dz + xz\,dzdx + xy\,dx\,dy)$ *over the surface of the sphere* $x^2 + y^2 + z^2 = 1$ *in the positive octant.*

SOLUTION: $\vec{F}.\hat{n}ds = yzdydz + xzdzdx + xydxdy$

$$\therefore \quad \vec{F} = yzi + zxj + xyk.$$

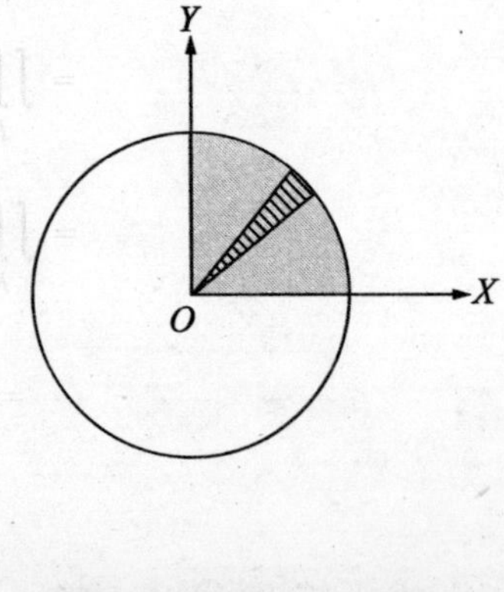

Given surface is $\phi \equiv x^2 + y^2 + z^2 - 1 = 0$

$$\therefore \quad \hat{n} = \frac{\nabla\phi}{|\nabla\phi|} = \frac{2xi + 2yj + 2zk}{\sqrt{4x^2+4y^2+4z^2}} = \frac{xi + yj + zk}{\sqrt{x^2+y^2+z^2}}$$

If R is the projection of the sphere $x^2+y^2+z^2=1$ on the XY-plane, then

$$dS = \frac{dxdy}{|\hat{n}.\hat{k}|} = 2xi + 2yj + 2zk$$

$$\iint_S yzdydz + xzdzdx + xydxdy = \iint_S (yzi + xzj + xyk).(xi + yj + zk)\frac{dxdy}{z}$$

$$= \iint_R 3xyz\frac{dxdy}{z} = 3\iint_R xydxdy$$

Putting $x = r\cos\theta,\ y = r\sin\theta, dxdy = rdrd\theta$

$$\iint yzdydz + zxdzdx + xydxdy = 3\int_0^{\pi/2} r\cos\theta.r\sin\theta.rdr\,d\theta = 3\int_0^{\pi/2}\frac{1}{2}\sin 2\theta\, d\theta\int_0^1 r^3 dr$$

$$= \frac{3}{2}\left|\frac{-\cos 2\theta}{2}\right|_0^{\pi/2}\left[\frac{r^4}{4}\right]_0^4 = \frac{3}{16}(-\cos\pi + \cos 0) = \frac{3}{8}.$$ **Ans.**

11. *Verify Stoke's theorem for* $\vec{F} = (x+y)i + (y+z)j - xk$ *and* S *is the surface of the plane* $2x + y + z = 2$ *in the first octant.*

SOLUTION: To show that $\oint_C \vec{F}.\vec{dr} = \iint_S (\nabla\times\vec{F}).\hat{n}dS.$

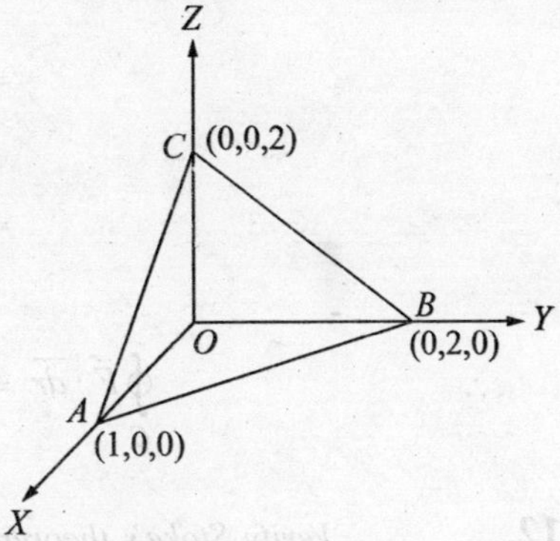

$$\nabla\times F = \begin{vmatrix} i & j & k \\ \frac{\partial}{\partial x} & \frac{\partial}{\partial y} & \frac{\partial}{\partial z} \\ x+y & y+z & -x \end{vmatrix} = i(-1) + j(1) + k(-1).$$

Let $\phi = 2x + y + z$ hence $\nabla\phi = 2i + j + k$

$$\hat{n} = \frac{\nabla\phi}{|\nabla\phi|} = \frac{2i+j+k}{\sqrt{4+1+1}} = \frac{2i+j+k}{\sqrt{6}}$$

$$dS = \frac{dxdy}{|\hat{n}.k|} = \sqrt{6}\,dxdy$$

$$\therefore \iint_S (\nabla\times F).\hat{n}dS = \iint(-i+j-k).\frac{(2i+j+k)}{\sqrt{6}}.\sqrt{6}\,dxdy$$

$$= \int_0^1\int_0^{2-2x}(-2+1-1)dxdy = -2\int_0^1 [y]_0^{2-2x}dx$$

$$= -2\int_0^1 (2-2x)dx = -2[2x-x^2]_0^1 = -2$$

Now for $\oint_C \vec{F}.\vec{dr} = \int_{\Delta ABC} (x+y)dx+(y+z)dy-xdz$

$$= \int_{AB} \vec{F}.\vec{dr} + \int_{BC} \vec{F}.\vec{dr} + \int_{CA} \vec{F}.\vec{dr}$$

Now $\int_{AB} (x+y)dx+(y+z)dy-xdz$. On AB, $z=0$, $y=2-2x$

$$= \int_1^0 (x+2-2x)dx+(2-2x+0)(-2dx)-0 = \int_1^0 (3x-2)\,dx$$

$$= \left[\frac{3x^2}{2}-2x\right]_1^0 = \frac{-3}{2}+2 = \frac{1}{2}.$$

$\int_{BC} (x+y)dx+(y+z)dy-xdz$ Here, on BC $x=0$, $y+z=2$

$$= \int_2^0 [0+2dy+0] = -2\int_0^2 dy = -4$$

$\int_{CA} (x+y)dx+(y+z)dy-xdz$ Here, on CA $y=0$, $2x+z=2$

$$= \int_0^1 x\,dx+(2-2x)\,(0)-x\,(-2dx)$$

$$= \left[\frac{3x^2}{2}\right]_0^1 = \frac{3}{2}.$$

$\therefore$ $\oint_C \vec{F}\cdot\vec{dr} = \frac{1}{2}-4+\frac{3}{2}=-2.$ **Hence the theorem is verified**

12. *Verify Stoke's theorem for* $\vec{F}=4yi-4xj+3k$, *where* S *is a disc of radius* 1 *lying on the plane* $z=1$ *and* C *is its boundary.*

SOLUTION: To verify $\iint_S (\nabla\times\vec{F}).\vec{ds} = \oint_C \vec{F}\cdot\vec{dr}$. Here S is the surface of the disk of radius 1 lying on the plane $z=1$ and C is circle $x^2+y^2=1, z=1$.

$$\nabla\times F = \begin{vmatrix} i & j & k \\ \frac{\partial}{\partial x} & \frac{\partial}{\partial y} & \frac{\partial}{\partial z} \\ 4y & -4x & 3 \end{vmatrix} = i(0-0)+j(0-0)+k(-4-4) = -8\hat{k}.$$

Here $\hat{n}=k$, $ds=\frac{dxdy}{|\hat{n}.k|}=dxdy$

$\therefore \quad \iint_S (\nabla\times\vec{F}).\hat{n}\,dS = \iint_S -8k.k\,dxdy = -8\iint_S dxdy$

$= -8 \times$ area of the disk $= -8\pi(1)^2 = -8\pi.$

Further, $\int_C \vec{F}.\vec{dr} = \int_C (4yi - 4xj + 3k).(idx + jdy + kdz) = \int_C (4ydx - 4xdy + 3dz)$

On the disk $z = 1$ hence $dz = 0$. Let us put $x = \cos\theta,\ y = \sin\theta$

$\therefore \quad \int_C \vec{F}.\vec{dr} = \int_0^{2\pi} 4\sin\theta(-\sin\theta)\,d\theta - 4\cos\theta(\cos\theta)d\theta = -4\int_0^{2\pi} d\theta = -8\pi$ **Hence verified.**

13. *Verify Stoke's theorem for* $\vec{F} = (x^2+y-4)\,i + 3xyj + (2xz+z^2)k$ *over the surface of the sphere* $x^2+y^2+z^2 = 16$ *above the XY-plane.*

SOLUTION: We are to verify $\iint_S (\nabla\times\vec{F}).\vec{dS} = \oint_C \vec{F}.\vec{dr}$ where S is the surface of the hemisphere $x^2+y^2+z^2=16$ above XY-plane and C is the circle $x^2+y^2=16,\ z=0.$

$$\nabla\times\vec{F} = \begin{vmatrix} i & j & k \\ \dfrac{\partial}{\partial x} & \dfrac{\partial}{\partial y} & \dfrac{\partial}{\partial z} \\ x^2+y-4 & 3xy & 2xz+z^2 \end{vmatrix} = i(0-0) + j(0-2z) + k(3y-1) = -2zj + k(3y-1).$$

Let $\phi = x^2+y^2+z^2$ then $\hat{n} = \dfrac{\nabla\phi}{|\nabla\phi|} = \dfrac{2xi+2yj+2zk}{\sqrt{4x^2+4y^2+4z^2}} = \dfrac{xi+yj+zk}{4}$ (since $x^2+y^2+z^2=16$)

$$dS = \frac{dxdy}{|\hat{n}.k|} = \frac{dxdy}{z/4} = \frac{4dxdy}{z}.$$

$$\therefore \quad \iint_S \nabla\times\vec{F}.\hat{n}ds = \iint_S [-2zj + (3y-1)k]\cdot\frac{(xi+yj+zk)}{4}\frac{4dxdy}{z}$$

$$= \iint_S [-2zy + z(3y-1)]\frac{dxdy}{z} = \iint_S (-2y+3y-1)\,dxdy = \iint_S (y-1)dxdy.$$

$$= \int_0^{2\pi}\int_0^4 (r\sin\theta - 1)\,r\,drd\theta$$

(On putting $x = r\cos\theta,\ y = r\sin\theta,\ r$ varying from 0 to 4 and θ from 0 to 2π)

$$= \left[\frac{r^3}{3}\right]_0^4 (-\cos\theta)_0^{2\pi} - \left[\frac{r^2}{2}\right]_0^4 [\theta]_0^{2\pi} = \frac{64}{3}(-\cos 2\pi + \cos 0) - \frac{16}{2}.2\pi = -16\pi$$

Next, $\int_C \vec{F}.\vec{dr} = \int_C \left[(x^2+y-4)i + 3xyj + (2xz+z^2)k\right]\cdot(i\,dx + j\,dy + k\,dz)$

$$= \int_C (x^2+y-4)\,dx + 3xy\,dy + (2xz+z^2)\,dz$$

Here $x = 4\cos\theta,\ y = 4\sin\theta,\ z = 0$ since C is $x^2+y^2 = 4^2,\ z = 0.$

$$= \int_0^{2\pi}\left[16\cos^2\theta + 4\sin\theta - 4)(-4\sin\theta\,d\theta) + 3.4.4\cos\theta\sin\theta\,(4\cos\theta\,d\theta) + 0\right]$$

$$= \int_0^{2\pi}(-64\cos^2\theta\sin\theta - 16\sin^2\theta + 16\sin\theta + 192\cos^2\theta\sin\theta)\,d\theta = \int_0^{2\pi} 16\sin^2\theta\,d\theta + 0 + 0$$

$= -16\pi$ **Hence verified.**

14. *Using Stoke's theorem find the work done in moving a particle once round the perimeter of the triangle with vertices at (2, 0, 0), (0, 3, 0) and (0, 0, 6),* $\vec{F}=(x+y)i+(2x-z)j+(y+z)k$

SOLUTION: Work done = $\oint_C \vec{F}.\vec{dr}$. By Stoke's theorem $\oint_C \vec{F}.\vec{dr}.=\iint_3 (\nabla\times\vec{F}).\hat{n}dS$ S is the surface of the plane ABC.

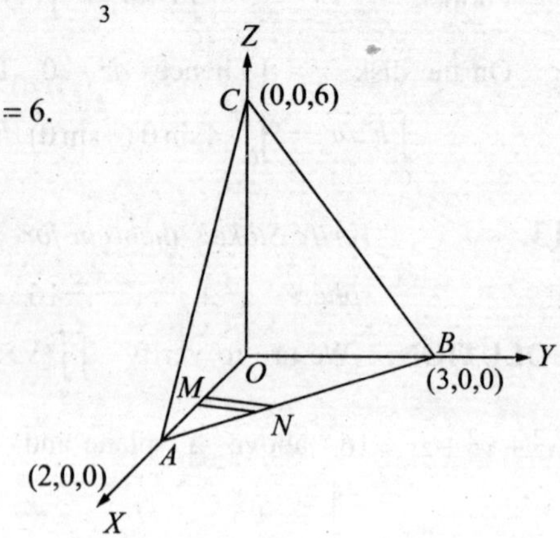

Equation of the plane is $\frac{x}{2}+\frac{y}{3}+\frac{z}{6}=1$ or $3x+2y+z=6$.

$$\nabla\times\vec{F} = \begin{vmatrix} i & j & k \\ \frac{\partial}{\partial x} & \frac{\partial}{\partial y} & \frac{\partial}{\partial z} \\ x+y & 2x-z & y+z \end{vmatrix} = 2i+0j+k.$$

Here $\phi = 3x + 2y + z$, hence

$$\hat{n} = \frac{\nabla\phi}{|\nabla\phi|}=\frac{3i+2j+k}{\sqrt{9+4+1}}$$

and $$dS = \frac{dxdy}{|\hat{n}.k|}=\frac{dxdy}{1/\sqrt{14}}=\sqrt{14}\,dxdy.$$

Let R be the region bounded by ΔOAB as projection of ΔABC on XY-plane. Along the vertical strip MN, y varies from 0 to $\frac{6-3x}{2}$ and x varies from 0 to 2.

$$\iint_S (\nabla\times\vec{F}).\hat{n}\,dS = \iint_R (2i+k).\left(\frac{3i+2j+k}{\sqrt{14}}\right)\sqrt{14}\,dxdy$$

$$= \int_0^2\int_0^{\frac{6-3x}{2}}(6+0+1)dxdy = 7\int_0^2 [y]_0^{\frac{6-3x}{2}}dx = 7\int_0^2 \frac{6-3x}{2}dx$$

$$= \frac{7}{2}\left[6x-\frac{3x^2}{2}\right]_0^2 = \frac{7}{2}(12-6)=21.$$ **Ans.**

15. *Evaluate* $\iint_S (x^3dydz+x^2ydzdx+x^2zdxdy)$ *where* S *is the closed surface consisting of the cylinder* $x^2+y^2=a^2$, $z=0$ *and* $z=b$.

SOLUTION: As S is a closed surface the Gauss divergence theorem is applicable.

By Gauss theorem $\iint_S \vec{F}.\hat{n}\,dS=\iiint_V \nabla.\vec{F}dxdydz$

or $$\iint_S F_1dydz+F_2dzdx+F_3dxdy = \iiint_V\left(\frac{\partial F_1}{\partial x}+\frac{\partial F_2}{\partial y}+\frac{\partial F_3}{\partial z}\right)dx\,dy\,dz$$

$\therefore$ $$F_1 = x^3,\ F_2=x^2y,\ F_3=x^2z$$

$\therefore$ $$\frac{\partial F_1}{\partial x}+\frac{\partial F_2}{\partial y}+\frac{\partial F_3}{\partial z} = 3x^2+x^2+x^2=5x^2$$

$$\therefore \quad \iiint_V \left(\frac{\partial F_1}{\partial x} + \frac{\partial F_2}{\partial y} + \frac{\partial F_3}{\partial z} \right) dxdydz = \iiint_V 5x^2 dxdydz$$

Converting to Cylindrical polar coordinates $x = r \cos \theta$, $y = r \sin \theta$, $z = z$, we get the above integral as

$$5\int_{z=0}^{b} \int_{\theta=0}^{2\pi} \int_{r=0}^{a} r^2 \cos^2 \theta \, rdr \, d\theta dz = 5[z]_0^b \int_0^a r^3 dr \int_0^{2\pi} \cos^2 \theta \, d\theta = 5b. \left[\frac{r^4}{4} \right]_0^a \int_0^{2\pi} \frac{(1+\cos 2\theta)}{2} d\theta$$

$$= \frac{5ba^4}{4} \cdot \frac{1}{2} \left[\theta + \frac{1}{2} \sin 2\theta \right]_0^{2\pi} = \frac{5}{4} \pi a^4 b. \qquad \textbf{Ans.}$$

16. *Verify Gauss divergence theorem for* $\vec{F} = 2xzi + yzj + z^2k$, *over the upper half of the sphere* $x^2 + y^2 + z^2 = a^2$.

SOLUTION: By Gauss theorem $\iiint_V \nabla . \vec{F} dV = \iint_S F . \hat{n} dS$

$$\nabla \cdot \vec{F} = \frac{\partial}{\partial x}(2xz) + \frac{\partial}{\partial y}(yz) + \frac{\partial}{\partial z}(z^2) = 2z + z + 2z = 5z$$

$\therefore$ The volume integral $\iiint_V \nabla . \vec{F} dV = \iiint_V 5z dxdydz$

We convert it into spherical polar coordinates $x = r \sin \theta \cos \phi$, $y = r \sin \theta \sin \phi$, $z = r \cos \theta$ we have $dxdydz = r^2 \sin \theta \, dr d\theta d\phi$. Equation of sphere is $r = a$. Here r varies from 0 to a, θ varies from 0 to $\pi/2$ and ϕ varies from 0 to 2π.

$$\therefore \quad \iiint \nabla . \vec{F} \, dv = 5\int_{\phi=0}^{2\pi} \int_{\theta=0}^{\pi/2} \int_{r=0}^{a} r \cos\theta . r^2 \sin\theta \, dr d\theta d\phi$$

$$= 5[\phi]_0^{2\pi} \left[\frac{r^4}{4} \right]_0^a \int_0^{\pi/2} \cos\theta \sin\theta \, d\theta = 5(2\pi) \frac{a^4}{4} \left[\frac{1}{2} \sin^2 \theta \right]_0^{\pi/2}$$

$$= \frac{5\pi}{4} a^4.$$

Next, in the surface integral $\iint_S \vec{F} . \hat{n} ds$ the surface S consists of S_1 the curved surface of the upper half of the sphere and S_2 the plane circular surface S_2 in the XY-plane.

$$\therefore \quad \iint_S \vec{F} . \hat{n} dS = \iint_{S_1} \vec{F} . \hat{n} dS + \iint_{S_2} \vec{F} . \hat{n} dS$$

Here S_1 is actually the curved surface of the upper half of the sphere.

Let $\phi = x^2 + y^2 + z^2$, $\hat{n} = \frac{\nabla \phi}{|\nabla \phi|} \frac{2xi + 2yj + 2zk}{\sqrt{4(x^2 + y^2 + z^2)}} = \frac{xi + yj + zk}{a}$

If R is the projection of S_1 on the XY-plane which is the circular plate $x^2 + y^2 = a^2$, then

$$dS = \frac{dxdy}{|\hat{n} . k|} = \frac{adxdy}{z}$$

$$\therefore \quad \iint_{S_1} \vec{F}.\hat{n}dS = \iint_R (2xzi + yzj + z^2k).\left(\frac{xi + yj + zk}{a}\right)\frac{adxdy}{z}$$

$$= \iint_R (2x^2 + y^2 + z^2)dxdy = \iint_R (2x^2 + y^2 + a^2 - x^2 - y^2)dxdy = \iint_R (x^2 + a^2)dxdy$$

(since $x^2 + y^2 + z^2 = a^2$)

$$= \int_0^{2\pi}\int_0^a (r^2 \cos^2\theta + a^2)\, rdrd\theta \quad \text{(On converting to polar coordinates)}$$

$$= \int_0^{2\pi}\left[\frac{r^4}{4}\right]_0^a \cos^2\theta\, d\theta + \int_0^{2\pi} a^2\left[\frac{r^2}{2}\right]_0^a d\theta$$

$$= \int_0^{2\pi}\left[\frac{a^4}{4}\left(\frac{1+\cos 2\theta}{2}\right) + \frac{a^4}{2}\right]d\theta = \frac{a^4}{8}.2\pi + 0 + \frac{a^4}{2}.2\pi = \frac{5\pi a^4}{4}$$

Further, consider the surface S_2 which is circular plate $x^2 + y^2 = a^2$ in XY-Plane where $\hat{n} = -k$ (outward drawn normal)

$$dS = \frac{dxdy}{|\hat{n}.k|} = dxdy$$

$$\therefore \quad \iint_{S_2} \vec{F}.\hat{n}dS = \iint_{S_2} (2xzi + yzj + z^2k)(-\hat{k})\, dxdy = 0 \text{ as } z = 0$$

$$\therefore \quad \iint_S F.\hat{n}dS = \frac{5}{4}\pi a^4 + 0.$$

Thus the Gauss divergence theorem is verified.

17. *Evaluate* $\iint_S (lx + my + nz)\, dS$ *where* l, m, n *are the direction cosines of the outer normal to the surface of sphere whose radius is 2.*

SOLUTION: Sphere $x^2 + y^2 + z^2 = 2^2$ is a closed surface hence Gauss divergence theorem can be applied here.

$$\iint_S \vec{F}.\hat{n}dS = \iiint_V \nabla.\vec{F}\, dV$$

$$\vec{F}.\hat{n} = lx + my + nz = (xi + yi + zk)\,.\,(li + mj + nk)$$

$$\therefore \quad \iiint_V \nabla.\vec{F}\, dV = \iiint 3\, dv = 3 \times \text{Volume of sphere with radius 2.}$$

$$= 3.\frac{4}{3}\pi.2^3 = 32\pi.$$

$$\Rightarrow \quad \int_S (lx + my + nz)\, dS = 32\pi.$$

Ans.

18. *Evaluate* $\iint \vec{F}.\overrightarrow{ds}$ *using Gauss divergence theorem where* $\vec{F} = 2xyi + yz^2 j + zxk$ *and* S *is the surface of the region bounded by* $x = 0,\ y = 0,\ z = 0,\ y = 3,\ x + 2z = 6.$

SOLUTION: S is the closed surface bounded by $x = 0,\ y = 0,\ z = 0,\ y = 3,\ x + 2z = 6.$

By Gauss theorem $\iint_S \vec{F}.\overrightarrow{ds} = \iiint_V (\nabla.\vec{F})\,dv$

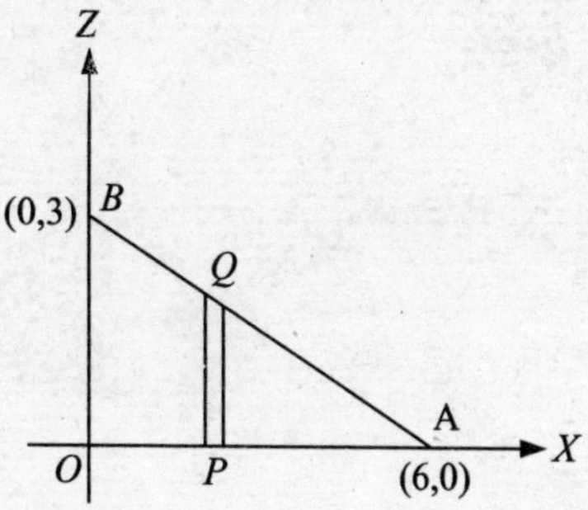

$\vec{F} = 2xy\,i + yz^2 j + zxk \quad \therefore \quad \nabla.\vec{F} = 2y + z^2 + x$

$$\therefore \quad \iiint_V \nabla.\vec{F}dv = \iiint (2y + z^2 + x)\,dxdydz$$

As per the given region y varies from $y = 0$ to $y = 3$ and in the XZ plane region is bounded by $x = 0,\ z = 0$ and $x + 2z = 6.$

Along the vertical strip PQ, Z varies from 0 to $\dfrac{6-x}{2}$ and the x varies from 0 to $6.$

$$\iiint \nabla.\vec{F}dv = \int_{x=0}^{6}\int_{z=0}^{\frac{6-x}{2}}\int_{y=0}^{3} (2y + z^2 + x)\,dydzdx$$

$$= \int_0^6\int_0^{\frac{6-x}{2}} \left[y^2 + yz^2 + xy\right]_{y=0}^{3} dzdx$$

$$= \int_0^6\int_0^{\frac{6-x}{2}} (9 + 3z^2 + 3x)\,dz\,dx = \int_0^6 \left[9z + z^3 + 3xz\right]^{\frac{6-x}{2}} dx$$

$$= \int_0^6 \left[\frac{9(6-x)}{2} + \frac{(6-x)^3}{8} + \frac{3x(6-x)}{2}\right] dx$$

$$= \left[\frac{-9}{4}(6-x)^2 - \frac{(6-x)^4}{32} + \frac{9x^2}{2} - \frac{x^3}{2}\right]_0^6$$

$$= 0 + 0 + \frac{9}{2}.36 - \frac{6.36}{2} + \frac{9}{4}.36 + \frac{36.36}{32}$$

$$= 162 - 108 + 81 + \frac{81}{2} = \frac{351}{2}.$$ **Ans.**

CHAPTER

16 Functions of a Complex Variable — Analyticity and Transformation

IMPORTANT FORMULAE AND USEFUL RESULTS

1. **Cauchy-Riemann Equations**

If $f(z) = u + iv$ is a function of z defined in domain D then for $f'(z)$ to exist at $z = z_0$, the pair of equations

$$\frac{\partial u}{\partial x} = \frac{\partial v}{\partial y} \text{ and } \frac{\partial u}{\partial y} = -\frac{\partial v}{\partial x} \quad \text{must be satisfied.}$$

Analytic Function: If $w = f(z)$ possesses a derivative at $z = z_0$ and at every point in some neighbourhood of z_0 then $f(z)$ is called analytic at $z = z_0$.

Canchy-Riemann Equations in Polar Form

$$\frac{\partial u}{\partial r} = \frac{1}{r}\frac{\partial v}{\partial \theta} \quad \text{and} \quad \frac{\partial v}{\partial r} = \frac{-1}{r}\frac{\partial v}{\partial \theta}$$

Harmonic Function and Laplace Equation

A function $f(x, y)$ which satisfies the Laplace equation

$$\nabla f = \frac{\partial^2 f}{\partial x^2} + \frac{\partial^2 f}{\partial y^2} = 0 \quad \text{is called harmonic function.}$$

If $f(z) = u + iv$ is analytic in a domain D then u and v both are harmonic functions. Here v is said to be conjugate of u and u is conjugate of v.

Milne-thonpson method: To get $f'(z)$ in terms of z, we replace x by z and y by 0 in $u(x, y)$ and $v(x, y)$ occuring in $\frac{dw}{dz} = \frac{\partial u(x, y)}{\partial x} + i\frac{\partial v(x, y)}{\partial x}$

and then integrating it gives $w = f(z)$.

2. **Mapping Conformal:** Let two curves C_1 and C_2 intersect in z plane at a point $z = z_0$ at an angle α. If their image curves s_1 and s_2 in the w plane intersect at the point $w = w_0$ at the same angle α then the mapping is called conformal.

3. **Some Standard Mappings:**

   (*i*) **Translation:** $w = z + c$ where c is complex . Here a figure in z-plane is shifted in the direction of c.

(ii) Rotation and Magnification: $w = AZ$ where A is complex constant. If $z = re^{i\theta}$, $w = \rho e^{i\phi}$ and $A = ae^{i\alpha}$ then $\rho e^{i\phi} = Are^{i(\theta+\alpha)}$

$\Rightarrow \rho = Ar, \quad \phi = \theta + \alpha$

(iii) Inversion and Reflection: $w = \dfrac{1}{z}$

$$\therefore \quad \rho e^{i\phi} = \frac{1}{r} e^{-e\theta} \quad \Rightarrow \quad \rho = \frac{1}{r} \text{ and } \phi = -\theta.$$

(iv) Bilinear Transformation: $w = \dfrac{az+b}{cz+d}$.

It consists of translation, rotation and magnification and inversion and refection. If z_1, z_2, z_3 in z-plane map into w_1, w_2, w_3 in w-plane, the bilinear mapping is

$$\frac{w-w_1}{w-w_3} \cdot \frac{w_2-w_3}{w_2-w_1} = \frac{(z-z_1)(z_2-z_3)}{(z-z_3)(z_2-z_1)}.$$

SOLUTIONS TO EXERCISE 16A

1. *If* $\omega = f(z) = u + iv$ *is analytic, show that*

$$|f'(z)|^2 = \begin{vmatrix} \dfrac{\partial u}{\partial x} & \dfrac{\partial u}{\partial y} \\ \dfrac{\partial v}{\partial x} & \dfrac{\partial v}{\partial y} \end{vmatrix}$$

SOLUTION: $f(z) = u + iv$ is analytic hence $\dfrac{\partial u}{\partial x} = \dfrac{\partial v}{\partial y}$ and $\dfrac{\partial u}{\partial y} = -\dfrac{\partial v}{\partial x}$

$$f'(z) = \frac{\partial u}{\partial x} + i\frac{\partial v}{\partial x} \qquad \therefore \qquad |f'(z)|^2 = \left(\frac{\partial u}{\partial x}\right)^2 + \left(\frac{\partial v}{\partial x}\right)^2$$

$$= \frac{\partial u}{\partial x}\cdot\frac{\partial v}{\partial y} + \frac{\partial v}{\partial x}\left(-\frac{\partial u}{\partial y}\right) = \begin{vmatrix} \dfrac{\partial u}{\partial x} & \dfrac{\partial u}{\partial y} \\ \dfrac{\partial v}{\partial x} & \dfrac{\partial v}{\partial y} \end{vmatrix}$$

Hence proved.

2. *Find* $\displaystyle \underset{z\to 1+i}{\text{limit}} \frac{z^2 - 2i}{z^2 - 2z + 2}$

SOLUTION: $\displaystyle \lim_{z\to 1+i} \frac{z^2 - 2i}{z^2 - 2z + 2}$. It is of the form $\dfrac{0}{0}$. Applying L' Hospital's rule,

$$\text{Given limit} = \lim_{z\to 1+i} \frac{2z}{(2z-2)} = \frac{1+i}{1+i-1} = 1 - i$$

Ans.

3. *A function* $f(x, y)$ *is given by*

$$f(x,y) = \begin{cases} \dfrac{x^3 - y^3}{x^3 + y^3}, & x, y \neq 0 \\ 0, & x = y = 0 \end{cases}$$

Test the continuity of $f(x, y)$ *at* $(0, 0)$.

SOLUTION:

$$f(x,y) = \frac{x^3 - y^3}{x^3 + y^3}$$

$$\underset{x\to 0}{\text{lt}}\ \underset{y\to 0}{\text{lt}}\ f(x,y) = \underset{x\to 0}{\text{lt}} \frac{x^3 - 0}{x^3 + 0} = 1 \quad \text{and} \quad \underset{y\to 0}{\text{lt}}\ \underset{x\to 0}{\text{lt}}\ f(x,y) = \underset{y\to 0}{\text{lt}} \frac{0 - y^3}{0 + y^3} = -1$$

$\therefore$ $f(x,y)$ is discontinuous at $(0,0)$. **Ans.**

4. *The function* $f(x, y) = \dfrac{xy}{x^2 + y^2}$. *Find* $\lim\limits_{(x,y)\to(0,0)} f(x, y)$.

SOLUTION: $f(x, y) = \dfrac{xy}{x^2 + y^2}$. $\therefore \underset{(x,y)\to(0,0)}{\text{Lt}} f(x, y) = \underset{x\to 0}{\text{Lt}} \dfrac{x^2}{x^2 + x^2} = \dfrac{1}{2}$ along $y = x$.

$$= \underset{x\to 0}{\text{Lt}} \frac{-x^2}{x^2 + x^2} = -\frac{1}{2} \quad \text{along } y = -x$$

Hence the limit does not exist. **Ans.**

5. *Show that the function* $f(z) = e^{-y} \cos x + i e^{-y} \sin x$ *is differentiable for all* z *and find its derivative.*

SOLUTION: $f(z) = u + iv$, $\therefore u = e^{-y} \cos x$, $v = e^{-y} \sin x$

There $\dfrac{\partial u}{\partial x} = -e^{-y} \sin x$, $\dfrac{\partial u}{\partial y} = -e^{-y} \cos x$, $\dfrac{\partial v}{\partial x} = e^{-y} \cos x$, $\dfrac{\partial v}{\partial y} = -e^{-y} \sin x$,

Since $\dfrac{\partial u}{\partial x} = \dfrac{\partial v}{\partial y}$ and $\dfrac{\partial u}{\partial y} = -\dfrac{\partial v}{\partial x}$ here, $f(z)$ is differentiable for all z.

$$f'(z) = \frac{\partial u}{\partial x} + i\frac{\partial v}{\partial x} = -e^{-y} \sin x + ie^{-y} \cos x = ie^{-y} (\cos x + i \sin x) \qquad \textbf{Ans.}$$

6. (*a*) *Show that* $u(x, y) = xy^3 - x^3y$ *is a harmonic function and find its conjugate harmonic function* $v(x, y)$.

(*b*) *Show that the function* $u = \dfrac{1}{2} \log(x^2 + y^2)$ *is harmonic, find its conjugate harmonic.*

SOLUTION: (*a*) $u(x, y) = xy^3 - x^3y$ hence $\dfrac{\partial u}{\partial x} = y^3 - 3x^2y$ and $\dfrac{\partial u}{\partial y} = 3xy^2 - x^3$.

Also $\dfrac{\partial^2 u}{\partial x^2} = -6xy$ and $\dfrac{\partial^2 u}{\partial y^2} = 6xy$ Hence $\dfrac{\partial^2 u}{\partial x^2} + \dfrac{\partial^2 u}{\partial y^2} = 0$ so the function $u(x, y)$ is harmonic.

To find its conjugate harmonic $v(x, y)$, it is implied that $f(z) = u + iv$ is analytic and hence C.R. equations are satisfied.

Therefore $\dfrac{\partial v}{\partial y} = y^3 - 3x^2y$ and $\dfrac{\partial v}{\partial x} = -3xy^2 + x^3$.

$\Rightarrow$ $v(x, y) = \dfrac{1}{4}y^4 - \dfrac{3x^2y^2}{2} + c(x)$ where $c(x)$ is a function of x.

Now $\dfrac{\partial v}{\partial x} = -3xy^2 + \dfrac{dc}{dx}$, But $\dfrac{\partial v}{\partial x} = -3xy^2 + x^3$, Hence $\dfrac{dc}{dx} = x^3$

$\therefore$ $c(x) = \dfrac{x^4}{4}$ Therefore, $v = \dfrac{1}{4}y^4 - \dfrac{3x^2y^2}{2} + \dfrac{x^4}{4} = \dfrac{1}{4}(x^4 + y^4) - \dfrac{3}{2}x^2y^2$ **Ans.**

(*b*) $u = \frac{1}{2}\log(x^2+y^2)$ hence $\frac{\partial u}{\partial x} = \frac{x}{x^2+y^2}$, $\frac{\partial u}{\partial y} = \frac{y}{x^2+y^2}$, $\frac{\partial^2 u}{\partial x^2} = \frac{y^2-x^2}{(y^2+x^2)}$ and

$$\frac{\partial^2 u}{\partial y^2} = \frac{x^2-y^2}{(x^2+y^2)^2}$$

$\Rightarrow \frac{\partial^2 u}{\partial x^2} + \frac{\partial^2 u}{\partial y^2} = 0$ hence u is a harmonic function.

Let $f(z) = u + iv$ be analytic and if v is the conjugate harmonic of u. C.R. equations are true.

Therefore $$\frac{\partial v}{\partial y} = \frac{x}{x^2+y^2} \qquad ...(1)$$

and $$\frac{\partial v}{\partial x} = \frac{-y}{x^2+y^2} \qquad ...(2)$$

From (1) we get $v = \tan^{-1}\left(\frac{y}{x}\right) + K(x)$ $\therefore$ $\frac{\partial v}{\partial x} = \frac{-y}{x^2+y^2}$ hence $K(x) = 0$.

Thus, $v \equiv \tan^{-1}\left(\frac{y}{x}\right)$. **Ans.**

7. (*i*) *Show that the function defined by* $f(z) = |z|$ *is not differentiable at the origin but is continuous everywhere.*

(*ii*) *Show that* $\frac{d}{dz}(\bar{z})^2$ *does not exist at any point except at* $z = 0$.

SOLUTION: $f(z) = |z| = \sqrt{x^2+y^2}$. It is continuous everywhere. $f(z)$ is differentiable at every point other than the origin.

At the origin $$f'(0) = \operatorname*{Lt}_{\delta x\to 0} \frac{f(0+\delta z) - f(0)}{\delta z} = \operatorname*{Lt}_{\delta x\to 0} \frac{|\delta z|}{\delta z} = \operatorname*{Lt}_{\substack{\delta x\to 0\\ \delta y\to 0}} = \frac{\sqrt{(\delta x)^2 + (\delta y)^2}}{\delta x + i\delta y}$$

Now $$f'(0) = \operatorname*{Lt}_{\delta x\to 0}\operatorname*{Lt}_{\delta y\to 0} \frac{\sqrt{(\delta x)^2 + (\delta y)^2}}{\delta x + i\delta y} = \operatorname*{Lt}_{\delta x\to 0} \frac{\delta x}{\delta x} = 1$$

But $\operatorname*{Lt}_{\delta y\to 0}\operatorname*{Lt}_{\delta x\to 0} \frac{\sqrt{(\delta x)^2 + (\delta y)^2}}{\delta x + i\delta y} = \operatorname*{Lt}_{\delta y\to 0} \frac{\delta y}{i\delta y} = \frac{1}{i}$. Hence $|z|$ is not differentiable at origin.

8. *Show that the functions defined by* (*i*) $f(z) = z^3 + 1 - iz^2$ (*ii*) e^z *are analytic everywhere.* **(GGSIPU IInd Sem. End Term 2005)**

SOLUTION: (*i*) $f(z) = u + iv = (x+iy)^3 + 1 - i(x+iy)^2$

$\therefore$ $u = x^3 - 3xy^2 + 1 + 2xy$ and $v = 3x^2y - y^3 - x^2 + y^2$

Next, $\frac{\partial u}{\partial x} = 3x^2 - 3y^2 + 2y$, $\frac{\partial u}{\partial y} = -6xy + 2x$, $\frac{\partial v}{\partial x} = 6xy - 2x$ and $\frac{\partial v}{\partial y} = 3x^2 - 3y^2 + 2y$.

$\therefore \quad \frac{\partial u}{\partial x} = \frac{\partial v}{\partial y}$ and $\frac{\partial u}{\partial y} = -\frac{\partial v}{\partial x}$. Hence $f(z)$ is analytic everywhere.

(*ii*) $f(z) = e^z = e^{(x+iy)} = u + iv$. Here $u = e^x \cos y$ and $v = e^x \sin y$

$$\therefore \quad \frac{\partial u}{\partial x} = e^x \cos y, \quad \frac{\partial u}{\partial y} = -e^x \sin y, \quad \frac{\partial v}{\partial x} = e^x \sin y \text{ and } \frac{\partial v}{\partial y} = e^x \cos y.$$

Hence C.R. equations are satisfied here so e^z is analytic everywhere. **Hence proved.**

9. *Obtain the singularities of the function*

$$f(z) = \frac{\cos z}{z^2 + 1}.$$

SOLUTION: $f(z) = \frac{\cos z}{z^2+1}$. Its singularities are given by $z^2 + 1 = 0$, *i.e.*, at $z = \pm i$.

Actually, $f(z)$ is analytic everywhere except at $z = \pm i$. **Ans.**

10. *Show that the function* $f(z) = \sqrt{|xy|}$ *satisfies the Cauchy-Riemann equations at the origin but is not analytic there.*

SOLUTION: $f(z) = \sqrt{|xy|} = u + iv$. Here $u = \sqrt{|xy|}$ and $v = 0$

$$\therefore \quad \frac{\partial v}{\partial x} = 0, \quad \frac{\partial v}{\partial y} = 0 \cdot \quad \frac{\partial u}{\partial x} = \frac{\partial}{\partial x}\sqrt{|xy|} = 0 \text{ at } (0, 0) \text{ and } \frac{\partial u}{\partial y} = \frac{\partial}{\partial y}\sqrt{|xy|} = 0 \text{ at } (0, 0)$$

Hence C.R. equations are satisfied at the origin. For analyticity of $f(z)$ at the origin

$$\underset{\delta z \to 0}{\text{Lt}} \frac{f(z+\delta z) - f(z)}{\delta z} = \underset{\delta z \to 0}{\text{Lt}} \frac{\sqrt{|\delta x\, \delta y|} - 0}{\delta x + i\delta y} = \begin{cases} \frac{1}{1+i} & \text{when } \delta z \to 0 \text{ along the line } y = x \\ \frac{1}{1-i} & \text{when } \delta z \to 0 \text{ along the line } y = -x \end{cases}$$

Hence $f(z)$ is not analytic at the origin.

11. *If* $u = \frac{x}{x^2 + y^2}$, *find its conjugate harmonic* v *such that* $u + iv$ *is analytic.*

SOLUTION: $u = \frac{x}{x^2 + y^2}$. To find v so that $u + iv$ is analytic, we have

$$\frac{\partial u}{\partial x} = \frac{(x^2+y^2)1 - x(2x)}{(x^2+y^2)^2} = \frac{y^2 - x^2}{(x^2+y^2)^2}, \quad \frac{\partial u}{\partial y} = \frac{-2xy}{(x^2+y^2)^2}$$

Thus $\frac{\partial v}{\partial y} = \frac{y^2 - x^2}{(y^2 + x^2)^2}$ and $\frac{\partial v}{\partial x} = \frac{2xy}{(x^2+y^2)^2}$.

From $\frac{\partial v}{\partial x} = \frac{2xy}{(x^2+y^2)^2}$ we get $v = \frac{-y}{(x^2+y^2)} + k(y)$

$\therefore \quad \frac{\partial v}{\partial y} = \frac{y^2 - x^2}{(x^2+y^2)} + \frac{dk}{dy}$. But we already have $\frac{\partial u}{\partial x} = \frac{\partial v}{\partial y} = \frac{y^2 - x^2}{(x^2+y^2)^2}$

Hence $\frac{dk}{dy} = 0$ where k is pure constant $= c$ $\quad \therefore \quad v = \frac{-y}{x^2+y^2} + c$ **Ans.**

12. *Find the values of* k *such that* $u = e^{kx}\cos y$ *is the real part of the analytic function* $u + iv$. *Also find this function.*

SOLUTION: $u = e^{kx}\cos y,\quad f(z) = u + iv$ is analytic

Here $\dfrac{\partial u}{\partial x} = ke^{kx}\cos y$ and $\dfrac{\partial v}{\partial y} = -ke^{kx}\sin y$. Since $\dfrac{\partial u}{\partial x} = \dfrac{\partial v}{\partial y}$ and $\dfrac{\partial u}{\partial y} = -\dfrac{\partial v}{\partial x}$,

We have $\dfrac{\partial v}{\partial y} = ke^{kx}\cos y$ and $\dfrac{\partial v}{\partial x} = e^{kx}\sin y$.

$\Rightarrow \quad v = ke^{kx}\sin y + k(x) \quad \therefore \quad \dfrac{\partial v}{\partial x} = k^2e^{kx}\sin y + \dfrac{dk}{dx} = e^{kx}\sin y$

Hence $\dfrac{dk}{dx} = 0 \quad \therefore \quad k$ is pure constant and $k^2 = 1 \quad \therefore \quad k = \pm 1$.

$\therefore \quad f(z) = u + iv = e^{kx}\cos y + ike^{kx}\sin y$

For $k = 1,\quad f(z) = e^x(\cos y + i\sin y) = e^z$ and for $k = -1,\quad f(z) = e^{-z}$. **Ans.**

13. *Let* $f(z) = u + iv$ *be an analytic function. If* $u = v^2$ *show that* $f(z)$ *is constant.*

SOLUTION: $f(z) = u + iv$ is analytic. Since $u = v^2$, we have

$\therefore \quad \dfrac{\partial u}{\partial x} = 2v\dfrac{\partial v}{\partial x}$ and $\dfrac{\partial u}{\partial y} = 2v\dfrac{\partial v}{\partial y}$, therefore, we get

$$\frac{\partial v}{\partial y} = 2v\frac{-\partial v}{\partial x} \quad \text{and} \quad \frac{-\partial v}{\partial x} = 2v\frac{\partial v}{\partial y}$$

$$\Rightarrow \quad \frac{-1}{2v}\frac{\partial v}{\partial x} = 2v\frac{\partial v}{\partial y} \quad \text{or} \quad (4v^2+1)\frac{\partial v}{\partial x} = 0$$

Therefore $\dfrac{\partial v}{\partial x} = 0$. Hence v is constant and since $u = v^2$ we have u also is constant. Thus $f(z)$ is constant. **Hence proved.**

14. *Determine the value(s) of 'p' such that the function.*

$$f(z) = r^2\cos 2\theta + i\,r^2\sin p\theta \quad \text{is analytic.}$$

SOLUTION: $f(z) = r^2\cos 2\theta + ir^2\sin p\theta$ is given to be analytic function.

$\therefore \quad u = r^2\cos 2\theta$ and $v = r^2\sin p\theta$.

The C.R. equations in polar form are $\dfrac{\partial u}{\partial r} = \dfrac{1}{r}\dfrac{\partial v}{\partial \theta}$ and $\dfrac{\partial v}{\partial r} = -\dfrac{1}{r}\dfrac{\partial u}{\partial \theta}$.

Here $\dfrac{\partial u}{\partial r} = 2r\cos 2\theta,\quad \dfrac{\partial u}{\partial \theta} = -2r^2\sin 2\theta,\quad \dfrac{\partial v}{\partial r} = 2r\sin p\theta,\quad \dfrac{\partial v}{\partial \theta} = pr^2\cos p\theta.$

For $f(z)$ to be analytic, by C.R. equations, we have

$$2r\cos 2\theta = \frac{1}{r}pr^2\cos p\theta \quad \text{and} \quad 2r\sin p\theta = -\frac{1}{r}(-2r^2\sin p\theta).$$

It follows that $p = 2$. **Ans.**

15. *Prove that $u = y^3 - 3x^2 y$ is a harmonic function. Find its harmonic conjugate and the corresponding analytic function.*

SOLUTION: $u = y^3 - 3x^2y$ is harmonic if $\dfrac{\partial^2 u}{\partial x^2} + \dfrac{\partial^2 u}{\partial y^2} = 0$

$$\therefore \quad \frac{\partial u}{\partial x} = -6xy, \quad \frac{\partial^2 u}{\partial x^2} = -6y \quad \text{and} \quad \frac{\partial u}{\partial y} = 3y^2 - 3x^2, \quad \frac{\partial^2 u}{\partial y^2} = 6y$$

$$\therefore \quad \frac{\partial^2 u}{\partial x^2} + \frac{\partial^2 u}{\partial y^2} = 0, \quad \text{hence } u \text{ is harmonic.}$$

Since $f(z) = u + iv$ is analytic, C.R. equations are satisfied.

Therefore $\dfrac{\partial v}{\partial x} = -\dfrac{\partial u}{\partial y} = -3y^2 + 3x^2$ and $\dfrac{\partial v}{\partial y} = \dfrac{\partial u}{\partial x} = -6xy$

$$\Rightarrow \quad v = -3xy^2 + x^3 + k(y) \quad \text{hence} \quad \frac{\partial v}{\partial y} = -6xy + \frac{dk}{dy} \quad \Rightarrow \quad \frac{dk}{dy} = 0.$$

So k is constant hence $v = -3xy^2 + x^3$

$\therefore \; f(z) = u + iv = y^3 - 3x^2y - 3ixy^2 + ix^3 \;\Rightarrow\; f(z) = iz^3.$

Next let us do it by Milne Thompson method.

$$f'(z) = \frac{\partial u}{\partial x} + i\frac{\partial v}{\partial x} = -6xy + i(-3y^2 + 3x^2)$$

Replacing x by z and y by 0 we get $f'(z) = 3iz^2$. Therefore $f(z) = iz^3$. **Ans.**

16. *Given* $f(z) = \dfrac{x^3(1+i) - y^3(1-i)}{x^2 + y^2}$, *when* $z \neq 0$

$= 0$ *when* $z = 0$

Test if $f(z)$ satisfies Cauchy-Riemann equations at the origin. Is $f(z)$ analytic at the origin?

SOLUTION: $f(z) = \dfrac{x^3(1+i) - y^3(1-i)}{x^2 + y^2}$ for $z \neq 0$ and $f(0) = 0$

Here $u = \dfrac{x^3 - y^3}{x^2 + y^2}, \quad v = \dfrac{x^3 + y^3}{x^2 + y^2}$

$f(z)$ is continuous everywhere and also at the origin. At the origin $(0, 0)$, we have

$$\frac{\partial v}{\partial x} = \lim_{x \to 0} \frac{u(x,0) - u(0,0)}{x - 0} = \operatorname*{Lt}_{x \to 0} \frac{x - 0}{x} = 1,$$

$$\frac{\partial u}{\partial y} = \lim_{y \to 0} \frac{u(0,y) - u(0,0)}{y - 0} = \operatorname*{Lt}_{y \to 0} \frac{-y - 0}{y} = -1$$

$$\frac{\partial v}{\partial x} = \lim_{x \to 0} \frac{v(x,0) - v(0,0)}{x - 0} = \lim_{x \to 0} \frac{x - 0}{x} = 1,$$

$$\frac{\partial v}{\partial y} = \lim_{y \to 0} \frac{v(0,y) - v(0,0)}{y - 0} = \operatorname*{Lt}_{y \to 0} \frac{y - 0}{y} = 1$$

Hence at the origin $\frac{\partial u}{\partial x} = \frac{\partial v}{\partial y}$ and $\frac{\partial u}{\partial y} = \frac{-\partial v}{\partial x}$. Thus C.R. equations are satisfied at the origin.

Now $f'(0) = \underset{z \to 0}{\text{Lt}} \frac{f(z) - f(0)}{z} = \underset{z \to 0}{\text{Lt}} \frac{x^3 - y^3 + i(x^3 + y^3) - 0}{(x^2 + y^2)(x + iy)}$

When $z \to 0$ along x-axis, $f'(0) = \underset{x \to 0}{\text{Lt}} \frac{x^3(1+i)}{x^3(1)} = 1 + i$

When $z \to 0$ along $y = x$, $f'(0) = \underset{x \to 0}{\text{Lt}} \frac{0 + i x^3 (2)}{2x^2(1+i)x} = \frac{i}{1+i} = \frac{1}{1-i}$

Since values of $f'(0)$ are different along the two paths, $f'(0)$ does not exist therefore $f(z)$ is not analytic at the origin. **Ans.**

17. *Find the potential due to two charged lines at the points $(a, 0)$ and $(-a, 0)$ having the same charge.*

SOLUTION: The electric field due to line charge $= \frac{\lambda}{2\epsilon_0} \frac{1}{r} = \frac{-d\phi}{dr}$ where ϕ is the potential field, r being complex, can be replaced by z. Therefore, $\phi = k\int_{-a}^{a} \frac{dr}{r}$.

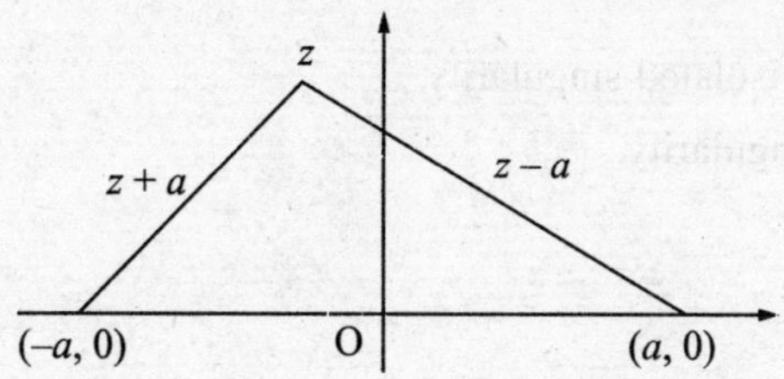

The potential fields due to the charge are $f_1(z) = k \log(z - a)$ and $f_2(z) = k \log(z + a)$ respectively.

The combined potential field is

$$f(z) = f_1(z) + f_2(z) = k \log(z^2 - a^2) = k[x^2 - y^2 - a^2 + 2ixy]$$

$$\therefore \quad \phi = \frac{k}{2} \log[(x^2 - y^2 - a^2)^2 + 4x^2y^2] = \frac{k}{2} \log[(x^2 + y^2)^2 - 2a^2(x^2 - y^2) + a^4].$$

Ans.

18. *If $f(z) = u + iv$ is analytic, show that u and v both satisfy the relation*

$$\frac{\partial^2 f}{\partial r^2} + \frac{1}{r}\frac{\partial f}{\partial r} + \frac{1}{r^2}\frac{\partial^2 f}{\partial \theta^2} = 0.$$

SOLUTION: $f(z) = u + iv$ is analytic hence the C.R. equations

$\frac{\partial u}{\partial r} = \frac{1}{r}\frac{\partial v}{\partial \theta}$ and $\frac{\partial v}{\partial r} = -\frac{1}{r}\frac{\partial u}{\partial \theta}$ should be staisfied.

$$\Rightarrow \quad \frac{\partial^2 u}{\partial r^2} = \frac{1}{r}\frac{\partial^2 v}{\partial r \partial \theta} - \frac{1}{r^2}\frac{\partial v}{\partial \theta} = \frac{1}{r}\frac{\partial}{\partial \theta}\left(\frac{\partial v}{\partial r}\right) - \frac{1}{r^2}\left(r\frac{\partial u}{\partial r}\right)$$

$$= \frac{1}{r}\frac{\partial}{\partial \theta}\left(-\frac{1}{r}\frac{\partial u}{\partial \theta}\right) - \frac{1}{r}\frac{\partial u}{\partial r} = \frac{1}{r}\left(-\frac{1}{r}\right)\frac{\partial^2 u}{\partial \theta^2} - \frac{1}{r}\frac{\partial u}{\partial r}$$

or $$\frac{\partial^2 u}{\partial r^2} + \frac{1}{r}\frac{\partial u}{\partial r} + \frac{1}{r^2}\frac{\partial^2 u}{\partial \theta^2} = 0 \qquad (1)$$

Next $$\frac{\partial^2 v}{\partial r^2} = \frac{1}{r^2}\frac{\partial u}{\partial \theta} - \frac{1}{r}\frac{\partial^2 u}{\partial r \partial \theta} = \frac{1}{r}\left(-\frac{\partial v}{\partial r}\right) - \frac{1}{r}\frac{\partial}{\partial \theta}\left(\frac{1}{r}\frac{\partial v}{\partial \theta}\right)$$

or $$\frac{\partial^2 v}{\partial r^2} + \frac{1}{r}\frac{\partial v}{\partial r} + \frac{1}{r^2}\frac{\partial^2 v}{\partial \theta^2} = 0 \qquad (2)$$

Equation (1) and (2) show that u and v both satisfy the equation

$$\frac{\partial^2 f}{\partial r^2} + \frac{1}{r}\frac{\partial f}{\partial r} - + \frac{1}{r^2}\frac{\partial^2 f}{\partial \theta^2} = 0.$$

Hence proved.

19. *What type of singularities, the following functions have*

(*i*) $\dfrac{\tan z}{z}$ (*ii*) e^{-z^3}

SOLUTION: (*i*) Removable isolated singularity.

(*ii*) Essential singularity. **Ans.**

SOLUTIONS TO EXERCISE 16B

1. *Obtain the bilinear transformation which maps the points $z = 1, i, -1$ onto the points $w = i, 0, -i$ and hence find* **(GGSIPU IInd Sem. IInd Term 2003)**

(a) the image of $|z| < 1$

(b) the invariant points of this transformation.

SOLUTION: If points z_1, z_2, z_3 in the z-plane are mapped to the points w_1, w_2, w_3 in the w-plane under the bilinear transformation, the bilinear transformation is then

$$\frac{w-w_1}{w-w_3}\cdot\frac{w_2-w_3}{w_2-w_1}=\frac{z-z_1}{w-z_3}\cdot\frac{z_2-z_3}{z_2-z_1}$$ Here $z_1=1,\ z_2=i,\ z_3=-1$ and $w_1=i,\ w_2=0,\ w_3=-i$

$$\therefore\quad \frac{w-i}{w+i}\cdot\frac{0+i}{0-i}=\frac{z-1}{z+1}\cdot\frac{i+1}{i-1}\quad\text{or}\quad \frac{w-i}{w+i}=\frac{iz-i}{z+1}$$

or $$\frac{w}{i}=\frac{(1+i)z+1-i}{(1-i)z+1+i}=i\cdot\frac{z-i}{z+i}$$

or $$w=\frac{i-z}{i+z}=\frac{1+iz}{1-iz}$$ is the required bilinear transformation **Ans.**

(*a*) To find image of $|z|<1$, $\dfrac{2iz}{2}=\dfrac{w-1}{w+1}$ or $iz=\dfrac{w-1}{w+1}$

$\therefore\quad |iz|=|z|=\left|\dfrac{w-1}{w+1}\right|$. Therefore $|z|<1\ \Rightarrow\ |w-1|<|w+1|$ **Ans.**

That is, w is nearer to 1 than to -1 hence $u>0$ is the image of $|z|<1$.

(*b*) The points in the z-plane which map into themselves are called fixed points or invariants. For that

$$z=\frac{i-z}{i+z}\quad\text{or}\quad z^2+(i+1)\,z-i=0\quad\therefore\quad z=\frac{-(1+i)\pm\sqrt{2i+4i}}{2}=\frac{-1}{2}\left[1+i\pm\sqrt{6i}\right]$$ are invariants.

Ans.

2. *Show that the image of the real axis in the z-plane on the w-plane by the transformation*

$$w=\frac{1}{z+i}$$ *is a circle. Find its radius and centre.*

SOLUTION: $w=\dfrac{1}{z+i}$ or $z=\dfrac{1}{w}-i$ or $x+iy=\dfrac{1}{u+iv}-i=\dfrac{u-iv}{u^2+v^2}-i$

$$\Rightarrow\quad x=\frac{u}{u^2+v^2}\quad\text{and}\quad y=\frac{-v}{u^2+v^2}-1.$$

$\therefore$ Real axis $y=0$ in z-plane is mapped into $u^2+v^2+v=0$

which is a circle with centre at $\left(0,\dfrac{-1}{2}\right)$ and radius $=\dfrac{1}{2}$. **Ans.**

3. *Given the bilinear transformation*

$w = \dfrac{2z+3}{z-4}$ *find the mapping of the circle* $x^2 + y^2 - 4x = 0$ *in the w-plane.*

SOLUTION: $w = \dfrac{2z+3}{z-4} \Rightarrow z = \dfrac{4w+3}{w-2} = \dfrac{4u+3+4iv}{u-2+iv}$

or $$x + iy = \frac{4u+3+4iv}{u-2+iv}\cdot\frac{u-2-iv}{u-2-iv}$$

$$= \frac{(4u+3)(u-2)+4v^2+i[4v(u-2)-v(4u+3)]}{(u-2)^2+v^2}$$

$\Rightarrow$ $$x = \frac{(4u+3)(u-2)+4v^2}{(u-2)^2+v^2},\quad y = \frac{4v(u-2)-v(4u+3)}{(u-2)^2+v^2} = \frac{-11v}{(u-2)^2+v^2}$$

Putting these in $x^2 + y^2 = 4x$, we get

$$\frac{[(4u+3)(u-2)+4v^2]^2+121v^2}{[(u-2)^2+v^2]^2} = \frac{4[(4u+3)(u-2)+4v^2]}{(u-2)^2+v^2}$$

or $$(4u+3)^2(u-2)^2+16v^4+8v^2(4u+3)(u-2)+121v^2 = [4(u-2)^2+4v^2][(4u+3)(u-2)+4v^2]$$

On simplification, we get $(u-2)^2(4u+3) + v^2(4u+3) = 0$

or $(4u+3)[(u-2)^2+v^2] = 0 \Rightarrow 4u+3 = 0$, since $(u-2)^2+v^2 \neq 0$. $\therefore\ u = -\dfrac{3}{4}$ **Ans.**

4. (a) *Discuss the mapping* $w = \dfrac{1}{z}$ *and show that a circle in the z-plane, in general, maps into another circle in the w-plane.*

(b) *Show that the function* $w = \dfrac{4}{z}$ *transforms the line* $x = c$ *in the z-plane into a circle in the w-plane.*

SOLUTION: (a) $w = \dfrac{1}{z}$ or $z = \dfrac{1}{w}$ or $x + iy = \dfrac{1}{u+iv} = \dfrac{u-iv}{u^2+v^2}$

$\Rightarrow$ $x = \dfrac{u}{u^2+v^2},\quad y = \dfrac{-v}{u^2+v^2}$

Let a circle is z-plane be $x^2 + y^2 + 2gx + 2fy + c = 0$

$\therefore$ Its image in w-plane is $\dfrac{u^2}{(u^2+v^2)^2} + \dfrac{v^2}{(u^2+v^2)^2} + \dfrac{2gu}{u^2+v^2} - \dfrac{2fv}{u^2+v^2} + c = 0,$

Multiplying throughout by $u^2 + v^2$, gives

$u^2 + v^2 + \dfrac{2g}{c}u - \dfrac{2f}{c}v + \dfrac{1}{c} = 0$ which is another circle in w-plane. **Hence proved.**

(b) $u + iv = \dfrac{4}{x+iy} = \dfrac{4(x-iy)}{x^2+y^2} \Rightarrow u = \dfrac{4x}{x^2+y^2},\quad v = \dfrac{-4y}{x^2+y^2} \quad \therefore\ \dfrac{y}{x} = \dfrac{-v}{u}$

Next, $x = c$ gives $y = -\dfrac{cv}{u}$.

Hence $u = \dfrac{4c}{c^2 + \dfrac{c^2v^2}{u^2}}$ or $u^2 + v^2 = \dfrac{4u}{c^2}$ a circle. **Ans.**

5. *Obtain the bilinear transformation which maps the points $z = 0, 1, \infty$ into the points $w = -3, -1, 1$ respectively. Also find the fixed points.*

SOLUTION: Let the bilmear transformation be $w = f(z)$

The points $z_1 = 0$, $z_2 = 1$, $z_3 = \infty$ in z-plane are mapped into the points $w_1 = -3$, $w_2 = -1$, $w_3 = 1$ in the w-plane respectively. The transformation is written as

$$\frac{w - w_1}{w - w_3} \cdot \frac{w_2 - w_3}{w_2 - w_1} = \frac{z - z_1}{z - z_3} \cdot \frac{z_2 - z_3}{z_2 - z_1}$$

or $$\frac{w+3}{w-1} \cdot \frac{-1-1}{-1+3} = \frac{z-0}{z-z_3} \cdot \frac{1-z_3}{1-0} = \frac{z\left(\frac{1}{z_3} - 1\right)}{\frac{z}{z_3} - 1} = z\left(\text{as } \frac{1}{z_3} = 0\right)$$

or $$-\left(\frac{w+3}{w-1}\right) = z, \qquad \text{or} \qquad w = \frac{z-3}{z+1}$$

which is the required bilinear mapping. For fixed points or invariants

$$z = \frac{z-3}{z+1} \quad \text{or} \quad z^2 = -3 \quad \text{or} \quad z = \pm i\sqrt{3}$$ **Ans.**

6. *Let $w = z^2$ define a transformation from z-plane to the w-plane. Consider a triangle in the z-plane with vertices at A (2, 1), B (4, 1), C (4, 3). Show that the image of this triangle is a curvilinear triangle in the w-plane.*

SOLUTION: $w = z^2$ gives $(u + iv) = (x + iy)^2 = x^2 - y^2 + 2ixy$.

Hence $u = x^2 - y^2 \quad v = 2ixy$ ...(1)

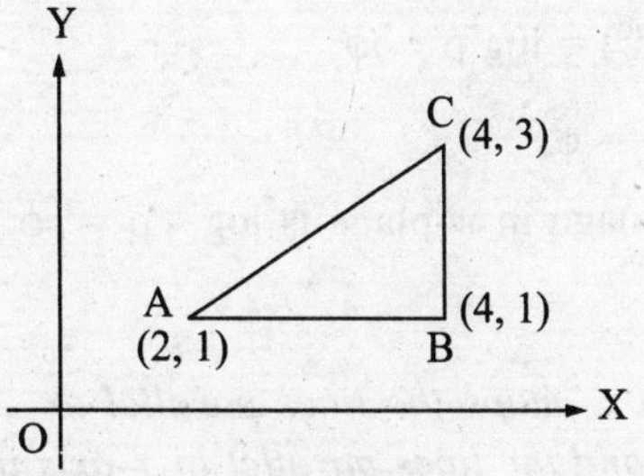

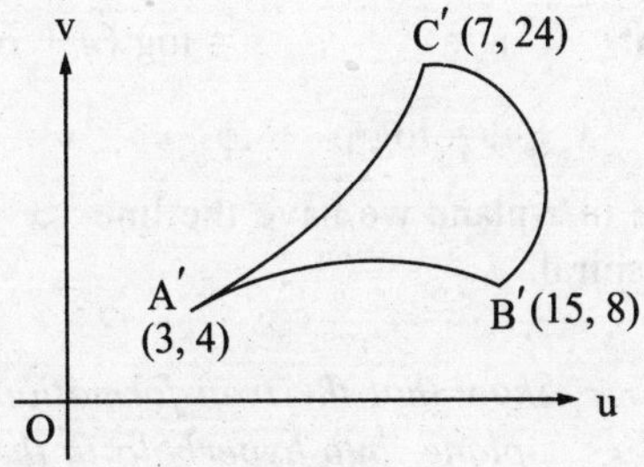

Point $A(2, 1)$ is imapped to $A'(3, 4)$ point $B(4, 1)$ is mapped to $B'(15, 8)$ and the point C (4, 3) is mapped to $C'(7, 24)$ in the w-plane

Line AB having equation $y = 1$ is mapped to $u = x^2 - 1$, $v = 2x$

or $u + 1 = \frac{v^2}{4}$ or $v^2 = 4(u + 1)$ a parabola in w-plane.

Line BC having equation $x = 4$ is mapped to $u = 16 - y^2$, $v = 8y$

or $\frac{v^2}{64} = 16 - u$ or $v^2 = -64(u - 16)$ another parabola in w-plane.

Line AC has the equation $y - 1 = \frac{3-1}{4-2}(x-2)$ or $x - y = 1$ ...(2)

Eliminating x and y in (1) and (2), gives $u^2 = 2v + 1$.

Thus we get as mapping a curvilinear line or triangle. **Ans.**

7. *Under the transformation* $z = \frac{4}{(w+1)^2}$, *find the region in the z-plane corresponding tp the interior of the circle* $|w| = 1$.

SOLUTION: $z = \frac{4}{(w+1)^2}$ or $w + 1 = \frac{2}{\sqrt{z}}$ converting to polar form

$$w = \frac{2}{\sqrt{re^{c\theta}}} - 1 = \frac{2}{\sqrt{r}}e^{-i\theta/2} - 1 = \frac{1}{\sqrt{r}}\left[2\left(\cos\frac{\theta}{2} - i\sin\frac{\theta}{2}\right) - \sqrt{r}\right]$$

$$= \frac{1}{\sqrt{r}}\left[2\cos\frac{\theta}{2} - \sqrt{r} - 2i\sin\frac{\theta}{2}\right]$$

Now $|w| < 1 \Rightarrow (2\cos\theta/2 - \sqrt{r})^2 + 4\sin^2\theta/2 < r$

or $4(\cos^2\theta/2 + \sin^2\theta/2) + r - 4\sqrt{r}\cos r\theta/2 < r$ or $\sqrt{r}\cos\theta/2 > 1$.

or $r > \frac{1}{\cos^2\frac{\theta}{2}} = \frac{2}{1+\cos\theta}$ $\therefore$ $r\cos\theta + r > 2$

or $r > (2 - x)$ or $x^2 + y^2 > 4 - 4x + x^2$ or $y^2 > 4 - 4x$

which is the exterior of the parabola $y^2 = -4x + 4$ **Ans.**

8. *Under the mapping* $w = e^z$ *determine the image of the lines* $x = cy$.

SOLUTION: $w = e^z$ $\therefore$ $z = \log(u + iv) = \log(\rho e^{i\phi}) = \log\rho + i\phi$

or $x + iy = \log\rho + i\phi \Rightarrow x = \log\rho$ and $y = \phi$.

Since in z-plane we have the line $x = cy$, its image in w-plane is $\log\rho = c\phi$ or $\rho = e^{c\phi}$ which is a spiral. **Ans.**

9. *Show that the transformation* $w = \cosh z$ *maps the lines parallel to X-axis in the z-plane into hyperbola in the w-plane and the lines parallel to Y-axis in the z-plane into ellipses in the w-plane.*

SOLUTION: $w = \cosh z$

or $2w = e^z + e^{-z} = e^x(\cos y + i\sin y) + e^{-x}(\cos y - i\sin y) = 2(u + iv)$

$\Rightarrow$ $2u = (e^x + e^{-x})\cos y$, $2v = (e^x - e^{-x})\sin y$.

Let the line parallel to X-axis in z-plane be $y = a$, hence

$$\frac{2u}{\cos a} = e^x + e^{-x} \quad \text{and} \quad \frac{2v}{\sin a} = e^x - e^{-x}.$$

Squaring and subtracting these, gives

$$\frac{4u^2}{\cos^2 a}-\frac{4v^2}{\sin^2 a}=4 \quad \text{or} \quad \frac{u^2}{\cos^2 a}-\frac{v^2}{\sin^2 a}=1 \quad \text{which is a hyperbola in } w\text{-plane.}$$

Similarly line parallel to y-axis in z-plane be $x = b$, hence

$$w = u + iv = \cos hb \cos y + i \sin hb \sin y$$

$\therefore \quad u = \cos hb \cos y \quad$ and $\quad v = \sin hb \sin y$

$$\Rightarrow \quad \frac{u^2}{\cos h^2 b}+\frac{v^2}{\sin h^2 b}=1 \quad \text{which is an ellipse in the } w\text{-plane.}$$

Hence proved.

10. *Show that the transformation* $w = z + \dfrac{a^2-b^2}{4z}$ *transforms the circle of radius* $(a + b)/2$, *centre at the origin, in the* z-*plane into an ellipse of semi axes* a *and* b *in the* w-*plane.*

SOLUTION:
$$w = z+\frac{a^2-b^2}{4z} \quad \text{or} \quad u+iv = x+iy+\left(\frac{a^2-b^2}{4}\right)\frac{x-iy}{x^2+y^2}$$

But $\quad x^2+y^2=\left(\dfrac{a+b}{2}\right)^2 \quad$ hence $\quad u+iv = x+iy+\dfrac{a^2-b^2}{(a+b)^2}(x-iy)$

$$\Rightarrow \quad u = x+\frac{a^2-b^2}{(a+b)^2}x, \qquad v = y-\frac{(a^2-b^2)}{(a+b)^2}y$$

$$\text{or} \quad x=\frac{u(a+b)^2}{(a+b)^2+a^2-b^2} \quad \text{and} \quad y=\frac{v(a+b)^2}{(a+b)^2-(a^2-b^2)}.$$

Therefore image of $x^2+y^2=\left(\dfrac{a+b}{2}\right)^2$ is

$$\frac{u^2(a+b)^4}{(2a^2+2ab)^2}+\frac{v^2(a+b)^4}{(2ab+2b^2)^2}=\left(\frac{a+b}{2}\right)^2$$

or $\quad \dfrac{u^2}{a^2}+\dfrac{v^2}{b^2}=1.$ which is ellipse with a and b as semi axes. **Hence proved.**

11. *Into what curves in the* w-*plane does the function* $f(z) = \cos z$ *transform the lines of the families* $x = c$ *and* $y = k$ *in the* z-*plane?*

SOLUTION: $f(z) = \cos z \quad$ or $\quad u + iv = \cos(x + iy) = \cos x \cos hy - i \sin x \sin hy$

$\Rightarrow \quad u = \cos x \cos hy \quad$ and $\quad v = -\sin x \sin hy$

For the image of $x = c$, we have $u = \cos c \cos hy$, $v = -\sin c \sin hy$

Eliminating y here, we get $\dfrac{u^2}{\cos^2 c}-\dfrac{v^2}{\sin^2 c}=1.$

which is a hyperbola in the w-plane.

For the image of $y = k$, we have $u = \cos x \cos hk$, $v = -\sin x \sin hk$

Eliminating x here, we get $\dfrac{u^2}{\cos h^2 k} + \dfrac{v^2}{\sin h^2 k} = 1$

Which is an ellipse in w-plane.

12. *Under the transformation* $w = e^z$ *find the image of*

(*i*) *the straight lines* $x = x_0$ *(constant) and* $y = y_0$ *(constant),*

(*ii*) *the rectangle* $0 \le x \le 1, \frac{1}{2} < y < 1,$

(*iii*) *the fundamental region* $-\pi < y \le \pi,$

(*iv*) *the horizontal strip* $0 \le y \le \pi.$

SOLUTION: (*i*) $w = e^z = e^{x+iy} = e^x(\cos y + i \sin y)$.

Line $x = x_0$ is mapped to $|\, w \,| = e^{x_0}$ which is a circle in w-plane with centre at the pole and raduis e^{x_0}.

Image of line $y = y_0$ in z-plane into w-plane.

For that $u + iv = e^x(\cos y_0 + i \sin y_0) \Rightarrow u = e^x \cos y_0,\ v = e^x \sin y_0$

or $\dfrac{v}{u} = \tan y_0$ is a line in w-plane passing through origin and making an angle y_0 with the u-axis.

(*ii*) $w = \rho e^{i\varphi} = e^x e^{iy} \Rightarrow \rho = e^x,\ \varphi = y$ region bounded by the portions of rays and circles

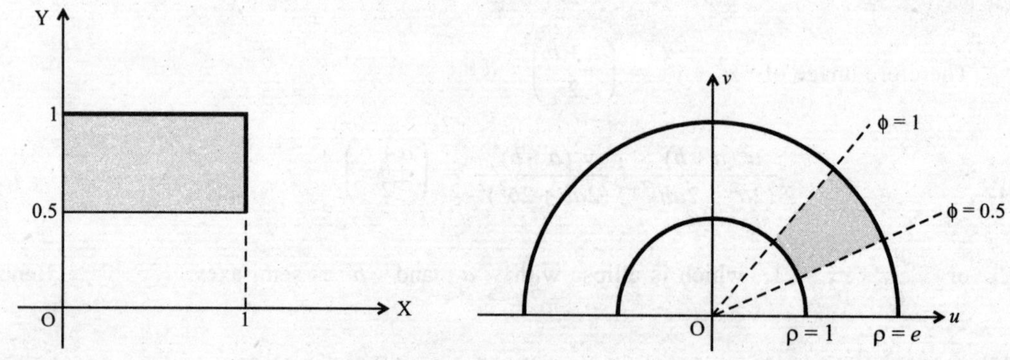

(*iii*) The fundamental region $-\pi < y < \pi$ is mapped onto the fulll w-plane (cut along the negative real axis) without the origin O. More generally, every horizontal strip bounded by two lines $y = c$ and $y = c + 2\pi$ is mapped onto the full ω-plane without the origin. This is so because $\omega = e^z$ is periodic with period $2\pi i$.

(*iv*) The strip $0 < y < \pi$ is mapped onto the upper half of the ω-plane. Its right half ($x > 0$) is mapped onto the exterior of the unit circle $|\omega| = 1$ and its left half ($x < 0$) onto the interior without the origin 0.

ADDITIONAL IMPORTANT PROBLEMS

1. *If $f(z)$ is an analytic function of z, then show that*

$$\left\{\frac{\partial}{\partial x}|f(z)|\right\}^2+\left\{\frac{\partial}{\partial y}|f(z)|\right\}^2=|f'(z)|^2.$$

SOLUTION: Let $f(z)=u+iv$, then $|f(z)|=(u^2+v^2)^{1/2}$

Consider $$\frac{\partial}{\partial x}|f(z)|=\frac{\partial}{\partial x}(u^2+v^2)^{1/2}=\frac{1}{2}(u^2+v^2)^{-1/2}\cdot\left(2u\frac{\partial u}{\partial x}+2v\frac{\partial v}{\partial x}\right)=\frac{u\dfrac{\partial u}{\partial x}+v\dfrac{\partial v}{\partial x}}{(u^2+v^2)^{1/2}}$$

or, $$\left\{\frac{\partial}{\partial x}|f(z)|\right\}^2=\frac{\left[u\dfrac{\partial u}{\partial x}+v\dfrac{\partial v}{\partial x}\right]^2}{u^2+v^2} \tag{1}$$

Similarly, $$\left\{\frac{\partial}{\partial y}|f(z)|\right\}^2=\frac{\left[u\dfrac{\partial u}{\partial y}+v\dfrac{\partial v}{\partial y}\right]^2}{u^2+v^2}.$$

Since $f(z)$ is analytic, using C.R. equations, it becomes

$$\left\{\frac{\partial}{\partial y}|f(z)|\right\}^2=\frac{\left\{v\dfrac{\partial u}{\partial x}-u\dfrac{\partial v}{\partial x}\right\}^2}{u^2+v^2} \tag{2}$$

From (1) and (2), we have

$$\left\{\frac{\partial}{\partial x}|f(z)|\right\}^2+\left\{\frac{\partial}{\partial y}|f(z)|\right\}^2=\frac{u^2\left(\dfrac{\partial u}{\partial x}\right)^2+v^2\left(\dfrac{\partial v}{\partial x}\right)^2+u^2\left(\dfrac{\partial v}{\partial x}\right)^2+v^2\left(\dfrac{\partial u}{\partial x}\right)^2}{u^2+v^2}$$

$$=\left(\frac{\partial u}{\partial x}\right)^2+\left(\frac{\partial v}{\partial x}\right)^2=|f'(z)|^2$$ **Hence Proved.**

2. *If $f(z)$ is an analytic function with constant modulus, then $f(z)$ is constant.*

SOLUTION: Let $f(z)=u+iv$, then $|f(z)|=\sqrt{u^2+v^2}$ = constant

Differentiating u^2+v^2 = constant w.r.t. x, we have $u\dfrac{\partial u}{\partial x}+v\dfrac{\partial v}{\partial x}=0$

Differentiating it again w.r.t. x, we have, $\left(\dfrac{\partial u}{\partial x}\right)^2+u\dfrac{\partial^2 u}{\partial x^2}+\left(\dfrac{\partial v}{\partial x}\right)^2+v\dfrac{\partial^2 v}{\partial x^2}=0$

Similarly, $\left(\frac{\partial u}{\partial y}\right)^2 + u\frac{\partial^2 u}{\partial y^2} + \left(\frac{\partial v}{\partial y}\right)^2 + v\frac{\partial^2 v}{\partial y^2} = 0$

Adding these two and using $\nabla^2 u = 0$ and $\nabla^2 v = 0$, (since u and v are harmonic, being real and imaginary parts of an analytic function $f(z) = u + iv$), we obtain

$$\left(\frac{\partial u}{\partial x}\right)^2 + \left(\frac{\partial v}{\partial x}\right)^2 + \left(\frac{\partial u}{\partial y}\right)^2 + \left(\frac{\partial v}{\partial y}\right)^2 = 0$$

which implies, $\frac{\partial u}{\partial x} = \frac{\partial v}{\partial x} = \frac{\partial u}{\partial y} = \frac{\partial v}{\partial y} = 0$

Thus u and v are independent of both x and y and hence $f(z)$ is constant.

3. *Show that the function $v(x, y) = ln(x^2 + y^2) + x - 2y$ is harmonic. Find its conjugate harmonic function $u(x, y)$ and the corresponding analytic function $f(z)$.*

SOLUTION: We have $v(x, y) = \ln(x^2 + y^2) + x - 2y$. This gives

$$\frac{\partial v}{\partial x} = -\frac{2x}{x^2 + y^2} + 1, \quad \frac{\partial v}{\partial y} = \frac{2y}{x^2 + y^2} - 2,$$

$$\frac{\partial^2 v}{\partial x^2} = \frac{2(y^2 - x^2)}{(x^2 + y^2)^2}, \quad \frac{\partial^2 v}{\partial y^2} = \frac{2(x^2 - y^2)}{(x^2 + y^2)^2}$$

Since $\frac{\partial^2 v}{\partial x^2} + \frac{\partial^2 v}{\partial y^2} = 0$, the function $v(x, y)$ is harmonic.

From the Cauchy-Reimann equation $\frac{\partial u}{\partial x} = \frac{\partial v}{\partial y}$, we get

$$\frac{\partial u}{\partial x} = \frac{2y}{x^2 + y^2} - 2$$

Integrating it w.r.t. x, we get $u(x, y) = 2\tan^{-1}\frac{x}{y} - 2x + \phi(y)$,

where $\phi(y)$ is an arbitrary function of y.

Differentiating w.r.t. y it gives, $\frac{\partial u}{\partial y} = \frac{-2x}{x^2 + y^2} + \phi'(y)$

Next, using the Cauchy-Riemann equation $\frac{\partial y}{\partial y} = -\frac{\partial v}{\partial x}$, we get

$$-\frac{2x}{x^2 + y^2} + \phi'(y) = -\frac{2x}{x^2 + y^2} - 1, \quad \text{or} \quad \phi'(y) = -1$$

Integrating with respect to y, we obtain $\phi(y) = -y + c$, where c is any arbitrary constant.

Hence, $u(x, y) = 2\tan^{-1}\frac{x}{y} - 2x - y + c$.

$$f(z) = \left(2\tan^{-1}\frac{x}{y} - 2x - y\right) + i\{\ln(x^2+y^2) + x - 2y\} + c$$

$$= 2i\left[\frac{1}{2}\ln(x^2+y^2) + i\tan^{-1}\frac{y}{x}\right] - (2-i)(x+iy) + c$$

$$= 2i \ln z - (2 - i)\, z + c.$$ **Ans.**

4. *Show that the function $v(x, y) = e^x \sin y$ is harmonic. Find its conjugate harmonic $u(x, y)$ and the corresponding analytic function.*

SOLUTION: We have $v(x, y) = e^x \sin y$. This gives

$$\frac{\partial v}{\partial x} = e^x \sin y, \quad \frac{\partial v}{\partial y} = e^x \cos y$$

and
$$\frac{\partial^2 v}{\partial x^2} = e^x \sin y, \quad \frac{\partial^2 v}{\partial y^2} = -e^x \sin y$$

Since $\dfrac{\partial^2 v}{\partial x^2} + \dfrac{\partial^2 v}{\partial y^2} = 0$, the given function $v(x, y)$ is harmonic.

Next, let $f(z) = u + iv$, then

$$f'(z) = \frac{\partial u}{\partial x} + i\frac{\partial v}{\partial x} = \frac{\partial v}{\partial y} + i\frac{\partial v}{\partial x} = e^x \cos y + ie^x \sin y.$$

Using Milne-Thomson method, replace x by z and y by zero, we get $f'(z) = e^z$

$\Rightarrow \quad f(z) = e^z + c$ where c is an arbitrary real constant.

5. *Determine the analytic function $f(z) = u + iv$, if $v = \ln[(x - 1)^2 + (y - 2)^2]$.*

SOLUTION: We have $v = \ln[(x - 1)^2 + (y - 2)^2]$. This gives

$$\frac{\partial v}{\partial x} = \frac{2(x-1)}{(x-1)^2 - (y-2)^2} \quad \text{and} \quad \frac{\partial v}{\partial y} = \frac{2(y-2)}{(x-1)^2 + (y-2)^2}$$

From the Cauchy-Reimann equation $\dfrac{\partial u}{\partial x} = \dfrac{\partial v}{\partial y}$, we get

$$\frac{\partial u}{\partial x} = \frac{2(y-2)}{(x-1)^2 + (y-2)^2}$$

Integrating w.r.t. x, we get $u(x, y) = 2\tan^{-1}\dfrac{x-1}{y-2} + \phi(y)$, where ϕ is an arbitrary function of y.

Differentiating it w.r.t. y, we obtain

$$\frac{\partial u}{\partial y} = \frac{-2(x-1)}{(x-1)^2 + (y-2)^2} + \phi'(y)$$

Using the Cauchy-Riemann equation $\dfrac{\partial u}{\partial y} = -\dfrac{\partial v}{\partial x}$, we get

$$\frac{-2(x-1)}{(x-1)^2+(y-2)^2}+\phi'(y) = \frac{-2(x-1)}{(x-1)^2+(y-2)^2}$$

which gives $\phi'(y) = 0$, or $\phi(y) = c$, where c is a constant.

Hence, $$u(x, y) = 2\tan^{-1}\frac{x-1}{y-2}+c = -2\tan^{-1}\frac{y-2}{x-1}+c.$$

Thus, $$f(z) = -2\tan^{-1}\frac{y-2}{x-1}+i\ln[(x-1)^2+(y-2)^2]+c$$

$$= 2i\left[\ln|(x-1)+i(y-2)|+i\tan^{-1}\frac{y-2}{x-1}\right]+c$$

$$= 2i\ln\{(x-1)+i(y-2)\}+c = 2i\ln(z-1-2i)+c,$$

is the required analytic function. **Ans.**

6. *Show that the function $v(r, \theta) = r^2\cos 2\theta - r\cos\theta + 2$ is harmonic. Find its conjugate harmonic function and the corresponding analytic function $f(z)$.*

SOLUTION: We have, $v(r, \theta) = r^2\cos 2\theta - r\cos\theta + 2$. This gives

$$v_r = 2r\cos 2\theta - \cos\theta,\quad v_{rr} = 2\cos 2\theta$$

and $$v_\theta = -2r^2\sin 2\theta + r\sin\theta,\quad v_{\theta\theta} = -4r^2\cos 2\theta + r\cos\theta.$$

Since, $v_{rr}+\frac{1}{r}v_r+\frac{1}{r^2}v_{\theta\theta} = 2\cos 2\theta+2\cos 2\theta-\frac{1}{r}\cos\theta-4\cos 2\theta+\frac{1}{r}\cos\theta = 0$, thus the function $v(r, \theta)$ satisfies the Laplace equation and, therefore, is harmonic.

From the Cauchy-Riemann equations in polar coordinates, $ru_r = v_\theta = -2r^2\sin 2\theta + r\sin\theta$,

hence $$u_r = -2r\sin 2\theta + \sin\theta. \tag{1}$$

Similarly, $-\frac{1}{r}u_\theta = v_r = 2r\cos 2\theta - \cos\theta$, gives

$$u_\theta = -2r^2\cos 2\theta + r\cos\theta \tag{2}$$

Integrating (1) with respect to r, we get

$u = -r^2\sin 2\theta + r\sin\theta + \phi(\theta)$, where $\phi(\theta)$ is an arbitrary function of θ.

Differentiating it partially w.r.t. θ, we have

$$u_\theta = -2r^2\cos 2\theta + r\cos\theta + \phi'(\theta) \tag{3}$$

From (2) and (3), we get $\phi'(\theta) = 0$, or $\phi(\theta) = c$

Thus, $$u = -r^2\sin 2\theta + r\sin\theta + c$$

Hence, $$f(z) = u + iv = (-r^2\sin 2\theta + r\sin\theta + c) + i(r^2\cos 2\theta - r\cos\theta + 2)$$

$$= r^2(-\sin 2\theta + i\cos 2\theta) + r(\sin\theta - i\cos\theta) + c + 2i$$

$$= i(r^2 e^{2iq} - re^{i\theta}) + c + 2i$$ **Ans.**

7. *Find the image of the infinite strip $1/4 \le y \le 1/2$ under the transformation $w = 1/z$.*

SOLUTION: If $z = x + iy$, $w = u + iv$, then under the given transformation $w = 1/z$, we have

$$x = \frac{u}{u^2 + v^2} \quad \text{and} \quad y = \frac{-v}{u^2 + v^2}$$

Thus $y = \frac{1}{4}$ gives $\frac{-v}{u^2+v^2} = \frac{1}{4}$, that is, $u^2 + v^2 + 4v = 0$, or $u^2 + (v + 2)^2 - 4 = 0$,

and $y = \frac{1}{2}$ gives $\frac{-v}{u^2+v^2} = \frac{1}{2}$, that is, $u^2 + v^2 + 2v = 0$, or $u^2 + (v + 1)^2 - 1 = 0$.

Hence the infinite strip $1/4 \le y \le 1/2$ in the z-plane, as shown in Fig., is transformed into the region $u^2 + (v + 2)^2 - 4 \le y \le u^2 + (v + 1)^2 - 1$ in the w-plane, that is, the region between the two circles both passing through (0, 0). Circles are: $u^2 + (v + 2)^2 = 4$ with center (0, –2) and radius 2 and, $u^2 + (v + 1)^2 = 1$ with center (0, –1) and radius 1 as shown in Fig.

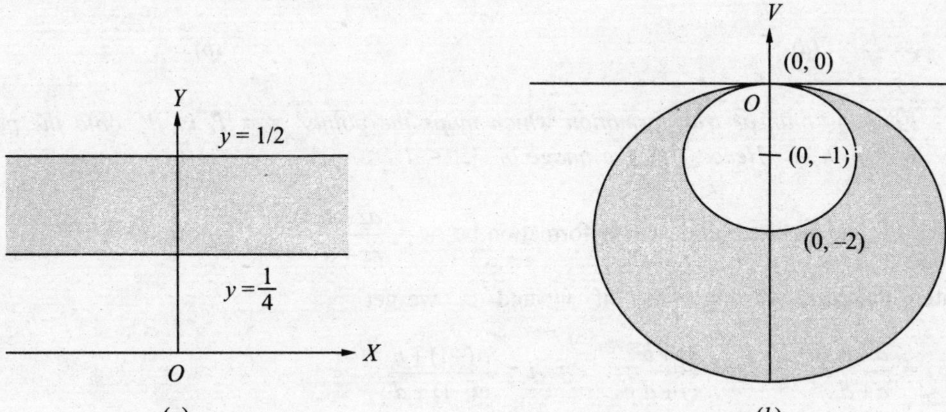

(a) (b)

8. *Find the image of the region $|z - i| < 2$ under the mapping $w = (1 + i)/(z + i)$.*

SOLUTION: The given mapping is

$$w = \frac{1+i}{z+i} \quad \text{or} \quad z = \frac{1+i(1-w)}{w}$$

Hence, $$|z - i| = \left|\frac{1+i(1-w)}{w} - i\right| = \left|\frac{1+i(1-2w)}{w}\right|$$

Thus, $|z - i| < 2$ gives $|1 + i(1 - 2w)| < 2\,|w|$

or, $|(1 + 2v) + i(1 - 2u)| < 2\,|u + iv|$, where $w = u + iv$

or, $(1 + 2v)^2 + (1 - 2u)^2 < 4(u^2 + v^2)$ or $2v - 2u + 1 < 0$

Therefore, the interior of the circular disk $|z - i| < 2$ as shown in Fig. is mapped as the half plane below the line $2v - 2u + 1 = 0$ and the boundary circle $|z - i| = 2$ is mapped as the line $2v - 2u + 1 = 0$, as shown in the figure.

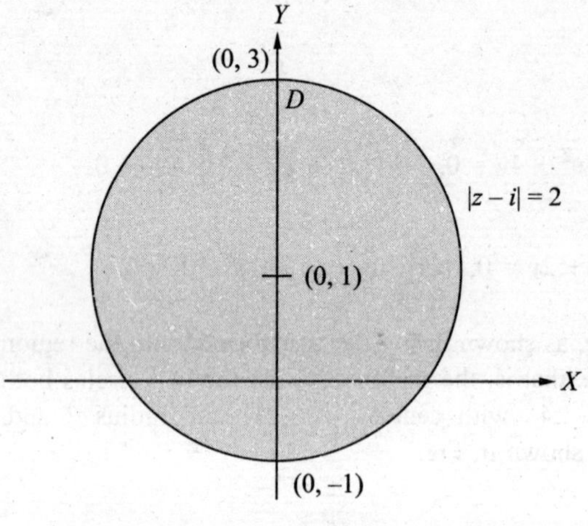

(a)

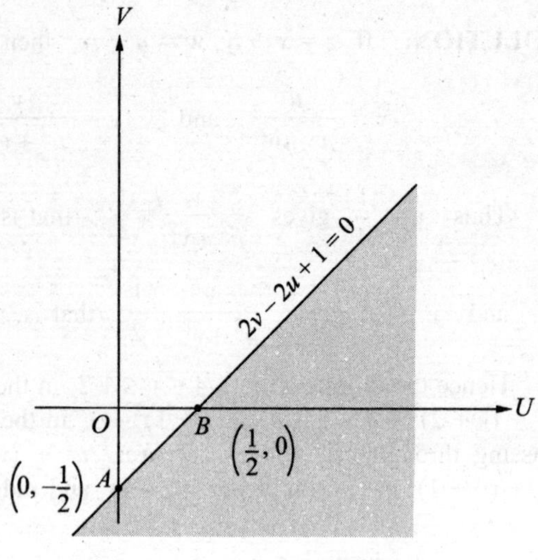

(b)

9. *Find the bilinear transformation which maps the points $z = 1, i, -1$ onto the points $w = i, 0, -i$. Hence, find the image of $|z| < 1$.*

SOLUTION: Let the required bilinear transformation be $w = \dfrac{az+b}{cz+d}$

Substituting the corresponding values of w and z, we get

$$i = \frac{a+b}{c+d}, \qquad 0 = \frac{ai+b}{ci+d}, \qquad -i = \frac{a(-1)+b}{c(-1)+d}.$$

which give respectively

$(a + b) - i\,(c + d) = 0, \quad b + ia = 0, \quad$ and $\quad (-a + b) + i(-c + d) = 0$

Solving these equations in term of a, we get $b = -ia, \quad c = -a, \quad d = -ia.$

Substituting for b, c and d, the required transformation becomes

$$w = \frac{az - ia}{-az - ia}, \quad \text{or} \quad w = \frac{i - z}{i + z}, \quad \text{or} \quad z = \frac{i(1-w)}{1+w}$$

Thus the region $|z| < 1$ is mapped onto the region

$$\left|\frac{i(1-w)}{1+w}\right| < 1, \quad \text{or} \quad \frac{|i|\,|1-w|}{|1+w|} < 1, \quad \text{or} \quad |1 - w| < |1 + w|$$

or, $\qquad |1 - u - iv| < |1 + u + iv|, \quad$ where $w = u + iv$

or, $\qquad (1 - u)^2 + v^2 < (1 + u)^2 + v^2 \quad$ or $\quad u > 0$

Hence, the interior of the circle $x^2 + y^2 = 1$ in the z-plane is mapped into the entire positive half plane $u > 0$ in the w-plane. Further we note that the circle $|z| = 1$ in the z-plane is mapped into the imaginary axis $u = 0$ in the w-plane.

10. *Find the image of the infinite strip $0 \le x \le \pi/2$ in the z-plane under the mapping $w = \tan^2 z/2$.*

SOLUTION: We have $w = \tan^2(z/2) = \dfrac{\sin^2(z/2)}{\cos^2(z/2)} = \dfrac{1-\cos z}{1+\cos z}$

Also $\cos z = \cos(x + iy) = \cos x \cosh y - i \sin x \sinh y$.

At $x = 0$, $\cos z = \cosh y$ and hence $w = \dfrac{1-\cosh y}{1+\cosh y}$ is purely real at $x = 0$ and thus

$$v = 0 \quad \text{and} \quad u = \frac{1-\cosh y}{1+\cosh y}.$$

At $y = 0$, $u = 0$ and for $0 < y < \infty$, u varies from 0 to -1. Also as $y \to -\infty$, u again tends to -1.

At $x = \dfrac{\pi}{2}$, $\cos z = -i \sinh y$ and hence $w = \dfrac{1+i\sinh y}{1-i\sinh y}$. It gives

$$|w| = \frac{|1+i\sin y|}{|1-i\sin y|} = \frac{\sqrt{1+\sinh^2 y}}{\sqrt{1-\sinh^2 y}} = 1, \text{ for all } y.$$

Thus the line $x = \dfrac{\pi}{2}$ in the z-plane is mapped into the unit circle $|w| = 1$ in the w-plane.

For any line $x = a$, when $0 < a < \pi/2$, we have

$$|w| = \left|\frac{1-\cos(a+iy)}{1+\cos(a+iy)}\right| = \frac{\sqrt{(1-\cos a\cosh y)^2 + (\sin a \sinh y)^2}}{\sqrt{(1+\cos a\cosh y)^2 + (\sin a \sinh y)^2}} < 1, \text{ for all } y.$$

Thus as x goes from 0 to a, $0 < a < \pi/2$, the interior of the circle $|w| = 1$ is mapped. The mapping is shown in Figure.

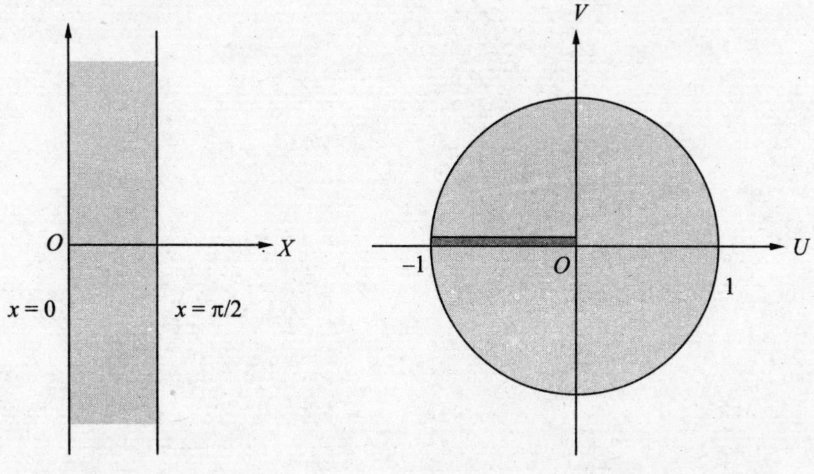

(a) (b)

CHAPTER

17 Complex Integration

IMPORTANT FORMULAE AND USEFUL RESULTS

1. **Cauchy's Integral Theorem :** If $f(z)$ is analytic and $f'(z)$ is continuous at all points inside and on a closed curve c then $\int_C f(z)dz = 0$ multiply connected

   Corollary : If $f(z)$ is analytic in a multiply connected region R between two curves C_1 and C_2 then $\int_{C_1} f(z)dz = \int_{C_2} f(z)dz$

2. **Cauchy's Integral Formula :** If $f(z)$ is analytic within and on a closed curve C and 'a' is any point within C, then $\int_C \frac{f(z)}{z-a}dz = 2\pi i f(a)$.

   Deductions : $\int_c \frac{f(z)dz}{(z-1)^2} = \frac{2\pi i}{1} f'(a)$, $\int_C \frac{f(z)dz}{(z-a)^{n+1}} = \frac{2\pi i}{n!} f^{(n)}(a)$, $n = 1, 2, 3, ...$

3. **Morera's Theorem :** If $f(z)$ is continuous in a region D bounded by a closed curve C, and $\int_c f(z)dz = 0$ around any closed curve C then $f(z)$ is analytic D.

4. **Taylor's series :** If $f(z)$ is analytic inside a circle C with centre at $z = a$ then $f(z) = f(a) + (z-a)f'(a) + \frac{(z-a)^2}{2!} f''(a) + ... + \frac{(z-a)^n}{n!} f^{(n)}(a) + ...$ which is convergent at every point inside C.

5. **Laurent's Series :** If $f(z)$ is analytic in an annular region R bounded by two concentric circles C_1 and C_2 of radii r_1 and r_2, we have

$$f(z) = a_0 + a_1(z-a) + a_2(z-a)^2 + ... + a_n(z-a)^n + ... \frac{a_{-1}}{(z-a)} + \frac{a_{-2}}{(z-a)^2} + ... + \frac{a_{-m}}{(z-a)^m} + ...$$

where $$a_n = \frac{1}{2\pi i}\int_C \frac{f(t)dt}{(t-a)^{n+1}}$$

6. **Singularities–poles :** If $f(z)$ is $= \sum_{n=1}^{\infty} a_n (z-a)^n + \frac{b_1}{z-a} + \frac{b_2}{(z-a)^2} + \frac{b_m}{(z-a)^m}$ then $f(z)$ is said to have pole of order m at $z = a$.

7. **Residue :** The coefficient of $\frac{1}{z-a}$ equal to b_1 in the Laurent's expansion of $f(z)$ about $z = a$, is called the residue of $f(z)$ at $z = a$. If $f(z)$ has a pole of order m at $z = a$ then residue of $f(z)$ at $z = a$ is equal to

$$b_1 = \frac{1}{(m-1)!} \lim_{z \to a} \frac{d^{m-1}}{dz^{m-1}} f(z)(z-a)^m.$$

8. **Cauchy's Residue Theorem :** If $f(z)$ is analytic it all points inside a closed curve C except at some points $z = z_1, z_2, \ldots z_n$ then $\int_C f(z) dz = 2\pi i$ (Sum of the residues of $f(z)$ at $z = z_1, z_2, \ldots$)

9. **Evaluation of real integrals :**

(*i*) of the form $\int_0^{2\pi} (\cos\theta, \sin\theta) d\theta$ take $C : |z| = 1$ or $z = e^{i\theta}$.

(*ii*) of the form $\int_{-\infty}^{\infty} f(x) dx$ take C as a semicircle $|z| = R$ upper half and diameter on X-axis from $(-R, 0)$ to $(R, 0)$ along X-axis.

SOLUTIONS TO EXERCISE 17A

1. *Evaluate* $\int_0^{1+i}(x-y+ix^2)\,dz$

(*a*) *along the straight line from* (0, 0) to (1, 1).

(*b*) *over the path along the lines* $y=0$ *and* $x=1$.

(*c*) *over the path along the lines* $x=0$ *and* $y=1$.

SOLUTION: (*a*) Line from (0, 0) to (1, 1) is $y=x$, so $dy=dx$

$$\therefore \quad \int_0^{1+i}(x-y+ix^2)dz = \int_{z=0}^{1+i}(x-y+ix^2)(dx+idy) = \int_{x=0}^{1}(0+ix^2)(1+i)dx$$

$$= (i-1)\left[\frac{x^3}{3}\right] = \frac{1}{3}(i-1).$$

(*b*) Path along the limes $y=0$ and $x=1$ is from 0 to A and from A to B.

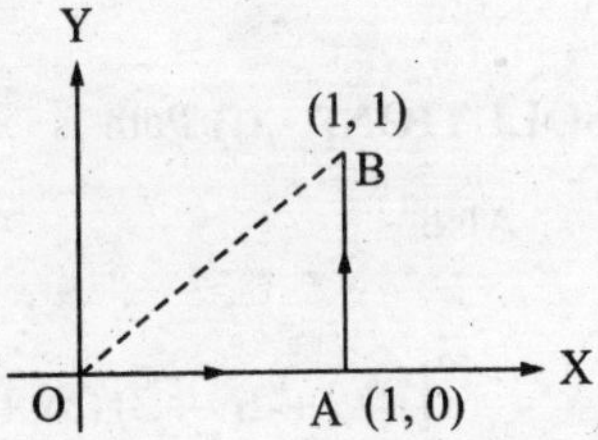

$\therefore$ Given integral along this path equals

$$\int_{x=0}^{1}(x+ix^2)dx + \int_{y=0}^{1}(1-y+i)idy$$

$$= \left[\frac{x^2}{2}+i\frac{x^3}{3}\right]_0^1 + \left[(1+i)y-\frac{y^2}{2}\right]_0^1 = -\frac{1}{2}+\frac{5i}{6}$$

(*c*) Path along the lines $x=0$ and $y=1$ is from O to C and then from C to B

$\therefore$ Given integral along this path equals

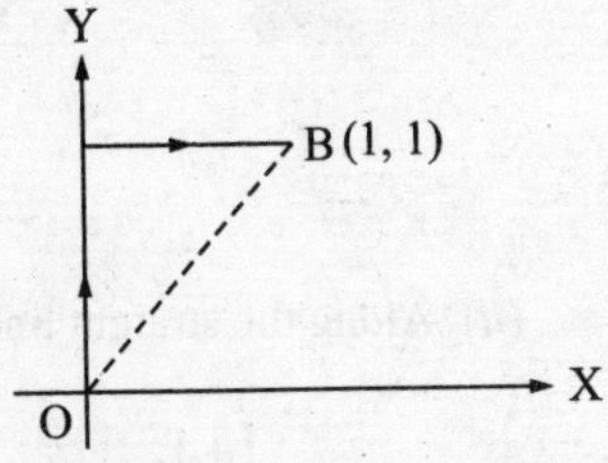

$$\int_{y=0}^{1}-y\,idy + \int_{x=0}^{1}(x-1+ix^2)dx$$

$$= -i\left[\frac{y^2}{2}\right]_0^1 + \left[\frac{x^2}{2}-x+i\frac{x^3}{3}\right]_0^1 = -\frac{1}{2}-\frac{i}{6}$$

2. *Evaluate* $\int_0^{2+i}(\bar{z})^2\,dz$ *along* (*i*) *the line* $y=x/2$ (*ii*) *the real axis to* 2 *and then vertically to* $2+i$.

SOLUTION: (*i*) The line is $y=\frac{x}{2}$, so $dy=\frac{1}{2}dx$

Therefore $\quad \int_0^{2+i}(\bar{z})dz = \int_0^{2+i}(x-iy)^2(dx+idy) = \int_0^2\left(x-i\frac{x}{2}\right)^2\left(dx+\frac{i}{2}dx\right)$

$$\int_0^2\left(1-\frac{i}{2}\right)^2\left(1+\frac{i}{2}\right)x^2dx = \left(1+\frac{1}{4}\right)\left(1+\frac{i}{2}\right)\left[\frac{x^3}{3}\right]_0^2 = \frac{10}{3}-\frac{5i}{3}$$

(*ii*) $$\int_0^{2+i} (x-iy)^2(dx+idy) = \int_0^2 x^2 dx + \int_0^1 (2-iy)^2 i dy$$

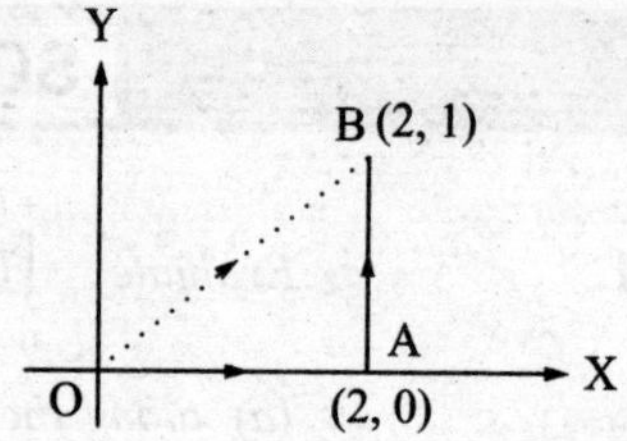

$$\left[\frac{x^3}{3}\right]_0^2 + i\int_0^1 (4-y^2-4iy)dy$$

$$= \frac{8}{3} + i\left[4 - \frac{1}{3} - 2i\right] = \frac{14}{3} + \frac{11i}{3}$$ **Ans.**

3. *Evaluate* $\int_{1-i}^{2+i} (2x+2iy+3)\, dz$ *along*

(*i*) *the path* $x = t+1, \quad y = 2t^2 - 1$

(*ii*) *the straight line from* $1-i$ *to* $2+i$.

SOLUTION: (*i*) Path is $x = t+1, \quad y = 2t^2 - 1 \quad \therefore \quad dx = dt, \; dy = 4tdt$

Also, $t=0$ at $z = 1-i$ and $t = 1$ at $z = 2+i$

$$\int_{1-i}^{2+i} (2x+2iy+3)(dx+idy) = \int_{t=0}^{1} [2(t+1) + 2i(2t^2-1) + 3][dt + 4itdt]$$

$$= \int_{t=0}^{1} [-16t^3 + 12it^2 + (10+20i)t + 5 + 2i]\, dt$$

$$= \left[-4t^4 + 4it^3 + (5+10i)t^2 + (5-2i)t\right]_0^1 = 6+12i.$$ **Ans.**

(*ii*) Along the straight line from $1-i$ to $2+i$, the equation of line being

$$y+1 = \frac{1+1}{2-1}(x-1) \quad \text{or} \quad y = 2x-3$$

$$\therefore \quad \int_{1-i}^{2+i} (2x+2iy+3)(dx+idy) = \int_{x=1}^{2} (2x+4ix-6i+3)(dx+2idx)$$

$$= \int_1^2 [(2+4i)x - 6i + 3](1+2i)dx = \left[(-3+4i)x^2 + 15x\right]_1^2 = 6+12i$$ **Ans.**

4. *Evaluate* $\int_C |z|\, dz$ *where* C *is the contour*

(*a*) *the straight line from* $z = -i$ *to* $z = i$

(*b*) *left half of the unit circle* $|z| = 1$ *from* $z = -i$ *to* $z = i$.

SOLUTION: (*a*) $\int |z|\, dz = \int \sqrt{x^2 + y^2}\, (dx + i dy)$

On straight line from $z = -i$ to $z = i$ we have $x = 0$ and y varies from -1 to 1.

Therefore $\int |z|\, dz = 2\int_0^1 iy\, dy = i$ **Ans.**

(*b*) On the left half of the circle $|z| = 1$ from $z = -i$ to $z = i$.

Parametric equation of the circle is $x = \cos\theta$, $y = \sin\theta$, θ varies from $-\frac{\pi}{2}$ to $\frac{\pi}{2}$.

Therefore $\int |z|\, dz = \int 1(dx + i dy) = \int_{-\pi/2}^{\pi/2} -\sin\theta\, d\theta + i\cos\theta\, d\theta$

$$= 0 + i\int_{-\pi/2}^{\pi/2} \cos\theta\, d\theta = 2i.$$ **Ans.**

5. *Evaluate* $\oint_C |z|^2\, dz$ *around the square with vertices at* (0, 0), (1, 0), (1, 1) *and* (0, 1).

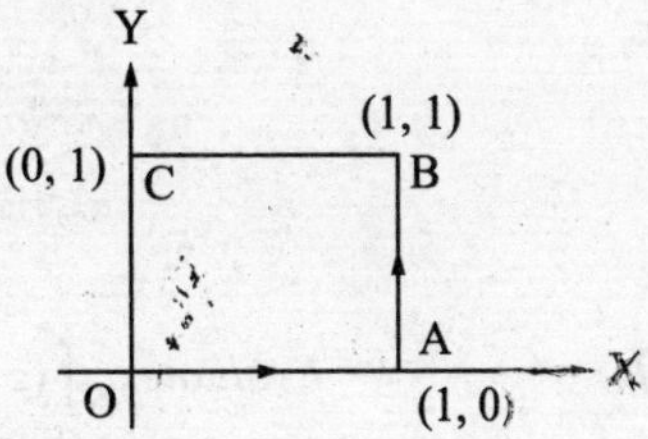

SOLUTION: $\int |z|^2\, dz = \int (x^2 + y^2)(dx + i dy) = \int_0^A + \int_A^B + \int_B^C + \int_C^0$

$$= \int_0^1 x^2 dx + \int_0^1 (1 + y^2) i dy + \int_1^0 (x^2 + 1) dx + \int_1^0 y^2 i dy$$

$$= \left[\frac{x^3}{3}\right]_0^1 + i\left[y + \frac{y^3}{3}\right]_0^1 - \left[\frac{x^3}{3} + x\right]_0^1 - i\left[\frac{y^3}{3}\right]_0^1$$

$$= \frac{1}{3} + \frac{4i}{3} - \frac{4}{3} - \frac{i}{3} = -1 + i$$ **Ans.**

6. *Evaluate* $\int_{1-2i}^{3+i} (2z + 3)\, dz$

(*a*) *along the path* $x = 2t + 1$, $y = 4t^2 - t - 2$, $0 \le t \le 1$.

(*b*) *along the straight line joining* $1 - 2i$ *and* $3 + i$.

(*c*) *along the straight lines from* $1 - 2i$ to $1 + i$ *and then to* $3 + i$.

SOLUTION: $I = \int_{1-2i}^{3+i} (2z + 3) dz = \int_{1-2i}^{3+i} (2x + 3 + 2iy)(dx + i dy)$

(*a*) Along the path $x = 2t + 1$, $y = 4t^2 - t - 2$, $0 \le t \le 1$

Hence, $$I = \int_0^1 [2(2t+1)+3+2i(4t^2-t-2)][2dt + i(8it-1)dt$$

$$= \int_0^1 [8it^2 + (4-2i)t + 5 - 4i]\,(2 - i + 8it)dt$$

$$= \int_0^1 [-64t^3 + t^2(32i+16+16i+8) + t(8-2-8i+40i+32) + 10 - 4 - 13i]dt$$

$$= [-16t^4 + t^3 + 16i + 8t^3 + t^2(16i+19) + (6-3i)t]_0^1 = 17 + 19i. \qquad \textbf{Ans.}$$

(*b*) Along the straight line from $1-2i$ to $3+i$. Equation of line is $y+2 = \frac{3}{2}(x-1)$ or $y = \frac{3}{2}x - \frac{7}{2}$

∴ Given integral $$I = \int_1^3 \left[2x+3+2i\left(\frac{3x}{2}-\frac{7}{2}\right)\right]\left(dx + \frac{3i}{2}dx\right)$$

$$= \int_1^3 [(2+3i)x + 3 - 7i]\left(1 + \frac{3i}{2}\right)dx$$

$$= (2+3i)\left(1+\frac{3i}{2}\right)\left[\frac{x^2}{2}\right]_1^3 + (3-7i)\left(1+\frac{3i}{2}\right)[x]_1^3 = 17 + 19i. \qquad \textbf{Ans.}$$

(*c*) Given integral $$I = \int_A^B + \int_B^C$$

$$= \int_{y=-2}^{1} (2+3+2iy)i\,dy + \int_{x=1}^{3} (2x+3+2i)dx$$

$$= [5iy - y^2]_{-2}^{1} + [x^2 + (3+2i)x]_1^3 = 17 + 19i. \qquad \textbf{Ans.}$$

7. *Evaluate* $\oint_C (z - z^2)\,dz$ *where* C *is the upper half of the circle* $|z| = 1$.

SOLUTION: On the curve C, $z = e^{i\theta}$, θ varying from 0 to π.

∴ Given integral $$= \int_0^\pi (e^{i\theta} - e^{2i\theta})\, ie^{i\theta} d\theta = i\int_0^\pi (e^{2i\theta} - e^{3i\theta})\, d\theta$$

$$= i\left[\frac{e^{2i\theta}}{2i} - \frac{e^{3i\theta}}{3i}\right]_0^\pi = \frac{1}{2}(e^{2\pi i} - 1) - \frac{1}{3}(e^{3\pi i} - 1)$$

$$= 0 - \frac{1}{3}(-1-i) = \frac{2}{3}. \qquad \textbf{Ans.}$$

8. *Evaluate* $\int_0^{3+i} z^2 dz$ *along the parabola* $y^2 = \frac{x}{3}$.

SOLUTION: $$I = \int_0^{3+i} z^2 dz = \int_{(0,0)}^{(3,1)} (x^2 - y^2 + 2i\,xy)\,(dx + i\,dy) \qquad (x = 3y^2 \;\therefore\; dx = 6y dy)$$

$$= \int_0^1 (9y^4 - y^2 + 6iy^3)\,(6y + i)\,dy = 6 + \frac{26i}{3}. \qquad \textbf{Ans.}$$

SOLUTIONS TO EXERCISE 17B

1. *Verify Cauchy's integral theorem for the function $f(z) = z^2$ taking C to be the circle $|z| = 3$.*

SOLUTION: $\int_C z^2 dz$ where $C : |z| = 3$. Here z^2 is analytic inside and on C.

On $|z| = 3$, $z = 3e^{i\theta}$ $\therefore$ $dz = 3ie^{i\theta}\, d\theta$

$$\therefore \qquad \int_C z^2 dz = \int_0^{2\pi} 9e^{2i\theta}\,.\,3ie^{i\theta}\, d\theta = \frac{27i}{3i}\left[e^{3i\theta}\right]_0^{2\pi} = 0$$

Hence Cauchys integral theorem is verified.

2. *Evaluate $\int_C \frac{z^2+1}{z^2-1}\, dz$ where C is the circle with centre at $z=1$ and unit radius.*

SOLUTION: $\int_c \frac{z^2+1}{z^2-1} dz$ where $C : |z-1| = 1$

There are two singularities $z = \pm 1$ but only $z = 1$ lies inside C

$\therefore$ By Cauchy's integral formula we have

$$\int_c \frac{z^2+1}{z^2-1} dz = \int_C \frac{\left(\frac{z^2+1}{z+1}\right)dz}{z-1} = 2\pi i\left[\frac{z^2+1}{z+1}\right]_{z=1} = 2\pi i \qquad \textbf{Ans.}$$

3. *Evaluate the integral $\int_C \left(\frac{2}{z} + \frac{3}{z^2}\right) dz$ where C is a path enclosing the origin.*

SOLUTION: $I = \int_c \left(\frac{2}{z} + \frac{3}{z^2}\right) dz$ where C is a path enclosing the origin. The singularity $z = 0$ lies inside C hence using Cauchy's integral formula

$$I = 2\pi i(2) + 2\pi i\left[\frac{d(3)}{dz}\right]_{z=1} = 4\pi i. \qquad \textbf{Ans.}$$

4. *Use Cauchy's integral formula to evaluate $\int_C \frac{e^{2z}}{(z+1)^4} dz$ where C is the circle $|z| = 2$.*

SOLUTION: $I = \int_C \frac{e^{2z}\, dz}{(z+1)^4}$ where $C : |z| = 2$.

The singularity is $z = -1$ which lies inside C, hence using Cauchy's integral formula, the given integral

$$I = \frac{2\pi i}{(4-1)!}\left[\frac{d^3}{dz^3} e^{2z}\right]_{z=-1} = \frac{2\pi i}{6}.[2^3 e^{2z}]_{z=-1} = \frac{8\pi i}{3e^2} \qquad \textbf{Ans.}$$

5. *Evaluate* $\int_C \frac{z^2+z+1}{z^2-7z+2}\,dz$ *where C is the ellipse* $4x^2+9y^2=1$.

SOLUTION: $\int_C \frac{z^2+z+1}{z^2-7z+2}\,dz$ where $C: 4x^2+9y^2=1$ or $\left(\frac{x}{1/2}\right)^2+\left(\frac{y}{1/3}\right)^2=1$

Since the singularities are given by $z^2-7z+2=0$

$\therefore \quad z=\frac{7\pm\sqrt{49-8}}{2}$ or $\frac{7+\sqrt{41}}{2}, \frac{7-\sqrt{41}}{2}$. Out of these, only $\frac{7-\sqrt{41}}{2}$ lies inside C.

$$\text{Given integral} = \int_C \frac{(z^2+z+1)dz}{\left[z-\left(\frac{7+\sqrt{41}}{2}\right)\right]\left[z-\left(\frac{7-\sqrt{41}}{2}\right)\right]} = 2\pi i\left[\frac{z^2+z+1}{\left[z-\left(\frac{7+\sqrt{41}}{2}\right)\right]}\right]_{z=\frac{7-\sqrt{41}}{2}}$$

$$= 2\pi i\,\frac{\left(\frac{7-\sqrt{41}}{2}\right)^2+\left(\frac{7-\sqrt{41}}{2}\right)+1}{-\sqrt{41}} = \frac{2\pi i}{-4\sqrt{41}}[49+41-14\sqrt{41}+14-2\sqrt{41}+4]$$

$$= \frac{\pi i}{\sqrt{41}}[54-8\sqrt{41}].$$ **Ans.**

6. *Evaluate* $\int_C \frac{\sin^2 z}{(z-\pi/6)^3}\,dz$ *where C is the circle* $|z|=1$.

SOLUTION: $\int_C \frac{f(z)dz}{(z-a)^3} = \frac{f''(a)}{2!}$ if $z=a$ lies in C.

$$\int_C \frac{\sin^2 z\,dz}{\left(z-\frac{\pi}{6}\right)^3} = 2\pi i\,.\,\frac{1}{2!}\left[\frac{d^2}{dz^2}\sin^2 z\right]_{z=\pi/6} \qquad \left(\text{Since } z=\frac{\pi}{6} \text{ lies in } C:|z|=1\right)$$

$$= \pi i\left[\frac{d}{dz}2\sin z\cos z\right]_{z=\pi/6} = \pi i[2\cos 2z]_{z=\pi/6} = \pi i.$$ **Ans.**

7. *Verify Cauchy's integral theorem for the function*

$f(z)=\frac{1}{z^2+2z+2}$ *when the contour C is the circle* $|z|=1$.

SOLUTION: $\int_C \frac{dz}{z^2+2z+2}$ where $C:|z|=1$

The singularites of $\frac{1}{z^2+2z+2}$ are given by $z^2+2z+2=0$ hence $z=-1\pm i$.

Since $|-1\pm i|=\sqrt{2}$, both the singularities lies outside C hence $\int_C \frac{dz}{z^2+2z+2}=0$ by Cauchy's integral theorem.

Let us evaluate this integral otherwise, taking $z = e^{i\theta}, 0 \le 2\pi$.

$$\therefore \quad \text{Integral} = \int_0^{2\pi} \frac{ie^{i\theta}\, d\theta}{e^{2i\theta} + 2e^{i\theta} + 2} = i\int_0^{2\pi} \frac{d\theta}{e^{i\theta} + 2 + 2e^{-i\theta}} = i\int_0^{2\pi} \frac{d\theta}{3\cos\theta - i\sin\theta + 2}$$

$$= i\int_0^{2\pi} \frac{d\theta}{3(\cos^2\theta/2 - \sin^2\theta/2) - 2i\sin\theta/2\cos\theta/2 + 2} = i\int_0^{2\pi} \frac{\sec^2\theta/2\, d\theta}{\tan^2\theta/2 - 2i\tan\theta/2 + 5}$$

Put $\tan\theta/2 = t$ and evaluate the integral which equals 0. **Hence proved.**

8. *Evaluate* $\int_C \frac{z\, dz}{9 + 9z - z^2 - z^3}$ *where* $C : |z| = 2$.

SOLUTION: The singularities of the integrand are given by $(z+1)(z^2 - 9) = 0$

$\therefore \quad z = -1, \pm 3$ out of these $z = \pm 3$ lie outside $c : |z| = 2$ and $z = -1$ lies in it.

$\therefore$ By Cauchy's integral formula

$$\int_C \frac{z\, dz}{(z+1)(9 - z^2)} = \int_C \frac{\left(\frac{z}{9 - z^2}\right) dz}{z+1} = 2\pi i\left[\frac{z}{9 - z^2}\right]_{z=-1} = -\frac{i\pi}{4}.$$ **Ans.**

9. *Evaluate* $\int_C \frac{\cos z}{z^2}\, dz$ *where* C *is any closed contour around the origin.*

SOLUTION: $\int_C \frac{\cos z\, dz}{z^2}$ where C is any closed curve enclosing the origin.

By Cauchy's integral formula $I = 2\pi i\left[\frac{d}{dz}\cos z\right]_{z=0} = 0$ **Ans.**

10. *Show that* $\int_C \frac{dz}{z^2(z-3)} = \frac{-2\pi i}{9}$ *where* C *is the circle* $|z| = 2$.

SOLUTION: Let $I = \int_C \frac{dz}{z^2(z-3)}$ where $C : |z| = 2$

The singularities are at $z = 0$ and $z = 3$, but $z = 3$ lies outside C hence $z = 0$ only lies inside C.

$\therefore$ By Cauchy's integral formula

$$I = \int_C \frac{\left(\frac{1}{z-3}\right) dz}{z^2} = 2\pi i\left[\frac{d}{dz}\left(\frac{1}{z-3}\right)\right]_{z=0} = 2\pi i\left[\frac{-1}{(z-3)^2}\right]_{z=0} = \frac{-2\pi i}{9}.$$ **Hence proved.**

11. *Evaluate* $f(2)$ *and* $f(3)$ *if* $f(a) = \int_C \frac{2z^2 - z - 2}{z - a}\, dz$ *where* C: $|z| = 2.5$

SOLUTION: $f(a) = 2\pi i\,[2z^2 - z - 2]_{z=a}$ if $z = a$ lies inside C.

$\therefore \quad f(2) = 8\pi i$ and $f(3) = 0$. **Ans.**

SOLUTIONS TO EXERCISE 17C

1. *Expand $\frac{1}{z}$ by Taylor's series about the point $z = 1$.*

SOLUTION: We know that Taylor's expansion of $f(z)$ about $z = a$, is

$$f(z) = f(a) + (z-a)\,f'(z) + \frac{(z-a)^2}{2!}f''(a) + \ldots$$

Therefore, the expansion of $\frac{1}{z}$ about $z = 1$, is

$$\frac{1}{z} = \frac{1}{z-1+1} = [1+(z-1)]^{-1} = 1-(z-1)+(z-1)^2-(z-1)^3+\ldots$$

which is the required expansion.

2. *Expand $\tan z$ as a Maclaurin's series in powers of z.*

SOLUTION: Singularities of $\tan z$ are at $z = \pm\frac{\pi}{2}, \pm\frac{3\pi}{2}, \ldots$. Consider $C : |z| < \frac{\pi}{2}$. Maclaurin's expansion of $f(z)$ in powers of z is the same as Taylor's expansion about the origin.

$$f(z) = \tan z = f(0) + zf'(0) + \frac{z^2}{2!}f''(0) + \frac{z^3}{3!}f'''(0) + \ldots$$

Here $f(0) = 0, \quad f'(z) = \sec^2 z \quad \therefore \quad f'(0) = \sec^2 0 = 1.$

$$f''(z) = 2\sec z(\sec z \tan z) = 2\sec^2 z \tan z \quad \text{hence } f''(0) = 0.$$

$$f'''(z) = 2\sec^2 z\,(\sec^2 z) + 4\sec^2 z\tan^2 z = 2\sec^4 z + 4\sec^4 z - 4\sec^2 z$$

$$= 6\sec^4 z - 4\sec^2 z$$

$\therefore \quad f'''(0) = 6 - 4 = 2$

Next $\quad f^{(iv)}(z) = 24\sec^3 z\,(\sec z\tan z) - 8\sec z\,(\sec z\tan z) = (24\sec^4 z + 8\sec^2 z)\tan z$

$\therefore \quad f^{(iv)}(0) = 0$

Similarly, we can obtain $f^{(v)}(0) = \frac{16}{15}$ and $f^{(vi)}(0) = 0$, etc.

Hence $\quad \tan z = z + \frac{z^3}{3} + \frac{2}{15}z^5 + \ldots$ where $|z| < \frac{\pi}{2}$. **Ans.**

3. *Determine the Laurent's expansion for $f(z) = \frac{1}{(1-z)(-z+2)}$ valid for the domain $1 < |z| < 2$.*

SOLUTION: $f(z) = \frac{1}{(1-z)(-z+2)}$, domain of $f(z)$ being $1 < |z| < 2$.

$$f(z) = \frac{1}{z-2} - \frac{1}{z-1} = -\frac{1}{2}\left(1-\frac{z}{2}\right)^{-1} - \frac{1}{z}\left(1-\frac{1}{z}\right)^{-1} \text{ as } |z|>1 \text{ and } |z|<2$$

$$= -\frac{1}{2}\left[1+\frac{z}{2}+\frac{z^2}{4}+\frac{z^4}{8}+...\right] - \frac{1}{z}\left[1+\frac{1}{z}+\frac{1}{z^2}+\frac{1}{z^3}+...\right]$$

$$= -\left[\frac{1}{2}+\frac{z}{2^2}+\frac{z}{2^3}+\frac{z^3}{2^4}+...\right] - \left[\frac{1}{z}+\frac{1}{z^2}+\frac{1}{z^3}+...\right]$$

$$= -\sum_{0}^{\infty}\frac{z^n}{2^{n+1}} - \sum_{1}^{\infty}\frac{1}{z^n}$$ **Ans.**

4. *Determine the order of the pole of the function*

$$f(z) = \frac{\sinh z}{z^5}.$$

SOLUTION:

$$f(z) = \frac{\sin hz}{z^5} = \frac{1}{z^5}\left[z+\frac{z^3}{3!}+\frac{z^5}{5!}+...\right]$$

$$= \frac{1}{z^4}+\frac{1}{3!z^2}+\frac{1}{5!}+\frac{1}{7!}z^2+...$$

Hence $f(z)$ has pole of order 4 at $z = 0$.

5. *Show that when* $|z+1|<1$,

$$\frac{1}{z^2} = 1+\sum_{n=1}^{\infty}(n+1)(z+1)^n.$$

SOLUTION: We want to expand $\frac{1}{z^2}$ in powers of $z+1$. Here $|z+1|<1$

$$f(z) = \frac{1}{z^2} = \frac{1}{(z+1-1)^2} = [1-(z+1)]^{-2}$$

$$= 1-(-2)(z+1)+\frac{(-2)(-3)}{2!}(z+1)^2 - \frac{(-2)(-3)(-4)}{3!}(z+1)^3 + ...$$

$$= 1+2(z+1)+3(z+1)^2+4(z+1)^3+5(z+1)^4+...$$

$$= 1+\sum_{n=1}^{\infty}(n+1)(z+1)^n$$

6. *Find the residue of* $f(z) = \frac{3z-4}{z(z-1)(z-2)}$ *at each of its poles.*

SOLUTION: Let $f(z) = \frac{3z-4}{z(z-1)(z-2)}$, it has simple poles at $z = 0, z = 1, z = 2$.

Therefore $\operatorname{Res}_{z=0} f(z) = \operatorname{Lt}_{z\to 0} \dfrac{3z-4}{z(z-1)(z-2)} = \dfrac{-4}{(-1)(-2)} = -2,$

and $\operatorname{Res}_{z=1} f(z) = \operatorname{Lt}_{z\to 1} \dfrac{3z-4}{z(z-2)} = \dfrac{3-4}{(1-2)} = 1$

and $\operatorname{Res}_{z=2} f(z) = \operatorname{Lt}_{z\to 2} \dfrac{3z-4}{z(z-1)} = \dfrac{6-4}{2(2-1)} = 1$

Thus residues of $f(z)$ at $z = 0, 1, 2$ are $-2, 1, 1$ respectively. **Ans.**

7. *Obtain the residue of* $f(z) = \left(\dfrac{z+1}{z-1}\right)^3$ *at its pole.*

SOLUTION: Let $f(z) = \left(\dfrac{z+1}{z-1}\right)^3$ has a pole at $z = 1$ of order 3.

$$\therefore \quad \operatorname{Res}_{z=1} f(z) = \frac{1}{(3-1)!} \operatorname{Lt}_{z\to 1} \frac{d^2}{dz^2}(z-1)^3\left(\frac{z+1}{z-1}\right)^3 = \frac{1}{2} \operatorname{Lt}_{z\to 1} \frac{d^2}{dz^2}(z+1)^3 = \frac{1}{2} \operatorname{Lt}_{z\to 1} \frac{d}{dz} 3(z+1)^2$$

$$= \frac{1}{2} \operatorname{Lt}_{z\to 1} 3\,.\,2(z+1) = 6$$ **Ans.**

8. *Evaluate the integral* $\int_C \dfrac{e^z - 1}{z(z+1)}\, dz$ where $C : |z| = 2$.

SOLUTION: Let $I = \int_C \dfrac{(e^z - 1)dz}{z(z+1)}$ where $C : |z| = 2$

The integrand has poles at $z = 0$ and $z = -1$, both lie inside C.

$\therefore$ By Cauchy's residue theorem

$$\int_C \frac{e^z - 1\, dz}{z(z+1)} = 2\pi i \left[\operatorname{Res}_{z=0} f(z) + \operatorname{Res}_{z=-1} f(z)\right]$$

$$= 2\pi i \left[\operatorname{Res}_{z=0} f(z) + \operatorname{Res}_{z=-1} f(z)\right] = 2\pi i \left[0 + \frac{e^{-1}-1}{-1}\right] = 2\pi i \left(1 - \frac{1}{e}\right)$$

9. *Find the value of the integral* $\int_C \frac{dz}{z^3(z+4)}$ *where* $C: |z| = 2.$

SOLUTION: Let $I = \int_C \frac{dz}{z^3(z+4)} = \int_C f(z)dz$ where $C: |z| = 2$

$f(z)$ has pole of order 3 at $z = 0$ and simple pole at $z = -4$. Here $z = 0$ lies inside C whereas $z = -4$ lies outside C. Therefore, by residue theorem, we have

$$\int_c f(z)dx = 2\pi i \operatorname*{Res}_{z=0} f(z) = 2\pi i \frac{1}{(3-1)!} \operatorname*{Lt}_{z\to 0} \frac{d^2}{dz^2}\left[\frac{z^3}{z^3(z+4)}\right]$$

$$= \frac{2\pi i}{2} \operatorname*{Lt}_{z\to 0} \frac{d}{dz}\left[\frac{-1}{(z+4)^2}\right] = \pi i \operatorname*{Lt}_{z\to 0} \frac{2}{(z+4)^3} = \frac{\pi i}{32}$$ **Ans.**

10. *Determine the residues of* $f(z) = \frac{z^2}{(z-2)(z^2+1)}$ *at its poles.*

SOLUTION: $f(z) = \frac{z^2}{(z-2)(z^2+1)}$ has simple poles at $z = 2, +i, -i.$

$$\therefore \quad \operatorname*{Res}_{z=2} f(z) = \operatorname*{Lt}_{z\to 2} \frac{z^2}{z^2+1} = \frac{4}{5}, \quad \operatorname*{Res}_{z=i} f(z) = \operatorname*{Lt}_{z\to i} \frac{z^2}{(z-2)(z+i)} = \frac{i^2}{(i-2)(2i)} = \frac{1-2i}{10}$$

$$\operatorname*{Res}_{z=-i} f(z) = \operatorname*{Lt}_{z\to -i} \frac{z^2}{(z-2)(z-i)} = \frac{-1}{-2+4i} = \frac{1+2i}{10}.$$

$\therefore$ Residues of $f(z)$ at $z = 2, i$ and $-i$ are $\frac{4}{5}, \frac{1-2i}{10}$ and $\frac{1+2i}{10}$ respectively.

11. *Find the residues of* $\frac{1}{z(z+2)^3}$ at $z = 0$ *and* $z = -2.$

SOLUTION: $f(z) = \frac{1}{z(z+2)^3}$ has simple pole at $z = 0$ and pole of order 3 at $z = -2.$

$$\operatorname*{Res}_{z=0} f(z) = \operatorname*{Lt}_{z\to 0} \frac{1}{(z+2)^3} = \frac{1}{8}$$

$$\operatorname*{Res}_{z=-2} f(z) = \frac{1}{(3-1)!} \operatorname*{Lt}_{z\to -2} \frac{d^2}{dz^2}\left[(z+2)^3 \frac{1}{z(z+2)^3}\right] = \frac{1}{2!} \operatorname*{Lt}_{z\to -2} \frac{d^2}{dz^2}\left(\frac{1}{z}\right)$$

$$= \frac{1}{2} \operatorname*{Lt}_{z\to -2} \frac{d}{dz}\left(-\frac{1}{z^2}\right) = \frac{1}{2} \operatorname*{Lt}_{z\to -2}\left[\frac{(-1)(-2)}{z^3}\right] = -\frac{1}{8}.$$

Thus, Residues of $f(z)$ at $z = 0$ and at $z = -2$ are $\frac{1}{8}$ and $-\frac{1}{8}$ respectively. **Ans.**

12. *Evaluate* $\oint_C \frac{e^z\,dz}{(z-1)(z+3)^2}$ *where (i)* $C : |z| = 3/2$. *(ii)* $C : |z| = 10$

SOLUTION: (*i*) $\int_C f(z)dz = \int_C \frac{e^z\,dz}{(z-1)(z+3)^2}$ where $C : |z| = \frac{3}{2}$.

$f(z)$ has simple pole at $z = 1$ and pole of order 2 at $z - 3$, but $z = -3$ lies outside C, and $Z = 1$ lies in C.

$$\therefore \qquad \int_C f(z)dz = 2\pi i \operatorname*{Res}_{z=1} f(z) = 2\pi i \operatorname*{Lt}_{z\to -2} \frac{e^z}{(z+3)^2} = \frac{i\pi e}{8} \qquad \textbf{Ans.}$$

(*ii*) Here $z = 1$ and $z = -3$ both lie inside $C: |z| = 10$.

$\therefore$ By residue theorem

$$\int_C f(z)\,dz = 2\pi i \quad \text{(sum of the residues of } f(z))$$

Here
$$\operatorname*{Res}_{z=1} f(z) = \left[\frac{e^z}{(z+3)^2}\right]_{z=1} = \frac{e}{16}$$

and
$$\operatorname*{Res}_{z=-3} f(z) = \operatorname*{Lt}_{z\to -3} \frac{d}{dz}\left(\frac{e^z}{z-1}\right) = \operatorname*{Lt}_{z\to -3}\left[\frac{(z-1)\,e^z - e^z.1}{(z-1)^2}\right]$$

$$= e^{-3}\left(\frac{-3-2}{16}\right)$$

$$\therefore \qquad \text{Value of integral} = 2\pi i\left(\frac{e}{16} - \frac{5e^{-3}}{16}\right) = \frac{\pi i}{8}(e - 5e^{-3}) \qquad \textbf{Ans.}$$

SOLUTIONS TO EXERCISE 17D

1. *Evaluate* $\displaystyle\int_0^{2\pi} \frac{d\theta}{1+\frac{1}{2}\cos\theta}$.

SOLUTION: $I = \displaystyle\int_0^{2\pi} \frac{d\theta}{1+\frac{1}{2}\cos\theta} = \int_0^{2\pi} \frac{d\theta}{1+\frac{1}{4}(e^{i\theta}+e^{-i\theta})} = \int_0^{2\pi} \frac{e^{i\theta}\,d\theta}{e^{2i\theta}+4e^{i\theta}+1}$

Consider $C:|z|=1$ so $z=e^{i\theta},\ 0\le\theta\le 2\pi,\quad dz = ie^{i\theta}\,d\theta$

$\therefore \quad \displaystyle\int_C f(z)dz = \int_C \frac{-4i\,dz}{z^2+4z+1}$ The poles of $f(z)$ are given by $z^2+4z+1=0$,

Hence $z = \dfrac{-4\pm\sqrt{16-4}}{2} = -2\pm\sqrt{3}$, of these only $z=-2+\sqrt{3}$ lies inside C.

$\therefore \quad \displaystyle\int_C f(z)dz = \int_C \frac{-4i}{[z-(-2+\sqrt{3})][z-(-2-\sqrt{3}]}$. Applying residue theorem, we get

$$= 2\pi i \operatorname*{Res}_{z=-2+\sqrt{3}} f(z) = 2\pi i(-4i) \operatorname*{Lt}_{z=-2+\sqrt{3}} \frac{1}{z-(-2-\sqrt{3})} = \frac{4\pi}{\sqrt{3}} \qquad \textbf{Ans.}$$

2. *Evaluate* $\displaystyle\int_{-\pi}^{\pi} \frac{d\theta}{1+\sin^2\theta}$.

SOLUTION: $I = \displaystyle\int_{-\pi}^{\pi} \frac{d\theta}{1+\sin^2\theta} = \int_{-\pi}^{\pi} \frac{d\theta}{1+\frac{1}{2}(1-\cos 2\theta)} = \int_{-\pi}^{\pi} \frac{2d\theta}{3-\frac{1}{2}(e^{2i\theta}+e^{-2i\theta})}$

or $I = \displaystyle\int_{-\pi}^{\pi} \frac{4e^{2i\theta}\,d\theta}{6e^{2i\theta}-e^{4i\theta}-1}$ Consider $C:|z|=1\ \ z=e^{i\theta} \quad \therefore \quad dz = i\,e^{i\theta}\,d\theta$

Then $I = \displaystyle\int_C f(z)dz = \int_C \frac{-4iz\,dz}{6z^2-z^4-1} = \int_C \frac{4iz\,dz}{z^4-6z^2+1} = \int_C \frac{4iz\,dz}{(z^2+2z-1)(z^2-2z-1)}$

Pole of $f(z)$ are given by $z^2+2z-1=0$ and $z^2-2z-1=0 \ \Rightarrow\ z = (\pm 1\pm\sqrt{2})$.

Out of these $z=\pm(1-\sqrt{2})$ lie inside c and $z=\pm(1+\sqrt{2})$ lie outside C.

Therefore $\qquad f(z) = \dfrac{4iz}{(z-1-\sqrt{2})(z-1+\sqrt{2})(z+1+\sqrt{2})(z+1-\sqrt{2})}$

and $\qquad \displaystyle\operatorname*{Res}_{z=1-\sqrt{2}} f(z) = \operatorname*{Lt}_{z=1-\sqrt{2}} \frac{4iz(z-1+\sqrt{2})}{(z-1-\sqrt{2})(z-1+\sqrt{2})(z+1+\sqrt{2})(z+1-\sqrt{2})}$

$$= \operatorname*{Lt}_{z=1-\sqrt{2}} \frac{4iz}{(z-1-\sqrt{2})(z+1+\sqrt{2})(z+1-\sqrt{2})}$$

$$= \operatorname*{Lt}_{z=1-\sqrt{2}} \frac{4iz(1-\sqrt{2})}{(-2\sqrt{2})\,2.2(1-\sqrt{2})} = \frac{-i}{2\sqrt{2}}$$

and
$$\operatorname*{Res}_{z=-(1-\sqrt{2})} f(z) = \operatorname*{Lt}_{z=-1+\sqrt{2}} \frac{4iz(z+1-\sqrt{2})}{(z-1-\sqrt{2})(z-1+\sqrt{2})(z+1+\sqrt{2})(z+1-\sqrt{2})}$$

$$= \operatorname*{Lt}_{z=-1+\sqrt{2}} \frac{4iz(-1+\sqrt{2})}{(-2)2(-1+\sqrt{2})(2\sqrt{2})} = \frac{-i}{2\sqrt{2}}$$

By Residue theorem

$$\therefore \quad \int_C f(z)dz = 2\pi i \times \text{Sum of the residues}$$

$$= -2\pi i\left[\frac{i}{2\sqrt{2}} + \frac{i}{2\sqrt{2}}\right] = \frac{2\pi}{\sqrt{2}} = \pi\sqrt{2}$$ **Ans.**

3. *Evaluate* $\displaystyle\int_0^{\pi} \frac{a\,d\theta}{a^2+\sin^2\theta}$ $(a > 0)$.

SOLUTION: $I = \displaystyle\int_0^{\pi} \frac{a\,d\theta}{a^2+\sin^2\theta} = \frac{a}{2}\int_0^{2\pi} \frac{d\theta}{a^2+\sin^2\theta}$ by property of definite integrals $(a > 0)$

or
$$I = \frac{a}{2}\int_0^{2\pi} \frac{d\theta}{a^2+\frac{1}{2}(1-\cos 2\theta)} = \frac{a}{2}\int_0^{2\pi} \frac{2d\theta}{2a^2+1-\cos 2\theta} = \int_0^{2\pi} \frac{a d\theta}{2a^2+1-\frac{1}{2}(e^{2i\theta}+e^{-2i\theta})}$$

$$= \int_0^{2\pi} \frac{2ae^{2i\theta}\,d\theta}{2(2a^2+1)e^{2i\theta}-e^{4i\theta}-1} = \int_0^{2\pi} \frac{-2ae^{2i\theta}\,d\theta}{e^{4i\theta}-2(2a^2+1)e^{2i\theta}+1}$$

Now, consider C as unit circle $|z|=1$ so $z=e^{i\theta}$ and $dz = ie^{i\theta}d\theta$

$$\therefore \quad I = \int_c f(z)dz = \int_c \frac{2az\,idz}{z^4-2(2a^2+1)z^2+1}$$

The poles of $f(z)$ are given by $z^4-2(2a^2+1)z^2+1=0$

$$\therefore \quad z^2 = 2a^2+1 \pm \sqrt{(2a^2+1)^2-1} = 2a^2+1 \pm \sqrt{2a^2(2a^2+2)}$$

$$= \left[a \pm \sqrt{a^2+1}\right]^2 \quad \text{hence} \quad z = \pm\left[a \pm \sqrt{a^2+1}\right]$$

Out of these the poles at $z = \pm(a+\sqrt{a^2+1})$ lie outside $C : |z| = 1$ while $z = \pm(\sqrt{a^2+1}-a)$ lie inside C.

Therefore, by Cauchy's residue theorem

$$\int_C f(z)\,dz = 2\pi i\ [\text{Sum of the residues at } z = \pm(\sqrt{a^2+1}-a)]$$

$$\underset{z=\sqrt{a^2+1}-a}{\text{Res}} f(z) = \underset{z=\sqrt{a^2+1}}{\text{Lt}} \frac{2aiz}{-a\left[z-\sqrt{a^2+1}+a\right]\left[z+\sqrt{a^2+1}+a\right]\left[z+\sqrt{a^2+1}-a\right]} = \frac{-i}{4\sqrt{a^2+1}}.$$

$$\underset{z=\sqrt{a^2+1}+a}{\text{Res}} f(z) = \underset{z=[z-(\sqrt{a+1}+a)]}{\text{Lt}} \frac{2aiz}{[z+\sqrt{a^2+a}][z+a-\sqrt{a^2-1}]} = \frac{--i}{4\sqrt{a^2+1}}.$$

$$\therefore \qquad \int_C f(z)dz = 2\pi i\left[\frac{-i}{4\sqrt{a^2+1}} - \frac{i}{4\sqrt{a^2+1}}\right] = \frac{\pi}{\sqrt{a^2+1}} \qquad \textbf{Ans.}$$

4. *Evaluate* $\int_0^{2\pi} \frac{\cos 2\theta \, d\theta}{1-2a\cos\theta+a^2}$ $(a^2 < 1)$.

SOLUTION: $I = \int_0^{2\pi} \frac{\cos 2\theta \, d\theta}{1-2a\cos\theta+a^2}$ (where $a < 1$) $= \int_0^{2\pi} \frac{(e^{2i\theta}+e^{-2i\theta})d\theta}{2[1+a^2-a(e^{i\theta}+e^{-i\theta})]}$

Consider $C: |z| = 1$ so $z = e^{i\theta}$ and $dz = izd\theta$

$$\therefore \qquad I = \int_C f(z)dz = \int_C \frac{\left(z^2+\frac{1}{z^2}\right)dz}{i2z\left[1-a\left(z+\frac{1}{z}\right)+a^2\right]} = \int_C \frac{(z^4+1)dz}{2z^2(z-az^2-a+a^2z)}$$

$$= \int_C \frac{(z^4+1)dz}{i2z^2(z-a)(1-az)} = \frac{-1}{2a}\int_C \frac{(z^4+1)dz}{z^2(z-a)\left(z-\frac{1}{a}\right)}$$

The poles of the $f(z)$ are at $z = 0$, $z = a, z = \frac{1}{a}$. Out of these $z = 0$ and $z = a$ lie inside C and $z = \frac{1}{a}$ lies outside C since $a^2 < 1$. $z = 0$ is pole of order two and $z = a$ is a simple pole.

$$\underset{z=0}{\text{Res}} f(z) = \underset{z\to 0}{\text{Lt}} \frac{d}{dz}\left[\frac{z^4+1}{2(z-a)(1-az)}\right]$$

$$= \underset{z\to 0}{\text{Lt}} \frac{(z-a)(1-az)\,4z^3-(z^4+1)\{-a(z-a)(1-az)\}}{2(z-a)^2(1-az)^2} = \frac{0-1(a^2+1)}{2\,.\,a^2\,.1} = \frac{-(a^2+1)}{2a^2}$$

and $\qquad \underset{z=a}{\text{Res}} f(z) = \underset{z\to a}{\text{Lt}} \frac{z^4+1}{2z^2(1-az)} = \frac{a^4+1}{2a^2(1-a^2)}$

By Cauchy's residue theorem

$$\therefore \qquad \int_C f(z)dz = \frac{2\pi i}{i}\left(\frac{-(a^2+1)}{2a^2} + \frac{a^4+1}{2a^2(1-a^2)}\right) = \frac{2\pi i}{2a^2(1-a^2)i}[(a^4-1)+a^4+1] = \frac{2\pi a^2}{1-a^2} \qquad \textbf{Ans.}$$

5. *Evaluate* $\int\limits_0^\infty \frac{dx}{x^2+1}$.

SOLUTION: Consider $\int_C f(z)dz$ where $f(z) = \frac{1}{z^2+1}$ and C consists of semi circle C_1 and from A to B on real axis. $C_1 : z = Re^{i\theta}, 0 \le \theta \le \pi$.

$$\therefore \quad \int_C \frac{1}{z^2+1} dz = \int_{C_1} \frac{dz}{z^2+1} + \int_{-R}^{R} \frac{dx}{x^2+1} = \int_0^\pi \frac{iRe^{i\theta}\, d\theta}{R^2 e^{2i\theta}+1} + \int_{-R}^{R} \frac{dx}{x^2+1}$$

We take limit when $R \to \infty$. $f(z)$ has simple poles at $z = \pm i$ of which $z = i$ lies inside C and $z = -i$ outside C.

$$\therefore \quad \int_C \frac{dz}{z^2+1} = 2\pi i \operatorname*{Res}_{z=i} \frac{1}{z^2+1} = 2\pi i \operatorname*{Lt}_{z\to i} \frac{z-i}{(z-i)(z+i)} = \frac{2\pi i}{2i} = \pi$$

and $$\int_{C_1} \frac{dz}{z^2+1} = \int_0^\pi \frac{Rie^{i\theta}\, d\theta}{R^2 e^{2i\theta}+1} \to 0 \text{ as } R \to \infty.$$

and $$\int_{-R}^{R} \frac{dx}{x^2+1} \to \int_{-\infty}^{\infty} \frac{dx}{x^2+1}$$

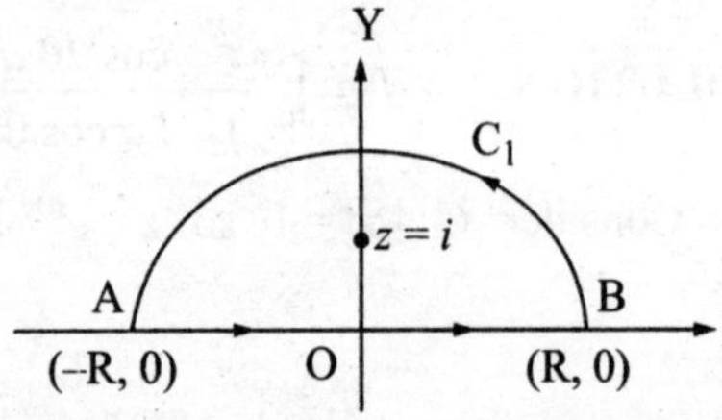

$$\therefore \quad 2\int_0^\infty \frac{dx}{x^2+1} = \pi \quad \text{or} \quad \int_0^\infty \frac{dx}{x^2+1} = \frac{\pi}{2}.$$ **Ans.**

6. *Evaluate* $\int\limits_{-\infty}^{\infty} \frac{x^2\, dx}{(x^2+a^2)(x^2+b^2)}$ *and hence find the value of* $\int\limits_{-\infty}^{\infty} \frac{x^2\, dx}{(x^2+1)(x^2+4)}$.

SOLUTION: Consider $\int\limits_C f(z)dz = \int\limits_C \frac{z^2 dz}{(z^2+a^2)(z^2+b^2)}$ where C is the upper half C_1 of the circle $|z| = R$ and line segment AB from $(R, 0)$ to $(R, 0)$ on real axis.

$$\therefore \quad \int\limits_C f(z)dz = \int\limits_{C_1} \frac{z^2 dz}{(z^2+a^2)(z^2+b^2)} + \int\limits_{-R}^{R} \frac{x^2\, dx}{(x^2+a^2)(x^2+b^2)}$$

$f(z)$ has simple poles at $z = \pm ai$ and $z = \pm bi$ out of which only $z = ai$ and $z = bi$ lie in C. On C_1, $z = Re^{i\theta}$, $0 \le \theta \le \pi$. We shall take limit as $R \to \infty$. Using Cauchy's reside theorem, above equation becomes

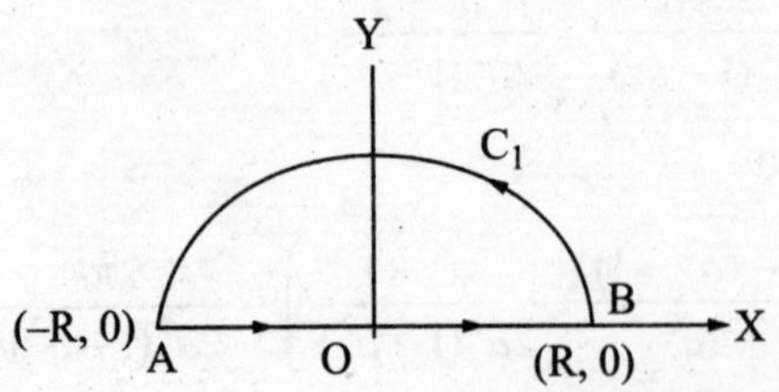

$$\int_C f(z)dz = 2\pi i\left(\begin{matrix}\text{Res } f(z) + \text{Res } f(z)\\ z = ai \qquad z = bi\end{matrix}\right) = \int_0^{\pi} \frac{R^2 e^{2i\theta} \cdot iRe^{i\theta} d\theta}{(R^2e^{2i\theta}+a^2)(R^2e^{2i\theta}+b^2)} + \int_{-R}^{R} \frac{x^2 dx}{(x^2+a^2)(x^2+b^2)}$$

$$\therefore \quad 2\pi i\left[\lim_{z\to ai} \frac{z^2(z-ai)}{(z^2+a^2)(z^2+b^2)} + \lim_{z\to bi} \frac{z^2(z-bi)}{(z^2+a^2)(z^2+b^2)}\right] = 0 + \int_{-\infty}^{\infty} \frac{x^2 dx}{(x^2+a^2)(x^2+b^2)} \qquad (\text{as } R\to\infty)$$

$$\therefore \quad \int_{-\infty}^{\infty} \frac{x^2 dx}{(x^2+a^2)(x^2+b^2)} = 2\pi i\left[\lim_{z\to ai} \frac{z^2}{(z+ai)(z^2+b^2)} + \lim_{z\to bi} \frac{z^2}{(z^2+a^2)(z+bi)}\right]$$

$$= 2\pi i\left[\frac{-a^2}{2ai(b^2-a^2)} + \frac{-b^2}{2bi(a^2-b^2)}\right] = \frac{\pi(a-b)}{a^2-b^2} = \frac{\pi}{a+b} \qquad \textbf{Ans.}$$

$$\therefore \quad \int_{-\infty}^{\infty} \frac{x^2 dx}{(x^2+1)(x^2+4)} = \frac{\pi}{1+2} = \frac{\pi}{3} \qquad \textbf{Ans.}$$

7. *Evaluate* (*i*) $\int_0^{\infty} \frac{\cos mx\, dx}{x^2+a^2}$. (*ii*) $\int_0^{\infty} \frac{\sin mx\, dx}{x^2+a^2}$

SOLUTION: Consider $\int_C f(z)dz$ where $f(z) = \frac{e^{miz}}{z^2+a^2}$ and C is semi circle $C_1 : |z| = R,\ 0 \le \theta \le \pi$ and the lime segment from $A(-R, 0)$ to $B(R, 0)$.

$$\therefore \quad \int_C \frac{e^{miz} dz}{z^2+a^2} = \int_{C_1} \frac{e^{miz} dz}{z^2+a^2} + \int_{-R}^{R} \frac{e^{mix} dx}{x^2+a^2} \quad \text{We shall take limit as } R\to\infty. \qquad ...(1)$$

$f(z)$ has simple poles at $z = \pm\, ai$ of which only $z = ai$ lies inside C.

$\therefore$ By Cauchy's residue theorem

$$\int_{C_1} \frac{e^{miz} dz}{z^2+a^2} = 2\pi i \operatorname*{Res}_{z=ai} \cdot \frac{e^{miz}}{z^2+a^2} = 2\pi i \operatorname*{Lt}_{z\to ai} \frac{e^{-am}(z-ai)}{z^2+a^2} = \frac{\pi}{a} e^{-am}$$

$$\int_{C_1} \frac{e^{miz} dz}{z^2+a^2} = \int_0^{\pi} \frac{e^{miRe^{i\theta}} \cdot Rie^{i\theta} d\theta}{R^2e^{2i\theta}+a^2} = \int_0^{\pi} \frac{e^{miR\cos\theta} \cdot e^{-mR\sin\theta}}{R^2e^{2C\theta}+a^2} \cdot Rie^{i\theta} \to \theta \text{ as } R\to\infty$$

and
$$\int_{-R}^{R} \frac{e^{mix} dx}{x^2+a^2} \longrightarrow \int_{-\infty}^{\infty} \frac{\cos mx + i\sin x}{x^2+a^2} dx$$

$\therefore$ (1) because $\frac{\pi}{a} e^{-am} = \int_{-\infty}^{\infty} \frac{\cos mix + i\sin x}{x^2+a^2} dx$. Equating real and imaginary parts on both sides, gives

$$\int_0^{\infty} \frac{\cos mx}{x^2+a^2} = \frac{\pi}{2a} e^{-am} \qquad \text{and} \qquad \int_{-\infty}^{\infty} \frac{\sin mx}{x^2+a^2} dx = 0 \qquad \textbf{Ans.}$$

8. *Evaluate* $\int_{-\infty}^{\infty} \frac{x \sin \pi x}{1-x^2} dx.$

SOLUTION: Consider $\oint_C f(z)\, dz = \oint_C \frac{z\, e^{i\pi z}}{1-z^2} dz$ where $f(z)$ has poles at $z = \pm 1$ which are indented by two small semi-circles C_1 and C_2. Hence C is the closed contour consisting of C' the upper half of the circle $|z| = R$, R being large, line segment A to B and upper half semi-circle $C_1 : |Z+1| = \in$, being small, line segment from E to F, then the upper half, semi-circle $C_2 : |z-1| = \in$ and line segment G to H.

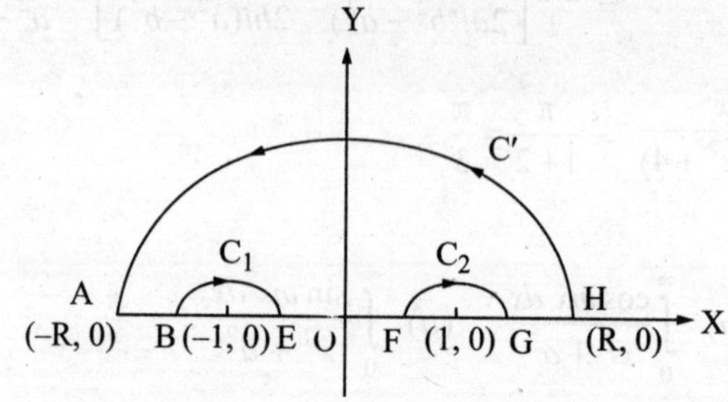

$$\int_C \frac{ze^{i\pi z}}{1-z^2} dz = \int_{C'} \frac{ze^{i\pi z} dz}{1-z^2} + \int_{-R}^{-1-\in} \frac{xe^{i\pi x} dx}{1-x^2} + \int_{C_1} \frac{ze^{i\pi z} dz}{1-z^2} + \int_{-1+\in}^{1-\in} \frac{xe^{i\pi x} dx}{1-x^2} + \int_{C_2} \frac{ze^{i\pi z} dz}{1-z^2} + \int_{1+\in}^{R} \frac{x\, e^{i\pi x}}{1-x^2} dx$$

By Cauchy's integral theorem $\int_C f(z)dz = 0$

$$\therefore \quad 0 = \int_0^{\pi} \frac{\text{R}e^{i\theta}\, e^{i\pi \text{R}} e^{i\theta}}{1-R^2 e^{2iQ}} iRe^{i\theta} d\theta + \int_{-R}^{-1-\in} \frac{xe^{i\pi x}}{1-x^2} dx + \int_{\pi}^{0} \frac{(-1+\in e^{i\theta}) e^{i\pi(-1+\in e^{i\theta})} i \in e^{i\theta} d\theta}{1-(-1+\in e^{i\theta})^2}$$

$$+ \int_{-1+\in}^{1+\in} \frac{xe^{i\pi x} dx}{1-x^2} + \int_{\pi}^{0} \frac{(1+\in e^{i\theta}) e^{i\theta(1+\in e^{i\theta})} i \in e^{i\theta} d\theta}{1-(1-\in e^{i\theta})^2} + \int_{i+\in}^{R} \frac{xe^{i\pi x} dx}{1-x^2}$$

Now let $R \to \infty$ and $\in \to 0$ and we have

$$0 = \underset{R\to\infty}{\text{Lt}}\ i \int_0^{\pi} \frac{e^{i\pi R}\, e^{i\theta}\, d\theta}{\frac{1}{R^2} - e^{2i\theta}} + \int_{-\infty}^{\infty} \frac{xe^{i\pi x}}{1-x^2} dx + \int_{\pi}^{0} \frac{i\in(-1+\in e^{i\theta})\, e^{-i\pi} e^{i\pi \in e^{i\theta}}\, e^{i\theta}\, d\theta}{2\in e^{2\theta} - \in^2 e^{2i\theta}}$$

$$= \underset{R\to\infty}{\text{Lt}}\ i \int_0^{\pi} \frac{e^{i\pi R \cos\theta} e^{-\pi R \sin\theta}\, d\theta}{\frac{1}{R^2} - e^{2i\theta}} + \int_{-\infty}^{\infty} \frac{x(\cos \pi x + i \sin \pi x) dx}{1-x^2} + \int_{\pi}^{0} \frac{i(-1+\in e^{i\theta}).1e^{i\theta} d\theta}{2e^{i\theta} - \in e^{2i\theta}}$$

$$= 0 + \int_{-\infty}^{\infty} \frac{x(\cos \pi x + i \sin \pi x) dx}{1-x^2} - \frac{i}{2} \int_0^{\pi} d\theta - \frac{i}{2} \int_0^{\pi} d\theta$$

$$\Rightarrow \quad \int_{-\infty}^{\infty} \frac{\sin \pi x dx}{1-x^2} = \pi$$ **Ans.**

9. *Evaluate* $\int_{-\infty}^{\infty} \frac{dx}{(x^2+1)^4}$.

SOLUTION: $\int_C f(z)dz = \int_C \frac{dz}{(1+z^2)^4}$ where C is consisting of C_1 the semi-circle $|Z| = R$ and line segment $(-R, 0)$ to $(R, 0)$, R being large.

$$\therefore \quad \oint_C \frac{dz}{(z^2+1)^4} = \int_{C_1} \frac{dz}{(z^2+1)^4} + \int_{-R}^{R} \frac{dx}{(x^2+1)^4}. \quad \text{Here } C_1 : |Z| = R,\ 0 \le \theta \le \pi.$$

$f(z)$ has poles of order 4 at $z = \pm i$. Out of these only $z = i$ lies inside C. Therefore by Cauchy's residue theorem, we have

$$\oint_C \frac{dz}{(z^2+1)^4} = 2\pi i \operatorname*{Res}_{z=i} f(z) = \frac{2\pi i}{3!} \operatorname*{Lt}_{z\to i} \frac{d^3}{dz^3} \frac{(z-i)^4}{(z^2+1)^2} = \frac{\pi i}{3} \operatorname*{Lt}_{z\to i} \frac{d^3}{dz^3} \frac{1}{(z+i)^4}$$

$$= \frac{\pi i}{3} \operatorname*{Lt}_{Z\to i} \frac{(-4)(-5)(-6)}{(Z+i)^7} = \frac{-\pi i 120}{3(2i)^7} = \pi \frac{40}{128} = \frac{5\pi}{16}$$

$$\int_{C_1} \frac{Rie^{i\theta}\, d\theta}{(R^2 e^{2i\theta}+1)^4} \to 0 \text{ as } R \to \infty \quad \text{and} \quad \int_{-R}^{R} \frac{dx}{(x^2+1)^4} \to \int_{-\infty}^{\infty} \frac{dx}{(x^2+1)^4}.$$

$$\therefore \quad \int_{-\infty}^{\infty} \frac{dx}{(x^2+1)^4} = \frac{5\pi}{16}$$

Ans.

10. *Evaluate* $\int_0^{\infty} \frac{\sin x}{x}\, dx$.

SOLUTION: Consider $\int_C f(z)dz = \int_C \frac{e^{iz}dz}{Z}$ where $f(z)$ has pole at $z = 0$ and C consists of C_1 upper half semi circle $|z| = R$, R being large, line segment BP and upper half of semi circle $|Z| = r$, r being small, then along line segment QA.

$$\text{Therefore} \quad \int_C \frac{e^{iz}dz}{z} = \int_{C_1} \frac{e^{iz}dz}{z} + \int_{-R}^{-r} \frac{e^{i.x}dx}{x} + \int_{C_2} \frac{e^{iz}dz}{z} + \int_r^R \frac{e^{i.x}dx}{x}$$

$$= \int_0^{\pi} \frac{e^{i\text{Re}^{i\theta}}\, i.\text{Re}^{i\theta}\, d\theta}{\text{Re}^{i\theta}} + \int_{-R}^{-r} \frac{e^{ix}dx}{x} + \int_{\pi}^{0} \frac{e^{ire^{i\theta}}\, .ire^{i\theta}\, d\theta}{re^{i\theta}} + \int_r^R \frac{e^{ix}dx}{x}$$

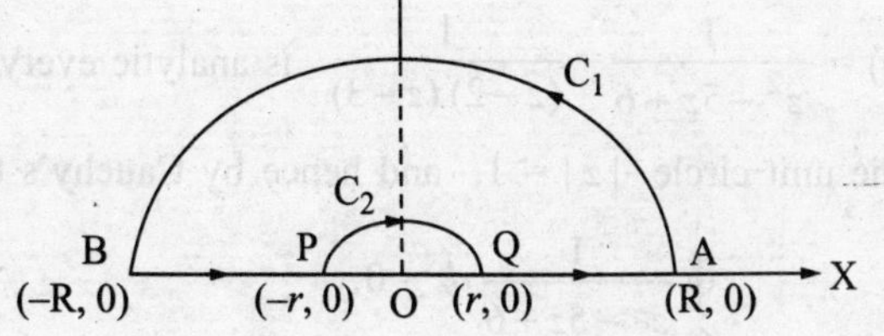

Here $R \to \infty$ and $r \to 0$. By Cauchy's integral theorem $\int_C f(z)dz = 0$

$$\therefore \qquad 0 = i\int_0^{\pi} e^{iRe^{i\theta}}\, d\theta + \int_{-\infty}^{0} \frac{e^{ix}dx}{x} + i\int_{\pi}^{0} e^{ire^{i\theta}}\, d\theta + \int_0^{\infty} \frac{e^{ix}dx}{x}$$

or
$$0 = i\int_0^{\pi} e^{+iR\cos\theta - R\sin\theta}.d\theta + \int_{-\infty}^{\infty} \frac{\cos x + i\sin x}{x} - i\int_0^{\pi} e^{ir\cos\theta} e^{-r\sin\theta} d\theta$$

Since $R \to \infty$ and $r \to 0$, we have

$$0 = 0 + \int_{-\infty}^{\infty} \frac{\cos x + i\sin x}{x} dx - i\int_0^{\pi} d\theta$$

Since $\frac{\cos x}{x}$ is an odd function of x $\int_{-\infty}^{\infty} \frac{\cos x}{x} dx = 0$, hence we have

$i\int_{-\infty}^{\infty} \frac{\sin x}{x} dx = i\pi$. Here $\frac{\sin x}{x}$ is an even function hence $2\int_0^{\infty} \frac{\sin x}{x} dx = \pi$

or $\int_0^{\infty} \frac{\sin x}{x} dx = \frac{\pi}{2}$ **Ans.**

ADDITIONAL IMPORTANT PROBLEMS

1. *Evaluate the following integrals by applying Cauchy's integral theorem, in case applicable*

(a) $\oint_C \cos z\, dz$ *(b)* $\oint_C \sec z\, dz$ *(c)* $\oint_C \frac{dz}{z^2 - 5z + 6} dz$ *(d)* $\oint_C \bar{z}\, dz$

where C *is the unit circle* $|z| = 1$.

SOLUTION: (*a*) The integrand $f(z) = \cos z$ is analytic for all z and also $f'(z) = \sin z$ is continuous everywhere, and hence on and inside C also. Thus by Cauchy's theorem

$$\oint_C \cos z\, dz = 0$$

(*b*) The integrand $f(z) = \sec z = \frac{1}{\cos z}$ is not analytic at the points $z = \pm\frac{\pi}{2}, \pm\frac{3\pi}{2}, ...$ but all these points lie outside the unit circle $|z| = 1$. Hence $f(z)$ is analytic and $f'(z)$ is continuous in and on C, and thus

$$\oint_C \sec z\, dz = 0.$$

(*c*) The integrand $f(z) = \frac{1}{z^2 - 5z + 6} = \frac{1}{(z-2)(z-3)}$ is analytic everywhere except at $z = 2, 3$, the points which lie outside the unit circle $|z| = 1$, and hence by Cauchy's theorem

$$\oint_C \frac{1}{z^2 - 5z + 6} dz = 0.$$

(*d*) The integrand $f(z) = \bar{z}$ is not analytic and hence the Cauchy's theorem is not applicable. In fact, about $C : |z| = 1$, we have

$$\oint_C \bar{z}\, dz = \int_0^{2\pi} e^{-i\theta}\, ie^{i\theta}\, d\theta = i\int_0^{2\pi} d\theta = 2\pi i. \qquad \textbf{Ans.}$$

2. *Evaluate* $I = \oint_C \dfrac{dz}{z^2(z-2)(z-4)}$, *where* C *is the rectangle joining the points (–1, –1), (3, –1), (3, 1) and (–1, 1) in the complex plane.*

SOLUTION: The curve C is the rectangle $ABCD$ as shown in figure. Resolving the integrand into partial fractions, we obtain

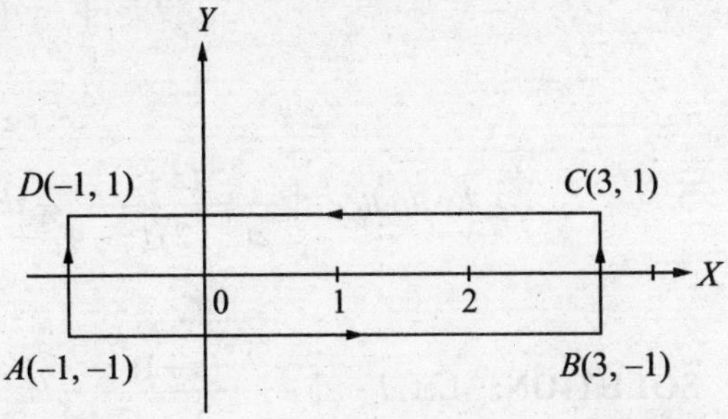

$$I = \frac{3}{32}\oint_C \frac{dz}{z} + \frac{1}{8}\oint_C \frac{dz}{z^2} - \frac{1}{8}\oint_C \frac{dz}{z-2} + \frac{1}{32}\oint_C \frac{dz}{z-4}$$

$$= \frac{3}{32}(2\pi i) + \frac{1}{8}(0) - \frac{1}{8}(2\pi i) + \frac{1}{32}(0) = -\frac{\pi i}{16}$$

Here the first three integrals above on the right side are evaluated by using the fact that $\int (z-a)^n\, dz$ $= \begin{cases} 2\pi i, & n = -1 \\ 0, & n \neq -1 \end{cases}$ and the last integral is zero by Cauchy's integral theorem. $\therefore \quad I = \dfrac{-\pi i}{16}$ **Ans.**

3. *Evaluate the integral* $\oint_C \dfrac{z^2+1}{z(2z-1)}\, dz$, *C*: $|z| = 1$.

SOLUTION: Let $I = \oint_C \dfrac{z^2+1}{z(2z-1)}\, dz$.

The integrand $(z^2 + 1)/z(2z - 1)$ is not analytic at the point $z = 0$ and $z = 1/2$ both of which lie inside C. Let us write it as

$$\frac{z^2+1}{z(2z-1)} = (z^2+1)\left[\frac{1}{(z-1/2)} - \frac{1}{z}\right]$$

Therefore, $I = \oint_C \dfrac{z^2+1}{z-1/2}\, dz - \oint_C \dfrac{z^2+1}{z}\, dz = 2\pi i\left[z^2+1\right]_{z=1/2} - 2\pi i\left[z^2+1\right]_{z=0}$

$$= \frac{5\pi i}{2} - 2\pi i = \frac{\pi i}{2}$$, using the Cauchy's integral formula. **Ans.**

4. *Evaluate the integral* $\oint_C \frac{e^z}{z^3}\,dz$, *C*: $|z| = 1$, *taken in counter-clockwise sense.*

SOLUTION: Let $I = \oint_C \frac{e^z}{z^3}\,dz$. Here $f(z) = e^z$ is analytic in the region bounded by the simple closed curve $|z| = 1$. The singular point $z = 0$ lies inside C: $|z| = 1$. Hence, applying the generalized Cauchy's integral formula

$$I = \oint_C \frac{e^z}{z^3}\,dz = \frac{2\pi i}{2!}\frac{d^2}{dz^2}(e^z)\Big|_{z=0} = \pi i.$$ **Ans.**

5. *Evaluate* $\oint_C \frac{(z+1)}{z(z-2)(z-4)^3}\,dz$, *C*: $|z-3| = 2$ *in the counter-clockwise sense.*

SOLUTION: Let $I = \oint_C \frac{(z+1)}{z(z-2)(z-4)^3}\,dz.$

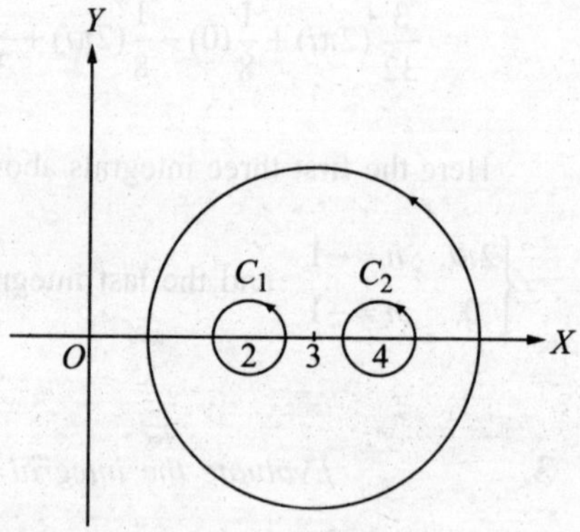

The integrand has singularities at $z = 0$, $z = 2$, and $z = 4$, out of these $z = 2$ and $z = 4$ lie inside C.

Consider two non-intersecting closed contour C_1 and C_2, as shown in Fig., lying completely within C, respectively about the point $z = 2$ and $z = 4$. Applying the principle of deformation the integral I becomes

$$I = \oint_C \frac{z+1}{z(z-2)(z-4)^3}\,dz$$

$$= \oint_{C_1}\left[\frac{z+1}{z(z-4)^3}\right]\frac{dz}{z-2} + \oint_{C_2}\left[\frac{z+1}{z(z-2)}\right]\frac{dz}{(z-4)^3} = I_1 + I_2, \text{ say.}$$

Now, $I_1 = \oint_{C_1}\left[\frac{z+1}{z(z-4)^3}\right]\frac{dz}{z-2} = 2\pi i\left[\frac{z+1}{z(z-4)^3}\right]_{z=2} = -\frac{3\pi i}{8}$ (using Cauchy's integral formula).

Similarly, $I_2 = \oint_{C_2}\left[\frac{z+1}{z(z-2)}\right]\frac{dz}{(z-4)^3} = \frac{2\pi i}{2!}\frac{d^2}{dz^2}\left[\frac{z+1}{z(z-2)}\right]_{z=4} = \frac{23\pi i}{64}$

Therefore, $I = \frac{-3\pi i}{8} + \frac{23\pi i}{64} = \frac{\pi i}{64}.$ **Ans.**

6. *Find the Taylor series expansion of*

$$f(z) = \frac{2z^2+9z+5}{z^3+z^2-8z-12} \quad \textit{about point } z = 1.$$

SOLUTION: We have

$$f(z) = \frac{2z^2+9z+5}{z^3+z^2-8z-12} = \frac{2z^2+9z+5}{(z+2)^2(z-3)} = \frac{1}{(z+2)^2} + \frac{2}{(z-3)}$$

Consider $\dfrac{1}{(z+2)^2} = \dfrac{1}{[3+(z-1)]^2} = \dfrac{1}{9}\left[1+\dfrac{z-1}{3}\right]^{-2}$

$$= \frac{1}{9}\left[1+(-2)\left(\frac{z-1}{3}\right)+\frac{(-2)(-3)}{2!}\left(\frac{z-1}{3}\right)^2+\frac{(-2)(-3)(-4)}{3!}\left(\frac{z-1}{3}\right)^3+\ldots\right], |z-1|<3,$$

$$= \frac{1}{9}\sum_{n=0}^{\infty}\binom{-2}{n}\left(\frac{z-1}{3}\right)^n, \quad |z-1|<3.$$

Also, $\dfrac{1}{z-3} = \dfrac{1}{-2+(z-1)} = -\dfrac{1}{2}\left[1-\dfrac{z-1}{2}\right]^{-1} = -\dfrac{1}{2}\sum_{n=0}^{\infty}\left[\dfrac{(z-1)}{2}\right]^n$, $|z-1|<2$

Hence $f(z) = \left[\dfrac{1}{9}\displaystyle\sum_{n=0}^{\infty}\binom{-2}{n}\left(\dfrac{z-1}{3}\right)^n - \sum_{n=0}^{\infty}\left(\dfrac{z-1}{2}\right)^n\right] = \displaystyle\sum_{n=0}^{\infty}\left[\binom{-2}{n}\dfrac{1}{3^{n+2}}-\dfrac{1}{2^n}\right](z-1)^n,$

with radius of convergence $R = \min\{2, 3\} = 2$.

7. *Obtain the Taylor series expansion of the function $f(z) = \ln z$ about the point $z = (-1+i)$. Obtain its radius of convergence.*

SOLUTION: We have, $f(z) = \ln z = \ln[(-1+i)+\{z-(-1+i)\}] = \ln(-1+i) + \ln\left[1+\dfrac{z-(-1+i)}{-1+i}\right]$

Using $\qquad \ln(1+z) = z - \dfrac{z^2}{2}+\dfrac{z^3}{3}-\dfrac{z^4}{z}+\ldots$

We obtain, $\qquad f(z) = \ln(-1+i) + \dfrac{z-(-1+i)}{-1+i} - \dfrac{1}{2}\left(\dfrac{z-(-1+i)}{-1+i}\right)^2 + \dfrac{1}{3}\left(\dfrac{z-(-1+i)}{-1+i}\right)^3 + \ldots$

as the Taylor series expansion about $(-1+i)$.

The series is convergent for $\left|\dfrac{z-(-1+i)}{-1+i}\right| < 1$, or $|z-(-1+i)| < \sqrt{2}$.

Hence the radius of convergence is $\sqrt{2}$.

8. *Find the Taylor series and Laurent series expansion of* $f(z) = \dfrac{1}{6-z-z^2}$ *about* $z = 0$.

SOLUTION: The function $f(z) = \dfrac{1}{6-z-z^2} = \dfrac{1}{(2-z)(z+3)}$ has singularities at the points $z = 2, -3$. So we find the expansions of $f(z)$ with center 0 in the regions

(*a*) $|z| < 2$, (*b*) $2 < |z| < 3$, and (*c*) $|z| > 3$.

We have, $$f(z) = \frac{1}{6-z-z^2} = \frac{1}{(2-z)(z+3)} = \frac{1}{5}\left[\frac{1}{2-z} + \frac{1}{z+3}\right]$$

(*a*) For $|z| < 2$, since $|z/2| < 1$ and $|z/3| < 1$, let us write

$$f(z) = \frac{1}{5}\left[\frac{1}{2}\left(1-\frac{z}{2}\right)^{-1} + \frac{1}{3}\left(1+\frac{z}{3}\right)^{-1}\right] = \frac{1}{5}\left[\frac{1}{2}\sum_{n=0}^{\infty}\left(\frac{z}{2}\right)^n + \frac{1}{3}\sum_{n=0}^{\infty}(-1)^n\left(\frac{z}{3}\right)^n\right]$$

$$= \sum_{n=0}^{\infty}\frac{1}{5}\left[\frac{1}{2^{n+1}} + \frac{(-1)^n}{3^{n+1}}\right]z^n.$$

This expansion contains no principal part and since $f(z)$ is analytic in $|z| < 2$, the expansion is just the Maclaurin series expansion of $f(z)$ in $|z| < 2$.

(*b*) For $2 < |z| < 3$, since $|2/z| < 1$ and $|z/3| < 1$, write

$$f(z) = \frac{1}{5}\left[-\frac{1}{z}\left(1-\frac{2}{z}\right)^{-1} + \frac{1}{3}\left(1+\frac{z}{3}\right)^{-1}\right] = \frac{1}{5}\left[-\frac{1}{z}\sum_{n=0}^{\infty}\left(\frac{2}{z}\right)^n + \frac{1}{3}\sum_{n=0}^{\infty}(-1)^n\left(\frac{z}{3}\right)^n\right]$$

$$= \frac{1}{5}\left[-\sum_{n=1}^{\infty}\frac{2^{n-1}}{z^n} + \sum_{n=1}^{\infty}(-1)^n\frac{z^n}{3^{n+1}}\right] = \sum_{n=1}^{\infty}\left(-\frac{2^{n-1}}{5}\right)\frac{1}{z^n} + \sum_{n=0}^{\infty}\left(\frac{(-1)^n}{5.3^{n+1}}\right)z^n$$

The first summation represents the principal part and the second summation the regular part of the Laurent series expansion of $f(z)$ in the domain $2 < |z| < 3$.

(*c*) For $|z| > 3$, since $|2/z| < 1$ and $|3/z| < 1$, let us write

$$f(z) = \frac{1}{5}\left[-\frac{1}{z}\left(1-\frac{2}{z}\right)^{-1} + \frac{1}{z}\left(1+\frac{3}{z}\right)^{-1}\right] = \frac{1}{5}\left[-\frac{1}{z}\sum_{n=0}^{\infty}\left(\frac{2}{z}\right)^n + \frac{1}{z}\sum_{n=0}^{\infty}(-1)^n\left(\frac{3}{z}\right)^n\right]$$

$$= \left[\sum_{n=1}^{\infty}\left(-\frac{2^{n-1}}{5}\right) + \left(\frac{(-1)^{n-1}\,3^{n-1}}{5}\right)\right]\frac{1}{z^n}$$

This expansion contains only principal part.

9. *Expand* $f(z) = \exp\left(z + \dfrac{1}{z}\right)$ *as a Laurent series about the origin.*

SOLUTION: The function $f(z)$ is analytic everywhere except at the origin which is a singular point and hence $f(z)$ can be expanded about the origin in Laurent series of the form

$$\exp\left(z+\frac{1}{z}\right) = \sum_{n=0}^{\infty}a_n z^n + \sum_{n=1}^{\infty}a_{-n}\frac{1}{z^n}, \text{ for } |z| > 0. \tag{1}$$

To determine the coefficients a_n and a_{-n}, we write

$$f(z) = (\exp z)\left(\exp\frac{1}{z}\right) = \left(1+z+\frac{z^2}{2!}+\frac{z^3}{3!}+\frac{z^4}{4!}+...\right)\left(1+\frac{1}{z}+\frac{1}{2!z^2}+\frac{1}{3!z^3}+\frac{1}{4!z^4}+...\right) \quad (2)$$

The constant term in (2) is

$$a_0 = 1 + 1 + \frac{1}{(2!)^2}+\frac{1}{(3!)^2}+\frac{1}{(4!)^2}+... = \sum_{k=0}^{\infty}\frac{1}{(k!)^2}$$

Also from (2) we observe that the coefficients a_n and a_{-n} are equal so it is sufficient to determine a_n only. Now a_n the coefficient of z^n in (2) is

$$a_n = \frac{1}{n!}+\frac{1}{1!(n+1)!}+\frac{1}{2!(n+2)!}+\frac{1}{3!(n+3)!}+... = \sum_{k=0}^{\infty}\frac{1}{k!(n+k)!}$$

Hence, Laurent expansion of $f(z)$ is (1), where

$$a_n = a_{-n} = \sum_{k=0}^{\infty}\frac{1}{k!(n+k)!}$$

and the region of convergence is $|z| > 0$. **Ans.**

10. *Expand* $f(z) = \dfrac{7z-2}{z(z+1)(z-2)}$ *as a Laurent series in the region* $1 < |z+1| < 3$.

SOLUTION: Set $z + 1 = w$, the function $f(z)$ becomes

$$f(z) = \frac{7(w-1)-2}{w(w-1)(w-3)} = \frac{7w-9}{w(w-1)(w-3)} = \frac{3}{w}+\frac{1}{w-1}+\frac{2}{w-3}.$$

The region is $1 < |w| < 3$. Since $|1/w| < 1$ and $|w/3| < 1$, we write

$$f(z) = -\frac{3}{w}+\frac{1}{w}\left(1-\frac{1}{w}\right)^{-1}-\frac{2}{3}\left(1-\frac{w}{3}\right)^{-1} = -\frac{3}{w}+\frac{1}{w}\sum_{n=0}^{\infty}\left(\frac{1}{w}\right)^n-\frac{2}{3}\sum_{n=0}^{\infty}\left(\frac{w}{3}\right)^n$$

$$= -\frac{2}{w}+\sum_{n=2}^{\infty}\left(\frac{1}{w}\right)^n-\sum_{n=0}^{\infty}\left(\frac{2}{3^{n+1}}\right)w^n$$

or,

$$f(z) = -\frac{2}{1+z}+\sum_{n=2}^{\infty}\frac{1}{(1+z)^n}-\sum_{n=0}^{\infty}\frac{2}{3^{n+1}}(1+z)^n$$

which is valid in the region $1 < |z+1| < 3$. **Ans.**

11. *Evaluate by residue method, the integral* $\oint_C z^{-4}\sin z\,dz$; C: $|z| = 1$ *taken in counter-clockwise sense.*

SOLUTION: Let $f(z) = \dfrac{\sin z}{z^4} = \dfrac{1}{z^4}\left(z-\dfrac{z^3}{3!}+\dfrac{z^5}{5!}-\dfrac{z^7}{7!}+...\right) = \dfrac{1}{z^3}-\dfrac{1}{3!z}+\dfrac{z}{5!}-\dfrac{z^3}{7!}+...$

This series is convergent for all $|z| > 0$ and $f(z)$ has a pole of third order at $z = 0$ and the residue of $f(z)$ at $z = 0$ is $b_1 = -1/3!$. Thus by

Therefore $\displaystyle\oint_C \frac{\sin z}{z^4}\,dz = 2\pi i b = -\frac{\pi i}{3}$. **Ans.**

12. *Determine the poles of function* $f(z) = \dfrac{e^z}{(z+2i)^3(z+i)}$ *and the residue at each pole. Hence calculate* $\oint_C f(z)\,dz$; *C:* $|z| = 2.5$, *taken in counter-clockwise sense.*

SOLUTION: Since $\lim_{z\to -i}[(z+i)f(z)] = \lim_{z\to -i}\dfrac{e^z}{(z+2i)^3} = ie^{-i}$ is finite and non-zero, $f(z)$ has *a* simple pole at $z = -i$ and $\operatorname{Res}_{z\to -i} f(z) = ie^{-i}$.

Since $\lim_{z\to -2i}[(z+2i)^3 f(z)] = \lim_{z\to -2i}\dfrac{e^z}{z+i} = \dfrac{e^{-2i}}{-i} = ie^{-2i}$ is finite and non-zero, thus $f(z)$ has a pole of order 3 at $z = -2i$, and

$$\operatorname*{Res}_{z=-2i} f(z) = \frac{1}{2!}\lim_{z\to -2i}\frac{d^2}{dz^2}\left(\frac{e^z}{z+i}\right)$$

$$= \frac{1}{2}\lim_{z\to -2i}[e^z\{(z+i)^{-1} - 2(z+i)^{-2} + 2(z+i)^{-3}\}] = \frac{1}{2}e^{-2i}(2-i).$$

To evaluate the integral $\oint_C f(z)\,dz$; *C:* $|z| = 2.5$, we observe that $f(z)$ is analytic on and inside $|z| = 2.5$ at all points except at the poles $z = -i,\ 2i$ and hence by residue theorem

$$\oint_C f(z)\,dz = 2\pi i\left[\operatorname*{Res}_{z=-i} f(z) + \operatorname*{Res}_{z=-i} f(z)\right] = 2\pi i\left[ie^{-i} + \frac{1}{2}e^{-2i}(2-i)\right].$$

13. *Evaluate* $\displaystyle\int_{-\infty}^{\infty}\frac{1}{x^6+64}\,dx.$

SOLUTION: Consider the contour integral $\oint_C \dfrac{1}{z^6+64}\,dz = \int_C f(z)\,dz$, where C is the contour consisting of the semicircle C_R of radius R and the segment of the real axis from $-R$ to R as in Fig. The integrand $f(z)$ has six simple poles at the points given by

$$z = (-64)^{1/6} = 2e^{i(\pi + 2n\pi)/6}, \quad n = 0, 1, 2, 3, 4, 5.$$

These are: $z_1 = 2e^{\pi i/6},\ z_2 = 2e^{3\pi i/6},\ z_3 = 2e^{5\pi i/6},\ z_4 = 2e^{7\pi i/6},\ z_5 = 2e^{9\pi i/6},\ z_6 = 2e^{11\pi i/6}.$

Out of these z_1, z_2 and z_3 lies only in the upper half-plane as shown in the adjoining figure.

We have $\displaystyle\operatorname*{Res}_{z=z_1} f(z) = \operatorname*{Res}_{z=2e^{\pi i/6}}\left(\frac{1}{z^6+64}\right) = \left[\frac{1}{6z^5}\right]_{z=2e^{\pi i/6}} = \frac{1}{192}e^{-5\pi i/6},$

Similarly, $\displaystyle\operatorname*{Res}_{z=z_2} f(z) = \operatorname*{Res}_{z=2e^{3\pi i/6}}\left(\frac{1}{z^6+64}\right) = \frac{1}{192}e^{-15\pi i/6} = \frac{1}{192}e^{-3\pi i/6}$

and $\displaystyle\operatorname*{Res}_{z=z_3} f(z) = \operatorname*{Res}_{z=2e^{5\pi i/6}}\left(\frac{1}{z^6+64}\right) = \frac{1}{192}e^{-25\pi i/6} = \frac{1}{192}e^{-\pi i/6}.$

Thus, by residue theorem

$$\int_C \frac{1}{z^6+64}\,dz = \frac{2\pi i}{192}\left[\cos\frac{5\pi}{6} - i\sin\frac{5\pi}{6} + \cos\frac{3\pi}{6} - i\sin\frac{3\pi}{6} + \cos\frac{\pi}{6} - i\sin\frac{\pi}{6}\right]$$

$$= \frac{\pi i}{96}\left[\left(\cos\frac{5\pi}{6} + \cos\frac{\pi}{6}\right) - i\left(\sin\frac{5\pi}{6} + \sin\frac{\pi}{6}\right) - i\right]$$

Using $\cos\frac{5\pi}{6} + \cos\frac{\pi}{6} = 0$ and $\sin\frac{5\pi}{6} + \sin\frac{\pi}{6} = 1$, we get

$$\oint_C \frac{1}{z^6+64}\,dz = \frac{\pi}{48}$$

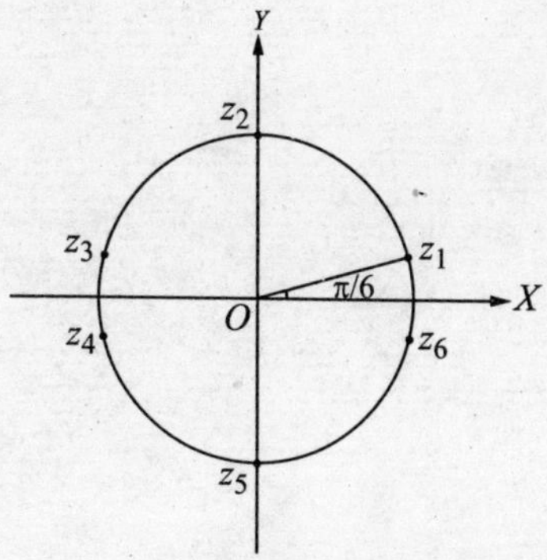

Also,

$$\oint_C \frac{1}{z^6+64}\,dz = \int_{-R}^{R} \frac{1}{x^6+64}\,dx + \int_{C_R} \frac{1}{z^6+64}\,dz$$

Now as $R \to \infty$, for any point on C_R, $|z| \to \infty$ and consequently the integrand in the second integral on the right side tends to zero and hence it vanishes and thus, we have

$$\int_{-\infty}^{\infty} \frac{1}{x^6+64}\,dx = \frac{\pi}{48}.$$

CHAPTER

18 Laplace Transformation and its Applications

IMPORTANT FORMULAE AND USEFUL RESULTS

1. $L[f(t); t \to s] = L(f(t)) = \overline{f}(s) = \int_0^{\infty} \overline{e}^{st} f(t)dt \; (s > 0)$ if it exists. here s is a dummy parameter.

(*i*) $L(e^{at}), a > 0 = \frac{1}{s-a} \quad \therefore L(1) = \frac{1}{s}.$

(*ii*) $L(\cos at) = \frac{s}{s^2+a^2}, \quad L(\sin at) = \frac{a}{s^2+a^2}.$

(*iii*) $L(\cosh at) = \frac{s}{s^2-a^2}, \quad L(\cosh at) = \frac{a}{s^2-a^2}.$

(*iv*) $L(t^n) = \frac{\overline{|n+1}}{s^{n+1}}$ and $L(e^{at} \cdot t^n) = \frac{\overline{|n+1}}{(s-a)^{n+1}}.$

(*v*) $L(e^{at} \cos bt) = \frac{s-a}{(s-a)^2+b^2}$ and $L(e^{at} \sin bt) = \frac{b}{(s-a)^2+b^2}$

2. Properties of Laplace Transform: If $L(f(t)) = \overline{f}(s)$ then

(*i*) Change of scale $L(f(at)) = \frac{1}{a}\overline{f}\left(\frac{s}{a}\right)$

(*ii*) $L(tf(t)) = -\frac{d}{ds}\overline{f}(s)$ and $L(t^n f(t)) = (-1)^n \frac{d^n}{ds^n}\overline{f}(s).$

(*iii*) $L\left(\frac{1}{t}f(t)\right) = \int_s^{\infty} \overline{f}(s)ds$ and $L\left(\frac{1}{t^2}f(t)\right) = \int_s^{\infty}\int_s^{\infty} \overline{f}(s)ds\,ds.$

(*iv*) $L(f'(t)) = s\overline{f}(s) - f(o)$ and $L(f^{(n)}(t)) = s^n\overline{f}(s) - s^{n-1}f(o) - s^{n-2}f'(o)... - f^{(n-1)}(o).$

(*v*) $L\left(\int_o^t f(u)du\right) = \frac{1}{s}\overline{f}(s)$

3. Inverse Laplace Transforms:

If $L(f(t) = \bar{f}(s)$ then $L^{-1}(\bar{f}(s)) = f(t)$.

(*i*) $L^{-1}\left(\frac{1}{s}\right) = 1$ (*ii*) $L^{-1}\left(\frac{1}{s^{n+1}}\right) = \frac{t^n}{n!}$, $n = 0, 1, 2, \ldots$

(*iii*) $L^{-1}\left(\frac{1}{s-a}\right) = e^{at}$ (*iv*) $L^{-1}\left(\frac{1}{(s-a)^{n+1}}\right) = \frac{e^{at}t^n}{n!}$, $n = 0, 1, 2, \ldots$

(*v*) $L^{-1}\left(\frac{1}{s^2+a^2}\right) = \frac{1}{a}\sin at$, $L^{-1}\left(\frac{s}{s^2+a^2}\right) = \cos at$

(*vi*) $L^{-1}\left(\frac{1}{s^2-a^2}\right) = \frac{1}{2}\sinh at$, $L^{-1}\left(\frac{s}{s^2-a^2}\right) = \cosh at$

(*vii*) $L^{-1}\left[\frac{1}{(s-a)^2+b^2}\right] = \frac{e^{at}}{b}\sin bt$, $L^{-1}\left\{\frac{s-a}{(s-a)^2+b^2}\right\} = e^{at}\cos bt$

(*viii*) $L^{-1}\left\{\frac{s}{(s^2+a^2)^2}\right\} = \frac{1}{2a}t\sin at$ (*ix*) $L^{-1}\left\{\frac{1}{(s^2+a^2)^2}\right\} = \frac{1}{2a}(\sin at - at\cos at)$.

4. Convolution Theorem.

Convolution of functions f_1 and $f_2 = f_1 \oplus f_2 = \int_o^t f_1(u) f_2(t-u)du = f_2 \oplus f_1$

If $L(f_1) = \bar{f}_1(s)$, $L(f_2) = \bar{f}_2(s)$ then

$$L^{-1}\{\bar{f}_1(s)\bar{f}_2(s)\} = f_1 \oplus f_2 = \int_o^t f_1(u) f_2(t-u)du$$

5. Laplace Transform of Periodic Functions

If $f(t)$ is T-periodic function, *i.e.*, $f(t + T) = f(t)$ then $L(f(t)) = \frac{1}{1-e^{-sT}}\int_o^T e^{-st} f(t)dt$.

6. Unit Step Function and Laplace Transform

$$U(t-a) = \begin{cases} 1 & \text{for } t > a \\ 0 & \text{for } t \le a \end{cases} \qquad L\{U(t-a)\} = \frac{e^{-as}}{s}.$$

$$L\{f(t-a)U(t-a)\} = e^{-as}\bar{f}(s).$$

7. Unit Impulse Function (or Direc-delta Function)

$$\delta(t-a) = \begin{cases} \frac{1}{\in}, & a \le t \le a + \in \\ 0, & \text{otherwise} \end{cases} \quad \text{where } \in \to 0. \text{ Here } \int_0^\infty \delta(t-a)dt = 1$$

$$L\{\delta(t-a)\} = e^{-as} \text{ and hence } L(\delta(t)) = 1.$$

Also $L\{f(t)\,\delta(t-a)\} = e^{-as}f(a)$.

SOLUTIONS TO EXERCISE 18A

1. *Find the Laplace transform of the function* (*i*) $3e^{4t} + 6t^2 - 4\sin 3t + \cos 2t$. (*ii*) te^{-t} (*iii*) $t^2 e^{-2t}$.

SOLUTION: (*i*) $L[3e^{4t} + 6t^2 - 4\sin 3t + \cos 2t] = \dfrac{3}{s-4} + \dfrac{6\cdot\Gamma 3}{s^3} - \dfrac{4\cdot 3}{s^2+9} + \dfrac{s}{s^2+4}$

$$= \frac{3}{s-4} + \frac{12}{s^3} - \frac{12}{s^2+9} + \frac{s}{s^2+4}$$ **Ans.**

(*ii*) $L(te^{-t}) = \dfrac{\Gamma 2}{(s+1)^2} = \dfrac{1}{(s+1)^2}$ (*iii*) $L(t^2 e^{-2t}) = (-1)^2 \dfrac{d^2}{ds^2}\dfrac{1}{(s-2)} = \dfrac{2}{(s-2)^3}$. **Ans.**

2. *Find the Laplace transform of* (*i*) $t\sin t$ (*ii*) $t^2\cos t$ (*iii*) $\cos(2t+3)$.

SOLUTION: (*i*) $L(t\sin t) = -\dfrac{d}{ds}\left(\dfrac{1}{s^2+1}\right) = \dfrac{2s}{(s^2+1)^2}$ **Ans.**

(*ii*) $L(t^2\cos t) = (-1)^2\dfrac{d^2}{ds^2}\left(\dfrac{s}{s^2+1}\right) = \dfrac{d}{ds}\dfrac{(s^2+1)1 - 2s^2}{(s^2+1)^2} = \dfrac{d}{ds}\dfrac{1-s^2}{(s^2+1)^2}$

$$= \frac{(s^2+1)^2(-2s) - (1-s^2)2(s^2+1)2s}{(s^2+1)^4} = \frac{-2s(s^2+1) - 4s(1-s^2)}{(s^2+1)^3}$$

$$= \frac{2s^3 - 6s}{(s^2+1)^3}$$ **Ans.**

(*iii*) $L[\cos(2t+3)] = L[\cos 2t\cos 3 - \sin 2t\cdot\sin 3]$

$$= \cos 3\cdot\frac{s}{s^2+2^2} - \sin 3\cdot\frac{2}{s^2+2^2} = \frac{s\cos 3 - 2\sin 3}{s^2+4}$$ **Ans.**

3. *Find the Laplace transform of* (*i*) $t^3 e^{4t}$ (*ii*) $e^{-2t}\sqrt{t}$.

SOLUTION: (*i*) $L(t^3 e^{4t}) = \left[\dfrac{\Gamma 4}{s^4}\right]_{s\to s-4} = \dfrac{\Gamma 4}{(s-4)^4} = \dfrac{6}{(s-4)^4}$

(*ii*) $L(e^{-2t}\sqrt{t}) = \left[\dfrac{\Gamma 3/2}{s^{3/2}}\right]_{s\to s+2} = \dfrac{\frac{1}{2}\Gamma\frac{1}{2}}{(s+2)^{3/2}} = \dfrac{\sqrt{\pi}}{2(s+2)^{3/2}}$.

4. *Find the Laplace transform of* (*i*) $\int_0^t x^2 e^x\,dx$ (*ii*) $\int_0^t \cos^2 u\,du$.

SOLUTION: (*i*) $L\left[\int_0^t x^2 e^x\,dx\right] = \dfrac{1}{s}L(x^2 e^x) = \dfrac{1}{s}\dfrac{\Gamma 3}{(s-1)^3} = \dfrac{2}{s(s-1)^3}$

(*ii*) $$L\left[\int_0^t \cos^2 u\,du\right] = \frac{1}{s}L(\cos^2 u) = \frac{1}{s}L\left[\frac{1+\cos 2u}{2}\right]$$

$$= \frac{1}{s}\left[\frac{1}{2s} + \frac{1}{2}\frac{s}{s^2+2^2}\right] = \frac{1}{2s^2} + \frac{1}{2(s^2+4)} = \frac{s^2+2}{s^2(s^2+4)}$$ **Ans.**

5. *Find the Laplace transform of the following functions*

(*i*) $f(t) = \begin{cases} 0, & 0 < t < 2 \\ 1, & t > 2 \end{cases}$ (*ii*) $f(t) = \begin{cases} 0, & t < 1 \\ t^2 - 2t + 2, & t \geq 1 \end{cases}$

SOLUTION: (*i*) $$L(f(t)) = \int_0^\infty e^{-st} f(t)dt = \int_0^2 oe^{-st}dt + \int_2^\infty 1\cdot e^{-st}dt = 0 + \left[\frac{e^{-st}}{-s}\right]_2^\infty = \frac{e^{-2s}}{s}$$

(*ii*) $$L(f(t)) = \int_0^\infty e^{-st} f(t)dt = \int_0^1 o^{-st}dt + \int_1^\infty (t^2 - 2t + 2)e^{-st}dt$$

$$= 0 + \left[(t^2 - 2t + 2)\frac{e^{-st}}{-s}\right]_1^\infty - \int_1^\infty (2t-2)\frac{e^{-st}}{-s}dt$$

$$= \frac{1}{s}(1 - 2 + 2)e^{-s} + 2\left[(t-1)\frac{e^{-st}}{-s^2}\right]_1^\infty - 2\int_1^\infty 1\frac{e^{-st}}{-s^2}dt$$

$$= \frac{1}{s}e^{-s} - 2\left[\frac{e^{-st}}{s^3}\right]_1^\infty = e^{-s}\left(\frac{1}{s} + \frac{2}{s^3}\right)$$ **Ans.**

6. *Find* (*i*) $L\left(\frac{\cos 2t - \cos 3t}{t}\right)$ (*ii*) $L\left(\frac{1-\cos t}{t^2}\right)$. (*iii*) L (sinh *at* sin *at*)

SOLUTION: (*i*) $$L\left(\frac{\cos 2t - \cos 3t}{t}\right) = \int_s^\infty L(\cos 2t - \cos 3t)dt$$

$$= \int_s^\infty \left(\frac{s}{s^2+2^2} - \frac{s}{s^2+3^2}\right)ds = \left[\frac{1}{2}\log(s^2+2^2) - \frac{1}{2}\log(s^2+3^2)\right]_s^\infty$$

$$= \frac{1}{2}\left[\log\frac{s^2+2^2}{s^2+3^2}\right] = \frac{1}{2}\left[\log 1 - \log\frac{s^2+2^2}{s^2+3^2}\right] = \frac{1}{2}\log\frac{s^2+9}{s^2+4}$$ **Ans.**

(*ii*) $$L\left(\frac{1-\cos t}{t^2}\right) = \int_s^\infty\int_s^\infty L(1-\cos t)ds\,ds$$

$$= \int_s^\infty\left[\int_s^\infty\left(\frac{1}{s} - \frac{s}{s^2+1}\right)ds\right]ds = \int_s^\infty\left[\log s - \frac{1}{2}\log(s^2+1)\right]_s^\infty ds$$

$$= \int_s^\infty \left[\log \frac{s}{\sqrt{s^2+1}}\right]_s^\infty ds = \int_s^\infty \left(\log 1 - \log \frac{s}{\sqrt{s^2+1}}\right) ds = \int_s^\infty \log\left(\frac{\sqrt{s^2+1}}{s}\right) ds$$

$$= \left[\log\left(\frac{\sqrt{s^2+1}}{s}\right)\cdot s\right]_s^\infty - \int_s^\infty \frac{s}{\sqrt{s^2+1}} \frac{d}{ds}\left(\frac{\sqrt{s^2+1}}{s}\right)\cdot s\, ds$$

$$= -s\log\left(\frac{\sqrt{s^2+1}}{s}\right) - \int_s^\infty \frac{s^2}{\sqrt{s^2+1}} \frac{s\cdot\dfrac{s}{\sqrt{s^2+1}} - \sqrt{s^2+1}}{s^2} ds.$$

$$= -s\log\frac{\sqrt{s^2+1}}{s} + \int_s^\infty \frac{ds}{s^2+1} = -s\log\sqrt{1+\frac{1}{s^2}} + [\tan^{-1} s]_s^\infty$$

$$= \frac{-s}{2}\log\left(1+\frac{1}{s^2}\right) + \frac{\pi}{2} - \tan^{-1} s = \cot^{-1} s - \frac{s}{2}\log\left(1+\frac{1}{s^2}\right)$$ **Ans.**

(iii) $\quad \mathrm{L}[\sin h\, at \sin at] = \mathrm{L}\left[\frac{e^{at}-e^{-at}}{2}\sin at\right] = \frac{1}{2}\mathrm{L}(e^{at}\sin at) - \frac{1}{2}L(e^{-at}\sin at)$

$$= \frac{1}{2}\frac{a}{(s-a)^2+a^2} - \frac{1}{2}\frac{a}{(s+a)^2+a^2} = \frac{a}{2}\frac{(s+a)^2+a^2-(s-a)^2-a^2}{[(s-a)^2+a^2][(s+a)^2+a^2]}$$

$$= \frac{a}{2}\frac{4as}{(s^2-a^2)^2+a^4+a^2\{(s-a)^2+(s+a)^2\}} = \frac{2a^2 s}{s^4+4a^4}$$ **Ans.**

7. *Evaluate* (*i*) $\int_0^\infty e^{-2t}\sin^3 t\, dt$ $\quad$ (*ii*) $\int_0^\infty \frac{\cos 6t - \cos 4t}{t} dt$

SOLUTION: (*i*) $\quad \mathrm{L}(\sin^3 t) = \int_0^\infty e^{-st}\sin^3 t\, dt$ by definition.

We know that $\quad \mathrm{L}(\sin^3 t) = \mathrm{L}\left[\frac{3\sin t - \sin 3t}{4}\right] = \frac{3}{4}\frac{1}{s^2+1} - \frac{1}{4}\frac{3}{s^2+9}$

$$\therefore \quad \int_0^\infty e^{-2t}\sin^3 t\, dt = \frac{3}{4}\left[\frac{1}{s^2+1} - \frac{1}{s^2+9}\right]_{s=2} = \frac{6}{65}$$ **Ans.**

(*ii*) $\quad \mathrm{L}\left(\frac{\cos 6t - \cos 4t}{t}\right) = \int_0^\infty e^{-st}\left(\frac{\cos 6t - \cos 4t}{t}\right) dt$ by defintion.

We know that $\quad L\left(\frac{\cos 6t - \cos 4t}{t}\right) = \int_s^\infty [L(\cos 6t) - L(\cos 4t)] ds$

$$= \int_s^\infty \left(\frac{s}{s^2+36} - \frac{s}{s^2+16}\right) ds = \left[\frac{1}{2}\log(s^2+36) - \frac{1}{2}\log(s^2+16)\right]_s^\infty$$

$$= \left[\frac{1}{2}\log\left(\frac{s^2+36}{s^2+16}\right)\right]_s^\infty = \log\sqrt{\frac{s^2+16}{s^2+36}}$$

$$\therefore \quad \int_0^\infty \frac{\cos 6t - \cos 4t}{t} dt = \left[\log\sqrt{\frac{s^2+16}{s^2+36}}\right]_{s=0} = \log\frac{4}{6} = \log\frac{2}{3}$$ **Ans.**

8. *If* $L(\cos^2 t) = \dfrac{s^2+2}{s(s^2+4)}$ *find* $L(\cos^2 at)$.

SOLUTION: We know that if $L(f(t)) = \bar{f}(s)$ then $L(f(at)) = \dfrac{1}{a}\bar{f}\left(\dfrac{s}{a}\right)$

$$\therefore \quad L(\cos^2 at) = \frac{1}{a}\left[\frac{s^2+2}{s(s^2+4)}\right]_{s \to \frac{s}{a}} = \frac{s^2+2a^2}{s(s^2+4a^2)}$$ **Ans.**

9. *If* $L(f(t)) = \bar{f}(s)$ *and* $g(t) = \begin{cases} f(t-a), & t > a \\ 0, & 0 < t < a \end{cases}$

then prove that $L(g(t)) = e^{-as}\bar{f}(s)$.

SOLUTION: By definition we have

$$L(g(t)) = \int_0^\infty e^{-st} g(t) dt = \int_0^a o\, e^{-st} dt + \int_a^\infty f(t-a) e^{-st} dt$$

$$= 0 + \int_0^\infty f(x) e^{-s(x+a)} dt \quad \text{(On putting } t-a=x)$$

$$= e^{-st}\int_0^\infty f(x) e^{-sx} dx = e^{-as}\bar{f}(s).$$ **Hence proved.**

10. (i) *Find* $L\left(\sin\sqrt{t}\right)$.

(ii) *If* $L\left(\sin\sqrt{t}\right) = \dfrac{\sqrt{\pi}}{2p^{3/2}} e^{-1/4p}$, *show that* $L\left(\dfrac{\cos\sqrt{t}}{\sqrt{t}}\right) = \left(\dfrac{\pi}{p}\right)^{1/2} e^{-1/4p}$.

SOLUTION: (i) $$L(\sin\sqrt{t}) = L\left[\sqrt{t} - \frac{1}{3!}(\sqrt{t})^3 + \frac{1}{5!}(\sqrt{t})^5 - \frac{1}{7!}(\sqrt{t})^7 + ...\right]$$

$$= L\left[t^{1/2} - \frac{1}{3!}t^{3/2} + \frac{1}{5!}t^{5/2} - \frac{1}{7!}t^{7/2} + ...\right]$$

$$= \frac{\overline{|3/2}}{s^{3/2}} - \frac{1}{3!}\frac{\overline{|5/2}}{s^{5/2}} + \frac{1}{5!}\frac{\overline{|7/2}}{s^{7/2}} - \frac{1}{7!}\frac{\overline{|9/2}}{s^{9/2}} + \ldots$$

$$= \frac{\overline{|3/2}}{s^{3/2}}\left[1 - \frac{3/2}{3!s} + \frac{(5/2)(3/2)}{5!s^2} - \frac{(7/2)\,(5/2)\,(3/2)}{7!s^3} + \ldots\right]$$

$$= \frac{1}{2s^{3/2}}\left[1 - \frac{1}{4s} + \frac{1}{2!}\left(\frac{1}{4s}\right)^2 - \frac{1}{3!}\left(\frac{1}{4s}\right)^3 + \ldots\right]$$

$$= \frac{\sqrt{\pi}}{2s^{3/2}}e^{-\frac{1}{4s}} \quad \text{since} \quad \overline{|1/2} = \sqrt{\pi}.$$ **Ans.**

(*ii*) We know that $L(f'(t)) = s\bar{f}(s) - f(0)$. where $\bar{f}(s) = L(f(t))$.

Taking $f(t) = \sin\sqrt{t}$ we get

$$L\left(\frac{d}{dt}\sin\sqrt{t}\right) = s\left[\frac{\sqrt{\pi}}{2\cdot s^{3/2}}e^{-\frac{1}{4s}}\right] - (\sin\sqrt{t})_{t=0}$$

or $$L\left(\frac{1\cos\sqrt{t}}{2\sqrt{t}}\right) = \frac{1}{2}\frac{\sqrt{\pi}}{\sqrt{s}}e^{-\frac{1}{4s}}$$

or $$L\left(\frac{1}{\sqrt{t}}\cos\sqrt{t}\right) = \sqrt{\frac{\pi}{s}}\,e^{-\frac{1}{4s}}$$ **Hence proved.**

11. *Find the Laplace transform of* $f(x) = \begin{cases} t^2, & 0 < t < 2 \\ t-1, & 2 < t < 3 \\ 7, & t > 3. \end{cases}$

(GGSIPU II Sem End Term 2009)

SOLUTION: $$L[f(t)] = \int_0^\infty e^{-st} f(t)\,dt = \int_0^2 t^2 e^{-st}\,dt + \int_2^3 (t-1)\,e^{-st}\,dt + \int_3^\infty 7\,e^{-st}\,dt$$

$$= \left[t^2\frac{e^{-st}}{-s}\right]_0^2 - \int_0^2 2t\frac{e^{-st}}{-s}\,dt + \left[(t-1)\frac{e^{-st}}{-s}\right]_2^3 - \int_2^3 \frac{e^{-st}}{-s}\,ds + 7\left[\frac{e^{-st}}{-s}\right]_3^\infty$$

$$= -\frac{4}{s}e^{-2s} + \frac{2}{s}\left[t\frac{e^{-st}}{-s}\right]_0^2 - \frac{2}{s}\int_0^2 \frac{e^{-st}}{-s}\,dt - \frac{1}{s}\left[2e^{-3s} - e^{-2s}\right]_2^3 - \left[\frac{e^{-st}}{s^2}\right]_2^3 + \frac{7}{s}e^{-3s}$$

$$= -\frac{4}{s}e^{-2s} - \frac{4}{s^2}e^{-2s} - \frac{2}{s^3}\left[e^{-st}\right]_0^2 - \frac{2}{s}e^{-3s} + \frac{1}{s}e^{-2s} - \frac{e^{-3s}}{s^2} + \frac{e^{-2s}}{s^2} + \frac{7}{s}e^{-3s}$$

$$= e^{-2s}\left(-\frac{2}{s^3}-\frac{3}{s^2}-\frac{3}{s}\right)+e^{-3s}\left(\frac{5}{s}-\frac{1}{s^2}\right)+\frac{2}{s^3}$$

$$= \frac{2}{s^3}(1-e^{-2s})-3e^{-2s}\left(\frac{1}{s}+\frac{1}{s^2}\right)+e^{-3s}\left(\frac{5}{s}-\frac{1}{s^2}\right).$$ **Ans.**

12. *Evaluate* $L\left[\int_0^t \frac{e^t \sin t}{t}\,dt\right]$

SOLUTION: $L\left[\int_0^t \frac{e^t \sin t}{t}\,dt\right] = \frac{1}{s}L\left[\frac{e^t \sin t}{t}\right]$

$$= \frac{1}{s}\int_s^\infty L(e^t \sin t)\,ds = \frac{1}{s}\int_s^\infty \frac{ds}{(s-1)^2+1}$$

$$= \frac{1}{s}\left[\tan^{-1}(s-1)\right]_s^\infty = \frac{1}{s}\left[\frac{\pi}{2}-\tan^{-1}(s-1)\right] = \frac{1}{s}\cot^{-1}(s-1).$$ **Ans.**

SOLUTIONS TO EXERCISE 18B

1. *Find the inverse Laplace transform of*

(*i*) $\frac{3s+1}{(s-1)(s^2+1)}$ (*ii*) $\frac{4s+5}{(s-1)^2(s+2)}$

SOLUTION: (*i*) $\frac{3s+1}{(s-1)(s^2+1)} = \frac{A}{s-1}+\frac{Bs+C}{s^2+1} \Rightarrow 3s+1 = A(s^2+1)+(s-1)(Bs+C)$

Hence $A + B = 0$, $A - C = 1$, $-B + C = 3$, $\therefore A = 2$, $B = -2$, $C = 1$.

$$\therefore\ L^{-1}\left[\frac{3s+1}{(s-1)(s^2+1)}\right] = L^{-1}\left(\frac{2}{(s-1)}\right)+L^{-1}\left(\frac{-2s}{s^2+1}\right)+L^{-1}\left(\frac{1}{s^2+1}\right)$$

$$= 2e^t - 2\cos t + \sin t.$$ **Ans.**

(*ii*) $\frac{4s+5}{(s-1)^2(s+2)} = \frac{A}{(s-1)}+\frac{B}{(s-1)^2}+\frac{C}{s+2}$ Here $B = 3$, $C = \frac{-1}{3}$

$\Rightarrow 4s + 5 = A(s-1)(s+2) + 3(s+2) - \frac{1}{3}(s-1)^2 \quad \therefore A = \frac{1}{3}$

$$\therefore\ L^{-1}\frac{4s+5}{(s-1)^2(s+2)} = L^{-1}\left[\frac{1}{3(s-1)}+\frac{3}{(s-1)^2}-\frac{1}{3(s+2)}\right] = \frac{1}{3}e^t + 3e^t\cdot t - \frac{1}{3}e^{-2t}$$ **Ans.**

2. *Find the inverse Laplace transform of*

(*i*) $\dfrac{4s^2+3s}{(s-1)(s^2+2s+5)}$ (*ii*) $\log\left(1+\dfrac{1}{s}\right)$

SOLUTION: (*i*) $\dfrac{4s^2+3s}{(s-1)(s^2+2s+5)} = \dfrac{7}{8(s-1)}+\dfrac{25(s+1)}{8[(s+1)^2+4]}+\dfrac{10}{8[(s+1)^2+2^2]}$

$$\therefore \quad L^{-1}\left[\frac{4s^2+3s}{(s-1)(s^2+2s+5)}\right] = \frac{7}{8}L^{-1}\left(\frac{1}{s-1}\right)+\frac{25}{8}L^{-1}\left[\frac{s+1}{(s+1)^2+2^2}\right]+\frac{5}{8}L^{-1}\frac{2}{(s+1)^2+2^2}$$

$$= \frac{7}{8}e^t+\frac{25}{8}e^{-t}\cos 2t+\frac{5}{8}e^{-t}\sin 2t \qquad \textbf{Ans.}$$

(*ii*) Let $\bar{f}(s) = L(f(t)) = \log\left(1+\dfrac{1}{s}\right)$

Therefore, $L[t\,f(t)] = \dfrac{-d}{ds}\bar{f}(s) = -\dfrac{d}{ds}[\log(s+1)-\log s] = \dfrac{1}{s}-\dfrac{1}{s+1}$

$\therefore \quad t\,f(t) = L^{-1}\left[\dfrac{1}{s}-\dfrac{1}{s+1}\right] = 1-e^{-t} \quad \therefore \; f(t) = \dfrac{1-e^{-t}}{t}.$ **Ans.**

3. *Find the inverse Laplace transform of*

(*i*) $\dfrac{s}{(s+7)^4}$ (*ii*) $\log\dfrac{s+a}{s+b}$

SOLUTION: (*i*) $L^{-1}\dfrac{s}{(s+7)^4} = L^{-1}\left[\dfrac{s+7-7}{(s+7)^4}\right] = L^{-1}\left[\dfrac{1}{(s+7)^3}-\dfrac{7}{(s+7)^4}\right]$

$$= \frac{e^{-7t}\cdot t^2}{2!}-7e^{-7t}\frac{t^3}{3!} = e^{-7t}\left(\frac{t^2}{2}-\frac{7t^3}{6}\right) \qquad \textbf{Ans.}$$

(*ii*) Let $L^{-1}\log\left(\dfrac{s+a}{s+b}\right) = f(t) \;\therefore\; L(f(t)) = \log\left(\dfrac{s+a}{s+b}\right) = \bar{f}(s)$, say.

We know that $L(t\,f(t)) = \dfrac{-d}{ds}\bar{f}(s) = \dfrac{-d}{ds}[\log(s+a)-\log(s+b)]$

or $L(t\,f(t)) = -\dfrac{1}{s+a}+\dfrac{1}{s+b} \quad \therefore \; t\,f(t) = L^{-1}\dfrac{1}{s+b}-L^{-1}\dfrac{1}{s+a} = e^{-bt}-e^{-at}$

$\therefore \quad f(t) = \dfrac{e^{-bt}-e^{-at}}{t}$ **Ans.**

4. *Find the inverse Laplace transform of*

(*i*) $\dfrac{3s-2}{s^2-2s+10}$ (*ii*) $\dfrac{1}{s^3-a^3}$.

SOLUTION: (*i*)
$$L^{-1}\frac{3s-2}{s^2-2s+10} = L^{-1}\left[\frac{3(s-1+1)-2}{(s-1)^2+3^2}\right]$$
$$= 3L^{-1}\frac{s-1}{(s-1)^2+3^2}+L^{-1}\frac{1}{(s-1)^2+3^2} = 3e^t\cos 3t+e^t\cdot\frac{1}{3}\sin 3t$$
$$= e^t\left[3\cos 3t+\frac{1}{3}\sin 3t\right]$$ **Ans.**

(*ii*)
$$\frac{1}{s^3-a^3} = \frac{1}{(s-a)(s^2+as+a^2)} = \frac{A}{s-a}+\frac{Bs+C}{s^2+as+a^2}$$
$$= \frac{1}{3a^2(s-a)}+\frac{Bs+C}{s^2+as+a^2}$$

or
$$1 = \frac{1}{3a}(s^2+as+a^2)+(s-a)(Bs+C) \quad \therefore\ B = \frac{-1}{3a^2} \text{ and } C = \frac{-2}{3a}$$

$\therefore$
$$L^{-1}\left(\frac{1}{s^3-a^3}\right) = \frac{1}{3a^2}L^{-1}\left(\frac{1}{s-a}\right)-\frac{1}{3a^2}L^{-1}\left[\frac{s+2a}{s^2+as+\frac{a^2}{4}+\frac{3a^2}{4}}\right]$$
$$= \frac{1}{3a^2}e^{at}-\frac{1}{3a^2}\left[\frac{s+\frac{a}{2}+\frac{3a}{2}}{\left(s+\frac{a}{2}\right)^2+\left(\frac{a\sqrt{3}}{2}\right)^2}\right]$$
$$= \frac{1}{3a^2}e^{at}-\frac{e^{-at/2}}{3a^2}\left[\cos\frac{a\sqrt{3}}{2}t+\sqrt{3}\sin\frac{a\sqrt{3}}{2}t\right]$$ **Ans.**

5. *Show that* $f*g=g*f$ *where* $f*g$ *is the convolution of* f *and* g.

SOLUTION: (*i*) We know that the convolution of f and $g = f*g = \int_0^t f(u)g(t-u)du$

hence $g*f = \int_0^t g(u)f(t-u)du$ Putting $t-u=x$ so $du = -dx$

$$= \int_t^0 g(t-x)f(x)(-dx) = \int_0^t g(t-x)f(x)dx = \int_0^t g(t-u)f(u)du.$$ **Hence proved.**

6. *Find $f * g(t)$ if $f(t)=1$ and $g(t)=1$.*

SOLUTION: (*i*) $g(t) = 1$ and $f(t) = 1$ hence

$$f * g(t) = \int_0^t f(u)g(t-u)du = \int_0^t 1\cdot 1\cdot du = [u]_0^t = t.$$ **Ans.**

7. *If $f_1(t)=t$ and $f_2(t)=e^t$ find their convolution $f_1 * f_2$.*

SOLUTION: $f_1(t) = t$ and $f_2(t) = e^t$, hence their convolution is given by

$$f_1 * f_2 = \int_0^t f_1(u) f_2(t-u)du = \int_0^t u\, e^{t-u} du = e^t \int_0^t u\, e^{-u} du$$

$$= e^t\left[\{-u\, e^{-u}\}_0^t - \int_0^t 1(-e^{-u})du\right] = e^t\left[(-t\, e^{-t}) + \int_0^t e^{-u} du\right]$$

$$= e^t\left[-t\, e^{-t} - \{e^{-u}\}_0^t\right] = -t - 1 + e^t$$ **Ans.**

8. *Applying the convolution, find the inverse Laplace transform of the function:*

(*i*) $\dfrac{1}{(s-2)(s-3)}$ (*ii*) $\dfrac{1}{(s+1)(s+9)^2}$

SOLUTION: (*i*) Let $\overline{f}(s) = \dfrac{1}{s-2}$ and $\overline{g}(s) = \dfrac{1}{s-3}$ and we know by convolution theorem that

$$L^{-1}[\overline{f}(s)\overline{g}(s)] = f * g = \int_0^t f(u)g(t-u)du$$

$$\therefore \quad L^{-1}\left(\frac{1}{s-2}\cdot\frac{1}{s-3}\right) = \int_0^t e^{2u}\, e^{3(t-u)} du = \int_0^t e^{3t}\, e^{-u} du = e^{3t}[-e^{-t}+1] = e^{3t} - e^{2t}$$ **Ans.**

(*ii*) Let $\overline{f}(s) = \dfrac{1}{s+1}, \overline{g}(s) = \dfrac{1}{(s+9)^2}$ so $f(t) = e^{-t}$ and $g(t) = e^{-9t}\cdot t$

By convolution theorem

$$L^{-1}\left[\frac{1}{s+1}\frac{1}{(s+9)^2}\right] = \int_0^t e^{-u}\, e^{-9(t-u)}\,(t-u)\, du = e^{-9t}\int_0^t e^{8u}(t-u)du$$

$$= e^{-9t}\left[(t-u)\frac{e^{8u}}{8}\right]_0^t - e^{-9t}\int_0^t -\frac{e^{8u}}{8} du = -\frac{t}{8}e^{-9t} + e^{-9t}\left[\frac{e^{8u}}{64}\right]_0^t$$

$$= \frac{-t}{8}e^{-9t} + \frac{1}{64}(e^{-t} - e^{-9t}) = \frac{e^{-t}}{64}\left[1 - e^{-8t}(1+8t)\right]$$ **Ans.**

9. *Show that*

(*i*) $L^{-1}\left(\frac{1}{s}\sin\frac{1}{s}; s\to t\right) = t - \frac{t^3}{(3!)^2} + \frac{t^5}{(5!)^2} - \frac{t^7}{(7!)^2} + \cdots$

(*ii*) $L\left[1 - \frac{t^2}{(2!)^2} + \frac{t^4}{(4!)^2} - \frac{t^6}{(6!)^2} + \cdots\right] = \frac{1}{s}\cos\frac{1}{s}.$

SOLUTION: (*i*)

$$\frac{1}{s}\sin\frac{1}{s} = \frac{1}{s}\left[\frac{1}{s} - \frac{1}{3!}\frac{1}{s^3} + \frac{1}{5!}\frac{1}{s^5} - \frac{1}{7!}\frac{1}{s^7} + \ldots\right]$$

$$= \frac{1}{s^2} - \frac{1}{3!}\frac{1}{s^4} + \frac{1}{5!}\frac{1}{s^6} - \frac{1}{7!}\frac{1}{s^8} + \ldots$$

$$\therefore \quad L^{-1}\left(\frac{1}{s}\sin\frac{1}{s}\right) = L^{-1}\left[\frac{1}{s^2} - \frac{1}{3!s^4} + \frac{1}{5!s^6}\ldots\right] = t - \frac{1}{3!}\frac{t^3}{3!} + \frac{1}{5!}\frac{t^5}{5!} - \frac{1}{7!}\frac{t^7}{7!} + \ldots$$ **Hence proved.**

(*ii*)

$$L\left[1 - \frac{t^2}{(2!)^2} + \frac{t^4}{(4!)^2} - \frac{t^6}{(6!)^2} + \ldots\right] = \frac{1}{s} - \frac{1}{(2!)^3}\cdot\frac{2!}{s^3} + \frac{1}{(4!)^2}\cdot\frac{4!}{s^5} - \frac{1}{(6!)^2}\cdot\frac{6!}{s^7} + \ldots$$

$$= \frac{1}{s}\left[1 - \frac{1}{2!}\frac{1}{s^2} + \frac{1}{4!}\frac{1}{s^4} - \frac{1}{6!}\frac{1}{s^6} + \ldots\right] = \frac{1}{s}\cos\frac{1}{s}.$$ **Hence proved.**

10. *Find* $L^{-1}\log\sqrt{\frac{s^2+a^2}{s^2+b^2}}$.

SOLUTION: Let $L[f(t)] = \log\sqrt{\frac{s^2+a^2}{s^2+b^2}} = \frac{1}{2}\left[\log(s^2+a^2) - \log(s^2+b^2)\right]$

$$\therefore \quad L[t f(t)] = \frac{-d}{ds}\left[\frac{1}{2}\log(s^2+a^2) - \frac{1}{2}\log(s^2+b^2)\right]$$

$$= \frac{-s}{s^2+a^2} + \frac{s}{s^2+b^2} = -L(\cos at) + L(\cos bt)$$

Hence $t f(t) = -\cos at + \cos bt \quad \therefore \; f(t) = \frac{\cos bt - \cos at}{t}.$ **Ans.**

11. *Find* $L^{-1}\tan^{-1}\left(\frac{1}{s}\right)$.

SOLUTION: Let $L[f(t)] = \tan^{-1}\frac{1}{s} \qquad \therefore \; L[t f(t)] = \frac{-d}{ds}\tan^{-1}\frac{1}{s}$

or $$L[t f(t)] = -\frac{1}{1+\frac{1}{s^2}}\left(\frac{-1}{s^2}\right) = \frac{1}{1+s^2} = L(\sin t)$$

$\Rightarrow \quad t f(t) = \sin t, \quad$ hence $f(t) = \frac{\sin t}{t}.$ **Ans.**

12. *Show that* $L^{-1}\left(\dfrac{s}{s^4+s^2+1}\right)=\dfrac{2}{\sqrt{3}}\sinh\left(\dfrac{t}{2}\right)\sin\left(\dfrac{\sqrt{3}\,t}{2}\right).$

SOLUTION: $$\frac{s}{s^4+s^2+1}=\frac{s}{\left(s^2+\frac{1}{2}\right)^2+\left(\frac{\sqrt{3}}{2}\right)^2}=\frac{s}{\left(s^2+\frac{1}{2}+\frac{i\sqrt{3}}{2}\right)\left(s^2+\frac{1}{2}-\frac{i\sqrt{3}}{2}\right)}$$

$$=\frac{s}{i\sqrt{3}}\left[\frac{1}{s^2+\frac{1}{2}-\frac{i\sqrt{3}}{2}}-\frac{1}{s^2+\frac{1}{2}+\frac{i\sqrt{3}}{2}}\right]$$

$$=\frac{1}{i\sqrt{3}}\left[\frac{s}{s^2+\left(\frac{-i+\sqrt{3}}{2}\right)^2}-\frac{s}{s^2+\left(\frac{-i-\sqrt{3}}{2}\right)^2}\right]$$

$$\therefore\quad L^{-1}\left(\frac{s}{s^4+s^2+1}\right)=\frac{1}{i\sqrt{3}}L^{-1}\left[\frac{s}{s^2+\left(\frac{-i+\sqrt{3}}{2}\right)^2}-\frac{s}{s^2+\left(\frac{i-\sqrt{3}}{2}\right)^2}\right]$$

$$=\frac{1}{i\sqrt{3}}\left[\cos\left(\frac{-i+\sqrt{3}}{2}\right)t-\cos\left(\frac{-i-\sqrt{3}}{2}\right)t\right]=\frac{2}{i\sqrt{3}}\sin\frac{t\sqrt{3}}{2}\sin i\frac{t}{2}$$

$$=\frac{2}{i\sqrt{3}}\sin\frac{t\sqrt{3}}{2}i\sin h\frac{t}{2}=\frac{2}{\sqrt{3}}\sin\frac{\sqrt{3}t}{2}\sinh\frac{t}{2}.$$ **Hence proved**

13. *Apply convolution theorem to prove that* $\beta(m,n)=\int_0^1 u^{m-1}(1-u)^{n-1}\,du=\dfrac{\overline{|m}\,\overline{|n}}{\overline{|m+n}}.$

SOLUTION: Let $f(x)=x^{m-1}$, $g(x)=x^{n-1}$ then $\bar{f}(s)=\dfrac{\overline{|m}}{s^m}$ and $\bar{g}(s)=\dfrac{\overline{|n}}{s^n}$

By convolution theorem

$$L^{-1}[\bar{f}(s)\bar{g}(s)]=\int_0^t f(u)g(t-u)du$$

$$\therefore\quad L^{-1}\left[\frac{\overline{|m}}{s^m}\frac{\overline{|n}}{s^n}\right]=\int_0^t u^{m-1}(t-u)^{n-1}du \qquad \text{(Put } u=tx\text{)}$$

or $$L^{-1}\frac{\overline{|m}\,\overline{|n}}{s^{m+n}}=\int_0^1 t^{m-1}x^{m-1}\cdot t^{n-1}(1-x)^{n-1}\,t\,dx=t^{m+n-1}\int_0^1 x^{m-1}(1-x)^{n-1}dx$$

$$\therefore\quad \frac{\overline{|m}\,\overline{|n}}{s^{m+n}}=L(t^{m+n-1})\int_0^1 x^{m-1}(1-x)^{n-1}dx \qquad \frac{\overline{|m}\,\overline{|n}}{s^{m+n}}=\frac{\overline{|m}\,\overline{|n}}{s^{m+n}}\beta(m,n)$$

$$\Rightarrow\quad \beta(m,n)=\int_0^1 x^{m-1}(1-x)^{n-1}dx=\frac{\overline{|m}\,\overline{|n}}{\overline{|m+n}}.$$ **Hence proved.**

14. *Find the inverse Laplace transform of* $\cos^{-1}\left(\dfrac{s}{2}\right)$.

SOLUTION: Let $L[f(t)] = \bar{f}(s) = \cot^{-1}\left(\dfrac{s}{2}\right)$

Then $$L[tf(t)] = -\frac{d}{d}\bar{f}(s) = \frac{1}{1+\left(\frac{s}{2}\right)^2} = \left(\frac{1}{2}\right)\frac{2}{s^2+4} = L(\sin 2t)$$

$\Rightarrow$ $tf(t) = \sin 2t$, hence $f(t) = \dfrac{1}{t}\sin 2t$. **Ans.**

SOLUTIONS TO EXERCISE 18C

1. *The* $\dfrac{2\pi}{\omega}$ *– periodic function* $f(t)$ *is given by*

$$f(t) = a\sin\omega t, \quad 0 < t < \frac{\pi}{\omega}$$
$$= 0, \quad \frac{\pi}{\omega} < t < \frac{2\pi}{\omega}.$$

Find its Laplace transform.

SOLUTION: $f(t)$ is $\dfrac{2\pi}{\omega}$ periodic function and we know that if $f(t)$ is T-periodic, then

$$L[f(t)] = \frac{\int_0^T e^{-st} f(t)\,dt}{1-e^{-sT}} = \frac{\int_0^{\pi/\omega} e^{-st}\, a\sin\omega t\,dt + \int_{\pi/\omega}^{2\pi/\omega} 0.\,e^{-st}\,dt}{[1-e^{-s\,2\pi/\omega}]}$$

or $$\bar{f}(s) = \frac{\left[a\dfrac{e^{-st}}{(\omega^2+s^2)}(-s\sin\omega t - \omega\cos\omega t)\right]_0^{\pi/\omega}}{1-e^{-2\pi s/\omega}}$$

$$= \frac{a}{(\omega^2+s^2)(1-e^{-2\pi s/\omega})}[e^{-\pi s/\omega}(-s\sin\pi - \omega\cos\pi) - e^0(0-\omega\cos 0)]$$

$$= \frac{a\omega(e^{-\pi s/\omega}+1)}{(\omega^2+s^2)(1-e^{-2\pi s/\omega})} = \frac{a\omega}{(\omega^2+s^2)(1-e^{-\pi s/\omega})}$$

2. *Find* $L\{f(t)\}$ *if* $f(t) = \dfrac{\lambda t}{T}$, $0 < t < T$ *and* $f(t+T) = f(t)$.

SOLUTION: Since $f(x)$ is given to be T-periodic function,

$\therefore$ $$L(f(t)) = \frac{\int_0^T e^{-st}\dfrac{\lambda t}{T}\,dt}{1-e^{-sT}} = \frac{\left[\dfrac{\lambda t}{T}\dfrac{e^{-st}}{-s}\right]_0^T - \int_0^T \dfrac{\lambda}{T}\cdot\dfrac{e^{-st}}{-s}\,dt}{1-e^{-sT}}$$

$$= \frac{\left(\frac{-\lambda}{s} e^{-sT} - 0\right) + \frac{\lambda}{Ts}\left[\frac{e^{-st}}{-s}\right]_0^T}{1 - e^{-sT}} = \frac{\frac{-\lambda}{s} e^{-sT}\left(1 + \frac{1}{Ts}\right) + \frac{\lambda}{Ts^2}}{1 - e^{-sT}}$$

$$= \frac{\lambda\,[1 - (1 + Ts)\,e^{-sT}]}{Ts^2\,(1 - e^{sT})}.$$ **Ans.**

3. *Find the Laplace transform of the 2l-periodic function $f(t)$ given by*

$$f(t) = \frac{t}{l}, \qquad 0 < t < l$$

$$= \frac{2l - t}{l}, \qquad l < t < 2l.$$

SOLUTION: $f(t)$ is $2l$–periodic, hence

$$L(f(t)) = \frac{\int_0^{2t} e^{-st} f(t)\,dt}{1 - e^{-2ls}} = \frac{\int_0^l e^{-st}\frac{t}{l}\,dt + \int_l^{2l}\frac{2l - t}{l} e^{-st}\,dt}{1 - e^{-2ls}}$$

$$= \frac{1}{l(1 - e^{2ls})}\left[t\frac{e^{-st}}{s}\bigg|_0^l - \int_0^l 1\cdot\frac{e^{-st}}{s}\,dt + (2lt)\frac{e^{-st}}{s}\bigg|_l^{2l} - \int_l^{2l}(-1)\frac{e^{-st}}{s}\,dt\right]$$

$$= \frac{1}{l\,(1 - e^{-2ls})}\left[-\frac{l}{s}e^{-ls} - \frac{1}{s^2}e^{-st}\bigg|_0^l + \frac{l}{s}e^{-ls} + \frac{1}{s^2}e^{-st}\bigg|_l^{2l}\right]$$

$$= \frac{1}{l\,(1 - e^{-2ls})}\left[\frac{-l}{s}e^{-ls} - \frac{1}{s^2}(e^{-ls} - 1) + \frac{l}{s}e^{-ls} + \frac{1}{s^2}(e^{-2ls} - e^{-ls})\right] = \frac{1 - e^{-ls}}{ls^2\,(1 + e^{-ls})}$$ **Ans.**

4. *Solve the differential equation using L -transform* $\frac{d^2x}{dt^2} + 4x = 5e^{-t}$, *given that* $x(0) = 2$ *and* $x'(0) = 3$.

SOLUTION: Equation is $(D^2 + 4)x = 5e^{-t}$ where $D \equiv \frac{d}{dt}$.

Taking Laplace transform throughout, we get

$$s^2\bar{x} - sx(0) - x'(0) + 4\bar{x} = \frac{5}{s+1}.$$ Here $x(0) = 2$, $x'(0) = 3$, therefore we have

$$(s^2 + 4)\,\bar{x} = \frac{5}{s+1} + 2s + 3 = \frac{2s^2 + 5s + 8}{s+1}$$

or

$$\bar{x} = \frac{2s^2 + 5s + 8}{(s^2 + 4)(s+1)} = \frac{1}{s+1} + \frac{s+4}{s^2+4}$$

Taking inverse Laplace transform, we get

$$x(t) = e^{-t} + \cos 2t + 2\sin 2t$$ **Ans.**

5. *Solve the differential equation using L -transform* $\dfrac{d^2y}{dt^2} + 2\dfrac{dy}{dt} + y = e^{-t}$, *given that* $y(0) = 0$ *and* $y'(0) = 1$.

SOLUTION: Equation is $(D^2 + 2D + 1)y = e^{-t}$ where $D \equiv \dfrac{d}{dt}$.

Taking L–transform on both sides, gives

$$s^2\bar{y} - sy(0) - y'(0) + 2[s\bar{y} - y(0)\} + \bar{y} = \frac{1}{s+1}. \quad \text{Here } y(0) = 0, \; y'(0) = 1$$

$$\therefore \quad (s^2 + 2s + 1)\,\bar{y} = 1 + \frac{1}{s+1} \quad \text{or} \quad \bar{y} = \frac{1}{(s+1)^2} + \frac{1}{(s+1)^3}$$

Taking inverse Laplace transform, we get

$$y(t) = te^{-t} + \frac{t^2}{2}e^{-t} = \frac{1}{2}e^{-t}(t^2 + 2t).$$ **Ans.**

6. *Solve the differential equation using L -transform* $\dfrac{d^3x}{dt^3} - \dfrac{dx}{dt} = 2\cos t$ *subject to the conditions* $x(0) = 3$, $x'(0) = 2$ *and* $x''(0) = 1$.

SOLUTION: The given equation is $(D^3 - D)x = 2\cos t$ where $D \equiv \dfrac{d}{dt}$.

Taking L–transform on both sides, gives

$$s^3\bar{x} - s^2x(0) - sx'(0) - x''(0) - [s\bar{x} - x(0)] = \frac{2s}{s^2+1}$$

or $$(s^3 - s)\bar{x} = \frac{2s}{s^2+1} + 3s^2 + 2s + 1 - 3 = \frac{3s^4 + 2s^3 + s^2 + 4s - 2}{s^2+1}$$

or $$\bar{x} = \frac{3s^4 + 2s^3 + s^2 + 4s - 2}{s(s^2-1)(s^2+1)} = \frac{2}{s} - \frac{1}{s+1} + \frac{2}{s-1} - \frac{1}{s^2+1}$$

Taking inverse L–transform on both sides, we get

$$x(t) = 2 - e^{-t} + 2e^{t} - \sin t.$$ **Ans.**

7. *Solve the differential equation using L -transform* $\dfrac{d^2x}{dt^2} + 2\dfrac{dx}{dt} + 2x = 2\cos 2t - 4\sin 2t$ *given that* $x = 0$ *and* $\dfrac{dx}{dt} = 1$ *at* $t = 0$.

SOLUTION: The given equation is $(D^2 + 2D + 2)x = 2\cos 2t - 4\sin 2t$ where $D \equiv \dfrac{d}{dt}$

Taking Laplace transform on both sides, we get

$$s^2\bar{x} - sx(0) - x'(0) + 2[s\bar{x} - x(0)] + 2\bar{x} = \frac{2s}{s^2+4} - \frac{8}{s^2+4}$$

or $\quad (s^2 + 2s + 2)\,\overline{x} = 1 + \dfrac{2s}{s^2+4} - \dfrac{8}{s^2+4} = \dfrac{s^2 + 2s - 4}{s^2+4}$

or $\quad \overline{x} = \dfrac{s^2 + 2s - 4}{(s^2+4)(s^2+2s+2)} = \dfrac{3s+8}{5(s^2+4)} + \dfrac{-3s-9}{5(s^2+2s+2)}$

$$= \frac{3s}{5(s^2+4)} + \frac{8}{5(s^2+4)} - \frac{3(s+1)}{5[(s+1)^2+1]} - \frac{6}{5[(s+1)^2+1]}$$

$\therefore \quad x(t) = \dfrac{3}{5}\cos 2t + \dfrac{4}{5}\sin 2t - \dfrac{3}{5}e^{-t}[\cos t + 2\sin t]$ **Ans.**

8. *Solve the differential equations using L -transform* $\dfrac{d^2y}{dt^2} + 9y = 18\,t$ *given that* $y(0) = 1$ *and* $y\left(\dfrac{\pi}{2}\right) = -1.$

SOLUTION: Equation can be written as $(D^2 + 9)y = 18t$ where $D \equiv \dfrac{d}{dt}$

Taking Laplace transform on both sides, we get $s^2\overline{y} - sy(0) - y'(0) + 9\overline{y} = \dfrac{18}{s^2}$

or $\quad (s^2+9)\overline{y} = \dfrac{18}{s^2} + s + k$ where $k = y'(0)$

or $\quad \overline{y} = \dfrac{s^3 + ks^2 + 18}{s^2(s^2+9)} = \dfrac{s}{s^2+9} + \dfrac{2}{s^2} + \dfrac{k-2}{s^2+9}$

Taking inverse L–transform on both sides, gives

$y(t) = \cos 3t + 2t + \dfrac{k-2}{3}\sin 3t$ Since $y(\pi/2) = -1$ we have $-1 = 0 + \pi - \dfrac{k-2}{3}$

$y(t) = \cos 3t + 2t + (\pi + 1)\sin 3t$ **Ans.**

9. *Solve the following system of simultaneous equations*

$$\frac{dx}{dt} - y = e^t$$

$\dfrac{dy}{dt} + x = \sin t,$ *given that* $x(0) = 1$ *and* $y(0) = 0.$

SOLUTION: Taking L–transform on both the equation, we get

$s\overline{x} - x(0) - \overline{y} = \dfrac{1}{s-1}$ and $s\overline{y} - y(0) + \overline{x} = \dfrac{1}{s^2+1}$

or $\quad s\overline{x} - \overline{y} = 1 + \dfrac{1}{s-1} = \dfrac{s}{s-1}$ and $\overline{x} + s\overline{y} = \dfrac{1}{s^2+1}$

Eliminating $\overline{x}$ in the above equations, we get

$$(s^2+1)\,\overline{y} = \frac{s}{s^2+1} - \frac{s}{s-1}$$

or $$\bar{y} = \frac{s}{(s^2+1)^2} - \frac{s}{(s-1)(s^2+1)} = \frac{s}{(s^2+1)^2} - \frac{1}{2(s-1)} + \frac{1}{2}\frac{(s-1)}{s^2+1}$$

$\therefore$ $$y(t) = \frac{1}{2}t\sin t - \frac{1}{2}e^t + \frac{1}{2}\cot t - \frac{1}{2}\sin t = \frac{1}{2}(t\sin t - e^t + \cos t - \sin t)$$

Putting this value of y in the second given equation, we get

$$x = \sin t - \frac{dy}{dt} = \sin t - \frac{1}{2}\frac{d}{dt}[t\sin t - e^t + \cos t - \sin t]$$
$$= \sin t - \frac{1}{2}[t\cos t + \sin t - e^t - \sin t - \cos t] = \frac{1}{2}[e^t + 2\sin t + (1-t)\cos t]$$

Thus, $$x = \frac{1}{2}[e^t + 2\sin t + (1-t)\cos t],\quad y = \frac{1}{2}[t\sin t - e^t + \cos t - \sin t]$$ **Ans.**

10. *Solve the simultaneous equations*

$$\frac{dx}{dt} + 3x + y = 0$$
$$\frac{dy}{dt} - x + y = 0 \quad \text{given that} \quad x(0) = y(0) = 1.$$

SOLUTION: Applying L–transform in both the equations, we get

$$s\bar{x} - x(0) + 3\bar{x} + \bar{y} = 0 \quad \text{and} \quad s\bar{y} - y(0) - \bar{x} + \bar{y} = 0$$

or $$(s+3)\,\bar{x} + \bar{y} = 1 \quad \text{and} \quad -\bar{x} + (s+1)\bar{y} = 1$$

Eliminating $\bar{x}$ in the above equations, we get

$$[(s+1)(s+3)+1]\,\bar{y} = 1 + (s+3) \quad \text{or} \quad \bar{y} = \frac{s+4}{s^2+4s+4} = \frac{s+2+2}{(s+2)^2}$$

or $$\bar{y} = \frac{1}{s+2} + \frac{2}{(s+2)^2} \quad \Rightarrow \quad y(t) = e^{-2t} + 2e^{-2t}t = (1+2t)\,e^{-2t}$$

$\therefore$ $$x(t) = y + \frac{dy}{dt} = e^{-2t} - 2te^{-2t} = (1-2t)e^{-2t}$$

Thus, the solution is $x = (1-2t)e^{-2t}$, $y = (1+2t)e^{-2t}$ **Ans.**

11. *Solve the following set of simultaneous equations*

$$\frac{dx}{dt} - \frac{dy}{dt} + 2y = \cos 2t$$
$$\frac{dx}{dt} + \frac{dy}{dt} - 2x = \sin 2t$$

subject to the conditions $x(0) = 0$, $y(0) = -1$.

SOLUTION: Taking L–transform on both the equations, we get

$$s\bar{x} - x(0) - s\bar{y} + y(0) + 2\bar{y} = \frac{s}{s^2+4} \quad \text{and} \quad s\bar{x} - x(0) + s\bar{y} - y(0) - 2\bar{x} = \frac{2}{s^2+4}$$

or $$s\bar{x} - (s-2)\,\bar{y} = \frac{s}{s^2+4} + 1 \quad \text{and} \quad (s-2)\bar{x} + s\bar{y} = \frac{2}{s^2+4} - 1$$

Eliminating $\bar{y}$ in the above equations, we get

$$[s^2 + (s-2)^2]\,\bar{x} = \frac{s^2}{s^2+4} + s + \frac{2(s-2)}{s^2+4} - (s-2) = \frac{3s^2+2s+4}{(s^2+4)}$$

or
$$\bar{x} = \frac{3s^2+2s+4}{2(s^2-2s+2)(s^2+4)} = \frac{s+1}{2(s^2-2s+2)} - \frac{s}{2(s^2+4)} = \frac{s-1+2}{2[(s-1)^2+1} - \frac{s}{2(s^2+4)}$$

which on inversion, gives

$$\therefore \quad x(t) = \frac{1}{2}[e^t \cos t + e^t \sin t] - \frac{1}{2}\cos 2t$$

Similarly, we can obtain $\bar{y}$ and then on inversion, get

$$y(t) = -\, e^t (\cos t - \sin t) - \sin 2t$$

Ans.

12. *An electron is projected into a uniform magnetic field which is perpendicular to its direction of motion. If x and y are the co-ordinates of its position at any time t, the equations of motion are given by*

$$m\frac{d^2x}{dt^2} = k\frac{dy}{dt} \quad \text{and} \quad m\frac{d^2y}{dt^2} = -k\frac{dx}{dt}$$

the initial conditions being $x = y = \frac{dy}{dt} = 0$ *and* $\frac{dx}{dt} = c$ *at* $t = 0$. *Use Laplace transform method to find x and y at time t.*

SOLUTION: The simultaneous equations are $m\frac{d^2x}{dt^2} = k\frac{dy}{dt}$ and $m\frac{d^2y}{dt^2} = -k\frac{dx}{dt}$

Taking L–transform on both of these, we get

$$m[s^2\bar{x} - sx(0) - x'(0)] - k[s\bar{y} - y(0)] = 0. \quad \text{and} \quad k[s\bar{x} - x(0)] + m[s^2\bar{y} - sy(0) - y'(0)] = 0$$

Given: $x(0) = y(0) = y'(0) = 0,\ x'(0) = c,$

Eliminating $\bar{x}$ in the above equations, we get

$$[m^2s^3 + k^2 s]\,\bar{y} = -\,mck \quad \text{or} \quad \bar{y} = \frac{-\,mck}{s(m^2s^2+k^2)} = \frac{-\,kc}{ms\left(s^2+\frac{k^2}{m^2}\right)} = \frac{mc}{k}\left[\frac{s}{s^2+\frac{k^2}{m^2}} - \frac{1}{s}\right]$$

On inversion, we get $y(t) = \frac{mc}{k}\left(\cos\frac{kt}{m} - 1\right)$

Putting this value of y in $m\frac{d^2y}{dt^2} = -k\frac{dx}{dt}$, we get

$$\frac{dx}{dt} = -\frac{m}{k}\frac{d^2y}{dt^2} = -\frac{m}{k}\frac{mc}{k}\frac{k^2}{m^2}\frac{kt}{m} = c\cos\frac{kt}{m}$$

$$\therefore \quad x(t) = \frac{mc}{k}\sin\frac{kt}{m}$$

Thus, $x = \frac{mc}{k}\sin\frac{Kt}{m},\quad y = \frac{mc}{k}\left(\cos\frac{kt}{m} - 1\right)$ is the solution. **Ans.**

13. *Solve the equation* $(D^4 + 2D^2 + 1)\,y = 0$ *where* $D \equiv \dfrac{d}{dx}$, *and* $y(0) = 0, y'(0) = 1$ *and* $y''(0) = 2$ *and* $y'''(0) = -3$.

SOLUTION: Taking L–transform on the given equation, we get

$$s^4\bar{y} - s^3y(0) - s^2y'(0) - sy''(0) - y'''(0) + 2[s^2\bar{y} - sy(0) - y'(0)] + \bar{y} = 0$$

or $$(s^4 + 2s^2 + 1)\bar{y} = s^2 + 2s - 1 \quad \text{as } y(0) = 0,\ y'(0) = 1,\ y''(0) = 2,\ y'''(0) = -3.$$

$\therefore$ $$\bar{y} = \frac{s^2 + 2s - 1}{(s^2+1)^2} = \frac{1}{s^2+1} + \frac{2s}{(s^2+1)^2} - \frac{2}{(s^2+1)^2}$$

$\therefore$ $$y(t) = \sin t + t\sin t - \frac{2}{2}(\sin t - t\cos t) = t(\sin t + \cos t)$$

since $L(t\sin at) = \dfrac{2as}{(s^2+a^2)}$ and $L(t\cos at) = \dfrac{1}{2a^3}(\sin at - at\cos at)$

Hence the solution is $y = t(\sin t + \cos t)$. **Ans.**

14. *Solve the equation* $(D^3 - 2D^2 + 5D)x = 0$ *where* $D \equiv \dfrac{d}{dt}$ *and* $x(0) = 0, x'(0) = 1$ *and* $x(\pi/8) = 1$.

SOLUTION: Taking L' transform on both sides of the given equation, we get

$$s^3\bar{x} - s^2x(0) - sx'(0) - x''(0) - 2[s^2\bar{x} - sx(0) - x'(0)] + 5[s\bar{x} - x(0)] = 0.$$

or $(s^3 - 2s^2 + 5s)\,\bar{x} = s + k - 2$ where $k = x''(0)$. and $x(0) = 0$, $x'(0) = 1$, $x(\pi/8) = 1$.

or $$\bar{x} = \frac{s+k-2}{s^3 - 2s^2 + 5s} = \frac{s+k-2}{s(s^2 - 2s + 5)} = \frac{k-2}{5s} + \frac{As+B}{s^2 - 2s + 5}$$

Here $$A = \frac{-(k-2)}{5} \text{ and } B = \frac{2k+1}{5}$$

$\therefore$ $$\bar{x} = \frac{k-2}{5s} - \left(\frac{k-2}{5}\right)\frac{s-1+1}{s^2-2s+5} + \left(\frac{2k-1}{5}\right)\frac{1}{(s^2-2s+5)} + \frac{k+3}{5[(s-1)^2+2^2]}$$

$\therefore$ $$x(t) = \frac{k-2}{5}(1 - e^t\cos 2t) + \frac{k+3}{10}e^t\sin 2t$$

Since $x(\pi/8) = 1$, we have $$1 = \frac{k-2}{5} + \frac{1\,e^{\pi/8}}{\sqrt{2}}\left(\frac{k+3}{10} - \frac{k-2}{5}\right)$$

This is possible when $k = 7$.

$\therefore$ $$x(t) = 1 - e^t(\cos 2t - \sin 2t)$$ **Ans.**

15. *Solve the equation* $\frac{d^2x}{dt^2} + x = t \cos 2t$, *where* $x = 0$ *and* $\frac{dx}{dt} = 0$ *at* $t = 0$.

SOLUTION: Taking Laplace transform on the equation, we get

$$s^2\bar{x} - sx(0) - x'(0) + \bar{x} = L(t \cos 2t) = \frac{-d}{ds}\frac{s}{s^2+4} = \frac{s^2-4}{(s^2+4)^2}$$

or $$(s^2+1)\,\bar{x} = \frac{s^2-4}{(s^2+4)^2} \quad \text{since} \quad x(0) = x'(0) = 0, \quad \text{given}$$

$$\therefore \quad \bar{x} = \frac{s^2-4}{(s^2+1)(s^2+4)^2} = \frac{-5}{9(s^2+1)} + \frac{8}{3(s^2+4)^2} + \frac{5}{9(s^2+4)}$$

$$\therefore \quad x(t) = \frac{-5}{9}\sin t + \frac{5}{18}\sin 2t + \frac{8}{3}\frac{1}{2.3^3}(\sin 2t - 2t\cos 2t)$$

$$= \frac{-5}{9}\sin t + \frac{4}{9}\sin 2t - \frac{t}{3}\cos 2t \left[\text{since } L(t\cos at) = \frac{1}{2a^3}(t - at\cos at)\right]$$ **Ans.**

16. *Using Laplace transform technique solve the differential equation* $\frac{d^2y}{dx^2} - 3\frac{dy}{dx} + 2y = 4x + e^{3x}$

where $y(0) = 1$ *and* $y'(0) = -1$.

SOLUTION: Taking Laplace transform throughout, we get

$$s^2\bar{y} - sy(0) - y'(0) - 3[s\bar{y} - y(0)] + 2\bar{y} = \frac{4}{s^2} + \frac{1}{s-3}$$

or $$(s^2 - 3s + 2)\bar{y} = s - 1 - 3 + \frac{4}{s^2} + \frac{1}{s-3} = \frac{s^2 + 4(s-3) + s^2(s-3)(s-4)}{s^2(s-3)}$$

Therefore $$\bar{y} = \frac{s^4 - 7s^3 + 13s^2 \; 4s - 12}{s^2(s-1)(s-2)(s-3)} = \frac{2}{s^2} + \frac{3}{s} - \frac{1}{2(s-1)} - \frac{2}{s-2} + \frac{1}{2(s-3)}$$

$$\therefore \quad y(x) = 2x + 3 - \frac{1}{2}e^x - 2e^{2x} + \frac{1}{2}e^{3x}$$ **Ans.**

17. *Solve* $y'' + y' - 2y = t$, $y(0 = 1$, $y'(0) = 0$ *using Laplace transformation.*

SOLUTION: Taking Laplace tranform on the given equation, we get

$$s^2\bar{y} - sy(0) - y'(0) + s\bar{y} - y(0) - 2\bar{y} = \frac{1}{s^2}$$

or $$(s^2 + s - 2)\,\bar{y} = s + 1 + \frac{1}{s^2} = \frac{s^3 + s^2 + 1}{s^2} \quad \text{since} \quad y(0) = 1 \quad \text{and} \quad y'(0) = 0$$

or $$\bar{y} = \frac{s^3 + s^2 + 1}{s^2\,(s^2 + s - 2)} = \frac{A}{s} + \frac{B}{s^2} + \frac{Cs + D}{s^2 + s - 2} = \frac{-1}{4s} - \frac{1}{2s^2} + \frac{5s + 7}{4(s^2 + s - 2)}$$

$$= -\frac{1}{4s} - \frac{1}{2s^2} + \frac{5}{4}\frac{\left(s + \frac{1}{2}\right) - \frac{1}{2}}{\left(s + \frac{1}{2}\right)^2 - \left(\frac{3}{2}\right)^2} + \frac{9/8}{\left(s + \frac{1}{2}\right)^2 - \left(\frac{3}{2}\right)^2}$$

$\therefore$ $$y = -\frac{1}{4} - \frac{t}{2} + \frac{5}{4} e^{-t/2} \cosh\left(\frac{3t}{2}\right) + \frac{9}{8} e^{-t/2} \frac{2}{3} \sinh\frac{3t}{2}$$

$$y = \frac{11}{8} e^{t} - \frac{1}{16} e^{-2t} - \frac{t}{2} - \frac{1}{4}$$ **Ans.**

18. *Solve $y'' + 4y' + 3y = e^{-t}$, $y(0) = y'(0) = 1$ using Laplace transformation.*

SOLUTION: Taking Laplace transform on the equation, we get

$$s^2\bar{y} - sy(0) - y'(0) + 4s\bar{y} - 4y(0) + 3\bar{y} = \frac{1}{s+1}$$

or $$(s^2 + 4s + 3)\,\bar{y} = \frac{1}{s+1} + s + 5 \quad \text{as} \quad y(0) = y'(0) = 1$$

or $$\bar{y} = \frac{s^2 + 6s + 6}{(s+1)\,(s^2 + 4s + 3)} = \frac{s^2 + 6s + 6}{(s+1)^2\,(s+3)} = \frac{1}{2}(s+1)^2 - \frac{3}{4(s+3)} + \frac{5}{4(s-1)}.$$

On applying inverse Laplace transform, we get

$$y(t) = t\frac{e^{-t}}{2} - \frac{3}{4} e^{-3t} + \frac{5}{4} e^{-t} = \frac{e^{-t}}{4}(5 + 2t) - \frac{3}{4} e^{-3t}$$ **Ans.**

19. *Solve using Laplace transform method,*

$y''(t) + 4y'(t) + 4y(t) = 6e^{-t}$, $y(0) = -2$, $y'(0) = 8$.

SOLUTION: Applying Laplace transoform on the given equation, we get

$$s^2\bar{y} - sy(0) - y'(0) + 4[s\bar{y} - y(0)] + 4\bar{y} = \frac{6}{s+1}$$

or $$(s^2 + 4s + 4)\bar{y} = -2s + 8 - 8 + \frac{6}{s+1} = \frac{-2s^2 - 2s + 6}{s+1} \quad \text{since} \quad y(0) = 2,\ y'(0) = 8.$$

or $$\bar{y} = \frac{-2s^2 - 2s + 6}{(s+1)\,(s+2)^2} = \frac{s}{s+1} - \frac{2}{(s+2)^2} - \frac{8}{s+2}$$

$\therefore$ $$y(t) = 6e^{-t} - 2e^{-2t}t - 8e^{-2t} = 6e^{-t} - 2e^{-2t}\,(t + 4)$$ **Ans.**

20. *Using Laplace transform find the solution of the initial value problem*

$$\frac{d^2y}{dt^2}+9y=6\cos 3t \quad \text{where } y(0)=2,\ y'(0)=0.$$

SOLUTION: Taking Laplace transform on the given equation, gives $s^2\bar{y}-sy(0)-y'(0)+9\bar{y}=\dfrac{6s}{s^2+9}$

or $$(s^2+9)\bar{y} = \frac{6s}{s^2+9}+2s \quad \text{since } y(0)=2 \text{ and } y'(0)=0.$$

or $$\bar{y} = \frac{2s^3+24s}{(s^2+9^2)}=\frac{2s\,(s^2+9+3)}{(s^2+9)^2}=\frac{2s}{(s^2+9)}+\frac{6s}{(s^2+9)^2}$$

$\Rightarrow$ $$y(t) = 2\cos 3t + t\sin 3t.$$ **Ans.**

21. *A function $f(t)$ obeys the equation*

$$f(t)+2\int_0^t f(t)\,dt=\cosh 2t.$$

Find the Laplace transform of $f(t)$ and hence $f(t)$.

SOLUTION: Taking Laplace transform on the given equation, we get

$$\bar{f}(s)+\frac{2}{s}\bar{f}(s) = \frac{s}{s^2-4} \quad \text{where } L(f(t))=\bar{f}(s).$$

or $$\bar{f}(s) = \frac{s^2}{(s+2)(s^2-4)}=\frac{s^2-4+4}{(s+2)(s^2-4)}=\frac{1}{s+2}+\frac{4}{(s+2)^2(s-2)}$$

$$= \frac{1}{s+2}+\frac{s+2+s-2}{(s+2)^2(s-2)}=\frac{1}{s+2}+\frac{1}{s^2+4}+\frac{1}{(s+2)^2}$$

$\therefore$ $$f(t) = e^{-2t}+\frac{1}{2}\sin h\,2t+e^{-2t}t=(1+t)e^{-2t}+\frac{1}{4}(e^{2t}-e^{-2t})$$

$$= \frac{1}{4}e^{2t}+e^{-2t}\left(t+\frac{3}{4}\right)$$ **Ans.**

22. *By using Laplace transform, solve the equation*

$$\frac{d^2x}{dt^2}+y=-5\cos 2t, \quad \frac{d^2y}{dt^2}+x=5\cos 2t,$$

where $x(0)=x'(0)=y'(0)=-1$ and $y(0)=1$.

SOLUTION: Taking Laplace transform on both the equations, we get

$$s^2\bar{x}-sx(0)-x'(0)+\bar{y} = \frac{-5s}{s^2+4} \quad \text{and} \quad s^2\bar{y}-sy(0)-y'(0)+\bar{x} = \frac{5s}{s^2+4}$$

Using $x(0) = x'(0) = y'(0) = -1$ and $y(0) = 1$ we get

$$s^2\bar{x} + \bar{y} = -s - 1 - \frac{5s}{s^2 + 4} \qquad ...(1)$$

$$\bar{x} + s^2\bar{y} = s - 1 + \frac{5s}{s^2 + 4} \qquad ...(2)$$

Adding (1) and (2), gives $(s^2 + 1)(\bar{x} + \bar{y}) = -2$ or $\bar{x} + \bar{y} = \frac{-2}{s^2 + 1}$

Taking inverse Laplace transform here we get

$$x + y = -2\sin t \qquad ...(3)$$

Subtracting (2) from (1) we get

$$(s^2 - 1)(\bar{x} - \bar{y}) = -2s - \frac{10s}{s^2 + 4} = -2s\frac{(s^2 + 9)}{s^2 + 4}$$

or $$\bar{x} - \bar{y} = \frac{-2s(s^2 + 9)}{(s^2 - 1)(s^2 + 4)} = \frac{-4s}{s^2 - 1} + \frac{2s}{s^2 + 4}$$

Taking inverse Laplace transform here, gives

$$x - y = -4\cosh t + 2\cos 2t \qquad ...(4)$$

From (3) and (4) we get

$$x = -2\cosh t - \sin t + \cos(2t)$$

and $$y = 2\cosh t - \sin t - \cos(2t)$$ **Ans.**

23. *Solve the simultaneous equations*

$$\frac{dx}{dt} - \frac{dy}{dt} - 2x + 2y = 1 - 2t, \quad \frac{d^2x}{dt^2} + 2\frac{dy}{dt} + x = 0$$

given that $x = 0,\ y = 0,\ \frac{dx}{dt} = 0$ *when* $t = 0$ *using Laplace transform*

SOLUTION: Taking Laplace transform on both the given equations, we get

$$s\bar{x} - x(0) - s\bar{y} + y(0) - 2\bar{x} + 2\bar{y} = \frac{1}{s} - \frac{2}{s^2} \quad \text{and} \quad s^2\bar{x} - sx(0) - x'(0) - 2s\bar{y} - 2y(0) + \bar{x} = 0$$

or $$s(\bar{x} - \bar{y}) - 2(\bar{x} - \bar{y}) = \frac{s - 2}{s^2} \qquad ...(1)$$

and $(s^2 + 1)\bar{x} + 2s\bar{y} = 0$. (On using initial conditions). ...(2)

or $$\bar{x} - \bar{y} = \frac{1}{s^2} \qquad ...(3)$$

which on taking inverse L' transform, gives

$$x - y = t \qquad ...(4)$$

Using (3) in (2), gives $(s^2+1)\,\bar{x} + 2s\left(\bar{x} - \frac{1}{s^2}\right) = 0.$ or $(s+1)^2\bar{x} = \frac{2}{s}$

or $\bar{x} = \frac{2}{s(s+1)^2} = \frac{2}{s} - \frac{2}{(s+1)^2} - \frac{2}{s+1}$ which on taking inverse L transform, gives

$$x = 2 - 2t\,e^{-t} - 2e^{-t}$$

$$\therefore \quad y = x - t = 2 - t - 2e^{-t}\,(t + 1)$$

Ans.

24. *Use Laplace transform to solve* $\frac{dx}{dt} + y = \sin t,\ \frac{dy}{dt} + x = \cos t$ *given that* $x = 2,\ y = 0$ *at* $t = 0.$

SOLUTION: Taking Laplace transform on both the given equations, we get

$$s\bar{x} - x(0) + \bar{y} = \frac{1}{s^2+1} \quad \text{and} \quad s\bar{y} - y(0) + \bar{x} = \frac{s}{s^2+1}$$

Since $x(0) = 2,\ y(0) = 0$ we have

$$s\bar{x} + \bar{y} = 2 + \frac{1}{s^2+1} \quad \text{and} \quad \bar{x} + s\bar{y} = \frac{s}{s^2+1}$$

Eliminating $\bar{y}$ in the above equation we get $\bar{x} = \frac{2s}{s^2-1}$

and hence $\bar{y} = \frac{1}{s^2+1} - \frac{2}{s^2-1}$

Taking inverse Laplace transform here, we get

$x = 2\cosh t$ and $y = \sin t - 2\sinh t.$ **Ans.**

25. *The coordinates* (x, y) *of a particle moving along a plane curve at any time* t *are given by*

$$\frac{dy}{dt} + 2x = \sin 2t,\quad \frac{dx}{dt} - 2y = \cos 2t,\quad t > 0$$

given that $x = 1, y = 0$ *at* $t = 0.$ *Show, using transform, that the particle moves along the curve* $4x^2 + 4xy + 5y^2 = 4.$

SOLUTION: Taking Laplace transform on both the given equations, we get

$$s\bar{y} - y(0) + 2\bar{x} = \frac{2}{s^2+4} \quad \text{and} \quad s\bar{x} - x(0) - 2\bar{y} = \frac{s}{s^2+4}$$

Using $x(0) = 1$ and $y(0) = 0,$ we have

$$2\bar{x} + s\bar{y} = \frac{2}{s^2+4} \quad \text{and} \quad s\bar{x} - 2\bar{y} = 1 + \frac{s}{s^2+4}$$

Eliminating $\overline{y}$ in the above equations, we get

$$(s^2+4)\,\overline{x} = \frac{4}{s^2+4}+s+\frac{s^2}{s^2+4} = s+1$$

Hence $\overline{x} = \frac{s+1}{s^2+4}$. Inverting the transform here, gives

$x = \cos 2t + \frac{1}{2}\sin 2t$. Putting this in the second given equation we get

$$y = \frac{1}{2}\left[\frac{dx}{dt} - \cos 2t\right] = \frac{1}{2}[-2\sin 2t + \cos 2t - \cos 2t] = -\sin 2t$$

$\therefore$ $4x^2 + 4xy + 5y^2 = (2\cos 2t + \sin 2t)^2 + 2\,(2\cos 2t + \sin 2t)\,(-\sin 2t) + 5\sin^2 2t$

$= 4$ (on simphication) **Hence Proved.**

26. *Solve by using Laplace transform* $\frac{d^2y}{dt^2}+2\frac{dy}{dt}+5y = \frac{17}{2}\sin 2t$, *given that* $y = 2$ *and* $\frac{dy}{dt} = -4$ *when* $t = 0$.

SOLUTION: Taking Laplace transform on both sides of the given equation, given

$$s^2\overline{y}(s) - sy(0) - y'(0) + 2s\overline{y}(s) - 2y(0) + 5\overline{y}(s) = \frac{17}{s^2+4}$$

or $(s^2+2s+5)\overline{y} = \frac{17}{s^2+4} + 2s + 0$ since $y(0) = 2,\ y'(0) = -4$

or
$$\overline{y} = \frac{2s^3+8s+17}{(s^2+4)\,(s^2+2s+5)} = \frac{-2s+1}{s^2+4} + \frac{4s+3}{s^2+2s+5}$$

$$= \frac{-2s}{s^2+4} + \frac{1}{s^2+4} + \frac{4(s+1)}{(s+1)^2+4} - \frac{1}{(s+1)^2+4}$$

$\therefore$ $y(t) = -2\cos 2t + \frac{1}{2}\sin 2t + 4e^{-t}\cos 2t - \frac{1}{2}e^{-t}\sin 2t$.

$= 2(2e^{-t}-1)\cos 2t + \frac{1}{2}(1-e^{-t})\sin 2t.$ **Ans.**

27. *Solve* $y'' + y' - 2y = 3\sin t - \cos t$, *if* $0 < t < 2\pi$

$= 3\sin 2t - \cos 2t$ *if* $t > 2\pi$

given that $y(0) = 1$ *and* $y'(0) = 0$.

SOLUTION: Taking L'transform on both sides of the given equation, we get

$$s^2\overline{y} - sy(0) - y'(0) + s\overline{y} - y(0) - 2\overline{y} = \int_0^{2\pi} e^{-st}\,(3\sin t - \cos t)\,dt + \int_{2\pi}^{\infty} e^{-st}\,(3\sin 2t - \cos 2t)\,dt.$$

$$\text{or} \quad (s^2+s-2)\,\bar{y} - s - 1 = \left[\frac{3e^{-st}}{s^2+1}(-s\sin t - \cos t) - \frac{e^{-st}}{s^2+1}(-s\cos t + \sin t)\right]_0^{2\pi}$$

$$+ \left[\frac{3e^{-st}}{s^2+4}(-s\sin 2t - 2\cos 2t) - \frac{e^{-st}}{s^2+4}(-s\cos 2t + 2\sin 2t)\right]_{2\pi}^{\infty}$$

$$= \frac{3e^{-2\pi s}(-1)}{s^2+1} - \frac{e^{-2\pi s}(-s)}{s^2+1} - \frac{3}{s^2+1}(-1) + \frac{1}{s^2+1}(-s) + 0 - \frac{3e^{-2\pi s}(-2)}{s^2+4} + \frac{e^{-2\pi s}(-s)}{s^2+4}$$

$$\text{or} \quad (s^2+s-2)\,\bar{y} = s+1+\frac{e^{-2\pi s}}{s^2+1}(s-3) + \frac{e^{-2\pi s}}{s^2+4}(6-s) + \frac{3-s}{s^2+1}$$

$$\text{or} \quad \bar{y} = \frac{s+1}{s^2+s-2} + \frac{e^{-2\pi s}(s-3)}{(s^2+1)(s^2+s-2)} + \frac{e^{-2\pi s}(6-s)}{(s^2+4)(s^2+s-2)} + \frac{-s+3}{(s^2+1)(s^2+s-2)}$$

$$\therefore \quad y = L^{-1}\left(\frac{s+1}{s^2+s-2}\right) + L^{-1}\frac{(s-3)(e^{-2\pi s}-1)}{(s^2+1)(s+2)(s+1)} + L^{-1}\frac{e^{-2\pi s}(6-s)}{(s^2+4)(s+2)(s-1)}$$

$$= \frac{2}{3}e^t + \frac{1}{3}e^{-2t} - \frac{1}{3}\cos t + \frac{1}{6}\sin t + \frac{1}{3}e^t + \frac{1}{3}e^{-2t} + \left[\frac{1}{3}\cos(t-2\pi) - \frac{1}{6}\sin(t-2\pi) - \frac{1}{3}e^{t-2\pi} - \frac{1}{3}e^{-2(t-2\pi)}\right] U(t-2\pi)$$

$$- \left[\frac{-1}{3}\cos 2(t-2\pi) + \sin 2(t-2\pi) + \frac{1}{3}e^{t-2\pi} - \frac{1}{3}e^{-2(t-2\pi)}\right] U(t-2\pi)$$

$$= e^t + \frac{2}{3}e^{-2t} - \frac{1}{3}\cos t + \frac{1}{6}\sin t + \left[\frac{1}{3}\cos t + \frac{1}{6}\sin t - \frac{1}{3}\cos 2t - \sin 2t - \frac{2}{3}e^{4\pi - 2t}\right] U(t-2\pi)$$ **Ans.**

SOLUTIONS TO EXERCISE 18D

1. *Find the Laplace transform of* $t^4\,U(t-2)$

SOLUTION: $t^4 = (t-2+2)^4 = 2^4 + {}^4C_1\, 2^3(t-2) + {}^4C_2.2^2(t-2)^2 + {}^4C_3.2(t-2)^3 + (t-2)^4$

$$\therefore \quad t^4\,U(t-2) = [2^4 + 4.2^3(t-2) + 6.2^3(t-2)^2 + 4.\,2(t-2)^3 + (t-2)^4]\,U(t-2)$$

Now using the formula $L[f(t-a)\,U(t-a)] = e^{-as}\bar{f}(s)$, we have

$$L[t^4\,U(t-2)] = e^{-2s}\left[2^4\,L(1) + 2^3 4.\frac{1}{s^2} + 2^2.6.\frac{2!}{s^3} + 4.2\frac{3!}{s^4} + \frac{4!}{s^5}\right]$$

$$= e^{-2s}\left[\frac{16}{s} + \frac{32}{s^2} + \frac{48}{s^3} + \frac{48}{s^4} + \frac{24}{s^5}\right]$$ **Ans.**

2. *A function $f(t)$ is given by*

$$f(t) = \begin{cases} t^2 & \text{for } 0 < t < 1 \\ 4t & \text{for } t > 1 \end{cases}$$

Express it in terms of unit step function and then find its L' transform.

SOLUTION:

$$f(t) = t^2 \quad \text{for} \quad 0 < t < 1$$
$$= 4t \quad \text{for} \quad t > 1$$

$$\therefore \quad f(t) = t^2\, U(t) + 4t\, U(t-1) - t^2\, U(t-1) = t^2\, [U(t) - U(t-1)] + 4t\, U(t-1)$$

To find $L(f(t))$ we write

$$f(t) = t^2\, U(t) - (t-1+1)^2\, U(t-1) + 4(t-1+1)\, U(t-1)$$
$$= t^2\, U(t) - (t-1)^2\, U(t-1) - 2\,(t-1)\, U(t-1) - U(t-1) + 4(t-1)\, U(t-1) + 4U(t-1)$$
$$= t^2\, U(t) - (t-1)^2\, U(t-1) + 2\,(t-1)\, U(t-1) + 3\, U(t-1)$$

$$\therefore \quad L(f(t)) = e^{-0s}\, L(t^2) - e^{-s}\, L(t^2) + 2e^{-s}\, L\,(t) + 3L\, U(t-1)$$

$$= \frac{2}{s^3} - e^{-s}\frac{2}{s^3} + 2e^{-s}\frac{1}{s^2} + \frac{3e^{-s}}{s} = \frac{2}{s^3} + e^{-s}\left[\frac{3}{s} + \frac{2}{s^2} - \frac{2}{s^3}\right]$$ **Ans.**

3. *Find the Laplace transform of function* $e^{-t}\,[1 - U\,(t-2)]$.

SOLUTION:

$$L[e^{-t}\,(1 - U(t-2)] = L\, e^{-t} - L[e^{-t}\, U(t-2)]$$
$$= \frac{1}{s+1} - L\,[e^{-(t-2)} e^{-2}\, U(t-2)]$$
$$= \frac{1}{s+1} - e^{-2}\, e^{-2s}\, L(e^{-t}) = \frac{1}{s+1} - e^{-2}\, e^{-2s}\, \frac{1}{s+1}$$
$$= \frac{1 - e^{-2(s+1)}}{s+1}$$ **Ans.**

4. *Find the inverse Laplace transform of* (*i*) $\dfrac{s\,e^{-\frac{s}{2}} + \pi\, e^{-s}}{s^2 + \pi^2}$ (*ii*) $\dfrac{e^{4-3s}}{(s+4)^{5/2}}$.

SOLUTION: (*i*)

$$L^{-1}\frac{se^{-s/2} + \pi e^{-s}}{s^2 + \pi^2} = L^{-1}\left[e^{-s/2}\left(\frac{s}{s^2+\pi^2}\right)\right] + L^{-1}\left[e^{-s}\frac{\pi}{s^2+\pi^2}\right]$$
$$= L^{-1}\,[e^{-s/2}\, L(\cos \pi\, t)] + L^{-1}[e^{-s}\, L(\sin \pi t)]$$
$$= \cos\left[\pi\left(t - \frac{1}{2}\right)\right] U\left(t - \frac{1}{2}\right) + \sin \pi\,(t-1)\, U\,(t-1)$$
$$= U\left(t - \frac{1}{2}\right)\sin \pi\, t - U(t-1)\sin \pi\, t = \left[U\left(t - \frac{1}{2}\right) - U(t-1)\right]\sin \pi t$$ **Ans.**

(*ii*) $L^{-1}\left(\dfrac{1}{(s+4)^{5/2}}\right) = e^{-4t}\dfrac{t^{3/2}}{\lceil\frac{5}{2}} = \dfrac{e^{-4t}\,t^{3/2}}{\frac{3}{2}\cdot\frac{1}{2}\sqrt{\pi}}.$

$$\therefore\quad L^{-1}\left(\frac{e^{4-3s}}{(s+4)^{5/2}}\right) = e^{4}\cdot\frac{4}{3\sqrt{\pi}}\left[e^{-4(t-3)}(t-3)^{3/2}U(t-3)\right] \text{ (since } L^{-1}[f(t-a)]\,U(t-a) = e^{-at}\,\bar{f}(s))$$

$$= \frac{4e^{16}}{3\sqrt{\pi}}\left[e^{-4t}(t-3)^{3/2}U(t-3)\right]$$ **Ans.**

5. *Find the inverse Laplace transform of* $\dfrac{1-\sqrt{s}}{s^{\frac{3}{2}}}e^{-s}$.

SOLUTION: $$L^{-1}\left[\frac{1-\sqrt{s}}{s^{3/2}}e^{-s}\right] = L^{-1}\left[e^{-s}\,L\left(\frac{2\sqrt{t}}{\sqrt{\frac{1}{2}}}-1\right)\right]$$

(On using the formula $f(t-a)\,U(t-a) = L^{-1}[e^{-as}\bar{f}(s)]$)

$$= \left[L^{-1}\,e^{-s}\,L\left(\frac{2\sqrt{t}}{\sqrt{\pi}}-1\right)\right] = \left(\frac{2\sqrt{t-1}}{\sqrt{\pi}}-1\right)U(t-1)$$ **Ans.**

6. *A function f (t) is given by*

$f(t) = \cos t$ *for* $0 < t < \pi$

$= \sin t$ *for* $t > \pi$

Express it in terms of Heaviside unit step function and find its Laplace transform.

SOLUTION: $f(t) = \cos t,\ 0 < t < \pi$

$= \sin t,\ t > \pi.$

In terms of unit step function (or Heawiside function)

$$\begin{aligned} f(t) &= \cos t\,[U(t) - U(t-\pi)] + \sin t\,U(t-\pi) \\ &= \cos t\,U(t) - \cos(t-\pi+\pi)\,U(t-\pi) + \sin(t-\pi+\pi)\,U(t-\pi) \\ &= \cos t\,U(t) + \cos(\pi-t)\,U(t-\pi) - \sin(t-\pi)\,U(t-\pi) \end{aligned}$$

$$\therefore\quad L(f(t)) = e^{0s}\frac{s}{s^2+1} + e^{-\pi s}\left(\frac{s}{s^2+1}\right) + \frac{e^{-\pi s}}{s^2+1}$$

$$= \frac{1}{s^2+1}[s + (s-1)\,e^{-\pi s}]$$ **Ans.**

7. *Find the Laplace transform of* $\sin 3t \cdot \delta\left(t-\frac{\pi}{4}\right) - t^2\,\delta\,(t-3)$.

SOLUTION: We know that $L[f(t)\,\delta(t-a)] = e^{-as} f(a)$

Hence $L\left[\sin 3t\,\delta\left(t-\frac{\pi}{4}\right) - t^2\,\delta\,(t-3)\right]$

$$= L\left[\sin 3t\,\delta\left(t-\frac{\pi}{4}\right)\right] - L[t^2\,\delta(t-3)]$$

$$= e^{-\pi s/4}\sin\left(\frac{3\pi}{4}\right) - e^{-3s}\,3^2 = e^{-\pi s/4}\,\frac{1}{\sqrt{2}} - 9e^{-3s}$$ **Ans.**

8. *Evaluate* $\int_0^\infty e^{-3t}\,\delta\,(t-2)\,dt$.

SOLUTION: We know that $L(f(t)) = \int_0^\infty e^{-st} f(t)\,dt$ and $L[f(t)\,\delta(t-a)] = e^{-as} f(a)$

$\therefore$ $L[1\,\delta(t-2)] = e^{-2s}.1 = \int_0^\infty e^{-st}\delta\,(t-2)\,dt$ (by definition).

Taking $s = 3$ we get $\int_0^\infty e^{-3t}\,\delta(t-2)\,dt = e^{-6}$ **Ans.**

9. *Solve the differential equation*

$$\frac{d^2x}{dt^2} + 3\frac{dx}{dt} + 2x = \delta\,(t) \quad \textit{given that} \quad x\,(0) = x'\,(0) = 0.$$

SOLUTION: Taking L–transform on the given equation, we get

$$s^2\bar{x} - sx(0) - x'(0) + 3\,s\bar{x} - 3x(0) + 2\bar{x} = L(\delta(t)) = 1$$

or $(s^2 + 3s + 2)\bar{x} = 1$ since $x(0) = x'(0) = 0$

or $$\bar{x} = \frac{1}{(s+2)(s+1)} = \frac{1}{s+1} - \frac{1}{s+2}$$

$\therefore$ $x(t) = e^t - e^{2t}$. **Ans.**

10. *Solve the differential equation*

$$\frac{d^2y}{dt^2} + 3\frac{dy}{dt} + 2y = f(x), \quad y\,(0) = y'\,(0) = 0$$

where $f(x) = 1$ *for* $0 < x < 1$

$= 0$ *for* $x > 1$.

SOLUTION: Taking L–transform on the given equation, we get

$$s^2\bar{y} - sy(0) - y'(0) + 3\,[s\bar{y} - y(0)] + 2\bar{y} = L(f(x))$$

where $f(x) = 1$ for $0 < x < 1$ and $f(x) = 0$ for $x > 1$, $y(0) = y'(0) = 0$

$$\therefore \quad L(f(x)) = \int_0^1 e^{-sx}.1\,dx + \int_1^\infty 0\,e^{-sx}dx = \frac{e^{-s}-1}{-s}$$

$$\therefore \quad \bar{y}\,(s^2+3s+2) = \frac{1-e^{-s}}{s}$$

$$\text{or} \quad \bar{y} = \frac{1}{s(s+1)\,(s+2)}(1-e^{-s}) = \left[\frac{1}{2s} - \frac{1}{s+1} + \frac{1}{2(s+2)}\right](1-e^{-s})$$

$$\therefore \quad y(x) = L^{-1}\left[\frac{1}{2s} - \frac{1}{s+1} + \frac{1}{2(s+2)}\right] - L^{-1}\left[\frac{1}{2s} - \frac{1}{s+1} + \frac{1}{2(s+2)}\right]e^{-s}$$

$$= \frac{1}{2} - e^{-t} + \frac{1}{2}e^{-2t} - \left[\frac{1}{2}U(t-1) - e^{-(t-1)}\,U(t-1) + \frac{1}{2}e^{-2(t-1)}\,U(t-1)\right]$$

$$= \left[\frac{1}{2} - e^{-t} + \frac{1}{2}e^{-2t}\right] - \left[\frac{1}{2} - e^{-(t-1)} + \frac{1}{2}e^{-2(t-1)}\right]U(t-1)$$ **Ans.**

11. *Solve the differential equation*

$$\frac{d^2y}{dt^2} + a^2y = f(t) \quad \text{with} \quad y(0) = y'(0) = 0$$

$$\text{where} \quad f(t) = 0 \quad \text{when} \quad 0 < t < \pi$$
$$= \sin t \quad \text{when} \quad \pi < t < 2\pi$$
$$= 0 \quad \text{when} \quad t > 2\pi.$$

SOLUTION: Taking L–transform on the given equation, we get

$$s^2\bar{y} - sy(0) - y'(0) + a^2\bar{y} = L(f(t)) \quad \text{where} \quad y(0) = y'(0) = 0$$

and $f(t) = 0$ when $0 < t < \pi$, $f(t) = \sin t$, $\pi < t < 2\pi$, $f(t) = 0$, $t > 2\pi$.

$$\therefore \quad (s^2+a^2)\,\bar{y} = \int_0^\pi e^{-st}.0\,dt + \int_\pi^{2\pi} e^{-st}\sin t\,dt + \int_{2\pi}^\infty e^{-st}.0\,dt$$

$$= \left[\frac{e^{-st}}{s^2+1}(-s\sin t - 1\cos t)\right]_\pi^{2\pi} = \frac{e^{-2\pi s}}{s^2+1}(-1) - \frac{e^{-\pi s}}{s^2+1}(1)$$

$$\text{or} \quad \bar{y} = \frac{(e^{-2\pi s} + e^{-\pi s})}{(s^2+1)\,(s^2+a^2)} = \frac{-1}{a^2-1}\left(\frac{1}{s^2+1} - \frac{1}{s^2+a^2}\right)(e^{-2\pi s} + e^{-\pi s})$$

$\therefore$ Taking inverse L' transform, we get

$$\therefore \quad y(t) = \frac{-1}{a^2-1}\left[\sin(t-2\pi) - \frac{1}{a}\sin a\,(t-2\pi)\right]U(t-2\pi) - \frac{1}{a^2-1}$$

$$\left[\sin(t-\pi) - \frac{1}{a}\sin a(t-\pi)\right]U(t-\pi)$$

For $0 < t < \pi$, $y(t) = 0$

For $\pi < t < 2\pi$, $y(t) = 0 - \frac{1}{a^2-1}\left[-\sin t - \frac{1}{a}(\sin at \cos a\pi - \cos at \sin a\pi)\right]$

$$= \frac{1}{a(a^2-1)}[a \sin t + \sin at \cos a\pi + \cos at \sin a\pi]$$

For $t > 2\pi$, $y(t) = \frac{1}{a(a^2-1)}[\sin at (\cos 2a\pi + \cos a\pi - \cos at (\sin 2a\pi + \sin a\pi)]$ **Ans.**

12. *A beam is simply supported at its one end x = 0 and is clamped at the other end x = l and is carrying a load W at x = l/4. Find the deflection of the beam at any point.*

SOLUTION: The differential equation for the deflection of a loaded beam is

$$\frac{d^4y}{dx^4} = \frac{w}{El}\delta(x - l/n) \quad \text{where} \quad y(0) = y''(0) = 0.$$

Taking L-transform of the differential equation on both sides, we get

$$s^4\bar{y} - s^3y(0) - s^2y'(0) - sy''(0) - y'''(0) = \frac{W}{El}e^{-sl/n}$$

or $$s^4\bar{y} = \frac{W}{El}e^{-sl/n} + s^2y'(0) + y'''(0) \quad \text{as} \quad y(0) = y''(0) = 0.$$

or $$\bar{y} = \frac{W}{El}\frac{e^{-ls/n}}{s^4} + -s^2 + s^4 \quad \text{where} \quad y'(0) = C_1 \quad \text{and} \quad y'''(0) = C_2$$

$\therefore$ $$y(x) = x + \frac{x^3}{3!} + El\,L^{-1}\left[\frac{e^{-ls/n}}{s^4}\right]$$

$$= 9x + \frac{C_2x^3}{6} + \frac{W}{6El}\left(x - \frac{l}{n}\right)^3 U\left(x - \frac{l}{n}\right)$$

$$= \begin{cases} c_1x + \frac{c_2}{6}x^3 & \text{when } 0 < x < \frac{l}{n} \\ c_1x + \frac{c_2}{6}x^3 & \text{when } \frac{l}{n} < x < l \end{cases}$$ **Ans.**

13. *Find the Laplace transform of the square wave function of period 'a' defined as*

$$f(t) = \begin{cases} 1 & \text{for} \quad 0 < t < a/2 \\ -1 & \text{for} \quad a/2 < t < a. \end{cases}$$

SOLUTION: $f(t)$ is 'a'-periodic function

$\therefore$ $$L(f(t)) = \frac{\int_0^a e^{-st}f(t)dt}{1-e^{-sa}} = \frac{1}{1-e^{-as}}\left[\int_0^{a/2} 1\cdot e^{-st}dt + \int_{a/2}^{a} -1e^{-st}dt\right]$$

$$= \frac{1}{1-e^{-as}}\left[\left\{\frac{e^{-st}}{-s}\right\}_0^{a/2} - \left\{\frac{e^{-st}}{-s}\right\}_{a/2}^{a}\right] = \frac{-1}{s(1-e^{-as})}\left[e^{-as/2} - 1 - e^{-as} + e^{-as/2}\right]$$

$$= \frac{1(1-e^{-as/2})^2}{s(1-e^{-as/2})(1+e^{-as/2})} = \frac{1-e^{-as/2}}{s(1+e^{-as/2})} = \frac{e^{+as/4} - e^{-as/4}}{s(e^{+as/4} + e^{-as/4})}$$

$$= \frac{1}{s}\tanh\left(\frac{as}{4}\right)$$ **Ans.**

14. *Find the Laplace transform of the function*

$$f(t) = \begin{cases} t, & 0 < t < c \\ 2c - t, & c < t < 2c \end{cases}$$

SOLUTION: $f(t) = t\,[U(t) - U(t-c)] + (2c - t)\,[U(t-c) - U(t-2c)]$

$= t\,U(t) - (t - c + c)\,U(t-c) + (t - 2c)\,U(t-2c) - (t - c - c)\,U(t-c)$

$= t\,U(t) - (t - c)\,U(t-c) - cU(t-c) + (t-2c)\,U(t-2c) - (t-c)\,U(t-c) + cu\,(t-c)$

$= t\,U(t) - 2(t-c)\,U(t-c) + (t-2c)\,U(t-2c)$

Using the property $L[f(t-a)\,U(t-a)] = e^{-as}\bar{f}(s)$, we get

$\therefore$ $$L(f(t)) = e^{-os}\frac{1}{s^2} - 2e^{-cs}\frac{1}{s^2} + e^{-2cs}\frac{1}{s^2} = \frac{1}{s^2}(1-e^{-cs})^2$$ **Ans.**

15. *An impulsive voltage $E\delta(t)$ is applied to a circuit consisting of L, R, C in series with zero initial conditions. If i be the current at any subsequent time t, find the limit of i as $t \to 0$.*

SOLUTION: In L, C, R circuit current i is given by

$$\frac{di}{dt} + \frac{R}{L}i + \frac{1}{LC}\int i\,dt = E\,\delta t$$

Taking Laplace transform on both sides of the above equation, we get

$$s\bar{i} - i(0) + 2\lambda i + n^2 \cdot \frac{1}{s}\bar{i} = E\,L(\delta t) = E$$

where $2\lambda = \frac{R}{L}$ and $n^2 = \frac{1}{LC}$.

or $(s^2 + 2\lambda s + n^2)\,\bar{i} = sE$ as $i(0) = 0$

or $$\bar{i} = \frac{s}{s^2 + 2\lambda s + n^2}E = \frac{(s + \lambda - \lambda)\,E}{(s+\lambda)^2 + n^2 - \lambda^2}$$

$\therefore$ $$i(t) = L^{-1}\frac{(s+\lambda-\lambda)\,E}{(s+\lambda)^2 + n^2 - \lambda^2} = E\left[e^{-\lambda t}\cos\sqrt{n^2-\lambda^2}\,t - \frac{\lambda e^{-\lambda t}}{\sqrt{n^2-\lambda^2}}\sin\sqrt{n^2-\lambda^2}\,t\right]$$ when $n > \lambda$.

$$= E\left[e^{-\lambda t}\cos\left(\sqrt{\lambda^2 - n^2}\right)t - \frac{\lambda}{\sqrt{\lambda^2-n^2}}e^{-\lambda t}\sinh\left(\sqrt{\lambda^2-n^2}\,t\right)\right]$$ when $\lambda > n$. **Ans.**

ADDITIONAL IMPORTANT PROBLEMS

1. *Find the Laplace transform of the functions*

(i) $f(t) = t,\ 0 < t < \dfrac{1}{2}$
$= t - 1,\ \dfrac{1}{2} < t < 1$
$= 0,\ t > 1$

(ii) $f(t) = 0,\ 0 < t < \pi$
$= \sin^2(t - \pi),\ t > \pi.$

SOLUTION: (i)

$$L\{f(t)\} = \int_0^\infty e^{-st} f(t)\,dt = \int_0^{1/2} te^{-st}dt + \int_{1/2}^{1} (t-1)\,e^{-st}dt + \int_1^\infty e^{-st}(0)\,dt$$

$$= \left[\frac{te^{-st}}{-s}\right]_0^{1/2} - \int_0^{1/2}\frac{e^{-st}}{-s}\,dt + \left[(t-1)\frac{e^{-st}}{-s}\right]_{1/2}^{1} - \int_{1/2}^{1}\frac{e^{-st}}{-s}\,dt$$

$$= \frac{-1}{2s}e^{-s/2} - \left[\frac{e^{-st}}{s^2}\right]_0^{1/2} - \frac{1}{2s}e^{-s/2} - \left[\frac{e^{-st}}{s^2}\right]_{1/2}^{1}$$

$$= -\frac{1}{s}e^{-s/2} - \frac{1}{s^2}(e^{-s/2} - 1 + e^{-s} - e^{-s/2})$$

$$= \frac{1}{s^2} - \frac{e^{-s}}{s^2} - \frac{e^{-s/2}}{s} = \frac{1-e^{-s}}{s^2} - \frac{e^{-s/2}}{s} \qquad \textbf{Ans.}$$

(ii)

$$L\{f(t)\} = \int_0^\infty e^{-st} f(t)\,dt = \int_0^\pi e^{-st}.0\,dt + \int_\pi^\infty e^{-st}\sin^2(t-\pi)\,dt$$

$$= \frac{1}{2}\int_\pi^\infty \{1 - \cos(2t - 2\pi)\}\,e^{-st}dt = \frac{1}{2}\int_\pi^\infty e^{-st}(1 - \cos 2t)\,dt$$

$$= \frac{1}{2}\int_\pi^\infty e^{-st}dt - \frac{1}{2}\int_\pi^\infty e^{-st}\cos 2t\,dt = \frac{-1}{2}\left[\frac{e^{-st}}{s}\right]_\pi^\infty - \frac{1}{2}\left[\frac{e^{-st}}{s^2+4}(-s\cos 2t + 2\sin 2t)\right]_\pi^\infty$$

$$= \frac{1}{2s}e^{-\pi s} + \frac{e^{-\pi s}}{2(s^2+4)}(-s) = \frac{e^{-\pi s}}{2}\left(\frac{1}{s} - \frac{s}{s^2+4}\right) \qquad \textbf{Ans.}$$

2. *If* $J_0(t) = \sum_{r=0}^{\infty}\dfrac{(-1)^r}{(r!)^2}\left(\dfrac{t}{2}\right)^{2r}$, *find* $L\{J_0(t)\}$.

SOLUTION:

$$J_0(t) = 1 - \left(\frac{t}{2}\right)^2 + \frac{1}{(2!)^2}\left(\frac{t}{2}\right)^4 - \frac{1}{(3!)^2}\left(\frac{t}{2}\right)^6 + \ldots$$

$$= 1 - \frac{t^2}{4} + \frac{1}{64}t^4 - \frac{1}{2304}t^6 + \ldots$$

$$\therefore \quad L\{J_0(t)\} = L(1) - \frac{1}{4}L(t^2) + \frac{1\,L(t^4)}{64} - \frac{1}{2304}L(t^6) + \ldots$$

$$= \frac{1}{s} - \frac{1}{4}\frac{2!}{s^3} + \frac{1}{64}\frac{4!}{s^5} - \frac{1}{2304}\frac{6!}{s^7} + \ldots$$

$$= \frac{1}{s}\left[1 - \frac{1}{2}\left(\frac{1}{s^2}\right) + \frac{3}{8}\left(\frac{1}{s^2}\right)^2 - \frac{15}{48}\left(\frac{1}{s^2}\right)^3 + \ldots\right]$$

$$= \frac{1}{s}\left(1 + \frac{1}{s^2}\right)^{-1/2} = \frac{1}{\sqrt{1+s^2}}$$

Ans.

3. *(i) If* $L\{f(t)\} = \log\frac{s+3}{s+1}$ *find* $L\{f(2t)\}$

(ii) If $L\{J_0(t)\} = \frac{1}{\sqrt{s^2+1}}$ *find* $L\{J_0(3t)\}$

SOLUTION: *(i)* $L\{f(t)\} = \log\frac{s+3}{s+1}$. We use the change of scale property that, if $L\{f(t)\} = F(s)$ then $L\{f(at)\} = \frac{1}{a}F\left(\frac{s}{a}\right)$.

$$\therefore \quad L\{f(2t)\} = \frac{1}{2}\log\frac{s/2+3}{s/2+1} = \frac{1}{2}\log\frac{s+6}{s+2}.$$

(ii) Since $L(J_0(t)) = \frac{1}{\sqrt{s^2+1}}$. Using change of scale property

$$L\{J_0(3t)\} = \frac{1}{3\sqrt{\left(\frac{s}{3}\right)^2+1}} = \frac{1}{\sqrt{s^2+9}}$$

4. *Find the Laplace transform of the following functions*

(i) $\frac{\cos 2t \sin t}{e^t}$ *(ii) sin 2t cos t cosh 2t*

SOLUTION: *(i)* $\frac{\cos 2t \sin t}{e^t} = e^{-t}\frac{\sin 3t - \sin t}{2} = \frac{1}{2}e^{-t}\sin 3t - \frac{1}{2}e^{-t}\sin t$

$$\therefore \quad L\left[\frac{\cos 2t \sin t}{e^t}\right] = \frac{1}{2}L(e^{-t}\sin 3t) - \frac{1}{2}L(e^{-t}\sin t)$$

$$= \frac{1}{2}\frac{3}{(s+1)^2+9} - \frac{1}{2}\frac{1}{(s+1)^2+1}$$

$$= \frac{s^2+2s-2}{(s^2+2s+10)(s^2+2s+2)}$$

(*ii*) $$f(t) = \sin 2t \cos t \cosh 2t = \left(\frac{\sin 3t + \sin t}{2}\right)\left(\frac{e^{2t}+e^{-2t}}{2}\right)$$

$\therefore$ $$L\{f(t)\} = \frac{1}{4} L\{e^{2t} \sin 3t + e^{2t} \sin t + e^{-2t} \sin 3t + e^{-2t} \sin t\}$$

$$= \frac{1}{4}\left[\frac{3}{(s-2)^2+9} + \frac{1}{(s-2)^2+1} + \frac{3}{(s+2)^2+9} + \frac{1}{(s+2)^2+1}\right]$$

$$= \frac{1}{4}\left[\frac{3(s^2+13)}{s^4+10s^2+169} + \frac{s^2+5}{s^4-6s^2+25}\right].$$ **Ans.**

5. *Find the Laplace transform of the following functions*

(*i*) $t\sqrt{1+\sin t}$ (*ii*) $t^2 e^t \sin 4t$

SOLUTION: (*i*) Let $f(t) = t\sqrt{1+\sin t} = t\left(\sin\frac{t}{2}+\cos\frac{t}{2}\right)$

Now $$L(\sqrt{1+\sin t}) = L\left[\sin^2\frac{t}{2} + \cos^2\frac{t}{2} + 2\sin\frac{t}{2}\cos\frac{t}{2}\right]^{1/2} = L\left(\sin\frac{t}{2}+\cos\frac{t}{2}\right)$$

$$= \frac{1/2}{s^2+\frac{1}{4}} + \frac{s}{s^2+\frac{1}{4}} = \frac{4s+2}{4s^2+1}$$

$\therefore$ $$L\{t\sqrt{1+\sin t}\} = \frac{-d}{ds}\frac{4s+2}{4s^2+1} = -\left[\frac{(4s^2+1)(4)-(4s+2)8s}{(4s^2+1)^2}\right]$$

$$= \frac{16s^2+16s-4}{(4s^2+1)^2} = 4\frac{(4s^2+4s-1)}{(4s^2+1)}$$ **Ans.**

(*ii*) $$L(\sin 4t) = \frac{4}{s^2+16}$$

$\therefore$ $$L(t^2 \sin 4t) = (-1)^2\frac{d^2}{ds^2}\frac{4}{s^2+16} = \frac{d}{ds}\left(\frac{-8s}{(4s^2+16)^2}\right)$$

$$= \frac{-8(s^2+16)^2+16s(s^2+16)(2s)}{(s^2+16)^4} = \frac{-8(s^2+16)+32s^2}{(s^2+16)^3}$$

$$= \frac{24s^2-128}{(s^2+16)^3} = \frac{8(3s^2-16)}{(s^2+16)^2}$$

Now applying first shifting theorem

$$L(t^2 e^t \sin 4t) = \frac{8\{3(s-1)^2-16\}}{\{(s-1)^2+16\}^3} = \frac{8(3s^2-6s-13)}{(s^2-2s+17)^3}$$ **Ans.**

6. *Find the Laplace transform of the following functions*

(*i*) $\dfrac{e^{-t}\sin t}{t}$ (*ii*) $\dfrac{\sin^2 t}{t^2}$

SOLUTION: (*i*) $L(\sin t) = \dfrac{1}{s^2+1} \quad \therefore \quad L(e^{-t}\sin t) = \dfrac{1}{(s+1)^2+1}$

Hence
$$L\left(\frac{e^{-t}\sin t}{t}\right) = \int_s^\infty \frac{ds}{(s+1)^2+1} = \left[\tan^{-1}(s+1)\right]_s^\infty$$

$$= \frac{\pi}{2} - \tan^{-1}(s+1) = \cot^{-1}(s+1) \quad \textbf{Ans.}$$

(*ii*)
$$L\left(\frac{\sin^2 t}{t^2}\right) = L\left[\frac{1-\cos 2t}{2t^2}\right] = \frac{1}{2}L\left[\frac{1-\cos 2t}{t^2}\right].$$

Since $L(1 - \cos 2t) = \dfrac{1}{s} - \dfrac{s}{s^2+4}$, we have

$$L\left(\frac{1-\cos 2t}{t}\right) = \int_s^\infty \left(\frac{1}{s} - \frac{s}{s^2+4}\right) ds = \left[\log s - \frac{1}{2}\log(s^2+4)\right]_s^\infty$$

$$= \left[\log\frac{s}{\sqrt{s^2+4}}\right]_s^\infty = \left[\log\frac{1}{\sqrt{\left(1+\frac{4}{s^2}\right)}}\right]_s^\infty = \log 1 - \log\frac{1}{\sqrt{\left(1+\frac{4}{s^2}\right)}} = \frac{1}{2}\log\frac{s^2+4}{s^2}$$

Hence
$$L\left(\frac{1-\cos 2t}{t^2}\right) = \int_s^\infty \frac{1}{2}\log\left(\frac{s^2+4}{s^2}\right) ds$$

$$= \frac{1}{2}\left[s\log\frac{s^2+4}{s^2}\right]_s^\infty - \int_s^\infty \frac{s.s^2}{s^2+4}\left[\frac{2s.s^2 - 2s(s^2+4)}{s^4}\right] ds$$

$$= \frac{1}{2}\left[-s\log\frac{s^2+4}{s^2} - \int_s^\infty \frac{-8}{s^2+4}\,ds\right] = \left[-s\log\frac{s^2+4}{s^2} + 8.\frac{1}{2}\left[\tan^{-1}\frac{s}{2}\right]_s^\infty\right]$$

$$= \frac{1}{2}\left[-s\log\frac{s^2+4}{s^2} + 4\left(\frac{\pi}{2} - \tan^{-1}\frac{s}{2}\right)\right]$$

$$\therefore \quad L\left(\frac{\sin^2 t}{t^2}\right) = \frac{1}{2}L\left(\frac{1-\cos 2t}{t^2}\right) = \frac{1}{4}\left[-s\log\frac{s^2+4}{s^2} + 4\cot^{-1}\frac{s}{2}\right] \quad \textbf{Ans.}$$

7. *If* $f(t) = t, \quad 0 \le t < 3$
$= 6, \quad t > 3$
find $L(f(t))$ *and* $L(f'(t))$.

SOLUTION: $$F(s) = L(f(t)) = \int_0^\infty e^{-st} f(t)\,dt = \int_0^3 te^{-st}\,dt + \int_3^\infty 6e^{-st}\,dt$$

or $$F(s) = \left[\frac{te^{-st}}{-s}\right]_0^3 - \int_0^3 1.\frac{e^{-st}}{-s}\,dt + 6\left[\frac{e^{-st}}{-s}\right]_3^\infty$$

$$= \frac{-3}{s}e^{-3s} - \left[\frac{e^{-st}}{s^2}\right]_0^3 + 6\left(0 + \frac{e^{-3s}}{s}\right) = \frac{1}{s^2} + e^{-3s}\left(\frac{3}{s} - \frac{1}{s^2}\right)$$ **Ans.**

$\therefore$ $$L(f'(t)) = s\,F(s) - f(0) = s\left[\frac{1}{s^2} + e^{-3s}\left(\frac{3}{s} - \frac{1}{s^2}\right)\right] - 0$$

$$= \frac{1}{s} + e^{-3s}\left(3 - \frac{1}{s}\right)$$ **Ans.**

8. *Find the Laplace transform of the following functions*

(*i*) $\int_0^t t\,e^{-4t} \sin 3t\,dt$ (*ii*) $e^{-t}\int_0^t \frac{\sin t}{t}\,dt.$

SOLUTION: (*i*) $$L(t \sin 3t) = \frac{-d}{ds} L(\sin 3t) = \frac{-d}{ds}\frac{3}{s^2+9} = \frac{6s}{(s^2+9)^2}$$

$\therefore$ $$L(t\,e^{-4t} \sin 3t) = \frac{6\,(s+4)}{[(s+4)^2+9]^2} = \frac{6(s+4)}{(s^2+8s+25)^2}$$

$\therefore$ $$L\int_0^t t\,e^{-4t} \sin 3t\,dt = \frac{1}{s} L(t\,e^{-4t} \sin 3t) = \frac{6\,(s+4)}{s(s^2+8s+25)^2}$$

(*ii*) $$L\left(\frac{\sin t}{t}\right) = \int_s^\infty L(\sin t)\,ds = \int_s^\infty \frac{1}{s^2+1}\,ds$$

$$= \left[\tan^{-1}\right]_s^\infty = \frac{\pi}{2} - \tan^{-1} s = \cot^{-1} s$$

$\therefore$ $$L\left[\int_0^t \frac{\sin t}{t}\,dt\right] = \frac{1}{s} L\left(\frac{\sin t}{t}\right) = \frac{1}{S}\cot^{-1} s$$

$\therefore$ $$L\left[e^{-t}\int_0^t \frac{\sin t}{t}\,dt\right] = \frac{1}{s+1}\cot^{-1}(s+1)$$ **Ans.**

9. *Show that* $\int_0^\infty e^{-3t}\, t \sin t \, dt = \frac{3}{50}$.

SOLUTION: $L(t \sin t) = \frac{-d}{ds}\frac{1}{s^2+1} = \frac{2s}{(s^2+1)^2}$

or $$\int_0^\infty e^{-st} t \sin t \, dt = \frac{2s}{(s^2+1)^2}$$

Putting $s = 3$ here, we get

$$\int_0^\infty e^{-3t} t \sin t \, dt = \frac{6}{(9+1)^2} = \frac{3}{50}$$ **Hence proved.**

10. *Show that* $\int_0^\infty e^{-t}\frac{\sin^2 t}{t}\, dt = \frac{1}{4}\log 5$.

SOLUTION: By definition $L\left(\frac{\sin^2 t}{t}\right) = \int_0^\infty e^{-st}\frac{\sin^2 t}{t}\, dt = L\left(\frac{1-\cos 2t}{2t}\right)$

$$= \frac{1}{2}\int_s^\infty \left(\frac{1}{s} - \frac{s}{s^2+4}\right) ds = \frac{1}{2}\left[\log s - \frac{1}{2}\log(s^2+4)\right]_s^\infty$$

$$= \frac{1}{2}\left[\log\frac{s}{\sqrt{s^2+4}}\right]_s^\infty = \frac{1}{2}\left[\lim_{s\to\infty}\log\frac{s}{\sqrt{s^2+4}} - \log\frac{s}{\sqrt{s^2+4}}\right]$$

$$= \frac{1}{2}\left[\log 1 - \log\frac{s}{\sqrt{s^2+4}}\right] = \frac{1}{2}\log\sqrt{\left(\frac{s^2+4}{s}\right)}$$

Putting $s = 1$, we get $$\int_0^\infty e^{-t}\frac{\sin^2 t}{t}\, dt = \frac{1}{2}\log\frac{\sqrt{5}}{1} = \frac{1}{4}\log 5.$$ **Ans.**

11. *Evaluate* $\int_0^\infty \frac{\cos 6t - \cos 4t}{t}\, dt$.

SOLUTION: By definition

$$L\left(\frac{\cos 6t - \cos 4t}{t}\right) = \int_0^\infty e^{-st}\frac{\cos 6t - \cos 4t}{t}\, dt$$

$$= \int_s^\infty L(\cos 6t - \cos 4t)\, ds = \int_s^\infty \left(\frac{s}{s^2+6^2} - \frac{s}{s^2+4^2}\right) ds$$

$$= \frac{1}{2}\left[\log\left(\frac{s^2+36}{s^2+16}\right)\right]_0^{\infty} = \frac{1}{2}\left[\lim_{s\to\infty}\log\frac{s^2+36}{s^2+36} - \log\frac{s^2+36}{s^2+16}\right]$$

$$= \frac{1}{2}\left[\lim_{s\to\infty}\log\frac{1+36/s^2}{1+16/s^2} - \log\frac{s^2+36}{s^2+16}\right]$$

$$= \frac{1}{2}\left[\log 1 - \log\frac{s^2+36}{s^2+16}\right] = \frac{1}{2}\log\frac{s^2+16}{s^2+36}$$

Putting $s = 0$, we get

$$\int_0^{\infty}\frac{\cos 6t - \cos 4t}{t}\,dt = \frac{1}{2}\log\frac{16}{36} = \frac{1}{2}\log\left(\frac{2}{3}\right)^2 = \log\frac{2}{3}.$$ **Ans.**

12. *Find the Laplace transformation of (i) $t^2 \cup (t-2)$ (ii) $e^{-t}\sin t \cup (t-\pi)$.*

SOLUTION: (*i*) By second shifting theorem $L[f(t-a)\cup(t-a)] = e^{-as}f(s)$.

Here

$$t^2\cup(t-2) = [(t-2)^2 + 4t - 4]\cup(t-2)$$

$$= (t-2)^2\cup(t-2) + 4(t-2)\cup(t-2) + 4\cup(t-2)$$

$$\therefore \quad L[t^2\cup(t-2)] = L[(t-2)^2\cup(t-2)] + 4\,L\{(t-2)\cup(t-2)] + 4\,L[\cup(t-2)]$$

$$= e^{-2s}L(t^2) + 4\,e^{-2s}L(t) + 4e^{-2s}L(1)$$

$$= e^{-2s}\left(\frac{2}{s^3}\right) + 4e^{-2s}\left(\frac{1}{s^2}\right) + 4\left(\frac{1}{s}\right)e^{-2s}$$

$$= e^{-2s}\left[\frac{2}{s^3} + \frac{4}{s^2} + \frac{4}{s}\right]$$ **Ans.**

(*ii*)

$$e^{-t}\sin t\cup(t-\pi) = e^{-t}\sin(t-\pi+\pi)\cup(t-\pi)$$

$$= -e^{-t}\sin(t-\pi)\cup(t-\pi)$$

$$\therefore \quad L[e^{-t}\sin t\cup(t-\pi)] = -L\,[e^{-t}\sin(t-\pi)\cup(t-\pi)]$$

$$= -\left[\frac{e^{-\pi s}}{s^2+1}\right]_{s\to s+1} = -\frac{e^{-\pi(s+1)}}{(s+1)^2+1}$$

$$= -\frac{e^{-\pi(s+1)}}{s^2+2s+2}.$$ **Ans.**

13. *Find the Laplace transform of the following functions*

(i) $f(t) = t^2,\ 0 < t < 1$
$= 4t,\ t > 1$

(ii) $f(t) = \sin 2t,\ 2\pi < t < 4\pi$
$= 0$ otherwise

SOLUTION: (*i*) Writing $f(t)$ in terms of unit step function

$$f(t) = t^2\cup(t) - t^2\cup(t-1) + 4t\cup(t-1)$$

$$= t^2\cup(t) - [(t-1)^2 + 2t - 1]\cup(t-1) + 4\,(t-1+1)\cup(t-1)$$

$$= t^2 \cup (t) - (t-1)^2 \cup (t-1) + 2(t-1) \cup (t-1) + \cup (t-1)$$
$$+ 4(t-1) \cup (t-1) + 4 \cup (t-1)$$

$$\therefore \quad L(f(t)) = \frac{2}{s^3} - e^{-s} \cdot \frac{2}{s^3} + 2e^{-s}\frac{1}{s^2} + \frac{e^{-s}}{s} + 4e^{-s}\frac{1}{s^2} + \frac{4e^{-s}}{s}$$

$$= \frac{2}{s^3} + e^{-s}\left(-\frac{2}{s^3} + \frac{2}{s^2} + \frac{5}{s}\right)$$

(*ii*)
$$f(t) = \sin 2t,\ 2\pi < t < 4\pi$$
$$= 0 \quad ,\ \text{otherwise}$$
$$f(t) = \sin 2t\,[\cup (t-2\pi) - \cup (t-4\pi)]$$
$$= \sin [2(t-2\pi) - 4\pi] \cup (t-2\pi)] - \sin [2(t-4\pi) + 8\pi] \cup (t-4\pi)$$
$$= \sin (2t - 4\pi) \cup (t-2\pi) - \sin [(2t-8\pi) \cup (t-4\pi)]$$

$$\therefore \quad L(f(t) = e^{-2\pi s} \cdot \frac{2}{s^2+4} - e^{-4\pi s}\frac{2}{s^2+4} = \frac{2}{s^2+4}(e^{-2\pi s} - e^{-4\pi s})$$ **Ans.**

14. *Find the Laplace transform of f(t) when*

(*i*) $f(t) = \sin 2t \cdot \delta\left(t - \frac{\pi}{4}\right) - t^2\delta(t-2)$ (*ii*) $f(t) = t \cup (t-4) + t^2\,\delta\,(t-4)$

SOLUTION: (*i*) We know that $L[f(t)\,\delta\,(t-a)] = e^{-as} f(a)$.

$$\therefore \quad L\left[\sin 2t\,\delta\left(t - \frac{\pi}{4}\right) - t^2\,\delta\,(t-2)\right] = e^{-\pi s/4}\sin\left(2 \cdot \frac{\pi}{4}\right) - 2^2 \cdot e^{-2s}$$

Therefore
$$L(f(t)) = e^{-\pi s/4} - 4e^{-2s}$$ **Ans.**

(*ii*)
$$L(t \cup (t-4) + t^2\,\delta\,(t-4)] = L[(t-4) \cup (t-4) + 4 \cup (t-4) + t^2\,\delta\,(t-4)]$$

$$= e^{-4s}\frac{1}{s^2} + \frac{4e^{-4s}}{s} + e^{-4s}.(4^2) = e^{-4s}\left[\frac{1}{s^2} + \frac{4}{s} + 16\right]$$ **Ans.**

15. *Find the Laplace transform of the following functions*

(*i*) $f(t) = t^2,\ 0 < t < 2$ (*ii*) $f(t) = e^t,\ 0 < t < 2\pi$

$f(t+2) = f(2)$ $f(t + 2\pi) = f(t)$

SOLUTION: (*i*) The function is periodic with period 2.

$$L[f(t)] = \frac{1}{1-e^{-2s}}\int_0^2 e^{-st} f(t)\,dt = \frac{1}{1-e^{-2s}}\int_0^2 t^2 e^{-st}\,dt$$

$$= \frac{1}{1-e^{-2s}}\left[t^2\left(\frac{e^{-st}}{-s}\right) - 2t\left(\frac{e^{-st}}{s^2}\right) + 2\left(\frac{e^{-st}}{-s^3}\right)\right]_0^2$$

$$= \frac{1}{1-e^{-2s}}\left[-\frac{4}{s}e^{-2s} - \frac{4}{s^2}e^{-2s} - \frac{2}{s^3}e^{-2s} + \frac{2}{s^3}\right]$$

$$= \frac{1}{1-e^{-2s}}\left[\frac{2}{s^3} - e^{-2s}\left(\frac{4}{s}+\frac{4}{s^2}+\frac{2}{s^3}\right)\right]$$ **Ans.**

(*ii*) The function is 2π-periodic function

$$\therefore \quad L[f(t)] = \frac{1}{1-e^{-2\pi s}}\int_0^{2\pi} e^{-st} f(t)\, dt = \frac{1}{1-e^{-2\pi s}}\int_0^{2\pi} e^{-st} e^t\, dt$$

$$= \frac{1}{1-e^{-2\pi s}}\int_0^{2\pi} e^{(1-s)t}\, dt = \frac{1}{1-e^{-2\pi s}}\left[\frac{e^{(1-s)t}}{1-s}\right]_0^{2\pi}$$

$$= \frac{1}{1-e^{-2\pi s}}\left[\frac{e^{(1-s)\,2\pi}}{1-s} - \frac{1}{1-s}\right] = \frac{e^{(1-s)2\pi}-1}{(1-s)\,(1-e^{-2\pi s})}.$$ **Ans.**

16. *Find the inverse Laplace transform of the following functions*

(*i*) $\frac{s}{(s^2+1)\,(s^2+4)}$ (*ii*) $\frac{s+2}{(s^2+4s+8)\,(s^2+4s+13)}$ (*iii*) $\frac{s}{s^4+s^2+1}$.

SOLUTION: (*i*) $F(s) = \frac{s}{(s^2+1)\,(s^2+4)} = \frac{s}{3}\left[\frac{1}{s^2+1} - \frac{1}{s^2+4}\right] = \frac{1}{3}\left[\frac{s}{s^2+1} - \frac{s}{s^2+4}\right]$

$$\therefore \quad L^{-1}[F(s)] = \frac{1}{3}[\cos t - \cos 2t]$$

(*ii*)
$$F(s) = \frac{s+2}{(s^2+4s+8)\,(s^2+4s+13)} = \frac{(s+2)}{5}\left[\frac{1}{s^2+4s+8} - \frac{1}{s^2+4s+13}\right]$$

$$= \frac{1}{5}\left[\frac{s+2}{(s+2)^2+4} - \frac{s+2}{(s+2)^2+9}\right]$$

$$\therefore \quad L^{-1}[F(s)] = \frac{1}{5}\left[L^{-1}\left(\frac{s+2}{(s+2)^2+4}\right) - L^{-1}\left(\frac{s+2}{(s+2)^2+9}\right)\right]$$

$$= \frac{e^{-2t}}{5} L^{-1}\left(\frac{s}{s^2+4}\right) - \frac{e^{-2t}}{5} L^{-1}\left(\frac{s}{s^2+9}\right)$$

$$= \frac{e^{-2t}}{5}(\cos 2t - \cos 3t).$$ **Ans.**

(*iii*)
$$F(s) = \frac{s}{s^4+s^2+1} = \frac{s}{s^4+2s^2+1-s^2} = \frac{s}{(s^2+1)^2-s^2}$$

$$= \frac{s}{(s^2+s+1)\,(s^2-s+1)} = \frac{1}{2}\left[\frac{1}{s^2+1-s} - \frac{1}{s^2+1+s}\right]$$

$$= \frac{1}{2}\left[\frac{1}{\left(s-\frac{1}{2}\right)^2+\frac{3}{4}} - \frac{1}{\left(s+\frac{1}{2}\right)^2+\frac{3}{4}}\right]$$

$$\therefore \quad L^{-1}[F(s)] = \frac{1}{2}L^{-1}\left[\frac{1}{\left(s-\frac{1}{2}\right)^2+\left(\frac{\sqrt{3}}{2}\right)^2} - \frac{1}{\left(s+\frac{1}{2}\right)^2+\left(\frac{\sqrt{3}}{2}\right)^2}\right]$$

$$= \frac{1}{2}e^{t/2}L^{-1}\left[\frac{1}{s^2+\left(\frac{\sqrt{3}}{2}\right)^2}\right] - \frac{1}{2}e^{-t/2}L^{-1}\left[\frac{1}{s^2+\left(\frac{\sqrt{3}}{2}\right)^2}\right]$$

$$= \frac{1}{2}e^{t/2}\frac{2}{\sqrt{3}}\sin\frac{\sqrt{3}t}{2} - \frac{1}{2}e^{-t/2}\frac{2}{\sqrt{3}}\sin\frac{\sqrt{3}t}{2}$$

$$= \frac{2}{\sqrt{3}}\sin\frac{\sqrt{3}t}{2}\left(\frac{e^{t/2}-e^{-t/2}}{2}\right) = \frac{2}{\sqrt{3}}\sin\frac{\sqrt{3}t}{2}\sinh\frac{t}{2}$$ **Ans.**

17. *Use convolution theorem to find the inverse Laplace transform of*

(*i*) $\dfrac{s(s+1)}{(s^2+1)(s^2+2s+2)}$ (*ii*) $\dfrac{(s+2)^2}{(s^2+4s+8)^2}$

SOLUTION: (*i*) Let $F_1(s) = \dfrac{s}{s^2+1}$ and $F_2(s) = \dfrac{s+1}{s^2+2s+2} = \dfrac{s+1}{(s+1)^2+1}$

$\therefore \quad f_1(t) = L^{-1}(F_1(s)) = \cos t$ and $f_2(t) = L^{-1}(F_2(s)) = e^{-t}\cos t.$

By convolution theorem, if $F(s) = F_1(s)\,F_2(s)$, we have

$$L^{-1}[F(s)] = \int_0^t e^{-u}\cos u\cos(t-u)\,du$$

$$= \frac{1}{2}\int_0^t e^{-u}[\cos t + \cos(2u-t)\,du$$

$$= \frac{1}{2}\cos t\int_0^t e^{-u}du + \frac{1}{2}\int_0^t e^{-u}\cos(2u-t)\,du$$

$$= \frac{1}{2}\cos t\,(1-e^{-t}) + \frac{1}{2}\int_0^t e^{-u}[\cos 2u\cos t + \sin 2u\sin t]\,du$$

$$= \frac{1}{2}\cos t(1-e^{-t}) + \frac{1}{2}\cos t\int_0^t e^{-u}\cos 2u\,du + \frac{1}{2}\sin t\int_0^t e^{-u}\sin 2u\,du$$

$$= \frac{1}{2}\cos t\,(1-e^{-t}) + \frac{1}{2}\cos t\left[\frac{e^{-u}}{1+4}(-\cos 2u + 2\sin 2u)\right]_0^t + \frac{1}{2}\sin t\left[\frac{e^{-u}}{1+4}(-\sin 2u - 2\cos 2u\right]_0^t$$

$$= \frac{1}{2}\cos t\,(1-e^{-t}) + \frac{1}{2}\cos t\left[\frac{e^{-t}}{1+4}(-\cos 2t + 2\sin 2t) + \frac{1}{5}\right] + \frac{1}{2}\sin t\left[\frac{e^{-t}}{1+4}(-\sin 2t - 2\cos 2t) + \frac{2}{5}\right]$$

$$= \frac{1}{2}\cos t(1-e^{-t}) + \frac{e^{-t}}{10}[-\cos t\cos 2t - \sin t\sin 2t + 2\sin 2t\cos t - 2\cos 2t\sin t + \cos t + 2\sin t]$$

$$= \frac{1}{2}\cos t(1-e^{-t}) + \frac{e^{-t}}{10}[-\cos t + 2\sin t + \cos t + 2\sin t]$$

$$= \frac{1}{2}\cos t\,(1-e^{-t}) + \frac{2}{5}e^{-t}\sin t. \qquad \textbf{Ans.}$$

(*ii*) Let $F(s) = F_1(s)\,F_2(s)$ where $F_1(s) = \dfrac{s+2}{s^2+4s+8}$, $F_2(s) = \dfrac{s+2}{s^2+4s+8} = F_1(s)$

$$L^{-1}(F_1(s)) = L^{-1}\frac{s+2}{s^2+4s+8} = L^{-1}\frac{s+2}{(s+2)^2+4} = e^{-2t}\cos 2t = L^{-1}(F_2(s))$$

By convolution theorem

$$L^{-1}[F(s)] = \int_0^t e^{-2u}\cos 2u\, e^{-2(t-u)}\cos 2(t-u)\,du$$

$$= \frac{e^{-2t}}{2}\int_0^t [\cos 2t + \cos(4u-2t)]\,du$$

$$= \frac{1}{2}e^{-2t}\left[u\cos 2t + \frac{1}{4}\sin(4u-2t)\right]_0^t$$

$$= \frac{1}{2}e^{-2t}\left[t\cos 2t + \frac{1}{4}\sin 2t + \frac{1}{4}\sin 2t\right]$$

$$= \frac{1}{2}e^{-2t}\left[t\cos 2t + \frac{1}{2}\sin 2t\right] = \frac{e^{-2t}}{4}[\sin 2t + 2t\cos 2t] \qquad \textbf{Ans.}$$

18. *Find the inverse Laplace transform of*

(*i*) $log\ \dfrac{s^2+a^2}{(s+b)^2}$ (*ii*) $cot^{-1}(s+1)$ (*iii*) $tan^{-1}\left(\dfrac{2}{s^2}\right)$

SOLUTION: (*i*) $F(s) = \log\dfrac{s^2+a^2}{(s+b)^2} = \log(s^2+a^2) - 2\log(s+b)$

$$\therefore \quad F'(s) = \frac{2s}{s^2+a^2} - \frac{2}{s+b}$$

Using the property that $L(t\,f(t)) = -\dfrac{d}{ds}F(s)$ where $F(s) = L(f(t))$

$$\Rightarrow \qquad t\,f(t) = -\,\mathrm{L}^{-1}[F'(s)] = -\,\mathrm{L}^{-1}\left(\frac{2s}{s^2+a^2} - \frac{2}{s+b}\right)$$

$$= -\,2\cos at + 2e^{-bt}$$

$$\Rightarrow \qquad f(t) = \frac{1}{t}[-2\cos at + 2e^{-bt}] = \frac{2}{t}[e^{-bt} - \cos at] \qquad \textbf{Ans.}$$

(ii) $$F(s) = \cot^{-1}(s+1) \quad \therefore \quad F'(s) = \frac{-1}{s^2+1}$$

$$\therefore \qquad \mathrm{L}^{-1}[F(s)] = -\frac{1}{t}\mathrm{L}^{-1}\left(\frac{-1}{s^2+1}\right) = \frac{1}{t}\sin t \qquad \textbf{Ans.}$$

(iii) $$F(s) = \tan^{-1}\left(\frac{2}{s^2}\right) \quad \therefore \quad F'(s) = \frac{1}{1+\dfrac{4}{s^4}}\left(\frac{-4}{s^3}\right) = \frac{-4s}{s^4+4}$$

$$\therefore \qquad \mathrm{L}^{-1}[F(s)] = \frac{-1}{t}\mathrm{L}^{-1}\left(\frac{-4s}{s^4+4}\right) = \frac{4}{t}\mathrm{L}^{-1}\left(\frac{s}{s^4+4}\right)$$

$$= \frac{4}{t}L^{-1}\left\{\frac{s}{(s^2+2)^2-(2s)^2}\right\} = \frac{4}{t}L^{-1}\left\{\frac{s}{(s^2-2s+2)\,(s^2+2s+2)}\right\}$$

$$= \frac{1}{t}\mathrm{L}^{-1}\left\{\frac{1}{s^2-2s+2} - \frac{1}{s^2+2s+2}\right\} = \frac{1}{t}\mathrm{L}^{-1}\left[\frac{1}{(s-1)^2+1} - \frac{1}{(s+1)^2+1}\right]$$

$$= \frac{1}{t}(e^t\sin t - e^{-t}\sin t) = \frac{2}{t}\sin t\sinh t\,. \qquad \textbf{Ans.}$$

19. *Find the inverse Laplace transform of*

(i) $\dfrac{e^{4-3s}}{(s+4)^{5/2}}$ *(ii)* $\dfrac{s\,e^{-2s}}{s^2+2s+2}$

SOLUTION: *(i)* Let $$F(s) = \frac{1}{(s+4)^{5/2}}$$

$$\therefore \qquad \mathrm{L}^{-1}(F(s)) = e^{-4t}\mathrm{L}^{-1}\left(\frac{1}{s^{5/2}}\right) = e^{-4t}\frac{t^{3/2}}{\left|\overline{\frac{5}{2}}\right.} = \frac{e^{-4t}\,t^{3/2}}{\frac{3}{2}\cdot\frac{1}{2}\left|\overline{\frac{1}{2}}\right.} = \frac{4e^{-4t}}{3\sqrt{\pi}}\,t^{3/2}$$

Hence $$\mathrm{L}^{-1}[e^{4-3s}F(s)] = \frac{e^4.4}{3\sqrt{\pi}}\,e^{-4(t-3)}\,(t-3)^{3/2} \cup (t-3) \qquad \textbf{Ans.}$$

(ii) Let $$F(s) = \frac{s}{s^2+2s+2} \quad \therefore \quad \mathrm{L}^{-1}[F(s)] = \mathrm{L}^{-1}\frac{s+1-1}{(s+1)^2+1} = e^{-t}(\cos t - \sin t)$$

$$\therefore \qquad \mathrm{L}^{-1}[e^{-2s}F(s)] = e^{-(t-2)}\,[\cos(t-2) - \sin(t-2)] \qquad \textbf{Ans.}$$

20. *Solve* $\dfrac{dy}{dt} + y = \cos 2t,\quad y(0) = 1$ *using Laplace transformation.*

SOLUTION: Taking Laplace transform on both sides, we get

$$s\bar{y} - y(0) + \bar{y} = \frac{s}{s^2+4} \quad \text{or} \quad (s+1)\bar{y} = \frac{s}{s^2+4} + 1 \qquad (\text{as } y(0) = 1)$$

$$\text{or} \quad \bar{y} = \frac{s^2+s+4}{(s+1)(s^2+4)} = \frac{4}{5(s+1)} + \frac{1}{5}\frac{s}{s^2+4} + \frac{4}{5}\cdot\frac{1}{s^2+4} \qquad (\text{using partial fractions})$$

Taking inverse Laplace transform, we get

$$y(t) = \frac{4}{5}e^{-t} + \frac{1}{5}\cos 2t + \frac{2}{5}\sin 2t\,. \qquad \textbf{Ans.}$$

21. *Solve, by Laplace transform,* $\dfrac{dy}{dt} + 2y + \int_0^t y\,dt = \sin t,\quad y(0) = 1.$

SOLUTION: Taking Laplace transform on both sides, we get

$$s\bar{y} - y(0) + 2\bar{y} + \frac{1}{s}\bar{y} = \frac{1}{s^2+1}$$

$$\text{or} \quad \bar{y}\left[s + \frac{1}{s} + 2\right] = 1 + \frac{1}{s^2+1} = \frac{s^2+2}{s^2+1}$$

$$\text{or} \quad \bar{y} = \frac{s(s^2+2)}{(s^2+1)(s+1)^2}$$

$$\text{or} \quad \bar{y} = \frac{1}{s+1} - \frac{3}{2}\frac{1}{(s+1)^2} + \frac{1}{2}\frac{1}{s^2+1} \qquad (\text{on resolving into partial fraction}).$$

Taking inverse Laplace transform, we get

$$y(t) = e^{-t} - \frac{3}{2}t\,e^{-t} + \frac{1}{2}\sin t \qquad \textbf{Ans.}$$

22. *Solve the equation* $y'' + 3y' + 2y = t\,\delta(t-1),\quad y(0) = 0,\quad y'(0) = 0.$

SOLUTION: Taking Laplace transform on both sides, we get

$$s^2\bar{y} - sy(0) - y'(0) + 3\{s\bar{y} - y(0)\} + 2\bar{y} = e^{-s}.[t]_{t=1} = e^{-s}.$$

$$\text{or} \quad \bar{y}(s^2+3s+2) = e^{-s} \quad (\text{as } y(0) = y'(0) = 0)$$

$$\text{or} \quad \bar{y} = \frac{e^{-s}}{s^2+3s+2} = \frac{e^{-s}}{(s+1)(s+2)} = e^{-s}\left[\frac{1}{s+1} - \frac{1}{s+2}\right]$$

Taking inverse Laplace transform, we get

$$y(t) = e^{-(t-1)} \cup (t-1) - e^{-2(t-1)} \cup (t-1). \qquad \textbf{Ans.}$$

23. *Solve, using Laplace transform, the equation*

$$\frac{d^2y}{dt^2}+4y=f(t) \quad \text{where} \quad f(t)=1, \; 0<t<1$$

$$= 0, \; t>1$$

Given that $y(0) = 0$ *and* $y'(0) = 1$.

SOLUTION: Here $f(t) = \cup(t) - \cup(t-1)$.

Taking Laplace transform on both sides, we get

$$s^2\bar{y} - sy(0) - y'(0) + 4\bar{y} = \frac{1}{s} - \frac{e^{-s}}{s}$$

or $$\bar{y}(s^2+4) = \frac{1}{s} - \frac{e^{-s}}{s} + 1 \quad \text{or} \quad \bar{y} = \frac{1}{s(s^2+4)} - \frac{e^{-s}}{s(s^2+4)} + \frac{1}{s^2+4}$$

or $$\bar{y} = \frac{1}{4s} - \frac{1}{4}\frac{s}{(s^2+4)} - e^{-s}\frac{1}{4s} + e^{-s}.\frac{1}{4}\frac{s}{(s^2+4)} + \frac{1}{s^2+4}$$

Taking inverse Laplace transform, we get

$$y(t) = \frac{1}{4} - \frac{1}{4}\cos 2t - \frac{1}{4}\cup(t-1) + \frac{1}{4}\cos 2(t-1)\cup(t-1) + \frac{1}{2}\sin 2t.$$ **Ans.**

24. *Solve the equation* $\frac{dx}{dt}+5x-2y=t$ *and* $\frac{dy}{dl}+2x+y=0$ *where* $x(0) = 0$ *and* $y(0) = 0$.

SOLUTION: Taking L'transform on both the equations, we get

$$s\bar{x} - x(0) + 5\bar{x} - 2\bar{y} = \frac{1}{s^2} \quad \text{or} \quad (s+5)\bar{x} - 2\bar{y} = \frac{1}{s^2}$$

and $$2\bar{x} + s\bar{y} - y(0) + \bar{y} = 0 \quad \text{or} \quad 2\bar{x} + (s+1)\bar{y} = 0$$

Eliminating $\bar{y}$ in the above equations, we get

$$\bar{x} = \frac{s+1}{s^2(s+3)^2} = \frac{1}{27s} + \frac{1}{9s^2} - \frac{1}{27(s+3)} - \frac{2}{9}\frac{1}{(s+3)^2}$$

$\Rightarrow$ $$x(t) = \frac{1}{27} + \frac{1}{9}t - \frac{1}{27}e^{-3t} - \frac{2}{9}te^{-3t}$$

Substituting for x in first given equation, we get

$$2y = \frac{dx}{dt} + 5x - t$$

$$= \frac{1}{9} + \frac{3}{27}e^{-3t} - \frac{2}{9}e^{-3t} + \frac{2}{3}te^{-3t} + \frac{5}{27} + \frac{5}{9}t - \frac{5}{27}e^{-3t} - \frac{10}{9}te^{-3t} - t$$

$$= \frac{8}{27} - \frac{4}{9}t - \frac{8}{27}e^{-3t} - \frac{4}{9}te^{-3t}$$

$\therefore$ $$y = \frac{4}{27} - \frac{2}{9}t - \frac{4}{27}e^{-3t} - \frac{2}{9}te^{-3t}.$$ **Ans.**

25. *Solve* $\frac{d^2x}{dt^2} - \frac{dy}{dt} = t\,e^{-t} - 2e^{-t} - 3$ *and* $\frac{dx}{dt} - 2y - x = -2t\,e^{-t} + e^{-t} - 6t$

using Laplace transform, given that $x(0) = 0,\ x'(0) = 1$ *and* $y(0) = 0.$

SOLUTION: Taking Laplace transform on both the equations, we get

$$s^2\bar{x} - sx(0) - x'(0) - [s\bar{y} - y(0)] = \frac{1}{(s+1)^2} - \frac{2}{s+1} - \frac{3}{s} \quad \text{or} \quad s^2\bar{x} - s\bar{y} = 1 - \frac{2s+1}{(s+1)^2} - \frac{3}{s}$$

and $$s\bar{x} - x(0) - 2\bar{y} - \bar{x} = \frac{-2}{(s+1)^2} + \frac{1}{s+1} - \frac{6}{s^2}$$

or $$s^2\bar{x} - s\bar{y} = \frac{s^2}{(s+1)^2} - \frac{3}{s} \quad \text{and} \quad (s-1)\bar{x} - 2\bar{y} = \frac{s-1}{(s+1)^2} - \frac{6}{s^2}$$

Eliminating $\bar{y}$ in the above equations, we get

$$\frac{s(s+1)}{2}\bar{x} = \frac{s^2+s}{2(s+1)^2} \quad \text{or} \quad \bar{x} = \frac{1}{(s+1)^2} \Rightarrow x(t) = t\,e^{-t}.$$

Substituting for x in the second given equation, we get

$$2y = \frac{dx}{dt} - x + 2t\,e^{-t} - e^{-t} + 6t$$

$$= -t\,e^{-t} + e^{-t} - t\,e^{-t} + 2\,t\,e^{-t} - e^{-t} + 6t = 6t$$

Therefore, $y(t) = 3t$.

Thus the solution is $x = t\,e^{-t}$ and $y = 3t$. **Ans.**

CHAPTER

19 Fourier Series and Harmonic Analysis

IMPORTANT FORMULAE AND USEFUL RESULTS

1. Fourier series of a 2π-periodic Function:

$f(x) = \frac{1}{2} a_0 + \sum_{n=1}^{\infty} (a_n \cos nx + b_n \sin nx)$ if $f(x)$ is defined in $(0, 2\pi)$ where $a_0 = \frac{1}{\pi}\int_0^{2\pi} f(x)\,dx$,

$a_n = \frac{1}{\pi}\int_0^{2\pi} f(x)\cos nx\,dx, \quad b_n = \frac{1}{\pi}\int_0^{2\pi} f(x)\sin nx\,dx.$

And, if $f(x)$ is defined in $(-\pi, \pi)$ then

$$a_0 = \frac{1}{\pi}\int_{-\pi}^{\pi} f(x)\,dx, \quad a_n = \frac{1}{\pi}\int_{-\pi}^{\pi} f(x)\cos nx\,dx, \quad b_n = \frac{1}{\pi}\int_{-\pi}^{\pi} f(x)\sin nx\,dx$$

If $f(x)$ is **even function,** in $(-\pi, \pi)$ *i.e.,* $f(-x) = f(x)$ then

$$f(x) = \frac{1}{2} a_0 + \sum_{1}^{\infty} a_n \cos nx \quad \text{where } a_0 = \frac{2}{\pi}\int_0^{\pi} f(x)\,dx \text{ and } a_n = \frac{2}{\pi}\int_0^{\pi} f(x)\cos nx\,dx$$

and if $f(x)$ is **odd** in $(-\pi, \pi)$, *i.e.,* $f(-x) = -f(x)$ then

$$f(x) = \sum_{1}^{\infty} b_n \sin nx \quad \text{where } b_n = \frac{2}{\pi}\int_0^{\pi} f(x)\sin nx\,dx$$

2. Fourier series of 2*l*-periodic function.

$f(x) = \frac{1}{2} a_0 + \sum_{n=1}^{\infty}\left(a_n \cos\frac{n\pi x}{l} + b_n \sin\frac{n\pi x}{l}\right)$ if $f(x)$ is defined in $(0, 2l)$ or in $(-l, l)$

where $a_0 = \frac{1}{l}\int_0^{2l} f(x)\,dx, \quad a_n = \frac{1}{l}\int_0^{2l} f(x)\cos\frac{n\pi x}{l}\,dx, \quad b_n = \frac{1}{l}\int_0^{2l} f(x)\sin\frac{n\pi x}{l}\,dx$

and if $f(x)$ is defined in $(-l, l)$ then

$$a_0 = \frac{1}{l}\int_{-l}^{l} f(x)\,dx, \quad a_n = \frac{1}{l}\int_{-l}^{l} f(x)\cos\frac{n\pi x}{l}\,dx, \quad b_n = \int_{-l}^{l} f(x)\sin\frac{n\pi x}{l}\,dx$$

3. Even or Odd Function

If $f(x)$ is even in $(-l, l)$ then $f(x) = \frac{1}{2} a_0 + \sum_{1}^{\infty} a_n \cos \frac{n\pi x}{l}$

where $a_0 = \frac{2}{l} \int_0^l f(x)\, dx,\ a_n = \frac{2}{l} \int_0^l f(x) \cos \frac{n\pi x}{l}\, dx$

and if $f(x)$ is odd in $(-l, l)$ then $f(x) = \sum_{1}^{\infty} b_n \sin \frac{n\pi x}{l}$

where $b_n = \frac{2}{l} \int_0^l f(x) \sin \frac{n\pi x}{l}\, dx$

4. Half-range series.

If $f(x)$ is $2l$-periodic or 2π-periodic but defined in half-range only, *i.e.*, in $(0, l)$ or $(0, \pi)$. Then

Half-range sine series of $f(x) = \sum_{1}^{\infty} b_n \sin \frac{n\pi x}{l}$ where $\frac{b}{n} = \frac{2}{l} \int_0^l f(x) \sin \frac{n\pi x}{l}\, dx$

or $f(x) = \Sigma b_n \sin nx$ where $b_n = \frac{2}{\pi} \int_0^\pi f(x) \sin nx\, dx$

Similarly, half-range consine series of $f(x)$ will be

$$f(x) = \frac{1}{2} a_0 + \sum a_n \cos nx \quad \text{where} \quad a_0 = \frac{2}{\pi} \int_0^\pi f(x)\, dx,\ a_n = \frac{2}{\pi} \int_0^\pi f(x) \cos nx\, dx$$

or $f(x) = \frac{1}{2} a_0 + \sum a_n \cos \frac{n\pi x}{l}$ where $a_0 = \frac{2}{l} \int_0^l f(x)\, dx,\ a_n = \frac{2}{l} \int_0^l f(x) \sin \frac{n\pi x}{l}\, dx$

5. Practical Harmonic Analysis

Let (x_i, y_i), $i = 1, 2 \ldots\ldots, k$ be the given set of values where x_i are equi-spaced. Let $y = f(x)$ be 2π-periodic in $(-\pi, \pi)$. Fourier series for $y = f(x)$ is obtained as

$$y = \frac{1}{2} a_0 + (a_1 \cos x + b_1 \sin x) + (a_2 \cos 2x + b_2 \sin 2x) + \ldots\ldots$$

where $a_0 = \frac{2}{k} \sum_{i=1}^{k} y_i,\ a_n = \frac{2}{k} \sum_{i=1}^{k} y_i \cos n\, x_i,\ b_n = \frac{2}{k} \sum_{k=1}^{k} y_i \sin nx_i.$

Here First harmonic $= a_1 \cos x + b_1 \sin x$

Second harmonic $= a_2 \cos 2x + b_2 \sin 2x$, etc.

SOLUTIONS TO EXERCISE 19A

1. *Find the Fourier series for the function*

$$f(x) = \cos kx \quad in \quad (-\pi, \pi) \quad (k > 0).$$

SOLUTION: $f(x) = \cos kx, \ -\pi < x < \pi$ is an even function of x hence will have cosine expansion only.

Let $\cos kx = \dfrac{1}{2} a_0 + \displaystyle\sum_{n=1}^{\infty} a_n \cos nx$

$$a_0 = \frac{2}{2\pi}\int_{-\pi}^{\pi} f(x)\,dx = \frac{2}{\pi}\int_0^{\pi} \cos kx\,dx = \frac{2}{\pi}\left[\frac{\sin kx}{k}\right]_0^{\pi} = \frac{2}{\pi k}\sin k\pi$$

and
$$a_n = \frac{2}{\pi}\int_0^{\pi} f(x)\cos nx\,dx = \frac{2}{\pi}\int_0^{\pi} \cos kx \cos nx\,dx$$

$$= \frac{2}{2\pi}\int_0^{\pi} [\cos (k+n)x + \cos (n-k)]\,dx$$

$$= \frac{2}{2\pi}\left[\frac{\sin (k+n)x}{k+n} + \frac{\sin (n-k)x}{n-k}\right]_0^{\pi} \quad \text{for all } x \text{ except when } k = n$$

$$= \frac{1}{\pi}\left[\frac{\sin (k+n)\pi}{k+n} + \frac{\sin (k-n)\pi}{k-n}\right]$$

$$= \frac{1}{\pi(k^2 - n^2)}[(k-n)\sin (k+n)\pi + (k+n)\sin (k-n)\pi]$$

$$= \frac{1}{\pi(k^2 - n^2)}[(k-n)\{\sin k\pi(-1)^n + \cos k\pi.0\} + (k+n)\{\sin k\pi(-1)^n - \cos k\pi.0\}$$

$$= \frac{(-1)^n}{\pi(k^2 - n^2)}[(k-n)\sin k\pi + (k+n)\sin k\pi] = \frac{(-1)^n \sin k\pi\,(2k)}{\pi(k^2 - n^2)}$$

$$\therefore \quad f(x) = \frac{\sin k\pi}{k\pi} + \sum_{n=1}^{\infty}\frac{2k(-1)^n \sin k\pi \cos nx}{\pi(k^2 - n^2)} = \frac{2k}{\pi}\sin k\pi\left[\frac{1}{2k^2} + \sum_{n=1}^{\infty}\frac{(-1)^n \cos nx}{k^2 - n^2}\right] \quad \textbf{Ans.}$$

2. *Find the Fourier expansion for the function $f(x)$ given by*

$$f(x) = 0, \qquad -\pi \le x \le 0$$
$$= x, \qquad 0 \le x \le \pi.$$

SOLUTION: Let $f(x) = \dfrac{1}{2} a_0 + \displaystyle\sum_{1}^{\infty}(a_n \cos nx + b_n \sin nx)$ where

$$a_0 = \frac{2}{2\pi}\int_{-\pi}^{\pi} f(x)\,dx = \frac{1}{\pi}\int_{-\pi}^{0} 0\,dx + \frac{1}{\pi}\int_0^{\pi} x\,dx = \frac{\pi}{2},$$

and $$a_n = \frac{2}{2\pi}\int_{-\pi}^{\pi} f(x)\cos nx\,dx = \frac{1}{\pi}\int_{-\pi}^{0} 0\cos nx\,dx + \frac{1}{\pi}\int_{0}^{\pi} x\cos nx\,dx$$

$$= 0 + \frac{1}{\pi}\left[\left.\frac{x\sin nx}{n}\right|_0^{\pi} - \int_0^{\pi}\frac{\sin nx}{n}\,dx\right]$$

$$= -\frac{1}{n\pi}\left[-\frac{\cos nx}{n}\right]_0^{\pi} = \frac{1}{\pi n^2}(\cos n\pi - 1) = \frac{1}{\pi n^2}[(-1)^n - 1]$$

and $$b_n = \frac{2}{2\pi}\int_{-\pi}^{\pi} f(x)\sin nx\,dx = \frac{1}{\pi}\int_{-\pi}^{0} 0\sin nx\,dx + \frac{1}{\pi}\int_{0}^{\pi} x\sin nx\,dx$$

$$= \frac{-1}{\pi}\left[\frac{x\cos nx}{n}\right]_0^{\pi} - \frac{1}{\pi}\int_0^{\pi}\frac{-\cos nx}{n}\,dx\,\Bigg] = \frac{-1}{n}\cos n\pi + \frac{1}{n\pi}\left[\frac{\sin nx}{n}\right]_0^{\pi}$$

$$= \frac{(-n)^{n+1}}{n} + 0$$

$$\therefore \quad f(x) = \frac{\pi}{4} + \frac{1}{\pi}\sum_{1}^{\infty}\frac{[(-1)^n - 1]\cos nx}{n^2} + \sum_{1}^{\infty}\frac{(-1)^{n+1}}{n}\sin nx$$

$$= \frac{\pi}{4} - \frac{2}{\pi}\sum_{0}^{\infty}\frac{1\cos(2n+1)\pi}{(2n+1)^2} + \sum_{1}^{\infty}\frac{1}{n}(-1)^{n+1}\sin nx$$ **Ans.**

3. *Obtain the Fourier series for* $\sqrt{1-\cos x}$, $0 \le x \le 2\pi$ *and hence find the sum of the series* $\sum_{1}^{\infty}\frac{1}{4n^2-1}$.

SOLUTION. $f(x) = \sqrt{1-\cos x} = \sqrt{2}\sin\frac{x}{2}, \quad 0 \le x \le 2\pi$

$$\text{F.S. for } f(x) = \frac{1}{2}a_0 + \sum_{1}^{\infty}(a_n\cos nx + b_n\sin x)$$

$$a_0 = \frac{1}{\pi}\int_0^{2\pi} f(x)\,dx = \frac{\sqrt{2}}{\pi}\int_0^{2\pi}\sin\frac{x}{2}\,dx = \frac{\sqrt{2}}{\pi}\left[-2\cos\frac{x}{2}\right]_0^{2\pi} = \frac{-2\sqrt{2}}{\pi}[-1-1] = \frac{4\sqrt{2}}{\pi}$$

$$a_n = \frac{1}{\pi}\int_0^{2\pi} f(x)\cos nx\,dx = \frac{1}{\pi}\int_0^{2\pi}\sqrt{2}\sin\frac{x}{2}\cos nx\,dx$$

$$= \frac{\sqrt{2}}{2\pi}\int_0^{2\pi}\left[\sin(2n+1)\frac{x}{2} - \sin(2n-1)\frac{x}{2}\right]dx$$

$$= \frac{\sqrt{2}}{2\pi}\left[-\frac{2}{2n+1}\cos(2n+1)\frac{x}{2} + \frac{2}{2n-1}\cos(2n-1)\frac{x}{2}\right]_0^{2\pi}$$

$$= \frac{\sqrt{2}}{\pi}\left[-\frac{\cos(2n+1)\pi}{2n+1} + \frac{1}{2n-1}\cos(2n-1)\pi + \frac{1}{2n+1} - \frac{1}{2n-1}\right]$$

$$= \frac{2\sqrt{2}}{\pi}\left(\frac{1}{2n+1} - \frac{1}{2n-1}\right) \qquad \text{(since } \cos(2n+1)\pi = \cos(2n-1)\pi = \cos\pi = -1\text{)}$$

$$= \frac{-4\sqrt{2}}{\pi(4n^2-1)}.$$

Since $f(2\pi - x) = \sqrt{2}\sin\left(\frac{2\pi - x}{2}\right) = \sqrt{2}\sin\left(\pi - \frac{x}{2}\right) = \sqrt{2}\sin\frac{x}{2}$, $f(x)$ is even hence will have cosine series only.

$$\therefore \quad f(x) = \frac{2\sqrt{2}}{\pi} - \frac{4\sqrt{2}}{\pi}\sum_{n=1}^{\infty}\frac{\cos nx}{4n^2-1} \qquad \textbf{Ans.}$$

Putting $x = 0$ we get $f(0) = 0 = \frac{2\sqrt{2}}{\pi} - \frac{4\sqrt{2}}{\pi}\sum_{1}^{\infty}\frac{1}{4n^2-1}$, therefore $\sum_{1}^{\infty}\frac{1}{4n^2-1} = \frac{1}{2}$.

4. *Find the Fourier expansion for the function*

$$f(x) = \frac{x}{12}(\pi - x)(2\pi - x) \quad \text{in} \quad (0, 2\pi).$$

SOLUTION: Let $f(x) = \frac{1}{2}a_0 + \sum_{n=1}^{\infty}(a_n \cos nx + b_n \sin x), \; 0 < x < 2\pi$

Here $f(2\pi - x) = \frac{2\pi - x}{12}[\pi - (2\pi - x)][2\pi - (2\pi - x)] = \frac{-x}{12}(\pi - x)(2\pi - x)$

$= -f(x)$ hence $f(x)$ is odd function if x and will have only sine terms

and $a_0 = a_n = 0.$

and $$b_n = \frac{1}{\pi}\int_0^{2\pi}\frac{x}{12}(\pi - x)(2\pi - x)\sin nx\,dx = \frac{1}{12\pi}\int_0^{2\pi}(x^3 - 3\pi x^2 + 2\pi^2 x)\sin nx\,dx$$

$$= \frac{1}{12\pi}\left[(x^3 - 3\pi x^2 + 2\pi^2 x)\left(\frac{-\cos nx}{n}\right)\right]_0^{2\pi} + \frac{1}{12\pi}\int_0^{2\pi}(3x^2 - 6\pi x + 2\pi^2)\frac{\cos nx}{n}\,dx$$

$$= \frac{-1}{12\pi n}(8\pi^3 - 12\pi^3 + 4\pi^3) + \frac{1}{12\pi}\left[(3x^2 - 6\pi x + 2\pi^2)\frac{\sin nx}{n^2}\right]_0^{2\pi}$$

$$- \frac{1}{12\pi}\int_0^{2\pi}(6x - 6\pi)\frac{\sin nx}{n^2}\,dx$$

$$= 0 + 0 + \frac{6}{12\pi}\left[(x - \pi)\frac{\cos nx}{n^3}\right]_0^{2\pi} - \frac{6}{12\pi}\int_0^{2\pi}1.\frac{\cos nx}{n^3}\,dx$$

$$= \frac{1}{2\pi.n^3}[\pi + \pi] - \frac{1}{2\pi}\left[\frac{\sin nx}{n^4}\right]_0^{2\pi} = \frac{1}{n^3} - 0 = \frac{1}{n^3}$$

$$\therefore \quad f(x) = \sum_{1}^{\infty}b_n \sin nx = \sum_{1}^{\infty}\frac{1}{n^3}\sin nx \qquad \textbf{Ans.}$$

5. *Determine the Fourier expansion for the function*

$$f(x) = \cos x, \qquad -\pi < x < 0$$
$$= \sin x, \qquad 0 < x < \pi.$$

SOLUTION: Let $f(x) = \frac{1}{2}a_0 + \sum_{n=1}^{\infty}(a_n \cos nx + b_n \sin nx)$ be the Fourier series.

Here, $$a_0 = \frac{2}{2\pi}\int_{-\pi}^{\pi} f(x)\,dx = \frac{1}{\pi}\int_{-\pi}^{0}\cos x\,dx + \frac{1}{\pi}\int_0^{\pi}\sin x\,dx = \frac{1}{\pi}[\sin x]_{-\pi}^{0} - \frac{1}{\pi}[\cos x]_0^{\pi} = \frac{2}{\pi}$$

and $$a_n = \frac{1}{\pi}\int_{-\pi}^{\pi} f(x)\cos nx\,dx = \frac{1}{\pi}\int_{-\pi}^{0}\cos x\cos nx\,dx + \frac{1}{\pi}\int_0^{\pi}\sin x\cos nx\,dx$$

$$= \frac{1}{2\pi}\int_{-\pi}^{0}(\cos(n+1)x + \cos(n-1)x]\,dx + \frac{1}{2\pi}\int_0^{\pi}[\sin(n+1)x - \sin(n-1)x]\,dx$$

$$= \frac{1}{2\pi}\left[\frac{\sin(n+1)x}{n+1} + \frac{\sin(n-1)x}{n-1}\right]_{-\pi}^{0} + \frac{1}{2\pi}\left[\frac{-\cos(n+1)x}{n+1} + \frac{\cos(n-1)x}{n-1}\right]_0^{\pi} \quad n \neq 1.$$

$$= 0 + \frac{1}{2\pi}\left[\frac{1-\cos(n+1)\pi}{n+1} + \frac{\cos(n-1)\pi - 1}{n-1}\right] n \neq 1$$

$$\therefore \quad a_n \begin{cases} = 0 & \text{if } n \text{ is odd } (n \neq 1) \\ = \frac{1}{\pi}\left(\frac{1}{n+1} - \frac{1}{n-1}\right) = \frac{-2}{\pi(n^2-1)} & \text{if } n \text{ is even.} \end{cases}$$

Next, $$b_n = \frac{1}{\pi}\int_{-\pi}^{\pi} f(x)\sin nx\,dx = \frac{1}{\pi}\int_{-\pi}^{0}\cos x\sin nx\,dx + \frac{1}{\pi}\int_0^{\pi}\sin x\sin nx\,dx$$

$$= \frac{1}{2\pi}\int_{-\pi}^{0}[\sin(n+1)x + \sin(n-1)x]\,dx + \frac{1}{2\pi}\int_0^{\pi}[\cos(n-1)x - \cos(n+1)x]dx$$

$$= -\frac{1}{2\pi}\left[\frac{\cos(n+1)x}{n+1} + \frac{\cos(n-1)x}{n-1}\right]_{-\pi}^{0} + \frac{1}{2\pi}\left[\frac{\sin(n-1)x}{n-1} - \frac{\sin(n+1)x}{n+1}\right]_0^{\pi} \quad n \neq 1$$

$$= -\frac{1}{2\pi}\left[\frac{1-\cos(n+1)\pi}{n+1} + \frac{1-\cos(n-1)\pi}{n-1}\right] + 0,\ n \neq 1$$

$$\therefore \quad b_n \begin{cases} = 0 & \text{if } n \text{ is odd } (n \neq 1) \\ = -\frac{1}{\pi}\left(\frac{1}{n+1} + \frac{1}{n-1}\right) = \frac{-2n}{\pi(n^2-1)} & \text{if } n \text{ is even} \end{cases}$$

and $$a_1 = \frac{1}{\pi}\int_{-\pi}^{\pi} f(x)\cos x\,dx = \frac{1}{\pi}\int_{-\pi}^{0}\cos^2 x\,dx + \frac{1}{\pi}\int_0^{\pi}\sin x\cos x\,dx$$

$$= \frac{1}{2\pi}\int_{-\pi}^{0}(1+\cos 2x)\,dx + \frac{1}{2\pi}[\sin^2 x]_0^{\pi}$$

$$= \frac{1}{2\pi}\left[x + \frac{1}{2}\sin 2x\right]_{-\pi}^{0} + 0 = \frac{1}{2}$$

$$b_1 = \frac{1}{\pi}\int_{-\pi}^{\pi} f(x)\sin x\,dx = \frac{1}{\pi}\int_{-\pi}^{0}\cos x\sin x\,dx + \frac{1}{\pi}\int_{0}^{\pi}\sin^2 x\,dx$$

$$= \frac{1}{2}[\sin^2 x]_{-\pi}^{0} + \frac{1}{2\pi}\int_0^{\pi}(1-\cos 2x)\,dx$$

$$= 0 + \frac{1}{2\pi}\left[x - \frac{1}{2}\sin 2x\right]_0^{\pi} = \frac{1}{2}$$

$$\therefore \quad f(x) = \frac{1}{\pi} + \frac{1}{2}(\cos x + \sin x) - \frac{2}{\pi}\sum_1^{\infty}\frac{1}{4n^2-1}\cos 2nx - \frac{2}{\pi}\sum_1^{\infty}\frac{2n}{4n^2-1}\sin 2nx \quad \textbf{Ans.}$$

6. *Draw the sketch of the saw tooth wave defined by the function*

$f(x) = x + \pi$ *for* $-\pi < x < \pi$

and find its Fourier series.

SOLUTION:

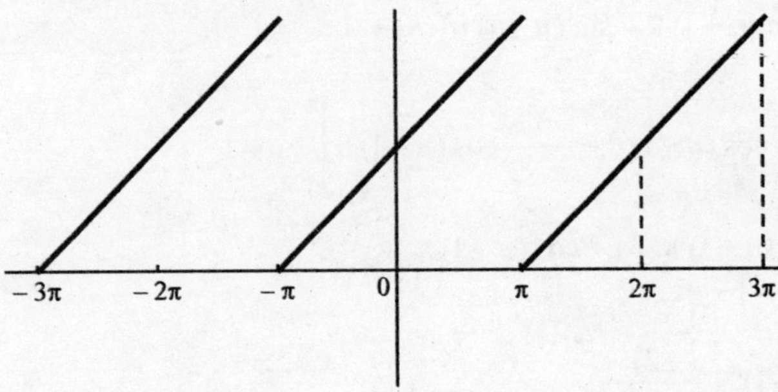

Let $\quad f(x) = \frac{1}{2}a_0 + \sum_{n=1}^{\infty}(a_n\cos nx + b_n\sin nx)$ be the Fourier series

Here, $\quad a_0 = \frac{1}{\pi}\int_{-\pi}^{\pi} f(x)\,dx = \frac{1}{\pi}\int_{-\pi}^{\pi}(x+\pi)\,dx = 0 + \frac{2}{\pi}\int_0^{\pi}\pi\,dx = 2\pi$

and $\quad a_n = \frac{1}{\pi}\int_{-\pi}^{\pi} f(x)\cos nx\,dx = \frac{1}{\pi}\int_{-\pi}^{\pi}(\pi + x)\cos nx\,dx$

$$= \frac{2}{\pi}\int_{\pi}^{\pi}\cos nx\,dx + 0 = 2\left[\frac{\sin nx}{n}\right]_0^{\pi} = 0$$

and $\quad b_n = \frac{1}{\pi}\int_{-\pi}^{\pi} f(x)\sin nx\,dx = \frac{1}{\pi}\int_{-\pi}^{\pi}(\pi + x)\sin nx\,dx = 0 + \frac{2}{\pi}\int_0^{\pi} x\sin nx\,dx$

$$= \frac{2}{\pi}\left[-x\frac{\cos nx}{n}\right]_0^{\pi} - \frac{2}{\pi}\int_0^{\pi}\left(\frac{-\cos nx}{n}\right)dx = -2\frac{\cos n\pi}{n} = \frac{2(-1)^{n+1}}{n}$$

$$\therefore \quad f(x) = \pi + 2\sum_1^{\infty}\frac{(-1)^{n+1}}{n}\sin 4x = \pi + 2\left(\frac{\sin x}{1} - \frac{\sin 2x}{2} + \frac{\sin 3x}{3} - \frac{\sin 4x}{4} + \ldots..\right)$$

Ans.

7. *An alternating current i after passing through a rectifier, is given by*

$$i = I_0 \sin\theta, \quad 0 < \theta < \pi$$
$$= 0, \quad \pi < \theta < 2\pi$$

where I_0 is the maximum current and the period is 2π. Find its Fourier series.

SOLUTION: $i = I_0 \sin\theta,\ \ 0 < \theta < \pi$ and $i = 0,\ \ \pi < \theta < 2\pi$

Let the Fourier series for i be $i = \dfrac{1}{2}a_0 + \sum_{n=1}^{\infty}(a_n \cos n\theta + b_n \sin n\theta)$

Here $$a_0 = \frac{1}{\pi}\int_0^{2\pi} f(\theta)\,d\theta = \frac{1}{\pi}\int_0^{\pi} i_0 \sin\theta\,d\theta + \frac{1}{\pi}\int_0^{\pi} 0\,d\theta = \frac{I_0}{\pi}[-\cos\theta]_0^{\pi} + 0 = \frac{2I_0}{\pi},$$

and $$a_n = \frac{1}{\pi}\int_0^{2\pi} f(\theta)\cos n\theta\,d\theta = \frac{1}{\pi}\int_0^{\pi} I_0 \sin\theta\cos n\theta\,d\theta + \frac{1}{\pi}\int_{\pi}^{2\pi} 0\cos n\theta\,d\theta$$

$$= \frac{I_0}{2\pi}\int_0^{\pi}[\sin(n+1)\theta - \sin(n-1)\theta]\,d\theta + 0$$

$$= \frac{I_0}{2\pi}\left[\frac{-1}{n+1}\cos(n+1)\theta + \frac{1}{n-1}\cos(n-1)\theta\right]_0^{\pi},\ n \neq 1$$

$$= \frac{I_0}{2\pi}\left[\frac{1-\cos(n+1)\pi}{n+1} - \frac{1-\cos(n-1)\pi}{n-1}\right],\ n \neq 1$$

$$= \frac{I_0}{2\pi}[1+(-1)^n]\frac{(-2)}{n^2-1}$$

and $$b_n = \frac{1}{\pi}\int_0^{2\pi} f(\theta)\sin n\theta\,d\theta = \frac{1}{\pi}\int_0^{\pi} I_0 \sin\theta\sin n\theta\,d\theta + \frac{1}{\pi}\int_{\pi}^{2\pi} 0\sin n\theta\,d\theta$$

$$= \frac{I_0}{2\pi}\int_0^{\pi}[\cos(n-1)\theta - \cos(n+1)\theta]\,d\theta + 0$$

$$= \frac{I_0}{2\pi}\left[\frac{1}{n-1}\sin(n-1)\theta - \frac{1}{n+1}\sin(n+1)\theta\right]_0^{\pi} = 0,\ (n \neq 1)$$

and $$b_1 = \frac{1}{\pi}\int_0^{2\pi} f(\theta)\sin\theta\,d\theta = \frac{1}{\pi}\int_0^{\pi} I_0 \sin^2\theta\,d\theta + \frac{1}{\pi}\int_{\pi}^{2\pi} 0\sin\theta\,d\theta$$

$$= \frac{I_0}{2\pi}\int_0^{\pi}(1-\cos 2\theta)\,d\theta + 0 = \frac{I_0}{2\pi}\left[\theta - \frac{1}{2}\sin 2\theta\right]_0^{\pi} = \frac{I_0}{2}$$

and $$a_1 = \frac{1}{\pi}\int_0^{2\pi} f(\theta)\cos\theta\,d\theta = \frac{1}{\pi}\int_0^{\pi} I_0 \sin\theta\cos\theta\,d\theta + \frac{1}{\pi}\int_{\pi}^{2\pi} 0\cos\theta\,d\theta$$

$$= \frac{I_0}{2}\int_0^{2\pi}\sin 2\theta\,d\theta + 0 = \frac{-I_0}{2}\cdot\frac{1}{2}[\cos 2\theta]_0^{\pi} = 0$$

$$\therefore \quad f(\theta) = i = \frac{I_0}{\pi} + \frac{I_0}{2}\sin\theta + \sum_{n=2}^{\infty}\frac{-I_0}{\pi}\frac{[1+(-1)^n]}{n^2-1}\cos n\theta$$ **Ans.**

8. *Find the Fourier series for the function* $f(x) = x^2$ *in* $(-l, l)$ *and show that*

$$\frac{1}{1^2} - \frac{1}{2^2} + \frac{1}{3^2} - \frac{1}{4^2} + \ldots = \frac{\pi^2}{12}.$$

SOLUTION: $f(x) = x^2$ is even and $2l$-perodic. Let $f(x) = \frac{1}{2} a_0 + \sum_1^{\infty} a_n \cos \frac{n\pi x}{l}$

where $a_0 = \frac{2}{2l} \int_{-l}^{l} f(x)\, dx = \frac{2}{l} \int_0^l x^2 dx = \frac{2}{l} \left[\frac{x^3}{3}\right]_0^l = \frac{2}{3} l^2$

and $a_n = \frac{2}{2l} \int_{-l}^{l} f(x) \cos \frac{n\pi x}{l}\, dx = \frac{2}{l} \int_0^l x^2 \cos \frac{n\pi x}{l}\, dx$

$$= \frac{2}{l}\left[x^2 \sin\left(\frac{n\pi x}{l}\right)\frac{l}{n}\right]_0^l - \frac{2}{l}\int_0^l 2x. \sin \frac{n\pi x}{l} \cdot \frac{l\, dx}{n\pi}$$

$$= 0 - \frac{4}{n\pi}\int_0^l x \sin \frac{n\pi x}{l}\, dx = \frac{-4}{n\pi}\left[-x \cos \frac{n\pi x}{l} \cdot \frac{l}{n\pi}\right]_0^l + \frac{4}{n\pi}\int_0^l -\cos \frac{n\pi x}{l} \cdot \frac{l}{n\pi}\, dx$$

$$= \frac{4l^2}{n^2\pi^2} \cos n\pi - \frac{4l}{n^2\pi}\left[\sin \frac{n\pi x}{l} \cdot \frac{l}{n\pi}\right]_0^l = \frac{4l^2}{n^2\pi^2}(-1)^n - 0$$

$\therefore$ $$f(x) = \frac{1}{3} l^2 + \sum_{n=1}^{\infty} \frac{4l^2 (-1)^n}{n^2\pi^2} \cos \frac{n\pi x}{l} = \frac{l^2}{3} + \frac{4l^2}{\pi^2}\sum_1^{\infty} \frac{(-1)^n}{n^2} \cos \frac{n\pi x}{l}$$ **Ans.**

9. *Determine the Fourier series for the function*

$$f(x) = \pi x, \quad 0 \le x < 1$$
$$= 0, \quad x = 1$$
$$= \pi (x - 2), \quad 1 < x \le 2.$$

SOLUTION: $f(2 - x) = \pi (2 - x)$ for $0 < (2 - x) < 1$ or $1 < x \le 2$

$= \pi(2 - x - 2) = -\pi x$ for $1 < 2 - x \le 2$ or $0 \le x < 1$

Hence $f(x)$ is an odd function hence $f(x)$ has Fourier sine expansion.

$\therefore$ $f(x) = \sum b_n \sin \frac{n\pi x}{l}$. It is periodic with period 2, hence $l = 1$

or $f(x) = \sum b_n \sin n\pi x$

where $b_n = \frac{2}{2l} \int_0^{2l} f(x) \sin \frac{n\pi x}{l}\, dx = \int_0^2 f(x) \sin n\pi x\, dx$

$$= \int_0^1 \pi x \sin n\pi x\, dx + \int_1^2 \pi(x - 2) \sin n\pi x\, dx$$

$$= \left[\frac{-\pi x \cos n\pi x}{n\pi}\right]_0^1 + \int_0^1 \frac{\pi \cos n\pi x\, dx}{n\pi} - \left[\pi(x - 2)\frac{\cos n\pi x}{n\pi}\right]_1^2 + \int_1^2 \frac{\pi \cos n\pi x}{n\pi}\, dx$$

$$= \frac{-1}{n}\cos n\pi + \frac{\pi}{n^2\pi^2}[\sin n\pi x]_0^1 - \frac{1}{n\pi}(0+\cos n\pi) + \left[\frac{\pi \sin n\pi x}{n^2\pi^2}\right]_1^2$$

$$= -\frac{2}{n}\cos n\pi$$

$$\therefore \quad f(x) = \sum_1^\infty \frac{-2}{n}\cos n\pi \sin n\pi x = -2\sum_1^\infty \frac{(-1)^n}{n}\sin n\pi x$$

10. *In the range* $(0, 2c)$ *the function* $f(x)$ *is defined as*

$$f(x) = 0, \qquad 0 < x < c$$
$$= k, \qquad c < x < 2c.$$

Expand $f(x)$ *in Fourier series of period* $2c$.

SOLUTION: $f(x)$ is $2c$-periodic function, hence

$$f(x) = \frac{1}{2}a_0 + \sum_1^\infty\left(a_n \cos\frac{n\pi x}{c} + b_n \frac{\sin n\pi x}{c}\right)$$

Here
$$a_0 = \frac{2}{2c}\int_0^{2c} f(x)\,dx = \frac{1}{c}\int_0^c 0\,dx + \frac{1}{c}\int_c^{2c} k\,dx = \frac{k}{c}[2c-c] = k,$$

$$a_n = \frac{2}{2c}\int_0^{2c} f(x)\frac{\cos n\pi x}{c}\,dx = \frac{1}{c}\int_0^c 0\cos\frac{n\pi x}{c}\,dx + \frac{1}{c}\int_c^{2c} k\cos\frac{n\pi x}{c}\,dx$$

$$= \frac{k}{c}\cdot\frac{c}{n\pi}\left[\sin\frac{n\pi x}{c}\right]_c^{2c}$$

$$= \frac{k}{n\pi}[\sin 2n\pi - \sin n\pi] = 0$$

and
$$b_n = \frac{2}{2c}\int_0^{2c} f(x)\sin\frac{n\pi x}{c}\,dx = \frac{1}{c}\int_0^c 0\sin\frac{n\pi x}{c}\,dx + \frac{1}{c}\int_c^{2c} k\sin\frac{n\pi x}{c}\,dx$$

$$= \frac{k}{c}\cdot\frac{c}{n\pi}\left[-\cos\frac{n\pi x}{c}\right]_c^{2c}$$

$$= \frac{k}{n\pi}[-\cos 2n\pi + \cos n\pi] = \frac{k}{n\pi}[-1+(-1)^n]$$

$$= \begin{cases} 0 & \text{, if } n \text{ is even} \\ \dfrac{-2k}{n\pi} & \text{, if } n \text{ is odd} \end{cases}$$

$$\therefore \quad f(x) = \frac{k}{2} - \frac{2k}{\pi}\sum_1^\infty \frac{\sin(2n-1)\frac{\pi x}{c}}{(2n-1)}$$

Ans.

11. *Determine the Fourier expansion of the function*

$$f(x) = 0, \qquad -3 < x < -1$$
$$= 1 + \cos \pi x, \quad -1 < x < 1$$
$$= 0, \qquad 1 < x < 3.$$

SOLUTION: $f(x)$ is periodic function with period = 6 in (– 3, 3), $\therefore$ $l = 3$.

It is an even function hence will have cosine series $= \frac{1}{2}a_0 + \sum_1^\infty a_n \cos \frac{n\pi x}{l}$

Here $a_0 = \frac{2}{2l}\int_{-l}^{l} f(x)\,dx = \frac{1}{3}\int_{-3}^{3} f(x)\,dx = \frac{1}{3}\int_{-3}^{-1} 0\,dx + \frac{1}{3}\int_{-1}^{1}(1+\cos \pi x) + \frac{1}{3}\int_{1}^{3} 0\,dx = \frac{2}{3}.$

and $a_n = \frac{2}{2l}\int_{-l}^{l} f(x)\cos\frac{n\pi x}{l}\,dx = \frac{1}{3}\int_{-3}^{-1} 0\cos\frac{n\pi x}{3}\,dx + \frac{1}{3}\int_{-1}^{1}(1+\cos \pi x)\cos\frac{n\pi x}{3}\,dx + \frac{1}{3}\int_{1}^{3} 0\cos\frac{n\pi x}{3}\,dx$

$$= \frac{2}{3}\int_0^1 \left[\cos\frac{n\pi x}{3} + \frac{1}{2}\left\{\cos\left(\frac{n+3}{3}\pi x\right) + \cos\left(\frac{n-3}{3}\pi x\right)\right\}\right]_0^1 dx$$

$$= \left[\frac{2}{n\pi}\sin\frac{n\pi x}{3} + \frac{1}{(n+3)\pi}\sin\left(\frac{n+3}{3}.\pi x\right) + \frac{1}{(n-3)\pi}\sin\left(\frac{n-3}{3}nx\right)\right]_0^1$$

$$= \frac{2}{n\pi}\sin\frac{n\pi}{3} + \frac{1}{(n+3)\pi}\sin(n+3)\frac{\pi}{3} + \frac{1}{(n-3)\pi}\sin(n-3)\frac{\pi}{3}$$

$$= \frac{2}{n\pi}\sin\frac{n\pi}{3} - \frac{1}{(n+3)\pi}\sin\frac{n\pi}{3} - \frac{1}{(n-3)\pi}\sin\frac{n\pi}{3}$$

$$= \frac{1}{\pi}\sin\frac{n\pi}{3}\left[\frac{2}{n} - \frac{1}{n+3} - \frac{1}{n-3}\right] = \frac{-18\sin\frac{n\pi}{3}}{\pi n(n^2-9)}$$

$$\therefore \quad f(x) = \frac{1}{3} - \frac{18}{\pi}\sum_1^\infty \frac{\sin\frac{n\pi}{3}\cos\frac{n\pi x}{3}}{n(n^2-9)}$$ **Ans.**

12. *Obtain the Fourier series for*

$$f(x) = \begin{cases} x & \text{for} \quad -1 \le x \le 0 \\ x+2 & \text{for} \quad 0 < x \le 1 \end{cases}$$

and hence deduce the sum of the series

$$1 - \frac{1}{3} + \frac{1}{5} - \frac{1}{7} + \ldots.$$

SOLUTION: $f(x)$ is periodic with period $= 2l = 2$ in (– 1, 1)

Let $f(x) = \frac{1}{2}a_0 + \sum_1^\infty \left(a_n \cos\frac{n\pi x}{l} + b_n \frac{n\pi x}{l}\right)$

Here $a_0 = \frac{2}{2l}\int_{-1}^{1} f(x)\,dx = \int_0^1 x\,dx + \int_{-1}^{0}(x+2)\,dx = -\frac{1}{2} + \frac{1}{2} + 2 = 2$

$$a_n = \frac{2}{2l}\int_{-1}^{1} f(x)\cos\frac{n\pi x}{l}\,dx = \int_{-1}^{0} x\cos n\pi x\,dx + \int_{0}^{1}(x+2)\cos n\pi x\,dx$$

$$= \left[x\frac{\sin n\pi x}{n\pi}\right]_{-1}^{0} - \int_{-1}^{0} 1.\frac{\sin n\pi x}{n\pi}\,dx + \left[(x+2)\frac{\sin n\pi x}{n\pi}\right]_{0}^{1} - \int_{0}^{1} 1.\frac{\sin n\pi x}{n\pi}\,dx$$

$$= 0 + \left[\frac{\cos n\pi x}{n^2\pi^2}\right]_{-1}^{0} + 0 + \left[\frac{\cos n\pi x}{n^2\pi^2}\right]_{0}^{1} = \frac{1-\cos n\pi}{n^2\pi^2} + \frac{\cos n\pi - 1}{n^2\pi^2} = 0$$

$$b_n = \frac{2}{2l}\int_{-1}^{1} f(x)\sin n\pi x\,dx = \int_{-1}^{0} x\sin n\pi x\,dx + \int_{0}^{1}(x+2)\sin n\pi x\,dx$$

$$= \left[-\frac{x\cos n\pi x}{n\pi}\right]_{-1}^{0} + \int_{-1}^{0} 1\frac{\cos n\pi x}{n\pi}\,dx - \left[(x+2)\frac{\cos n\pi x}{n\pi}\right]_{0}^{1} + \int_{0}^{1} 1.\frac{\cos n\pi x}{n\pi}\,dx$$

$$= \frac{-\cos n\pi}{n\pi} + \left[\frac{\sin n\pi x}{n^2\pi^2}\right]_{-1}^{0} - \frac{3}{n\pi}\cos n\pi + \frac{2.1}{n\pi} + \left[\frac{\sin n\pi x}{n^2\pi^2}\right]_{0}^{1} = \frac{2}{n\pi}[1-2\cos n\pi]$$

$$\therefore \quad f(x) = 1 + \sum_{1}^{\infty}\frac{2}{n\pi}[1-2\cos n\pi]\sin n\pi x = 1 + \frac{2}{\pi}\sum_{1}^{\infty}\frac{1-2(-1)^n}{n}\sin n\pi x$$

At $x = \frac{1}{2}$, $f(x) = \frac{5}{2}$ hence $\frac{5}{2} = 1 + \frac{2}{\pi}\sum_{1}^{\infty}\frac{1-2(-1)^n}{n}\sin\frac{n\pi}{2}$

or $$\frac{3\pi}{4} = 3\sin\frac{\pi}{2} + \frac{3}{3}\sin\frac{3\pi}{2} + \frac{3}{5}\sin\frac{5\pi}{2} + \frac{3}{7}\sin\frac{7\pi}{2} + \ldots\ldots$$

or $$1 - \frac{1}{3} + \frac{1}{5} - \frac{1}{7} + \ldots\ldots = \frac{\pi}{4}$$ **Ans.**

13. *If* $f(x) = 0, \; -\pi \le x \le 0$

$= \sin x, \quad 0 \le x \le \pi,$ *prove that*

$$f(x) = \frac{1}{\pi} + \frac{1}{2}\sin x - \frac{2}{\pi}\sum_{n=1}^{\infty}\frac{\cos nx}{4n^2-1}$$

SOLUTION: Let $$f(x) = \frac{1}{2}a_0 + \sum_{1}^{\infty} a_n\cos nx + b_n\sin nx, \; -\pi \le x \le \pi$$

$$a_0 = \frac{1}{\pi}\int_{-\pi}^{\pi} f(x)\,dx = \frac{1}{\pi}\int_{-\pi}^{0} 0\,dx + \frac{1}{\pi}\int_{0}^{\pi}\sin x\,dx = 0 + \frac{1}{\pi}[-\cos x]_0^{\pi} = \frac{2}{\pi}$$

$$a_n = \frac{1}{\pi}\int_{-\pi}^{\pi} f(x)\cos nx\,dx = \frac{1}{\pi}\int_{-\pi}^{0} 0\cos nx\,dx + \frac{1}{\pi}\int_{0}^{\pi}\sin x\cos nx\,dx$$

$$= \frac{1}{2\pi}\int_{0}^{\pi}[\sin(n+1)x - \sin(n-1)x]\,dx \quad \frac{1}{2\pi}\left[\frac{1-\cos(n+1)\pi}{n+1} + \frac{\cos(n-1)\pi - 1}{n-1}\right]$$

$$= \frac{1}{2\pi}\left[\frac{1+\cos n\pi}{n+1} - \frac{1+\cos n\pi}{n-1}\right] = -\frac{2}{2\pi}\frac{1+(-1)^n}{(n^2-1)}, \quad n \ne 1$$

Hence $a_n = 0$ when n is odd and $a_n = \dfrac{-2}{\pi(n^2-1)}$ when n is even, $n \neq 1$.

$$b_n = \frac{1}{\pi}\int_{-\pi}^{\pi} f(x) \sin nx \, dx = \frac{1}{\pi}\int_{-\pi}^{0} 0 \sin nx \, dx + \frac{1}{\pi}\int_{0}^{\pi} \sin x \sin nx \, dx$$

$$= \frac{1}{2\pi}\int_0^{\pi} [\cos(n-1)x - \cos(n+1)x] \, dx$$

$$= \frac{1}{2\pi}\left[\frac{\sin(n-1)x}{n-1} - \frac{\sin(n+1)x}{n+1}\right]_0^{\pi} = 0 \quad \text{except for } n = 1$$

$$b_1 = \frac{1}{\pi}\int_0^{\pi} \sin^2 x \, dx = \frac{1}{2\pi}\int_0^{\pi} (1-\cos 2x) \, dx = \frac{1}{2\pi}\left[x - \frac{1}{2}\sin 2x\right]_0^{\pi} = \frac{1}{2}$$

Hence $\quad f(x) = \dfrac{1}{\pi} + \dfrac{1}{2}\sin x - \dfrac{2}{\pi}\displaystyle\sum_{n=1}^{\infty} \dfrac{\cos 2nx}{4n^2-1}$ **Hence proved.**

14. *Prove that* $\quad x^2 = \dfrac{\pi^2}{3} + 3\displaystyle\sum_{1}^{\infty} (-1)^n \dfrac{\cos nx}{n^2}, \quad -\pi < x < \pi$

Hence show that $\quad \dfrac{\pi^2}{12} = \dfrac{1}{1^2} - \dfrac{1}{2^2} + \dfrac{1}{3^2} - \dfrac{1}{4^2} + \ldots$

SOLUTION: $f(x) = x^2$ is an even function in $(-\pi, \pi)$ hence $f(x) = \dfrac{1}{2}a_0 + \displaystyle\sum_{1}^{\infty} a_n \cos nx$

where $\quad a_0 = \dfrac{2}{\pi}\displaystyle\int_0^{\pi} x^2 \, dx = \dfrac{2}{\pi}\left[\dfrac{x^3}{3}\right]_0^{\pi} = \dfrac{2}{3}\pi^2$

$$a_n = \frac{2}{\pi}\int_0^{\pi} x^2 \cos nx \, dx = \frac{2}{\pi}\left[x^2 \frac{\sin nx}{n}\right]_0^{\pi} - \frac{2}{\pi}\int_0^{\pi} \frac{2x \sin nx}{n} \, dx = 0 - \frac{4}{n\pi}\int_0^{\pi} x \sin nx \, dx$$

$$= -\frac{4}{n\pi}\left[\frac{-x\cos nx}{n}\right]_0^{\pi} - \frac{4}{n\pi}\int_0^{\pi} 1.\frac{\cos nx}{n} \, dx = \frac{4\pi\cos n\pi}{n^2\pi} + \frac{4}{\pi n^2}\left[\frac{\sin nx}{n}\right]_0^{\pi} = \frac{4}{n^2}(-1)^n$$

$\therefore \quad f(x) = \dfrac{\pi^2}{3} + 4\displaystyle\sum_{1}^{\infty} \dfrac{(-1)^n}{n^2} \cos nx, \; -\pi < x < \pi$

At $\; x = 0, \; f(x) = 0$

$\therefore \quad 0 = \dfrac{\pi^2}{3} + 4\displaystyle\sum_{1}^{\infty} \dfrac{(-1)^n}{n^2}$

or $\quad \dfrac{\pi^2}{12} = \dfrac{1}{1^2} - \dfrac{1}{2^2} + \dfrac{1}{3^2} - \dfrac{1}{4^2} + \ldots\ldots\ldots$ **Hence proved.**

15. *Find the Fourier series for the function* $f(x) = x + x^2, \quad -\pi < x < \pi.$

Hence show that $\dfrac{\pi^2}{6} = 1 + \dfrac{1}{2^2} + \dfrac{1}{3^2} + \dfrac{1}{4^2} + \ldots$

SOLUTION: Let $f(x) = \dfrac{1}{2} a_0 + \sum_1^{\infty} (a_n \cos nx + b_n \sin nx)$

Here $$a_0 = \frac{1}{\pi}\int_{-\pi}^{\pi} (x + x^2)\, dx = 0 + \frac{2}{\pi}\int_0^{\pi} x^2 dx = \frac{2}{\pi}\left[\frac{x^3}{3}\right]_0^{\pi} = \frac{2\pi^2}{3},$$

$$a_n = \frac{1}{\pi}\int_{-\pi}^{\pi} (x + x^2) \cos nx\, dx = 0 + \frac{2}{\pi}\int_0^{\pi} x^2 \cos nx\, dx = \frac{2}{\pi}\left[\frac{x^2 \sin nx}{n}\right]_0^{\pi} - \frac{2}{\pi}\int_0^{\pi} \frac{2x}{n} \sin nx\, dx$$

$$= 0 - \frac{4}{n\pi}\left[-x\frac{\cos nx}{n}\right]_0^{\pi} + \frac{4}{n\pi}\int_0^{\pi} -1.\frac{\cos nx}{n}\, dx$$

$$= \frac{4}{n^2}\cos n\pi - \frac{4}{\pi n^2}\left[\frac{\sin nx}{n}\right]_0^{\pi} = \frac{4}{n^2}(-1)^n - 0$$

and $$b_n = \frac{1}{\pi}\int_{-\pi}^{\pi} (x + x^2) \sin nx\, dx = 0 + \frac{2}{\pi}\int_0^{\pi} x \sin nx\, dx$$

$$= \frac{2}{\pi}\left[-\frac{x \cos nx}{n}\right]_0^{\pi} - \frac{2}{\pi}\int_0^{\pi} \frac{-\cos nx}{n}\, dx$$

$$= \frac{-2}{\pi}\frac{\pi}{n}\cos n\pi + \frac{2}{n\pi}\left[\frac{\sin nx}{n}\right]_0^{\pi} = -\frac{2}{n}(-1)^n$$

$$\therefore \quad f(x) = \frac{\pi^2}{3} + \sum_{n=1}^{\infty}\left[\frac{4}{n^2}(-1)^n \cos nx - \frac{2}{n}(-1)^n \sin nx\right] \qquad \ldots(1)$$ **Ans.**

In (1) put $x = \pi$, we get $\pi + \pi^2 = \dfrac{\pi^2}{3} + \sum_1^{\infty} \dfrac{4}{n^2}(-1)^n \cos n\pi = \dfrac{\pi^2}{3} + 4\sum_1^{\infty}\dfrac{1}{n^2}$

and putting $x = -\pi$, we get $-\pi + \pi^2 = \dfrac{\pi^2}{3} + 4\sum \dfrac{1}{n^2}$

Adding the above relations, gives $2\pi^2 = \dfrac{2\pi^2}{3} + 8\sum_1^{\infty}\dfrac{1}{n^2}$ or $\sum \dfrac{1}{n^2} = \dfrac{\pi^2}{6}$ **Hence proved.**

16. *Find the Fourier series of the function defined as*

$$f(x) = \begin{cases} x + \pi & \text{for} \quad 0 \le x \le \pi \\ -x - \pi & \text{for} \quad -\pi \le x < 0 \end{cases}$$

SOLUTION: Let $f(x) = \dfrac{1}{2} a_0 + \sum_1^{\infty} (a_n \cos nx + b_n \sin nx)$

where $$a_0 = \frac{1}{\pi}\int_{-\pi}^{\pi} f(x)\,dx = \frac{1}{\pi}\int_{-\pi}^{0} (-x-\pi)\,dx + \frac{1}{\pi}\int_{0}^{\pi} (x+\pi)\,dx$$

$$= \frac{1}{\pi}\left[\frac{-x^2}{2} - \pi x\right]_{-\pi}^{0} + \frac{1}{\pi}\left[\frac{x^2}{2} + \pi x\right]_{0}^{\pi}$$

$$= 0 - \frac{1}{\pi}\left(\frac{-\pi^2}{2} + \pi^2\right) + \frac{1}{\pi}\left(\frac{\pi^2}{2} + \pi^2\right) = \pi$$

and $$a_n = \frac{1}{\pi}\int_{-\pi}^{\pi} f(x)\cos nx\,dx = \frac{1}{\pi}\int_{-\pi}^{0} -(x+\pi)\cos nx\,dx + \frac{1}{\pi}\int_{0}^{\pi}(x+\pi)\cos nx\,dx$$

$$= \frac{1}{\pi}\left[-(x+\pi)\frac{\sin nx}{n}\right]_{-\pi}^{0} + \frac{1}{\pi}\int_{-\pi}^{0}\frac{\sin nx\,dx}{n} + \frac{1}{\pi}\left[(x+\pi)\frac{\sin nx}{n}\right]_{0}^{\pi} - \frac{1}{\pi}\int_{0}^{\pi}\frac{\sin nx\,dx}{n}$$

$$= 0 - \frac{1}{n\pi}\left[\frac{\cos nx}{n}\right]_{-\pi}^{0} + 0 + \frac{1}{n\pi}\left[\frac{\cos nx}{n}\right]_{0}^{\pi} = \frac{2}{n^2\pi}(\cos n\pi - 1)$$

and $$b_n = \frac{1}{\pi}\int_{-\pi}^{\pi} f(x)\sin nx\,dx = \frac{1}{\pi}\int_{-\pi}^{0} -(x+\pi)\sin nx\,dx + \frac{1}{\pi}\int_{0}^{\pi}(x+\pi)\sin nx\,dx$$

$$= \frac{1}{\pi}\left[(x+\pi)\frac{\cos nx}{n}\right]_{-\pi}^{0} - \frac{1}{\pi}\int_{-\pi}^{0}\frac{\cos nx\,dx}{n} - \frac{1}{\pi}\left[(x+\pi)\frac{\cos nx}{n}\right]_{0}^{\pi} + \frac{1}{\pi}\int_{0}^{\pi}\frac{\cos nx}{x}\,dx$$

$$= \frac{1}{n\pi}[\pi - 0] - \frac{1}{\pi n^2}[\sin nx]_{-\pi}^{0} - \frac{1}{n\pi}\left[(x+\pi)\frac{\cos nx}{n}\right]_{0}^{\pi} + \frac{1}{\pi}\int_{0}^{\pi}\frac{\cos nx}{n}\,dx$$

$$= \frac{1}{n} - 0 - \frac{1}{n}[2\cos n\pi - 1] + 0 = \frac{2}{n}(1 - \cos n\pi)$$

$\therefore$ $$f(x) = \frac{\pi}{2} + \sum_{1}^{\infty}\frac{2}{\pi n^2}(\cos n\pi - 1)\cos nx + \sum_{1}^{\infty}\frac{2}{n}(1 - \cos n\pi)\sin nx$$

$$= \frac{\pi}{2} - \frac{4}{\pi}\left[\frac{\cos x}{1^2} + \frac{\cos 3x}{3^2} + \frac{\cos 5x}{5^2} + \ldots\ldots\right] + 4\left(\frac{\sin x}{1} + \frac{\sin 3x}{3} + \frac{\sin 5x}{5} + \ldots\ldots\right)$$

Ans.

17. *If* $-\pi < x < \pi$, *prove that*

$$x\sin x = 1 - \frac{1}{2}\cos x - \frac{2\cos 2x}{1\cdot 3} + \frac{2\cos 3x}{2\cdot 4} - \frac{2\cos 4x}{3\cdot 5} + \ldots$$

and hence show that

$$\frac{\pi}{4} = \frac{1}{2} + \frac{1}{1\cdot 3} + \frac{1}{3\cdot 5} + \frac{1}{5\cdot 7} + \ldots$$

SOLUTION: $f(x) = x\sin x$ in $(-\pi, \pi)$. It is even functions of x,

$\therefore$ $$f(x) = \frac{1}{2}a_0 + \sum_{1}^{\infty} a_n \cos nx.$$

$$a_0 = \frac{2}{\pi}\int_0^{\pi} x \sin x \, dx = \frac{2}{\pi}[-x\cos x]_0^{\pi} - \frac{2}{\pi}\int_0^{\pi} -\cos x \, dx = \frac{2}{\pi}[-\pi(-1)] + \frac{2}{\pi}[\sin x]_0^{\pi} = 2$$

$$a_n = \frac{2}{\pi}\int_0^{\pi} x \sin x \cos nx \, dx = \frac{1}{\pi}\int_0^{\pi} x[\sin(n+1)x - \sin(n-1)x]\, dx$$

$$= \frac{1}{\pi}\left[x\left\{\frac{-\cos(n+1)x}{n+1} + \frac{\cos(n-1)x}{n-1}\right\}\right]_0^{\pi} - \frac{1}{\pi}\int_0^{\pi}\left[\frac{-\cos(n+1)x}{n+1} + \frac{\cos(n-1)x}{n-1}\right] dx, n \neq 1$$

$$= \left[\frac{-\cos(n+1)\pi}{n+1} + \frac{\cos(n-1)\pi}{n-1}\right] + \frac{1}{\pi}\left[\frac{-\sin(n+1)\pi}{(n+1)^2} + \frac{\sin(n-1)\pi}{(n-1)^2}\right], n \neq 1$$

$$= \frac{\cos n\pi}{n+1} - \frac{\cos n\pi}{n-1} + 0 = (-1)^n\left(\frac{1}{n+1} - \frac{1}{n-1}\right) = \frac{2(-1)^{n+1}}{n^2-1}, n \neq 1$$

and $$a_1 = \frac{2}{\pi}\int_0^{\pi} x \sin x \cos x \, dx = \frac{1}{\pi}\int_0^{\pi} x \sin 2x \, dx = \frac{1}{\pi}\left[\frac{-x\cos 2x}{2}\right]_0^{\pi} - \frac{1}{\pi}\int_0^{\pi}\frac{-\cos 2x}{2}\, dx$$

$$= \frac{-\pi}{2\pi}\cos 2\pi + \left[\frac{\sin 2x}{4}\right]_0^{\pi} = -\frac{1}{2}$$

$$\therefore \quad f(x) = 1 - \frac{1}{2}\cos x - \frac{2}{2^2-1}\cos 2x + \frac{2}{3^2-1}\cos 3x - \frac{2}{4^2-1}\cos 4x + \ldots\ldots$$

$$= 1 - \frac{1}{2}\cos x - \frac{2}{1.3}\cos 2x + \frac{2}{2.4}\cos 3x - \frac{2}{3.5}\cos 4x + \ldots\ldots$$ **Hence proved.**

Let us now put $x = \frac{\pi}{2}$ at which $f(x) = \frac{\pi}{2}$, we have

$$\frac{\pi}{2} = 1 + \frac{2}{1.3} - \frac{2}{3.5} + \frac{2}{5.7} - \ldots\ldots$$

or $$\frac{\pi}{4} = \frac{1}{2} + \frac{1}{1.3} - \frac{1}{3.5} + \frac{1}{5.7} - \ldots$$ **Hence proved.**

SOLUTIONS TO EXERCISE 19B

1. *Obtain the half-range sine series for the function*
$f(x) = \sinh ax$ in $0 < x < \pi$.

SOLUTION: $f(x) = \sinh ax$ in half range $(0, \pi)$. To find sine series $\sum b_n \sin nx$

where
$$b_n = \frac{2}{\pi}\int_0^{\pi} \sinh ax \sin nx = \frac{1}{\pi}\int_0^{\pi}(e^{ax} - e^{-ax}) \sin nx\, dx$$

$$= \frac{1}{\pi}\left[\frac{e^{ax}}{a^2+n^2}(a \sin nx - n \cos nx)\right]_0^{\pi} - \frac{1}{\pi}\left[\frac{e^{-ax}}{a^2+n^2}(-a \sin nx - n \cos nx)\right]_0^{\pi}$$

$$= \frac{1}{\pi(a^2+n^2)}[e^{a\pi}(a \sin n\pi - n \cos n\pi) - (0-n) - e^{-a\pi}(-a \sin n\pi - \cos n\pi) - (0-n)]$$

$$= \frac{1}{\pi(a^2+n^2)}[-n(-1)^n(e^{a\pi} - e^{-a\pi})] = \frac{2n(-1)^{n+1}}{\pi(a^2+n^2)} \sinh a\pi$$

$\therefore$ Half range series $= \sinh ax = \dfrac{2}{\pi}\sum \dfrac{(-1)^{n+1}\, n \sinh a\pi}{a^2+n^2} \sin nx$ **Ans.**

2. *Expand* $f(x) = 1$ *in a sine series in* $0 < x < \pi$.

SOLUTION: $f(x) = 1$ as half range sine series in $(0, \pi) = \sum b_n \sin nx$

where
$$b_n = \frac{2}{\pi}\int_0^{\pi} 1 \sin nx\, dx = \frac{2}{\pi}\left[\frac{-\cos nx}{n}\right]_0^{\pi} = \frac{2}{n\pi}(1 - \cos n\pi)$$

$\therefore$ $f(x) = 1 = \dfrac{2}{\pi}\sum \dfrac{1-(-1)^n}{n} \sin nx = \dfrac{4}{\pi}\left[\sin x + \dfrac{\sin 3x}{3} + \dfrac{\sin 5x}{5} + \ldots\ldots\right]$ **Ans.**

3. *Given that* $f(x) = \cos x$ *for* $0 < x < \pi$, *find the half-range sine Fourier series of* $f(x)$.

SOLUTION: $f(x) = \cos x$ in $0 < x < \pi$ we are to find its half-range sine series.

Let $f(x) = \sum b_n \sin nx$. Here

$$b_n = \frac{2}{\pi}\int_0^{\pi} \cos x \sin nx\, dx = \frac{1}{\pi}\int_0^{\pi}[\sin (n+1)x + \sin (n-1)\, x]\, dx$$

$$= \frac{1}{\pi}\left[\frac{-\cos (n+1)x}{n+1} - \frac{\cos (n-1)x}{n-1}\right]_0^{\pi}, \quad (n \neq 1)$$

$$= \frac{-1}{\pi}\left[\frac{(-1)^{n+1}-1}{n+1} + \frac{(-1)^{n-1}-1}{n-1}\right] = \frac{-1}{\pi}[(-1)^{n+1}-1]\left(\frac{1}{n+1} + \frac{1}{n-1}\right)$$

$$= \frac{[1+(-1)^n]\, 2n}{\pi(n^2-1)} = \frac{2}{\pi}\frac{[1+(-1)^n]n}{n^2-1}$$

$$\therefore \quad f(x) = \frac{2}{\pi}\sum \frac{[1+(-1)^n]\,n}{n^2-1}\sin nx \qquad n \neq 1. \text{ Also, } b_1 = \frac{2}{\pi}\int_0^{\pi} \cos x \sin x \, dx = 0.$$

$$= \frac{2}{\pi}\sum_{m=1}^{\infty} \frac{(1+1)\,2m}{4m^2-1}\sin 2mx = \frac{8}{\pi}\sum_{m=1}^{\infty} \frac{m}{4m^2-1}\sin 2mx$$

$$= \frac{8}{\pi}\sum_{n=1}^{\infty} \frac{n}{4n^2-1}\sin 2nx$$ **Ans.**

4. *Obtain the half-range Fourier cosine series for* $x \sin x$ *in the interval* $(0, \pi)$.

SOLUTION: $f(x) = x \sin x$ in $(0, \pi)$. The half-range cosine series is $\frac{1}{2}a_0 + \sum_{1}^{\infty} a_n \cos nx$

where $$a_0 = \frac{2}{\pi}\int_0^{\pi} x \sin x \, dx = \frac{2}{\pi}[-x\cos x]_0^{\pi} - \frac{2}{\pi}\int_0^{\pi} -\cos x\, dx = \frac{2}{\pi}(\pi - 0) + \frac{2}{\pi}[\sin x]_0^{\pi} = 2$$

and $$a_n = \frac{2}{\pi}\int_0^{\pi} x \sin x \cos nx \, dx = \frac{1}{\pi}\int_0^{\pi} x\,[\sin(n+1)x - \sin(n-1)x]\,dx$$

$$= \frac{1}{\pi}\left[x\left\{\frac{-\cos(n+1)x}{n+1} + \frac{\cos(n-1)x}{n-1}\right\}\right]_0^{\pi} - \frac{1}{\pi}\int \left[\frac{-\cos(n+1)x}{n+1} + \frac{\cos(n-1)x}{n-1}\right] dx, \; n \neq 1$$

$$= \frac{-\cos(n+1)\pi}{n+1} + \frac{\cos(n-1)\pi}{n-1} - \frac{1}{\pi}\left[\frac{-\sin(n+1)x}{(n+1)^2} + \frac{\sin(n-1)x}{(n-1)^2}\right]_0^{\pi}, \; n \neq 1$$

$$= \frac{(-1)^n}{n+1} - \frac{(-1)^n}{n-1} = \frac{(-1)^n(-2)}{n^2-1}, \; n \neq 1$$

Here $$a_1 = \frac{2}{\pi}\int_0^{\pi} x \sin x \cos x \, dx = \frac{1}{\pi}\int_0^{\pi} x \sin 2x \, dx = \frac{1}{\pi}\left[\frac{-x\cos 2x}{2}\right]_0^{\pi} - \frac{1}{\pi}\int_0^{\pi} -\frac{1}{2}\cos 2x \, dx$$

$$= -\frac{1}{2} + \frac{1}{2\pi}\left[\frac{\sin 2x}{2}\right]_0^{\pi} = \frac{-1}{2}$$

$$\therefore \quad f(x) = 1 - \frac{1}{2}\cos x + \sum_{n=2}^{\infty} \frac{2(-1)^{n+1}}{n^2-1}\cos nx$$

$$= 1 - \frac{1}{2}\cos x - \frac{2}{1.3}\cos 2x + \frac{2}{2.4}\cos 3x - \frac{2}{3.5}\cos 4x + \ldots\ldots$$ **Ans.**

5. *Find the half-range sine series for* $f(x)$ *given by*

$$f(x) = \begin{cases} x/2, & 0 \le x \le c \\ c/2, & c \le x \le \pi - c \\ (\pi - x)/2, & \pi - c \le x \le \pi \end{cases}$$

SOLUTION: $f(x)$ is 2π-periodic defined in half-range $(0, \pi)$. Let the sine series be $f(x) = \sum b_n \sin nx$

where $$b_n = \frac{2}{\pi}\int_0^{\pi} f(x)\sin nx\,dx = \frac{2}{\pi}\left[\int_0^c \frac{x}{2}\sin nx\,dx + \int_c^{\pi-c}\frac{c}{2}\sin nx\,dx + \int_{\pi-c}^{\pi}\frac{\pi-x}{2}\sin nx\,dx\right]$$

$$= \frac{2}{\pi}\left[\frac{-x}{2}\frac{\cos nx}{n}\right]_0^c + \frac{2}{\pi}\int_0^c \frac{1}{2}\frac{\cos nx}{n}\,dx - \frac{c}{\pi}\left[\frac{\cos nx}{n}\right]_c^{\pi-c}$$

$$+ \frac{2}{\pi}\left[-\frac{\pi-x}{2}\frac{\cos nx}{n}\right]_{\pi-c}^{\pi} - \frac{2}{\pi}\int_{\pi-c}^{\pi}\frac{1}{2}\frac{\cos nx}{n}\,dx$$

$$= \frac{-c}{\pi n}\cos nc + \frac{1}{\pi}\left[\frac{\sin nx}{n^2}\right]_0^c - \frac{c}{\pi n}[\cos(n\pi - nc) - \cos nc] + \frac{c}{n\pi}\cos(n\pi - nc) + \frac{1}{n\pi}\left[\frac{\sin nx}{n}\right]_{\pi-c}^{\pi}$$

$$= \frac{-c}{\pi n}\cos nc + \frac{1}{\pi n^2}\sin nc + \frac{c}{n\pi}\cos nc + \frac{1}{\pi n^2}\sin(n\pi - nc) + 0$$

$$= \frac{1}{\pi n^2}[\sin nc + \sin n\pi\cos nc - \cos n\pi\sin nc] = \frac{(1-\cos n\pi)}{\pi n^2}\sin nc$$

$\therefore$ $$f(x) = \sum_1^{\infty} b_n \sin nx = \sum_1^{\infty}\frac{(1-\cos n\pi)}{\pi n^2}\sin nc\sin nx$$

or $$f(x) = \frac{2}{\pi}\sum\frac{1}{(2n+1)^2}\sin(2n+1)c\sin(2n+1)x$$ **Ans.**

6. *Find the half-period sine series for $f(x)$ given by*

$$f(x) = kx, \qquad 0 \le x \le \frac{\pi}{2}$$
$$= k(\pi - x), \qquad \frac{\pi}{2} \le x \le \pi.$$

SOLUTION: $f(x)$ is defined in half-range $(0, \pi)$. The half-range since series of $f(x) = \sum_1^{\infty} b_n \sin nx$

where $$b_n = \frac{2}{\pi}\int_0^{\pi} f(x)\sin nx\,dx = \frac{2}{\pi}\left[\int_0^{\pi/2} kx\sin nx\,dx + \int_{\pi/2}^{\pi} k(\pi - x)\sin nx\,dx\right.$$

$$= \frac{2k}{\pi}\left[\frac{-x\cos nx}{n}\right]_0^{\pi/2} + \frac{2k}{\pi}\int_0^{\pi/2}\frac{\cos nx}{n}\,dx + \frac{2k}{\pi}\left[-(\pi - x)\frac{\cos nx}{n}\right]_{\pi/2}^{\pi} - \frac{2k}{\pi}\int_{\pi/2}^{\pi}\frac{\cos nx}{n}\,dx$$

$$= -\frac{2k}{\pi}\frac{\pi}{2\pi}\cos\frac{n\pi}{2} + \frac{2k}{\pi n^2}[\sin nx]_0^{\pi/2} + \frac{2k}{n\pi}\frac{\pi}{2}\cos\frac{n\pi}{2} - \frac{2k}{\pi n^2}[\sin nx]_{\pi/2}^{\pi}$$

$$= \frac{2k}{\pi n^2}\sin\frac{n\pi}{2} + \frac{2k}{\pi n^2}\sin\frac{n\pi}{2} = \frac{4k}{\pi n^2}\sin\frac{n\pi}{2}$$

Thus, half-range since series of $f(x)$ is

$$f(x) = \frac{4k}{\pi}\sum_1^{\infty}\frac{1}{n^2}\sin\frac{n\pi}{2}\sin x = \frac{4k}{\pi}\left[\frac{1}{1^2}\sin x - \frac{1}{3^2}\sin 3x + \frac{1}{5^2}\sin 5x - \frac{\sin 7x}{7^2} + ...\right]$$ **Ans.**

7. *Obtain the half-range sine series for* x^2 *in* $(0, \pi)$.

SOLUTION: We want sine series for $f(x) = x^2$ with half-range (o, π).

$$f(x) = \sum b_n \sin nx$$

where
$$b_n = \frac{2}{\pi}\int_0^{\pi} x^2 \sin nx\, dx = \frac{2}{\pi}\left[\frac{-x^2\cos nx}{n}\right]_0^{\pi} + \frac{2}{\pi}\int_0^{\pi}\frac{2x\cos nx}{n}\,dx$$

$$= \frac{-2\pi}{n}\cos n\pi + 0 + \frac{4}{n\pi}\left[x\frac{\sin nx}{n}\right]_0^{\pi} - \frac{4}{n\pi}\int_0^{\pi}\frac{\sin nx}{n}\,dx$$

$$= \frac{-2\pi}{n}\cos n\pi + \frac{4}{\pi n^2}\left[\frac{\cos nx}{n}\right]_0^{\pi} = \frac{-2\pi}{n}\cos n\pi - \frac{4}{\pi n^3}(1-\cos n\pi)$$

$$= \begin{cases} \dfrac{2\pi}{n} - \dfrac{8}{\pi n^3} & \text{when } n \text{ is odd} \\ \dfrac{-2\pi}{n} & \text{when } n \text{ is even} \end{cases}$$

$$\therefore \quad f(x) = \sum_1^{\infty} b_n \sin nx = \left(2\pi - \frac{8}{\pi}\right)\sin x - \pi\sin 2x + \left(\frac{2\pi}{3} - \frac{8}{\pi . 3^3}\right)\sin 3x - \frac{\pi}{2}\sin 4x + \ldots$$

or Half-range sine series is

$$f(x) = \frac{2}{\pi}\left[\left(\frac{\pi^2}{1} - \frac{4}{1^3}\right)\sin x - \frac{\pi^2}{2}\sin 2x + \left(\frac{\pi^2}{3} - \frac{4}{3^3}\right)\sin 3x - \frac{\pi^2}{4}\sin 4x + \ldots\right]$$ **Ans.**

8. *Find the half-range cosine series for*

$f(x) = e^x$ *in* $0 < x < 1$

SOLUTION: Half-range as is (0, 1) cosine series of $f(x) = e^x$ is to be obtained, period $= 2l = 2$

$$f(x) = \frac{1}{2}a_0 + \sum_1^{\infty} a_n \cos\frac{n\pi x}{l} = \frac{1}{2}a_0 + \sum a_n \cos n\pi x$$

where
$$a_0 = \frac{2}{l}\int_0^l f(x)\,dx = 2\int_0^1 e^x\,dx = 2[e^x]_0^1 = 2(e-1)$$

and
$$a_n = \frac{2}{l}\int_0^l f(x)\cos\frac{n\pi x}{l}\,dx = 2\int_0^1 e^x \cos n\pi x\,dx$$

$$= \left[\frac{2e^x}{1+n^2\pi^2}\{1.\cos n\pi x + n\pi\sin n\pi x\}\right]_0^1$$

$$= \frac{2e}{1+n^2\pi^2}(\cos n\pi + 0) - \frac{2}{1+n^2\pi^2}(1+0) = \frac{2}{1+n^2\pi^2}[e\cos n\pi - 1]$$

$\therefore$ Half-range cosine series of $f(x) = e - 1 + \sum_1^{\infty}\frac{2}{1+n^2\pi^2}\{e(-1)^n - 1\}\cos n\pi x$ **Ans.**

9. *If* $f(x) = cx, \ 0 \le x \le \dfrac{l}{2}$

$= c\,(l - x), \ \dfrac{l}{2} \le x \le l$

find the half-range cosine series for $f(x)$ *in* $(0, l)$

SOLUTION: Half-range is $(0, l)$. We are to find half-range cosine series of $f(x)$.

Let $f(x) = \dfrac{1}{2}a_0 + \sum_{1}^{\infty} a_n \cos n\pi \dfrac{x}{l}$ where

$$a_0 = \frac{2}{l}\int_0^l f(x)\,dx = \frac{2}{l}\int_0^{l/2} cx\,dx + \frac{2}{l}\int_{l/2}^{l} c\,(l-x)\,dx = \frac{2c}{l}\left[\frac{x^2}{2}\right]_0^{l/2} + \frac{2c}{l}\left[lx - \frac{x^2}{2}\right]_{\frac{l}{2}}^{l}$$

$$= \frac{cl}{4} + \frac{2c}{l}\left(l^2 - \frac{l^2}{2} - \frac{l^2}{2} + \frac{l^2}{8}\right) = \frac{cl}{2}$$

and $$a_n = \frac{2}{l}\int_0^l f(x)\cos n\pi x/l\,dx = \frac{2}{l}\int_0^{l/2} cx\cos n\pi x/l\,dx + \frac{2}{l}\int_{l/2}^{l} c(l-x)\cos n\pi x/l\,dx$$

$$= \frac{2c}{l}\left[\frac{x\sin n\pi x/l}{n\pi/l}\right]_0^{l/2} - \frac{2c}{l}\int_0^{l/2}\frac{\sin n\pi x/l}{n\pi/l}\,dx + \frac{2c}{l}\left[(l-x)\frac{\sin n\pi x/l}{n\pi/l}\right]_{l/2}^{l}$$

$$+ \frac{2c}{l}\int_{l.2}^{l}\frac{\sin n\pi x/l}{n\pi/l}\,dx$$

$$= \frac{2c}{n\pi}\left[\frac{l}{2}\sin\frac{n\pi}{2} - 0\right] + \frac{2c}{n\pi}\left[\frac{\cos n\pi x/l}{n\pi/l}\right]_0^{l/2} - \frac{cl}{n\pi}\sin\frac{n\pi}{2} - \frac{2c}{n\pi}\left[\frac{\cos n\pi x/l}{n\pi/l}\right]_{l/2}^{l}$$

$$= \frac{2cl}{n^2\pi^2}\left(\cos\frac{n\pi}{2} - 1\right) - \frac{2cl}{n^2\pi^2}\left(\cos n\pi - \cos\frac{n\pi}{2}\right)$$

$$= \frac{2cl}{n^2\pi^2}\left[2\cos\frac{n\pi}{2} - (1 + \cos n\pi)\right] = \frac{4cl}{n^2\pi^2}\left[\cos\frac{n\pi}{2} - \cos^2\frac{n\pi}{2}\right]$$

$$= \frac{4cl}{n^2\pi^2}\cos\frac{n\pi}{2}\left(1 - \cos\frac{n\pi}{2}\right) = \frac{8cl}{n^2\pi^2}\cos\frac{n\pi}{2}\sin^2\frac{n\pi}{4}$$

$\therefore$ Half-range cosine seires of $f(x)$ is

$$\frac{cl}{4} + \frac{8cl}{\pi^2}\sum_{1}^{\infty}\frac{1}{n^2}\cos\frac{n\pi}{2}\sin^2\frac{n\pi}{2}\cos\frac{n\pi x}{l}$$ **Ans.**

10. *Determine the half-period cosine series for*

$f(x) = e^x, \qquad 0 < x < l.$

SOLUTION: $f(x) = \dfrac{1}{2}a_0 + \sum_{n=1}^{\infty} a_n \cos\dfrac{n\pi x}{l}, \ 0 < x < l,$ it is half-range cosine series of $f(x)$.

Here $a_0 = \dfrac{2}{l}\int_0^l f(x)\,dx, \ a_n = \dfrac{2}{l}\int_0^l f(x)\cos n\pi x/l\,dx$

$$\therefore \quad a_0 = \frac{2}{l}\int_0^l e^x dx = \frac{2}{l}[e^x]_0^l = \frac{2}{l}[e^l - 1]$$

and
$$a_n = \frac{2}{l}\int_0^l e^x \cos\frac{n\pi x}{l}\,dx = \frac{2}{l}\left[\frac{e^x}{1^2 + n^2\pi^2/l^2}\left(1.\cos\frac{n\pi x}{l} + \frac{n\pi}{l}\sin\frac{n\pi x}{l}\right)\right]_0^l$$

$$= \frac{2}{l}\left[\frac{e^l}{1+n^2\pi^2/l^2}(\cos n\pi) - \frac{1}{1+n^2\pi^2/l^2}(1+0)\right]$$

$$= \frac{2}{l(1+n^2\pi^2/l^2)}[e^l \cos n\pi - 1] = \frac{2l}{l^2+n^2\pi^2}[e^l(-1)^n - 1]$$

$\therefore$ Half-range cosine series of $f(x)$ is

$$\frac{1}{l}(e^l - 1) + 2l\sum\frac{[e^l(-1)^n - 1]}{l^2+n^2\pi^2}\cos\frac{n\pi x}{l}. \qquad \textbf{Ans.}$$

11. *If* $f(x) = \dfrac{x}{c}$ *for* $0 < x < c$

$$= \frac{l-x}{l-c} \quad \textit{for } c < x < l$$

show that the half-range sine series for $f(x)$ *is*

$$f(x) = \frac{2l^2}{c(l-c)\pi^2}\sum_1^\infty \frac{1}{n^2}\sin\frac{n\pi c}{l}\sin\frac{n\pi x}{l}.$$

SOLUTION: We need half-range sine series of $f(x)$ is $(0, l)$. Let $f(x) = \sum b_n \sin n\pi x/l$

where $b_n = \dfrac{2}{l}\displaystyle\int_0^l f(x)\sin\frac{n\pi x}{l}dx = \frac{2}{l}\int_0^c \frac{x}{c}\sin\frac{n\pi x}{l}dx + \frac{2}{l}\int_c^l \frac{l-x}{l-c}\sin\frac{n\pi x}{l}dx$

$$= \frac{2}{lc}\left[-x\frac{\cos n\pi x/l}{n\pi/l}\right]_0^c + \frac{2}{lc}\int_0^c \frac{\cos n\pi x/l}{n\pi/l}dx$$

$$+ \frac{2}{l(l-c)}\left[-(l-x)\frac{\cos n\pi x/l}{n\pi/l}\right]_c^l - \frac{2}{l(l-c)}\int_c^l \frac{\cos n\pi x/l}{n\pi/l}dx$$

$$= \frac{-2}{cn\pi}c\cos n\pi c/l - \frac{2}{\pi nc}\left[\frac{\sin n\pi x/l}{n\pi/l}\right]_0^l + \frac{2}{n\pi(l-c)}(l-c)\cos n\pi c/l - \frac{2}{n\pi(l-c)}\left[\frac{\sin n\pi x/l}{n\pi/l}\right]_c^l$$

$$= \frac{-2}{n\pi}\cos n\pi c/l + \frac{2}{n\pi}\cos n\pi c/l + \frac{2l}{n^2\pi^2(l-c)}\sin n\pi c/l = \frac{2l}{n^2\pi^2(l-c)}\sin n\pi c/l$$

$$\therefore \quad f(x) = \sum\frac{2l}{n^2\pi^2(l-c)}\sin\frac{n\pi c}{l}\sin n\pi x/l = \frac{2l}{\pi^2(l-c)}\sum\frac{1}{n^2}\sin n\pi c/l\sin n\pi x/l \qquad \textbf{Ans.}$$

12. *If $f(x)$ is represented by the Fourier series*

$$f(x) = \frac{1}{2}a_0 + \sum_{n=1}^{\infty} a_n \cos\frac{n\pi x}{c}, \qquad 0 < x < c$$

then prove that $\frac{2}{c}\int_0^c [f(x)]^2\,dx = \frac{1}{2}a_0^2 + \sum_{n=1}^{\infty} a_n^2$.

SOLUTION: $f(x) = \frac{1}{2}a_0 + \sum_{n=1}^{\infty} a_n \cos\frac{n\pi x}{c}, \quad 0 < x < c$

where $a_0 = \frac{2}{c}\int_0^c f(x)\,dx, \quad a_n = \frac{2}{c}\int_0^c f(x)\cos\frac{n\pi x}{c}\,dx$

$$[f(x)]^2 = \frac{1}{4}{a_0}^2 + \sum_{n=1}^{\infty} a_n^2\cos^2\frac{n\pi x}{c} + 2.\frac{a_0}{2}\sum_{n=1}^{\infty} a_n\cos\frac{n\pi x}{c} + 2\sum_{\substack{m,n=1\\ m\neq n}}^{\infty} a_m a_n \cos\frac{n\pi x}{c}\cos\frac{m\pi x}{c}$$

$$\int_0^c [f(x)]^2\,dx = \frac{1}{4}a_0^2\int_0^c dx + \sum_{n=1}^{\infty}\frac{1}{2}a_n^2\int_0^c\left(1+\cos\frac{2n\pi}{c}\right)dx + a_0\sum_{n=1}^{\infty} a_n\int_0^c \cos\frac{n\pi x}{c}\,dx$$

$$+ 2\sum_{\substack{m,n=1\\ m\neq n}}^{\infty} a_m a_n \int_0^c \cos\frac{m\pi x}{c}\cos\frac{n\pi x}{c}\,dx$$

$$= \frac{c\,a_0^2}{4} + \sum_{n=1}^{\infty}\frac{1}{2}a_n^2\left[x + \frac{c}{2n\pi}\sin\frac{2n\pi x}{c}\right]_0^c + a_0\sum_{n=1}^{\infty} a_n\frac{c}{n\pi}\sin\frac{n\pi x}{c}\Big|_0^c$$

$$+ \sum_{m,n=1}^{\infty} a_m a_n \int_0^c\left[\cos(m+n)\frac{\pi x}{c} + \cos(m-n)\frac{\pi x}{c}\right]_0^c$$

$$= \frac{c\,a_0^2}{4} + \sum_{n=1}^{\infty}\frac{1}{4}a_n^2(c+0) + a_0\sum_1^{\infty} a_n(0)$$

$$+ \sum_{m,n=1}^{\infty} a_m a_n\left[\frac{c}{m+n}\sin(m+n)\frac{\pi x}{c} + \frac{c}{m-n}\sin(m-n)\frac{\pi x}{c}\right]_0^c$$

$$= \frac{c}{4}a_0^2 + \frac{1}{2}\sum a_n^2 + 0 + 0$$

$$\therefore \quad \frac{2}{c}\int_0^c f(x)^2\,dx = \frac{1}{2}a_0^2 + \sum_1^{\infty} a_n^2$$

Hence proved.

13. *Express* $f(x) = \begin{cases} \frac{1}{4} - x, & 0 < x < \frac{1}{2} \\ x - \frac{3}{4}, & \frac{1}{2} < x < 1 \end{cases}$

as Fourier series of sine terms.

SOLUTION: Half-range Fourier sine series of $f(x)$ is to be obtained.

Let $\quad f(x) = \sum b_n \sin \frac{n\pi x}{l} \quad$ Here $\quad l = 1$

when $\quad b_n = \frac{2}{l}\int_0^1 f(x)\sin\frac{n\pi x}{l}\,dx = 2\int_0^{1/2}\left(\frac{1}{4}-x\right)\sin n\pi x\,dx + 2\int_{1/2}^1\left(x-\frac{3}{4}\right)\sin n\pi x\,dx$

$$= -2\left[\left(\frac{1}{4}-x\right)\frac{\cos n\pi x}{n\pi}\right]_0^{1/2} - 2\int_0^{1/2}(-1)^2\frac{\cos n\pi x}{n\pi}\,dx - 2\left[\left(x-\frac{3}{4}\right)\frac{\cos n\pi x}{n\pi}\right]_{1/2}^1 - 2\int_{1/2}^1\frac{(-1)\cos n\pi x}{n\pi}\,dx$$

$$= -2\left[-\frac{1}{4n\pi}\cos\frac{n\pi}{2} - \frac{1}{4n\pi}\right] - 2\left[\frac{\sin n\pi x}{n^2\pi^2}\right]_0^{1/2} - 2\left[\frac{1}{4n\pi}\cos n\pi + \frac{1}{4n\pi}\cos\frac{n\pi}{2}\right] + 2\left[\frac{\sin n\pi x}{n^2\pi^2}\right]_{1/2}^1$$

$$= \frac{1}{2n\pi}\left[\cos\frac{n\pi}{2}+1\right] - \frac{2}{n^2\pi^2}\sin\frac{n\pi}{2} - \frac{1}{2n\pi}\left[\cos n\pi + \cos\frac{n\pi}{2}\right] - \frac{2}{n^2\pi^2}\sin\frac{n\pi}{2}$$

$$= \frac{1}{2n\pi}(1-\cos n\pi) - \frac{4}{n^2\pi^2}\sin\frac{n\pi}{2} = 0 \quad \text{when } x \text{ is even}$$

$$= \frac{1}{n\pi} - \frac{4}{n^2\pi^2}\sin\frac{n\pi}{2} \quad \text{when } n \text{ is odd.}$$

$$\therefore \quad f(x) = \sum_{n=1,3,5,\ldots}\left(\frac{1}{n\pi} - \frac{4}{n^2\pi^2}\sin\frac{n\pi}{2}\right)\sin n\pi x$$ **Ans.**

14. *Prove that for* $0 \le x \le \pi$

$$x(\pi - x) = \frac{\pi^2}{6} - \left[\frac{\cos 2x}{1^2} + \frac{\cos 4x}{2^2} + \frac{\cos 6x}{3^2} + \ldots\right]$$

SOLUTION: Half-range Fourier cosine series for $f(x) = x(\pi - x)$ is to be obtained

Let $\quad f(x) = \frac{1}{2}a_0 + \sum_1^\infty a_n \cos nx \quad$ Here $\quad a_0 = \frac{2}{\pi}\int_0^\pi x(\pi - x)\,dx = \frac{\pi^2}{3}.$

and $\quad a_n = \frac{2}{\pi}\int_0^\pi (\pi x - x^2)\cos nx\,dx = \frac{2}{\pi}\left[(\pi x - x^2)\frac{\sin nx}{n}\right]_0^\pi - \frac{2}{\pi}\int_0^\pi(\pi - 2x)\frac{\sin nx}{n}\,dx$

$$= 0 - \frac{2}{n\pi}\left[-(\pi - 2x)\frac{\cos nx}{n}\right]_0^\pi + \frac{2}{n\pi}\int_0^\pi (2)\frac{\cos nx}{n}\,dx$$

$$= \frac{2}{\pi n^2}[-\pi\cos n\pi - \pi] + \frac{4}{n\pi^2}\left[\frac{\sin nx}{n}\right]_0^\pi = \frac{-2}{n^2}(1+\cos n\pi)$$

$$= \begin{cases} -\dfrac{4}{n^2} & \text{when } n \text{ is even} \\ 0 & \text{when } n \text{ is odd.}\end{cases}$$

$$\therefore \quad f(x) = \frac{\pi^2}{6} - 4\sum_{n\ \text{even}}\frac{\cos nx}{n^2} = \frac{\pi^2}{6} - 4\sum_{m=1}^\infty \frac{\cos 2mx}{4m^2}$$

or $\quad f(x) = \frac{\pi^2}{6} - \left(\frac{\cos 2x}{1^2} + \frac{\cos 4x}{2^2} + \frac{\cos 6x}{3^2} + \ldots\right)$ **Hence proved.**

SOLUTIONS TO EXERCISE 19C

1. *The following Table gives* 12 *values of* y *corresponding to equispaced values of the angle* x *in the range* (0º, 360º):

x	0º	30º	60º	90º	120º	150º	180º	210º	240º	270º	300º	330º
y	10.5	20.2	26.4	29.3	27.0	21.5	12.8	1.6	–11.2	–18.0	–15.8	–0.4

Obtain the Fourier series for y *in terms of* x *upto third harmonics.*

SOLUTION: $y = \frac{1}{2}a_0 + a_1 \cos x + b_1 \sin x + a_2 \cos 2x + b_2 \sin 2x + ...$

Here $a_0 = \frac{2}{12}\sum_{l=1}^{12} y_i = \frac{1}{6}(16.8) = 8.4$

$a_1 = \frac{2}{12}\sum y_i \cos x_i = 23.1$ $\qquad b_1 = \frac{2}{12}\sum y_i \sin x_i = -1.7$

$a_2 = \frac{2}{12}\sum y_i \cos 2x_i = 1.1$ $\qquad b_2 = \frac{2}{12}\sum y_i \sin 2x_i = 3.1$

$a_3 = \frac{2}{12}\sum y_i \cos 3x_i = -0.6$ $\qquad b_3 = \frac{2}{12}\sum y_i \sin 3x_i = 0.45$

$\therefore\ y = 8.4 + 23.1 \cos x - 1.7 \sin x + 1.1 \cos 2x + 3.1 \sin 2x - 0.6 \cos 3x + 0.45 \sin 3x$ **Ans.**

2. *The values of current* i *are connected with the equispaced values of the deflection as per following Tabular data:*

θ	0º	30º	60º	90º	120º	150º	180º	210º	240º	270º	300º	330º
i	0	24.0	33.5	27.5	18.2	13.0	0	–24.0	–33.5	–27.5	–18.2	–13.0

Express the current i *in terms of* θ *by Fourier harmonics upto third order.*

SOLUTION: Let $i = \frac{1}{2}a_0 + a_1 \cos\theta + b_1 \sin\theta + a_2 \cos 2\theta + b_2 \sin 2\theta + a_3 \cos 3\theta + b_3 \sin 3\theta + ...$

Here $a_0 = \frac{2}{12}\sum_{k=1}^{12} i_k = 0$

$a_1 = \frac{2}{12}\sum i_k \cos\theta_k = 5.7,$ $\qquad b_1 = \frac{2}{12}\sum i_k \sin\theta_k = 30.3$

$a_2 = \frac{2}{12}\sum i_k \cos 2\theta_k = 0,$ $\qquad b_2 = \frac{2}{12}\sum i_k \sin 2\theta_k = 0$

$a_3 = \frac{2}{12}\sum i_k \cos 3\theta_k = -5.1,$ $\qquad b_3 = \frac{2}{12}\sum i_k \sin 3\theta_k = 3.2$

$a_4 = \frac{2}{12}\sum i_k \cos 4\theta_k = 0,$ $\qquad b_4 = \frac{2}{12}\sum i_k \sin 4\theta_k = 0$

$a_5 = \frac{2}{12}\sum i_k \cos 5\theta_k = -0.6,$ $\qquad b_5 = \frac{2}{12}\sum i_k \sin 5\theta_k = 0.4$

$\therefore\ i = 5.7 \cos\theta + 30.3 \sin\theta - 5.1 \cos 3\theta + 3.2 \sin 3\theta - 0.6 \cos 5\theta + 0.4 \sin 5\theta$ **Ans.**

3. *The values of y corresponding to certain equispaced values of x are given in the range $0 \le x \le \pi$ as follows:*

x	0	15	30	45	60	75	90	105	120	135	150	165
y	0	3.1	4.9	5.8	6.9	8.2	12.8	17.0	22.0	21.6	14.9	7.6

Prove that y can be expressed approximately, as the following Fourier expansion.

$$y = 16.8 \sin (x - 20)^{o} + 5.9 \sin (3x + 60^{o}) + \sin (5x + 155^{o})$$

under the assumption that the graph of y for x between 180° and 360° is exactly similar to that between $x = 0$ and $x = 180^{o}$ but lies below the axis of x.

SOLUTION: Let $y = \frac{1}{2} a_0 + a_1 \cos x + b_1 \sin x + a_2 \cos 2x + b_2 \sin 2x + a_3 \cos 3x + b_3 \sin 3x + ...$

Here $a_0 = \frac{2}{24} \sum y_i = 0$

$a_1 = \frac{2}{24} \sum y_i \cos x_i \qquad b_1 = \frac{2}{24} \sum y_i \sin x_i$

$a_2 = 0, \; b_2 = 0$

$a_3 = \frac{2}{24} \sum y_i \cos 3x_i \qquad b_3 = \frac{2}{24} \sum y_i \sin 3x_i$

$a_4 = 0, \; b_4 = 0$

$a_5 = \frac{2}{24} \sum y_i \cos 5x_i \qquad b_5 = \frac{2}{24} \sum y_i \sin 5x_i$

On comptutation and adjustment we can show that

$$y = 16.8 \sin (x - 20^{\circ}) + 5.9 \sin (3x + 60^{\circ}) + \sin (5x + 155^{\circ})$$

4. *The displacement x of a certain part of a machine corresponding to the movement θ of the crank are given in the following Table:*

θ	0°	30°	60°	90°	120°	150°	180°	210°	240°	270°	300°	330°	360°
x	1.8	1.1	0.3	0.16	0.5	1.5	2.16	1.88	1.25	1.30	1.76	2.00	1.8

Expand x in Fourier series approximately upto fifth harmonics.

SOLUTION: Let $x = \frac{1}{2} a_0 + a_1 \cos \theta + b_1 \sin \theta + a_2 \cos 2\theta + b_2 \sin 2\theta + a_3 \cos 3\theta + b_3 \sin 3\theta + a_4 \cos 4\theta + b_4 \sin 4\theta + a_5 \cos 5\theta + b_5 \sin 5\theta + ...$

Here $a_0 = \frac{2}{12} \sum x_i = 2.62$

$a_1 = \frac{2}{12} \sum x_i \cos \theta_i = -0.07, \qquad b_1 = \frac{2}{12} \sum x_i \sin \theta_i = -0.62$

$a_2 = \frac{2}{12} \sum x_i \cos 2\theta_i = 0.64, \qquad b_2 = \frac{2}{12} \sum x_i \sin 2\theta_i = 0.18$

$a_3 = \frac{2}{12} \sum x_i \cos 3\theta_i = -0.11, \qquad b_3 = \frac{2}{12} \sum x_i \sin 3\theta_i = -0.02$

$$a_4 = \frac{2}{12}\sum x_i \cos 4\theta_i = 0.05, \qquad b_4 = \frac{2}{12}\sum x_i \sin 4\theta_i = 0.03$$

$$a_5 = \frac{2}{12}\sum x_i \cos 5\theta_i = -0.01, \qquad b_5 = \frac{1}{12}\sum x_i \sin 5\theta_i = -0.02$$

$$\therefore \quad x = 1.31 - 0.07\cos\theta - 0.62\sin\theta + 0.64\cos 2\theta - 0.18\sin 2\theta - 0.11\cos 3\theta - 0.02\sin 3\theta + 0.05\cos 4\theta + 0.03\sin 4\theta - 0.01\cos 5\theta - 0.02\sin 5\theta$$ **Ans.**

5. *Following Table connects the values of θ and x, expand x in terms of θ in Fourier series upto third harmonics.*

θ	0°	30°	60°	90°	120°	150°	180°	210°	240°	270°	300°	330°
x	0	1.0	1.732	2.0	1.732	1.0	0	0	0	0	0	0

SOLUTION: Let $x = \frac{1}{2}a_0 + a_1\cos\theta + b_1\sin\theta + a_2\cos 2\theta + b_2\sin 2\theta$ (upto second harmonics)

Here $a_0 = \frac{2}{12}\sum x_i = \frac{1}{6}(7.64) = 1.244$

$$a_1 = \frac{2}{12}\sum x_i\cos\theta_i = \frac{1}{6}\left[1\left(\frac{\sqrt{3}}{2}\right) + 1.732\left(\frac{1}{2}\right) + 0 + \sqrt{3}\left(\frac{-1}{2}\right) + 1.732\left(\frac{-1}{2}\right)\right] = 0$$

$$b_1 = \frac{2}{12}\sum x_i\cos\theta_i = \frac{1}{6}\left[1\left(\frac{1}{2}\right) + 1.732\frac{\sqrt{3}}{2} + 2(1) + 1.732\frac{\sqrt{3}}{2} + \frac{1}{2}(1)\right] = \frac{6}{6} = 1$$

$$a_2 = \frac{2}{12}\sum x_i\cos 2\theta_i = \frac{1}{6}[1\cos 60° + 1.732\cos 120° + 2\cos 180° + 1.732\cos 240° + 1\cos 300°]$$

$$= \frac{1}{6}\left[1\left(\frac{1}{2}\right) + 1.732\left(-\frac{1}{2}\right) + 2(-1) + 1.732\left(-\frac{1}{2}\right) + 1\left(\frac{1}{2}\right)\right]$$

$$= \frac{1}{12}[1 - 1.732 - 4 - 1.732 + 1] = \frac{-5.464}{12} = \frac{-1.366}{3} = \frac{-4}{3\pi}$$ **approx.**

$$b_2 = \frac{2}{12}\sum x_i\sin 2\theta_i = \frac{1}{6}[1\sin 60° + 1.732\sin 120° + 2\sin 180° + 1.732\sin 240° + 1\sin 300°]$$

$$= \frac{1}{6}\left[\frac{\sqrt{3}}{2} + 1.732\frac{\sqrt{3}}{2} + 0 - 1.732\frac{\sqrt{3}}{2} - \frac{\sqrt{3}}{2}\right] = 0$$

$$\therefore \quad x = \frac{2}{\pi}\left[1 + \frac{\pi}{2}\sin\theta - \frac{2}{3}\cos 2\theta\right]$$ **approx.**

6. *Variations of a periodic current i (in Amperes) against time t over a period T are given in the following Table:*

t	0	$\frac{T}{6}$	$\frac{T}{3}$	$\frac{T}{2}$	$\frac{2T}{3}$	$\frac{5T}{6}$	T
A (amps)	1.98	1.30	1.05	1.30	−0.88	−0.25	1.98

Show that there is a current of 0.75 *ampears. Also, find the amplitudes of first harmonic.*

SOLUTION: Let $A = \frac{1}{2} a_0 + a_1 \cos t + b_1 \sin t + a_2 \cos 2t + b_2 \sin 2t + ...$

where $a_0 = \frac{2}{6} \sum A_i = \frac{1}{3} [1.98 + 1.30 + 1.05 + 1.30 - 0.88 - 0.25] = \frac{3.45}{3} = 1.15,$

$$a_1 = \frac{2}{6} \sum A_i \cos t_i,$$

$$b_1 = \frac{2}{6} \sum A_i \sin t_i.$$

Amplitude of first harmonic = $\sqrt{a_1^2 + b_1^2} = 1.071$ **Ans.**

7. *Compute the Fourier coefficients* a_0, a_1, a_2, b_1, b_2 *in the Fourier expansion of the function* $y(x)$ *Tabulated below:*

x	0	1	2	3	4	5
y	9	18	24	28	26	20

Find also the amplitude of the first harmonic.

SOLUTION:

x	y	$\cos \pi x/3$	$\sin \pi x/3$
0	9	1	0
1	18	$\frac{1}{2}$	$\frac{\sqrt{3}}{2}$
2	24	$\frac{-1}{2}$	$\frac{\sqrt{3}}{2}$
3	28	-1	0
4	26	$\frac{-1}{2}$	$-\frac{\sqrt{3}}{2}$
5	20	$\frac{1}{2}$	$-\frac{\sqrt{3}}{2}$

Here $a_0 = \frac{2}{6} \sum y_i = \frac{1}{3}(125) = 41.67$

$$a_1 = \frac{2}{6} \sum y_i \cos \frac{\pi x}{3} = \frac{2}{6} \left[9 - 28 + \frac{1}{2}(18 - 24 - 26 + 20) \right] = \frac{1}{3} (-19 - 6) = -8.33$$

$$b_1 = \frac{2}{6} \sum y_i \sin \frac{\pi x}{3} = \frac{2}{6} \left[\frac{\sqrt{3}}{2} (18 + 24 - 26 - 20) \right] = \frac{\sqrt{3}}{2} (-4) = \frac{-2\sqrt{3}}{3} = -1.15$$

$\therefore$ Amplitude = $\sqrt{a_1^2 + b_1^2} = \sqrt{8.33^2 + 1.15^2}$ **Ans.**

ADDITIONAL IMPORTANT PROBLEMS

1. *Find the Fourier series of* $f(x) = \frac{1}{2}(\pi - x)$ *in the internal* $(0, 2\pi)$.

Hence deduce that $\frac{\pi}{4} = 1 - \frac{1}{3} + \frac{1}{5} - \frac{1}{7} + \ldots$

SOLUTION: The Fourier series of $f(x)$ with period 2π, is

$$f(x) = \frac{1}{2}a_0 + \sum_{n=1}^{\infty}(a_n \cos nx + b_n \sin nx)$$

where
$$a_0 = \frac{1}{\pi}\int_0^{2\pi} f(x)\,dx = \frac{1}{\pi}\int_0^{2\pi}\frac{1}{2}(\pi - x)dx = \frac{1}{2\pi}\left[\pi x - \frac{x^2}{2}\right]_0^{2\pi} = \frac{1}{2\pi}[2\pi^2 - 2\pi^2] = 0$$

and
$$a_n = \frac{1}{\pi}\int_0^{2\pi} f(x)\cos nx\,dx = \frac{1}{2\pi}\int_0^{2\pi}(\pi - x)\cos nx\,dx$$

$$= \frac{1}{2\pi}\left[(\pi - x)\frac{\sin nx}{n}\right]_0^{2\pi} - \frac{1}{2\pi}\int_0^{2\pi}(-1)\frac{\sin nx}{n}dx = \frac{1}{2\pi}(0) - \frac{1}{2\pi n^2}[\cos nx]_0^{2\pi} = 0$$

and
$$b_n = \frac{1}{\pi}\int_0^{2\pi} f(x)\sin nx\,dx = \frac{1}{2\pi}\int_0^{2\pi}(\pi - x)\sin nx\,dx$$

$$= \frac{-1}{2\pi}\left[(\pi - x)\frac{\cos nx}{n}\right]_0^{2\pi} - \frac{1}{2\pi}\int_0^{2\pi}\frac{(-1)(-\cos nx)}{n}dx$$

$$= -\frac{1}{2n\pi}[-\pi\cos 2n\pi - \pi] - \frac{1}{2n\pi}\left[\frac{\sin nx}{n^2}\right]_0^{2\pi} = \frac{1}{2n}(2) + 0 = \frac{1}{n}$$

Hence
$$f(x) = \sum_1^{\infty}\frac{1}{n}\sin nx$$ **Ans.**

Putting $x = \frac{\pi}{2}$ here we get

$$f\left(\frac{\pi}{2}\right) = \sum_1^{\infty}\frac{1}{n}\sin\frac{n\pi}{2}$$

or
$$\frac{\pi}{4} = 1 - \frac{1}{3} + \frac{1}{5} - \frac{1}{7} + \ldots$$ **Hence proved.**

2. *Find the Fourier series of* $f(x) = \dfrac{3x^2 - 6\pi x + 2\pi^2}{12}$ *in the interval* $(0, 2\pi)$ *and hence deduce that* $\dfrac{\pi^2}{6} = 1 + \dfrac{1}{2^2} + \dfrac{1}{3^2} + \ldots$

SOLUTION: $f(x) = \dfrac{1}{2}a_0 + \sum_1^{\infty}(a_n \cos nx + b_n \sin nx)$ where

$$a_0 = \frac{1}{\pi}\int_0^{2\pi}\frac{1}{12}(3x^2 - 6\pi x + 2\pi^2)\,dx = \frac{1}{12\pi}\Big[x^3 - 3\pi x^2 + 2\pi^2 x\Big]_0^{2\pi} = \frac{1}{12\pi}[8\pi^3 - 12\pi^3 + 4\pi^3] = 0$$

$$a_x = \frac{1}{\pi}\int_0^{2\pi}\frac{1}{12}(3x^2 - 6\pi x + 2\pi^2)\cos nx\,dx$$

$$= \frac{1}{12\pi}\left[(3x^2 - 6\pi x + 2\pi^2)\frac{\sin nx}{n}\right]_0^{2\pi} - \frac{1}{12\pi}\int_0^{2\pi}(6x - 6\pi)\frac{\sin nx}{n}\,dx$$

$$= 0 - \frac{1}{12\pi}\left[6(x-\pi)\frac{(-\cos nx)}{n^2}\right]_0^{2\pi} + \frac{1}{12\pi}\int_0^{2\pi}\frac{6(-\cos nx)}{n^2}\,dx$$

$$= \frac{1}{2n^2}(\cos 2n\pi + \cos 0) + \frac{1}{2\pi}\left[\frac{-\sin nx}{n^3}\right]_0^{2\pi} = \frac{1}{n^2} + 0$$

$$b_n = \frac{1}{\pi}\int_0^{2\pi}\frac{1}{12}(3x^2 - 6\pi x + 2\pi^2)\sin nx\,dx$$

$$= \frac{1}{12\pi}\left[(3x^2 - 6\pi x + 2\pi^2)\left(\frac{-\cos nx}{n}\right)\right]_0^{2\pi} - \frac{1}{12\pi}\int_0^{2\pi}6(x-\pi)\left(\frac{-\cos nx}{n}\right)dx$$

$$= \frac{-1}{12\pi n}\Big[(12\pi^2 - 12\pi^2 + 2\pi^2)\cos 2n\pi - 2\pi^2\cos 0\Big] + \frac{1}{2n\pi}\left[(x-\pi)\frac{\sin nx}{n}\right]_0^{2\pi} - \frac{1}{2n\pi}\int_0^{2\pi}\frac{\sin nx}{n}\,dx$$

$$= 0 + 0 + \frac{1}{2\pi n^3}[\cos nx]_0^{2\pi} = 0$$

$$\therefore\ f(x) = \sum_1^{\infty}\frac{1}{n^2}\cos nx$$

Putting $x = 0$ here, we get

$$\frac{2\pi^2}{12} = \sum_1^{\infty}\frac{1}{n^2}\cos 0 \quad\text{or}\quad \frac{\pi^2}{6} = \sum\frac{1}{n^2} = 1 + \frac{1}{2^2} + \frac{1}{3^2} + \frac{1}{4^2} + \ldots$$

Hence proved.

3. *Find the Fourier series of* $f(x) = x, \quad \dfrac{-\pi}{2} < x < \dfrac{\pi}{2}$

$$= \pi - x, \quad \frac{\pi}{2} < x < \frac{3\pi}{2}.$$

SOLUTION: Fourier series of $f(x)$ in $\left(\dfrac{-\pi}{2}, \dfrac{3\pi}{2}\right)$ is

$$f(x) = \frac{1}{2}a_0 + \sum_{n=1}^{\infty}(a_n \cos nx + b_n \sin nx)$$

where
$$a_0 = \frac{1}{\pi}\int_{-\pi/2}^{3\pi/2} f(x)\,dx = \frac{1}{\pi}\int_{-\pi/2}^{\pi/2} x\,dx + \frac{1}{\pi}\int_{\pi/2}^{3\pi/2}(\pi - x)\,dx = 0 + \frac{1}{\pi}\left[\pi x - \frac{x^2}{2}\right]_{\pi/2}^{3\pi/2}$$

$$= \frac{1}{\pi}\left[\frac{3\pi^2}{2} - \frac{9\pi^2}{8} - \frac{\pi^2}{2} + \frac{\pi^2}{8}\right] = 0$$

and
$$a_n = \frac{1}{\pi}\int_{-\pi/2}^{3\pi/2} f(x)\cos nx\,dx = \frac{1}{\pi}\int_{-\pi/2}^{\pi/2} x\cos nx\,dx + \frac{1}{\pi}\int_{\pi/2}^{3\pi/2}(\pi - x)\cos nx\,dx$$

Here
$$\int_{-\pi/2}^{\pi/2} x\cos nx\,dx = 0, \text{ by property of definite integrals}$$

and
$$\int_{\pi/2}^{3\pi/2}(\pi - x)\cos nx\,dx = \left[(\pi - x)\frac{\sin x}{n}\right]_{\pi/2}^{3\pi/2} - \int_{\pi/2}^{3\pi/2}(-1)\frac{\sin nx}{n}\,dx$$

$$= \frac{-\pi}{2n}\sin\frac{3n\pi}{2} - \frac{\pi}{2n}\sin\frac{n\pi}{2} - \left[\frac{\cos nx}{n^2}\right]_{\pi/2}^{3\pi/2}$$

$$= \frac{-\pi}{2n}\left[\sin\left(2n\pi - \frac{n\pi}{2}\right) + \sin\frac{n\pi}{2}\right] - \frac{1}{n^2}\left[\cos\frac{3n\pi}{2} - \cos\frac{n\pi}{2}\right]$$

$$= \frac{-\pi}{2}\left(-\frac{\sin n\pi}{2} + \frac{\sin n\pi}{2}\right) - \frac{1}{n^2}\left[\cos\left(2n\pi - \frac{n\pi}{2}\right) - \frac{\cos n\pi}{2}\right]$$

$$= 0 - \frac{1}{n^2}\left[\frac{\cos n\pi}{2} - \frac{\cos n\pi}{2}\right] = 0$$

$\therefore \quad a_n = 0$

$$b_n = \frac{1}{\pi}\int_{-\pi/2}^{3\pi/2} f(x)\sin nx\,dx = \frac{1}{\pi}\int_{-\pi/2}^{\pi/2} x\sin nx\,dx + \frac{1}{\pi}\int_{\pi/2}^{3\pi/2}(\pi - x)\sin nx\,dx$$

$$= \frac{1}{\pi}\left[\frac{-x\cos nx}{n}\right]_{-\pi/2}^{\pi/2} - \frac{1}{\pi}\int_{-\pi/2}^{\pi/2}\frac{1(-\cos nx)}{n}\,dx + \frac{1}{\pi}\left[\frac{(\pi - x)(-\cos nx)}{n}\right]_{\pi/2}^{3\pi/2} - \frac{1}{\pi}\int_{\pi/2}^{3\pi/2}(-1)\frac{(-\cos nx)}{n}\,dx$$

$$= \frac{1}{n\pi}\left[\frac{-\pi}{2}\cos\frac{n\pi}{2} - \frac{\pi}{2}\cos\frac{n\pi}{2}\right] + \frac{1}{\pi n^2}[\sin nx]_{-\pi/2}^{\pi/2} + \frac{1}{n\pi}\left[\frac{\pi}{2}\cos\frac{3n\pi}{2} + \frac{\pi}{2}\cos\frac{n\pi}{2}\right] - \left[\frac{\sin nx}{\pi n^2}\right]_{\pi/2}^{3\pi/2}$$

$$= -\frac{1}{n}\cos\frac{n\pi}{2} + \frac{2}{\pi n^2}\sin\frac{n\pi}{2} + \frac{1}{2n}\left[\cos\frac{n\pi}{2} + \cos\left(2n\pi - \frac{n\pi}{2}\right)\right] - \frac{1}{\pi n^2}\left[\sin\frac{3n\pi}{2} - \sin\frac{n\pi}{2}\right]$$

$$= 0 + \frac{2}{\pi n^2}\sin\frac{n\pi}{2} - \frac{1}{\pi n^2}\left[\sin\left(2n\pi - \frac{n\pi}{2}\right) - \sin\frac{n\pi}{2}\right] = \frac{1}{\pi n^2}\left[2\sin\frac{n\pi}{2} + 2\sin\frac{n\pi}{2}\right]$$

$$= \frac{4}{\pi n^2}\sin\frac{n\pi}{2}$$

$$\therefore \quad f(x) = \frac{1}{\pi}\sum_{1}^{\infty}\frac{4}{\pi n^2}\sin nx \sin\frac{n\pi}{2}$$ **Ans.**

4. *Find the Fourier series of $f(x) = x^2$ in the interval (0, 4). Hence deduce that*

$$\frac{\pi^2}{6} = \frac{1}{1^2} + \frac{1}{2^2} + \frac{1}{3^2} + \ldots$$

SOLUTION: Fourier series of $f(x)$ with period $2l = 4$ is

$$f(x) = \frac{1}{2}a_0 + \sum_{n=1}^{\infty}\left(a_n\cos\frac{n\pi x}{l} + b_n\sin\frac{n\pi x}{l}\right) = \frac{1}{2}a_0 + \sum_{1}^{\infty}\left(a_n\cos\frac{n\pi x}{2} + b_n\sin\frac{n\pi x}{2}\right)$$

$$a_0 = \frac{1}{l}\int_0^{2l} f(x)\,dx = \frac{1}{2}\int_0^4 x^2\,dx = \frac{1}{2}\left[\frac{x^3}{3}\right]_0^4 = \frac{32}{3}.$$

$$a_n = \frac{1}{l}\int_0^{2l} f(x)\cos\frac{nx\pi}{l}\,dx = \frac{1}{2}\int_0^4 x^2\cos\frac{n\pi x}{2}\,dx$$

$$= \frac{1}{2}\left[x^2\left(\frac{2}{n\pi}\sin\frac{n\pi x}{2}\right) - 2x\left(\frac{-4}{n^2\pi^2}\cos\frac{n\pi x}{2}\right) + 2\left(\frac{-8}{n^3\pi^3}\sin\frac{n\pi x}{2}\right)\right]_0^4$$

$$= \frac{1}{2}\left[8\left(\frac{4}{n^2\pi^2}\right)\right] + 0 = \frac{16}{n^2\pi^2}.$$

$$b_n = \frac{1}{l}\int_0^{2l} f(x)\sin\frac{n\pi x}{l}\,dx = \frac{1}{2}\int_0^4 x^2\sin\frac{n\pi x}{2}\,dx$$

$$= \frac{1}{2}\left[x^2\left(\frac{-2}{n\pi}\cos\frac{n\pi x}{2}\right) - 2x\left(\frac{-4}{n^2\pi^2}\sin\frac{n\pi x}{2}\right) + 2\left(\frac{8}{n^3\pi^3}\cos\frac{n\pi x}{2}\right)\right]_0^4$$

$$= \frac{1}{2}\left(\frac{-32}{n\pi}\right) + 0 + 0 = \frac{-16}{n\pi}$$

$$\therefore \quad f(x) = \frac{16}{3} + \frac{16}{\pi^2}\sum_{1}^{\infty}\frac{1}{n^2}\cos\frac{n\pi x}{2} - \frac{16}{\pi}\sum_{1}^{\infty}\frac{1}{n}\sin\frac{n\pi x}{2} \quad \ldots(1)$$

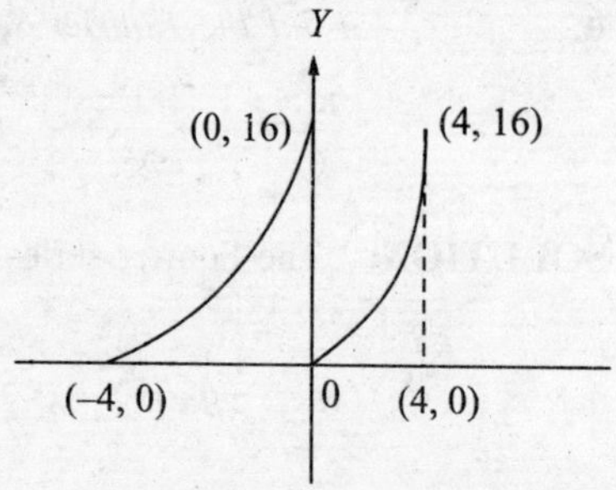

The graph of $f(x)$ of period 4 is as shown in the figure.

Here $f(0-) = 16$ and $f(0+) = 0$, hence

we can take $f(0) = \frac{1}{2}[f(0+) + f(0-)]$.

Putting $x = 0$ in (1), we get

$$\frac{0+16}{2} = \frac{16}{3} + \frac{16}{\pi^2}\sum\frac{1}{n^2} \quad \text{or} \quad 8 - \frac{16}{3} = \frac{16}{\pi^2}\sum\frac{1}{n^2}$$

or
$$\frac{\pi^2}{6} = \sum\frac{1}{n^2}$$

Hence Proved.

5. *Find the Fourier series of $f(x) = 4 - x^2$ in the interval $(0, 2)$ and hence deduce that*
$$\frac{\pi^2}{6} = \frac{1}{1^2} + \frac{1}{2^2} + \frac{1}{3^2} + \ldots$$

SOLUTION: Fourier series with period $2l = 2$, is

$$f(x) = \frac{1}{2}a_0 + \sum_1^\infty\left(a_n \cos\frac{n\pi x}{l} + b_n \sin\frac{n\pi x}{l}\right) = \frac{1}{2}a_0 + \sum_1^\infty(a_n \cos n\pi x + b_n \sin n\pi x)$$

$$a_0 = \frac{1}{l}\int_0^{2l} f(x)\,dx = \frac{1}{1}\int_0^2 (4-x^2)dx = \left[4x - \frac{x^3}{3}\right]_0^2 = 8 - \frac{8}{3} = \frac{16}{3}.$$

$$a_n = \frac{1}{l}\int_0^{2l} f(x)\cos\frac{n\pi x}{l}\,dx = \frac{1}{1}\int_0^2 (4-x^2)\cos n\pi x\,dx$$

$$= \left[(4-x^2)\frac{\sin n\pi x}{n\pi} - (-2x)\left(\frac{-\cos n\pi x}{n^2\pi^2}\right) + (-2)\left(\frac{-\sin n\pi x}{n^3\pi^3}\right)\right]_0^2 = -\frac{4}{n^2\pi^2}$$

$$b_n = \frac{1}{l}\int_0^{2l} f(x)\sin\frac{n\pi x}{l}\,dx = \int_0^2 (4-x^2)\sin n\pi x\,dx$$

$$= \left[(4-x^2)\left(\frac{-\cos n\pi x}{n\pi}\right) - (-2x)\left(\frac{-\sin n\pi x}{n^2\pi^2}\right) + (-2)\left(\frac{\cos n\pi x}{n^3\pi^3}\right)\right]_0^2 = \frac{4}{n\pi}$$

$$\therefore \quad f(x) = \frac{8}{3} - \frac{4}{\pi^2}\sum_1^\infty \frac{1}{n^2}\cos n\pi x + \frac{4}{\pi}\sum_1^\infty \frac{1}{n}\sin n\pi x \qquad \ldots(1) \quad \textbf{Ans.}$$

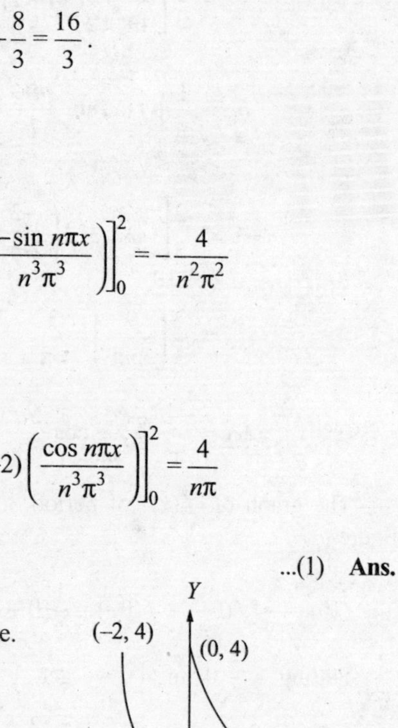

The graph of $f(x)$ of period 2 is shown in the adjacent figure.

Here $f(x)$ 4 and $f(0-) = 0$ hence

we can take $f(0) = \frac{1}{2}[f(0+) + f(0-) = \frac{1}{2}(4+0) = 2.$

Hence proved.

6. *Find the Fourier series of* $f(x) = 2x - x^2$ *in the interval* $(0, 3)$. *Hence deduce that*

$$\frac{\pi^2}{6} = \frac{1}{1^2} + \frac{1}{2^2} + \frac{1}{3^2} + \ldots$$

SOLUTION: The Fourier series of $f(x)$ with period $2l = 3$ is given by

$$f(x) = \frac{1}{2}a_0 + \sum_1^{\infty} a_n \cos\frac{n\pi x}{l} + b_n \sin\frac{n\pi x}{l} = \frac{1}{2}a_0 + \sum_1^{\infty}\left(a_n \cos\frac{2n\pi x}{3} + b_n \sin\frac{2n\pi x}{3}\right)$$

$$a_0 = \frac{1}{l}\int_0^{2l} f(x)\, dx = \frac{2}{3}\int_0^3 (2x - x^2)\, dx = \frac{2}{3}\left[x^2 - \frac{x^3}{3}\right]_0^3 = \frac{2}{9}\left[9 - \frac{27}{3}\right] = 0.$$

$$a_n = \frac{1}{l}\int_0^{2l} f(x) \cos\frac{n\pi x}{l}\, dx = \frac{2}{3}\int_0^3 (2x - x^2) \cos\frac{2n\pi x}{3}\, dx$$

$$= \frac{2}{3}\left[(2x - x^2)\left(\frac{3}{2n\pi}\sin\frac{2n\pi x}{3}\right) - (2 - 2x)\left(\frac{-9}{4n^2\pi^2}\cos\frac{2n\pi x}{3}\right) + (-2)\left(\frac{-27}{8n^3\pi^3}\sin\frac{2n\pi x}{3}\right)\right]_0^3$$

$$= \frac{2}{3}\left[\frac{9}{4n^2\pi^2}(-4 - 2)\right] = \frac{-9}{n^2\pi^2}$$

$$b_n = \frac{1}{l}\int_0^{2l} f(x) \sin\frac{2n\pi x}{l}\, dx = \frac{2}{3}\int_0^3 (2x - x^2) \sin\frac{2n\pi x}{3}\, dx$$

$$= \frac{2}{3}\left[(2x - x^2)\left(\frac{-3}{2n\pi}\cos\frac{2n\pi x}{3}\right) - (2 - 2x)\left(\frac{-9}{4n^2\pi^2}\sin\frac{2n\pi x}{3}\right) + (-2)\left(\frac{27}{8n^3\pi^3}\cos\frac{2n\pi x}{3}\right)\right]_0^3$$

$$= \frac{2}{3}\left(\frac{9}{2n\pi}\right) = \frac{3}{n\pi}$$

$$\therefore \quad f(x) = \frac{-9}{\pi^2}\sum\frac{1}{n^2}\cos\frac{2n\pi x}{3} + \frac{3}{\pi}\sum\frac{1}{n}\sin\frac{2n\pi x}{3} \qquad \ldots(1) \quad \textbf{Ans.}$$

The graph of $f(x)$ of period 3, is shown in the adjoining figure.

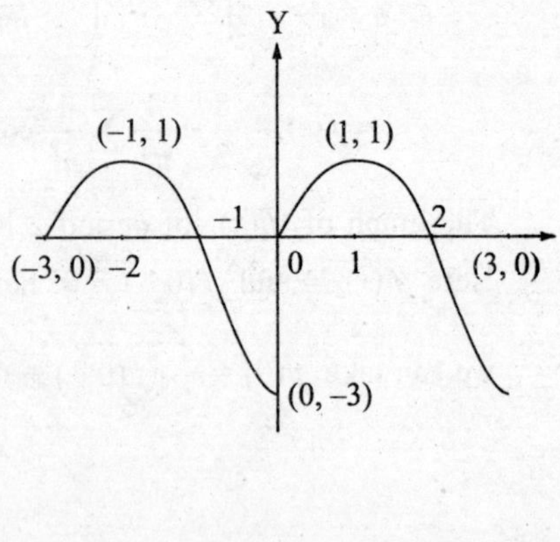

$$f(0) = \frac{1}{2}[f(0+) + f(0-)] = \frac{1}{2}[0 + (-3)] = -\frac{3}{2}$$

Putting $x = 0$ in (1), we get

$$\frac{-3}{2} = \frac{-9}{\pi^2}\sum\frac{1}{n^2} + 0 \quad \text{or} \quad \frac{\pi^2}{6} = \sum\frac{1}{n^2}. \qquad \textbf{Hence Proved.}$$

7. *Find the Fourier series of* $f(x) = 4 - x,\ 3 < x < 4$
$= x - 4,\ 4 < x < 5$

SOLUTION: Fourier series of $f(x)$ with period $2l = 5 - 3 = 2$. $\therefore$ $l = 1$, is

$$f(x) = \frac{1}{2}a_0 + \sum\left(a_n \cos\frac{n\pi x}{l} + b_n \sin\frac{n\pi x}{l}\right) = \frac{1}{2}a_0 + \sum(a_n \cos n\pi x + b_n \sin n\pi x)$$

$$a_0 = \frac{1}{l}\int_c^{c+2l} f(x)\,dx = \int_3^4 (4-x)\,dx + \int_4^5 (x-4)\,dx = \left[4x - \frac{x^2}{2}\right]_3^4 + \left[\frac{x^2}{2} - 4x\right]_4^5 = 1$$

$$a_n = \frac{1}{l}\int_c^{c+2l} f(x)\cos\frac{n\pi x}{l}\,dx = \int_3^4 (4-x)\cos n\pi x\,dx + \int_4^5 (x-4)\cos n\pi x\,dx$$

$$= \left[(4-x)\frac{\sin n\pi x}{n\pi} - (-1)^2\frac{\cos n\pi x}{n^2\pi^2}\right]_3^4 + \left[(x-4)\frac{\sin n\pi x}{n\pi} - (-1)\frac{\cos n\pi x}{n^2\pi^2}\right]_4^5$$

$$= \frac{-1}{n^2\pi^2}(\cos 4n\pi - \cos 3n\pi) + \frac{1}{n^2\pi^2}(\cos 5n\pi - \cos 4n\pi)$$

$$= \frac{1}{n^2\pi^2}[-1 + (-1)^n + (-1)^n - 1] = \frac{2}{n^2\pi^2}[(-1)^n - 1].$$

$$b_n = \frac{1}{l}\int_c^{c+2l} f(x)\sin\frac{n\pi x}{l}\,dx = \int_3^4 (4-x)\sin n\pi x\,dx + \int_4^5 (x-4)\sin n\pi x\,dx$$

$$= \left[(4-x)\left(\frac{-\cos n\pi x}{n\pi}\right) - (-1)\left(\frac{-\sin n\pi x}{n^2\pi^2}\right)\right]_3^4 + \left[(x-4)\left(\frac{-\cos n\pi x}{n\pi}\right) - (+1)\left(\frac{-\sin n\pi x}{n^2\pi^2}\right)\right]_4^5$$

$$= \frac{+1}{n\pi}\cos 3n\pi - \frac{1}{n\pi}\cos 5n\pi = 0$$

$$\therefore\ f(x) = \frac{1}{2} + \sum_1^\infty \frac{2}{n^2\pi^2}[(-1)^n - 1]\cos n\pi x = \frac{1}{2} + \frac{2}{\pi^2}\sum_1^\infty \frac{(-1)^n - 1}{n^2}\cos n\pi x.$$ **Ans.**

8. *Obtain the Fourier series of* $f(x) = \cos x,\ -\pi < x < 0$
$= -\cos x,\ 0 < x < \pi.$

SOLUTION: $f(-x) = \cos(-x),\quad -\pi < -x < 0$
$= -\cos(-x),\quad 0 < -x < \pi$

that is, $f(-x) = \cos x,\quad 0 < x < \pi$
$= -\cos x,\quad -\pi < x < 0$

Thus $f(-x) = -f(x)$ hence $f(x)$ is odd function of period 2π.

$\therefore$ $f(x) = \sum b_n \sin nx$

$$b_n = \frac{2}{\pi}\int_0^\pi f(x)\sin nx\,dx = \frac{2}{\pi}\int_0^\pi(-\cos x)\sin nx\,dx$$

$$= \frac{-1}{\pi}\int_0^{\pi}[\sin(n+1)x+\sin(n-1)x]\,dx = \frac{-1}{\pi}\left[\frac{-\cos(n+1)x}{n+1}-\frac{\cos(n-1)x}{n-1}\right]_0^{\pi} \quad (n\neq 1)$$

$$= \frac{1}{\pi}\left[\frac{\cos(n+1)\pi}{n+1}+\frac{\cos(n-1)\pi}{n-1}-\frac{1}{n-1}-\frac{1}{n+1}\right] \qquad (n\neq 1)$$

$$= \frac{1}{\pi}\left[\left(\frac{-\cos n\pi-1}{n+1}\right)-\frac{1+\cos n\pi}{n-1}\right] \quad (n\neq 1) = \frac{-2n}{\pi(n^2-1)}[1+(-1)^n] \qquad (n\neq 1)$$

and $$b_1 = \frac{2}{\pi}\int_0^{\pi} f(x)\sin x\,dx = \frac{2}{\pi}\int_0^{\pi} -\cos x\sin x\,dx = \left[\frac{\cos 2x}{2\pi}\right]_0^{\pi} = 0$$

$$\therefore \quad f(x) = \frac{-2}{\pi}\sum_2^{\infty}\frac{n}{n^2-1}[1+(-1)^n]\sin nx$$

$$= \frac{-2}{\pi}\left[\frac{4}{1.3}\sin 2x+\frac{8}{3.5}\sin 4x+\frac{6}{5.7}\sin 6x+\ldots\right]$$ **Ans.**

9. *Find the Fourier series of $f(x) = x|x|$ in the interval $(-1, 1)$.*

SOLUTION: Clearly $f(x)$ is odd in $(-1, 1)$.

$$\therefore \quad f(x) = -x^2, \quad -1<x<0$$
$$= x^2, \quad 0<x<1$$

Fourier series of $f(x)$ which is odd in $(-1, 1)$, with period $2l = 2$, is

$$f(x) = \sum b_n \sin\frac{n\pi x}{l} = \sum b_n \sin n\pi x$$

where $$b_n = \frac{2}{l}\int_0^1 f(x)\sin\frac{n\pi x}{l}\,dx = 2\int_0^1 x^2\sin n\pi x\,dx$$

$$= 2\left[x^2\left(\frac{-\cos n\pi x}{n\pi}\right)-2x\left(\frac{-\sin n\pi x}{n^2\pi^2}\right)+2\left(\frac{\cos n\pi x}{n^3\pi^3}\right)\right]_0^1$$

$$= 2\left[-\frac{\cos n\pi}{n\pi}+\frac{2\cos n\pi}{n^3\pi^3}-\frac{2}{n^3\pi^3}\right]$$

$$\therefore \quad f(x) = 2\sum_1^{\infty}\left[-\frac{(-1)^n}{n\pi}+\frac{2(-1)^n-2}{n^3\pi^3}\right]\sin n\pi x.$$ **Ans.**

10. *Find the half-range sine series of $f(x) = e^{ax}$ in $(0, \pi)$.*

SOLUTION: Half-range sine series of $f(x)$ defined in $(0, \pi)$ is

$$f(x) = \sum_1^{\infty} b_n \sin nx \quad \text{where } b_n = \frac{2}{\pi}\int_0^{\pi} f(x)\sin nx\,dx$$

or $$b_n = \frac{2}{\pi}\int_0^{\pi} e^{ax} \sin nx\, dx = \frac{2}{\pi}\left[\frac{e^{ax}}{a^2+n^2}(a \sin nx - n \cos nx)\right]_0^{\pi}$$

$$= \frac{2}{\pi}\left[\frac{e^{a\pi}}{a^2+n^2}(a \sin n\pi - n \cos n\pi) - \frac{1}{a^2+n^2}(0-n)\right]$$

$$= \frac{2n}{\pi(a^2+n^2)}[1-(-1)^n e^{a\pi}]$$

$$\therefore \quad f(x) = \frac{2}{\pi}\sum_1^{\infty}\frac{n}{a^2+n^2}[1-(-1)^n e^{a\pi}] \sin nx \qquad \textbf{Ans.}$$

11. *Find the half range cosine series of f(x) where*

$$f(x) = x, \qquad 0 < x < \frac{\pi}{2}$$

$$= \pi - x, \qquad \frac{\pi}{2} < x < \pi$$

SOLUTION: $f(x) = \frac{1}{2}a_0 + \sum a_n \cos nx$ where

$$a_0 = \frac{2}{\pi}\int_0^{\pi} f(x)\, dx = \frac{2}{\pi}\left[\int_0^{\pi/2} x\, dx + \int_{\pi/2}^{\pi}(\pi - x)\, dx\right] = \frac{2}{\pi}\left[\left(\frac{x^2}{2}\right)_0^{\pi/2} + \left(\pi x - \frac{x^2}{2}\right)_{\pi/2}^{\pi}\right] = \frac{\pi}{2}$$

$$a_n = \frac{2}{\pi}\int_0^{\pi} f(x) \cos nx\, dx = \frac{2}{\pi}\left[\int_0^{\pi/2} x \cos nx\, dx + \int_{\pi/2}^{\pi}(\pi - x)\cos nx\, dx\right]$$

$$= \frac{2}{\pi}\left[x\frac{\sin nx}{n} - (1)\left(-\frac{\cos nx}{n^2}\right)\right]_0^{\pi/2} + \frac{2}{\pi}\left[(\pi - x)\frac{\sin nx}{n} - (-1)\left(-\frac{\cos nx}{n^2}\right)\right]_{\pi/2}^{\pi}$$

$$= \frac{2}{\pi n^2}\left[2\cos\frac{n\pi}{2} - 1 - (-1)^n\right]$$

$$\therefore \quad f(x) = \frac{\pi}{4} + \frac{2}{\pi}\sum\frac{1}{n^2}\left[2\cos\frac{n\pi}{2} - 1 - (-1)^n\right]\cos nx$$

$$= \frac{\pi}{4} - \frac{2}{\pi}\left(\frac{1}{1^2}\cos 2x + \frac{1}{3^2}\cos 6x + \frac{1}{5^2}\cos 10x + \ldots\right) \qquad \textbf{Ans.}$$

12. *Find the half-range sine series of* $f(x) = x, \; 0 < x < 1$

$= 2 - x, \; 1 < x < 2$

and hence deduce that $\dfrac{\pi^2}{8} = \dfrac{1}{1^2} + \dfrac{1}{3^2} + \dfrac{1}{5^2} + \ldots$

SOLUTION: Half-range Fourier sine series of $f(x)$ with period $2l = 4$, is

$$f(x) = \sum b_n \sin\frac{n\pi x}{l} = \sum b_n \sin\frac{n\pi x}{2}.$$

where
$$b_n = \frac{2}{l}\int_0^l f(x)\sin\frac{n\pi x}{l}\,dx = \frac{2}{2}\left[\int_0^1 x\sin\frac{n\pi x}{2}\,dx + \int_1^2 (2-x)\sin\frac{n\pi x}{2}\,dx\right]$$

$$= \left[-\frac{2x}{n\pi}\cos\frac{n\pi x}{2} - (1)\left(\frac{-4}{n^2\pi^2}\sin\frac{n\pi x}{2}\right)\right]_0^1$$

$$+\left[(2-x)\left(\frac{-2}{n\pi}\cos\frac{n\pi x}{2}\right) - (-1)\left(\frac{-4}{n^2\pi^2}\sin\frac{n\pi x}{2}\right)\right]_1^2$$

$$= \frac{-2}{n\pi}\cos\frac{n\pi}{2} + \frac{4}{n^2\pi^2}\sin\frac{n\pi}{2} + 0 - \frac{4}{n^2\pi^2}\sin n\pi + \frac{2}{n\pi}\cos\frac{n\pi}{2} + \frac{4}{n^2\pi^2}\sin\frac{n\pi}{2}$$

$$= \frac{8}{n^2\pi^2}\sin\frac{n\pi}{2}$$

$$\therefore \quad f(x) = \frac{8}{\pi^2}\sum_1^\infty \frac{1}{n^2}\sin\frac{n\pi}{2}\sin\frac{n\pi x}{2}$$

At $x = 1$, $f(1)$ is not defined but can be taken as $\dfrac{f(1-)+f(1+)}{2}$

or $f(1) = \dfrac{1+(2-1)}{2} = 1$. Putting $x = 1$ in the above Fourier series,

$$f(1) = 1 = \frac{8}{\pi^2}\sum_1^\infty \frac{1}{n^2}\sin^2\frac{n\pi}{2}$$

or
$$\frac{\pi^2}{8} = \sum_1^\infty \frac{1}{n^2}\sin^2\frac{n\pi}{2} = \frac{1}{1^2} + \frac{1}{3^2} + \frac{1}{5^2} + \frac{1}{7^2} + \ldots$$

Hence proved.

CHAPTER

20 BETA, GAMMA Functions and Series Solution of Differential Equations

IMPORTANT FORMULAE AND USEFUL RESULTS

1. Gamma Function: $\Gamma n = \int_0^\infty e^{-x} x^{n-1}\, dx,\ n > 0,\ \Gamma\frac{1}{2} = \sqrt{\pi}$

$\Gamma 1 = 1,\ \Gamma 2 = 1,\ \Gamma 3 = 1.2,...,\qquad \Gamma(n+1) = n\Gamma n.$

2. Beta Function: $\beta(m, n) = \int_0^1 x^{m-1}(1-x)^{n-1}\, dx$

Here $\beta(m, n) = \beta(n, m).$

$$\beta(m, n) = \int_0^\infty \frac{t^{n-1}\, dt}{(1+t)^{m+n}} \quad \text{and} \quad \beta(m, n) = 2\int_0^{\pi/2} \sin^{2m-1}\theta \cos^{2n-1}\theta\, d\theta$$

3. Relation between Beta and Gamma Function

$$\beta(m, n) = \frac{\Gamma m\, \Gamma n}{\Gamma(m+n)}$$

$$\int_0^{\pi/2} \sin^p\theta \cos^q\theta\, d\theta = \frac{\Gamma\frac{p+1}{2}\, \Gamma\frac{q+1}{2}}{2\,\Gamma\frac{p+q+2}{2}}.$$

4. SOLUTION-IN SERIES (FROBENIUS METHOD)

Frobenius method is applied to second order differential equations of the form

$$P(x)\frac{d^2y}{dx^2} + Q(x)\frac{dy}{dx} + R(x)y = 0$$

or $\dfrac{d^2y}{dx^2} + p(x)\dfrac{dy}{dx} + q(x)y = 0$ where $p(x) = \dfrac{Q(x)}{P(x)}$ and $q(x) = \dfrac{R(x)}{P(x)}$.

In general, solution is taken as $y = \Sigma a_r x^{m+r} = x^m (a_0 + a_1 x + a_2 x^2 + ...)$

We substitute for y, $\dfrac{dy}{dx}$ and $\dfrac{d^2y}{dx^2}$ in the given equation. When the coefficient of the lowest degree term in x is equated to zero, we get indicial equation as it determines m.

The general solution of the given equation is of the form $y = A y_1(x) + B y_2(x)$. where A and B are arbitrary constants and y_1, y_2 are independent solutions.

Case I: Roots of indicial equation are distinct but not differing by an integer, we get $y = A y_1(x) + B y_2(x)$ where $y_1(x)$ corresponds to $m = m_1$ and $y_2(x)$ to $m = m_2$.

Case II: Roots equal $m_1 = m_2$.

Here both y_1, y_2 become identical. After getting y_1 we get $y_2 = \left(\dfrac{\partial y}{\partial m}\right)_{m=m_1, \text{ or } m_2}$

Case III: Roots differ by an integer, the solution is

$$y = A[y(x)]_{m=m_1} + B\left(\frac{\partial y}{\partial m}\right)_{m=m_1}$$

SOLUTIONS TO EXERCISE 20A

1. *Evaluate* $\int_0^\infty x\, e^{-ax} \cos bx\, dx,\ (a > 0)$

SOLUTION: Consider $I = \int_0^\infty x\, e^{-ax}\, e^{ibx} dx = \int_0^\infty x\, e^{(-a+ib)x} dx$

$$= \left[x \frac{e^{(-a+ib)x}}{-a+ib}\right]_0^\infty - \int_0^\infty \frac{1\, e^{(-a+ib)x}}{-a+ib} dx = 0 - \left[\frac{e^{(-a+ib)x}}{(-a+ib)^2}\right]_0^\infty$$

$$= 0 - 0 + \frac{1}{(-a+ib)^2} = \frac{(a+ib)^2}{(a^2+b^2)^2} = \frac{a^2-b^2+2iab}{(a^2+b^2)^2}$$

or $$I = \frac{a^2-b^2}{(a^2+b^2)^2} + \frac{2i\, ab}{(a^2+b^2)^2}$$

Equating real parts on both sides, gives

$$\int_0^\infty x\, e^{-ax} \cos bx\, du = \frac{a^2-b^2}{(a^2+b^2)^2}$$ **Ans.**

2. *Evaluate* $\int_0^\infty \frac{x^4\, dx}{4^x}$.

SOLUTION: $I = \int_0^\infty \frac{x^4\, du}{4^x}$ Putting $4^x = e^t$ or $x \log 4 = t$, we have

$$I = \int_0^\infty \frac{(t)^4}{(\log 4)^4 e^t} \frac{dt}{\log 4} = \int_0^\infty \frac{e^{-t} \cdot t^4}{(\log 4)^5} dt = \frac{\overline{|5}}{(\log 4)^5} = \frac{24}{(\log 4)^5}$$ **Ans.**

3. *Show that* $\int_0^1 (x \log x)^3\, dx = \frac{-3}{128}$.

SOLUTION: $I = \int_0^1 (x \log x)^3\, dx$. Put $x = e^{-t}$ or $\log x = -t$ so $dx = -e^{-t}\, dt$.

$$\therefore \quad I = -\int_\infty^0 [e^{-t}(-t)^3]\, e^{-t}\, dt = -\int_0^\infty t^3 e^{-4t}\, dt.$$ Now putting $4t = y$

$$= -\int_0^\infty \frac{y^3}{4^3} e^{-y} \frac{dy}{4} = -\frac{1}{4^4}\overline{|4} = \frac{-3}{128}$$ **Ans.**

4. *Evaluate* $\int_0^1 x^m (\log x)^n \, dx$.

SOLUTION: $I = \int_0^1 x^m (\log x)^n \, dx$. Putting $x = e^{-t}$ or $\log x = -t$, so $dx = -e^{-t} \, dt$

$$\therefore \quad I = \int_\infty^0 e^{-mt} (-t)^n (-e^{-t}) \, dt = \int_0^\infty (-1)^n e^{-(m+1)t} t^n \, dt.$$

Now put $(m+1)t = y$ then

$$I = (-1)^n \int_0^\infty e^{-y} \frac{y^n \, dy}{(m+1)^{n+1}} = \frac{(-1)^n \boxed{n+1}}{(m+1)^{n+1}}$$ **Ans.**

5. *Show that* $\beta(m, n+1) + \beta(m+1, n) = \beta(m, n)$.

SOLUTION: We know that $\beta(m, n) = \dfrac{\boxed{m}\boxed{n}}{\boxed{m+n}}$ therefore

$$\beta(m, n+1) + \beta(m+1, n) = \frac{\boxed{m}\boxed{n+1}}{\boxed{m+n+1}} + \frac{\boxed{m+1}\boxed{n}}{\boxed{m+n+1}} = \frac{\boxed{m} \cdot n\boxed{n} + m\boxed{m}\boxed{n}}{\boxed{m+n+1}}$$

$$= \frac{\boxed{m}\boxed{n}(m+n)}{(m+n)\boxed{m+n}} = \frac{\boxed{m}\boxed{n}}{\boxed{m+n}} = \beta(m, n)$$ **Hence proved.**

6. *Evaluate* $\int_0^\infty \frac{x^4 (1+x^5) \, dx}{(1+x)^{15}}$.

SOLUTION: Let $$I = \int_0^\infty \frac{x^4(1+x^5) \, dx}{(1+x)^{15}} = \int_0^\infty \frac{x^{5-1} \, dx}{(1+x)^{15}} + \int_0^\infty \frac{x^{10-1} \, dx}{(1+x)^{15}}$$

We know that $$\beta(m, n) = \int_0^1 x^{m-1} (1-x)^{n-1} \, dx = \int_0^\infty \frac{t^{n-1}}{(1+t)^{m+1}} \quad \left(\text{on putting } x = \frac{t}{1+t}\right)$$

and since $$\beta(n, m) = \beta(m, n) = \int_0^\infty \frac{t^{m-1} \, dt}{(1+t)^{m+n}}$$

$$\therefore \quad I = \beta(5, 10) + \beta(10, 5) = \frac{2\boxed{5}\boxed{10}}{\boxed{15}} = \frac{1}{7 \cdot 13 \cdot 11 \cdot 5} = \frac{1}{5005}$$ **Ans.**

7. *Show that* $\int_0^1 \frac{x^{m-1} (1-x)^{n-1} \, dx}{(a+x)^{m+n}} = \frac{\beta(m,n)}{a^n (a+1)^m}$.

SOLUTION: Let $I = \int_0^1 \frac{x^{m-1} (1-x)^{n-1} \, dx}{(a+x)^{m+n}}$. Put $\frac{x}{a+x} = \frac{t}{a+1}$ so $x = \frac{at}{a+1-t}$

and $$(1-x) = \frac{(a+1)(1-t)}{a+1-t} \text{ and } x+a = \frac{a(a+1)}{a+1-t} \text{ and } dx = \frac{a(a+1)\,dt}{(a+1-t)^2}$$

$$\therefore \quad I = \int_0^1 \frac{(at)^{m-1}(a+1)^{n-1}(1-t)^{n-1}(a+1-t)^{m+n}}{(a+1-t)^{m-1}(a+1-t)^{n-1}a^{m+n}(a+1)^{m+n}} \frac{a(a+1)\,dt}{(a+1-t)^2}$$

$$= \int_0^1 \frac{t^{m-1}(1-t)^{n-1}\,dt}{a^n(a+1)^m} = \frac{\beta(m,n)}{a^n(a+1)^m}$$ **Hence proved.**

8. *Evaluate* $\int_0^1 \frac{x^{m-1}(1-x)^{n-1}}{(1+x)^{m+n}} dx.$

SOLUTION: In last question if we replace a by 1, we get

$$\int_0^1 \frac{x^{m-1}(1-x)^{n-1}\,du}{(1+x)^{m+n}} = \frac{\beta(m,n)}{2^m}$$ **Ans.**

9. *Show that* $\int_1^\infty \frac{dx}{x^{m+1}(x-1)^n} = \beta(m+n, 1-n), -m < n < 1.$

SOLUTION: Let $I = \int_1^\infty \frac{dx}{x^{m+1}(x-1)^n}$, Putting $x = \frac{1}{t}$, we get

$$I = -\int_1^0 \frac{\frac{dt}{t^2}}{\left(\frac{1}{t}\right)^{m+1}\left(\frac{1}{t}-1\right)^n} = \int_0^1 \frac{t^{m+1+n-2}\,dt}{(1-t)^n} = \int_0^1 t^{m+n-1}(1-t)^{-n+1-1}\,dt$$

$$= \beta(m+n, 1-n) \text{ where } -m < n < 1.$$ **Ans.**

10. *If* $\Gamma n\,\Gamma(1-n) = \frac{\pi}{\sin n\pi}$, show that $\int_0^\infty \frac{x^{n-1}\,dx}{(1+x)} = \frac{\pi}{\sin n\pi}.$

SOLUTION: Let $I = \int_0^\infty \frac{x^{n-1}\,dx}{1+x}$ then $I = \int_0^\infty \frac{x^{n-1}\,dx}{(1+x)^{n+1-n}}$

We know that $\beta(m,n) = \int_0^\infty \frac{t^{n-1}\,dt}{(1+t)^{m+n}}$ hence

$$I = \int_0^\infty \frac{x^{n-1}\,dx}{(1+x)^{n+1-n}} = \beta(1-n, n) = \frac{\Gamma(1-n)\,\Gamma n}{\Gamma 1} = \frac{\pi}{\sin n\pi}$$ **Hence proved.**

11. *Show that* $\beta(m+1, n) = \dfrac{m}{m+n}\beta(m,n)$.

SOLUTION: $$\beta(m+1, n) = \frac{\Gamma(m+1)\,\Gamma(n)}{\Gamma(m+n+1)} = \frac{m\,\Gamma(m)\,\Gamma(n)}{(m+n)\,\Gamma(m+n)} = \frac{m}{m+n}\,\beta(m, n)$$ **Hence proved.**

12. *Prove that* $\beta(n, n) = 2\displaystyle\int_0^{1/2} t^{n-1}(1-t)^{n-1}\,dt$.

SOLUTION: We know that $\beta(m, n) = \displaystyle\int_0^1 x^{m-1}(1-x)^{n-1}\,dx \quad \therefore\ I = \beta(n, n) = \int_0^1 x^{n-1}(1-x)^{n-1}\,dx$

or $$I = \int_0^{1/2} x^{n-1}(1-x)^{n-1}\,dx + \int_{1/2}^{1} x^{n-1}(1-x)^{n-1}\,dx = \int_0^{1/2} x^{n-1}(1-x)^{n-1}\,dx + I'$$

where $I' = \displaystyle\int_{1/2}^{1} x^{n-1}(1-x)^{n-1}\,dx$ Putting $x = 1-y$ so $y = 1-x$

$$= -\int_{1/2}^{0}(1-y)^{n-1}y^{n-1}\,dy = \int_0^{1/2} y^{n-1}(1-y)^{n-1}\,dy = \int_0^{1/2} x^{n-1}(1-x)^{n-1}\,dx$$

Hence $\beta(n, n) = 2\displaystyle\int_0^{1/2} x^{n-1}(1-x)^{n-1}\,dx.$ **Hence proved.**

13. *If* $I_n = \dfrac{\sqrt{\pi}\,\Gamma(n+1)}{2\,\Gamma(n/2+1)}$, *show that* $I_{n+2} = (n+1)\,I_n$.

SOLUTION: $$I_{n+2} = \frac{\sqrt{\pi}\,\Gamma(n+3)}{2\,\Gamma\left(\frac{n+2}{2}+1\right)} = \frac{\sqrt{\pi}\,(n+2)(n+1)\,\Gamma(n+1)}{2\,\Gamma\left(\frac{n}{2}+2\right)}$$

$$= \frac{\sqrt{\pi}\,(n+2)(n+1)\,\Gamma(n+1)}{2\left(\frac{n}{2}+1\right)\Gamma\left(\frac{n}{2}+1\right)} = (n+1)\,I_n$$ **Hence proved.**

14. *Evaluate* $\displaystyle\int_0^1 \left(1-x^{1/n}\right)^{m-1}dx$

SOLUTION: Put $x = \sin^{2n}\theta$ so $dx = 2n\sin^{2n-1}\theta\cos\theta\,d\theta$

$$\therefore\ I = \int_0^1\left[1-x^{1/n}\right]^m dx = \int_0^{\pi/2}(1-\sin^2\theta)^{m-1}\,2n\sin^{2n-1}\theta\cos\theta\,d\theta = 2n\int_0^{\pi/2}\sin^{2n-1}\theta\cos^{2m-1}\theta\,d\theta$$

$$= 2n\,\frac{\Gamma(m)\,\Gamma(n)}{\Gamma(m+n)}$$ **Ans.**

15. *Prove the following*

$$\int_0^\infty x^{n-1}e^{-ax}\cos bx\,dx=\frac{\Gamma n}{(a^2+b^2)^{n/2}}\cos\left(n\tan^{-1}\frac{b}{a}\right)$$

SOLUTION: Consider $I=\int_0^\infty x^{n-1}\,e^{-ax-ibx}dx$ whose real part is the given integral.

$$=\int_0^\infty x^{n-1}\,e^{-(a+ib)x}\,dx \quad \text{Putting } (a+ib)x=t, \text{ we get}$$

$$=\int_0^\infty \frac{t^{n-1}}{(a+ib)^{n-1}}\,e^{-t}\,\frac{dt}{a+ib}=\frac{1}{(a+ib)^n}\int_0^\infty t^{n-1}\,e^{-t}\,dt$$

$$=\frac{\Gamma n}{(a+ib)^n}.\quad \text{If we write } a=r\cos\alpha,\ b=r\sin\alpha$$

or

$$I=\frac{\Gamma n}{r^n\,(\cos\alpha+\sin\alpha)^n}=\frac{\Gamma n}{r^n}(\cos n\alpha-i\sin n\alpha)$$

Equating real parts on both sides, we get

$$\int_0^\infty x^{n-1}\,e^{-ax}\cos bx\,dx=\frac{\Gamma n\cos n\alpha}{(a^2+b^2)^{n/2}}$$

where $\alpha=\tan^{-1}\left(\frac{b}{a}\right)$. **Hence proved.**

SOLUTIONS TO EXERCISE 20B

1. *Solve in series* $\frac{d^2y}{dx^2} - x\frac{dy}{dx} + x^2 y = 0.$

SOLUTION: Let $y = a_0 + a_1 x + a_2 x^2 + a_3 x^3 + ... = \sum_0^\infty a_n x^n$

then $\frac{dy}{du} = \sum_{n=1}^\infty na_n x^{n-1}$ and $\frac{d^2y}{dx^2} = \sum_{n=2}^\infty n(n-1)a_n x^{n-2}$. Hence the given equation becomes an identity as $\sum_2^\infty n(n-1)a_n x^{n-2} - \sum_1^\infty na_n x^n + \sum_0^\infty a_n x^{n+2} = 0$

Equating to zero the coefficients of various degree terms in x, we get

$2 \cdot 1\, a_2 = 0,\ \ 3 \cdot 2\, a_3 - 1\, a_1 = 0$ Thus $a_2 = 0$ and $a_3 = \frac{1}{6}a_1$.

and in general, $(n+2)(n+1)a_{n+2} - na_n + a_{n-2} = 0$

For $n = 2$, $\quad 4 \cdot 3\, a_4 - 2a_2 + a_0 = 0 \quad \Rightarrow a_4 = -\frac{a_0}{12}$

For $n = 3$, $\quad 5 \cdot 4\, a_5 - 3a_3 + a_1 = 0 \quad \Rightarrow a_5 = \frac{1}{20}\left(\frac{1}{2} - 1\right)a_1 = \frac{-a_1}{40}$

For $n = 4$, $\quad 6 \cdot 5 \cdot a_6 - 4a_4 + a_2 = 0 \quad \Rightarrow a_6 = \frac{-2}{15}\frac{a_0}{12} = \frac{-a_0}{90}$

and For $n = 5$, $\quad 7 \cdot 6 \cdot a_7 - 5a_5 + a_3 = 0 \quad \Rightarrow a_7 = \frac{1}{7 \cdot 6}\left[\frac{a_1}{8} - \frac{a_1}{6}\right] = \frac{-a_1}{1008}$, etc.

$\therefore \quad y = a_0\left(1 - \frac{x^4}{12} - \frac{x^6}{90} - ...\right) + a_1\left(x + \frac{x^3}{6} - \frac{x^5}{40} - \frac{x^7}{1008} - ...\right)$ **Ans.**

2. *Find the series solution of* $4x^2\frac{d^2y}{dx^2} + 4x\frac{dy}{dx} - y = 0.$

SOLUTION: Let $\quad y = \sum_{n=0}^\infty a_n x^{m+n} \quad \therefore\ y' = \Sigma a_n (m+n) x^{m+n-1}$

and $y'' = \Sigma a_n (m+n)(m+n-1) x^{m+n-2}$. Putting these in the given equation, we get

$$4\sum_0^\infty a_n (m+n)(m+n-1)x^{m+n} + 4\sum_0^\infty a_n (m+n)x^{m+n} - \sum_0^\infty a_n x^{m+n} = 0$$

or $\quad \sum_{n=0}^\infty a_n[4(m+n)(m+n-1) + 4(m+n) - 1]\, x^{m+n} = 0$

or $\quad \sum_{n=0}^\infty a_n[4(m+n)^2 - 1]\, x^{m+n} = 0$

Equating to zero the coefficient of lowest degree term in x we get the indicial equation as $4m^2 - 1 = 0$ or $m = \pm\frac{1}{2}$

Equating to zero the coefficient of x^{m+n}, we get

$$a_n [4(m+n)^2 - 1] = 0 \text{ or } a_n \left[4\left(n \pm \frac{1}{2}\right)^2 - 1\right] = 0 \Rightarrow a_n = 0 \text{ for } n = 1, 2, 3, \ldots$$

$\therefore$ Solution is $y = Ax^{1/2} + Bx^{-1/2}$ where A and B are arbitrary constants. **Ans.**

3. *Find the solution in series of* $4x\frac{d^2y}{dx^2} + 2\frac{dy}{dx} + y = 0.$

SOLUTION: Let $y = \sum_{0}^{\infty} a_n x^{m+n}$ hence $y' = \Sigma(m+n)\, a_n x^{m+n-1}$,

$y'' = \Sigma(m+n)(m+n-1)\, a_n\, x^{m+n-2}$ Putting these in the given equation, we get

$$4\sum_{0}^{\infty} (m+n)(m+n-1)\, a_n\, x^{m+n-1} + 2\sum_{0}^{\infty} (m+n) a_n x^{m+n-1} + \sum a_n\, x^{m+n} = 0$$

Equating to zero the coefficient of the lowest degree term x^{m-1}, we get the indicial equation as

$$2m(2m-1)\, a_0 = 0 \Rightarrow m = 0, \frac{1}{2} \quad \text{since} \quad a_0 \neq 0.$$

Next, equating to zero the coefficient of x^{m+n}, we get $2(m+n+1)(2m+2n+1)\, a_{n+1} + a_n = 0$

or
$$a_{n+1} = \frac{-a_n}{(2m+2n+2)(2m+2n+1)}$$

For $m = 0$,
$$a_{n+1} = \frac{-a_n}{(2n+2)(2n+1)} \quad \text{hence } a_1 = \frac{-a_0}{2\cdot 1},\ a_2 = \frac{-a_1}{4\cdot 3} = \frac{a_0}{4\cdot 3\cdot 2\cdot 1} = \frac{a_0}{4!},$$

$$a_3 = \frac{-a_2}{6\cdot 5} = \frac{a_0}{6!},\quad a_4 = \frac{-a_0}{8!}, \ldots$$

For $m = \frac{1}{2}$
$$a_{n+1} = \frac{-a_n}{(2n+3)(2n+2)} \quad \text{hence} \quad a_1 = \frac{-a_0}{3\cdot 2},$$

$$a_2 = \frac{-a_1}{5\cdot 4} = \frac{a_0}{3\cdot 2\cdot 5\cdot 4},\quad a_3 = \frac{-a_2}{7\cdot 6} = \frac{-a_0}{7!}, \ldots$$

$$\therefore \quad y_1 = a_0\left[1 - \frac{x^1}{2!} + \frac{x^2}{4!} - \ldots\right] \quad \text{and} \quad y_2 = a_0\left[x^{1/2} - \frac{x^{3/2}}{3!} + \frac{x^{5/2}}{5!} - \ldots\right]$$

$$\therefore \quad y_1 = a_0 \cos\sqrt{x} \quad \text{and} \quad y_2 = a_0 \sin\sqrt{x}$$

or $\quad y = A\cos\sqrt{x} + B\sin\sqrt{x}$ **Ans.**

4. *Solve in series the differential equation* $\dfrac{d^2y}{dx^2} + xy = 0.$

SOLUTION: Let $y = \sum_0^\infty a_n x^n,\ y' = \sum_1^\infty n a_n x^{n-1},\ y'' = \sum n(n-1)x^{n-2}.$

Putting these in the given equation we have $\sum_2^\infty n(n-1)a_n x^{n-2} + \sum_1^\infty a_n x^{n+1} = 0.$

Equating to zero the coefficient of the lowest degree term, we get $2 \cdot 1\, a_2 = 0$, and equating to zero the coefficient of x^n, we get

$$(n+3)(n+2)a_{n+3} + a_n = 0 \quad \text{or} \quad a_{n+3} = \frac{-a_n}{(n+3)(n+2)} \Rightarrow a_2 = a_5 = a_8 = a_{11} = 0 \ldots$$

$$\Rightarrow \quad a_3 = \frac{-a_0}{2 \cdot 3},\quad a_4 = \frac{-a_1}{4 \cdot 3},\quad a_5 = 0,\quad a_6 = \frac{-a_3}{6 \cdot 5} = \frac{(-1)^2 a_0}{2 \cdot 3 \cdot 5 \cdot 6},\quad a_7 = \frac{-a_4}{7 \cdot 6} = \frac{(-1)^2 a_1}{3 \cdot 4 \cdot 6 \cdot 7},\quad a_8 = 0,$$

$$a_9 = \frac{-a_6}{9 \cdot 8} = \frac{(-1)^3 a_0}{2 \cdot 3 \cdot 5 \cdot 6 \cdot 89},\quad a_{10} = \frac{-a_7}{10 \cdot 9} = \frac{(-1)^3 a_1}{3 \cdot 4 \cdot 6 \cdot 7 \cdot 9 \cdot 10},\quad a_{11} = 0, \ldots$$

$$\therefore \quad y = a_0\left[1 - \frac{1}{3!}x^3 + \frac{1 \cdot 4}{6!}x^6 - \frac{1 \cdot 4 \cdot 7}{9!}x^9 + \ldots\right] + a_1\left[x - \frac{1 \cdot 2}{4!}x^4 + \frac{2 \cdot 5}{7!}x^7 - \ldots\right]$$ **Ans.**

5. *Solve the differential equation* $9x(1-x)\dfrac{d^2y}{dx^2} - 12\dfrac{dy}{dx} + 4y = 0$ *in power series.*

SOLUTION: Let $y = \sum_0^\infty a_n x^{m+n} \quad \therefore\ y' = \sum (m+n) a_n x^{m+n-1},$

and $$y'' = \sum (m+n)(m+n-1) a_n x^{m+n-2}$$

Putting these in the given equation, we get

$$9(x - x^2)\sum (m+n)(m+n-1)a_n x^{m+n-2} - 12 \sum (m+n) a_n x^{m+n-1} + 4\sum a_n x^{m+n} = 0$$

or $$\sum [-9(m+n)(m+n-1) + 4] a_n x^{m+n} + \sum [9(m+n)(m+n-1) - 12(m+n)] a_n x^{m+n-1} = 0$$

Equating to zero the coefficient of the lowest degree term in x, we get the indicial equation as $9m(m-1) - 12m = 0$ whose roots are $m = 0, \dfrac{7}{3}.$

Next, equating to zero the coefficient of x^{m+n}, we get

$$[-9(m+n)(m+n-1) + 4] a_n + [9(m+n+1)(m+n) - 12(m+n+1)] a_{n+1} = 0$$

or $$a_{n+1} = \frac{9(m+n)^2 - 9(m+n) - 4}{3(m+n+1)(3m+3n-4)} a_n$$

$$= \frac{(3m+3n+1)(3m+3n-4)}{(3m+3n+3)(3m+3n-4)} a_n = \frac{3m+3n+1}{3m+3n+3} a_n$$

For $m = 0$ we have $a_{n+1} = \dfrac{(3n+1)}{(3n+3)} a_n$

$$\therefore \quad a_1 = \frac{a_0}{3}, \quad a_2 = \frac{4}{6} a_1 = \frac{1\cdot 4}{3\cdot 6} a_0, \quad a_3 = \frac{7}{9} a_2 = \frac{1\cdot 4\cdot 7}{3\cdot 6\cdot 9} a_0, \text{ etc.}$$

$\therefore$ One solution is $y_1 = \left[1 + \dfrac{1}{3}x + \dfrac{1\cdot 4}{3\cdot 6}x^2 + \dfrac{1\cdot 4\cdot 7}{3\cdot 6\cdot 9}x^3 + ...\right]$

For $m = \dfrac{7}{3}$ or $3m = 7$, we have $a_{n+1} = \dfrac{3n+8}{3n+10} a_n$

$$\therefore \quad a_1 = \frac{8}{10} a_0, \quad a_2 = \frac{11}{13} a_1 = \frac{8\cdot 11}{10\cdot 13} a_0, \quad a_3 = \frac{14}{16} a_2 = \frac{8\cdot 11\cdot 14}{10\cdot 13\cdot 16} a_0, \text{ etc.}$$

$\therefore$ Second solution is $y_2 = a_0 \left[1 + \dfrac{8}{10}x + \dfrac{8\cdot 11}{10\cdot 13}x^2 + \dfrac{8\cdot 11\cdot 14}{10\cdot 13\cdot 16}x^3 + ...\right]$

$\therefore$ Finally we have the complete solution as

$$y = A\left[1 + \frac{1}{3}x + \frac{1\cdot 4}{3\cdot 6}x^2 + \frac{1\cdot 4\cdot 7}{3\cdot 6\cdot 9}x^3 + ...\right]$$

$$+ Bx^{7/3}\left[1 + \frac{8}{10}x + \frac{8\cdot 11}{10\cdot 13}x^2 + \frac{8\cdot 11\cdot 14}{10\cdot 13\cdot 16}x^3 + ...\right]$$ **Ans.**

6. *Solve in series* $x(1-x)\dfrac{d^2y}{dx^2} - (1+3x)\dfrac{dy}{dx} - y = 0.$

SOLUTION: Let $y = \Sigma a_n x^{m+n}$, $y' = \Sigma (m+n) a_n x^{m+n-1}$, $y'' = \Sigma (m+n)(m+n-1) a_n x^{m+n-2}$

Putting these in the given equation, we get an identity in x as

$$(x - x^2)\,\Sigma (m+n)(m+n-1) a_n x^{m+n-2} - (1+3x)\,\Sigma (m+n) a_n x^{m+n-1} - \Sigma a_n x^{m+n} = 0$$

or $\sum [(m+n)(m+n-1) - (m+n)] a_n x^{m+n-1} + \sum [-(m+n)(m+n-1) - 3(m+n) - 1] a_n x^{m+n} = 0$

or $\sum (m+n)(m+n-2) a_n x^{m+n-1} - \sum (m+n+1)^2 a_n x^{m+n} = 0$

Equating to zero the coefficient of the lowest degree term in x, we get indicial equation

$$m(m-2)a_0 = 0 \quad \text{hence} \quad m = 0, 2 \text{ as } a_0 \neq 0.$$

Now, in general, equating to zero the coefficeint of x^{m+n}, we get

$$(m+n+1)(m+n-1)a_{n+1} - (m+n+1)^2 a_n = 0 \quad \text{or} \quad a_{n+1} = \frac{m+n+1}{m+n-1} a_n$$

$$\therefore \quad y = x^m a_0 \left[1 + \frac{m+1}{m-1}x + \frac{(m+2)(m+1)}{m(m-1)}x^2 \right.$$

$$\left. + \frac{(m+3)(m+2)(m+1)}{(m+1)m(m-1)}x^3 + \frac{(m+4)(m+3)(m+2)(m+1)x^4}{(m+2)(m+1)m(m-1)} + ...\right]$$

It does not exist for $m = 0$ $\therefore$ we take $a_0 = b_0\, m$, then

$$y = x^m b_0\left[m + \frac{m(m+1)}{m-1}x + \frac{(m+2)(m+1)}{(m-1)}x^2\right.$$

$$\left. + \frac{(m+3)(m+2)(m+1)}{(m+1)m(m-1)}x^3 + \frac{(m+4)(m+3)(m+2)(m+1)x^4}{(m+2)(m+1)m(m-1)} + \ldots\right]$$

Now for $m = 0$ we have one solution as

$$y_1 = b_0\left[0 + 0 + \frac{2\cdot 1}{-1}x^2 + \frac{3\cdot 2\cdot 1}{1(-1)}x^3 + \frac{4\cdot 3\cdot 2\cdot 1}{2\cdot 1(-1)}x^4 + \ldots\right]$$

or $$y_1 = -2b_0 x^2\left[1 + 3x + \frac{4\cdot 3}{2}x^2 + \ldots\right]$$

The other solution is obtained as $y = \left(\dfrac{\partial y}{\partial m}\right)_{m=0}$

or $$y_2 = b_0\, x^m \log x\left[m + \frac{m(m+1)}{m-1}x + \frac{(m+2)(m+1)}{m-1}x^2 + \frac{(m+3)(m+2)(m+1)}{(m+1)(m-1)}x^3 + \ldots\right]$$

$$+ b_0\, x^m\left[1 + \frac{m^2 - 2m - 1}{(m-1)^2}x + \frac{m^2 - 2m - 5}{(m-1)^2}x^2 + \ldots\right]$$

Putting $m = 0$, we get

$$y_2 = -b_0 \log x\left[\frac{2\cdot 1}{-1}x^2 + \frac{3\cdot 2\cdot 1}{1(-1)}x^3 + \ldots\right] + b_0\,[1 - x + 5x^2 - \ldots]$$

$\therefore$ Complete solution is

$$y = (C'_1 + C'_2 \log x)\,(2x^2 + 2\cdot 3\,x^3 + 3\cdot 4\,x^4 + \ldots) + C'_2\,(-1 + x + 5x^2 - 11\,x^3 + \ldots)$$ **Ans.**

7. *Solve in series the differential equation* $x^2\dfrac{d^2y}{dx^2} + x\dfrac{dy}{dx} + (x^2 - 4)\,y = 0.$

SOLUTION: Since $x = 0$ is a regular singularity of the equation, let its solution be

$$y = x^m(a_0 + a_1 x + a_2 x^2 + \ldots) = \sum_{k=0}^{\infty} a_k\, x^{m+k}$$

then $\dfrac{dy}{dx} = \Sigma\,(m + K)\, a_K\, x^{m+K-1}$ and $\dfrac{d^2y}{dx^2} = \sum\,(m + K)(m + K - 1)\, a_K\, x^{m+K-2}$

Putting these in the given equation, we get

$$\Sigma\,(m + K)\,(m + K - 1)\, a_K\, x^{m+K} + \Sigma(m + K)\, a_K\, x^{m+K-1} + \Sigma\, a_K\, x^{m+K+2} - 4\,\Sigma a_K x^{m+K} = 0$$

or $$\Sigma\,[(m + K)^2 - 4]\, a_K\, x^{m+K} + \Sigma\, a_K\, x^{m+K+2} = 0$$

Equating to zero the coefficient of term of lowest power of x, that is, x^m, gives $(m^2 - 4)\, a_0 = 0$

$\therefore$ $m = 2, -2$ as $a_0 \neq 0$. Roots differ by integer. Equating to zero the coefficient of x^{m+1}, we get

$$[(m + 1)^2 - 4]\, a_1 = 0 \Rightarrow a_1 = 0$$

Equating to zero the coefficient of x^{m+2}, gives

$$(m^2+4m)\,a_2+a_0=0 \quad \therefore \quad a_2=\frac{-a_0}{m(m+4)}$$

Similarly, $$a_3=\frac{-a_1}{(m+1)(m+5)}=0 \quad \text{as } a_1=0, \quad a_4=\frac{-a_2}{(m+2)(m+6)}, \text{ etc.}$$

$\therefore$ $$a_1=a_3-a_5=a_7=\ldots=0$$

and $$a_4=\frac{a_0}{m(m+2)(m+4)(m+6)}, \quad a_6=\frac{-a_0}{m(m+2)(m+4)^2(m+6)(m+8)}, \text{ etc.}$$

$\therefore$ Solution is $$y=a_0x^m\left[1-\frac{x^2}{m(m+4)}+\frac{x^4}{m(m+2)(m+4)(m+6)}-\frac{x^6}{m(m+2)(m+4)^2(m+6)(m+8)}+\ldots\right] \quad \ldots(1)$$

It does not exist for $m=-2$. To overcome it we take $a_0=b_0(m+2)$

then $$y=b_0x^m\left[(m+2)-\frac{(m+2)x^2}{m(m+4)}+\frac{x^4}{m(m+4)(m+6)}-\frac{x^6}{m(m+2)(m+4)^2(m+6)(m+8)}+\ldots\right] \quad \ldots(2)$$

Put $m=-2$, $$y_1=b_0x^{-2}\left[\frac{x^4}{(-2)(2)(4)}-\frac{x^6}{(-2)(2)(4)(8)}+\ldots\right]$$

Differentiating (2) partially w.r.t. m, we get

$$\frac{\partial y}{\partial m}=b_0x^m\log x\left[(m+2)-\frac{(m+2)x^2}{m(m+4)}+\frac{x^4}{m(m+4)(m+6)}-\ldots\right]$$
$$+\,b_0x^m\left[1-\frac{m+2}{m(m+4)}\left(\frac{1}{m+2}-\frac{1}{m}-\frac{1}{m+4}\right)x^2+\ldots\right]$$

$\therefore$ Second solution is $$y_2=\left(\frac{\partial y}{\partial m}\right)_{m=-2}=b_0x^{-2}\log x\left[\frac{x^4}{(-2)(2)(4)}-\frac{x^6}{(-2)(2)^2(4)(6)}+\ldots\right]$$
$$+\,b_0x^{-2}\left[1-\frac{x^2}{(-2)(2)}+\frac{1}{(-2)(2)(4)}\left(\frac{1}{2}-\frac{1}{2}-\frac{1}{4}\right)x^4-\ldots\right]$$

$\therefore$ Complete solution is $y=c_1y_1+c_2y_2$

or $$y=Ax^2\left[1-\frac{x^2}{2\cdot 6}+\frac{x^4}{2\cdot 4\cdot 6\cdot 8}-\frac{x^6}{2\cdot 4\cdot 6^2\cdot 8\cdot 10}+\ldots\right]$$
$$+\,B\left[x^2\cdot\log x\left(-\frac{1}{2^2\cdot 4}+\frac{x^4}{2^3\cdot 4\cdot 6}-\ldots\right)+x^{-2}\left(1+\frac{x^2}{2^2}+\frac{x^4}{2^2\cdot 4^2}+\ldots\right)\right]$$

where $$A=\frac{c_1b_0}{(-2)(2)(4)}, \quad B=c_2b_0.$$ **Ans.**

8. *Obtain the general solution in power series of the equation*

$$(x - x^2)\, y'' - 3y' + 2y = 0.$$

SOLUTION: Let $y = \Sigma a_n x^{m+n}$, $y' = \Sigma(m + n)\, a_n x^{m+n-1}$

and $y'' = \Sigma(m + n)(m + n - 1)\, a_n x^{m+n-2}$

Putting these in the given equation, we get the identity in x as

$$(x - x^2)\, \Sigma(m + n)(m + n - 1)\, a_n x^{m+n-2} - 3\Sigma(m + n)\, a_n x^{m+n-1} + 2\Sigma a_n x^{m+n} = 0$$

or $$\Sigma\, [(m + n)(m + n - 1) - 3(m + n)]\, a_n x^{m+n-1} - \Sigma[(m + n)(m + n - 1) - 2]\, a_n x^{m+n} = 0$$

or $$\Sigma\, (m + n)(m + n - 4)\, a_n x^{m+n-1} - \Sigma\, (m + n - 2)(m + n + 1)\, a_n x^{m+n} = 0$$

Equating to zero the coefficient of the lowest degree term, we get the indicial equation as $m(m - 4) = 0$ $\therefore$ $m = 0, 4.$ as $a_0 \neq 0.$

Next, equating to zero the coefficient of x^{m+n}, we get

$$(m + n + 1)(m + n - 3)\, a_{n+1} - (m + n - 2)(m + n + 1)\, a_n = 0$$

$$\therefore \quad a_{n+1} = \frac{(m+n-2)\, a_n}{m+n-3}$$

$$\therefore \quad a_1 = \frac{m-2}{m-3} a_0, \quad a_2 = \frac{m-1}{m-2} a_1 = \frac{m-1}{m-2} \cdot \frac{m-2}{m-3} a_0 = \frac{m-1}{m-3} a_0,$$

$$a_3 = \frac{m}{m-1} a_2 = \frac{m}{m-1} \cdot \frac{m-1}{m-3} a_0 = \frac{m}{m-3} a_0, \ldots, a_4 = \frac{m+1}{m} a_3 = \frac{m+1}{m} \cdot \frac{m}{m-3} a_0 = \frac{m+1}{m-3} a_0, \ldots$$

$$\therefore \quad y = a_0 x^m \left[1 + \frac{m-2}{m-3} x + \frac{m-1}{m-3} x^2 + \frac{m}{m-3} x^3 + \frac{m+1}{m-3} x^4 + \ldots\right]$$

$$= \frac{a_0 x^m}{m-3} [(m-3) + (m-2)x + (m-1)x^2 + mx^3 + (m+1)x^4 + \ldots]$$

For $m = 0$, $$y = a_0 \left[1 + \frac{2}{3}x + \frac{1}{3}x^2 - \frac{x^4}{3} + \ldots\right]$$

and for $m = 4$, $$y = a_0 x^4\, [1 + 2x + 3x^2 + 4x^3 + \ldots]$$

$\therefore$ Complete solution is

$$y = C_1 \left(1 + \frac{2x}{3} + \frac{1}{3}x^2 - \frac{x^4}{3} + \ldots\right) + C_2 x^4 (1 + 2x + 3x^2 + 4x^3 + \ldots)$$ **Ans.**

9. *Obtain the series solution of the equation*

$$(x^2 - 1)x^2y'' - (x^2 + 1)\,xy' + (x^2 + 1)\,y = 0.$$

SOLUTION: Let $y = \Sigma a_n\, x^{m+n}$, $y' = \Sigma a_n\,(m + n)x^{m+n-1}$ and $y'' = \Sigma a_n(m + n)\,(m + n - 1)x^{m+n-2}$

Hence the given equation becomes an identity in x as

$$(x^4 - x^2)\,\Sigma a_n(m + n)\,(m + n - 1)\,x^{m+n-2} - (x^3 + x)\,\Sigma\, a_n(m + n)\,x^{m+n-1} + (x^2 + 1)\,\Sigma a_n\,x^{m+n} = 0$$

or $\Sigma a_n[(m + n)\,(m + n - 1) - (m + n + 1)]\,x^{m+n+2} - \Sigma a_n[(m + n)\,(m + n - 1) + (m + n) - 1]\,x^{m+n} = 0$

or $\Sigma a_n(m + n - 1)^2\,x^{m+n+2} - \Sigma\, a_n(m + n + 1)\,(m + n - 1)\,x^{m+n} = 0$

Equating to zero the coefficient of the lowest degree term x^m, we get the indicial equation as

$$-\,a_0\,(m + 1)\,(m - 1) = 0 \;\Rightarrow\; m = 1, -1;$$ case of roots differing by an integer.

Next, equating to zero the coefficient of x^{m+n}, we get

$$a_{n-2}\,(m + n - 3)^2 - a_n\,(m + n + 1)\,(m + n - 1) = 0$$

$$\therefore \qquad a_n = \frac{(m+n-3)^2}{(m+n+1)\,(m+n-1)}\,a_{n-2} = \frac{(m+n-3)^2}{(m+n)^2-1}\,a_{n-2}$$

Equating to zero the coefficient of x^{m+1}, we get

$$-a_1\,(m + 2)\,(m) = 0 \;\Rightarrow\; a_1 = 0 \text{ Hence } a_3 = a_5 = \dots = 0$$

$$\therefore \quad \frac{a_2}{a_0} = \frac{(m-1)^2}{(m^2-1)},$$ Let us choose $a_0 = (m^2 - 1)\,b_0$ then $a_2 = (m - 3)^2\,b_0$

and $$\frac{a_4}{a_2} = \frac{(m+1)^2}{(m+4)^2-1} \quad \therefore \quad a_4 = \frac{(m+1)^2\,(m-3)^2}{[(m+4)^2-1]}\,b_0$$

and $$\frac{a_6}{a_4} = \frac{(m+3)^2}{(m+6)^2-1} \quad \therefore \quad a_6 = \frac{(m+3)^2\,(m+1)^2\,(m-3)^2}{[(m+4)^2-1][(m+6)^2-1]}\,b_0$$

$\therefore$ The solution is

$$y = b_0\,x^m\left[1+(m-3)^2\,x^2+\frac{(m+1)^2\,(m-3)^2\,x^4}{(m+4)^2-1}+\frac{(m+3)^2\,(m+1)^2\,(m-3)^2\,x^6}{[(m+4)^2-1][(m+6)^2-1]}+\dots\right]$$

For $m=1$ $$y = y_1 = b_0\,x^{-1}\,[1 + 16x^2] = b_0\left[\frac{1}{x}+16x\right]$$

For the other solution $$y_2 = \left(\frac{\partial y}{\partial m}\right)_{m=-1}.$$

$$\frac{\partial y}{\partial m} = b_0\,x^m\log x\left[1+(m-3)^2\,x^2+\frac{(m+1)^2\,(m-3)^2}{(m+4)^2-1}\,x^4+\dots\right] + b_0\,x^m\,[0+2(m-3)\,x^2+\dots]$$

$$\left(\frac{\partial y}{\partial m}\right)_{m=-1} = \frac{b_0}{x}\log x\,[1+16x^2+0]+\frac{b_0}{x}[2(4)x^2+\dots] \qquad y_2 = b_0\,\log x\left[\frac{1}{x}+\frac{1}{16x}\right]+b_0\,(8x+\dots)$$

$\therefore$ Complete solution is of the form

$$y = ax + b\left(x\log x+\frac{1}{2x}\right)$$ **Ans.**

ADDITIONAL IMPORTANT PROBLEMS

1. *Evaluate the following integrals interms of Gamma functions*

$$(a)\ \int_0^\infty x^{2/3} e^{-\sqrt{x}}\, dx \quad (b)\ \int_0^1 \frac{dx}{\sqrt{1-x^4}} \quad (c)\ \int_0^\infty a^{-bx^2} dx \quad (d)\ \int_0^1 x^{3/2}(1-\sqrt{x})^{1/2}\, dx.$$

SOLUTION: (*a*) Consider $I = \int_0^\infty x^{2/3} e^{-\sqrt{x}}\, dx$ (setting $\sqrt{x} = t$)

$$I = \int_0^\infty (t^2)^{2/3} e^{-t}.2t\, dt = 2\int_0^\infty e^{-t} t^{7/3}\, dt = 2\Gamma\left(\frac{10}{3}\right) = 2.\frac{7}{3}\cdot\frac{4}{3}\cdot\frac{1}{3}\cdot\Gamma\left(\frac{1}{3}\right) = \frac{56}{27}\Gamma\left(\frac{1}{3}\right).$$

(*b*) Consider $I = \int_0^1 \frac{dx}{\sqrt{1-x^4}}$. (putting $x^2 = \sin\theta$)

$$I = \int_0^{\pi/2} \frac{\frac{1}{2}(\sin\theta)^{-\frac{1}{2}} \cos\theta\, d\theta}{\sqrt{1-\sin^2\theta}} = \frac{1}{2}\int_0^{\pi/2} (\sin\theta)^{-\frac{1}{2}}\, d\theta,$$

$$= \frac{1}{2}\frac{\Gamma\left(\frac{1}{4}\right)\Gamma\left(\frac{1}{2}\right)}{\Gamma\left(\frac{3}{4}\right)} = \frac{\sqrt{\pi}}{2}\cdot\frac{\Gamma\left(\frac{1}{4}\right)}{\Gamma\left(\frac{3}{4}\right)}.$$

(*c*) Consider $I = \int_0^\infty a^{-bx^2 dx} = \int_0^\infty e^{-(b\ln a)x^2}\, dx.$

Set $(b \ln a)\, x^2 = t$, so that, $2(b \ln a)\, x dx = dt$, then

$$I = \frac{1}{(b\log a)^{3/2}} \int_0^\infty e^{-t} t^{-1/2}\, dt = \frac{\Gamma(1/2)}{(b\log a)^{3/2}} = \frac{\sqrt{\pi}}{(b\log a)^{3/2}}$$

(*d*) Consider $I = \int_0^1 x^{3/2}(1-\sqrt{x})^{1/2} dx.$ (putting $\sqrt{x} = t$)

$$I = 2\int_0^1 t^3(1-t)^{1/2}\, t\, dt = 2\int_0^1 t^4(1-t)^{1/2}\, dt = 2B\left(5, \frac{3}{2}\right)$$

$$= 2\frac{\Gamma(5)\,\Gamma\left(\frac{3}{2}\right)}{\Gamma\left(\frac{13}{2}\right)} = 2\frac{4!.\left(\frac{1}{2}\right).\sqrt{\pi}}{\frac{11}{2}.\frac{9}{2}.\frac{7}{2}.\frac{5}{2}.\frac{3}{2}.\frac{1}{2}.\sqrt{\pi}} = \frac{512}{3465}.$$ **Ans.**

2. *Show that* $\int_0^1 \frac{x^2 dx}{\sqrt{(1-x^4)}} \times \int_0^1 \frac{dx}{\sqrt{(1+x^4)}} = \frac{\pi}{4\sqrt{2}}$.

SOLUTION: (*a*) Consider, $I_1 = \int_0^1 \frac{x^2 dx}{\sqrt{(1-x^4)}}$, (putting $x^2 = \sin\theta$)

$$I_1 = \frac{1}{2}\int_0^{\pi/2} \frac{\sin\theta\cos\theta}{\cos\theta\sqrt{\sin\theta}}\,d\theta = \frac{1}{2}\int_0^{\pi/2}\sqrt{\sin\theta}\,d\theta = \frac{1}{4}B\left(\frac{3}{4},\frac{1}{2}\right).$$

Next consider, $I_2 = \int_0^1 \frac{dx}{\sqrt{(1+x^4)}}$ (putting $x^2 = \tan\theta$)

$$I_2 = \frac{1}{2}\int_0^{\pi/4} \frac{\sec^2\theta\,d\theta}{\sqrt{\tan\theta}\sec\theta} = \frac{1}{2}\int_0^{\pi/4}\frac{d\theta}{\sqrt{\sin\theta\cos\theta}}$$

$$= \frac{1}{\sqrt{2}}\int_0^{\pi/4}\frac{d\theta}{\sqrt{\sin 2\theta}} = \frac{1}{2\sqrt{2}}\int_0^{\pi/2}\frac{d\phi}{\sqrt{\sin\phi}} \text{ (on putting } 2\theta = \phi) = \frac{1}{4\sqrt{2}}B\left(\frac{1}{4},\frac{1}{2}\right).$$

Thus $\int_0^1 \frac{x^2 dx}{\sqrt{(1-x^4)}} \times \int_0^1 \frac{dx}{\sqrt{(1+x^4)}} = \frac{1}{16\sqrt{2}}B\left(\frac{3}{4},\frac{1}{2}\right)B\left(\frac{1}{4},\frac{1}{2}\right)$

$$= \frac{1}{16\sqrt{2}}\frac{\Gamma\left(\frac{3}{4}\right)\Gamma\left(\frac{1}{2}\right)}{\Gamma\left(\frac{5}{4}\right)}\cdot\frac{\Gamma\left(\frac{1}{4}\right)\Gamma\left(\frac{1}{2}\right)}{\Gamma\left(\frac{3}{4}\right)} = \frac{\pi}{4\sqrt{2}}.$$

3. *Show that*

(*a*) $\int_0^\infty xe^{-ax}\cos bx\,dx = \frac{a^2-b^2}{(a^2+b^2)^2}$ (*b*) $\int_0^\infty xe^{-ax}\sin bx\,dx = \frac{2ab}{(a^2+b^2)^2}$

SOLUTION: Let $I = \int_0^\infty xe^{-ax}\cos bx\,dx + i\int_0^\infty xe^{-ax}\sin bx\,dx$

$$= \int_0^\infty xe^{-ax}(\cos bx + i\sin bx)\,dx = \int_0^\infty xe^{-ax}e^{ibx}\,dx$$

$$= \int_0^\infty xe^{-(a-ib)x}dx = \frac{\Gamma(2)}{(a-ib)^2}, \text{ since } \int_0^\infty xe^{-mx}dx = \frac{\Gamma(2)}{m^2} \text{ (integrating by parts)}$$

$$= \frac{1}{(a-ib)^2} = \frac{(a+ib)^2}{(a-ib)^2(a+ib)^2}$$

$$= \frac{a^2-b^2+2iab}{(a^2+b^2)^2} = \frac{a^2-b^2}{(a^2+b^2)^2} + i\frac{2ab}{(a^2+b^2)^2}$$

Equating the imaginary parts on both sides, we obtain

$$\int_0^\infty xe^{-ax} \cos bx\, dx = \frac{a^2 - b^2}{(a^2+b^2)^2}, \quad \text{and} \quad \int_0^\infty xe^{-ax} \sin bx\, dx = \frac{2ab}{(a^2+b^2)^2}.$$

4. *If l and m are positive real numbers then*

$$\iint_D x^{l-1} y^{m-1}\, dx\, dy = \frac{\Gamma(l)\,\Gamma(m)}{\Gamma(l+m+1)},$$

where D is the domain $x \geq 0,\ y \geq 0$ and $x + y \leq 1$.

SOLUTION: The region of integration is $D = \{(x, y);\ 0 \leq y \leq 1 - x,\ \ 0 \leq x \leq 1\}$.

Let, $$I = \iint_D x^{l-1} y^{m-1} dx\, dy = \int_0^1 \left(\int_0^{1-x} x^{l-1}\, y^{m-1}\, dy \right) dx = \frac{1}{m} \int_0^1 x^{l-1} (1-x)^m\, dx = \frac{1}{m} B(l, m+1)$$

$$= \frac{1}{m} \frac{\Gamma(l)\,\Gamma(m+1)}{\Gamma(l+m+1)} = \frac{\Gamma(l)\,\Gamma(m)}{\Gamma(l+m+1)}.$$

The above integral is known as **DIRICHLET'S INTEGRAL FOR TWO VARIABLES.** The integral can be extended to three variables called **DIRICHLET'S INTEGRAL FOR THREE VARIABLES** considered in the next example.

5. *If l, m, n are positive number, then*

$$\iiint_D x^{l-1} y^{m-1} z^{n-1} dx\, dy\, dz = \frac{\Gamma(l)\,\Gamma(m)\,\Gamma(n)}{\Gamma(l+m+n+1)},$$

where D is the region $x \geq 0,\ y \geq 0,\ z \geq 0,\ x + y + z \leq 1$.

SOLUTION: The region of integration is

$$D = \{(x, y, z) : 0 \leq z \leq 1 - x - y,\ 0 \leq y \leq 1 - x,\ 0 \leq x \leq 1\}.$$

Let $$I = \iiint_D x^{l-1} y^{m-1} z^{n-1} dx\, dy\, dz = \int_0^1 \int_0^{1-x} \int_0^{1-x-y} x^{l-1} y^{m-1} z^{n-1} dz\, dy\, dx.$$

Consider $I' = \int_0^{1-x} \int_0^{1-x-y} y^{m-1} z^{n-1} dz\, dy$. Set $1 - x = p$, then

$$I' = \int_0^h \int_0^{h-y} y^{m-1} z^{n-1} dz\, dy = \frac{1}{n} \int_0^h y^{m-1} (h-y)^n\, dy.$$

Now setting $y = hY$, we have

$$I' = \frac{1}{n} \int_0^1 (hY)^{m-1} h^n (1-Y)^n\, h\, dY = \frac{h^{m+n}}{n} \int_0^1 Y^{m-1} (1-Y)^n\, dY$$

$$= \frac{h^{m+n}}{n} B(m, n+1) = \frac{(1-x)^{m+n}}{n} B(m, n+1).$$

Thus
$$I = \frac{B(m, n+1)}{n}\int_0^1 x^{l-1}(1-x)^{m+n}\,dx$$

$$= \frac{B(m, n+1)}{n}.B(l, m+n+1) = \frac{\Gamma(l)\,\Gamma(m)\,\Gamma(n)}{\Gamma(l+m+n+1)}.$$

Dirichlet's integrals are used to evaluate some specific area integrals and volume integrals, as shown in the examples considered below.

6. *Evaluate* $\iiint_D xyz\,dx\,dy\,dz$, *where* D *is the region enclosed by the ellipsoid* $\frac{x^2}{a^2}+\frac{y^2}{b^2}+\frac{z^2}{c^2} = 1.$

SOLUTION: Since the region D of integration $\frac{x^2}{a^2}+\frac{y^2}{b^2}+\frac{z^2}{c^2} \le 1$ is symmetrical in all the eight octagons, therefore the given integral can be written as

$$I = 8\iiint_D xyz\,dx\,dy\,dz,$$

where D is given as $x \ge 0,\ y \ge 0,\ z \ge 0$ and $\frac{x^2}{a^2}+\frac{y^2}{b^2}+\frac{z^2}{c^2} \le 1.$

Set $\frac{x^2}{a^2} = u,\ \frac{y^2}{b^2} = v$ and $\frac{z^2}{c^2} = w$, so that $2xdx = a^2du$, $2ydy = b^2dv$ and $2zdz = c^2dw$, then

$$I = a^2b^2c^2\iiint_{D'} du\,dv\,dw,$$

where D' is given as $u \ge 0,\quad v \ge 0,\quad w \ge 0$ and $u + v + w \le 1.$

or
$$I = a^2b^2c^2\iiint_{D''} u^{1-1}\,v^{1-1}\,w^{1-1}du\,dv\,dw = a^2b^2c^2\frac{\Gamma(1)\,\Gamma(1)\,\Gamma(1)}{\Gamma(4)} = \frac{1}{6}a^2b^2c^2.$$
Ans.

7. *Express* $\int_0^2 (8-x^3)^{-1/3}dx$ *as a Beta function.*

SOLUTION: Putting $x^3 = 8t$, or $x = 2t^{1/3}$ $\therefore$ $dx = \frac{2}{3}t^{-2/3}dt$, we get

$$\int_0^2 (8-x^3)^{-1/3}dx = \frac{2}{3}\cdot\frac{1}{2}\int_0^1 t^{-2/3}(1-t)^{-1/3}dt$$

$$= \frac{1}{3}B\left(\frac{-2}{3}+1, \frac{-1}{3}+1\right) = \frac{1}{3}B\left(\frac{1}{3}, \frac{2}{3}\right).$$
Ans.

8. *Prove that* $\dfrac{B(m+1,n)}{B(m,n)} = \dfrac{m}{m+n}$, $m > 0,\ n > 0.$

SOLUTION: $B(m+1, n) = \int_0^1 x^m (1-x)^{n-1}\,dx$ (integrating by parts)

$$= \left[-x^m \frac{(1-x)^n}{n}\right]_0^1 - \int_0^1 m\,x^{m-1}\frac{(1-x)^n}{-n}\,dx$$

$$= 0 + \frac{m}{n}\int_0^1 x^{m-1}(1-x)^{n-1}(1-x)\,dx$$

$$= \frac{m}{n}\int_0^1 x^{m-1}(1-x)^{n-1}\,dx - \frac{m}{n}\int_0^1 x^m(1-x)^{n-1}\,dx$$

$$= \frac{m}{n}B(m,n) - \frac{m}{n}B(m+1,n)$$

or $$\left(1+\frac{m}{n}\right)B(m+1,n) = \frac{m}{n}B(m,n)$$

or $$B(m+1, n) = \frac{m}{m+n}B(m,n)$$ **Hence Proved**

9. *Evaluate* $\displaystyle\int_0^\infty \frac{x^8(1-x^6)}{(1+x)^{24}}\,dx.$

SOLUTION: $$\int_0^\infty \frac{x^8(1-x^6)}{(1+x)^{24}}\,dx = \int_0^\infty \frac{x^8}{(1+x)^{24}}\,dx - \int_0^\infty \frac{x^{14}}{(1+x)^{24}}\,dx$$

$$= \int_0^1 \frac{x^8\,dx}{(1+x)^{9+15}} - \int_0^1 \frac{x^{14}\,dx}{(1+x)^{9+15}}$$

$$= \beta(9, 15) - \beta(15, 9) = 0$$ **Ans.**

10. *Show that* $\Gamma\left(\frac{-1}{2}\right) = -2\sqrt{\pi}$ *and* $\Gamma\left(\frac{-3}{2}\right) = \frac{4\sqrt{\pi}}{3}$.

SOLUTION: We know that $\Gamma(n)\,\Gamma(1-n) = \dfrac{\pi}{\sin n\pi}$ ...(1)

Putting $n = \dfrac{-1}{2}$, we get $\Gamma\left(\frac{-1}{2}\right)\Gamma\left(1+\frac{1}{2}\right) = \dfrac{\pi}{\sin\left(\frac{-\pi}{2}\right)} = -\pi.$

or $\Gamma\left(\frac{-1}{2}\right)\Gamma\left(\frac{3}{2}\right) = -\pi \quad\Rightarrow\quad \Gamma\left(\frac{-1}{2}\right) = \dfrac{-\pi}{\frac{1}{2}\Gamma\left(\frac{1}{2}\right)} = -2\sqrt{\pi}\,.$

Putting $n = \dfrac{-3}{2}$ in (1), we get

$$\Gamma\left(\frac{-3}{2}\right)\Gamma\left(1+\frac{3}{2}\right) = \frac{\pi}{\sin\left(\frac{-3\pi}{2}\right)} = \frac{\pi}{\sin\frac{\pi}{2}} = \pi$$

or $$\Gamma\left(\frac{-3}{2}\right)\left(\frac{3}{2}\right)\Gamma\left(\frac{3}{2}\right) = \pi \quad \text{or} \quad \Gamma\left(\frac{-3}{2}\right) = \frac{2\pi}{3\cdot\frac{1}{2}\Gamma\left(\frac{1}{2}\right)} = \frac{4\pi}{3\sqrt{\pi}} = \frac{4}{3}\sqrt{\pi}.$$ **Hence proved.**

11. *Solve the differential equation $4xy'' + 2y' + y = 0$ using the Frobenius method.*

SOLUTION: It is easy to verify that $x = 0$ is a regular singular point of the given equation. Let the series solution be

$$y(x) = \sum_{n=0}^{\infty} c_n x^{n+r}, \quad c_0 \neq 0. \qquad ...(1)$$

Differentiating (1), we obtain

$$y'(x) = \sum_{n=0}^{\infty}(n+r)c_n x^{n+r-1} \quad \text{and} \quad y''(x) = \sum_{n=0}^{\infty}(n+r)(n+r-1)c_n x^{n+r-2}$$

Substituting these in the given differential equation, we obtain

$$4\sum_{n=0}^{\infty}(n+r)(n+r-1)c_n x^{n+r-1} + 2\sum_{n=0}^{\infty}(n+r)c_n x^{n+r-1} + \sum_{n=0}^{\infty}c_n x^{n+r} = 0.$$

or $$\sum_{n=0}^{\infty}(n+r)(4n+4r-2)c_n x^{n+r-1} + \sum_{n=0}^{\infty}c_n x^{n+r} = 0$$

or $$2r(2r-1)c_0 x^{r-1} + \sum_{n=1}^{\infty}(n+r)(4n+4r-2)c_n x^{n+r-1} + \sum_{n=0}^{\infty}c_n x^{n+r} = 0$$

or $$2r(2r-1)c_0 x^{r-1} + \sum_{n=0}^{\infty}(n+r+1)(4n+4r+2)c_{n+1}x^{n+r} + \sum_{n=0}^{\infty}c_n x^{n+r} = 0 \qquad ...(2)$$

Equating to zero the coefficient of x^{r-1}, the lowest power of x, in (2) we obtain $2r(2r-1)c_0 = 0$. Since $c_0 \neq 0$, the indicial equation is $r(2r - 1) = 0$, which gives, $r = 0, 1/2$.

Equating to zero the coefficient of x^{n+r} in (2), we obtain

$$2(n + r + 1)(2n + 2r + 1)c_{n+1} + c_n = 0,$$

which gives $$c_{n+1} = \frac{-c_n}{2(n+r+1)(2n+2r+1)}, \quad n = 0, 1, 2, ... \qquad ...(3)$$

as the recurrence relation. It follows that

$$C_1 = \frac{-C_0}{(2r+1)(2r+2)},\ C_2 = \frac{-C_1}{(2r+3)(2r+4)} = \frac{(-1)^2 C_0}{(2r+1)(2r+2)(2r+3)(2r+4)},$$

$$C_3 = \frac{-C_2}{(2r+5)(2r+6)} = \frac{(-1)^3 C_0}{(2r+1)(2r+2)(2r+3)(2r+4)(2r+5)(2r+6)}, \ldots$$

Thus, $y = 2r(2r-1)\,C_0 x^{r-1} + C_0 x^r \left[1 - \frac{x}{(2r+1)(2r+2)} + \frac{x^2}{(2r+1)(2r+2)(2r+3)(2r+4)} - \frac{x^3}{(2r+1)(2r+2)(2r+3)(2r+4)(2r+5)(2r+6)} + \ldots\right]$

For $r = 0$ we have $y_1 = C_0\left[1 - \frac{x}{1\cdot 2} + \frac{x^2}{1\cdot 2\cdot 3\cdot 4} - \frac{x^3}{1\cdot 2\cdot 3\cdot 4\cdot 5\cdot 6} + \ldots\right]$

$$= C_0\left[1 - \frac{x}{2!} + \frac{x^2}{4!} - \frac{x^3}{6!} + \frac{x^4}{8!} - \ldots\right]$$

For $r = \frac{1}{2}$ we have $y_2 = C_0 x^{1/2}\left[1 - \frac{x}{2\cdot 3} + \frac{x^2}{2\cdot 3\cdot 4\cdot 5} - \frac{x^3}{2\cdot 3\cdot 4\cdot 5\cdot 6\cdot 7} + \ldots\right]$

$$= C_0 x^{1/2}\left[1 - \frac{x}{3!} + \frac{x^2}{5!} - \frac{x^3}{7!} + \ldots\right]$$

Thus, the solution is $y = Ay_1 + By_2$ where A and B are arbitrary constants.

12. *Solve the differential equation $x^2y'' - (x + x^2)\,y' + y = 0$ using the Frobenius method.*

SOLUTION: It is easy to verify that $x = 0$ is a regular singular point of the given equation $x^2y'' - (x + x^2)y' + y = 0$

Let the series solution be $y(x) = \sum_{n=0}^{\infty} c_n x^{n+r}$, $c_0 \neq 0$.

Substituting this in the given equation, we obtain

$$\sum_{n=0}^{\infty}(n+r)(n+r-1)c_n x^{n+r} - \sum_{n=0}^{\infty}(n+r)c_n x^{n+r} - \sum_{n=0}^{\infty}(n+r)c_n x^{n+r+1} + \sum_{n=0}^{\infty} c_n x^{n+r} = 0$$

To shift the indices in the third summation on the left side of the above equation, replace $n + 1$ by n, we obtain

$$\sum_{n=0}^{\infty}(n+r)(n+r-1)c_n x^{n+r} - \sum_{n=0}^{\infty}(n+r)c_n x^{n+r} - \sum_{n=1}^{\infty}(n+r-1)c_{n-1} x^{n+r} + \sum_{n=0}^{\infty} c_n x^{n+r} = 0$$

or, $$(r^2 - 2r + 1)c_0 x^r + \sum_{n=1}^{\infty}[\{(n+r)(n+r-2)+1\}c_n - \{n+r-1\}\,c_{n-1}]\,x^{n+r} = 0 \qquad \ldots(1)$$

Equating the coefficient of x^r on left side of Eq. (1) to zero, we obtain $(r^2 - 2r + 1)c_0 = 0$, which gives $r^2 - 2r + 1 = 0$ as the indicial equation since $c_0 \neq 0$. The roots are repeated $r = 1, 1$.

Equating the coefficient of x^{n+r} to zero, we obtain the recurrence relation as

$$c_n = \frac{(n+r-1)}{(n+r)(n+r-2)+1} c_{n-1}, \quad n = 1, 2, \ldots$$

It gives
$$c_1 = \frac{1}{r} c_0, \quad c_2 = \frac{1}{r(r+1)} c_0, \quad c_3 = \frac{1}{r(r+1)(r+2)} c_0, \text{ etc.}$$

Hence the solution, taking $c_0 = 1$, is

$$y(x) = x^r \left[1 + \frac{x}{r} + \frac{x^2}{r(r+1)} + \frac{x^3}{r(r+1)(r+2)} + \ldots \right] \quad \ldots(2)$$

One Frobenius solution corresponding to $r = 1$ is taken as

$$y_1(x) = x \left[1 + x + \frac{x^2}{2!} + \frac{x^3}{3!} + \ldots \right]$$

Since the indicial roots are equal, the second linearly independent Frobenius solution obtained from (2) as

$$y_2(x) = \left.\frac{\partial y}{\partial r}\right|_{r=1} = \left[x^r (\ln x) \left\{ 1 + \frac{x}{r} + \frac{x^2}{r(r+1)} + \frac{x^3}{r(r+1)(r+2)} + \ldots \right\} \right]_{r=1}$$

$$+ \left[x^r \left\{ -\frac{x}{r^2} - \frac{x^2}{r(r+1)} \left(\frac{1}{r} + \frac{1}{r+1} \right) - \frac{x^3}{r(r+1)(r+2)} \left(\frac{1}{r} + \frac{1}{r+1} + \frac{1}{r+2} \right) - \ldots \right\} \right]_{r=1}$$

or,
$$y_2(x) = (\ln x)\, y_1(x) - x^2 \left[1 + \frac{x}{2!} \left(1 + \frac{1}{2} \right) + \frac{x^3}{3!} \left(1 + \frac{1}{2} + \frac{1}{3} \right) + \ldots \right]$$

Hence the general solution is $y(x) = Ay_1(x) + By_2(x)$, where A and B are two arbitrary constants.

13. *Solve $4x^2y'' + 4xy' - y = 0$ using the Frobenius method.*

SOLUTION: The point $x = 0$ is a regular singular point of the equation. Let the series solution be

$$y(x) = \sum_{n=0}^{\infty} c_n x^{n+r}, \quad c_0 \neq 0, \text{ where } r \text{ is any real number.} \quad \ldots(1)$$

Substituting for y'', y' and y in the given equation, we obtain

$$4\sum_{n=0}^{\infty} (n+r)(n+r-1) c_n x^{n+r} + 4\sum_{n=0}^{\infty} (n+r) c_n x^{n+r} - \sum_{n=0}^{\infty} c_n x^{n+r} = 0$$

or,
$$\sum_{n=0}^{\infty} [4(n+r)^2 - 1]\, c_n x^{n+r} = 0 \quad \ldots(2)$$

Equating the coefficient of the least power of x, that is, of x^r, on the left side of (2) to zero, we obtain, $c_0(4r^2 - 1) = 0$. Since $c_0 \neq 0$, hence the indicial equation is $(4r^2 - 1) = 0$, which gives, $r = \pm\frac{1}{2}$, as its roots.

Next, equating the coefficient of x^{n+r} on the left side of Eq. (2) to zero, we obtain

$$[4(n + r)^2 - 1]c_n = 0, \quad n = 1, 2, \ldots \qquad \text{...(3)}$$

as the recurrence relation.

For $r = \frac{1}{2}$, (3) becomes $4n(n + 1)c_n = 0$, which gives, $c_n = 0$, for all n, except when $n = 0$, or -1.

Since $n = -1$ is not admissible, therefore one Frobenius solution from (1) is $y_1(x) = c_0x^{1/2}$. ...(4)

For $r = \frac{-1}{2}$, (3) becomes $4n(n - 1) c_n = 0$ which gives $c_n = 0$ for all n except when $n = 0$ or -1.

$$y_2(x) = c_0x^{-\frac{1}{2}} + c_1x^{\frac{1}{2}} \qquad \text{...(5)}$$

We note that the solution (4) is contained in (5) as its part and hence the general solution is given by (5) itself, where c_0 and c_1 are two arbitrary constants.

14. *Solve $x^2y'' + x(x + 2)y' - 2y = 0$ by the Frobenius method.*

SOLUTION: It is easy to check that $x = 0$ is a regular singular point of the given equation. Let the series solution be $y(x) = \sum_{n=0}^{\infty} c_nx^{n+r}$, $c_0 \neq 0$.

Substituting for y, y' and y'' in the given equation, we obtain

$$\sum_{n=0}^{\infty}(n+r)(n+r-1)c_nx^{n+r} + \sum_{n=0}^{\infty}(n+r)c_nx^{n+r+1} + 2\sum_{n=0}^{\infty}(n+r)c_nx^{n+r} - 2\sum_{n=0}^{\infty}c_nx^{n+r} = 0.$$

Replacing $(n + 1)$ by n in the second summation on the left, we obtain

$$\sum_{n=0}^{\infty}(n+r)(n+r-1)c_nx^{n+r} + \sum_{n=1}^{\infty}(n+r-1)c_{n-1}x^{n+r} + 2\sum_{n=0}^{\infty}(n+r)c_nx^{n+r} - 2\sum_{n=0}^{\infty}c_nx^{n+r} = 0 \qquad \text{...(1)}$$

Equating to zero the coefficient of x^r, on the left side of Eq. (1), we obtain $(r^2 + r - 2)c_0 = 0$. Since $c_0 \neq 0$, it gives $r^2 + r - 2 = 0$ as the indicial equation with $r_1 = 1$ and $r_2 = -2$ as its two roots. Here $r_1 - r_2 = 3$, hence the case of roots differing by an integer.

Equating to zero the coefficient of x^{n+r} on the left side of Eq. (1), the recurrence relation obtained is

$$[(n + r)^2 - (n + r) + 2(n + r) - 2]\ c_n + (n + r - 1)\ c_{n-1} = 0$$

or $$(n + r - 1)\ (n + r + 2)\ c_n + (n + r - 1)\ c_{n-1} = 0$$

or $$c_n = -\frac{c_{n-1}}{(n+r+2)}$$

For $r = -2$,

$$c_n = -\frac{1}{n}c_{n-1}.$$

Hence, $c_1 = -c_0,\quad c_2 = -\frac{1}{2}c_1 = \frac{1}{2!}c_0,\quad c_3 = -\frac{1}{3}c_2 = -\frac{1}{3!}c_0,\quad c_4 = -\frac{1}{4}c_3 = -\frac{1}{4!}c_0, \ldots$

Thus, one Frobenius solution is

$$y_1(x) = c_0 x^{-2}\left[1 - x + \frac{1}{2!}x^2 - \frac{1}{3!}x^3 + \frac{1}{4!}x^4 - \ldots\right]$$

or,

$$y_1(x) = c_0\left[\frac{1}{x^2} - \frac{1}{x} + \frac{1}{2!}\right] - c_0\left[\frac{1}{3!}x - \frac{1}{4!}x^2 + \frac{1}{5!}x^3 + \ldots\right] \qquad \ldots(3)$$

For $r = 1$, (2) becomes $c_n = -\dfrac{1}{n+3}c_{n-1}$

$$c_1 = -\frac{c_0}{4},\quad c_2 = \frac{c_0}{4\cdot5},\quad c_3 = \frac{c_0}{4\cdot5\cdot6},\quad c_4 = \frac{c_0}{4\cdot5\cdot6\cdot7}, \ldots$$

Thus, the second solution is

$$y_2(x) = c_0 x\left[1 - \frac{1}{4}x + \frac{1}{4\cdot5}x^2 - \frac{1}{4\cdot5\cdot6}x^3 + \frac{1}{4\cdot5\cdot6\cdot7}x^4 - \ldots\right] = 3!c_0\left[\frac{1}{3!}x - \frac{1}{4!}x^2 + \frac{1}{5!}x^3 \ldots\right] \qquad \ldots(4)$$

It is clear from (3) and (4) that $y_1(x)$ includes the solution $y_2(x)$.

The general solution can be given as

$$y(x) = A\left[\frac{1}{x^2} - \frac{1}{x} + \frac{1}{2}\right] + Bx\left[\frac{1}{3!} - \frac{x}{4!} + \frac{x^2}{5!} - \ldots\right],$$

where A and B are two arbitrary constants.

15. *Solve the differential equation $x(1 + x)y'' + 3xy' + y = 0$, about $x = 0$, by Frobenius method.*

SOLUTION: Obviously $x = 0$ is a regular singular point of the given equation.

Let the series solution be $y(x) = \sum_{n=0}^{\infty} c_n x^{n+r},\ c_0 \neq 0.$

Substituting for y, y' and y'' in the given equation, we obtain

$$\sum_{n=0}^{\infty}(n+r)(n+r-1)c_n x^{n+r-1} + \sum_{n=0}^{\infty}(n+r)(n+r-1)c_n x^{n+r} + 3\sum_{n=0}^{\infty}(n+r)c_n x^{n+r} + \sum_{n=0}^{\infty}c_n x^{n+r} = 0$$

Collecting the terms of the least power of x, that is x^{r-1}; replacing $n-1$ with n in the first summation on left side and then regrouping the terms of x^{n+r}, we obtain

$$r(r-1)c_0 x^{r-1} + \sum_{n=0}^{\infty}[(n+r+1)(n+r)c_{n+1} + \{(n+r)(n+r+2)+1\}c_n]x^{n+r} = 0 \qquad \ldots(1)$$

Equating to zero the coefficient of x^{r-1} on the left side of (1), the indicial equation, since $c_0 \neq 0$, is $r(r-1) = 0$, which gives, $r = 0, 1$. Here $r_1 - r_2$ equals an integer.

Next, equating to zero the coefficient of x^{n+r} on the left side of (1), the recurrence relation obtained is

$$c_{n+1} = -\frac{[(n+r)(n+r+2)+1]}{(n+r+1)(n+r)}c_n = -\frac{(n+r+1)^2}{(n+r+1)(n+r)}c_n = -\frac{(n+r+1)}{n+r}c_n$$

For $n = 0, 1, 2, \ldots$ this gives

$$c_1 = -\frac{r+1}{r}c_0,\quad c_2 = -\frac{r+2}{r+1}c_1 = \frac{r+2}{r}c_0,\quad c_3 = -\frac{r+3}{r+2}c_2 = -\frac{r+3}{r}c_0, \text{ etc.}$$

The solution can be written as

$$y(x) = c_0 x^r\left[1 - \frac{r+1}{r}x + \frac{r+2}{r}x^2 - \frac{r+3}{r}x^3 + \ldots\right] \qquad \ldots(2)$$

We observe that the coefficients c_1 onward in (2) become infinite when $r = 0$. We modify this by putting $c_0 = a_0(r-0) = a_0 r$, $a_0 \neq 0$, so that (2) becomes

$$y(x) = a_0 x^r[r - (r+1)x + (r+2)x^2 - (r+3)x^3 + \ldots] \qquad \ldots(3)$$

For $r = 0$, it gives $y_1(x) = a_0[-x + 2x^2 - 3x^3 + \ldots] = -a_0 x(1-x)^{-2}$ as one Frobenius solution.

The second Frobenius solution is obtained from (3) as

$$y_2(x) = \left[\frac{\partial y}{\partial r}\right]_{r=0} = [(\ln x)\, y(x) + a_0 x^r\{1 - x + x^2 - x^3 + \ldots\}]_{r=0}$$

$$= (\ln x)y_1(x) + \frac{a_0}{1+x} = -a_0 x(\ln x)(1-x)^{-2} + \frac{a_0}{1+x}$$

The general solution is $y(x) = Ay_1(x) + By_2(x)$ where A and B are two arbitrary constants.

We can verify that solution corresponding to the larger root $r = 1$ is linearly dependent on $y_1(x)$.

We know that if a series solution is derived about an analytic point of a differential equation, then it is a power series. Now we attempt Frobenius method to find a series solution about an analytic point. We expect the roots of the indicial equation to be non-negative integer, since the Frobenius method should also lead to the same power series.

We illustrate this by the following example.

16. *Solve the differential equation $y'' + xy = 0$ in series about $x = 0$ by Frobenius method.*

SOLUTION: Here $x = 0$ is an analytic point for the given equation.

Let the solution be $y(x) = \sum_{n=0}^{\infty} c_n x^{n+r}$

Substituting in the given equation, we obtain

$$\sum_{n=0}^{\infty}(n+r)(n+r-1)c_n x^{n+r-2} + \sum_{n=0}^{\infty}c_n x^{n+r+1} = 0.$$

Rewriting this as

$$r(r-1)c_0x^{r-2}+(r+1)rc_1x^{r-1}+(r+2)(r+1)c_2x^r+\sum_{n=3}^{\infty}(n+r)(n+r-1)c_nx^{n+r-2}+\sum_{n=0}^{\infty}c_nx^{n+r+1}=0.$$

Replacing $(n-2)$ by n in the first summation and replacing $(n+1)$ by n in the second summation on the left side and regrouping the terms of x^{n+r}, we obtain

$$r(r-1)c_0x^{r-2}+(r+1)rc_1x^{r-1}+(r+2)(r+1)c_2x^r+\sum_{n=1}^{\infty}[(n+r+2)(n+r+1)c_{n+2}+c_{n-1}]x^{n+r}=0. \quad ...(1)$$

Equating to zero the coefficient of the least power of x, that is, x^{r-2} in (1) gives $r(r-1)=0$, since $c_0 \neq 0$. Thus, indicial roots are $r = 0, 1$.

Next, equating the coefficient of x^{r-1} in (1) to zero, gives $(r+1)rc_1 = 0$.

It gives $c_1 = 0$, for $r = 1$, and c_1 as arbitrary constant for $r = 0$.

Next, equating to zero the coefficient of x^r, gives $(r+1)(r+2)c_2 = 0$, which implies $c_2 = 0$, for $r = 0, 1$ both.

Equating to zero the coefficient of x^{n+r} in (1) the recurrence relation is

$$c_{n+2} = -\frac{c_{n-1}}{(n+r+2)(n+r+1)}, \quad n = 1, 2, ...$$

For $r = 0$, it becomes $c_{n+2} = -\dfrac{c_{n-1}}{(n+1)(n+2)}, \quad n = 1, 2, ...$

Hence $$c_3 = -\frac{c_0}{2.3}, \quad c_4 = -\frac{c_1}{3.4} \quad c_5 = -\frac{c_2}{4.5} = 0,$$

$$c_6 = -\frac{c_3}{5.6} = \frac{c_0}{2.3.5.6} \qquad c_7 = -\frac{c_4}{6.7} = \frac{c_1}{3.4.6.7}, \quad c_8 = 0, \text{ etc.}$$

Hence, one Frobenius solution corresponding to $r = 0$, is

$$y_1(x) = \left[c_0 + c_1x - \frac{c_0}{2.3}x^3 - \frac{c_1}{3.4}x^4 + \frac{c_0}{2.3.5.6}x^6 + \frac{c_1}{3.4.6.7}x^7 + ...\right]$$

$$= c_0\left[1 - \frac{1}{6}x^3 + \frac{1}{180}x^6 - ...\right] + c_1x\left[1 - \frac{1}{12}x^3 + \frac{1}{504}x^6 - ...\right],$$

where, c_0 and c_1 are two arbitrary constants.

Since it involves two arbitrary constants, this can be considered as the complete solution of the given differential equation.

CHAPTER

21 Bessel's Equation and Bessel Functions

IMPORTANT FORMULAE AND USEFUL RESULTS

1. Equation is $x^2\ y'' + xy' + (x^2 - n^2)\ y = 0$

$$J_n(x) = \sum_0^{\infty} \frac{(-1)^n}{r!\,\lceil n+r+1}\left(\frac{x}{2}\right)^{n+2r} \quad \text{Bessel's Function of first kind.}$$

Solution is $\quad y = A\ J_n(x) + B\ J_{-n}(K).$

2. When n is integer, J_n and J_{-n} are linearly dependent and we have $J_{-n}(x) = (-1)^n J_n(x)$.

3. **Bessel Function of Half-order**

$$J_{1/2}(x) = \sqrt{\frac{2}{\pi x}} \sin x \quad \text{and} \quad J_{-1/2}(x) = \sqrt{\frac{2}{\pi x}} \cos x$$

4. **Recurrence Relations**

(i) $\dfrac{d}{dx}(x^n J_n) = x^n J_{n-1}(x)$

(ii) $\dfrac{d}{dx}(x^{-n} J_n) = -x^{-n} J_{n+1}(x).$

(iii) $(J'_n(x)) = J_{n-1}(x) - \dfrac{n}{x} J_n(x).$

(iv) $J'_n = -J_{n+1} + \dfrac{n}{x} J_n$

(v) $J'_n = \dfrac{1}{2}[J_{n-1} - J_{n+1}]$

(vi) $\dfrac{2n}{x} J_n = J_{n-1} + J_{n+1}$

5. **Generating Function for $J_n(x)$**

$$e^{x/2}\left(t-\frac{1}{t}\right) = \sum_{-\infty}^{\infty} t^n J_n(x)$$

Jacobi series are

$$\cos(x\sin\theta) = J_0 + 2[J_2\cos 2\theta + J_4\cos 4\theta + ...]$$

and $$\sin(x\sin\theta) = 2[J_1\sin\theta + J_3\sin 3\theta + ...]$$

Integral Representation of J_n is

$$J_n = \frac{1}{\pi}\int_0^{\pi}\cos(x\sin\theta - n\theta)\,d\theta.$$

6. **Orthogonality of Bessel's Functions:**

If α and β are the roots of the equation $J_n(x) = 0$, *i.e.*, if $J_n(\alpha) = J_n(\beta) = 0$ then

$$\int_0^1 x\,J_n(\alpha x)\,J_n(\beta x)\,du = \begin{cases} 0 & \text{if } \alpha \neq \beta \\ \frac{1}{2}J_{n+1}^2(\alpha) & \text{if } \alpha = \beta \end{cases}$$

7. **Fourier-Bessel Expansion of a Function:**

If $\alpha_1, \alpha_2, ..., \alpha_n$ are roots of $J_n(x) = 0$ in $(0, a)$ then we can write $f(x) = \sum_{K=1}^{\infty} C_K J_n(\alpha_K x)$

where $$C_K = \frac{2}{a^2 J_{n+1}^2(a\alpha_K)}\int_0^a x\,f(x)\,J_n(\alpha_K x)\,dx$$

Ber and Bei Functions

Consider the equation $x\frac{d^2y}{dx^2} + \frac{dy}{dx} - ixy = 0$

whose solution is $y = J_0(i^{3/2}x)$

or $$y = \sum_{n=0}^{\infty}\frac{(-1)^r}{r!\,\lceil 0+r+1}\left[\frac{i^{3/2}x}{2}\right]^{0+2r} = \text{Ber}(x) + i\,\text{Bei}(x)$$

where $$\text{Ber}(x) = 1 - \frac{x^4}{2^2\cdot4^2} + \frac{x^8}{2^2\cdot4^2\cdot6^2\cdot8^2} - \frac{x^{12}}{2^2\cdot4^2\cdot6^2\cdot8^2\cdot10^2\cdot12^2} + ...$$

and $$\text{Bei}(x) = \frac{x^2}{2^2} - \frac{x^6}{2^2\cdot4^2\cdot6^2} + \frac{x^{10}}{2^2\cdot4^2\cdot6^2\cdot8^2\cdot10^2} - ...$$

Thus, $J_0(i^{3/2}x) = \text{Ber}(x) + i\,\text{Bei}(x)$

SOLUTIONS TO EXERCISE 22A

1. *Express $P_4(x)$ as an algebraic function of x using Rodrigue's formula.*

SOLUTION: Rodrigue's formula is $P_n(x) = \frac{1}{2^n.n!}\frac{d^n}{dx^n}(x^2-1)^n$

$$P_4(x) = \frac{1}{2^4.4!}\frac{d^4}{dx^4}(x^2-1)^4 = \frac{1}{16(24)}\frac{d^3}{dx^3}\left[8x(x^2-1)^3\right]$$

$$= \frac{1}{48}\frac{d^2}{dx^2}\left[(x^2-1)^3 + 3x(x^2-1)^2.2x\right] = \frac{1}{48}\frac{d^2}{dx^2}\left[(x^2-1)^2(7x^2-1)\right]$$

$$= \frac{1}{48}\frac{d}{dx}[14x(x^2-1)^2 + 2(7x^2-1)(x^2-1)2x] = \frac{1}{24}\frac{d}{dx}[x(x^2-1)(7x^2-7+14x^2-2)]$$

$$= \frac{1}{24}[(3x^2-1)(21x^2-9) + (x^3-x)(42x)]$$

$$= \frac{1}{24}[63x^4 + 42x^4 - 48x^2 - 42x^2 + 9] = \frac{1}{8}[35x^4 - 30x^2 + 3]$$ **Ans.**

2. *Show that $P'_5(x) = 9P_4(x) + 5P_2(x) + P_0(x)$.*

SOLUTION: As $P_5(x) = \frac{1}{2^5.5!}\frac{d^5}{dx^5}(x^2-1)^5$

$$P'_5(x) = \frac{1}{32(120)}\frac{d^6}{dx^6}(x^2-1)^5 = \frac{1}{32(120)}\frac{d^5}{dx^5}\left[5(x^2-1)^4.2x\right]$$

$$= \frac{1}{32(12)}\frac{d^4}{dx^4}[4(x^2-1)^3\cdot 2x^2 + (x^2-1)^4]$$

$$= \frac{1}{32(12)}\frac{d^4}{dx^4}[(x^2-1)^4 + 8(x^2-1)^4 + 8(x^2-1)^3]$$

$$= \frac{1}{32(12)}\left[9\frac{d^4}{dx^4}(x^2-1)^4 + 8\frac{d^3}{dx^3}\{3(x^2-1)^2.2x\}\right]$$

$$= \frac{9}{32(12)}2^4(4!)P_4(x) + \frac{1}{8}\frac{d^3}{dx^3}[x(x^2-1)^2]$$

$$= 9P_4(x) + \frac{1}{8}\frac{d^2}{dx^2}[1(x^2-1)^2 + 4x^2(x^2-1)]$$

$$= 9\,P_4(x) + \frac{1}{8}\frac{d^2}{dx^2}[5\,(x^2-1)^2 + 4(x^2-1)]$$

$$= 9\,P_4(x) + \frac{5}{8}\cdot\frac{d^2}{dx^2}(x^2-1)^2 + \frac{1}{2}\frac{d^2}{dx^2}(x^2-1) = 9\,P_4(x) + \frac{5}{8}.2^2.\,2!P_2(x) + \frac{1}{4}\,(4)$$

$$= 9\,P_4(x) + 5\,P_2(x) + 1 = 9\,P_4(x) + 5\,P_2(x) + P_0(x) \quad (\text{as } P_0(x) = 1)$$ **Hence proved.**

3. *Evaluate* $\int_{-1}^{1} x^2 P_n^2(x)\,dx$.

SOLUTION: The recurrence realation $(n+1)\,P_{n+1} = (2n+1)x\,P_n - nP_{n-1}$

can be written as $xP_n = \frac{1}{2n+1}[(n+1)P_{n+1} + nP_{n-1}]$

$$\therefore \quad \int_{-1}^{1} x^2 P_x^2 = \int_{-1}^{1} \frac{1}{(2n-1)^2}[(n+1)P_{n+1} + n\,P_{n-1}]^2\,dx$$

$$= \frac{1}{(2n+1)^2}\int_{-1}^{1}[(n+1)^2\,P_{n+1}^2 + n^2\,P_{n-1}^2 + 2n\,(n+1)\,P_{n-1}P_{n+1}]\,dx$$

$$= \frac{(n+1)^2}{(2n+1)^2}\int_{-1}^{1} P_{n+1}^2\,dx + \frac{n^2}{(2n+1)^2}\int_{-1}^{1} P_{n-1}^2\,dx + \frac{2n(n+1)}{(2n+1)^2}\int_{-1}^{1} P_{n-1}\,P_{n+1}\,dx$$

$$= \frac{(n+1)^2}{(2n+1)^2}\,\frac{2}{2(n+1)+1} + \frac{n^2}{(2n+1)^2}\,\frac{2}{2n+1} + \frac{2n(n+1)}{(2n+1)^2}(0)$$

(by the orthgonality property of Lengendres polynominals)

$$= \frac{2}{(2n+1)^2}\left[\frac{(n+1)^2}{2n\cdot 3} + \frac{n^2}{2n+1}\right] = \frac{8n^3 + 12n^2 - 2}{(2n+1)^2\,(2n+3)\,(2n-1)}$$

$$= \frac{1}{8\,(2n+3)} + \frac{1}{8(2n-1)} + \frac{3}{4(2n+1)}$$ (On resolving into partial fractions) **Ans.**

4. *If* $f(x) = 0, \quad -1 < x \le 0$

$\quad\quad = x, \quad 0 < x < 1$

find the Fourier Legendre expansion of $f(x)$ *in terms of Legendre polynomials of different orders.*

SOLUTION: Let $f(x) = \sum_{m=0}^{\infty} a_m\,P_m(x), -| \le x \le |$.

To find a_n we multiply both sides by $P_n(x)$ and integrate in $(-1, 1)$.

$$\int_{-1}^{1} f(x)\,P_n(x)\,dx = \sum_{m=0}^{\infty} a_m \int_{-1}^{1} P_n(x)\,P_m(x)\,dx = 0 + a_n \int_{-1}^{1} Pn^2(x)\,dx$$

or $$\int_{-1}^{0} 0\, P_n(x)\, dx + \int_{0}^{1} x\, P_n(x)\, dx = a_n \cdot \frac{2}{2n+1}$$

Hence $$a_n = \frac{2n+1}{2} \int_0^1 x\, P_n(x)\, du$$

For $n = 0,$ $$a_0 = \frac{1}{2}\int_0^1 x\, P_0\, dx = \frac{1}{2}\int_0^1 x\, dx = \frac{1}{4} \quad \text{as } P_0(x) = 1$$

For $n = 1,$ $$a_1 = \frac{3}{2}\int_0^1 x\, P_1\, dx = \frac{3}{2}\int_0^1 x^2\, dx = \frac{3}{2}\cdot\frac{1}{3} = \frac{1}{2} \quad \text{as } P_1(x) = x$$

For $n=2,$ $$a_2 = \frac{5}{2}\int_0^1 x\, P_2\, dx = \frac{5}{2}\int_0^1 x\cdot\frac{1}{2}(3x^2-1)\, dx = \frac{5}{4}\left[\frac{x^4}{2} - \frac{x^2}{4}\right]_0^1 = \frac{5}{16} \quad \textit{as } P_2(x) = \frac{1}{3}(3x^2-1)$$

For $n = 3,$ $$a_3 = \frac{7}{2}\int_0^1 x\, P_3\, dx = \frac{7}{2}\int_0^1 x\cdot\frac{1}{2}(5x^3-3x)\, dx = \frac{7}{4}\int_0^1 (5x^4-3x^2)\, dx$$

$$= \frac{7}{4}[x^5 - x^3]_0^1 = 0 \quad \text{as } P_3(x) = \frac{1}{2}(5x^3-3x)$$

For $n = 4,$ $$a_4 = \frac{-3}{32}, \text{ etc.}$$

$$\therefore \quad f(x) = \frac{1}{4}P_0(x) + \frac{1}{2}P_1(x) + \frac{5}{16}P_2(x) - \frac{3}{32}P_4(x) + \ldots$$ **Ans.**

5. *Prove that* $x^5 = \dfrac{8}{63}\left[\mathrm{P}_5 + \dfrac{7}{2}\mathrm{P}_3 + \dfrac{27}{8}\mathrm{P}_1\right]$

SOLUTION: Using Rodrigue's formula $P_n(x) = \dfrac{1}{2^n.n!}\dfrac{d^n}{dx^4}(x^2-1)^n$, we have

$$P_0(x) = 1,\ P_1(x) = x,\ P_2(x) = \frac{1}{2}(3x^2-1),\ P_3(x) = \frac{1}{2}(5x^3-3x), \ldots$$

Using the recurrence relation $(n+1)P_{n+1} = (2x+1)x\ \mathrm{P}_n - nP_{n-1}$ ...(1)

and taking $n = 3$, we get $4P_4 = 7xP_3 - 3P_2$

or $$P_4(x) = \frac{1}{4}[7xP_3 - 3P_2] = \frac{1}{4}\left[\frac{7x}{2}(5x^3-3x) - \frac{3}{2}(3x^2-1)\right] = \frac{1}{8}\left[35x^4 - 30x^2 + 3\right]$$

and taking $n = 4$ is (1), we get $5P_5 = 9xP_4 - 4P_3$

$$\therefore \quad P_5 = \frac{1}{5}\left[\frac{9x}{8}(35x^4-30x^2+3) - \frac{4}{2}(5x^3-3x)\right]$$

$$= \frac{1}{5}\left[\frac{9.35x^5}{8} - \frac{9.30x^3}{8} + \frac{27}{8}x - 10x^3 + 6x\right]$$

$$= \frac{63}{8}x^5 - \left(\frac{54}{8} + 2\right)x^3 + \frac{75x}{8.5} = \frac{63}{8}x^5 - \frac{70}{8}x^3 + \frac{15}{8}x$$

or $\qquad 8P_5 = 63x^5 - 70x^3 + 15x$

Hence $\qquad x^5 = \frac{1}{63}\left[8P_5 + \frac{70}{5}(2P_3 + 3P_1) - 15P_1\right] = \frac{8}{63}\left[P_5 + \frac{7}{2}P_3 + \frac{27}{8}P_1\right]$ **Hence proved.**

6. *Express (i)* $f(x) = 10x^3 + 6x^2 - 2x + 1$ *(ii)* $f(x) = x^4 + 3x^3 - x^2 + 5x - 2$ *in terms of Legendre polynomials.*

SOLUTION: (*i*) Since $P_0(x) = 1, P_1(x) = x, P_2(x) = \frac{1}{2}(3x^2 - 1), P_3(x) = \frac{1}{2}(5x^3 - 3x)$.

$\therefore \qquad x^3 = \frac{1}{5}[2P_3(x) + 3P_1(x)],\ x^2 = \frac{1}{3}(2P_2 + P_0)$

$\therefore \qquad f(x) = 10x^3 + 6x^2 - 2x + 1 = 4P_3(x) + 6P_1(x) + 4P_2 + 2P_0 - 2P_1 + P_0$

$= 4P_3 + 4P_2 + 4P_1 + 3P_0.$ **Ans.**

(*ii*) In the last problem we had

$$P_4 = \frac{1}{8}\left[35x^4 - 30x^2 + 3\right], \quad P_3 = \frac{1}{2}(5x^3 - 3x), \quad P_2 = \frac{1}{2}(3x^2 - 1), \quad P_1(x) = x, P_0(x) = 1$$

$\therefore \qquad x^2 = \frac{1}{3}(2P_2 + P_0),\ x^3 = \frac{1}{5}[2P_3 + 3P_1],\ x^4 = \frac{1}{35}[8P_4 + 30x^2 + 3]$

or $\qquad x^4 = \frac{1}{35}[8P_4 + 10(2P_2 + P_0) - 3P_0] = \frac{1}{35}[8P_4 + 20P_2 + 7P_0]$

Therefore $\quad f(x) = x^4 + 3x^3 - x^2 + 5x - 2$

$$= \frac{1}{35}[8P_4 + 20P_2 + 7P_0] + \frac{3}{5}[2P_3 + 3P_1] - \frac{1}{3}(2p_2 + p_0) + 5P_1 - 2P_0$$

$$= \frac{8}{35}P_4 + \frac{6}{5}P_3 - \frac{2}{21}P_2 + \frac{34}{5}P_1 - \frac{32}{15}P_0$$ **Ans.**

7. *Evaluate* $\int_{-1}^{1} P_n(x)(1 - 2hx + h^2)^{-1/2}\,dx$.

SOLUTION: We know that $(1 - 2hx + h^2)^{-1/2} = \sum_{r=0}^{\infty} P_r(x)\,h^r$

$$\therefore \quad \int_{-1}^{1} P_n(x)(1 - 2hx + h^2)^{-1/2}\,dx = \int_{-1}^{1} P_n(x)\sum_{r=0}^{\infty} P_r(x)\,h^r\,dx = \int_{-1}^{1} h^n P_n^2(x)\,dx + 0$$

(by orthogonality of Legendres polynomials) $= h^n\left(\frac{2}{2n+1}\right)$ **Ans.**

8. *Show that* $\int_{-1}^{1} x \, P_n \, P_{n-1} \, dx = \dfrac{2n}{4n^2 - 1}$.

SOLUTION: By the recurrence relation $(n + 1) \, P_{n+1} = (2n + 1) \, x \, P_n - n \, P_{n-1}$ we have

$$x \, P_n = \frac{1}{2n+1}[(n+1) \, P_{n+1} + n \, P_{n-1}]$$

$$\therefore \quad \int_{-1}^{1} x \, P_n \, P_{n-1} \, dx = \int_{-1}^{1} \frac{P_{n-1}}{2n+1} \{(n+1) \, P_{n+1} + n \, P_{n-1}\}$$

$$= \frac{n+1}{2n+1} \int_{-1}^{1} P_{n-1} \, P_{n+1} \, dx + \frac{n}{2n+1} \int_{-1}^{1} P_{n-1}^2 \, dx = 0 + \frac{n}{2n+1} \cdot \frac{2}{2(n-1)+1}$$

(by the orthogonality property of Legendre polynomials)

$$= \frac{2n}{(2n+1)(2n-1)} = \frac{2n}{4n^2 - 1}$$ **Hence proved.**

9. *Find the Legendre polynomial expansion of the function* $f(x)$ *given by*

$$f(x) = -1, \; -1 < x < 0$$
$$= 1, \quad [0 < x < 1$$

SOLUTION: Let $f(x) = \sum_{0}^{\infty} a_k \, P_k(x)$ as Fourier Legendre expansion.

Multiplying both sides by P_n and integrating in $(-1, 1)$, we get

$$\int_{-1}^{1} f(x) \, P_n(x) \, dx = \int_{-1}^{1} P_n(x) \sum_{k=0}^{\infty} a_k \, P_k(x) \, dx$$

$$= \sum_{\substack{k=0 \\ k \neq n}}^{\infty} a_k \int_{-1}^{1} P_n(x) \, P_k(x) \, dx + a_n \int_{-1}^{1} P_n^2(x) \, dx = 0 + a_n \cdot \frac{2}{2n+1}$$

by orthogonality property of Legendre polynomials.

Next, $$\int_{-1}^{1} f(x) \, P_n(x) \, dx = \int_{-1}^{0} -1 \, P_n(x) \, dx + \int_{0}^{1} 1 \, P_n(x) \, dx = \int_{0}^{1} P_n(x) \, dx + I_1$$

where $$I_1 = \int_{-1}^{0} -P_n(x) \, dx = \int_{1}^{0} P_n(-y) \, dy = -\int_{0}^{1} (-1)^n \, P_n(y) \, dy = (-1)^{n+1} \int_{0}^{1} P_n(x) \, dx$$

(Putting $x = -y$ so $dx = -dy$)

Therefore, $\int_{-1}^{1} f(x) \, P_n(x) \, dx = \left[1 + (-1)^{n+1}\right] \int_{0}^{1} P_n(x) \, dx$ and, in turn, we have

$$a_n = \frac{2n+1}{2} [1 + (-1)^{n+1}] \int_{0}^{1} P_n(x) \, dx .$$

Also know that $$\int_0^1 P_n(x)\,dx = \frac{1}{n+1}P_{n-1}(0)$$

hence $$a_n = \frac{2n+1}{2(n+1)}[1+(-1)^{n+1}]\,P_{n-1}(0) = 0 \quad \text{when } n \text{ is even.}$$

And when n is odd, $$a_1 = \frac{3}{2}P_0(0) = \frac{3}{2},\quad a_3 = \frac{7}{4}P_2(0) = \frac{7}{4}(-1)\frac{1}{2} = \frac{-7}{8}, \text{ etc.}$$

Thus, $$f(x) = \frac{3}{2}P_1(x) - \frac{7}{8}P_3(x) - \ldots = \frac{3x}{2} - \frac{7}{8}\left(\frac{5x^3-3x}{2}\right) + \ldots$$ **Ans.**

10. *Show that* $\int_x^1 \mathrm{P}_n(x)\,dx = \frac{1}{2n+1}[\mathrm{P}_{n-1}(x) - \mathrm{P}_{n+1}(x)]$.

Solution: By recurrence relation, we have

$$(2n+1)\,P_n(x) = P'_{n+1}(x) - P'_{n-1}(x)$$

$$\therefore \int_x^1 (2n+1)\,P_n(x)\,dx = [P_{n+1}(x) - P_{n-1}(x)]_x^1$$

or $$\int_x^1 P_n(x)\,dx = \frac{1}{2n+1}[\{P_{n+1}(1) - P_{n-1}(1)\} - \{P_{n+1}(x) - P_{n-1}(x)\}]$$

$$= \frac{1}{2n+1}[(1-1) - P_{n+1}(x) + P_{n-1}(x)]$$

or $$\int_x^1 P_n(x)\,dx = \frac{1}{2n+1}[P_{n-1}(x) - P_{n+1}(x)]$$ **Hence proved.**

ADDITIONAL IMPORTANT PROBLEMS

1. *If $P_n(x)$ denotes the nth order Legendre polynomial show that*

$$\frac{1-z^2}{(1-2xz+z^2)^{3/2}} = \sum_{n=0}^{\infty}(2n+1)\,P_n(x)\,z^n.$$

SOLUTION: We know that $(1-2xz+z^2)^{-1/2} = \sum_{n=0}^{\infty} z^n P_n(x)$ ...(1)

Differentiating w.r.t. z on both sides, we get

$$-\frac{1}{2}(1-2xz+z^2)^{-3/2}(-2x+2z) = \sum nz^{n-1}P_n(x)$$

or $$(x-z)(1-2xz+z^2)^{-3/2} = \sum nz^{n-1}P_n(x).$$

Multiplying throughout by $2z$, gives

$$(2xz - 2z^2)\,(1 - 2xz + z^2)^{-3/2} = \sum 2nz^n\, P_n(x) \qquad ...(2)$$

Adding (1) and (2), gives

$$\frac{2xz - 2z^2}{(1 - 2xz + z^2)^{3/2}} + \frac{1}{(1 - 2xz + z^2)^{1/2}} = \sum (2n+1)\, z^n\, P_n(x)$$

or $$(2xz - 2z^2 + 1 - 2xz + z^2)\,(1 - 2xz + z^2)^{-3/2} = \sum (2n+1) z^n\, P_n(x)$$

or $$\frac{1 - z^2}{(1 - 2xz + z^2)^{3/2}} = \sum (2n+1)\, z^n\, P_n(x)$$ **Hence proved.**

2. *If $P_n(x)$ stands for Legendre's polynomial of order n, show that*

(a) $P'_n(1) = \dfrac{n(n+1)}{2}$

(b) $P'_n(-1) = (-1)^{n-1} \cdot \dfrac{n(n+1)}{2}.$

SOLUTION: (a) Since $P_n(x)$ is the solution of the Legendre's equation

$$(1 - x^2)\, P''_n(x) - 2x\, P'_n(x) + n(n + 1)\, P_n(x) = 0 \qquad ...(1)$$

Putting $x = 1$ in (1), we get $-2P'_n(1) + n(n + 1)\, P_n(1) = 0$

Since $P_n(1) = 1$ we have $P'_n(1) = \dfrac{n(n+1)}{2}.$

(b) Putting $x = -1$ in (1), we get $-2(-1)\, P'_n(-1) + n(n + 1)\, P_n(-1) = 0$

Since $P(-1) = (-1)^n$ we get $P_n' = \dfrac{-n(n+1)}{2}(-1)^n = \dfrac{n(n+1)}{2}(-1)^{n-1}$ **Hence proved.**

3. *Express x^4 in terms of Legendre polynomials of various orders*

SOLUTION: We know that $P_4(x) = \dfrac{1}{8}(35x^4 - 30x^2 + 3),\ P_2(x) = \dfrac{1}{2}(3x^2 - 1),\ P_0(x) = 1$

$$\therefore \quad x^4 = \frac{1}{35}\left[8P_4(x) + 30x^2 - 3\right] = \frac{1}{35}\left[8P_4(x) + \frac{30}{3}\{2P_2(x) + P_0\} - 3\right]$$

$$= \frac{1}{35}[8P_4(x) + 20P_2(x) + 10P_0 - 3P_0]$$

or $$x^4 = \frac{1}{35}[8P_4(x) + 20P_2(x) + 7P_0(x)]$$ **Ans.**

4. *Prove the Beltrami's result that*

$$(2n + 1)\,(x^2 - 1)\,P'_n(x) = n(n + 1)\,\{P_{n+1}(x) - P_{n-1}(x)\}$$

SOLUTION: The required result can be obtained from the following two recurrence relations:

$$(1 - x^2)\,P'_n(x) = n[P_{n-1}(x) - xP_n(x)] \quad \text{and} \quad (1 - x^2)\,P'_n(x) = (n + 1)\,[xP_n(x) - P_{n+1}(x)]$$

or $\quad \frac{1}{n}(1-x^2)\,P_n'(x) = P_{n-1}(x) - xP_n(x)$ and $\frac{1}{n+1}(1-x^2)\,P_n'(x) = xP_n(x) - P_{n+1}(x)$

Adding the above two relations, we get

$$\left(\frac{1}{n}+\frac{1}{n+1}\right)(1-x^2)\,P_n'(x) = P_{n-1}(x) - P_{n+1}(x)$$

or $\quad (2n + 1)(1 - x^2)\,P'_n(x) = n(n + 1)\,[P_{n-1}(x) - P_{n+1}(x)]$ **Hence Proved**

5. If n^{th} derivative of a function $f(x)$ is continous in $(-1, 1)$ then prove that

$$\int_{-1}^{1} f(x)\,P_n(x)\,dx = \frac{(-1)^n}{2^n \cdot n!}\int_{-1}^{1}(x^2-1)^n\, f^{(n)}(x)\,dx$$

SOLUTION: $\displaystyle\int_{-1}^{1} f(x)\,P_n(x)\,dx = \int_{-1}^{1} f(x)\cdot\frac{1}{2^n\cdot n!}\frac{d^n}{dx^n}(x^2-1)^n\,dx$ (using Rodrigue's formula)

$$= \frac{1}{2^n\cdot n!}\left[\left\{f(x)\frac{d^{n-1}}{dx^{n-1}}(x^2-1)^n\right\}_{-1}^{1} - \int_{-1}^{1} f'(x)\frac{d^{n-1}}{dx^{n-1}}(x^2-1)^n dx\right]$$

(on integrating by parts)

$$= \frac{1}{2^n\cdot n!}\left[0 - \int_{-1}^{1} f'(x)\frac{d^{n-1}}{dx^{n-1}}(x^2-1)^n\,dx\right] = \frac{(-1)'}{2^n\cdot n!}\left[\int_{-1}^{1} f'(x)\frac{d^{n-1}}{dx^{n-1}}(x^2-1)^n\,dx\right]$$

(on integrating by parts again)

$$= \frac{(-1)^2}{2^n\cdot n!}\left[\int_{-1}^{1} f''(x)\frac{d^{n-2}}{dx^{n-2}}(x^2-1)^n\,dx\right] = \frac{(-1)^n}{2^n\cdot n!}\int_{-1}^{1} f^{(n)}(x)\,(x^2-1)^n\,dx$$

(On integrating by parts $(n - 2)$ times) **Hence proved.**

6. *Prove that* $\displaystyle\int_{-1}^{1}(x^2-1)\,P_{n+1}\,P_n'\,dx = \frac{2n(n+1)}{(2n+1)(2n+3)}.$

SOLUTION: As a recurrence relation we have $(x^2 - 1)\,P'_n = n(xP_n - P_{n-1})$

$$\therefore \quad \int_{-1}^{1}(x^2-1)\,P_{n+1}\,P_n'\,dx = \int_{-1}^{1} n\,P_{n+1}\,(xP_n - P_{n-1})\,dx$$

$$= n\int_{-1}^{1} xP_n\,P_{n+1}\,dx - n\int_{-1}^{1} P_{n+1}\,P_{n-1}\,dx = n\int_{-1}^{1} x\,P_n\,P_{n+1}\,dx - 0$$

(by othogonality property of Legendre polynomials)

Using here another recurrence relation $(2n + 1)\,x\,P_n = (n + 1)\,P_{n+1} + n\,P_{n-1}$ we get

$$\int_{-1}^{1}(x^2-1)\,P_{n+1}P_n'dx = \frac{n}{2n+1}\int_{-1}^{1}\{(n+1)\,P_{n+1} + nP_{n-1}\}\,P_{n+1}\,dx$$

$$= \frac{n(n+1)}{2n+1}\int_{-1}^{1} P_{n+1}^2 dx + \frac{n^2}{2n+1}\int_{-1}^{1} P_{n-1}P_{n+1}\, dx$$

$$= \frac{n(n+1)}{2n+1}\int_{-1}^{1} P_{n+1}^2\, dx + 0 \qquad \text{(by orthogonality property)}$$

$$= \frac{n(n+1)}{2n+1}\,\frac{2}{2(n+1)+1} \qquad \left(\text{Since } \int_{-1}^{1} P_n^2\, dx = \frac{2}{2n+1}\right)$$

$$= \frac{2n(n+1)}{(2n+1)(2n+3)}$$

Hence proved.

7. *Show that* $\sum_{n=0}^{\infty} \frac{x^{n+1}}{n+1} P_n(x) = \frac{1}{2}\log\left(\frac{1+x}{1-x}\right)$.

SOLUTION: We know that $\sum_{0}^{\infty} h^n P_n(x) = (1-2xh+h^2)^{-1/2}$. Integrating both sides w.r.t. h from 0 to h, we get

$$\sum_{0}^{\infty} \frac{h^{n+1}}{n+1} P_n(x) = \int_0^h (1-2xh+h^2)^{-1/2}\, dh = \int_0^h \frac{dh}{\sqrt{(1-x^2)+(h-x)^2}} \quad \text{if } |x| < 1.$$

$$= \left[\log \frac{(h-x)+\sqrt{h^2-2hx+1}}{\sqrt{1-x^2}}\right]_0^h = \log\left[\frac{h-x+\sqrt{h^2-2hx+1}}{1-x}\right]$$

Replacing h by x on both sides; we get

$$\sum_{0}^{\infty} \frac{x^{n+1}}{n+1} P_n(x) = \log \frac{0+\sqrt{(1-x^2)}}{1-x} = \log\sqrt{\frac{1+x}{1-x}}$$

$$= \frac{1}{2}\log\frac{1+x}{1-x}$$

Hence proved.

8. *Using the recurrence relation* $(n+1)\,P_{n+1}(x) = (2n+1)\,xP_n(x) - nP_{n-1}(x)$ *generate the Legendre polynomials* P_2, P_3, P_4 *given that* $P_0(x) = 1$ *and* $P_1(x) = x$.

SOLUTION: The recurrence relation is

$$(n+1)P_{n+1}(x) = (2n+1)\,xP_n(x) - nP_{n-1}(x)$$

For $n = 1$, it gives, $\quad 2P_2(x) = 3x\,P_1(x) - P_0(x) = 3x^2 - 1$, (using $P_0(x) = 1$ and $P_1(x) = x$)

or, $P_2(x) = (3x^2 - 1)/2.$

For $n = 2$, we have $3P_3(x) = 5x\,P_2(x) - 2P_1(x) = 5x(3x^2 - 1)/2 - 2x = 3(5x^3 - 3x)/2$

or, $P_3(x) = (5x^3 - 3x)/2.$

For $n = 3$, we have $4P_4(x) = 7x\,P_3(x) - 3P_2(x) = 7x(5x^3 - 3x)/2 - 3(3x^2 - 2)/2$

$$= (35x^4 - 30x^2 + 3)/2$$

or $P_4(x) = (35x^4 - 30x^2 + 3)/8.$

9. *Let $f(x)$ be a polynomial of degree m and let $n > m$, then show that*

$$\int_{-1}^{1} f(x)\,P_n(x)\,dx = 0.$$

SOLUTION: Since $f(x)$ is a polynomial of degree m, we have

$$f(x) = c_0\,P_0(x) + c_1\,P_1(x) + \ldots + c_m\,P_m(x), \text{ for some } c\text{'s}.$$

Multiplying it with $P_n(x)$ and integrating w.r.t. x over the interval $[-1, 1]$, we get

$$\int_{-1}^{1} f(x)\,P_n(x)\,dx = \sum_{k=0}^{m} c_k \int_{-1}^{1} P_k(x)\,P_n(x)\,dx = 0$$

using the orthogonality property of the Legendre polynomials as $k \neq n$.

10. *Show that*

$$\int_{-1}^{1} x^2 P_{n-1}(x)\,P_{n+1}(x)\,dx = \frac{2n(n+1)}{(2n-1)(2n+1)(2n+3)}$$

SOLUTION: Consider the recurrence relation $(2n + 1)\,xP_n = (n + 1)\,P_{n+1} + nP_{n-1}$.

Replacing n by $(n - 1)$ and $(n + 1)$, respectively, we get

$$xP_{n-1} = \frac{1}{2n-1}[nP_n + (n-1)\,P_{n-2}]$$

and, $$xP_{n+1} = \frac{1}{2n+3}[(n+2)\,P_{n+2} + (n+1)\,P_n].$$

Multiplying these two and integrating w.r.t. x over $[-1, 1]$ and using orthogonality property, we get

$$\int_{-1}^{1} x^2 P_{n-1}(x)\,P_{n+1}(x)\,dx = \frac{1}{(2n-1)(2n+3)} \cdot n(n+1) \int_{-1}^{1} P_n^2(x)\,dx,$$

Thus,

$$\int_{-1}^{1} x^2 P_{n-1}(x)\,P_{n+1}(x)\,dx = \frac{2n(n+1)}{(2n-1)(2n+1)(2n+3)}.$$

Hence Proved

11. *Show that* $\int_{-1}^{1} f(x)\, P_n(x)\, dx = \frac{(-1)^n}{2^n \cdot n!} \int_{-1}^{1} f^{(n)}(x) \cdot (x^2-1)^n\, dx.$

Hence, deduce that $\int_{-1}^{1} x^m\, P_n(x)\, dx = \begin{cases} 0, & \text{if} \quad m < n \\ \frac{2^{n+1}(n!)^2}{(2n+1)!}, & \text{if} \quad m = n \end{cases}$

SOLUTION: Using Rodrigue's formula, we have

$$\int_{-1}^{1} f(x)\, P_n(x)\, dx = \frac{1}{2^n\, n!} \int_{-1}^{1} f(x) \frac{d^n}{dx^n}(x^2-1)^n\, dx \qquad \text{(Integrating by parts)}$$

$$= \frac{1}{2^n\, n!}\left[\left\{f(x)\frac{d^{n-1}}{dx^{n-1}}(x^2-1)^n\right\}_{-1}^{1} - \int_{-1}^{1} f'(x)\frac{d^{n-1}}{dx^{n-1}}(x^2-1)^n\, dx\right]$$

$$= \frac{-1}{2^n \cdot n!}\int_{-1}^{1} f'(x)\cdot\frac{d^{n-1}}{dx^{n-1}}(x^2-1)^n\, dx, \text{ since the first term is zero at } x = \pm 1.$$

Integrating by parts further $(n-1)$ times, we obtain

$$\int_{-1}^{1} f(x)\, P_n(x) = \frac{(-1)^n}{2^n \cdot n!}\int_{-1}^{1} f^{(n)}(x)\,(x^2-1)^n\, dx.$$

Writing $f(x) = x^m$, so that $f^{(n)}(x) = 0$, when $m < n$ and, $f^{(n)}(x) = n!$, when $m = n$.

Thus, $$\int_{-1}^{1} x^m P_n(x)\, dx = 0 + \frac{(-1)^n}{2^n\, n!}\int_{-1}^{1} n!(x^2-1)^n\, dx = \frac{2}{2^n}\int_{0}^{1}(1-x^2)^n\, dx$$

$$= \frac{2}{2^n}\int_{0}^{\pi/2} \cos^{(2n+1)}\theta\, d\theta, \text{ (using } x = \sin\theta)$$

$$= \frac{2}{2^n}\left[\frac{(2n)(2n-2)\dots 4.2.}{(2n+1)(2n-1)\dots 3.1}\right] = \frac{2}{2^n}\cdot\left[\frac{2^{2n}(n!)^2}{(2n+1)^n}\right]$$

$$= \frac{2^{n+1}(n!)^2}{(2n+1)!}$$

Hence Proved

12. *Expand* $f(x) = \begin{cases} 0, & -1 < x \le 0 \\ x, & 0 < x < 1 \end{cases}$ *as a Fourier-Legendre series.*

SOLUTION: Let $f(x) = \sum_{n=0}^{\infty} c_n P_n(x)$, where c_i's are Fourier-Legendre coefficients, given by

$$c_n = \frac{2n+1}{2} \int_{-1}^{1} f(x)\, P_n(x)\, dx, \quad n = 0, 1, 2, \ldots$$

For given $f(x)$, $$c_n = \frac{2n+1}{2} \int_{0}^{1} x P_n(x)\, dx.$$

Thus, $$c_0 = \frac{1}{2} \int_{0}^{1} x P_0(x)\, dx = \frac{1}{2} \int_{0}^{1} x dx = \frac{1}{4}, \quad c_1 = \frac{3}{2} \int_{0}^{1} x P_1(x)\, dx = \frac{3}{2} \int_{0}^{1} x^2\, dx = \frac{1}{2},$$

$$c_2 = \frac{5}{2} \int_{0}^{1} x P_2(x)\, dx = \frac{5}{2} \int_{0}^{1} x \left(\frac{3x^2 - 1}{2} \right) dx = \frac{5}{16},$$

$$c_3 = \frac{7}{2} \int_{0}^{1} x P_3(x)\, dx = \frac{7}{2} \int_{0}^{1} x \left(\frac{5x^3 - 3x}{2} \right) dx = 0$$

and $$c_4 = \frac{9}{2} \int_{0}^{1} x P_4(x)\, dx = \frac{9}{2} \int_{0}^{1} x \left(\frac{35x^4 - 30x^2 + 3}{8} \right) dx = -\frac{3}{32}$$ and so on.

Hence, $$f(x) = \frac{1}{4} P_0(x) + \frac{1}{2} P_1(x) + \frac{5}{16} P_2(x) - \frac{3}{32} P_4(x) + \ldots$$

13. *Expand* $f(x) = \cos \frac{\pi x}{2}$ *as a Fourier-Legendre series.*

SOLUTION: Let $f(x) = \sum_{n=0}^{\infty} c_n P_n(x)$, where c_i's are Fourier-Legendre coefficients, given as

$$c_n = \frac{2n+1}{2} \int_{-1}^{1} f(x)\, P_n(x)\, dx, \quad n = 0, 1, 2, \ldots$$

For $f(x) = \cos \frac{\pi x}{2}$, $$c_n = \frac{2n+1}{2} \int_{-1}^{1} \cos \frac{\pi x}{2} P_n(x)\, dx$$

CHAPTER

23 First Order Partial Differential Equations

IMPORTANT FORMULAE AND USEFUL RESULTS

1. **Linear Lagrangian Form:** $Pp + Qq = R$ where $p = \dfrac{\partial z}{\partial x}$, $q = \dfrac{\partial z}{\partial y}$. and P, Q, R are functions of x, y, z. It is called linear in p and q. To solve it we form auxiliary equation $\dfrac{dx}{P} = \dfrac{dy}{Q} = \dfrac{dz}{R}$.

   If $u = a$ and $v = b$ are two independent solutions then the solution of the above equation is $\phi(u, v) = 0$.

2. **Non-Linear Equation:** $F(x, y, z, p, q) = 0$

   Type I: $F(p, q) = 0$. Its solution is $z = ax + by + c$ where $F(a, b) = 0$.

   Type II: $z = px + qy + f(p, q)$

   Its solution is $z = ax + by + f(a, b)$.

   Type III: $F(p, q, z) = 0$ setting $t = x + ay$. we get $F\left(\dfrac{dz}{dt}, a\dfrac{dz}{dt}, z\right) = 0$

   $\therefore$ Solution $z = \phi(x + ay, b)$.

   Type IV: $F_1(x, p) = F_2(y, q) \Rightarrow F_1(x, p) = F_2(y, q) = a$

   Solve these for p and q as $p = \phi_1(x, a)$, $q = \phi_2(y, a)$

   Put them in $dz = p\,dx + q\,dy$ to get $z = \int \phi_1(x, a)\,dx + \int \phi_2(y, a)\,dy + b$

   Type V: Charpit's Method.

   Equation is $f(x, y, z, p, q) = 0$. Hence we have an auxiliary system of equation as

$$\frac{dp}{\dfrac{\partial f}{\partial x} + p\dfrac{\partial f}{\partial z}} = \frac{dq}{\dfrac{\partial f}{\partial y} + q\dfrac{\partial f}{\partial z}} = \frac{dz}{-p\dfrac{\partial f}{\partial p} - q\dfrac{\partial f}{\partial q}} = \frac{dx}{\dfrac{-\partial f}{\partial p}} = \frac{dy}{\dfrac{-\partial f}{\partial q}} = \frac{\partial F}{0}.$$

SOLUTIONS TO EXERCISE 23A

1. *Form a differential equation of all planes which are at a constant distance 'a' from the origin.*

SOLUTION: Equation of the plane which is at a constant distance 'a' from the origin is

$$lx + my + nz = a \quad ...(1) \qquad \text{where } l^2 + m^2 + n^2 = 1$$

or $\quad lx + my + z\sqrt{1-l^2-m^2} = a.$

Differentiating it partially w.r.t. x and also y, we get

$$l + \sqrt{1-l^2-m^2}\; p = 0 \quad \text{and} \quad m + \sqrt{1-l^2-m^2}\; q = 0$$

or $\quad l^2 = (1 - l^2 - m^2)\, p^2 \quad$ and $\quad m^2 = (1 - l^2 - m^2)\, q^2$

On adding these, we get $\quad l^2 + m^2 = (1 - l^2 - m^2)\,(p^2 + q^2)$

or $\quad l^2 + m^2 = \dfrac{p^2+q^2}{1+p^2+q^2} \quad$ and $\quad 1 - l^2 - m^2 = \dfrac{1}{1+p^2+q^2}$

$\Rightarrow \quad l = \dfrac{p}{\sqrt{1+p^2+q^2}} \quad$ and $\quad m = \dfrac{q}{\sqrt{1+p^2+q^2}}$

Substituting for l, m and $n = \sqrt{1-l^2-m^2}$ in (1), gives $\dfrac{px+qy}{\sqrt{1+p^2+q^2}} + \dfrac{z}{\sqrt{1+p^2+q^2}} = a$

or $\quad px + qy + z = a\sqrt{1+p^2+q^2} \quad$ is the required differential equation. **Ans.**

2. *Solve* $\quad px + qy = nz$

SOLUTION: It is of the Lagrangian form $Pp + Qq = R$ where P, Q, R are all functions of x, y, z. Its subsidiary equation is

$$\frac{dx}{P} = \frac{dy}{Q} = \frac{dz}{R} \quad \text{or} \quad \frac{dx}{x} = \frac{dy}{y} = \frac{dz}{nz}$$

Equation $\dfrac{dx}{x} = \dfrac{dy}{y}$ gives $\log x = \log y - \log c_1 \quad$ or $\quad \dfrac{y}{x} = c_1$

Also $\quad \dfrac{dz}{z} = \dfrac{n\,dx}{x}$ gives $\log z = n \log x + \log c_2 \quad$ or $\quad z = c_2\, x^n$

$\therefore$ We can write $\quad z = x^n \cdot \phi\left(\dfrac{y}{x}\right)$ **Ans.**

3. *Solve* $y^2p - xyq = x(z - 2y)$

SOLUTION: Given equation is of Lagrangian form $Pp + Qq = R$. Its subsidiary equation is

$$\frac{dx}{P} = \frac{dy}{Q} = \frac{dz}{R} \quad \text{or} \quad \frac{dx}{y^2} = \frac{dy}{-xy} = \frac{dz}{x(z-2y)}$$

One solution is obtained from $\dfrac{dx}{y} = \dfrac{dy}{-x}$ or $xdx + ydy = 0$ as $x^2 + y^2 = C_1$

Second solution is obtained from $\dfrac{dy}{-y} = \dfrac{dz}{z-2y}$ or $ydz + zdy = 2y\,dy$

as $yz = y^2 + C_2$

$\therefore$ Complete solution is $yz - y^2 = f(x^2 + y^2)$ where f is an arbitrary function. **Ans.**

4. *Solve* $px(z - 2y^2) = (z - qy)(z - y^2 - 2x^3)$

Solution: Equation can be written as $Pp + Qq = R$

where $P = x(z - 2y^2)$, $Q = y(z - y^2 - 2x^3)$ and $R = z(z - y^2 - 2x^3)$.

Substidiary equation is $\dfrac{dx}{P} = \dfrac{dy}{Q} = \dfrac{dz}{R}$

or $$\frac{dx}{x(z-2y^2)} = \frac{dy}{y(z-y^2-2x^3)} = \frac{dz}{z(z-y^2-2x^3)} \quad ...(1)$$

From $\dfrac{dy}{Q} = \dfrac{dz}{R}$ we get $\dfrac{dy}{y} = \dfrac{dz}{z}$

Hence we get one solution as $\log\dfrac{y}{z} = \log C_1$ or $\dfrac{y}{z} = C_1$

Now using multipliers in the first two and last two in (1), gives

$$\frac{z\,dz - x\,dx}{z - y^2 - 2x^3 - z + 2y^2} = \frac{y\,dy - x\,dx}{z - y^2 - 2x^3 - z + 2y^2}$$

or $\dfrac{z\,dz - x\,dx}{y^2 - 2x^3} = \dfrac{y\,dy - x\,dx}{y^2 - 2x^3}$ or $z\,dz - x\,dx = y\,dy - x\,dx$

or $z\,dz = y\,dy = 0$ $\therefore$ $z^2 - y^2 = c_2$. This is the other solution.

$\therefore$ Complete solution is $f\left(\dfrac{y}{z}, z^2 - y^2\right) = 0$ or $z^2 - y^2 = \phi\left(\dfrac{y}{z}\right)$ **Ans.**

5. *Solve* $z = px + qy + c\sqrt{1 + p^2 + q^2}$

SOLUTION: The given equation is of the form $z = px + qy + f(p, q)$.

where $p = \dfrac{\partial z}{\partial x}$ and $q = \dfrac{\partial z}{\partial y}$. Its solution is $z = ax + by + f(a, b)$

Therefore, the solution of the given partial differential equation, is

$z = ax + by + c\sqrt{1 + a^2 + b^2}$ where c is the given constant and a and b are arbitrary constants of integration.

6. *Solve* $p^2 + q^2 = npq$

SOLUTION: The equation is non-linear of the form $f(p, q) = 0$.

Let its solution be taken as $z = ax + by + c$

$\therefore\ p = a,\ q = b$ hence $f(p, q) = a^2 + b^2 - nab = 0$

Hence the solution is $z = ax + by + c$ where $a^2 + b^2 = nab$

And this has only two arbitrary constants a and b. **Ans.**

7. *Solve* $z = pq$.

SOLUTION: Let z be a function of a variable t where $t = x + ay$.

We write $z = f(t)$. Now $p = \dfrac{\partial z}{\partial x} = \dfrac{dz}{dt}\cdot 1$ and $q = \dfrac{\partial z}{\partial y} = \dfrac{dz}{dt}\cdot a$

Therefore we have $z = \dfrac{dz}{dt}\cdot a\dfrac{dz}{dt}$ or $\dfrac{dz}{dt} = \sqrt{\dfrac{z}{a}}$

or $\dfrac{dz}{\sqrt{z}} = \dfrac{1}{\sqrt{a}}dt$ which on integration, gives

$2\sqrt{a}\sqrt{z} = t + b$ or $4az = (x + ay + b)^2$ where a and b are arbitrary cosntants. **Ans.**

8. *Solve* $p^2 - pq = 1 - z^2$

SOLUTION: The equation is of the form $F(p, q, z) = 0$. We take z as a function of t where $t = x + ay$. Therefore $\dfrac{\partial z}{\partial x} = p = \dfrac{dz}{dt}$ and $\dfrac{\partial z}{\partial y} = q = a\dfrac{dz}{dt}$. Then the given equation becomes

$\left(\dfrac{dz}{dt}\right)^2 - a\left(\dfrac{dz}{dt}\right)^2 = 1 - z^2$ or $\dfrac{dz}{\sqrt{1-z^2}} = \dfrac{dt}{\sqrt{1-a}}$ which on integration gives

$\sin^{-1} z = \dfrac{t}{\sqrt{1-a}} + b$ or $z = \sin\left(\dfrac{x+ay}{\sqrt{1-a}} + b\right)$ **Ans.**

9. *Solve* $q = -px + p^2$

SOLUTION: The given equation is of the form $F_1(x, p) = F_2(y, q)$.

We can write $F_1(x, p) = a,\ F_2(y, q) = a$

or $q = a$ and $p^2 - px = a$ where a is an arbitrary constant.

$$p^2 - px - a = 0 \quad \text{gives} \quad p = \frac{x \pm \sqrt{x^2 + 4a}}{2}, \quad q = a.$$

Also, $dz = \dfrac{\partial z}{\partial x}dx + \dfrac{\partial z}{\partial y}dy = p\,dx + q\,dy = \dfrac{x}{2} \pm \dfrac{1}{2}\sqrt{x^2 + 4a}\,dx + a\,dy$

SOLUTIONS TO EXERCISE 22A

1. *Express $P_4(x)$ as an algebraic function of x using Rodrigue's formula.*

SOLUTION: Rodrigue's formula is $P_n(x) = \frac{1}{2^n.n!}\frac{d^n}{dx^n}(x^2-1)^n$

$$P_4(x) = \frac{1}{2^4.4!}\frac{d^4}{dx^4}(x^2-1)^4 = \frac{1}{16(24)}\frac{d^3}{dx^3}\left[8x(x^2-1)^3\right]$$

$$= \frac{1}{48}\frac{d^2}{dx^2}\left[(x^2-1)^3 + 3x(x^2-1)^2.2x\right] = \frac{1}{48}\frac{d^2}{dx^2}\left[(x^2-1)^2(7x^2-1)\right]$$

$$= \frac{1}{48}\frac{d}{dx}[14x(x^2-1)^2 + 2(7x^2-1)(x^2-1)2x] = \frac{1}{24}\frac{d}{dx}[x(x^2-1)(7x^2-7+14x^2-2)]$$

$$= \frac{1}{24}[(3x^2-1)(21x^2-9) + (x^3-x)(42x)]$$

$$= \frac{1}{24}[63x^4 + 42x^4 - 48x^2 - 42x^2 + 9] = \frac{1}{8}[35x^4 - 30x^2 + 3]$$ **Ans.**

2. *Show that $P'_5(x) = 9P_4(x) + 5P_2(x) + P_0(x)$.*

SOLUTION: As $P_5(x) = \frac{1}{2^5.5!}\frac{d^5}{dx^5}(x^2-1)^5$

$$P'_5(x) = \frac{1}{32(120)}\frac{d^6}{dx^6}(x^2-1)^5 = \frac{1}{32(120)}\frac{d^5}{dx^5}\left[5(x^2-1)^4.2x\right]$$

$$= \frac{1}{32(12)}\frac{d^4}{dx^4}[4(x^2-1)^3\cdot 2x^2 + (x^2-1)^4]$$

$$= \frac{1}{32(12)}\frac{d^4}{dx^4}[(x^2-1)^4 + 8(x^2-1)^4 + 8(x^2-1)^3]$$

$$= \frac{1}{32(12)}\left[9\frac{d^4}{dx^4}(x^2-1)^4 + 8\frac{d^3}{dx^3}\{3(x^2-1)^2.2x\}\right]$$

$$= \frac{9}{32(12)}2^4(4!)P_4(x) + \frac{1}{8}\frac{d^3}{dx^3}[x(x^2-1)^2]$$

$$= 9P_4(x) + \frac{1}{8}\frac{d^2}{dx^2}[1(x^2-1)^2 + 4x^2(x^2-1)]$$

$$= 9\, P_4(x) + \frac{1}{8}\frac{d^2}{dx^2}[5\,(x^2-1)^2 + 4(x^2-1)]$$

$$= 9\, P_4(x) + \frac{5}{8}.\frac{d^2}{dx^2}(x^2-1)^2 + \frac{1}{2}\frac{d^2}{dx^2}(x^2-1) = 9\, P_4(x) + \frac{5}{8}.2^2.\, 2!P_2(x) + \frac{1}{4}\ (4)$$

$$= 9\, P_4(x) + 5\, P_2(x) + 1 = 9\, P_4(x) + 5\, P_2(x) + P_0(x) \quad (\text{as } P_0(x) = 1)$$ **Hence proved.**

3. *Evaluate* $\int_{-1}^{1} x^2 P_n^2(x)\, dx$.

SOLUTION: The recurrence realation $(n+1)\, P_{n+1} = (2n+1)x\, P_n - nP_{n-1}$

can be written as $xP_n = \frac{1}{2n+1}[(n+1)P_{n+1} + nP_{n-1}]$

$$\therefore \int_{-1}^{1} x^2 P_x^2 = \int_{-1}^{1} \frac{1}{(2n-1)^2}[(n+1)P_{n+1} + n\, P_{n-1}]^2\, dx$$

$$= \frac{1}{(2n+1)^2}\int_{-1}^{1}[(n+1)^2\, P_{n+1}^2 + n^2\, P_{n-1}^2 + 2n\,(n+1)\, P_{n-1}P_{n+1}]\, dx$$

$$= \frac{(n+1)^2}{(2n+1)^2}\int_{-1}^{1} P_{n+1}^2\, dx + \frac{n^2}{(2n+1)^2}\int_{-1}^{1} P_{n-1}^2\, dx + \frac{2n(n+1)}{(2n+1)^2}\int_{-1}^{1} P_{n-1}\, P_{n+1}\, dx$$

$$= \frac{(n+1)^2}{(2n+1)^2}\frac{2}{2(n+1)+1} + \frac{n^2}{(2n+1)^2}\frac{2}{2n+1} + \frac{2n(n+1)}{(2n+1)^2}(0)$$

(by the orthgonality property of Lengendres polynominals)

$$= \frac{2}{(2n+1)^2}\left[\frac{(n+1)^2}{2n \cdot 3} + \frac{n^2}{2n+1}\right] = \frac{8n^3 + 12n^2 - 2}{(2n+1)^2\,(2n+3)\,(2n-1)}$$

$$= \frac{1}{8\,(2n+3)} + \frac{1}{8(2n-1)} + \frac{3}{4(2n+1)}$$ (On resolving into partial fractions) **Ans.**

4. *If* $f(x) = 0, \quad -1 < x \le 0$

$= x, \quad 0 < x < 1$

find the Fourier Legendre expansion of $f(x)$ *in terms of Legendre polynomials of different orders.*

SOLUTION: Let $f(x) = \sum_{m=0}^{\infty} a_m\, P_m(x), -| \le x \le |.$

To find a_n we multiply both sides by $P_n(x)$ and integrate in $(-1, 1)$.

$$\int_{-1}^{1} f(x)\, P_n(x)\, dx = \sum_{m=0}^{\infty} a_m \int_{-1}^{1} P_n(x)\, P_m(x)\, dx = 0 + a_n \int_{-1}^{1} Pn^2(x)\, dx$$

or $$\int_{-1}^{0} 0\, P_n(x)\, dx + \int_{0}^{1} x\, P_n(x)\, dx = a_n \cdot \frac{2}{2n+1}$$

Hence $$a_n = \frac{2n+1}{2} \int_{0}^{1} x\, P_n(x)\, du$$

For $n = 0$, $$a_0 = \frac{1}{2} \int_{0}^{1} x\, P_0\, dx = \frac{1}{2} \int_{0}^{1} x\, dx = \frac{1}{4} \quad \text{as } P_0(x) = 1$$

For $n = 1$, $$a_1 = \frac{3}{2} \int_{0}^{1} x\, P_1\, dx = \frac{3}{2} \int_{0}^{1} x^2\, dx = \frac{3}{2} \cdot \frac{1}{3} = \frac{1}{2} \quad \text{as } P_1(x) = x$$

For $n = 2$, $$a_2 = \frac{5}{2} \int_{0}^{1} x\, P_2\, dx = \frac{5}{2} \int_{0}^{1} x \cdot \frac{1}{2}(3x^2 - 1)\, dx = \frac{5}{4}\left[\frac{x^4}{2} - \frac{x^2}{4}\right]_0^1 = \frac{5}{16} \quad \textit{as } P_2(x) = \frac{1}{3}(3x^2 - 1)$$

For $n = 3$, $$a_3 = \frac{7}{2} \int_{0}^{1} x\, P_3\, dx = \frac{7}{2} \int_{0}^{1} x \cdot \frac{1}{2}(5x^3 - 3x)\, dx = \frac{7}{4} \int_{0}^{1} (5x^4 - 3x^2)\, dx$$

$$= \frac{7}{4}[x^5 - x^3]_0^1 = 0 \quad \text{as } P_3(x) = \frac{1}{2}(5x^3 - 3x)$$

For $n = 4$, $$a_4 = \frac{-3}{32}, \text{ etc.}$$

$\therefore$ $$f(x) = \frac{1}{4} P_0(x) + \frac{1}{2} P_1(x) + \frac{5}{16} P_2(x) - \frac{3}{32} P_4(x) + \ldots$$ **Ans.**

5. *Prove that* $x^5 = \dfrac{8}{63}\left[P_5 + \dfrac{7}{2} P_3 + \dfrac{27}{8} P_1\right]$

SOLUTION: Using Rodrigue's formula $P_n(x) = \dfrac{1}{2^n . n!} \dfrac{d^n}{dx^4} (x^2 - 1)^n$, we have

$$P_0(x) = 1,\ P_1(x) = x,\ P_2(x) = \frac{1}{2}(3x^2 - 1),\ P_3(x) = \frac{1}{2}(5x^3 - 3x), \ldots$$

Using the recurrence relation $(n + 1)P_{n+1} = (2x + 1)x\, P_n - nP_{n-1}$ ...(1)

and taking $n = 3$, we get $4P_4 = 7xP_3 - 3P_2$

or $$P_4(x) = \frac{1}{4}[7xP_3 - 3P_2] = \frac{1}{4}\left[\frac{7x}{2}(5x^3 - 3x) - \frac{3}{2}(3x^2 - 1)\right] = \frac{1}{8}\left[35x^4 - 30x^2 + 3\right]$$

and taking $n = 4$ is (1), we get $5P_5 = 9xP_4 - 4P_3$

$\therefore$ $$P_5 = \frac{1}{5}\left[\frac{9x}{8}(35x^4 - 30x^2 + 3) - \frac{4}{2}(5x^3 - 3x)\right]$$

$$= \frac{1}{5}\left[\frac{9.35x^5}{8} - \frac{9.30x^3}{8} + \frac{27}{8}x - 10x^3 + 6x\right]$$

$$= \frac{63}{8}x^5 - \left(\frac{54}{8}+2\right)x^3 + \frac{75x}{8.5} = \frac{63}{8}x^5 - \frac{70}{8}x^3 + \frac{15}{8}x$$

or $\quad 8P_5 = 63x^5 - 70x^3 + 15x$

Hence $\quad x^5 = \frac{1}{63}\left[8P_5 + \frac{70}{5}(2P_3 + 3P_1) - 15P_1\right] = \frac{8}{63}\left[P_5 + \frac{7}{2}P_3 + \frac{27}{8}P_1\right]$ **Hence proved.**

6. *Express (i) $f(x) = 10x^3 + 6x^2 - 2x + 1$ (ii) $f(x) = x^4 + 3x^3 - x^2 + 5x - 2$ in terms of Legendre polynomials.*

SOLUTION: (*i*) Since $P_0(x) = 1, P_1(x) = x, P_2(x) = \frac{1}{2}(3x^2 - 1), P_3(x) = \frac{1}{2}(5x^3 - 3x)$.

$\therefore \quad x^3 = \frac{1}{5}[2P_3(x) + 3P_1(x)],\ x^2 = \frac{1}{3}(2P_2 + P_0)$

$\therefore \quad f(x) = 10x^3 + 6x^2 - 2x + 1 = 4P_3(x) + 6P_1(x) + 4P_2 + 2P_0 - 2P_1 + P_0$

$= 4P_3 + 4P_2 + 4P_1 + 3P_0.$ **Ans.**

(*ii*) In the last problem we had

$$P_4 = \frac{1}{8}\left[35x^4 - 30x^2 + 3\right],\quad P_3 = \frac{1}{2}(5x^3 - 3x),\ P_2 = \frac{1}{2}(3x^2 - 1),\ P_1(x) = x, P_0(x) = 1$$

$\therefore \quad x^2 = \frac{1}{3}(2P_2 + P_0),\ x^3 = \frac{1}{5}[2P_3 + 3P_1],\ x^4 = \frac{1}{35}[8P_4 + 30x^2 + 3]$

or $\quad x^4 = \frac{1}{35}[8P_4 + 10(2P_2 + P_0) - 3P_0] = \frac{1}{35}[8P_4 + 20P_2 + 7P_0]$

Therefore $\quad f(x) = x^4 + 3x^3 - x^2 + 5x - 2$

$$= \frac{1}{35}[8P_4 + 20P_2 + 7P_0] + \frac{3}{5}[2P_3 + 3P_1] - \frac{1}{3}(2p_2 + p_0) + 5P_1 - 2P_0$$

$$= \frac{8}{35}P_4 + \frac{6}{5}P_3 - \frac{2}{21}P_2 + \frac{34}{5}P_1 - \frac{32}{15}P_0$$ **Ans.**

7. *Evaluate* $\int_{-1}^{1} P_n(x)(1 - 2hx + h^2)^{-1/2}\,dx$.

SOLUTION: We know that $(1 - 2hx + h^2)^{-1/2} = \sum_{r=0}^{\infty} P_r(x)\,h^r$

$$\therefore \quad \int_{-1}^{1} P_n(x)(1 - 2hx + h^2)^{-1/2}\,dx = \int_{-1}^{1} P_n(x)\sum_{r=0}^{\infty} P_r(x)\,h^r\,dx = \int_{-1}^{1} h^n P_n^2(x)\,dx + 0$$

(by orthogonality of Legendres polynomials) $= h^n\left(\frac{2}{2n+1}\right)$ **Ans.**

8. *Show that* $\int_{-1}^{1} x\, P_n\, P_{n-1}\, dx = \dfrac{2n}{4n^2 - 1}$.

SOLUTION: By the recurrence relation $(n+1)\, P_{n+1} = (2n+1)\, x\, P_n - n\, P_{n-1}$ we have

$$x\, P_n = \frac{1}{2n+1}[(n+1)\, P_{n+1} + n\, P_{n-1}]$$

$$\therefore \quad \int_{-1}^{1} x\, P_n\, P_{n-1}\, dx = \int_{-1}^{1} \frac{P_{n-1}}{2n+1}\{(n+1)\, P_{n+1} + n\, P_{n-1}\}$$

$$= \frac{n+1}{2n+1}\int_{-1}^{1} P_{n-1}\, P_{n+1}\, dx + \frac{n}{2n+1}\int_{-1}^{1} P_{n-1}^2\, dx = 0 + \frac{n}{2n+1}\cdot\frac{2}{2(n-1)+1}$$

(by the orthogonality property of Legendre polynomials)

$$= \frac{2n}{(2n+1)(2n-1)} = \frac{2n}{4n^2 - 1}$$

Hence proved.

9. *Find the Legendre polynomial expansion of the function* $f(x)$ *given by*

$$f(x) = -1, \ -1 < x < 0$$
$$= 1, \quad [0 < x < 1$$

SOLUTION: Let $f(x) = \sum_{0}^{\infty} a_k\, P_k(x)$ as Fourier Legendre expansion.

Multiplying both sides by P_n and integrating in $(-1, 1)$, we get

$$\int_{-1}^{1} f(x)\, P_n(x)\, dx = \int_{-1}^{1} P_n(x) \sum_{k=0}^{\infty} a_k\, P_k(x)\, dx$$

$$= \sum_{\substack{k=0 \\ k \neq n}}^{\infty} a_k \int_{-1}^{1} P_n(x)\, P_k(x)\, dx + a_n \int_{-1}^{1} P_n^2(x)\, dx = 0 + a_n \cdot \frac{2}{2n+1}$$

by orthogonality property of Legendre polynomials.

Next, $\int_{-1}^{1} f(x)\, P_n(x)\, dx = \int_{-1}^{0} -1\, P_n(x)\, dx + \int_{0}^{1} 1\, P_n(x)\, dx = \int_{0}^{1} P_n(x)\, dx + I_1$

where $I_1 = \int_{-1}^{0} -P_n(x)\, dx = \int_{1}^{0} P_n(-y)\, dy = -\int_{0}^{1} (-1)^n\, P_n(y)\, dy = (-1)^{n+1}\int_{0}^{1} P_n(x)\, dx$

(Putting $x = -y$ so $dx = -dy$)

Therefore, $\int_{-1}^{1} f(x)\, P_n(x)\, dx = \left[1 + (-1)^{n+1}\right]\int_{0}^{1} P_n(x)\, dx$ and, in turn, we have

$$a_n = \frac{2n+1}{2}[1 + (-1)^{n+1}]\int_{0}^{1} P_n(x)\, dx\,.$$

Also know that $\int_0^1 P_n(x)\,dx = \frac{1}{n+1} P_{n-1}(0)$

hence $a_n = \frac{2n+1}{2(n+1)}[1+(-1)^{n+1}]\,P_{n-1}(0) = 0$ when n is even.

And when n is odd, $a_1 = \frac{3}{2} P_0(0) = \frac{3}{2}$, $a_3 = \frac{7}{4} P_2(0) = \frac{7}{4}(-1)\frac{1}{2} = \frac{-7}{8}$, etc.

Thus, $f(x) = \frac{3}{2} P_1(x) - \frac{7}{8} P_3(x) - \ldots = \frac{3x}{2} - \frac{7}{8}\left(\frac{5x^3-3x}{2}\right) + \ldots$ **Ans.**

10. *Show that* $\int_x^1 \mathrm{P}_n(x)\,dx = \frac{1}{2n+1}[\mathrm{P}_{n-1}(x) - \mathrm{P}_{n+1}(x)]$.

Solution: By recurrence relation, we have

$$(2n+1)\,P_n(x) = P'_{n+1}(x) - P'_{n-1}(x)$$

$$\therefore \int_x^1 (2n+1)\,P_n(x)\,dx = [P_{n+1}(x) - P_{n-1}(x)]_x^1$$

or $$\int_x^1 P_n(x)\,dx = \frac{1}{2n+1}[\{P_{n+1}(1) - P_{n-1}(1)\} - \{P_{n+1}(x) - P_{n-1}(x)\}]$$

$$= \frac{1}{2n+1}[(1-1) - P_{n+1}(x) + P_{n-1}(x)]$$

or $$\int_x^1 P_n(x)\,dx = \frac{1}{2n+1}[P_{n-1}(x) - P_{n+1}(x)]$$ **Hence proved.**

ADDITIONAL IMPORTANT PROBLEMS

1. *If $P_n(x)$ denotes the nth order Legendre polynomial show that*

$$\frac{1-z^2}{(1-2xz+z^2)^{3/2}} = \sum_{n=0}^{\infty} (2n+1)\,P_n(x)\,z^n.$$

SOLUTION: We know that $(1 - 2xz + z^2)^{-1/2} = \sum_{n=0}^{\infty} z^n\,P_n(x)$ ...(1)

Differentiating w.r.t. z on both sides, we get

$$-\frac{1}{2}(1-2xz+z^2)^{-3/2}(-2x+2z) = \sum nz^{n-1}\,P_n(x)$$

or $$(x-z)(1-2xz+z^2)^{-3/2} = \sum nz^{n-1}\,P_n(x).$$

Multiplying throughout by 2z, gives

$$(2xz - 2z^2)\,(1 - 2xz + z^2)^{-3/2} = \sum 2nz^n\, P_n(x) \qquad ...(2)$$

Adding (1) and (2), gives

$$\frac{2xz - 2z^2}{(1-2xz+z^2)^{3/2}} + \frac{1}{(1-2xz+z^2)^{1/2}} = \sum (2n+1)\, z^n\, P_n(x)$$

or $$(2xz - 2z^2 + 1 - 2xz + z^2)\,(1-2xz+z^2)^{-3/2} = \sum (2n+1) z^n\, P_n(x)$$

or $$\frac{1-z^2}{(1-2xz+z^2)^{3/2}} = \sum (2n+1)\, z^n\, P_n(x)$$ **Hence proved.**

2. *If $P_n(x)$ stands for Legendre's polynomial of order n, show that*

(*a*) $P'_n(1) = \dfrac{n(n+1)}{2}$

(*b*) $P'_n(-1) = (-1)^{n-1} \cdot \dfrac{n(n+1)}{2}$.

SOLUTION: (*a*) Since $P_n(x)$ is the solution of the Legendre's equation

$$(1 - x^2)\, P''_n(x) - 2x\, P'_n(x) + n(n+1)\, P_n(x) = 0 \qquad ...(1)$$

Putting $x = 1$ in (1), we get $-2P'_n(1) + n(n+1)\, P_n(1) = 0$

Since $P_n(1) = 1$ we have $P'_n(1) = \dfrac{n(n+1)}{2}$.

(*b*) Putting $x = -1$ in (1), we get $-2(-1)\, P'_n(-1) + n(n+1)\, P_n(-1) = 0$

Since $P(-1) = (-1)^n$ we get $P'_n = \dfrac{-n(n+1)}{2}(-1)^n = \dfrac{n(n+1)}{2}(-1)^{n-1}$ **Hence proved.**

3. *Express x^4 in terms of Legendre polynomials of various orders*

SOLUTION: We know that $P_4(x) = \dfrac{1}{8}(35x^4 - 30x^2 + 3)$, $P_2(x) = \dfrac{1}{2}(3x^2 - 1)$, $P_0(x) = 1$

$$\therefore \quad x^4 = \frac{1}{35}\left[8P_4(x) + 30x^2 - 3\right] = \frac{1}{35}\left[8P_4(x) + \frac{30}{3}\{2P_2(x) + P_0\} - 3\right]$$

$$= \frac{1}{35}[8P_4(x) + 20P_2(x) + 10P_0 - 3P_0]$$

or $$x^4 = \frac{1}{35}[8P_4(x) + 20P_2(x) + 7P_0(x)]$$ **Ans.**

4. *Prove the Beltrami's result that*

$$(2n+1)\,(x^2 - 1)\, P'_n(x) = n(n+1)\,\{P_{n+1}(x) - P_{n-1}(x)\}$$

SOLUTION: The required result can be obtained from the following two recurrence relations:

$$(1 - x^2)\, P'_n(x) = n[P_{n-1}(x) - xP_n(x)] \quad \text{and} \quad (1 - x^2)\, P'_n(x) = (n+1)\, [xP_n(x) - P_{n+1}(x)]$$

or $\quad \frac{1}{n}(1-x^2)\,P_n'(x) = P_{n-1}(x) - xP_n(x)$ and $\frac{1}{n+1}(1-x^2)\,P_n'(x) = xP_n(x) - P_{n+1}(x)$

Adding the above two relations, we get

$$\left(\frac{1}{n}+\frac{1}{n+1}\right)(1-x^2)\,P_n'(x) = P_{n-1}(x) - P_{n+1}(x)$$

or $\quad (2n + 1)(1 - x^2)\,P'_n(x) = n(n + 1)\,[P_{n-1}(x) - P_{n+1}(x)]$ **Hence Proved**

5. If n^{th} derivative of a function $f(x)$ is continous in (–1, 1) then prove that

$$\int_{-1}^{1} f(x)\,P_n(x)\,dx = \frac{(-1)^n}{2^n \cdot n!}\int_{-1}^{1}(x^2-1)^n\,f^{(n)}(x)\,dx$$

SOLUTION: $\displaystyle\int_{-1}^{1} f(x)\,P_n(x)\,dx = \int_{-1}^{1} f(x)\cdot\frac{1}{2^n \cdot n!}\frac{d^n}{dx^n}(x^2-1)^n\,dx$ (using Rodrigue's formula)

$$= \frac{1}{2^n \cdot n!}\left[\left\{f(x)\frac{d^{n-1}}{dx^{n-1}}(x^2-1)^n\right\}_{-1}^{1} - \int_{-1}^{1} f'(x)\frac{d^{n-1}}{dx^{n-1}}(x^2-1)^n dx\right]$$

(on integrating by parts)

$$= \frac{1}{2^n \cdot n!}\left[0-\int_{-1}^{1} f'(x)\frac{d^{n-1}}{dx^{n-1}}(x^2-1)^n\,dx\right] = \frac{(-1)'}{2^n \cdot n!}\left[\int_{-1}^{1} f'(x)\frac{d^{n-1}}{dx^{n-1}}(x^2-1)^n\,dx\right]$$

(on integrating by parts again)

$$= \frac{(-1)^2}{2^n \cdot n!}\left[\int_{-1}^{1} f''(x)\frac{d^{n-2}}{dx^{n-2}}(x^2-1)^n\,dx\right] = \frac{(-1)^n}{2^n \cdot n!}\int_{-1}^{1} f^{(n)}(x)\,(x^2-1)^n\,dx$$

(On integrating by parts (n – 2) times) **Hence proved.**

6. *Prove that* $\displaystyle\int_{-1}^{1}(x^2-1)\,P_{n+1}\,P_n'\,dx = \frac{2n(n+1)}{(2n+1)(2n+3)}$.

SOLUTION: As a recurrence relation we have $(x^2 - 1)\,P'_n = n(xP_n - P_{n-1})$

$$\therefore \quad \int_{-1}^{1}(x^2-1)\,P_{n+1}\,P_n'\,dx = \int_{-1}^{1} n\,P_{n+1}\,(xP_n - P_{n-1})\,dx$$

$$= n\int_{-1}^{1} xP_n\,P_{n+1}\,dx - n\int_{-1}^{1} P_{n+1}\,P_{n-1}\,dx = n\int_{-1}^{1} x\,P_n\,P_{n+1}\,dx - 0$$

(by othogonality property of Legendre polynomials)

Using here another recurrence relation $(2n + 1)\,x\,P_n = (n + 1)\,P_{n+1} + n\,P_{n-1}$ we get

$$\int_{-1}^{1}(x^2-1)\,P_{n+1}P_n'dx = \frac{n}{2n+1}\int_{-1}^{1}\{(n+1)\,P_{n+1} + nP_{n-1}\}\,P_{n+1}\,dx$$

$$= \frac{n(n+1)}{2n+1}\int_{-1}^{1} P_{n+1}^2 dx + \frac{n^2}{2n+1}\int_{-1}^{1} P_{n-1}P_{n+1}\, dx$$

$$= \frac{n\,(n+1)}{2n+1}\int_{-1}^{1} P_{n+1}^2\, dx + 0 \qquad \text{(by orthogonality property)}$$

$$= \frac{n(n+1)}{2n+1}\,\frac{2}{2(n+1)+1} \qquad \left(\text{Since } \int_{-1}^{1} P_n^2\, dx = \frac{2}{2n+1}\right)$$

$$= \frac{2n\,(n+1)}{(2n+1)\,(2n+3)}$$

Hence proved.

7. *Show that* $\sum_{n=0}^{\infty} \frac{x^{n+1}}{n+1} P_n(x) = \frac{1}{2}\log\left(\frac{1+x}{1-x}\right)$.

SOLUTION: We know that $\sum_{0}^{\infty} h^n\, P_n(x) = (1-2xh+h^2)^{-1/2}$. Integrating both sides w.r.t. h from 0 to h, we get

$$\sum_{0}^{\infty} \frac{h^{n+1}}{n+1} P_n(x) = \int_0^h (1-2xh+h^2)^{-1/2}\, dh = \int_0^h \frac{dh}{\sqrt{(1-x^2)+(h-x)^2}} \quad \text{if } |x|<1.$$

$$= \left[\log \frac{(h-x)+\sqrt{h^2-2hx+1}}{\sqrt{1-x^2}}\right]_0^h = \log\left[\frac{h-x+\sqrt{h^2-2hx+1}}{1-x}\right]$$

Replacing h by x on both sides; we get

$$\sum_{0}^{\infty} \frac{x^{n+1}}{n+1} P_n(x) = \log \frac{0+\sqrt{(1-x^2)}}{1-x} = \log\sqrt{\frac{1+x}{1-x}}$$

$$= \frac{1}{2}\log\frac{1+x}{1-x}$$

Hence proved.

8. *Using the recurrence relation* $(n+1)\,P_{n+1}(x) = (2n+1)\,xP_n(x) - nP_{n-1}(x)$ *generate the Legendre polynomials* P_2, P_3, P_4 *given that* $P_0(x) = 1$ *and* $P_1(x) = x$.

SOLUTION: The recurrence relation is

$$(n+1)P_{n+1}(x) = (2n+1)\,xP_n(x) - nP_{n-1}(x)$$

For $n = 1$, it gives, $2P_2(x) = 3x\,P_1(x) - P_0(x) = 3x^2 - 1$, (using $P_0(x) = 1$ and $P_1(x) = x$)

or, $P_2(x) = (3x^2 - 1)/2.$

For $n = 2$, we have $3P_3(x) = 5x\,P_2(x) - 2P_1(x) = 5x(3x^2 - 1)/2 - 2x = 3(5x^3 - 3x)/2$

or, $P_3(x) = (5x^3 - 3x)/2.$

For $n = 3$, we have $4P_4(x) = 7x\,P_3(x) - 3P_2(x) = 7x(5x^3 - 3x)/2 - 3(3x^2 - 2)/2$

$$= (35x^4 - 30x^2 + 3)/2$$

or $P_4(x) = (35x^4 - 30x^2 + 3)/8.$

9. *Let $f(x)$ be a polynomial of degree m and let $n > m$, then show that*

$$\int_{-1}^{1} f(x)\,P_n(x)\,dx = 0.$$

SOLUTION: Since $f(x)$ is a polynomial of degree m, we have

$$f(x) = c_0\,P_0(x) + c_1\,P_1(x) + \dots + c_m\,P_m(x), \text{ for some } c\text{'s}.$$

Multiplying it with $P_n(x)$ and integrating w.r.t. x over the interval $[-1, 1]$, we get

$$\int_{-1}^{1} f(x)\,P_n(x)\,dx = \sum_{k=0}^{m} c_k \int_{-1}^{1} P_k(x)\,P_n(x)\,dx = 0$$

using the orthogonality property of the Legendre polynomials as $k \neq n$.

10. *Show that*

$$\int_{-1}^{1} x^2 P_{n-1}(x)\,P_{n+1}(x)\,dx = \frac{2n(n+1)}{(2n-1)(2n+1)(2n+3)}$$

SOLUTION: Consider the recurrence relation $(2n + 1)\,xP_n = (n + 1)\,P_{n+1} + nP_{n-1}$.

Replacing n by $(n - 1)$ and $(n + 1)$, respectively, we get

$$xP_{n-1} = \frac{1}{2n-1}[nP_n + (n-1)\,P_{n-2}]$$

and, $$xP_{n+1} = \frac{1}{2n+3}[(n+2)\,P_{n+2} + (n+1)\,P_n].$$

Multiplying these two and integrating w.r.t. x over $[-1, 1]$ and using orthogonality property, we get

$$\int_{-1}^{1} x^2 P_{n-1}(x)\,P_{n+1}(x)\,dx = \frac{1}{(2n-1)(2n+3)}\cdot n(n+1)\int_{-1}^{1} P_n^2(x)\,dx,$$

Thus,

$$\int_{-1}^{1} x^2 P_{n-1}(x)\,P_{n+1}(x)\,dx = \frac{2n(n+1)}{(2n-1)(2n+1)(2n+3)}.$$ **Hence Proved**

11. *Show that* $\int_{-1}^{1} f(x)\, P_n(x)\, dx = \dfrac{(-1)^n}{2^n \cdot n!} \int_{-1}^{1} f^{(n)}(x) \cdot (x^2-1)^n\, dx.$

Hence, deduce that $$\int_{-1}^{1} x^m P_n(x)\, dx = \begin{cases} 0, & \text{if} \quad m < n \\ \dfrac{2^{n+1}(n!)^2}{(2n+1)!}, & \text{if} \quad m = n \end{cases}$$

SOLUTION: Using Rodrigue's formula, we have

$$\int_{-1}^{1} f(x)\, P_n(x)\, dx = \frac{1}{2^n\, n!} \int_{-1}^{1} f(x) \frac{d^n}{dx^n} (x^2-1)^n\, dx \qquad \text{(Integrating by parts)}$$

$$= \frac{1}{2^n\, n!} \left[\left\{ f(x) \frac{d^{n-1}}{dx^{n-1}} (x^2-1)^n \right\}_{-1}^{1} - \int_{-1}^{1} f'(x) \frac{d^{n-1}}{dx^{n-1}} (x^2-1)^n\, dx \right]$$

$$= \frac{-1}{2^n \cdot n!} \int_{-1}^{1} f'(x) \cdot \frac{d^{n-1}}{dx^{n-1}} (x^2-1)^n\, dx, \text{ since the first term is zero at } x = \pm 1.$$

Integrating by parts further $(n-1)$ times, we obtain

$$\int_{-1}^{1} f(x)\, P_n(x) = \frac{(-1)^n}{2^n \cdot n!} \int_{-1}^{1} f^{(n)}(x)\, (x^2-1)^n\, dx.$$

Writing $f(x) = x^m$, so that $f^{(n)}(x) = 0$, when $m < n$ and, $f^{(n)}(x) = n!$, when $m = n$.

Thus, $$\int_{-1}^{1} x^m P_n(x)\, dx = 0 + \frac{(-1)^n}{2^n\, n!} \int_{-1}^{1} n!(x^2-1)^n\, dx = \frac{2}{2^n} \int_{0}^{1} (1-x^2)^n\, dx$$

$$= \frac{2}{2^n} \int_{0}^{\pi/2} \cos^{(2n+1)} \theta\, d\theta, \text{ (using } x = \sin\theta)$$

$$= \frac{2}{2^n} \left[\frac{(2n)(2n-2)\ldots 4.2.}{(2n+1)(2n-1)\ldots 3.1} \right] = \frac{2}{2^n} \cdot \left[\frac{2^{2n}(n!)^2}{(2n+1)^n} \right]$$

$$= \frac{2^{n+1}(n!)^2}{(2n+1)!}$$

Hence Proved

12. *Expand* $f(x) = \begin{cases} 0, & -1 < x \le 0 \\ x, & 0 < x < 1 \end{cases}$ *as a Fourier-Legendre series.*

SOLUTION: Let $f(x) = \sum_{n=0}^{\infty} c_n P_n(x)$, where c_i's are Fourier-Legendre coefficients, given by

$$c_n = \frac{2n+1}{2}\int_{-1}^{1} f(x)\,P_n(x)\,dx, \quad n = 0, 1, 2, \ldots$$

For given $f(x)$, $$c_n = \frac{2n+1}{2}\int_0^1 xP_n(x)\,dx.$$

Thus, $$c_0 = \frac{1}{2}\int_0^1 xP_0(x)\,dx = \frac{1}{2}\int_0^1 x\,dx = \frac{1}{4}, \quad c_1 = \frac{3}{2}\int_0^1 xP_1(x)\,dx = \frac{3}{2}\int_0^1 x^2\,dx = \frac{1}{2},$$

$$c_2 = \frac{5}{2}\int_0^1 xP_2(x)\,dx = \frac{5}{2}\int_0^1 x\left(\frac{3x^2-1}{2}\right)dx = \frac{5}{16},$$

$$c_3 = \frac{7}{2}\int_0^1 xP_3(x)\,dx = \frac{7}{2}\int_0^1 x\left(\frac{5x^3-3x}{2}\right)dx = 0$$

and $$c_4 = \frac{9}{2}\int_0^1 xP_4(x)\,dx = \frac{9}{2}\int_0^1 x\left(\frac{35x^4-30x^2+3}{8}\right)dx = -\frac{3}{32}$$ and so on.

Hence, $$f(x) = \frac{1}{4}P_0(x) + \frac{1}{2}P_1(x) + \frac{5}{16}P_2(x) - \frac{3}{32}P_4(x) + \ldots$$

13. *Expand* $f(x) = \cos\frac{\pi x}{2}$ *as a Fourier-Legendre series.*

SOLUTION: Let $f(x) = \sum_{n=0}^{\infty} c_n P_n(x)$, where c_i's are Fourier-Legendre coefficients, given as

$$c_n = \frac{2n+1}{2}\int_{-1}^{1} f(x)\,P_n(x)\,dx, \quad n = 0, 1, 2, \ldots$$

For $f(x) = \cos\frac{\pi x}{2}$, $$c_n = \frac{2n+1}{2}\int_{-1}^{1}\cos\frac{\pi x}{2}\,P_n(x)\,dx$$

CHAPTER

23 First Order Partial Differential Equations

IMPORTANT FORMULAE AND USEFUL RESULTS

1. **Linear Lagrangian Form:** $Pp + Qq = R$ where $p = \frac{\partial z}{\partial x}$, $q = \frac{\partial z}{\partial y}$. and P, Q, R are functions of x, y, z. It is called linear in p and q. To solve it we form auxiliary equation $\frac{dx}{P} = \frac{dy}{Q} = \frac{dz}{R}$.

   If $u = a$ and $v = b$ are two independent solutions then the solution of the above equation is $\phi(u, v) = 0$.

2. **Non-Linear Equation:** $F(x, y, z, p, q) = 0$

   Type I: $F(p, q) = 0$. Its solution is $z = ax + by + c$ where $F(a, b) = 0$.

   Type II: $z = px + qy + f(p, q)$

   Its solution is $z = ax + by + f(a, b)$.

   Type III: $F(p, q, z) = 0$ setting $t = x + ay$. we get $F\left(\frac{dz}{dt}, a\frac{dz}{dt}, z\right) = 0$

   $\therefore$ Solution $z = \phi(x + ay, b)$.

   Type IV: $F_1(x, p) = F_2(y, q) \Rightarrow F_1(x, p) = F_2(y, q) = a$

   Solve these for p and q as $p = \phi_1(x, a)$, $q = \phi_2(y, a)$

   Put them in $dz = p\,dx + q\,dy$ to get $z = \int \phi_1(x, a)\,dx + \int \phi_2(y, a)\,dy + b$

   Type V: Charpit's Method.

   Equation is $f(x, y, z, p, q) = 0$. Hence we have an auxiliary system of equation as

$$\frac{dp}{\frac{\partial f}{\partial x} + p\frac{\partial f}{\partial z}} = \frac{dq}{\frac{\partial f}{\partial y} + q\frac{\partial f}{\partial z}} = \frac{dz}{-p\frac{\partial f}{\partial p} - q\frac{\partial f}{\partial q}} = \frac{dx}{\frac{-\partial f}{\partial p}} = \frac{dy}{\frac{-\partial f}{\partial q}} = \frac{\partial F}{0}.$$

SOLUTIONS TO EXERCISE 23A

1. *Form a differential equation of all planes which are at a constant distance 'a' from the origin.*

SOLUTION: Equation of the plane which is at a constant distance '*a*' from the origin is

$$lx + my + nz = a \quad ...(1) \quad \text{where } l^2 + m^2 + n^2 = 1$$

or $lx + my + z\sqrt{1-l^2-m^2} = a.$

Differentiating it partially w.r.t. x and also y, we get

$$l + \sqrt{1-l^2-m^2}\; p = 0 \quad \text{and} \quad m + \sqrt{1-l^2-m^2}\; q = 0$$

or $\quad l^2 = (1 - l^2 - m^2)\, p^2 \quad$ and $\quad m^2 = (1 - l^2 - m^2)\, q^2$

On adding these, we get $\quad l^2 + m^2 = (1 - l^2 - m^2)\,(p^2 + q^2)$

or $\quad l^2 + m^2 = \dfrac{p^2+q^2}{1+p^2+q^2} \quad$ and $\quad 1 - l^2 - m^2 = \dfrac{1}{1+p^2+q^2}$

$\Rightarrow \quad l = \dfrac{p}{\sqrt{1+p^2+q^2}} \quad$ and $\quad m = \dfrac{q}{\sqrt{1+p^2+q^2}}$

Substituting for l, m and $n = \sqrt{1-l^2-m^2}$ in (1), gives $\dfrac{px+qy}{\sqrt{1+p^2+q^2}} + \dfrac{z}{\sqrt{1+p^2+q^2}} = a$

or $\quad px + qy + z = a\sqrt{1+p^2+q^2}\quad$ is the required differential equation. **Ans.**

2. *Solve* $\quad px + qy = nz$

SOLUTION: It is of the Lagrangian form $Pp + Qq = R$ where P, Q, R are all functions of x, y, z. Its subsidiary equation is

$$\frac{dx}{P} = \frac{dy}{Q} = \frac{dz}{R} \quad \text{or} \quad \frac{dx}{x} = \frac{dy}{y} = \frac{dz}{nz}$$

Equation $\dfrac{dx}{x} = \dfrac{dy}{y}$ gives $\log x = \log y - \log c_1 \quad$ or $\quad \dfrac{y}{x} = c_1$

Also $\quad \dfrac{dz}{z} = \dfrac{n\,dx}{x}$ gives $\log z = n \log x + \log c_2 \quad$ or $\quad z = c_2\, x^n$

$\therefore$ We can write $\quad z = x^n \cdot \phi\left(\dfrac{y}{x}\right)$ **Ans.**

3. *Solve* $y^2p - xyq = x(z - 2y)$

SOLUTION: Given equation is of Lagrangian form $Pp + Qq = R$. Its subsidiary equation is

$$\frac{dx}{P} = \frac{dy}{Q} = \frac{dz}{R} \quad \text{or} \quad \frac{dx}{y^2} = \frac{dy}{-xy} = \frac{dz}{x(z-2y)}$$

One solution is obtained from $\dfrac{dx}{y} = \dfrac{dy}{-x}$ or $xdx + ydy = 0$ as $x^2 + y^2 = C_1$

Second solution is obtained from $\dfrac{dy}{-y} = \dfrac{dz}{z-2y}$ or $ydz + zdy = 2y\,dy$

as $$yz = y^2 + C_2$$

$\therefore$ Complete solution is $yz - y^2 = f(x^2 + y^2)$ where f is an arbitrary function. **Ans.**

4. *Solve* $px(z - 2y^2) = (z - qy)(z - y^2 - 2x^3)$

Solution: Equation can be written as $Pp + Qq = R$

where $P = x(z - 2y^2)$, $Q = y(z - y^2 - 2x^3)$ and $R = z(z - y^2 - 2x^3)$.

Substidiary equation is $\dfrac{dx}{P} = \dfrac{dy}{Q} = \dfrac{dz}{R}$

or $$\frac{dx}{x(z-2y^2)} = \frac{dy}{y(z-y^2-2x^3)} = \frac{dz}{z(z-y^2-2x^3)} \qquad ...(1)$$

From $\dfrac{dy}{Q} = \dfrac{dz}{R}$ we get $\dfrac{dy}{y} = \dfrac{dz}{z}$

Hence we get one solution as $\log\dfrac{y}{z} = \log C_1$ or $\dfrac{y}{z} = C_1$

Now using multipliers in the first two and last two in (1), gives

$$\frac{z\,dz - x\,dx}{z - y^2 - 2x^3 - z + 2y^2} = \frac{y\,dy - x\,dx}{z - y^2 - 2x^3 - z + 2y^2}$$

or $$\frac{z\,dz - x\,dx}{y^2 - 2x^3} = \frac{y\,dy - x\,dx}{y^2 - 2x^3} \quad \text{or} \quad z\,dz - x\,dx = y\,dy - x\,dx$$

or $z\,dz = y\,dy = 0$ $\therefore$ $z^2 - y^2 = c_2$. This is the other solution.

$\therefore$ Complete solution is $f\left(\dfrac{y}{z}, z^2 - y^2\right) = 0$ or $z^2 - y^2 = \phi\left(\dfrac{y}{z}\right)$ **Ans.**

5. *Solve* $z = px + qy + c\sqrt{1 + p^2 + q^2}$

SOLUTION: The given equation is of the form $z = px + qy + f(p, q)$.

where $p = \dfrac{\partial z}{\partial x}$ and $q = \dfrac{\partial z}{\partial y}$. Its solution is $z = ax + by + f(a, b)$

Therefore, the solution of the given partial differential equation, is

$z = ax + by + c\sqrt{1 + a^2 + b^2}$ where c is the given constant and a and b are arbitrary constants of integration.

6. *Solve* $p^2 + q^2 = npq$

SOLUTION: The equation is non-linear of the form $f(p, q) = 0$.

Let its solution be taken as $z = ax + by + c$

$\therefore\ p = a,\ q = b$ hence $f(p, q) = a^2 + b^2 - nab = 0$

Hence the solution is $z = ax + by + c$ where $a^2 + b^2 = nab$

And this has only two arbitrary constants a and b. **Ans.**

7. *Solve* $z = pq.$

SOLUTION: Let z be a function of a variable t where $t = x + ay$.

We write $z = f(t)$. Now $p = \dfrac{\partial z}{\partial x} = \dfrac{dz}{dt}\cdot 1$ and $q = \dfrac{\partial z}{\partial y} = \dfrac{dz}{dt}\cdot a$

Therefore we have $$z = \frac{dz}{dt}\cdot a\frac{dz}{dt} \quad \text{or} \quad \frac{dz}{dt} = \sqrt{\frac{z}{a}}$$

or $$\frac{dz}{\sqrt{z}} = \frac{1}{\sqrt{a}}\,dt \quad \text{which on integration, gives}$$

$$2\sqrt{a}\sqrt{z} = t + b \quad \text{or} \quad 4az = (x + ay + b)^2$$ where a and b are arbitrary cosntants. **Ans.**

8. *Solve* $p^2 - pq = 1 - z^2$

SOLUTION: The equation is of the form $F(p, q, z) = 0$. We take z as a function of t where $t = x + ay$. Therefore $\dfrac{\partial z}{\partial x} = p = \dfrac{dz}{dt}$ and $\dfrac{\partial z}{\partial y} = q = a\dfrac{dz}{dt}$. Then the given equation becomes

$$\left(\frac{dz}{dt}\right)^2 - a\left(\frac{dz}{dt}\right)^2 = 1 - z^2 \quad \text{or} \quad \frac{dz}{\sqrt{1-z^2}} = \frac{dt}{\sqrt{1-a}} \quad \text{which on integration gives}$$

$$\sin^{-1} z = \frac{t}{\sqrt{1-a}} + b \quad \text{or} \quad z = \sin\left(\frac{x + ay}{\sqrt{1-a}} + b\right)$$ **Ans.**

9. *Solve* $q = -px + p^2$

SOLUTION: The given equation is of the form $F_1(x, p) = F_2(y, q)$.

We can write $F_1(x, p) = a,\ F_2(y, q) = a$

or $q = a$ and $p^2 - px = a$ where a is an arbitrary constant.

$$p^2 - px - a = 0 \quad \text{gives} \quad p = \frac{x \pm \sqrt{x^2 + 4a}}{2}, \quad q = a.$$

Also, $$dz = \frac{\partial z}{\partial x}dx + \frac{\partial z}{\partial y}dy = p\,dx + q\,dy = \frac{x}{2} \pm \frac{1}{2}\sqrt{x^2 + 4a}\,dx + a\,dy$$

On integration we get $z = \int \left(\frac{x}{2} \pm \frac{1}{2}\sqrt{x^2 + 4a} \right) dx + ay + c$

or $z = \frac{x^2}{4} + \frac{1}{4}\left\{ x\sqrt{x^2 + 4a} + a \log\left(x + \sqrt{x^2 + 4a} \right) \right\} + ay + c$ **Ans.**

10. *Solve* $pq + qx = y$

SOLUTION: The gives equation is $p + x = \frac{y}{q}$ which is of the form $F_1(p, x) = F_2(q, y)$.

We write $F_1(p, x) = F_2(q, y) = a$ constant.

or $p + x = a$ and $\frac{y}{q} = a.$

Using $dz = p\,dx + q\,dy$ we get $dz = (a - x)\,dx + \frac{y}{a}\,dy$

which on integration gives the solution as $z = \frac{-1}{2}(a - x)^2 + \frac{y^2}{2a} + b$

or $2az + a(a - x)^2 - y^2 = 2ab$ where a and b are arbitrary constants. **Ans.**

11. *Solve the equation* $p(1 + q) = qz.$

SOLUTION: The equation is of the form $f(p, q, z) = 0$

We take $t = x + ay$ and let z be a function of t where a is constant.

$\therefore \quad p = \frac{\partial z}{\partial x} = \frac{dz}{dt} \cdot \frac{\partial t}{\partial x} = \frac{dz}{dt}$ and $q = \frac{\partial z}{\partial y} = \frac{dz}{dt} \cdot \frac{\partial t}{\partial y} = a\frac{dz}{dt}$

Putting these values of p and q in the given equation, we get

$$\frac{dz}{dt}\left(1 + a\frac{dz}{dt}\right) = az\frac{dz}{dt}$$

Here $\frac{dz}{dt} \neq 0$ hence $1 + a\frac{dz}{dt} = az$ or $dt = \frac{a\,dz}{az - 1}$ $\therefore \quad t = \log(az - 1) - b$

where a and b are arbitrary constants.

$\therefore \quad x + ay = \log(az - 1) - b$ or $\log(az - 1) = x + ay + b$ **Ans.**

12. *Solve* $z^2(p^2 + q^2) = x^2 + y^2$ *where* $p = \frac{\partial z}{\partial x}, q = \frac{\partial z}{\partial y}$

SOLUTION: The given equation is

$\left(z\frac{\partial z}{\partial x}\right)^2 + \left(z\frac{\partial z}{\partial y}\right)^2 = x^2 + y^2.$ Let us write $Z = \frac{z^2}{2}$ then $P = \frac{\partial Z}{\partial x} = z\frac{\partial z}{\partial x}$ and $Q = \frac{\partial Z}{\partial y} = z\frac{\partial z}{\partial y}$

Hence the given equation can be written as

$$P^2 + Q^2 = x^2 + y^2 \quad \text{or} \quad P^2 - x^2 = -Q^2 + y^2$$

which is of the form $F_1(x, P) = F_2(y, Q)$

$\Rightarrow \qquad P^2 - x^2 = -Q^2 + y^2 = a,$ where a is an arbitrary constant.

or $\qquad P = \sqrt{x^2 + a}, \qquad Q = \sqrt{y^2 - a}.$

Substituting these in $dZ = P\,dx + Q\,dy$, we get $dZ = \sqrt{a + x^2}\;dx + \sqrt{y^2 - a}\;dy$

This on integration, gives

$$Z = \frac{x}{2}\sqrt{x^2 + a} + \frac{a}{2}\log\left(x + \sqrt{x^2 + a}\right) + \frac{y}{2}\sqrt{y^2 - a} - \frac{a}{2}\log\left(y + \sqrt{y^2 - a}\right)$$

or $$z^2 = x\sqrt{x^2 + a} + y\sqrt{y^2 - a} + a\log\left(\frac{x + \sqrt{x^2 + a}}{y + \sqrt{y^2 - a}}\right) + C$$ **Ans.**

13. *Solve* $q^2y^2 = z(z - px)$

SOLUTION: Let us put $X = \log x$, $Y = \log y$ then

$$p = \frac{\partial z}{\partial x} = \frac{\partial z}{\partial X}\frac{dX}{dx} = \frac{1}{x}\frac{\partial z}{\partial X} \quad \therefore \quad \frac{\partial z}{\partial X} = px \text{ and similarly } \frac{\partial z}{\partial Y} = qy.$$

Thus the given equation becomes $\qquad Q^2 = z^2 - Pz \qquad$ ...(1)

which is of the form $F(P, Q, Z) = 0$.

Let z be a function of t where $t = X + aY$ where a is an arbitrary constant.

then $\quad P = \dfrac{\partial z}{\partial X} = \dfrac{dz}{dt}\cdot\dfrac{\partial t}{\partial X} = \dfrac{dz}{dt}\cdot 1 \qquad \therefore \quad Q = \dfrac{\partial z}{\partial Y} = \dfrac{dz}{dt}\cdot\dfrac{\partial t}{\partial Y} = a\dfrac{dz}{dt}$

$\therefore$ (1) becomes

$$a^2\left(\frac{dz}{dt}\right)^2 = z^2 - z\frac{dz}{dt} \quad \text{or} \quad a^2\left(\frac{dz}{dt}\right)^2 + z\left(\frac{dz}{dt}\right) - z^2 = 0$$

$$\therefore \quad \frac{dz}{dt} = \frac{-z \pm \sqrt{z^2 + 4a^2 z^2}}{2a^2} \quad \text{or} \quad \frac{dz}{z} = \frac{-1 \pm \sqrt{1 + 4a^2}}{2a^2}\,dt$$

$\therefore \quad K\log z = t \pm \log b$ where $K = \dfrac{2a^2}{-1 \pm \sqrt{4a^2 + 1}}$ and b is constant.

or $\quad K\log z = X + aY + \log b = \log x + a\log y + \log b$

or $\qquad z^K = b\,xy^a$ is the solution where a and b are arbitrary constants. **Ans.**

14. *Solve* $\quad p^2 - q^2 = \dfrac{x - y}{z}$

SOLUTION: Substituting $Z = z^{3/2}$ or $z = Z^{2/3}$, we get

$$p = \frac{\partial z}{\partial x} = \frac{\partial}{\partial x}Z^{2/3} = \frac{2}{3}\frac{1}{Z^{1/3}}\frac{\partial Z}{\partial x} = \frac{2}{3\sqrt{z}}\frac{\partial Z}{\partial x} \quad \text{and} \quad q = \frac{2}{3\sqrt{z}}\frac{\partial Z}{\partial y}$$

$$\therefore \quad z(p^2 - q^2) = \frac{4}{9}\left[\left(\frac{\partial Z}{\partial x}\right)^2 - \left(\frac{\partial Z}{\partial y}\right)^2\right] = x - y$$

or $$P^2 - Q^2 = \frac{9}{4}(x-y) \quad \text{where } P = \frac{\partial Z}{\partial x} \text{ and } Q = \frac{dZ}{\partial y}$$

or $$P^2 - \frac{9}{4}x = Q^2 - \frac{9}{4}y = a, \quad \text{where } a \text{ is a constant.}$$

Thus, $$P = \sqrt{\frac{9}{4}x + a} \quad \text{and} \quad Q = \sqrt{\frac{9}{4}y + a}.$$

Next, since $dZ = P\,dx + Q\,dy$, we get

$$dZ = \sqrt{\frac{9}{4}x + a}\,dx + \sqrt{\frac{9}{4}y + a}\,dy \quad \text{which, on integration, gives}$$

$$Z = \frac{1}{2}\cdot\frac{2}{3}(9x+4a)^{3/2} + \frac{1}{2}\cdot\frac{2}{3}(9y+4a)^{3/2} + b \quad \text{where } b \text{ is an arbitrary constant.}$$

or $$z^{3/2} = \frac{1}{3}(9x+4a)^{3/2} + \frac{1}{3}(9y+4a)^{3/2} + b$$ **Ans.**

15. *Solve, using Charpits method,* $z = p^2x + q^2y$ *or otherwise*

SOLUTION: Let us put $x = X^2$ and $y = Y^2$ then

$$p = \frac{\partial z}{\partial x} = \frac{\partial z}{\partial X}\frac{dX}{dx} = \frac{1}{2\sqrt{x}}\frac{\partial z}{\partial X} = \frac{P}{2\sqrt{x}} \quad \text{where } P = \frac{\partial z}{\partial X}$$

and $$q = \frac{\partial z}{\partial y} = \frac{\partial z}{\partial Y}\cdot\frac{dY}{dy} = \frac{1}{2\sqrt{y}}\frac{\partial z}{\partial Y} = \frac{Q}{2\sqrt{y}} \quad \text{where } Q = \frac{\partial z}{\partial Y}$$

$\therefore$ Given equation becomes $\dfrac{P^2}{4} + \dfrac{Q^2}{4} = z$ which is of the form $F(P, Q, z) = 0$ ...(1)

Now, Let $t = X + aY$ where z is a function of t, then

$$P = \frac{\partial z}{\partial X} = \frac{dz}{dt}\frac{\partial t}{\partial X} = \frac{dz}{dt} \quad \text{and} \quad Q = \frac{\partial z}{\partial y} = \frac{dz}{dt}\frac{\partial t}{\partial y} = a\frac{dz}{dt}$$

Putting these values of P and Q in (1), we get

$$\left(\frac{dz}{dt}\right)^2 + a^2\left(\frac{dz}{dt}\right)^2 = 4z$$

or $$\frac{dz}{2\sqrt{z}} = \frac{dt}{\sqrt{1+a^2}} \quad \text{which on integration, gives}$$

$$\sqrt{z} = \frac{t}{\sqrt{1+a^2}} + b = \frac{X+aY}{\sqrt{1+a^2}} + b$$

or $$\sqrt{z} = \frac{\sqrt{x} + a\sqrt{y}}{\sqrt{1+a^2}} + b$$ **Ans.**

The given problem can be solved by charpit's method as well.

16. *Solve, using Charpits method,* $p^2 + qy = z$

SOLUTION: The given equation is $f \equiv p^2 + qy - z = 0$

Charpit's auxiliary equations are

$$\frac{dp}{\frac{\partial f}{\partial x} + p\frac{\partial f}{\partial z}} = \frac{dq}{\frac{\partial f}{\partial y} + q\frac{\partial f}{\partial z}} = \frac{dz}{-p\frac{\partial f}{\partial p} - q\frac{\partial f}{\partial q}} = \frac{dx}{\frac{-\partial f}{\partial p}} = \frac{dy}{\frac{-\partial f}{\partial q}} = \frac{df}{0}$$

or $$\frac{dp}{0 + p(-1)} = \frac{dq}{q + q(-1)} = \frac{dz}{-p(2p) - q(y)} = \frac{dx}{-2p} = \frac{dy}{-y} = \frac{df}{0}$$

or $$\frac{dp}{-p} = \frac{dq}{0} = \frac{dz}{-2p^2 - qy} = \frac{dx}{-2p} = \frac{dy}{-y}$$

$\Rightarrow\ dq = 0 \quad \therefore \quad q = a,$ an arbitrary constant.

Then given equation becomes $p^2 = z - ay$ or $p = \sqrt{z - ay}$

Now $$dz = pdx + qdy = \sqrt{z - ay}\,dx + ady$$

or $$\frac{dz - a\,dy}{\sqrt{z - ay}} = dx \quad \text{or} \quad \frac{d(z - ay)}{\sqrt{z - ay}} = dx$$

On integration, we get $x = 2\sqrt{z - ay} + b$ or $z = ay + \left(\frac{x - b}{2}\right)^2$ **Ans.**

17. *Solve* $2z + p^2 + qy + 2y^2 = 0$

SOLUTION: We apply Charpit's method here to solve it.

Let $$f(x, y, z, p, q) = 2z + p^2 + qy + 2y^2.$$

The subsidiary equations are

$$\frac{dx}{\frac{-\partial f}{\partial p}} = \frac{dy}{\frac{-\partial f}{\partial q}} = \frac{dz}{-p\frac{\partial f}{\partial p} - q\frac{\partial f}{\partial q}} = \frac{dp}{\frac{\partial f}{\partial x} + p\frac{\partial f}{\partial z}} = \frac{dq}{\frac{\partial f}{\partial y} + q\frac{\partial f}{\partial z}}$$

or $$\frac{dx}{-2p} = \frac{dy}{-y} = \frac{dz}{-2p^2 - qy} = \frac{dp}{2p} = \frac{dq}{2q + 4y}.$$

$\Rightarrow\ dp = -dx$ or $p = -x + a$ where a is arbitrary constant. Putting this value of p in the given equation, we get $$q = \frac{1}{y}[-2z - 2y^2 - (a - x)^2]$$

$\therefore$ $dz = p\,dx + q\,dy$ gives $dz = (a - x)\,dx + \frac{1}{y}[-2z - 2y^2 - (a - x)^2]\,dy$

Multiplying throughout by $2y^2$, gives

$$2y^2\,dz + 4yz\,dy = 2y^2(a - x)\,dx - 4y^3\,dy - 2y(a - x)^2\,dy$$

On integration, we get $2y^2 z = -y^2(a - x)^2 - y^4 + b$ where a, b are constants. **Ans.**

18. *Solve* $(x + y)(p + q)^2 + (x - y)(p - q)^2 = 1$

SOLUTION: Let us put $x + y = X^2$ and $x - y = Y^2$, then

$$p = \frac{\partial z}{\partial x} = \frac{\partial z}{\partial X}\frac{\partial X}{\partial x} + \frac{\partial z}{\partial Y}\frac{\partial Y}{\partial x} = \frac{\partial z}{\partial X}\frac{1}{2X} + \frac{\partial z}{\partial Y}\cdot\frac{1}{2Y}$$

and
$$q = \frac{\partial z}{\partial y} = \frac{\partial z}{\partial X}\frac{\partial X}{\partial y} + \frac{\partial z}{\partial Y}\frac{\partial Y}{\partial y} = \frac{\partial z}{\partial X}\frac{1}{2X} - \frac{\partial z}{\partial Y}\frac{1}{2Y}$$

$\therefore$
$$p + q = \frac{1}{X}\frac{\partial z}{\partial X} \quad \text{and} \quad p - q = \frac{1}{Y}\frac{\partial z}{\partial Y}.$$

Putting these values in the given equation, we get

$$X^2 \cdot \frac{1}{X^2}\left(\frac{\partial z}{\partial X}\right)^2 + Y^2 \cdot \frac{1}{Y^2}\left(\frac{\partial z}{\partial Y}\right)^2 = 1 \quad \text{or} \quad \left(\frac{\partial z}{\partial X}\right)^2 + \left(\frac{\partial z}{\partial Y}\right)^2 = 1$$

$\therefore$ The complete solution is $z = aX + bY + c$ where $a^2 + b^2 = 1$ $\therefore$ $b = \sqrt{1-a^2}$

$\therefore$
$$z = aX + \sqrt{1-a^2}\, Y + c = a\sqrt{x+y} + \sqrt{1-a^2}\,\sqrt{x-y} + c.$$
Ans.

19. *Solve* $4xyz = pq + 2px^2y + 2qxy^2$

SOLUTION: Putting $x = \sqrt{X}$, $y = \sqrt{Y}$ we have

$$p = \frac{\partial z}{\partial x} = \frac{\partial z}{\partial X}\frac{\partial X}{\partial x} = 2\sqrt{X}\frac{\partial z}{\partial X}, \quad \text{and} \quad q = \frac{\partial z}{\partial y} = \frac{\partial z}{\partial Y}\cdot\frac{\partial Y}{\partial y} = 2\sqrt{Y}\frac{\partial z}{\partial y}$$

$\therefore$ Equation becomes

$$z = \frac{pq}{4xy} + \frac{px}{2} + \frac{qy}{2} = \frac{\partial z}{\partial X}\frac{\partial z}{\partial Y} + X\frac{\partial z}{\partial X} + Y\frac{\partial z}{\partial Y}$$

or
$$z = PQ + PX + QY \quad \text{where} \quad P = \frac{\partial z}{\partial X},\ Q = \frac{\partial z}{\partial Y}$$

Its solution is $z = aX + bY + ab$ or $z = ax^2 + by^2 + ab$ **Ans.**

20. *Use Charpit's method to solve* $16p^2z^2 + 9q^2z^2 + 4z^2 - 4 = 0$

SOLUTION: $f(p, q, x, y, z) = 16p^2z^2 + 9q^2z^2 + 4z^2 - 4$

By Charpit's method, we have

$$\frac{dp}{\frac{\partial f}{\partial x} + p\frac{\partial f}{\partial z}} = \frac{dq}{\frac{\partial f}{\partial y} + q\frac{\partial f}{\partial z}} = \frac{dz}{-p\frac{\partial f}{\partial p} - q\frac{\partial f}{\partial q}} = \frac{\partial x}{\frac{-\partial f}{\partial p}} = \frac{dy}{\frac{-\partial f}{\partial q}} = \frac{\partial f}{0}$$

or
$$\frac{dp}{0 + 2zp\,(16p^2 + 9q^2)} = \frac{dq}{0 + 2zq\,(16p^2 + 9q^2)} = \frac{dz}{-p(32\,pz^2) - q(18q\,z^2)} = \frac{dx}{0} = \frac{dy}{0}$$

$$\Rightarrow \quad \frac{dp}{p} = \frac{dq}{q} \Rightarrow \log p = \log q + \log a \quad \text{or} \quad p = aq$$
where a is an arbitrary constant.

Putting this in the given equation we get

$$16a^2 q^2 z^2 + 9q^2 z^2 + 4z^2 = 4 \qquad \text{or} \qquad q = \sqrt{\frac{4(1-z^2)}{(16a^2+9)z^2}} = \frac{2}{\sqrt{16a^2+9}}\sqrt{\frac{1-z^2}{z^2}}$$

and $p = aq$. Substituting these in $dz = pdx + qdy$, we get

$$\frac{dz}{q} = adx + dy \qquad \text{or} \qquad adx + dy = \frac{z\sqrt{16a^2+9}}{2\sqrt{1-z^2}}\,dz$$

or
$$ax + y = \frac{1}{2}\sqrt{16a^2+9}\int\frac{z\,dz}{\sqrt{1-z^2}}\,dz = \frac{-1}{2}\sqrt{16a^2+9}\,\sqrt{1-z^2} + b$$

$\therefore$ Solution is $ax + y + \frac{1}{2}\sqrt{16a^2+9}\,\sqrt{1-z^2} = b$ where a and b are arbitrary constants. **Ans.**

21. *Apply Charpit's method to solve* $2(pq + py + qx) + x^2 + y^2 = 0$

SOLUTION: Let $f(p, q, x, y, z) = 2(pq + py + qx) + x^2 + y^2$

By Charpits method

$$\frac{dp}{2q+2x} = \frac{dq}{2p+2y} = \frac{dz}{-p(2q+2y)-q(2p+2x)} = \frac{dx}{-(2q+2y)} = \frac{dy}{-(2p+2x)} = \frac{df}{0}$$

$$= \frac{dp+dq+dx+dy}{0} \quad \Rightarrow \quad p + q + x + y = a \quad \text{where } a \text{ is an arbitrary constant}$$

or $p + y = a - (q + x)$ The given equation can be written as

$$(p + y)^2 + (q + x)^2 = (p - q)^2 \qquad ...(2)$$

Substituting for $p + y$ from (1) in (2), we get $[a - (q + x)]^2 + (q + x)^2 = [a - (q + x) - y - q]^2$

Solve for q and substitute the values of p and q in $dz = pdx + qdy$. Integrate it and get

$$2z = ax - x^2 + ay - y^2 + \frac{1}{2}(x-y)\sqrt{(x-y)^2 + a^2} + \frac{a^2}{2\sqrt{2}}\log\left[\sqrt{2(x-y)} + \sqrt{2(x-y)^2 + a^2}\right] + b \quad \textbf{Ans.}$$

CHAPTER

24 Partial Differential Equations of Higher Order and Applications

IMPORTANT FORMULAE AND USEFUL RESULTS

1. Equation (*Homogeneous*) of nth degree is of the form.

$F(D, D')Z = (D^n + a_1 D^{n-1} D' + a_2 D^{n-2} D'^2 + \ldots + D'^n) Z = f(x, y)$

A.E. is $m^n + a_1 m^{n-1} + a_2 m^{n-2} + \ldots + a_n = 0$. roots being $m_1, m_2, \ldots, m_n$.

C.F. corresponding to $(D - m D')z = 0$ is $z = \phi(y + mx)$.

$\therefore$ C.F. $= \phi_1(y + m_1 x) + \phi_2(y + m_2 x) + \ldots + \phi_n(y + m_n x)$

For $m = m_2 = m_1$ C.F. $= x\,\phi_1(y + m_1 x) + \phi_2(y + m_1 x)$

2. **P.I.** To find $\dfrac{1}{D - mD'} f(x, y)$, first change y to $y - mx$, integrate it w.r.t. x treating y as constant,, then in the integral obtained replace y by $y + mx$ and that is the value.

3. **Short Cut Methods for P.I.**

(*i*) $$\frac{1}{F(D, D')} e^{ax + by} = \frac{1}{F(a, b)} e^{ax + by} \quad \text{provided } F(a, b) \neq 0$$

(*ii*) $$\frac{1}{F(D^2, DD^1, D^2)} \begin{matrix}\sin\\ \cos\end{matrix}(ax + by) = \frac{1}{F(-a^2, -ab, -b^2)} \begin{matrix}\sin\\ \cos\end{matrix}(ax + by) \quad \text{provided } F(-a^2, -ab, -b^2) \neq 0$$

(*iii*) $$\frac{1}{F(D, D')} x^m y^n = [F(D, D')]^{-1} x^m y^n$$

4. **Non-Homogeneous Linear Equations**

Equation is of the form $F(D, D')z = f(x, y)$

For C.F. $F(D, D')z = (a_1 D + b_1 D' + C_1)(a_2 D + b_2 D' + C_2) \ldots (a_n D + b_n D' + c_n) z = 0$

C.F. $= \phi_1(a_1 y - b_1 x) e^{-C_1 x/a_1} + \phi_2(a_1 y - b_2 x) e^{-C_2 x/a_2} + \ldots + \phi_n(a_n y - b_n x) e^{-C_n x/a_n}$

For P.I. (*i*) $\frac{1}{F(D, D')} e^{ax+by} = \frac{1}{F(a, b)} e^{ax+by}$

(*ii*) $\frac{1}{F(D^2, DD', D'^2, D, D')} \frac{\sin (ax+by)}{\cos} = \frac{1}{F(-a^2, -ab, -b^2, D, D')} \frac{\sin (ax+by)}{\cos}$

then rationalise, etc.

(*iii*) $\frac{1}{F(D, D')} x^m y^n = [F(D, D')]^{-1} x^m y^n$

(*iv*) $\frac{1}{F(D, D')} [e^{ax+by} V(x, y)] = e^{ax+by} \frac{1}{F(D+a, D'+b)} V(x, y)$

5. Method of Separation of Variables.

SOLUTIONS TO EXERCISE 24A

1. *Solve* $(D^2 - 3DD' + 2D'^2)\, z = e^{2x+3y} + e^{x+y} + \sin(x - 2y)$

SOLUTION: For complementary function the auxiliary equation is $m^2 - 3m + 2 = 0$ or $(m - 1)(m - 2) = 0$, roots being $m = 1, 2$.

$\therefore$ $\quad C.F = \phi_1(y + 1x) + \phi_2(y + 2x)$

and
$$\text{P.I.} = \frac{1}{D^2 - 3DD' + 2D'^2}[e^{2x+3y} + e^{x+y} + \sin(x-2y)]$$

$$= \frac{1}{2^2 - 3(2)(3) + 2(3)^2} e^{2x+3y} + \frac{1}{(D-D')(D-2D')} e^{x+y}$$

$$+ \frac{1}{-1^2 + 3(1)(-2) + 2(-2^2)} \sin(x-2y)$$

$$= \frac{1}{4} e^{2x+3y} + \frac{1}{(D-D')}\left[\frac{1}{1-2(1)} e^{x+y}\right] - \frac{1}{13}\sin(x-2y)$$

$$= \frac{1}{4} e^{2x+3y} - \frac{1}{D-D'} e^{x+y} - \frac{1}{13}\sin(x-2y)$$

$$= \frac{1}{4} e^{2x+3y} - \frac{x}{\frac{\partial}{\partial D}(D-D')} e^{x+y} - \frac{1}{15}\sin(x-2y)$$

$$= \frac{1}{4} e^{2x+3y} - \frac{x}{1} e^{x+y} - \frac{1}{15}\sin(x-2y).$$ **Ans.**

2. *Find the particular integral of* $p + 3q = \cos(2x + y)$ *where* $p = \dfrac{\partial z}{\partial x}$ *and* $q = \dfrac{\partial z}{\partial y}$.

SOLUTION: The equation is $(D + 3D')\, z = \cos(2x + y)$ where $D = \dfrac{\partial}{\partial x}$, $D' \equiv \dfrac{\partial}{\partial y}$.

$$\therefore \quad \text{P.I.} = \frac{1}{D+3D'}\cos(2x+y) = \frac{D-3D'}{D^2-9D'^2}\cos(2x+y)$$

$$= (D-3D')\frac{1}{-4-9(-1)}\cos(2x+y) = \frac{1}{5}(D-3D')\cos(2x+y)$$

$$= \frac{1}{5}[D\cos(2x+y) - 3D'\cos(2x+y)]$$

$$= \frac{1}{5}[-2\sin(2x+y) + 3\sin(2x+y)] = \frac{1}{5}\sin(2x+y).$$

Ans.

3. *Solve* $(D - 2D')^2 (D + 3D') z = e^{2x+y}$

SOLUTION: The A.E. is $(m - 2)^2 (m + 3) = 0$ whose roots are $m = 2, 2, -3$

$\therefore$ C.F. $= \phi_1 (y + 2x) + x\,\phi_2 (y + 2x) + \phi_3 (y - 3x)$

and
$$\text{P.I.} = \frac{1}{(D-2D')^2 (D+3D')} e^{2x+y} = \frac{1}{(D-2D')^2}\left[\frac{1}{D+3D'} e^{2x+y}\right]$$

$$= \frac{1}{(D-2D')^2}\frac{1}{2+3} e^{2x+y} = \frac{1}{5}\frac{1}{(D-2D')^2} e^{2x+y} \quad \text{(case of double failure)}$$

$$= \frac{1}{5}\frac{x^2}{\frac{\partial^2}{\partial D^2}(D-2D')^2} e^{2x+y} = \frac{x^2}{2\cdot 5} e^{2x+y}$$

$\therefore$ Complete solution is $z = \phi_1 (y + 2x) + x\,\phi_2 (y + 2x) + \phi_3 (y - 3x) + \frac{x^2}{10} e^{2x+y}$ **Ans.**

4. *Solve* $(D^3 - 7DD'^2 - 6D'^3) z = \cos (x - y) + x^2 + xy^2 + y^3.$

SOLUTION: The Auxiliary equation to find C.F, is $m^3 - 7m - 6 = 0$

or $(m + 1)(m^2 - m - 6) = 0$ whose roots are $m = -1, -2, 3.$

$\therefore$ C.F $= \phi_1 (y - x) + \phi_2 (y - 2x) + \phi_3 (y + 3x)$

and
$$\text{P.I} = \frac{1}{D^3 - 7DD'^2 - 6D'^3}\cos(x-y) + \frac{1}{D^3 - 7DD'^2 - 6D'^3}(x^2 + xy^2 + y^3)$$

Here
$$\frac{1}{D^3 - 7DD'^2 - 6D'^3}\cos(x-y)$$

$$= \frac{1}{D(-1^2) - 7D'(1) - 6D'(-1)}\cos(x-y)$$

$$= \frac{1}{-D-D'}\cos(x-y) = \frac{-(D-D')}{D^2 - D'^2}\cos(x-y)$$

$$= \frac{-x(D-D')}{\frac{d}{dD}\left(D^2 - D'^2\right)}\cos(x-y) = -\frac{x}{2D}[D\cos(x-y) - D'\cos(x-y)]$$

$$= -\frac{x}{2D}[-\sin(x-y) - \sin(x-y)] = \frac{2x}{2}\int \sin(x-y)dx = -x\cos(x-y)$$

and
$$\frac{1}{D^3 - 7DD'^2 - 6D'^3}(x^2 + xy^2 + y^3)$$

$$= \frac{1}{D^3}\left[1 - \left(7\frac{D'^2}{D^2} + \frac{6D'^3}{D^3}\right)\right]^{-1}(x^2 + xy^2 + y^3)$$

$$= \frac{1}{D^3}\left[1+\left(\frac{7D'^2}{D^2}+\frac{6D'^3}{D^3}\right)+\left(7\frac{D'^2}{D^2}+\frac{6D'^3}{D^3}\right)^2+...\right](x^2+xy^2+y^3)$$

$$= \frac{1}{D^3}\left[(x^2+xy+y^3)+\frac{7}{D^2}(0+2x+6y)+\frac{6}{D^3}(0+0+6)\right]$$

$$= \frac{1}{D^3}\left[x^2+xy^2+y^3+7\left(\frac{x^3}{3}+6y\frac{x^2}{2}\right)+36\frac{x^3}{3.2}\right]$$

$$= \frac{1}{D^3}\left[x^2+xy^2+21x^2y+\frac{25}{3}x^3+y^3\right] = \frac{1}{D^2}\left[\frac{x^3}{3}+\frac{x^2y^2}{2}+7x^3y+\frac{25}{12}x^4+xy^3\right]$$

$$= \frac{1}{D}\left[\frac{x^4}{12}+\frac{x^3y^2}{6}+\frac{7x^4y}{4}+\frac{5x^5}{12}+\frac{x^2y^3}{2}\right] = \frac{x^5}{60}+\frac{x^4y^2}{24}+\frac{7x^5y}{20}+\frac{5x^6}{72}+\frac{x^3y^3}{6}$$

$\therefore$ Complete solution is

$$z = \phi_1(y-x)+\phi_2(y-2x)+\phi_3(y+3x)$$

$$+\frac{x^5}{60}+\frac{x^4y^2}{24}+\frac{7x^5y}{20}+\frac{5x^6}{72}+\frac{x^3y^3}{6}-x\cos(x-y)$$ **Ans.**

5. *Solve* $(D-D')^2\ z = x+\phi(x+y)$

SOLUTION: The auxiliary equation is $(m-1)^2=0 \quad \therefore \quad m=1, 1$

$\therefore$ $\quad$ $C.F = \phi_1(y+x)+x\,\phi_2(y+x)$

and $\quad$ $$P.I. = \frac{1}{(D-D')^2}[x+\phi(x+y)] = \frac{1}{(D-D')^2}x+\frac{1}{(D-D')^2}\phi(y+x)$$

Now $$\frac{1}{(D-D')^2}x = \frac{1}{D^2}\left(1-\frac{D'}{D}\right)^{-2}x = \frac{1}{D^2}\left[1+2\frac{D'}{D}+\frac{(-2)(-3)}{2!}\frac{D'^2}{D^2}+...\right]x$$

$$= \frac{1}{D^2}[x+0+0] = \frac{1}{D}\left(\frac{x^2}{2}\right) = \frac{x^3}{6}$$

and $$\frac{1}{(D-D')^2}\phi(y+x) = \frac{1}{D-D'}\left[\frac{1}{D-D'}\phi(y+x)\right] = \frac{x}{D-D},\ \phi(y+x) = \frac{x^2}{2}\phi(y+x)$$

$\therefore$ Complete solution is $Z = \phi_1(y+x)+x\phi_2(y+x)+\dfrac{x^3}{6}+\dfrac{x^2}{2}\phi(y+x)$ **Ans.**

6. *Solve* $r+s-6t = y\cos x$

SOLUTION: The equation is $\dfrac{\partial^2 z}{\partial x^2}+\dfrac{\partial^2 z}{\partial x\partial y}-6\dfrac{\partial^2 z}{\partial y^2} = y\cos x$

or $\quad$ $(D^2+DD'-6D'^2)\,z = y\cos x$ $\quad$ where $\quad D = \dfrac{\partial}{dx},\quad D' = \dfrac{\partial}{dy}.$

The auxiliary equation is $m^2 + m - 6 = 0$ or $(m + 3)(m - 2) = 0$ Its roots are $m = 2, -3$.

$\therefore$ C.F. $= \phi_1(y + 2x) + \phi_2(y - 3x)$

$$\text{P.I.} = \frac{1}{D^2 + DD' - 6D'^2} y\cos x = \frac{1}{D^2\left(1 + \frac{D'}{D} - \frac{6D'^2}{D^2}\right)} y\cos x$$

$$= \frac{1}{D^2}\left[1 + \left(\frac{D'}{D} - \frac{6D'^2}{D^2}\right)\right]^{-1} y\cos x = \frac{1}{D^2}\left[1 - \frac{D'}{D} + \frac{6D'^2}{D^2}\right] y\cos x$$

$$= \frac{1}{D^2}\left[y\cos x - \frac{1}{D}\cos x + 0 + \ldots\right] = \frac{1}{D^2}(y\cos x - \sin x) = -y\cos x + \sin x$$

$\therefore$ Complete solution is $z = \phi_1(y + 2x) + \phi_2(y - 3x) - y\cos x + \sin x$ **Ans.**

7. *Solve* $(D^2 + 2DD' + D'^2)\, z = 2\cos y - x\sin y$.

SOLUTION: The given equation has auxiliary equation as $m^2 + 2m + 1 = 0$ hence $m = -1, -1$

$\therefore$ C.F $= \phi_1(y - x) + x\,\phi_2(y - x)$

and
$$\text{P.I.} = \frac{1}{(D^2 + 2DD' + D'^2)}(2\cos y - x\sin y)$$

$$= \frac{2}{D^2 + 2DD' + D'^2}\cos(0x + y) - \frac{1}{(D + D')^2} x\sin y$$

$$= \frac{2}{0 + 0 - 1}\cos y - \frac{1}{D'^2}\left(1 + \frac{D}{D'}\right)^{-2}(x\sin y)$$

$$= -2\cos y - \frac{1}{D'^2}\left[1 - \frac{2D}{D'} + \frac{6}{2}\frac{D^2}{D'^2} + \ldots\right] x\sin y$$

$$= -2\cos y - \frac{1}{D'^2}\left[x\sin y - \frac{2}{D'}\sin y\right] = -2\cos y - \frac{1}{D'^2}[x\sin y + 2\cos y]$$

$$= -2\cos y - [-x\sin y - 2\cos y] = x\sin y$$

$\therefore$ Complete solution $z = \phi_1(y - x) + x\phi_2(y - x) + x\sin y$ **Ans.**

8. *Solve* $(D^2 - 2DD')\, z = e^{2x} + x^2 y$

SOLUTION: The A.E is $m^2 - 2m = 0$ whose roots are $m = 0, 2$

$\therefore$ C.F. $= \phi_1(y) + \phi_2(y + 2x)$

and
$$\text{P.I} = \frac{1}{D^2 - 2DD'}(e^{2x} + x^2 y) = \frac{1}{2^2 - 2(0)} e^{2x} + \frac{1}{D^2}\left(1 - \frac{2D'}{D}\right)^{-1}(x^2 y)$$

$$= \frac{1}{4}e^{2x} + \frac{1}{D^2}\left[1 + \frac{2D'}{D} + \frac{4D'^2}{D^2} + \ldots\right](x^2 y) = \frac{1}{4}e^{2x} + \frac{1}{D^2}\left[x^2 y + \frac{2D'}{D}(x^2 y)\right]$$

$$= \frac{1}{4} e^{2x} + \frac{1}{D^2}\left(x^2 y + \frac{2x^3}{3}\right) = \frac{1}{4} e^{2x} + \frac{x^4 y}{6} + \frac{x^5}{30}$$

Hence the complete solution is

$$Z = \phi_1 (y) + \phi_2 (y + 2x) + \frac{1}{4} e^{2x} + \frac{1}{6} x^4 y + \frac{1}{30} x^5$$ **Ans.**

9. *Solve* $(D^3 - 2D^2D')\, z = 2e^{2x} + 3x^2y$

SOLUTION: The A.E. is $m^3 - 2m^2 = 0$ whose roots are $m = 0, 0, 2$

$\therefore$ C.F. $= \phi_1 (y) + x\, \phi_2 (y) + \phi_3 (y + 2x)$

and $$\text{P.I.} = \frac{1}{D^3 - 2D^2 D'}(2e^{2x} + 3x^2 y) = \frac{1}{D^2 (D - 2D')} 2e^{2x} + \frac{1}{D^2 (D - 2D')} 3x^2 y$$

$$= \frac{1}{D^2}\left(\frac{2}{2-0}\right) e^{2x} + \frac{1}{D^3}\left(1 - \frac{2D'}{D}\right)^{-1} (3x^2 y) = \frac{1}{D^2}(e^{2x}) + \frac{1}{D^3}\left[1 + \frac{2D'}{D} + \frac{4D'^2}{D^2} + \ldots\right](3x^2 y)$$

$$= \frac{1}{4} e^{2x} + \frac{1}{D^3}\left[3x^2 y + \frac{6}{D}(x)^2 + 0\right] = \frac{e^{2x}}{4} + \frac{1}{D^3}(3x^2 y) + \frac{6}{D^4}(x^2) = \frac{x}{2} e^{2x} + \frac{x^5 y}{4.5} + \frac{x^6}{60}$$

$\therefore$ Complete solution is $Z = \phi_1 (y) + x\, \phi_2 (y) + \phi_3 (y + 2x) + \frac{1}{4} e^{2x} + \frac{x^5 y}{20} + \frac{x^6}{60}$

10. Solve $(D^3 - 3DD'^2 - 2D'^3)\, z = \cos (x + 2y) - e^y (y + 2x)$

SOLUTION: The A.E. is $m^3 - 3m - 2 = 0$ Its roots are $m = -1, -1, 2.$

$\therefore$ C.F. $= \phi_1 (y - x) + x\, \phi_2 (y - x) + \phi_3 (y + 2x)$

and $$\text{P.I.} = \frac{1}{D^3 - 3DD'^2 - 2D'^3}[\cos(x + 2y) - e^y (y + 2x)]$$

$$= \frac{1}{(-1)D + 12D + 8D'} \cos(x + 2y) - e^y \frac{1}{D^3 - 3D(D' + 1)^2 - 2(D' + 1)^3}(y + 2x)$$

$$= \frac{11D - 8D'}{121D^2 - 64D'^2} \cos(x + 2y)$$

$$- e^y \frac{1}{2 + 6D' + 6D'^2 + 3DD'^2 + 6DD' + 3D - D^3 - 2D'^3}(y + 2x)$$

$$= \frac{11D - 8D'}{121(-1) - 64(-4)} \cos(x + 2y) + \frac{e^y}{2}$$

$$\left[1 + \left(3D' - 3D'^2 + D'^3 + \frac{3}{2} DD'^2 + 3DD' - \frac{3}{2} D - \frac{1}{2} D^3\right)\right]^{-1} (y + 2x)$$

$$= \frac{11D - 8D'}{135}\cos(x+2y) + \frac{1}{2}e^y\left[1 - 3D' - 3D'^2 - D'^3 - \frac{3}{2}DD' + \frac{3}{2}DD' - \frac{3}{2}D\right](y+2x)$$

$$= \frac{5}{135}\sin(x+2y) + \frac{1}{2}e^y\, 2(y+2x-6)$$

$\therefore$ Complete solution is $z = \phi_1(y-x) + x\phi_2(y-x) + \phi_3(y+2x)$

$$+\frac{1}{27}\sin(x+2y) + \frac{1}{2}e^y\,(y+2x-6)$$

11. *Solve* $\dfrac{\partial^3 u}{\partial x^3} + \dfrac{\partial^3 u}{\partial x^2 \partial y} + 4\dfrac{\partial^3 u}{\partial y^3} = e^{x+2y}$

SOLUTION: Given equation is $(D^3 + D^2D' + 4D'^3)\,u = e^{x+2y}$ where $D \equiv \dfrac{\partial}{\partial x}, D' \equiv \dfrac{\partial}{\partial y}$

For C.F., Auxiliary is $m^3 + m^2 + 4 = 0$ or $(m+2)(m^2 - m + 2) = 0$

whose roots are $m = -2$ and $m = \dfrac{1 \pm i\sqrt{7}}{2}$.

$$\therefore \quad \text{C.F.} = \phi_1(y-2x) + \phi_2\left(y + \frac{1+i\sqrt{7}}{2}x\right) + \phi_3\left(y + \frac{1-i\sqrt{7}}{2}x\right)$$

and $$\text{P.I.} = \frac{1}{D^3 + D^2D' + 4D'^3}e^{x+2y} = \frac{1}{1^3 + 1^2.2 + 4.2^3}e^{x+2y} = \frac{1}{35}e^{x+2y}$$

$\therefore$ Complete Solution is $u = \phi_1(y-2x) + \phi_2\left(y + \dfrac{1+i\sqrt{7}}{2}x\right) + \phi_3\left(y + \dfrac{1-i\sqrt{7}}{2}x\right) + \dfrac{e^{x+2y}}{35}$ **Ans.**

12. *Solve the partial differential equation*

$$\frac{\partial^2 z}{\partial x^2} - 2\frac{\partial^2 z}{\partial x \partial y} + \frac{\partial^2 z}{\partial y^2} = \sin x$$

SOLUTION: The given equation is $(D^2 - 2DD' + D'^2)z = \sin x$ where $D \equiv \dfrac{\partial}{\partial x}, D' \equiv \dfrac{\partial}{\partial y}$.

The A.E. is $m^2 - 2m + 1 = 0$ hence $m = 1, 1$.

Therefore C.F. $= \phi_1(y+x) + x\phi_2(y+x)$

and $$\text{P.I.} = \frac{1}{D^2 - 2DD' + D'^2}\sin x = \frac{\sin x}{-1+0+0}$$

$\therefore$ Complete solution is $z = \phi_1(y+x) + x\phi_2(y+x) - \sin x$ **Ans.**

13. *Solve* $\dfrac{\partial^2 z}{\partial x^2} - \dfrac{\partial^2 z}{\partial x \partial y} = \cos x \cos 2y$

SOLUTION: Given equation is $(D^2 - DD')z = \cos x \cos 2y$ where $D = \dfrac{\partial}{\partial x}$, $D' = \dfrac{\partial}{\partial y}$.

The A.E. is $m^2 - m = 0$ hence $m = 0, 1$. $\therefore$ C.F. $= \phi_1(y) + \phi_2(y + x)$

and
$$\text{P.I.} = \frac{1}{D^2 - DD'}(\cos x \cos 2y) = \frac{1}{2(D^2 - DD')}\cos(x + 2y) + \frac{1}{2(D^2 - DD')}\cos(x - 2y)$$

$$= \frac{1}{2[-1-(-2)]}\cos(x+2y) + \frac{1}{2[-1-(2)]} = \frac{1}{2}\cos(x+2y) - \frac{1}{6}\cos(x-2y)$$

$\therefore$ Complete solution is $z = \phi_1(y) + \phi_2(y+x) + \frac{1}{2}\cos(x+2y) - \frac{1}{6}\cos(x-2y)$ **Ans.**

14. *Solve* $2\dfrac{\partial^2 z}{\partial x^2} - 5\dfrac{\partial^2 z}{\partial x \partial y} + 2\dfrac{\partial^2 z}{\partial y^2} = 5\sin(2x + y)$

SOLUTION: Given equations is $(2D^2 - 5DD' + 2D'^2)z = 5\sin(2x + y)$.

A.E. is $2m^2 - 5m + 2 = 0$ whose roots are $m = 2$ and $\frac{1}{2}$

$\therefore$ C.F. $= \phi_1(y + 2x) + \phi_2(y + \frac{1}{2}x) = \phi_1(y+2x) + \phi_2(2y+x)$

and
$$\text{P.I.} = \frac{5}{2D^2 - 5DD' + 2D'^2}\sin(2x+y) = \frac{1}{2(-4) - 5(-2) + 2(-1)}5\sin(2x+y) \text{ (case of faliure)}$$

$$= \frac{5x}{\frac{\partial}{\partial D}[2D^2 - 5DD' + 2D'^2]}\sin(2x+y) = \frac{5x}{[4D - 5D']}\sin(2x+y)$$

$$= 5x\frac{(4D + 5D')}{16D^2 - 25D'^2}\sin(2x+y) = \frac{5x(4D+5D')}{16(-4) - 25(-1)}\sin(2x+y)$$

$$= \frac{5x}{-39}[8\cos(2x+y) + 5\cos(2x+y)] = -\frac{5x}{3}\cos(2x+y)$$

$\therefore$ Complete solution is $z = \phi_1(y+2x) + \phi_2(2y+x) - \frac{5x}{3}\cos(2x+y)$ **Ans.**

SOLUTIONS TO EXERCISE 24B

1. *Solve* $\dfrac{\partial^2 z}{\partial x^2} - \dfrac{\partial^2 z}{\partial x \partial y} - 2\dfrac{\partial^2 z}{\partial x^2} + 2\dfrac{\partial z}{\partial x} - 4\dfrac{\partial z}{\partial y} = 0.$

SOLUTION: The equation is $(D^2 - DD' - 2D'^2 + 2D - 4D')z = 0$ where $D \equiv \dfrac{\partial}{\partial x}$, $D' \equiv \dfrac{\partial}{\partial y}$

or $(D + D' + 2)(D - 2D')z = 0$

$\therefore$ C.F. $= \phi_1(y - x)\, e^{-2x} + \phi_2(y + 2x)$ and P.I. $= 0$

$\therefore$ Solution is $z = e^{-2x}\, \phi_1(y - x) + \phi_2(y + 2x)$ **Ans.**

2. *Solve* $\dfrac{\partial^2 z}{\partial x^2} + \dfrac{\partial^2 z}{\partial x \partial y} + \dfrac{\partial z}{\partial y} - z = e^x + e^{-x}.$

SOLUTION: The equation is $(D^2 + DD' + D' - 1)\, z = e^x + e^{-x}$ where $D \equiv \dfrac{\partial}{\partial x}$, $D' \equiv \dfrac{\partial}{\partial y}$

or $(D + 1)(D + D' - 1)\, z = e^x + e^{-x}$ $\therefore$ C.F. $= \phi_1(y)e^{-x} + \phi_2(y - x)e^x$

and P.I. $= \dfrac{1}{D^2 + DD' + D' - 1}(e^{-x} + e^x)$

$$= \frac{1}{[D^2 + DD' + D' - 1]_{D=-1,\, D'=0}}\, e^{-x} + \frac{1}{[D^2 + DD' + D' - 1]_{D=1,\, D'=0}}\, e^x$$

$$= \frac{x}{\left[\dfrac{\partial}{\partial D}(D^2 + DD' + D' - 1)\right]_{D=-1,\, D'=0}}\, e^{-x} + \frac{x}{\left[\dfrac{\partial}{\partial D}(D^2 + DD' + D' - 1)\right]_{D=1,\, D'=0}}\, e^x$$

$$= \frac{x}{[2D + D']_{D=-1,\, D'=0}}\, e^{-x} + \frac{x}{[2D + D']_{D=1,\, D'=0}}\, e^x = \frac{-x}{2} e^{-x} + \frac{x}{2} e^x.$$

$\therefore$ Complete solution is $z = e^{-x}\, \phi_1(y) + e^x\, \phi_2(y - x) + \dfrac{x}{2}(e^x - e^{-x})$ **Ans.**

3. *Solve* $\dfrac{2\partial^2 z}{\partial x \partial y} + \dfrac{\partial^2 z}{\partial y^2} - 3\dfrac{\partial z}{\partial y} = 3\cos(3x - 2y)$

SOLUTION: The equation is $D'(D' - 2D - 3)\, z = 3\cos(3x - 2y)$

Here C.F. $= \phi_1(x) + e^{3y}\, \phi(x - 2y)$

and P.I. $= \dfrac{1}{D'^2 + 2DD' - 3D'}\, 3\cos(3x - 2y)$

$$= \frac{3}{-4 - 2(3)(-2) - 3D'}, \cos(3x - 2y) = \frac{3}{8 - 3D'} \cos(3x - 2y)$$

$$= \frac{3(8+3D')}{64-9D'^2}\cos(3x-2y) = \frac{3[8\cos(3x-2y) + 6\sin(3x-2y)]}{64-9(-4)}$$

$$= \frac{3}{30}[4\cos(3x-2y)+3\sin(3x-2y)]$$

$\therefore$ Complete Solution is $z = \phi_1(x) + e^{3y}\phi_2(x-2y) + \frac{3}{50}[4\cos(3x-2y)+3\sin(3x-2y)]$ **Ans.**

4. *Solve* $\frac{\partial^2 z}{\partial x^2} + \frac{\partial^2 z}{\partial y^2} - 2\frac{\partial^2 z}{\partial x\partial y} - 3\frac{\partial z}{\partial x} + 3\frac{\partial z}{\partial y} + 2z = e^{2x-y}$

SOLUTION: The equation is $(D^2 + D'^2 - 2DD' - 3D' + 3D' + 2)z = e^{2x-y}$

or $(D - D' - 2)(D - D' - 1)z = e^{2x-y}$

$\therefore$ C.F. $= e^{2x}\phi_1(y+x) + e^x\phi_2(y+x)$.

and P.I. $= \frac{1}{D^2 + D'^2 - 2DD' - 3D + 3D' + 2}e^{2x-y} = \frac{1}{2^2 + 1 - 2(2)(-1) - 3(2) + 3(-1) + 2}e^{2x-y}$

$\therefore$ Solution is $z = e^{2x}\phi_1(y+x) + e^x\phi_2(y+x) + \frac{1}{2}e^{2x-y}$. **Ans.**

5. *Solve* $\frac{\partial^2 z}{\partial x^2} + 3\frac{\partial^2 z}{\partial x\partial y} + 2\frac{\partial^2 z}{\partial y^2} - 4\frac{\partial z}{\partial x} - 5\frac{\partial z}{\partial y} + 3z = 2x + 3y$

SOLUTION: The given equation is $[D^2 + 3DD' + 2D'^2 - 4D - 5D' + 3]\ z = 2x + 3y$

or $(D + D' - 1)(D + 2D' - 3)Z = 2x + 3y$

$\therefore$ C.F. $= \phi_1(y-x)e^{+x} + \phi_2(y-2x)e^{3x}$

and P.I. $= \frac{1}{3}\left[1 - \frac{4}{3}D - \frac{5}{3}D' + \frac{2D'^2}{3} + DD' + \frac{D^2}{3}\right]^{-1}(2x+3y)$

$$= \frac{1}{3}\left[1 + \frac{4D}{3} + \frac{5D'}{3} - \frac{2D'^2}{3} - DD' - \frac{D^2}{3} + \ldots\right](2x+3y)$$

$$= \frac{1}{3}\left[(2x+3y) + \frac{4}{3}(2) + \frac{5}{3}(3) + 0\right](2x+3y) = \frac{1}{3}(2x+3y) + \frac{23}{3}$$

$\therefore$ Complete solution is $z = \phi_1(y-x)e^x + \phi_2(y-2x)e^{3x} + \frac{2}{3}x + y + \frac{23}{9}$. **Ans.**

6. *Solve* $\frac{\partial^2 z}{\partial x^2} + \frac{\partial^2 z}{\partial x\partial y} + \frac{\partial z}{\partial y} - z = x^2 y$

SOLUTION: The given equation is $(D^2 + DD' + D' - 1)z = x^2 y$

or $(D+1)(D + D' - 1)z = x^2 y$ $\therefore$ C.F. $= e^{-x}\phi_1(y) + e^x\phi_2(y+x)$

and P.I. $= \frac{1}{D^2 + DD' + D' - 1}x^2 y = -[1 - (D' + DD' + D^2)]^{-1}x^2 y$

$$= -[1 + (D' + DD' + D^2) + (D' + DD' + D^2)^2 + \ldots]\,x^2 y$$

$$= -[x^2 y + (x^2 + 2x + 2y) + 2D^2 D' x^2 y + 0] = -[x^2 y + x^2 + 2x + 2y + 4]$$

$\therefore$ Complete solution is $z = e^{-x}\phi_1(y) + e^x\phi_2(y+x) - (x^2 y + x^2 + 2x + 2y + 4)$ **Ans.**

7. *Solve* $\dfrac{\partial^2 z}{\partial x^2} - \dfrac{\partial^2 z}{\partial y^2} + \dfrac{\partial z}{\partial x} + 3\dfrac{\partial z}{\partial y} - 2z = e^{x-y} - x^2 y$

SOLUTION: The given equation is $(D^2 - D'^2 + D + 3D' - 2)\, z = e^{x-y} - x^2 y$

For C.F., we have $(D + D' - 1)(D - D' + 2)\, Z = 0$ $\therefore$ C.F. $= e^x\, \phi_1(y - x) + e^{-2x}\, \phi_2(y + x)$

and

$$\text{P.I.} = \frac{1}{D^2 - D'^2 + D + 3D' - 2} e^{x-y} - \frac{1}{D^2 - D'^2 + D + 3D' - 2} x^2 y$$

$$= \frac{1}{1^2 - (-1)^2 + 1 + 3(-1) - 2} e^{x-y} + \frac{1}{2}\left[1 - \left(\frac{3}{2}D' + \frac{D}{2} - \frac{1}{2}D'^2 + \frac{1}{2}D^2\right)\right]^{-1} x^2 y$$

$$= -\frac{1}{4} e^{x-y} + \frac{1}{2}\left[1 + \left(\frac{3}{2}D' + \frac{D}{2} - \frac{1}{2}D'^2 + \frac{1}{2}D^2\right) + \left(\frac{3}{2}D' + \frac{D}{2} - \frac{1}{2}D'^2 + \frac{1}{2}D^2\right)^2 + \ldots\right] x^2 y$$

$$= -\frac{1}{4} e^{x-y} + \frac{1}{2}\left[x^2 y + \frac{3}{2}x^2 + xy + y + \frac{y}{2} + 3x\right] + \frac{3}{2}$$

$\therefore$ Complete solution is

$$z = e^x\, \phi_1(y - x) + e^{-2x}\, \phi_2(y + x) - \frac{1}{4} e^{x-y} + \frac{1}{2}x^2 y + \frac{3}{4}x^2 + \frac{1}{2}xy + \frac{3y}{4} + \frac{3x}{2} + \frac{3}{2}. \qquad \textbf{Ans.}$$

8. *Solve* $(D - 3D' - 2)^2 z = 2e^{2x} \tan(y + 3x)$ where $D = \dfrac{\partial}{\partial x}$ and $D' = \dfrac{\partial}{\partial y}$.

SOLUTION: C.F., as the solution to $(D - 3D' - 2)^2 z = 0$ is given by

$$\text{C.F.} = e^{2x}[\phi_1(y + 3x) + x\phi_2(y + 3x)]$$

and

$$\text{P.I.} = \frac{1}{(D - 3D' - 2)^2} 2e^{2x} \tan(y + 3x)$$

$$= 2e^{2x} \frac{1}{(D + 2 - 3D' - 2)^2} \tan(y + 3x) = 2e^{2x} \frac{1}{(D - 3D')^2} \tan(y + 3x)$$

Now, as we know in case of homogeneous equations, to evaluate $\dfrac{1}{D - mD'} f(x, y)$ we first change y to $y - mx$ in $f(x, y)$ and then integrate it w.r.t. x treating y as constant, and then in the integral thus obtained change y to $y + mx$ and get the desired value.

Therefore,

$$\frac{1}{(D - 3D')}\left[\frac{1}{(D - 3D')} \tan(y + 3x)\right]$$

$$= \frac{1}{D - 3D'} \int \tan(y - 3x + 3x)\, dx = \frac{1}{D - 3D'} [x \tan y]\,(y \to y + 3x)$$

(repeating the process)

$$= \left[\int x \tan(y - 3x + 3x)\, dx\right]_{y \to y+3x} = \left[\int x \tan y\, dx\right]_{y \to y+3x} = \frac{x^2}{2} \tan(y + 3x)$$

$\therefore$ P.I. $= x^2 e^{2x} \tan(y + 3x)$

Thus the complete solution is

$$z = e^{2x}[\phi_1(y + 3x) + x\, \phi_2(y + 3x)] + x^2 e^{2x} \tan(y + 3x) \qquad \textbf{Ans.}$$

SOLUTIONS TO EXERCISE 24C

1. *Obtain the form for* Y *as a function of* y *alone if* $u = Y \sin mx$ *is a solution of the Laplace equation* $\dfrac{\partial^2 u}{\partial x^2} + \dfrac{\partial^2 u}{\partial y^2} = 0$ *and then find the solution of the equation such that* $u = 0$ *when* $y = -1$ *and* $u = \sin x$ *when* $y = 1$.

SOLUTION: Given $u = Y \sin mx$ where $Y = Y(y)$.

$$\therefore \quad \frac{\partial^2 u}{\partial x^2} = Y \frac{d^2}{dx^2} \sin mx \quad \text{and} \quad \frac{\partial^2 u}{\partial y^2} = \sin mx \frac{d^2 Y}{dy^2}.$$

$\therefore$ Laplace equation becomes $-m^2 Y \sin mx + \sin mx \dfrac{d^2 Y}{dy^2} = 0$

$\Rightarrow$ either $\sin mx = 0$ so that $mx = n\pi,\ n \in I$.

or $\dfrac{d^2 Y}{dy^2} - m^2 Y = 0$, whose solution is $Y = Ae^{my} + Be^{-my}$ where A and B are arbitrary constants.

$\therefore$ The solution is $u = \sin mx\,(A e^{my} + Be^{-my})$.

We are give that $u = 0$ when $y = -1$ and $u = \sin x$ when $y = 1$

Therefore, $A e^{-m} + B e^{m} = 0$ and $(A e^{m} + B e^{-m}) \sin mx = \sin x \Rightarrow m = 1$.

Thus we have $Be + \dfrac{A}{e} = 0$ and $Ae + \dfrac{B}{e} = 1$

when solved for A and B gives $B = \dfrac{e}{1 - e^4}$ and $A = \dfrac{-e^3}{1 - e^4}$

Therefore, the solution is $u = \sin x \left[\dfrac{-e^3}{1 - e^4} e^{y} + \dfrac{e}{1 - e^4} e^{-y} \right] = \dfrac{\sin x}{1 - e^4} (e^{1-y} - e^{3+y})$. **Ans.**

2. *Solve the Laplace equation* $\dfrac{\partial^2 u}{\partial x^2} + \dfrac{\partial^2 u}{\partial y^2} = 0$ *subject to the conditions*

(*i*) $u(0, y) = 0$, (*ii*) $u(1, y) = 0$,

(*iii*) $u(x, \infty) = 0$ and (*iv*) $u(x, 0) = \sin^3 \pi x$ for $0 < x < 1$.

SOLUTION: By the method of separation of variables, let $u = X(x)\, Y(y)$.

Hence $Y \dfrac{d^2 X}{dx^2} + X \dfrac{d^2 Y}{dy^2} = 0$ or $\dfrac{X''}{X} = \dfrac{-Y''}{Y} = -k^2$, say or $$\frac{d^2X}{dx^2}+k^2X=0 \text{ and } \frac{d^2Y}{dy^2}-k^2Y=0$$

$\Rightarrow$ $X = c_1 \cos kx + c_2 \sin kx$ and $Y = c_3 e^{ky} + c_4 e^{-ky}$

$\therefore$ $u = (c_1 \cos kx + c_2 \sin kx)(c_3 e^{ky} + c_4 e^{-ky})$.

Since $u(o, y) = 0$ we have $c_1 = 0$ and since $u(x, \infty) = 0$ we have $c_3 = 0$.

Thus, we have $u = c_2 \sin kx \cdot c_4 e^{-ky}$ or $u = c \sin kx\, e^{-ky}$.

Next, since $u(1, y) = 0$ we have $\sin k = 0 \Rightarrow k = n\pi,\ n \in I$

Hence we can write $u(x, y) = \sum_{1}^{\infty} c_n \sin n\pi x\, e^{-n\pi y}$.

Further, we are given that $u(x, 0) = \sin^3 \pi x,\ 0 < x < 1$

$\therefore$ $\sin^3\pi x = \sum_{1}^{\infty} c_n \sin n\pi x$. Also, $\sin^3\pi x = \frac{3}{4}\sin \pi x - \frac{1}{4}\sin 3\pi x$

$\Rightarrow$ $c_1 = \frac{3}{4}$ and $c_2 = 0,\ c_3 = \frac{-1}{4},\ c_4 = c_5 = \ldots = 0$.

$\therefore$ The solution is $u(x, y) = \frac{3}{4}\sin \pi x\, e^{-\pi y} - \frac{1}{4}\sin 3\pi x e^{-3\pi y}$. **Ans.**

3. *A long rectangular plate of width a cm. with insulated surfaces has its temperature equal to zero on both long sides and on one of the short sides so that $u(0, y) = 0$, $u(a, y) = 0$, and $u(x, \infty) = 0$ and $u(x, 0) = kx$. Find the expression for the temperature at any point of the plate in the steady state.*

SOLUTION: Steady state temperature $u(x, y)$ is given by $\frac{\partial^2 u}{\partial x^2}+\frac{\partial^2 u}{\partial y^2}=0$

B. Cs. are $u(0, y) = 0$ for $0 < y < \infty$, $u(a, y) = 0$, $0 < y < \infty$ $u(x, \infty) = 0$ for $0 < x < a$ and $u(x, 0) = kx$ for $0 < x < \pi$ The possible solution is $u = (A \cos px + B \sin px)(C e^{ky} + De^{-ky})$ because we are given that $u = 0$ as $y \to \infty$. Also, A = 0 since $u(0, y) = 0$

Further, $u(a, y) = 0 \Rightarrow \sin ap = 0$ hence $ap = n\pi,\ n \in I$

Next, since $u(x, \infty) = 0$ for $0 < x < a$, we have $C = 0$.

Thus, $u_n(x, y) = D_n \sin \frac{n\pi x}{a} e^{-ky}$

$\therefore$ $$u(x, y) = \sum_{n=1}^{\infty} D_n \sin\frac{n\pi x}{a} e^{-ky}$$

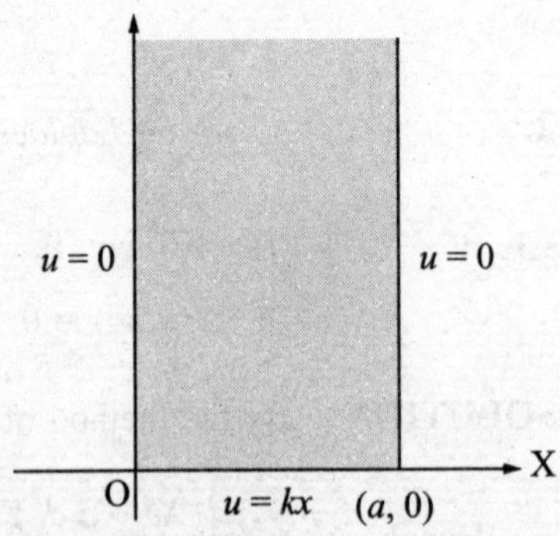

Also since $u(x, 0) = kx$ we have $kx = \sum D_n \sin\frac{n\pi x}{a}$

which is a half range sine series for $f(x) = kx$ in $(0, a)$

$$\therefore \quad D_n = \frac{2}{a}\int_0^a kx \sin\frac{n\pi x}{a}\,dx = \frac{2}{a}\left[\frac{-kx}{n\pi/a}\cos\frac{n\pi x}{a}\right]_0^a + \frac{2}{a}\int_0^a \frac{k}{n\pi/a}\cos\frac{n\pi x}{a}\,dx$$

$$= \frac{-2ka(-1)^n}{\pi n} + \frac{2k}{n\pi}\left[\sin\frac{n\pi x}{a}\cdot\frac{a}{n\pi}\right]_0^a = (-1)^{n+1}\frac{2ak}{n\pi} + 0.$$

$$\therefore \quad u(x, y) = \sum_{n=1}^{\infty} (-1)^{n+1}\frac{2ak}{n\pi}\sin\frac{n\pi x}{a}\cdot e^{-ky} \quad \text{is the required solution.}$$ **Ans.**

4. *A square plate of width 'a' is bounded by the lines $x = 0,\ x = a,\ y = 0,\ y = a$. The edges $x = 0,\ y = a$ are kept at zero temperature. The edge $y = 0$ is insulated and the edge $x = a$ is kept at the temperature u_0. Find the steady state temperature u at any point of the plate.*

SOLUTION: We wish to solve the Laplace equation $\dfrac{\partial^2 u}{\partial x^2} + \dfrac{\partial^2 u}{\partial y^2} = 0$

The boundary conditions are $u(0, y) = 0,\ u(x, a) = 0,\ u(a, y) = u_0$

and $\left(\dfrac{\partial u}{\partial y}\right) = 0$ since the edge $y = 0$ is insulated.

The possible solution is $u = (c_1 \cosh px + c_2 \sinh px)(c_3 \cos py + c_4 \sin py)$

Since $u(0, y) = 0$ we have $c_1 = 0 \quad \therefore \quad u = \sinh px\,(A \cos py + B \sin py)$

and $\dfrac{\partial u}{\partial y} = \sinh px\,(-Ap \sin py + Bp \cos py).$

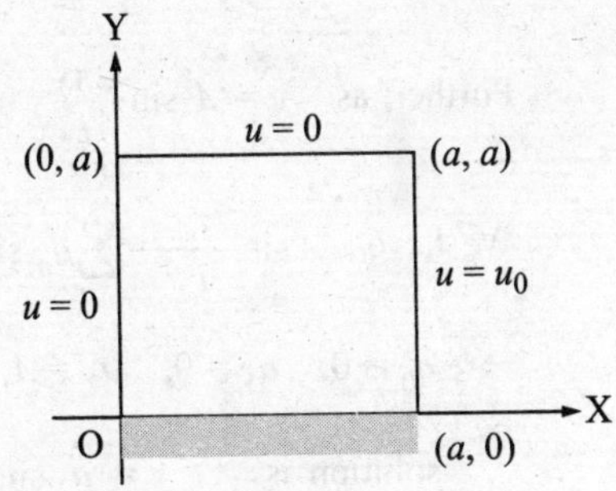

Since $\dfrac{\partial u}{\partial y} = 0$ on $y = 0$, we get $B = 0$.

Therefore $u = A \sinh px \cos py.$

The condition $u(x, a) = 0$ gives $\cos pa = 0$

$$\Rightarrow \quad pa = (2r + 1)\frac{\pi}{2}, \quad r = 0, 1, 2 \ldots$$

$$\therefore \quad u = (-1)^{r+1} A \sinh (2r+1)\frac{\pi}{2} x \cos (2r+1)\frac{\pi}{2} y, \quad r = 0, 1, 2, \ldots$$

The general solution is $$u = \sum_{r=0}^{\infty} (-1)^{r+1} A_r \sinh\left[\frac{2r+1}{2}\frac{\pi x}{a}\right]\cos\left[\frac{(2r+1)\pi y}{2a}\right]$$

Next, $u(a, y) = u_0, \quad 0 < y < a$

Let us have the Fourier sine series of u defined above, as

$$u = \sum b_n \sin\frac{n\pi y}{a} \quad \text{where} \quad b_n = \frac{2}{a}\int_0^a u_0 \sin\frac{n\pi y}{a}\,dy$$

or $$b_n = \frac{2}{a}u_0 \cdot \frac{a}{n\pi}\left[-\cos\frac{n\pi y}{a}\right]_0^a = \frac{-2u_0}{n\pi}[\cos n\pi - 1] = \begin{cases} 0, & \text{when } n \text{ is even} \\ \dfrac{4u_0}{n\pi} & \text{when } n \text{ is odd.} \end{cases}$$

Thus, the solution becomes

$$u = \frac{4u_0}{\pi}\sum_{r=0}^{\infty}\frac{(-1)^{r+1}}{(2r+1)}\sinh\frac{(2r+1)\pi x}{2a}\cos\frac{(2r+1)\pi y}{2a}.$$

5. *A tightly stretched string of length l is attached at $x = 0$ and at $x = l$. Find the expression for y the displacement of the string at a distance x, given that $y = A\sin\frac{2\pi x}{l}$ at $t = 0$.*

SOLUTION: The wave equation is $\frac{\partial^2 y}{\partial t^2} = c^2\frac{\partial^2 y}{\partial x^2}$. Its possible solution is

$$y = (c_1\cos px + c_2\sin px)\,(c_3\cos p\,ct + c_4\sin pct).$$

Since $y(0, t) = 0$ we have $c_1 = 0$ and since $\frac{\partial y}{\partial t} = 0$ at $t = 0$ we have $c_4 = 0$

$\therefore$ $y = K\sin px\cos cpt.$

Also since $y(l, t) = 0$ we have $\sin pl = 0 \Rightarrow pl = n\pi$ or $p = \frac{n\pi}{l}$ $n = 1, 2, 3, \ldots$

$\therefore$ we can write $y = \sum_{n=1}^{\infty} a_n \sin\frac{n\pi x}{l}\cos\frac{cn\pi t}{l}$ we have $A\sin\frac{2\pi x}{l} = \sum_{n=1}^{\infty} a_n\sin\frac{n\pi x}{l}\cos cm\pi$

Further, as $y = A\sin\frac{2\pi x}{l}$ at $t = 0$

We have $A\sin\frac{2\pi x}{l} = \sum_{n=1}^{\infty} a_n\sin\frac{n\pi x}{l}$

$\Rightarrow a_0 = 0,\ a_1 = 0,\ a_2 = A,\ a_3 = a_4 = \ldots = 0$

$\therefore$ solution is $y = a_2\sin\frac{2\pi x}{l}\cos\frac{2c\pi t}{l} = A\sin\frac{2\pi x}{l}\cos\frac{2\pi ct}{l}$ **Ans.**

6. *An elastic string of length l is stretched. One end is taken as the origin and at a distance $2l/3$ from this end the string is displaced through a distance 'a' transversely and the released from rest. Obtain $y(x, t)$ the vertical displacement at any x if y satisfies the equation $\frac{\partial^2 y}{\partial t^2} = c^2\frac{\partial^2 y}{\partial x^2}$.*

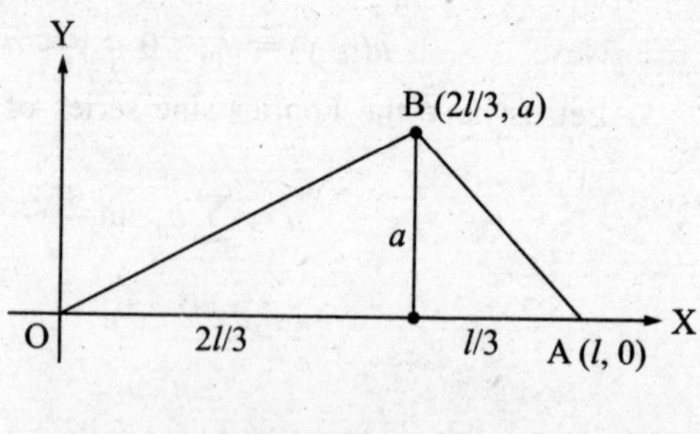

SOLUTION: Equation of OB is $y = \frac{3ax}{2l}$.

Equation of AB is $y = \frac{3a}{l}(l - x)$.

The boundary conditions are

$$y(0, t) = 0, \ \ y(l, t) = 0$$

and that initial conditions are

$$\left(\frac{\partial y}{\partial t}\right)_{t=0} = 0 \quad \text{and} \quad y(x, 0) = \begin{cases} \dfrac{3ax}{2l}, & 0 < x < \dfrac{2l}{3} \\ \dfrac{3a}{l}(l-x), & \dfrac{2l}{3} < x < l. \end{cases}$$

The most general solution is

$$y(x, t) = (c_1 \cos px + c_2 \sin px)(c_3 \cos cpt + c_4 \sin cpt).$$

The condition $y(0, t) = 0$ gives $c_1 = 0$

and the condition $\left(\dfrac{\partial y}{\partial t}\right)_{t=0} = 0$ gives $c_4 = 0.$

$\therefore$ we can write $y(x, t) = c \sin px \cos cpt$

The condition $y(l, t) = 0$ gives $\sin pl = 0 \ \therefore \ p = \dfrac{n\pi}{l}.$

$$\therefore \qquad y(x, t) = c \sin\left(\frac{n\pi x}{l}\right)\cos\left(\frac{cn\pi t}{l}\right), \quad n = 1, 2, 3, ...$$

Combining all these solutions, we can write

$$y(x, t) = \sum_{n=1}^{\infty} b_n \sin\frac{n\pi x}{l}\cos\frac{cn\pi t}{l}.$$

Next we obtain the Fourier sine series of $y(x, 0) = f(x) = \sum b_n \sin\dfrac{n\pi x}{l}$

where $f(x) = \dfrac{3ax}{2l}, \ 0 < x < \dfrac{2l}{3}$

and $\qquad = \dfrac{3a}{l}(l-x), \ \dfrac{2l}{3} < x < l.$

$$\therefore \qquad b_n = \frac{2}{l}\int_0^l f(x)\sin\frac{n\pi x}{l}\,dx = \frac{2}{l}\int_0^{2l/3}\frac{3ax}{2l}\sin\frac{n\pi x}{l}\,dx + \frac{2}{l}\int_{2l/3}^{l}\frac{3a}{l}(l-x)\sin\frac{n\pi x}{l}\,dx$$

$$= \frac{6a}{l^2}\left[\left\{\frac{x}{2}\left(-\frac{l}{n\pi}\cos\frac{n\pi x}{l}\right) - \left(\frac{1}{2}\right)\left(\frac{-l^2}{n^2\pi^2}\sin\frac{n\pi x}{l}\right)\right\}_0^{2l/3}\right.$$

$$\left. + \left\{(l-x)\left(\frac{-l}{n\pi}\cos\frac{n\pi x}{l}\right) - (-1)\left(\frac{-l^2}{n^2\pi^2}\sin\frac{n\pi x}{l}\right)\right\}_{2l/3}^{l}\right]$$

$$= \frac{6a}{l^2}\left[-\frac{l^2}{3n\pi}\cos\frac{2n\pi}{3} + \frac{l^2}{2n^2\pi^2}\sin\frac{2n\pi}{3} + \frac{l^2}{3n\pi}\cos\frac{2n\pi}{3} + \frac{l^2}{n^2\pi^2}\sin\frac{2n\pi}{3}\right]$$

$$= \frac{6a}{l^2}\left[\frac{3}{2}\frac{l^2}{n^2\pi^2}\sin\frac{2n\pi}{3}\right] = \frac{9a}{n^2\pi^2}\sin\left(\frac{2n\pi}{3}\right)$$

$$\therefore \qquad y(x, t) = \frac{9a}{\pi^2}\sum_{1}^{\infty}\frac{1}{n^2}\sin\frac{2n\pi}{3}\sin\frac{n\pi x}{l}\cos\frac{n\pi ct}{l}$$ is the required solution. **Ans.**

7. *A string is stretched tightly between $x = 0$ and $x = l$ and both the ends are given the displacement $y = a \sin pt$ perpendicular to the string. The displacement y in the string satisfies the equation $\frac{\partial^2 y}{\partial x^2} = \frac{1}{c^2} \cdot \frac{\partial^2 y}{\partial t^2}$. Obtain the expression for the displacement in the string at any time t.*

SOLUTION: $\frac{\partial^2 y}{\partial t^2} = c^2 \frac{\partial^2 y}{\partial x^2}$ whose suitable solution is

$$y = (c_1 \cos cx + c_2 \sin cx)(c_3 \cos ckt + c_4 \sin ckt)$$

Given that $y(0, t) = a \sin pt$ gives $c_1(c_3 \cos ckt + c_4 \sin ckt) = a \sin pt$

$\Rightarrow \quad c_1 \cdot c_3 = 0$ and $ck = p \quad \therefore \quad k = \frac{p}{c}$ and $c_1 c_4 = a$

and we can write $y = \left(a \cos \frac{px}{c} + c_2 \sin \frac{px}{c}\right) \sin pt$

Further, since $y(l, t) = a \sin pt$ we get

$$a \cos \frac{pl}{c} + c_2 \sin \frac{pl}{c} = a \quad \therefore \quad c_2 = \frac{a\left(1 - \cos \frac{pl}{c}\right)}{\sin \frac{pl}{c}} = a \tan\left(\frac{pl}{2c}\right)$$

Therefore $\quad y = \left[a \cos \frac{px}{c} + a \tan\left(\frac{pl}{2c}\right) \sin \frac{px}{c}\right] \sin pt$

$$= \frac{a}{\cos\left(\frac{pl}{2c}\right)} \left[\cos \frac{px}{c} \cos \frac{pl}{2c} + \sin \frac{px}{c} \sin \frac{pl}{2c}\right] \sin pt$$

or $\quad y = a \sec\left(\frac{pl}{2c}\right) \cos\left(\frac{px}{c} - \frac{pl}{2c}\right) \sin pt$ is the solution. **Ans.**

8. *Solve the differential equation $\frac{\partial^2 u}{\partial t^2} = 4 \frac{\partial^2 u}{\partial x^2}$ subject to the conditions $u = \sin t$ at $x = 0$ and $\frac{\partial u}{\partial x} = \sin t$ at $x = 0$.*

SOLUTION: Let $\quad u = X(x)\, T(t) \quad$ hence $\quad XT'' = uTX''$

or $\quad \frac{X''}{X} = \frac{1}{4} \frac{T''}{T} = -p^2$, say, a constant suited here.

$\Rightarrow \quad X'' + p^2 X = 0$ and $T'' + 4p^2 T = 0$

$\therefore \quad u = (c_1 \cos px + c_2 \sin px)(c_3 \cos 2pt + c_4 \sin 2pt)$

From the condition $\quad u(0, t) = \sin t$ we get $c_1(c_3 \cos 2pt + c_4 \sin 2pt) = \sin t$

$\Rightarrow \quad c_1 c_3 = 0$ hence $c_3 = 0$ and $c_1 c_4 = 1$ and $p = \frac{1}{2}$

$$\therefore \quad u = \left(c_1 \cos\frac{x}{2} + c_2 \sin\frac{x}{2}\right) c_4 \sin t = \left(\cos\frac{x}{2} + A \sin\frac{x}{2}\right)\sin t$$

Further $\dfrac{\partial u}{\partial x} = \left(\dfrac{-1}{2}\sin\dfrac{x}{2} + \dfrac{A}{2}\cos\dfrac{x}{2}\right)\sin t$ since we are given that $\dfrac{\partial u}{\partial x} = \sin t$ at $x = 0$ hence $A = 2$

$\therefore \quad u = \left(\cos\dfrac{x}{2} + 2\sin\dfrac{x}{2}\right)\sin t$ is the solution. **Ans.**

9. *A rod* l cm *long with insulated lateral surface is initially at zero temperature throughout. If one end is kept at* 0ºC *and the other at* 100º C *find the temperature function* $u(x, t)$ *where* x *is the distance measured along the rod from the end which is at* 0º C.

SOLUTION: This is one-dimensional heat flow and therefore the governing equation is $\dfrac{\partial u}{\partial t} = a^2 \dfrac{\partial^2 u}{\partial x^2}$ where the boundary conditions are $u(x, 0) = 0$, $u(0, t) = 0$, $u(l, t) = 100$

Applying the method of separation of variables we get

$X T' = a^2 X'' T$ *i.e.*, $\dfrac{T'}{T} = \dfrac{a^2 X''}{X} = -p^2$, a constant.

Hence $T = A e^{-p^2 t}$ and $X = B\cos\dfrac{px}{a} + C\sin\dfrac{px}{a}$.

Since $u(0, t) = 0$ we have $B = 0$ $\therefore$ $u(x, t) = c' e^{-p^2 t}\sin\dfrac{px}{a}$ where c' is constant.

Next, since $u(l, t) = 0$ we have $\sin\dfrac{pl}{a} = 0$ hence $\dfrac{pl}{a} = n\pi$ or $p = \dfrac{an\pi}{l}$

$$\therefore \quad u(x, t) = c' e^{-a^2 n^2 \pi^2 t/l^2} \sin\left(\frac{an\pi}{l}\cdot\frac{x}{a}\right).$$

The most general solution is $u = A + Bx + \sum c_n e^{-n^2 a^2 \pi^2 t/l^2} \sin\dfrac{n\pi x}{l}$

since $u = A + Bx$ also satisfies the given equation.

Now since $u(x, 0) = 0$ we have $0 = A + Bx + \sum c_n \sin\dfrac{n\pi x}{l}$ and since $u(0, t) = 0$ we have $A = 0$.

Also since $u(l, t) = 100$ we have $100 = A + Bl$ $\therefore$ $B = 100/l$.

Thus $$\frac{-100}{l} x = \sum_{n=1}^{\infty} c_n \sin\frac{n\pi x}{l}.$$

To determine c_n we expand $\dfrac{-100x}{l}$ as half-range Fourier sine series

and obtain $$c_n = \frac{2}{l}\int_0^l \frac{-100x}{l}\sin\frac{n\pi x}{l}\,dx = \frac{-200}{l^2}\int_0^l x\sin\frac{n\pi x}{l}\,dx = \frac{-200}{l^2}\left[\frac{-l^2}{n\pi}\cos n\pi\right] = \frac{200}{n\pi}\cos n\pi$$

$$\therefore \quad u(x, t) = \frac{100x}{l} + \sum_{1}^{\infty} \frac{200}{n\pi} \cos n\pi \, e^{-n^2 a^2 \pi^2 t/l^2} \sin \frac{n\pi x}{l}$$

$$= \frac{100x}{l} + \left[\frac{-200}{\pi} e^{-a^2\pi^2 t/l^2} \sin \frac{\pi x}{l} + \frac{200}{2\pi} e^{-4a^2 t/l^2} \sin \frac{2\pi x}{l} + \ldots \right]$$

or $$u = \frac{100x}{l} - \frac{200}{\pi} \left[e^{-\pi^2 a^2 t/l^2} \sin \frac{\pi x}{l} - \frac{1}{2} e^{-4\pi a^2 t/l^2} \sin \frac{2\pi x}{l} + \ldots \right]$$ **Ans.**

10. *Solve the equation* $\frac{\partial u}{\partial t} = K \frac{\partial^2 u}{\partial x^2}$ *for the conduction of heat along a thin rod without radiation subject to the following boundary and initial conductions*

(*i*) u remains finite for any lapse of time

(*ii*) $\frac{\partial u}{\partial x} = 0$ for $x = 0$ and $x = l$

(*iii*) $u = x(l - x)$ at $t = 0$ between $x = 0$ and $x = l$.

SOLUTION: Substituting $u = XT$ the given equation becomes $\frac{X''}{X} = \frac{T''}{KT} = -p^2$ (say)

$$\therefore \quad \frac{d^2 X}{dx^2} + p^2 x^2 = 0 \quad \text{and} \quad \frac{dT}{dx} + p^2 KT = 0.$$

Whose solution is $X = (c_1 \cos px + c_2 \sin px)$ and $T = c_3 e^{-p^2 Kt}$

$$\therefore \quad u = (c_1 \cos px + c_2 \sin px) \, c_3 e^{-p^2 Kt} \quad \text{when } t = 0, \; X = Ax + B, \; T = c_3$$

and $$\frac{\partial u}{\partial x} = (-c_1 p \sin px + c_2 p \cos px) \, c_3 e^{-p^2 Kt}$$

Since $\frac{\partial u}{\partial x} = 0$ for $x = 0$ we have $c_2 = 0$ and $A = 0$

and $\frac{\partial u}{\partial x} = 0$ for $x = l$ we get $\sin pl = 0 \quad \therefore \; p = \frac{n\pi}{l}$, $n = 1, 2, 3, \ldots$

$$\therefore \quad u = c_1 \cos px \, e^{-p^2 Kt} = a_n \cos \frac{n\pi x}{l} e^{\frac{-n^2\pi^2}{l^2} Kt}$$

Therefore the general solution becomes $u = a_0 + \sum a_n \cos \frac{n\pi x}{l} e^{\frac{-n^2\pi^2 Kt}{l^2}}$

Now using the condition $u = lx - x^2$ for $t = 0, 0 < x < l$.

$$\therefore \quad lx - x^2 = a_0 + \sum a_n \cos \frac{n\pi x}{l}.$$

This being the expansion of $lx - x^2$ as a half-range Fourier consine series in $(0, l)$, we have

$$a_0 = \frac{2}{l} \int_0^l (lx - x^2) \cos \frac{n\pi x}{l} \, dx$$

$$= \frac{2}{l}\left[(lx - x^2)\sin\frac{n\pi x}{l}\cdot\frac{l}{n\pi} - (l-2x)\left(-\frac{l^2}{n^2\pi^2}\cos\frac{n\pi x}{l}\right) + (-2)\left(\frac{-l^3}{n^3\pi^3}\sin\frac{n\pi x}{l}\right)\right]_0^l$$

$$= \frac{2}{l}\left[0 - \frac{l^3}{\pi^2 n^2}(\cos n\pi + 1) + 0\right] = \begin{cases} -\dfrac{4l^2}{n^2\pi^2} & \text{when } x \text{ is even} \\ 0 & \text{otherwise.} \end{cases}$$

Thus, taking $n = 2m$, the solution can be written as

$$u = \frac{l^2}{6} - \frac{l^2}{\pi^2}\sum_{m=1}^{\infty}\frac{1}{m^2}\cos\left(\frac{2m\pi x}{l}\right)e^{-4m^2\pi^2 Kt/l^2}$$ **Ans.**

11. *Using the method of separation of variables, solve*

$$\frac{\partial u}{\partial x} = 2\frac{\partial u}{\partial t} + u \quad \text{where} \quad u(x, 0) = 6e^{-3x}.$$

SOLUTION: Let $u = X(x)\,T(t)$. Putting it in the given equation, we get

$$\frac{dX}{dx}T = 2X\frac{dT}{dt} + XT \quad \text{or} \quad \frac{X'}{X} = \frac{2T'}{T} + 1 = K, \text{ a pure constant.}$$

Therefore, we have $\dfrac{dX}{dx} - KX = 0$ or $\dfrac{dX}{X} = K dx \Rightarrow X = c_1 e^{Kx}$.

And $\dfrac{2dT}{dt} = (K-1)T$ or $\dfrac{dT}{T} = \dfrac{K-1}{2}dt \Rightarrow T = c_2 e^{(K-1)t/2}$

Thus, the solution is $u = XT = c_1 e^{Kx}\, c_2 e^{(K-1)t/2} = c\, e^{Kx} e^{(K-1)t/2}$

Since $u(x, 0) = 6e^{-3x}$ we get $c = 6$ and $K = -3$

$\therefore$ The solution is $u(x, t) = 6e^{-3x}e^{-2t} = 6e^{-(3x+2t)}$. **Ans.**

12. *Solve the equation* $\dfrac{\partial^2 z}{\partial x^2} - 2\dfrac{\partial z}{\partial x} + \dfrac{\partial z}{\partial y} = 0$ *by the method of separation of variables.*

SOLUTION: Let $z = X(x)\,Y(y)$, then the equation can be written as

$$Y\frac{d^2X}{dx^2} - 2Y\frac{dX}{dx} + X\frac{dY}{dy} = 0 \quad \text{or} \quad \frac{1}{X}\frac{d^2X}{dx^2} - \frac{2}{X}\frac{dX}{dx} + \frac{1}{Y}\frac{dY}{dy} = 0$$

or
$$\frac{1}{X}\frac{d^2X}{dx^2} - \frac{2}{X}\frac{dX}{dx} = \frac{-1}{Y}\frac{dY}{dy} = K, \text{ a pure constant}$$

$\Rightarrow$
$$\frac{d^2X}{dx^2} - 2\frac{dX}{dx} - KX = 0 \quad \text{and} \quad \frac{dY}{dy} + KY = 0$$

or
$$(D^2 - 2D - K)X = 0 \text{ and } (D' + K)Y = 0 \text{ where } D \equiv \frac{\partial}{\partial x},\ D' \equiv \frac{\partial}{\partial y}.$$

The auxiliary equations are $m^2 - 2m - K = 0 \therefore m = 1 \pm \sqrt{K+1}$ and $n + K = 0 \therefore n = -K$

$$\therefore \quad X = c_1 e^{(1+\sqrt{K+1})x} + c_2 e^{(1-\sqrt{K+1})x} \quad \text{and} \quad Y = c_3 e^{-Ky}$$

$$\therefore \quad z = XY = \left[c_1 e^{(1+\sqrt{K+1})x} + c_2 e^{(1-\sqrt{K+1})x}\right] c_3 e^{-Ky}$$

or
$$z = \left[A e^{(1+\sqrt{K+1})x} + Be^{(1-\sqrt{K+1})x}\right] e^{-Ky}$$
Ans.

13. *An insulated rod of length l has its ends A and B maintained at $0\,°C$ and $100\,°C$ respectively until steady state conditions prevail. If B is suddenly reduced to $0\,°C$ and maintained at $0\,°C$, find the temperature at a distance x from A at time t.*

SOLUTION: If $u(x, t)$ is the temperature distribution at a distance x from the end A at any time t then it is one dimensional heat equation

$$\frac{\partial u}{\partial t} = c^2 \frac{\partial^2 u}{\partial x^2}, \quad 0 < x < l, \; t > 0. \qquad ...(1)$$

Before the temperature change at the end B the heat flow was independent of time, so the system was in steady state hence

$\dfrac{\partial^2 u}{\partial x^2} = 0$ which gives $u = ax + b$ where a and b are arbitrary constants.

As per the conditions $u(0, 0) = 0$ and $u(l, 0) = 100$

$$\Rightarrow \; b = 0, \; a = 100/l \quad \text{hence} \quad u(x, 0) = \frac{100}{l} x \qquad ...(2)$$

The boundary conditions for the subsequent flow are $u(0, t) = 0$ and $u(l, t) = 0$

As discussed in earlier problems of one dimensional heat equation, we have

$$u(x, t) = \sum_{n=1}^{\infty} B_n \sin \frac{n\pi x}{l} e^{-\frac{c^2\pi^2 t}{l^2}} \qquad ...(3)$$

Using (2)
$$\frac{100x}{l} = \sum B_n \sin \frac{n\pi x}{l}.$$

which is the half-range sine series of $f(x) = \dfrac{100x}{l}$ in $(0, l)$.

Hence
$$B_n = \frac{2}{l}\int_0^l \frac{100x}{l} \sin\frac{n\pi x}{l}\, dx = \frac{200}{l^2}\left[-\frac{xl}{n\pi}\cos\frac{n\pi x}{l} + \frac{l^2}{(n\pi)^2}\sin\frac{n\pi x}{l}\right]_0^l = \frac{200}{-n\pi}\cos n\pi = \frac{200}{n\pi}(-1)^{n+1}$$

Substituting for B_n in (3) we get the solution as

$$u(x, t) = \frac{200}{\pi}\sum \frac{(-1)^{n+1}}{n} \sin \frac{n\pi x}{l} e^{\frac{-c^2 n^2 \pi^2 t}{l^2}}, \; 0 < x < l, \; t \geq 0.$$
Ans.

❑❑❑

CHAPTER

25 Fourier Transforms and Applications

IMPORTANT FORMULAE AND USEFUL RESULTS

1. Fourier Integral Theorem

$$f(x) = \frac{1}{\pi}\int_0^\infty\int_0^\infty f(t)\cos\lambda(t-x)\,dt\,d\lambda = \text{Fourier integral of } f(x)$$

$$f(x) = \frac{2}{\pi}\int_0^\infty\int_0^\infty f(t)\cos\lambda t\cos\lambda x\,dt\,d\lambda = \text{Fourier cosine integral}$$

$$f(x) = \frac{2}{\pi}\int_0^\infty\int_0^\infty f(t)\sin\lambda t\sin\lambda x\,dt\,d\lambda = \text{Fourier sine Integral}$$

Fourier transform of $f(x)$ is $F(\lambda) = \dfrac{1}{\sqrt{2\pi}}\displaystyle\int_{-\infty}^{\infty} e^{i\lambda x} f(t)\,dt = \overline{f}(\lambda)$

Hence $\quad f(x) = \dfrac{1}{\sqrt{2\pi}}\displaystyle\int_{-\infty}^{\infty} e^{-i\pi x} F(\lambda)\,d\lambda$

Fourier Cosine transform of $f(x) = F_C(\lambda) = \sqrt{\dfrac{2}{\pi}}\displaystyle\int_{-\infty}^{\infty} f(x)\cos\lambda x\,dx$

Hence $\quad f(x) = \sqrt{\dfrac{2}{\pi}}\displaystyle\int_{-\infty}^{\infty} F_C(\lambda)\cos\lambda x\,d\lambda$

Fourier Sine transform of $f(x) = F_5(\lambda) = \sqrt{\dfrac{2}{\pi}}\displaystyle\int_0^{\infty} f(x)\sin\lambda x\,dx$

Hence $\quad f(x) = \sqrt{\dfrac{2}{\pi}}\displaystyle\int_0^{\infty} F_5(\lambda)\sin\lambda x\,d\lambda$

2. Properties of Fourier Transform

Change of scale: $F[f(ax)] = \frac{1}{a} F\left(\frac{\lambda}{a}\right)$ if $F(\lambda) = F(f(x))$

Shifting Property: $F[f(x-a)] = e^{i\lambda a} F(\lambda)$

Modulation Property: If $F[f(x)] = F(\lambda)$ **then** $F[f(x)\, e^{-aix}] = F(\lambda - a)$

Convolution Theorem: Convolution of $f(x)$ and $g(x)$ $f * g$ is given by

$$f(x) * g(x) = \int_{-\infty}^{\infty} f(u)\, g(x-u)\, du$$

If $F[f(x)] = F(\lambda)$ and $F(g(x)) = G(\lambda)$, then

$$F[f(x) * g(x)] = F(\lambda)\, G(x)$$

3. Fourier Transform of Derivatives Let $u = u(x, t)$, then

$$F\left(\frac{\partial u}{\partial x}\right) = 0 - i\lambda \bar{u}(\lambda, t)$$

$$F\left(\frac{\partial^2 u}{\partial x^2}\right) = 0 - \lambda^2 \bar{u}(\lambda, t)$$

$$F\left(\frac{\partial u}{\partial t}\right) = \frac{d}{dt} \bar{u}(\lambda, t)$$

$$F_5\left(\frac{\partial^2 u}{\partial x^2}\right) = \lambda u(0, t) - \lambda^2\, \bar{u}_s(\lambda, t)$$

$$F_C\left(\frac{\partial^2 u}{\partial x^2}\right) = -\left[\frac{\partial u}{\partial x}\right]_{x=0} - \lambda^2 F_C(u(x, t))$$

SOLUTIONS TO EXERCISE 25A

1. *Show that* $\int_0^{\infty} \frac{\sin\alpha\cos\alpha x}{\alpha}\, d\alpha = \pi/2$ *when* $0 \le x < 1$

$$= \pi/4 \quad \textit{when} \quad x = 1$$

$$= 0 \quad \textit{when} \quad x > 1.$$

SOLUTION: Consider $f(x) = 1$ for $0 < x < 1$

$$= 0 \text{ for } x > 1$$

then $F_c(s) = \int_{-\infty}^{\infty} f(x)\cos sx\, dx = \int_0^1 1\cos sx\, dx + \int_1^{\infty} 0\cos sx\, dx = \left[\frac{\sin sx}{s}\right]_0^1 = \frac{\sin s}{s}$

Inverting the Fourier cosine transform above, gives

$$f(x) = \frac{2}{\pi}\int_0^{\infty} \frac{\sin s}{s}\cos sx\, ds = 1 \text{ for } 0 < x < 1$$

or $$= 0 \quad \text{for } x > 1$$

or $$\int_0^{\infty} \frac{\sin s}{s}\cos sx\, ds = \frac{\pi}{2} \quad \text{for } 0 < x < 1$$

$$= 0 \quad \text{for } x > 1$$

At $x = 1$ $\int_0^{\infty} \frac{\sin s}{s}\cos sx\, ds =$ Mean of $\frac{\pi}{2}$ and $0 = \frac{\pi}{4}$. **Hence proved.**

2. *Find the Fourier transform of the function given by* $f(x) = e^{-|x|}$.

SOLUTION: $F(f(x)) = F(\lambda) = \int_{-\infty}^{\infty} e^{-|x|}\, e^{i\lambda x}\, dx$

$$= \int_{-\infty}^{0} e^{+x}\, e^{i\lambda x} dx + \int_0^{\infty} e^{-x}\, e^{i\lambda x}\, dx = I_1 + \int_0^{\infty} e^{-x}\, e^{i\lambda x}\, dx$$

In the first integral I_1 on the RHS, we put $x = -y$

$\therefore$ $$I_1 = -\int_{\infty}^{0} e^{-y}\, e^{-i\lambda y}\, dy = \int_0^{\infty} e^{-y}\, e^{-i\lambda y} dy = \int_0^{\infty} e^{-x}\, e^{-i\lambda x}\, dx$$

Therefore $$F(\lambda) = \int_0^{\infty} e^{-x}\, e^{-i\lambda x}\, dx + \int_0^{\infty} e^{-x}\, e^{i\lambda x}\, dx = 2\int_0^{\infty} e^{-x}\cos\lambda x\, dx$$

$$= \left[\frac{2e^{-x}}{1+\lambda^2}(-1\cos\lambda x + \lambda\sin\lambda x)\right]_0^{\infty} = \frac{2}{1+\lambda^2}$$ **Ans.**

3. *Find the Fourier sine transform of* $f(x) = \frac{1}{x} e^{-ax}$, $a > 0$ *and use it to evaluate*

$$\int_0^{\infty} \tan^{-1}\left(\frac{x}{a}\right) \sin x \, dx$$

SOLUTION: $f(x) = \frac{1}{x} e^{-ax}$, its Fourier sine transform is

$$F_s(f(x)) = \int_0^{\infty} \frac{1}{x} e^{-ax} \sin \lambda x \, dx = F_s(\lambda) \quad \text{...(1)}$$

$$\therefore \quad \frac{d}{d\lambda} F_s(\lambda) = \int_0^{\infty} \frac{1}{x} e^{-ax} x \cos \lambda x \, . \, dx = \int_0^{\infty} e^{-ax} \cos \lambda x \, dx$$

$$= \left[\frac{e^{-ax}}{a^2 + \lambda^2} (-a \cos \lambda x + \lambda \sin \lambda x) \right]_0^{\infty} = 0 + \frac{a}{a^2 + \lambda^2}$$

On integration we get $F_s(\lambda) = \int \frac{a \, d\lambda}{a^2 + \lambda^2} = \tan^{-1} \frac{\lambda}{a} + C$

Putting here $\lambda = 0$ gives $F_s(0) = 0 + C$ But $F_s(0) = 0$ from (1) $\therefore$ $C = 0$

Therefore $F_s(\lambda) = \tan^{-1}\left(\frac{\lambda}{a}\right)$

Taking inverse Fourier sine transform, we get

$$f(x) = \frac{2}{\pi} \int_0^{\infty} \tan^{-1}\left(\frac{\lambda}{a}\right) \sin \lambda x \, d\lambda = \frac{1}{x} e^{-ax}$$

Taking $x = 1$ we get $f(1) = \int_0^{\infty} \tan^{-1} \frac{\lambda}{a} \sin \lambda \, d\lambda = \frac{\pi}{2} e^{-a}$. **Ans.**

4. *Find the Fourier cosine integral representation of*

$$f(x) = \begin{cases} x^2, & 0 < x < a \\ 0, & x > a \end{cases}.$$

SOLUTION: Since $f(x) = \begin{cases} x^2, & 0 < x < a \\ 0, & x > a \end{cases}$ we have

$$F_c[f(x)] = F_c(\lambda) = \int_0^a x^2 \cos \lambda x \, dx + \int_a^{\infty} 0 \cos \lambda x \, dx = \int_0^a x^2 \cos \lambda x \, dx$$

$$= \left[x^2 \frac{\sin \lambda x}{\lambda} \right]_0^a - \int_0^a \frac{2x}{\lambda} \sin \lambda x \, dx = \frac{a^2 \sin a\lambda}{\lambda} - \frac{2}{\lambda} \int_0^a x \sin \lambda x \, dx$$

$$= \frac{a^2}{\lambda}\sin a\lambda - \frac{2}{\lambda}\left[-x\frac{\cos\lambda x}{\lambda}\right]_0^a - \frac{2}{\lambda}\int_0^a 1\frac{\cos\lambda x\, dx}{\lambda}$$

$$= \frac{a^2}{\lambda}\sin a\lambda + \frac{2a}{\lambda^2}\cos a\lambda - \frac{2}{\lambda^2}\left[\frac{\sin\lambda x}{\lambda}\right]_0^a = \left(\frac{a^2}{\lambda} - \frac{2}{\lambda^3}\right)\sin a\lambda + \frac{2a}{\lambda^2}\cos a\lambda$$

On inverting the Fourier cosine transform on both sides, we get

$$f(x) = \frac{2}{\pi}\int_0^\infty \left[\left(\frac{a^2}{\lambda} - \frac{2}{\lambda^3}\right)\sin a\lambda + \frac{2a}{\lambda^2}\cos a\lambda\right]\cos\lambda x\, d\lambda$$ **Ans.**

5. *Find the Fourier cosine transform of the function*

$$f(x) = \begin{cases}\cos x & \text{for } 0 < x < a \\ 0 & \text{for } x > a\end{cases}$$

SOLUTION: Taking Fourier cosine transform of $f(x)$, we get

$$F_c(f(x)) = F_c(\lambda) = \int_0^a \cos x\cos\lambda x\, dx + \int_a^0 0.\cos\lambda x\, dx = \frac{1}{2}\int_0^a [\cos(\lambda+1)x + \cos(\lambda-1)x]dx$$

$$= \frac{1}{2}\left[\frac{\sin(\lambda+1)x}{\lambda+1} + \frac{\sin(\lambda-1)x}{\lambda-1}\right]_0^a = \frac{1}{2}\left[\frac{\sin(\lambda+1)a}{\lambda+1} + \frac{\sin(\lambda-1)a}{\lambda-1}\right]$$ **Ans.**

6. *Find the Fourier sine and cosine transform of $f(x) = e^{-x}$ and hence show that*

$$\int_0^\infty \frac{\cos mx}{1+x^2}dx = \frac{\pi}{2}e^{-m} = \int_0^\infty \frac{x\sin mx}{1+x^2}dx.$$

SOLUTION: $f(x) = e^{-x}$ hence $F_s(f(x)) = \int_0^\infty e^{-x}\sin sx\, dx = \left[\frac{e^{-x}}{1+s^2}(-\sin sx - s\cos sx)\right]_0^\infty = \frac{s}{1+s^2}$ ...(1)

Similarly $F_C(f(x)) = \int_0^\infty e^{-x}\cos sx\, dx = \left[\frac{e^{-x}}{1+s^2}(-1\cos sx + s\sin sx)\right]_0^\infty = \frac{1}{1+s^2}$ ...(2)

Taking now the inverse sine transform in (1), gives

$$e^{-x} = \frac{2}{\pi}\int_0^\infty \frac{s}{1+s^2}\sin sx\, ds \quad \text{or} \quad \int_0^\infty \frac{s\sin sx}{1+s^2}ds = \frac{\pi}{2}e^{-x}$$

Replacing s by x and x by m we get $\int_0^\infty \frac{x\sin mx\, dx}{1+x^2} = \frac{\pi}{2}e^{-m}$

Next taking inverse cosine transform in (2), gives $f(x) = e^{-x} = \frac{2}{\pi}\int_0^\infty \frac{1}{1+s^2}\cos sx\, dx.$

Now replacing s by x and x by m, we get $e^{-m}\frac{\pi}{2} = \int_0^\infty \frac{\cos mx\, dx}{1+x^2}$ **Hence proved.**

7. *Show that the Fourier sine and cosine transforms of* x^{n-1} are $\frac{\Gamma n}{s^n}\sin\frac{n\pi}{2}$ *and* $\frac{\Gamma n}{s^n}\cos\frac{n\pi}{2}$ *respectively.*

SOLUTION: We know that $\int_0^\infty e^{-x}x^{n-1}\,dx = \Gamma n$

Putting $x = i\lambda y$ so that $dx = i\lambda dy$, we get

$$\int_0^\infty e^{-i\lambda y}(i\lambda y)^{n-1}\,i\lambda dy = \Gamma n \quad \text{or} \quad \int_0^\infty \lambda^n y^{n-1}(\cos\lambda y - i\sin\lambda y)\,dy = \Gamma n\, i^{-n}$$

or $$\int_0^\infty y^{n-1}\cos\lambda y\,dy - i\int_0^\infty y^{n-1}\sin\lambda y\,dy = \frac{\Gamma n\, i^{-n}}{\lambda^n} = \frac{\Gamma n}{\lambda^n}e^{-i\,n\pi/2} = \frac{\Gamma n}{\lambda^n}\left(\cos\frac{n\pi}{2} - i\sin\frac{n\pi}{2}\right)$$

Equating real and Imaginary parts on both sides, gives,

$$\int_0^\infty y^{n-1}\cos\lambda y\,dy = \frac{\Gamma n}{\lambda^n}\cos\frac{n\pi}{2}$$

and $$\int_0^\infty y^{n-1}\sin\lambda y\,dy = \frac{\Gamma n}{\lambda^n}\sin\frac{n\pi}{2}$$

Hence proved.

SOLUTIONS TO EXERCISE 25B

1. *The temperature $u(x, t)$ at any point of an infinite bar satisfies the equation*

$$\frac{\partial u}{\partial t} = \frac{\partial^2 u}{\partial x^2}, \quad -\infty < x < \infty, \quad t > 0$$

and the initial temperature along the length of the bar is given by

$$u(x, 0) = \begin{cases} 2 & \text{for} \quad |x| < 1 \\ 0 & \text{for} \quad |x| > 1 \end{cases}$$

Determine the expression for $u(x, t)$.

SOLUTION: Taking Fourier transform on the given equation w.r.t. x, we get

$$\int_{-\infty}^{\infty} \frac{\partial u}{\partial t} e^{i\lambda x}\, dx = \int_{-\infty}^{\infty} \frac{\partial^2 u}{\partial x^2} e^{i\lambda x}\, dx$$

or $$\frac{d}{dt}\int_{-\infty}^{\infty} u\, e^{i\lambda x}\, dx = \left[\frac{\partial u}{\partial x} e^{i\lambda x} - \int i\lambda \frac{\partial u}{\partial x} e^{i\lambda x} dx\right]_{-\infty}^{\infty} = 0 - i\lambda \int_{-\infty}^{\infty} \frac{\partial u}{\partial x} e^{i\lambda x} dx = -i\lambda\left[u e^{i\lambda x} - \int i\lambda u\, e^{i\lambda x} dx\right]_{-\infty}^{\infty}$$

or $$\frac{d}{dt}\bar{u}(\lambda, t) = (i\lambda)^2\, \bar{u}(\lambda, t) \quad \text{since } u \text{ and } \frac{\partial u}{\partial x} \text{ are 0 at } x = \pm\infty.$$

or $$\frac{d\bar{u}}{\bar{u}} = -\lambda^2 dt \quad \text{which on integration gives}$$

$$\log \bar{u} = -\lambda^2 t + \log A \quad \text{or} \quad \bar{u} = Ae^{-\lambda^2 t} \quad \text{where A is an arbitrary constant.} \qquad ...(1)$$

Next, Since $u(x, 0) = 2$ for $|x| < 1$ and 0 for $|x| > 1$, taking inverse Fourier transform, we get

$$\bar{u}(\lambda, 0) = \int_{-\infty}^{\infty} u(x, 0)\, e^{i\lambda x}\, dx = \int_{-1}^{1} 2\, e^{i\lambda x}\, dx + 0 = \frac{2}{i\lambda}\left[e^{i\lambda x}\right]_{-1}^{1} = \frac{2}{i\lambda}(e^{i\lambda} - e^{-i\lambda}) = \frac{4}{\lambda}\sin\lambda$$

From (1) $\bar{u}(\lambda, 0) = A = \frac{4}{\lambda}\sin\lambda$, hence $\bar{u} = \frac{4}{\lambda}\sin\lambda\, e^{-\lambda^2 t}$

Taking inverse Fourier transform, we get

$$u(x, t) = \frac{1}{2\pi}\int_{-\infty}^{\infty} \frac{4}{\lambda}\sin\lambda\, e^{-\lambda^2 t} e^{-i\lambda x}\, d\lambda = \frac{4}{2\pi}\int_{-\infty}^{\infty} \frac{\sin\lambda}{\lambda} e^{-\lambda^2 t}(\cos\lambda x - i\sin\lambda x)\, d\lambda$$

$$= \frac{2.2}{\pi}\int_{0}^{\infty} \frac{1}{\lambda} e^{-\lambda^2 t} \sin\lambda\cos\lambda x + 0 \qquad \text{(By property of definite integrals)}$$

$$= \frac{2}{\pi}\int_{0}^{\infty} e^{-\lambda^2 t}\left[\frac{\sin(1+x)\lambda + \sin(1-x)\lambda}{\lambda}\right] d\lambda$$

Thus $$u(x, t) = \frac{2}{\pi}\int_{0}^{\infty} \frac{e^{-\lambda^2 t}}{\lambda}[\sin(1+x)\lambda + \sin(1-x)\lambda\, d\lambda]$$ **Ans.**

2. *Using Fourier sine transform, solve the equation*

$$\frac{\partial u}{\partial t} = 2\,\frac{\partial^2 u}{\partial x^2},\quad 0 < x < \infty,\ t > 0 \quad \textit{subject to the following conditions}$$

(*i*) $u(0, t) = 0, \quad t > 0$ (*ii*) $u(x, 0) = e^{-x}, \quad x > 0$

(*iii*) *u and* $\dfrac{\partial u}{\partial x}$ *both tend to zero as* $x \to \infty$,

SOLUTION: Taking Fourier sine transform on both sides w.r.t x, we get

$$F_s\left(\frac{\partial u}{\partial t}\right) = 2F_s\left(\frac{\partial^2 u}{\partial x^2}\right) \Rightarrow \frac{d}{dt}[\bar{u}_s(\lambda, t)] = 0 - 2\lambda^2\,\bar{u}_s(\lambda, t)$$

On using the initial condition that u and $\dfrac{\partial u}{\partial x}$ both tend 0 as $x \to \infty$. and $u(0, t) = 0,\ t > 0$.

we get $\dfrac{d}{dt}\bar{u} = -2\lambda^2\bar{u}$ Hence $\bar{u}_s(\lambda, t) = Ae^{-2\lambda^2 t}$ where A is an arbitrary constant.

Taking Fourier sine transform on the condition $u(x, 0) = e^{-x},\ x > 0$, we get

$$\bar{u}_s(\lambda, 0) = \int_0^\infty e^{-x}\sin\lambda x\,dx = \left[\frac{e^{-x}}{1+\lambda^2}\{-1\sin x - \lambda\cos\lambda x\}\right]_0^\infty$$

$$= 0 - \frac{1}{1+\lambda^2}(0-\lambda) = \frac{\lambda}{1+\lambda^2}$$

$$\Rightarrow\ A = \frac{\lambda}{1+\lambda^2}.\ \text{Therefore,}\quad \bar{u}(\lambda, t) = \frac{\lambda}{1+\lambda^2}e^{-2\lambda^2 t}$$

Taking inverse Fourier sine transform on both sides, we get the solution as

$$u(x, t) = \frac{2}{\pi}\int_0^\infty \frac{\lambda}{1+\lambda^2}e^{-2\lambda^2 t}\sin\lambda x\,d\lambda = \frac{2}{\pi}\int_0^\infty \frac{s}{s^2+1}e^{-2s^2 t}\sin sx\,ds.$$ **Ans.**

3. *Solve the equation* $\dfrac{\partial u}{\partial t} = K\dfrac{\partial^2 u}{\partial x^2},\quad 0 < x < \infty,\quad t > 0$ *subject to the boundary condition* $u(0, t) = f(t),\quad t > 0$ *and the initial condition* $u(x,0) = 0,\quad x > 0$.

SOLUTION: Taking Fourier sine transform on boths sides of the given equation w.r.t. x, we get

$$F_s\left(\frac{\partial u}{\partial t}\right) = KF_s\left(\frac{\partial^2 u}{\partial x^2}\right) \quad\text{or}\quad \frac{d}{dt}\bar{u}(\lambda, t) = K\int_0^\infty \frac{\partial^2 u}{\partial x^2}\sin\lambda x\,dx$$

or $$\frac{d}{dt}\bar{u}_s = KF_s\left(\frac{\partial^2 u}{\partial x^2}\right) = K(-\lambda^2)\,\bar{u}_s(\lambda, t)\ \text{using boundary and initial conditions.}$$

or $$\frac{d\bar{u}_s}{\bar{u}_s} = -K\lambda^2\,dt \quad\Rightarrow\quad \bar{u}(\lambda, t) = Ae^{-K\lambda^2 t}. \qquad ...(1)$$

Applying Fourier sine transform on the boundary condition $u(0, t) = f(t)$, gives

$$\bar{u}(0,\lambda) = \int_0^\infty f(t)\sin\lambda t\,dt = \int_0^\infty f(x)\sin\lambda x\,dx = F_s(\lambda)$$

Taking $t = 0$ in (1), we get $\bar{u}(0,\lambda) = A = F_s(\lambda)$.

Now inverting the Fourier sine transform, gives the solution as

$$u(x, t) = \frac{2}{\pi}\int_0^\infty F_s(\lambda)\,e^{-K\lambda^2 t}\sin\lambda x\,d\lambda$$

where $F_s(\lambda)$ is the Fourier sine transform of $f(t)$. **Ans.**

4. *Find the bounded harmonic function* $V(x, y)$ *in the semi-infinite strip* $x > 0,\ 0 < y < 1$ *that satisfies the boundary conditions*
$V_x(0, y) = 0, \quad V_y(x, 0) = 0, \quad V(x, 1) = f(x).$

SOLUTION: Here we are to solve the two dimensional Laplace equation $\dfrac{\partial^2 V}{\partial x^2} + \dfrac{\partial^2 V}{\partial y^2} = 0$, $x > 0,\ 0 < y < 1$, conditions being $\dfrac{\partial V}{\partial x} = 0$ for $x = 0$ and all y, $\dfrac{\partial V}{\partial y} = 0$ for $y = 0$ and all x, and $V(x, 1) = f(x)$.

By the method of separation of variables, solution should be of the form

$$V(x, y) = (C_1 \cos\lambda x + C_2 \sin\lambda x)(C_3 \cosh\lambda y + C_4 \sinh\lambda y)$$

Here $\dfrac{\partial V}{\partial x} = (-C_1\lambda\sin\lambda x + C_2\lambda\cos\lambda x)(C_3\cosh\lambda y + C_4\sinh\lambda y)$

Since $\dfrac{\partial V}{\partial x} = 0$ at $x = 0$, all y, we get $C_2 = 0$,

and $\dfrac{\partial V}{\partial y} = (C_1\cos\lambda x + C_2\sin\lambda x)(C_3\lambda\sinh\lambda y + C_4\lambda\cosh\lambda y)$

Since $\dfrac{\partial V}{\partial y} = 0$ at $y = 0$, for all x, we get $C_4 = 0$

Hence the solution is $V(x, y) = C_1 C_3 \cosh\lambda y\cos\lambda x = A\cosh\lambda y\cos\lambda x$ where A is a constant.

Since the bar is semi-infinite and no restriction on λ we can replace A by A (λ)

hence the most general solution can be taken as $V(x, y) = \displaystyle\int_0^\infty A(\lambda)\cosh\lambda y\cos\lambda x\,d\lambda$

Now to find A(λ) we use the condition $V(x, 1) = f(x)$ and get

$$V(x, 1) = \int_0^\infty A(\lambda)\cosh\lambda\cos\lambda x\,d\lambda = f(x)$$

Here $A(\lambda)\cosh\lambda$ is the cosine transform of $f(x)$ hence $A(\lambda)\cosh\lambda = \dfrac{2}{\pi}\displaystyle\int_0^\infty f(x)\cos\lambda x\,dx$

Thus the required solution is

$$V(x, y) = \int_0^\infty A(\lambda)\cosh\lambda y\cos\lambda x\,d\lambda \quad \text{where } A(\lambda) = \frac{2}{\pi\cosh\lambda}\int_0^\infty f(x)\cos\lambda x\,dx.$$ **Ans.**

5. *Find the bounded harmonic function* $V(x, y)$ *in the semi-infinite strip* $y > 0,\ 0 < x < 1$ *which satisfies* $\frac{\partial V}{\partial y}(x, 0) = 0,\quad V(0, y) = e^{-y}$ *and* $V(1, y) = 0.$

SOLUTION: We are to solve here the two dimension Laplace equation $\frac{\partial^2 V}{\partial x^2} + \frac{\partial^2 V}{\partial y^2} = 0$, $y > 0,\ 0 < x < 1$, conditions being $\frac{\partial V}{\partial y} = 0$ for $y = 0$ and all x, $V(0, y) = e^{-y}$ and $V(1, y) = 0$.

Using the method of separation of variables, the solution can be of the form

$$V(x, y) = (C_1 \cosh \lambda x + C_2 \sinh \lambda x)\,(C_3 \cos \lambda y + C_4 \sin \lambda y).$$

From the condition $\frac{\partial V}{\partial y} = 0$ at $y = 0$ we get $C_4 = 0$.

$$\therefore \quad V(x, y) = (C_1 \cosh \lambda x + C_2 \sinh \lambda x)\,(C_3 \cos \lambda y).$$

and from $V(1, y) = 0$ we get $C_1 \cosh \lambda + C_2 \sinh \lambda = 0$ or $C_2 = -C_1 \frac{\cosh \lambda}{\sinh \lambda}$

$$\therefore \quad V(x, y) = C_1 C_3 \cos \lambda y \left[\cosh \lambda x - \frac{\cosh \lambda}{\sinh \lambda} \sinh \lambda x\right]$$

$$= A \cos \lambda y \,\frac{\sinh \lambda (1 - x)}{\sinh \lambda} \quad \text{where } A \text{ is an arbitrary constant.}$$

Since the strip is semi-infinite as $0 < y < \infty$, let $A = A(\lambda)$, $0 < \lambda < \infty$, the most general solution is

$$V(x, y) = \int_0^\infty A(\lambda) \frac{\sinh \lambda (1 - x)}{\sinh \lambda} \cos \lambda y \, d\lambda.$$

Now, since $V(0, y) = e^{-y}$ we have $\int_0^\infty A(\lambda) \cos \lambda y \, d\lambda = e^{-y}$.

$$\Rightarrow \quad A(\lambda) = \frac{2}{\pi} \int_0^\infty e^{-y} \cos \lambda y \, dy = \frac{2}{\pi}\left[\frac{e^{-y}}{1 + \lambda^2}(-\cos \lambda y + \lambda \sin \lambda y)\right]_0^\infty = \frac{2 \cdot 1}{\pi (1 + \lambda^2)}$$

Substituting this value of $A(\lambda)$, the above result becomes

$$V(x, y) = \frac{2}{\pi} \int_0^\infty \frac{1}{1 + \lambda^2} \frac{\sinh \lambda (1 - x)}{\sinh \lambda} \cos \lambda y \, d\lambda$$ and this is the required solution. **Ans.**

6. *Use Fourier sine transform to solve the equation*

$$\frac{\partial V}{\partial t} = K \frac{\partial^2 V}{\partial x^2}, \qquad x > 0,\ t > 0$$

subject to the conditions

$V = V_0$ *when* $x = 0,\ t > 0$ *and* $V = 0$ *when* $t = 0,\ x > 0.$

SOLUTION: Taking Fourier sine transform on both sides of equation, we get

$$F_s\left(\frac{\partial V}{\partial t}\right) = K F_s\left(\frac{\partial^2 V}{\partial x^2}\right)$$

or $$\frac{d}{dt}F_s(V) = K\int_0^\infty \frac{\partial^2 V}{\partial x^2}\sin\lambda x\, dx = K\left[\sin\lambda x\frac{\partial V}{\partial x}\right]_0^\infty - K\int_0^\infty \lambda\cos\lambda x\frac{\partial V}{\partial x}\,dx$$

or $$\frac{d}{dt}\bar{V}(\lambda, t) = 0 - K\lambda[\cos\lambda x\, V(x,t)]_0^\infty - K\lambda^2\int_0^\infty V(x,t)\sin\lambda x\, dx = K\lambda V_0 - K\lambda^2\bar{V}(\lambda, t)$$

using the conditions V and $\frac{\partial V}{\partial x}\to 0$ as $x\to\infty$ and $V = V_0$ at $x = 0,\ t > 0$.

or $$\frac{d\bar{V}}{dt} + K\lambda^2\bar{V} = K\lambda V_0 \quad \text{which is linear in } \bar{V}, \quad \text{I.F.} = e^{\int K\lambda^2 dt} = e^{K\lambda^2 t}$$

$\therefore$ Its solution is $\bar{V}e^{K\lambda^2 t} = \int K\lambda V_0\, e^{k\lambda^2 t}\,dt + C$ where C is constant

$$= \frac{K\lambda V_0}{K\lambda^2}e^{K\lambda^2 t} + C \quad \text{or} \quad \bar{V}(\lambda, t) = \frac{V_0}{\lambda} + C\,e^{-K\lambda^2 t}.$$

Next taking Fourier sine transform of the initial condition $V(x, 0) = 0$, we get $\bar{V}(\lambda, 0) = 0$.

Therefore, $\frac{V_0}{\lambda} + C = 0 \quad \therefore\ C = \frac{-V_0}{\pi}$

Thus we have $\bar{V}(\lambda, t) = \frac{V_o}{\lambda}[1 - e^{-K\lambda^2 t}]$.

Taking here the inverse Fourier sine transform, gives

$$V(x, t) = \frac{2}{\pi}\int_0^\infty \frac{V_0}{\lambda}(1 - e^{-K\lambda^2 t})\sin\lambda x\, d\lambda = \frac{2V_0}{\pi}\left[\int_0^\infty \frac{\sin\lambda x\, d\lambda}{\lambda} - \int_0^\infty \frac{e^{-K\lambda^2 t}\sin\lambda x}{\lambda}\,d\lambda\right]$$

Next, putting $\lambda x = y$ so that $x d\lambda = dy$ we have $\int_0^\infty \frac{\sin\lambda x}{\lambda}\,d\lambda = \int_0^\infty \frac{\sin y}{y} = \frac{\pi}{2}$ (well known).

Hence $$V(x, t) = \frac{2V_0}{\pi}\left[\frac{\pi}{2} - \int_0^\infty \frac{\sin\lambda x}{\lambda}e^{-K\lambda^2 t}\,d\lambda\right] \quad \text{or} \quad V(x, t) = V_0 - \frac{2V_0}{\pi}\int_0^\infty \frac{\sin\lambda x}{\lambda}e^{-K\lambda^2 t}\,d\lambda$$ **Ans.**

7. *Solve the partial differential equation* $\frac{\partial u}{\partial t} = c^2\frac{\partial^2 u}{\partial x^2}$ *for* $x > 0,\ t > 0$

under the boundary condition $u(0, t) = f(t)$

and the initial condition $u(x, 0) = 0$.

SOLUTION: The problem is identical to question No. 3 where K is to be replaced by C^2. **Ans.**

ADDITIONAL IMPORTANT PROBLEMS

1. *Find the Fourier transform of* $f(x) = e^{-ax}, \; 0 < x < \infty$
$= 0, \; x < 0$

SOLUTION: $F(s)$ the complex fourier transform of $f(x)$, is given by

$$F(s) = \int_{-\infty}^{\infty} f(x)\, e^{isx}\, dx = \int_{-\infty}^{0} 0\, e^{isx}\, dx + \int_{0}^{\infty} e^{-ax}\, e^{isx}\, dx$$

$$= \int_{0}^{\infty} e^{-x(a-is)}\, dx = \left[\frac{e^{-x(a-is)}}{-(a-is)}\right]_0^{\infty} = 0 + \frac{1}{a-is}$$

or $$F(s) = \frac{a+is}{a^2+s^2}$$ **Ans.**

2. *Obtain the Fourier transform of* $f(x) = xe^{-x}, \; x > 0$
$= 0, \; x < 0.$

SOLUTION: The complex Fourier transform of $f(x)$ is given by

$$F(s) = \int_{-\infty}^{\infty} f(x)\, e^{isx}\, dx = \int_{-\infty}^{0} 0\, e^{isx}\, du + \int_{0}^{\infty} x e^{-x}\, e^{isx}\, dx = \int_{0}^{\infty} x\, e^{-(1-is)x}\, dx$$

$$= \left[\frac{x\, e^{-(1-is)x}}{-(1-is)}\right]_0^{\infty} - \int_{0}^{\infty} \frac{e^{-(1-is)x}\, dx}{-(1-is)} = 0 + \frac{1}{(1-is)^2} = 0 - \left[\frac{e^{-(1-is)x}}{(1-is)^2}\right]_0^{\infty}$$

or $$F(s) = \frac{(1+is)^2}{(1+s^2)^2}$$ **Ans.**

3. *Find the Fourier transform of* $f(t) = \sin t, \; 0 < t < \pi$
$= 0, \; t > \pi$ *and* $t < 0.$

and hence deduce that $\displaystyle\int_0^{\infty} \frac{\cos \pi s/2}{1-s^2}\, dx = \frac{\pi}{2}.$

SOLUTION: $$F(s) = \int_0^{\infty} f(t)\, e^{ist}\, dt = 0 + \int_0^{\pi} e^{ist} \sin t\, dt$$

$$= \left[\frac{e^{ist}}{1+(is)^2}(is \sin t - 1 \cos t)\right]_0^{\pi} = \frac{e^{is\pi}+1}{1-s^2}$$

Taking inverse Fourier transform throughout, we get

$$f(t) = \frac{1}{2\pi}\int_{-\infty}^{\infty} F(s)\, e^{-ist}\, ds = \frac{1}{2\pi}\int_{-\infty}^{\infty} \frac{1+e^{is\pi}}{1-s^2}\, e^{-ist}\, ds$$

$$= \frac{1}{2\pi}\int_{-\infty}^{\infty}\frac{e^{-ist}+e^{-is(t-\pi)}}{1-s^2}ds = \frac{1}{2\pi}\int_{-\infty}^{\infty}\frac{\cos st+\cos s(t-\pi)-i\sin st-i\sin s(t-\pi)}{1-s^2}ds$$

$$= \frac{2}{2\pi}\int_{0}^{\infty}\frac{\cos st+\cos s(t-\pi)}{1-s^2}ds+0 \quad \text{(by property of definite integrals.)}$$

Now putting $t = \pi/2$ on both sides, we get

$$f\left(\frac{\pi}{2}\right) = \frac{1}{\pi}\int_{0}^{\infty}\frac{\cos s\pi/2+\cos s\pi/2}{1-s^2}ds \quad \text{but } f\left(\frac{\pi}{2}\right) = 1, \text{ given}$$

$$\therefore \quad 1 = \frac{2}{\pi}\int_{0}^{\infty}\frac{\cos s\pi/2}{1-s^2}ds \quad \text{or} \quad \int_{0}^{\infty}\frac{\cos s\pi/2}{1-s^2} = \frac{\pi}{2}.$$

Hence Proved

4. *Find the Fourier transform of* $f(x) = a - |x|, \; |x| \le a$
$= 0, \; |x| > a$

Hence show that $\int_{0}^{\infty}\frac{\sin^2 x}{x^2}dx = \frac{\pi}{2}.$

SOLUTION: Here $f(x)$ is an even functioin of x as $f(-x) = f(x)$. $F(s)$ the complex Fourier transform of $f(x)$ is given by

$$F(s) = \int_{-\infty}^{\infty} f(x)\,e^{isx}\,dx = \int_{-\infty}^{\infty} f(x)\cos sx\,dx + i\int_{-\infty}^{\infty} f(x)\sin sx\,dx = 2\int_{0}^{\infty} f(x)\cos sx\,dx + 0$$

$$= 2\int_{0}^{a}(a-x)\cos sx\,dx = 2\left[\left((a-x)\frac{\sin sx}{s}\right)_0^a - \int_{0}^{a}(-1)\frac{\sin sx}{s}dx\right]$$

$$= 0+\frac{2}{s}\int_{0}^{a}\sin sx\,dx = \frac{2}{s}\left[\frac{-\cos sx}{s}\right]_0^a = \frac{2(1-\cos as)}{s^2}.$$

Ans.

Taking inverse Fourier transform

$$f(x) = \frac{1}{2\pi}\int_{-\infty}^{\infty}F(s)e^{-isx}ds = \frac{1}{2\pi}\int_{-\infty}^{\infty}\frac{2(1-\cos as)}{s^2}(\cos sx - i\sin sx)\,ds$$

$$= \frac{2}{\pi}\int_{0}^{\infty}\frac{(1-\cos as)\cos sx}{s^2}ds \quad \text{(by property of defnite integrals).}$$

Putting $x = 0$ we get $f(0) = \frac{2}{\pi}\int_{0}^{\infty}\frac{1-\cos as}{s^2}ds = a$ as $f(0) = a$

Now taking $a = 2$, gives $2 = \frac{2}{\pi}\int_{0}^{\infty}\frac{1-\cos 2s}{s^2}ds$

or $\int_{0}^{\infty}\frac{\sin^2 s}{s^2}ds = \frac{\pi}{2}.$ Changing the variable of integration, we get

$$\int_{0}^{\infty}\frac{\sin^2 x\,dx}{x^2} = \frac{\pi}{2}.$$

Hence Proved.

5. *Find the Fourier transform of* $f(x) = 1+\frac{x}{a}, \ -a < x < 0$

$$= 1-\frac{x}{a}, \ 0 < x < a$$
$$= 0, \ \text{otherwise}$$

SOLUTION: Clearly $f(x)$ is an even function of x. Fourier transform of $f(x)$ is

$$F(s) = \int_{-\infty}^{\infty} f(x)\, e^{isx}\, dx = \int_{-\infty}^{\infty} f(x)\,(\cos sx + i \sin sx)\, dx = 2\int_{0}^{\infty} f(x) \cos sx\, dx + 0$$

(by property of definite integrals)

$$= 2\int_{0}^{\infty}\left(1-\frac{x}{a}\right)\cos sx\, dx = 2\left[\left(1-\frac{x}{a}\right)\frac{\sin sx}{s}\right]_0^a - 2\int_0^a\left(\frac{-1}{a}\right)\frac{\sin sx}{s}\,dx$$

$$= 0 + 2\left[\frac{-\cos sx}{a\,s^2}\right]_0^a = \frac{2(1-\cos as)}{as^2} = \frac{4}{as^2}\sin^2(as/2).$$

6. *Determine the Fourier transform of* $f(t) = e^{-at}, \ t > 0$

$$= -e^{at}, \ t < 0$$

SOLUTION: $f(-t) = -f(t)$ hence $f(t)$ is an odd function of t.

$\therefore$ Fourier transform of $f(t)$ is given by

$$F(\omega) = \int_{-\infty}^{\infty} f(t)\, e^{i\omega t}\, dt = \int_{-\infty}^{\infty} f(t)\,(\cos \omega t + i \sin \omega t)\, dt$$

$$= 0 + 2i\int_0^{\infty} f(t) \sin \omega t\, dt = 2i\int_0^{\infty} e^{-at} \sin \omega t\, dt$$

$$= 2i\left[\frac{e^{-at}}{a^2+\omega^2}(-a \sin \omega t - \omega \cos \omega t)\right]_0^{\infty} = \frac{2i\omega}{a^2+\omega^2}$$

Thus, $$F(\omega) = \frac{2i\,\omega}{a^2+\omega^2}$$ **Ans.**

7. *Find Fourier cosine and sine transforms of* $f(x) = e^{-ax}, \ a > 0.$

SOLUTION: Fourier cosine transform of $f(x)$, $F_C(s)$ is given by

$$F_C(s) = \int_0^{\infty} f(x) \cos sx\, dx = \int_0^{\infty} e^{-ax} \cos sx\, dx = \left[\frac{e^{-ax}}{a^2+s^2}(-a \cos sx + s \sin sx)\right]_0^{\infty}$$

$$= 0 - \frac{1}{a^2+s^2}(-a) = \frac{a}{s^2+a^2}$$

and Fourier sine transform of $f(x)$, $F_S(s)$ is given by

$$F_S(s) = \int_0^\infty f(x)\sin sx\,dx = \int_0^\infty e^{-ax}\sin sx\,dx = \left[\frac{e^{-ax}}{a^2+s^2}(-a\sin sx - s\cos sx)\right]_0^\infty$$

$$= 0 - \frac{1}{a^2+s^2}(-s) = \frac{s}{s^2+a^2}$$

Thus $F_C(s) = \dfrac{1}{s^2+a^2}$ and $F_S(s) = \dfrac{s}{s^2+a^2}$. **Ans.**

8. *Find the Fourier sine transform of* $\dfrac{e^{-at}}{t}$.

SOLUTION: Fourier sine transform of $f(t) = \dfrac{e^{-at}}{t}$, as $F_S(s)$ is given by

$$F_S(s) = \int_0^\infty f(t)\sin st\,dt = \int_0^\infty \frac{e^{-at}}{t}\sin st\,dt$$

Differentiating both sides w.r.t. s, we get

$$F'_S(s) = \int_0^\infty \frac{e^{-at}}{t}.t\cos st\,dt = \int_0^\infty e^{-at}\cos st\,dt$$

$$= \left[\frac{e^{-at}}{a^2+s^2}(-a\cos st + s\sin st)\right]_0^\infty = 0 - \frac{1}{a^2+s^2}(-a) = \frac{a}{s^2+a^2}.$$

Integrating both sides w.r.t. s, we get

$$F_S(s) = \int \frac{a\,ds}{s^2+a^2} = \tan^{-1}\frac{s}{a} + c, \quad c \text{ being constant.}$$

When $s = 0$ $F_S(0) = 0$ hence $c = 0$ $\therefore$ $F_S(s) = \tan^{-1}\left(\dfrac{s}{a}\right)$. **Ans.**

9. *Find $f(t)$ if its Fourier sine transform is* $\dfrac{1}{\omega}e^{-a\omega}$. *Hence find* $F_S^{-1}\left(\dfrac{1}{\omega}\right)$.

SOLUTION: The inverse Fourier sine transform of $F_S(\omega)$, is given by

$$f(t) = \frac{2}{\pi}\int_0^\infty F_S(\omega)\sin\omega t\,d\omega = \frac{2}{\pi}\int_0^\infty \frac{1}{\omega}e^{-a\omega}\sin\omega t\,d\omega$$

Differentiating on both sides w.r.t. t, we get

$$f'(t) = \frac{2}{\pi}\int_0^\infty \frac{1}{\omega} e^{-a\omega} \cos \omega t . \omega d\omega = \frac{2}{\pi}\int_0^\infty e^{-a\omega} \cos \omega t \, d\omega$$

$$= \frac{2}{\pi}\left[\frac{e^{-a\omega}}{a^2+t^2}(-a \cos \omega t + \omega \sin \omega t)\right]_0^\infty = 0 - \frac{2}{\pi}\frac{1}{a^2+t^2}(-a) = \frac{2}{\pi}\frac{a}{a^2+t^2}$$

Integrating both sides w.r.t. t, we get

$$f(t) = \frac{2}{\pi}\int \frac{a\,dt}{a^2+t^2} = \frac{2}{\pi}\tan^{-1}\frac{t}{a} + c \quad \text{where } c \text{ is arbitrary constant}$$

At $t = 0$, $f(0) = 0$ Hence $c = 0$ $\quad \therefore \quad f(t) = \dfrac{2}{\pi}\tan^{-1}\dfrac{t}{a} = F_S^{-1}\left(\dfrac{1}{\omega}e^{-a\omega}\right)$

Putting here $a = 0$, we get $F_S^{-1}\left(\dfrac{1}{\omega}\right) = \dfrac{2}{\pi}\tan^{-1}(\infty) = \dfrac{2}{\pi}.\dfrac{\pi}{2} = 1.$ **Ans.**

10. *Find $f(t)$ if its Fourier sine transform is* $\dfrac{\omega}{\omega^2+1}$.

SOLUTION: We need to find $F_S^{-1}\left(\dfrac{\omega}{\omega^2+1}\right) = f(t).$

$$\therefore \; f(t) = \frac{2}{\pi}\int_0^\infty F_S(\omega)\sin \omega t \, d\omega = \frac{2}{\pi}\int_0^\infty \frac{\omega}{\omega^2+1}\sin \omega t \, d\omega$$

$$= \frac{2}{\pi}\int_0^\infty \frac{\omega^2}{\omega(\omega^2+1)}\sin \omega t \, d\omega = \frac{2}{\pi}\int_0^\infty \frac{\omega^2+1-1}{\omega(\omega^2+1)}\sin \omega t \, d\omega$$

$$= \frac{2}{\pi}\int_0^\infty \frac{\sin \omega t \, d\omega}{\omega} - \frac{2}{\pi}\int_0^\infty \frac{\sin \omega t \, d\omega}{\omega(\omega^2+1)} = \frac{2}{\pi}\left(\frac{\pi}{2}\right) - \frac{2}{\pi}\int_0^\infty \frac{\sin \omega t \, d\omega}{\omega(\omega^2+1)} \quad \text{...(1)}$$

Now differentiating both sides w.r.t. t, on both sides, we get

$$f'(t) = 0 - \frac{2}{\pi}\int_0^\infty \frac{\omega \cos \omega t \, d\omega}{\omega(\omega^2+1)} = \frac{-2}{\pi}\int_0^\infty \frac{\cos \omega t \, d\omega}{\omega^2+1} \quad \text{...(2)}$$

Again differentiating w.r.t. t, gives $f''(t) = \dfrac{2}{\pi}\displaystyle\int_0^\infty \frac{\omega}{\omega^2+1}\sin \omega t \, d\omega = f(t)$

or $\quad D^2 f(t) - f(t) = 0$ where $D \equiv \dfrac{d}{dt}$. A.E. is $m^2 - 1 = 0 \;\therefore\; m = \pm 1$

Therefore $f(t) = c_1 e^t + c_2 e^{-t}$ and $f'(t) = c_1 e^t - c_2 e^{-t}$

Putting $t = 0$ in (1), gives $f(0) = 1 \Rightarrow c_1 + c_2 = 1$ ...(3)

and (2) gives $$f'(0) = -\frac{2}{\pi}\int_0^\infty \frac{d\omega}{\omega^2+1} = \frac{-2}{\pi}\left[\tan^{-1}\omega\right]_0^\infty = \frac{-2}{\pi}\cdot\left(\frac{\pi}{2}-0\right) = -1$$

$\therefore$ $$c_1 - c_2 = -1 \qquad ...(4)$$

Hence from (3) and (4) we get $c_1 = 0$ and $c_2 = 1$ which gives $f(t) = 1e^{-t}$ **Ans.**

11. *Solve the integral equation* $\int_0^\infty f(x)\sin\omega x\, dx = 1, \; 0 \le \omega < 1$
$= 2, \; 1 \le \omega < 2$
$= 0, \; \omega > 2$

SOLUTION: Fourier sine transform of $f(x)$ as $F_S(\omega)$ is given by

$$F_S(\omega) = \int_0^\infty f(x)\sin\omega x\, dx = \begin{cases} 1, & 0 \le \omega < 1 \\ 2, & 1 \le \omega < 2 \\ 0, & \omega > 2 \end{cases}$$

Taking inverse Fourier sine transform, gives

$$f(x) = \frac{2}{\pi}\int_0^\infty F_S(\omega)\sin\omega x\, d\omega = \frac{2}{\pi}\left[\int_0^1 \sin\omega x\, d\omega + \int_1^2 2\sin\omega x\, d\omega + 0\right]$$

$$= \frac{2}{\pi}\left[-\left[\frac{\cos\omega x}{x}\right]_0^1 - 2\left[\frac{\cos\omega x}{x}\right]_1^2\right] = \frac{2}{\pi x}[(1-\cos x) + 2(\cos x - \cos 2x)]$$

Thus $$f(x) = \frac{2}{\pi x}(1 + \cos x - 2\cos x)$$ **Ans.**

12. *Find the Fourier cosine transform of* $f(x) = \dfrac{1}{1+x^2}$.

Hence derive the Fourier sine transform of $\dfrac{x}{1+x^2}$.

SOLUTION: $$F_C(\omega) = \int_0^\infty f(x)\cos\omega x\, dx = \int_0^\infty \frac{\cos\omega x\, dx}{1+x^2} \qquad ...(1)$$

Differentiating w.r.t. ω, we get

$$F_C'(\omega) = \int_0^\infty \frac{-x\sin\omega x\, dx}{1+x^2} = -\int_0^\infty \frac{x^2\sin\omega x\, dx}{x(1+x^2)} = -\int_0^\infty \frac{(x^2+1-1)\sin\omega x\, dx}{x(1+x^2)}$$

$$= -\int_0^\infty \frac{\sin\omega x\, dx}{x} + \int_0^\infty \frac{\sin\omega x\, dx}{x(1+x^2)} = -\frac{\pi}{2} + \int_0^\infty \frac{\sin\omega x\, dx}{x(1+x^2)} \quad \left(\text{since} \int_0^\infty \frac{\sin\omega x}{x}dx = \frac{\pi}{2}\right) \qquad ...(2)$$

Again differentiation w.r.t. ω, gives

$$F''_C(\omega) = 0 + \int_0^\infty \frac{x \cos \omega x dx}{x(1+x^2)} dx = \int_0^\infty \frac{\cos \omega x dx}{x^2+1} = F_C(\omega)$$

or $(D^2 - 1) F_C(\omega) = 0$ where $D \equiv \dfrac{d}{d\omega}$. It's solution is

$$F_C(\omega) = Ae^{\omega} + Be^{-\omega} \text{ where } A, B \text{ are arbitrary constants.}$$

$$\Rightarrow \quad F'_C(\omega) = Ae^{\omega} - Be^{-\omega} \Rightarrow F''_C(\omega) = Ae^{\omega} + Be^{-\omega}.$$

Writing $\omega = 0$ in (1) gives $F_C(0) = \displaystyle\int_0^\infty \frac{dx}{1+x^2} = \frac{\pi}{2} \Rightarrow A+B = \frac{\pi}{2}$.

Putting $\omega = 0$ from eqn (2) we get $F'_C(0) = \dfrac{-\pi}{2} \quad \therefore \quad A - B = -\dfrac{\pi}{2}$.

$\Rightarrow A = 0,\ B = \pi/2$ hence $F_C(\omega) = \dfrac{\pi}{2} e^{-\omega}$ $F_S(\omega) = \displaystyle\int_0^\pi \frac{x \sin \omega x\, dx}{1+x^2} dx = -F'_C(\omega) = \frac{\pi}{2} e^{-\omega}$. **Ans.**

13. *Obtain* $f(x)$ *satisfying the integral equation* $\displaystyle\int_0^\infty f(x) \cos \omega x\, dx = e^{-\omega}, \omega > 0.$

SOLUTION: The given equation is nothing but $F_C(\omega) = e^{-\omega}$.
Inverting the Fourier cosine transform, we get

$$f(x) = \frac{2}{\pi}\int_0^\infty e^{-\omega} \cos \omega x\, dx = \frac{2}{\pi}\left[\frac{e^{-\omega}}{1+x^2}(-1 \cos \omega x + \omega \sin \omega x)\right]_0^\infty$$

$$= \frac{2}{\pi(1+x^2)}[0-(-1)] = \frac{2}{\pi(1+x^2)}.$$ **Ans.**

CHAPTER 26

Error Analysis and Numerical Solution of Equations

IMPORTANT FORMULAE AND USEFUL RESULTS

1. Absolute Error in $X = E_A = |X - X_1|$

   Relative Error $= E_R = \dfrac{E_A}{X}$

   Percentage Error $= E_p = 100\, E_R$.

2. **Solution of Algebraic and Transcendental Equations**

   (*i*) **Bisection Method :** If $f(x) = 0$ and $f(a)$ and $f(b)$ are of opposite signs then $x_0 = \dfrac{a+b}{2}$ is approximate root. Check for the sign of $f(x_0)$ and process can be continued.

   (*ii*) **Regula Falsi Method :** If $f(a)\, f(b) < 0$ then first approximate root $x_1 = \dfrac{a\, f(b) - b\, f(a)}{f(b) - f(a)}$. Check for the sign of $f(x_1)$ and process can be continued.

   (*iii*) **Newton-Raphson Method :** If $x = x_0$ is the starting value of a root of $f(x) = 0$ then $x_1 = x_0 - \dfrac{f(x_0)}{f'(x_0)}$ is the first approximation. Check for the sign of $f(x_1)$ and process can be continued. In general $x_{n+1} = x_n - \dfrac{f(x_n)}{f'(x_n)}$.

   Newton's method is of second order convergence.

   (*iv*) **Direct Iteration Method :** If $f(x) = 0$ can be written as $x = \phi(x)$ and $|\phi'(x_0)| < 1$ then better approximations are given by

   $x_1 = \phi(x_0), \quad x_2 = \phi(x_1), \quad x_3 = \phi(x_2), \; ..., \; x_{n+1} = \phi(x_n).$

3. Solution of Linear Simultaneous Equations :

Direct Methods :

1. Determinant Method
2. Matrix Inversion Method
3. Gauss Elimination
4. Gauss Jordan Method
5. Crout's Method

Indirect Methods :

1. Gauss Jacobi Method
2. Gauss-seidal Method
3. Relaxation Method

SOLUTIONS TO EXERCISE 26A

1. *Find the sum of the numbers each correct to indicated significant digits*

136.426, 28.3, 321, 68.243 and 17.482

Also find the total error.

SOLUTION : Since the number 321 is known to only three significant digits, we round off the other numbers to 0.1 and get 136.4, 28.3, 321, 68.2 and 17.5.

The total of these numbers = 571.4 = 571

To determine the total error of the result we find the individual errors :

(*a*) The sum of the errors of the original data :

$$\in_1 = 10^{-3} + 10^{-1} + 10^0 + 10^{-3} + 10^{-3}$$

$$= 0.001 + 0.1 + 1 + 0.001 + 0.001 = 1.103$$

(*b*) The absolute value of the sum of the rounding errors

$= \in_2 = 1.026 + 0 + 0 + 0.043 + 0.0181 = 0.051$ and finally.

(*c*) The round off error in the sum $= \in_3 = 0.5$

Hence the total error $\in = \in_1 + \in_2 + \in_3 = 1.103 + 0.051 + 0.5 = 1.654 < 2$

The required sum is therefore 571 ± 2. **Ans.**

2. *Find the difference* $\sqrt{10} - \pi$ *correct to five significant digits.*

SOLUTION : Since $\sqrt{10} = 3.1622776 \ldots$ and $\pi = 3.1415926 \ldots$

The required result is $D = 3.1622776 \ldots - 3.1415926 \ldots$

$= 0.0206850$

Which is correct to five significant digits. **Ans.**

3. *Evaluate* $\sqrt{3.01} - \sqrt{3}$ *correct to three significant digits.*

SOLUTION : To evaluate $\sqrt{3.01} - \sqrt{3}$ correct to three significant digits.

Let $D = \sqrt{3.01} - \sqrt{3} = \dfrac{(\sqrt{3.01} - \sqrt{3})(\sqrt{3.01} + \sqrt{3})}{\sqrt{3.01} + \sqrt{3}}$

$$= \frac{3.01 - 3}{\sqrt{3.01} + \sqrt{3}} = \frac{0.01}{1.73 + 1.73} = \frac{0.01}{3.46} = 0.00289 = 2.89 \times 10^{-3}$$ **Ans.**

4. *Add the numbers 83.72 and 1.529 in a decimal computer with a fixed word length of four. Also find the relative error.*

SOLUTION : We assume that the machine does 'ehopping', the sum $S = 83.72 + 1.52 = 85.24$. **Ans.**

The absolute error $\in_s$ is given by $\in_s = 0.009$

and the relative error r_s is given by $r_s = \dfrac{0.009}{85.24} = 0.0001055 = 0.106 \times 10^{-4}$. **Ans.**

5. *Find the product of two approximate numbers $a = 2.5$ and $b = 72.398$ correct to the number of significant digits given.*

SOLUTION : We take $a = 2.5$ and $b = 72.4$ (after rounding)

$\therefore$ Product $P = ab = 2.5 \times 72.4 = 181.0 = 1.8\ 1 \times 10^2$. **Ans.**

6. *Find the product of the numbers 512.7 and 0.115756 assuming that the computer is a decimal one chopping after 4 digits in the mantissa. Also find the relative error.*

SOLUTION : True value $= P = 512.7 \times 0.11576 = 59.350152$

Approximate value of the product $P = 0.5127 \times 10^3 \times 0.1157$

$= 0.05932 \times 10^3 = 59.32 = 59$

$\therefore$ The absolute error $= 0.350152$ and the relative error $= \dfrac{0.350152}{59.350152} = 5.93 \times 10^{-3}$ **Ans.**

7. *Find the number of significant figures in the quotient of 786.5/289.3 assuming that both the numbers are approximate and correct only to the number of digits given.*

SOLUTION : Maximum relative error is given by $r_Q = \dfrac{0.05}{289.3} = 0.0001728 = \in$

and quotient $\dfrac{a}{b} = \dfrac{786.5}{289.3} = 2.722$

$\therefore$ Using the formula $\in_Q \leq \dfrac{2a}{b} . \in$, $\in_Q = 2(2.7220)\ (0.0001728) = 0.0009406$

Since the error affects the fourth decimal place only, we can coulude that the quotient is correct to three decimal places and that the quotient Q is equal to $Q = 2.722$ **Ans.**

SOLUTIONS TO EXERCISE 26B

1. *Obtain a root of the equation* $x^3 - 4x - 9 = 0$ *employing bisection method in four stages.*

SOLUTION : $f(x) = x^3 - 4x - 9$, $f(2) < 0$ and $f(3) > 0$ hence a root lies in (2, 3).

As first approximation $x_1 = \frac{1}{2}(2 + 3) = 2.5.$

Therefore $f(x_1) = (2.5)^3 - 4(2.5) - 9 = -3.375 < 0$

$\therefore$ Second approximation x_2 lies between x_1 and 2 $\therefore$ $x_2 = \frac{1}{2}(2.5 + 3) = 2.75.$

Then $f(x_2) = (2.75)^3 - 4(2.75) - 9 = 0.7369 > 0$

$\therefore$ Root lies between x_1 and x_2 $\therefore$ $x_3 = \frac{1}{2}(x_1 + x_2) = 2.625$

Then $f(x_3) = (2.625)^3 - 4(2.625) - 9 = -1.4121 < 0$

$\therefore$ Root lies between x_2 and x_3 and fourth approximation to the root is $x_4 = \frac{1}{2}(x_2 + x_3) = 2.6875.$ **Ans.**

2. *Find the positive real root of the equation* $4x^4 - x - 8 = 0$ *correct to three places of decimal using method of false position.*

SOLUTION : $f(x) = 4x^4 - x - 8 = 0$. Here $f(1) = -5, f(2) = 54$, Also, $f(1.5) = 4(1.5)^4 - 1.5 - 8 = 10.75$

Hence one root lies between 1 and 1.5 $\therefore$ $a = 1$, $b = 1.5$, $f(a) = -5$, $f(b) = 10.75$

By Regula Falsi method

$$x_1 = \frac{af(b) - bf(a)}{f(b) - f(a)} = \frac{1(10.75) - 1.5(-5)}{10.75 - (-5)} = 1.1587$$

$f(x_1) = -1.948 < 0$ hence root lies between 1.1587 and 1.5.

$$\therefore \quad x_2 = \frac{1.1587(10.75) - 1.5(-1.948)}{10.75 - (-1.948)} = 1.2123$$ **Ans.**

3. *Use Regula Falsi method to find the root of the equation* $2x - \log_{10} x = 7$ *lying between* 3.5 *and* 4 *correct to three decimal places.*

SOLUTION : Here $f(x) = 2x - \log_{10} x - 7 = 0$

$\therefore$ $f(3.5) = 2(3.5) - \log_{10} 3.5 - 7 = -\log_{10} 3.5 = -0.544068$

and $f(4.0) = 2(4) - \log_{10}(4) - 7 = 1 - \log_{10} 4 = 1 - 0.60206 = 0.398$

By Regula Falsi Method first approximation is

$$x_1 = \frac{3.5(0.398) - 4(-0.544)}{0.398 + 0.544} = 3.7885$$

$\therefore$ $f(x_1) = 2(3.7885) - 7 - \log_{10} 3.7885 = -0.00147 < 0$

Hence root lies between 3.7885 and 4

$\therefore$ second approximation $x_2 = \dfrac{3.7885(0.398) - 4(-0.00147)}{0.398 - (-0.00147)} = \dfrac{1.5137}{0.39947} = 3.7893$

$$f(x_2) = 2(3.7893) - 0.5785 - 7 = 0.00035 > 0$$

Third approximation when calculated = 3.7893 correct to four decimal places. **Ans.**

4. *Find a real root of the equation $x = e^{-x}$, using Newton Raphson method correct to three places of decimal.*

SOLUTION : $f(x) = x - e^{-x} = 0$. Initially $x = x_0 = 0$, $f'(x) = 1 + e^{-x}$

By N.R. method $x_1 = x_0 + h$ where $h = \dfrac{-f(x_0)}{f'(x_0)} = \dfrac{e^{-x} - x}{1 + e^{-x}} = \dfrac{1}{2} = 0.5$

$\therefore$ First approximation $x_1 = 0 + 0.5 = 0.5$

For x_2, $x_2 = x_1 + h$, $h = \dfrac{-f(x_1)}{f'(x_1)} = \dfrac{e^{-x_1} - x_1}{1 + e^{-x_1}} = \dfrac{e^{-0.5} - 0.5}{1 + e^{-0.5}} = 0.06$

$\therefore$ $x_2 = 0.5 + 0.06 = 0.56$

Next, $x_3 = x_2 + h$ and $h = \dfrac{e^{-0.56} - 0.56}{1 + e^{-0.56}} = \dfrac{0.571 - 0.56}{1.571} = 0.007$

$\therefore$ $x_3 = 0.56 + 0.007 = 0.567$. **Ans.**

5. *Find the root of the equation $x^3 - 3x - 5 = 0$ lying between 2 and 3.*

SOLUTION : $f(x) = x^3 - 3x - 5 = 0$, $f(2) = -3$, $f(3) = 13$

Applying Regula Falsi method, as first approximation we get

$$x_1 = \frac{af(b) - bf(a)}{f(b) - f(a)} = \frac{2(13) - 3(-3)}{13 - (-3)} = \frac{35}{16} = 2.1875$$

$\therefore$ $f(x_1) = (2.1875)^3 - (2.1875) - 5 = 10.4675 - 11.5625 = -1.09497 < 0$

Hence root lies between 2.1875 and 3.

$\therefore$ Second approximation x_2 is given by

$$x_2 = \frac{2.1875(13) - 3(-1.09497)}{13 - (-1.09497)} = \frac{28.4375 + 3.2849}{14.09497} = 2.251$$

and $f(x_2) = (2.251)^3 - 3(2.251) - 5 = -0.346 < 0$

Hence root lies between 2.251 and 3

$\therefore$ Third approximation is $x_3 = \dfrac{2.251(13) - 3(-0.346)}{13 - (-0.346)} = 2.2704$. **Ans.**

6. *Find the approximate value of* $\sqrt{28}$ *correct to four decimal places using Newton -Raphson method*

SOLUTION : $x = \sqrt{28}$ then we have $f(x) = x^2 - 28$.

By N.R. method $x_{n+1} = x_n - \dfrac{f(x_n)}{f'(x_n)} = x_n - \dfrac{x_n}{2} + \dfrac{28}{2x_n} = \dfrac{1}{2}\left(x_n + \dfrac{28}{x_n}\right)$

We take $x_0 = 5$ then $x_1 = \dfrac{1}{2}\left(x_0 + \dfrac{28}{x_0}\right) = \dfrac{1}{2}\left(5 + \dfrac{28}{5}\right) = 5.3$

Similarly, $x_2 = \dfrac{1}{2}\left(x_1 + \dfrac{28}{x_1}\right) = \dfrac{1}{2}\left(5.3 + \dfrac{28}{5.3}\right) = \dfrac{1}{2}(5.3 + 5.283) = 5.291$

and $x_3 = \dfrac{1}{2}\left(5.291 + \dfrac{28}{5.291}\right) = 5.2915.$ **Ans.**

7. *Find real root of the following system of non linear equations*

$$x^2 - y^2 = 4 \quad \text{and} \quad x^2 + y^2 = 16$$

using Newton Raphson method, starting values being (2.8, 2.8).

SOLUTION : $x^2 - y^2 = 4$, $x^2 + y^2 = 16$. On solving these we can get

$$x = \sqrt{10} = 3.162 \quad \text{and} \quad y = \sqrt{6} = 2.449$$

To apply Newton-Raphson method, we need initial values x_0, y_0.

Let us take $x_0 = y_0 = 2\sqrt{2} = 2.828$ and $f(x, y) = x^2 - y^2 - 4$ and $g(x, y) = x^2 + y^2 - 16$

$$\therefore \quad \frac{\partial f}{\partial x} = 2x, \quad \frac{\partial f}{\partial y} = -2y \quad \text{and} \quad \frac{\partial g}{\partial x} = 2x, \quad \frac{\partial g}{\partial y} = 2y.$$

Next, let the correct values of x and y be $x = x_0 + h, y = y_0 + k$

$\therefore f(x_0 + h, y_0 + k) = 0$ and $g(x_0 + h, y_0 + k) = 0$

By Taylor's expansion we get $f_0 + h\left(\dfrac{\partial f}{\partial x}\right)_{(0,0)} + k\left(\dfrac{\partial f}{\partial y}\right)_{(0,0)} = 0$ and $g_0 + h\left(\dfrac{\partial g}{\partial x}\right)_{(0,0)} + k\left(\dfrac{\partial g}{\partial y}\right)_{(0,0)} = 0$

where higher degree terms of h and k are neglected.

$$\therefore \quad h(2x_0) + k(-2y_0) = x_0^2 - y_0^2 + 4 \quad \text{and} \quad h(2x_0) + k(2y_0) = x_0^2 + y_0^2 - 16.$$

or $h - k = \dfrac{8-8-4}{4\sqrt{2}} = +\dfrac{1}{\sqrt{2}} = 0.707$ and $h + k = \dfrac{8+8-16}{4\sqrt{2}} = 0$

$\Rightarrow \quad x_1 = x_0 + h = 3.182$ and $y_1 = y_0 + k = 2.474.$

Repeating the process we get $h = -0.0204$ and $k = -0.0242$

$\therefore \quad x_2 = 3.162$ and $y_2 = 2.450$ **Ans.**

8. *Solve the equation* $5x - \sin x = 2$ *correct to four places of decimal.*

SOLUTION : $f(x) = 5x - \sin x - 2 = 0, \quad x = x_0 = \pi/6$ and $x_1 = x_0 + h$

then
$$h = -\frac{f(x)}{f'(x)} = \left[\frac{\sin x + 2 - 5x}{5 - \cos x}\right]_{x=\frac{\pi}{6}} = \frac{2 + \frac{1}{2} - \frac{5\pi}{6}}{5 - \cos\frac{\pi}{6}} = \frac{\frac{1}{2} - \frac{\pi}{6}}{1 - \frac{\sqrt{3}}{10}} = \frac{1 - \frac{3.14}{3}}{2 - \frac{1.732}{5}} = -0.0285$$

$\therefore$
$$x_1 = \frac{\pi}{6} - 0.0285 = 0.52358 - 0.0285 = 0.4950.$$
Ans.

9. *Obtain the root of the equation* $x^3 - 2x^2 + 3x - 5 = 0$ *lying between* 1 *and* 2 *correct to four places of decimal, using*

(*a*) *Regula Falsi method* (*b*) *Newton-Raphson method.*

SOLUTION : Let $f(x) = x^3 - 2x^2 + 3x - 5 = 0$, root lying between 1 and 2.

$f(1) = -3, \quad f(2) = 1$

(*i*) By Regula Falsi Method. $a = 1, b = 2, \quad f(a) = -3, \quad f(b) = 1.$

$\therefore$
$$x_1 = \frac{af(b) - bf(a)}{f(b) - f(a)} = \frac{1(1) - 2(-3)}{1 - (-3)} = \frac{7}{4} = 1.75$$

and
$$f(x_1) = 1.75^3 - 2(1.75)^2 + 3(1.75) - 5 = -6.125 + 5.36 + 5.25 - 5 = -0.495$$

Hence root lies between 1.75 and 2

$\therefore$
$$x_2 = \frac{1.75(1) - 2(-0.495)}{1 - (-0.495)} = \frac{2.74}{1.495} = 1.833$$

and
$$f(x_2) = 1.833^3 - 2(1.833)^2 + 3(1.833) - 5 = 6.158 - 6.72 + 5.499 - 5 = -0.62$$

Hence the root lies between 1.833 and 2.0

$$\therefore \quad x_3 = \frac{1.833(1) - 2(-0.62)}{1 - (-0.62)} = \frac{3.113}{1.64} = 1.89$$

(*ii*) By N.R. method we take $x_0 = 1.5$ and $x_1 = x_0 + h$,

Here
$$h = \frac{-f(x_0)}{f'(x_0)} = -\left[\frac{x^3 - 2x^2 + 3x - 5}{3x^2 - 4x + 3}\right]_{x=1.5} = -\left[\frac{3.375 - 4.5 + 4.5 - 5}{6.75 - 6 + 3}\right] = \frac{1.625}{3.75} = 0.433$$

$\therefore$
$$x_1 = 1.5 + 0.433 = 1.933$$

Next
$$x_2 = x_1 + h, \quad h = -\left[\frac{x^3 - 2x^2 + 3x - 5}{3x^2 - 4x + 3}\right]_{1.933} = -\left[\frac{7.223 - 7.473 + 5.799 - 5}{11.21 - 7.732 + 3}\right] = -0.0847$$

$\therefore$
$$x_2 = 1.933 - 0.0847 = 1.848$$
Ans.

10. *Use bisection method to find the real root of the equation* $x^3 - 18 = 0$ *correct to three decimal places.*

SOLUTION : $f(x) = x^3 - 18 = 0$, $f(2) = -10$, $f(3) = 9$ $\therefore$ roots lies in $(2, 3)$,

let $x_1 = \dfrac{2+3}{2} = 2.5$ $\therefore$ $f(2.5) = 2.5^3 - 18 = 15.625 - 18 = -2.375 < 0$

Thus, root lies between 2.5 and 3 $\therefore$ $x_2 = \dfrac{2.5+3}{2} = 2.75$

and $f(2.75) = 2.75^3 - 18 = 20.79687 - 18 = 2.7968 > 0$

Thus, root lies between 2.5 and 2.75 $\therefore$ $x_3 = \dfrac{2.5+2.75}{2} = 2.625.$

Similarly, we get $x_4 = 2.5625$

and $f(2.5625) = (2.5625)^3 - 18 = -1.1735$

Thus, root lies between 2.5625 and 2.625

$\therefore$ $x_5 = \dfrac{2.5625+2.625}{2} = 2.5937$ and $f(x_5) = f(2.5937) = -0.55078 < 0.$

Hence root lies between 2.5937 and 2.625

$\therefore$ $x_6 = \dfrac{2.5937+2.625}{2} = 2.609.$ Further $x_7 = 2.62$ **Ans.**

11. *Using Regula Falsi method, find the roots of the equation.* $2x - 3\sin x - 5 = 0$ *correct to three decimal places.*

SOLUTION : This is a transcendental equation. We can write it as $3\sin x = 2x - 5.$

By drawing the graph of $y = 3\sin x$ and $y = 2x - 5$ we locate the root between 2 and 3.

Let $f(x) = 2x - 3\sin x - 5$ $\therefore$ $f(2) = -3.7279$ and $f(3) = 0.5766$

As first approximate root by Regula Falsi method, we get

$$x_1 = \frac{2(0.5766) - 3(-3.7279)}{0.5766 - (-3.7279)} = \frac{12.3369}{4.3045} = 2.8660$$

For second iteration $f(x_1) = -0.0844 < 0.$ Hence root lies between 2.8660 and 3.

$$x_2 = \frac{2.8660(0.5766) - 3(-0.0844)}{0.5766 - (-0.0844)} = \frac{1.9057}{0.861} = 2.8830$$

Similarly we can get third approximation as $x_3 = 2.8831$ **Ans.**

12. *Perform five iterations of the bisection method to obtain the smallest positive root of the equation* $x^3 - 5x + 1 = 0$

SOLUTION : $f(x) = x^3 - 5x + 1 = 0$ $\therefore$ $f(0) = 1$, $f(1) = -3$, $f(2) = -1$, $f(3) = 13$, $f(-2) = 3$, $f(-3) = -11.$

Hence roots lies in $(0, 1)$, $(2, 3)$ and $(-3, -2)$. Thus smallest positive root lies in $(0, 1)$.

By bisection method, first approximation gives $x_1 = (0+1)/2 = 0.5$

$f(0.5) = 0.125 - 2.5 + 1 = -1.375$

Since $f(0)\,f(0.5) < 0$, one root lies between 0 and 0.5

As second approximation $x_2 = 0.25$, hence $f(0.25) = 0.25^3 - 5(0.25) + 1 = 0.0156 - 1.25 + 1 = -0.2344$

Thus root lies between 0 and 0.25.

$\therefore\ x_3 = 0.125$ and $f(0.125) = 0.00195 - 0.625 + 1 = 0.377$

$\therefore$ Root lies between 0.125 and 0.25

After two more iterations we get the root as 0.2031. **Ans.**

13. *Using Newton-Raphson method, find the root of the equation* $2x^3 - 3x + 4 = 0$ *lying between* -2 *and* -1 *correct to four decimal places.*

SOLUTION : Let $f(x) = 2x^3 - 3x + 4 = 0$, $f(-2) = -6$ and $f(-1) = 5$

Let the initial value of the root be $\dfrac{-1-2}{2} = -1.5$

$\therefore f(-1.5) = -6.75 + 4.5 + 4 = -1.75$. Also $f'(x) = 6x^2 - 3 \quad \therefore \quad f'(1.5) = 10.5$

By N.R. Method, the first approximate root $x_1 = -1.5 - \dfrac{-1.75}{10.5} = -1.333$

Next $f(x_1) = -4.737 + 4 + 4 = 3.2620$

Therefore, $x_2 = -1.333 - \dfrac{f(x_1)}{f'(x_1)} = -1.333 - \dfrac{3.2620}{7.661} = -1.7588$

$\therefore \quad f(x_2) = -10.881 + 5.2767 + 4 = -1.6045$ and $f'(x_2) = 15.5624$

Similarly, $x_3 = -1.7589 - \left(\dfrac{-1.6045}{15.5624}\right) = -1.655$

Continuing the iterations we get the required root correct to four decimal places as -1.6474. **Ans.**

14. *Employing Newton Raphson method, find the zeros of* $f(x) = 1 - 3x + \frac{1}{2}xe^x$ *acurate to five places of decimal.*

SOLUTION : $f(x) = 1 - 3x + \frac{1}{2}xe^x$. Here $f(0) = 1$ and $f(1) = -2 + \dfrac{e}{2} < 0$

$\therefore$ One root lies in (0, 1) we take $x_0 = 0.5$. Using N.R. method

$$x_1 = x_0 - \frac{f(x_0)}{f'(x_0)} \qquad \text{Here } f'(x) = -3 + \frac{e^x}{2}(1+x)$$

$f(0.5) = 1 - 1.5 + 0.25\,(1.6487) = -0.0878$ and $f'(0.5) = -1.76346$

$\therefore \quad x_1 = 0.5 - \dfrac{0.0878}{(1.76346)} = 0.4502.$

Next $f(0.4502) = 0.3531 + 1 - 1.3506 = 0.0024977$ and $f'(0.4502) = -3 + \dfrac{1.4502}{2}(1.5686) = -1.8626$

$\therefore \quad x_2 = 0.4502 - \dfrac{0.0024977}{(-1.8626)} = 0.45154$ **Ans.**

Also, $f(1) = -0.6408$ and $f(2) = 2.389$ therefore one root lies in $(1, 2)$.

Taking $x_0 = 1.5$ we can follow N.R. method as above and get the root = 1.54955. **Ans.**

15. *Compute $23^{1/3}$ by solving for the zero of $f(x) = 23 - x^3$ using Newton's method correct to four decimal places.*

SOLUTION : $f(x) = x^3 - 23 = 0$, $f'(x) = 3x^2$. Using N.R. method we get

$$x_1 = x_0 - \frac{f(x_0)}{f'(x_0)} = 3 - \frac{3^3 - 23}{3.9} = 3 - \frac{4}{27} = 2.85$$

$$x_2 = 2.85 - \frac{2.85^3 - 23}{3(2.85)^2} = 2.85 - \frac{0.149}{24.3675} = 2.844$$

$$x_3 = 2.844 - \frac{2.844^3 - 23}{3(2.844)^2} = 2.844 - \frac{0.00323}{24.265} = 2.84387$$ **Ans.**

16. *Approximate the largest root of the equation $x^4 - 4x^3 + x^2 + 1.2 = 0$ using Newton's method correct to four places of decimal.*

SOLUTION : $f(x) = x^4 - 4x^3 + x^2 + 1.2 = 0$ or $x^3(x - 4) + x^2 + 1.2 = 0$

Obviously no root lies in $(4, \infty)$. Here $f'(x) = 4x^3 - 12x^2 + 2x$.

$$f(3) = -27 + 9 + 1.2 < 0 \quad \text{and} \quad f(4) = 0 + 16 + 1.2 > 0$$

$\therefore$ Largest positive root lies in (3, 4). We take $x_0 = 3.5$

Hence $f(x_0) = -21.4375 + 13.45 = -7.9875$ and $f'(x_0) = 45.5$.

Therefore $$x_1 = x_0 - \frac{f(x_0)}{f'(x_0)} = 3.5 - \frac{(-7.9875)}{45.5} = 3.6755.$$

Next, $f(3.6755) = -1.4032$ and $f'(3.6755) = 198.6137 - 162.1116 + 7.351 = 43.8531$

Here $$x_2 = 3.6755 + \frac{1.4032}{43.8531} = 3.7095$$

Doing one more iteration same way, we get $x_3 = 3.7066$ **Ans.**

17. *Devise a fixed point iteration scheme to find the root of the quadratic equation $2x^2 - 24x + 41 = 0$ near $x = 2$ and find the root.*

SOLUTION : $f(x) = 2x^2 - 24x + 41 = 0$. We are to find root near $x = 2$. By direct iteration method.

$$x = \frac{2x^2 + 41}{24} = \frac{x^2}{12} + \frac{41}{24} = \phi(x)$$

$\therefore$ $\phi'(x) = \frac{x}{6}$ and $\left|\frac{x}{6}\right| < 1$ near $x = 2$. Hence by direct iteration.

$$x_1 = \phi(x_0) = \phi(2) = \frac{49}{24} = 2.0416$$

$$\Rightarrow \quad x_2 = \phi(x_1) = \phi(2.0416) = 2.0556$$

$$x_3 = \phi(x_2) = \phi(2.0556) = 2.0605$$

$$x_4 = \phi(x_3) = \phi(2.0605) = 2.0621$$

$$x_5 = \phi(x_4) = \phi(2.0621) = 2.0627$$
$$x_6 = \phi(x_5) = \phi(2.0627) = 2.06289$$
$$x_7 = \phi(x_6) = \phi(2.06289) = 2.0629$$

18. *Find the root of the equation $x^3 - 4x + 2 = 0$ near $x = 0.6$ using direct iteration method correct to three places of decimal.*

SOLUTION : $f(x) = x^3 - 4x + 2 = 0$. For the root near $x = 0.6$, by direct iteration

we have $x = \dfrac{x^3+2}{4} = \phi(x)$. Here $x_0 = 0.6$

Also $|\phi'(x)| = \left|\dfrac{3x^2}{4}\right| < 1$ near $x = 0.6$.

$\therefore\ x_1 = \phi(x_0) = \left(\dfrac{0.6^3+2}{4}\right) = 0.554$ and $x_2 = \phi(x_1) = \phi(0.554) = 0.542$

$$x_3 = \phi(x_2) = \phi(0.542) = \frac{(0.542)^3+2}{4} = 0.5398$$
$$x_4 = \phi(x_3) = \phi(0.5398) = \frac{(0.5398)^3+2}{4} = 0.5393$$
$$x_5 = \phi(x_4) = \phi(0.5393) = \frac{(0.5393)^3+2}{4} = 0.5392$$
$$x_6 = \phi(x_5) = \phi(0.5392) = \frac{(0.5392)^3+2}{4} = 0.539191$$

$\therefore$ desired root = 0.5392 **Ans.**

19. *Find the root of the equation $2x = \cos x + 3$ correct to three decimal places using direct iteration method.*

SOLUTION : Given equation can be written as $x = \dfrac{1}{2}(\cos x + 3)$

Therefore $\phi(x) = \dfrac{1}{2}(\cos x + 3)$ and $\phi'(x) = -\dfrac{1}{2}\sin x$

and $|\phi'(x)| < 1$ hence direct iteration process is applicable.

We take initial value as $x_0 = \dfrac{\pi}{2}$ and successive iterations give as follows :

$$x_1 = \phi(\pi/2) = 1.5, \quad x_2 = \phi(1.5) = 1.535,$$
$$x_3 = \phi(1.535) = 1.518, \quad x_4 = \phi(1.518) = 1.526$$
$$x_5 = \phi(1.526) = 1.522, \quad x_6 = \phi(1.522) = 1.524$$
$$x_7 = \phi(1.524) = 1.523, \quad x_7 = \phi(1.523) = 1.524$$

Therefore root correct to three decimal places is 1.524. **Ans.**

SOLUTIONS TO EXERCISE 26C

1. *Solve the following system by Gauss elimination*

$$x + y + z = 3$$
$$4x + 2y - z = 5$$
$$9x + 5y - z = 13$$

SOLUTION : System of equations is consistent because rank (D) = rank (A) = 2 < 3 hence many solutions. One of these is $x = y = z = 1$

2. *Solve the following linear system of equations by Gauss elimination*

$$5x + 10y - 2z = -0.30$$
$$2x - y + z = 1.91$$
$$3x + 4y = 1.16$$

SOLUTION : $D = \left[\begin{array}{ccc|c} 5 & 10 & -2 & -0.3 \\ 2 & -1 & 1 & 1.91 \\ 3 & 4 & 0 & 1.16 \end{array}\right]$ Applying $R_1 \to \frac{1}{5} R_1$

$$\sim \begin{bmatrix} 1 & 2 & -0.4 & . & -0.06 \\ 2 & -1 & 1 & . & 1.91 \\ 3 & 4 & 0 & . & 1.16 \end{bmatrix} \text{ Applying } R_2 \to R_2 - 2R_1 \text{ and } R_3 \to R_3 - 3R_1$$

$$\sim \begin{bmatrix} 1 & 2 & -0.4 & . & -0.06 \\ 0 & -5 & 1.8 & . & 2.03 \\ 0 & -2 & 1.2 & . & 1.34 \end{bmatrix} \text{ Applying } R_2 \to -\frac{1}{5} R_2$$

$$\sim \begin{bmatrix} 1 & 2 & -0.4 & . & -0.06 \\ 0 & 1 & -0.36 & . & -0.406 \\ 0 & -2 & 1.2 & . & 1.34 \end{bmatrix} \text{ Applying } R_3 \to R_3 + 2R_2$$

$$\sim \begin{bmatrix} 1 & 2 & -0.4 & . & -0.06 \\ 0 & 1 & -0.36 & . & -0.4.06 \\ 0 & 0 & 0.48 & . & 0.528 \end{bmatrix}$$

$\Rightarrow$ $0.48\,z = 0.528$ $\therefore\ z = 1.1$

and $y - 0.36\,z = -0.406$ $\therefore\ y = -0.406 + 0.36\,(1.1) = -0.01$

and $x + 2y - 0.4z = -0.06,$ $x = -0.06 + 0.4\,(1.1) - 2\,(-0.01) = 0.4$

thus $x = 0.4,$ $y = -0.01$ and $z = 1.1$ **Ans.**

3. *Solve by using Gauss Jordan scheme, the following system of equations*

$$5x + 3y + 7z = 4$$
$$x + 3y + 3z = 2$$
$$7x + 2y + 10z = 5$$

SOLUTION: Augmented matrix = $[A : B] = \left[\begin{array}{ccc|c} 5 & 3 & 7 & 4 \\ 1 & 3 & 3 & 2 \\ 7 & 2 & 10 & 5 \end{array}\right] \sim \left[\begin{array}{ccc|c} 1 & 0 & 0 & 1/5 \\ 0 & 1 & 0 & 3/10 \\ 0 & 0 & 1 & 3/10 \end{array}\right]$

$\Rightarrow \quad x = \frac{1}{5}, \; y = z = \frac{3}{10}.$ **Ans.**

4. *Solve, by using Gauss Jordan elimination method, the following system of equations*

$$2x_1 + 2x_2 + 4x_3 - 2x_4 = 10$$
$$x_1 + 3x_2 + 2x_3 + x_4 = 17$$
$$3x_1 + x_2 + 3x_3 + x_4 = 18$$
$$x_1 + 3x_2 + 4x_3 + 2x_4 = 27$$

SOLUTION : $AX = B$. The augmented matrix D is given by

$$D = \begin{bmatrix} 2 & 2 & 4 & -2 & . & 10 \\ 1 & 3 & 2 & 1 & . & 17 \\ 3 & 1 & 3 & 1 & . & 18 \\ 1 & 3 & 4 & 2 & . & 27 \end{bmatrix} \text{ applying } R_1 \to \frac{1}{2}R_1 \text{ and } R_2 - 1R_1,\ R_4 - R_1 \text{ and } R_3 - 3R_1$$

$$\sim \begin{bmatrix} 1 & 1 & 2 & -1 & . & 5 \\ 0 & 2 & 0 & 2 & . & 12 \\ 0 & -2 & -3 & 4 & . & 3 \\ 0 & 2 & 2 & 3 & . & 22 \end{bmatrix} \text{ Applying } R_2 \to \frac{1}{2}R_2,\ R_3 \to R_3 + R_2,\ R_4 \to R_4 - R_2$$

$$\sim \begin{bmatrix} 1 & 1 & 2 & -1 & . & 5 \\ 0 & 1 & 0 & 1 & . & 6 \\ 0 & 0 & -3 & 6 & . & 15 \\ 0 & 0 & 2 & 1 & . & 10 \end{bmatrix} \text{ Applying } R_3 \to -\frac{1}{3}R_3 \text{ and } R_4 - 2R_3$$

$$\sim \begin{bmatrix} 1 & 1 & 2 & -1 & . & 5 \\ 0 & 1 & 0 & 1 & . & 6 \\ 0 & 0 & 1 & -2 & . & -5 \\ 0 & 0 & 2 & 1 & . & 10 \end{bmatrix} \text{ Applying } R_4 \to R_4 - 2R_3$$

$$\sim \begin{bmatrix} 1 & 1 & 2 & -1 & . & 5 \\ 0 & 1 & 0 & 1 & . & 6 \\ 0 & 0 & 1 & -2 & . & -5 \\ 0 & 0 & 0 & 5 & . & 20 \end{bmatrix} \text{ Now applying } R_4 \to \frac{1}{5}R_4 \text{ and } R_1 \to R_1 - R_2$$

$$\sim\left[\begin{array}{cccc|c} 1 & 0 & 2 & -2 & -1 \\ 0 & 1 & 0 & -2 & 6 \\ 0 & 0 & 0 & 1 & -5 \\ 0 & 0 & 0 & 1 & 4 \end{array}\right]$$ Next applying $R_3 \to R_3 + 2R_4$

$$\sim\left[\begin{array}{cccc|c} 1 & 0 & 2 & -2 & -1 \\ 0 & 1 & 0 & 1 & 6 \\ 0 & 0 & 1 & 0 & 3 \\ 0 & 0 & 0 & 1 & 4 \end{array}\right]$$ Finally applying $R_2 \to R_2 - R_4$, $R_1 - 2R_3$ and $R_1 \to R_1 + 2R_4$ we get

$$\sim\left[\begin{array}{cccc|c} 1 & 0 & 0 & 0 & 1 \\ 0 & 1 & 0 & 0 & 2 \\ 0 & 0 & 1 & 0 & 3 \\ 0 & 0 & 0 & 1 & 4 \end{array}\right]$$ Hence $x_1 = 1$ $x_2 = 2$, $x_3 = 3$ x_4, $= 4$ **Ans.**

5. *Employ Crout's method for solving the following system of algebraic simultaneous equations*

$$2x + 3y + 2z = 2$$
$$10x + 3y + 4z = 16$$
$$3x + 6y + z = -6$$

SOLUTION : The matrix equation is $AX = B$

where $A = \begin{bmatrix} 2 & 3 & 2 \\ 10 & 3 & 4 \\ 3 & 6 & 1 \end{bmatrix}$, $X = \begin{bmatrix} x \\ y \\ z \end{bmatrix}$, $B = \begin{bmatrix} 2 \\ 16 \\ -6 \end{bmatrix}$

Let $$A = LU = \begin{bmatrix} 1 & 0 & 0 \\ l_2 & 1 & 0 \\ l_{31} & l_{32} & 1 \end{bmatrix}\begin{bmatrix} u_{11} & u_{12} & u_{13} \\ 0 & u_{22} & u_{23} \\ 0 & 0 & u_{33} \end{bmatrix} = \begin{bmatrix} 1 & 0 & 0 \\ 5 & 1 & 0 \\ \frac{3}{2} & -\frac{1}{8} & 1 \end{bmatrix}\begin{bmatrix} 2 & 3 & 2 \\ 0 & -12 & -6 \\ 0 & 0 & -\frac{11}{7} \end{bmatrix}$$

To solve the given equations we have $L(UX) = B$ and then $LY = B$ where $UX = Y$.

Here $LY = B$ gives $\begin{bmatrix} 1 & 0 & 0 \\ 5 & 1 & 0 \\ \frac{3}{2} & \frac{-1}{8} & 1 \end{bmatrix}\begin{bmatrix} y_1 \\ y_2 \\ y_3 \end{bmatrix} = \begin{bmatrix} 2 \\ 16 \\ -6 \end{bmatrix}$ Hence $y_1 = 2$, $y_2 = 6$ and $y_3 = -\frac{-33}{4}$.

Further, $UX = Y$ gives $\begin{bmatrix} 2 & 3 & 2 \\ 0 & -12 & -6 \\ 0 & 0 & \frac{-11}{4} \end{bmatrix}\begin{bmatrix} x_1 \\ x_2 \\ x_3 \end{bmatrix} = \begin{bmatrix} 2 \\ 6 \\ -33/4 \end{bmatrix}$

Hence $x_3 = 3$, $x_2 = -2$ and $x_1 = 1$. Thus, the solution is $x_1 = 1$, $x_2 = -2$, $x_3 = 3$. **Ans.**

6. *Apply Crout's method to solve the following system of equations*

$$5x - y - 2z = 142$$
$$x - 3y - z = -30$$
$$2x - y - 3z = 5$$

SOLUTION : The matrix equation is $AX = B$.

Now, $$A = LU = \begin{bmatrix} 5 & -1 & -2 \\ 1 & -3 & -1 \\ 2 & -1 & -3 \end{bmatrix} = \begin{bmatrix} 1 & 0 & 0 \\ 1/5 & 1 & 0 \\ 2/5 & 3/4 & 1 \end{bmatrix} \begin{bmatrix} 5 & -1 & -2 \\ 0 & -\dfrac{14}{5} & -3/5 \\ 0 & 0 & 29/14 \end{bmatrix}$$

Also, $LUX = B$, If $UX = Y$ then $LY = B$.

$$LY = B \text{ or } \begin{bmatrix} 1 & 0 & 0 \\ 1/5 & 1 & 0 \\ 2/5 & 3/14 & 1 \end{bmatrix} \begin{bmatrix} y_1 \\ y_2 \\ y_3 \end{bmatrix} = \begin{bmatrix} 142 \\ -30 \\ 5 \end{bmatrix}$$ which gives $y_1 = 142$, $y_2 = 58.4$, $y_3 = 39.2857$.

Next, $$UX = Y \text{ or } \begin{bmatrix} 5 & -1 & -2 \\ 0 & -14/5 & -3/4 \\ 0 & 0 & 29/14 \end{bmatrix} \begin{bmatrix} x_1 \\ x_2 \\ x_3 \end{bmatrix} = \begin{bmatrix} 142 \\ 58.4 \\ 39.2857 \end{bmatrix}$$

gives $x_1 = 39.34$, $x_2 = 16.78$, $x_3 = 18.96 = 19$. **Ans.**

7. *Use Crouts method to solve the following set of simultaneous equations*

$$10x_1 - 7x_2 + 3x_3 + 5x_4 = 6$$
$$-6x_1 + 8x_2 - x_3 - 4x_4 = 5$$
$$3x_1 + x_2 + 4x_3 + 11x_4 = 2$$
$$5x_1 - 9x_2 - 2x_3 + 4x_4 = 7$$

SOLUTION : $AX = B$ let $A = \text{LU}$ then $L\,(\text{UX}) = B$. If $\text{UX} = Y$ then $L(Y) = B$.

$$A = LU = \begin{bmatrix} 1 & 0 & 0 & 0 \\ -0.6 & 1 & 0 & 0 \\ .3 & 0.816 & 1 & 0 \\ 0.5 & -1.447 & -0.957 & 1 \end{bmatrix} \begin{bmatrix} 10 & -7 & 3 & 5 \\ 0 & 3.8 & 0.8 & -1.0 \\ 0 & 0 & 2.447 & 10.316 \\ 0 & 0 & 0 & 9.925 \end{bmatrix}$$

Now $$LY = B \text{ or } \begin{bmatrix} 1 & 0 & 0 & 0 \\ -0.6 & 1 & 0 & 0 \\ 0.3 & 0.816 & 1 & 0 \\ 0.5 & -1.447 & -0.957 & 1 \end{bmatrix} \begin{bmatrix} y_1 \\ y_2 \\ y_3 \\ y_4 \end{bmatrix} = \begin{bmatrix} 6 \\ 5 \\ 2 \\ 7 \end{bmatrix}$$

which gives $y_1 = 6$, $y_2 = 8.6$, $y_3 = -6.8158$, $y_4 = 9.9247$

Next $$UX = Y \text{ or } \begin{bmatrix} 10 & -7 & 3 & 5 \\ 0 & 3.8 & 0.8 & -1.0 \\ 0 & 0 & 2.447 & 10.316 \\ 0 & 0 & 0 & 9.925 \end{bmatrix} \begin{bmatrix} x_1 \\ x_2 \\ x_3 \\ x_4 \end{bmatrix} = \begin{bmatrix} 6 \\ 8.6 \\ -6.8158 \\ 9.9247 \end{bmatrix}$$

$\therefore \quad x_4 = 1, 2.447\ x_3 + 10.316\,(1) = -6.8156 \quad \therefore \quad x_3 = -7$

$3.8\,x_2 + 0.8\,x_3 - 1.0x_4 = 8.6$ hence $x_2 = 4$ and $10\,x_1 - 7\,x_2 + 5x_4 = 6$ gives $x_1 = 5$

$\therefore$ solution is $x_1 = 5, \quad x_2 = 4, \quad x_3 = -7, \quad x_4 = 1$ **Ans.**

8. *Apply Gauss-Seidel method (two steps) to solve the following system of equations*

$$13x_1 + 5x_2 - 3x_3 + x_4 = 18$$
$$2x_1 + 12x_2 + x_3 - 4x_4 = 13$$
$$3x_1 - 4x_2 + 10x_3 + x_4 = 29$$
$$2x_1 + x_2 - 3x_3 + 9x_4 = 31$$

SOLUTION : Since the conditions of Gauss-Seidal method are satisfied we can apply this process.

Take $x_1^{(0)} = x_2^{(0)} = x_3^{(0)} = x_4^{(0)} = 0$. The iteration formula is

$$x_1^{(n+1)} = \frac{1}{13}\left[18 - 5x_2^{(n)} + 3x_3^{(n)} - x_4^{(n)}\right]$$

$$x_2^{(n+1)} = \frac{1}{12}\left[13 - 2x_1^{(n+1)} - x_x^{(n)} + 4x_4^{(n)}\right]$$

$$x_3^{(n+1)} = \frac{1}{10}\left[29 - 3x_1^{(n+1)} + 4x_2^{(n+1)} - x_4^{(n)}\right]$$

$$x_4^{(n+1)} = \frac{1}{9}\left[31 - 2x_1^{(n+1)} - x_2^{(n+1)} + 3x_3^{(n+1)}\right]$$

For $n = 0,\ x_1^{(1)} = \frac{18}{13} = 1.3846,\quad x_2^{(1)} = \frac{1}{12}\left[13 - 2x^{(1)} - x_3^{(0)} + 4x_4^{(0)}\right] = \frac{13.7692}{12} = 0.8525$

$$x_3^{(1)} = \frac{1}{10}\left[29 - 3x_1^{(1)} + 4_1^{(1)} + 4x_2^{(1)} - x_4^{(0)}\right] = 2.8256$$

$$x_4^{(1)} = \frac{1}{9}\left[31 - 2x_1^{(1)} - x_2^{(1)} + 3x_3^{(1)}\right] = \frac{1}{9}[31 - 2.7692 - 0.8525 + 8.4768] = 3.9839$$

Next, for $n = 1$, we have

$$x_1^{(2)} = \frac{1}{13}[18 - 5(0.82525) + (2.8256) - 3.9839] = 1.4023$$

$$x_2^{(2)} = \frac{1}{12}\left[13 - 2, x_1^{(2)} - x_3^{(1)} + 4x_4^{(1)}\right] = \frac{1}{12}[13 - 2(1.4023) - 2.8256 + 4(3.9839)] = 1.942$$

$$x_3^{(2)} = \frac{1}{10}\left[29 - 3x_1^{(2)} + 4x_2^{(2)} - x_4^{(1)}\right] = \frac{1}{10}[29 - 3\,(1.4023) + 4\,(1.942) - 3.9839) = 3.1979$$

$$x_4^{(2)} = \frac{1}{9}\left[31 - 2x_1^{(2)} - x_2^{(2)} + 3x_3^{(2)}\right] = \frac{1}{9}[31 - 2(1.4023) - 1.942 + 3\,(3.1979) = 3.9784$$

Therefore, after two steps of Gauss Seidal method, we get

$x_1 = 1.4, \quad x_2 = 1.94, \quad x_3 = 3.2$ and $x_4 = 3.97 = 4$ approx. **Ans.**

9. *Solve by Crout's method the following system of equations.*

$$x + 2y + 3z = 7$$
$$2x + 7y + 15z = 26$$
$$3x + 15y + 41z = 62$$

SOLUTION : $AX = B, \quad A = LU \quad \Rightarrow \quad UX = Y \quad$ where $\quad LY = B.$

$$A = LU \quad \text{or} \quad \begin{bmatrix} 1 & 2 & 3 \\ 2 & 7 & 15 \\ 3 & 15 & 41 \end{bmatrix} = \begin{bmatrix} 1 & 0 & 0 \\ 2 & 1 & 0 \\ 3 & 3 & 1 \end{bmatrix} \begin{bmatrix} 1 & 2 & 3 \\ 0 & 3 & 9 \\ 0 & 0 & 5 \end{bmatrix}$$

$$\text{Next,} \quad LY = B \quad \text{or} \quad \begin{bmatrix} 1 & 0 & 0 \\ 2 & 1 & 0 \\ 3 & 3 & 1 \end{bmatrix} \begin{bmatrix} y_1 \\ y_2 \\ y_3 \end{bmatrix} = \begin{bmatrix} 7 \\ 26 \\ 62 \end{bmatrix}$$

which gives $y_1 = 7, \quad y_2 = -2\,(7) + 26 = 12 \quad$ and $\quad y_3 = 62 - 3(7) - 3(12) = 5.$

$$\text{Therefore,} \quad UX = Y \quad \text{becomes} \quad \begin{bmatrix} 1 & 2 & 3 \\ 0 & 3 & 9 \\ 0 & 0 & 5 \end{bmatrix} \begin{bmatrix} x_1 \\ x_2 \\ x_3 \end{bmatrix} = \begin{bmatrix} 7 \\ 12 \\ 5 \end{bmatrix}$$

which gives $x_3 = 1, \quad x_2 = 1 \quad$ and $\quad x_1 = 2$

$\therefore$ solution is $x_1 = 2, \quad x_2 = 1, \quad x_3 = 1.$ **Ans.**

10. *Use Gauss–Seidal procedure to solve the following set of equations.*

$$5x + 2y + z = 12$$
$$x + 4y + 2z = 15$$
$$x + 2y + 5z = 20$$

SOLUTION :

$$x = \frac{1}{5}\,[12 - 2y - z]$$
$$y = \frac{1}{4}\,[15 - x - 2z]$$
$$z = \frac{1}{5}\,[20 - x - 2y]$$

By Gauss Seidal method

$$x^{(n+1)} = \frac{1}{5}\,[12 - 2y^{(n)} - z^{(n)}]$$
$$y^{(n+1)} = \frac{1}{4}\,[15 - x^{(n+1)} - 2z^{(n)}]$$
$$z^{(n+1)} = \frac{1}{5}\,[20 - x^{(n+1)} - 2y^{(n+1)}]$$

Let $x^{(0)} = y^{(0)} = z^{(0)} = 0$ then

$$x^{(1)} = \frac{1}{5}\,[12 - 0 - 0] = 2.4$$

$$y^{(1)} = \frac{1}{4}\ [15 - 2.4 - 2\ (0)] = 3.15$$

$$z^{(1)} = \frac{1}{5} = [20 - 2.4 - 2\ (3.15)] = 2.26$$

$$x^{(2)} = \frac{1}{5}\ [12 - 2\ (3.15) - 2.26] = 0.688$$

$$y^{(2)} = \frac{1}{4} = [15 - 0.688 - 2\ (2.26)] = 2.448$$

$$z^{(2)} = \frac{1}{5} = [20 - 0.688 - 2\ (2.448)] = 2.883$$

$$x^{(3)} = \frac{1}{5}\ [12 - 2\ (2.448) - 2.883] = 0.8442$$

$$y^{(3)} = \frac{1}{4}\ [15 - 0.8442 - 2(2.883)] = 2.0974$$

$$z^{(3)} = \frac{1}{5}\ [20 - 0.8442 - 2(2.0974)] = 2.9922$$

$$x^{(4)} = \frac{1}{5}\ [12 - 2\ (2.0974) - 2.9922] = 0.9622$$

$$y^{(4)} = \frac{1}{4}\ [15 - 0.9626 - 2(2.99222)] = 2.0132$$

$$z^{(4)} = \frac{1}{5}\ [20 - 0.9626 - 2\ (2.0132)] = 3.0022$$

Therefore, after four iterations, we hrve $x = 0.9626$, $y = 2.0132$, $z = 3.0022$.

Further proceeding the same scheme we can finally get $x = 1$, $y = 2$, $x = 3$. **Ans.**

11. *Apply Gauss elimination process to solve the following system of equations*

$$4.7x + 1.3y - 1.6z = 1.3$$
$$x - 4.1y + 1.1z = 4.6$$
$$2.1x + 1.4y + 6.2z = 5.2$$

SOLUTION: Equations can be written as

$$x + 0.2766\,y - 0.3404\,z = 0.2766$$
$$x - 4.1\,y + 1.1\,z = 4.6$$
$$x + 0.6666\,y + 2.9524\,z = 2.4762$$

Eliminating x with last two equations, we get

$$4.3766\,y - 1.4404\,z = -4.3234 \quad \text{or} \quad y - 0.329\,z = -1.0123$$

and $\quad 4.7666\,y + 1.8\,524\,z = -2.1238 \quad \text{or} \quad y + 0.3886\,z = -0.4455$

Eliminating y here, we get $0.07176\,z = 0.56667 \quad \therefore \quad z = 0.75556$

In turn $y = -0.73918$ and $x = 0.73826$

Solution is $x = 0.73826$, $y = 0.73918$, $z = 0.75556$. **Ans.**

12. *Use Guauss-Seidal iterative process to find the solution of the following system of simultaneous equations*

$$1.2x_1 + 4.4x_2 - 1.9x_3 = -4.2$$
$$5.1x_1 - 1.3x_2 + 2.4x_3 = 2.7$$
$$-2.6x_1 + 1.7x_2 - 6.3x_3 = 9.6$$

correct to three places of decimal.

SOLUTION: On interchanging first two equations the system becomes diagonally dominant.

Take $x_1^{(0)} = x_2^{(0)} = x_3^{(0)} = 1$. Scheme of iteration is

$$x_1^{(n+1)} = \frac{1}{5.1}(2.7 + 1.3x_2^{(n)} - 2.4x_3^{(n)})$$

$$x_2^{n+1} = \frac{1}{4.4}(-4.2 - 1.2x_1^{(n+1)} + 1.9x_3^{(n)})$$

$$x_3^{(x+1)} = \frac{1}{6.3}(-9.6 - 2.6x_1^{(n+1)} + 1.7x_2^{(n+1)})$$

Iteration No.	0	1	2	3	4	5
x_1	1	0.313726	1.229617	1.219913	1.175631	1.162857
x_2	1	–0.608289	–2.074693	–2.406137	–2.430951	–2.422740
x_3	1	–1.817425	–2.591108	–2.676541	–2.664962	–2.657887
	6	7	8			
x_1	1.162621	1.162815	1.162924			
x_2	–2.419348	–2.418785	–2.418809			
x_3	–2.656461	–2.656389	–2.656450			

Ans.

13. *Solve the following equations using Gauss elimination method upto three decimal places*

$$5x_1 - x_2 + x_3 = 10$$
$$2x_1 + 4x_2 = 12$$
$$x_1 + x_2 + 5x_3 = -1$$

SOLUTION : Given equations can be written as

$$x_1 + x_2 + 5x_3 = -1 \quad ...(1)$$

$$x_1 + 2x_2 = 6 \quad ...(2)$$

$$5x_1 - x_2 + x_3 = 10 \quad ...(3)$$

Eliminating x_3 in (1) and (3) we get $-24x_1 + 6x_2 = -51$ or $-4x_1 + x_2 = -8.5$ ...(4)

From (2) and (4), we get $9x_1 = 23$ $\therefore$ $x_1 = 2.555$

and hence $x_2 = 1.723$ and lastly $x_3 = 10 + x_2 - 5x_1 = -$ **1.043** **Ans.**

14. *Solve the following equations using Gauss-Jordan method upto three decimal places*

$$2x + 2y + z = 12$$
$$3x + 2y + 2z = 8$$
$$5x + 10y - 8z = 10$$

SOLUTION : Eliminating z in given equations, we get $x + 2y = 16$ and $17x + 18y = 42$

which give $x = -12.75$ and $y = 14.37$

$\therefore$ $z = 12 - 2x - 2y = 8.76$

Thus, solution is $x = -12.75,\ y = 14.37,\ z = 8.76.$ **Ans.**

ADDITIONAL IMPORTANT PROBLEMS

1. *Solve $x^3 - 9x + 1 = 0$ for the root lying between 2 and 3, correct to three significant figures.*

SOLUTION: We have $f(x) = x^3 - 9x + 1 \quad \therefore \quad f(2) = -9, \; f(3) = 1$ and $f(2)f(3) < 0.$

Therefore root of $f(x) = 0$ lies between 2 and 3.

Let $x_0 = 2, \; x_1 = 3$ and x_2 be the root of the equation.

Then, we have $x_2 = \dfrac{x_0 + x_1}{2} = \dfrac{2+3}{2} = 2.5 \quad \therefore \quad f(x_2) = (2.5)^2 - 9(2.5) + 1 = -5.8$

Since $f(x_2)$ and $f(x_1)$ have opposite signs $\quad \therefore$ the root lies between x_2 and x_1.

The next approximation is $x_3 = \dfrac{x_2 + x_1}{2} = \dfrac{2.5+3}{2} = 2.75 \quad \therefore \quad f(x_3) = (2.75)^3 - 9(2.75) + 1 = -2.9$

Since $f(x_3) \neq 0$ and $f(x_3) f(x_1) < 0,$ the root lies in the interval $[x_3, x_1]$, which is given by

$$x_4 = \frac{x_3 + x_1}{2} = \frac{2.75+3}{2} = 2.88 \quad \therefore \quad f(x_4) = (2.88)^3 - 9(2.88) + 1 = -1.03$$

We have $f(x_4) \neq 0$ and $f(x_4) f(x_1) < 0,$ therefore root lies in the interval $[x_4, x_1]$, hence

$$x_5 = \frac{x_4 + x_1}{2} = \frac{2.88+3}{2} = 2.94 \quad \therefore \quad f(x_5) = (2.94)^3 - 9(2.24) + 1 = -0.05$$

Again, since $f(x_5) \neq 0$ and $f(x_5) f(x_1) < 0,$ therefore, we get

$$x_6 = \frac{x_5 + x_1}{2} = \frac{2.94+3}{2} = 2.97 \quad \therefore \quad f(x_6) = 0.47 \quad f(x_6) \neq 0 \quad \text{and} \quad f(x_5) f(x_6) < 0$$

root lies between x_5 and x_6. Therefore, we get

$$x_7 = \frac{x_5 + x_6}{2} = \frac{2.94+2.97}{2} = 2.955$$

We have $f(x_7) = 0.21$ Since $f(x_7) \neq 0,$ and $f(x_7) f(x_5) < 0,$
therefore, root of (1) is in the interval (x_5, x_7). Hence

$$x_8 = \frac{x_5 + x_7}{2} = \frac{2.94+2.955}{2} = 2.9475$$

We have $f(x_8) = 0.08.$ $f(x_8) \neq 0,$ and $f(x_5) f(x_8) < 0,$
therefore, root of (1) lies between x_5 and x_8. Therefore

$$x_9 = \frac{x_5 + x_8}{2} = \frac{2.94+2.9475}{2} = 2.9438 \quad \therefore \quad f(x_9) = 0.017$$

Since $f(x_9) \neq 0$ and $f(x_5) f(x_9) < 0.$ The next approximation of the root is

$$x_{10} = \frac{2.94+2.9438}{2} = 2.9419$$

$f(x_{10})\,f(x_9) < 0$. Therefore, the root lies between x_{10} and x_9.

The next approximation of the root is

$$x_{11} = \frac{x_{10}+x_9}{2} = \frac{2.9419+2.9438}{2} = 2.9419$$

x_9, x_{10} and x_{11} are approximately equal upto three significant figures hence, we can take 2.94 as a root upto 3 significant figures.

2. *Use the method of bisection to find a root of the equation*

$$f(x) = x^4 + 2x^3 - x - 1 = 0$$

lying in the interval [0, 1] at the end of fifth interation.

SOLUTION: We have $f(x) = x^4 + 2x^3 - x - 1$. Here $x_0 = 0$, $x_1 = 1$.

$f(0) = -1$ and $f(1) = 1 + 2 - 1 - 1 = 1$ $\therefore$ $f(x_0)\,f(x_1) = f(0)\,f(1) < 0$

The root lies between $x_0 = 0$, and $x_1 = 1$. Using bisection method, we get

$$x_2 = \frac{0+1}{2} = 0.5 \quad \text{we have, } f(x_2) = f(0.5) = -1.1875.$$

and $f(x_2)\,f(x_1) < 0$, root lies in [0.5, 1], therefore we get

$$x_3 = \frac{x_2+x_1}{2} = \frac{0.5+1}{2} = 0.75 \quad \therefore \quad f(x_3) = (0.75)^4 + 2.(0.75)^3 - 0.75 - 1 = -0.5898$$

Since $f(x_3)\,f(x_1) < 0$, root lies between x_3 and x_1, we get next approximation as

$$x_4 = \frac{0.75+1}{2} = 0.875. \quad \therefore \quad f(x_4) = f(0.875) = 0.051$$

Since $f(x_3)\,f(x_4) < 0$, the root lies between x_3 and x_4 and is given by

$$x_5 = \frac{0.75+0.875}{2} = 0.8125 \quad \therefore \quad f(x_5) = f(0.8125) = -0.3039$$

We have $f(x_5)\,f(x_4) < 0$ the root of the given equation lies in [0.8125, 0.875]. **Ans.**

3. *Find an approximate value of the root of the equation $x^3 + x - 1 = 0$ near $x = 1$, by the method of Regula-falsi using the formula twice.*

SOLUTION: Here, we have $f(x) = x^3 + x - 1$. $\therefore$ $f(0.5) = -0.375$, $f(1) = 1$

Hence, one root of $f(x) = 0$ lies between 0.5 and 1.

We take $x_0 = 0.5$ and $x_1 = 1$ such that

$$x_2 = \frac{x_0\,f(x_1) - x_1 f(x_0)}{f(x_1) - f(x_0)} = \frac{0.5(1) - 1.(-0.375)}{1-(-0.375)} = 0.6363 \approx 0.64.$$

Now $f(x_2) = f(0.64) = -0.0979 < 0$.

Since $f(x_2)\,f(x_1) < 0$, the root of $f(x) = 0$, lies in the between x_2 and x_1.

The next approximation is $x_3 = \dfrac{x_2 f(x_1) - x_1 f(x_2)}{f(x_1) - f(x_2)} = \dfrac{(0.64)\,(1) - (1)\,(-0.0979)}{1-(-0.0979)}$

$$= \frac{0.7379}{1.0997} = 0.6721$$

The approximate value of the root is 0.6721.

4. Using Regula-falsi method find the real root of the equation $e^x - 4x = 0$.

SOLUTION: We have $f(x) = e^x - 4x = 0$.

Here $f(0) = e^0 - 4.(0) = 1$, $f(1) = e - 4 = -1.28$, $f(0)f(1) < 0$.

Therefore, root of (1) lies between 0 and 1. $x_0 = 0$ and $x_1 = 1$, $f(x_0) = 1, f(x_1) = -1.28$

The next approximation of the root is x_2, then

$$x_2 = \frac{x_0 f(x_1) - x_1 f(x_0)}{f(x_1) - f(x_0)} = \frac{0.(-1.128) - 11}{-1.128 - 1} = 0.439.$$

$\therefore$ $f(x_2) = f(0.439) = e^{0.439} - 4.(0.439) = -0.2048.$

We have $f(x_0) > 0$ and $f(x_2) < 0$ **hence** $f(x_0)\, f(x_2) < 0$

$\therefore$ **The root lies between** x_0 **and** x_2. **By Regula-falsi method we have**

$$x_3 = \frac{x_0 f(x_2) - x_2 f(x_0)}{f(x_2) - f(x_0)} = \frac{0.(-0.2048) - 0.439 \times 1}{-0.2041 - 1} = 0.3643$$

$\therefore$ $f(x_3) = f(0.3643) = e^{0.3664} - 4.(0.3643) = -0.0177$

Since $f(x_0)\ f(x_3) < 0$, **root lies between** x_0 **and** x_3, **which is given by**

$$x_4 = \frac{x_0 f(x_3) - x_3 f(x_0)}{f(x_0) - f(x_0)} = \frac{0.(-0.00177) - 0.36431.1}{-0.0177 - 1} = 0.35796.$$

$\therefore$ $f(x_4) = f(0.35796) - e^{0.35796} - 4.(0.35796) = -0.00143$

We get $f(x_0)\ f(x_4) < 0$, **the root lies in between** x_0 **and** x_4, **which is given by**

$$x_5 = \frac{x_0 f(x_4) - x_4 f(x_0)}{f(x_4) - f(x_0)} = \frac{0.(-0.00143) - (0.35796).1}{-0.0143 - 1} = 0.35745.$$

$\therefore$ $f(x_5) = e^{0.35745} - 4.(0.35745) = -0.000121.$

Since $f(x_0)\, f(x_5) < 0$, the root of (1) lies between x_0 and x_5, and is given by

$$x_6 = \frac{x_0 f(x_5) - x_5 f(x_0)}{f(x_5) - f(x_0)} = \frac{0.(-0.000121) - (0.35745).1}{-0.000121 - 1} = 0.35741.$$

$\therefore$ $f(x_6) = f(0.35741) = -0.0000181.$

The values of x_5 and x_6 slightly differ. Therefore, the root of the given equation is 3.5741. **Ans.**

5. *Find the roots of the equation* $x^2 - 5x + 1 = 0$ *using iteration method.*

SOLUTION: Let $f(x) = x^2 - 5x + 1 = 0$ We have, $f(0) = 1$, and $f(1) = 1^2 - 5.1 + 1 = -3$.
$f(0)f(1) < 0$.

$\therefore$ there is a real root of $f(x) = 0$ between 0 and 1.

Also $f(4) = 4^2 - 5.4 + 1 = -3$ and $f(5) = 5^2 - 5.5 + 1 = 1$

We have $f(4)\, f(5) < 0$, hence, there is another root of $f(x) = 0$ between 4 and 5.

The given equation can be written as $x = \frac{1}{5}(x^2 + 1) = \phi(x)$ (say) ...(2)

(i) Let $x_0 = 0$ be the initial approximation. The next approximation is given by

$$x_1 = \phi(x_0) = \frac{1}{5}(x_0^2 + 1) = \frac{1}{5}(0^2 + 1) + 0.2$$

The successive approximations are

$$x_2 = \phi(x_1) = \frac{1}{5}(x_1^2+1) = \frac{1}{5}[(0.2)^2+1] = 0.208$$

$$x_3 = \phi(x_2) = \frac{1}{5}(x_2^2+1) = \frac{1}{5}[(0.208)^2+1] = 0.2086528$$

$$x_4 = \phi(x_3) = \frac{1}{5}(x_3^2+1) = \frac{1}{5}[(0.2086578)^2+1] = 0.208707$$

We have $x_3 \simeq x_4 = 0.2087$. $\therefore$ one root of the given equation is 0.2087. **Ans.**

(*ii*) Consider $x = \phi(x)$ in [4, 5]

let $x_0 = 4$ be the initial approximation of the root in (4, 5). The next approximation is

$$x_1 = \phi(x_0) = \frac{1}{5}(x_0^2+1) = \frac{1}{5}(4^2+1) = 3.4$$

and $$x_2 = \phi(x_1) = \frac{1}{5}(x_1^2+1) = \frac{1}{5}[(3.4)^2+1] = 2.512$$

The value $2.512 \notin [4, 5]$

Therefore, we express $x^2 - 5x + 1 = 0$ as $x = \sqrt{5x-1} = \phi(x)$ (say)

let $x_0 = 4.8$ be the initial approximation to the root. Then proceed to get the root is 4.79. **Ans.**

6. *Find the smallest positive root of* $x^3 - 5x + 3 = 0.$

SOLUTION: We have $f(x) = x^3 - 5x + 3 = 0.$

$f(-2) = (-2)^3 - 5(-2) + 3 = 5$, $f(-3) = (-3)^3 - 5(-3) + 3 = -9.$

Here $f(-2)\ f(-3) < 0$ there is a root between -2 and -3.

Next, $f(1) = -1$, $f(2) = 1$

Since $f(1) f(2) < 0$ there is a root between 1 and 2

Also $f(0) = 3$, $f(1) = -1$ $\Rightarrow$ $f(0)\ f(1) < 0$ Hence, there is a root between 0 and 1

Thus the required (i.e. the smallest positive) root lies between 0 and 1.

We have $f(x) = x^3 - 5x + 3$; $f'(x) = 3x^2 - 5.$

Let $x_0 = 1$ be the initial approximation, then by Newton-=Raphson formula

$$x_1 = x_0 - \frac{f(x_0)}{f'(x_0)} = 1 - \frac{x_0^3 - 5x_0 + 3}{3x_0^2 - 5} = 1 - \frac{(-1)}{(-2)} = 1 - \frac{1}{2} = 0.5$$

Therefore $f(x_1) = f(0.5) = (0.5)^3 - 5(0.5) + 3 = 0.625.$

$$x_2 = x_1 - \frac{f(x_1)}{f'(x_2)} = 0.5 - \frac{0.625}{(-4.25)} = 0.5 - 0.14 = 0.64$$

$$x_3 = x_2 - \frac{f(x_2)}{f'(x_2)} = 0.64 - \frac{(0.062144)}{(-3.7712)} = 0.64 + 0.0165 = 0.6565$$

$$x_4 = 0.6565 - \frac{(0.000446412125)}{(-3.70702325)} = 0.6565 + 0.000120 = 0.656620$$

$$x_6 = 0.656620 - \frac{0.0000015976975}{(-3.7065505268)} = 0.656620 + 0.000000431047$$

$$= 0.656620431047 \simeq .6566$$

The root of equation at the fifth iteration is 0.6566. **Ans.**

7. *Find the negative root of $x^3 - 2x - 5 = 0$, correct to three places of decimal by Newton-Raphson method.*

SOLUTION: Let $\phi(x) = x^3 - 2x + 5$, then $\phi(-x) = -x^3 + 2x + 5 = -(x^3 - 2x - 5)$ ...(1)

Now, let $f(x) = x^3 - 2x - 5$, then $f'(x) = 3x_2 - 2$.

To obtain the negative root of given equation we find the positive root of $f(x) = 0$.

we have $f(2) = 23 - 2.22 - 5 = -1$ and $f(3) = 33 - 2.3 - 5 = 16$.

Since $f(2)\, f(3) < 0$ the root of $f(x) = 0$ lies between 2 and 3.

Let $x_0 = 2$ be the initial approximation since $|f(2)| < |f(3)|$ $\therefore$ $f(2)$ is nearer to zero than $f(3)$ is.

Applying Newton-Raphson formula, we have

$$x_1 = x_0 - \frac{f(x_0)}{f'(x_0)} = 2 - \frac{f(2)}{f'(2)} = 2 - \frac{(-1)}{3(2)^2 - 2} = 2.1$$

Since $f(x_1) = (2.1)^3 - 2(2.1) - 5 = 0.061$ and $f'(x_1) = 3(2.1)^2 - 2 = 11.23$

We get $$x_2 = x_1 - \frac{f(x_1)}{f'(x_1)} = 2.1 - \frac{(0.061)}{11.23} = 2.0946.$$

Since $f(x_2) = f(2.0946) = (2.0946)^3 - 2(2.0946) - 5 = 0.0005$ and $f(x_2) = 3(2.0946)^2 - 2 = 11.1619$

The next approximation is $$x_3 = x_2 - \frac{f(x_2)}{f'(x_2)} = 2.0946 - \frac{(0.0005)}{11.1619} = 2.09456.$$

Since $x_2 \simeq x_3$ upto 3 decimal places, therefore, the root of $f(x) = 0$ is 2.095 (correct to three decimal places).

Hence, the negative root of $\phi(x) = 0$ is -2.095. **Ans.**

8. *Using Newton's interative formula establish the iterative formula* $x_{n+1} = \frac{1}{3}\left[2x_n + \frac{N}{x_n^2}\right]$ *to calculate the cube root of N.*

SOLUTION: Let $x = 3\sqrt{N}$ or $x^3 - N = 0$

We have $f(x) = x^3 - N$ such that $f'(x) = 3x^2$

By Newton-Raphson rule if x_n denotes the nth iterate then

$$x_{n+1} = x_n - \left[\frac{f(x_n)}{f'(x_n)}\right]$$

i.e. $$x_{n+1} = x_n - \frac{x_n^3 - N}{3x_n^2} = \frac{3x_n^3 - x_n^3 + N}{3x_n^2} = \frac{2x_n^3 + N}{3x_n^2}$$

$$= \frac{1}{3}\left[2x_n + \frac{N}{x_n^2}\right], \quad n = 0, 1, 2, ...$$

Thus, the iteration formula for the cube root of N is

$$x_{n+1} = \frac{1}{3}\left[2x_n + \frac{N}{x_n^2}\right], \quad n = 0, 1, 2, ...$$ **Hence Proved.**

CHAPTER

27 Finite Differences and Interpolation

IMPORTANT FORMULAE AND USEFUL RESULTS

1. **Forward Finite Differences:** (First order) $\Delta y_i = y_{i+1} - y_i$ or $\Delta f(x) = f(x+h) - f(x)$
   (Second Order) $\Delta^2 y_i = \Delta y_{i+1} - \Delta y_i, \ldots\ \Delta^n y_i = \Delta(\Delta^{n-1} y_i),\ n = 1, 2, 3, \ldots$
2. **Backward Finite Differences:**
   (First Order) $\nabla y_i = y_i - y_{i-1}$ or $\nabla f(x) = f(x) - f(x-h)$
   (Second Order) $\nabla^2 y_i = \nabla(\nabla y_i) = \nabla y_i - y_{i-1}$
   (n^{th} order) $\nabla^n y_i = \nabla(\nabla^{n-1} y_i),\ n = 1, 2, 3, \ldots$
3. **Central Finite Differences:** (First Order) $\delta y_i = y_{i+1/2} - y_{i-1/2}$ or $\delta f(x) = f(x+h/2) - f(x-h/2)$
   (second order) $\delta^2 y_i = \delta(\delta y_i) = \delta y_{i+1/2} - \delta y_{i-1/2}, \ldots$
   (n^{th} order) $\delta^n y_i = \delta(\delta^{n-1} y_i),\quad n = 1, 2, 3, \ldots$
4. **Shift Operator and Averages:**

$$Ey_i = y_{i+1} \quad\text{and}\quad \frac{1}{E} y_i = y_{i-1}, \quad\text{or}\quad Ef(x) = f(x+h) \quad\text{and}\quad \frac{1}{E} f(x) = f(x-h).$$

$$\mu y_i = \frac{1}{2}[y_{i+1/2} + y_{i-1/2}] \quad\text{or}\quad \mu f(x) = \frac{1}{2}[f(x+h/2) + f(x-h/2)]$$

$$E \equiv 1 + \Delta,\ Ef(x) = e^{hD} f(x) \quad\text{where}\quad D \equiv \frac{d}{dx}$$

$$\mu\delta = \frac{1}{2}\left(E - \frac{1}{E}\right), \quad \mu^2 \equiv 1 + \delta^2/4.$$

5. **Newton's Divided Differences:**

(First Order) $$f(x_0, x_1) = [x_0, x_1] = \frac{f(x_1) - f(x_0)}{x_1 - x_0}$$

$$f(x_1, x_2) = [x_1, x_2] = \frac{f(x_2) - f(x_1)}{x_2 - x_1}, \text{ etc.}$$

(Second Order) $$f(x_0, x_1, x_2) = \frac{f(x_1, x_2) - f(x_0, x_1)}{x_2 - x_0} = [x_0, x_1, x_2]$$

(Third Order) $$f(x_0, x_1, x_2, x_3) = \frac{f(x_1, x_2, x_3) - f(x_0, x_1, x_2)}{x_3 - x_0} = [x_0, x_1, x_2, x_3], \text{ etc.}$$

INTERPOLATION

1. Newton's Divided Differences Interpolation Formula:

$$f(x) = P_n(x) = f(x_0) + (x - x_0) f(x_0, x_1) + (x - x_0)(x - x_1) f(x_0, x_1, x_2) + \ldots$$
$$+ (x - x_0)(x - x_1) \ldots (x - x_{n-1}) f(x_0, x_1, x_2, \ldots, x_n)$$

2. Newton-Gregory Forward Differences Interpolation Formula:

$$y(x_0 + rh) = y_0 + r\Delta y_0 + \frac{r(r-1)}{2!}\Delta^2 y_0 + \frac{r(r-1)(r-2)}{3!}\Delta^3 y_0 + \ldots + \frac{r(r-1)-(r-n+1)}{n!}\Delta^n y_0$$

3. Newton-Gregory Backward Differences Interpolation Formula:

$$y(x_0) = y_n + r\nabla y_n + \frac{r(r+1)}{2!}\nabla^2 y_n + \frac{r(r-1)(r-2)}{3!}\nabla^3 y_0 + \ldots + \frac{r(r+1)(r+2)\ldots(r+n-1)}{n!}\nabla^n y_n$$

4. Gauss Central Difference Formulae:

(a) Gauss Forward Difference Formula:

$$y_p = y_0 + p\Delta y_0 + \frac{p(p-1)}{2!}\Delta^2 y_{-1} + \frac{(p+1)p(p-1)}{3!}\Delta^3 y_{-1} + \frac{p(p-1)(p-2)(p+1)}{4!}\Delta^3 y_{-2} + \ldots$$

(b) Gauss Backward Difference Formula:

$$y_p = y_0 + p\,\Delta y_{-1} + \frac{p(p+1)}{2!}\Delta^2 y_{-1} + \frac{(p+1)p(p-1)}{3!}\Delta^3 y_{-2} + \frac{(p+2)(p+1)(p-1)}{4!}\Delta^4 y_{-2} + \ldots$$

5. Sterling's Central Difference Interpolation Formula

$$y_p = y_0 + p\frac{\Delta y_0 + \Delta y_{-1}}{2} + \frac{p^2}{2}\Delta^2 y_{-1} + \frac{p(p^2-1)}{3!}\left(\frac{\Delta^3 y_{-1} + \Delta^3 y_{-2}}{2}\right) + \frac{p^2(p^2-1)}{4!}\Delta^4 y_{-2} + \ldots$$

or $$y_p = y_0 + p\mu\delta y_0 + \frac{p^2}{2}\delta^2 y_0 + \frac{p(p^2-1)}{3!}\mu\delta^3 y_0 + \frac{p^2(p^2-1)}{4!}\delta^4 y_0 + \ldots$$

6. (a) Lagrange's Interpolation Formula for Unqual Intervals:

$$y = \frac{(x-x_1)(x-x_2)\ldots(x-x_n)}{(x_0-x_1)(x_0-x_2)\ldots(x_0-x_n)}y_0 + \frac{(x-x_0)(x-x_2)(x-x_3)\ldots(x-x_n)}{(x_1-x_0)(x_1-x_2)(x_1-x_3)\ldots(x-x_n)}y_1 + \ldots$$
$$+ \frac{(x-x_0)(x-x_1)\ldots(x-x_{n-1})y_n}{(x_n-x_0)(x_n-x_1)(x_n-x_2)\ldots(x_n-x_{n-1})}$$

(b) Lagrange's Inverse Interpolation Formula:

$$x = \frac{(y-y_1)(y-y_2)\ldots(y-y_n)}{(y_0-y_1)(y_0-y_2)\ldots(y_0-y_n)}x_0 + \frac{(y-y_0)(y-y_2)\ldots(y-y_n)}{(y_1-y_0)(y_1-y_2)\ldots(y_1-y_n)}x_1 + \ldots$$
$$+ \frac{(y-y_0)(y-y_1)\ldots(y-y_{n-1})}{(y_n-y_0)(y_n-y_1)\ldots(y_n-y_{n-1})}$$

7. The Factorial Polynomial: If the interval of differencing is 1, we have

$x^{(n)} = x(x-1)(x-2)\ldots(x-n+1)$ so $x^{(1)} = x$, $x^{(2)} = x(x-1)$, $x^{(3)} = x(x-1)(x-2), \ldots$

If the interval of differencing is h then

$$x^{(n)} = x(x-h)(x-2h)\ldots(x-\overline{n-1}\,h)$$
$$\Delta x^{(n)} = (x+1)^{(n)} - x^{(n)} = n\,x^{(n-1)}$$
$$\Delta^2 x^{(n)} = \Delta[nx^{(n-1)}] = n(n-1)\,x^{(n-2)}, \ \ldots\ \Delta^n x^{(n)} = n!.$$

SOLUTIONS TO EXERCISE 27A

1. *Given that* $y_5 = 4, y_6 = 3, y_7 = 4, y_8 = 10$ *and* $y_9 = 24$, *find the value of* $\Delta^4 y_5$ *using the difference Table.*

SOLUTION:

x	y	Δy	$\Delta^2 y$	$\Delta^3 y$	$\Delta^4 y$
$x_0 = 5$	4				
		−1			
6	3		2		
		1		3	
7	4		5		0
		6		3	
8	10		8		
		14			
9	24				

$\therefore \ \Delta^4 y_5 = 0$ **Ans.**

2. *Find a cubic polynomial for* y *in terms of* x *given that* $y(0) = 1$, $y(1) = 0$, $y(2) = 1$ and $y(3) = 10$ and hence find $y(4)$.

SOLUTION:

x	y	Δy	$\Delta^2 y$	$\Delta^3 y$
0	1			
		−1		
1	0		2	
		1		6
2	1		8	
		9		
3	10			

$$y = 1 + x(-1) + \frac{x(x-1)}{2!}(2) + \frac{x(x-1)(x-2)}{3!}(6) = x^3 - 2x^2 + 1 \quad \therefore \quad y(4) = 33$$ **Ans.**

3. *Using Newton's interpolation formula find the* 7th *term of the sequence*

2, 9, 28, 65, 126, 217.

SOLUTION:

x	y	Δy	$\Delta^2 y$	$\Delta^3 y$	$\Delta^4 y$	$\Delta^5 y$
0	2					
		7				
1	9		12			
		19		6		
2	28		18		0	
		37		6		0
3	65		24		0	
		61		6		
4	126		30			
		91				
5	217					

$$y = 2 + 7x + 12\,\frac{x(x-1)}{2!} + \frac{6x(x-1)(x-2)}{3!} + 0 + 0$$

$$= x^3 + 3x^2 + 3x + 2$$

$\therefore$ 7th Term $= y(6) = 6^3 + 3.6^2 + 3.6 + 2 = 344$ **Ans.**

4. *For the factorial polynomial* $x^{(n)}$ *show that*

$$\Delta x^{(n)} = nx^{(n-1)} \text{ and } \Delta^n x^{(n)} = n!$$

SOLUTION : $\Delta x^{(n)} = (x+1)^{(n)} - x^{(n)} = nx^{(n-1)}, \quad \Delta^2 x^{(n)} = n(n-1)\,x^{(n-2)}, \ldots. \quad \Delta^n x^{(n)} = n!$

5. *Express the polynomial* $f(x) = 2x^3 + 3x^2 - 5x + 4$ *in factorial notation and find its successive differences. Also, obtain a function whose first order finite difference is* $f(x)$.

SOLUTION : $f(x) = 2x^3 + 3x^2 - 5x + 4$. To convert into factorial notation, we apply synthetic division.

0	2	3	–5	4
		0	0	0
1	2	3	–5	4
		2	5	
2	2	5	0	
		4		
	2	9		
	2			

$\therefore \quad f(x) = 2x^{(3)} + 9\,x^{(2)} + 0\,x^{(1)} + 4\,x^{(0)}$

and $\quad \Delta f(x) = 2.3x^{(2)} + 9.2\,x^{(1)}, \quad \Delta^2 f(x) = 6.2\,x^{(1)} + 18\,x^{(0)}, \quad \Delta^3 f(x) = 12.$

If $\quad \Delta\,\phi(x) = f(x) \quad$ then $\quad \phi(x) = \dfrac{2x^{(4)}}{4} + \dfrac{9x^{(3)}}{3} + 4x^{(1)} = \dfrac{1}{2}(x^4 - 7x^2 + 14x) + C$ **Ans.**

6. *From the steam Table, we have the following data*

Temp in °C	140	150	160	170	180
Pressure kg/Cm^2	3.685	4.854	6.303	8.076	10.225

Find pressures at temperatures 142° C *and* 175° C.

SOLUTION:

T	P	ΔP	$\Delta^2 P$	$\Delta^3 P$	$\Delta^4 P$
140	3685				
		1169			
150	4854		280		
		1449		44	
160	6303		324		8
		1773		52	
170	8076		376		
		2149			
180	10225				

When $\quad T = 142°, \quad r = \dfrac{142-140}{10} = 0.2.$

$$P = P_0 + r\,\Delta\,P_0 + \frac{r(r-1)}{2!}\,\Delta^2 P_0 + \frac{r(r-1)(r-2)}{3!}\,\Delta^3 P_0 + \frac{r(r-1)(r-2)(r-3)}{4!}\,\Delta^4 P_0$$

$$= 3685 + 0.2\,(1169) + \frac{(0.2)(0.2-1)}{2}\,(280)$$

$$+ \frac{(0.2)(0.2-1)(0.2-2)}{6}(44) \quad + \frac{(0.2)(-0.8)(-1.8)(-2.8)(8)}{4.3.2}$$

$$= 3685 + 233.8 \; - (0.08)\,(280) + 0.048(44) - 0.2688 \; = \; 3917.2$$

When $\quad T = 175°, \quad r = \frac{(175-180)}{10} \; = -0.5$

$$P = P_n + r\,\nabla\,P_n + \frac{r(r+1)}{2}\;\nabla^2\;P_n \;\frac{r(r+1)(r+2)}{6} = \nabla^3\,P_n + ...$$

$$= 10225 - 0.5\,(2149) + \frac{(-0.5)(0.5)}{2}\,(376) + \frac{(-0.5)(0.5)(1.5)(52)}{3!}$$

$$+ \frac{(0-5)(0.5)(1.5)(2.5)}{4!}\,(8) = 9099.9375 = 9100.$$ **Ans.**

7. *From the following Table, find $f(3.75)$ using Gauss forward interpolation formula*

x	2.5	3.0	3.5	4.0	4.5	5.0
$f(x)$	24.145	22.043	20.225	18.644	17.262	16.047

SOLUTION:

x	$f(x)$	Δf	$\Delta^2 f$	$\Delta^3 f$	$\Delta^4 f$	$\Delta^5 f$
2.5	24.145					
		–2.102				
3.0	22.043		0.284			
		–1.818		–0.047		
3.5	20.225		0.237		0.009	
		–1.581		–0.038		–0.003
4.0	18.644		0.199		0.006	
		–1.382		0.32		
4.5	17.262		0.167			
		1.215				
5.00	16.047					

Take $x_0 = 3.5$ since we are to find $f(3.75)$, $\quad r = \frac{x - x_0}{h} = \frac{3.75 - 3.5}{0.5} = 0.5$

By Gauss forward difference formula

$$f(x) = f_0 + r\,\Delta f_0 + \frac{r(r-1)}{2!}\;\Delta^2 f_{-1} + \frac{r(r-1)(r+1)}{3!}\;\Delta^3 f_{-1} + \frac{r(r-1)(r-2)(r+1)}{4!}\;\Delta^4 f_{-2} + ...$$

$$\therefore \qquad f(3.75) = 20.225 + 0.5\,(-1.581) + \frac{0.5(-0.5)}{2}\,(0.237) + \frac{0.5(-0.5)(1.5)}{6}\,(-0.038)$$

$$+ \frac{0.5(-0.5)(-1.5)(1.5)}{24}\,(0.009)$$

$$= 20.225 - 0.7905 - 0.02962 + 0.00375 + 0.000211 = 19.408$$ **Ans.**

8. *From the following Table, find cos 51°42' using Gauss backward interpolation formula*

x	50°	51°	52°	53°	54°
cos x	0.6428	0.6293	0.6157	0.6018	0.5878

SOLUTION:

x	$y = \cos x$	Δy	$\Delta^2 y$	$\Delta^3 y$
50°	0.6428			
		−0.0135		
51°	0.6293		− 0.0001	
		−0.0136		− 0.0002
52°	0.6157		− 0.0003	
		− 0.139		+ 0.0002
53°	0.6018		−0.0001	
		−0.0140		
54°	0.5878			

To compute cos (51° 42′) we take $x_0 = 52°$, $x = 51° 42'$, $h = 1°$

$$\therefore \quad r = \frac{57° 42' - 52°}{1°} = \frac{-18'}{60'} = -0.3$$

By Gauss backward difference formula

$$y(x) = y_0 + r\,\Delta y_{-1} + \frac{r(r+1)}{2}\,\Delta^2 y_{-1} + \frac{(r+1)(r)(r-1)}{31}\,\Delta^3 y_{-2} + \ldots\ldots$$

or
$$y = 0.6157 + (-0.3)\,(-0.0136) + \frac{(0.3)(0.7)}{2}\,(-0.0003) + \frac{(-0.3)(0.7)(-1.3)}{6}\,(-0.002)$$

$$= 0.6157 + 0.00408 + 0.0000315 - 0.00009 = 0.61972$$ **Ans.**

9. *From the following Table estimate $e^{1.17}$ using Gauss forward interpolation formula*

x	1.00	1.05	1.10	1.15	1.20	1.25	1.30
e^x	2.7183	2.8577	3.0042	3.1582	3.3201	3.4903	3.6693

SOLUTION:

x	$y = e^x$	Δy	$\Delta^2 y$	$\Delta^3 y$	$\Delta^4 y$	$\Delta^5 y$
1.00	2.7183					
		0.1394				
1.05	2.8577		0.0071			
		0.1465		0.0004		
1.10	3.0042		0.0075		0.0	
		0.1540		0.0004		0.0
1.15	3.1582		0.0079		0.0	
		0.1169		0.0004		0.0001
1.20	3.3201		0.0083		0.0001	
		0.1702		0.0005		
1.25	3.4903		0.0088			
		0.1790				
1.30	3.6693					

We shall be using Gauss forward formula. $x_0 = 1.15$, $x = 1.17$ and $h = 0.5$

$$\therefore \quad r = \frac{1.17-1.15}{0.5} = \frac{.02}{.05} = 0.4$$

$$y = e^x = y_0 + r\,\Delta y_0 + \frac{r(r-1)}{2}\Delta^2 y_{-1} + \frac{r(r-1)(r+1)}{3!}\,\Delta^3 y_{-1} + \frac{r(r-1)(r-2)(r+1)}{6}\,\Delta^4 y_{-2} + \ldots$$

$$= 3.1582 + (0.4)\,(0.1619) + \frac{0.4(-0.6)}{2}\,(0.0079) + \frac{(0.4)(-0.6)(1.4)}{6}\,(0.0004) + 0$$

$$= 3.1582 + 0.06476 - 0.000948 - .0000224 = 3.222$$ **Ans.**

10. *Following data has been extracted from the census*

Year	1901	1911	1921	1931	1941	1951
Population (in thousand)	12	15	20	27	39	52

Estimate the population for the year 1936 *employing Gauss backward interpolation formula.*

SOLUTION:

x	y	Δy	$\Delta^2 y$	$\Delta^3 y$	$\Delta^4 y$	$\Delta^5 y$
1901	12					
		3				
1911	15		2			
		5		0		
1921	20		2		3	
		7		3		–10
1931	27		5		–7	
		12		–4		
1941	39		1			
		13				
1951	52					

We are to find population in the year 1936. Let us take $x_0 = 1931$, $h = 10$ and $x = x_0 + rh$

$$\therefore \quad r = \frac{1936-1931}{10} = 0.5.$$ Using Gauss backward difference formula

$$y(x) = y_0 + r\,\Delta y_{-1} + \frac{r(r+1)}{2}\,\Delta^2 y_{-1} + \frac{r(r+1)(r-1)}{6}\,\Delta^3 y_{-2} + \frac{r(r+1)(r+2)(r-1)}{4!}\,\Delta^4 y_{-2} + \ldots$$

$$\therefore \quad y(1936) = 27 + 0.5\,(7) + \frac{.5(1.5)}{2}\,(5) + \frac{.5(1.5)(-.5)}{6}\,(3) + \frac{(.5)(1.5)(2.5)(-0.5)}{24}\,(-7)$$

$$= 27 + 3.5 + 1.875 - 0.1375 + 0.2734 = 32.461$$ **Ans.**

11. *Employing Stirling's interpolation formula approximate* tan 16° *from the following data.*

x	0°	5°	10°	15°	20°	25°	30°
tan x	0.0	0.0875	0.1763	0.2679	0.3640	0.4663	0.5774

SOLUTION:

x	$\tan x = y$	Δy	$\Delta^2 y$	$\Delta^3 y$	$\Delta^4 y$	$\Delta^5 y$	$\Delta^6 y$
0°	0.0						
		0.0875					
5°	0.0875		0.0013				
		0.0888		0.0015			
10°	0.1763		0.0028		0.0002		
		0.0916		0.0017		− 0.0002	
15°	0.2679		0.0045		0.0000		0.0011
		0.0961		0.0017		0.0009	
20°	0.3640		0.00062		0.0009		
		0.1023		0.0026			
25°	0.4663		0.0088				
		0.1111					
30°	0.5774						

To compute $y = \tan 16°,\ x_0 = 15°,\ h = 1°$ and $r = \dfrac{16°-15°}{5} = 0.2$

By Sterling's formula $y_r = y_0 + r\,\mu\delta y_0 + \dfrac{1}{2!}\,r^2\Delta^2 y_0 + \dfrac{1}{3!}\,r\,(r^2-1)\,\mu\delta^3 y_0 + \dfrac{r^2(r^2-1)}{4!}\,\delta^4 y_0 + \ldots$

$$y(r=0.2) = 0.2679 + 0.2\,\frac{(0.0916+0.0961)}{2} + \frac{1}{2!}\,0.2^2\,(0.0045)$$

$$+\frac{(0.2)(0.04-1)}{6}\;\frac{(0.0017+0.0017)}{2} + \frac{-0.04(.04-1)}{24}\,(0)$$

$$= 0.2679 + 0.00877 + 0.00009 = 0.2867$$

Therefore $\tan 16° = 0.2867$ **Ans.**

12. *Employ Bessel's interpolation formula to estimate* $y(35)$ *from the following Table*

x	20	30	40	50
y	512	439	346	243

SOLUTION:

x	y	Δy	$\Delta^2 y$	$\Delta^3 y$
20	512			
		–73		
30	439		–20	
		–93		10
40	346		–10	
		–103		
50	243			

To Compute $y(35)$ by Bessel's formula, we take $x_0 = 30,\ x = 35,\ x_0 + rh \quad \therefore \quad r = 0.5.$

$$y = y_0 + r\,\delta y_{1/2} + \frac{r(r-1)}{2!}\,\mu\delta^2\,y_{1/2} + \frac{r(r-1)}{3!}\left(r-\frac{1}{2}\right)\delta^3 y_{1/2}$$

or
$$y = y_0 + r\,\Delta y_0 + \frac{r(r-1)}{2}\;\frac{\Delta^2 y_{-1} + \Delta^2 y_0}{2} + \frac{\left(r-\frac{1}{2}\right) r(r-1)}{3!}\,\Delta^3 y_{-1} + \ldots$$

$$= 439 + 0.5\,(-93) + \frac{0.5(-0.5)}{2}\;\frac{(-20-10)}{2} + 0 = 439 - 46.5 + 1.875 = 394.375$$ **Ans.**

13. *Obtain* y at $x = 10$ *from the following Table using Lagrange's interpolation formula*

x	5	6	9	11
y	12	13	14	16

SOLUTION : It is a case of interpolation in unequispaced values. We apply Lagrange's interpolation formula

$$y = \frac{(x-x_1)(x-x_2)(x-x_3)}{(x_0-x_1)(x_0-x_2)(x_0-x_3)} \cdot y_0 + \frac{(x-x_0)(x-x_2)(x-x_3)}{(x_1-x_0)(x_1-x_2)(x_1-x_3)} y_1 + \frac{(x-x_0)(x-x_1)(x-x_3)}{(x_2-x_0)(x_2-x_1)(x_2-x_3)} y_2$$

$$+ \frac{(x-x_0)(x-x_1)(x-x_2)}{(x_3-x_0)(x_3-x_1)(x_3-x_2)} y_3 = \frac{(x-6)(x-9)(x-11)}{(5-6)(5-9)(5-11)} (12) + \frac{(x-5)(x-9)(x-11)}{(6-5)(6-9)(6-11)} (13)$$

$$+ \frac{(x-5)(x-6)(x-11)}{(9-5)(9-6)(9-11)} (14) + \frac{(x-5)(x-6)(x-9)}{(11-5)(11-6)(11-9)} (16)$$

$$\therefore\ y(10) = \frac{4.1.(-1)}{(-2)(-4)(-6)} 12 + \frac{5.1.(-1)}{1(-3)(-5)} 13 + \frac{5.4.(-1)}{4.3.(-2)}.14 + \frac{5.4.1}{6.5.2} (16)$$

$$= 2 - 4.333 + 11.666 + 5.333 = 14.666.$$ **Ans.**

14. *Use Lagrange's inverse interpolation formula to approximate* x *when* $y = 13.5$ *from the following data*

x	93.0	96.2	100.0	104.2	108.7
y	11.38	12.80	14.70	17.07	19.91

SOLUTION : To calculate x when $y = 13.5$ we apply Lagrange's inverse interpolation formula as

$$x = \frac{(y-y_1)(y-y_2)(y-y_3)(y-y_4)}{(y_0-y_1)(y_0-y_2)(y_0-y_3)(y_0-y_2)} x_0 + \frac{(y-y_0)(y-y_2)(y-y_3)(y-y_4)}{(y_1-y_0)(y_1-y_2)(y_1-y_3)(y_1-y_4)} x_1$$

$$+ \frac{(y-y_0)(y-y_1)(y-y_3)(y-y_4)}{(y_2-y_0)(y_2-y_1)(y_2-y_3)(y_2-y_4)} x_2 + \frac{(y-y_0)(y-y_1)(y-y_2)(y-y_4)}{(y_3-y_1)(y_3-y_1)(y_3-y_2)(y_3-y_4)} x_3$$

$$+ \frac{(y-y_0)(y-y_1)(y-y_2)(y-y_3)}{(y_4-y_0)(y_4-y_1)(y_4-y_2)(y_4-y_3)} x_4$$

$$\therefore\quad x(13.5) = \frac{(13.5-12.80)(13.5-14.70)(13.5-17.07)(13.5-19.91)(93.0)}{(11.38-12.80)(11.38-14.70)(11.38-17.07)(11.38-19.91)}$$

$$+ \frac{(13.5-11.38)(13.5-14.70)(13.5-17.07)(13.5-19.91)(96.2)}{(12.80-11.38)(12.80-14.70)(12.80-17.07)(12.80-19.91)}$$

$$+ \frac{(13.5-11.38)(13.5-12.80)(13.5-17.07)(13.5-19.91)(100.0)}{(14.70-11.38)(14.70-12.80)(14.70-17.07)(14.70-19.91)}$$

$$+ \frac{(13.5-11.38)(13.5-12.80)(13.5-14.70)(13.5-19.91)(104.2)}{(17.07-11.38)(17.07-12.80)(17.07-14.70)(17.07-19.91)}$$

$$+ \frac{(13.5-11.38)(13.5-12.80)(13.5-14.70)(13.5-17.07)(108.7)}{(19.91-11.38)(19.91-12.80)(19.91-14.70)(19.91-17.07)}$$

$$= 97.65575 \text{ (on. simplification)}$$ **Ans.**

15. *Using Newton's divided difference interpolation formula estimate $f(8)$ and $f(15)$, given that*

x	4	5	7	10	11	13
y	48	100	294	900	1210	2028

SOLUTION:

x	$y = f(x)$	*Ist order difference*	*IInd order difference*	*3rd order dfference*	*4th order difference*
4	48				
		52			
5	100		15		
		97		1	
7	294		21		0
		202		1	
10	900		27		0
		310		1	
11	1210		33		
		409			
13	2028				

By Newton's divided difference interpolation formula

$$P_n(x) = f(x_0) + (x - x_0)\, f(x_0, x_1) + (x - x_0)(x - x_1)\, f(x_0, x_1, x_2)$$
$$+ (x - x_0)(x - x_1)(x - x_2)\, f(x_0, x_1, x_2, x_3) - \ldots$$

To calculate $P_n(8)$ we take $x_0 = 7$

$$\therefore \quad P_n(8) = 294 + (8-7)(202) + (8-7)(8-10)(27) + (8-7)(8-10)(8-11)\,1$$
$$= 294 + 202 - 54 + 6 = 448$$

Ans.

To calculate $P_n(15)$ we take $x_0 = 13$

$$\therefore \quad P_n(x) = 2028 + (x-4)(52) + (x-4)(x-5)(15) + (x-4)(x-5)(x-7)(1)$$
$$= 2028 + 52(x-4) + 15(x^2 - 9x + 20) + x^3 - 16x^2 - 53x - 140$$
$$= x^3 - x^2 - 136\,x + 100.$$

$$\therefore \quad P_n(15) = 15^3 - 15^2 - 136(15) + 1980 = 3090.$$

Ans.

16. *If the interval of differencing is unity, find* $\Delta(e^{2x} \log 3x)$

SOLUTION:

$$\Delta(e^{2x} \log 3x) = e^{2(x+1)} \log(3x+3) - e^{2x} \log 3x$$
$$= e^{2x}[e^2 \log(3x+3) - e^2 \log 3x + e^2 \log 3x - \log 3x]$$
$$= e^{2x}\left[e^2 \log \frac{x+1}{x} + (1 + e^2) \log 3x\right]$$

17. *Find* (*i*) $\Delta[\cot(a^x)]$ (*ii*) $\Delta^2(\cos 2x)$.

SOLUTION: (*i*) $\Delta \cot(a^x) = \cot a^{x+h} - \cot a^x = \dfrac{\cos a^{x+h} \sin a^x - \cos a^x \sin a^{x+h}}{\sin a^{x+h} . \sin a^x} = \dfrac{\sin(a^x - a^{x+h})}{\sin a^x . \sin a^{x+h}}$

(*ii*) $\Delta^2 \cos^2(2x) = \Delta[\cos 2(x+h)] - \Delta \cos 2x$

$$= \cos 2(x+2h) - \cos(2x+2h) - \cos(2x+2h) + \cos 2x$$
$$= \cos(2x+4h) + \cos 2x - 2\cos(2x+2h) = 2\cos(2x+2h)\cos 2h - 2\cos(2x+2h)$$
$$= 2\cos(2x+2h)[\cos 2h - 1] = -4\sin^2 h \cos(2x+2h)$$

where h is the interval of differencing.

Ans.

ADDITIONAL IMPORTANT PROBLEMS

1. *Prove the following* (a) $(1+\Delta)(1-\nabla) \equiv 1$ (b) $\Delta\nabla \equiv \Delta - \nabla$ (c) $\nabla \equiv E^{-1}\Delta$.

SOLUTION: (a) $(1+\Delta)(1-\nabla)\,f(x) = EE^{-1}f(x) = f(x)$ $\therefore$ $(1+\nabla)(1-\nabla) \equiv 1$.

(b) $\nabla\Delta f(x) = (E-1)(1-E^{-1})\,f(x) = (E-1)[f(x)-f(x-h)]$

$$= Ef(x)-f(x)-Ef(x-h)+f(x-h) = f(x+h)-f(x)-f(x)+f(x-h)$$

$\therefore$ $\nabla\Delta \equiv \Delta - \nabla$.

(c) $E^{-1}\Delta\, f(x) = E^{-1}[f(x+h)-f(x)] = E^{-1}f(x+h)-E^{-1}f(x) = f(x)-f(x-h) = \nabla f(x)$.

2. *Use the method of seperation of symbols and prove that*

$$u_x = u_{x-1}+\Delta u_{x-2}+\Delta^2 u_{x-3}+\ldots+\Delta^n u_{x-n}$$

SOLUTION:

$$\text{R.H.S.} = u_{x-1}+\Delta u_{x-2}+\Delta^2 u_{x-3}+\ldots+\Delta^n u_{x-n}$$

$$= E^{-1}u_x+\Delta E^{-2}u_{x-2}+\ldots+\Delta^{n-1}E^{-n}u_x+\Delta^n u_{x-n}$$

$$= E^{-1}[1+\Delta E^{-1}+\Delta^2E^{-2}+\ldots+\Delta^{n-1}E^{-(n-1)}]u_x+\Delta^n u_{x-n}$$

$$= E^{-1}\left[\frac{1-\Delta^nE^{-n}}{1-\Delta E^{-1}}\right]u_x+\Delta^n u_{x-n} = \frac{1}{E}\left[\frac{1-\Delta^nE^{-n}}{1-\Delta E^{-1}}\right]u_x+\Delta^n u_{x-n}$$

$$= \left[\frac{1-\Delta^nE^{-n}}{E-\Delta}\right]u_x+\Delta^n u_{x-n} = \left[\frac{1-\Delta^nE^{-n}}{1}\right]u_x+\Delta^n u_{x-n} \qquad (\because E = 1+\Delta)$$

$$= u_x-\Delta^n u_{x-n}+\Delta^n u_{x-n} = u_x = \text{L.H.S.}$$

Hence Proved

3. *The following data gives the melting point of an alloy of lead and zinc.*

Percentage of lead in the alloy x	50	60	70	80
Temperature T°C	205	225	248	274

Find the melting point of the alloy containing 54% of lead using Newton's forward interpolation formula.

SOLUTION: The difference Table for the given data is

x	y	Δ	Δ^2	Δ^3
50	205			
		20		
60	225		3	
		23		0
70	248		3	
		26		
80	274			

Here we have $x = 54,\ x_0 = 50,\ h = 10,\ p = \dfrac{x - x_0}{h} = \dfrac{54-50}{10} = 0.4$

Applying Newton's forward interpolation formula $T(x) = y_0 + p\,\Delta y_0 + \dfrac{p(p-1)}{2!}\Delta^2 y_0 + ...$

$$\therefore \quad T(54) = 205\ (0.4)\ (20) + \frac{(0.4)\ (0.4-1).3}{2!} = 205 + 8 + (0.2)\ (-0.6).3 = 212.64$$

Thus, the melting point of the alloy containing 54% of lead is 212.64 degrees

4. *Interpolate the value of y for $x = 1.5$ from the following data:*

x	0	1	2	3
y	2	5	13	35

using Newton's backward interpolation formula.

SOLUTION: We have $x = 1.5,\ x_n = 3,\ h = 1$ and $p = \dfrac{x - x_n}{h} = \dfrac{1.5-3}{1} = -1.5$

x	y	∇	∇^2	∇^3
0	2			
		3		
1	5		5	
		8		9
2	13		14	
		22		
3	35			

Newton's backward interpolation formula for equal interval is

$$y(x) = y_n + p\nabla y_n + \frac{p(p+1)}{2}\nabla^2 y_n + \frac{p(p+1)\,(p+2)}{6}\nabla^3 y_n + ...$$

$$\therefore \quad y(1.5) = 35 + (-1.5)\ (22) + \frac{(-1.5)\ (-1.5+1)}{6}.(14) + \frac{(-1.5\ (-1.5+1)\ (-1.5+2)}{6}\times 9$$

$$= 35 - 33 + 5.25 + 0.5625 = 7.8125$$

hence the value of y at $x = 1.5$ is 7.8125. **Ans.**

5. *Find the value of f(2.5) using Gauss forward interpolation formula, for the following data.*

x	1	2	3	4
$f(x)$	1	8	27	64

SOLUTION: Taking the origin at 2 we have

x	$y = f(x)$	Δy	$\Delta^2 y$	$\Delta^3 y$
1	1			
		7		
2	8		12	
		19		6
			18	
3	27			
		37		
4	64			

Here we have $x = 2$, $h = 1$, $p = \dfrac{x - x_0}{h} = \dfrac{2.5 - 2}{1} = 0.5$

Using Gauss forward interpolation formuia

$$y_p = y_0 + p\Delta y_0 + \frac{p(p-1)}{2!}\Delta^2 y_{-1} + \frac{(p+1)\,p(p-1)}{3!}\Delta^3 y_{-1} + \ldots -$$

We have $$y_{0.5} = 8 + (0.5)\,.\,(19) + \frac{(0.50)\,(0.5-1)}{2}.(12) + \frac{(0.5+1)\,(0.5)\,(0.5-1)}{6}.(6)$$

or $$y_{0.5} = 8 + 9.5 - 1.5 - 0.375 = 15.625$$

Hence the value of $f(x)$ at $x = 2.5$ is 15.625. **Ans.**

6. *Given that $f(25) = 0.2707$, $f(30) = 0.3027$, $f(35) = 0.3386$ and $f(40) = 0.3794$. Use Gauss backward interpolation formula to find $f(32)$.*

SOLUTION: We have $h = 5$, $x = 32$. Taking the origin at $x = 35$, $p = \dfrac{32-35}{5} = \dfrac{-3}{5} = -0.6$

The difference Table is

x	p	y	Δy	$\Delta^2 y$	$\Delta^3 y$
25	−2				
		0.2707			
30	−1		0.320		
		0.3027		0.0039	
35	0		0.0359		0.0010
		0.3386		0.0049	
40	1		0.408		
		0.3794			

Apply Gauss backward interpolation formula

$$y_p = y_0 + p\Delta y_{-1} + \frac{p(p-1)}{2!}\Delta^2 y_{-1} + \frac{(p+1)\,p(p-1)}{3!}\Delta^3 y_{-2} + \ldots$$

$$\therefore \quad y_{-0.6} = 0.3386 + (-0.6)\,(0.0359) + \frac{(-0.6)\,(-0.6+1)}{2}.(0.004) + \frac{(-0.6+1)\,(-0.6)\,(-0.6-1)}{6}.(0.0010)$$

$$= 0.3165.$$ **Ans.**

7. *Given that $f(x) = 168, 192, 336$ at $x = 1, 7, 15$ respectively, use Lagrange's formula and find the value of $f(10)$.*

SOLUTION: We have $x_0 = 1,\ x_1 = 7,\ x_2 = 15$ and $y_0 = 168,\ y_1 = 192,\ y_2 = 336$ and $x = 10$

Substituting in the Lagrange's formula

$$y = f(x) = \frac{(x-x_1)(x-x_2)}{(x_0-x_1)(x_0-x_2)} y_0 + \frac{(x-x_0)(x-x_2)}{(x_1-x_0)(x_1-x_2)} y_1 + \frac{(x-x_0)(x-x_1)}{(x_2-x_0)(x_2-x_1)} y_2$$

we get

$$f(10) = \frac{(10-7)(10-15)}{(1-7)(1-15)}.(168) + \frac{(10-1)(10-15)}{(7-1)(7-15)}.(192) + \frac{(10-1)(10-7)}{(15-1)(15-7)}.(336)$$

$$= \frac{(3)(-5)}{(-6)(-14)}.(168) + \frac{(9)(-5)}{(6)(-8)}.(192) + \frac{(9)(3)}{(14)(8)}.(336)$$

$$= -30.005 + 180 + 81.01 = 231.005.$$

Ans.

CHAPTER

28 Numerical Differentiation and Integration

IMPORTANT FORMULAE AND RESULTS

NUMERICAL DIFFERENTIATION

1. Derivatives in terms of Forward Differences

$$\frac{dy}{dx} = \frac{1}{h}\left[\Delta y_0 + \frac{2p-1}{2}\Delta^2 y_0 + \frac{3p^2-6p+2}{6}\Delta^3 y_0 + \frac{4p^3-18p^2+22p-6}{24}\Delta^4 y_0 + \ldots\ldots\right]$$

where $p = (x - x_0)/h$.

$$\therefore \quad \left(\frac{dy}{dx}\right)_{x=x_0} = \frac{1}{h}\left[\Delta y_0 - \frac{1}{2}\Delta^2 y_0 + \frac{1}{3}\Delta^3 y_0 - \frac{1}{4}\Delta^4 y_0 - \ldots\ldots\right]$$

and $$\left(\frac{d^2y}{dx^2}\right) = \frac{1}{h^2}\left[\Delta^2 y_0 + \frac{p-1}{1}\Delta^3 y_0 + \frac{6p^2-18p+11}{12}\Delta^4 y_0 + \ldots\ldots\right]$$

$$\left(\frac{d^2y}{dx^2}\right)_{x=x_0} = \frac{1}{h}\left[\Delta^2 y_0 - \Delta^3 y_0 + \frac{11}{12}\Delta^4 y_0 + \ldots\ldots\right]$$

2. Derivatives in terms of Backward Differences

$$\frac{dy}{dx} = \frac{1}{h}\left[\nabla y_n + \frac{2p+1}{2}\nabla^2 y_n + \frac{3p^2+6p+2}{6}\nabla^3 y_n + \ldots\ldots\right]$$

$$\left(\frac{dy}{dx}\right)_{x=x_n} = \frac{1}{h}\left[\nabla y_n + \frac{1}{2}\nabla^2 y_n + \frac{1}{3}\nabla^3 y_n + \ldots\ldots\right]$$

$$\left(\frac{d^2y}{dx^2}\right)_{x=x_n} = \frac{1}{h^2}\left[\nabla^2 y_n + \nabla^3 y_n + \frac{11}{12}\nabla^4 y_n + \ldots\ldots\right]$$

3. Derivatives in terms of Central Differences

$$\frac{dy}{dx} = \frac{1}{h}\left[\mu\delta y_0 + p\delta^2 y_0 + \frac{3p^2-1}{6}\mu\delta^3 y_0 + \frac{2p^2-p}{12}\delta^4 y_0 + ...\right]$$

$$\left(\frac{d^2y}{dx^2}\right) = \frac{1}{h}\left[\delta^2 y_0 + p\mu\delta^3 y_0 + \frac{6p^2-p}{12}\delta^4 y_0 + ...\right]$$

$$\therefore \quad \left(\frac{dy}{dx}\right)_{x=x_0} = \frac{1}{h}\left[\mu\delta y_0 - \frac{1}{6}\mu\delta^3 y_0 + ...\right]$$

$$\left(\frac{d^2y}{dx^2}\right)_{x=x_0} = \frac{1}{h^2}\left[\delta^2 y_0 - \frac{1}{12}\delta^4 y_0 + ...\right]$$

NUMERICAL INTEGRATION

1. Trapezoidal Rule

$$\int_a^b y\,dx = h\left[\frac{y_0+y_n}{2} + y_1 + y_2 + ... + y_{n-1}\right]$$

2. Simpson's one-third Rule

$$\int_a^b y\,dx = \frac{h}{3}[y_0 + y_n + 4(y_1 + y_3 + ... + y_{n-1}) + 2(y_2 + y_4 + y_6 + ... + y_{n-2})]$$

3. Simpson's Three-eight rule

$$\int_a^b y\,dx = \frac{3h}{8}[y_0 + y_n + 3(y_1 + y_2 + y_4 + y_5 + y_7 + ...) + 2(y_3 + y_6 + y_7 + ...)]$$

4. Boole's Rule: $\int_{x_0}^{x_4} y\,dx = \frac{2h}{45}[7y_0 + 32y_1 + 12y_2 + 32y_3 + 7y_4]$

5. Weedels' Rule: $\int_{x_0}^{x_6} y\,dx = \frac{3h}{10}[y_0 + 5y_1 + y_2 + 6y_3 + y_4 + 5y_5 + y_6]$

SOLUTIONS TO EXERCISE 28A

1. *The population of a town as per the census, is given below. Compute the rate of growth of population in* 1941 *and in* 1971.

year (x)	1931	1941	1951	1961	1971
Population (y) (in thousand)	40.62	60.80	79.95	103.56	132.65

SOLUTION:

x	y	Δy	$\Delta^2 y$	$\Delta^3 y$	$\Delta^4 y$
1931	40.62				
		20.18			
1941	60.80		−1.03		
		19.15		5.49	
1951	79.95		4.46		−4.47
		23.61		1.02	
1961	103.56		5.48		
		29.09			
1971	132.65				

Here $x_0 = 1931$, Take $x = x_0 + rh$ *i.e.*, $1941 = 1931 + 10r$ $\therefore\ r = 1$

By forward difference Interpolation formula

$$\left(\frac{dy}{dx}\right) = \frac{1}{h}\left[\Delta y_0 + \frac{2r-1}{2} + \Delta^2 y_0 + \frac{3r^2 - 6r + 2}{6}\Delta^3 y_0 + \frac{4r^3 - 18r^2 + 22r - 6}{24}\Delta^2 y_0\right]$$

$$\therefore\ \left(\frac{dy}{dx}\right)_{1941} = \frac{1}{10}\left[20.18 + \frac{1}{2}(-1.03) - \frac{1}{6}(5.49) + \frac{2}{24}(-4.47)\right]$$

$$= \frac{1}{10}[20.18 - 0.515 - 0.915 - 0.3725] = 1.83775.$$ **Ans.**

And by Backward interpolation Formula

$$\left(\frac{dy}{dx}\right)_{y_n} = \frac{1}{h}\left[\nabla y_n + \frac{1}{2}\nabla^2 y_n + \frac{1}{3}\nabla^3 y_n + \frac{1}{4}\nabla^4 y_n\right]$$

$$\left(\frac{dy}{dx}\right)_{1971} = \frac{1}{10}\left[29.07 + \frac{1}{2}(5.48) + \frac{1}{3}(1.02) + \frac{1}{4}(-4.47)\right]$$

$$= \frac{1}{10}[29.07 + 2.74 + 0.34 - 1.117] = 3.10525$$ **Ans.**

2. *In an observation in which* θ *is the temperature in degree celsius of a vessel of cooling water and* t *is the time in minutes from the begining of observation, the results are give below.*

t	1	3	5	7	9
θ	85.3	74.5	67.0	60.5	54.3

Find the approximate rate of cooling at $t = 3$ and at $t = 3.5$.

SOLUTION:

t	θ	$\Delta\theta$	$\Delta^2\theta$	$\Delta^3\theta$	$\Delta^4\theta$
1	85.3				
		– 10.8			
3	74.5		3.3		
		– 7.5		– 2.3	
5	67.0		1.0		1.6
		–6.5		–0.7	
7	60.5		0.3		
		– 6.2			
9	54.3				

We want to calculate $\frac{d\theta}{dt}$ at $t = 3$ and $t = 3.5$ Here $t_0 = 1$, $t = t_0 = rh$ $\therefore\ r = \frac{3-2}{2} = 1$

$$\therefore\ \frac{d\theta}{dt} = \frac{1}{h}\left[\Delta\theta_0 + \frac{2r-1}{2}\Delta^2\theta_0 + \frac{3r^2-6r+2}{6}\Delta^3\theta_0 + \frac{4r^3-18r^2+22r-6}{24}\Delta^4\theta_0\right]$$

$$= \frac{1}{2}\left[-10.8 + \frac{1}{2}(3.3) - \frac{1}{6}(-2.3) + \frac{2}{24}(1.6)\right] = \frac{1}{2}[-10.8 + 1.65 + 0.383 + 0.133] = -4.317$$

To compute $\frac{d\theta}{dt}$ at $t = 3.5$ we use Stirling's central difference formula

Here, $t_0 = 5$, $r = \frac{3.5-5}{2} = -0.75$

$$\frac{d\theta}{dt} = \frac{1}{h}\left[\mu\delta\theta_0 + r\delta^2\theta_0 + \frac{3r^2-1}{6}\mu\delta^3\theta_0 + \frac{2r^3-r}{12}\delta^4\delta_0 + ...\right]$$

$$= \frac{1}{h}\left[\frac{(\Delta\theta_0 + \Delta\theta_{-1})}{2} + r\Delta^2\theta_{-1} + \frac{3r^2-1}{6}\left(\frac{\Delta^3\theta_1 + \Delta^3\theta_{-2}}{2}\right) + \frac{2r^3-r}{12}\Delta^4\theta_{-2} + ...\right]$$

$$= \frac{1}{2}\left[\frac{-7.5-6.5}{2} - 0.75\,(1.0) + \frac{[3(-0.75)^2-1]}{6}\left(\frac{-2.3-0.7}{2}\right) + ...\right]$$

$$= \frac{1}{2}\left[-7 - 0.75 + \frac{.6875}{6}(-1.5)\right] = -3.96$$ **Ans.**

3. *Following Table gives values of x and their corresponding cube roots*

x	50	51	52	53	54	55	56
$y\ (= x^{1/3})$	3.6840	3.7084	3.7325	3.7563	3.7798	3.8030	3.8259

Estimate the values of first and second order derivatives of $x^{1/3}$ at $x = 50$ and at $x = 56$.

SOLUTION:

x	$y (= x^{1/3})$	Δy	$\Delta^2 y$	$\Delta^3 y$
50	3.6840			
		0.0244		
51	3.7084		– 0.0003	0
		– 0.0241		
52	3.7325		–0.0003	
		0.0238		0
53	3.7563		– 0.0003	
		0.0235		0
54	3.7798		– 0.0003	
		0.0232		0
55	3.8030		– 0.0003	
		0.0229		
56	3.8259			

Here $x_0 = 50$, and $x_n = 56$, $h = 1$.

$$\left(\frac{dy}{dx}\right)_{x=x_0} = \frac{1}{h}\left[\Delta y_0 - \frac{1}{2}\Delta^2 y_0 + \frac{1}{3}\Delta^3 y_0 - \frac{1}{4}\Delta^4 y_0 + ...\right]$$

$$= 1\left[0.0244 - \frac{1}{2}(-0.0003) + 0\right] = 0.02455$$ **Ans.**

$$\left(\frac{dy}{dx}\right)_{x=x_n} = \left(\frac{dy}{dx}\right)_{56} = \frac{1}{h}\left[\nabla y_n + \frac{1}{2}\nabla^2 y_n + \frac{1}{3}\nabla^3 y_n + ...\right]$$

$$= 1\left[0.0229 - \frac{1}{2}0.0003 + 0\right] = 0.02275$$ **Ans.**

4. *Following Table gives the velocity v of a body for various values of time t. Find its acceleration at $t = 1.1$*

t	1.0	1.1	1.2	1.3	1.4
v	43.1	47.7	52.1	56.4	60.8

SOLUTION:

t	V	ΔV	$\Delta^2 V$	$\Delta^3 V$	$\Delta^4 V$
1.0	43.1				
		4.6			
1.1	47.7		− 0.2		
		4.4		0.1	
1.2	52.1		− 0.1		0.1
		4.3		0.2	
1.3	56.4		+ 0.1		
		4.4			
1.4	60.8				

To compute $\left(\frac{dV}{dt}\right)_{t=1.1}$ we take $t_0 = 1.0$, $r = \frac{1.1 - 1.0}{0.1} = 1$

$$\frac{dV}{dt} = \frac{1}{h}\left[\Delta V_0 + \frac{2r-1}{2}\Delta^2 V_0 + \frac{3r^2 - 6r + 2}{6}\Delta^3 V_0 + \frac{4r^3 - 18r^2 + 2r - 6}{24}\Delta^4 V_0\right]$$

$$= \frac{1}{0.1}\left[4.6 + \frac{1}{2}(-0.2) - \frac{1}{6}(0.1) + \frac{2}{24}(0.1)\right]$$

$$= 10\,[4.6 - 0.1 - 0.0166 + 0.0083] = 44.917$$ **Ans.**

5. *A rod is rotating in a plane. The Table below gives various values of angle θ (in radians) through which the rod has turned, for various values of time t (in seconds). Find approximately the angular velocity and angular acceleration of the rod at $t = 0.6$ seconds.*

t	0	0.2	0.4	0.6	0.8	1.0	1.2
θ	0	0.12	0.49	1.12	2.02	3.20	4.67

SOLUTION:

t	θ	$\Delta\theta$	$\Delta^2\theta$	$\Delta^3\theta$	$\Delta^4\theta$	$\Delta^5\theta$	$\Delta^6\theta$
0	0						
		0.12					
0.2	0.02		0.25				
		0.37		0.01			
0.4	0.49		0.26		0.0		
		0.63		0.01		0	
0.6	1.12		0.27		0.0		0
		0.90		0.01		0	
0.8	2.02		0.28		0.0		
		1.18		0.01			
1.0	3.20		0.29				
		1.47					
1.2	4.67						

To find $\dfrac{d\theta}{dt}$ and $\dfrac{d^2\theta}{dt^2}$ at $t = 0.6$ we take up the Stirling's central difference formula

$$\left(\frac{d\theta}{dt}\right)_{t=0.6} = \frac{1}{h}\left[\mu\delta y_0 - \frac{1}{6}\mu\delta^3 y_0 + ...\right] = \frac{1}{h}\left[\frac{\Delta y_0 + \Delta y_{-1}}{2} - \frac{1}{6}\frac{\Delta^3 y_{-1} + \Delta^3 y_{-2}}{2} + ...\right]$$

$$= \frac{1}{0.2}\left[\frac{0.90 + 0.63}{2} - \frac{1}{6}\frac{0.01 + 0.01}{2} + ...\right] = 3.81665$$ **Ans.**

and $$\left(\frac{d^2\theta}{dt^2}\right)_{t=0.6} = \frac{1}{h^2}\left[\delta^2 y_0 - \frac{1}{12}\delta^4 y_0 + ...\right] = \frac{1}{h^2}\left[\Delta^2 y_{-1} - \frac{1}{12}\Delta^2 y_{-2} + ...\right]$$

$$= \frac{1}{(0.2)^2}\left[0.27 - \frac{1}{12}(0)\right] = \frac{0.27}{0.04} = 6.75$$ **Ans.**

6. *Following Table gives the values of y for different equispaced values of x*

x	3	4	5	6	7	8
y	0.205	0.240	0.259	0.262	0.250	0.224

Find the approximate value of x for which y is maximum and also the maximum value y.

SOLUTION:

x	y	Δy	$\Delta^2 y$	$\Delta^3 y$
3	0.205			
		0.035		
4	0.240		– 0.016	
		0.019		0.000
5	0.259		– 0.016	
		0.003		0.001
6	0.262		0.015	
		– 0.012		0.001
7	0.250		0.014	
		– 0.026		
8	0.224			

Here $x_0 = 3$, $y_0 = 0.205$, $\Delta y_0 = 0.035$, $\Delta^2 y_0 = -\,0.016$, $\Delta^3 y_0 = 0$.

By Newton's forward difference formula, we get

$$y = 0.205 + r(0.035) + \frac{r(r-1)}{2}(-0.016) + \frac{r(r-1)(r-2)}{3!}(0). \qquad ...(1)$$

$\therefore \quad \dfrac{dy}{dr} = 0.035 + \dfrac{2r-1}{2}(-0.016)$. For y to be minimum $\dfrac{dy}{dr} = 0$

$\therefore \quad 0.035 - 008\,(2r-1) = 0$ which gives $r = 2.6875$

Therefore $x = x_0 + rh = 3 + 2.6875 \times 1 = 5.6875.$ **Ans.**

Putting $r = 2.6875$ in (1) the minimum value of y

$$= 0.205 + 2.6875 \times 0.035 + \frac{1}{2}(2.6875 \times 1.6875)(-0.016) = 0.2628. \qquad \textbf{Ans.}$$

7. *The following Table gives ordinates of a few points for different abscissas:*

x	7.47	7.48	7.49	7.50	7.51	7.52
y	1.93	1.95	1.98	2.01	2.03	2.06

Estimate the area bounded by the curve passing through the given points and the axis of x.

SOLUTION: Here we can apply Trapezoidal rule only as $n = 5$ here.

$$\int_a^b y\,dx = h\left[\frac{y_0 + y_n}{2} + y_1 + y_2 + \ldots.. + y_{n-1}\right]$$

$$= 1\left[\frac{1}{2}(1.93 + 2.06) + 1.95 + 1.98 + 2.01 + 2.03\right] = 9.965. \qquad \textbf{Ans.}$$

8. *Evaluate the integral* $\int_0^6 \frac{1}{1+x}\,dx$ *approximately by (i) Trapezoidal rule (ii) Simpson's one-third rule (iii) Simpson's three-eighth rule (iv) Weedle's rule. Also compare your results with the actual value.*

SOLUTION: $\int_0^6 \frac{1}{1+x}\,dx = \int_0^6 y\,dx$ is to be approximated. Actual value = log 7 = 1.94591.

x	0	1	2	3	4	5	6
y	1	0.5	0.3333	0.25	0.2	0.1666	0.14286

By Trapezoidal Rule $\int_0^6 y\,dx = h\left[\frac{y_0 + y_6}{2} + y_1 + y_2 + y_3 + y_4 + y_5\right]$

$$= 1\left[\frac{1}{2}(1 + 1.14286) + 0.5 + 0.3333 + 0.25 + 0.2 + 0.1666\right] = 2.521428. \qquad \textbf{Ans.}$$

By Simpson's $\frac{1}{3}$ rule $\int_0^6 y\,dx = \frac{h}{3}[y_0 + y_6 + 4(y_1 + y_3 + y_5) + 2(y_2 + y_4)]$

$$= \frac{1}{3}[1 + 0.14286 + 4(0.5 + 0.25 + 0.1666) + 2(0.3333 + 0.2)]$$

$$= \frac{1}{3}[1.14286 + 3.6664 + 1.0666] = 1.95872 \qquad \textbf{Ans.}$$

By Simpson's $\frac{3}{8}$ rule $\int_0^6 y\,dx = \frac{3h}{8}[y_0 + y_6 + 3(y_1 + y_2 + y_4 + y_5) + 2(y_3)]$

$$= \frac{3}{8}[1 + 0.14286 + 3(0.5 + 0.3333 + 0.2 + 0.1666) + 2(0.25)]$$

$$= \frac{3}{8}[1.14286 + 3.5997 + 0.5] = 1.96697$$ **Ans.**

By Weedles rule $\int_0^6 y\,dx = \frac{3h}{10}[y_0 + 5y_1 + y_2 + 6y_3 + y_4 + 5y_5 + y_6]$

$$= \frac{3}{10}[1 + 0.14286 + 5(0.5) + 0.3333 + 6(0.25) + 0.2 + 5(0.1666)]$$

$$= \frac{3}{10}[1.14286 + 2.5 + 0.3333 + 1.5 + 0.2 + .8333] = 1.95284$$ **Ans.**

9. *The speed v of a racing car which starts from rest, is given in the following Table at equal interval of time t.*

t (in min)	2	4	6	8	10	12	14	16	18	20
v (in km/min)	10	18	25	29	32	20	11	5	2	0

Find the approximate distance covered by the car in 20 min.

SOLUTION: The mopped starts from rest hence $v = 0$ at $t = 0$, $h = 2$, $v_0 = 0$.

Distance covered $= \int_0^{20} v\,dt$ Applying Simpson's one-third rule

$$= \frac{h}{3}[X + 4(O) + 2(E)] \quad \text{where } X = v_0 + v_{10} = 0 + 0 = 0$$

$$O = v_1 + v_3 + v_5 + v_7 + v_9 = 10 + 25 + 32 + 11 + 2 = 80$$
$$E = v_2 + v_4 + v_6 + v_8 = 18 + 29 + 20 + 5 = 72$$

$$\int_0^{20} v\,dt = \frac{2}{3}[0 + 4(80) + 2(72)] = 309.33 \text{ km}$$ **Ans.**

10. *Approximate the value of the integral* $\int_0^1 \sqrt{1 - x^2}\, dx$ *for number of sub-intervals* $n = 10, 20$ *and* 30 *using (i) Trapezoidal rule (ii) Simpson's* $1/3$ *rule.*

SOLUTION: Using 10, 20 and 30 subintervals and employing trapezoidal and simpson's $\frac{1}{3}$ rule the value of the integral is Tabulated below.

N	Trapezoidal Rule	Simpson's rule
10	0.776129582	0.781752040
20	0.782116220	0.784111766
30	0.783610789	0.784698434

Ans.

11. *A curve passes through the points with following coordinates:*

x	0.00	0.25	0.50	0.75	1.00
y	1.0000	0.9896	0.9589	0.9089	0.8415

The area between the x-axis, the ordinates at $x = 0$ and $x = 1$ and above curve is rotated about the x-axis. Estimate the volume of the solid of resolution thus obtained correct to three decimal places.

SOLUTION: Here $y_0 = 1.0000$, $y_1 = 0.9896$, $y_2 = 0.9589$, $y_3 = 0.9089$, $y_4 = 0.8415$ and $h = 0.25$, The required volume of the solid of revolution

$$= \int_0^1 \pi y^2 dx = \frac{\pi h}{3}[y_0^2 + y_4^2 + 4(y_1^2 + y_3^2) + 2(y_2^2)] \qquad \text{(using Simpson's 1/3 rule)}$$

$$= \frac{\pi}{12}[1 + 0.8415^2 + 4(0.9896^2 + 0.9089^2) + 2(0.9589^2)]$$

$$= \frac{\pi}{12}[1 + 0.70812 + 4(0.9793) + 2(0.91949)]$$

$$= \frac{3.1416}{12}[1.70812 + 7.22164 + 1.83898] = 2.8193$$ **Ans.**

12. *A rocket is launched from the ground. Its acceleration f is noted for first 80 seconds as given in the following Table*

t (sec)	0	10	20	30	40	50	60	70	80
f (cm/sec^2)	30	31.63	33.34	35.47	37.75	40.33	43.25	46.69	50.67

Estimate the velocity of the rocket at $t = 80$ sec employing Simpson's one-third rule.

SOLUTION: Velocity V at $t = 80$ is $\int_0^{80} f\,dt$. Using Simpson's $\frac{1}{3}$ rule

$$V = \frac{h}{3}[f_0 + f_9 + 4(f_1 + f_3 + f_5 + f_7) + 2(f_2 + f_4 + f_6)]$$

$$= \frac{10}{3}[30 + 50.67) + 4\,(31.63 + 35.47 + 40.33 + 46.69) + 2(33.34 + 37.75 + 43.25)]$$

$$= \frac{10}{3}[80.67 + 4(154.12) + 2(114.34)] = \frac{10}{3}[80.67 + 616.48 + 228.68]$$

$= 3086.1$ cm/sec $= 30.861$ m/sec **Ans.**

13. *Employing Trapezoidal and Simpson's one-third rule, evaluate the integral*

$$\int_0^5 \frac{\sin 3x}{\sqrt{x^2 + x + 1}} dx \quad \text{with} \quad n = 10.$$

SOLUTION: Here $y = \frac{\sin 3x}{\sqrt{x^2 + x + 1}}$ we take $n = 10$ hence $h = 0.5$.

Using Simpson's one-third rule

$$\int_0^5 y\,dx = 0.5\,[y_0 + y_{10} + 4\,(y_1 + y_3 + y_5 + y_7 + y_9) + 2(y_2 + y_4 + y_6 + y_8)]$$

$$= 0.376738.$$ **Ans.**

Using Trapezoidal rule

$$\int_0^5 y\,dx = 0.5\left[\frac{y_0 + y_{10}}{2} + (y_1 + y_2 + \ldots.. + y_9)\right] = 0.290422$$ **Ans.**

14. *Use Trapezoidal rule and Simpson's one-third rule with* 10 *subintervals to estimate the integral* $I = \int_0^{\pi} \frac{\sin x}{x}\,dx$

and compare your results with $I = 1.851937$ *which is exact to six places of decimal.*

SOLUTION: $I = \int_0^{\pi} \frac{\sin x}{x}\,dx = \int_0^{\pi} y\,dx, \quad y = \frac{\sin x}{x}, \quad n = 10 \quad \therefore\ h = \frac{\pi}{10}$

x	0	$\frac{\pi}{10}$	$\frac{2\pi}{10}$	$\frac{3\pi}{10}$	$\frac{4\pi}{10}$	$\frac{5\pi}{10}$	$\frac{6\pi}{10}$	$\frac{7\pi}{10}$	$\frac{8\pi}{10}$	$\frac{9\pi}{10}$	π
$y = \frac{\sin x}{x}$	$y_0 = 1$	y_1	y_2	y_3	y_4	y_5	y_6	y_7	y_8	y_9	0

Using Trapezoidal Rule $\left(h = \frac{\pi}{10}\right)$

$$\int_0^{\pi} \frac{\sin x}{x}\,dx = \frac{h}{2}[y_0 + y_{10} + 2\,(y_1 + y_2 + \ldots.. + y_9)] = 1.849317$$ **Ans.**

Using Simpson's rule $\left(h = \frac{\pi}{10}\right)$

$$\int_0^{\pi} \frac{\sin x}{x}\,dx = \frac{h}{3}[y_0 + y_{10} + 4(y_1 + y_3 + \ldots\ldots + y_9) + 2(y_2 + y_4 + \ldots\ldots + y_8)]$$

$$= 1.8519944$$ **Ans.**

15. *Using the well known integral representation of Bessel's function*

$$J_1(x) = \frac{1}{\pi}\int_0^{\pi} \cos(x \sin\theta - \theta)\,d\theta$$

estimate $J_1(4)$ *by employing Simpson's rule with step length* $h = \pi/10$ *and compare your result with* $J_1(4) = -0.066043$ *which is correct to six decimal places.*

SOLUTION: $J_1(x) = \frac{1}{\pi}\int_0^{\pi} \cos\,(x \sin\theta - \theta)\,d\theta$. Problem is similar to that in 14.

Here also divide $(0, \pi)$ into 10 equal parts and obtain the result by Simpson's one-third rule as $J_1(4) = -0.065743.$ **Ans.**

ADDITIONAL IMPORTANT PROBLEMS

1. *Find* $\dfrac{dy}{dx}, \dfrac{d^2y}{dx^2}$ *at* $x = 0$; *where the function* $y = f(x)$ *is as per the following data.*

x:	0.00	0.05	0.10	0.15	0.20
y:	0.00000	0.10017	0.20134	0.30452	0.41075

SOLUTION: The difference table is

x	y	Δy	$\Delta^2 y$	$\Delta^3 y$	$\Delta^4 y$
0.00	0.00000				
		0.1007			
0.05	0.10017		0.00100		
		0.10117		0.00101	
0.10	0.20134		0.00201		0.00003
		0.10318		0.00104	
0.15	0.30452		0.00305		
		0.10623			
0.20	0.41075				

we have $h = 0.05, \quad x_0 = 0.00, \quad y_0 = 0.00000, \quad \Delta y_0 = 0.10017$

$$\Delta^2 y_0 = 0.00100, \quad \Delta^3 y_0 = 0.00101, \quad \Delta^4 y_0 = 0.00003$$

Using Netwon's formula, we get

$$\left(\frac{dy}{dx}\right)_{x=x_0} = \frac{1}{h}\left[\Delta y_0 - \frac{1}{2}\Delta^2 y_0 + \frac{1}{3}\Delta^3 y_0 - \frac{1}{4}\Delta^4 y_0 + ...\right]$$

$$= \frac{1}{0.05}\left[0.10017 - \frac{1}{2}(0.00100) + \frac{1}{3}(0.00101) - \frac{1}{4}(0.00003)\right]$$

Thus, $\left(\dfrac{dy}{dx}\right)_{x=0.00} = 1.99985.$ **Ans.**

Further, $$\left(\frac{d^2y}{dx^2}\right)_{x=x_0} = \frac{1}{h^2}\left(\Delta^2 y_0 - \Delta^3 y_0 + \frac{11}{12}\Delta^4 y_0 + ...\right)$$

$\therefore$ $$\left(\frac{d^2y}{dx^2}\right)_{x=x_0} = \frac{1}{(0.05)^2}\left[0.001 - 0.00101 + \frac{11}{12}(0.00003)\right] = 6.99$$ **Ans.**

2. *From the Table of values below, compute* $\dfrac{dy}{dx}$ and $\dfrac{d^2y}{dx^2}$ *for* $x = 1$.

x	1	2	3	4	5	6
y	1	8	27	64	125	216

SOLUTION: The difference Table is

x	y	Δy	$\Delta^2 y$	$\Delta^3 y$	$\Delta^4 y$
1	1				
		7			
2	8		12		
		19		6	
3	27		18		0
		37		6	
4	64		24		0
		61		6	
5	125		30		
		91			
6	216				

we have here $x_0 = 1,\ h = 1,\quad x = 1$ is at the beginning of the Table.

∴ we use Newton's forward formula and get

$$\left[\frac{dy}{dx}\right]_{x=x_0} = \frac{1}{h}\left[\Delta y_0 = \frac{1}{2}\Delta^2 y_0 + \frac{1}{3}\Delta^3 y_0 - \frac{1}{4}\Delta^4 y_0 + \ldots\right]$$

or
$$\left[\frac{dy}{dx}\right]_{x=1} = \frac{1}{h}\left[7 - \frac{1}{2}12 + \frac{1}{3}6 - 0 + \ldots\right] = 7 - 6 + 2 = 3$$

and
$$\left[\frac{d^2y}{dx^2}\right]_{x=x_0} = \frac{1}{h^2}\left[\Delta^2 y_0 - \Delta^3 y_0 + \frac{11}{12}\Delta^4 y_0 - \ldots\right] \quad \text{or} \quad \left[\frac{d^2y}{dx^2}\right]_{x=1} = \frac{1}{1^2}[12 - 6] = 0$$

Hence
$$\left[\frac{dy}{dx}\right]_{x=1} = 3,\quad \left[\frac{d^2y}{dx^2}\right]_{x=1} = 6.$$

3. *Find* $\dfrac{dy}{dx}$ *at* $x = 2.03$ *using the data given below*

x	1.96	1.98	2.00	2.02	2.04
y	0.7825	0.7739	0.7651	0.7563	0.7473

SOLUTION: $X = 2.03$ is at the end of the Table, therefore, we use Newton's backward interpolation formula for finding the derivative.

The backward difference table is

x	y	∇y	$\nabla^2 y$	$\nabla^3 y$	$\nabla^4 y$
1.96	0.7825				
		–0.0086			
1.98	0.7739		–0.0002		
		–0.0088		0.0002	
2.00	0.7651		0.0000		–0.0004
		–0.0088		–0.0002	
2.02	0.7563		–0.0002		
		–0.0090			
2.04	0.7473				

Here, $h = 0.02,\ x = 2.03,\ x_n = 2.04\quad \therefore\quad p = \dfrac{x - x_n}{h} = \dfrac{2.03 - 2.04}{0.02} = -0.5.$

$\therefore\quad \nabla y_n = -0.0090,\quad \nabla^2 y_n = -0.0002\quad \nabla^3 y_n = -0.0002,$ and $\nabla^4 y_n = -0.0004$

from Newton's formula, we have $\dfrac{dy}{dx} = \dfrac{1}{h}\left[\nabla y_n + \dfrac{(2p+1)}{2!}\nabla^2 y_n + \dfrac{(3p^2+6p+2)}{3!}\nabla^3 y_n + \ldots\right]$

$$\therefore \quad \left(\frac{dy}{dx}\right)_{x=2.03} = \frac{1}{0.02}\left[-0.0009 + \frac{[2(-0.5)+1}{2}\times(-0.0002) + \frac{(3\times 0.25 - 6\times 0.5 + 2}{6}\times(-0.0002)\right.$$
$$\left. + \frac{4\times(-0.125)+18\times(0.25)+22\times(-0.5)+6}{24}\times(-0.0004)\right]$$

hence, we have $\left(\dfrac{dy}{dx}\right)_{x=2.03} = -0.44876.$ **Ans.**

4. *The population of a city is given in the following Table. Find the rate of growth in the population in the year 1995.*

Year (x) :	1961	1971	1981	1991	2001
Population (y) :	40.62	60.80	79.95	103.56	132.65

SOLUTION: Here $h = 10$, $x = 132.65$, $y_n = 1995$.

Hence we use Newton's backward interpolation formula,

Here, we have $p = \dfrac{x - x_n}{h} = \dfrac{1995 - 2001}{10} = -0.6$. The difference table is

x	y	∇y	$\nabla^2 y$	$\nabla^3 y$	$\nabla^4 y$
1961	40.62				
		20.18			
1971	60.80		−1.03		
		19.15		5.49	
1981	79.95		4.46		−4.47
		23.61		1.02	
1991	103.56		5.48		
		29.09			
2001	132.65				

For rate of growth of population, we use here Newtons backward difference formula

$$\frac{dy}{dx} = \frac{1}{h}\left[\nabla y_n + \frac{2p+1}{2!}\nabla^2 y_n + \frac{(3p^2+6p+2)}{3!}\nabla^3 y_n + \ldots\right] \text{ and get}$$

$$\left(\frac{dy}{dx}\right)_{x=1995} = \frac{1}{10}\left[29.09 + \frac{1}{2}\{2(-0.6+1)\}(5.48) + \frac{1}{6}\{3(-0.6)^2 + 6(-0.6) + 2\}(1.02)\right]$$

$$= \frac{1}{10}[29.09 - 0.548 - 0.0918] = 2.845.$$ **Ans.**

5. *Calculate the value* $\displaystyle\int_0^6 \frac{x}{1+x}\,dx$ *correct up to three significant figures taking six intervals by Trapezoidal rule.*

SOLUTION: Here we have $f(x) = \dfrac{x}{1+x}$. $a = 0$, $b = 1$, $n = 6$, and $h = \dfrac{b-a}{n} = \dfrac{1}{6}$.

x	0	1/6	2/6	3/6	4/6	5/6	6/6 = 1
$y=f(x)$	0.00000	0.14286	0.25000	0.33333	0.40000	0.45454	0.50000
	y_0	y_1	y_2	y_3	y_4	y_5	y_6

By trapezoidal rule $$I=\int_a^b f(x)\,dx = \frac{h}{2}[(y_0+y_6)+2(y_1+y_2+y_3+y_4+y_5)]$$

$$= \frac{1}{12}[(0.00000+0.50000+2(0.14286+0.25000$$

$$+0.3333+0.4000+0.45454)] = 0.30512.$$

$\therefore$ $I = 0.0305$, correct to three significance figure. **Ans.**

6. *Find the value* $\int_0^{0.6} e^x dx$, *taking* $n = 6$, *correct to five significant figures by Simpson's one-third rule.*

SOLUTION: We have $f(x) = e^x$, $a = 0$, $b = 0.6$, $n = 6$. $h=\dfrac{b-a}{n}=\dfrac{0.6-0}{6}=0.1.$

x	0.0	0.1	0.2	0.3	0.4	0.5	0.6
$y=f(x)$	1.0000	1.10517	1.22140	1.34986	1.49182	1.64872	1.82212
	y_0	y_1	y_2	y_3	y_4	y_5	y_6

The Simpson's rule is $$\int_a^b f(x)\,dx = \frac{h}{3}[(y_0+y_6)+4(y_1+y_3+y_5)+2(y_2+y_4)]$$

$$= \frac{0.1}{3}[(1.00000+1.82212)+4(1.10517+1.34986+1.64872)$$

$$+2(1.22140+1.49182)]$$

$$= \frac{0.1}{3}[2.82213+4\,(4.1035+2(2.71322)]$$

or $I = 0.82212.$ **Ans.**

7. *The velocity of a train which starts from rest is given by the following Table, the time being reckoned in minutes from the start and the speed in km/hour.*

t (minutes)	*2*	*4*	*6*	*8*	*10*	*12*	*14*	*16*	*18*	*20*
v (km/hr)	*16*	*28.8*	*40*	*46.4*	*51.2*	*32.0*	*17.6*	*8*	*3.2*	*0*

Estimate approximately the total distance run in 20 minutes.

SOLUTION: $S = \int_0^{20} v.dt$ The train starts from rest, $\therefore$ the velocity $v = 0$ when $t = 0$

t	0	2	4	6	8	10	12	14	16	18	20
v	0	16	28.8	40	46.4	51.2	32.0	17.6	8	3.2	0
	y_0	y_1	y_2	y_3	y_4	y_5	y_6	y_7	y_8	y_9	y_{10}

$\therefore \quad h = \frac{2}{60}$ hrs $= \frac{1}{30}$ *hrs.* The Simpson's rule is

$$S = \int_0^{20} v.dt = \frac{h}{3}[(y_0 + y_{10}) + 4(y_1 + y_3 + y_7 + y_9) + 2(y_2 + y_4 + y_6 + y_8)]$$

$$= \frac{1}{30 \times 3}[(0+0)\ 4(16+40+51.2+17.6+3.2)+2(28.8+46.4+32.0+8)]$$

$$= \frac{1}{90}[0+4\times128+2\times115.2] = 8.25 \text{ km}.$$

Hence, the distance run by the train in 20 minutes = 8.25 km. **Ans.**

8. *Evaluate* $\int_{0.1}^{0.7} \frac{dx}{x}$, *with* $h = 0.1$, *using Simpson's* $\frac{3}{8}$ *rule.*

SOLUTION: We have $y = f(x) = \frac{1}{x}$, $h = 0.1$, $a = 0.1$, $b = 0.7$

x	0.1	0.2	0.3	0.4	0.5	0.6	0.7
$y = f(c) = \frac{1}{x}$	10.0	5.0	3.33	2.5	2.0	1.67	1.43
	y_0	y_1	y_2	y_3	y_4	y_5	y_6

Applying Simpson's three-eight rule

$$\int_{0.1}^{0.7} \frac{1}{x}\,dx = \frac{3h}{8}[(y_0 + y_n) + 3(y_1 + y_2 + y_4 + y_5 + \ldots) + 2(y_3 + y_6 + \ldots)]$$

$$= \frac{3\times(0.1)}{8}\ [(10-1.43)+3(5+3.33+2.0+1.67)+2(2.5)]$$

$$= 1.97$$ **Ans.**

9. *Evaluate* $\int_{-3}^{3} x^4 d\,x$ *by Weddle's rule.*

SOLUTION: Taking $n = 6$, we get $h = \frac{b-a}{n} = \frac{3-(-3)}{6} = \frac{6}{6} = 1$

x :	–3	–2	–1	0	1	2	3
$y = x^4$:	81	16	1	0	1	16	81
	y_0	y_1	y_2	y_3	y_4	y_5	y_6

By Weddle's rule $\int_{-3}^{3} f(x)\,dx = \int_{-3}^{3} x^4\,dx = \frac{3h}{10}[y_0 + 5y_1 + y_2 + 6y_3 + y_4 + 5y_5 + y_6]$

$$= \frac{3\times1}{10}[81+5\times16+1+6\times0+1+5\times16+81] = 97.2$$ **Ans.**

CHAPTER

29 Difference Equation and Numerical Solution of Ordinary Differential Equations

IMPORTANT FORMULAE AND USEFUL RESULTS

DIFFERENCE EQUATIONS

These are of the form $a_0\, y_{n+r} + a_1\, y_{n+r-1} + a_2\, y_{n+r-2} + \ldots\ldots + a_n\, y_n = \phi(n)$.

For complementary Function. Auxiliary equation is

$$m^n(a_0\, m^r + a_1 m^{r-1} + a_2\, m^{r-2} + \ldots\ldots + a_n) = 0. \quad \text{If its roots are } m_1, m_2, \ldots\ldots m_r$$

then $\quad$ C.F. $= c_1 m_1^n + c_2 m_2^n + c_3 m_3^n + \ldots\ldots + c_r m_r^n$ where $c_1, c_2, \ldots\ldots$ are arbitrary constants.

Particular Integral. For the equation $f(E)\, y = \phi(n)$

$$\text{P.I} = \frac{1}{f(E)}\, \phi(n).$$

DIFFERENTIAL EQUATIONS

1. **Taylor series method:** Equation is $\dfrac{dy}{dx} = f(x, y), \quad y(x_0) = y_0$

$$y = y_0 + (x - x_0) y_0' + \frac{(x - x_0)^2}{2!} y_0'' + \ldots$$

$$= y_0 + (x - x_0)\, f(x_0\, y_0) + \frac{(x - x_0)^2}{2!} f'(x_0, y_0) + \frac{(x - x_0)^3}{3!} f''(x_0\, y_0) + \ldots$$

2. **Pieard's method:** Equation is $\dfrac{dy}{dx} = f(x, y)$ where $y(x_0) = y_0$

$$\therefore \quad y = y_0 + \int_{x_0}^{x} f(x, y)\, dx,$$

$$y^{(1)} = y_0 + \int_{x_0}^{x} f(x, y_0)\, dx, \quad y^{(2)} = y_0 + \int_{x_0}^{x} f(x, y^{(1)})\, dx, \ldots\ldots, \quad y^{(n)} = y_0 + \int_{x_0}^{x} f(x, y^{(n-1)})\, dx$$

3. Euler's method: Equation is $\frac{dy}{dx} = f(x, y)$, $y(x_0) = y_0$

$$y_1 = y(x_0 + h) = y_0 + hf(x_0, y_0),\quad y_2 = y(x_0 + 2h) = y_1 + hf(x_1, y_1), \ldots$$

$$y_{n+1} = y_n + hf(x_n, y_n),\quad n = 0, 1, 2, \ldots$$

4. Modified Euler's Method: $y_1 = y_0 + hf(x_0, y_0)$

We modify it as $y_1^{(1)} = y_0 + \frac{h}{2}[f(x_0, y_0) + f(x_0 + h, y_1)]$. Further modifying it, gives

$y_1^{(2)} = y_0 + \frac{h}{2}[f(x_0, y_0) + f(x_0 + h, y_1^{(1)})]$, and so on till the desired accuracy is achieved.

Next, $y_2 = y_1 + hf(x_0 + h, y_1)$

modifying it we get $y_2^{(1)} = y_1 + \frac{h}{2}[f(x_0 + h, y_1) + f(x_0 + 2h, y_2)]$. Again modifying it, gives

$y_2^{(2)} = y_1 + \frac{h}{2}[f(x_0 + h, y_1) + f(x_0 + 2h, y_2^{(1)}]$ and so on till the desired accuracy is achieved.

5. Runge-Kutta method of Fourth order:

Equation is $\frac{dy}{dx} = f(x, y)$, $y(x_0) = y_0$

$$K_1 = hf(x_0, y_0)$$

$$K_2 = hf\left(x_0 + \frac{h}{2}, y_0 + \frac{K_1}{2}\right)$$

$$K_3 = hf\left(x_0 + \frac{h}{2}, y_0 + \frac{K_2}{2}\right)$$

$$K_4 = hf(x_0 + h, y_0 + K_3)$$

then $y(x_0 + h) = y_0 + \frac{1}{6}(K_1 + 2K_2 + 2K_3 + K_4)$.

SOLUTIONS TO EXERCISE 29A

1. *Solve the equation* $y_{x+2} + y_{x+1} - 6y_x = 0.$

SOLUTION: Equation is $(E^2 + E - 6)y_x = 0$ where $Ey_i = y_{i+1}$. Auxiliary equation is $m^2 + m - 6 = 0$ or $(m + 3)(m - 2) = 0$ $\therefore$ $m = 2, -3.$

$\therefore$ Solution is $y_x = c_1 2^x + c_2(-3)^x$ **Ans.**

2. *Solve* $y_{n+2} - 4y_{n+1} + 3y_n = 0$ *given that* $y_0 = 0, y_4 = 10.$

SOLUTION: Equation can be written as $(E^2 - 4E + 3)y_n = 0$

Here the A.E. is $m^2 - 4m + 3 = 0$ or $(m - 3)(m - 1) = 0$ $\therefore$ $m = 1, 3.$

$\therefore$ Solution is $y_n = c_1(1)^n + c_2(3)^n = c_1 + c_2(3^n)$

Since $y_0 = 0$ and $y_4 = 10$ we have $c_1 + c_2 = 0$ and $c_1 + 81c_2 = 10$

which give $c_2 = \frac{1}{8}$ and $c_1 = -\frac{1}{8}$. Therefore the solution is

$$y_n = -\frac{1}{8} + \frac{1}{8}3^n = \frac{1}{8}(3^n - 1).$$ **Ans.**

3. *Solve the equation* $y_{n+3} - 8y_{n+2} + 21y_{n+1} - 18y_n = 0.$

SOLUTION: Equation can be written as $(E^3 - 8E^2 + 21E - 18)y_n = 0$ where $Ey_i = y_{i+1}$.

The A.E. is $m^3 - 8m^2 + 21m - 18 = 0$ or $(m - 3)^2(m - 2) = 0$

$\therefore$ Solution is $y_n = c_1(2)^n + (c_2 + n.c_3)(3)^n$. **Ans.**

4. *Solve* $y_{n+2} - y_{n+1} + y_n = 0$ *given that* $y_0 = 1,\ y_1 = \frac{\sqrt{3}+1}{2}.$

SOLUTION: Equation can be written as $(E^2 - E + 1)y_n = 0$ where $Ey_i = y_{i+1}$.

There A.E. is $m^2 - m + 1 = 0$ $\therefore$ $m = \frac{1 \pm i\sqrt{3}}{2}$

$\therefore$ Solution is $y_n = r^n(c_1 \cos n\theta + c_2 \sin n\theta]$

where $r = \sqrt{\left(\frac{1}{2}\right)^2 + \left(\frac{\sqrt{3}}{2}\right)^2} = 1$ and $\theta = \pi/3$ $\therefore$ $y_n = 1^n\left[c_1 \cos\frac{n\pi}{3} + c_2 \sin\frac{n\pi}{3}\right].$

Given that $y_0 = 1$ and $y_1 = (\sqrt{3}+1)/2.$ $\Rightarrow$ $c_1 = c_2 = 1$

$\therefore$ solution is $y_n = \cos\frac{n\pi}{3} + \sin\frac{n\pi}{3}.$ **Ans.**

5. *Solve* $y_{x+2} - 4y_{x+1} + 3y_x = 4^x$

SOLUTION: Equation is $(E^2 - 4E + 3)y_x = 4^x$ where $Ey_x = y_{x+1}$.

Here A.E. is $m^2 - 4m + 3 = 0$ $\therefore$ $m = 1, 3$ hence C.F $= c_1\, 1^x + c_2\, 3^x$.

And P.I. $= \dfrac{1}{E^2 - 4E + 3} 4^x = \dfrac{1}{4^2 - 4(4) + 3} 4^x = \dfrac{1}{3} . 4^x$

$\therefore$ Solution is $y_x = c_1 + c_2(3^x) + \dfrac{1}{3}(4^x)$ **Ans.**

6. *Solve* $y_{n+2} - 4y_{n+1} + 3y_n = 3^n$.

SOLUTION: Difference equation can be written as $(E^2 - 4E + 3)y_n = 3^n$ where $Ey_n = y_{n+1}$.

The A.E. is $m^2 - 4m + 3 = 0 \therefore$ $m = 1, 3$

Hence C.F. $= c_1(1^n) + c_2(3^n) = c_1 + c_2 \cdot 3^n$

and P.I. $= \dfrac{1}{E^2 - 4E + 3} 3^n = \dfrac{1}{(E-1)(E-3)} 3^n = \dfrac{1}{E-3} \dfrac{1}{(3-1)}(3^n)$

$= \dfrac{1}{2} \dfrac{1}{E-3}(3^n) = \dfrac{1}{2} . n(3^n)$ (case of failure)

$\therefore$ Solution is $y_n = c_1 + \left(c_2 + \dfrac{n}{2}\right)(3^n)$ **Ans.**

7. *Solve the difference equation* $y_{n+2} - 2y_{n+1} + y_n = 3n + 4$.

SOLUTION: Difference equation can be written as $(E^2 - 2E + 1)y_n = 3n + 4$.

The A.E. is $m^2 - 2m + 1 = 0$ $\therefore$ $m = 1, 1$ $\therefore$ C.F. $= (c_1 + c_2 n)(1)^n = c_1 + nc_2$

and P.I $= \dfrac{1}{(E-1)^2}(3n + 4) = \dfrac{1}{\Delta^2}(3n^{(1)} + 4n^{(0)}]$ (since $E = 1 + \Delta$)

$= \dfrac{1}{\Delta}\left[\dfrac{3n^{(2)}}{2} + 4n^{(1)}\right] = \dfrac{3}{2} . \dfrac{n^{(3)}}{3} + \dfrac{4n^{(2)}}{2} = \dfrac{1}{2} n^{(3)} + 2n^{(2)}$

$= \dfrac{1}{2} n(n-1)(n-2) + 2n(n-1) = \dfrac{n}{2}(n-1)(n+2)$

$\therefore$ Solution is $y_n = c_1 + nc_2 + \dfrac{n}{2}(n-1)(n+2)$. **Ans.**

8. *Solve* $y_{x+2} - 6y_{x+1} + 8y_x = 2^x + 6x$.

SOLUTION: Difference equation can be written as $(E^2 - 6E + 8)y_x = 2^x + 6x$ where $Ey_x = y_{x+1}$.

A.E. is $m^2 - 6m + 8 = 0$ hence $m = 2, 4$ $\therefore$ C.F. $= c_1 \times 2^x + c_2 \times 4^x$.

and $$\text{P.I} = \frac{1}{(E-2)(E-4)}2^x + \frac{1}{(E-2)(E-4)}6x$$

$$= \frac{1}{E-2}.\left(\frac{1}{2-4}\right)(2^x) + \frac{1}{(1+\Delta-2)(1+\Delta-4)}6x \qquad (\text{as } E = 1+\Delta)$$

$$= -\frac{1}{2}x(2^x) + \frac{1}{(\Delta-1)(\Delta-3)}6x = \frac{-x}{2}(2^x) + \frac{1}{3-4\Delta+\Delta^2}6x \qquad ...(1)$$

$$= \frac{-x}{2}(2^x) + \frac{1}{3}\left[1 - \left(\frac{4}{3}\Delta - \frac{\Delta^2}{3}\right)\right]^{-1} 6x^{(1)} \qquad (\text{Since } x = x^{(1)} \text{ in factorial notation})$$

$$= \frac{-x}{2}.2^x + \frac{6}{3}\left[1+\frac{4}{3}\Delta - \frac{\Delta^2}{3}\right]x^{(1)} = \frac{-x}{2}\cdot 2^x + 2\left[x^{(1)} + \frac{4}{3}\right] = \frac{-x}{2}(2^x) + 2x + \frac{8}{3}$$

$\therefore$ Solution is $y_x = \left(c_1 - \dfrac{x}{2}\right)2^{(x)} + c_2(4^x) + 2x + \dfrac{8}{3}$ **Ans.**

SOLUTIONS TO EXERCISE 29B

1. *Using Picard's method, solve* $\dfrac{dy}{dx} = 1 + xy,\ y(2) = 0$

for $x = 2.1,\ 2.2,\ 2.3$ *correct to four places of decimal.*

SOLUTION: $dy = (1+xy)dx,\ \ y(2) = 0 \quad \therefore \quad x_0 = 2,\ \ y_0 = 0$

$\therefore$ $$\int_{y_0}^{y} dy = \int_{x_0}^{x}(1+xy)\,dx \qquad \text{or} \quad y_1 = y_0 + \int_2^x (1+x.\,y_0)\,dx = x-2$$

$\therefore$ $$y_2 = y_0 + \int_2^x (1+xy_1)\,dx = 0 + \int_2^x [1+x(x-2]dx = \left[x - x^2 + \frac{x^3}{3}\right]_2^x$$

or $$y_2 = x - x^2 + \frac{x^3}{3} - \left(2 - 4 + \frac{8}{3}\right) = \frac{-2}{3} + x - x^2 + \frac{x^3}{3}.$$

$$y_3 = y_0 + \int_2^x (1+xy_2)dx = 0 + \int_2^x \left[1 + x\left(\frac{-2}{3} + x - x^2 + \frac{x^3}{3}\right)\right]dx$$

$$= \int_2^x \left(1 - \frac{2}{3}x + x^2 - x^3 + \frac{x^4}{3}\right)dx$$

$$= x - \frac{x^2}{3} + \frac{x^3}{3} - \frac{x^4}{4} + \frac{x^5}{15} - \left(2 - \frac{4}{3} + \frac{8}{3} - 4 + \frac{32}{15}\right)$$

$$= -\frac{22}{15} + x - \frac{x^2}{3} + \frac{x^3}{3} - \frac{x^4}{4} + \frac{x^5}{15}$$

Next $$y_4 = y_0 + \int_2^x (1 + xy_3)\,dx = 0 + \int_2^x 1 + x\left(-\frac{22}{15} + x - \frac{x^2}{3} + \frac{x^3}{3} - \frac{x^4}{4} + \frac{x^5}{15}\right)dx$$

$$= x - \frac{11x^2}{15} + \frac{x^3}{3} - \frac{x^4}{12} + \frac{x^5}{15} - \frac{x^6}{24} + \frac{x^7}{105} - \left(2 - \frac{44}{15} + \frac{8}{3} - \frac{4}{3} + \frac{32}{15} - \frac{8}{3} + \frac{128}{105}\right)$$

$$= \frac{-114}{105} + x - \frac{11x^2}{15} + \frac{x^3}{3} - \frac{x^4}{12} + \frac{x^5}{15} - \frac{x^6}{24} + \frac{x^7}{105}$$

Putting $x = 2.1,\ 2.2.\ 2.3$ we get $y(2.1) = 0.1110$, $y(2.2) = 0.249$, $y(2.3) = 0.424$. **Ans.**

2. *Solve the differential equation* $\frac{dy}{dx} = y - x^2$, $y(0) = 1$ *by Picard's method upto third approximation and hence find* $y(0.1)$ *and* $y(0.2)$.

SOLUTION: $dy = (y - x^2)$ where $y(0) = 1$ $\therefore$ $x_0 = 0$, $y_0 = 1$

$$\therefore\ y = y_0 + \int_{x_0}^x (y - x^2)\,dx = 1 + \int_0^1 (y_0 - x^2)\,dx = 1 + \int_0^1 (1 - x^2)\,dx = 1 + x - \frac{x^3}{3} = y_1$$

Next $$y = y_0 + \int_{x_0}^x (y_1 - x^2)\,dx = 1 + \int_0^x \left(1 + x - \frac{x^3}{3} - x^2\right)dx$$

$$= \left[1 + x + \frac{x^2}{2} - \frac{x^4}{12} - \frac{x^3}{3}\right]_0^x = 1 + x + \frac{x^2}{2} - \frac{x^3}{3} - \frac{x^4}{12} = y_2$$

Next $$y = y_0 + \int_{x_0}^x (y_2 - x^2)\,dx = 1 + \int_0^x \left(1 + x + \frac{x^2}{2} - \frac{x^3}{3} - \frac{x^4}{12} - x^2\right)dx$$

The solution is $y = 1 + x + \frac{x^2}{2} - \frac{x^3}{6} - \frac{x^4}{12} - \frac{x^5}{60}$ **Ans.**

$\therefore$ $y(0.1) = 1.1048249$ and $y(02) = 1.218528$ **Ans.**

3. *Solve* $\frac{dy}{dx} = x^2 + y^2$, $y(0) = 1$ *using Picard's method.*

SOLUTION: We have $dy = (x^2 + y^2)dx$, $y(0) = y_0 = 1$. By Pieard's method

$$y = y_1 = y_0 + \int_0^x (x^2 + y_0^2)\,dx = 1 + \int_0^x (1 + x^2)\,dx = 1 + x + \frac{x^3}{3}$$

Next $$y = y_2 = y_0 + \int_0^x (x^2 + y_1^2)\,dx = 1 + \int_0^x \left[x^2 + \left(1 + x + \frac{x^3}{3}\right)^2\right]dx$$

$$= 1 + \int_0^x \left[x^2 + 1 + x^2 + \frac{x^6}{9} + 2x + \frac{2x^3}{3} + \frac{2x^4}{3}\right]dx$$

or $$y = 1 + x + x^2 + \frac{2x^3}{3} + \frac{2x^4}{12} + \frac{2x^5}{15} + \frac{x^7}{63}.$$ **Ans.**

4. *Solve* $\frac{dy}{dx} = x + y$, *given that* $y(1) = 0$, *to get* $y(1.1)$ *and* $y(1.2)$ *by using Taylor series method. Compare your results with those by analytical method.*

SOLUTION: $\frac{dy}{dx} = f(x, y) = x + y, \quad y(1) = 0 \quad \therefore \quad x_0 = 1, \quad y_0 = 0.$

At $x = x_0 + h$ we could calculate y by Taylor's method as

$$y = y_0 + h\left(\frac{dy}{dx}\right)_{(x_0, y_0)} + \frac{h^2}{2!}\left(\frac{d^2y}{dx^2}\right)_{(x_0, y_0)} + \frac{h^3}{3!}\left(\frac{d^3y}{dx^3}\right)_{(x_0, y_0)} + \ldots..$$

$$\therefore \quad y = y(1.1) = 0 + 0.1\, f(x_0, y_0) + \frac{h^2}{2}[1 + f(x_0, y_0)] + \frac{h^3}{3!}[1 + f(x_0, y_0)] + \ldots.$$

$$= 0 + 0.1\,(1 + 0) + \frac{0.01}{2}[1 + 1 + 0] + \frac{0.001}{3!}[1 + 1 + 0] + \ldots..$$

$$= 0 + 0.1 + 0.01 + \frac{0.001}{3} + \ldots... = 0.11033 = y(1.1)$$

Now take $x_0 = 1.1$ and $y_0 = 0.11033$, then

$$y(1.2) = 0.11033 + 0.1\,(x_0 + y_0) + \frac{0.01}{2!}(1 + x_0 + y_0) + \frac{0.001}{3!}(1 + x_0 + y_0)$$

$$= 0.11033 + 0.1\,(1.1 + 0.11033) + \frac{0.01}{2!}(1 + 1.1 + 0.11033) + \frac{0.001}{6}(1 + 1.1 + 0.11033) + \ldots.$$

$$= 0.11033 + 0.12103 + 0.0110516 + 0.000368 = 0.2424 \text{ approx.}$$ **Ans.**

5. *Use Taylor series method to approximate* $y(0.2)$ *and* $y(0.4)$, *given that*

$$\frac{dy}{dx} = 1 - 2xy, \quad y(0) = 0.$$

SOLUTION: $\frac{dy}{dx} = f(x, y) = 1 - 2xy, \quad y(0) = 0, \quad \therefore \quad x_0 = 0, \quad y_0 = 0, \quad h = 0.2. \quad \left(\frac{dy}{dx}\right)_{(0,\,0)} = 1$

$$\frac{d^2y}{dx^2} = -2x\frac{dy}{dx} - 2y = -2x\,(1 - 2xy) - 2y = 0 \text{ at } (0, 0)$$

$$\frac{d^3y}{dx^3} = \frac{d}{dx}[-2x - 2y + 4x^2y] = -2 - 2\frac{dy}{dx} + 8xy + 4x^2\frac{dy}{dx}$$

$$= -2 + 8xy + 2(1 - 2xy) + 4x^2(1 - 2xy) = -4 \quad \text{at} \quad (0, 0)$$

$\therefore$ Taylor's expansion is

$$y = y_0 + h\left(\frac{dy}{dx}\right)_{(0,\,0)} + \frac{h^2}{2!}\left(\frac{d^2y}{dx^2}\right)_{(0,\,0)} + \frac{h^3}{3!}\left(\frac{d^3y}{dx^3}\right)_{(0,\,0)} + \frac{h^4}{4!}\left(\frac{d^4y}{dx^4}\right)_{(0,\,0)} + \ldots$$

$$= 0 + 0.2\,(1) + \frac{0.04}{2}(0) + \frac{0.008}{3!}(-4) + \ldots.. = 0.2 - 0.0053 + \ldots.. = 0.1947 \text{ approx.}$$ **Ans.**

Similarly, taking $x_0 = 0.2$, $y_0 = 0.1947$ we can get

$y(0.4) = 0.35988$ approximately. **Ans.**

6. *Solve the differential equation* $\frac{dy}{dx} = x \cdot y^{1/3}$, $y(1) = 1$ *for* $y(1.1)$ *and* $y(1.2)$ *correct to four decimal places by employing Taylor series method.*

SOLUTION: $\frac{dy}{dx} = x\,y^{1/3} = f(x, y)$, $y(1) = 1$. To find $y(1.1)$ we take $h = 0.1$ $x_0 = 1$, $y_0 = 1$.

$$\left(\frac{dy}{dx}\right)_{(1,1)} = 1, \quad \frac{d^2y}{dx^2} = \frac{x}{3}\,y^{-2/3}\frac{dy}{dx} + y^{1/3} = \frac{x}{3}\,y^{-2/3}\,(xy^{1/3}) + y^{1/3} = \frac{x^2}{3}\,y^{-1/3} + y^{1/3}$$

$$\therefore \quad \left(\frac{d^2y}{dx^2}\right)_{(1,1)} = \frac{1}{3} + 1 = \frac{4}{3}$$

Similarly,
$$\frac{d^3y}{dx^3} = \frac{x^2}{3}\cdot\left(\frac{-1}{3}\right)y^{-4/3}\frac{dy}{dx} + \frac{2x}{3}\,y^{-1/3} + \frac{1}{3}\,y^{-2/3}\frac{dy}{dx}$$

$$= \frac{-x^2}{9}\,y^{-4/3}\,.\,xy^{1/3} + \frac{2x}{3}\,y^{-1/3} + \frac{1}{3}\,y^{-2/3}\,.\,xy^{1/3}$$

$$= -\frac{x^3}{9}\,y^{-1} + xy^{-1/3} = 1 - \frac{1}{9} = \frac{8}{9} \quad \text{at } (1,1).$$

$$\therefore \quad y = 1 + 0.1\left(\frac{dy}{dx}\right)_{(1,1)} + \frac{0.1^2}{2!}\left(\frac{d^2y}{dx^2}\right)_{(1,1)} + \frac{0.1^3}{3!}\left(\frac{d^3y}{dx^3}\right)_{(1,1)} + \ldots..$$

$$= 1 + 0.1\,(1) + \frac{0.01}{2}\left(\frac{4}{3}\right) + \frac{0.001}{6}\left(\frac{8}{9}\right) + \ldots.. = 1.1068 \text{ approx.}$$ **Ans.**

Similarly, taking $x_0 = 1.1$, $y_0 = 1.1068$ we can get

$y(1.2) = 1.2277$ approx. **Ans.**

7. *Given* $\frac{dy}{dx} = -y$, $y(0) = 1$, *approximate the value of* y *at* $x = 0.01$, 0.02, 0.03 *and* 0.04 *using Euler's Method.*

SOLUTION: $y(0) = 1$, $\frac{dy}{dx} = -y = f(x, y)$, $x_0 = 0$, $y_0 = 1$,

$\therefore$ By Euler's method $y = y_0 + h\,f(x_0, y_0)$. Taking $h = 0.01$ successively, gives

$y(0.01) = 1 + 0.01\,(-1) = 0.99$

$y(0.02) = 0.99 + 0.01\,(-0.99) = 0.9801$

$y(0.03) = 0.9801 + 0.01\,(-0.9801) = 0.9703$

and $y(0.04) = 0.9703 + 0.01\,(-0.9703) = 0.9606$. **Ans.**

8. *Consider* $\frac{dy}{dx} = x(y + x) - 2$, $y(0) = 2$. *Use Euler's method with stepsize* $h = 0.3$ *to compute approximation to* $y(0.6)$ *correct to five decimal places.*

SOLUTION: $\frac{dy}{dx} = x(y + x) = 2$, $y(0) = 2$. Take $h = 0.3$

By Euler's method (modified)

$$y = y_0 + hf(x_0, y_0) = 2 + 0.3[0 - 2] = 1.40 = y_1 = y(0.3)$$

$$y_1^{(1)} = 2 + \frac{h}{2}[f(x_0, y_0) + f(x_0 + h, y_1)]$$

$$= 2 + \frac{0.3}{2}[-2 + 0.3\,(1.4 + 0.3) - 2] = 2 - 0.5235 = 1.4765$$

$$y_1^{(2)} = 2 + \frac{0.3}{2}[-2 + 0.3\,(1.4765 + 0.3) - 2] = 1.47994 = 1.48 \text{ approx} = y(0.3)$$

Next, taking $y_0 = 1.48, \quad x_0 = 0.3, \quad h = 0.3$

We get $y(0.6) = y_0 + hf(x_0, y_0) = 1.48 + 0.3\,[0.3(1.48 + 0.3) - 2] = 1.016 = y_2 = y(0.6).$

$$y_2^{(1)} = 1.48 + \frac{0.3}{2}[0.3\,(1.48 + 0.3) - 2 + 0.6\,(1.016 + 0.6) - 2]$$

$$= 1.48 - 0.3666 = 1.1134 \text{ approx.}$$

Similarly, $y_2^{(2)} = 1.48 + \dfrac{0.3}{2}[0.3\,(1.48 + 0.3) - 2 + 0.6\,(1.1134 + 0.6) - 2] = 1.10215$ approx. **Ans.**

9. *Solve the differential equation* $\dfrac{dy}{dx} = x + y$, $y(0) = 1$, *for* $y(0.1)$, $y(0.2)$ *and* $y(0.3)$ *using modified Euler's method.*

SOLUTION: Taking $h = 0.1$, $\dfrac{dy}{dx} = x + y = f(x, y)$, $y(0) = 1$,

We arrange the calculation in tabular form:

x	$f(x, y)$	Mean slope	Old y + 0.1 (mean slope) = y
0.0	1		$1 + 0.1\,(1) = 1.1$
0.1	1.11	$\frac{1}{2}(1 + 1.2)$	$1 + 0.1\,(1.1) = 1.11$
0.1	1.21	$\frac{1}{2}(1 + 1.21)$	$1 + 0.1\,(1.105) = 1.1105$
0.1	1.2105	$\frac{1}{2}(1 + 1.2105)$	$1 + 0.1\,(1.11052) = 1.1105$
0.1	1.2105		$1.1105 + 0.1(1.2105) = 1.2316$
0.2	1.4316	$\frac{1}{2}(1.1215 + 1.4316)$	$1.1105 + 0.1\,(1.3211) = 1.2426$
0.2	1.4426	$\frac{1}{2}(1.2105 + 1.4426)$	$1.1105 + 0.1\,(1.3266) = 1.2432$
0.2	1.4432	$\frac{1}{2}(1.2105 + 1.4432)$	$1.1105 + 0.1\,(1.3268) = 1.2432$
0.2	1.4432		$1.2432 + 0.1\,(1.4432) = 1.3875$
0.3	1.6875	$\frac{1}{2}(1.4432 + 1.6875)$	$1.2432 + 0.1\,(1.5654) = 1.3997$
0.3	1.7003	$\frac{1}{2}(1.4432 + 1.7003)$	$1.2432 + 0.1(1.5718) = 1.4004$
0.3	1.7004	$\frac{1}{2}(1.4432 + 1.7004)$	$1.2432 + 0.1(1.5718) = 1.4004$

10. *Solve* $\frac{dy}{dx} = y^2 - \frac{y}{x}$, $y(1) = 1$ *for* $x = 1\ (0.1)\ 1.5$ *by employing*

(*i*) Euler's method (*ii*) Modified Euler's method.

SOLUTION: $\frac{dy}{dx} = y^2 - \frac{y}{x}$, $y(1) = 1$, $h = 0.1$. To calculate $y(1.1)$, $y(1.2)$, $y(1.3)$, $y(1.4)$, $y(1.5)$.

We proceed as in Q.9 above.

(*a*) By Euler's method

$y(1.1) = 1.0091$, $y(1.2) = 1.0269$, $y(1.3) = 1.0534$, $y(1.4) = 1.0891$ and $y(1.5) = 1.1357$

(*b*) By Modified Euler's method

$y(1.1) = 1.0142$, $y(1.2) = 1.0376$, $y(1.3) = 1.0704$, $y(1.4) = 1.1141$ and $y(1.5) = 1.1747$ **Ans.**

11. *Employing Runge-Kutta method of fourth order, estimate the value of y for x = 0.2 given that* $\frac{dy}{dx} = x + y$, $y(0) = 1$.

SOLUTION: $\frac{dy}{dx} = f(x, y) = x + y$, $y(0) = 1$, $x_0 = 0$, $y_0 = 1$, $h = 0.2$

$$k_1 = h f(x_0, y_0) = 0.2\ (0 + 1) = 0.2$$

$$k_2 = hf\left(x_0 + \frac{h}{2}, y_0 + \frac{k_1}{2}\right) = 0.2\,[0 + 0.1 + 1 + 0.1] = 0.24$$

$$k_3 = hf\left(x_0 + \frac{h}{2}, y_0 + \frac{k_2}{2}\right) = 0.2\,[0 + 0.1 + 1 + 0.12] = 0.244$$

$$k_4 = hf(x_0 + n, y_0 + k_3) = 0.2[0 + 0.2 + 1 + 0.244] = 0.2888$$

$$\therefore \quad y = y_0 + \frac{1}{6}(k_1 + 2k_2 + 2k_3 + k_4) = 1 + \frac{1}{6}(0.2 + 0.48 + 0.488 + 0.2888)$$

$$= 1.2428 \quad \therefore \ y(0.2) = 1.2428.$$ **Ans.**

12. *Compute y(0.3) given that* $\frac{dy}{dx} + y + xy^2 = 0$, $y(0) = 1$, *taking h = 0.1 and using Runge-Kutta method.*

SOLUTION: $\frac{dy}{dx} = -y(1 + xy)$, $y(0) = 1$, $h = 0.1$ we are two find $y(0.1)$, $y(0.2)$, $y(0.3)$

Adopting the procedure of Runge-Kutta we can get

$y(.1) = 0.9006$, $y(.2) = 0.8046$, $y(.3) = 0.7144$. **Ans.**

13. *Apply Runge-Kutta method of fourth order to solve*

$$\frac{dy}{dx} = \frac{y^2 - x^2}{y^2 + x^2}, \quad y(0) = 1 \quad \text{for} \quad x = 0.2,\ 0.4.$$

SOLUTION: Here $f(x, y) = \frac{y^2 - x^2}{y^2 + x^2}$ To find $y(0.2)$ and $y(0.4)$, $h = 0.2$, $x_0 = 0$, $y_0 = 1$

$\therefore \quad k_1 = hf(x_0, y_0) = 0.2\, f(0, 1) = 0.2$

$$k_2 = hf\left(x_0+\frac{h}{2}, y_0+\frac{k_1}{2}\right) = 0.2\, f(0.1, 1.1) = 0.19672$$

$$k_3 = hf\left(x_0+\frac{h}{2}, y_0+\frac{k_2}{2}\right) = 0.2\, f(0.1, 1.09836) = 0.1967$$

$$k_4 = hf(x_0+h, y_0+k_3) = 0.2\, f(0.2, 1.967) = 0.1891.$$

$$\therefore \quad y(0.2) = 1+\frac{1}{6}(k_1+2k_2+2k_3+k_4) = \frac{1}{6}[0.2+2(0.19672)+2(0.1967)+0.1891]$$

$$= 1.19599 = 1.196.$$

To find $y(0.4)$, take $x_0 = 0.2$, $y_0 = 1.196$, $h = 0.2$. Then

$$k_1 = hf(x_0, y_0) = 0.1891$$

$$k_2 = hf\left(x_0+\frac{h}{2}, y_0+\frac{k_1}{2}\right) = 0.1795$$

$$k_3 = hf\left(x_0+\frac{h}{2}, y_0+\frac{k_2}{2}\right) = 0.1793$$

$$k_4 = hf(x_0+h, y_0+k_3) = 0.1688$$

$$\therefore \quad y(0.4) = 1.196+\frac{1}{6}[1.189+2(0.1795)+2(0.1793)+0.1688] = 1.3752$$ **Ans.**

14. *Obtain the approximate values of* $y(0.1)$, *and* $z(0.1)$ *from the system of simultaneous equations* $\frac{dy}{dx} = x+z$, $\frac{dz}{dx} = x-y^2$ given that $y(0) = 2$, $z(0) = 1$ *using Runge-Kutta method of fourth order.*

SOLUTION: $y(0) = 2$, $z(0) = 1$, $h = 0.1$. To find $y(0.1)$ and $z(0.1)$. $\frac{dz}{dx} = x - y^2$.

$$\frac{dy}{dx} = f_1(x, y, z) = x+z,$$

$k_1 = 0.1\,(0+1) = 0.1$	$L_1 = 0.1\,(0-2^2) = -0.4$
$k_2 = 0.1\,(0.05+1-0.2) = 0.085$	$L_2 = 0.1\,[0.05+2.\,085^2] = -4.1525$
$k_3 = 0.1\,(0.05+1-0.207625) = 0.084237$	$L_3 = 0.1\,[0.05-(2+0.0425)^2] = 0.41218$
$k_4 = 0.1(0.1+1-0.41218) = 0.068702$	$L_4 = 0.1\,[0.1-(2.0687)^2] = -0.41632177$
$\therefore\ k = \frac{1}{6}[0.1+0.17+.168474+0.068702$	$L = \frac{1}{6}[-0.4-0.8305-0.82436-4.16332]$
$= 0.08453$	$= -0.41186$
	$\therefore\ z = 1-0.41186 = 0.58813$ **Ans.**

Thus, $y(0.1) = y_0 + k = 2.08453.$

15. *Employing Runge-Kutta method, solve* $\dfrac{d^2y}{dx^2} = x\left(\dfrac{dy}{dx}\right)^2 - y^2$ *for* $x = 0.2$ *correct to four decimal places, given that* $y = 1$ *and* $\dfrac{dy}{dx} = 0$ *when* $x = 0$.

SOLUTION: $\dfrac{dy}{dx} = z = f_1(x, y, z)$ and $\dfrac{dz}{dx} = xz^2 - y^2 = f_2(x, y, z)$

Here $x_0 = 0$, $y_0 = 1$, $z_0 = 0$, $h = 0.2$. Next as per R.K. Scheme

$k_1 = hf_1(x_0, y_0, z_0) = 0.2(0) = 0$ $\qquad L_1 = hf_2(x_0, y_0, z_0) = 0.2(-1) = -0.2$

$k_2 = hf_1\left(x_0 + \dfrac{h}{2}, y_0 + \dfrac{k_1}{2}, z_0 + \dfrac{L_1}{2}\right) = 0.2(-0.1) = -0.02$

$L_2 = hf_2\left(x_0 + \dfrac{h}{2}, y_0 + \dfrac{K_1}{2}, z_0 + \dfrac{L_1}{2}\right) = 0.2(-0.999) = -0.1998$

$k_3 = hf_1\left(x_0 + \dfrac{h}{2}, y_0 + \dfrac{k_2}{2}, z_0 + \dfrac{L_2}{2}\right) = 0.2(-0.0977) = -0.02$

$L_3 = hf_2\left(x_0 + \dfrac{h}{2}, y_0 + \dfrac{k_2}{2}, z_0 + \dfrac{L_2}{2}\right) = 0.2(-0.9791) = -0.1958.$

$k_4 = hf_1(x_0 + h, y_0 + k_2, z_0 + L_3) = -0.03916$

$L_4 = hf_2(x_0 + h, y_0 + k_3, z_0 + L_3) = -0.2(0.9527) = -0.1905.$

$\therefore\ k = \dfrac{1}{6}[k_1 + 2k_2 + 2k_3 + k_4] = -0.0199$ $\qquad L = \dfrac{1}{6}(L_1 + 2L_2 + 2L_3 + L_4) = -0.1970$

$\therefore\ y(0.2) = y_0 + k = 1 - 0.0199 = 0.9801$ $\qquad \therefore\ y' = z(0.2) = z_0 + L = -0.1920$ **Ans.**

16. *Evaluate* $y(1.1)$, $z(1.1)$, *given that* $\dfrac{dy}{dx} = xyz$, $\dfrac{dz}{dx} = \dfrac{xy}{z}$, $y(1) = \dfrac{1}{3}$, $z(1) = 1$

SOLUTION: The simultaneous equations are $\dfrac{dy}{dx} = xyz$, $\dfrac{dz}{dx} = \dfrac{xy}{z}$

Given that $y(1) = \dfrac{1}{3}$, $z(1) = 1$, $h = 0.1$, $x_0 = 1$, $y_0 = \dfrac{1}{3}$, $z_0 = 1$.

$\dfrac{dy}{dx} = f(x, y, z) = xyz$ $\qquad \dfrac{dz}{dx} = g(x, y, z) = \dfrac{x_0 y_0}{z}$

$k_1 = hf(x_0, y_0, z_0) = hx_0 y_0 z_0 = 0.1(1)\left(\dfrac{1}{3}\right)(1) = \dfrac{0.1}{3} = 0.033$

$L_1 = hg(x_0, y_0, z_0) = h\dfrac{xy}{z_0} = \dfrac{0.1}{1}(1)\left(\dfrac{1}{3}\right) = \dfrac{0.1}{3} = 0.0333$

$k_2 = hf\left(x_0 + \dfrac{h}{2}, y_0 + \dfrac{k_1}{2}, z_0 + \dfrac{L_1}{2}\right) = h\left(x_0 + \dfrac{h}{2}\right)\left(y_0 + \dfrac{k_1}{2}\right)\left(z_0 + \dfrac{L_1}{2}\right)$

$= 0.1(1 + 0.05)\left(\dfrac{1}{3} + \dfrac{0.1}{6}\right)\left(1 + \dfrac{0.1}{6}\right) = 0.037$

$L_2 = hg\left(x_0 + \dfrac{h}{2}, y_0 + \dfrac{k_1}{2} z_0 + \dfrac{L_1}{2}\right) = h\left(x_0 + \dfrac{h}{2}\right)\left(y_0 + \dfrac{k_1}{2}\right) \Big/ \left(z_0 + \dfrac{L_1}{2}\right)$

$= 0.1(1 + 0.5)\left(\dfrac{1}{3} + \dfrac{0.1}{6}\right) \Big/ \left(1 + \dfrac{0.1}{6}\right) = 0.036$

$$k_3 = hf\left(x_0 + \frac{h}{2},\ y_0 + \frac{k_2}{2},\ z_0 + \frac{L_2}{2}\right) \qquad L_3 = hg\left(x_0 + \frac{h}{2},\ y_0 + \frac{k_2}{2},\ z_0 + \frac{L_2}{2}\right)$$

$$= h\left(x_0 + \frac{h}{2}\right)\left(y_0 + \frac{k_2}{2}\right)\left(z_0 + \frac{L_2}{2}\right) \qquad = h\left(x_0 + \frac{h}{2}\right)\left(y_0 + \frac{k_2}{2}\right)\Big/\left(z_0 + \frac{L_2}{2}\right)$$

$$= 0.1\,(1 + 0.05)\left(\frac{1}{3} + \frac{0.037}{2}\right)\left(1 + \frac{0.036}{2}\right) \qquad = 0.1\,(1 + 0.05)\left(\frac{1}{3} + \frac{0.037}{2}\right)\Big/\left(1 + \frac{0.37}{2}\right)$$

$$= 0.0376 \qquad = 0.03637$$

$$k_4 = hf(x_0 + h,\ y_0 + k_3,\ z_0 + L_3) \qquad L_4 = hf(x_0 + h,\ y_0 + k_3,\ z_0 + L_3)$$

$$= h\,(x_0 + h)\,(y_0 + k_3)\,(z_0 + L_3) \qquad = h\,(x_0 + h)\,(y_0 + k_3)/(z_0 + L_3).$$

$$= 0.1\,(1 + 0.1)\left(\frac{1}{3} + 0.0376\right)(1 + 0.03637) \qquad = 0.1(1 + 0.1)\left(\frac{1}{3} + 0.0376\right)\Big/1 + 0.03637$$

$$\therefore\ y(1.1) = \frac{1}{3} + \frac{1}{6}[0.0333 + 0.037 \times 2 + 0.0376 \times 3 \qquad z(1.1) = 1 + \frac{1}{6}[0.0333 + 0.72 + 0.0/27 + 0.0393]$$

$$+ 0.04228] = 0.37076 \textbf{ Ans.} \qquad = 1.0862 \quad \textbf{Ans.}$$

17. *Solve the second order differential equation* $\dfrac{d^2y}{dx^2} = y^3$ for $y(0.1)$ *given that* $y(0) = 10,\ y'(0) = 5,$ *by using the Runge-Kutta method.*

SOLUTION: The given equation can be written as simultaneous equations $\dfrac{dz}{dx} = y^3$ where $z = \dfrac{dy}{dx}$.

Here $y(0) = 10,\ z(0) = 5,$ to find $y(0.1),\ h = 0.1$. Applying R.K. method of fourth order, we get

$$k_1 = hf_1(x_0, y_0, z_0) = hz_0 \qquad L_1 = hf_2(x_0, y_0, z_0) = hy^3$$

$$= 0.1(5) = 0.5 \qquad = 0.1(10)^3 = 100.$$

$$k_2 = hf_1\left(x_0 + \frac{h}{2},\ y_0 + \frac{k_1}{2},\ z_0 + \frac{L_1}{2}\right) \qquad L_2 = hf_2\left(x_0 + \frac{h}{2},\ y_0 + \frac{k_1}{2},\ z_0 + \frac{L_1}{2}\right) = h\left(y_0 + \frac{k_1}{2}\right)^3$$

$$= h\left(z_0 + \frac{L_1}{2}\right) = 0.1\,(5 + 50) = 5.5 \qquad = 0.1(10 + 0.25)^3 = 107.69$$

$$k_3 = hf_1\left(x_0 + \frac{h}{2},\ y_0 + \frac{k_2}{2},\ z_0 + \frac{L_2}{2}\right) \qquad L_3 = hf_2\left(x_0 + \frac{h}{2}, y_0 + \frac{k_2}{2}, z_0 + \frac{L_2}{2}\right)$$

$$= h\left(z_0 + \frac{L_2}{2}\right) = 0.1\,(5 + 53.84) \qquad = h\left(y_0 + \frac{k_2}{2}\right)^3 = 0.1\,(10 + 2.75)^3$$

$$= 5.884 \qquad = 207.27$$

$$k_4 = hf_1(x_0 + h), y_0 + k_3, z_0 + L_3) \qquad L_4 = hf_2(x_0 + h,\ y_0 + k_3,\ z_0 + L_3)$$

$$= 0.1\,(z_0 + L_3) = 0.1\,(5 + 207.27) \qquad = h\,(y_0 + k_3)^3 = 0.1\,(10 + 5.884)^3$$

$$= 21.227 \qquad = 400.755$$

Finally, $y(0.1) = y_0 + \dfrac{1}{6}(k_1 + 2k_2 + 2k_3 + k_4)$

$$= 10 + \frac{1}{6}[0.5 + 11 + 11.768 + 21.227] = 10 + \frac{44.495}{6} = 17.416 \quad \textbf{Ans.}$$

18. *Using Mine's predictor-corrector scheme find* $y(2.0)$ *as the solution to* $\frac{dy}{dx} = \frac{1}{2}(x + y)$, given that $y(0) = 2$, $y(0.5) = 2.636$, $y(1.0) = 3.595$, $y(1.5) = 4.968$.

SOLUTION: $\frac{dy}{dx} = f(x, y) = \frac{1}{2}(x + y)$, $x_0 = 0$, $y_0 = 2$, $h = 0.5$

$$f_0 = f(x_0, y_0) = \frac{1}{2}(0 + 2) = 1, \quad f_1 = f(x_1, y_1) = \frac{1}{2}(x_1 + y_1) = \frac{1}{2}(0.5 + 2.636) = 1.568$$

$$f_2 = \frac{1}{2}(x_2 + y_2) = \frac{1}{2}(1.0 + 3.595) = 2.2975 \quad f_3 = \frac{1}{2}(x_3 + y_3) = \frac{1}{2}(1.5 + 4.968) = 3.234$$

$$\therefore \quad y_4 = y(2.0) = y_0 + \frac{4h}{3}(2f_1 - f_2 + 2f_3) = 2 + \frac{4(0.5)}{3}[2(1.568) - 2.2975 + 2(3.234)]$$

$$(y_4)_p = 2 + \frac{2}{3}[3.136 - 2.2975 + 6.468] = 7.306 \quad \text{(predicted value)}$$

Now for corrected value $f_4 = \frac{1}{2}(x_4 + y_4) = \frac{1}{2}(2 + 7.306) = 4.653$

$$\therefore \quad y_4^{(1)} = y_2 + \frac{h}{3}(f_2 + 4f_3 + f_4) = 3.595 + \frac{1}{6}[2.2975 + 4(3.234) + 4.653] = 6.91.$$

Then further corrected value of $f_4 = \frac{1}{2}(2 + 6.91) = 4.455$

Therefore $y_4^{(2)} = 3.595 + \frac{1}{6}[2.2975 + 4(3.234) + 4.455] = 6.876$ **Ans.**

19. *Solve the differential equation* $\frac{dy}{dx} = \sin x - y$ with $y(0) = 1$ *in the interval* $0 \le x \le 1$ *using Euler's method and modified Euler's method. Also compare the result with exact results.*

SOLUTION: Let $h = 0.2$ so $n = 5$, $f(x, y) = \sin x - y$. Tabular results are:

n	x_n	y_n	$0.2 f(x_n, y_n)$	$y_{n+1} = y_0 + 02. f(x_n, y_n)$
0	0	1	− 0.2	0.8
1	0.2	0.8	− 0.1203	0.6797
2	0.4	0.6797	− 0.0581	0.6217
3	0.6	0.6217	− 0.0114	0.6103
4	0.8	0.6103	+ 0.0214	0.6317
5	1.0	0.6317	+ 0.0420	0.6236

Ans.

SOLUTIONS TO EXERCISE 29C

1. *Solve the boundary value problem*

$\frac{d^2y}{dx^2} = y$ with the boundary conditions $y(0) = 0$, $y(2) = 3.627$, taking $n = 4$.

SOLUTION: In terms of finite differences we know that

$$\frac{d^2y}{dx^2} = \frac{y(x+h) - 2y(x) + y(x-h)}{h^2} + 0(h^2)$$

or $y'' = \frac{y_{i+1} - 2y_i + y_{i-1}}{h^2}$. Here $y_0 = 0$, $y_4 = 3.627$, $n = 4$ $\therefore$ $h = \frac{1}{2}$

Given equation is $\frac{d^2y}{dx^2} = y$ hence $h^2 y_i = y_{i+1} - 2y_i + y_{i-1}$,

or $y_i = 4y_{i+1} - 8y_i + 4y_{i-1}$ or $4y_{i+1} + 4y_{i-1} - 9y_i = 0$

Taking $i = 1$, $4y_2 + 4y_0 - 9y_1 = 0$ or $y_2 = \frac{9}{4}y_1$ (as $y_0 = 0$). ...(1)

Taking $i = 2$, $4y_3 + 4y_1 - 9y_2 = 0$ or $4y_3 = 9y_2 - 4y_1$ hence $y_3 = \frac{65}{16}y_1$ ...(2)

Taking $i = 3$, $4y_4 + 4y_2 - 9y_3 = 0$, but $y_4 = 3.627$ (given)

$\therefore$ $4(3.627) + 4y_2 - 9\frac{(65)}{16}y_1 = 0.$ Using (1) and (2) we get

$$y_1 = 0.5263, \quad y_2 = \frac{9}{4}y_1 = 1.1843 \text{ and } y_3 = \frac{65}{4}y_1 = 2.1381.$$

Thus, $y(0.5) = 0.5263$, $y(1.0) = 1.1843$ and $y(1.5) = 2.1381$ **Ans.**

2. *Solve the boundary value problem* $\frac{d^2y}{dx^2} - 64y + 10 = 0$

with the boundary conditions $y(0) = y(1) = 0$.

Compute $y(0.5)$ *and compare it with the exact value.*

SOLUTION: Since $\frac{d^2y}{dx^2} = \frac{1}{h^2}\left[y_{i+1} - 2y_i + y_{i-1}\right]$ we have $\frac{y_{i+1} - 2y_i + y_{i-1}}{h^2} - 64y_i + 10 = 0$,

$y(0) = 0$, $y(1) = 0$.

Let us take first $n = 2$ then $h = \frac{1}{2}$ $\therefore$ $y_0 = y_2 = 0$. Thus, we have

$4y_{i+1} - 8y_i + 4y_{i-1} - 64y_i + 10 = 0$ or $y_{i+1} - 18y_i + y_{i-1} + 2.5 = 0.$

For $i = 1$ we get $y_2 - 18y_1 + y_0 + 2.5 = 0$. But $y_2 = y_0 = 0$, therefore

we have $y_1 = y(0.5) = \dfrac{2.5}{18} = 0.1389$ when $n = 2$. **Ans.**

Next, let us take $n = 4$ so that $h = \dfrac{1}{4}$ then we have

$$16y_{i+1} - 32y_i + 16y_{i-1} - 64y_i + 10 = 0 \quad \text{or} \quad y_{i+1} - 6y_i + y_{i-1} + \frac{10}{16} = 0$$

For $i = 1$ we have $y_2 - 6y_1 + y_0 + \dfrac{10}{16} = 0$ but $y_0 = 0$ $\therefore$ $y_2 = 6y_1 - \dfrac{10}{16}$ ...(1)

for $i = 2$, $y_3 - 6y_2 + y_1 + \dfrac{10}{16} = 0$ or $y_3 = 6y_2 - y_1 - \dfrac{10}{16}$ ...(2)

and for $i = 3$, $y_4 - 6y_3 + y_2 + \dfrac{10}{16} = 0$ but $y_4 = 0$ $\therefore$ $6y_3 = y_2 + \dfrac{10}{16}$ ...(3)

From equations (1), (2) and (3) we get

$y_1 = y(0.25) = 0.1287$, $y_2 = y(0.5) = 0.1470$ and $y_3 = y(0.75) = 0.1285$. **Ans.**

3. *Solve the boundary value problem* $\dfrac{d^2y}{dx^2} = x + y$,
given the boundary conditions $y(0) = y(1) = 0$, *taking* $n = 4$.

SOLUTION: $\dfrac{y_{i+1} - 2y_i + y_{i-1}}{h^2} = x_i + y_i = y_i + \dfrac{i}{4}$ as $n = 4$, $h = \dfrac{1}{4}$, $x_i = \dfrac{i}{4}$. $y_0 = y_4 = 0$

or $$16y_{i+1} - 33y_i + 16y_{i-1} = \frac{i}{4} \quad \text{or} \quad y_{i+1} - \frac{33}{16}y_i + y_{i-1} = \frac{i}{64}$$

For $i = 1$, $y_2 - 2.0625y_1 + y_0 = \dfrac{1}{64}$ or $y_2 = 2.0625\,y_1 + \dfrac{1}{64}$ (as $y_0 = 0$) ...(1)

For $i = 2$, $y_3 - 2.0625\,y_2 + y_1 = \dfrac{2}{64}$ or $y_3 = 2.0625\,y_2 - y_1 + \dfrac{2}{64}$

Using (1) here $y_3 = (0.0625)^2\,y_1 + \dfrac{2.0625}{64} + \dfrac{2}{64} - y_1 = [2.0625^2 - 1]y_1 + \dfrac{4.0625}{64}$ ...(2)

and for $i = 3$, $y_4 - 2.0625y_3 + y_2 = \dfrac{3}{64}$ but $y_4 = 0$

$$\therefore \quad y_3 = \frac{1}{2.0625}y_2 - \frac{3}{64 \times 2.0625} = \frac{1}{2.0625}\left[2.0625\,y_1 + \frac{1}{64}\right] - \frac{3}{64 \times 2.0625}$$

or $$= y_1 - \frac{2}{64 \times 2.0625}$$ ...(3)

From (2) and (3) we get $[(2.0625)^2 - 2]y_1 = \frac{4.0625}{64} - \frac{2}{64 \times 2.0625}$

or $2.2539\, y_1 = -0.0635 - 0.0151$ or $y_1 = -0.03489$

$\therefore$ $y_2 = -2.0625\,(0.03489) + 0.015625 = -0.05634$

and $y_3 = -0.03489 - \frac{1}{32 \times 2.0625} = -0.05003$

Therefore, $y(0.5) = -0.03489,\ y(1) = -0.05634,\ y(1.5) = -0.05003$ **Ans.**

4. *Solve the boundary value problem*

$$x\,\frac{d^2y}{dx^2} + y = 0 \quad \textit{with} \quad y(1) = 1 \ \textit{and} \ y(2) = 2, \ \textit{taking} \ n = 4.$$

SOLUTION: Using forward finite differences, the given equation becomes

$$x_i\,\frac{(y_{i+1} - 2y_i + y_{i-1})}{h^2} + y_i = 0. \quad \text{Here } x_i = 1 + \frac{i}{4}, \quad h = \frac{1}{4}.$$

$$\therefore \quad 16\left(1 + \frac{i}{4}\right)(y_{i+1} - 2y_i + y_{i-1}) + y_i = 0$$

$$\text{or} \quad y_{i+1} + y_{i-1} = y_i\left[2 - \frac{1}{16 + 4i}\right] = \frac{31 + 8i}{16 + 4i}\,y_i, \quad i = 1, 2, 3.$$

For $i = 1$, $y_2 + 1 = \frac{39}{20}\,y_i$ or $y_2 = 1.95y_1 - 1$, as $y_0 = 1$ ...(1)

For $i = 2$, $y_3 + y_1 = \frac{47}{24}\,y_2$ or $y_3 = 1.9583y_2 - y_1$ ...(2)

and for $i = 3$, $2 + y_2 = \frac{55}{28}\,y_3$ or $1.9643\,y_3 = y_2 + 2$ as $y_4 = 2.$ ...(3)

Eliminating y_3 in (2) and (3) we get $2.846y_2 = 1.9643y_1 + 2$

Solving (1) and (4) we get $y_1 = 1.3513$ and $y_2 = 1.635$ hence $y_3 = 1.8505$

Thus, $y(1.25) = 1.3513,\ y(1.5) = 1.635$ and $y(1.75) = 1.8505$ **Ans.**

5. *Solve the boundary value problem*

$$\frac{d^2y}{dx^2} + x\,\frac{dy}{dx} + y = 3x^2 + 2, \ \textit{given the boundary conditions} \quad y(0) = y(1) = 1.$$

SOLUTION: Using finite difference formulae, given equation can be written as

$$\left(\frac{y_{i+1} - 2y_i + y_{i-1}}{h^2}\right) + x_i\,\frac{y_{i+1} - y_{i-1}}{2h} + y_i = 3x_i^2 + 2, \ \ n = 4, \ \ h = \frac{1}{4}, \ \ y(0) = y(1) = 1 \ \therefore \ x_i = \frac{i}{4}.$$

or $$16(y_{i+1}-2y_i+y_{i-1})+\frac{i}{4}\frac{(y_{i+1}-y_{i-1})}{2}+y_i=\frac{3i^2}{16}+2,\ i=1,2,3.$$

or $$\left(16+\frac{i}{2}\right)y_{i+1}+\left(16-\frac{i}{2}\right)y_{i-1}-31y_i=\frac{3i^2}{16}+2$$

For $i = 1$ we have $\frac{33}{2}y_2+\frac{31}{2}y_0-31y_1=\frac{35}{16}$. But $y_0 = 1$

$$\therefore \quad 33y_2-62y_1=2\left(\frac{35}{16}-\frac{31}{2}\right)=-26.6250 \Rightarrow y_2=1.8788y_1-0.8068 \quad ...(1)$$

For $i = 2$ we have $17y_3 + 15y_1 - 31y_2 = 2.75$

or $$y_3 = 1.8235\,y_2 - 0.8823\,y_1 + 0.16176 \quad ...(2)$$

For $i = 3$ we have $\frac{35}{2}y_4+\frac{29}{2}y_2-31y_3=\frac{27}{16}+2=3.6875$ but $y_4 = 1$

or $$31y_3 = 14.5\,y_2 + 13.8125 \quad \text{or} \quad y_3 = 0.4677\,y_2 + 0.4455 \quad ...(3)$$

From (2) and (3) we get $1.3558\,y_2 - 0.8823\,y_1 = 0.28374$ or $y_2 = 0.65076\,y_1 + 0.20928$ ...(4)

From (1) and (2) we get $1.2280\,y_1 = 1.01608$ or $y_1 = 0.8274$

and $y_2 = 0.747$ and $y_3 = 0.7952$

Thus, we have $y(0.25) = 0.8274$, $y(0.50) = 0.7477$, $y(0.75) = 0.7952$. **Ans.**

6. *The boundary value problem governing the deflection of a beam of length* 3 *metres is given by* $\frac{d^4y}{dx^4}+2y=\frac{1}{9}x^2+\frac{2}{3}x+4$ with $y(0) = y'(0) = y(3) = y'(3) = 0$.

The beam is fixed at $x = 0$ *and simply supported at* $x = 3$. *Determine the deflection* y at $x = 1$ and $x = 2$.

SOLUTION: In terms of finite differences the given equation can be written as

$$\frac{1}{h^4}(y_{i+2}-4y_{i+1}+6y_i-4y_{i-1}+y_{i-2})+2y_i=\frac{1}{9}x_i^2+\frac{2}{3}x_i+4$$

Here beam is from $x = 0$ to $x = 3$, and $n = 3$, $h = 1$, $x_i = i$, $y(0) = y_0 = 0$, $y(3) = y_3 = 0$
$i = 0, 1, 2, 3$, $x_0 = 0$, $x_3 = 3$

$$\therefore \quad y_{i+2}-4y_{i+1}+8y_i-4y_{i-1}+y_{i-2}=\frac{i^2}{9}+\frac{2i}{3}+4$$

For $i = 1$, we have $y_3 - 4y_2 + 8y_1 - 4y_0 + y_{-1} = \frac{1}{9}+\frac{2}{3}+4=\frac{43}{9}$

and for $i = 2$, we have $y_4 - 4y_3 + 8y_2 - 4y_1 + y_0 = \frac{4}{9}+\frac{4}{3}+4=\frac{52}{9}$

Since $y_0 = y_3 = 0$ above equations become $-4y_2 + 8y_1 + y_{-1} = \frac{43}{9}$ ...(1)

and $$y_4 + 8y_2 - 4y_1 = \frac{52}{9}. \qquad ...(2)$$

Further $$\left(\frac{dy}{dx}\right)_{x = x_i} = \frac{y_{i+1} - y_{i-1}}{2h} \qquad ...(3)$$

Since $y'(0) = 0$ we put $i = 0$ in (3), to get $y_1 - y_{-1} = 0 \quad \therefore \quad y_{-1} = y_1$

and since $y'(3) = 0$ we put $i = 3$ in (3), to get $y_4 - y_2 = 0 \quad \therefore \quad y_4 = y_2$

In turn, equations (1) and (2) become $-4y_2 + 8y_1 + y_1 = \frac{43}{9}$ and $y_2 + 8y_2 - 4y_1 = \frac{52}{9}$

Which on solving for y_1 and y_2, gives $y_1 = 1.0171$ and $y_2 = 1.094$

that is $y(1) = 1.0171$ and $y(2) = 1.094$. **Ans.**

ADDITIONAL IMPORTANT PROBLEMS

1. *Solve* $\frac{dy}{dx} = 1 - y,\ y(0) = 0$ *using Euler's method and find* $y(0.1),\ y(0.2)$ *and* $y(0.3)$.

SOLUTION: $\frac{dy}{dx} = f(x, y) = 1 - y,\ y(0) = 0,$ that is, $x_0 = 0,\ y_0 = 0$ and $h = 0.1\ \therefore x_1 = 0.1,$ $x_2 = 0.2,\ x_3 = 0.3$.

The Euler's algorithm is $y_{n+1} = y_n + hf(x_n, y_n),\ n = 0, 1, 2, \ldots$

$$\therefore \quad y_1 = y_0 + hf(x_0, y_0) = 0 + 0.1(1 - 0) = 0.1,$$

$$y_2 = y_1 + hf(x_1, y_1) = 0.1 + 0.1(1 - 0.1) = 0.19$$

and $y_3 = y_2 + hf(x_2, y_2) = 0.19 + 0.1\,(1 - 0.19) = 0.271.$

Thus, $y(0.1) = 0.1,\ y(0.2) = 0.19$ and $y(0.3) = 0.279.$ **Ans.**

2. *Solve the equation* $\frac{dy}{dx} = x^2 + y$ *where* $y(0) = 1$ *using Modified Euler's method for* $x = 0.02$ *in two steps.*

SOLUTION: Here $f(x, y) = x^2 + y,\ x_0 = 0,\ y_0 = 1,\ h = 0.01,\ x_1 = 0.01,\ x_2 = 0.02.$

We have $y_1^{(0)} = y_0 + h\,f(x_0, y_0) = 1 + 0.1(x_0^2 + y_0) = 1 + 0.01(0^2 + 1) = 1.01.$

then $y_1^{(1)} = y_0 + \frac{h}{2}\left[f(x_0, y_0) + f(x_1, y_1^{(0)})\right] = 1 + \left(\frac{0.01}{2}\right)\left[0 + 1 + (0.01)^2 + (1.01)\right] = 1.01005$

$$y_1^{(2)} = y_0 + \frac{h}{2}\left[f(x_0, y_0) + f(x_1, y_1^{(1)})\right] = 1 + \left(\frac{0.01}{2}\right)[0 + 1 + (0.01)^2 + 1.01005] = 1.01005$$

Thus, finally $y(0.1) = 1.01005.$

Next, $y_2^{(0)} = y_1 + hf\,[(x_1, y_1) = 1.01005 + 0.1\,[(0.01)^2 + 1.01005] = 1.02015$

then $y_2^{(1)} = y_1 + \frac{h}{2}[f(x_1, y_1) + f(x_2, y_2^{(0)})]$

$$= 1.01005 + \frac{(0.01)}{2}[(0.01)^2 + 1.01005 + (0.02)^2 + 1.02015] = 1.020204$$

and $y_2^{(2)} = y_1 + \frac{h}{2}[f(x_1, y_1) + f(x_2, y_2^{(1)})]$

$$= 1.01005 + \frac{(0.01)}{2}[(0.01)^2 + 1.01005 + (0.02)^2 + 1.020204] = 1.020204$$

Thus, finally $y(0.02) = 1.020204.$ **Ans.**

3. *Solve the differential equation* $\frac{dy}{dx} = \frac{y-x}{y+x}$, $y(0) = 1$ *for* $x = 1$ *correct to four decimal places using Euler's method.*

SOLUTION: $\frac{dy}{dx} = f(x, y) = \frac{y-x}{y+x}$ given that $y = 1$ at $x = 0$ then $x_0 = 0$, $y_0 = 1$

Let us take $h = 0.02$ $x_1 = 0.02$, $x_2 = 0.04$, $x_3 = 0.06$, $x_4 = 0.08$, $x_5 = 0.$

Using Euler's method, we get $y_1 = y(0.02) = y_0 + h\frac{y_0 - x_0}{y_0 + x_0} = 1 + (0.02)\left(\frac{1-0}{1+0}\right) = 1.02$

Next, $y_2 = y(0.04) = y_1 + hf(x_1, y_1) = 1.02 + 0.02\left(\frac{y_1 - x_1}{y_1 + x_1}\right) = 1.02 + 0.02\left(\frac{1.02 - 0.02}{1.02 + 0.02}\right) = 1.0392$

$$y_3 = y(0.06) = y_2 + hf(x_2, y_2) = 1.0392 + 0.02\left(\frac{1.0392 - 0.04}{1.0392 + 0.04}\right) = 1.0577.$$

$$y_4 = y(0.08) = y_3 + hf(x_3, y_3) = 1.0577 + 0.02\left(\frac{1.0577 - 0.06}{1.0577 + 0.06}\right) = 1.0756$$

and $y_5 = y(1.0) = y_4 + hf(x_4, y_4) = 1.0756 + 0.02\left(\frac{1.0756 - 0.08}{1.0756 + 0.08}\right).$

Thus, $y(1.0) = 1.0928.$ **Ans.**

4. *Solve by Numerical method the equation* $\frac{dy}{dx} = 0.2x + 0.1y$ *for* $x = 0.2$ *given that* $y(0) = 2$, $y(0.05) = 2{:}0103$, $y(0.10) = 2.0211$, $y(0.15) = 2.0323$ *using Milne's method.*

SOLUTION: Here $h = 0.05$, $f(x, y) = 0.2x + 0.1y$, $x_0 = 0$, $x_1 = 0.05$, $x_2 = 0.10$, $x_3 = 0.15$, $x_4 = 0.20.$

$y_0 = 2$, $y_1 = 2.0103$, $y_2 = 2.0211$, $y_3 = 2.0323$, and $y_4(0.2)$ is to be obtained.

$$\therefore \quad f_0 = f(x_0, y_0) = 0.2(0) + 0.1(2.0) = 0.2$$
$$f_1 = f(x_1, y_1) = 0.2(0.05) + 0.1(2.0103) = 0.21103$$
$$f_2 = f(x_2, y_2) = 0.2(0.10) + 0.1(2.0211) = 0.22211$$
$$f_3 = f(x_3, y_3) = 0.2(0.15) + 0.1(2.0323) = 0.23323$$

Using the predictor formula $(y_4)_p = y_0 + \frac{4h}{3}(2f_1 - f_2 + 2f_3)$, we have

$$(y_4)_p = 2 + \frac{4(0.05)}{3}[2(0.21103) - 0.22211 + 2(0.3323)]$$

$$= 2 + \frac{0.2}{3}[0.42206 - 0.22211 + 0.46646] = 2.44427$$

$$\therefore \quad f_4 = f(x_4, (y_4)_p) = 0.2\,x_4 + 0.1\,(y_4)_p = 0.2\,(0.20) + 0.1\,(2.044427) = 0.2444427$$

Now applying the corrector formula $(y_4)_c = y_2 + \frac{4}{3}[f_2 + 4f_3 + f_4]$ we get

$$(y_4)_c = 20211 + \frac{0.05}{3}[0.22211 + 4(0.23323) + 0.2444427]$$

$$= 2.0211 + 0.0233245 = 2.0444245.$$

Therefore $y(0.2) = 2.0444245.$ **Ans.**

5. *Solve* $\frac{dy}{dx} = x + y$, for $x = 0.02$ and 0.25 *using Milne's predictor-corector method, initial conditions being* $x_0 = 0,\ y_0 = 1;\ x_1 = 0.05,\ y_1 = 1.1026;\ x_2 = 0.10,\ y_2 = 1.2104;\ x_3 = 0.15,\ y_3 = 1.3237.$

SOLUTION: Here $h = 0.05$, $f(x, y) = x + y \therefore f_0 = f(x_0, y_0) = x_0 + y_0 = 0 + 1 = 1$

$f_1(x, y) = x_1 + y_1 = 0.05 + 1.1026 = 1.1526$

$f_2(x, y) = x_2 + y_2 = 0.10 + 1.2104 = 1.3104$

$f_3(x, y) = x_3 + y_3 = 0.15 + 1.3237 = 1.4737$

Using the predictor formula $(y_4)_p = y_0 + \frac{4h}{3}(2f_1 - f_2 + 2f_3)$ we get

$$(y_4)_p = 1 + \frac{4(0.05)}{3}[2(1.1526) - 1.3104 + 2\,(1.4737)] = 1 + \frac{0.20}{3}[2.3052 - 1.3104 + 2.9474]$$

$$= 1.2628 \qquad \therefore (f_4)_p = 0.20 + 1.2628 = 1.4628$$

Now applying the corrector formula $(y_4)_c = y_2 + \frac{h}{3}(f_2 + 4f_3 + f_4)$, we get

$$(y_4)_c = 1.2104 + \frac{0.05}{3}[1.3104 + 4(1.4737) + 1.4628] = 12104 + 0.1445 = 1.3549$$ **Ans.**

Thus, $y(0.20) = 1.3549$

6. *Employing Runge-Kutta method of fourth order find y(0, 2) if* $\frac{dy}{dx} = x + y^2$, *y(0) = 1.*

SOLUTION: We do it in two steps so $h = 0.1$. Here $x_0 = 0$, $y_0 = 1$ and $f(x, y) = x + y^2$.

Now $K_1 = hf(x_0, y_0) = 0.1(0 + 1^2) = 0.1$

$$K_2 = hf\left(x_0 \frac{h}{2}, y_0 + \frac{K_1}{2}\right) = 0.1\left[005 + (0 + (1.05^2)\right] = 0.11525$$

$$K_3 = hf\left(x_0 \frac{h}{2}, y_0 + \frac{K_2}{2}\right) = 0.1\left[005 + (1.0576)^2\right] = 0.11685$$

$$K_4 = hf(x_0 + y, y_0 + K_3) = 0.1\left[0.1 + (1.11685)^2\right] = 0.13474$$

Then $$K = \frac{1}{6}[K_1 + 2K_2 + 2K_3 + K_4] = \frac{1}{6}[0.1 + 2(0.11525) + 2(0.11685) + 0.13474] = 0.1165$$

$\therefore \quad y(0.1) = y_0 + k = 1 + 0.1165 = 1.1165$

Next, we take the step two in which $x_0 = 0.1$, $y_0 = 1.1165$.

$$\therefore \quad K_1 = 0.1\left[0.1+(1.1165)^2\right] = 0.1347$$

$$K_2 = 0.1\left[0.15+(1.0673)^2\right] = 0.1288$$

$$K_3 = 0.1\left[0.15+(1.0644)^2\right] = 0.12833$$

$$K_4 = 0.1\left[0.2+(1.1283)^2\right] = 0.1473$$

$$\therefore \quad K = \frac{1}{6}[K_1 + 2K_2 + 2K_3 + K_4] = 0.1571$$

$\therefore \quad y(0.2) = 1.116 + 0.1327 = 1.2487$ **Ans.**

7. *Using Runge-Kutta method of fourth order find* $y(0.1)$, *and* $y(0.2)$ *given that*

$$\frac{dy}{dx} = 1 + xy, \quad y(0) = 2.$$

SOLUTION: $\dfrac{dy}{dx} = f(x, y) = 1 + xy.$ Here $x_0 = 0$, $y_0 = 2$, $h = 0.1$, $x_1 = 0.1$

Then by R.K. method $K_1 = hf(x_0, y_0) = 0.1[1 + 0(2)] = 0.1$

$$K_2 = hf\left(x_0 + \frac{h}{2},\ y_0 + \frac{K_1}{2}\right) = 0.1\left[1+\left(0+\frac{0.1}{2}\right)\left(2+\frac{0.1}{2}\right)\right]$$

$$= 0.1[1 + 0.05 \times 2.05] = 0.11025$$

$$K_3 = hf\left(x_0 + \frac{h}{2},\ y_0 + \frac{K_2}{2}\right) = 0.1\left[1+(0+0.05)\left(2+\frac{0.1105}{2}\right)\right]$$

$$= 0.1[1 + 0.05 \times 2.05525] = 0.1106$$

$$K_4 = hf(x_0 + h, y_0 + K_3) = 0.1[1 + (0 + 0.1)(2 + 0.1106)]$$

$$= 0.1[1 + 0.1 \times 2.106] = 0.1106$$

$$\therefore \quad y_1 = y_0 + \frac{1}{6}(K_1 + 2K_2 + 2K_3 + K_4) = 2 + \frac{1}{6}(0.1 + 0.2205 + 0.2212 + 0.1106) = 2.1086$$

To find $y(0.2)$ we proceed as follows

$$K_1 = hf(x_1, y_1) = h(1 + x_1 y_1) = 0.1(1 + 0.1 \times 2.086) = 0.1211$$

$$K_2 = hf\left(x_1 + \frac{h}{2}, y_1 + \frac{K_1}{2}\right) = 0.1\left[1 + (0.1 + 0.05)\,(2.1086 + 0.0606)\right] = 0.1325$$

$$K_3 = hf\left(x_1 + \frac{h}{2}, y_1 + \frac{K_2}{2}\right) = 0.1\left[1 + (0.1 + 0.05)\,(2.1086 + 0.0663)\right] = 0.1326$$

$$K_4 = hf(x_1 + h, y_1 + K_3) = 0.1[1 + (0.1 + 0.1)\,(2.1086 + 0.1326) = 0.1464$$

$$\therefore \quad y_2 = y(0.2) = y_1 + \frac{1}{6}(K_1 + 2K_2 + 2K_3 + K_4)$$

$$= 2.1086 + \frac{1}{6}\left[0.1211 + 2(0.1325) + 2(0.1326) + 0.1464\right] = 2.2416$$

Thus, $y(0.1) = 2.1086$ and $y(02) = 2.2416$. **Ans.**

❑❑❑

Win Prizes!

ATTENTION READERS

We request you for your frank assessment regarding some of the aspects of the book given as under:

QUESTION BANK IN ENGINEERING MATHEMATICS

By

Dr. V.P. Mishra

Please fill up the given space in neat capital letters.

(i) What topics of your syllabus that are important from you examination point of view are not covered in the book ?

....

....

....

....

(ii) What are the chapters wherein the treatment of the subject matter is not systematic or organised or updated ?

....

....

....

....

(iii) Have you come across misprints/mistakes/factual inaccuracies in the book? Please specify the chapters and the page numbers.

....

....

....

....

(iv) What other books on the same subject have you found/heard better that the present book? Please specify in terms of price and quality.

....

....

....

....

(v) Further suggestions and comments for the improvement of the book:

....

....

....

....

(PS: If need be please attach a separate sheet to write your views.)

Other Details :

(i) Who recommended you the book ? (Please tick in the box near the option relevant to you.)

☐ Teachers ☐ Friends ☐ Bookseller

(ii) Name of the recommending teacher, his designation and address.

....
....
....
....

(iii) Name and address of the bookseller you purchased the book form :

....
....
....
....

(iv) Name and address of your institution (Please mention the University or Board, as the case may be)

....
....
....
....

(v) Your name and complete postal address :

....
....
....
....

(vi) Course you are enrolled in :

....
....
....
....

The best assessment will be awarded each month. The award will be in the form of our publications as decided by the Editorial Board.

Please mail the filled up coupon at your earliest to:

S.K. Kataria & Sons®

4760-61/23, Ansari Road, Daryaganj Delhi-110002
Ph. : 23243489, 23269324
Fax : 011-23243489
e-mail: katariabooks@yahoo.com
katson_sanjeev@yahoo.co.in